Index of Study

Geometry Formulas and Definitions

- Perimeter of a triangle with sides s_1, s_2, and s_3:
  $P = s_1 + s_2 + s_3$

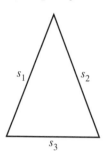

- Perimeter of a square with side s: $P = 4s$

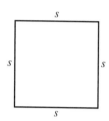

- Perimeter of a rectangle with length L and width W:
  $P = 2L + 2W$

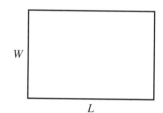

- **Circumference** (distance around the outside) of a circle with radius r: $C = 2\pi r$

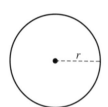

- Area of a triangle with base b and height h:

  $A = \dfrac{1}{2}bh$

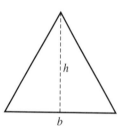

- Area of a square with side s: $A = s^2$

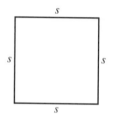

- Area of a rectangle with length L and width W: $A = LW$

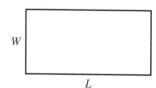

- Area of a circle with radius r:
  $A = \pi r^2$

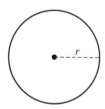

- An **equilateral triangle** is a triangle that has three equal sides.

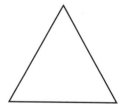

- An **isosceles triangle** is a triangle that has at least two equal sides.

ELEMENTARY AND INTERMEDIATE ALGEBRA

ELEMENTARY AND INTERMEDIATE ALGEBRA

George Woodbury
College of the Sequoias

PEARSON

Addison
Wesley

Boston San Francisco New York
London Toronto Sydney Tokyo Singapore Madrid
Mexico City Munich Paris Cape Town Hong Kong Montreal

Publisher	Greg Tobin
Executive Editor	Jennifer Crum
Executive Project Manager	Kari Heen
Project Editor	Lauren Morse
Editorial Assistants	Elizabeth Bernardi and Emily Ragsdale
Managing Editor	Ron Hampton
Senior Production Supervisor	Jeffrey Holcomb
Cover and Interior Designer	Dennis Schaefer
Photo Researcher	Beth Anderson
Media Producer	Sharon Smith
Marketing Manager	Jay Jenkins
Marketing Assistant	Alexandra Waibel
Senior Author Support/ Technology Specialist	Joe Vetere
Senior Prepress Supervisor	Caroline Fell
Senior Manufacturing Buyer	Evelyn Beaton
Production Coordination	Pat McCutcheon
Composition	WestWords, Inc.
Illustrations	Jim McLaughlin
Cover Image	© Nora Good/Masterfile. The wings of a Monarch butterfly.

For permission to use copyrighted material, grateful acknowledgment is made to the copyright holders listed on page P-1, which is hereby made part of this copyright page.

Many of the designations used by manufacturers and sellers to distinguish their products are claimed as trademarks. Where those designations appear in this book, and Addison-Wesley was aware of a trademark claim, the designations have been printed in initial caps or all caps.

Library of Congress Cataloging-in-Publication Data
Woodbury, George, 1967–
  Elementary and intermediate algebra / George Woodbury.
    p. cm.
  Includes index.
  ISBN 0-321-16640-X—ISBN 0-321-32167-7
    1. Algebra—Textbooks. I. Title.

QA152.3.W66 2005
512.9—dc22

2004062479

1 2 3 4 5 6 7 8 9 10—VH—09 08 07 06 05

To Tina, Dylan, and Alycia
You make everything meaningful and worthwhile—
yesterday, today, and tomorrow.

Contents

Preface

The goal of this text is to help students understand the topics within elementary and intermediate algebra and to better prepare them for future courses. The material is written specifically to move students through elementary and intermediate algebra seamlessly, continually building on topics throughout the text. Based on his experience teaching developmental-mathematics students, George Woodbury has included an abundance of clearly and completely worked-out examples. He understands that students often turn first to these examples for help. Most examples are followed by a Quick Check exercise, allowing students to assess their understanding of a similar problem. Practicing Quick Check exercises helps students build their confidence before approaching the exercise sets. You will find that Woodbury's text has more exercises than most other texts, offering students ample opportunity to develop their skills and increase their understanding. In addition, a great deal of attention has been focused on reading and interpreting graphs, graphing equations, and understanding and working with functions. Students often have difficulty with these topics, so the extra attention will help them fully understand the concepts. Finally, the text contains a large assortment of interesting applications, so your students will be better able to use mathematics and apply the skills they have learned to their daily lives.

Textbook Elements

Focusing on Taking Students Further

The book is written as a combined algebra text from the ground up; it is not a merging of separate elementary algebra and intermediate algebra texts. The focus is to create a textbook that flows smoothly from topic to topic, relating elementary algebra concepts to intermediate algebra concepts. The unnecessary overlap of material often found between an elementary and an intermediate algebra text has been eliminated. This approach takes students further into algebra and better prepares them for their next collegiate math course.

Following the Blueprint

Each **chapter opener** introduces students to the chapter's contents and presents a new study tip that is revisited and expanded upon throughout the chapter. (See pages 1, 65, and 127.)

Building Good Study Skills

A unique *Study Tip* is introduced in each chapter opener. Each **themed *Study Tip*** is

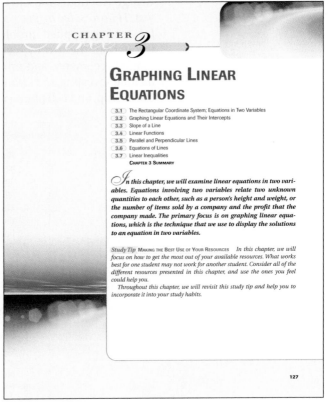

revisited after each section's exercise set and again at the end of the chapter. These helpful *Study Tips* outline good study habits that equip students for success as they progress through the textbook. Some of the *Study Tip* themes include Study Groups, Using Your Textbook Effectively, Test Taking, and Overcoming Math Anxiety. (See pages 127, 221, and 283.)

Study Tip Revisited   As soon as you receive your test, write down any formulas, rules, or procedures that will help you during the exam. Once you have written these down, you can refer to them as you work through the test; for example, a table or formula that you use to solve a particular type of word problem. Write it down on the test while it is still fresh in your memory. That way, when you reach the appropriate problem on the test, you will not need to worry about not being able to remember the table or formula.

By writing all of this information on your test, you will eliminate memorization difficulties that arise when taking a test. However, if you do not understand the material or how to use what you have written down, then you will struggle on the test. There is no substitute for understanding what you are doing.

Study Tip Revisited   When you take a test, work on the easiest problems first, saving the more difficult problems for later. One benefit to this approach is that you will gain confidence as you progress through the test, so you are confident when you attempt to solve a difficult problem. Students who struggle on a problem on the test will lose confidence, and this may cause students to miss later problems that they know how to do. By completing the easier problems quickly first, you will save time to work on the few difficult problems.

One exception to this practice is to begin the test by first attempting the most difficult type of problem for you. For example, if a particular type of word problem has been difficult for you to solve throughout your preparation for the exam, you may want to focus on how to solve that type of problem just before arriving for the test. With the solution of this type of problem committed to your short-term memory, your best opportunity for success comes at the beginning of the test prior to solving any other problems.

Study Tip Revisited   Try to leave yourself enough time to review the test at the end of the test period. Check for careless errors, which can cost you a fair amount of points. Also be sure that your answers make sense within the context of the problem. For example, if the question asks how tall a person is and your answer is 68 feet tall, then chances are that something has gone astray.

Check for problems, or parts of problems, that you may have skipped and left blank. There is no worse feeling in a math class than getting a test back and realizing that you simply forgot to do one or more of the problems.

Take all of the allotted time to review your test. There is no reward for turning in a test early, and the more that you work on the test, the more likely it is that you will find a mistake or a problem where you can gain points. Keep in mind that a majority of the students who turn in tests early do so because they are not prepared for the test and cannot do several of the problems.

Let's Solve Problems

The ability to solve problems is a skill that is required daily in the real world and in math. George Woodbury presents a **six-step problem-solving strategy** based on George Pólya's text, *How to Solve It*. (See page 85.) In Chapter 2, Section 3, Woodbury lays the foundation for solving applied problems. He then expands on this problem-solving strategy throughout the text by incorporating hundreds of applied problems on topics such as motion, geometry, and mixture problems. Interesting themes in the applied problems include investing and saving money, understanding sports statistics, landscaping, home ownership, and cell phone usage. (See pages 94–97, 259–262, and 604–607.)

Solving Applied Problems

1. **Read the problem.** This step is often overlooked, but misreading or misinterpreting the problem essentially guarantees an incorrect solution. Read the problem once quickly to get a rough idea of what is going on, and then read it more carefully a second time to gather all of the important information.

2. **List all of the important information.** Identify all known quantities presented in the problem, and determine which quantities we need to find. Creating a table to hold this information or a drawing to represent the problem are good ideas.

3. **Assign a variable to the unknown quantity.** If there is more than one unknown quantity, express them in terms of the same variable, if possible.

4. **Find an equation relating the known values to the variable expressions representing the unknown quantities.** Sometimes, this equation can be translated directly from the wording of the problem. Other times the equation will depend on general facts that are known outside the statement of the problem, such as geometry formulas.

5. **Solve the equation.** Solving the equation is not the end of the problem, as the value of the variable is often not what we were originally looking for. If we were asked for the length and width of a rectangle and we reply "x equals 4," we have not answered the question. Check your solution to the equation.

6. **Present the solution.** Take the value of the variable from the solution to the equation and use it to figure out all unknown quantities. Check these values to be sure that they make sense in the context of the problem. For example, the length of a rectangle cannot be negative. Finally, present your solution in a complete sentence, using the proper units.

Examples and Exercises that Drive Student Learning

The most distinguishing characteristic of George Woodbury's *Elementary and Intermediate Algebra* is its **extensive number of examples and exercises**—more examples than students normally see in a college-level math text. In conjunction with the many useful features described in this preface, these important examples and exercises ensure students' success in elementary and intermediate algebra, as well as in more advanced math courses. (See pages 98–108, 155–169, and 310–316.)

EXAMPLE 11 The number of Americans (y), in millions, who have a cellular phone can be approximated by the equation $y = 13x + 77$, where x is the number of years after 1999. Interpret the slope and y-intercept of this line. (*Source:* Cellular Telecommunications Industry Association)

Solution

The slope of this line is 13, and this tells us that each year we can expect an additional 13 million Americans to have a cellular phone. (The number of Americans who have a cellular phone is increasing because the slope is positive.) The y-intercept at (0, 77) tells us that approximately 77 million Americans had a cellular phone in 1999.

Quick Check **11** The number of women (y) accepted to medical school in a given year can be approximated by the equation $y = 218x + 7485$, where x is the number of years after 1997. Interpret the slope and y-intercept of this line. (*Source:* Association of American Medical Colleges)

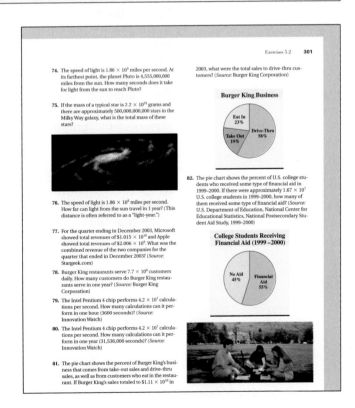

Exercises 5.2   **301**

74. The speed of light is 1.86×10^5 miles per second. At its farthest point, the planet Pluto is 4,555,000,000 miles from the sun. How many seconds does it take for light from the sun to reach Pluto?

75. If the mass of a typical star is 2.2×10^{33} grams and there are approximately 500,000,000,000 stars in the Milky Way galaxy, what is the total mass of these stars?

76. The speed of light is 1.86×10^5 miles per second. How far can light from the sun travel in 1 year? (This distance is often referred to as a "light-year.")

77. For the quarter ending in December 2003, Microsoft showed total revenues of $\$1.015 \times 10^{10}$ and Apple showed total revenues of $\$2.006 \times 10^9$. What was the combined revenue of the two companies for the quarter that ended in December 2003? (*Source:* Stargeek.com)

78. Burger King restaurants serve 7.7×10^6 customers daily. How many customers do Burger King restaurants serve in one year? (*Source:* Burger King Corporation)

79. The Intel Pentium 4 chip performs 4.2×10^7 calculations per second. How many calculations can it perform in one hour (3600 seconds)? (*Source:* Innovation Watch)

80. The Intel Pentium 4 chip performs 4.2×10^7 calculations per second. How many calculations can it perform in one year (31,536,000 seconds)? (*Source:* Innovation Watch)

81. The pie chart shows the percent of Burger King's business that comes from take-out and drive-thru sales, as well as from customers who eat in the restaurant. If Burger King's sales totaled to $\$1.11 \times 10^{10}$ in

2003, what were the total sales to drive-thru customers? (*Source:* Burger King Corporation)

Burger King Business

Eat In 23%

Drive-Thru 58%

Take Out 19%

82. The pie chart shows the percent of U.S. college students who received some type of financial aid in 1999–2000. If there were approximately 1.67×10^7 U.S. college students in 1999–2000, how many of them received some type of financial aid? (*Source:* U.S. Department of Education, National Center for Educational Statistics, National Postsecondary Student Aid Study, 1999–2000)

College Students Receiving Financial Aid (1999 –2000)

No Aid 45%

Financial Aid 55%

Students Can Function with Graphs

Woodbury takes an *early-and-often* approach to graphing and functions. These primary algebraic concepts are introduced early (Chapter 3) and are then consistently incorporated throughout the text, providing optimal opportunity for their use and review. Helping students become comfortable with reading and interpreting graphs and function notation establishes a basis for understanding that will prepare them for future math courses.

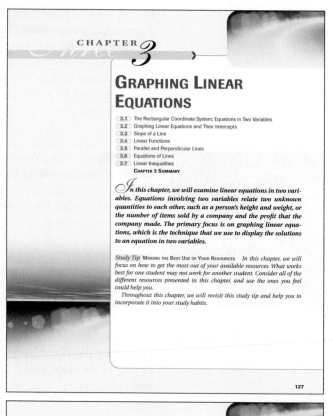

Making the Transition from Elementary to Intermediate Algebra

Chapter 8 (see page 460) is representative of the author's direct approach to teaching elementary and intermediate algebra with purpose and consistency. Serving as the transition between the two courses, this chapter is designed to begin the intermediate algebra course by **reviewing** and **extending** essential **elementary algebra concepts** in order to **introduce new intermediate algebra topics.**

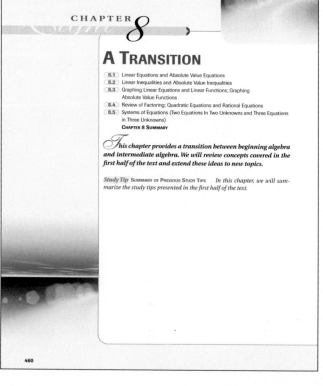

Immediate Feedback from Quick Check Exercises
Quick Check exercises appear after nearly every example in the text. Each of the confidence-building **Quick Check** exercises mirrors the example that precedes it, allowing students to practice what they are learning as they read through each section. The answers to the **Quick Check** exercises are provided in the back of the book so that students can immediately assess their understanding. (See pages 99, 160, and 305.)

EXAMPLE 10   Admission to the county fair is $8 per person, so the admission price for a group of x people can be represented by $8x$. If a Cub Scout group paid a total of $208 for admission to the county fair, how many people were in the group?

Solution

We can express this relationship in an equation by using the idea that the total cost of admission is equal to $208. Since the total cost for x people can be represented by $8x$, our equation is $8x = 208$.

$$8x = 208$$
$$\frac{8x}{8} = \frac{208}{8} \qquad \text{Divide both sides by 8.}$$
$$x = 26 \qquad \text{Divide.}$$

You can verify that this value checks as a solution. Whenever we work on an applied problem, we should present our solution as a complete sentence. There were 26 people in the Cub Scout group at the county fair.

Quick Check   9 Josh spent a total of $26.50 to take a date to the movies. This left him with only $38.50 in his pocket. How much money did he have with him before going to the movies?

"You *Will* Use This Information Someday!"
The many **real-world applications** encourage student understanding of algebra in a practical context. These fun and interesting applications give instructors an opportunity to answer the often-asked question, "When am I ever going to use this?" They also provide a springboard for classroom discussions or the synthesis of skills. (See pages 40, 163, and 301.)

36. The number of doctoral degrees conferred on women, y, in a particular year can be approximated by the equation $y = 473x + 17{,}811$, where x represents the number of years after 1995. The number of doctoral degrees conferred on men, y, in a particular year can be approximated by the equation $y = -423x + 26{,}841$, where x represents the number of years after 1995. In what year will the number of doctoral degrees conferred on women equal the number of doctoral degrees conferred on men? (*Source:* U.S. Department of Education, National Center for Educational Statistics)

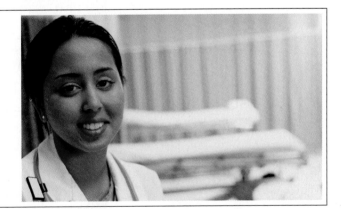

Practicing What Students Have Already Learned

Appearing once per chapter, **Quick Review** exercises are a short selection of review exercises geared toward helping students maintain skills they have previously learned and preparing them for upcoming concepts. (See pages 108, 262, and 363.)

QUICK REVIEW EXERCISES

Section 8.1

Evaluate.

1. $f(x) = 3x - 14, f(-6)$

2. $f(x) = x^2 + 17x - 60, f(-5)$

3. $f(x) = \dfrac{x^2 - 6x + 25}{3x + 7}, f(2)$

4. $f(x) = |x - 7| + 3, f(0)$

Taking It to Another Level

Appearing at the end of each chapter, **Stretch Your Thinking** exercises ask students to go the extra mile by taking what they have learned in the chapter and applying it to a problem that requires some additional thought. These exercises can be done individually or in groups. (See pages 63, 126, and 391.)

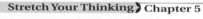

Stretch Your Thinking Chapter 5

At a recent school carnival, George agreed to be launched out of a cannon to become the world's first flying mathematician. He landed after 3.4625 seconds. George's height, in feet, after t seconds is given by the function $h(t) = -16t^2 + 55.4t$.

a) Complete the following chart listing George's height at 0.5-second intervals:

Time (in seconds)	Height (in feet)
0	
0.5	
1	
1.5	
2	
2.5	
3	
3.5	

b) Create a coordinate system in which the horizontal axis represents time (in seconds) and the vertical axis represents height (in feet). Plot these eight points and connect them to see George's course of flight through the air. (Your graph should be a smooth curve.)

Use the table and graph to answer the following questions.

c) What is George's height when he was launched out of the cannon?

d) What is George's height after 0.5 seconds?

e) How long did it take George to reach a height of 39.4 feet?

f) What is George's height when he lands on the ground? How long did it take (from the time of being launched) for George to land on the ground?

g) When does it appear that George reaches his maximum height? What is his maximum height?

"Who Thought of This?"

The **Mathematicians in History** activities provide a structured opportunity for students to learn about the rich and diverse history of mathematics. These short research projects, which ask students to investigate the life of a prominent mathematician, can be assigned as independent work or used as a collaborative learning activity. A list of directed questions is provided for each activity. Some of the art pieces associated with the *Mathematicians in History* features were created by Murleen Ray, one of the author's students. (See pages 63, 125, and 278.)

Mathematicians in History
Hypatia of Alexandria

Hypatia of Alexandria was one of the first women known to make major contributions to the field of mathematics. She was also active in the fields of astronomy and philosophy. One quote that has been attributed to her is, "Reserve your right to think, for even to think wrongly is better than not to think at all."

Write a one-page summary (*or* make a poster) of the life of Hypatia of Alexandria and her accomplishments.

Interesting issues:

- Where and when was Hypatia born?
- Who was Hypatia's father, and what fields was he involved in?
- Hypatia lectured on the philosophy of Neoplatonism. What are the principle ideas of Neoplatonism?
- What was Hypatia's relationship to Cyril and Orestes?
- In the year A.D. 415, Hypatia was murdered. Describe the circumstances of her death.

Pushing the Buttons When Appropriate

Optional *Using Your Calculator* boxes are presented throughout the text, giving students guided calculator instruction and screen shots, as appropriate, to complement the material presented. (See pages 30, 143, and 225.)

Watching Out for Common Mistakes and Miscalculations

Throughout the text, **A Word of Caution** helps students to avoid misconceptions by pointing out typical errors that students often make. (See pages 160, 287, and 342.)

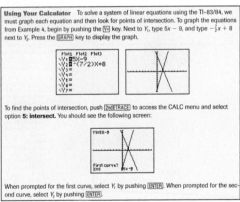

Using Your Calculator To solve a system of linear equations using the TI–83/84, we must graph each equation and then look for points of intersection. To graph the equations from Example 4, begin by pushing the [Y=] key. Next to Y_1, type $5x - 9$, and type $-\frac{7}{2}x + 8$ next to Y_2. Press the [GRAPH] key to display the graph.

To find the points of intersection, push [2nd][TRACE] to access the CALC menu and select option **5: intersect.** You should see the following screen:

When prompted for the first curve, select Y_1 by pushing [ENTER]. When prompted for the second curve, select Y_2 by pushing [ENTER].

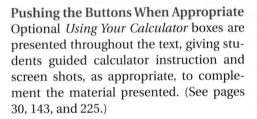

A Word of Caution When raising a quotient to a power, raise both the numerator and the denominator to that power. Do not just raise the numerator to that power.

$$\left(\frac{r}{s}\right)^3 = \frac{r^3}{s^3}, \text{not } \frac{r^3}{s} \qquad (\text{Assume } s \neq 0.)$$

Preparing Students for the Test
Each chapter concludes with a **Chapter Summary,** a **summary** of the **Study Tips, Review Exercises,** and a **Chapter Test,** which together are an excellent resource for extra practice and test preparation. (See pages 213–218, 273–277, and 387–390.)

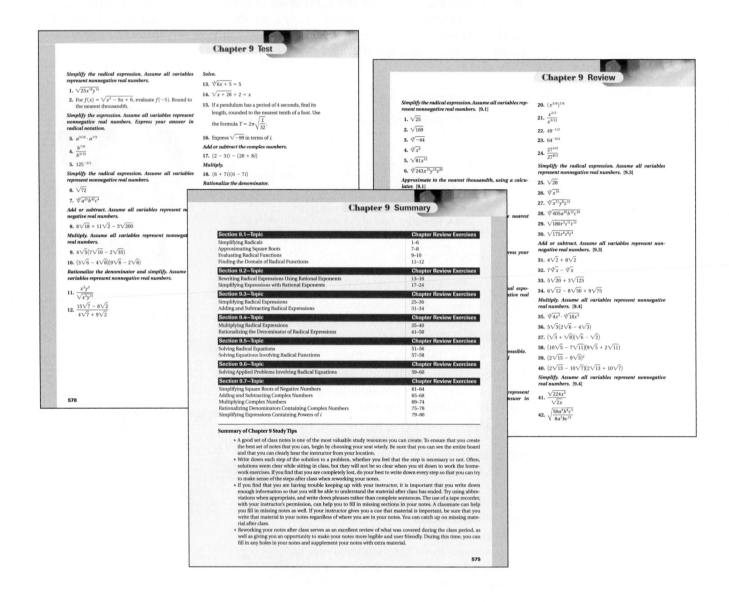

Putting It All Together
A set of **Cumulative Review** exercises can be found after Chapters 4, 7, 11, and 14. In both the elementary and intermediate algebra sections of the text, these exercises are strategically placed to help students review before midterm and final exams. (See pages 280–282 and 458–459.)

Cumulative Review   Chapters 5–7

Simplify the expression. Write the result without using negative exponents. (Assume all variables represent nonzero real numbers.) [5.1/5.2]

1. $\dfrac{x^{11}}{x^4}$

2. $12m^{10}n^7 \cdot 9m^6n^{14}$

3. $\left(\dfrac{5a^6b^{11}}{c^7}\right)^4$

4. $\dfrac{t^{12}}{t^{-9}}$

5. $(a^{-6}b^7)^{-8}$

6. $x^{-16} \cdot x^{-19}$

Rewrite in scientific notation. [5.2]

7. 330,000,000,000

8. 0.000000000000714

9. If a computer can perform a calculation in 0.000000000005 second, how long would it take to perform 200,000,000 calculations? [5.2]

Evaluate the polynomial for the given value of the variable. [5.3]

10. $x^2 + 11x - 16$ for $x = -7$

Add or subtract. [5.3]

11. $(x^2 + 6x - 20) + (x^2 - 13x - 39)$

12. $(x^2 - 15x - 9) - (2x^2 - 3x - 31)$

Evaluate the given function. [5.3]

13. $f(x) = x^2 + 5x - 37, f(8)$

Multiply. [5.4]

14. $3x(5x^2 - 8x + 12)$

15. $(4x - 7)(3x - 8)$

Divide. [5.5]

16. $\dfrac{10x^{12} + 24x^9 - 18x^6}{2x^5}$

17. $\dfrac{x^2 + 13x + 39}{x + 6}$

Factor completely. [6.1–6.5]

18. $x^3 - 4x^2 + 6x - 24$

19. $2x^2 - 26x + 80$

20. $x^2 + 5x - 36$

21. $2x^2 - x - 10$

22. $x^3 - 125$

23. $x^2 - 81$

Solve. [6.6]

24. $x^2 - 49 = 0$

25. $x^2 - 11x + 24 = 0$

26. $x^2 + 14x + 49 = 0$

27. $x(x + 8) = 20$

Find a quadratic equation with integer coefficients that has the given solution set. [6.6]

28. $\{-2, 7\}$

For the given function $f(x)$, find all values x for which $f(x) = 0$. [6.7]

29. $f(x) = x^2 - 100$

30. $f(x) = x^2 + 2x - 35$

31. A man is standing on a cliff above a beach. He throws a rock upward from a height of 192 feet above the beach with an initial velocity of 64 feet/second. The rock's height above the beach, in feet, after t seconds is given by the function $h(t) = -16t^2 + 64t + 192$, which is graphed as follows:

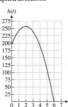

Use the function and its graph to answer the following questions: [6.7]

458

...e how high above ...seconds.

...rock to land on the

...s the rock reach its ...ach?

...e the greatest height ...k reaches.

...ntegers have a prod-... [6.8]

...ters more than twice ...gle is 133 square ...e width of the rec-

...ational expression is ...undefined. [7.1]

34. $\dfrac{x^2 + 11x + 24}{x^2 - 14x + 45}$

Simplify the given rational expression. (Assume all denominators are nonzero.) [7.1]

35. $\dfrac{x^2 + 3x - 54}{x^2 + 16x + 63}$

Evaluate the given rational function. [7.1]

36. $r(x) = \dfrac{x + 12}{x^2 + 7x + 5}, r(3)$

Multiply. [7.2]

37. $\dfrac{x^2 + 3x - 18}{x^2 + 15x + 44} \cdot \dfrac{x^2 + 2x - 8}{x^2 + 8x + 12}$

Divide. [7.2]

38. $\dfrac{x^2 + x - 90}{x^2 + 14x + 49} \div \dfrac{x^2 - 8x - 9}{x^2 + 5x - 14}$

Add or subtract. [7.3/7.4]

39. $\dfrac{5x + 7}{x^2 - 2x - 8} - \dfrac{2x + 19}{x^2 - 2x - 8}$

40. $\dfrac{6}{x^2 + x - 2} + \dfrac{2}{x^2 - 6x + 5}$

41. $\dfrac{x - 5}{x^2 - 10x + 21} - \dfrac{3}{x^2 - 9}$

42. $\dfrac{x + 4}{x^2 + 13x + 42} + \dfrac{6}{x^2 + 12x + 35}$

Simplify the complex fraction. [7.5]

43. $\dfrac{\dfrac{x - 5}{x + 3} - \dfrac{1}{x - 6}}{\dfrac{x^2 + 5x - 24}{x^2 - 3x - 18}}$

44. $\dfrac{1 + \dfrac{1}{x} - \dfrac{20}{x^2}}{1 + \dfrac{13}{x} + \dfrac{40}{x^2}}$

Solve. [7.6]

45. $6 + \dfrac{4}{x - 3} = \dfrac{3x + 1}{x - 3}$

46. $\dfrac{2x - 3}{x + 4} = \dfrac{x + 3}{x - 4}$

47. $\dfrac{5}{x + 7} + \dfrac{4}{x - 3} = \dfrac{x - 11}{x^2 + 4x - 21}$

Solve for the specified variable. [7.6]

48. $\dfrac{1}{x} + \dfrac{1}{y} = 2$ for x

49. The sum of the reciprocal of a number and $\frac{4}{15}$ is $\frac{11}{30}$. Find the number. [7.7]

50. Two pipes working together can fill a tank in 3 hours. Working alone, the smaller pipe would take 8 hours longer than the larger pipe to fill the tank. How long would it take the smaller pipe alone to fill the tank? [7.7]

❮ Student Supplements	**Instructor Supplements ❯**

Student's Solutions Manual

- By Mark Tom, *College of the Sequoias.*
- Contains complete solutions to the odd-numbered section exercises and solutions to all Quick Check, Review, Chapter Test, and Cumulative Review exercises.
  ISBN: 0-321-32175-8

Digital Video Tutor

- Complete set of digitized video lessons on CD-ROM for student use at home or on campus.
- Ideal for distance learning or supplemental instruction.
  ISBN: 0-321-32170-7

MathXL® Tutorials on CD

- Provide algorithmically generated practice exercises correlated with the exercises in the textbook.
- Every exercise is accompanied by an example and a guided solution, and selected exercises may also include a video clip to help students visualize concepts.
- The software presents helpful feedback for incorrect answers and can generate printed summaries of students' progress.
  ISBN: 0-321-35638-1

Addison-Wesley Math Tutor Center

- Staffed by qualified mathematics instructors.
- Provides tutoring on examples and odd-numbered exercises from the textbook through a registration number with a new textbook or purchased separately.
- Accessible via toll-free telephone, toll-free fax, e-mail, or the Internet.
- Whiteboard technology allows tutors and students to actually see problems worked while they "talk" in real time over the Internet during tutoring sessions.
  www.aw-bc.com/tutorcenter

Annotated Instructor's Edition

- Provides answers to all text exercises, including Quick Check exercises, in color next to the corresponding problems.
  ISBN: 0-321-32167-7

Instructor's Solutions Manual

- By Mark Tom, *College of the Sequoias.*
- Contains complete solutions to all section, Quick Check, Review, Chapter Test, and Cumulative Review exercises.
  ISBN: 0-321-32168-5

Instructor and Adjunct Support Manual

- Includes resources designed to help both new and experienced faculty with course preparation and classroom management.
- Offers helpful teaching tips correlated with the chapters of the text.
- Includes over 1500 additional practice exercises from every section of the text, along with answers.
  ISBN: 0-321-39670-7

Printed Test Bank

- By Don Rose, *College of the Sequoias.*
- Provides three free-response tests per chapter and two multiple-choice tests per chapter.
- Contains two free-response final exams and two multiple-choice final exams.
- Includes short quizzes of five to six questions for every chapter that can be used in class, for individual practice, or for group work.
  ISBN: 0-321-32169-3

Videotapes

- Presented by *Student's Solutions Manual* author Mark Tom and author George Woodbury.
- Include a series of lectures correlated directly with the section content of the text.
- Include a pause-the-video feature that encourages students to stop the videotape, work through an example, and resume play to watch the video instructor work through the same example.
  ISBN: 0-321-32171-5

TestGen®

- Enables instructors to build, edit, print, and administer tests.
- Features a computerized bank of questions developed to cover all text objectives.
- Available on a dual-platform Windows/Macintosh CD-ROM.
  ISBN: 0-321-32172-3

Addison-Wesley Math Adjunct Support Center

The Addison-Wesley Math Adjunct Support Center is staffed by qualified mathematics instructors with over 50 years of combined experience at both the community college and university level. Assistance is provided for faculty in the following areas:

- Syllabus consultation
- Tips on using materials packaged with your book
- Book-specific content assistance
- Teaching suggestions, including advice on classroom strategies

For more information, visit *www.aw-bc.com/tutorcenter/math-adjunct.html.*

MyMathLab: *www.mymathlab.com*

MyMathLab is a series of text-specific, easily customizable online courses for Addison-Wesley textbooks in mathematics and statistics. Powered by CourseCompass™ (Pearson Education's online teaching and learning environment) and MathXL® (our online homework, tutorial, and assessment system), MyMathLab gives you the tools you need to deliver all or a portion of your course online, whether your students are in a lab setting or working from home. MyMathLab provides a rich and flexible set of course materials, featuring free-response exercises that are algorithmically generated for unlimited practice and mastery. Students can also use online tools, such as video lectures, animations, and a multimedia textbook, to improve their understanding and performance independently. Instructors can use MyMathLab's homework and test managers to select and assign online exercises correlated directly with the textbook, and they can also create and assign their own online exercises and import TestGen tests for added flexibility. MyMathLab's online grade book—designed specifically for mathematics and statistics—automatically tracks students' homework and test results and gives the instructor control over how to calculate final grades. Instructors can also add offline (paper-and-pencil) grades to the gradebook. MyMathLab is available to qualified adopters. For more information, visit our Web site at *www.mymathlab.com* or contact your Addison-Wesley sales representative.

MathXL®: *www.mathxl.com*

MathXL® is a powerful online homework, tutorial, and assessment system that accompanies Addison-Wesley textbooks in mathematics or statistics. With MathXL, instructors can create, edit, and assign online homework and tests using algorithmically generated exercises correlated with the textbook at the objective level. They can also create and assign their own online exercises and import TestGen tests for added flexibility. All student work is tracked in MathXL's online gradebook. Students can take chapter tests in MathXL and receive personalized study plans based on their test results. The study plan diagnoses weaknesses and links students directly to tutorial exercises for the objectives they need to study and retest. Students can also access supplemental animations and video clips directly from selected exercises. MathXL is available to qualified adopters. For more information, visit our Web site at *www.mathxl.com* or contact your Addison-Wesley sales representative.

InterAct Math Tutorial Web site: *www.interactmath.com*

Get practice and tutorial help online! This interactive tutorial Web site provides algorithmically generated practice exercises that correlate directly with the exercises in the textbook. Students can retry an exercise as many times as they like, with new values each time, for unlimited practice and mastery. Every exercise is accompanied by an interactive guided solution that provides helpful feedback for incorrect answers, and students can also view a worked-out sample problem that steps them through an exercise similar to the one they're working on.

Acknowledgments

Writing a textbook such as this is a monumental task, and I would like to take the opportunity to thank everyone who has helped me along the way. The following reviewers provided thoughtful suggestions and were instrumental in the development of *Elementary and Intermediate Algebra:*

Richard Avery	*Dakota State University*
Kerry Bailey	*Laramie County Community College*
Teresa Betkowski	*Gordon College*
Pat Bezona	*Valdosta State University*
Stacey Boggs	*Allegany College of Maryland*
Michael Bowen	*Oxnard College*
Susan Bradley	*Angelina College*
Tim Britt	*Jackson State Community College*
Debra Bryant	*Tennessee Technological University*
Jared Burch	*College of the Sequoias*
Carolyn Case	*Vincennes University*
Patrick Cross	*University of Oklahoma*
Deric Davenport	*Pikes Peak Community College*
Stephan DeLong	*Front Range Community College*
Marlene Demerjian	*College of the Canyons*
Jeanne Draper	*College of the Sequoias*
Dawn Draus	*Lower Columbia College*
Noelle Eckley	*Lassen Community College*
Audrey Evans	*Hopkinsville Community College*
Carol Flakus	*Lower Columbia College*
Theresa Gallo	*San Diego City College*
Corinna Goehring	*Jackson State Community College*
Annette Hamilton	*Iowa Western Community College*
Margret Hathaway	*Kansas City Kansas Community College*
Bobbie Hill	*Coastal Bend College*
Linda Houston	*Tulsa Community College*
Charles Johnson	*College of the Canyons*
Tracey Johnson	*University of Georgia*
Judy Jones	*Valencia Community College*
Ed Kavanaugh	*Schoolcraft College*
David Keller	*Kirkwood Community College*
Mike Kirby	*Tidewater Community College–Virginia Beach*
Mary Landers	*Weatherford College*
Wing Lee	*Valdosta State University*
Frederick Lippman	*Shasta College*
Kirsten Lollis	*San Diego City College*

Patricia Lord	*Bainbridge College*
Frank Ma	*Los Angeles Harbor College*
Margaret Messinger	*Southwest Texas Junior College*
Bambi Moise	*De Anza College*
Debbie Moran	*Greenville Technical College*
Michael Oppedisano	*Morrisville State College*
Linda Padilla	*Joliet Junior College*
Shahrokh Parvini	*San Diego Mesa College*
Alvin Paul	*Prairie State College*
Faith Peters	*Miami-Dade College*
Patrick Riley	*Hopkinsville Community College*
Sherry Schulz	*College of the Canyons*
Jack Sharp	*Georgia Highlands College–Floyd campus*
Ronald Shepler	*Ferris State University*
Linda Shoesmith	*Scott Community College*
Mouwafac Sidaoui	*University of San Francisco*
Eric Smith	*Southeast Community College*
James Sousa	*Phoenix College*
Carrol Spears	*Amarillo College*
Sandra Stokes	*Baton Rouge Community College*
Paul Swatzel	*California State University, San Bernardino*
Janet Teeguarden	*Ivy Tech Community College of Indiana*
John Thomason	*Austin Community College*
John Thoo	*Yuba College*
Curt Vander Vere	*Pennsylvania College of Technology*
Leigh Ann Wheeler	*Greenville Technical College*
Bill Witte	*Montgomery College*
Bella Zamansky	*University of Cincinnati*

I have truly enjoyed working with the team at Addison-Wesley. My editor, Jenny Crum, has encouraged and inspired me constantly and helped me turn a great idea into an outstanding textbook. Her passion is contagious. (Thanks to Gretchen and Lucy as well!) Thanks to Susan Winslow for getting this all started—you are the best! I appreciate all the logistical help I received from Lauren Morse, Greg Erb, and Elizabeth Bernardi. They kept me on schedule and coordinated all reviews and supplements. Thanks are also due to Ron Hampton, Jeff Holcomb, and Dennis Schaefer for their efforts during production. Dona Kenly and Jay Jenkins provided great advice from the beginning and helped me to stay true to my goals. I do owe special thanks to Greg Tobin and Maureen O'Connor for believing in my vision and taking a chance on me. Anne Scanlan-Rohrer is an outstanding developmental editor, and her hard work and suggestions have had a tremendous impact on this book. I look forward to working with her again. Linda Murphy, Janis Cimperman, Perian Herring, Cathy Ferrer, and Sudhir Goel also deserve credit for their help in checking the manuscript and pages for accuracy.

My colleagues at College of the Sequoias have been helpful. They have given some great advice and were constantly listening to my ideas. Thanks to Vineta Harper for her help with the Mathematicians in History feature; she is a great resource. Also, thanks to Mark Tom (and his family) for the long hours put in on the student's and instructor's solutions manuals and to Don Rose for his work on the *Printed Test Bank*. Thanks to Stephanie Logan for creating the Stretch Your Thinking exercises. I owe thanks to Ross Rueger for class testing the text and providing great ideas at an early stage. I would also like to thank my students for keeping my fires burning. Truly, this book is all about the students.

My friends and family have been very supportive throughout this process. Gehrig38, Papi, and Bruce—you have always been there for me, and I do appreciate it. Thanks to my parents for all of their encouragement and especially to my grandmother Angela Woodbury, who remains one of my best friends.

Most importantly, thanks to my wife Tina and our wonderful children Dylan and Alycia. They are truly my greatest blessing, and I love them more than words can say. The process of writing a textbook is long and difficult, and they have been supportive and understanding at every turn.

Finally, this book is dedicated to the memory of my brother-in-law Scott Mello. We miss him tremendously, but he remains alive in our thoughts and actions every day.

George Woodbury

REVIEW OF REAL NUMBERS

This chapter reviews properties of real numbers and arithmetic that are necessary for success in algebra. The chapter also introduces several algebraic properties.

Study Tip STUDY GROUPS   *Throughout this book there are study tips to help you learn and be successful in this course. In this chapter, we will focus on getting involved in a study group.*

Working with a study group is an excellent way to learn mathematics, improve your confidence and level of interest, and improve your performance on tests and quizzes. When working with a group, you will be able to work through questions about the material you are studying. Also, by being able to explain how to solve a particular type of problem to another person in your group, you will increase your ability to retain this knowledge.

We will revisit this study tip throughout this chapter and help you incorporate it into your study habits. See the end of Section 1.1 for tips on how to get a study group started.

1.1

INTEGERS, OPPOSITES, AND ABSOLUTE VALUE

Objectives

1 Graph whole numbers on a number line.
2 Determine which is the greater of two whole numbers.
3 Graph integers on a number line.
4 Find the opposite of an integer.
5 Determine which is the greater of two integers.
6 Find the absolute value of an integer.

A **set** is a collection of objects, such as the set consisting of the numbers 1, 4, 9, and 16. This set can be written as {1, 4, 9, 16}. The braces, {}, are used to denote a set, and the values listed inside are said to be **elements**, or members, of the set. A set with no elements is called the **empty set**. A **subset** of a set is a collection of some or all of the elements of the set. For example, {1, 9} is a subset of the set {1, 4, 9, 16}. A subset can also be an empty set.

Whole Numbers

Objective 1 Graph whole numbers on a number line. This text, for the most part, deals with the set of real numbers.

Real Numbers

> A number is a **real number** if it can be written as a fraction or as a decimal number.

One subset of the set of real numbers is the set of natural numbers.

Natural Numbers

> The set of **natural numbers** is the set {1, 2, 3, . . . }.

If we include the number 0 with the set of natural numbers, we have the set of **whole numbers**.

Whole Numbers

> The set of **whole numbers** is the set {0, 1, 2, 3, . . . }. This set can be displayed on a number line as follows:
>
>
>
> 0 1 2 3 4 5 6 7 8 9 10

The arrow on the right-hand side of the number line indicates that the values continue to increase in this direction. There is no largest whole number, but we say that the values approach infinity (∞).

To graph any particular number on a number line, we place a point at that location on the number line.

EXAMPLE 1   Graph the number 6 on a number line.

Solution

To graph any number on a number line, we place a point at that number's location.

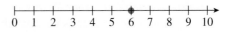

Quick Check **1**
Graph the number 4 on a number line.

Inequalities

Objective 2   Determine which is the greater of two whole numbers. When comparing two whole numbers a and b, we say that a is **greater than** b, denoted $a > b$, if the number a is to the right of the number b on the number line. The number a is **less than** b, denoted $a < b$, if a is to the left of b on the number line. The statements $a > b$ and $a < b$ are called **inequalities**.

EXAMPLE 2   Write the appropriate symbol, either $<$ or $>$, between the following:
6 _____ 4

Solution

Let's take a look at the two values graphed on a number line.

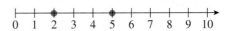

Since the number 6 is to the right of the number 4 on the number line, we say that 6 is greater than 4. So $6 > 4$.

EXAMPLE 3   Write the appropriate symbol, either $<$ or $>$, between the following:
2 _____ 5

Solution

Since 2 is to the left of 5 on the number line, $2 < 5$.

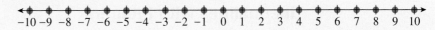

Quick Check **2**
Write the appropriate symbol, either $<$ or $>$, between the following:

a) 8 _____ 3
b) 19 _____ 23

Integers

Objective 3   Graph integers on a number line. Another important subset of the real numbers is the set of integers.

Integers

> The set of **integers** is the set $\{ \ldots, -3, -2, -1, 0, 1, 2, 3, \ldots \}$. We can display the set of integers on a number line as follows:
>
> ```
> ←┼─┼─┼─┼─┼─┼─┼─┼─┼─┼─┼─┼─┼─┼─┼─┼─┼─┼─┼─┼─┼→
>  -10 -9 -8 -7 -6 -5 -4 -3 -2 -1  0  1  2  3  4  5  6  7  8  9  10
> ```

The arrow on the left side indicates that the values continue to decrease in this direction, and they are said to approach negative infinity $(-\infty)$.

Opposites

Objective 4 **Find the opposite of an integer.** The set of integers is the set of whole numbers together with the opposites of the natural numbers. The **opposite** of a number is a number on the other side of 0 on the number line and is the same distance from 0 as that number. We denote the opposite of a real number a as $-a$. For example, -5 and 5 are opposites because they are both 5 units away from 0 and one is to the left of 0 while the other is to the right of 0.

Numbers to the left of 0 on the number line are called **negative numbers**. Negative numbers represent a quantity less than 0. For example, if you have written checks that the balance in your checking account cannot cover, your balance would be a negative number. A temperature that is below 0° F, a golf score that is below par, and an elevation that is below sea level are other examples of quantities that can be represented by negative numbers.

EXAMPLE 4   What is the opposite of 7?

Solution

The opposite of 7 is -7, since -7 is also 7 units away from 0 but is on the opposite side of 0.

EXAMPLE 5   What is the opposite of -6?

Solution

The opposite of -6 is 6:

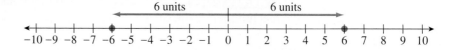

The opposite of 0 is 0 itself. Zero is the only number that is its own opposite.

Inequalities with Integers

Objective 5 **Determine which is the greater of two integers.** Inequalities for integers follow the same guidelines as they do for whole numbers. If we are given two integers a and b, the number that is greater is the number that is to the right on the number line.

EXAMPLE 6   Write the appropriate symbol, either $<$ or $>$, between the following:
-3 ____ 5

Solution

Looking at the number line, we can see that -3 is to the left of 5, so $-3 < 5$.

EXAMPLE 7   Write the appropriate symbol, either $<$ or $>$, between the following:
-2 ____ -7

Solution

On the number line, -2 is to the right of -7, so $-2 > -7$.

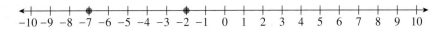

Quick Check 4
Write the appropriate
symbol, either $<$ or $>$,
between the following:

a) -14 ____ -11
b) 6 ____ -20

Absolute Values

Objective 6   **Find the absolute value of an integer.**

Absolute Value

> The **absolute value** of a number a, denoted $|a|$, is the distance between a and 0 on the number line.

Distance cannot be negative, so the absolute value of a number a will always be 0 or higher.

EXAMPLE 8   Find the absolute value of 6.

Solution

The number 6 is 6 units away from 0 on the number line, so $|6| = 6$.

EXAMPLE 9   Find the absolute value of -4.

Solution

The number -4 is 4 units away from 0 on the number line, so $|-4| = 4$.

Quick Check 5
Find the absolute value
of -9.

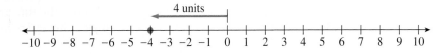

EXERCISES 1.1

Graph the following whole numbers on a number line.

1. 7
2. 3
3. 6
4. 9
5. 2
6. 13

Write the appropriate symbol, either $<$ or $>$, between the following whole numbers.

7. 3 ___ 13
8. 7 ___ 9
9. 8 ___ 6
10. 12 ___ 5
11. 45 ___ 27
12. 36 ___ 38

Graph the following integers on a number line.

13. -4
14. -7
15. 3
16. -9
17. -12
18. 6

Find the opposite of the following integers.

19. -7
20. 5
21. 14
22. -20
23. 0
24. -39

Write the appropriate symbol, either $<$ or $>$, between the following integers:

25. -7 ___ -9
26. -5 ___ -2
27. -13 ___ -11
28. -8 ___ -14
29. -16 ___ 0
30. 5 ___ -3
31. -4 ___ 3
32. -105 ___ -129

Find the following absolute values.

33. $|-15|$
34. $|9|$
35. $|0|$
36. $|-6|$
37. $|25|$
38. $|-13|$

Find the missing number, if possible. There may be more than one number that works; so find as many as possible. There may be no number that works.

39. $|?| = 7$
40. $|?| = 14$
41. $|?| = 0$
42. $|?| = -3$
43. $|?| + 5 = 11$
44. $3 \cdot |?| + 8 = 20$

Answer in complete sentences.

45. A fellow student tells you that to find the absolute value of any number, just make the number positive. Is this always true? Explain why or why not in your own words.

46. True or false: The opposite of the opposite of a number is the number itself. Explain your reasoning.

47. If the opposite of a nonzero integer is equal to the absolute value of that integer, is the integer positive or negative? Explain your reasoning.

48. If an integer is less than its opposite, is the integer positive or negative? Explain your reasoning.

Study Tip Revisited   To form a study group, you must begin with the question, "Who do I want to work with?" Look for students who are serious about learning, such as students who seem prepared for each class or students who ask good questions during class. If a student does not appear to do their homework, or is constantly late for class, then this student may not be the best person to work with.

Look for students you feel you can get along with. You are about to spend a long period working with this group, sometimes under stressful conditions. Working with group members who get on your nerves will not help you to learn.

If you take advantage of tutorial services provided by your college, keep an eye out for classmates who do the same. There is a strong chance that classmates who use the tutoring center are serious about learning mathematics and earning good grades.

For Extra Help

MyMathLab
MathXL
Interactmath.com
MathXL Tutorials on CD
DVT CD Videotape
Addison-Wesley Math Tutor Center
Student Solutions Manual

1.2

OPERATIONS WITH INTEGERS

Objectives

1 Add integers.
2 Subtract integers.
3 Multiply integers.
4 Divide integers.

Addition and Subtraction of Integers

Objective 1 **Add integers.** We can use the number line to help us learn how to add and subtract integers. Suppose we were trying to add the integers 3 and -7, which could be written as $3 + (-7)$. On a number line, we will start at the first number in the sum, 3. Adding -7 tells us to move 7 units in the negative, or left, direction. (Note that if we were adding a positive number to 3, we would have moved to the *right*.)

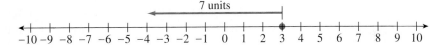

Since we end up at -4, this tells us that $3 + (-7) = -4$.

One important property of opposites is that the sum of two opposites is equal to 0.

Sum of Two Opposites

> For any real number a, $a + (-a) = 0$.

Suppose that we wanted to add the number 4 and its opposite, -4. Using the number line, we would begin at 4 and then move 4 units to the left, ending at 0.

We could also see that $3 + (-7) = -4$ through the use of manipulatives, which are hands-on tools used to demonstrate mathematical properties. Suppose that we had a bag of green and red candies. We could let each piece of green candy represent a positive 1 and each piece of red candy represent a negative 1. To add $3 + (-7)$, we would begin by combining 3 green candies (positive 3) with 7 red candies (negative 7). Combining 1 red candy with 1 green candy has a net result of 0, as the sum of two opposites is equal to 0. So, each time we can make a pair of a green candy and a red candy, these two cancel each other's effect and can be discarded. After doing this, we would be left with 4 red candies. Our answer would be -4.

Adding a Positive Number and a Negative Number

Whenever we are adding a positive number and a negative number, we can begin by subtracting the absolute values of the two numbers.

In the example $3 + (-7)$, $|3| = 3$ and $|-7| = 7$. The difference between the absolute values is 4. The sign of our answer is the same as the sign of the number that has the larger absolute value. In this case, -7 has the larger absolute value, so our result is negative.

EXAMPLE 1   Find the following sum: $12 + (-8)$.

Solution

$|12| = 12; \quad |8| = 8$ Find the absolute value of each number.
$12 - 8 = 4$ The difference between the absolute values is 4.
$12 + (-8) = 4$ Since the number with the larger absolute value is positive, the result is positive.

Quick Check 1 Find the sum $14 + (-6)$.

Notice that in the previous example $12 + (-8)$ is equivalent to $12 - 8$, which also equals 4. What the two expressions share in common is that there is one number (12) that contributes to the total in a positive fashion and a second number (8) that contributes to the total in a negative fashion.

Totaling a Positive Number and a Negative Number

1. Take the absolute value of each number, and find the difference between these two absolute values. (This is the difference between the two numbers' contributions to the total.)
2. The sign of the result is the same as the sign of the number that has the largest absolute value.

EXAMPLE 2   Find the sum $3 + (-11)$.

Solution

We can rewrite $3 + (-11)$ as $3 - 11$. Again we have one number (3) contributing to the total in a positive way and a second number (11) contributing in a negative way. The difference between their contributions is 8, and since the number making the larger contribution is negative, our result is -8.

Quick Check 2
Find the sum $4 + (-17)$.

$$3 + (-11) = -8$$

Note that $-11 + 3$ also equals -8. The rules for totaling a positive integer and a negative integer still apply when the first number is negative and the second number is positive.

Totaling Two (or More) Negative Numbers

1. Total the negative contributions of each number.
2. The sign of the result is negative.

EXAMPLE 3   Find the sum $-3 + (-7)$.

Solution

We can begin by rewriting $-3 + (-7)$ as $-3-7$. Both values contribute to the total in a negative fashion. Totaling the negative contributions of 3 and 7 gives us 10, and our result is negative because both of our numbers were negative.

Quick Check 3
Find the sum $-2 + (-9)$.

$$-3 + (-7) = -10$$

Objective 2   **Subtract integers.**   To subtract a negative integer from another integer, we will use the following property:

Subtraction of Real Numbers

For any real numbers a and b, $a - b = a + (-b)$.

This property tells us that we can add the opposite of b to a rather than subtracting b from a. Suppose that we were subtracting a negative integer, such as in the example $-8 - (-19)$. The property for subtraction of real numbers tells us that subtracting -19 is the same as adding its opposite (19), so we would then convert this subtraction to $-8 + 19$. Remember that subtracting a negative number is equivalent to adding a positive number.

EXAMPLE 4   Subtract $6 - (-27)$.

Solution

Quick Check 4
Subtract $11 - (-7)$.

$$6 - (-27) = 6 + 27 \qquad \text{Subtracting } -27 \text{ is the same as adding } 27.$$
$$= 33 \qquad \text{Add.}$$

General Strategy for Adding/Subtracting Integers

- Rewrite "double signs." *Adding a negative number* $(4 + (-5))$ *can be rewritten as subtracting a positive number* $(4-5)$. *Subtracting a negative number* $(-2 - (-7))$ *can be rewritten as adding a positive number* $(-2 + 7)$.
- Look at each integer and determine whether it is contributing positively or negatively to the total.
- Add up any integers contributing positively to the total, if there are any, resulting in a single positive integer. In a similar fashion, add up all integers that are contributing to the total negatively, resulting in a single negative integer. Finish by finding the sum of these two integers.

Rather than being told to add or subtract, we may be asked to simplify a numerical expression. To **simplify** an expression, perform all arithmetic operations.

EXAMPLE 5   Simplify $17 - (-11) - 6 + (-13) - (-21) + 3$.

Solution

We begin by working on the *double signs*. This produces the following:

$$17 - (-11) - 6 + (-13) - (-21) + 3$$
$$= 17 + 11 - 6 - 13 + 21 + 3$$
$$= 52 - 19$$

Rewrite double signs.
The four integers that contribute in a positive fashion (17, 11, 21, and 3) have a total of 52. The two integers that contribute in a negative fashion have a total of -19.

$$= 33$$

Subtract.

Quick Check   **5** Simplify $14 - 9 - (-22) - 6 + (-30) + 5$.

Using Your Calculator   When using your calculator, it is important to be able to distinguish between the subtraction key and the key for a negative number. On the TI-83/84, the subtraction key ⊟ is listed above the addition key on the right side of the calculator, while the negative key ⊡ is located to the left of the ENTER key at the bottom of the calculator. Here are two different ways to simplify the expression from the previous example on the TI-83/84:

```
17-(-11)-6+(-13)
-(-21)+3
```

```
17+11-6-13+21+3
```

In either case, pressing the ENTER key produces the result 33.

Multiplication and Division of Integers

Objective 3   Multiply integers. The result obtained when multiplying two numbers is called the **product** of the two numbers. The numbers that are multiplied are called **factors**. When we multiply two positive integers, their product is also a positive integer. For example, the product of the two positive integers 4 and 7 is the positive integer 28. This can be written as $4 \cdot 7 = 28$. The product $4 \cdot 7$ can also be written as $4(7)$ or $(4)(7)$.

The product $4 \cdot 7$ is another way to represent the repeated addition of 7 four times.

$$4 \cdot 7 = 7 + 7 + 7 + 7$$
$$= 28$$

We will use this concept to show that the product of a positive integer and a negative integer is a negative integer. Suppose we want to multiply 4 by -7. We can rewrite this as -7 being added four times or $(-7) + (-7) + (-7) + (-7)$. From our work earlier in this section we know that this total is -28, so $4(-7) = -28$. Anytime we multiply a positive integer by a negative integer, the result is negative. The same is true when we multiply a negative integer by a positive integer. So $(-7)(4)$ is also equal to -28.

Products of Integers

$$(\text{Positive}) \cdot (\text{Negative}) = \text{Negative}$$
$$(\text{Negative}) \cdot (\text{Positive}) = \text{Negative}$$

EXAMPLE 6   Multiply $5(-8)$.

Solution

We begin by multiplying 5 and 8, which equals 40. The next step is to determine the sign of our result. Whenever we multiply a positive integer by a negative integer the result is negative.

$$5(-8) = -40$$

Quick Check 6
Multiply $10(-6)$.

A Word of Caution   Note the difference between $5 - 8$ (a subtraction) and $5(-8)$ (a multiplication). A set of parentheses *without* a sign in front of them is used to imply multiplication.

Product	Result
$(3)(-5)$	-15
$(2)(-5)$	-10
$(1)(-5)$	-5
$(0)(-5)$	0
$(-1)(-5)$	?
$(-2)(-5)$	?

The product of two negative integers is a positive integer. Let's try to understand why this is true by considering the example $(-2)(-5)$. Examine the table to the left, which shows the products of some integers and -5.

Notice the pattern in the table. Each time the integer multiplied by -5 decreases by 1, the product increases by 5. As we go from $0(-5)$ to $(-1)(-5)$, the product should increase by 5. So $(-1)(-5) = 5$ and by the same reasoning $(-2)(-5) = 10$.

Product of Two Negative Integers

$$(\text{Negative}) \cdot (\text{Negative}) = \text{Positive}$$

EXAMPLE 7   Multiply $(-9)(-8)$.

Solution

We begin by multiplying 9 and 8, which equals 72. The next step is to determine the sign of our result. Whenever we multiply a negative integer by a negative integer the result is positive.

$$(-9)(-8) = 72.$$

Quick Check 7
Multiply $(-7)(-9)$.

Products of Integers

- If a product contains an *odd number* of negative factors, then the result is negative.
- If a product contains an *even number* of negative factors, then the result is positive.

The main idea behind this principle is that every two negative factors multiply to be positive.

EXAMPLE 8 Multiply $7(-2)(-5)(-3)$.

Solution

Since there are three negative factors, our product will be negative.

$$7(-2)(-5)(-3) = -210$$

Quick Check **8** Multiply $-4(-10)(5)(-2)$.

Using Your Calculator Here is what the screen should look like when using the TI-83/84 to multiply the expression in the previous example:

```
7(-2)(-5)(-3)
               -210
```

Notice that parentheses can be used to indicate multiplication without using the ⊠ key.

Before continuing on to division, we should mention multiplication by 0. Because any real number multiplied by 0 will be 0, we can state the multiplication property of 0.

Multiplication by 0

For any real number x,

$$0 \cdot x = 0$$
$$x \cdot 0 = 0$$

Objective 4 **Divide integers.** When dividing one number called the **dividend** by another number called the **divisor**, the result obtained is called the **quotient** of the two numbers:

The statement "6 divided by 3 is equal to 2" is true because the product of the quotient and the divisor, $2 \cdot 3$, is equal to the dividend 6.

$$6 \div 3 = 2 \longleftrightarrow 2 \cdot 3 = 6$$

When we divide two integers that have the same sign (both positive or both negative), the quotient is positive. When we divide two integers that have different signs (one negative, one positive), the quotient is negative. Note that this is consistent with our rules for multiplication.

EXAMPLE 9   Divide $(-54) \div (-6)$.

Solution

When we divide a negative number by another negative number, the result is positive.

$$(-54) \div (-6) = 9 \qquad \text{Note that } -6 \cdot 9 = -54.$$

EXAMPLE 10   Divide $(-33) \div 11$.

Solution

When we divide a negative number by a positive number, the result is negative.

$$(-33) \div 11 = -3$$

Quick Check 9
Divide $72 \div (-8)$.

Whenever 0 is divided by any integer (except 0), the quotient is 0. For example, $0 \div 16 = 0$. We can check that this quotient is correct by multiplying the quotient by the divisor. Since $0 \cdot 16 = 0$, the quotient is correct.

Division by Zero

Whenever an integer is divided by 0, the quotient is said to be **undefined**.

We use the word **undefined** to state that an operation cannot be performed or is meaningless. For example, $41 \div 0$ is undefined. Suppose that $41 \div 0 = a$ for some real number a. In that case, the product $a \cdot 0$ would be equal to 41. Since we know that the product of 0 and any real number is equal to 0, there does not exist such a number a.

EXERCISES *1.2*

Add.

1. $3 + (-5)$

2. $4 + (-2)$

3. $7 + (-25)$

4. $64 + (-105)$

5. $(-4) + 5$

6. $-9 + 2$

7. $-14 + 22$

8. $(-35) + 50$

9. $-5 + (-6)$

10. $-9 + (-9)$

Subtract.

11. $8 - 6$

12. $13 - 9$

13. $5 - 11$

14. $4 - 12$

15. $(-5) - 3$

16. $(-9) - 6$

17. $-7 - 14$

18. $-29 - 17$

19. $15 - (-12)$

20. $41 - (-35)$

Simplify.

21. $8 + 13 - 6$

22. $7 - 15 - 4$

23. $-9 + 7 - 4$

24. $-5 - 8 + 23$

25. $6 - (-16) + 5$

26. $18 - 21 - (-62)$

27. $4 + (-15) - 13 - (-25)$

28. $-13 + (-12) - (-1) - 29$

29. A mother with $30 in her purse paid $22 for her family to get into a movie. How much money did she have remaining?

30. A student had $60 in his checking account prior to writing an $85 check to the bookstore for books and supplies. What is his account's new balance?

31. The temperature at 6 A.M. in Fargo, North Dakota, was $-8°$ C. By 3 P.M., the temperature had risen by $12°$C. What was the temperature at 3 P.M.?

32. If a golfer completes a round at 3 strokes under par, her score is denoted -3. A professional golfer had rounds of -4, -2, 3, and -6 in a recent tournament. What was her total score for this tournament?

33. Dylan drove from a town located 400 feet below sea level to another town located 1750 feet above sea level. What was the change in elevation traveling from one town to another?

34. After withdrawing $80 from her bank using an ATM card, Alycia had $374 remaining in her savings account. How much money did Alycia have in the account prior to withdrawing the money?

Multiply.

35. $8(-6)$

36. $-3(5)$

37. $-4 \cdot 9$

38. $-7(-8)$

39. $-11(-13)$

40. $-6 \cdot 15$

41. $82(-1)$

42. $-1 \cdot 19$

43. $-6 \cdot 0$

44. $0(-240)$

45. $-6(-3)(5)$

46. $-2(-4)(-8)$

47. $5 \cdot 3(-2)(-6)$

48. $-7 \cdot 2(-7)(-2)$

Divide, if possible.

49. $36 \div (-6)$

50. $54 \div (-9)$

51. $-63 \div 7$

52. $-48 \div 12$

53. $-21 \div (-3)$

54. $-100 \div (-5)$

55. $126 \div (-9)$

56. $-420 \div 14$

57. $0 \div (-13)$

58. $0 \div 11$

59. $29 \div 0$

60. $-15 \div 0$

Simplify.

61. $-11(-12)$

62. $126 \div (-6)$

63. $5 - 13$

64. $5(-13)$

65. $17 - (-11) - 49$

66. $8(-7)(-6)$

67. $-432 \div 3$

68. $-5 \cdot 17$

69. $9(-24)$

70. $9 + (-24)$

71. $5 \cdot 3(-17)(-29)(0)$

72. $-16 + (-11) - 42 - (-58)$

73. Kevin owns 8 baseball cards that are each worth $12. What is the total worth of these cards?

74. An Internet startup company had an operating loss of $10,000 last month. If the four partners decide to divide up the loss equally among themselves, what is the loss for each person?

75. Tina owns 400 shares of a stock that dropped in value by $3 per share last month. She also owns 500 shares of a stock that went up by $2 per share last month. What is Tina's net income on these two stocks for last month?

76. Mario took over as the CEO for a company that lost $20 million dollars in the year 2000. The company lost three times as much in 2001. The company went on to lose $13 million more in 2002 than they had lost in 2001. How much money did Mario's company lose in 2002?

True or False (If false, give an example that shows why the statement is false.)

77. The sum of two integers is always an integer.

78. The difference of two integers is always an integer.

79. The sum of two whole numbers is always a whole number.

80. The difference of two whole numbers is always a whole number.

Explain each of the following in your own words.

81. Explain why subtracting a negative integer from another integer is the same as adding the opposite of that integer to it. Use the example $11 - (-5)$ in your explanation.

82. Explain why a positive integer times a negative integer produces a negative integer.

83. Explain why a negative integer times another negative integer produces a positive integer.

84. Explain why division by 0 is undefined.

85. When a certain integer is added to -11, the result is 15. What is that integer?

86. Fourteen less than a certain integer is -23. What is that integer?

87. When a certain integer is divided by -7, the result is 12. What is that integer?

88. When a certain integer is multiplied by -3, the result is -417. What is that integer?

Study Tip REVISITED   Once you have formed a study group, you must determine where and when to meet. It is a good idea to meet at least twice a week, for at least an hour per session. Consider a location where quiet discussion is allowed, such as the library or tutorial center.

Clearly, the group must meet at times when each person is available. Some groups like to meet during the hour before class, using the study group as a way to prepare for class. Other groups prefer to meet during the hour after class, allowing them to go over material while it is fresh in their minds. Another suggestion is to meet at a time when your instructor is holding office hours. This way, if the entire group is struggling with the same question, one or more members can visit the instructor for assistance.

1.3
FRACTIONS

Objectives

1 **Find the factor set of a natural number.**
2 **Determine whether a natural number is prime.**
3 **Find the prime factorization of a natural number.**
4 **Simplify a fraction to lowest terms.**
5 **Change an improper fraction to a mixed number.**
6 **Change a mixed number to an improper fraction.**

Factors

Objective   1   Find the factor set of a natural number. To factor a natural number, we express it as the product of two natural numbers. For example, one way to factor 12 is to rewrite it as $3 \cdot 4$. In this example, 3 and 4 are said to be factors of 12. The collection of all factors of a natural number is called its **factor set**. The factor set of 12 can be written as $\{1, 2, 3, 4, 6, 12\}$, since $1 \cdot 12 = 12, 2 \cdot 6 = 12$ and $3 \cdot 4 = 12$.

EXAMPLE   1   Write the factor set for 18.

Solution

Quick Check **1**
Write the factor set for 36.

Since 18 can be factored as $1 \cdot 18, 2 \cdot 9$ and $3 \cdot 6$, its factor set is $\{1, 2, 3, 6, 9, 18\}$.

Some numbers, such as 21, have only a few factors. The only factors of 21 other than 1 and 21 are 3 and 7, since $3 \cdot 7 = 21$. The factor set of 21 is $\{1, 3, 7, 21\}$. Other numbers, such as 13, have only two factors. The only factors of 13 are 1 and 13 itself, so the factor set of 13 is $\{1, 13\}$.

Prime Numbers

Objective   2   Determine whether a natural number is prime.

Prime Numbers

A natural number is **prime** if it is greater than 1 and its only two factors are 1 and itself.

For instance, the number 13 is prime because its only two factors are 1 and 13. The first 10 prime numbers are 2, 3, 5, 7, 11, 13, 17, 19, 23, and 29. The number 8 is not prime because it has factors other than 1 and 8, namely, 2 and 4. A natural number greater than 1 that is not prime is called a **composite** number. The number 1 is considered to be neither prime nor composite.

EXAMPLE 2   Determine whether the following numbers are prime or composite:
a) 26   **b)** 37

Solution

a) The factor set for 26 is {1, 2, 13, 26}. Since 26 has factors other than 1 and itself, it is a composite number.
b) Since it has no factors other than 1 and 37 itself, 37 is a prime number.

Quick Check 2 Determine whether the following numbers are prime or composite.

a) 57   **b)** 47   **c)** 48

Prime Factorization

Objective 3 **Find the prime factorization of a natural number.** When we rewrite a natural number as a product of prime factors, we obtain the **prime factorization** of the number. The prime factorization of 12 is $2 \cdot 2 \cdot 3$, because 2 and 3 are prime numbers and $2 \cdot 2 \cdot 3 = 12$. A **factor tree** is a useful tool for finding the prime factorization of a number. Here is an example of a factor tree for 72.

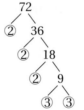

$72 = 2 \cdot 36$, 2 is prime.

$36 = 2 \cdot 18$, 2 is prime.

$18 = 2 \cdot 9$, 2 is prime.

$9 = 3 \cdot 3$, both factors are prime.

It should be noted that we could have begun by rewriting 72 as $8 \cdot 9$, and then factored these two numbers. The process for creating a factor tree for a natural number is not unique, although the prime factorization for the number is unique.

EXAMPLE 3   Find the prime factorization for 51.

Solution

Quick Check 3
Find the prime factorization for 63.

This number has only two factors: 3 and 17. The prime factorization is $3 \cdot 17$.

Fractions

Objective 4 **Simplify a fraction to lowest terms.**

Rational Numbers

A **rational number** is a real number that can be written as the quotient (or ratio) of two integers, the second of which is not zero. An **irrational number** is a real number that cannot be written in this way, such as the number π.

Rational numbers are often expressed using fraction notation such as $\frac{3}{7}$. Whole numbers, such as 7, can be written as a fraction whose denominator is 1, such as $\frac{7}{1}$. The number on the top of the fraction is called the **numerator**, and the number on the bottom of the fraction is called the **denominator**.

$$\frac{\text{numerator}}{\text{denominator}}$$

If the numerator and denominator do not have any common factors other than 1, then the fraction is said to be in **lowest terms**.

To simplify a fraction to lowest terms, we can begin by finding the prime factorization of both the numerator and denominator. Then, divide both the numerator and denominator by their common factors.

EXAMPLE 4   Simplify $\frac{18}{30}$ to lowest terms.

Solution

$$\frac{18}{30} = \frac{2 \cdot 3 \cdot 3}{2 \cdot 3 \cdot 5}$$ Find the prime factorization of the numerator and denominator. $18 = 2 \cdot 3 \cdot 3$, $30 = 2 \cdot 3 \cdot 5$.

$$= \frac{\overset{1}{\cancel{2}} \cdot \overset{1}{\cancel{3}} \cdot 3}{\underset{1}{\cancel{2}} \cdot \underset{1}{\cancel{3}} \cdot 5}$$ Divide out common factors.

$$= \frac{3}{5}$$ Simplify.

EXAMPLE 5   Simplify $\frac{4}{24}$ to lowest terms.

Solution

$$\frac{4}{24} = \frac{2 \cdot 2}{2 \cdot 2 \cdot 2 \cdot 3}$$ Find the prime factorization of the numerator and denominator.

$$\frac{\overset{1}{\cancel{2}} \cdot \overset{1}{\cancel{2}}}{\underset{1}{\cancel{2}} \cdot \underset{1}{\cancel{2}} \cdot 2 \cdot 3}$$ Divide out common factors.

$$= \frac{1}{6}$$ Simplify.

Quick Check 4
Simplify to lowest terms:

a) $\frac{45}{210}$ b) $\frac{24}{384}$

A Word of Caution   It is customary to leave off the denominator of a fraction if it is equal to 1. For example, rather than writing $\frac{19}{1}$, we usually write 19. However, we cannot omit a numerator that is equal to 1. In the previous example, it would have been a mistake if we had written 6 instead of $\frac{1}{6}$.

Mixed Numbers and Improper Fractions

Objective 5   **Change an improper fraction to a mixed number.** An **improper fraction** is a fraction whose numerator is greater than or equal to its denominator, such as $\frac{7}{4}, \frac{400}{150}, \frac{8}{8}$, and $\frac{35}{7}$. An improper fraction is often converted to a **mixed number**, which is the sum of a whole number and a fraction. For example, the improper fraction $\frac{14}{3}$ can be represented by the mixed number $4\frac{2}{3}$, which is equivalent to $4 + \frac{2}{3}$.

To convert an improper fraction to a mixed number, begin by dividing the denominator into the numerator. The quotient is the whole number portion of the mixed number, and the remainder becomes the numerator of the fractional part.

$$
\begin{array}{r}
\text{Denominator} \rightarrow 3\overline{)14} \quad \begin{array}{l} \leftarrow \text{Whole-number portion} \\ \text{of mixed number} \end{array}\\
\underline{12} \quad \begin{array}{l} \leftarrow \text{Numerator of fractional} \\ \text{portion of mixed number} \end{array}\\
2
\end{array}
$$

EXAMPLE 6 Convert the improper fraction $\frac{71}{9}$ to a mixed number.

Solution

We begin by dividing 9 into 71, which divides in 7 times with a remainder of 8.

$$
\begin{array}{r}
7\\
9\overline{)71}\\
\underline{63}\\
8
\end{array}
$$

Quick Check 5

Convert the improper fraction $\frac{121}{13}$ to a mixed number.

The mixed number for $\frac{71}{9}$ is $7\frac{8}{9}$.

Objective 6 **Change a mixed number to an improper fraction.** Often we have to convert a mixed number such as $2\frac{7}{15}$ into an improper fraction before proceeding with arithmetic operations.

Rewriting a Mixed Number as an Improper Fraction

- Multiply the whole number part of the mixed number by the denominator of the fractional part of the mixed number.
- Add this product to the numerator of the fractional part of the mixed number.
- The sum is the numerator of the improper fraction. The denominator stays the same.

$$
2\frac{7}{15} \quad \begin{array}{l} \leftarrow \text{Add product to} \\ \text{numerator} \end{array} \qquad\qquad 2\frac{7}{15} = \frac{37}{15}
$$
$$
\text{Multiply} \nearrow
$$

EXAMPLE 7 Convert the mixed number $5\frac{4}{7}$ to an improper fraction.

Solution

Quick Check 6

Convert the mixed number $8\frac{1}{6}$ to an improper fraction.

We begin by multiplying $5 \cdot 7 = 35$. We add this product to 4 to produce a numerator of 39.

$$
5\frac{4}{7} = \frac{39}{7}
$$

1. Is 7 a factor of 247?

2. Is 13 a factor of 273?

3. Is 6 a factor of 4836?

4. Is 9 a factor of 32,057?

5. Is 15 a factor of 2835?

6. Is 103 a factor of 1754?

Write the factor set for the following numbers.

7. 27   8. 15

9. 20   10. 16

11. 31   12. 48

13. 60

14. 21   15. 49

16. 25   17. 69

18. 103

19. 221

20. 209

Write the prime factorization of the following numbers. (If the number is prime, state this.)

21. 18   22. 12   23. 42

24. 28   25. 33   26. 50

27. 27   28. 32

29. 125   30. 49   31. 29

32. 38   33. 99   34. 90

35. 31   36. 91   37. 119

38. 53   39. 360

40. 126   41. 104

42. 109

Simplify the following fractions to lowest terms.

43. $\dfrac{10}{16}$ 44. $\dfrac{30}{36}$ 45. $\dfrac{9}{45}$

46. $\dfrac{38}{2}$ 47. $\dfrac{126}{294}$ 48. $\dfrac{60}{84}$

49. $\dfrac{27}{64}$ 50. $\dfrac{66}{154}$ 51. $\dfrac{160}{176}$

52. $\dfrac{56}{45}$

Convert the following mixed numbers to improper fractions.

53. $3\dfrac{4}{5}$ 54. $7\dfrac{2}{9}$

55. $2\dfrac{16}{17}$ 56. $6\dfrac{5}{14}$

57. $22\dfrac{13}{19}$ 58. $37\dfrac{6}{25}$

Convert the following improper fractions to whole numbers or mixed numbers.

59. $\dfrac{39}{5}$ 60. $\dfrac{56}{8}$

61. $\dfrac{101}{7}$ 62. $\dfrac{12}{4}$

63. $\dfrac{61}{11}$ 64. $\dfrac{85}{13}$

Answer in complete sentences.

65. Describe an example of a real-world situation involving fractions. Describe an example of a real-world situation involving mixed numbers.

66. Describe an example of a situation in which you should convert an improper fraction to a mixed number.

> **Study Tip** REVISITED
>
> - Some study groups prefer to meet off campus in the evening. One good place to meet would be at a coffee shop with tables large enough for everyone to work on, provided that the noise is not too distracting. The food court at a local mall would work as well.
> - Some groups take advantage of study rooms at public libraries.
> - Other groups like to meet at each other's homes. This typically provides a comfortable, relaxing atmosphere to work in. If your group is invited to meet at someone's house, try to be a perfect guest. Help to set up and clean up if you can, or bring beverages and healthy snacks.

Objectives

1 Multiply fractions and mixed numbers.
2 Divide fractions and mixed numbers.
3 Add and subtract fractions and mixed numbers with the same denominator.
4 Find the least common multiple (LCM) of two natural numbers.
5 Add and subtract fractions and mixed numbers with different denominators.

Multiplying Fractions

Objective 1 **Multiply fractions and mixed numbers.** To multiply fractions, we multiply the numerators together and multiply the denominators together. When multiplying fractions, we may simplify any individual fraction, as well as dividing a common factor from a numerator and a different denominator. Dividing out a common factor in this fashion is often referred to as **cross-canceling**.

EXAMPLE 1   Multiply $\frac{4}{11} \cdot \frac{5}{6}$.

Solution

The first numerator (4) and the second denominator (6) have a common factor of 2 that we can eliminate through division.

$$\frac{4}{11} \cdot \frac{5}{6} = \frac{\overset{2}{\cancel{4}}}{11} \cdot \frac{5}{\underset{3}{\cancel{6}}} \qquad \text{Divide out the common factor 2.}$$

$$= \frac{2}{11} \cdot \frac{5}{3} \qquad \text{Simplify.}$$

$$= \frac{10}{33} \qquad \text{Multiply the two numerators and the two denominators.}$$

Quick Check 1
Multiply $\frac{10}{63} \cdot \frac{9}{16}$.

EXAMPLE 2   Multiply $3\frac{1}{7} \cdot \frac{14}{55}$.

Solution

When multiplying a mixed number by another number, we should convert the mixed number to an improper fraction before proceeding.

$$3\frac{1}{7} \cdot \frac{14}{55} = \frac{22}{7} \cdot \frac{14}{55} \qquad \text{Convert } 3\frac{1}{7} \text{ to the improper fraction } \frac{22}{7}.$$

$$= \frac{\overset{2}{\cancel{22}}}{\underset{1}{\cancel{7}}} \cdot \frac{\overset{2}{\cancel{14}}}{\underset{5}{\cancel{55}}} \qquad \text{Divide out common factors of 11 and 7.}$$

Quick Check 2
Multiply $2\frac{2}{3} \cdot 8\frac{5}{8}$.

$$= \frac{4}{5} \qquad \text{Multiply.}$$

Dividing Fractions

Objective 2 **Divide fractions and mixed numbers.**

Reciprocal

When we invert a fraction, such as $\frac{3}{5}$ to $\frac{5}{3}$, the resulting fraction is called the **reciprocal** of the original fraction.

Consider the fraction $\frac{a}{b}$, where a and b are nonzero real numbers. The reciprocal of this fraction is $\frac{b}{a}$. Take note that if we multiply a fraction by its reciprocal, such as $\frac{a}{b} \cdot \frac{b}{a}$, then the result is 1. This property will be important in Chapter 2.

Reciprocal Property

For any nonzero real numbers a and b, $\frac{a}{b} \cdot \frac{b}{a} = 1$.

To divide a fraction by another fraction, we invert the divisor and then multiply the resulting fractions.

EXAMPLE 3 Divide $\frac{16}{25} \div \frac{22}{15}$.

Solution

$$\frac{16}{25} \div \frac{22}{15} = \frac{16}{25} \cdot \frac{15}{22} \qquad \text{Invert the divisor and multiply.}$$

$$= \frac{\overset{8}{\cancel{16}}}{\underset{5}{\cancel{25}}} \cdot \frac{\overset{3}{\cancel{15}}}{\underset{11}{\cancel{22}}} \qquad \text{Divide out the common factors 2 and 5.}$$

$$= \frac{24}{55} \qquad \text{Multiply.}$$

***Quick Check* 3**
Divide $\frac{12}{25} \div \frac{63}{10}$.

A Word of Caution When dividing a fraction by another fraction, we must invert the divisor before dividing out a common factor from a numerator and a denominator.

Adding and Subtracting Fractions

Objective 3 **Add and subtract fractions and mixed numbers with the same denominator.** To add or subtract fractions that have the same denominator, we add or subtract the numerators, placing the result over the common denominator. Be sure to simplify this fraction to lowest terms.

EXAMPLE 4 Add $\frac{1}{12} + \frac{5}{12}$.

Solution

$$\frac{1}{12} + \frac{5}{12} = \frac{6}{12} \qquad \text{Since the denominators are the same, add the numerators, place the sum over 12}$$

Quick Check 4
Add $\frac{7}{16} + \frac{3}{16}$.

$$= \frac{1}{2} \qquad \text{Simplify to lowest terms.}$$

EXAMPLE 5   Subtract $\frac{3}{8} - \frac{9}{8}$.

Solution

The two denominators are both the same (8), so we subtract the numerators. When we subtract $3 - 9$ the result is -6. Although we may leave the negative sign in the numerator, it is often brought out in front of the fraction itself.

$$\frac{3}{8} - \frac{9}{8} = -\frac{6}{8} \qquad \text{Subtract the numerators.}$$

$$= -\frac{3}{4} \qquad \text{Simplify to lowest terms.}$$

Quick Check 5
Subtract $\frac{17}{20} - \frac{5}{20}$.

Objective 4 Find the least common multiple (LCM) of two natural numbers.
Two fractions are said to be **equivalent fractions** if they have the same numerical value and can both be simplified to be the same fraction when simplifying to lowest terms. In order to add or subtract two fractions with different denominators, we must first convert them to equivalent fractions with the same denominator. To do this, we find the **least common multiple (LCM)** of the two denominators. This is the smallest number that is a multiple of both denominators. For example, the LCM of 4 and 6 is 12 because 12 is the smallest multiple of both 4 and 6.

To find the LCM for two numbers, we can begin by factoring them into their prime factorizations.

Finding the LCM of 2 or More Numbers

- Find the prime factorization of each number.
- Find the common factors of the numbers.
- Multiply the common factors by the remaining factors of the numbers.

EXAMPLE 6   Find the LCM of 24 and 30.

Solution

We begin with the prime factorizations of 24 and 30.

$$24 = 2 \cdot 2 \cdot 2 \cdot 3$$
$$30 = 2 \cdot 3 \cdot 5$$

The factors that the two numbers share in common are a 2 and a 3.

$$24 = ②\cdot 2 \cdot 2 \cdot ③$$
$$30 = ②\cdot③\cdot 5$$

Quick Check 6
Find the least common multiple of 18 and 42.

Additional factors are a pair of 2's, as well as a 5. So to find the LCM, we multiply the common factors (2 and 3) by the additional factors (2, 2, and 5).

$$2 \cdot 3 \cdot 2 \cdot 2 \cdot 5 = 120$$

The least common multiple of 24 and 30 is 120.

Another technique for finding the LCM for two numbers is to start listing the multiples of the larger number until we find a multiple that is also a multiple of the smaller number. For example, the first few multiples of 6 are

$$6: 6, 12, 18, 24, 30, \ldots$$

The first multiple listed that is also a multiple of 4 is 12, so the LCM of 4 and 6 is 12.

Objective 5 Add and subtract fractions and mixed numbers with different denominators. When adding or subtracting two fractions that do not have the same denominator, we first find a common denominator by finding the LCM of the two denominators. We then convert each fraction to an equivalent fraction whose denominator is that common denominator. Once we rewrite the two fractions so they have the same denominator, we can then add (or subtract) as done previously in this section.

Adding or Subtracting Fractions with Different Denominators

- Find the LCM of the denominators.
- Rewrite each fraction as an equivalent fraction whose denominator is the LCM of the original denominators.
- Add or subtract the numerators, placing the result over the common denominator.
- Simplify to lowest terms, if possible.

EXAMPLE 7   Add $\frac{5}{12} + \frac{9}{14}$.

Solution

The prime factorization of 12 is $2 \cdot 2 \cdot 3$, and the prime factorization of 14 is $2 \cdot 7$. The two denominators have a common factor of 2. If we multiply this common factor by the other factors of these two numbers, 2, 3, and 7, we see that the LCM of these two denominators is 84. We begin by rewriting each fraction as a fraction whose denominator is 84. We will multiply the first fraction by $\frac{7}{7}$ and the second fraction by $\frac{6}{6}$. Since $\frac{7}{7}$ and $\frac{6}{6}$ are both equal to 1, we are not changing the value of either fraction.

$$\frac{5}{12} + \frac{9}{14} = \frac{5}{12} \cdot \frac{7}{7} + \frac{9}{14} \cdot \frac{6}{6}$$

Multiply the first fraction's numerator and denominator by 7. Multiply the second fraction's numerator and denominator by 6.

$$= \frac{35}{84} + \frac{54}{84}$$

Multiply.

$$= \frac{89}{84}$$

Add.

Quick Check 7   This fraction is already in lowest terms.

Add $\frac{4}{9} + \frac{2}{15}$.

EXAMPLE 8   Simplify $\frac{5}{6} - \frac{3}{8} + \frac{2}{9}$.

Solution

Here is the prime factorization for each denominator:

$$6 = 2 \cdot 3 \qquad 8 = 2 \cdot 2 \cdot 2 \qquad 9 = 3 \cdot 3$$

The LCM for these three denominators is 72. We now begin by converting each fraction to an equivalent fraction whose denominator is 72.

$$\frac{5}{6} - \frac{3}{8} + \frac{2}{9} = \frac{5}{6} \cdot \frac{12}{12} - \frac{3}{8} \cdot \frac{9}{9} + \frac{2}{9} \cdot \frac{8}{8}$$

Multiply the first fraction by $\frac{12}{12}$, the second fraction by $\frac{9}{9}$, and the third fraction by $\frac{8}{8}$.

$$= \frac{60}{72} - \frac{27}{72} + \frac{16}{72}$$

Multiply.

$$= \frac{49}{72}$$

Simplify. $60 - 27 + 16 = 49$.

Quick Check **8**

Simplify $\frac{3}{4} + \frac{2}{3} - \frac{7}{12}$.

This fraction is in lowest terms.

EXERCISES *1.4*

Multiply. Your answer should be in lowest terms.

1. $\frac{5}{8} \cdot \frac{6}{25}$ **2.** $\frac{3}{4} \cdot \frac{18}{23}$

3. $\frac{9}{10} \cdot \frac{14}{15}$ **4.** $\frac{21}{10} \cdot \frac{30}{49}$

5. $4\frac{2}{7} \cdot \frac{14}{25}$ **6.** $3\frac{5}{9} \cdot 2\frac{1}{6}$

7. $5 \cdot 6\frac{3}{10}$ **8.** $8 \cdot \frac{7}{12}$

9. $\frac{2}{3} \cdot \frac{8}{9}$ **10.** $\frac{12}{35} \cdot \frac{14}{99}$

Divide. Your answer should be in lowest terms.

11. $\frac{3}{4} \div \frac{9}{16}$ **12.** $\frac{2}{5} \div \frac{7}{15}$

13. $\frac{1}{9} \div \frac{4}{3}$ **14.** $\frac{5}{6} \div \frac{13}{14}$

15. $\frac{17}{40} \div \frac{1}{2}$ **16.** $\frac{7}{30} \div \frac{1}{5}$

17. $\frac{4}{11} \div 3\frac{1}{5}$ **18.** $3\frac{3}{8} \div 6$

19. $2\frac{4}{5} \div 6\frac{2}{3}$ **20.** $7\frac{1}{9} \div 13\frac{3}{8}$

Add or subtract.

21. $\frac{7}{15} + \frac{4}{15}$ **22.** $\frac{1}{8} + \frac{5}{8}$

23. $\frac{5}{9} + \frac{4}{9}$ **24.** $\frac{3}{10} + \frac{9}{10}$

25. $\frac{4}{5} - \frac{2}{5}$ **26.** $\frac{17}{18} - \frac{5}{18}$

27. $\frac{9}{16} - \frac{3}{16}$ **28.** $\frac{13}{42} - \frac{29}{42}$

Find the LCM of the given numbers.

29. 12, 18

30. 6, 8

31. 10, 14

32. 9, 15

33. 9, 36

34. 9, 10

35. 4, 6, 9

36. 8, 10, 12

Add. Your answer should be in lowest terms.

37. $\frac{1}{2} + \frac{1}{3}$ **38.** $\frac{2}{3} + \frac{1}{5}$

39. $\frac{4}{5} + \frac{3}{4}$ **40.** $\frac{4}{7} + \frac{1}{4}$

41. $\frac{7}{10} + \frac{5}{8}$ **42.** $\frac{3}{4} + \frac{5}{6}$

43. $\frac{21}{63} + \frac{1}{5}$ **44.** $\frac{1}{4} + \frac{60}{80}$

45. $6\frac{1}{5} + 5$ **46.** $3 + 8\frac{3}{7}$

47. $6\frac{2}{3} + 5\frac{1}{6}$ **48.** $11\frac{4}{9} + 5\frac{1}{3}$

Subtract. Your answer should be in lowest terms.

49. $\frac{5}{6} - \frac{1}{3}$ **50.** $\frac{7}{12} - \frac{3}{4}$

51. $\frac{2}{3} - \frac{7}{15}$ **52.** $\frac{1}{2} - \frac{7}{9}$

53. $\frac{3}{4} - \frac{2}{7}$ **54.** $\frac{5}{8} - \frac{5}{6}$

55. $\frac{11}{20} - \frac{5}{12}$ **56.** $\frac{5}{7} - \frac{3}{14}$

57. $7\frac{1}{2} - 3\frac{1}{4}$ **58.** $12\frac{2}{3} - 6\frac{2}{5}$

59. $12\frac{3}{10} - 9$ **60.** $6 - 4\frac{3}{4}$

Simplify.

61. $\frac{8}{9} \cdot \frac{3}{5}$ **62.** $\frac{19}{20} - \frac{7}{20}$

63. $\frac{5}{12} + \frac{11}{12}$ **64.** $\frac{3}{4} + \frac{7}{10}$

65. $\frac{7}{30} \div \frac{35}{48}$ **66.** $\frac{12}{35} \cdot \frac{14}{27}$

67. $\frac{1}{6} - \frac{7}{8}$ **68.** $\frac{7}{24} - \frac{29}{40}$

69. $\frac{19}{30} + \frac{11}{18}$ **70.** $3\frac{1}{5} \cdot 4\frac{3}{8}$

71. $13 \div \frac{1}{8}$ **72.** $\frac{3}{5} - \frac{2}{3} - \frac{7}{10}$

73. $3\frac{4}{7} + 6\frac{3}{5} - 8$ **74.** $12\frac{1}{3} + 7\frac{1}{6} - 5\frac{1}{2}$

Find the missing number.

75. $\frac{11}{24} + \frac{5}{?} = \frac{13}{12}$ **76.** $\frac{?}{10} - \frac{1}{3} = \frac{1}{6}$

77. $\frac{10}{21} \cdot \frac{?}{75} = \frac{4}{45}$ **78.** $\frac{11}{40} + \frac{?}{40} = \frac{9}{10}$

79. Bruce is fixing a special dinner for his girlfriend. The three recipes he is preparing call for $\frac{1}{2}$ cup, $\frac{3}{4}$ cup, and $\frac{1}{3}$ cup of olive oil, respectively. In total, how much olive oil does Bruce need to make these three recipes?

80. Sue gave birth to twins. One of the babies weighed $4\frac{7}{8}$ pounds at birth, and the other baby weighed $5\frac{1}{4}$ pounds. Find the total weight of the twins at birth.

81. A chemist has $\frac{23}{40}$ fluid ounces of a solution. If she needs $\frac{1}{8}$ fluid ounces of the solution for an experiment, how much of the solution will remain?

82. A popular weed spray concentrate recommends using $1\frac{1}{4}$ tablespoons of concentrate for each quart of water. How much concentrate needs to be mixed with 6 quarts of water?

83. A board that is $4\frac{1}{5}$ feet long needs to be cut into 6 pieces of equal length. How long will each piece be?

84. Ross makes a batch of hot sauce that will be poured into bottles that hold $5\frac{3}{4}$ fluid ounces. If Ross has 115 fluid ounces of hot sauce, how many bottles can he fill?

85. A craftsman is making a rectangular picture frame. Two of the sides will each be $\frac{5}{6}$ of a foot long, while the other two sides will each be $\frac{2}{3}$ of a foot long. If the craftsman has one board that is $2\frac{3}{4}$ feet long, is this enough to make the picture frame? Explain why or why not.

86. A pancake recipe calls for $1\frac{1}{3}$ cups of whole-wheat flour to make 12 pancakes. How much flour would be needed in order to make 48 pancakes?

87. A pancake recipe calls for $1\frac{1}{3}$ cups of whole-wheat flour to make 12 pancakes. How much flour would be needed to make 6 pancakes?

88. Explain, using your own words, the difference between dividing a number in half and dividing a number by one half.

Answer in complete sentences.

89. Explain why we cannot divide a common factor of 2 from the numbers 4 and 6 in the expression $\frac{4}{7} \cdot \frac{6}{11}$.

90. Explain why it would be a bad idea to rewrite fractions with a common denominator before multiplying them.

Study Tip **REVISITED**   A study group can go over homework assignments together. It is important that each group member work on the assignment before arriving at the study session. If you struggled with a problem, or couldn't do it at all, ask for help or suggestions from your group members.

If there was a problem that you seem to understand better than your group, be sure to share your knowledge; explaining how to do a certain problem increases your chances of retaining that knowledge until the exam and beyond.

At some point, try several even-numbered problems from the text that are representative of the problems on the homework assignment. Check your solutions with other members of your group.

At the end of each session, quickly review what the group has accomplished. Finish the meeting by planning your next session, including the time and location of the meeting and what you plan to work on at that meeting.

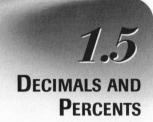

1.5

DECIMALS AND PERCENTS

Objectives

1 Perform arithmetic operations with decimals.
2 Rewrite a fraction as a decimal number.
3 Rewrite a decimal number as a fraction.
4 Rewrite a fraction as a percent.
5 Rewrite a decimal as a percent.
6 Rewrite a percent as a fraction.
7 Rewrite a percent as a decimal.

Decimals

Rational numbers can also be represented using **decimal notation**. The decimal 0.23 is equivalent to the fraction $\frac{23}{100}$ or twenty-three hundredths. The digit 2 is said to be in the tenths place, and the digit 3 is in the hundredths place. Here is a chart showing several place values for decimals:

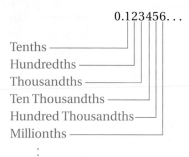

$$0.123456\ldots$$

Tenths ———
Hundredths ———
Thousandths ———
Ten Thousandths ———
Hundred Thousandths ———
Millionths ———
⋮

Objective 1 Perform arithmetic operations with decimals. Here is a brief summary of arithmetic operations using decimals:

- To add or subtract two decimal numbers, align the decimal points and add or subtract as you would with integers.

$$3.96 + 12.072$$

$$
\begin{array}{r}
3.96 \\
+\ 12.072 \\
\hline
16.032
\end{array}
$$

If the two decimals do not have the same number of decimal places, we can extend the number with fewer places by adding 0's to the end of that number. For instance, 3.96 can be written as 3.960 in the preceding example.

- To multiply two decimal numbers, multiply them as you would integers. The total number of decimal places in the two factors tell us how many decimal places are in the product.

$$-2.09 \cdot 3.1$$

In this example, the two factors have a total of three decimal places, so our product must have three decimal places. We multiply these two numbers as if they were 209 and 31 and then insert the decimal point in the appropriate place, leaving three digits to the right of the decimal point.

$$
\begin{array}{cc}
-209 & -2.09 \\
\underline{\times 31} & \underline{\times 3.1} \\
-6479 & -6.479
\end{array}
$$

3 places

- To divide two decimal numbers, we move the decimal point in the divisor to the right so that it becomes an integer. We must then move the decimal point in the other number (dividend) to the right by the same number of spaces. The decimal point in the answer will be aligned with this new location of the decimal point in the dividend.

$8.24 \div 0.4$

$0.4\overline{)8.24}$ We begin by moving each decimal point one place to the right.

$0.4\overline{)8.24}$

$\begin{array}{r} 20.6 \\ 4\overline{)82.4} \end{array}$

Rewriting Fractions as Decimals and Decimals as Fractions

Objective 2 Rewrite a fraction as a decimal number. To rewrite any fraction as a decimal, we divide its numerator by its denominator. The fraction line is simply another way to write "÷".

EXAMPLE 1   Rewrite the fraction $\frac{5}{8}$ as a decimal.

Solution

To rewrite this fraction as a decimal, we will divide 5 by 8. Since 8 does not divide into 5, we will place a decimal point after the 5 and add a 0 after it.

$8\overline{)5.0}$

We can now begin the division, adding 0's to the end of the dividend until there is no remainder.

$$
\begin{array}{r}
0.625 \\
8\overline{)5.000} \\
\underline{-4\,8} \\
20 \\
\underline{-16} \\
40 \\
\underline{-40} \\
0
\end{array}
$$

Quick Check 1
Rewrite the fraction $\frac{3}{4}$ as a decimal.

$\dfrac{5}{8} = 0.625.$

EXAMPLE 2   Rewrite the fraction $\frac{23}{30}$ as a decimal.

Solution

When we divide 23 by 30, our result is a decimal that does not terminate (0.76666 . . .). The pattern continues repeating the digit 6 forever. This is an example of a **repeating decimal**. We may place a bar over the repeating digit(s) to denote a repeating decimal.

$$\frac{23}{30} = 0.7\overline{6}$$

Quick Check **2**
Rewrite the fraction $\frac{5}{18}$ as a decimal.

Objective 3   **Rewrite a decimal number as a fraction.** Suppose we wanted to rewrite a decimal number such as 0.48 as a fraction. This decimal is read as "forty-eight hundredths" and is equivalent to the fraction $\frac{48}{100}$. Simplifying this fraction shows us that $0.48 = \frac{12}{25}$.

EXAMPLE 3   Rewrite the decimal 0.8 as a fraction in lowest terms.

Solution

The decimal ends in the tenths place, so we will begin with a fraction whose denominator is 10.

$$0.8 = \frac{8}{10} \qquad \text{Remove the decimal point to find the numerator.}$$
$$= \frac{4}{5} \qquad \text{Simplify to lowest terms.}$$

Quick Check **3**
Rewrite the decimal 0.04 as a fraction in lowest terms.

Using Your Calculator   To rewrite a decimal as a fraction using the TI-83/84, we type the decimal, press the [MATH] key, and select option 1. The following screens show how to rewrite the decimal 0.8 as a fraction using the TI-83/84 as in Example 3:

EXAMPLE 4   Rewrite the decimal 0.164 as a fraction in lowest terms.

Solution

This decimal ends in the thousandths place, so we start with a fraction of $\frac{164}{1000}$.

Quick Check **4**
Rewrite the decimal 0.425 as a fraction in lowest terms.

$$0.164 = \frac{164}{1000} \qquad \text{Rewrite as a fraction whose denominator is 1000.}$$
$$= \frac{41}{250} \qquad \text{Simplify to lowest terms.}$$

Percents

Objective **4** **Rewrite a fraction as a percent.** **Percents** (%) are used to represent numbers as parts of one hundred. One percent, which can be written as 1%, is equivalent to 1 part of 100 or $\frac{1}{100}$ or 0.01. The fraction $\frac{27}{100}$ is equivalent to 27%. Percents, decimals, and fractions are all ways to write a rational number. We will begin by learning to convert back and forth between percents and fractions, as well as between percents and decimals.

Rewriting Fractions and Decimals as Percents

To rewrite a fraction or a decimal as a percent, we multiply it by 100%.

EXAMPLE ▸ **5**   Rewrite $\frac{2}{5}$ as a percent.

Solution

We begin by multiplying by 100% and simplifying.

Quick Check **5**
Rewrite $\frac{7}{10}$ as a percent.

$$\frac{2}{\cancel{5}} \cdot \cancel{100}\,\overset{20}{\%} = 40\%$$

When rewriting a fraction as a percent, it may be helpful to multiply by $\frac{100}{1}\%$.

EXAMPLE ▸ **6**   Rewrite $\frac{3}{8}$ as a percent.

Solution

Again, we multiply by 100%. Occasionally, as in this example, we will end up with an improper fraction, which can be changed to a mixed number.

$$\frac{3}{\cancel{8}} \cdot \cancel{100}\,\overset{25}{\%} = \frac{75}{2}\%, \text{ which can be rewritten as } 37\frac{1}{2}\%.$$

Quick Check **6**
Rewrite $\frac{21}{40}$ as a percent.

Objective **5** **Rewrite a decimal as a percent.**

EXAMPLE ▸ **7**   Rewrite 0.3 as a percent.

Solution

When we multiply a decimal by 100, the result is the same as moving the decimal point two places to the right.

$$0.30$$

Quick Check **7**
Rewrite 0.42 as a percent.

$$0.3 \cdot 100\% = 30\%$$

Rewriting Percents as Fractions and Decimals

Objective **6** **Rewrite a percent as a fraction.** To rewrite a percent as a fraction or a decimal, we can divide it by 100 and omit the percent sign. When rewriting a percent as a fraction, we may choose to multiply by $\frac{1}{100}$ rather than dividing by 100.

EXAMPLE 8   Rewrite 44% as a fraction.

Solution

We will begin by multiplying by $\frac{1}{100}$ and omitting the percent sign.

$$\overset{11}{\cancel{44}} \cdot \frac{1}{\underset{25}{\cancel{100}}} = \frac{11}{25}$$

Quick Check **8** Rewrite 35% as a fraction.

EXAMPLE 9   Rewrite $16\frac{2}{3}$% as a fraction.

Solution

To begin, we rewrite the mixed number $16\frac{2}{3}$ as an improper fraction $\left(\frac{50}{3}\right)$ and then simplify.

$$\frac{\overset{1}{\cancel{50}}}{3} \cdot \frac{1}{\underset{2}{\cancel{100}}} = \frac{1}{6}$$

Quick Check **9** Rewrite $11\frac{2}{3}$% as a fraction.

Objective 7 **Rewrite a percent as a decimal.** In the next examples, we will rewrite percents as decimals, rather than as fractions.

EXAMPLE 10   Rewrite 56% as a decimal.

Solution

We start by dropping the percent sign and dividing by 100. Keep in mind that dividing a decimal number by 100 is the same as moving the decimal point two places to the left.

$$.56$$

$$56 \div 100 = 0.56$$

Quick Check **10** Rewrite 8% as a decimal.

EXAMPLE 11   Rewrite 143% as a decimal.

Solution

When a percent is greater than 100%, its equivalent decimal must be greater than 1.

$$143 \div 100 = 1.43$$

Quick Check **11** Rewrite 240% as a decimal.

Simplify the following decimal expressions.

1. $4.23 + 3.62$

2. $13.89 - 2.54$

3. 7×5.2

4. $69.54 \div 6$

5. $8.4 - 3.7$

6. $7.9 + 4.5$

7. $13.568 \div 0.4$

8. $3.6(4.7)$

9. $-2.2 \cdot 3.65$

10. $6.2 - 15.9$

11. $13.47 + 21.562$

12. $5.283 \div 0.25$

13. $-6.3(3.9)(-2.25)$

14. $-4.84 \div (-0.016)$

15. $37.278 + 56.722$

16. $109.309 - 27.46 - 52.3716$

Rewrite the following fractions as decimal numbers.

17. $\dfrac{3}{5}$ 18. $\dfrac{7}{10}$

19. $\dfrac{7}{8}$ 20. $\dfrac{33}{4}$

21. $\dfrac{16}{25}$ 22. $\dfrac{9}{16}$

23. $24\dfrac{29}{50}$ 24. $7\dfrac{3}{20}$

Rewrite the following decimal numbers as a fraction in lowest terms.

25. 0.2 26. 0.5

27. 0.44 28. 0.35

29. 0.75 30. 0.68

31. 0.375 32. 0.204

F O R   E X T R A   H E L P

MyMathLab
MyMathLab

MathXL
MathXL

Interactmath.com

MathXL
Tutorials on CD

DVT CD
Videotape

Tutor Center

Addison-Wesley
Math Tutor Center

Student Solutions
Manual

33. The normal body temperature for humans is 98.6° F. If Melody has a temperature that is 2.8° F above normal, what is her temperature?

34. Paul spent the following amounts on five gifts for his wife's birthday: $32.95, $16.99, $47.50, $12.37, and $285. How much did Paul spend on gifts for his wife?

35. The balance of Jessica's checking account is $293.42. If she writes a check for $47.47, what will her new balance be?

36. The S&P 500 Index is one method used to measure the performance of the stock market. Over a three-day period, the S&P 500 lost 7.26 points, gained 3.17 points, and lost 12.69 points. If the index was at 905.36 before these three days, what was the index at after these three days?

37. An office manager bought 12 cases of paper. If each case cost $21.47, what was the total cost for the 12 cases?

38. Jean gives Chris a $20 bill and tells him to go to the grocery store and buy as many hot dogs as he can. If each package of hot dogs costs $2.65, how many packages can Chris buy? How much change will Chris have?

Rewrite as percents.

39. $\dfrac{1}{2}$ 40. $\dfrac{1}{4}$

41. $\dfrac{3}{4}$ 42. $\dfrac{3}{5}$

43. $\dfrac{4}{5}$

44. $\dfrac{5}{8}$

45. $\dfrac{7}{8}$

46. $\dfrac{11}{12}$

47. $\dfrac{27}{4}$

48. $\dfrac{12}{5}$

49. 0.4

50. 0.6

51. 0.15

52. 0.87

53. 0.09

54. 0.03

55. 3.2

56. 2.75

Rewrite as fractions.

57. 53%

58. 71%

59. 84%

60. 80%

61. 7%

62. 2%

63. $11\dfrac{1}{9}\%$

64. $18\dfrac{2}{11}\%$

65. 520%

66. 275%

Rewrite as decimals.

67. 32%

68. 57%

69. 16%

70. 29%

71. 4%

72. 1%

73. 0.3%

74. 61.3%

75. 400%

76. 320%

Answer in complete sentences.

77. Until recently, stock prices at the New York Stock Exchange were reported as fractions. Now prices are reported as decimals. Do you feel that this was a good idea? Explain why.

78. Describe a real-world application involving decimals.

Study Tip REVISITED   Another way to structure a group study session is to have each member bring a list of three questions to the meeting. The questions could be about specific homework problems or they could be about topics or procedures that have been covered in class. Begin by having each person read one of their questions. Once the questions have been read, the group should attempt to come up with answers that each member understands. Repeat this process until all questions have been answered. If the group cannot answer a question, ask your instructor at the beginning of the next class or during office hours.

Objectives

1 **Create a pie chart.**
2 **Create a bar graph.**
3 **Display data over time using a line graph.**

In today's data-driven society, we constantly see graphs presenting **data** or information on the television news, as well as in newspapers and magazines. Not only is it important to be able to read and interpret graphs, we must be able to create them as well. This section focuses on the creation and interpretation of pie charts, bar graphs, and line graphs that display data over a period of time.

Pie Charts

Objective 1 Create a pie chart. A **pie chart** is a circle used to display how a set of data can be divided into categories by percent. The entire circle represents 100%, or the total. Each category is displayed as a wedge of the circle, showing the percentage of the total that the category represents. Here is an example of a pie chart, showing the 2003 market share of leading pizza makers:

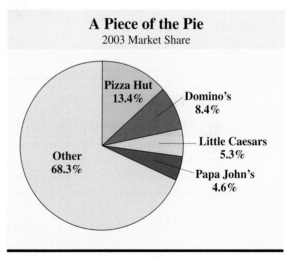

A Piece of the Pie
2003 Market Share

Pizza Hut 13.4%
Domino's 8.4%
Little Caesars 5.3%
Papa John's 4.6%
Other 68.3%

(*Source: Advertising Age*)

From the pie chart, we can see that Pizza Hut (13.4%) had the highest market share of any one chain. We also see that the category labeled as Other, which includes independent pizzerias and some minor chains, has a majority of the market share (68.3%).

EXAMPLE 1   A survey of 200 registered voters regarding an upcoming bond measure showed that 90 voters were in favor of the bond, 72 were opposed, and the rest were undecided. Create a pie chart that shows the percent of these voters who were in favor of the bond, opposed to it, or undecided.

Solution

We begin by finding what percent of the total is represented by each group. Since 90 of the 200 voters were in favor of the bond, this category is $\frac{90}{200}$ of the total. To rewrite this fraction as a percent, we multiply by 100%.

$$\frac{90}{200} \cdot 100\% = 45\% \qquad \text{Multiply by 100\% and simplify.}$$

45% of these voters were in favor of the bond. To find the percent of these voters who opposed the bond, we need to rewrite the fraction $\frac{72}{200}$ as a percent, which is 36%.

Since the three categories must total to 100%, we know that the percent of these voters who were undecided must be 19% because $100 - 45 - 36 = 19$. Here is the pie chart:

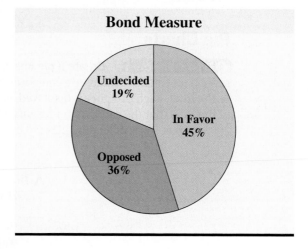

Bond Measure

Quick Check **1**
A survey of 150 students were asked whether there was adequate parking on their campus. Thirty-three of the students said yes, 87 said no, and the remaining 30 were unsure. Create a pie chart showing the percent of each response.

Bar Graphs

Objective 2 **Create a bar graph.** A **bar graph** is another graph that can be used to compare categories. To construct a bar graph, each category must be associated with a number. A bar is drawn above each category's name, and the height of the bar is determined by the number associated with the category. For example, the following graph shows the number of drive-in theaters in five different states:

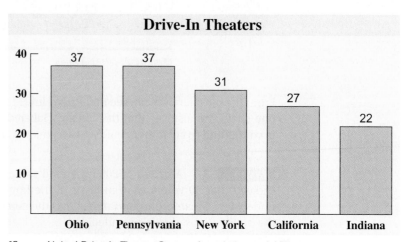

Drive-In Theaters

(*Source:* United Drive-In Theatre Owners Association and AP)

EXAMPLE 2   The following table shows the average tuition and fees in 2002–2003 for four different types of colleges and universities:

Type	Average Tuition and Fees
Private Four-Year	$18,273
Private Two-Year	$9890
Public Four-Year	$4081
Public Two-Year	$1735

(*Source:* Annual Survey of Colleges, The College Board)

Create a bar graph for this set of data.

Solution

We will use a scale of $2000 for the average tuition and fees.

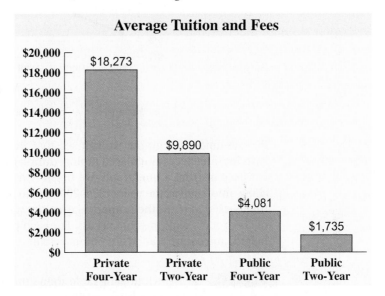

Quick Check   2 The following table shows the unemployment rates in 2001 for persons 25 years old and over, by highest level of education:

Level of Education	Percent Unemployed
Less than high school graduate	7.3
High school graduate, no college	4.2
Some college, no degree	3.5
Associate degree	2.9
Bachelor's or higher degree	2.3

(*Source:* U.S. Department of Labor, Current Population Survey, 2001)

Create a bar graph for this set of data.

Line Graphs

Objective 3 **Display data over time using a line graph.** The final graph introduced in this section is a **line graph**, which is used to show how a numerical quantity changes over time. The following line graph shows how the percentage of public school classrooms with Internet access changed during the years 1996–2001:

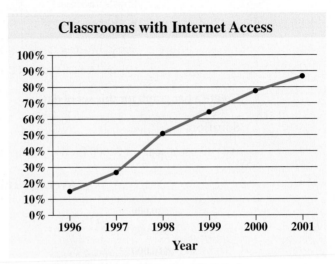

Classrooms with Internet Access

(*Source:* National Center for Educational Statistics)

By examining this graph, we can see that the percentage of public school classrooms with Internet access increased rapidly during this period.

To construct a line graph, we begin with a horizontal number line that is used to show time, such as the years 1996–2001 in the previous graph. We also draw a vertical line that is used to show the numerical quantity we are investigating, such as the percentage of classrooms with Internet access. For each period, we place a point at the appropriate height, and connect adjacent points with lines.

EXAMPLE 3   The following table shows that the revenue generated by DVD sales and rentals has grown steadily since 1998:

Year	1998	1999	2000	2001	2002	2003
Revenue (in billions)	$0.4	$1.5	$3.9	$7.1	$11.9	$17.5

(*Source:* Adams Media Research)

Create a line graph for this set of data.

Solution

The horizontal line used for showing the year needs to be labeled from 1998 to 2003. The vertical line used for showing the revenue must extend at least to 17.5, so we will use a scale of 2 and label the line up to 20.

The point for the year 1998 needs to be placed at a height of 0.4. Next we plot the remainder of the points, connecting them with lines.

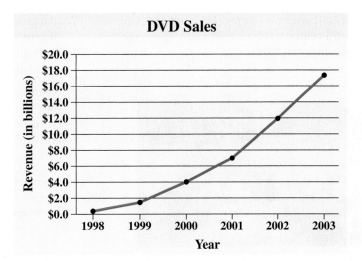

DVD Sales

Quick Check **3** Here are the number of reported collisions between wildlife (mostly birds) and civilian aircraft for selected years from 1990 to 2002:

Year	1990	1995	2000	2002
Number of Collisions	1990	2775	6323	6556

(*Source:* U.S. Air Force, Federal Aviation Administration)

Create a line graph for this set of data.

EXERCISES *1.6* ❭

For Exercises 1–10, construct a pie chart to represent the given set of data.

1. In 1999–2000, 75% of public school teachers were women and 25% were men. (*Source:* National Center for Education Statistics)

2. A 1999 survey of 14-year-old students from the United States showed that 85% of them said they are very likely or likely to vote in national elections once they turn 18, while 15% said that they were not likely to vote. (*Source:* International Association for the Evaluation of Educational Achievement (IEA), IEA Civic Education Study, 2001)

3. Thirty-one percent of fourth-grade public school students in the United States performed at or above the "proficient" level in the 2003 National Assessment of Educational Progress math test, while 69% performed below this level. (*Source:* National Center for Education Statistics)

F O R

E X T R A

H E L P

MyMathLab
MyMathLab

MathXL
MathXL

Interactmath.com

MathXL
Tutorials on CD

DVT CD
Videotape

Tutor Center
Addison-Wesley
Math Tutor Center

Student Solutions
Manual

4. In a March 2001 study, it was revealed that 34.3% of all persons 25 years and older held a college degree, and the remaining 65.7% did not. (*Source:* U.S. Department of Commerce, Bureau of the Census, Current Population Survey)

5. Participants in a survey were asked for the main reason that they shop for holiday gifts online. Here are the results:

Free shipping	34%
Convenience	33%
Comparison shopping	23%
Larger selection	6%
Speed	4%

(*Source:* CoolSavings 2003 Holiday Survey, October 2003)

6. Here is a table showing how long Americans planned to take to pay off their holiday debt:

Will not have any debt	Less than a month	One month to 5 months	More than 5 months	Not sure
57%	10%	22%	7%	4%

(*Source:* Maritz Poll survey)

7. In a group of 500 Americans who were asked whether they believe that only a few or no companies operate fairly and honestly, 170 respondents said yes. (*Source:* Wirthlin/LRN Corporate Ethics Study.)

8. A sample of 1400 workers who were asked whether they would not mind using electronic fingerprint or hand-recognition technology to record their time and attendance at work yielded the following results:

Would not mind	Privacy invasion	Not sure
728	280	392

(*Source:* American Payroll Association survey)

9. A sample of 400 high school freshmen who were asked when they planned to attend college produced the following results:

Right after graduation	256
After working for at least 2 years	48
Never	68
Unsure	28

10. A community college math department is offering 15 prealgebra classes, 21 elementary algebra classes, and 24 intermediate algebra classes this semester.

For Exercises 11–12, construct two pie charts to represent the given sets of data and illustrate the difference between them.

11. A 2003 survey showed that 44% of men shop online, while only 38% of women shop online. (*Source:* Blue Martini/Harris Interactive survey)

12. In 2000–2001, 48% of the bachelor's degrees awarded in mathematics were earned by females. In the same year, 29% of the master's degrees awarded in mathematics were earned by females. (*Source:* U.S. Department of Education, National Center for Educational Statistics, Integrated Postsecondary Education Data System (IPEDS), "Completions" survey)

For Exercises 13–18, construct a bar chart to represent the given set of data.

13. Here are the top personals sites by users in October 2003: (*Source:* Nielsen/NetRatings)

Site	Users (in millions)
Yahoo! Personals	4.9
Match	3.9
American Singles	3.7
MSN Dating & Personals	1.9
Netscape Love & Personals	1.5

14. Here are the percentages of children 19 to 24 months who consume these foods at least once a day:

Food	%
Hot dogs	25%
Sweetened beverages	23%
French fries	21%
Pizza	11%
Candy	10%

(*Source:* Mathematica Policy Research for Gerber Products Co.)

15. Here are the percentages of public school eighth-graders in certain cities who scored at or above "proficient" on national assessments of reading skills in 2003:

City	% Proficient
Atlanta	11%
Boston	22%
Chicago	15%
San Diego	20%
Washington, DC	10%

(*Source:* National Center for Educational Statistics)

16. Here are the percentages of public school eighth-graders in certain cities who scored at or above "proficient" on national assessments of math skills in 2003:

City	% Proficient
Atlanta	6%
Boston	17%
Chicago	9%
San Diego	18%
Washington, DC	6%

(*Source:* National Center for Educational Statistics)

17. Here is a list of the companies that were granted the most patents in 2003:

Company	Number of Patents
IBM	3415
Canon	1992
Hitachi	1893
Matsushita Electric	1786
Hewlett-Packard	1759

(*Source:* United States Patent and Trademark Office, IBM)

18. Here are the biggest initial public offerings (IPOs) as of 2003:

IPO	$ (in billions)
ENEL	$16.5
Deutsche Telekom	$13.0
AT&T Wireless	$10.6
Kraft	$8.7
France Telecom	$7.3
Telstra	$5.6
Swisscom	$5.6
UPS	$5.5
Infineon	$5.2
China Unicom	$4.9

(*Source:* Renaissance Capital)

For Exercises 19–26, create a line graph for the given set of data.

19. Here are the percentages of farms that own or lease computers, for selected years:

1997	1999	2001	2003
31%	40%	50%	54%

(*Source:* National Agriculture Statistics Service)

20. The following table shows that the percentage of U.S. households that have at least one mutual fund has grown steadily since the early 1990s:

1992	1994	1996	1998	2000	2002
26%	30%	37%	44%	52%	54%

(*Source:* Investment Company Institute)

21. Here are the number of worldwide spam messages sent daily, in billions, for selected years:

1999	2000	2001	2002	2003
1.0	2.3	4.0	5.6	7.3

(*Source:* IDC)

22. Here are the number of mobile homes in the United States since 1960, in millions:

1960	1970	1980	1990	2000
0.8	2.1	4.7	7.4	8.8

(*Source:* U.S. Bureau of the Census)

23. The following table shows that the average size of a new home, in square feet, has grown over time:

Year	1950	1960	1970	1980	1990	2000
Average Size	1100	1400	1500	1740	2080	2270

(*Source:* National Association of Home Builders)

24. Here are the number of U.S. cell phone subscribers, in millions, for selected years:

1998	1999	2000	2001	2002
69	86	109	128	141

(*Source:* Cellular Telecommunications, Internet Associates)

25. The following table shows that the number of obese people having weight-loss surgery (primarily gastric bypass surgery) has increased dramatically:

1997	2002	2003
23,100	63,100	103,200

(*Source:* American Society for Bariatric Surgery)

26. Here is the total amount of online retail sales in the United States, in billions, for the years 1999–2003:

1999	2000	2001	2002	2003
$26	$42	$51	$76	$96

(*Source:* Shop.org, Forrester Research)

27. Explain why it is important that each bar in a bar graph have the same width.

28. Explain why the following bar graph is misleading.

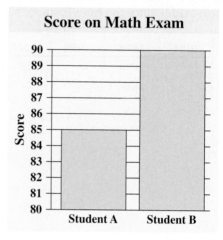

Score on Math Exam

QUICK REVIEW EXERCISES

Section 1.6

Simplify.

1. $-19 + 12$

2. $(-4)(-5)(-6)$

3. $-9 + 21 - (-6) - 38$

4. $\dfrac{4}{9} \div \left(-\dfrac{14}{165}\right)$

Study Tip REVISITED   It is important for group members to share phone numbers and email addresses. This will allow you to contact other group members if you misplace the homework assignment for that day. You can also contact others in your group if you have to miss class. This allows you to find out what was covered in class and which problems were assigned for homework. The person you contact can also give you advice for certain problems.

Some members of study groups agree to call each other if they are having trouble with homework problems. Calling a group member for this purpose should be a last resort. Be sure that you have used all available resources (examples in the text, notes, etc.) and have given the problem your fullest effort. Otherwise, some group members will be calling for help on all of the problems.

1 **Simplify exponents.**
2 **Use the order of operations to simplify arithmetic expressions.**

Exponents

Objective 1 **Simplify exponents.** Using the same number repeatedly as a factor can be represented using **exponential notation**. For example, $3 \cdot 3 \cdot 3 \cdot 3 \cdot 3$ can be written as 3^5.

Base, Exponent

> For the expression 3^5, the number being multiplied (3) is called the **base**. The **exponent** (5) tells us how many times that the base is used as a factor.

We read 3^5 as "*three raised to the fifth power*" or simply as "*three to the fifth power.*" When raising a base to the second power, we usually say that the base is being **squared**. When raising a base to the third power, we usually say that the base is being **cubed**. Exponents of 4 or higher do not have a special name.

EXAMPLE 1   Simplify 2^7.

Solution

In this example, we want to use 2 as a factor seven times.

$$2^7 = 2 \cdot 2 \cdot 2 \cdot 2 \cdot 2 \cdot 2 \cdot 2 \qquad \text{Write 2 as a factor seven times.}$$
$$= 128 \qquad \text{Multiply.}$$

Quick Check 1
Simplify 4^3.

Using Your Calculator Many calculators have a key that can be used to simplify expressions with exponents. When using the TI-83/84, we use the $\boxed{\wedge}$ key. Other calculators may have a key that is labeled $\boxed{y^x}$ or $\boxed{x^y}$. Here is the screen that you should see when using the TI-83/84 to simplify the expression in Example 1:

```
2^7
           128
```

EXAMPLE 2   Simplify $\left(\frac{2}{3}\right)^3$.

Solution

When the base is a fraction, the same rules apply. We use $\frac{2}{3}$ as a factor three times.

Quick Check 2
Simplify $\left(\frac{1}{8}\right)^4$.

$$\left(\frac{2}{3}\right)^3 = \frac{2}{3} \cdot \frac{2}{3} \cdot \frac{2}{3} \qquad \text{Write as a product.}$$
$$= \frac{8}{27} \qquad \text{Multiply.}$$

Consider the expression $(-3)^2$. The base is -3; so we multiply $(-3) \cdot (-3)$, and the result is positive 9. However, in the expression -2^4, the negative sign is not included in a set of parentheses with the 2. So the base of this expression is 2 and not -2. We will use 2 as a factor 4 times and then take the opposite of our result: $-2^4 = -(2 \cdot 2 \cdot 2 \cdot 2) = -16$.

Order of Operations

Objective 2 **Use the order of operations to simplify arithmetic expressions.**
Suppose we were asked to simplify the expression $2 + 4 \cdot 3$. We could obtain two different results, depending on whether we performed the addition or the multiplication first. Performing the addition first would give us $6 \cdot 3 = 18$, while performing the multiplication first would give us $2 + 12 = 14$. Only one result can be considered correct. The **order of operations agreement** is a standard order in which arithmetic operations are performed, ensuring a single correct answer.

Order of Operations

1. **Remove grouping symbols.** We begin by simplifying all expressions in parentheses, brackets, and absolute value bars. We also perform any operations in the numerator or denominator of a fraction. This is done by steps 2–4, presented next.
2. **Perform any operations involving exponents.** After all grouping symbols have been removed from the expression, we simplify any exponential expressions.
3. **Multiply and divide.** These two operations have equal priority. We perform multiplications or divisions in the order they appear from left to right.
4. **Add and subtract.** At this point, the only remaining operations should be additions and subtractions. Again these operations are of equal priority, and we perform them in the order that they appear from left to right. We can also use the strategy we used in totaling integers from Section 1.2.

Considering this information, we find that the correct result when we simplify $2 + 4 \cdot 3$ is 14, because the multiplication takes precedence over the addition.

$$2 + 4 \cdot 3 = 2 + 12 \qquad \text{Multiply } 4 \cdot 3.$$
$$= 14 \qquad \text{Add.}$$

EXAMPLE 3   Simplify $4 + 3 \cdot 5 - 2^6$.

Solution

In this example, the operation with the highest priority is 2^6, since simplifying exponents takes precedence over addition or multiplication. Be sure to note that the base is 2, and not -2, since the negative sign is not grouped with 2 inside a set of parentheses.

Quick Check **3**
Simplify
$-2 \cdot 7 + 11 - 3^4.$

$$4 + 3 \cdot 5 - 2^6 = 4 + 3 \cdot 5 - 64 \qquad \text{Raise 2 to the 6}^{\text{th}} \text{ power.}$$
$$= 4 + 15 - 64 \qquad \text{Multiply } 3 \cdot 5.$$
$$= -45 \qquad \text{Simplify.}$$

EXAMPLE 4 Simplify $(-5)^2 - 4(-2)(6)$.

Solution

We begin by squaring negative 5, which equals 25.

$$
\begin{aligned}
(-5)^2 - 4(-2)(6) &= 25 - 4(-2)(6) && \text{Square } -5. \\
&= 25 - (-48) && \text{Multiply } 4(-2)(6). \\
&= 25 + 48 && \text{Write as a sum, eliminating the} \\
& && \text{double signs.} \\
&= 73
\end{aligned}
$$

Quick Check 4
Simplify
$9^2 - 4(-2)(-10)$.

A Word of Caution When we square a negative number, such as $(-5)^2$ in the previous example, the result is a positive number.

EXAMPLE 5 Simplify $8 \div 2 + 3(7 - 4 \cdot 5)$.

Solution

The first step is to simplify the expression inside the set of parentheses. Here the multiplication takes precedence over the subtraction. Once we have simplified the expression inside of the parentheses, we proceed to multiply and divide. We finish by subtracting.

$$
\begin{aligned}
8 \div 2 + 3(7 - 4 \cdot 5) &= 8 \div 2 + 3(7 - 20) && \text{Multiply } 4 \cdot 5. \\
&= 8 \div 2 + 3(-13) && \text{Subtract } 7 - 20. \\
&= 4 + 3(-13) && \text{Divide } 8 \div 2. \\
&= 4 - 39 && \text{Multiply } 3(-13). \\
&= -35 && \text{Subtract.}
\end{aligned}
$$

Using Your Calculator When using your calculator to simplify an expression using the order of operations, you may want to perform one operation at a time. However, by carefully using parentheses, you can enter the entire expression at one time. Here is how we can simplify the expression in Example 5 using the TI-83/84:

```
8/2+3(7-4*5)
              -35
```

Quick Check 5
Simplify
$20 \div 5 \cdot 10(3 \cdot 6 - 9)$.

Occasionally, an expression will have one set of grouping symbols inside another set, such as the expression $3[7 - 4(9 - 3)] + 5 \cdot 4$. We call this **nesting** grouping symbols. We begin by simplifying the innermost set of grouping symbols and work our way out from there.

EXAMPLE 6   Simplify $3[7 - 4(9 - 3)] + 5 \cdot 4$.

Solution

We begin by simplifying the expression inside the set of parentheses. Once that has been done, we simplify the expression inside the square brackets.

$$3[7 - 4(9 - 3)] + 5 \cdot 4$$
$$= 3[7 - 4 \cdot 6] + 5 \cdot 4$$ Subtract $9 - 3$, to simplify the expression inside the parentheses. Now we turn our attention to simplifying the expression inside the square brackets.

$$= 3[7 - 24] + 5 \cdot 4$$ Multiply $4 \cdot 6$.
$$= 3[-17] + 5 \cdot 4$$ Subtract $7 - 24$.
$$= -51 + 20$$ Multiply $3[-17]$ and $5 \cdot 4$.
$$= -31$$ Simplify.

Quick Check 6
Simplify
$4[3^2 + 5(2 - 8)]$.

EXERCISES 1.7

Rewrite the given expression using exponential notation.

1. $2 \cdot 2 \cdot 2$

2. $9 \cdot 9 \cdot 9 \cdot 9 \cdot 9$

3. $\dfrac{2}{9} \cdot \dfrac{2}{9} \cdot \dfrac{2}{9} \cdot \dfrac{2}{9} \cdot \dfrac{2}{9} \cdot \dfrac{2}{9}$

4. $\dfrac{1}{7} \cdot \dfrac{1}{7} \cdot \dfrac{1}{7}$

5. $(-2)(-2)(-2)(-2)$

6. $(-6)(-6)(-6)(-6)(-6)$

7. $-3 \cdot 3 \cdot 3 \cdot 3 \cdot 3$

8. $-\dfrac{3}{5} \cdot \dfrac{3}{5} \cdot \dfrac{3}{5} \cdot \dfrac{3}{5}$

9. "five to the third power"

10. "seven squared"

Simplify the given expression.

11. 2^5 **12.** 5^2 **13.** 7^4 **14.** 6^3

15. 10^3 **16.** 10^4 **17.** 1^{723}

18. 0^{2364} **19.** $\left(\dfrac{3}{4}\right)^3$ **20.** $\left(\dfrac{1}{5}\right)^5$

21. 0.2^5 **22.** 0.5^4 **23.** $(-3)^4$

24. -3^4 **25.** -2^7 **26.** $(-2)^7$

27. $2^3 \cdot 3^2$ **28.** $4^2 \cdot 5^3$ **29.** $-3^2 \cdot (-4)^3$

30. $-\left(\dfrac{2}{3}\right)^2 \cdot \left(\dfrac{3}{5}\right)^3$

Simplify the given expression.

31. $1 + 2 \cdot 3$ **32.** $5 - 4 \cdot 3$

33. $12 - 8 \div 4$

34. $6 + 2 \cdot 11 - 27$

35. $4 + 6^2 \div 3$ **36.** $7 \cdot 6 - 3 \cdot 13$

37. $(-3)^2 - 4(-5)(-4)$

38. $3^2 + 4^2$ **39.** $(3 + 4)^2$

40. $5(7 - 2)$

41. $-4 - 5(7 - 3 \cdot 6)$

42. $3 - 6(5^2 - 4 + 2 \cdot 7)$

43. $\dfrac{1}{2} + \dfrac{1}{2} \cdot \dfrac{4}{7}$

44. $\dfrac{3}{5} \cdot \dfrac{2}{3} + \dfrac{15}{7} \cdot \dfrac{21}{20}$

45. $\dfrac{2}{9} \div \dfrac{5}{3} \cdot \dfrac{33}{50}$

46. $4 \cdot \dfrac{3}{8} - \left(\dfrac{7}{6}\right)^2$ **47.** $\dfrac{1}{9}\left(\dfrac{19}{28} - \dfrac{1}{4}\right)$

48. $\dfrac{8}{25} \div \left(\dfrac{4}{15} - \dfrac{1}{3} + \dfrac{3}{5}\right) \cdot \dfrac{7}{18}$ **49.** $\dfrac{2^2 - 9}{3^3 - 7}$

50. $\dfrac{3 + 5 \cdot 7 - 4 \cdot 2}{1 + 2 \cdot 17}$ **51.** $\dfrac{(3 + 5) \cdot 6 - 8}{1 + 3^2 + 2}$

52. $\dfrac{10^3 + 9^3}{1^3 + 12^3}$ **53.** $2 - 3[5 - (3 + 6)]$

54. $6[14 - 7(3 - 5) - 5^2]$

Use arithmetic operation signs such as +, −, ·, and ÷ between the values to produce the desired results. (You may use parentheses as well.)

55. 3   5   9   8   = 14

56. 4   7   2   6   = 14

57. 3   7   9   2   = 33

58. 8   2   3   14   4   7   = 2

Construct an "order-of-operations problem" of your own that involves at least four numbers and produces the given result.

59. 41

60. 0

61. −13

62. −19

The size of a computer's memory is measured by the number of bytes that it can store. The following table lists how many bytes there are in commonly used storage units:

1 kilobyte (KB)	2^{10} bytes
1 megabyte (MB)	2^{20} bytes
1 gigabyte (GB)	2^{30} bytes
1 terabyte (TB)	2^{40} bytes

Calculate the number of bytes in each of the following.

63. 1 kilobyte

64. 1 megabyte

65. 1 gigabyte

66. 1 terabyte

Find the missing number.

67. $2^? = 64$

68. $7^? = 16,807$

69. $?^4 = 1296$

70. $?^5 = -243$

71. $\left(\dfrac{3}{?}\right)^4 = \dfrac{81}{625}$

72. $\left(\dfrac{7}{4}\right)^? = \dfrac{2401}{256}$

73. Explain the difference between $(-2)^6$ and -2^6.

74. Explain how to use the order of operations to simplify expressions.

Study Tip **REVISITED**   For any group or team to be effective, it must be made up of members who are committed to giving their fullest effort to success. Here are some pointers for being a productive study group member:

- **Arrive at each session fully prepared.** Be sure that you have completed all required homework assignments. Bring a list of any questions that you have.
- **Stay focused during study sessions.** Do not spend your time socializing. It is all right to be friendly and begin each session with small talk, but be sure to shift the focus to math after a few minutes.
- **Be open minded.** During study sessions, you may be told that you are wrong. Keep in mind that the goal is to learn mathematics, not to always be correct initially.
- **Consider the feelings of others.** When a person has made a mistake or does not know what to do with a particular problem, be supportive and encouraging.
- **Know when to speak and when to listen.** A study group works best when it is a collaborative body. One person should not be asking all of the questions, and one person should not be answering all of the questions.

1.8
INTRODUCTION TO ALGEBRA

Objectives

1. **Build variable expressions.**
2. **Evaluate algebraic expressions.**
3. **Understand and use the commutative and associative properties of real numbers.**
4. **Use the distributive property.**
5. **Identify terms and their coefficients.**
6. **Simplify variable expressions.**

Variables

The number of students absent from a particular English class changes from day to day, as does the closing price for a share of Microsoft stock and the daily high temperature in Providence, Rhode Island. Quantities that change, or vary, are often represented by variables. A **variable** is a letter or symbol that is used to represent a quantity that either changes or whose value is unknown.

Variable Expressions

Objective 1   Build variable expressions. Suppose a buffet restaurant charges $7 per person to eat. To determine the bill for a family to eat in the restaurant, we multiply the number of people by $7. The number of people can change from family to family, so we can represent this quantity by a variable such as x. The bill for a family with x people can be written as $7 \cdot x$, or simply $7x$. This expression for the bill, $7x$, is known as a variable expression. A **variable expression** is a combination of one or more variables with numbers or arithmetic operations. When the operation is multiplication such as $7 \cdot x$, we often omit the multiplication dot. Here are some other examples of variable expressions:

$$3a + 5, \qquad x^2 + 3x - 10, \qquad \frac{y + 5}{y - 3}$$

EXAMPLE 1   Write an algebraic expression for "the sum of a number and 17."

Quick Check 1
Write an algebraic expression for "9 more than a number."

Solution

We must choose a variable to represent the unknown number. Let x represent the number. The expression is $x + 17$.
    Other phrases to look for that suggest addition are *plus, increased by, more than,* and *total.*

EXAMPLE 2   Write an algebraic expression for "five less than a number."

Solution

Let x represent the number. The expression is $x - 5$.

Quick Check 2
Write an algebraic expression for "a number decreased by 25."

Be careful with the order of subtraction when the phrase *less than* is used. Five less than a number tells us that we need to subtract 5 from that number. A common error is to write the subtraction in the opposite order.

Other phrases that suggest subtraction are *difference, minus,* and *decreased by.*

A Word of Caution   Five less than a number is written as $x - 5$, not $5 - x$.

EXAMPLE 3   Write an algebraic expression for "the product of 3 and two different numbers."

Quick Check 3
Write an algebraic expression for "twice a number."

Solution

Since we are building an expression involving two different unknown numbers, we need to introduce two variables. Let x and y represent the two numbers. The expression is $3xy$.

Other phrases that suggest multiplication are *times, multiplied by, of,* and *twice.*

EXAMPLE 4   Four friends decide to rent a fishing boat for the day. If all 4 friends decide to split evenly the cost of renting the boat, write a variable expression for the amount each friend would pay.

Quick Check 4
Write an algebraic expression for "the quotient of a number and 20."

Solution

Let c represent the cost of the boat. The expression is $c \div 4$ or $\frac{c}{4}$.

Other phrases that suggest division are *quotient, divided by,* and *ratio.*

Here are some different phrases that translate to $x + 5$, $x - 10$, $8x$, and $\frac{x}{6}$:

$x + 5$	The sum of a number and 5
	A number plus 5
	A number increased by 5
	5 more than a number
	The total of a number and 5
$x - 10$	10 less than a number
	The difference of a number and 10
	A number minus 10
	A number decreased by 10
$8x$	The product of 8 and a number
	8 times a number
	8 multiplied by a number
$\frac{x}{6}$	The quotient of a number and 6
	A number divided by 6
	The ratio of a number and 6

Evaluating Variable Expressions

Objective 2 **Evaluate algebraic expressions.** We will often have to evaluate variable expressions for particular values of variables. To do this, we substitute the appropriate numerical value for each variable and then simplify the resulting expression using the order of operations.

EXAMPLE ▸ **5**   Evaluate $2x - 7$ for $x = 6$.

Solution

The best first step when evaluating a variable expression is to rewrite the expression, replacing each variable by a set of parentheses. For example, rewrite $2x - 7$ as $2(\ ) - 7$. Then we can substitute the appropriate value for each variable and simplify.

$$2x - 7$$
$$2(6) - 7 \qquad \text{Substitute 6 for x.}$$
$$= 12 - 7 \qquad \text{Multiply.}$$
$$= 5 \qquad \text{Subtract.}$$

Quick Check ◂ **5**
Evaluate $5x + 2$ for
$x = 11$.

The expression $2x - 7$ is equal to 5 for $x = 6$.

EXAMPLE ▸ **6**   Evaluate $x^2 - 5x + 6$ for $x = 3$.

Solution

$$x^2 - 5x + 6$$
$$(3)^2 - 5(3) + 6 \qquad \text{Substitute 3 for } x.$$
$$= 9 - 5(3) + 6 \qquad \text{Square 3.}$$
$$= 9 - 15 + 6 \qquad \text{Multiply.}$$
$$= 0 \qquad \text{Simplify.}$$

Quick Check ◂ **6**
Evaluate $x^2 - 13x - 40$
for $x = -8$.

EXAMPLE ▸ **7**   Evaluate $b^2 - 4ac$ for $a = 3$, $b = -2$, and $c = -10$.

Solution

$$b^2 - 4ac$$
$$(-2)^2 - 4(3)(-10) \qquad \text{Substitute 3 for a, } -2 \text{ for b, and } -10 \text{ for c.}$$
$$= 4 - 4(3)(-10) \qquad \text{Square negative 2.}$$
$$= 4 - (-120) \qquad \text{Multiply.}$$
$$= 4 + 120 \qquad \text{Rewrite without double signs.}$$
$$= 124 \qquad \text{Total.}$$

Quick Check ◂ **7**
Evaluate $b^2 - 4ac$ for
$a = -1$, $b = 5$, and
$c = 18$.

Properties of Real Numbers

Objective **3**   **Understand and use the commutative and associative properties of real numbers.** We will now examine three properties of real numbers. The first is the **commutative property**, which is used for addition and multiplication.

Commutative Property

For all real numbers a and b,

$$a + b = b + a \qquad \text{and} \qquad a \cdot b = b \cdot a.$$

This property states that changing the order of the numbers in a sum or in a product does not change the result. For example, $3 + 9 = 9 + 3$ and $7 \cdot 8 = 8 \cdot 7$. Note that this property does not work for subtraction or for division. Changing the order of the numbers in a subtraction or a division will generally change the result.

The second property that we will cover is the **associative property**, which also holds for addition and multiplication.

Associative Property

For all real numbers a, b, and c,

$$(a + b) + c = a + (b + c) \quad \text{and} \quad (ab)c = a(bc).$$

This property states that changing the grouping of the numbers in a sum or in a product does not change the result. Check for yourself; does $(2 + 7) + 3$ equal $2 + (7 + 3)$? The first expression simplifies to $9 + 3$, while the second becomes $2 + 10$, both of which equal 12.

EXAMPLE ▶ 8   Simplify $5(12x)$ by applying the associative property.

Solution

By the associative property we know that $5(12x) = (5 \cdot 12)x$. We can now multiply $5 \cdot 12$ to produce a result of $60x$.

Quick Check **8**

Simplify $8(3a)$ by applying the associative property.

EXAMPLE ▶ 9   Simplify $(3x + 7) + 15$ by applying the associative property.

Solution

By the associative property we know that $(3x + 7) + 15 = 3x + (7 + 15)$. This allows us to add 7 and 15. The simplified expression is $3x + 22$.

Quick Check **9**

Simplify $(9x + 20) + 14$ by applying the associative property.

Objective 4 Use the distributive property. The third property of real numbers is the **distributive property**.

Distributive Property

For all real numbers a, b, and c,

$$a(b + c) = ab + ac.$$

This property tells us that we can distribute the factor outside of the parentheses to each number being added in the parentheses, perform the multiplications, and then add. (This property also holds true when the operation inside the parentheses is subtraction.) Consider the expression $3(5 + 4)$. The order of operations tells us that this is equal to $3 \cdot 9$, or 27. Here is how we simplify the expression using the distributive property:

$$
\begin{aligned}
3(5 + 4) &= 3 \cdot 5 + 3 \cdot 4 && \text{Apply the distributive property.} \\
&= 15 + 12 && \text{Multiply } 3 \cdot 5 \text{ and } 3 \cdot 4. \\
&= 27 && \text{Add.}
\end{aligned}
$$

Either way we get the same result.

EXAMPLE 10 Simplify $2(4 + 3x)$ using the distributive property.

Solution

Quick Check 10
Simplify $7(5x - 4)$ using the distributive property.

$$2(4 + 3x) = 2 \cdot 4 + 2 \cdot 3x \qquad \text{Distribute the 2.}$$
$$= 8 + 6x \qquad \text{Multiply. Recall from the example regarding the}$$

associative property that $2 \cdot 3x$ equals $6x$.

A Word of Caution The expression $8 + 6x$ is not equal to $14x$.

EXAMPLE 11 Simplify $7(2 + 3a - 4b)$ by using the distributive property.

Quick Check 11
Simplify $12(x - 2y + 3z)$ using the distributive property.

Solution

When there are more than two terms inside the parentheses, we distribute the factor to each term. Also, it is a good idea to be able to distribute the factor and multiply mentally.

$$7(2 + 3a - 4b) = 14 + 21a - 28b$$

EXAMPLE 12 Simplify $-4(2x - 5)$ using the distributive property.

Solution

When the factor outside the parentheses is negative, the negative number must be distributed to each term inside the parentheses. This will change the sign of each term inside the parentheses.

Quick Check 12
Simplify $-6(4x + 11)$ using the distributive property.

$$-4(2x - 5) = (-4) \cdot 2x - (-4) \cdot 5 \qquad \text{Distribute the } -4.$$
$$= -8x - (-20) \qquad \text{Multiply.}$$
$$= -8x + 20 \qquad \text{Rewrite without double signs.}$$

Simplifying Variable Expressions

Objective 5 Identify terms and their coefficients. In an algebraic expression, a **term** is either a number, a variable, or a product of a number and variables. Terms in an algebraic expression are separated by addition or subtraction. In the expression $7x - 5y + 3$, there are three terms: $7x$, $-5y$, and 3. The numerical factor of a term is its **coefficient**. The coefficients of these terms are 7, -5, and 3.

EXAMPLE 13 For the expression $-5x + y + 3xy - 19$, determine the number of terms, list them, and state the coefficient for each term.

Solution

This expression has four terms: $-5x$, y, $3xy$, and -19.

What is the coefficient for the second term? Although it does not appear to have a coefficient, its coefficient is 1. This is because y is the same as $1 \cdot y$. So the four coefficients are -5, 1, 3, and -19.

Quick Check 13 For the expression $x^3 - x^2 + 23x - 59$, determine the number of terms, list them, and state the coefficient for each term.

Objective **6** **Simplify variable expressions.** Two terms that have the same variable factors, or two constants, are called **like terms**. Consider the expression $9x + 8y + 6x - 3y + 8z - 5$. There are two sets of like terms in this expression: $9x$ and $6x$, as well as $8y$ and $-3y$. There are no like terms for $8z$, since no other term has z as its sole variable factor. Similarly, there are no like terms for the constant term -5.

Suppose we had to simplify the expression $8y - 3y$. This expression could be rewritten as $(8 - 3)y$ using the distributive property, or simply as $5y$.

Combining Like Terms

> When simplifying variable expressions, we can combine like terms into a single term by adding or subtracting the coefficients of the like terms.

EXAMPLE **14** Simplify $4x + 11x$ by combining like terms.

Solution

These two terms are like terms, since they both have the same variable factors. We can simply add the two coefficients together to produce the expression $15x$.

Quick Check **14**

Simplify
$3x + 7y + y - 5x$ by combining like terms.

$$4x + 11x = 15x.$$

A general strategy for simplifying algebraic expressions is to begin by performing any distributive multiplications. After all multiplications have been performed, all that remains should be addition and subtraction. At this point, we can combine any like terms.

EXAMPLE **15** Simplify $8(3x - 5) - 4x + 7$.

Solution

The first step is to use the distributive property by distributing the 8 to each term inside the parentheses. Then we will be able to combine like terms.

Quick Check **15**

Simplify
$5(2x + 3) + 9x - 8$.

$$
\begin{aligned}
8(3x - 5) - 4x + 7 &= 24x - 40 - 4x + 7 &&\text{Distribute the 8.} \\
&= 20x - 33 &&\text{Combine like terms.}
\end{aligned}
$$

EXAMPLE **16** Simplify $5(9 - 7x) - 10(3x + 4)$.

Solution

We must be sure to distribute the *negative* 10 into the second set of parentheses.

$$5(9 - 7x) - 10(3x + 4) = 45 - 35x - 30x - 40$$

Distribute the 5 to each term in the first set of parentheses and distribute the -10 to each term in the second set of parentheses.

$$= -65x + 5$$

Combine like terms.

Quick Check 16
Simplify
$3(2x - 7) - 8(x - 9)$.

Usually we will write our simplified result with variable terms preceding constant terms, but it would also be correct to write $5 - 65x$, because of the commutative property.

A Word of Caution   If a factor in front of a set of parentheses is negative or has a subtraction sign in front of it, we must distribute the negative sign along with the factor.

EXERCISES *1.8*

Build a variable expression for the following phrases.

1. The sum of a number and 11

2. A number decreased by 9

3. Five times a number

4. A number divided by 14

5. Thirteen minus a number

6. Twice a number

7. Four less than three times a number

8. Five times a number, increased by 17

9. The sum of two different numbers

10. One number divided by another

11. Seven times the difference of two numbers

12. Half of the sum of a number and 25

13. A college charges \$200/unit for tuition. If we let u represent the number of units that a student is taking, build a variable expression for the student's tuition.

14. Thomas Magnum, a private investigator from the 1980s TV show *Magnum, P. I.,* charged \$200 per day to take a case. If we let d represent the number of days it took Magnum to crack the case, build a variable expression for the amount of money he would charge for the case.

15. A professional baseball player is appearing at a baseball card convention. The promoter agreed to pay the player a flat fee of \$25,000 plus \$22 per autograph signed. If we let a represent the number of autographs signed, build a variable expression for the amount of money the player will be paid.

16. Jim Rockford, a private investigator from the 1970s TV show *The Rockford Files,* charged \$200 per day plus expenses to take a case. If we let d represent the number of days that Rockford worked on a case, and if he had \$425 in expenses, build a variable expression for the amount of money he would charge for the case.

Evaluate the following algebraic expressions under the given conditions.

17. $4x - 9$ for $x = 2$

18. $7 + 6x$ for $x = 5$

19. $3a - 2b$ for $a = 2$ and $b = -7$

20. $-5w + 4z$ for $w = -1$ and $z = 3$

21. $3(x - 5)$ for $x = 7$

22. $(x + 3)(2x - 5)$ for $x = -2$

23. $x^2 + 9x + 18$ for $x = 5$

24. $a^2 - 7a - 30$ for $a = 3$

25. $y^2 + 4y - 17$ for $y = -3$

26. $x^2 - 5x - 9$ for $x = -4$

27. $m^2 - 9$ for $m = 3$

28. $5 - x^2$ for $x = -2$

29. $x^7 + 5x^6 + 303x^3 - 754$ for $x = 0$

30. $a^3 - 4a - 13$ for $a = -7$

31. $b^2 - 4ac$ for $a = 2$, $b = 5$ and $c = -3$

32. $b^2 - 4ac$ for $a = -5$, $b = 7$ and $c = -4$

33. $b^2 - 4ac$ for $a = -1$, $b = -2$ and $c = 10$

34. $b^2 - 4ac$ for $a = 15$, $b = -6$ and $c = 9$

35. $5(x + h) - 17$ for $x = -3$ and $h = 0.01$

36. $-3(x + h) + 4$ for $x = -6$ and $h = 0.001$

37. $(x + h)^2 - 5(x + h) - 14$ for $x = -3$ and $h = 0.1$

38. $(x + h)^2 + 4(x + h) - 28$ for $x = 6$ and $h = 1$

39. The commutative property works for addition but does not work in general for subtraction.

a) Give an example of two numbers a and b such that $a - b \neq b - a$.

b) Can you find two numbers a and b such that $a - b = b - a$?

40. The commutative property works for multiplication but does not work in general for division.

a) Give an example of two numbers a and b such that $\frac{a}{b} \neq \frac{b}{a}$.

b) b) Can you find two numbers a and b such that $\frac{a}{b} = \frac{b}{a}$?

Simplify, where possible.

41. $3(x - 9)$

42. $4(2x + 5)$

43. $5(3 - 7x)$

44. $12(5x - 7)$

45. $7x + 9x$

46. $14x + 3x$

47. $3x - 8x$

48. $22a - 15a$

49. $3x - 7 + 4x + 11$

50. $15x + 32 - 19x - 57$

51. $32 - 49a + 87a - 95$

52. $52 + 32a - 17a + 45$

53. $4a + 5b$

54. $13x - 12y + 357$

55. $5x - 3y - 7x - 19y$

56. $15m - 11n - 6m + 22n$

57. $3(2x - 5) + 4x + 7$

58. $8 - 19k + 6(3k - 5)$

59. $-2(4x - 9) - 11$

60. $3(2a + 11b) - 13a$

61. $6y - 5(3y - 17)$

62. $5 - 9(3 - 4x)$

63. $3(4z - 7) + 9(2z + 3)$

64. $-6(5x - 9) - 7(13 - 12x)$

65. $-2(5a + 4b - 13c) + 3b$

66. $-7(-2x + 3y - 17z) - 5(3x - 11)$

67. Verify that the expression $2(3x - 9) + 5(3x - 9)$ is equal to the expression $7(3x - 9)$.

68. Use the result from the previous problem to find a shortcut for simplifying $13(5a - 4) + 7(5a - 4)$ without distributing the 13 and the 7.

69. Find a shortcut for simplifying $6(2a + 11) - 2(2a + 11)$ without distributing the 6 and the -2.

70. Find a shortcut for simplifying $3(9a - 19) - 10(9a - 19)$ without distributing the 3 and the -10.

For the following expressions,
a) determine the number of terms;
b) write down each term; and
c) write down the coefficient for each.

(Be sure to simplify each expression before answering.)

71. $5x^3 + 3x^2 - 7x - 15$

72. $-3a^2 - 7a - 10$

73. $3x - 17$

74. $9x^4 - 10x^3 + 13x^2 - 17x + 329$

75. $5(3x^2 - 7x + 11) - 6x$

76. $4(3a - 5b - 7c - 11) - 2(6b - 5c)$

77. $5(-7a - 3b + 5c) - 3(6b - 9c - 23)$

78. $2(-4x + 7y) - (3x + 9y)$

Answer in complete sentences.

79. Give an example of a real-world situation that can be described by the variable expression $60x$. Explain why the expression fits your situation.

80. Give an example of a real-world situation that can be described by the variable expression $20x + 35$. Explain why the expression fits your situation.

Study Tip REVISITED   Occasionally, a group will have a member who is not productive or is even disruptive. You should not allow one person to prevent your group from being successful. If you have a group member who is not contributing to the group in a positive way, try talking to that person one-on-one, discreetly. Once the person knows how important this study group is in helping you to learn mathematics, he or she may change their ways and start being an effective group member.

Some groups prefer to sit down as a group with the nonproductive or disruptive member and air their grievances. Always be honest with the person, but try not to be offensive. Try to find solutions that would be acceptable to the group and the group member in question. Seek advice from your instructor. Undoubtedly, your instructor has seen a group in a similar situation and may have valuable advice for you.

If you have worked on improving the situation but have seen no improvement, then it may be time to remove the person from the group. Try to be professional about it rather than adversarial. Remember, you will continue to see this person in class each day, and it would be best to separate on good terms.

Chapter 1 Summary

Section 1.1—Topic	Chapter Review Exercises
Determining Inequalities Involving Integers	1–2
Finding Absolute Values	3–4
Finding Opposites of Integers	5–6

Section 1.2—Topic	Chapter Review Exercises
Finding Sums and Differences of Integers	7–12
Finding Products and Quotients of Integers	13–14
Solving Applied Problems Involving Integers	47–49

Section 1.3—Topic	Chapter Review Exercises
Finding the Factor Set of a Number	15–16
Finding the Prime Factorization of a Number	17–18
Simplifying a Fraction to Lowest Terms	19–22
Rewriting Mixed Numbers as Improper Fractions	23–24
Rewriting Improper Fractions as Mixed Numbers	25–26

Section 1.4—Topic	Chapter Review Exercises
Multiplying and Dividing Fractions	27–30, 35, 37
Adding and Subtracting Fractions	31–34, 36, 38
Solving Applied Problems Involving Fractions	50

Section 1.5—Topic	Chapter Review Exercises
Performing Arithmetic Operations With Decimals	39–42
Rewriting Fractions as Decimals	43–44
Rewriting Decimals as Fractions	45–46

Section 1.7—Topic	Chapter Review Exercises
Simplifying Expressions Involving Exponents	51–54
Simplifying Expressions Using the Order of Operations	55–60

Section 1.8—Topic	Chapter Review Exercises
Building Variable Expressions	61–66
Evaluating Variable Expressions	67–72
Simplifying Variable Expressions	73–78
Identifying Terms and Their Coefficients	79–80

Summary of Chapter 1 Study Tips

- Creating a study group is one of the best ways to help you learn mathematics and improve your performance on quizzes and exams.
- Sometimes, a concept will be easier for you to understand if it is explained to you by a peer. One of your group members may have some insight into a particular topic, and that person's explanation may be just the trick to turn on the "lightbulb" in your head.
- When you explain a certain topic to a fellow student, or show a student how to solve a particular problem, you increase your chances of retaining this knowledge. Because you are forced to choose your words in such a way that the other student will completely understand, you have demonstrated that you truly understand.
- You will also find it helpful to have a support group: students who can lean on each other when times are bad. By being supportive of each other, you are increasing the chances that you will all learn and be successful in your class.

Write the appropriate symbol, either < or >, between the following integers. **[1.1]**

1. -10 ___ -7 **2.** 3 ___ -12

Find the following absolute values. **[1.1]**

3. $|8|$ **4.** $|-13|$

Find the opposite of the following integers. **[1.1]**

5. -6 **6.** 12

Add or subtract. **[1.2]**

7. $9 + (-16)$ **8.** $-10 + 7$

9. $-22 - 19$ **10.** $4 - (-23)$

Simplify. **[1.2]**

11. $3 - 16 - 24$ **12.** $8 - (-19) - 7$

Multiply or divide. **[1.2]**

13. $9(-6)$ **14.** $-144 \div (-9)$

Write the factor set for the following numbers. **[1.3]**

15. 42

16. 108

Write the prime factorization of the following numbers. (If prime, state this.) **[1.3]**

17. 32 **18.** 60

Simplify the following fractions to lowest terms. **[1.3]**

19. $\dfrac{24}{42}$ **20.** $\dfrac{9}{72}$ **21.** $\dfrac{40}{234}$ **22.** $\dfrac{98}{7}$

Rewrite the following mixed numbers as improper fractions. **[1.3]**

23. $5\dfrac{2}{3}$ **24.** $12\dfrac{7}{25}$

Rewrite the following improper fractions as mixed numbers. **[1.3]**

25. $\dfrac{38}{10}$ **26.** $\dfrac{55}{6}$

Multiply or divide. **[1.4]**

27. $\dfrac{9}{16} \cdot \dfrac{28}{75}$ **28.** $5\dfrac{1}{2} \cdot \dfrac{14}{33}$

29. $\dfrac{8}{15} \div \dfrac{20}{21}$ **30.** $3\dfrac{1}{5} \div 7\dfrac{1}{2}$

Add or subtract. **[1.4]**

31. $\dfrac{5}{8} + \dfrac{7}{12}$ **32.** $\dfrac{13}{20} + \dfrac{5}{6}$

33. $\dfrac{11}{18} - \dfrac{4}{9}$ **34.** $\dfrac{7}{36} - \dfrac{13}{42}$

Simplify. **[1.4]**

35. $\dfrac{30}{49} \cdot \dfrac{35}{66}$ **36.** $\dfrac{7}{16} - \dfrac{11}{48}$

37. $\dfrac{9}{40} \div \dfrac{39}{35}$ **38.** $\dfrac{7}{20} + \dfrac{1}{15}$

Simplify the following decimal expressions. **[1.5]**

39. $8.7 + 3.92$ **40.** $24.308 - 15.49$

41. 8.4×3.6 **42.** $40.92 \div 4.65$

Rewrite the following fractions as decimal numbers. **[1.5]**

43. $\dfrac{8}{25}$ **44.** $\dfrac{15}{16}$

Rewrite the following decimal numbers as fractions in lowest terms. **[1.5]**

45. 0.75 **46.** 0.28

47. Jeff bought a book at a yard sale for $23. If he had $40 prior to buying the book, how much money does Jeff have now? [1.2]

48. Gray had $78 in his checking account prior to writing a $125 check to the bookstore for books and supplies. What is his account's new balance? [1.2]

49. Three investors plan to start a new company. If start-up costs are $37,800, how much will each person have to invest? [1.2]

50. If one recipe calls for $1\frac{1}{2}$ cups of flour and a second recipe calls for $2\frac{2}{3}$ cups of flour, how much flour is needed to make both recipes? [1.4]

Simplify the given expression. **[1.7]**

51. 4^3 **52.** $\left(\dfrac{2}{5}\right)^3$

53. -2^6 **54.** $3^5 \cdot 5^2$

Simplify the given expression. **[1.7]**

55. $5 + 8 \cdot 4$ **56.** $25 - 15 \div 5$

57. $3 + 13 \cdot 5 - 20$ **58.** $(3 + 13) \cdot 5 - 20$

59. $54 - 27 \div 3^2$ **60.** $\dfrac{3}{4} + \dfrac{1}{4} \cdot \dfrac{12}{25}$

Build a variable expression for the following phrases.
[1.8]

61. The sum of a number and 14

62. A number decreased by 20

63. Eight less than twice a number

64. Nine more than 6 times a number

65. A coffeehouse charges $3.55 for a cup of coffee. If we let c represent the number of cups of coffee that a coffeehouse sells on a particular day, build a variable expression for the revenue from coffee sales. [1.8]

66. A rental company rents moving vans for $20, plus $0.15 per mile. If we let m represent the number of miles, build a variable expression for the cost to rent a moving van from this rental company. [1.8]

Evaluate the following algebraic expressions under the given conditions. **[1.8]**

67. $3x + 17$ for $x = 9$ **68.** $9 - 8x$ for $x = -2$

69. $10a - 4b$ for $a = 2$ and $b = -9$

70. $(8x - 9)(2x - 11)$ for $x = 4$

71. $x^2 - 7x - 30$ for $x = -3$

72. $b^2 - 4ac$ for $a = -1$, $b = -8$ and $c = 5$

Simplify. **[1.8]**

73. $5(x + 7)$

74. $6x + 21x$

75. $8x - 25 - 3x + 17$

76. $8y - 6(4y - 21)$

77. $15 - 23k + 7(4k - 9)$

78. $-8(2x + 25) - (103 - 19x)$

For the following expressions,
a) determine the number of terms;
b) write down each term; and
c) write down the coefficient for each.

79. $x^3 - 4x^2 - 10x + 41$

80. $-x^2 + 5x - 30$

Write the appropriate symbol, either < or >, between the following integers.

1. -15 ___ -18

Find the following absolute values.

2. $|-17|$

Simplify.

3. $7 + (-13)$

4. $-7(-9)$

5. Write the factor set for 45.

6. Write the prime factorization of 108. (If the number is prime, state this.)

7. Simplify $\dfrac{60}{84}$ to lowest terms.

8. Rewrite $\dfrac{67}{18}$ as a mixed number.

Simplify.

9. $\dfrac{11}{63} \cdot \dfrac{15}{44}$

10. $\dfrac{2}{9} \div \dfrac{8}{21}$

11. $\dfrac{3}{5} + \dfrac{11}{12}$

12. $\dfrac{5}{24} - \dfrac{4}{9}$

13. Simplify 8.05×2.27.

14. Rewrite 0.36 as a fraction. Your answer should be in lowest terms.

15. Eleanor bought 13 computers for $499 each. What was the total cost for the 13 computers?

16. After a deposit of $407.83, the balance in Lindsay's checking account was $1203.34. What was the balance before the deposit?

Simplify the given expression.

17. $16 - 8 \cdot 5$

18. $-9 + 4 \cdot 13 - 6 \cdot 3$

19. $\dfrac{4^2 + 3^2}{3 + 4 \cdot 13}$

20. Build a variable expression for the phrase *seven less than four times a number.*

21. A landscaper charges $50, plus $20 per hour, for yard maintenance. If we let h represent the number of hours spent working on a particular yard, build a variable expression for the landscaper's charge.

Evaluate the following algebraic expressions under the given conditions:

22. $16 - 5x$ for $x = -9$

23. $x^2 + 6x - 17$ for $x = -8$

Simplify.

24. $5(2x - 13)$

25. $7y - 8(2y - 30)$

Mathematicians in History
Srinivasa Ramanujan

Srinivasa Ramanujan was a self-taught Indian mathematician, viewed by many to be one of the greatest mathematical geniuses in history. As a student, he became so totally immersed in his work with mathematics that he ignored his other subjects and failed his college exams. Ramanujan's life is chronicled in the biography *The Man Who Knew Infinity: A Life of the Genius Ramanujan.*

Write a one-page summary (or make a poster) of the life of Srinivasa Ramanujan and his accomplishments.

Interesting issues:

- Where and when was Srinivasa Ramanujan born?
- At age 16, Ramanujan borrowed a mathematics book that strongly influenced his life as a mathematician. What was the name of the book?
- Ramanujan got married on July 14, 1909. The marriage was arranged by his mother. How old was his bride at the time?
- What jobs did Ramanujan hold in India?
- Which renowned mathematician invited Ramanujan to England in 1914?
- Ramanujan's health in England was poor. What was the cause of his poor health?
- What is the significance of the taxicab number 1729?
- What were the circumstances that led to Ramanujan's death, and what was his age when he died?

Stretch Your Thinking ⟩ Chapter 1

Smoothie Superb makes an orange juice and strawberry-banana smoothie called the Berry Ana. The store charges $3.78 for a regular size and gives the customer one free nutritional boost. For each additional nutritional boost, they charge 43 cents. Let b represent the number of nutritional boosts in a drink.

a) Build a variable expression for the total cost of a Berry Ana smoothie with *b* nutritional boosts.

b) What would be the total cost for a customer to order a Berry Ana smoothie with two nutritional boosts?

During December, Smoothie Superb makes a specialty drink called the Cold Terminator. The Cold Terminator is the exact same smoothie as the Berry Ana, but contains four nutritional boosts. A regular-size Cold Terminator sells for $4.99.

c) If a customer wanted to order a Berry Ana smoothie with four nutritional boosts, would it be more cost effective to buy a Cold Terminator? Explain your answer.

d) Set up an inequality statement (using either < or >) comparing the costs of the Cold Terminator and a Berry Ana with four boosts.

2

LINEAR EQUATIONS

In this chapter, we will learn to solve linear equations and investigate applications of this type of equation. We will also discuss applications involving percents and proportions. The chapter concludes with a section on linear inequalities.

Study Tip **USING YOUR TEXTBOOK EFFECTIVELY**   *In this chapter, we will focus on how to get the most out of your textbook. Students who treat their books solely as a source of homework exercises are turning their backs on one of their best resources.*

Throughout this chapter, we will revisit this study tip and help you incorporate it into your study habits.

Objectives

1. **Identify linear equations.**
2. **Determine whether a value is a solution of an equation.**
3. **Solve linear equations using the multiplication property of equality.**
4. **Solve linear equations using the addition property of equality.**
5. **Solve applied problems using the multiplication property of equality or the
    addition property of equality.**

Linear Equations

Objective 1   Identify linear equations. An **equation** is a mathematical statement of equality between two expressions. It is a statement that asserts that the value of the expression on the left side of the equation is equal to the value of the expression on the right side. Here are a few examples of equations:

$$2x = 8 \qquad x + 17 = 20 \qquad 3x - 8 = 2x + 6 \qquad 5(2x - 9) + 3 = 7(3x + 16)$$

All of these are examples of linear equations. A **linear equation** has a single variable, and the exponent for that variable is 1. For example, if the variable in a linear equation is x, then the equation cannot have terms containing x^2 or x^3. The variable in a linear equation cannot appear in a denominator either. Here are some examples of equations that are not linear equations:

$$x^2 - 5x - 6 = 0 \qquad \text{Variable is squared.}$$
$$m^3 - m^2 + 7m = 7 \qquad \text{Variable has exponents greater than 1.}$$
$$\frac{5x + 3}{x - 2} = -9 \qquad \text{Variable is in denominator.}$$

Solutions of Equations

Objective 2   Determine whether a value is a solution of an equation. A **solution** of an equation is a value that, when substituted for the variable in the equation, produces a true statement, such as $5 = 5$.

EXAMPLE 1   Is $x = 3$ a solution of $9x - 7 = 20$?

Solution

To check whether a particular value is a solution of an equation, we substitute that value for the variable in the equation. If, after simplifying both sides of the equation, we have a true mathematical statement, then the value is a solution.

$$9x - 7 = 20$$
$$9(3) - 7 = 20 \qquad \text{Substitute 3 for } x.$$
$$27 - 7 = 20 \qquad \text{Multiply.}$$
$$20 = 20 \qquad \text{Subtract.}$$

Since 20 is equal to 20, we know that $x = 3$ is a solution.

EXAMPLE 2   Is $x = -2$ a solution of $3 - 4x = -5$?

Solution

Again, we begin by substituting -2 for x and simplify both sides of the equation.

$$3 - 4x = -5$$
$$3 - 4(-2) = -5 \qquad \text{Substitute } -2 \text{ for } x.$$
$$3 + 8 = -5 \qquad \text{Multiply. } -4(-2) = 8$$
$$11 = -5 \qquad \text{Add.}$$

This statement, $11 = -5$, is not true because 11 is not equal to -5. $x = -2$ is not a solution of this equation.

Quick Check 1
Is $x = -7$ a solution of $4x + 23 = -5$?

The set of all solutions of an equation is called its **solution set**. The process of finding an equation's solution set is called **solving the equation**. When we find all of the solutions to an equation, we write these values using set notation inside braces { }.

When solving a linear equation, our goal is to convert it to an equivalent equation that has the variable isolated on one side with a number on the other side; for example, $x = 3$ or $-5 = y$. The value that is on the opposite side of the equation from the variable after it has been isolated is the solution of the equation.

Multiplication Property of Equality

Objective 3   Solve linear equations using the multiplication property of equality.   Our first tool for solving linear equations is the **multiplication property of equality**. It tells us that for any equation, if we multiply both sides of the equation by the same nonzero number, then both sides remain equal to each other.

Multiplication Property of Equality

For any algebraic expressions A and B, and any nonzero number n,

$$\text{if } A = B \text{ then } n \cdot A = n \cdot B.$$

Think of a scale that holds two weights in balance. If we double the amount of weight on each side of the scale, will the scale still be balanced? Of course it will.

We can think of an equation as a scale, and the expressions on each side as the weights that are balanced. Multiplying both sides by a nonzero number leaves both sides still balanced and equal to each other.

Why is it so important that the number we multiply both sides of the equation by not be 0? If we multiply both sides of an equation by 0, the resulting equation will be $0 = 0$. Since this equation no longer contains a variable, we will not be able to isolate the variable on one side of the equation.

EXAMPLE ▸ 3   Solve $\frac{x}{3} = 4$ by using the multiplication property of equality.

Solution

The goal when solving this linear equation is to isolate the variable x on one side of the equation. We begin by multiplying both sides of this equation by 3. The expression $\frac{x}{3}$ is equivalent to $\frac{1}{3}x$, so when we multiply by 3, we are multiplying by the reciprocal of $\frac{1}{3}$. The product of reciprocals is equal to 1, so the resulting expression on the left side of the equation is $1x$, or x.

$$\frac{x}{3} = 4$$

$$3 \cdot \frac{x}{3} = 3 \cdot 4 \qquad \text{Multiply both sides by 3.}$$

$$\overset{1}{\cancel{3}} \cdot \frac{x}{\underset{1}{\cancel{3}}} = 3 \cdot 4 \qquad \text{Divide out common factors.}$$

$$x = 12 \qquad \text{Multiply.}$$

The solution that we have found is $x = 12$. Before moving on, we must check this value to be sure that we have made no mistakes and that it is actually a solution.

Check

$$\frac{x}{3} = 4$$

$$\frac{12}{3} = 4 \qquad \text{Substitute 12 for } x.$$

$$4 = 4 \qquad \text{Divide.}$$

Quick Check 2
Solve $\frac{x}{8} = -2$ by using the multiplication property of equality.

This is a true statement, so $x = 12$ is a solution and the solution set is $\{12\}$.

The multiplication property of equality also allows us to divide both sides of an equation by the same nonzero number without affecting the equality of the two sides. This is because dividing both sides of an equation by a nonzero number, n, is equivalent to multiplying both sides of the equation by the reciprocal of the number, $\frac{1}{n}$.

EXAMPLE ▸ 4   Solve $5y = 40$ by using the multiplication property of equality.

Solution

We begin by dividing both sides of the equation by 5, which will isolate the variable y.

$$5y = 40$$

$$\frac{5y}{5} = \frac{40}{5} \qquad \text{Divide both sides by 5.}$$

$$\frac{\overset{1}{\cancel{5}}y}{\underset{1}{\cancel{5}}} = \frac{40}{5} \qquad \text{Divide out common factors on the left side.}$$

$$y = 8 \qquad \text{Simplify.}$$

Again, we check our solution before writing it in solution set notation.

Check

$$5y = 40$$
$$5(8) = 40 \qquad \text{Substitute 8 for } y.$$
$$40 = 40 \qquad \text{Multiply.}$$

$y = 8$ is indeed a solution and the solution set is $\{8\}$.

Quick Check **3** Solve $4a = 56$ by using the multiplication property of equality.

If the coefficient of the variable term is negative, we must divide both sides of the equation by that negative number.

EXAMPLE ▶ 5   Solve $-7n = -56$ by using the multiplication property of equality.

Solution

In this example, the coefficient of the variable term is -7. To solve this equation, we need to divide both sides by -7.

$$-7n = -56$$
$$\frac{-7n}{-7} = \frac{-56}{-7} \qquad \text{Divide both sides by } -7.$$
$$n = 8 \qquad \text{Simplify.}$$

The check of this solution is left to the reader. The solution set is $\{8\}$.

Quick Check **4** Solve $-9a = 144$ by using the multiplication property of equality.

A Word of Caution   When dividing both sides of an equation by a negative number, keep in mind that this will change the sign of the number on the other side of the equation.

Consider the equation $-x = 16$. The coefficient of the variable term is -1. To solve this equation we can either multiply both sides of the equation by -1, or divide both sides by -1. Either way, the solution is $x = -16$. We could also have solved the equation by inspection by reading the equation $-x = 16$ as *"the opposite of x is 16."* If the opposite of x is 16, then we know that x must be equal to -16.

We will now learn how to solve an equation in which the coefficient of the variable term is a fraction.

EXAMPLE 6   Solve $\frac{3}{8}a = -\frac{5}{2}$ by using the multiplication property of equality.

Solution

In this example, we have a variable multiplied by a fraction. In such a case, we can multiply both sides of the equation by the reciprocal of the fraction. When we multiply a fraction by its reciprocal, the result is 1. This will leave the variable isolated.

$$\frac{3}{8}a = -\frac{5}{2}$$

$$\frac{\overset{1}{\cancel{8}}}{\underset{1}{\cancel{3}}} \cdot \frac{\overset{1}{\cancel{3}}}{\underset{1}{\cancel{8}}}a = \frac{\overset{4}{\cancel{8}}}{3}\left(-\frac{5}{\underset{1}{\cancel{2}}}\right) \qquad \text{Multiply by reciprocal of the fraction } \tfrac{3}{8} \text{ and divide out common factors.}$$

$$a = -\frac{20}{3} \qquad \text{Simplify.}$$

Quick Check 5
Solve $\frac{9}{16}x = \frac{21}{8}$ by using the multiplication property of equality.

The check of this solution is left to the reader. The solution set is $\left\{-\frac{20}{3}\right\}$.

Addition Property of Equality

Objective 4 Solve linear equations using the addition property of equality.
The **addition property of equality** tells us that we can add the same number to both sides of an equation, or subtract the same number from both sides of an equation, without affecting the equality of the two sides.

Addition Property of Equality

> For any algebraic expressions A and B, and any number n,
>
> $$\text{if} \quad A = B \quad \text{then} \quad A + n = B + n,$$
> $$\text{and} \quad A - n = B - n.$$

This property helps us to solve equations in which we have a number either added to or subtracted from a variable on one side of an equation.

EXAMPLE 7   Solve $x + 4 = 11$ using the addition property of equality.

Solution

In this example, the number 4 is being added to the variable x. To isolate the variable, we use the addition property of equality to subtract 4 from both sides of the equation.

$$x + 4 = 11$$
$$x + 4 - 4 = 11 - 4 \qquad \text{Subtract 4 from both sides.}$$
$$x = 7 \qquad \text{Simplify.}$$

Checking this solution, we will substitute 7 for x in the original equation.

Check
$$x + 4 = 11$$
$$7 + 4 = 11 \qquad \text{Substitute 7 for } x.$$
$$11 = 11 \qquad \text{Add.}$$

We have a true statement, so the solution set is $\{7\}$.

Quick Check   6 Solve $a + 22 = -8$ by using the addition property of equality.

EXAMPLE 8   Solve $13 = y - 9$ using the addition property of equality.

Solution

When a value is subtracted from a variable, we isolate the variable by adding that value to both sides of the equation. In this example, we will add 9 to both sides.

$$13 = y - 9$$
$$13 + 9 = y - 9 + 9 \qquad \text{Add 9 to both sides.}$$
$$22 = y \qquad \text{Add.}$$

Check
$$13 = y - 9$$
$$13 = 22 - 9 \qquad \text{Substitute 22 for } y.$$
$$13 = 13 \qquad \text{Subtract.}$$

This is a true statement, so the solution set is $\{22\}$.

Quick Check   7 Solve $x - 28 = -13$ by using the addition property of equality.

EXAMPLE 9   Solve $m + \frac{5}{2} = \frac{2}{3}$ using the addition property of equality.

Solution

Although there are fractions in this equation, we proceed as we did in the previous examples. We begin by subtracting $\frac{5}{2}$ from both sides.

$$m + \frac{5}{2} = \frac{2}{3}$$

$$m + \frac{5}{2} - \frac{5}{2} = \frac{2}{3} - \frac{5}{2} \qquad \text{Subtract } \tfrac{5}{2} \text{ from both sides.}$$

$$m = \frac{4}{6} - \frac{15}{6} \qquad \begin{array}{l}\text{Simplify the left side. On the right side, rewrite the} \\ \text{fractions as equivalent fractions with a common} \\ \text{denominator of 6.}\end{array}$$

$$m = -\frac{11}{6} \qquad \text{Subtract.}$$

The check of this solution is left to the reader. The solution set is $\left\{ -\frac{11}{6} \right\}$.

Quick Check   8 Solve $x - \frac{7}{4} = \frac{5}{6}$ by using the addition property of equality.

In the next section, we will learn how to solve equations requiring us to use both the multiplication and addition properties of equality.

Applications

Objective **5** **Solve applied problems using the multiplication property of equality or the addition property of equality.** We finish this section with an example of an applied problem that can be solved with linear equations.

EXAMPLE ▶**10** Admission to the county fair is $8 per person, so the admission price for a group of x people can be represented by $8x$. If a Cub Scout group paid a total of $208 for admission to the county fair, how many people were in the group?

Solution

We can express this relationship in an equation by using the idea that the total cost of admission is equal to $208. Since the total cost for x people can be represented by $8x$, our equation is $8x = 208$.

$$8x = 208$$
$$\frac{8x}{8} = \frac{208}{8} \qquad \text{Divide both sides by 8.}$$
$$x = 26 \qquad \text{Divide.}$$

You can verify that this value checks as a solution. Whenever we work on an applied problem, we should present our solution as a complete sentence. There were 26 people in the Cub Scout group at the county fair.

Quick Check **9** **Josh spent a total of $26.50 to take a date to the movies. This left him with only $38.50 in his pocket. How much money did he have with him before going to the movies?**

EXERCISES *2.1* ▶

Which of the following equations are linear equations and which are not? If an equation is not a linear equation, explain why.

1. $5x^2 - 7x = 3x + 8$

2. $4x - 9 = 17$

3. $3x - 5(2x + 3) = 8 - x$

4. $\dfrac{7}{x^2} - \dfrac{5}{x} - 13 = 0$

5. $y = 5$

6. $x^4 - 1 = 0$

7. $\dfrac{3x}{11} - \dfrac{5}{4} = \dfrac{2x}{7}$

8. $x \cdot 4 - 1 = 0$

9. $x + \dfrac{3}{x} + 18 = 0$

10. $3(4x - 9) + 7(2x + 5) = 15$

Check to determine whether the given value is a solution of the equation.

11. $x = 7$, $\quad 5x - 9 = 26$

12. $x = 3$, $\quad 2x - 11 = x + 2$

13. $a = 8$, $\quad 3 - 2a = a + 2a - 11$

14. $m = 4$, $\quad 15 - 8m = 3m - 29$

15. $z = \dfrac{1}{4}$, $\quad \dfrac{2}{3}z + \dfrac{11}{6} = 2$

16. $t = \dfrac{5}{3},$ $\dfrac{1}{10}t + \dfrac{1}{3} = \dfrac{1}{2}$

Solve by using the multiplication property of equality.

17. $3x = 54$

18. $7a = -56$

19. $4y = -40$

20. $13x = 91$

21. $6b = 10$

22. $8z = -42$

23. $5a = 0$

24. $0 = 12x$

25. $-5t = 35$

26. $-11x = -44$

27. $-2x = -28$

28. $-9h = 54$

29. $-t = 45$

30. $-y = 31$

31. $\dfrac{x}{3} = 7$

32. $\dfrac{x}{4} = 3$

33. $-\dfrac{t}{8} = 12$

34. $-\dfrac{g}{13} = -7$

35. $\dfrac{2}{3} = \dfrac{x}{12}$

36. $\dfrac{3}{8} = \dfrac{x}{4}$

37. $-\dfrac{2}{5}x = 4$

38. $-\dfrac{5}{6}x = 15$

Solve by using the addition property of equality.

39. $a + 7 = 10$

40. $b + 5 = 13$

41. $x + 8 = 4$

42. $x + 1 = -15$

43. $x + 13 = 30$

44. $n - 19 = 11$

45. $a + 3.2 = 5.7$

46. $x - 14.9 = -21.2$

47. $12 = x + 3$

48. $4 = m + 10$

49. $b - 7 = 13$

50. $a - 9 = 99$

51. $t - 7 = -4$

52. $n - 3 = -18$

53. $-4 + x = 19$

54. $-15 + x = -7$

55. $x + 9 = 0$

56. $b - 17 = 0$

57. $a + \dfrac{5}{12} = \dfrac{5}{9}$

58. $x - \dfrac{7}{10} = \dfrac{4}{15}$

59. $a + 5 + 6 = 7$

60. $x + 4 - 9 = -3$

Solve the equation.

61. $60 = 5a$

62. $a - \dfrac{3}{4} = \dfrac{7}{6}$

63. $x - 27 = -11$

64. $\dfrac{3}{14}x = \dfrac{5}{2}$

65. $-\dfrac{b}{10} = -3$

66. $b + 39 = 30$

67. $x + 24 = -17$

68. $\dfrac{x}{5} = 13$

69. $11 = -\dfrac{n}{7}$

70. $\dfrac{5}{3} = n + \dfrac{7}{2}$

71. $y + \dfrac{2}{15} = -\dfrac{5}{6}$

72. $47 = y + 35$

73. $0 = x + 56$

74. $x - 38 = -57$

75. $t - \dfrac{17}{8} = -\dfrac{7}{4}$

76. $\dfrac{t}{15} = -7$

77. $\dfrac{4}{9}x = -\dfrac{14}{15}$

78. $x - \dfrac{3}{5} = -\dfrac{9}{10}$

79. $0 = 45m$

80. $40 = -6m$

The values given are solutions of a linear equation. Give an equation that can be solved using the multiplication property of equality and that has the given solution.

81. $x = 7$

82. $x = -13$

83. $n = \dfrac{5}{2}$

84. $m = -\dfrac{3}{10}$

The values given are solutions of a linear equation. Give an equation that can be solved using the addition property of equality and that has the given solution.

85. $x = -6$

86. $x = 14$

87. $b = \dfrac{1}{6}$

88. $a = -\dfrac{9}{14}$

Set up a linear equation and solve it for the following problems.

89. Zoe has only nickels in her pocket. If she has $1.35 in her pocket, how many nickels does she have?

90. Carter sold lemonade in front of his house to raise money for a new bike. He charged 25 cents per cup. If his total sales from yesterday were $23.75, how many cups of lemonade did he sell?

91. A local garage band, the Grease Monkeys, held a rent party. They charged $3 per person for admission, with the proceeds to be used to pay their rent. If their rent is $425, and they ended up with an extra $52 after paying the rent, how many people came to see them play?

92. Ross organized a tour of a local winery. Attendees paid Ross $12 to go on the tour, plus another $5 for lunch. If Ross collected $493, how many people came on the tour?

93. An insurance company hired 8 new employees. This brought their total to 174 employees. How many employees did the company have before these 8 people were hired?

94. Geena scored 17 points higher than Jared on the last math exam. If Geena's score was 85, find Jared's score.

95. As a cold front was moving in, the temperature in Visalia dropped by 19° F in a two-hour period. If the temperature dropped to 37° F, what was the temperature before the cold front moved in?

96. Stephanie switched to a vegan diet and lost 13 pounds in three months. If her new weight is 134 pounds, what was her weight prior to switching to a vegan diet?

97. Explain why the equation $0x = 15$ cannot be solved.

98. Explain why we cannot multiply both sides of an equation by 0 when applying the multiplication property of equality.

99. Find a value for x such that the expression $x + 21$ is less than -39.

100. Find a value of x such that the expression $-8x$ is greater than 95.

Study Tip **REVISITED**   One effective way to use your textbook is to read a section in the text before it is covered in class. This will give you an idea about the main concepts covered in the section, and these concepts can be clarified by your instructor in class.

When reading ahead, you should scan the section. Look for definitions that are introduced in the section, as well as procedures developed in the section. Pay closer attention to the examples. If you find a step in the examples that you cannot understand, you can ask your instructor about it in class.

2.2

SOLVING LINEAR EQUATIONS: A GENERAL STRATEGY

1 Solve linear equations using both the multiplication property of equality and the addition property of equality.

2 Solve linear equations containing fractions.

3 Solve linear equations using the five-step general strategy.

4 Identify linear equations with no solution.

5 Identify linear equations with infinitely many solutions.

6 Solve literal equations for a specified variable.

In the previous section, we solved equations that required the use of only one operation to isolate the variable. In this section, we will learn how to solve equations requiring the use of both the multiplication and addition properties of equality.

Solving Linear Equations

Objective 1   Solve linear equations using both the multiplication property of equality and the addition property of equality. Suppose we needed to solve the equation $4x - 7 = 17$. Should we divide both sides by 4 first? Should we add 7 to both sides first? We must first determine which of the two numbers is most closely connected to the variable. The order of operations tells us that in the expression $4x - 7$, we first multiply 4 by x and then subtract 7 from the result. To isolate the variable x, we undo these operations in the opposite order. We will first add 7 to both sides to undo the subtraction, and then we will divide both sides by 4 to undo the multiplication.

Solution	Check
$4x - 7 = 17$	$4(6) - 7 = 17$
$4x - 7 + 7 = 17 + 7$	$24 - 7 = 17$
$4x = 24$	$17 = 17$
$\dfrac{4x}{4} = \dfrac{24}{4}$	
$x = 6$	

Since our solution $x = 6$ checks, the solution set is $\{6\}$.

EXAMPLE 1   Solve the equation $3x + 41 = 8$.

Solution

To solve this equation, we will begin by subtracting 41 from both sides. This will isolate the term $3x$. We can then divide both sides of the equation by 3 to isolate the variable x.

$$3x + 41 = 8$$
$$3x + 41 - 41 = 8 - 41 \qquad \text{Subtract 41 from both sides to isolate } 3x.$$
$$3x = -33 \qquad \text{Subtract.}$$
$$\frac{3x}{3} = -\frac{33}{3} \qquad \text{Divide both sides by 3 to isolate } x.$$
$$x = -11 \qquad \text{Divide.}$$

Now we check our solution.

$$3x + 41 = 8$$
$$3(-11) + 41 = 8 \qquad \text{Substitute } -11 \text{ for } x.$$
$$-33 + 41 = 8 \qquad \text{Multiply.}$$
$$8 = 8 \qquad \text{Simplify.}$$

Quick Check **1** Since $x = -11$ produced a true statement, the solution set is $\{-11\}$.

Solve the equation
$5x - 2 = 33$.

EXAMPLE **2**   Solve the equation $-8x - 19 = 13$.

Solution

$$-8x - 19 = 13$$
$$-8x - 19 + 19 = 13 + 19 \qquad \text{Add 19 to both sides to isolate } -8x.$$
$$-8x = 32 \qquad \text{Add.}$$
$$\frac{-8x}{-8} = \frac{32}{-8} \qquad \text{Divide by } -8 \text{ to isolate } x.$$
$$x = -4 \qquad \text{Divide.}$$

Now we check our solution.

$$-8x - 19 = 13$$
$$-8(-4) - 19 = 13 \qquad \text{Substitute } -4 \text{ for } x.$$
$$32 - 19 = 13 \qquad \text{Multiply.}$$
$$13 = 13 \qquad \text{Simplify.}$$

Quick Check **2** The solution set is $\{-4\}$.

Solve the equation
$6 - 4x = 38$.

Solving Linear Equations Containing Fractions

Objective **2** **Solve linear equations containing fractions.** If an equation contains fractions, it can be helpful to convert it to an equivalent equation that does not contain fractions before solving it. This can be done by multiplying both sides of the equation by the LCM of the denominators.

EXAMPLE **3**   Solve the equation $\frac{2}{3}x - 5 = \frac{3}{4}$.

Solution

This equation contains two fractions, and the denominators are 3 and 4. The LCM of these two denominators is 12, so we will begin by multiplying both sides of the equation by 12.

$$\frac{2}{3}x - 5 = \frac{3}{4}$$
$$12\left(\frac{2}{3}x - 5\right) = 12\left(\frac{3}{4}\right) \qquad \text{Multiply both sides by the LCM 12.}$$
$$\overset{4}{\cancel{12}} \cdot \frac{2}{\underset{1}{\cancel{3}}}x - 12 \cdot 5 = \overset{3}{\cancel{12}} \cdot \frac{3}{\underset{1}{\cancel{4}}} \qquad \text{Distribute and divide out common factors.}$$

$$8x - 60 = 9 \qquad \text{Multiply.}$$
$$8x - 60 + 60 = 9 + 60 \qquad \text{Add 60 to both sides to isolate } 8x.$$
$$8x = 69 \qquad \text{Add.}$$
$$\frac{8x}{8} = \frac{69}{8} \qquad \text{Divide by 8 to isolate } x.$$
$$x = \frac{69}{8} \qquad \text{Simplify.}$$

Quick Check 3

Solve the equation
$\frac{2}{7}x + \frac{1}{2} = \frac{4}{3}$.

The check of this solution is left to the reader. The solution set is $\left\{\frac{69}{8}\right\}$.

A Word of Caution   When multiplying both sides of an equation by the LCM of the denominators, be sure to multiply each term by the LCM, including any terms that do not contain fractions.

A General Strategy for Solving Linear Equations

Objective 3   Solve linear equations using the five-step general strategy. We will now examine a process that can be used to solve any linear equation. This process works not only for types of equations we have already learned to solve but also for more complicated equations, such as $7x + 4 = 3x - 20$ and $5(2x - 9) + 3x = 7(3 - 8x)$.

Solving Linear Equations

1. **Simplify each side of the equation completely.**
   • Use the distributive property to clear any parentheses.
   • If there are fractions in the equation, multiply both sides of the equation by the LCM of the denominators to clear the fractions from the equation.
   • Combine any like terms that are on the same side of the equation. After completing this step, the equation should contain at most one variable term and at most one constant term on each side.
2. **Collect all variable terms on one side of the equation.** If there is a variable term on each side of the equation, use the addition property of equality to place both variable terms on the same side of the equation. It is a good idea to move the term whose coefficient is the lesser of the two coefficients. For instance, if one side of the equation has a variable term $4x$ and the other side has a variable term $7x$, then we will subtract $4x$ from both sides of the equation as 4 is less than 7. This will prevent having to divide both sides of the equation by a negative number, which could lead to a careless sign error.
3. **Collect all constant terms on the other side of the equation.** If there is a constant term on each side of the equation, use the addition property of equality to isolate the variable term.
4. **Divide both sides of the equation by the coefficient of the variable term.** At this point our equation should be of the form $ax = b$. We use the multiplication property of equality to find our solution.
5. **Check your solution.**

In the next example, we will solve equations that have variable terms and constant terms on both sides of the equation.

EXAMPLE ▶ 4   Solve the equation $3x + 8 = 7x - 6$.

Solution

We have no fractions to clear in this equation, there are no multiplications to be performed, and there are no like terms to be combined. We begin by gathering the variable terms on one side of the equation.

$$3x + 8 = 7x - 6$$

$$3x + 8 - 3x = 7x - 6 - 3x$$ Subtract $3x$ from both sides to gather the variable terms on the right side of the equation.

$$8 = 4x - 6$$ Subtract.

$$8 + 6 = 4x - 6 + 6$$ Add 6 to both sides to isolate $4x$.

$$14 = 4x$$ Add.

$$\frac{14}{4} = \frac{4x}{4}$$ Divide both sides by 4 to isolate x.

$$\frac{7}{2} = x$$ Simplify.

Quick Check ◀ 4   The check of this solution is left to the reader. The solution set is $\left\{\frac{7}{2}\right\}$.

Solve the equation
$6x + 19 = 3x - 8.$

Using Your Calculator   We can use a calculator to check the solutions to an equation. Substitute the value for the variable and simplify the expressions on each side of the equation. Here is how the check of the solution $x = \frac{7}{2}$ would look on the TI-83/84.

```
3(7/2)+8
                18.5
7(7/2)-6
                18.5
```

Since each expression is equal to 18.5 when $x = \frac{7}{2}$, this solution checks.

In the next example, we will solve an equation containing like terms on the same side of the equation.

EXAMPLE ▶ 5   Solve the equation $13x - 43 - 9x = 6x + 37 + 5x - 66$.

Solution

This equation has like terms on each side of the equation, so we begin by combining these like terms.

$$13x - 43 - 9x = 6x + 37 + 5x - 66$$

$$4x - 43 = 11x - 29$$ Combine like terms on each side.

$$4x - 43 - 4x = 11x - 29 - 4x$$ Subtract $4x$ from both sides to gather variable terms on one side of the equation.

$$-43 = 7x - 29$$ Subtract.

$$-43 + 29 = 7x - 29 + 29$$ Add 29 to both sides to isolate $7x$.

$$-14 = 7x$$ Add.

$$-\frac{14}{7} = \frac{7x}{7}$$ Divide both sides by 7 to isolate x.

$$-2 = x$$ Divide.

The check of this solution is left to the reader. The solution set is $\{-2\}$.

Quick Check 5 Solve the equation $6x - 9 + 4x = 3x + 13 - 8$.

EXAMPLE ▶ 6   Solve the equation $6(3x - 8) + 14 = x + 3(5 - x) + 1$.

Solution

We begin by distributing the 6 on the left side of the equation and the 3 on the right side of the equation. Then we combine like terms before solving.

$$6(3x - 8) + 14 = x + 3(5 - x) + 1$$
$$18x - 48 + 14 = x + 15 - 3x + 1 \qquad \text{Distribute.}$$
$$18x - 34 = -2x + 16 \qquad \text{Combine like terms.}$$
$$18x - 34 + 2x = -2x + 16 + 2x \qquad \text{Add } 2x \text{ to both sides to gather variable}$$
$$\text{terms on the left side of the equation.}$$

$$20x - 34 = 16 \qquad \text{Add.}$$
$$20x - 34 + 34 = 16 + 34 \qquad \text{Add 34 to both sides to isolate } 20x.$$
$$20x = 50 \qquad \text{Add.}$$
$$\frac{20x}{20} = \frac{50}{20} \qquad \text{Divide by 20 to isolate } x.$$

$$x = \frac{5}{2} \qquad \text{Simplify.}$$

The check of this solution is left to the reader. The solution set is $\left\{\frac{5}{2}\right\}$.

Quick Check 6 Solve the equation $3(2x - 7) + x = 2(x - 9) - 18$.

EXAMPLE ▶ 7   Solve the equation $8x - 7(3x - 5) = 6(4 - x) + x - 7$.

Solution

When applying the distributive property, we must be careful to distribute -7 on the left side. This changes the signs of the terms inside the parentheses.

$$8x - 7(3x - 5) = 6(4 - x) + x - 7$$
$$8x - 21x + 35 = 24 - 6x + x - 7 \qquad \text{Distribute. (Be careful with signs.)}$$
$$-13x + 35 = -5x + 17 \qquad \text{Combine like terms.}$$
$$-13x + 35 + 13x = -5x + 17 + 13x \qquad \text{Add } 13x \text{ to both sides to gather variable terms on the right side of the}$$
$$\text{equation.}$$

$$35 = 8x + 17 \qquad \text{Add.}$$
$$35 - 17 = 8x + 17 - 17 \qquad \text{Subtract 17 from both sides to isolate } 8x.$$

$$18 = 8x \qquad \text{Subtract.}$$

$$\frac{18}{8} = \frac{8x}{8}$$

Divide both sides by 8 to isolate x.

$$\frac{9}{4} = x$$

Simplify.

The check of this solution is left to the reader. The solution set is $\left\{\frac{9}{4}\right\}$.

Quick Check 7 Solve the equation $5 - 4(3x - 8) + 7x = 8 - (x - 13)$.

In the next example, we will clear the equations of fractions before solving.

EXAMPLE 8   Solve the equation $\frac{3}{5}x - \frac{2}{3} = \frac{3}{4}x + \frac{5}{6} + 2x$.

Solution

We begin by finding the LCM of the four denominators (5, 3, 4, and 6), which is 60. We then multiply both sides of the equation by 60 to clear the equation of fractions.

$$\frac{3}{5}x - \frac{2}{3} = \frac{3}{4}x + \frac{5}{6} + 2x$$

$$60\left(\frac{3}{5}x - \frac{2}{3}\right) = 60\left(\frac{3}{4}x + \frac{5}{6} + 2x\right)$$

Multiply both sides by LCM (60).

$$\overset{12}{\cancel{60}} \cdot \frac{3}{\cancel{5}}x - \overset{20}{\cancel{60}} \cdot \frac{2}{\cancel{3}} = \overset{15}{\cancel{60}} \cdot \frac{3}{\cancel{4}}x + \overset{10}{\cancel{60}} \cdot \frac{5}{\cancel{6}} + 60 \cdot 2x$$

Distribute and divide out common factors.

$$36x - 40 = 45x + 50 + 120x$$

Multiply.

$$36x - 40 = 165x + 50$$

Combine like terms.

$$36x - 40 - 36x = 165x + 50 - 36x$$

Subtract $36x$ from both sides to gather variable terms on the right side of the equation.

$$-40 = 129x + 50$$

Subtract.

$$-40 - 50 = 129x + 50 - 50$$

Subtract 50 from both sides to isolate $129x$.

$$-90 = 129x$$

Subtract.

$$-\frac{90}{129} = \frac{129x}{129}$$

Divide both sides by 129 to isolate x.

$$-\frac{30}{43} = x$$

Simplify.

Quick Check 8

Solve the equation
$\frac{1}{10}x - \frac{2}{5} = \frac{1}{20}x - \frac{7}{10}$.

The check of this solution is left to the reader. The solution set is $\left\{-\frac{30}{43}\right\}$.

Contradictions and Identities

Objective 4 Identify linear equations with no solution. In each equation that we have solved to this point, there has been exactly one solution. This will not always be the case. We now turn our attention to two special types of equations: contradictions and identities.

A **contradiction** is an equation that will always be false, regardless of the value that we substitute for the variable. A contradiction has no solution, so its solution set is the empty set { }. The empty set is also known as the *null set* and is denoted by the symbol $\varnothing$.

EXAMPLE ▶ 9   Solve the equation $5x + 5 = 5x + 6$.

Solution

We begin solving this equation by gathering both variable terms on the same side of the equation. This can be done by subtracting $5x$ from both sides of the equation.

$$5x + 5 = 5x + 6$$
$$5x + 5 - 5x = 5x + 6 - 5x \qquad \text{Subtract } 5x \text{ from both sides.}$$
$$5 = 6 \qquad \text{Simplify.}$$

We are left with an equation that is a false statement, since 5 can never be equal to 6. This equation is a contradiction and has no solutions. Its solution set is $\varnothing$.

Quick Check 9
Solve the equation
$3x - 4 = 3x + 4.$

Objective 5 Identify linear equations with infinitely many solutions. An **identity** is an equation that is always true. If we substitute any real number for the variable in an identity, it will produce a true statement. The solution set for an identity is the set of all real numbers. We will denote the set of all real numbers as $\Re$. An identity has infinitely many solutions rather than a single solution.

EXAMPLE ▶ 10   Solve the equation $3x + 8 = 3x + 8$.

Solution

We will begin to solve this equation by attempting to gather all variable terms on one side of the equation.

$$3x + 8 = 3x + 8$$
$$3x + 8 - 3x = 3x + 8 - 3x \qquad \text{Subtract } 3x \text{ from both sides.}$$
$$8 = 8$$

Since 8 is always equal to 8, regardless of the value of x, this equation is an identity. Its solution set is the set of all real numbers $\Re$.

Quick Check 10
Solve the equation
$5a + 4 = 4 + 5a.$

Literal Equations

Objective 6 Solve literal equations for a specified variable. A **literal equation** is an equation that contains two or more variables.

Perimeter of a Rectangle

The **perimeter** of a rectangle is a measure of the distance around the rectangle. The formula for the perimeter (P) of a rectangle with length L and width W is $P = 2L + 2W$.

$$P = 2L + 2W$$

W

L

The equation $P = 2L + 2W$ is a literal equation with three variables. Often we will be asked to solve literal equations for one variable in terms of the other variables in the equation. In the perimeter example, if we solved for the width W in terms of the length L and the perimeter P, then we would have a formula for the width of a rectangle if we knew its length and perimeter.

We solve literal equations by isolating the specified variable. We will use the same general strategy that we used for solving linear equations. We treat the other variables in the equation as if they were constants.

EXAMPLE ▶11 Solve the literal equation $P = 2L + 2W$ (perimeter of a rectangle) for W.

Solution

We want to gather all terms containing the variable W on one side of the equation and gather all other terms on the other side. This can be done by subtracting $2L$ from both sides.

$$P = 2L + 2W$$
$$P - 2L = 2L + 2W - 2L \qquad \text{Subtract } 2L \text{ to isolate } 2W.$$
$$P - 2L = 2W \qquad \text{Subtract.}$$
$$\frac{P - 2L}{2} = \frac{2W}{2} \qquad \text{Divide by 2 to isolate } W.$$
$$\frac{P - 2L}{2} = W$$

Quick Check 11
Solve the literal equation
$x + 2y = 5$ for y.

We usually rewrite the equation so that the variable we solved for appears on the left side: $W = \frac{P - 2L}{2}$.

EXAMPLE ▶12 Solve the literal equation $A = \frac{1}{2}bh$ (area of a triangle) for h.

Solution

Since this equation has a fraction, we can begin by multiplying both sides by 2.

$$A = \frac{1}{2}bh$$

$$2 \cdot A = 2 \cdot \frac{1}{2}bh \qquad \text{Multiply both sides by 2 to clear fractions.}$$

$$2 \cdot A = \overset{1}{\cancel{2}} \cdot \frac{1}{\underset{1}{\cancel{2}}}bh \qquad \text{Divide out common factors.}$$

$$2A = bh \qquad \text{Multiply.}$$

$$\frac{2A}{b} = \frac{bh}{b} \qquad \text{Divide both sides by } b \text{ to isolate } h.$$

$$\frac{2A}{b} = h \quad \text{or} \quad h = \frac{2A}{b}.$$

Quick Check 12
Solve the literal equation
$\frac{1}{5}xy = z$ for x.

Some geometry formulas contain the symbol π, such as $C = 2\pi r$ and $A = \pi r^2$. The symbol π is the Greek letter pi and is used to represent an irrational number that is approximately equal to 3.14. When solving literal equations containing π, do not replace it by its approximate value.

Solve.

1. $3x + 16 = 7$

2. $2x - 9 = 15$

3. $5 - 4x = 3$

4. $16 - 7x = -12$

5. $6 = 2a + 11$

6. $18 = 8 - 6b$

7. $9x + 24 = 24$

8. $-4x + 30 = -34$

9. $\dfrac{1}{4}x - \dfrac{1}{3} = \dfrac{5}{12}$

10. $\dfrac{3}{11}x + \dfrac{5}{2} = 8$

11. $\dfrac{x}{12} - \dfrac{11}{6} = \dfrac{5}{4}$

12. $\dfrac{2}{5} - \dfrac{3}{8}x = -\dfrac{11}{10}$

13. $7a + 11 = 5a - 9$

14. $3m - 11 = 9m + 5$

15. $16 - 4x = 2x + 61$

16. $6n + 14 = 27 + 6n$

17. $-6t + 17 = 3t - 34$

18. $-31 + 11n = 4n + 60$

19. $\dfrac{1}{2}x - 3 = \dfrac{11}{5} - \dfrac{3}{4}x$

20. $\dfrac{3}{5} - \dfrac{5}{6}x = 2x + \dfrac{4}{3}$

21. $27x - 16 - 6x = 14 + 21x - 30$

22. $-5n + 11 - 4n + 9 = n - 45$

23. $9n - 17 - 5n - 32 = 23 + 4n + 5$

24. $5t + 72 + 8t = 17t - 28 + 11t$

25. $2(2x - 5) - 7 = x - 12$

26. $10 + 4(3x + 1) = 5x - 7 - 2x$

27. $13(3x + 4) - 7(2x - 5) = 5x - 13$

28. $3(2x - 8) - 5(15 - 6x) = 9(4x - 11)$

Find a linear equation that has the given solution. (Answers may vary.)

29. $x = 9$

30. $x = 6$

31. $a = \dfrac{1}{9}$

32. $b = \dfrac{5}{7}$

33. Write a linear equation that is a contradiction.

34. Write a linear equation that is an identity.

35. Write an equation that has infinitely many solutions.

36. Write an equation that has no solutions.

37. Write an equation whose single solution is negative.

38. Write an equation whose single solution is a fraction.

Solve the following literal equations for the specified variable.

39. $5x + y = -2$ for y

40. $-6x + y = 9$ for y

41. $8x + 2y = 4$ for y

42. $12x + 4y = 20$ for y

43. $6x - 3y = 9$ for y

44. $15x - 5y = -20$ for y

45. $P = a + b + c$ for b

46. $P = a + b + 2c$ for c

47. $d = r \cdot t$ for t

48. $d = r \cdot t$ for r

49. $C = 2\pi r$ for r

50. $S = 2\pi r h$ for r

Temperatures are usually reported using either the Celsius scale or the Fahrenheit scale. To convert a Celsius temperature to a Fahrenheit temperature, we multiply the Celsius temperature by $\frac{9}{5}$ and add 32 to the result.

51. If the temperature outside is 68° F, find the Celsius temperature.

52. If the normal body temperature for a person is 98.6° F, find the Celsius equivalent.

53. A number is doubled and then added to 7. If the result is 715, find the number.

54. If three-fourths of a number is added to 11, the result is 20. Find the number.

55. It costs $200 to rent a booth at a craft fair. Tina wants to sell homemade kites at the fair. It cost Tina $4 in material to make each kite.

    a) If she has $520 available to buy material and pay expenses, how many kites could she make to sell at the craft fair after paying the $200 rental charge?

    b) If she is able to sell all of the kites, how much would she need to charge for each kite in order to break even?

    c) How much should she charge for each kite in order to make a profit of $500?

56. A churro is a Mexican dessert pastry. Dan is able to buy churros for $0.75 each, which he plans to sell at a high school baseball game. Dan has to donate $50 to the high school team in order to be able to sell the churros at the game.

    a) If Dan has $110 to invest in this venture, how many churros will he be able to buy after paying $50 to the team?

    b) If Dan charges $2 for each churro, how many must he sell in order to break even?

    c) If Dan charges $2 for each churro and he is able to sell all of his churros, what will his profit be?

57. Find the missing value such that $x = 3$ is a solution of $5x - ? = 11$.

58. Find the missing value such that $x = -2$ is a solution of $3(2x - 5) + ? = 5(3 - x)$.

59. Find the missing value such that $x = -\frac{3}{2}$ is a solution of $2x + 9 = 6x + ?$.

60. Find the missing value such that $x = \frac{2}{5}$ is a solution of $4(2x + 7) = 7(3x + ?)$.

61. Solve the linear equation $14 = 2x - 9$, showing all steps. Next to this, show all of the steps necessary to solve the literal equation $y = mx + b$ for x. In your own words, write a paragraph explaining how these two processes are similar and how they are different.

62. Write a word problem whose solution can be found from the equation $4.75x + 225 = 795$. Explain how the equation fits your problem.

> **Study Tip** REVISITED   The Quick Check exercises following most examples in this text are designed to be similar to the examples of the book. After reading through, and possibly reworking an example, try the corresponding Quick Check exercise. After checking your answer in the back of the book, you will be able to determine whether you are ready to move forward in the section or whether further practice is needed.

2.3

PROBLEM SOLVING; APPLICATIONS OF LINEAR EQUATIONS

Objectives

1 Understand the six steps for solving applied problems.
2 Solve problems involving an unknown number.
3 Solve problems involving geometric formulas.
4 Solve problems involving consecutive integers.
5 Solve problems involving motion.
6 Solve other applied problems.

Introduction to Problem Solving

Although the ability to perform mathematical computations and abstract algebraic manipulations is important and valuable, one of the most important skills developed in math classes is the skill of problem solving. Every day we are faced with making important decisions and predictions, and the thought process required in decision-making is quite similar to the process of solving mathematical problems.

When faced with a problem to solve in the real world, we first take inventory of the facts that we know. We also determine exactly what it is that we are trying to figure out. Then we develop a plan for taking what we already know and using it to help us figure out how to solve the problem. Once we have solved the problem, we reflect on the route that we took to solve it in order to make sure that that route was a logical way to solve the problem. After examining our solution for correctness and practicality, we often finish by presenting our solution to others for their consideration, information, or approval.

An example of one such real-world problem would be determining how early to leave for school on the first day of classes. Suppose your first class is at 9 A.M. The problem to solve is figuring out what time to leave your house so that you will not be late. Now think about the facts that you know: It's normally a 20-minute drive to school, and the classroom is a 10-minute walk from the parking lot. Add information specific to the first day of classes: Traffic near the school is more congested than normal, and it takes longer to find a parking space in the parking lot. Based on your past experience, you figure an extra 15 minutes for traffic and another 10 minutes for parking. Adding up all of these times tells us that you need 55 minutes to get to your class. Adding an extra 15 minutes, just to be sure, you decide that you need to leave home by 7:50 A.M.

Objective 1   Understand the six steps for solving applied problems. In the previous example, we identified the problem to be solved, gathered our facts, and used them to find a solution to the problem. Solving applied math problems will require that we follow a similar procedure. George Pólya, this chapter's Mathematician in History, was a leader in the study of problem solving. Here is a general plan for solving applied math problems, based on the work of Pólya's text *How to Solve It*.

Solving Applied Problems

1. **Read the problem.** This step is often overlooked, but misreading or misinterpreting the problem essentially guarantees an incorrect solution. Read the problem once quickly to get a rough idea of what is going on, and then read it more carefully a second time to gather all of the important information.

continued

2. **List all of the important information.** Identify all known quantities presented in the problem, and determine which quantities we need to find. Creating a table to hold this information or a drawing to represent the problem are good ideas.
3. **Assign a variable to the unknown quantity.** If there is more than one unknown quantity, express them in terms of the same variable, if possible.
4. **Find an equation relating the known values to the variable expressions representing the unknown quantities.** Sometimes, this equation can be translated directly from the wording of the problem. Other times the equation will depend on general facts that are known outside the statement of the problem, such as geometry formulas.
5. **Solve the equation.** Solving the equation is not the end of the problem, as the value of the variable is often not what we were originally looking for. If we were asked for the length and width of a rectangle and we reply "x equals 4," we have not answered the question. Check your solution to the equation.
6. **Present the solution.** Take the value of the variable from the solution to the equation and use it to figure out all unknown quantities. Check these values to be sure that they make sense in the context of the problem. For example, the length of a rectangle cannot be negative. Finally, present your solution in a complete sentence, using the proper units.

We will now put this strategy to use. All of the applied problems in this section will lead to linear equations.

Objective 2   Solve problems involving an unknown number.

EXAMPLE 1   Seven more than twice a number is 39. Find the number.

Solution

This is not a very exciting problem, but it provides us with a good opportunity to practice setting up applied problems.

There is one unknown in this problem. Let's use n to represent the *unknown number.*

Unknown
Number: n

Now we translate the sentence "Seven more than twice a number is 39" into an equation. In this case, twice a number is $2n$, so seven more than twice a number is $2n + 7$:

$$\underbrace{\text{Seven more than twice a number}}_{2n + 7} \underbrace{\text{is}}_{=} \underbrace{39}_{39}$$

Now we solve the equation.

$$
\begin{aligned}
2n + 7 &= 39 \\
2n &= 32 \qquad \text{Subtract 7 from both sides.} \\
n &= 16 \qquad \text{Divide both sides by 2.}
\end{aligned}
$$

The solution of the equation is $n = 16$. Now we refer back to the table containing our unknown quantity. Since our unknown number was n, the solution to the problem is the same as the solution of the equation. The number is 16.

It is a good idea to check our solution. Twice our number is 32, and seven more than that is 39.

Quick Check **1**

Three less than four times a number is 29. Find the number.

Geometry Problems

Objective **3** **Solve problems involving geometric formulas.** The next example involves the perimeter of a rectangle, which is a measure of the distance around the outside of the rectangle. The perimeter P of a rectangle whose length is L and whose width is W is given by the formula $P = 2L + 2W$.

EXAMPLE **2**   Mark's vegetable garden is in the shape of a rectangle, and he can enclose it with 96 feet of fencing. If the length of his garden is 16 feet more than 3 times the width, find the dimensions of his garden.

Solution

The unknown quantities are the length and the width. Since the length is given in terms of the width, it is a good idea to pick a variable to represent the width. We can then write the length in terms of this variable. We will let w represent the width of the rectangle. The length is 16 feet more than 3 times the width, so we can express the length as $3w + 16$. One piece of information that we are given is that the perimeter is 96 feet. This information can be recorded in the following table or drawing:

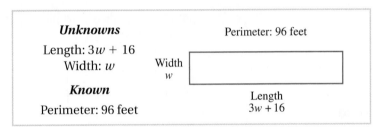

We start with the formula for perimeter and substitute the appropriate values and expressions.

$$P = 2L + 2W$$
$$96 = 2(3w + 16) + 2w \qquad \text{Substitute 96 for } P, 3w + 16 \text{ for } L, \text{ and } w \text{ for } W.$$
$$96 = 6w + 32 + 2w \qquad \text{Distribute.}$$
$$96 = 8w + 32 \qquad \text{Combine like terms.}$$
$$64 = 8w \qquad \text{Subtract 32 from both sides.}$$
$$8 = w \qquad \text{Divide both sides by 8.}$$

We now take this solution and substitute 8 for w in the expressions for length and width, which can be found in the table where we listed our unknowns:

Length: $3w + 16 = 3(8) + 16 = 40$
Width: $w = 8$

The length of Mark's garden is 40 feet, and the width is 8 feet. It checks that the perimeter of this rectangle is 96 feet.

Quick Check   2 A rectangular living room has a perimeter of 80 feet. The length of the room is 8 feet less than twice the width of the room. Find the dimensions of the room.

Here is a summary of useful formulas and definitions from geometry:

Geometry Formulas and Definitions

- Perimeter of a triangle with sides s_1, s_2, and s_3: $P = s_1 + s_2 + s_3$

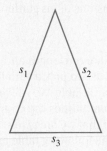

- Perimeter of a square with side s: $P = 4s$

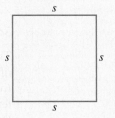

- Perimeter of a rectangle with length L and width W: $P = 2L + 2W$

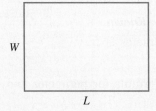

- **Circumference** (distance around the outside) of a circle with radius r: $C = 2\pi r$

- Area of a triangle with base b and height h: $A = \dfrac{1}{2}bh$

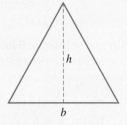

- Area of a square with side s: $A = s^2$

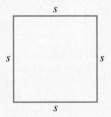

- Area of a rectangle with length L and width W: $A = LW$

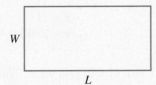

- Area of a circle with radius r: $A = \pi r^2$

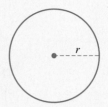

- An **equilateral triangle** is a triangle that has three equal sides.

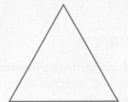

- An **isosceles triangle** is a triangle that has at least two equal sides.

EXAMPLE ▸ 3   An isosceles triangle has a perimeter of 30 cm. The third side is 3 cm shorter than each of the sides that have equal lengths. Find the lengths of the three sides.

Solution

In this problem, there are three unknowns, which are the lengths of the three sides. Because the triangle is an isosceles triangle, the first two sides have equal lengths. We can represent the length of each side by the variable x. The third side is 3 cm shorter than the other two sides and its length can be represented by $x - 3$. Here is a summary of this information, together with the known perimeter:

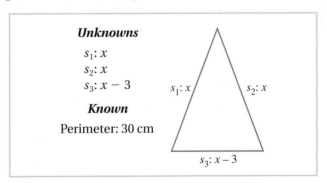

Unknowns
s_1: x
s_2: x
s_3: $x - 3$

Known
Perimeter: 30 cm

We now set up the equation and solve for x.

$$P = s_1 + s_2 + s_3$$
$$30 = x + x + x - 3 \qquad \text{Substitute 30 for } P, x \text{ for } s_1 \text{ and } s_2, \text{ and } x - 3 \text{ for } s_3.$$
$$30 = 3x - 3 \qquad \text{Combine like terms.}$$
$$33 = 3x \qquad \text{Add 3 to both sides.}$$
$$11 = x \qquad \text{Divide both sides by 3.}$$

Looking back to the table, we substitute 11 for x.

$$s_1: x = 11$$
$$s_2: x = 11$$
$$s_3: x - 3 = 11 - 3 = 8$$

The three sides are 11 cm, 11 cm, and 8 cm. The perimeter of this triangle checks to be 30 cm.

Quick Check   **3**   The perimeter of a triangle is 112 inches. The longest side is 8 inches longer than 3 times the shortest side. The other side of the triangle is 20 inches longer than twice the shortest side. Find the lengths of the 3 sides.

Consecutive Integers

Objective   4   Solve problems involving consecutive integers. The following example involves consecutive integers, which are integers that are 1 unit apart from each other on the number line.

EXAMPLE 4   The sum of three consecutive integers is 87. Find them.

Solution

In this problem, the three consecutive integers are our three unknowns. We will let x represent the first integer. Since consecutive integers are 1 unit apart from each other we can let $x + 1$ represent the second integer. Adding another 1, we can let $x + 2$ represent the third integer. Here is a table of the unknowns:

> *Unknowns*
>
> First: x
> Second: $x + 1$
> Third: $x + 2$

We know that the sum of these three integers is 87, which leads to our equation.

$$x + (x + 1) + (x + 2) = 87$$
$$3x + 3 = 87 \qquad \text{Combine like terms.}$$
$$3x = 84 \qquad \text{Subtract 3 from both sides.}$$
$$x = 28 \qquad \text{Divide both sides by 3.}$$

We now substitute this solution to find our three unknowns.

> First: $x = 28$
> Second: $x + 1 = 28 + 1 = 29$
> Third: $x + 2 = 28 + 2 = 30$

Quick Check 4

The sum of three consecutive integers is 213. Find them.

The three integers are 28, 29, and 30. The three integers do add up to be 87.

Suppose that a problem involved consecutive *odd* integers rather than consecutive integers. Consider any string of consecutive odd integers, such as 15, 17, 19, 21, 23, ... How far apart are these consecutive odd integers? Each odd integer is 2 away from the previous one. When working with consecutive odd integers, or consecutive even integers for that matter, we will let x represent the first integer and then will add 2 to find the next consecutive integer of that type. Here is a table showing the pattern of unknowns for consecutive integers, consecutive odd integers, and consecutive even integers:

Consecutive Integers	Consecutive Odd Integers	Consecutive Even Integers
First: x	First: x	First: x
Second: $x + 1$	Second: $x + 2$	Second: $x + 2$
Third: $x + 2$	Third: $x + 4$	Third: $x + 4$
⋮	⋮	⋮

Motion Problems

Objective 5 **Solve problems involving motion.**

Distance Formula

When an object, such as a car, moves at a constant rate of speed, r, for an amount of time, t, then the distance traveled, d, is given by the formula $d = r \cdot t$.

Suppose that a person drove 238 miles in 3.5 hours. To determine the car's rate of speed for this trip, we can substitute 238 for d and 3.5 for t in the equation $d = r \cdot t$.

$$d = r \cdot t$$
$$238 = r \cdot 3.5 \qquad \text{Substitute 238 for } d \text{ and 3.5 for } t.$$
$$\frac{238}{3.5} = \frac{r \cdot 3.5}{3.5} \qquad \text{Divide both sides by 3.5.}$$
$$68 = r \qquad \text{Simplify.}$$

The car's rate of speed for this trip was 68 mph.

EXAMPLE 5   Tina drove her car at a rate of 60 mph from her home to Rochester. Her mother Linda made the same trip at a rate of 80 mph, and it took her 2 hours less than Tina to make the trip. How far is it from Tina's home to Rochester?

Solution

For this problem, we can begin by setting up a table showing the relevant information. We will let t represent the time it took for Tina to make the trip. Since it took Linda 2 hours less to make the trip, we can represent her time by $t - 2$. We multiply each person's rate of speed by the time she traveled to find the distance she traveled.

	Rate	Time	Distance ($d = r \cdot t$)
Tina	60	t	$60t$
Linda	80	$t - 2$	$80(t - 2)$

Since we know that both trips were exactly the same distance, we get our equation by setting the expression for Tina's distance equal to the expression for Linda's distance, and we then solve for t.

$$60t = 80(t - 2)$$
$$60t = 80t - 160 \qquad \text{Distribute.}$$
$$-20t = -160 \qquad \text{Subtract } 80t \text{ from both sides.}$$
$$t = 8 \qquad \text{Divide both sides by } -20.$$

At this point, we must be careful to answer the appropriate question. We were asked to find the distance that was traveled. We can substitute 8 for t in either of the expressions for distance. Using the expression for the distance traveled by Tina, $60t = 60(8) = 480$. Using the expression for the distance traveled by Linda, $80(t - 2)$, produces the same result. It is 480 miles from Tina's home to Rochester.

Quick Check **5** Jake drove at a rate of 85 mph for a certain amount of time. His brother Elwood then took over as the driver. Elwood drove at a rate of 90 mph for 3 hours longer than Jake had driven. If the two brothers drove a total of 970 miles, how long did Elwood drive?

Other Problems

Objective 6 Solve other applied problems.

EXAMPLE 6 One number is 5 more than twice a second number. If the sum of the two numbers is 80, find the two numbers.

Solution

In this problem, there are two unknown numbers. We will let one of the numbers be x. Since we know that one number is 5 more than twice the second number, we can represent this number as $2x + 5$.

> ***Unknowns***
>
> First: x
> Second: $2x + 5$

Finally, we know that the sum of these two numbers is 80, which leads to the equation $x + 2x + 5 = 80$.

$$x + 2x + 5 = 80$$
$$3x + 5 = 80 \qquad \text{Combine like terms.}$$
$$3x = 75 \qquad \text{Subtract 5 from both sides.}$$
$$x = 25 \qquad \text{Divide both sides by 3.}$$

We now substitute this solution to find the two unknown numbers:

> First: $x = 25$
> Second: $2x + 5 = 2(25) + 5 = 50 + 5 = 55$

The two numbers are 25 and 55. The sum of these two numbers is 80.

Quick Check   6 **A number is one more than five times a second number. If the sum of the two numbers is 103, find the two numbers.**

We conclude this section with a problem involving coins.

EXAMPLE ▸ 7   Ernie has nickels and dimes in his pocket. He has 5 more dimes than nickels. If Ernie has $2.30 in his pocket, how many nickels and dimes does he have?

Solution

We know the amount of money Ernie has in nickels and dimes totals $2.30. Since the number of dimes is given in terms of the number of nickels, we will let n represent the number of nickels. Since we know that Ernie has 5 more dimes than nickels, we can represent the number of dimes by $n + 5$. A table can be quite helpful in organizing our information, as well as in finding our equation.

	Number of Coins	Value of Coin	Amount of Money
Nickels	n	0.05	$0.05n$
Dimes	$n + 5$	0.10	$0.10(n + 5)$

The amount of money that Ernie has in nickels and dimes is equal to $2.30.

$$0.05n + 0.10(n + 5) = 2.30$$
$$0.05n + 0.10n + 0.50 = 2.30 \qquad \text{Distribute and multiply.}$$
$$0.15n + 0.50 = 2.30 \qquad \text{Combine like terms.}$$
$$0.15n = 1.80 \qquad \text{Subtract 0.50 from both sides.}$$
$$n = 12 \qquad \text{Divide both sides by 0.15.}$$

The number of nickels is 12. The number of dimes $(n + 5)$ is $12 + 5$ or 17. Ernie has 12 nickels and 17 dimes. It is left to the reader to verify that these coins are worth a total of $2.30.

Quick Check   7 **Bert has a pocketful of nickels and quarters in his pocket. He has 13 more nickels than quarters. If Bert has $3.35 in his pocket, how many nickels and quarters does he have?**

1. One number is 17 more than another. If the sum of the two numbers is 71, find the two numbers.

2. One number is 20 more than another. If the sum of the two numbers is 106, find the two numbers.

3. One number is 8 less than another number. If the sum of the two numbers is 96, find the two numbers.

4. One number is 27 less than another number. If the sum of the two numbers is 153, find the two numbers.

5. One number is 5 more than another number. If the smaller number is doubled and added to 3 times the larger number, the sum is 80. Find the two numbers.

6. One number is 11 more than another number. If 3 times the smaller number is added to 5 times the larger number, the sum is 191. Find the two numbers.

7. The width of a rectangle is 5 meters longer that its length, and the perimeter is 26 meters. Find the length and the width of the rectangle.

8. A rectangle has a perimeter of 62 feet. If its length is 9 feet longer than its width, find the dimensions of the rectangle.

9. A schoolyard sandbox is in the shape of a rectangle. It has 54 feet of boards around the border that enclose the sandbox. If the length of the sandbox is 3 feet less than twice the width, find the dimensions of the sandbox.

10. A rectangular sheet of paper has a perimeter of 39 inches. The length of the paper is 6 inches less than twice its width. Find the dimensions of the sheet of paper.

11. A rectangle has a perimeter of 130 cm. If the width of the rectangle is 4 times its length, find the dimensions of the rectangle.

12. A homeowner has built a rectangular shuffleboard court in her backyard. The length of the court is 7 times as long as the width, and the perimeter is 96 feet. Find the dimensions of the court.

13. A square has a perimeter of 220 feet. What is the length of one of its sides?

14. The bases in a Major League Baseball infield are set out in the shape of a square. When Manny Ramirez hits a home run, he has to run 360 feet to make it all the way around the bases. How far is the distance from home plate to first base?

15. An equilateral triangle has a perimeter of 96 cm. Find the length of each side of this triangle.

16. The third side of an isosceles triangle is 2 inches shorter than each of the other two sides. If the perimeter of this triangle is 67 inches, find the length of each side.

17. One side of a triangle is 4 inches longer than the side that is the base of the triangle. The third side is 7 inches longer than the base. If the perimeter is 26 inches, find the length of each side.

18. Two sides of a triangle are 3 feet and 6 feet longer than the third side of the triangle, respectively. If the perimeter of the triangle is 36 feet, find the length of each side.

19. A circle has a circumference of 69.08 inches. Use $\pi \approx 3.14$ to find the radius of the circle.

20. A circle has a circumference of 48π inches. Find its radius.

Two positive angles are said to be **complementary** if their measures add up to 90°.

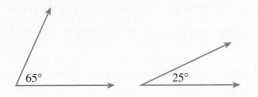

21. Angles *A* and *B* are complementary angles. If the measure of angle *A* is 30° more than the measure of angle *B*, find the measures of the two angles.

22. Angles *A* and *B* are complementary angles. Angle *A* is 15° less than angle *B*. Find the measures of the two angles.

23. An angle is 10° more than 3 times its complementary angle. Find the measures of the two angles.

24. An angle is 6° less than twice its complementary angle. Find the measures of the two angles.

Two positive angles are said to be **supplementary** if their measures add up to 180°.

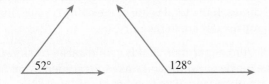

25. An angle is 12° more than 3 times its supplementary angle. Find the measures of the two angles.

26. An angle is 39° less than twice its supplementary angle. Find the measures of the two angles.

The measures of the three angles inside a triangle total 180°.

27. One angle in a triangle is 10° more than the smallest angle in the triangle, while the other angle is 20° more than the smallest angle. Find the measures of the three angles.

28. A triangle contains three angles: *A*, *B*, and *C*. Angle *B* is twice angle *A*, and angle *C* is 20° more than angle *A*. Find the measures of the three angles.

29. If the perimeter of a rectangle is 104 inches and the length of the rectangle is 6 inches less than its width, find the *area* of the rectangle.

30. If the perimeter of a rectangle is 76 inches and the width of the rectangle is 4 inches more than its length, find the *area* of the rectangle.

31. A rectangle has a perimeter of 46 inches. The width of the rectangle is 7 inches shorter than twice its length. If a square's side is as long as the width of this rectangle, find the perimeter of the square.

32. A rectangular pen with a perimeter of 200 feet is divided up into four square pens by building three fences inside the rectangular pen as follows:

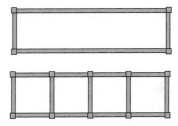

a) What are the original dimensions of the rectangular pen?

b) What is the total length of the three sections of fencing that were added to the inside of the rectangular pen?

c) If the additional fencing cost $2.85 per foot, find the cost of the fencing required to make the four rectangular pens.

33. The sum of two consecutive integers is 305. Find the two integers.

34. The sum of three consecutive integers is 186. Find the three integers.

35. If the sum of three consecutive integers is 237, what is the sum of the two smaller integers?

36. If the sum of five consecutive integers is 755, what is the product of the smallest integer and the largest integer?

37. The sum of three consecutive odd integers is 225. Find the three odd integers.

38. The sum of three consecutive even integers is 96. Find the three even integers.

39. The sum of four consecutive odd integers is 360. Find the largest of the four odd integers.

40. The sum of five odd integers is 575. Find the average of these five odd integers.

41. Arthur has a book open and the two page numbers that he is looking at have a sum of 277. What are the page numbers of the pages he is looking at?

42. Tyler's dance recitals are held on three consecutive days in the month of June. The three dates add up to be 72. What is the date of the first recital?

43. The smallest of three consecutive integers is 18 less than the sum of the two larger integers. Find the three integers.

44. There are three consecutive odd integers, and the sum of the smaller two integers is 49 less than 3 times the larger integer. Find the three odd integers.

45. A. J. drove his race car at a rate of 125 miles per hour for 4 hours. How far did he drive?

46. Angela made the 300-mile drive to Las Vegas in 6 hours. What was her rate for the trip?

47. If Mario drives at a rate of 68 miles per hour, how long will it take him to drive 374 miles?

48. A bullet train travels 240 kilometers per hour. How far can it travel in 45 minutes?

49. Janet swam a 50-meter race in 28 seconds. What was her speed in meters per second?

50. In 1927, Charles Lindbergh flew his airplane the *Spirit of St. Louis* 3500 miles from New York City to Paris in 33.5 hours. What was his speed for the trip, to the nearest tenth of a mile per hour?

51. Susan averaged 80 miles per hour on the way to Phoenix to make a sales call. On the way home, she averaged 70 miles per hour and it took her 1 hour longer to drive home than it did to drive to Phoenix.

   a) What was the total driving time for Susan's trip?

   b) How far does Susan live from Phoenix?

52. Don started driving east at 6 A.M. at a rate of 75 miles per hour. If Dennis leaves the same place heading east at 8 A.M. at a rate of 80 miles per hour, how long will it take him to catch up to Don?

53. After saving quarters for a month, Larry has $13.75 in quarters. How many quarters does Larry have?

54. A roll of nickels is worth $2. How many nickels are in a roll?

55. Chris has a jar with dimes and quarters in it. The jar has 7 more quarters in it than it has dimes. If the total value of the coins is $5.60, how many quarters are in the jar?

56. Kay's change purse has nickels and dimes in it. The number of nickels is 5 more than twice the number of dimes. If the total value of the coins is $4.05, how many nickels are in the purse?

57. The film department holds a fund-raiser by showing the movie π. Admission was $4 for students and $7 for nonstudents. The number of students who attended was 10 more than 4 times the number of nonstudents who attended. If the department raised $500, how many students attended the movie?

58. A movie theater charges $4.50 for children to see a matinee and $6.75 for adults. At today's matinee there were 20 more children than adults, and the total receipts were $405. How many children were at today's matinee?

Answer in complete sentences.

59. Write a word problem involving a rectangle whose length is 50 feet and whose width is 20 feet. Explain how you created your problem.

60. Write a word problem whose solution is "There were 70 children and 40 adults in attendance." Explain how you created your problem.

61. Write a word problem for the given table and equation. Explain how you created your problem.
Equation: $55t = 70(t - 3)$

	Rate	Time	Distance $(d = r \cdot t)$
Steve	55	t	$55t$
Ross	70	$t-3$	$70(t-3)$

62. Write a word problem for the given table and equation. Explain how you created your problem.
Equation: $18(t - 2) = 5t + 9$

	Rate	Time	Distance $(d = r \cdot t)$
?	5	t	$5t$
?	18	$t-2$	$18(t-2)$

Study Tip REVISITED   When rewriting your classroom notes, your textbook can be helpful in supplementing your notes. If your instructor gave you a set of definitions in class, look up the corresponding definition in the textbook. You can also look for examples similar to the ones provided by your instructor during class.

The textbook should also be kept available while you are working on your homework. If you are stuck on a particular problem, look through the section for an example that may give you an idea about how to proceed.

2.4

APPLICATIONS INVOLVING PERCENTS; RATIO AND PROPORTION

Objectives

1. Use the basic percent equation to find an unknown amount, base, or percent.
2. Solve applied problems using the basic percent equation.
3. Solve applied problems involving percent increase or percent decrease.
4. Solve problems involving interest.
5. Solve mixture problems.
6. Solve proportions.
7. Solve applied problems involving proportions.

Basic Percent Equation

Objective 1 Use the basic percent equation to find an unknown amount, base, or percent. We know that 40 is one half of 80. Since the fraction $\frac{1}{2}$ is the same as 50%, we could also say that 40 is 50% of 80. In this example, the number 40 is referred to as the **amount** and the number 80 is referred to as the **base**. The basic equation relating these two quantities reflects that the amount is a percentage of the base, or

$$\text{Amount} = \text{Percent} \cdot \text{Base}$$

In this equation, it is important to express the percent as a decimal or a fraction rather than as a percent. Obviously, it is crucial to identify the amount, the percent, and the base correctly. It is a good idea to write the information in the following form:

$$\underset{\text{(Amount)}}{\underline{\hspace{2cm}}} \text{ is } \underset{\text{(Percent)}}{\underline{\hspace{2cm}}} \text{ \% of } \underset{\text{(Base)}}{\underline{\hspace{2cm}}}$$

EXAMPLE 1   What number is 40% of 45?

Solution

Let n represent the unknown number. We can now write the information as

$$\underset{\text{(Amount)}}{\underline{\hspace{1.5cm}n\hspace{1.5cm}}} \text{ is } \underset{\text{(Percent)}}{\underline{\hspace{1cm}40\hspace{1cm}}} \text{ \% of } \underset{\text{(Base)}}{\underline{\hspace{1cm}45\hspace{1cm}}}$$

We see that the amount is n, the percent is 40%, or 0.4, and the base is 45.

We now translate this information to an equation, making sure that the percent is written as a decimal rather than as a percent.

$$n = 0.4(45)$$
$$n = 18 \qquad \text{Multiply.}$$

We have found that 18 is 40% of 45.

Quick Check 1
25% of 44 is what number?

> **A Word of Caution**   When using the basic percent equation, be sure to rewrite the percent as a decimal number or a fraction before solving for an unknown amount or base.

EXAMPLE 2   Eight percent of what number is 12?

Solution

Letting n represent the unknown number, we can rewrite this information as

$$\underset{\text{(Amount)}}{\underline{\quad 12 \quad}} \text{ is } \underset{\text{(Percent)}}{\underline{\quad 8 \quad}} \% \text{ of } \underset{\text{(Base)}}{\underline{\quad n \quad}}$$

The amount is 12, the percent is 8%, and the base is unknown. We must be sure to write the percent as a decimal before working with the equation.

$$12 = 0.08n$$
$$\frac{12}{0.08} = \frac{0.08n}{0.08} \qquad \text{Divide both sides by 0.08.}$$
$$150 = n \qquad \text{Simplify.}$$

Quick Check **2**
39 is 60% of what number?

Eight percent of 150 is 12.

EXAMPLE 3   What percent of 56 is 35?

Solution

In this problem, the percent is unknown. If we let p represent the percent, then we can write this information as

$$\underset{\text{(Amount)}}{\underline{\quad 35 \quad}} \text{ is } \underset{\text{(Percent)}}{\underline{\quad p \quad}} \% \text{ of } \underset{\text{(Base)}}{\underline{\quad 56 \quad}}$$

We can now translate directly to the basic percent equation and solve it.

$$35 = p \cdot 56$$
$$\frac{35}{56} = \frac{p \cdot 56}{56} \qquad \text{Divide both sides by 56.}$$
$$\frac{5}{8} = p \qquad \text{Simplify.}$$

When the percent is unknown, we must convert our result from a fraction or a decimal to a percent by multiplying by 100%.

$$p = \frac{5}{8} \cdot 100\% \qquad \text{Multiply by 100\%.}$$
$$p = \frac{5}{\overset{}{\underset{2}{8}}} \cdot \overset{25}{\cancel{100}}\% \qquad \text{Divide out common factors.}$$
$$p = \frac{125}{2}\% \qquad \text{Simplify.}$$
$$p = 62\frac{1}{2}\% \qquad \text{Write as a mixed number.}$$

Quick Check **3**
56 is what percent of 80?

$62\frac{1}{2}\%$ of 56 is 35.

A Word of Caution   After using the basic percent equation to solve for an unknown percent, be sure to rewrite the solution as a percent by multiplying by 100%.

When using the basic percent equation to solve for an unknown percent, keep in mind that the base is not necessarily the larger of the two given numbers. For example, if we are trying to determine what percent of 80 is 200, the base is the smaller number (80).

Applications

Objective 2 Solve applied problems using the basic percent equation. We will now use our basic percent equation to solve applied problems. We will continue to use the strategy developed in Section 2.3.

EXAMPLE 4   There are 7107 female students at a certain community college. If 60% of all the students at the college are female, what is the total enrollment of the college?

Solution

We know that 60% of the students are female, so the number of female students (7107) is 60% of the total enrollment. If we let n represent the unknown total enrollment, then our information can be written as

$$\underline{\quad 7107 \quad} \text{ is } \underline{\quad 60 \quad} \% \text{ of } \underline{\quad n \quad}$$
$$\text{(Amount)} \qquad \text{(Percent)} \qquad \text{(Base)}$$

We translate this sentence to an equation and solve.

$$7107 = 0.6n$$
$$11{,}845 = n \qquad \text{Divide both sides by 0.6.}$$

Quick Check 4   There are 11,845 students at the college.

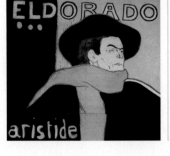

Quick Check 4
Of the 500 children who attend a particular elementary school, 28% buy their lunch at school. How many children buy their lunch at this school?

Percent Increase and Percent Decrease

Objective 3 Solve applied problems involving percent increase or percent decrease. **Percent increase** is a measure of how much a quantity has increased from its original value. The amount of increase is the amount in the basic percent equation, while the original value is the base. **Percent decrease** is a similar measure of how much a quantity has decreased from its original value.

EXAMPLE 5   An art collector bought a lithograph for $2500. After three years, the lithograph was valued at $3800. Find the percent increase in the value of the lithograph over the three years.

Solution

The amount of increase in value is $1300, because $3800 - 2500 = 1300$. We need to find out what percent of the original value is the increase in value. If we let p represent the unknown percent, then we can write our information as

$$\underline{\quad 1300 \quad} \text{ is } \underline{\quad p \quad} \% \text{ of } \underline{\quad 2500 \quad}$$
$$\text{(Amount)} \qquad \text{(Percent)} \qquad \text{(Base)}$$

This translates to the equation $1300 = p \cdot 2500$, which we can then solve.

$$1300 = p \cdot 2500$$
$$0.52 = p \qquad \text{Divide both sides by 2500.}$$

Converting this decimal result to a percent, we see that the lithograph increased in value by 52%.

Quick Check 5
Lee bought a new car for $30,000 three years ago, and now it is worth $13,800. Find the percent decrease in the value of this car.

> **A Word of Caution**   When solving problems involving percent increase or percent decrease, the base is always the *original amount.*

Percents are involved in a great number of business applications. One such problem is determining the sale price of an item after a percent discount has been applied.

EXAMPLE 6   A department store is having a "20% off" sale, so all prices have been reduced by 20% of the original price. If a robe was originally priced at $37.50, what is the sale price after the 20% discount?

Solution

We begin by finding the amount by which the original price has been discounted. Let d represent the amount of the discount. Since the amount of the discount is 20% of the original price ($37.50), we can write our information as

$$\underset{\text{(Amount)}}{\underline{\quad d \quad}} \text{ is } \underset{\text{(Percent)}}{\underline{\quad 20 \quad}} \% \text{ of } \underset{\text{(Base)}}{\underline{\quad \$37.50 \quad}}.$$

This translates to the equation $d = 0.2(37.50)$, which we solve for d.

$$d = 0.2(37.50)$$
$$d = 7.50 \qquad \text{Multiply.}$$

Quick Check 6   The discount is $7.50, so the sale price is $37.50 − $7.50 or $30.00.

A department store bought a shipment of CD players for $42 wholesale. If the store marks the CD players up by 45%, what is the selling price of a CD player?

Interest

Objective 4   **Solve problems involving interest.**   **Interest** is an amount of money paid by a borrower for the privilege of borrowing the money. Banks also pay interest to customers for depositing money in their bank. Simple interest is calculated as a percentage of the **principal**, which is the amount borrowed or deposited. If r is the annual interest rate, then the simple interest (I) owed on a principal (P) is given by the formula $I = P \cdot r \cdot t$, where t is time in years.

   If an investor deposited $3000 in an account that paid 4% annual interest for one year, we could determine how much money would be in the account at the end of one year by substituting 3000 for P, 0.04 for r, and 1 for t in the formula $I = P \cdot r \cdot t$.

$$I = P \cdot r \cdot t$$
$$I = 3000(0.04)(1) \qquad \text{Substitute 3000 for } P, 0.04 \text{ for } r, \text{ and 1 for } t.$$
$$I = 120 \qquad \text{Multiply.}$$

The interest earned in one year is $120. There is $3000 plus the $120 in interest, or $3120, in the account after 1 year.

Here is an applied problem involving interest that will require more work to set it up.

EXAMPLE ▶ 7   Martha had some spare cash and decided to invest it in an account that paid 6% annual interest. Her friend Stuart found an account that paid 7% interest, and invested $5000 more in this account than Martha did in her account. If the pair earned a total of $1650 in interest from the two accounts in one year, how much did Martha invest? How much did Stuart invest?

Solution

If we let x represent the amount Martha invested at 6%, then the amount Stuart invested at 7% can be represented by $x + 5000$. The equation for this problem comes from the fact that the interest earned in each account must add up to $1650. We will use a table to hold the important information for this problem.

<table>
<tr><td></td><td>**Principal (P)**</td><td>**Rate (r)**</td><td>**Time (t)**</td><td>**Interest $I = P \cdot r \cdot t$**</td></tr>
<tr><td>6%</td><td>x</td><td>0.06</td><td>1</td><td>$0.06x$</td></tr>
<tr><td>7%</td><td>$x + 5000$</td><td>0.07</td><td>1</td><td>$0.07(x + 5000)$</td></tr>
<tr><td>**Total**</td><td></td><td></td><td></td><td>1650</td></tr>
</table>

Quick Check ▶ 7
Mark invested money in two accounts. One account paid 3% annual interest, and the other account paid 5% annual interest. He invested $2000 more in the account that paid 5% interest than in the account that paid 3% interest. If Mark earned a total of $500 in interest from the two accounts in one year, how much did he invest in each account?

The interest from the first account is $0.06x$, and the interest from the second account is $0.07(x + 5000)$, so their sum must be $1650. Essentially, we can read this equation in the final column in the table.

$$0.06x + 0.07(x + 5000) = 1650$$

$0.06x + 0.07x + 350 = 1650$	Distribute.
$0.13x + 350 = 1650$	Combine like terms.
$0.13x = 1300$	Subtract 350 from both sides.
$x = 10,000$	Divide both sides by 0.13.

Since $x = 10,000$, the amount Martha invested at 6% interest (x) is $10,000 and the amount Stuart invested at 7% interest $(x + 5000)$ is $15,000. The reader may verify that Martha earned $600 in interest and Stuart earned $1050 in interest, which totals to $1650.

Mixture Problems

Objective 5   Solve mixture problems. The next example is an example of a **mixture problem**. In this problem, we will be mixing two solutions that have different concentrations of alcohol to produce a mixture whose concentration of alcohol is somewhere in between the two individual solutions. We will determine how much of each solution should be used in order to produce a mixture of the desired specifications.

If two quantities total to some number V, and we let x represent one of the quantities, then the other quantity can be expressed as $V - x$. For example, if two numbers add up to 75 and we let x represent one of the other numbers, then the other number can be expressed as $75 - x$.

EXAMPLE ▶ 8   A chemist has two solutions. The first is 20% alcohol, and the second is 30% alcohol. How many milliliters (ml) of each solution should she use if she wants to make 40 ml of a solution that is 24% alcohol?

Solution

There are two unknown quantities in this problem: the amount of the 20% alcohol solution that she should use and the amount of the 30% alcohol solution that she should use. We will let x represent the amount of the 20% alcohol solution in ml. We need to express the amount of the 30% solution in terms of x as well. Since the two quantities must add up to be 40 ml, the amount of the 30% solution can be represented by $40 - x$.

The equation that we will use to solve this problem is based on the amount of alcohol in each solution, as well as the amount of alcohol in the mixture. When we add the amount of alcohol that comes from the 20% solution to the amount of alcohol that comes from the 30% solution, it should equal the amount of alcohol in the mixture.

Since we want to end up with 40 ml of a solution that is 24% alcohol, this solution should contain $0.24(40) = 9.6$ ml of alcohol. The following table, which is quite similar to the one used in the previous example involving interest, presents all of the information:

Solution	Amount of Solution (ml)	% Alcohol	Amount of Alcohol (ml)
Solution 1 (20%)	x	0.2	$0.2x$
Solution 2 (30%)	$40 - x$	0.3	$0.3(40 - x)$
Mixture	40	0.24	$0.24(40) = 9.6$

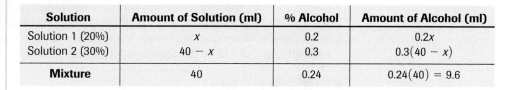

Alcohol from solution 1 + Alcohol from solution 2 = Alcohol in mixture

$$0.2x \qquad + \qquad 0.3(40 - x) \qquad = \qquad 9.6$$

Now we solve the equation for x.

$$
\begin{aligned}
0.2x + 0.3(40 - x) &= 9.6 \\
0.2x + 12 - 0.3x &= 9.6 && \text{Distribute.} \\
-0.1x + 12 &= 9.6 && \text{Combine like terms.} \\
-0.1x &= -2.4 && \text{Subtract 12 from both sides.} \\
x &= 24 && \text{Divide both sides by } -0.1.
\end{aligned}
$$

Since x represented the amount of the 20% alcohol solution that the chemist should use, she should use 24 ml of the 20% solution. Subtracting this amount from 40 ml tells us that she should use 16 ml of the 30% alcohol solution.

Ratio and Proportion

A **ratio** is a comparison of two quantities using division and is usually written as a fraction. Suppose a first-grade student has 20 minutes to eat her lunch and this is followed by a 30-minute recess. The ratio of the time for eating to the time for recess is $\frac{20}{30}$, which can be simplified to $\frac{2}{3}$. This means that for every 2 minutes of eating time, there are 3 minutes of recess time.

Objective 6 **Solve proportions.** Ratios are important tools for solving some applied problems involving proportions. A **proportion** is a statement of equality between two ratios. In other words, a proportion is an equation in which two ratios are equal to each other. Here are a few examples of proportions:

$$\frac{5}{8} = \frac{n}{56} \qquad \frac{8}{11} = \frac{5}{n} \qquad \frac{9}{2} = \frac{4x + 5}{x}$$

Quick Check **8**

Marie had one solution that was 30% alcohol and a second solution that was 42% alcohol. How many milliliters of each solution should be mixed to make 80 milliliters of a solution that is 39% alcohol?

The **cross products** of a proportion are the two products obtained when we multiply the numerator of one fraction by the denominator of the other fraction. The two cross products for the proportion $\frac{a}{b} = \frac{c}{d}$ are $a \cdot d$ and $b \cdot c$.

When we solve a proportion, we are looking for the value of the variable that produces a true statement. We will use the fact that the cross products of a proportion are equal if the proportion is indeed true.

$$\text{If } \frac{a}{b} = \frac{c}{d}, \text{ then } a \cdot d = b \cdot c.$$

Consider the two equal fractions $\frac{3}{6}$ and $\frac{5}{10}$. If we write these two fractions in the form of an equation $\frac{3}{6} = \frac{5}{10}$, the two cross products ($3 \cdot 10$ and $6 \cdot 5$) are both equal to 30. This holds true for any proportion.

To solve a proportion, we will multiply each numerator by the denominator on the other side of the equation, and we will set these products equal to each other. We then solve the resulting equation for the variable in the problem. This process is called **cross multiplying** and will be demonstrated in the next example.

EXAMPLE ▸ 9   Solve $\frac{3}{4} = \frac{n}{68}$.

Solution

We begin by cross multiplying. We multiply the numerator on the left by the denominator on the right ($3 \cdot 68$), and we multiply the denominator on the left by the numerator on the right ($4 \cdot n$).

Now we write an equation that states that the two products are equal to each other, and we solve this equation.

$$\frac{3}{4} = \frac{n}{68}$$

$3 \cdot 68 = 4 \cdot n$	Cross multiply.
$204 = 4n$	Multiply.
$51 = n$	Divide both sides by 4.

Quick Check 9   The solution set is $\{51\}$.

Solve $\frac{3}{17} = \frac{n}{187}$.

> ***A Word of Caution***   When solving a proportion, we cross multiply. *Do not "cross cancel."*

Applications of Proportions

Objective 7   **Solve applied problems involving proportions.** Now we turn our attention to applied problems involving proportions.

EXAMPLE ▸ 10   Studies show that one out of every nine people is left handed. In a sample of 216 people, how many would be left handed according to these studies?

Solution

We begin by setting up a ratio based on the known information. Since there is one person out of every nine who is left handed, the ratio would be $\frac{1}{9}$. Notice that the numerator represents the number of left-handed people while the denominator represents the number of all people. When we set up a proportion it is crucial to keep this ordering consistent. This given ratio will be the left side of the proportion. To set up the right side of the proportion, we need to write a second ratio relating the unknown quantity to the given quantity. The unknown quantity would be the number of left-handed people in the group of 216 people. Let n represent the number of left-handed people. The proportion would then be $\frac{1}{9} = \frac{n}{216}$. Again, be very careful that the ratio on the left side $\left(\frac{\text{left handed}}{\text{total}}\right)$ is consistent with the ratio on the right side. Thus,

$$\frac{1}{9} = \frac{n}{216}$$
$$216 = 9n \qquad \text{Cross multiply.}$$
$$24 = n \qquad \text{Divide both sides by 9.}$$

According to these studies, there should be 24 left-handed people in the group of 216 people.

> ***Quick Check* 10**
>
> **Studies show that 4 out of 5 dentists recommend sugarless gum for their patients that chew gum. In a group of 85 dentists, how many would recommend sugarless gum to their patients according to these studies?**

EXERCISES *2.4*

1. Fourteen percent of 50 is what number?

2. What percent of 80 is 24?

3. Forty percent of what number is 56?

4. Seventeen is 4% of what number?

5. What percent of 190 is 133?

6. What number is 29% of 62?

7. What is $37\frac{1}{2}$% of 104?

8. Fifteen is what percent of 36?

9. What is 240% of 68?

10. What percent of 24 is 108?

11. Fifty-five is 125% of what number?

12. What is 600% of 53?

13. Thirty percent of the M&M's in a bowl are brown. If there are 280 M&M's in the bowl, how many are brown?

14. A doctor helped 320 women deliver their babies last year. Eight of these women had twins. What percent of the doctor's patients had twins?

15. Forty percent of the registered voters in a certain precinct are registered as Democrats. If 410 of the registered voters in the precinct are registered as Democrats, how many registered voters are there in the precinct?

16. An online retailer adds a 15% charge to all items for shipping and handling. If a shipping-and-handling fee of $96 is added to the cost of a computer, what was the original price of the computer?

17. A certain brand of rum is 40% alcohol. How many milliliters of alcohol are there in a 750-milliliter bottle of rum?

18. A certificate of deposit (CD) pays 3.08% interest. Find the amount of interest earned in one year on a $5000 CD.

19. Dylan bought a baseball card for $11.50. The card increased in value by 60% in one year. How much was the card worth after one year?

F
O   MyMathLab
R   MathXL
     MathXL
E
X   Interactmath.com
T
R
A   MathXL
     Tutorials on CD
H
E   DVT CD
L   Videotape
P
     Tutor Center
     Addison-Wesley
     Math Tutor Center

     Student Solutions
     Manual

20. Kristy bought shares of a stock valued at $48.24. If the stock price soared to $66.33, find the percent increase in the value of the stock.

21. The price of a gallon of milk increased by $0.26. This represented a price increase of 8%. What was the original price of the milk?

22. A bookstore charges its customers 32% over the wholesale cost of the book. How much does the bookstore charge for a book that has a wholesale price of $64?

23. A store is discounting all of its prices by 15%. If a sweater's original price was $24.95, what is the price after the discount?

24. The average score for a class on test 1 was 88. The average score for test 2 was 77. What was the percent decrease in the average score from test 1 to test 2?

25. A community college has 1600 parking spots on campus. Tracy has designed a new library, but to build it, the college will lose 120 of its parking spots. What is the percent decrease in the number of available parking spots?

26. A community college had its budget slashed by $3,101,000. This represents a cut of 7% of its total budget from last year. What was last year's budget?

27. Last year, the average SAT math score for students at a certain high school was 500. This year the average score decreased by 20%. By what percent do scores need to increase next year to bring the average score back up to 500? (*Hint:* The answer is not 20%.)

28. During the year 2003, an auto plant produced 25% fewer cars than it had in the year 2002. Production fell by another 20% during 2004. By what percent will production have to increase in 2005 to make the same number of cars as the plant produced in 2002?

29. Janet invested her savings in two accounts. One of the accounts paid 2% interest, and the other paid 5%. Janet put twice as much money in the account that paid 5% than she put in the account that paid 2%. If she earned a total of $84 in interest from the two accounts in one year, find the amount invested in each account.

30. One of Jaleel's mutual funds made a 10% profit last year, while the other mutual fund made an 8% profit. Jaleel had invested $1400 more in the fund that made an 8% profit than he did in the fund that made a 10% profit. If he made $45 more from the 8% fund than he did from the 10% fund, how much was invested in each fund?

31. Kamiran deposits a total of $6500 in savings accounts at two different banks. The first bank pays 4% interest, while the second bank pays 5% interest. If he earns a total of $300 in interest in one year, how much was deposited at each bank?

32. Dianne was given $35,000 when she retired. She invested some at 7% interest and the rest at 9% interest. If she earned $2910 in interest in one year, how much was invested in each account?

33. A mechanic has two antifreeze solutions: one is 70% antifreeze and the other is 40% antifreeze. How much of each solution should be mixed to make 60 gallons of a solution that is 52% antifreeze?

34. A chemist has two saline solutions: one is 2% salt and the other is 6% salt. How much of each solution should be mixed to make 2 liters of a saline solution that is 3% salt?

35. A dairyman has some milk that is 5% butterfat, as well as some lowfat milk that is 2% butterfat. How much of each needs to be mixed together to make 1000 gallons of milk that is 3.2% butterfat?

36. Patti has two solutions: one is 27% acid and the other is 39% acid. How much of each solution should be mixed together to make 9 liters of a solution that is 29% acid?

37. Paula has 400 milliliters of a solution that is 80% alcohol. She plans to dilute it so that it is only 50% alcohol. How much water does she have to add in order to do this?

38. A bartender has rum that is 40% alcohol. How much rum and how much cola need to be mixed together to make 5 liters of rum-and-cola that is 16% alcohol?

Solve the proportion.

39. $\dfrac{2}{3} = \dfrac{n}{48}$

40. $\dfrac{5}{8} = \dfrac{n}{32}$

41. $\dfrac{20}{n} = \dfrac{45}{81}$

42. $\dfrac{22}{30} = \dfrac{55}{n}$

43. $\dfrac{2}{5} = \dfrac{n}{126}$

44. $\dfrac{3}{8} = \dfrac{n}{200}$

45. $\dfrac{7}{6} = \dfrac{n}{75}$

46. $\dfrac{3}{100} = \dfrac{n}{412}$

Set up a proportion and solve it to solve the following problems.

47. At a certain community college, three out of every five students are female. If the college has 9200 students, how many are female?

48. A day care center has a policy that there will be at least 4 staff members present for every 18 children. How large of a staff is necessary to accommodate 63 children?

49. The directions for a powdered plant food state that 3 tablespoons of the food need to be mixed with 2 gallons of water. How many tablespoons need to be mixed with 15 gallons of water?

50. Eight out of every nine people are right handed. If a factory has 352 right-handed workers, how many left-handed workers are there at the factory?

51. If 5 out of every 6 teachers in a city meet the minimum qualifications to be teaching and the city has 864 teachers, how many do not meet the minimum qualifications?

52. A survey showed that 7 out of every 10 registered voters in a county are in favor of a bond measure to raise money for a new college campus. At this rate, if there are 140,000 registered voters in the county, how many are not in favor of the bond measure?

53. Two hundred people were asked about their habits when traveling internationally. The following pie chart shows the responses by category:

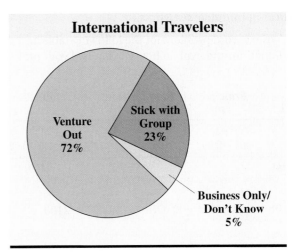

International Travelers

(*Source:* Harris Interactive survey for InterContinental Hotels & Resorts)

Approximately how many of these people venture out on their own?

54. In 2002, 5524 people were killed in work-related incidents in the United States. The following pie chart shows the deaths by category:

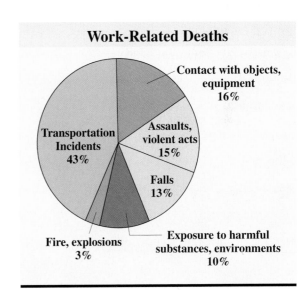

Work-Related Deaths

(*Source:* Bureau of Labor Statistics, Census of Fatal Occupational Injuries)

Approximately how many deaths were due to transportation incidents?

Answer in complete sentences.

55. Write a word problem for the following table and equation, and explain how you created your problem.

	Principal (**P**)	Rate (**r**)	Time (**t**)	Interest $I = P \cdot r \cdot t$
	x	0.03	1	$0.03x$
	$x + 10{,}000$	0.05	1	$0.05(x + 10{,}000)$
Total				2100

Equation: $0.03x + 0.05(x + 10{,}000) = 2100$

56. Write a word problem for the following table and equation, and explain how you created your problem.

Solution	Amount of Solution (ml)	% ?	Amount of Alcohol (ml)
Solution 1	x	0.38	$0.38x$
Solution 2	$60 - x$	0.5	$0.5(60 - x)$
Mixture	60	0.4	24

Equation: $0.38x + 0.5(60 - x) = 24$

QUICK REVIEW EXERCISES

Section 2.4

Solve.

1. $4x + 17 = 41$

2. $5x - 19 = 3x - 51$

3. $2x + 2(x + 8) = 68$

4. $x + 3(x + 2) = 2(x + 4) + 30$

Study Tip **REVISITED**   Your textbook can be used to create a series of note cards for studying new terms and procedures or difficult problems. Each time you find a new term in the textbook, write the term on one side of a note card and its definition on the other side. Collect all of these cards and cycle through them, reading the term and trying to recite the definition. Check your definition on the other side of the card. The same thing can be done for each new procedure introduced in the textbook.

If there is a particular type of problem you are struggling with, find an example that has been worked out in the text. Write the problem on the front of the card and write the complete solution on the back of the card. Collect all of the cards containing difficult problems and select one at random. Try to solve the problem and then check the solution given on the back of the card. Repeat this for the other cards.

Some students only have trouble recalling what the first step is for solving particular problems. If you are in this group, you can write a series of problems on the front of some note cards and the first step on the back of the cards. You can then cycle through these cards in the same way. It is easy to carry a set of note cards with you. Anytime you have 5 to 15 free minutes, cycle through them.

1　**Present the solutions of an inequality on a number line.**
2　**Present the solutions of an inequality using interval notation.**
3　**Solve linear inequalities.**
4　**Solve compound inequalities involving the union of two linear inequalities.**
5　**Solve compound inequalities involving the intersection of two linear inequalities.**
6　**Solve applied problems using linear inequalities.**

2.5
LINEAR INEQUALITIES

Suppose that you went to a doctor and she told you that your temperature was normal. This would mean that your temperature was 98.6° F. However, if the doctor told you that you had a fever, could you tell what your temperature was? No, all you would know is that it was above 98.6° F. If we let the variable t represent your temperature, this relationship could be written as $t > 98.6$. The expression $t > 98.6$ is an example of an inequality. An inequality is a mathematical statement comparing two quantities using the symbols $<$ (*less than*) or $>$ (*greater than*). It states that one quantity is smaller than (or larger than) the other quantity.

Two other symbols that may be used in an inequality are $\leq$ and $\geq$. The symbol $\leq$ is read as *"less than or equal to,"* and the symbol $\geq$ is read as *"greater than or equal to."* The inequality $x \leq 7$ is used to represent all real numbers that are either less than 7 or are equal to 7. Inequalities involving the symbols $\leq$ or $\geq$ are often called **weak inequalities**, while inequalities involving the symbols $<$ or $>$ are called **strict inequalities** because they involve numbers that are strictly greater than (or less than) a given value.

Presenting Solutions of Inequalities Using a Number Line

Objective　1　**Present the solutions of an inequality on a number line.**

Linear Inequality

A **linear inequality** is an inequality containing linear expressions.

A linear inequality often has infinitely many solutions, and it is not possible to list every solution. Consider the inequality $x < 3$. There are infinitely many solutions of this inequality, but the one thing that all of the solutions share in common is that they are all to the left of 3 on the number line. Values located to the right of 3 on the number line are greater than 3. All shaded numbers to the left are solutions of the inequality and can be represented on a number line as follows:

An **endpoint** of an inequality is a point on a number line that separates values that are solutions from values that are not. The open circle at $x = 3$ tells us that the endpoint is not included in the solution, because 3 is not less than 3. If the inequality were $x \le 3$, then we would fill in the circle at $x = 3$.

Here are three more inequalities, with their solutions graphed on a number line:

Inequality	Solution
$x \le -2$	
$x > 5$	
$x \ge -4$	

Saying that a is less than b is equivalent to saying that b is greater than a. In other words, the inequality $a < b$ is equivalent to the inequality $b > a$. This is an important tool, as we will want to rewrite an inequality so that the variable is on the left side of the inequality before graphing it on a number line.

EXAMPLE ▶ 1   Graph the solutions of the inequality $1 > x$ on a number line.

Solution

Before we graph the solutions, we should rewrite the inequality so that the variable is on the left side in order to avoid confusion about which direction should be shaded on the number line. Saying that 1 is greater than x is the same as saying that x is less than 1, or $x < 1$. The values that are less than 1 are to the left of 1 on the number line.

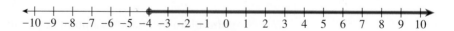

Quick Check ◀ **1**
Graph the solutions of the inequality $8 < x$ on a number line.

Interval Notation

Objective 2   **Present the solutions of an inequality using interval notation.**
Another way to present the solutions of an inequality is by using **interval notation**. A range of values on a number line, such as the solutions of the inequalities $x < -2$ and $x \ge 4$, is called an **interval**. Inequalities typically have one or more intervals as their solutions. Interval notation presents an interval that is a solution by listing its left and right endpoints. We use parentheses around the endpoints if the endpoints are not included as solutions, and we use brackets if the endpoints are included as solutions.

When an interval continues on indefinitely to the right on a number line, we will use the symbol ∞ (infinity) in the place of the right endpoint and follow it with a parenthesis. Let's look again at the number line associated with the solutions of the inequality $x \ge -4$.

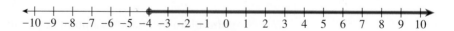

The solutions begin at -4 and include any number that is greater than -4. The interval is bounded by -4 on the left side and extends without bound on the right side, which can be written in interval notation as $[-4, \infty)$. The endpoint -4 is included in the interval, so we write a square bracket in front of it, because it is a solution, while we will always write a parenthesis after ∞, as it is not an actual number that ends the interval at that point.

If the inequality had been $x > -4$, instead of $x \geq -4$, then we would have written a parenthesis rather than a square bracket before -4. In other words, the solutions of this inequality can be expressed in interval notation as $(-4, \infty)$.

For intervals that continue on indefinitely to the left on a number line, we will use $-\infty$ in the place of the left endpoint.

Here is a summary of different inequalities with the solutions presented on a number line and in interval notation:

Inequality	Number Line	Interval Notation
$x > 4$		$(4, \infty)$
$x \geq 4$		$[4, \infty)$
$x < 4$		$(-\infty, 4)$
$x \leq 4$		$(-\infty, 4]$

Solving Linear Inequalities

Objective **3**   **Solve linear inequalities.**   Solving a linear inequality, such as $7x - 11 \leq -32$, is very similar to solving a linear equation. In fact, there is only one difference between the two procedures: Whenever we multiply both sides of an inequality by a negative number, or divide both sides by a negative number, the direction of the inequality changes. Why is this? Consider the inequality $3 < 5$, which is a true statement. If we multiply each side by -2, is the left side $(-2 \cdot 3 = -6)$ still less than the right side $(-2 \cdot 5 = -10)$? No, $-6 > -10$ because -6 is located to the right of -10 on the number line. Changing the direction of the inequality after multiplying (or dividing) by a negative number produces an inequality that is still a true statement.

EXAMPLE 2   Solve the inequality $7x - 11 \leq -32$. Present your solutions on a number line, as well as using interval notation.

Solution

Think about the steps that you would take to solve the equation $7x - 11 = -32$. The same steps will be followed in solving this inequality.

$$7x - 11 \leq -32$$
$$7x \leq -21 \qquad \text{Add 11 to both sides.}$$
$$x \leq -3 \qquad \text{Divide both sides by 7.}$$

Now we present our solution on a number line.

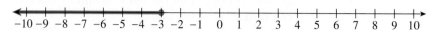

This can be expressed in interval notation as $(-\infty, -3]$.

Quick Check   2 Solve the inequality $2x - 9 < 5$. Present your solutions on a number line, as well as using interval notation.

EXAMPLE   3   Solve the inequality $-2x - 5 > 9$. Present your solutions on a number line, as well as using interval notation.

Solution

We begin by solving the inequality.

$$-2x - 5 > 9$$
$$-2x > 14 \qquad \text{Add 5 to both sides.}$$
$$\frac{-2x}{-2} < \frac{14}{-2} \qquad \text{Divide both sides by } -2 \text{ to isolate } x. \text{ Notice that the direction of the inequality must change because we are dividing by a negative number.}$$
$$x < -7 \qquad \text{Simplify.}$$

Now we present our solutions.

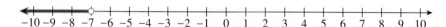

In interval notation this is $(-\infty, -7)$.

Quick Check   3 Solve the inequality $7 - 4x \le -13$. Present your solutions on a number line, as well as using interval notation.

EXAMPLE   4   Solve the inequality $5x - 12 > 8 - 3x$. Present your solutions on a number line, as well as using interval notation.

Solution

We need to collect all variable terms on one side of the inequality and all constant terms on the other side.

$$5x - 12 > 8 - 3x$$
$$8x - 12 > 8 \qquad \text{Add } 3x \text{ to both sides.}$$
$$8x > 20 \qquad \text{Add 12 to both sides.}$$
$$x > \frac{5}{2} \qquad \text{Divide both sides by 8 and simplify.}$$

Here is our solution on a number line:

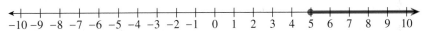

In interval notation, this is written as $\left(\frac{5}{2}, \infty\right)$.

Quick Check **4** Solve the inequality $3x + 8 < x + 2$. Present your solutions on a number line, as well as using interval notation.

EXAMPLE ▶ **5**   Solve the inequality $3(7 - 2x) + 6x \leq 5x - 4$. Present your solutions on a number line, as well as using interval notation.

Solution

As with equations, we begin by simplifying each side completely.

$$3(7 - 2x) + 6x \leq 5x - 4$$

$21 - 6x + 6x \leq 5x - 4$	Distribute.
$21 \leq 5x - 4$	Combine like terms.
$25 \leq 5x$	Add 4 to both sides.
$5 \leq x$	Divide both sides by 5.

We can rewrite this solution as $x \geq 5$, which will help us when we display our solutions on a number line.

This can be expressed in interval notation as $[5, \infty)$.

Quick Check **5** Solve the inequality $(11x - 3) - (2x - 13) \geq 4(2x + 1)$. Present your solutions on a number line, as well as using interval notation.

Compound Inequalities

Objective **4** **Solve compound inequalities involving the union of two linear inequalities.** A **compound inequality** is made up of two or more individual inequalities. One type of compound inequality involves the word "or." In this type of inequality, we are looking for real numbers that are solutions of one inequality or the other. For example, the solutions of the compound inequality $x < 2$ or $x > 4$ are real numbers that are either less than 2 or greater than 4. The solutions of this compound inequality can be displayed on a number line as follows:

To express the solutions using interval notation, we write both intervals with the symbol for **union** ($\cup$) between them. The interval notation for $x < 2$ or $x > 4$ is $(-\infty, 2) \cup (4, \infty)$.

To solve a compound inequality involving "or," we simply solve each individual inequality separately.

EXAMPLE 6   Solve $9x + 20 < -34$ or $5 + 4x \geq 19$.

Solution

We must solve each inequality separately. We begin with $9x + 20 < -34$.

$$9x + 20 < -34$$
$$9x < -54 \qquad \text{Subtract 20 from both sides.}$$
$$x < -6 \qquad \text{Divide both sides by 9.}$$

Now we solve the second inequality, $5 + 4x \geq 19$.

$$5 + 4x \geq 19$$
$$4x \geq 14 \qquad \text{Subtract 5 from both sides.}$$
$$x \geq \frac{7}{2} \qquad \text{Divide both sides by 4 and simplify.}$$

We next combine our solutions on a single number line.

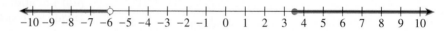

Our solutions can be expressed in interval notation as $(-\infty, -6) \cup [\frac{7}{2}, \infty)$.

Quick Check 6   Solve $3x - 2 \leq 10$ or $4x - 13 \geq 15$.

Objective 5   **Solve compound inequalities involving the intersection of two linear inequalities.**   Another type of compound inequality involves the word "and," and the solutions of this compound inequality must be solutions of each individual inequality. For example, the solutions of the inequality $x > -5$ and $x < 1$ are real numbers that are, at the same time, greater than -5 and less than 1. This type of inequality is generally written in the condensed form $-5 < x < 1$. You can think of the solutions as being between -5 and 1.

To solve a compound inequality of this type, we need to isolate the variable in the middle part of the inequality between two real numbers. A key aspect of our approach will be to work on all three parts of the inequality at once.

EXAMPLE 7   Solve $-5 \leq 2x - 1 < 9$.

Solution

We are trying to isolate x in the middle of this inequality. The first step will be to add 1 to all three parts of this inequality, after which we will divide by 2.

$$-5 \leq 2x - 1 < 9$$
$$-4 \leq 2x < 10 \qquad \text{Add 1 to each part of the inequality.}$$
$$-2 \leq x < 5 \qquad \text{Divide each part of the inequality by 2.}$$

The solutions are presented on the following number line:

These solutions can be expressed in interval notation as $[-2, 5)$.

Quick Check   7 Solve $-11 < 4x + 1 < 7$.

Applications

Objective 6   **Solve applied problems using linear inequalities.** Many applied problems involve inequalities rather than equations. Here are some key phrases and their translations into inequalities:

Key Phrases for Inequalities

x is greater than a x is more than a x is higher than a x is above a	$x > a$
x is at least a x is a or higher	$x \geq a$
x is less than a x is lower than a x is below a	$x < a$
x is at most a x is a or lower	$x \leq a$
x is between a and b x is more than a but less than b	$a < x < b$
x is between a and b, inclusive x is at least a but no more than b	$a \leq x \leq b$

EXAMPLE 8   A sign next to an amusement park ride says that you must be at least 48″ tall in order to get on the ride. Set up an inequality that shows the heights of people who are able to get on the ride.

Solution

Let h represent the height of a person. If a person must be at least 48″ tall, then his or her height must be 48″ or above. In other words, the person's height must be greater than or equal to 48″. The inequality is $h \geq 48$.

A Word of Caution   The phrase "at least" means "greater than or equal to" ($\geq$) and does not mean "less than."

EXAMPLE ▸ 9   An instructor tells her class that a score of 70 or higher out of a possible 100 on the final exam is needed to pass. Set up an inequality that shows the scores that are not passing.

Solution

Let s represent a student's score. Since a score of 70 or higher will pass, any score lower than 70 will not pass. This inequality can be written as $s < 70$. If we happen to know that the lowest possible score is 0, we could also write the inequality as $0 \leq s < 70$.

Quick Check   8 The Brainiac Club has a bylaw that states a person must have an IQ of at least 130 in order to be admitted to the club. Set up an inequality that shows the IQs of people who are unable to be admitted to the club.

We will now set up and solve applied problems involving inequalities.

EXAMPLE ▸ 10   If a student averages 90 or higher on the five tests given in a math class, then the student will earn a grade of A. Sean's scores on the first four tests are 82, 87, 93, and 92. What scores on the fifth test will give Sean an A?

Solution

To find the average of five test scores, we add the five scores and divide by 5. Let x represent the score of the fifth test. The average can then be expressed as $\frac{82 + 87 + 93 + 92 + x}{5}$. We are interested in which scores on the fifth test give an average that is 90 or higher.

$$\frac{82 + 87 + 93 + 92 + x}{5} \geq 90$$

$$\frac{354 + x}{5} \geq 90 \qquad \text{Simplify the numerator.}$$

$$5 \cdot \frac{354 + x}{5} \geq 5 \cdot 90 \qquad \text{Multiply both sides by 5 to clear the fraction.}$$

$$354 + x \geq 450 \qquad \text{Simplify.}$$

$$x \geq 96 \qquad \text{Subtract 354 from both sides.}$$

Sean must score at least 96 on the fifth test to earn an A.

Quick Check   9 If a student averages 70 or lower on the four tests given in a math class, then the student will fail the class. Bobby's scores on the first three tests are 74, 78, and 80. What scores on the fourth test will result in Bobby failing the class?

Graph each inequality on a number line, and write it in interval notation.

1. $x < 3$

2. $x > -6$

3. $x \geq -1$

4. $x \leq 13$

5. $-2 < x < 8$

6. $3 < x \leq 12$

7. $x > \dfrac{9}{2}$

8. $x \geq -\dfrac{7}{3}$

9. $x > 8$ or $x \leq 2$

10. $x < -5$ or $x > 0$

11. $x < \dfrac{3}{2}$ or $x > \dfrac{13}{4}$

12. $-5 \leq x \leq 5$

Solve. Graph your solution on a number line, and express it in interval notation.

19. $x + 9 < 5$

20. $x - 7 > 3$

21. $2x > 12$

22. $4x \geq -20$

23. $-5x > 15$

24. $-8x < -24$

25. $3x + 2 < 14$

26. $2x - 7 > 11$

27. $5 + 13x \leq 31$

28. $-3 + 8x \geq -37$

29. $-2x + 9 \leq 29$

Write an inequality associated with the given graph or interval notation.

13.

14.
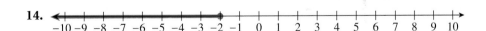

15.

16. $(-3, -2)$ **17.** $(-\infty, -4)$

18. $[-8, \infty)$

30. $7 - 4x > 19$

31. $8 < 3x - 13$

32. $1 \geq 5x + 16$

33. $\dfrac{2}{3}x > \dfrac{10}{9}$

34. $\dfrac{5}{9}x - \dfrac{4}{3} < 7$

35. $7(x + 4) \leq -21$

36. $3(x - 5) > 18$

37. $4(2x - 7) + 10 < 38$

38. $5(x - 3) - 1 \geq -31$

39. $5x + 13 > 2x - 11$

40. $-2x + 51 < 9 - 8x$

41. $5(x + 3) + 2 > 3(x + 4) - 6$

42. $5(2x - 3) - 3 < 2(2x + 9) - 4$

43. $2x < -8$ or $x + 7 > 5$

44. $x - 3 \leq -5$ or $6x > -6$

45. $9x - 7 < -25$ or $3x + 11 > 25$

46. $4x - 9 < -17$ or $4x - 9 > 17$

47. $-2x + 17 < 5$ or $10 - x > 7$

48. $-4x + 13 \leq 9$ or $6 - 5x \geq 21$

49. $\dfrac{1}{4}x + \dfrac{7}{24} \leq \dfrac{2}{3}$ or $\dfrac{1}{4}x + \dfrac{9}{20} \geq \dfrac{6}{5}$

50. $\dfrac{1}{8}x - \dfrac{1}{5} < \dfrac{1}{12}x - \dfrac{3}{40}$ or $\dfrac{1}{2}x - \dfrac{3}{20} > \dfrac{1}{4}x + \dfrac{11}{10}$

51. $-4 < x - 3 < 1$

52. $2 \leq x + 7 \leq 5$

53. $-12 \leq 8x \leq 18$

54. $-6 < 2x < 0$

55. $-2 < 5x - 7 < 28$

56. $18 \leq 4x - 10 \leq 50$

57. $6 < 3x + 15 \leq 33$

58. $-12 \le 4x - 18 < 7$

59. $\dfrac{1}{5} \le \dfrac{1}{2}x - \dfrac{1}{3} \le \dfrac{7}{4}$

60. $-\dfrac{3}{10} < \dfrac{1}{5}x - \dfrac{1}{2} < \dfrac{3}{5}$

61. Explain, in your own words, why the inequality $8 < 9x + 7 < 3$ has no solutions.

62. Explain, in your own words, why solving the compound inequality $x + 7 < 9$ or $x + 7 < 13$ is the same as solving the inequality $x + 7 < 13$ only.

63. Adrian needs to score at least 2 goals in the final game of the season to set a new scoring record. Write an inequality that shows the number of goals that will set a new record.

64. If Eddie sells fewer than seven cars this week, he will be fired. Write an inequality that shows the number of cars that will make Eddie unemployed.

65. A certain type of shrub is tolerant to 30° F, which means that it can survive at temperatures down to 30° F. Write an inequality that shows the temperatures at which the shrub cannot survive.

66. If Chad scores below 40% on the final exam, he will not pass the class. Write an inequality that shows the scores at which Chad will pass the class.

67. The manager of a winery tells the owner that he has lost at least 12,500 grapevines due to Pierce's disease spread by the insect, the glassy-winged sharpshooter.

Write an inequality for the number of grapevines that have been lost.

68. A community college must have 11,200 or more students this semester in order to be eligible for growth funding from the state. Write an inequality that shows the number of students that makes the college eligible for growth funding.

69. John is paid to get signatures on petitions. He gets paid 10 cents for each signature. How many signatures does he need to gather today in order to earn at least $30?

70. Carlo attends a charity wine-tasting festival. There is a $10 admission fee. In addition, there is a $4 charge for each variety tasted. If Carlo brings $40 with him, how many varieties can he taste?

71. Angelica is going out of town on business. Her company will reimburse her up to $85 for a rental car. She rents a car for $19.95 plus 7 cents per mile. How many miles can Angelica drive without going over the amount that her company will reimburse?

72. An elementary school's booster club is holding a fundraiser by selling cookie dough. Each tub of cookie dough sells for $11, and the booster club gets to keep half of the proceeds. If the booster club wants to raise at least $12,000, how many tubs of cookie dough do they need to sell?

73. Charles told his son Harry that he should always give at least 15% of his earnings to charity. If Harry earned $37,500 last year, how much should he have given to charity?

74. Karina is looking for a formal dress to wear to a ball, and she has $400 to spend. If the store charges 7% sales tax on each dress, what price range should Karina be considering?

75. Students in a real-estate class must have an average score of at least 80% on their exams in order to pass. If Jacqui has scored 92, 93, 85, and 96 on the first four exams, what scores on the fifth exam would allow her to pass the course? (Assume the highest possible score is 100.)

76. Students with an average test score below 70 after the third test will be sent an Early Alert warning. Robert scored 62 and 59 on the first two tests. What scores on the third exam will save Robert from receiving an Early Alert warning? (Assume the highest possible score is 100.)

77. Explain, in your own words, why we must change the direction of an inequality when we divide both sides of that inequality by a negative number.

Study Tip **REVISITED**   Two features of this textbook that you may find helpful are labeled A Word of Caution and Using Your Calculator. Each Word of Caution warns you about common student errors for certain problems, and tells you how to avoid repeating the same mistake. After a section is covered in class, look through the section for this feature before attempting the homework exercises.

Using Your Calculator shows how you can use the Texas Instruments TI-83/84 calculator to help with selected examples in the text. If you own one of these calculators, look for this feature before beginning your homework. If you own a different calculator, you should be able to interpret the directions for your calculator.

Section 2.1—Topic	Chapter Review Exercises
Determining Whether a Value Is a Solution	1–2
Solving Equations Using the Multiplication and Addition Properties of Equality	3–6
Solving Applied Problems Involving Linear Equations	7–8

Section 2.2—Topic	Chapter Review Exercises
Solving Linear Equations	9–18
Solving Literal Equations	19–22

Section 2.3—Topic	Chapter Review Exercises
Solving Applied Problems Involving Linear Equations	23–27

Section 2.4—Topic	Chapter Review Exercises
Converting Fractions and Decimals to Percents	28–31
Converting Percents to Fractions and Decimals	32–39
Solving Percent Equations	40–42
Solving Applied Problems Involving Percents	43–47
Solving Proportions	48–51
Solving Applied Problems Involving Proportions	52–53

Section 2.5—Topic	Chapter Review Exercises
Solving Linear Inequalities	54–57
Solving Compound Linear Inequalities	58–61
Solving Applied Problems Involving Inequalities	62–64

Summary of Chapter 2 Study Tips

Using your textbook as more than a source of the homework exercises and answers to the odd problems will help you to learn mathematics.

- Read the section the night before your instructor covers it to familiarize yourself with the material. If you find a step in one of the examples that you do not understand, then you can have questions prepared for your instructor in class.
- Reread the section after it is covered in class. After you read through an example, attempt the Quick Check exercise that follows it. The answers to these exercises can be found at the back of the textbook. As you work through the homework exercises, refer back to the examples found in that section for assistance.
- Use the textbook to create a series of note cards for each section. Note cards can be used to memorize new terms or procedures, and can help you remember how to solve particular problems.
- In each section, look for the feature labeled A Word of Caution. These will point out common student mistakes and offer advice for not making the same mistake yourself.
- Finally, the feature labeled Using Your Calculator will show you how to use the TI-83/84 calculator to help with the problems in that section.

Chapter 2 Review

Check to determine whether the given value is a solution of the equation. [2.1]

1. $x = 7, 4x + 9 = 19$

2. $a = -9, (4a - 15) - (2a + 7) = -2(11 - a)$

Solve by using the multiplication property of equality. [2.1]

3. $2x = 32$ 4. $-9x = 54$

Solve by using the addition property of equality. [2.1]

5. $x - 3 = -9$ 6. $a + 14 = 8$

For each of the following problems, set up a linear equation and solve it. [2.1]

7. A local club charged an $8 cover charge to see a "Battle of the Bands." After paying the winning band a prize of $250, the club made a profit of $694. How many people came to see the "Battle of the Bands?"

8. Thanks to proper diet and exercise, Randy's cholesterol dropped by 23 points in the last two months. If his new cholesterol level is 189 points, what was his cholesterol level two months ago?

Solve. [2.2]

9. $5x - 8 = 27$ 10. $-11 = 3a + 16$

11. $-4x + 15 = 7$ 12. $51 - 2x = -21$

13. $\dfrac{x}{6} - \dfrac{5}{12} = \dfrac{5}{4}$ 14. $\dfrac{1}{5}x - \dfrac{1}{2} = \dfrac{5}{4}$

15. $4x - 19 = 5x + 14$

16. $2m + 23 = 17 - 8m$

17. $2n - 11 + 3n + 4 = 7n - 39$

18. $3(2x - 7) - (x - 9) = 2x - 33$

Solve the following literal equations for the specified variable. [2.2]

19. $7x + y = 8$ for y

20. $15x + 10y = 30$ for y

21. $C = \pi \cdot d$ for d

22. $P = 3a + b + 2c$ for a

23. One number is 26 more than another. If the sum of the two numbers is 98, find the two numbers. [2.3]

24. The width of a rectangle is 7 feet longer than its length, and the perimeter is 50 feet. Find the length and the width of the rectangle. [2.3]

25. The sum of three consecutive even integers is 204. Find the three even integers. [2.3]

26. If Jason drives at a speed of 72 miles per hour, how long will it take him to drive 342 miles? [2.3]

27. Alycia's piggy bank has dimes and quarters in it. The number of dimes is four less than three times the number of quarters. If the total value of the coins is $7.85, how many dimes are in the piggy bank? [2.3]

Rewrite as a percent. [2.4]

28. $\dfrac{2}{5}$ 29. $\dfrac{7}{25}$

30. 0.9 31. 0.45

Rewrite as a fraction. [2.4]

32. 30% 33. 55% 34. 24% 35. 6%

Rewrite as a decimal. [2.4]

36. 15% 37. 90%

38. 4% 39. 340%

40. Twenty percent of 95 is what number? [2.4]

41. What percent of 136 is 51? [2.4]

42. Thirty-two percent of what number is 1360? [2.4]

43. Fifty-five percent of the students at a college are female. If the college has 10,520 students, how many are female? [2.4]

44. In a group of 320 college graduates, 204 decided to pursue a master's degree. What percent of these students decided to pursue a master's degree? [2.4]

45. Jerry's bank account earned $553.50 in interest last year. This account pays 4.5% interest. How much money did Jerry have in the account at the start of the year? [2.4]

46. A store is discounting all of its prices by 20%. If a sweater's original price was $59.95, what is the price after the discount? [2.4]

122

47. Karl deposits a total of $2000 in savings accounts at two different banks. The first bank pays 6% interest, while the second bank pays 5% interest. If he earns a total of $117 in interest in the first year, how much was deposited at each bank? [2.4]

Solve the proportion. [2.4]

48. $\dfrac{n}{6} = \dfrac{18}{27}$ **49.** $\dfrac{4}{9} = \dfrac{n}{72}$

50. $\dfrac{32}{13} = \dfrac{192}{n}$ **51.** $\dfrac{38}{n} = \dfrac{57}{75}$

Set up a proportion and solve it to solve the following problems. [2.4]

52. At a certain community college, 3 out of every 7 students plan to transfer to a four-year university. If the college has 14,952 students, how many plan to transfer to a four-year university?

53. Six tablespoons of a concentrated weed killer must be mixed with 1 gallon of water before spraying. How many tablespoons of the weed killer need to be mixed with $2\frac{1}{2}$ gallons of water?

Solve. Graph your solution on a number line, and express it in interval notation. [2.5]

54. $x + 5 < 2$

55. $x - 3 > -8$

56. $2x - 7 \geq 3$

57. $3x + 4 \leq -8$

58. $3x < -18$ or $x + 3 > 10$

59. $x + 5 \leq -3$ or $4x \geq -12$

60. $-21 \leq 3x - 6 \leq 24$

61. $-11 < 2x - 9 < 11$

62. Erin needs to make at least seven sales this week to qualify for a bonus. Write an inequality that shows the number of sales that will qualify Erin to receive a bonus. [2.5]

63. A high school yearbook staff is selling advertisements for $45. If the yearbook staff needs to raise at least $3000 in advertisement revenue, how many advertisements do they need to sell? [2.5]

64. If Steve's test average is at least 70, but lower than 80, then his final grade will be a C. His scores on the first four tests were 58, 72, 66, and 75. What scores on the fifth exam will give him a C for the class? (Assume that the highest possible score is 100.) [2.5]

1. Solve $-8x = 56$ by using the multiplication property of equality.

2. Solve $x + 17 = -22$ by using the addition property of equality.

Solve.

3. $7x - 20 = 36$

4. $72 = 30 - 8a$

5. $3x + 34 = 5x - 21$

6. $3(5x - 16) - 9x = x - 63$

7. $\dfrac{1}{8}x - \dfrac{2}{3} = \dfrac{5}{6}$

8. $\dfrac{n}{100} = \dfrac{22}{40}$

9. Solve the literal equation $4x + 2y = 28$ for y.

10. Rewrite 72% as a fraction.

11. Rewrite 6% as a decimal.

12. What number is 16% of 175?

13. What percent of 660 is 99?

Solve. Graph your solution on a number line, and express it in interval notation.

14. $6x + 8 \le -7$

15. $x + 4 \le -5$ or $2x + 3 \ge 0$

16. $-23 < 3x + 7 < 23$

17. The width of a rectangle is 3 feet less than twice its length, and the perimeter is 72 feet. Find the length and the width of the rectangle.

18. Barry's wallet contains $5 and $10 bills. He has seven more $5 bills than $10 bills. If Barry has $185 in his wallet, how many $5 bills are in the wallet?

19. At a certain university, 2 out of every 9 professors earned their degree outside of the United States. If the university has 396 professors, how many of them earned their degrees outside of the United States?

20. Wayne earns a commission of $500 for every car he sells. How many cars does he need to sell in order to earn at least $17,000 in commissions this month?

Mathematicians in History
George Pólya

George Pólya was a Hungarian mathematician who spent much of his career on problem solving and wrote the landmark text *How to Solve It*. Alan Schoenfeld has said of this book, "For mathematics education and the world of problem solving it marked a line of demarcation between two eras, problem solving before and after Pólya."

Write a one-page summary (OR make a poster) of the life of George Pólya and his accomplishments. Also, look up Pólya's most famous quotes and list your favorite quote.

Interesting issues:

- Where and when was George Pólya born?
- What circumstances led to Pólya leaving Göttingen after Christmas in 1913?
- The political situation in Europe in 1940 forced Pólya to leave Zürich for the United States. Which university did Pólya work at upon arriving in the United States?
- When was Pólya's monumental book *How to Solve It* published?
- Summarize Pólya's strategy for solving problems.
- At which American university did Pólya spend most of his career?
- What was Pólya's mnemonic for the first fourteen digits of π?
- How old was Pólya when he taught his last class, and what was the subject?

Step 1: Form groups of no more than four students. Each group should have a regular-sized bag of plain M&M's.

Step 2: Sort the M&M's into the six colors. Count and record the number of M&M's you have of each color. Write these values in the "Number" column of the following table:

Color	Number	Variable Representation	Observed Percent	Expected Percent	Expected Number
Brown		n		13%	
Yellow				14%	
Red				13%	
Blue				24%	
Orange				20%	
Green				16%	

Step 3: Create a variable representation for the number you have of each color. Let the number of brown M&M's be n and use this variable in the creation of the representations for the other colors. For example, if you have 13 brown M&M's and 20 yellow M&M's, you could represent the number of yellow M&M's by $n + 7$ or $2n - 6$. Be creative in how you describe the number of yellow, red, blue, orange, and green M&M's.

Step 4: Using the variable representations of all six colors, create an equation that relates these variables to your total number of M&M's.

Step 5: Once all the groups have created an equation, pass yours on to another group and have them solve it. Did they get the correct numbers for all six colors? If not, first check to make sure that your variable representations are correct. If you believe your representations are correct, have the other group check their solving process. If the group still cannot solve your equation correctly, have your instructor check your representations and the other groups' process of solving the equation to see where the error may have occurred. If the group found solving your equation an easy process, see if you can create more elaborate variable representations to try and stump them.

Step 6: Each bag of plain M&M's is supposed to have a certain percentage of each color present. (See the "Given Percent" column for those percentages.) Using the numbers you counted, calculate the percentage of each color that is actually present in your bag of M&M's and list these values in the "Observed Percent" column. Remember that the total of your "Observed Percent" column should be very close to 100%. Are your percentages close to the expected percents?

Step 7: Calculate the expected number of each color of M&M's that you should have found in your bag and write these values in the "Expected Number" column. How would you calculate this number? Discuss possible processes with your group. Here's a hint: It has to do with the expected percentages the candy company claims to have in each bag and your total number of M&M's.

Step 8: Along with your completed chart, you should submit your equation(s) to your instructor for review.

GRAPHING LINEAR EQUATIONS

In this chapter, we will examine linear equations in two variables. Equations involving two variables relate two unknown quantities to each other, such as a person's height and weight, or the number of items sold by a company and the profit that the company made. The primary focus is on graphing linear equations, which is the technique that we use to display the solutions to an equation in two variables.

Study Tip **MAKING THE BEST USE OF YOUR RESOURCES** *In this chapter, we will focus on how to get the most out of your available resources. What works best for one student may not work for another student. Consider all of the different resources presented in this chapter, and use the ones you feel could help you.*

Throughout this chapter, we will revisit this study tip and help you to incorporate it into your study habits.

3.1
THE RECTANGULAR COORDINATE SYSTEM; EQUATIONS IN TWO VARIABLES

Objectives

1. Determine whether an ordered pair is a solution of an equation in two variables.
2. Plot ordered pairs on a rectangular coordinate plane.
3. Complete ordered pairs for a linear equation in two variables.
4. Graph linear equations in two variables by plotting points.

Equations in Two Variables and Their Solutions; Ordered Pairs

Objective 1   Determine whether an ordered pair is a solution of an equation in two variables. In this section, we will begin to examine equations in two variables, which relate two different quantities to each other. Here are some examples of equations in two variables:

$$y = 4x + 7, \, 2a + 3b = -6, \, y = 3x^2 - 2x - 9, \, t = \frac{4n + 9}{3n - 7}$$

As in the last chapter, we will be looking for solutions of these equations. A solution of an equation in two variables is a pair of values that, when substituted for the two variables, produce a true statement. Consider the equation $3x + 4y = 10$. One solution of this equation is $x = 2$ and $y = 1$. We see that this is a solution by substituting 2 for x and 1 for y into the equation:

$$
\begin{aligned}
3x + 4y &= 10 \\
3(2) + 4(1) &= 10 \qquad \text{Substitute 2 for } x \text{ and 1 for } y. \\
6 + 4 &= 10 \qquad \text{Multiply.} \\
10 &= 10 \qquad \text{Add.}
\end{aligned}
$$

Since this produces a true statement, our pair of values together form a solution. Solutions to this type of equation are written as **ordered pairs** (x, y). In our example, the ordered pair $(2, 1)$ is a solution. We call it an ordered pair because the two values are listed in the specific order of x first and y second. The values are often referred to as **coordinates.**

EXAMPLE 1   Is the ordered pair $(-3, 2)$ a solution of the equation $y = 3x + 11$?

Solution

To determine whether this ordered pair is a solution, we need to substitute -3 for x and 2 for y:

$$
\begin{aligned}
y &= 3x + 11 \\
(2) &= 3(-3) + 11 \qquad \text{Substitute } -3 \text{ for } x \text{ and 2 for } y. \\
2 &= -9 + 11 \qquad \text{Multiply.} \\
2 &= 2 \qquad \text{Simplify.}
\end{aligned}
$$

Since this is a true statement, the ordered pair $(-3, 2)$ is a solution of the equation $y = 3x + 11$.

EXAMPLE **2**   Is $(4, 0)$ a solution of the equation $y - 2x = 8$?

Solution

Again, we check to see whether this ordered pair is a solution by substituting the given values for x and y. We must be careful to substitute 4 for x and 0 for y, not vice versa:

$$y - 2x = 8$$
$$(0) - 2(4) = 8 \qquad \text{Substitute 4 for } x \text{ and 0 for } y.$$
$$0 - 8 = 8 \qquad \text{Multiply.}$$
$$-8 = 8 \qquad \text{Subtract.}$$

Since this is a false statement, $(4, 0)$ is not a solution of the equation $y - 2x = 8$.

Quick Check **1**

Is $(0, -6)$ a solution of the equation $3x + 2y = -12$?

The Rectangular Coordinate Plane

Objective **2** **Plot ordered pairs on a rectangular coordinate plane.** Ordered pairs can be displayed graphically using the **rectangular coordinate plane.** The rectangular coordinate plane is also known as the **Cartesian plane,** named after the famous French philosopher and mathematician René Descartes.

The rectangular coordinate plane is made up of two number lines, known as **axes.** The axes are drawn at right angles to each other, intersecting at 0 on each axis. The accompanying graph shows how it looks.

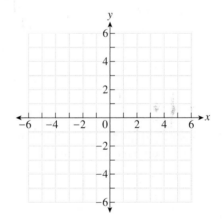

x-axis, *y*-axis, Origin

> The **horizontal axis,** or **x-axis,** is used to show the first value of an ordered pair. The **vertical axis,** or **y-axis,** is used to show the second value of an ordered pair. The point where the two axes cross is called the **origin** and represents the ordered pair $(0, 0)$.

Positive values of x are found to the right of the origin, and negative values of x are found to the left of the origin. Positive values of y are found above the origin, and negative values of y are found below the origin.

We can use the rectangular coordinate plane to represent ordered pairs that are solutions of an equation. To plot an ordered pair (x, y) on a rectangular coordinate plane, we begin at the origin. The x-coordinate tells us how far to the left or right of the origin to move. The y-coordinate then tells us how far up or down to move from there.

Suppose we wanted to plot the ordered pair $(-3, 2)$ on a rectangular coordinate plane. The x-coordinate is -3, and that tells us to move three units to the left of the origin. The y-coordinate is 2, so we move up two units to find the location of our ordered pair, and we place a point at this location. We will often refer to an ordered pair as a **point** when we plot it on a graph.

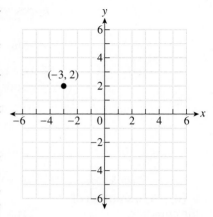

One final note about the rectangular coordinate plane is that the two axes divide the plane into four sections called **quadrants.** The four quadrants are labeled I, II, III, and IV. A point that lies on one of the axes is not considered to be in one of the quadrants.

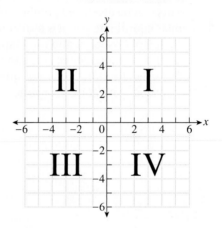

EXAMPLE 3   Plot the ordered pairs $(5, 1)$, $(4, -2)$, $(-2, -4)$, and $(-4, 0)$ on a rectangular coordinate plane.

Quick Check 2

Plot the ordered pairs $(3, 8)$, $(5, -4)$, $(-3, 6)$, and $(0, 7)$ on a rectangular coordinate plane.

Solution

The four ordered pairs are plotted on the following graph:

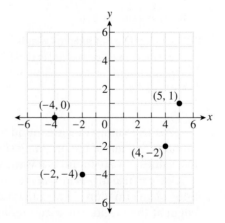

Choosing the appropriate scale for a graph is an important skill. To plot the ordered pair $\left(\frac{7}{2}, -4\right)$ on a rectangular coordinate plane, we could use a scale of $\frac{1}{2}$ on the axes. A rectangular plane with axes showing values ranging from -10 to 10 is not large enough to plot the ordered pair $(-45, 25)$. A good choice would be to label the axes showing values that range from -50 to 50. Since this is a wide range of values, it makes sense to increase the scale. Since each coordinate is multiple of 5, we will use a scale of 5 units.

It is important to be able to read the coordinates of an ordered pair from a graph.

EXAMPLE 4   Find the coordinates of points A through G on the following graph:

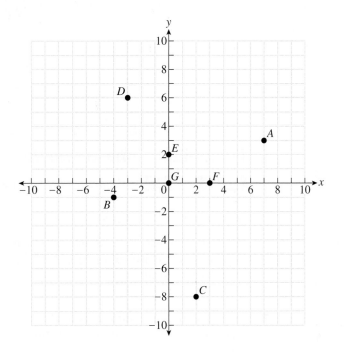

Solution

When determining coordinates for a point on a rectangular coordinate plane, we begin by finding the x-coordinate. We do this by determining how far to the left or right of the origin the point is. If the point is to the right, then the x-coordinate is positive. If the point is to the left, then the x-coordinate is negative. If the point is neither to the right nor to the left of the origin, then the x-coordinate is 0.

We next determine the y-coordinate of the point by measuring how far above or below the origin the point is. If the point is above the origin, then the y-coordinate is positive. If the point is below the origin, then the y-coordinate is negative. If the point is neither above nor below the origin, then the y-coordinate is 0.

Here is a table containing the coordinates of each point:

Point	A	B	C	D	E	F	G
Coordinates	$(7, 3)$	$(-4, -1)$	$(2, -8)$	$(-3, 6)$	$(0, 2)$	$(3, 0)$	$(0, 0)$

Quick Check **3** Find the coordinates of points *A* through *E* on the following graph:

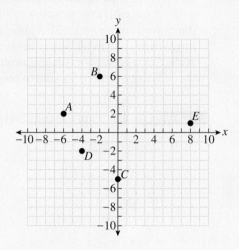

Linear Equations in Two Variables

Objective **3** **Complete ordered pairs for a linear equation in two variables.**

Standard Form of a Linear Equation

A linear equation in two variables is an equation that can be written in the form
$Ax + By = C$, where *A*, *B*, and *C* are real numbers and *A* and *B* are not both 0. This is
called the **standard form** of a linear equation in two variables.

To find an ordered pair that is a solution of a linear equation, we can arbitrarily select a value for one of the variables, substitute this value into the equation, and solve for the remaining variable. The next example gives us practice with finding solutions of linear equations in two variables.

EXAMPLE **5**   Find the missing coordinate such that the ordered pair $(2, y)$ is a solution of the equation $y = 5x - 4$.

Solution

Quick Check **4**
Find the missing coordinate such that the ordered pair $(3, y)$ is a solution of the equation $y = -3x + 5$.

We are given an *x*-coordinate of 2, so we substitute this for *x* in the equation.

$y = 5(2) - 4$ Substitute 2 for *x* and solve for *y*.
$y = 10 - 4$ Multiply.
$y = 6$ Subtract.

The ordered pair $(2, 6)$ is a solution of this equation.

The solutions of any linear equation in two variables follow a pattern. We will now begin to explore this pattern.

EXAMPLE 6   Complete the following table of ordered pairs with solutions of the linear equation $x + y = 5$:

x	3		0
y		-2	

Plot the solutions on a rectangular coordinate plane.

Solution

The first ordered pair has an x-coordinate of 3. Substituting this for x gives us the equation $(3) + y = 5$, which has a solution of $y = 2$. $(3, 2)$ is a solution.

The second ordered pair has a y-coordinate of -2. Substituting for y, this produces the equation $x + (-2) = 5$. The solution of this equation is $x = 7$, so $(7, -2)$ is a solution.

The third ordered pair has a x-coordinate of 0. When we substitute this for x, we get the equation $(0) + y = 5$. This equation has a solution of $y = 5$, so $(0, 5)$ is a solution. Completing the table gives us

x	3	7	0
y	2	-2	5

When we plot these three points on a rectangular coordinate plane, it looks like the following graph. Notice that the three points appear to fall on a straight line.

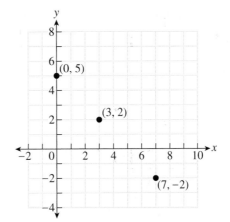

Quick Check 5

Complete the table of ordered pairs with solutions of the linear equation $x - y = 6$:

x	4		-1
y		3	

Plot the solutions on a rectangular coordinate plane.

Graphing Linear Equations by Plotting Points

Objective 4   Graph linear equations in two variables by plotting points.

When we plot the points for each set of solutions, we should notice that the three points appear to be in line with each other. In each case, we can draw a straight line that passes through each of the three points. In addition to the three points on the graph, every point that lies on the line is a solution of the equation. Every ordered pair that is a solution must lie on this line.

We can sketch a graph of the solutions of a linear equation by first plotting ordered pairs that are solutions and then drawing the straight line through these points. The line we sketch continues indefinitely in both directions. Although only two points are needed to determine a line, we will find three points. The third point is used as a check to make sure that the first two points are accurate. We will arbitrarily pick three values for one of the variables and find the other coordinates.

EXAMPLE 7   Graph the linear equation $y = 3x - 4$.

Solution

When the equation has y isolated on one side of the equals sign, it is a good idea to select values for x and find the corresponding y-values. We find three points on the line by letting $x = 0$, $x = 1$, and $x = 2$. We did not need to use 0, 1, and 2 for x; they were chosen arbitrarily. If you start with other values, your ordered pairs will still be on the same line. The following table shows the calculations:

$x = 0$	$x = 1$	$x = 2$
$y = 3(0) - 4$	$y = 3(1) - 4$	$y = 3(2) - 4$
$y = 0 - 4$	$y = 3 - 4$	$y = 6 - 4$
$y = -4$	$y = -1$	$y = 2$

Our three ordered pairs are $(0, -4)$, $(1, -1)$, and $(2, 2)$. We plot these three points on a rectangular coordinate plane and draw a line through them.

Quick Check 6

Graph the linear equation $y = 2x + 1$.

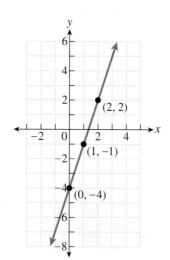

Is the ordered pair a solution of the given equation?

1. $(9, 3)$, $2x + y = 21$

2. $(5, 4)$, $3x + 2y = 23$

3. $(1, -4)$, $x + 2y = 7$

4. $(3, -3)$, $4x - 3y = 21$

5. $(-6, 2)$, $5x - 4y = -38$

6. $(-4, 3)$, $6x - 5y = 9$

7. $(3, 2)$, $-2x + 5y = 4$

8. $(5, -2)$, $-x + 7y = -9$

9. $(-3, 0)$, $-4x + y = 12$

10. $(0, 4)$, $2x - 5y = -20$

Plot the points on a rectangular coordinate plane.

11. $(5, 2)$ $(3, 4)$ $(2, -6)$ $(-4, 1)$

12. $(-3, -5)$ $(-5, 3)$ $(4, -3)$ $(-2, -1)$

13. $(0, 4)$ $(-2, 0)$ $(-4, 2)$ $(2, 4)$

14. $(-4, 0)$ $(-5, 0)$ $(0, -3)$ $(0, 2)$

15. $\left(\frac{7}{2}, 5\right)\left(-2, \frac{13}{4}\right)$ $\left(-\frac{23}{8}, -1\right)\left(\frac{3}{2}, -\frac{5}{2}\right)$

16. $(-70, 30)$ $(20, -90)$ $(45, 25)$ $(-50, -68)$

Find the coordinates of the labeled points A, B, C, and D.

17.

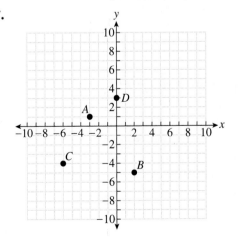

18.

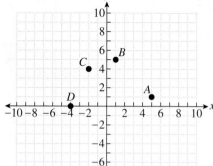

19.

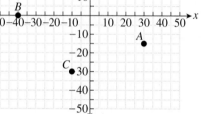

20.

Which quadrant (I, II, III, or IV) is the given point in?

21. $(5, -1)$ **22.** $(-2, 8)$ **23.** $(3, 4)$

24. $(-6, -4)$ **25.** $(2, 5)$ **26.** $(-3, -3)$

27. $\left(\dfrac{1}{2}, -\dfrac{3}{5}\right)$ **28.** $(-700, 1000)$

Using the given coordinate, find the other coordinate that makes the ordered pair a solution of the equation.

29. $3x - 2y = 8, (2, y)$

30. $x + 4y = 7, (x, 3)$

31. $-2x + 5y = -6, (x, -4)$

32. $5x + y = 3, (3, y)$

33. $-x - 3y = -4, (10, y)$

34. $-3x - 2y = 5, (x, -4)$

35. $7x - 11y = -22, (0, y)$

36. $4x + 3y = -20, (x, 0)$

37. $3x + \dfrac{1}{2}y = -2, (x, 8)$

38. $\dfrac{1}{5}x + 3y = 20, (10, y)$

Complete each table with ordered pairs that are solutions.

39. $2x + y = 7$ **40.** $x + 3y = 8$

x	y
1	
	1
0	

x	y
2	
	3
0	

41. $5x - 3y = -12$ **42.** $-2x + 4y = -16$

x	y
−3	
0	
3	

x	y
−2	
0	
2	

43. $y = 4x - 9$

x	y
0	
1	
4	

44. $y = -3x + 2$

x	y
-1	
0	
3	

45. $y = \dfrac{2}{3}x - 2$

x	y
-3	
0	
6	

46. $y = \dfrac{7}{5}x + 1$

x	y
-5	
0	
5	

51. $y = 3x$

52. $y = -5x$

53. $y = \dfrac{1}{4}x + 3$

54. $y = \dfrac{1}{2}x - 3$

Find three ordered pairs that are solutions of the given linear equation, and use them to graph the equation.

47. $y = 4x + 1$

48. $y = 2x - 3$

55. $y = -\dfrac{3}{5}x + 6$

56. $y = -\dfrac{5}{2}x + 4$

49. $y = -2x + 7$

50. $y = -3x + 6$

57. $3x + y = 9$

58. $2x + 3y = 6$

Answer in complete sentences.

59. Explain, in your own words, why the order of the coordinates in an ordered pair is important. Is there a difference between plotting the ordered pair (a, b) and the ordered pair (b, a)?

60. If the ordered pair (a, b) is in quadrant II, which quadrant is (b, a) in? Which quadrant is $(-a, b)$ in? Explain how you determined your answers.

> **Study Tip** **REVISITED**   One of the most important resources available to you is your instructor. Take advantage of your instructor's experience and knowledge. If you have a question during the lecture, be sure to ask. The question that you have is probably on the minds of several other students as well.
>
> Learn where your instructor's office is located and when office hours are held. Visiting your instructor during office hours is a great way to get one-on-one help. There are some questions that are difficult for the instructor to answer completely during the class period that are more easily handled in your instructor's office. Your instructor may also be able to give you extra problems to work on. In addition to answering your questions during class, some instructors are available for questions either right before or right after class.

3.2

GRAPHING LINEAR EQUATIONS AND THEIR INTERCEPTS

Objectives

1. Find the *x*- and *y*-intercepts of a line from its graph.
2. Find the *x*- and *y*-intercepts of a line from its equation.
3. Graph linear equations by using their intercepts.
4. Graph linear equations that pass through the origin.
5. Graph horizontal lines.
6. Graph vertical lines.
7. Interpret the graph of an applied linear equation.

In the previous section we learned that we can use a rectangular coordinate plane to display the solutions of an equation in two variables. We also learned that the solutions of a linear equation in two variables were all on a straight line. In this section, we will learn to graph the line associated with a linear equation in two variables in a more systematic fashion.

Intercepts

Objective **1**   Find the *x*- and *y*-intercepts of a line from its graph.

A point at which a graph crosses the *x*-axis is called an ***x*-intercept,** and a point at which a graph crosses the *y*-axis is called a ***y*-intercept.**

EXAMPLE 1   Find the coordinates of any *x*-intercepts and *y*-intercepts.

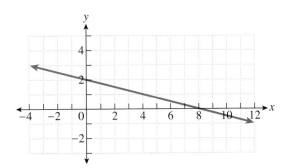

Solution

An *x*-intercept is a point at which the graph crosses the *x*-axis. We see that this graph crosses the *x*-axis at the point $(8, 0)$, so the *x*-intercept is the point $(8, 0)$.

This graph crosses the *y*-axis at the point $(0, 2)$, so its *y*-intercept is $(0, 2)$.

Quick Check **1** Find the coordinates of any *x*-intercepts and *y*-intercepts.

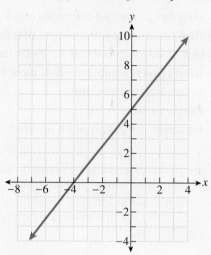

Some lines do not have both an *x*- and *y*-intercept. This horizontal line does not cross the *x*-axis, so it does not have an *x*-intercept. Its *y*-intercept is at the point $(0, -6)$.

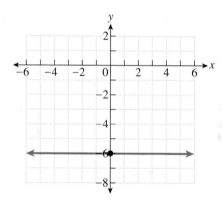

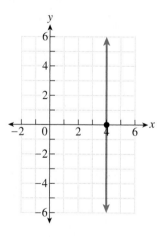

The *x*-intercept of this vertical line is the point $(4, 0)$. This graph does not cross the *y*-axis, so it does not have a *y*-intercept.

Objective **2** **Find the *x*- and *y*-intercepts of a line from its equation.** Any point that lies on the *x*-axis has a *y*-coordinate of 0. Similarly, any point that lies on the *y*-axis has an *x*-coordinate of 0. We can use these facts to find the intercepts of an equation's graph algebraically. To find an *x*-intercept, we substitute 0 for *y* and solve for *x*. To find a *y*-intercept, we substitute 0 for *x* and solve for *y*.

EXAMPLE ▸ 2 Find the x- and y-intercepts for the equation $5x - 2y = 10$.

Solution

Let's begin with the x-intercept by substituting 0 for y and solving for x.

$$5x - 2y = 10$$
$$5x - 2(0) = 10 \qquad \text{Substitute 0 for } y.$$
$$5x = 10 \qquad \text{Simplify the left side of the equation.}$$
$$x = 2 \qquad \text{Divide both sides by 5.}$$

The x-intercept is at $(2, 0)$. To find the y-intercept, we substitute 0 for x and solve for y.

Quick Check **2**

Find the x- and y-intercepts for the equation $-3x + 4y = -12$.

$$5x - 2y = 10$$
$$5(0) - 2y = 10 \qquad \text{Substitute 0 for } x.$$
$$-2y = 10 \qquad \text{Simplify the left side of the equation.}$$
$$y = -5 \qquad \text{Divide both sides by } -2.$$

The y-intercept is $(0, -5)$.

EXAMPLE ▸ 3 Find the x- and y-intercepts for the equation $y = -3x - 7$.

Solution

We will begin by finding the x-intercept. To do this, we substitute 0 for y and solve for x.

$$y = -3x - 7$$
$$0 = -3x - 7 \qquad \text{Substitute 0 for } y.$$
$$3x = -7 \qquad \text{Add } 3x \text{ to both sides.}$$
$$x = -\frac{7}{3} \qquad \text{Divide both sides by 3.}$$

Quick Check **3**

Find the x- and y-intercepts for the equation $y = 2x - 9$.

The x-intercept is $\left(-\frac{7}{3}, 0\right)$. To find the y-intercept, we substitute 0 for x and solve for y.

$$y = -3x - 7$$
$$y = -3(0) - 7 \qquad \text{Substitute 0 for } x.$$
$$y = -7 \qquad \text{Simplify.}$$

The y-intercept is at $(0, -7)$.

Graphing a Linear Equation by Using Its Intercepts

Objective 3 **Graph linear equations by using their intercepts.** When graphing a linear equation, we will begin by finding any intercepts. This will often give us two points for our graph. Although only two points are necessary to graph a straight line, we will plot at least three points to be sure that our first two points have been found accurately. In other words, we will use a third point as a "check" for the first two points we find. We find this third point by selecting a value for x, substituting it into the equation, and solving for y.

EXAMPLE ▸ 4 Graph $3x + y = 6$. Label any intercepts.

Solution

We begin by finding the x- and y-intercepts.

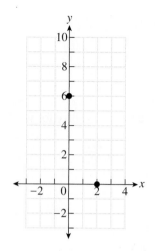

	x-intercept			**y-intercept**	
$3x + (0) = 6$	Substitute 0 for y.		$3(0) + y = 6$	Substitute 0 for x.	
$3x = 6$	Simplify.		$y = 6$	Simplify.	
$x = 2$	Divide both sides by 3.				

The x-intercept is $(2, 0)$. The y-intercept is $(0, 6)$.

Before looking for a third point, let's plot the two intercepts on a graph. This can be helpful in selecting a value to substitute for x when finding a third point. It is a good idea to select a value for x that is somewhat near the x-coordinates of the other two points. For example, $x = 1$ would be a good choice for this example. If we choose a value of x that is too far away from the x-coordinates of the other two points, then we may have to change the scale of the axes or extend them to show this third point.

$$3(1) + y = 6 \qquad \text{Substitute 1 for } x.$$
$$3 + y = 6 \qquad \text{Simplify the left side of the equation.}$$
$$y = 3 \qquad \text{Subtract 3.}$$

The point $(1, 3)$ is our third point. Keep in mind that the third point can vary with our choice for x, but the three points should still be on the same line. After plotting the third point, we finish by drawing a straight line that passes through all three points.

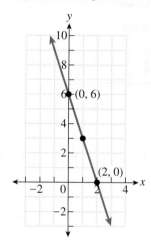

Quick Check 4

Graph $2x - 3y = 12$.
Label any intercepts.

EXAMPLE 5   Graph $y = 2x + 5$. Label any intercepts.

Solution

We will begin by finding the x- and y-intercepts.

x-intercept $(y = 0)$	**y-intercept** $(x = 0)$
$0 = 2x + 5$	$y = 2(0) + 5$
$-5 = 2x$	$y = 5$
$-\dfrac{5}{2} = x$	$(0, 5)$
$\left(-\dfrac{5}{2}, 0\right)$	

Choosing $x = 1$ to find our third point is a good choice: It is close to the two intercepts, and it will be easy to substitute 1 for x in the equation.

$$y = 2(1) + 5 \qquad \text{Substitute 1 for } x \text{ in the equation.}$$
$$y = 7 \qquad\qquad \text{Simplify.}$$

Our third point is $(1, 7)$. Here is the graph:

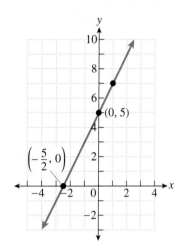

Quick Check 5
Graph $y = -3x - 9$.
Label any intercepts.

Using Your Calculator   We can use the TI-83/84 to graph a line. To enter the equation of the line that we graphed in Example 5, press the [Y=] key. Enter 2x + 5 next to Y_1, using the key labeled [X,T,Θ,n] to type the variable x. Now press the key labeled [GRAPH] to graph the line.

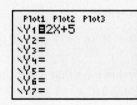
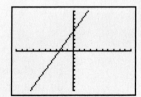

Graphing a Line by Using Its Intercepts

- Find the *x*-intercept by substituting 0 for *y* and solving for *x*.
- Find the *y*-intercept by substituting 0 for *x* and solving for *y*.
- Select a value of *x*, substitute it into the equation, and solve for *y* to find the coordinates of a third point on the line.
- Graph the line that passes through these three points.

Graphing Linear Equations That Pass through the Origin

Objective **4** **Graph linear equations that pass through the origin.** Some lines do not have two intercepts. For example, a line that passes through the origin has its x- and y-intercepts at the same point. In this case, we will need to find two additional points.

EXAMPLE **6**   Graph $2x - 8y = 0$. Label any intercepts.

Solution

Let's start by finding the x-intercept.

$$
\begin{aligned}
2x - 8(0) &= 0 \qquad \text{Substitute 0 for } x.\\
2x &= 0 \qquad \text{Simplify.}\\
x &= 0 \qquad \text{Divide both sides by 2.}
\end{aligned}
$$

The x-intercept is at the origin $(0, 0)$. This point is also the y-intercept, which you can verify by substituting 0 for x and solving for y. We will need to find two more points before we graph the line. Suppose that we choose $x = 1$. Then we have

$$
\begin{aligned}
2(1) - 8y &= 0 \qquad \text{Substitute 1 for } x.\\
2 - 8y &= 0 \qquad \text{Simplify.}\\
2 &= 8y \qquad \text{Add } 8y.\\
\frac{1}{4} &= y \qquad \text{Divide both sides by 8 and simplify.}
\end{aligned}
$$

The point $\left(1, \frac{1}{4}\right)$ is on the line. However, it may not be easy for us to put this point on the graph because of its fractional y-coordinate. We can try to avoid this problem by solving the equation for y before choosing a value for x.

$$
\begin{aligned}
2x - 8y &= 0\\
2x &= 8y \qquad\qquad\qquad \text{Add } 8y \text{ to both sides.}\\
\frac{1}{4}x &= y \;\; \text{or} \;\; y = \frac{1}{4}x. \qquad \text{Divide both sides by 8 and simplify.}
\end{aligned}
$$

If we choose a value for x that is a multiple of 4, then we will have a y-coordinate that is an integer. We will use $x = 4$ and $x = -4$, although there are many other choices that would work.

$x = 4$	$x = -4$
$y = \dfrac{1}{4}(4)$	$y = \dfrac{1}{4}(-4)$
$y = 1$	$y = -1$
$(4, 1)$	$(-4, -1)$

Here is the graph:

Quick Check 6

Graph $y = 3x$. Label any intercepts.

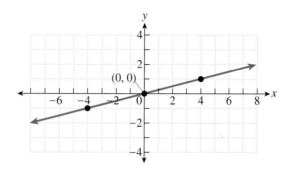

Graphing a Line Passing through the Origin

- Determine that the x- and y-intercepts are both at the origin.
- Find a second point on the line by selecting a value of x and substituting it into the equation. Solve the equation for y.
- Find a third point on the line by selecting another value of x and substituting it into the equation. Solve the equation for y.
- Graph the line that passes through these three points.

Horizontal Lines

Objective 5 **Graph horizontal lines.** Can we draw a line that does not cross the x-axis? Yes, a horizontal line can be drawn that does not cross the x-axis. Here's an example:

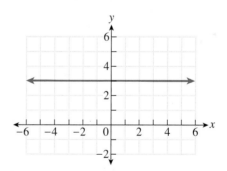

What is the equation of the line in this graph? Notice that each point has the same y-coordinate (3). The equation for the line is $y = 3$. The equation for a horizontal line is always of the form $y = b$, where b is a real number. To graph the line $y = b$, plot the y-intercept $(0, b)$ on the y-axis and draw a horizontal line through this point.

EXAMPLE 7   Graph $y = -2$. Label any intercepts.

Solution

This graph will be a horizontal line, as its equation is of the form $y = b$. If an equation does not have a term containing the variable x, then its graph will be a horizontal line.

Quick Check 7

Graph $y = 7$. Label any intercepts.

We begin by plotting the point $(0, -2)$, which is the y-intercept. We then draw a horizontal line through this point.

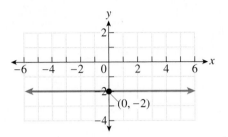

Vertical Lines

Objective 6 **Graph vertical lines.** Vertical lines have equations that are of the form $x = a$, where a is a real number. Just as the equation of a horizontal line does not have any terms containing the variable x, the equation of a vertical line does not have any terms containing the variable y. We begin to graph an equation of this form by plotting its x-intercept at $(a, 0)$. We then draw a vertical line through this point.

EXAMPLE 8   Graph $x = 6$. Label any intercepts.

Solution

The x-intercept for this line is at $(6, 0)$. We plot this point and then draw a vertical line through it.

Quick Check 8

Graph $x = -15$. Label any intercepts.

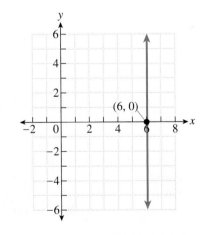

Horizontal Lines
Equation: $y = b$
- Plot the y-intercept at $(0, b)$.
- Graph the horizontal line passing through this point.

Vertical Lines
Equation: $x = a$
- Plot the x-intercept at $(a, 0)$.
- Graph the vertical line passing through this point.

Applications of the Graphs of Linear Equations

Objective 7 **Interpret the graph of an applied linear equation.** The intercepts of a graph have important interpretations within the context of an applied problem. The y-intercept often gives us an initial condition, such as how much money is owed on a loan, or the fixed costs to attend college before adding in the number of units that a student is taking. The x-intercept often shows us a break-even point, such as when a business goes from a loss to a profit.

EXAMPLE 9 Chris borrowed $3600 from his friend Mack, promising to pay him $50 per month until he had paid off the loan. The amount of money (y) that Chris owes Mack after x months have passed is given by the equation $y = 3600 - 50x$.

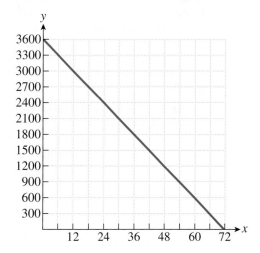

Use the graph of this equation to answer the following questions (since the number of months cannot be negative, the graph begins at $x = 0$):

a) What does the y-intercept represent in this situation?

Solution

The y-intercept is at $(0, 3600)$. It tells us that after borrowing the money from Mack, Chris owes Mack $3600.

b) What does the x-intercept represent in this situation?

Solution

The x-intercept is at $(72, 0)$. It tells us that Chris will pay off his debt in 72 months.

Using Your Calculator   When graphing a line using the TI-83/84, we will occasionally need to resize the window. In Example 9, the x-intercept is at the point $(72, 0)$ and the y-intercept is at $(0, 3600)$. Our graph should show these important points. After entering $3600 - 50x$ for Y_1, press the [WINDOW] key. Fill in the screen as follows, and press the [GRAPH] key to display the graph:

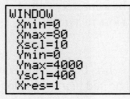

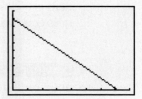

Quick Check   **9**   Nancy invests \$10,000 in a new business and expects to make a \$2000 profit each month. Nancy's net profit (y) after x months have passed is given by the equation $y = 2000x - 10,000$.

   Use the graph of this equation to answer the following questions (since the number of months cannot be negative, the graph begins at $x = 0$):

a) What does the y-intercept represent in this situation?

b) What does the x-intercept represent in this situation?

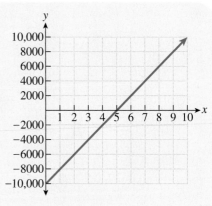

EXERCISES 3.2 ❯

Find the x- and y- intercepts from the graph.

1.

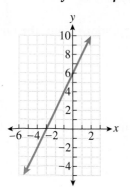

2.

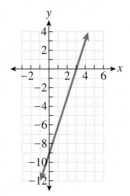

3.

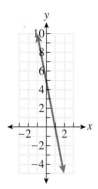

4.

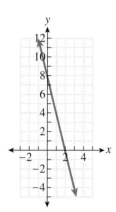

5.

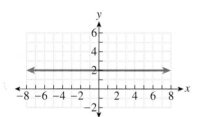

No *x*-int.,

6.

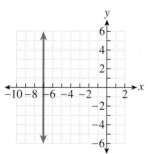

No *y*-int.

Find the x- and y- intercepts.

7. $x + y = 6$

8. $x + y = -3$

9. $x - y = -8$

10. $x - y = 4$

11. $3x + y = 9$

12. $2x + y = -10$

13. $-3x + 5y = 45$

14. $-2x - 6y = 18$

15. $-4x + 6y = 14$

16. $5x + 8y = -60$

17. $\dfrac{6}{5}x + \dfrac{3}{10}y = 10$

18. $\dfrac{1}{3}x + \dfrac{1}{2}y = 2$

19. $\dfrac{3}{5}x + \dfrac{1}{2}y = 3$

20. $\dfrac{2}{9}x - \dfrac{1}{12}y = \dfrac{2}{3}$

21. $y = -x + 8$

22. $y = x - 5$

23. $y = 3x - 12$

24. $y = -2x + 16$

25. $y = 3x$

26. $y = -2x$

27. $y = \dfrac{3}{5}x - 2$

28. $y = \dfrac{4}{7}x - 4$

29. $y = 2$ No *x*-int.,

30. $x = 10$ No *y*-int.

Find the intercepts, and then graph the line.

31. $x + y = 4$

32. $x + y = -2$

33. $2x - 5y = 10$

34. $-4x + 2y = 8$

35. $-3x + 4y = -24$

36. $6x + 5y = -30$

37. $3x + y = 9$

38. $x + 2y = 8$

39. $-4x + 7y = 14$

40. $3x - 4y = 18$

41. $y = x - 4$

42. $y = x + 6$

43. $y = 5x$

44. $y = -4x$

49. $y = 4$

50. $y = -6$

45. $y = -2x + 7$

46. $y = 4x - 2$

51. $x = 3$

47. $y = \dfrac{7}{2}x - 7$

52. $x = -8$

48. $y = -\dfrac{1}{3}x + 2$

53. A lawyer bought a new copy machine for her office in the year 2005, paying $12,000. For tax purposes, the copy machine depreciates at $1500 per year. If we let x represent the number of years after 2005, then the value of the copier (y) after x years is given by the equation $y = 12{,}000 - 1500x$.

   a) Find the y-intercept of this equation. Explain, in your own words, what this intercept signifies.

b) Find the x-intercept of this equation. Explain, in your own words, what this intercept signifies.

c) In 2008, what will be the value of the copy machine?

54. Jerry was ordered to perform 400 hours of community service as a reading tutor. Jerry spends eight hours each Saturday teaching people to read. The equation that tells the number of hours (y) that remain on the sentence after x Saturdays is $y = 400 - 8x$.

a) Find the y-intercept of this equation. Explain, in your own words, what this intercept signifies.

b) Find the x-intercept of this equation. Explain, in your own words, what this intercept signifies.

c) After 12 Saturdays, how many hours of community service does Jerry have remaining?

55. At an exclusive country club, members pay an annual fee of $5000. In addition, they must pay $75 for each round of golf that they pay. The equation that gives the cost (y) to play x rounds of golf per year is $y = 5000 + 75x$.

a) Find the y-intercept of this equation. Explain, in your own words, what this intercept signifies.

b) In your own words, explain why this equation has no x-intercept within the context of this problem.

c) If a person plays one round of golf per week, how much will she pay to the country club in one year?

56. In addition to paying $24 per unit, a community college student pays a registration fee of $100 per semester. The equation that gives the cost (y) to take x units is $y = 24x + 100$.

a) Find the y-intercept of this equation. Within the context of this problem, is this cost possible? Explain why or why not.

b) In your own words, explain why this equation has no x-intercept within the context of this problem.

c) A full load for a student is 12 units. How much will a full-time student pay per semester?

Study Tip **REVISITED**   Most college campuses have a tutorial center that provides free tutoring in math and other subjects. Some tutorial centers offer walk-in tutoring, others schedule appointments for one-on-one or group tutoring, and some combine these approaches.

Walk-in tutoring allows you to come in whenever you want to and ask questions as they arise. One-on-one tutoring is designed for students who will require more assistance than simple questions and answers, such as if you need someone to help explain the topic to you. If you do not completely understand a tutor's answer or directions, do not be afraid to ask for further explanation.

While it is a good idea to ask a tutor if you are doing a problem correctly or to give you an idea of a starting point for a problem, it is not a good idea to ask a tutor to do a problem for you. We learn by doing, not by watching. It may look easy while the tutor is working a problem, but this will not help you to do the next problem.

3.3

SLOPE OF A LINE

Objectives

1. Understand the slope of a line.
2. Find the slope of a line from its graph.
3. Find the slope of a line passing through two points by using the slope formula.
4. Find the slopes of horizontal and vertical lines.
5. Find the slope and *y*-intercept of a line from its equation.
6. Find the equation of a line given its slope and *y*-intercept.
7. Graph a line by using its slope and *y*-intercept.
8. Interpret the slope and *y*-intercept in real-world applications.

Slope of a Line

Objective 1   **Understand the slope of a line.** Here are four different lines that pass through the point $(0, 2)$:

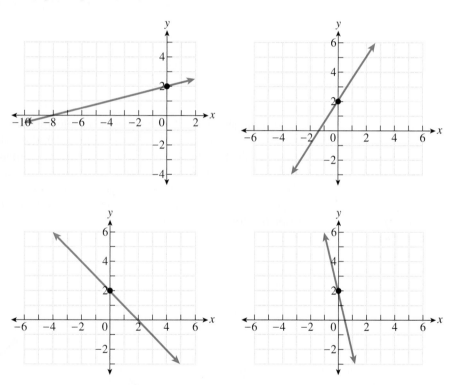

Notice that some of the lines are rising to the right, while others are falling to the right. Also, some are rising or falling more steeply than the others. The characteristic that distinguishes these lines is their slope.

Slope

> The **slope** of a line is a measure of how steeply a line rises or falls as it moves to the right. We use the letter *m* to stand for the slope of a line.

If a line rises as it moves to the right, then its slope is positive. If a line falls as it moves to the right, then its slope is negative.

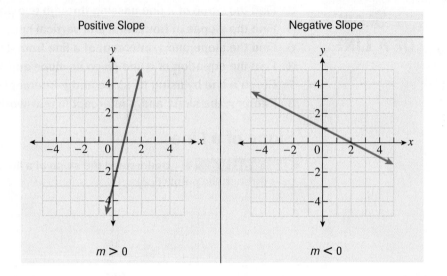

Positive Slope | Negative Slope

$m > 0$ | $m < 0$

To find the slope of a line, we begin by selecting two points that are on the line. As we move from the point on the left to the point on the right, we measure how much the line rises or falls. The slope is equal to this distance divided by the distance traveled from left to right. This is often referred to as "rise over run," or as the change in y divided by the change in x. Sometimes the change in y is written as Δy, and the change in x is written as Δx. The Greek letter Δ (delta) is often used to represent change in a quantity. The slope of a line is represented as

$$m = \frac{\text{rise}}{\text{run}} \text{ or } m = \frac{\Delta y}{\Delta x}.$$

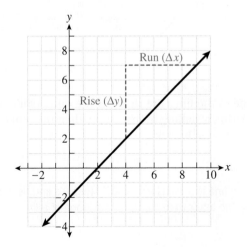

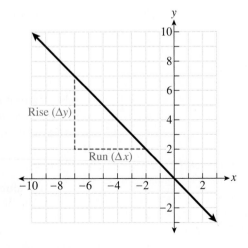

Finding the Slope of a Line from Its Graph

Objective 2 **Find the slope of a line from its graph.** Consider the following line that passes through the points $(1, 2)$ and $(3, 8)$. To move from the point on the left to the point on the right, the line rises by six units as it moves two units to the right.

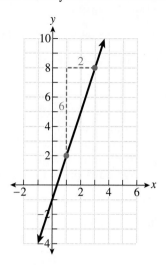

The slope m is $\frac{6}{2}$ or 3. When $m = 3$, this means that every time the line moves up by three units (y increases by 3), it moves one unit to the right (x increases by 1). Look at the graph again. Starting at the point $(1, 2)$, move three units up and one unit to the right. This places you at the point $(2, 5)$. This point is also on the line. We can continue this pattern to find other points on the line.

EXAMPLE 1   Find the slope of the line that passes through the points $(4, 7)$ and $(6, 1)$.

Solution

We will begin by plotting the two points on a graph. Notice that this line falls from left to right, so its slope is negative. The line drops by six units as it moves two units to the right, so its slope is $-\frac{6}{2}$ or -3.

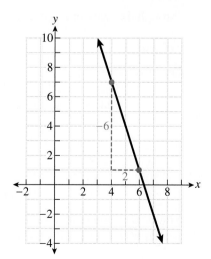

Quick Check 1
Find the slope of the line that passes through $(-2, 5)$ and $(4, -7)$.

Slope Formula

Objective 3 **Find the slope of a line passing through two points by using the slope formula.**

Slope Formula

> If a line passes through two points (x_1, y_1) and (x_2, y_2), then we can calculate its
> slope using the formula $m = \dfrac{y_2 - y_1}{x_2 - x_1}$.

This is consistent with the technique that we used to find slope in the previous example. The numerator $(y_2 - y_1)$ represents the vertical change, while the denominator $(x_2 - x_1)$ represents the horizontal change. Using this formula saves us from having to plot the points on a graph, but keep in mind that you can continue to use that technique.

We have previously seen that the slope of a line that passes through the points $(1, 2)$ and $(3, 8)$ is 3. We will now use the formula $m = \dfrac{y_2 - y_1}{x_2 - x_1}$ to find the slope of the line through these two points. We will use the point $(1, 2)$ as (x_1, y_1) and the point $(3, 8)$ as (x_2, y_2).

$$m = \frac{8 - 2}{3 - 1}$$
$$= \frac{6}{2}$$
$$= 3$$

Notice that we get the same result as we did before. We could change the ordering of the points and still calculate the same slope. In other words, we could use the point $(3, 8)$ as (x_1, y_1) and the point $(1, 2)$ as (x_2, y_2) and still find that $m = 3$. We can subtract in either order as long as we are consistent. Whichever order we subtract the y-coordinates in, we must subtract the x-coordinates in the same order.

EXAMPLE 2 Use the formula $m = \dfrac{y_2 - y_1}{x_2 - x_1}$ to find the slope of the line that passes through the two points $(0, 7)$ and $(3, 1)$.

Solution

One way to think of the numerator in this formula is "*the second* y-*coordinate minus the first* y-*coordinate,*" which, in our example, would be $1 - 7$. Using a similar approach for the x-coordinates in the denominator, we would begin with a denominator of $3 - 0$.

Quick Check 2
Use the formula
$m = \dfrac{y_2 - y_1}{x_2 - x_1}$ to find the
slope of the line that
passes through the two
points $(1, 2)$ and $(5, 14)$.

$$m = \frac{1 - 7}{3 - 0} \qquad \text{Substitute into the formula.}$$
$$= \frac{-6}{3} \qquad \text{Simplify the numerator and denominator.}$$
$$= -2 \qquad \text{Simplify.}$$

The slope of the line that passes through these two points is -2. This line falls 2 units for each unit it moves to the right.

A Word of Caution   It does not matter which point is labeled as (x_1, y_1) and which point is labeled as (x_2, y_2). It *is* important that the order in which the y-coordinates are subtracted in the numerator is also the order in which the x-coordinates are subtracted in the denominator.

EXAMPLE   3   Use the formula $m = \dfrac{y_2 - y_1}{x_2 - x_1}$ to find the slope of the line that passes through the two points $(-5, -2)$ and $(-1, 4)$.

Solution

In this example, we learn to apply the formula to points that have negative x- or y-coordinates.

$$m = \frac{4 - (-2)}{-1 - (-5)} \qquad \text{Substitute into the formula.}$$

$$= \frac{4 + 2}{-1 + 5} \qquad \text{Eliminate "double signs."}$$

$$= \frac{6}{4} \qquad \text{Simplify the numerator and denominator.}$$

$$= \frac{3}{2} \qquad \text{Simplify.}$$

The slope of this line is $\frac{3}{2}$. The line rises by 3 units for every 2 units that it moves to the right.

Quick Check   3 Use the formula $m = \dfrac{y_2 - y_1}{x_2 - x_1}$ to find the slope of the line that passes through the two points $(-3, 5)$ and $(1, -5)$.

Horizontal and Vertical Lines

Objective   4   **Find the slopes of horizontal and vertical lines.** Let's look at the line that passes through the points $(1, 3)$ and $(4, 3)$.

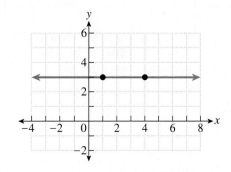

This is a horizontal line. What is its slope?

$$m = \frac{3 - 3}{4 - 1}$$

$$= \frac{0}{3}$$

$$= 0$$

The slope of this line is 0. The same is true for any horizontal line. The vertical change between any two points on a horizontal line is always equal to 0, and when we divide 0 by any nonzero number the result is equal to 0. Thus, the slope of any horizontal line is equal to 0.

Here is a brief summary of the properties of horizontal lines.

Horizontal Lines

Equation: $y = b$, where b is a real number
  y-intercept: $(0, b)$
  Slope: $m = 0$

Look at the following vertical line:

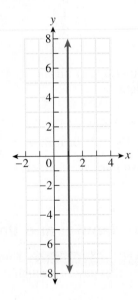

Notice that each point on this line has an x-coordinate of $x = 1$. We can attempt to find the slope of this line by selecting any two points on the line, such as $(1, 2)$ and $(1, 5)$, and using the slope formula.

$$m = \frac{5 - 2}{1 - 1}$$

$$= \frac{3}{0}$$

The fraction $\frac{3}{0}$ is undefined. So the slope of a vertical line is said to be undefined.

Here is a brief summary of the properties of vertical lines.

Vertical Lines

> Equation: $x = a$, where a is a real number
> x-intercept: $(a, 0)$
> Slope: Undefined

EXAMPLE 4 Find the slope, if it exists, of the line $x = -2$.

Solution

Quick Check 4
Find the slope, if it exists, of the line $x = 5$.

This line is a vertical line. The slope of a vertical line is undefined, so this line has undefined slope.

EXAMPLE 5 Find the slope, if it exists, of the line $y = 4$.

Solution

This line is a horizontal line. The slope of any horizontal line is 0, so the slope of this line is 0.

Quick Check 5
Find the slope, if it exists, of the line $y = -2$.

A Word of Caution Avoid stating that a line has "no slope." This is a vague phrase; some may take it to mean that the slope is equal to 0, while others may interpret it as meaning that the slope is undefined.

Slope–Intercept Form of a Line

Objective 5 Find the slope and *y*-intercept of a line from its equation. If we solve a given equation for y so that it is in the form $y = mx + b$, this is said to be the **slope–intercept form** of a line. The number being multiplied by x is the slope m, while b represents the y-coordinate of the y-intercept.

Slope–Intercept Form of a Line

> $y = mx + b$
> m: Slope of the line
> b: y-coordinate of the y-intercept

This means that if we were graphing the line $y = 2x + 5$, then the line would have a slope of 2 and a y-intercept of $(0, 5)$. Let's verify that this is true, beginning with the slope. Every time we increase the value of x by 1, y will increase by 2 because x is being multiplied by 2. If y increases by 2 (vertical change) every time x increases by 1 (horizontal change), this means that the slope of the line is equal to 2. As for the y-intercept, we can verify that this is true by substituting 0 for x in the equation.

$$y = 2x + 5$$
$$y = 2(0) + 5$$
$$y = 5$$

This shows that the y-intercept is at $(0, 5)$. Repeating this for the general form $y = mx + b$, we see that the y-coordinate of the y-intercept is b.

EXAMPLE 6   Find the slope and y-intercept for the line $y = -3x + 7$.

Solution

Since this equation is already solved for y, we can read the slope and the y-intercept directly from the equation. The slope is -3, which is the coefficient of the term containing x in the equation. The y-intercept is at $(0, 7)$, since 7 is the constant in the equation.

EXAMPLE 7   Find the slope and y-intercept for the line $2x + 2y = 11$.

Solution

We begin by solving for y.

$$2x + 2y = 11$$
$$2y = -2x + 11 \qquad \text{Subtract } 2x \text{ from both sides.}$$
$$\frac{2y}{2} = \frac{-2x}{2} + \frac{11}{2} \qquad \text{Divide each term by 2.}$$
$$y = -x + \frac{11}{2} \qquad \text{Simplify.}$$

The slope of this line is -1. When we see $-x$ in the equation, we need to remember that this is the same as $-1x$. The y-intercept is $(0, \frac{11}{2})$.

> **A Word of Caution**   We cannot determine the slope and y-intercept of a line from its equation unless the equation has been solved for y first.

Objective 6   Find the equation of a line given its slope and y-intercept.

EXAMPLE 8   Find the equation of a line which has a slope of 2 and a y-intercept of $(0, -15)$.

Solution

We will use the slope–intercept form $(y = mx + b)$ to help us find the equation of this line. Since the slope is 2, we can replace m by 2. Also, since the y-intercept is $(0, -15)$, we can replace b by -15. The equation of this line is $y = 2x - 15$.

Graphing a Line by Using Its Slope and y-Intercept

Objective 7   **Graph a line by using its slope and y-intercept.** Once we have an equation in slope–intercept form, we can use this information to graph the line. We can begin by plotting the y-intercept as our first point. We can then use the slope of the line to find a second point. To find a third point, we can find the x-intercept by letting $y = 0$ and solving for x.

EXAMPLE ▸ **9**   Graph the line $y = -4x + 8$. Label any intercepts.

Solution

We will start at the y-intercept, which is at $(0, 8)$. Since the slope is -4, this tells us that the line moves down four units as it moves one unit to the right. This would give us a second point at $(1, 4)$.

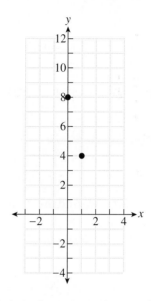

We can find the x-intercept by substituting 0 for y and solving for x.

$$0 = -4x + 8$$
$$4x = 8 \qquad \text{Add } 4x \text{ to both sides.}$$
$$x = 2 \qquad \text{Divide both sides by 4.}$$

The x-intercept is at $(2, 0)$. Below is the graph, showing the three points.

Quick Check **9**

Graph the line $y = 2x + 6$. Label any intercepts.

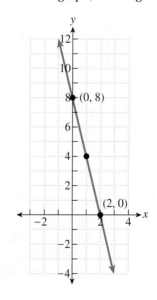

EXAMPLE 10   Graph the line $y = \frac{1}{2}x - 3$. Label any intercepts.

Solution

We will start by plotting the y-intercept at $(0, -3)$. The slope is $\frac{1}{2}$, which tells us that the line rises by one unit as it moves two units to the right. This will give us a second point at $(2, -2)$.

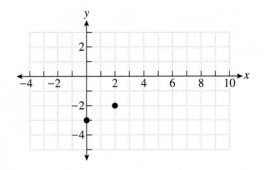

Now we find the x-intercept by substituting 0 for y.

$$0 = \frac{1}{2}x - 3$$

$$2 \cdot 0 = 2\left(\frac{1}{2}x - 3\right) \qquad \text{Multiply both sides by 2 to clear fractions.}$$

$$0 = \overset{1}{\cancel{2}} \cdot \frac{1}{\underset{1}{\cancel{2}}}x - 6 \qquad \text{Distribute and divide out common factors.}$$

$$0 = x - 6 \qquad \text{Simplify.}$$

$$6 = x \qquad \text{Add 6 to both sides.}$$

The x-intercept is at $(6, 0)$. The graph is shown below.

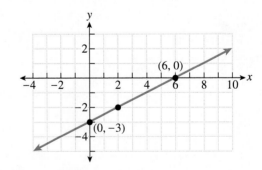

Quick Check 10
Graph the line
$y = -\dfrac{2}{3}x + 4$. Label any intercepts.

Graphing a Line by Using Its Slope and y-intercept

- Plot the y-intercept $(0, b)$.
- Use the slope m to find another point on the line.
- Find the x-intercept by substituting 0 for y and solving for x.
- Graph the line that passes through these 3 points.

Applications

Objective 8 Interpret the slope and *y*-intercept in real-world applications.
The concept of slope becomes much more important to us when we apply it to real-world situations.

EXAMPLE 11 The number of Americans (*y*), in millions, who have a cellular phone can be approximated by the equation $y = 13x + 77$, where *x* is the number of years after 1999. Interpret the slope and *y*-intercept of this line. (*Source:* Cellular Telecommunications Industry Association)

Solution

The slope of this line is 13, and this tells us that each year we can expect an additional 13 million Americans to have a cellular phone. (The number of Americans who have a cellular phone is increasing because the slope is positive.) The *y*-intercept at $(0, 77)$ tells us that approximately 77 million Americans had a cellular phone in 1999.

Quick Check **11** The number of women (*y*) accepted to medical school in a given year can be approximated by the equation $y = 218x + 7485$, where *x* is the number of years after 1997. Interpret the slope and *y*-intercept of this line. (*Source:* Association of American Medical Colleges)

EXERCISES 3.3 ❯

Determine whether the given line has a positive or negative slope.

1.

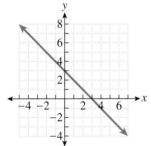

2.

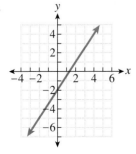

3.

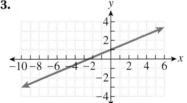

4.

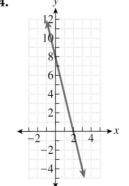

Find the slope of the given line. If the slope is undefined, state this.

5.

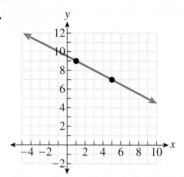

6.

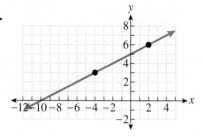

7.

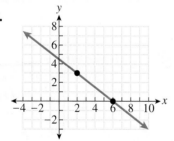

8.

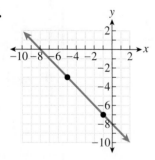

9.

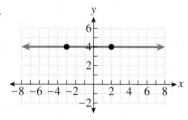

10.

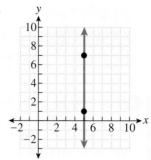

Find the slope of a line that passes through the given two points. If the slope is undefined, state this.

11. $(3, 5)$ and $(4, 7)$

12. $(2, 6)$ and $(3, 1)$

13. $(-2, 3)$ and $(-5, -6)$

14. $(5, -4)$ and $(-1, 5)$

15. $(3, 6)$ and $(-4, 6)$

16. $(-2, -6)$ and $(2, 6)$

17. $(0, 4)$ and $(0, 7)$

18. $(0, 0)$ and $(4, 7)$

Graph. Label any intercepts and determine the slope of the line. If the slope is undefined, state this.

19. $y = 4$

20. $x = -2$

21. $x = 0$

22. $y = -\dfrac{17}{5}$

26.

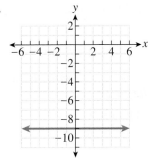

27.

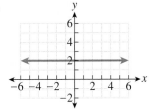

23. $x = 8$

28.

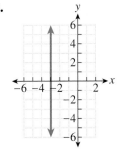

24. $y = 0$

Find the slope and the y-intercept of the given line.

29. $y = 6x - 7$ **30.** $y = 4x + 11$

31. $y = -2x + 3$

32. $y = -5x - 8$

33. $6x + 4y = -10$

34. $8x - 6y = 12$

Determine the equation of the given line, as well as the slope of the line. If the slope is undefined, state this.

35. $x - 5y = 8$

36. $x + 4y = 14$

25.

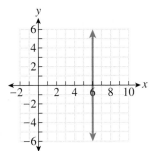

Find the equation of a line with the given slope and y-intercept.

37. Slope -2, y-intercept $(0, 5)$

38. Slope 4, y-intercept $(0, 3)$

39. Slope 3, y-intercept $(0, -6)$

40. Slope -5, y-intercept $(0, -2)$

41. Slope 0, y-intercept $(0, -4)$

42. Slope 0, y-intercept $(0, 1)$

Graph using the slope and y-intercept. Find the x-intercept and label it on the graph.

43. $y = 3x + 6$ **44.** $y = 2x - 8$

45. $y = -2x - 6$ **46.** $y = -5x + 10$

47. $y = x - 3$ **48.** $y = -x - 1$

49. $y = 4x$ **50.** $y = -2x$

51. $y = \dfrac{7}{2}x - 7$ **52.** $y = -\dfrac{5}{4}x + 10$

Graph. Label any intercepts.

53. $y = -x + 2$

54. $y = -\dfrac{1}{4}x + 2$

55. $y = -2x + 8$

56. $y = -3x + 7$

62. $x = \dfrac{9}{2}$

63. $y = 4$

57. $y = 6x - 9$

58. $y = -2$

64. $y = x + 7$

65. $y = \dfrac{3}{4}x + 3$

59. $y = -x - 5$

60. $y = -5x$

66. $x = -3$

67. $3x + y = 1$

61. $y = \dfrac{4}{5}x$

68. $y = \dfrac{2}{5}x + 1$

69. $5x - y = -5$ **70.** $-8x + 2y = -10$

a) Find the slope of the equation. In your own words, explain what this slope signifies.

b) Find the y-intercept of the equation. In your own words, explain what this y-intercept signifies.

c) Use this equation to predict the value of the card in the year 2015.

74. The number of nursing students in college (y) at a time x years after 1999 is approximated by the equation $y = -3600x + 60{,}000$. (Based on data from the years 1995–1999. *Source:* American Association of Colleges of Nursing)

Graph the two lines and find the coordinates of the point of intersection.

71. $y = 3x - 6$ and $y = -x - 2$

a) Find the slope of the equation. In your own words, explain what this slope signifies.

72. $5x - 4y = 20$ and $y = \dfrac{1}{2}x + 1$

b) Find the y-intercept of the equation. In your own words, explain what this y-intercept signifies.

c) Use this equation to predict the number of nursing students in the year 2025. Does this number seem possible? Explain why or why not.

75. On a 1000-foot stretch of highway through the mountains, the road rises by a total of 50 feet. Find the slope of the road.

76. George walked 1 mile (5280 feet) on a treadmill. The grade was set to 4%, which means that the slope of the treadmill is 4% or 0.04. Over the course of his workout, how many feet did George climb (vertically)?

73. Over a 10-year period from 1990 to 2000, the value of a baseball card increased by \$40. If we let x represent the number of years after 1990, the value (y) of the card is given by the equation $y = 4x + 50$.

77. In general, the larger the animal, the slower its heart rate. Would an equation relating heart rate (y) to an animal's weight (x) have a positive or negative slope? Explain why.

78. The governor of a state predicts unemployment rates will increase for the next five years. Would an equation relating unemployment rate (y) to the number of years after the statement was made (x) have a positive or negative slope? Explain why.

79. Consider the equations $y = -\frac{3}{8}x + 7$ and $20x - 8y = 320$. For each equation, do you feel that graphing by finding the x- and y-intercepts or by using the slope and y-intercept is the most efficient way to graph the equation? Explain your choices.

> ***Study Tip*** **REVISITED**   Some colleges run short-term study skills courses or seminars, and these courses can be quite helpful to students who are learning mathematics. These courses cover everything from note taking and test taking to overcoming math anxiety. The tips and suggestions that you pick up can only help to improve your understanding of mathematics. To find out whether there are any such courses or seminars at your school, ask an academic counselor.

3.4
LINEAR
FUNCTIONS

Objectives

1 Define *function, domain,* and *range.*

2 Evaluate functions.

3 Graph linear functions.

4 Interpret the graph of a linear function.

5 Determine the domain and range of a function from its graph.

A coffeehouse sells coffee for $2 per cup. We know that it would cost $4 to buy 2 cups, $6 for 3 cups, $8 for 4 cups, and so on. We also know that the general formula for the cost of x cups in dollars is $2 \cdot x$. The cost depends on the number of cups bought. We say that the cost is a function of the number of cups bought.

Functions; Domain and Range

Objective 1 Define *function, domain,* and *range.* A **function** is a rule that takes an input value and assigns a particular output value to it. For a rule to define a function, it is necessary that each input value be assigned one and only one output value. In the example about coffee, the input value is the number of cups of coffee bought, and the output value is the cost. For each number of cups bought, there is only one possible cost. The rule for determining the cost is to multiply the number of cups by $2.

Cups	1	2	3	4	. . .	x	. . .
	↓	↓	↓	↓	. . .	↓	
Cost	$2	$4	$6	$8		$2 · x	

The set of input values for a function is called the **domain** of the function. The domain for the example about coffee is the set of natural numbers $\{1, 2, 3, \dots\}$. The set of output values for a function is called the **range** of the function. The range of the function in the coffee example is $\{2, 4, 6, \dots\}$.

Function, Domain, and Range

> A **function** is a rule that takes an input value and assigns one and only one output value to it. The **domain** of the function is the set of input values, and the **range** is the set of output values.
>
>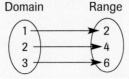

If an input value can be associated with more than one output value, then the rule is not a function. For example, a rule that took a month of the year as its input and listed the people at your school who were born in that month as its output would not be a function because each month has more than one person born in that month.

EXAMPLE ▶1   A mail-order company is selling holiday ornaments for $4 each. There is an additional $4.95 charge for shipping and handling per order. Find the function for the total cost of an order, as well as the domain and range of the function.

Solution

To determine the cost of an order, we begin by multiplying the number of ornaments by $4. To this we still need to add $4.95 for shipping and handling.

Function: If n is the number of ornaments ordered, then the cost is equal to $4n + 4.95$.
Domain: Set of possible number of ornaments ordered $\{1, 2, 3, \dots\}$
Range: Set of possible costs of the orders $\{\$8.95, \$12.95, \$16.95, \dots\}$

Quick Check   **1** ▶ A rental agency rents small moving trucks for $19.95 plus $0.15 per mile. Find the function for the total cost to rent a truck, as well as the domain and range of the function.

Function Notation

The perimeter of a square with side x can be found using the formula $P = 4x$. This is a function that can also be expressed using function notation as $P(x) = 4x$. **Function notation** is a way to present the output value of a function for the input x. The notation on the left side, $P(x)$, tells us the name of the function, P, as well as the input variable, x. The expression on the right side, $4x$, is the formula for the function. The parentheses on the left side are used to identify the input variable and are not used to indicate multiplication. We read $P(x)$ as "the function P evaluated at x" or simply as "P of x." We used P as the name of the function (P for perimeter), but we could have used any letter that we wanted. The letters f and g are frequently used for function names.

Evaluating Functions

Objective 2 **Evaluate functions.** Suppose we wanted to find the perimeter of a square with a side of 3 inches. We are looking to evaluate our perimeter function for an input of 3, or, in other words, $P(3)$. When we find the output value of a function for a particular value of x, this is called **evaluating** the function. To evaluate a function for a particular value of the variable, we substitute that value for the variable in the function's formula, and then we simplify the resulting expression. To evaluate $P(3)$, we substitute 3 for x in the formula and simplify the resulting expression.

$$P(x) = 4x$$
$$P(3) = 4(3)$$
$$= 12$$

Since $P(3) = 12$, the perimeter is 12 inches.

EXAMPLE ▸ 2   Let $f(x) = x + 7$. Find $f(4)$.

Solution

We need to replace x in the function's formula by 4 and simplify the resulting expression.

$$f(4) = 4 + 7 \qquad \text{Substitute 4 for } x.$$
$$= 11 \qquad \text{Add.}$$

$f(4) = 11$. This means that when the input is $x = 4$, the output of the function is 11.

EXAMPLE ▸ 3   Let $g(x) = 3x - 8$. Find $g(-2)$.

Solution

Quick Check ◂ 2

Let $g(x) = 2x + 9$. Find $g(-6)$.

In this example, we need to replace x by -2. As our functions become more complicated, it is a good idea to use parentheses when we substitute our input value.

$$g(-2) = 3(-2) - 8 \qquad \text{Substitute } -2 \text{ for } x.$$
$$= -6 - 8 \qquad \text{Multiply.}$$
$$= -14 \qquad \text{Simplify.}$$

EXAMPLE ▸ 4   Let $g(x) = 7x + 11$. Find $g(0)$.

Solution

Quick Check ◂ 3

Let $g(x) = 9x - 23$. Find $g(0)$.

Often, we will evaluate a function at $x = 0$. In this example, the term $7x$ is equal to 0 when $x = 0$, leaving the function equal to 11.

$$g(0) = 7(0) + 11 \qquad \text{Substitute 0 for } x.$$
$$= 11 \qquad \text{Simplify.}$$

EXAMPLE ▸ 5   Let $g(x) = 8 - 3x$. Find $g(a + 3)$.

Solution

Quick Check ◂ 4

Let $g(x) = 7x - 12$. Find $g(a + 8)$.

In this example, we are substituting a variable expression for x in the function. After we replace x by $a + 3$, we will need to simplify the resulting variable expression.

$$g(a + 3) = 8 - 3(a + 3) \qquad \text{Replace } x \text{ by } a + 3.$$
$$= 8 - 3a - 9 \qquad \text{Distribute } -3.$$
$$= -3a - 1 \qquad \text{Combine like terms.}$$

Linear Functions and Their Graphs

Objective 3 **Graph linear functions.**

> A **linear function** is a function of the form $f(x) = mx + b$, where m and b are real numbers.

Some examples of linear functions are $f(x) = x - 9$, $f(x) = 5x$, $f(x) = 3x + 11$, and $f(x) = 6$. We now turn our attention towards graphing linear functions. We graph any function $f(x)$ by plotting points of the form $(x, f(x))$. The output value of the function

$f(x)$ is treated as the variable y was when we were graphing linear equations in two variables. When we graph a function $f(x)$, the vertical axis is used to represent the output values of the function.

We can begin to graph a linear function by finding the y-intercept. As with a linear equation that is in slope–intercept form, the y-intercept for the graph of a linear function $f(x) = mx + b$ is the point $(0, b)$. For example, the y-intercept for the graph of the function $f(x) = 5x - 8$ is the point $(0, -8)$. In general, to find the y-intercept of any function $f(x)$, we can find $f(0)$. After plotting the y-intercept, we can then use the slope m to find other points. Finally, we can find the x-intercept of the graph by setting the formula for $f(x)$ equal to 0 and solving for x.

EXAMPLE 6 Graph the linear function $f(x) = \frac{1}{2}x + 2$.

Solution

We can start with the y-intercept, which is at $(0, 2)$. The slope of the line is $\frac{1}{2}$, so beginning at the point $(0, 2)$, we move up one unit and two units to the right. This leads to a second point at $(2, 3)$.

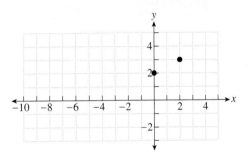

To find the x-intercept, we set $f(x)$ equal to 0 and solve for x.

$$f(x) = 0 \qquad \text{Set } f(x) \text{ equal to 0.}$$
$$\frac{1}{2}x + 2 = 0 \qquad \text{Replace } f(x) \text{ by its formula.}$$
$$\frac{1}{2}x = -2 \qquad \text{Subtract 2.}$$
$$2 \cdot \frac{1}{2}x = 2(-2) \qquad \text{Multiply by 2.}$$
$$x = -4$$

The x-intercept is at $(-4, 0)$. Here is the graph of the function $f(x) = \frac{1}{2}x + 2$.

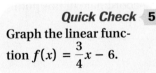

Quick Check 5
Graph the linear function $f(x) = \frac{3}{4}x - 6$.

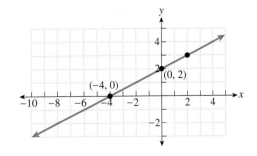

EXAMPLE 7 Graph the linear function $f(x) = -3$.

Solution

Quick Check 6
Graph the linear function $f(x) = 4$.

This function is known as a **constant function.** The function is constantly equal to -3, regardless of the input value x. Its graph is a horizontal line with a y-intercept at $(0, -3)$.

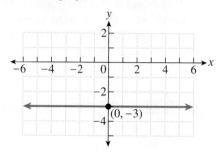

Interpreting the Graph of a Linear Function

Objective 4 Interpret the graph of a linear function. Here is the graph of a function $f(x)$:

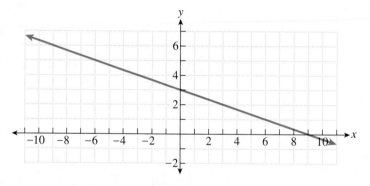

The ability to read an interpret a graph is an important skill. This line has an x-intercept at the point $(9, 0)$, so $f(9) = 0$. The y-intercept is at the point $(0, 3)$, so $f(0) = 3$.

Suppose that we wanted to find $f(3)$ for this particular function. We can do so by finding a point on the line that has an x-coordinate of 3. The y-coordinate of this point is $f(3)$. In this case, $f(3) = 2$.

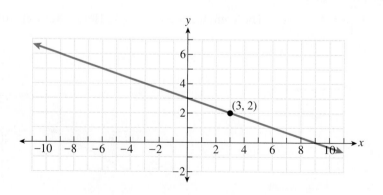

We can also use this graph to solve the equation $f(x) = 6$. Look for the point on the graph that has a y-coordinate of 6. The x-coordinate of this point is -9, so $x = -9$ is the solution of the equation $f(x) = 6$.

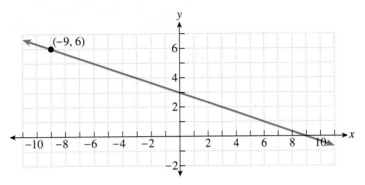

EXAMPLE ▶ 8   Consider the following graph of a function.

a) Find the x-intercept and the y-intercept.

Solution

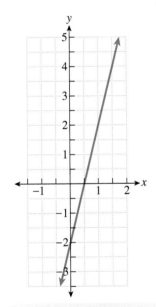

Note the scale on this graph, measured in units equal to $\frac{1}{2}$. The x-intercept is $\left(\frac{1}{2}, 0\right)$. The y-intercept is $(0, -2)$.

b) Find $f\left(1\frac{1}{2}\right)$.

Solution

We are looking for a point on the line that has an x-coordinate of $1\frac{1}{2}$. The point is $\left(1\frac{1}{2}, 4\right)$, so $f\left(1\frac{1}{2}\right) = 4$.

c) Find a value x such that $f(x) = 2$.

Solution

We are looking for a point on the line that has a y-coordinate of 2, and this point is $(1, 2)$. The value that satisfies the equation $f(x) = 2$ is $x = 1$.

***Quick Check*   7**   Consider the following graph of a function:

a) Find the x-intercept and the y-intercept.

b) Find $f(6)$.
c) Find a value x such that $f(x) = 6$.

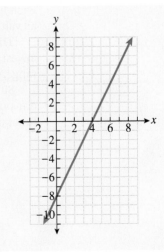

Objective 5 Determine the domain and range of a function from its graph.
The domain and range of a function can also be read from a graph. Recall that the domain of a function is the set of all input values. This corresponds to all of the *x*-coordinates of the points on the graph. The domain is the interval of values on the *x*-axis for which the graph exists. We read the domain from left to right on the graph.

The domain of a linear function is the set of all real numbers, which can be written in interval notation as $(-\infty, \infty)$. The graphs of linear functions continue on to the left and to the right. Look at the following three graphs of linear functions.

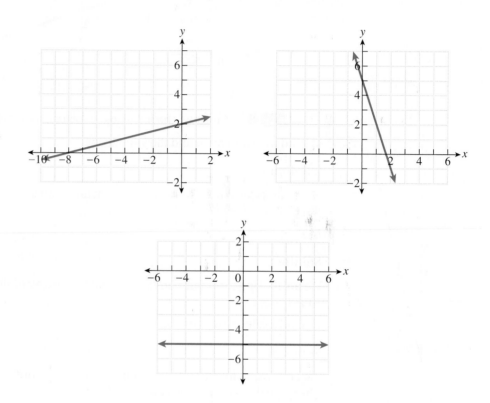

Each line continues to the left as well as to the right. This tells us that the graph exists for all values of *x* in the interval $(-\infty, \infty)$.

The range of a function can be read vertically from a graph. The range goes from the lowest point on the graph to the highest point. All linear functions have $(-\infty, \infty)$ as their range, except constant functions.

Since a constant function of the form $f(x) = c$, where *c* is a real number, has only one possible output value, its range would be $\{c\}$. For example, the range of the constant function $f(x) = 3$ is $\{3\}$.

The following table lists the names of the 11 professional basketball players who played on the "Dream Team" in the 1992 Barcelona Olympics, along with the NBA team they played on.

Player	NBA Team
Patrick Ewing	New York Knicks
David Robinson	San Antonio Spurs
Karl Malone	Utah Jazz
Charles Barkley	Philadelphia 76ers
Larry Bird	Boston Celtics
Scottie Pippen	Chicago Bulls
Chris Mullin	Golden State Warriors
Michael Jordan	Chicago Bulls
Clyde Drexler	Portland Trail Blazers
Magic Johnson	Los Angeles Lakers
John Stockton	Utah Jazz

1. Would a rule that took a player's name from the "Dream Team" as an input and listed his NBA team as an output be a function? Explain why or why not.

2. Could a function be defined in the opposite direction, with the name of the NBA team as the input and the "Dream Team" player's name as the output? Explain why or why not.

3. Would a rule that took a person as an input and listed that person's mother as an output be a function? Why or why not?

4. Could a function be defined with a woman as an input and her child as an output? Why or why not?

For Exercises 5–8, determine whether a function exists with
a) set A as the input and set B as the output, and
b) set B as the input and set A as the output.

5. High Temperatures on December 16

Set A City	Set B High Temp.
Boston, MA	39° F
Orlando, FL	75° F
Providence, RI	39° F
Rochester, NY	26° F
Visalia, CA	57° F

6.

Set A Person	Set B Last 4 Digits of SSN
Jenny Crum	1234
Maureen O'Connor	5283
Dona Kenly	6405
Susan Winslow	5555
Lauren Morse	9200

7.

Set A Person	Set B Birthday
Greg Erb	December 15
Karen Guardino	December 17
Jolene Lehr	October 15
Siméon Poisson	June 21
Lindsay Skay	May 28
Sharon Smith	June 21

8.

Set *A* Football Player	Set *B* Uniform Number
Jeff Garcia	5
Garrison Hearst	20
Terrell Owens	81
Bryant Young	97
Julian Peterson	98

For the given set of ordered pairs, determine whether a function could be defined for which the input would be an x-coordinate and the output would be the corresponding y-coordinate. If a function cannot be defined in this manner, explain why.

9. $\{(-2, 4), (-1, 1), (0, 0), (1, 1), (2, 4), (3, 9)\}$

10. $\{(2, -2), (1, -1), (0, 0), (1, 1), (2, 2)\}$

11. $\{(5, 3), (2, 7), (-4, -6), (5, -2), (0, 4)\}$

12. $\{(-6, 3), (-2, 3), (1, 3), (5, 3), (11, 3)\}$

13. $\{(2, -5), (2, -1), (2, 0), (2, 3), (2, 5)\}$

14. $\{(1, 1), (2, 2), (3, 3), (4, 4), (5, 5)\}$

15. A Celsius temperature can be converted to a Fahrenheit temperature by multiplying it by 1.8 and then adding 32.

   a) Create a function $F(x)$ that converts a Celsius temperature *x* to a Fahrenheit temperature.

   b) Use the function $F(x)$ from part a) to convert the following Celsius temperatures to Fahrenheit temperatures.

   $0°\,C$ $100°\,C$ $30°\,C$ $-10°\,C$ $-40°\,C$

16. To convert a Fahrenheit temperature to a Celsius temperature we first subtract 32 from it, and then multiply that difference by $\frac{5}{9}$. Create a function $C(x)$ that converts a Fahrenheit temperature *x* into a Celsius temperature.

17. A college student takes a summer job selling newspaper subscriptions door-to-door. She is paid $36 for a 4-hour shift. She also earns $7 for each subscription sold.

   a) Create a function $f(x)$ for the amount she earns on a night when she sells *x* subscriptions.

   b) Use the function $f(x)$ to determine how much she earns on a night that she sells 12 subscriptions.

18. An elementary school holds a carnival. Attendees are charged $8 for dinner, plus 25¢ for each game played.

   a) Create a function $f(x)$ for the amount that a person pays if he eats dinner and plays *x* games.

   b) Use the function $f(x)$ to determine how much a person pays if he eats dinner and plays 45 games.

19. Create a linear function whose graph has a slope of 4 and a *y*-intercept at $(0, 3)$.

20. Create a linear function whose graph has a slope of 5 and a *y*-intercept at $(0, -9)$.

21. Create a linear function whose graph has a slope of -3 and a *y*-intercept at $(0, -4)$.

22. Create a linear function whose graph has a slope of $-\frac{1}{2}$ and a *y*-intercept at $(0, \frac{2}{3})$.

23. Create a linear function whose graph has a slope of 0 and a *y*-intercept at $(0, 6)$.

24. Create a linear function whose graph has a slope of 0 and a *y*-intercept at $(0, 0)$.

Evaluate the given function.

25. $g(x) = x + 2,\ g(-5)$

26. $h(x) = -5x,\ h(4)$

27. $f(x) = 2x - 3,\ f(5)$

28. $f(x) = 4x + 1,\ f(-2)$

29. $f(x) = -7x + 11,\ f(-3)$

30. $f(x) = 15 - 4x,\ f(4)$

31. $f(x) = 9x - 25,\ f(0)$

32. $f(x) = 6x + 13,\ f(0)$

33. $g(x) = 3x - 1,\ g\left(\dfrac{2}{3}\right)$

34. $g(x) = 5x + 7, g\left(\dfrac{9}{5}\right)$

35. $f(x) = 3x + 4, f(a)$

36. $f(x) = 2x - 3, f(b)$

37. $f(x) = 7x - 2, f(a + 3)$

38. $f(x) = 5x + 9, f(a - 7)$

39. $f(x) = 16 - 3x, f(2a - 5)$

40. $f(x) = 5 - 6x, f(3a - 4)$

41. $f(x) = 6x + 4, f(x + h)$

42. $f(x) = 3x + 11, f(x + h)$

Graph the linear function. Label any intercepts.

43. $f(x) = 6x - 6$　　**44.** $f(x) = 2x + 6$

45. $f(x) = -3x + 3$　　**46.** $f(x) = -x - 8$

47. $f(x) = \dfrac{4}{3}x + 4$　　　**48.** $f(x) = \dfrac{2}{5}x - 2$

49. $f(x) = 7x$　　　　**50.** $f(x) = -9x$

51. $f(x) = \dfrac{8}{3}x$

52. $f(x) = -\dfrac{3}{7}x$

53. $f(x) = 4$

Find the domain and range of the given function.

57.

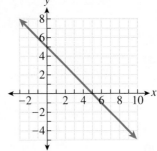

54. $f(x) = -6$

58.

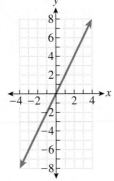

55. Refer to the graph of the function $f(x)$.

   a) Find the x-intercept.

   b) Find the y-intercept.

   c) Find $f(5)$.

   d) Find a value a such that $f(a) = -8$.

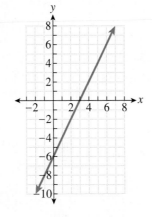

59.

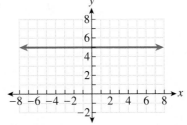

56. Refer to the graph of the function $g(x)$.

   a) What is the x-intercept?

   b) What is the y-intercept?

   c) Find $g(-4)$.

   d) Find a value a such that $g(a) = -2$.

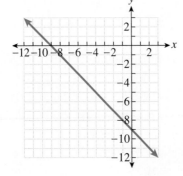

60.

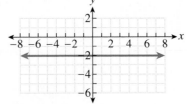

Answer in complete sentences.

61. Give an example of two sets *A* and *B* and a rule that is a function from set *A* to set *B*. Explain why your rule meets the definition of a function. Give another rule that would not be a function from set *B* to set *A*. Explain why your rule does not meet the definition of a function.

Study Tip **Revisited**   The student solutions manual for a mathematics textbook can be a valuable resource when used properly but it can be detrimental to learning if used improperly. You should refer to the solutions manual to check your work or to see where you went wrong with your solution. The detailed solution can be helpful in this case.

Some students do their homework with the solutions manual sitting open in front of them, looking at what they should write for the next line. If you do nothing except essentially copy the solutions manual onto your own paper, you will most likely have tremendous difficulty on the next exam.

3.5

PARALLEL AND PERPENDICULAR LINES

Objectives

1 Determine whether two lines are parallel.
2 Determine whether two lines are perpendicular.

Parallel Lines

Objective 1   Determine whether two lines are parallel. In this section, we will examine the relationship between two lines. Consider the following pair of lines.

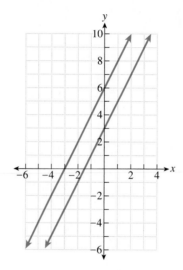

We notice that these two lines do not intersect. The lines have the same slope and are called **parallel lines.**

Parallel Lines

- Two nonvertical lines are **parallel** if they have the same slope. In other words, if we denote the slope of one line as m_1 and the slope of the other line as m_2, the two lines are parallel if $m_1 = m_2$.
- If two lines are both vertical lines, they are parallel.

Following are some examples of lines that are parallel:

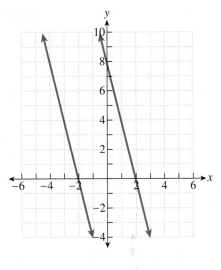

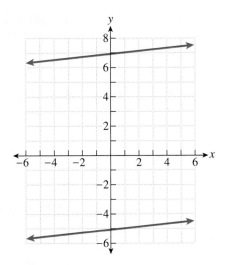

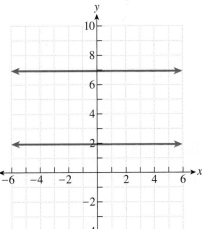

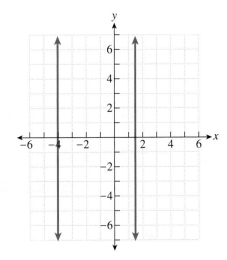

EXAMPLE 1   Are the two lines $y = 2x - 5$ and $y = 3x + 1$ parallel?

Quick Check 1

Are the two lines
$y = 4x + 3$ and $y = -4x$
parallel?

Solution

These two equations are in slope–intercept form, so we can see that the slope of the first line is 2 and the slope of the second line is 3. Since the slopes are not equal, these two lines are not parallel.

EXAMPLE 2   Are the two lines $6x + 3y = -9$ and $y = -2x + 2$ parallel?

Solution

To find the slope of the first line, we solve the equation for y:

$$6x + 3y = -9$$
$$3y = -6x - 9 \qquad \text{Subtract } 6x \text{ from both sides.}$$
$$y = -2x - 3 \qquad \text{Divide by 3.}$$

The slope of the first line is -2. Since the second line is already in slope–intercept form, we see that the slope of that line is also -2. Since the two slopes are equal, the two lines are parallel.

Quick Check **2** Are the two lines $8x + 6y = 36$ and $y = -\dfrac{4}{3}x + 5$ parallel?

Any two horizontal lines, such as $y = 4$ and $y = 1$, are parallel to each other, since their slopes are both 0. Any two vertical lines, such as $x = 2$ and $x = -1$, are parallel to each other as well.

EXAMPLE 3 Find the slope of a line that is parallel to the line $4x + 3y = 8$.

Solution

For a line to be parallel to $4x + 3y = 8$, it must have the same slope as this line. To find the slope of this line, we solve the equation for y.

$$4x + 3y = 8$$
$$3y = -4x + 8 \qquad \text{Subtract } 4x \text{ from both sides.}$$
$$y = -\frac{4}{3}x + \frac{8}{3} \qquad \text{Divide by 3.}$$

Quick Check **3**
Find the slope of a line that is parallel to the line $2x - 9y = 36$.

The slope of the line $4x + 3y = 8$ is $-\frac{4}{3}$. A line that is parallel to $4x + 3y = 8$ has slope $-\frac{4}{3}$.

Perpendicular Lines

Objective 2 Determine whether two lines are perpendicular.

Two distinct lines that are not parallel intersect at one point. One special type of intersecting lines are **perpendicular lines.** Here is an example of two lines that are perpendicular. Perpendicular lines intersect at right angles. Notice that one of these lines has a positive slope and the other line has a negative slope.

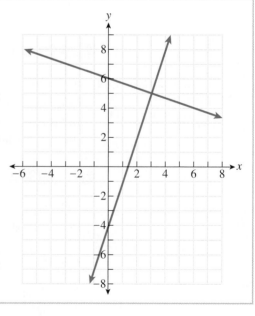

Perpendicular Lines

- Two nonvertical lines are **perpendicular** if their slopes are **negative reciprocals.** In other words, if we denote the slope of one line as m_1 and the slope of the other line as m_2, the two lines are perpendicular if $m_1 = -\dfrac{1}{m_2}$. (This is equivalent to saying that two lines are perpendicular if the product of their slopes is -1.)

- A vertical line is perpendicular to a horizontal line.

Two numbers are **negative reciprocals** if they are reciprocals that have opposite signs, such as 5 and $-\frac{1}{5}$, -2 and $\frac{1}{2}$, and $-\frac{3}{10}$ and $\frac{10}{3}$.

EXAMPLE 4   Are the two lines $y = 4x + 5$ and $y = -4x + 5$ perpendicular?

Solution

> **Quick Check 4**
> Are the two lines
> $y = x + 7$ and
> $y = -x + 3$
> perpendicular?

The slopes of these two lines are 4 and -4, respectively. While the signs of these two slopes are opposite, the slopes are not reciprocals. Therefore, the two lines are not perpendicular.

EXAMPLE 5   Are the two lines $y = 3x + 7$ and $2x + 6y = 5$ perpendicular?

Solution

The slope of the first line is 3. To find the slope of the second line, we solve the equation for y.

$$2x + 6y = 5$$
$$6y = -2x + 5 \qquad \text{Subtract } 2x.$$
$$y = -\frac{2}{6}x + \frac{5}{6} \qquad \text{Divide by 6.}$$

> **Quick Check 5**
> Are the two lines
> $y = \dfrac{2}{3}x + 5$ and
> $6x + 4y = 12$
> perpendicular?

$$y = -\frac{1}{3}x + \frac{5}{6} \qquad \text{Simplify.}$$

The slope of the second line is $-\frac{1}{3}$. The two slopes are negative reciprocals, so the lines are perpendicular.

Any vertical line, such as $x = 2$, is perpendicular to any horizontal line, such as $y = -2$.

EXAMPLE 6   Find the slope of a line that is perpendicular to the line $5x + 2y = 8$.

Solution

We begin by finding the slope of this line. This can be done by solving the equation $5x + 2y = 8$ for y.

$$5x + 2y = 8$$
$$2y = -5x + 8 \qquad \text{Subtract } 5x \text{ from both sides.}$$
$$y = -\frac{5}{2}x + \frac{8}{2} \qquad \text{Divide by 2.}$$
$$y = -\frac{5}{2}x + 4 \qquad \text{Simplify.}$$

The slope of this line is $-\frac{5}{2}$. To find the slope of a line, perpendicular to this line, we take the reciprocal of this slope and change its sign from negative to positive. A line is perpendicular to $5x + 2y = 8$ if it has a slope of $\frac{2}{5}$.

Quick Check **6**

Find the slope of a line that is perpendicular to the line $7x - 3y = 21$.

Determining Whether Two Lines Are Parallel, Perpendicular, or Neither

To summarize, nonvertical lines are parallel if and only if they have the same slope and are perpendicular if and only if their slopes are negative reciprocals. Horizontal lines are parallel to other horizontal lines, and vertical lines are parallel to other vertical lines. Finally, a horizontal line and a vertical line are perpendicular to each other. These concepts are summarized in the following table:

If . . .	the lines are parallel if . . .	the lines are perpendicular if . . .
two nonvertical lines have slopes m_1 and m_2	$m_1 = m_2$	$m_1 = -\dfrac{1}{m_2}$
one of the two lines is horizontal	the other line is horizontal	the other line is vertical
one of the two lines is vertical	the other line is vertical	the other line is horizontal

EXAMPLE 7   Are the two lines $y = 6x + 2$ and $y = \frac{1}{6}x - 3$ parallel, perpendicular, or neither?

Solution

The slope of the first line is 6 and the slope of the second line is $\frac{1}{6}$. The slopes are not equal, so the lines are not parallel.

The slopes are not negative reciprocals, so the lines are not perpendicular either.

The two lines are neither parallel nor perpendicular.

Quick Check **7**

Are the two lines $y = 2x - 5$ and $y = -2x$ parallel, perpendicular, or neither?

EXAMPLE 8   Are the two lines $-3x + y = 7$ and $-6x + 2y = -4$ parallel, perpendicular, or neither?

Solution

We find the slope of each line by solving each equation for y.

$$-3x + y = 7$$
$$y = 3x + 7 \qquad \text{Add } 3x \text{ to both sides.}$$

The slope of the first line is 3. Now we find the slope of the second line.

$$-6x + 2y = -4$$
$$2y = 6x - 4 \qquad \text{Add } 6x \text{ to both sides.}$$
$$y = 3x - 2 \qquad \text{Divide by 2.}$$

The slope of the second line is also 3.

Since the two slopes are equal, the lines are parallel.

Quick Check **8**

Are the two lines $y = 4x + 3$ and $8x - 2y = 12$ parallel, perpendicular, or neither?

EXAMPLE ▶ 9 Are the two lines $10x + 2y = 0$ and $x - 5y = 3$ parallel, perpendicular, or neither?

Solution

We begin by finding the slope of each line.

$$10x + 2y = 0$$
$$2y = -10x \qquad \text{Subtract } 10x \text{ from both sides.}$$
$$y = -5x \qquad \text{Divide by 2.}$$

The slope of the first line is -5. Now we find the slope of the second line.

Quick Check 9
Are the two lines
$y = -\frac{2}{5}x + 9$ and
$5x - 2y = -8$ parallel,
perpendicular, or
neither?

$$x - 5y = 3$$
$$-5y = -x + 3 \qquad \text{Subtract } x \text{ from both sides.}$$
$$y = \frac{1}{5}x - \frac{3}{5} \qquad \text{Divide by } -5.$$

The slope of the second line is $\frac{1}{5}$. Since the two slopes are negative reciprocals, the lines are perpendicular.

EXAMPLE ▶ 10 Find the equation of the line that is perpendicular to $y = \frac{1}{3}x + 2$ and has a y-intercept at $(0, 6)$.

Quick Check 10
Find the equation of the
line that is parallel to
$6x + 4y = 21$ and has a
y-intercept at $(0, -7)$.

Solution

The slope of the line $y = \frac{1}{3}x + 2$ is $\frac{1}{3}$. Since we are looking for the equation of a line that is perpendicular to $y = \frac{1}{3}x + 2$, the slope of this line is -3. In addition to the slope of the line being -3, we know that the y-coordinate of the y-intercept is 6. We can now find the equation of this line using the point–slope form ($y = mx + b$). The equation of the line is $y = -3x + 6$.

EXERCISES *3.5* ❯

Are the two given lines parallel?

1. $y = 3x + 5,\ y = 3x - 2$

2. $y = 5x - 7,\ y = -5x + 3$

3. $4x + 2y = 9,\ 3y = 6x + 7$

4. $x + 3y = -4,\ 3x + 9y = 8$

5. $y = 6,\ y = -6$

6. $x = 2,\ x = 7$

Are the two given lines perpendicular?

7. $y = 4x,\ y = \dfrac{1}{4}x - 3$

8. $y = -\dfrac{3}{2}x + 2,\ y = \dfrac{2}{3}x + 1$

9. $15x + 3y = 11,\ x - 5y = -4$

10. $x + y = 6,\ x - y = -3$

11. $y = -7,\ y = \dfrac{1}{7}$

12. $x = 3,\ y = 4$

Are the two given lines parallel, perpendicular, or neither?

13. $y = 6x - 2,\ y = -6x + 5$

14. $y = 7x - 9,\ y = \dfrac{1}{7}x + 3$

15. $y = 4x + 3,\ y = -\dfrac{1}{4}x - 6$

16. $y = 8x - 16, y = 8x - 1$

17. $10x + 2y = 7, y = -5x + 12$

18. $9x + 3y = 15, 4x - 12y = 8$

19. $y = 4, y = \dfrac{1}{4}$

20. $x - 4y = 11, 8y = 2x + 3$

21. $30x + 10y = 17, 12x - 4y = 9$

22. $x = 3, x = 0$

23. $x = 5, y = -2$

24. $x + y = 16, y = x - 7$

Are the lines associated with the given functions parallel, perpendicular, or neither?

25. $f(x) = 7x - 5, g(x) = 7x + 3$

26. $f(x) = -2x - 5, g(x) = -\dfrac{1}{2}x + 4$

27. $f(x) = 6, g(x) = 6x + 4$

28. $f(x) = 4x + 9, g(x) = -\dfrac{1}{4}x + 3$

29. $f(x) = x + 7, g(x) = x + 2$

30. $f(x) = 5, g(x) = -5$

Find the slope of a line that is parallel to the given line. If the slope is undefined, state this.

31. $y = -6x + 7$

32. $y = 3x + 2$

33. $y = 5x$

34. $y = -3$

35. $12x + 4y = 8$

36. $2x + 6y = -7$

Find the slope of a line that is perpendicular to the given line. If the slope is undefined, state this.

37. $y = \dfrac{1}{4}x + 1$

38. $y = -3x + 14$

39. $y = \dfrac{3}{5}x - 6$

40. $14x - 6y = 17$

41. $15x + 24y = 39$

42. $x = -63$

43. Find the slope of a line that is parallel to the line $Ax + By = C$. (A, B, and C are real numbers, $A \neq 0$, and $B \neq 0$.)

44. Find the slope of a line that is perpendicular to the line $Ax + By = C$. (A, B, and C are real numbers, $A \neq 0$, and $B \neq 0$.)

45. Is the line that passes through $(3, 7)$ and $(5, -1)$ parallel to the line $y = -4x + 11$?

46. Is the line that passes through $(-6, 2)$ and $(4, 8)$ parallel to the line $-3x + 5y = 15$?

47. Is the line that passes through $(4, 7)$ and $(-1, 9)$ perpendicular to the line $-10x + 4y = 8$?

48. Is the line that passes through $(-3, -8)$ and $(2, 7)$ perpendicular to the line that passes through $(-5, 3)$ and $(7, 7)$?

49. Find the equation of a line that is parallel to $y = 2x + 9$ and has a y-intercept at $(0, -6)$. Graph the line and label any intercepts.

50. Find the equation of a line that is parallel to $3x + 2y = 7$ and has a y-intercept at $(0, 9)$. Graph the line and label any intercepts.

51. Find the equation of a line that is parallel to $y = 13$ and has a y-intercept at $(0, -4)$. Graph the line and label any intercepts.

52. Find the equation of a line that is perpendicular to $y = -3x + 12$ and has a y-intercept at $(0, 3)$. Graph the line and label any intercepts.

53. Find the equation of a line that is perpendicular to $5x - 3y = -9$ and has a y-intercept at $(0, -3)$. Graph the line and label any intercepts.

54. Find the equation of a line that is perpendicular to $x = 3$ and has a y-intercept at $(0, 7)$. Graph the line and label any intercepts.

55. Describe three real-world examples of parallel lines.

56. Describe three real-world examples of perpendicular lines.

57. Explain the process for determining whether two lines of the form $Ax + By = C$ are parallel.

58. Explain the process for determining whether two lines of the form $Ax + By = C$ are perpendicular.

> **Study Tip** REVISITED   Some frequently overlooked resources are the students who are taking the class with you.
>
> - If you have a quick question about a particular problem, a classmate can help you.
> - If your notes are incomplete for a certain day, a classmate may allow you to use their notes to fill in the holes in your own notes.
> - If you are forced to miss class, a quick call to a classmate can help you find out what was covered in class and what the homework assignment is.
> - You can form a study group with your fellow classmates.
> - A classmate can also help to motivate you when you are feeling low.

3.6

EQUATIONS OF LINES

Objectives

1 **Find the equation of a line by using the point–slope form.**
2 **Find the equation of a line given two points on the line.**
3 **Find a linear equation to describe real data.**
4 **Find the equation of a parallel or perpendicular line.**

We are already familiar with the slope–intercept form of the equation of a line: $y = mx + b$. When written in this form, we know both the slope of the line (m) and the y-coordinate of the y-intercept (b). This form is convenient for graphing lines. In this section, we will look at another form of the equation of a line. We will also learn how to find the equation of a line if we know the slope of a line and any point which the line passes through.

Point–Slope Form of the Equation of a Line

Objective 1 **Find the equation of a line by using the point–slope form.** In Section 3.3, we learned that if we know the slope of a line and its y-intercept, we can write the equation of the line using the slope–intercept form of a line $y = mx + b$, where m is the slope of the line and b is the y-coordinate of the y-intercept. If we know the slope of a line and the coordinates of any point on that line, not just the y-intercept, we can write the equation of the line using the point–slope form of an equation.

Point–Slope Form of the Equation of a Line

> The **point–slope form** of the equation of a line with slope m that passes through the point (x_1, y_1) is
>
> $$y - y_1 = m(x - x_1)$$

This form can be derived directly from the slope formula $m = \dfrac{y_2 - y_1}{x_2 - x_1}$. If we let (x, y) represent an arbitrary point on the line, this formula becomes $m = \dfrac{y - y_1}{x - x_1}$. Multiplying both sides of that equation by $(x - x_1)$ produces the point–slope form of the equation of a line.

If a line has slope 2 and passes through the point $(3, 1)$, we find its equation by substituting 2 for m, 3 for x_1 and 1 for y_1. Its equation is $y - 1 = 2(x - 3)$. This equation can be left in the form it is, or converted to slope–intercept form or standard form.

EXAMPLE 1 Find the equation for a line whose slope is 4 and passes through the point $(-2, 5)$. Write the equation in slope–intercept form.

Solution

We will substitute 4 for m, -2 for x_1, and 5 for y_1 in the point–slope form. We will then solve the equation for y in order to write the equation in slope–intercept form.

$$y - 5 = 4(x - (-2))$$ Substitute into point–slope form.
$$y - 5 = 4(x + 2)$$ Eliminate double signs.
$$y - 5 = 4x + 8$$ Distribute.
$$y = 4x + 13$$ Add 5 to isolate y.

The equation for a line with slope 4 that passes through $(-2, 5)$ is $y = 4x + 13$. Below is the graph of the line, showing that it passes through the point $(-2, 5)$.

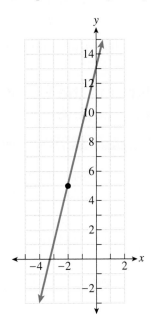

Quick Check 1

Find the equation for a line whose slope is 2 and passes through the point $(1, 5)$. Write the equation in slope–intercept form.

EXAMPLE 2   Find the equation for a line whose slope is -1 and passes through the point $(4, -3)$. Write the equation in slope–intercept form.

Solution

We begin by substituting -1 for m, 4 for x_1, and -3 for y_1 in the point–slope form. After substituting, we solve the equation for y.

$$y - (-3) = -1(x - 4)$$ Substitute into point–slope form.
$$y + 3 = -1(x - 4)$$ Eliminate double signs.
$$y + 3 = -x + 4$$ Distribute -1.
$$y = -x + 1$$ Subtract 3 to isolate y.

Quick Check 2

Find the equation for a line whose slope is -3 and passes through the point $(4, -6)$. Write the equation in slope–intercept form.

The equation for a line with slope -1 that passes through $(4, -3)$ is $y = -x + 1$.

In the previous example, we converted the equation from point–slope form to slope–intercept form. This is a good idea in general. We use the point–slope form because it is a convenient form for finding the equation of a line if we know its slope and the coordinates of a point on the line. We convert the equation to slope–intercept form because it is easier to graph a line whose equation is in this form.

EXAMPLE 3   The line associated with the linear function $f(x)$ has a slope of 2. If $f(-3) = -7$, find the function $f(x)$.

Solution

A linear function is of the form $f(x) = mx + b$. In this example, we know that $m = 2$, so $f(x) = 2x + b$. We will now use the fact that $f(-3) = -7$ to find b:

$$f(-3) = -7$$
$$2(-3) + b = -7 \qquad \text{Substitute } -3 \text{ for } x \text{ in the function } f(x).$$
$$-6 + b = -7 \qquad \text{Multiply.}$$
$$b = -1 \qquad \text{Add 6.}$$

> **Quick Check** **3**
> The line associated with the linear function $f(x)$ has a slope of -4. If $f(-2) = 13$, find the function $f(x)$.

Now that we have found b, we know that $f(x) = 2x - 1$.

Finding the Equation of a Line Given Two Points on the Line

Objective **2** **Find the equation of a line given two points on the line.** Another use of the point–slope form is to find the equation of a line that passes through two given points. We begin by finding the slope of the line passing through those two points using the slope formula $m = \dfrac{y_2 - y_1}{x_2 - x_1}$. Then we use the point–slope form with this slope and either of the two points we were given. We will finish by rewriting the equation in slope–intercept form.

EXAMPLE ▸ **4**   Find the equation of a line that passes through the two points $(-1, 6)$ and $(2, 0)$.

Solution

We begin by finding the slope of the line that passes through these two points.

$$= \frac{0 - 6}{2 - (-1)} \qquad \text{Substitute into the formula } m = \frac{y_2 - y_1}{x_2 - x_1}.$$
$$= \frac{-6}{3} \qquad \text{Simplify numerator and denominator.}$$
$$= -2 \qquad \text{Simplify.}$$

We can now substitute into the point–slope form using either of the two points with $m = -2$. We will use $(2, 0)$:

$$y - 0 = -2(x - 2) \qquad \text{Substitute in point–slope form.}$$
$$y = -2x + 4 \qquad \text{Distribute.}$$

> **Quick Check** **4**
> Find the equation of a line that passes through the two points $(-2, 7)$ and $(6, -5)$.

The equation of the line that passes through these two points is $y = -2x + 4$. If we had used the point $(-1, 6)$ instead of $(2, 0)$, our result would have been the same.

At the right is the graph of the line, along with the points $(-1, 6)$ and $(2, 0)$.

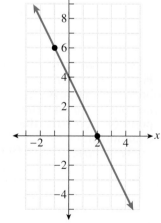

EXAMPLE ▶ 5   Find the equation of a line that passes through the two points $(8, 2)$ and $(8, 9)$.

Solution

We begin by attempting to find the slope of the line that passes through these two points.

$$m = \frac{9 - 2}{8 - 8} \qquad \text{Substitute into the slope formula.}$$

$$= \frac{7}{0} \qquad \text{Simplify the numerator and denominator.}$$

Quick Check 5
Find the equation of a line that passes through the two points $(-6, 4)$ and $(3, 4)$.

The slope is undefined, so this line is a vertical line. (We could have discovered this by plotting these two points on a graph.) The equation for this line is $x = 8$.

Finding a Linear Equation to Describe Linear Data

Objective 3   Find a linear equation to describe real data.

EXAMPLE ▶ 6   In 1995, there were approximately 75,000 nursing students in college in the United States. By 1999, the number had dropped to approximately 60,000 students. Find a linear equation that describes the number (y) of nursing students x years after 1995. (*Source:* American Association of Colleges of Nursing)

Solution

Since x represents the number of years after 1995, then $x = 0$ for the year 1995 and $x = 4$ for the year 1999. This tells us that two points on the line are $(0, 75{,}000)$ and $(4, 60{,}000)$. We begin by calculating the slope of the line.

$$m = \frac{60{,}000 - 75{,}000}{4 - 0} \qquad \text{Substitute into the slope formula.}$$

$$= \frac{-15{,}000}{4} \qquad \text{Simplify the numerator and denominator.}$$

$$= -3750 \qquad \text{Divide.}$$

The slope is -3750, which tells us that the number of nursing students in college decreases at a rate of 3750 students per year. In this example, we know the y-intercept is $(0, 75{,}000)$, so we can write the equation directly in slope–intercept form. The equation is $y = -3750x + 75{,}000$. (If we did not know the y-intercept of the line, we would find the equation of the line by substituting into the point–slope form.)

If we found the x-intercept of the line in the previous nursing example, we would see that the number of nursing students in college would be 0 only 20 years after 1995. Common sense tells us this is very unlikely to occur. Although there was a trend of decreased enrollment between 1995 and 1999, there is no way we can guarantee that this will remain true in the future. Chances are excellent that enrollment will level off or begin to increase in the near future, thanks to a high demand for nurses and a shortage of supply due to nurses retiring.

Quick Check **6** In 2000, approximately 84.6 million people traveled to a theme or amusement park. By 2002, this number had increased to approximately 92.4 million people. Find a linear equation that tells the number (y) of people, in millions, who traveled to a theme or amusement park x years after 2000. (*Source:* Domestic Travel Market Report, Travel Industry Association of America)

Finding the Equation of a Parallel or Perpendicular Line

Objective **4** **Find the equation of a parallel or perpendicular line.** To find the equation of a line, we must know the slope of the line and the coordinates of a point on the line. In the previous examples, either we were given the slope or we calculated the slope using two points which were on the line. Sometimes, the slope of the line is given in terms of another line. We could be given the equation of a line either parallel or perpendicular to the line for which we are trying to find the equation.

EXAMPLE ▶**7** Find the equation of a line which is parallel to the line $y = -\frac{3}{4}x + 15$ and passes through $(-8, 5)$.

Solution

Since the line is parallel to $y = -\frac{3}{4}x + 15$, its slope must be $-\frac{3}{4}$. We will substitute this slope, along with the point $(-8, 5)$, into the point–slope form to find the equation of this line:

$$y - 5 = -\frac{3}{4}(x - (-8)) \qquad \text{Substitute into point–slope form.}$$

$$y - 5 = -\frac{3}{4}(x + 8) \qquad \text{Eliminate double signs.}$$

$$y - 5 = -\frac{3}{4}x - \frac{3}{\overset{1}{\cancel{4}}} \cdot \overset{2}{\cancel{8}} \qquad \text{Distribute and divide out common factors.}$$

$$y - 5 = -\frac{3}{4}x - 6 \qquad \text{Simplify.}$$

$$y = -\frac{3}{4}x - 1 \qquad \text{Add 5.}$$

Quick Check ▶**7**
Find the equation of a line parallel to the line $y = \frac{2}{5}x - 9$ and passing through $(5, 4)$.

The equation of the line parallel to $y = -\frac{3}{4}x + 15$ and passing through $(-8, 5)$ is $y = -\frac{3}{4}x - 1$.

EXAMPLE ▶**8** Find the equation of a line perpendicular to the line $y = -\frac{1}{2}x + 7$ and passing through $(5, 3)$.

Solution

Quick Check ▶**8**
Find the equation of a line perpendicular to the line $9x - 12y = 4$ and passing through $(6, -2)$.

The slope of the line $y = -\frac{1}{2}x + 7$ is $-\frac{1}{2}$, so the slope of a perpendicular line must be the negative reciprocal of $-\frac{1}{2}$, or 2. We now can substitute into the point–slope form:

$$y - 3 = 2(x - 5) \qquad \text{Substitute into point–slope form.}$$
$$y - 3 = 2x - 10 \qquad \text{Distribute.}$$
$$y = 2x - 7 \qquad \text{Add 3.}$$

The equation of the line is $y = 2x - 7$.

If we are given . . .	We find the equation by . . .
The slope and the y-intercept	Substituting m and b into the slope–intercept form $y = mx + b$
The slope and a point on the line	Substituting m and the coordinates of the point into the point–slope form $y - y_1 = m(x - x_1)$
Two points on the line	Calculating m using the slope formula $m = \dfrac{y_2 - y_1}{x_2 - x_1}$ and then substituting the slope and the coordinates of one of the points into the point–slope form $y - y_1 = m(x - x_1)$
A point on the line and the equation of a parallel line	Substituting the slope of that line and the coordinates of the point into the point–slope form $y - y_1 = m(x - x_1)$
A point on the line and the equation of a perpendicular line	Substituting the negative reciprocal of the slope of that line and the coordinates of the point into the point–slope form $y - y_1 = m(x - x_1)$

Keep in mind that if the slope of the line is undefined, then the line is vertical and its equation is of the form $x = a$, where a is the x-coordinate of the given point. We do not use the point–slope form or the slope–intercept form to find the equation of a vertical line.

EXERCISES 3.6

Write the following equations in slope–intercept form.

1. $4x + 2y = 8$

2. $-3x + 6y = 9$

3. $x - 5y = 10$

4. $-12x - 16y = 10$

5. $7x - y = 3$

6. $6x + 4y = 0$

Find the equation of a line with the given slope and y-intercept.

7. Slope -3, y-intercept $(0, 5)$

8. Slope 2, y-intercept $(0, -4)$

9. Slope $\dfrac{2}{3}$, y-intercept $(0, -3)$

10. Slope 0, y-intercept $(0, 9)$

11. Slope 5, y-intercept $(0, 8)$

12. Slope $-\dfrac{3}{5}$, y-intercept $(0, 2)$

Find the slope–intercept form of the equation of a line with the given slope that passes through the given point. Graph the line and label any intercepts.

13. Slope 1, through $(3, 2)$

14. Slope -2, through $(1, 6)$

15. Slope -3, through $(-2, 9)$

20. Undefined slope, through $(8, 5)$

21. Find a linear function $f(x)$ with slope -2 such that $f(-4) = 23$.

22. Find a linear function $f(x)$ with slope 5 such that $f(3) = 12$.

16. Slope -1, through $(-6, -3)$

17. Slope $\dfrac{3}{2}$, through $(-6, -9)$

23. Find a linear function $f(x)$ with slope $\dfrac{3}{5}$ such that $f(-15) = -17$.

24. Find a linear function $f(x)$ with slope 0 such that $f(316) = 228$.

Find the slope–intercept form of the equation of a line that passes through the given points. Graph the line and label any intercepts.

25. $(2, -3), (7, 2)$

18. Slope $\dfrac{2}{5}$, through $(-5, -4)$

26. $(4, 6), (8, 10)$

19. Slope 0, through $(8, 5)$

27. $(-3, -3), (1, 9)$ **28.** $(-6, 9), (-2, -3)$

For Exercises 33–38, find the slope–intercept form of the equation of a line whose x-intercept and y-intercept is given.

	33.	34.	35.	36.	37.	38.
x-intercept	$(6, 0)$	$(-9, 0)$	$(4, 0)$	$(-10, 0)$	$(6, 0)$	$(10, 0)$
y-intercept	$(0, 2)$	$(0, 3)$	$(0, -4)$	$(0, -2)$	$(0, -6)$	$(0, 4)$

39. Jamie sells newspaper subscriptions door-to-door to help pay her tuition. She is paid a certain salary each night, plus a commission on each sale she makes. On Monday, she sold 3 subscriptions and was paid $66. On Tuesday she sold 8 subscriptions and was paid $116.

a) Find a linear equation that calculates Jamie's nightly pay on a night that she sells x subscriptions.

b) How much will Jamie be paid on a night that she makes no sales?

29. $(-10, -12), (6, 20)$ **30.** $(-7, 4), (2, 4)$

40. Luis contracted with a landscaper to install a brick patio in his backyard. The landscaper charged $10,000. Luis paid the landscaper a deposit the first month, and now makes a fixed monthly payment until the balance has been paid in full. After three months, Luis owed $6800. After seven months, Luis still owed $5200.

31. $(-2, -9), (-2, -3)$ **32.** $(8, -4), (8, 2)$

a) Find a linear equation for Luis's balance (y) after x months.

b) How much was the deposit that Luis paid?

c) How many months will it take to pay off the entire balance?

41. Members of a country club pay an annual membership fee. In addition, they pay a greens fee each time they play a round of golf. Last year Dave played golf 37 times and paid the country club a total of $6790 (membership fee and greens fees). Last year, Vijay played golf 15 times and paid the country club $5250.

    a) Find a linear equation for the amount owed (y) by a member who plays x rounds of golf in a year.

    b) What is the annual membership fee at this country club?

    c) What is the cost for a single round of golf?

42. A computer repairman charges a service fee in addition to an hourly rate to fix a computer. To fix Frank's computer, the repairman took two hours and charged a total of $225 (service fee plus hourly rate for two hours work). Bundy's computer was in deeper trouble, and took six hours to fix. The charge to Bundy was $505.

    a) Find a linear equation for the charge (y) for a repair that takes x hours to perform.

    b) How much is the repairman's service fee?

    c) How much is the repairman's hourly rate?

Find the slope–intercept form of the equation of a line parallel to the given line and passing through the given point.

43. Parallel to $y = -2x + 13$, through $(-6, -3)$

44. Parallel to $y = 4x - 11$, through $(3, 8)$

45. Parallel to $-3x + y = 9$, through $(4, -6)$

46. Parallel to $10x + 2y = 40$, through $(-2, 9)$

47. Parallel to $y = 5$, through $(2, 7)$

48. Parallel to $x = -5$, through $(-6, -4)$

Find the slope–intercept form of the equation of a line parallel to the graphed line and passing through the point plotted on the graph.

49.

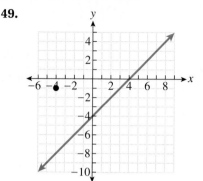

50.

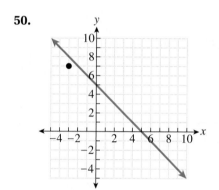

Find the slope–intercept form of the equation of a line perpendicular to the given line and passing through the given point.

51. Perpendicular to $y = 3x - 7$, through $(9, 4)$

52. Perpendicular to $y = -\dfrac{1}{2}x + 5$, through $(1, -3)$

53. Perpendicular to $4x + 5y = 9$, through $(-4, 2)$

54. Perpendicular to $3x - 2y = 8$, through $(-6, -7)$

55. Perpendicular to $y = 3$, through $(-5, -3)$

56. Perpendicular to $x = -8$, through $(6, 1)$

Find the slope–intercept form of the equation of a line perpendicular to the graphed line and passing through the point plotted on the graph.

57.

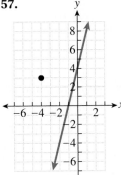

58.

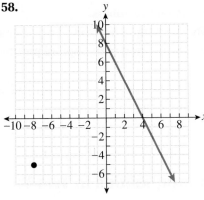

59. Explain why the equation of a vertical line has the form $x = a$ and the equation of a horizontal line has the form $y = b$.

60. Explain why a horizontal line has a slope of 0 and a vertical line has undefined slope.

QUICK REVIEW EXERCISES

Section 3.6

Graph. Label any x- and y-intercepts.

1. $5x - 2y = 10$

2. $y = -3x - 8$

3. $y = \dfrac{2}{3}x - 4$

4. $y = -8$

Study Tip **REVISITED**   Consider using the Internet as a resource that will allow you to conduct further research on topics you are learning. There are a great number of Web sites dedicated to mathematics. These sites have alternative explanations, examples, and practice problems. If you find a Web site that helps you with one particular topic, check that site when you are researching another topic. Keep in mind that anyone can create a Web site, so there is no guarantee that the information is correct. If you have any questions, be sure to ask your instructor.

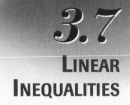

LINEAR INEQUALITIES

Objectives

1 Determine whether an ordered pair is a solution of a linear inequality in two variables.

2 Graph a linear inequality in two variables.

3 Graph a linear inequality involving a horizontal or vertical line.

4 Graph linear inequalities associated with applied problems.

In this section, we will learn how to solve **linear inequalities** in two variables. Here are some examples of linear inequalities in two variables:

$$2x + 3y \leq 6 \qquad 5x - 4y \geq -8 \qquad -x + 9y < -18 \qquad -3x - 4y > 7$$

Solutions of Linear Inequalities in Two Variables

Objective **1** Determine whether an ordered pair is a solution of a linear inequality in two variables.

Solutions of Linear Inequalities

> A solution of a linear inequality in two variables is an ordered pair (x, y) such that when the coordinates are substituted into the inequality, a true statement results.

For example, consider the linear inequality $2x + 3y \leq 6$. The ordered pair $(2, 0)$ is a solution because, when we substitute these coordinates into the inequality, it produces the following result:

$$2x + 3y \leq 6$$
$$2(2) + 3(0) \leq 6$$
$$4 + 0 \leq 6$$
$$4 \leq 6$$

The last inequality is a true statement, so $(2, 0)$ is a solution. The ordered pair $(3, 4)$ is not a solution, because $2(3) + 3(4)$ is not less than or equal to 6. Any ordered pair (x, y) for which $2x + 3y$ evaluates to be less than or equal to 6 is a solution of this inequality, and there are infinitely many solutions to this inequality. We will display our solutions on a graph.

Graphing Linear Inequalities in Two Variables

Objective **2** Graph a linear inequality in two variables. Ordered pairs that are solutions of the inequality $2x + 3y \leq 6$ are of one of two types. Ordered pairs (x, y) for which $2x + 3y = 6$ are solutions, and ordered pairs (x, y) for which $2x + 3y < 6$ are also solutions. Points satisfying $2x + 3y = 6$ lie on a line, so we begin by graphing this line. Since the equation is in standard form, a quick way to graph this line is by finding its x- and y-intercepts.

x-intercept (*y* = 0)	*y*-intercept (*x* = 0)
$2x + 3(0) = 6$	$2(0) + 3y = 6$
$2x + 0 = 6$	$0 + 3y = 6$
$2x = 6$	$3y = 6$
$x = 3$	$y = 2$
$(3, 0)$	$(0, 2)$

Here is the graph of the line:

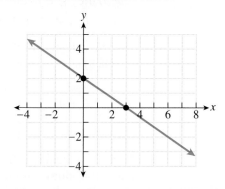

This line divides the plane into two half-planes and is the dividing line between ordered pairs for which $2x + 3y < 6$ and ordered pairs for which $2x + 3y > 6$. To finish graphing our solutions, we must determine which half-plane contains the ordered pairs for which $2x + 3y < 6$. To do this we will use a **test point,** which is a point that is not on the graph of the line, whose coordinates are used for determining which half-plane contains the solutions of the inequality. We substitute the test point's coordinates into the original inequality, and if the resulting inequality is true, this point and all other points on the same side of the line are solutions, and we will shade that half-plane. If the resulting inequality is false, then the solutions are on the other side of the line, and we shade that half-plane instead. A wise choice for the test point is the origin $(0, 0)$ if it is not on the line which has been graphed, since its coordinates are easy to work with when substituting into the inequality. Since $(0, 0)$ is not on the line, we will use it as a test point.

$$\text{Test Point: } (0, 0)$$
$$2(0) + 3(0) \leq 6$$
$$0 + 0 \leq 6$$
$$0 \leq 6$$

Since the last line is a true inequality, $(0, 0)$ is a solution, and we will shade the half-plane containing this point.

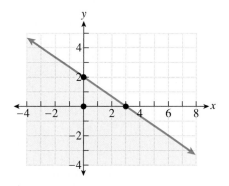

EXAMPLE 1   Graph $y \geq 4x + 3$.

Solution

We begin by graphing the line $y = 4x + 3$. This equation is in slope–intercept form, so we can graph it by plotting its y-intercept $(0, 3)$ and using the slope (up four units, one unit to the right) to find other points on the line. We can also verify that the x-intercept is at $\left(-\dfrac{3}{4}, 0 \right)$.

Since the line does not pass through the origin, we can use $(0, 0)$ as a test point.

$$\text{Test Point: } (0, 0)$$
$$0 + 4(0) \geq 3$$
$$0 + 0 \geq 3$$
$$0 \geq 3$$

The last inequality is false, so the solutions are in the half-plane that does not contain the origin. The graph of the inequality is shown at the right.

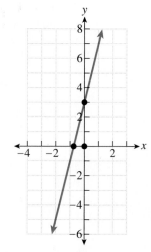

The first two inequalities we graphed were **weak linear inequalities,** which are inequalities involving the symbols $\leq$ or $\geq$. We now turn our attention to **strict linear inequalities,** which involve the symbols $<$ or $>$. An example of a strict linear inequality would be $x - 5y < 5$. Ordered pairs for which $x - 5y = 5$ are not solutions to this inequality, so the points on the line are not included as solutions. We denote this on the graph by graphing the line as a dashed or broken line. We still pick a test point and shade the appropriate half-plane in the same way that we did in the previous examples.

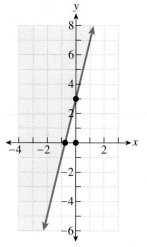

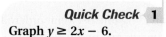

Quick Check **1**
Graph $y \geq 2x - 6$.

EXAMPLE 2   Graph $x - 5y < 5$.

Solution

We start by graphing the line $x - 5y = 5$ as a dashed line. This equation is in standard form, so we can graph the line by finding its intercepts.

x-intercept $(y = 0)$	y-intercept $(x = 0)$
$x - 5(0) = 5$	$0 - 5y = 5$
$x - 0 = 5$	$0 - 5y = 5$
$x = 5$	$-5y = 5$
	$y = -1$
$(5, 0)$	$(0, -1)$

At the right is the graph of the line, with an *x*-intercept at $(5, 0)$ and a *y*-intercept at $(0, -1)$. Again, notice that the line is a dashed line.

Since the line does not pass through the origin, we will use $(0, 0)$ as a test point.

$$\text{Test Point: } (0, 0)$$
$$0 - 5(0) < 5$$
$$0 - 0 < 5$$
$$0 < 5$$

This inequality is true, so the origin is a solution. We shade the half-plane containing the origin.

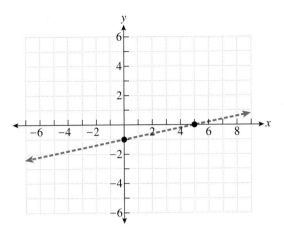

Quick Check **2**
Graph $x + 2y < 6$.

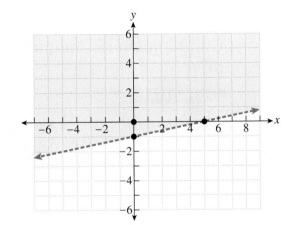

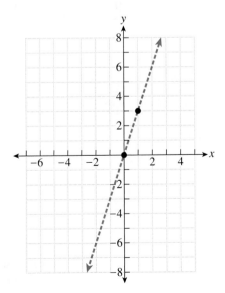

EXAMPLE **3**   Graph $y > 3x$.

Solution

This is a strict inequality, so we begin by graphing the line $y = 3x$ as a dashed line.

The equation is in slope–intercept form, with a slope of 3 and a *y*-intercept at $(0, 0)$. After plotting the *y*-intercept, we can use the slope to find a second point at $(1, 3)$. The graph of the line is shown at the right.

Since the line passes through the origin, we cannot use $(0, 0)$ as a test point. We will try to choose a point that is clearly not on the line, such as $(4, 0)$, which is to the right of the line.

Quick Check 3

Graph $y \leq -\dfrac{4}{3}x.$

Test Point: $(4, 0)$

$0 > 3(4)$

$0 > 12$

This inequality is false, so we will shade the half-plane that does not contain the test point $(4, 0)$. At the right is the graph.

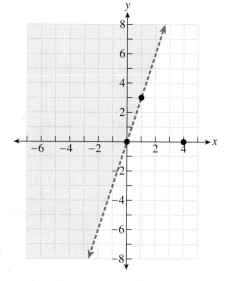

Graphing Linear Inequalities Involving Horizontal or Vertical Lines

Objective 3 Graph a linear inequality involving a horizontal or vertical line.
A linear inequality involving a vertical line or a horizontal line has only one variable but can still be graphed on a plane rather than on a number line. After graphing the related line, we find that it is not necessary to use a test point. Instead, we can use reasoning to determine where to shade. However, we may continue to use test points if we choose.

To graph the inequality $x > 5$, we begin by graphing the vertical line $x = 5$, using a dashed line. The values of x that are greater than 5 are to the right of this line, so we shade the half-plane to the right of $x = 5$. If we had used the origin as a test point, the resulting inequality $(0 > 5)$ would be false, so we would shade the half-plane that does not contain the origin, producing the same graph.

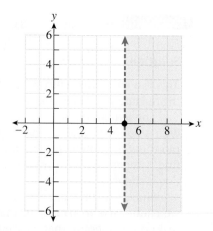

To graph the inequality $y \leq -2$, we begin by graphing the horizontal line $y = -2$, using a solid line. The values of y that are less than -2 are below this line, so we shade the half-plane below the line.

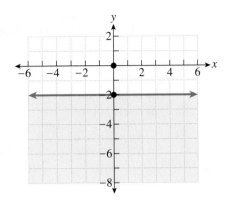

Here is a brief summary of how to graph linear inequalities in two variables:

Graphing Linear Inequalities in Two Variables

- Graph the line related to the inequality, which is found by replacing the inequality symbol with an equals sign.
- If the inequality symbol includes equality ($\leq$ or $\geq$), graph the line as a solid line.
- If the inequality symbol does not include equality ($<$ or $>$), graph the line as a dashed or broken line.
- Select a test point that is not on the line. Use the origin if possible. Substitute the coordinates of the point into the original inequality. If the resulting inequality is true, then shade the region on the side of the line that contains the test point. If the resulting inequality is false, shade the region on the side of the line that does not contain the test point.

Applications

Objective 4 **Graph linear inequalities associated with applied problems.** We turn our attention to an application problem to end the section.

EXAMPLE 4   A movie theater has 120 seats. The number of adults and children that can be admitted cannot exceed the number of seats. Set up and graph the appropriate inequality.

Solution

There are two unknowns in this problem: the number of adults and the number of children. We will let x represent the number of adults and y represent the number of children. Since the total number of adults and children cannot exceed 120, we know that $x + y \leq 120$. (We also know that $x \geq 0$ and $y \geq 0$, since we cannot have a negative number of children or adults. Therefore, we are restricted to the first quadrant, the positive x-axis and the positive y-axis.)

To graph this inequality, we begin by graphing $x + y = 120$ as a solid line. Since this equation is in standard form, we graph the line using its x-intercept $(120, 0)$ and its y-intercept $(0, 120)$.

The origin is not on this line, so we can use $(0, 0)$ as a test point.

$$x + y \leq 120$$
$$0 + 0 \leq 120 \qquad \text{Substitute } 0 \text{ for } x \text{ and } 0 \text{ for } y.$$
$$0 \leq 120 \qquad \text{True.}$$

This is a true statement, so we will shade on the side of the line that contains the origin.

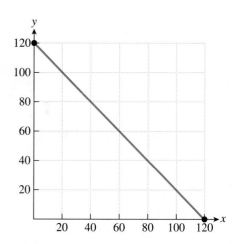

Quick Check 4
To make a fruit salad, Irv needs a total of at least 60 pieces of fruit. If Irv decides to buy only apples and pears, set up and graph the appropriate inequality.

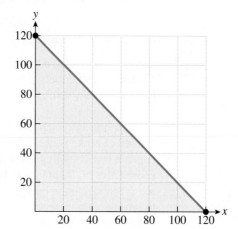

EXERCISES 3.7 ❯

Determine whether the ordered pair is a solution to the given linear inequality.

1. $5x + 3y \leq 22$

  a) $(0, 0)$ **b)** $(8, -4)$
  c) $(2, 4)$ **d)** $(-3, 9)$

2. $2x - 4y > 6$

  a) $(4, 1)$ **b)** $(2, -3)$
  c) $(7, 2)$ **d)** $(5, 1)$

3. $y < 6x - 11$

  a) $(5, 8)$ **b)** $(0, 0)$
  c) $(2, 1)$ **d)** $(-4, -13)$

4. $x - 7y \geq -3$

  a) $(4, 1)$ **b)** $(9, 2)$
  c) $(0, 0)$ **d)** $(-16, -2)$

5. $y < 7$

  a) $(0, 0)$ **b)** $(8, 6)$
  c) $(-3, 10)$ **d)** $(6, 7)$

6. $x \leq -2$

  a) $(0, 0)$ **b)** $(-1, 3)$
  c) $(-2, 19)$ **d)** $(-5, -4)$

Complete the solution of the linear inequality by shading the appropriate region.

7. $4x + y \geq 7$

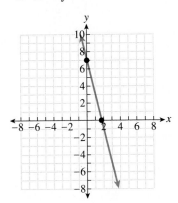

8. $-3x + 2y \leq -6$

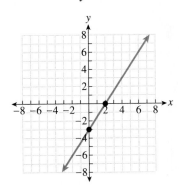

9. $2x + 8y < -4$

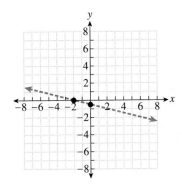

10. $10x - 2y > 0$

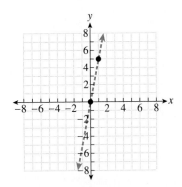

Which graph, A or B, represents the solution to the linear inequality? Explain how you chose your answer.

11. $3x + 4y \geq 24$

A)

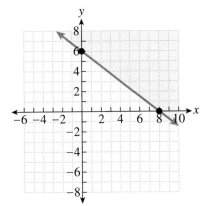

B)

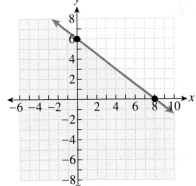

12. $y \leq \dfrac{1}{8}x$

A)

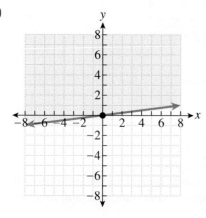

B)

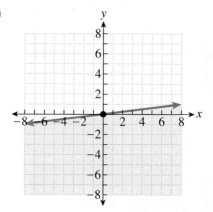

13. $3x - 2y < -18$

A)

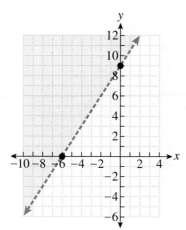

B)

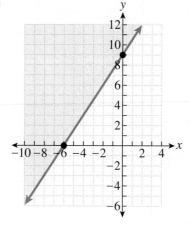

14. $5x - 4y > 20$

A)

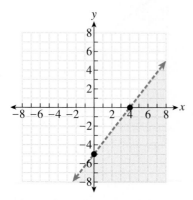

B)

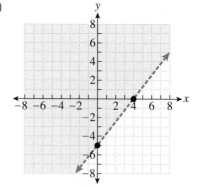

Determine the missing inequality sign ($<$, $>$, $\leq$, or $\geq$) for the linear inequality based on the given graph.

15. $8x - 3y$ _____ 24

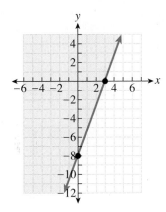

16. $3x + 6y$ _____ -27

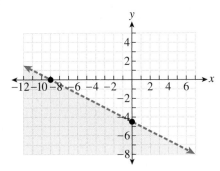

17. $-5x + 8y$ _____ 0

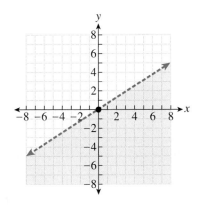

18. y _____ $4x - 6$

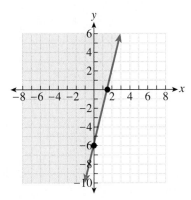

Graph the linear inequality on a plane.

19. $6x - 5y < -30$ **20.** $2x + 4y \leq 16$

21. $-x + 3y \geq 9$

22. $7x - 2y > -14$

23. $-9x + 4y \leq 0$

24. $y < -\dfrac{1}{4}x$

29. $y \leq -2x - 4$

30. $y < -5$

31. $y > -\dfrac{3}{2}x - \dfrac{3}{2}$

32. $y \geq \dfrac{3}{5}x$

25. $y > 3$

26. $x \geq 5$

33. $x \leq -2$

34. $y < x$

27. $y \geq \dfrac{2}{3}x - 6$

28. $y > 4x - 8$

35. $y > 2x - 30$ **36.** $y \geq -x + 15$

39.

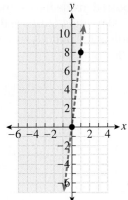

Determine the linear inequality associated with the given solution. (First, find the equation of the line. Then rewrite this equation as an inequality with the appropriate inequality sign: $<$, $>$, $\leq$, or $\geq$.)

37.

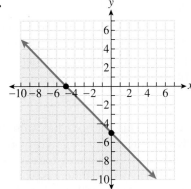

40.

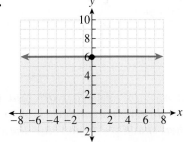

41. A charity basketball game was attended by more than 50 people. Some of the people were faculty members and the rest were students. Set up and graph an inequality involving the number of faculty members and the number of students in attendance.

38.

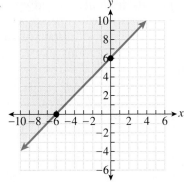

42. An elevator has a warning posted inside the car that the maximum capacity is 1200 pounds. Suppose that the average weight of an adult is 160 pounds and the average weight of a child is 60 pounds.

    a) Set up and graph an inequality involving the number of adults and the number of children that can safely ride on the elevator.

    b) From the graph from part a), can two adults and 16 children ride in the elevator at the same time?

Find the region that contains ordered pairs that are solutions to both inequalities. First graph each inequality separately, and then shade the region that the two graphs have in common.

43. $x + y < 3$ and $y > x - 7$

44. $2x + 3y \leq 18$ and $2x + y \geq 8$

Answer in complete sentences.

45. True or false: every ordered pair is a solution of either $3x - 8y < 24$ or $3x - 8y > 24$. Explain your answer.

46. Explain why we used a dashed line when graphing inequalities involving the symbols $<$ and $>$ and why we use a solid line when graphing inequalities involving the symbols $\leq$ and $\geq$.

Chapter 3 Summary

Section 3.1—Topic	Chapter Review Exercises
Finding the Coordinates of a Point on a Graph	1
Determining which Quadrant a Point Is in	2–3
Determining Whether an Ordered Pair Is a Solution of an Equation	4–6

Section 3.2—Topic	Chapter Review Exercises
Finding the x- and y-Intercepts of a Line	7–10
Graphing a Line by Using Its Intercepts	11–16, 51–56

Section 3.3—Topic	Chapter Review Exercises
Finding the Slope of a Line Passing through Two Points	17–20
Finding the Slope and y-Intercept of a Line from Its Equation	21–26, 36–37
Graphing a Line by Using Its Slope and y-Intercept	27–32, 51–56

Section 3.4—Topic	Chapter Review Exercises
Solving Applications Involving Linear Functions	38
Evaluating Linear Functions	39–41
Interpreting the Graph of a Function	42

Section 3.5—Topic	Chapter Review Exercises
Determining Whether Two Lines Are Parallel, Perpendicular, or Neither	33–35

Section 3.6—Topic	Chapter Review Exercises
Finding the Equation of a Line Given Its Slope and a Point on the Line	43–45
Finding the Equation of a Line Given Two Points on the Line	46–48
Finding the Equation of a Line from Its Graph	49–50

Section 3.7—Topic	Chapter Review Exercises
Graphing Linear Inequalities	57–60

Summary of Chapter 3 Study Tips

You have a great number of resources at your disposal to help you learn mathematics, and your success may depend on how well you use them.

- Your instructor is a valuable resource, who can answer your questions and provide advice. In addition to being available during the class period and office hours, some instructors take questions from students right before class begins or just after class ends.
- The tutorial center is another valuable resource. Keep in mind that the goal is for you to understand the material and to be able to solve the problems yourself.
- Many colleges offer short courses or seminars in study skills. Ask an academic counselor about these courses and whether they would be helpful.
- Some other resources discussed in this chapter were the student solutions manual, your classmates, and the Internet.
- Finally, the online supplement to this textbook at MyMathLab.com can be quite helpful. At this site you will find video clips, tutorial exercises, and much, much more.

Find the coordinates of the labeled points A, B, C, and D.
[3.1]

1.

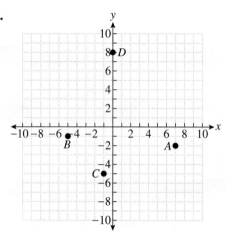

Which quadrant (I, II, III, or IV) is the given point located in? [3.1]

2. $(-3, -3)$　　　**3.** $(-9, 4)$

Is the ordered pair a solution to the given equation? [3.1]

4. $(5, -2), 4x - 2y = 16$

5. $\left(\dfrac{7}{2}, \dfrac{3}{5}\right), 4x + 5y = 17$

6. $(-6, -4), x - 3y = 6$

Find the x-intercept and y-intercept, if possible. [3.2]

7. $2x - 8y = -40$

8. $5x - y = -15$

9. $3x - 7y = 0$

10. $4x - 7y = 14$

Find the x- and y-intercepts, and use them to graph the equation. [3.2]

11. $-3x + 2y = 12$

12. $x - 4y = -8$

13. $2x + 3y = 0$

14. $4x + 3y = -6$

15. $y = -\dfrac{3}{2}x + 6$　　　　**16.** $y = 2x - 4$

Find the slope of the line that passes through the given points. [3.3]

17. $(-5, 3)$ and $(-3, 9)$

18. $(6, -2)$ and $(3, 10)$

19. $(3, -8)$ and $(8, -3)$

20. $(2, -7)$ and $(7, -7)$

Find the slope and the y-intercept of the given line. **[3.3]**

21. $y = -2x + 7$

22. $y = 4x - 6$

23. $y = -\dfrac{2}{3}x$

24. $4x - 2y = 9$

25. $5x + 3y = 18$

26. $y = -7$

Graph using the slope and y-intercept. Find the x-intercept and label it on the graph. **[3.3]**

27. $y = 3x - 6$

28. $y = -5x - 5$

29. $y = \dfrac{2}{5}x + 2$

30. $y = x$

31. $y = -\dfrac{7}{3}x + 3$

32. $y = \dfrac{1}{2}x + \dfrac{3}{2}$

Are the two given lines parallel, perpendicular, or neither? **[3.5]**

33. $y = 2x - 9$
    $y = -2x + 9$

34. $4x + y = 11$
    $y = \dfrac{1}{4}x + \dfrac{5}{2}$

35. $12x - 9y = 17$
    $-8x + 6y = 10$

Find the equation of a line with the given slope and y-intercept. **[3.3]**

36. Slope -4, y-intercept $(0, -2)$

37. Slope $\dfrac{2}{5}$, y-intercept $(0, -6)$

38. Monique's vacation fund has $720 in it. She plans to add $50 to this fund each month. **[3.4]**

   a) Create a function $f(x)$ for the amount of money in Monique's vacation fund after x months.

   b) Use the function from part a) to determine how much money will be in the vacation fund after 11 months.

   c) Use the function from part a) to determine how many months it will take until Monique has enough money to take a trip to Hawaii, which will cost $2750.

Evaluate the given function. **[3.4]**

39. $f(x) = 9x + 7, f(-2)$

40. $g(x) = 3 - 8x$, $g(-5)$

41. $f(x) = 3x + 7$, $f(5a - 1)$

42. Consider the following graph of a function $f(x)$. [3.4]

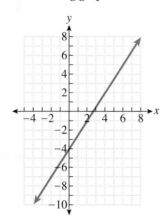

   a) Find $f(-2)$.

   b) Find all values x for which $f(x) = 5$.

   c) Find the domain of $f(x)$.

   d) Find the range of $f(x)$.

Find the slope–intercept form of the equation of a line with the given slope that passes through the given point. **[3.6]**

43. Slope 1, through $(4, -2)$

44. Slope -5, through $(1, 3)$

45. Slope $-\dfrac{3}{2}$, through $(-4, 9)$

Find the slope–intercept form of the equation of a line that passes through the two given points. **[3.6]**

46. $(-2, 1)$ and $(2, -7)$

47. $(-6, -5)$ and $(2, 7)$

48. $(5, 3)$ and $(-9, 3)$

Find the equation of the line that has been graphed. **[3.6]**

49.

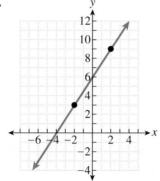

50.

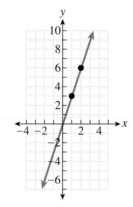

Graph. Label any intercepts. **[3.2/3.3]**

51. $y = 4x + 8$

52. $y = -\dfrac{1}{2}x + 3$

53. $y = -5x + 2$ **54.** $x = 5$ **59.** $y \leq -4$

60. $y > \dfrac{3}{2}x + 5$

55. $3x - 4y = 24$

56. $y = -6$

Graph the inequality on a plane. [3.7]

57. $3x + y < -9$ **58.** $2x + 7y \geq 14$

Which quadrant (I, II, III, or IV) is the given point located in?

1. $(-3, 7)$

Is the ordered pair a solution to the given equation?

2. $(4, -2), -3x - 7y = 2$

Find the x-intercept and y-intercept, if possible.

3. $5x - 6y = -15$

4. $y = -\dfrac{7}{2}x + 21$

Find the x- and y-intercepts, and use them to graph the equation.

5. $4x - y = 6$

6. $y = \dfrac{2}{3}x - 4$

Find the slope of the line that passes through the given points.

7. $(2, 8)$ and $(-3, 9)$

Find the slope and the y-intercept of the given line.

8. $y = 7x - 8$

9. $3x + 5y = 45$

Graph using the slope and y-intercept. Find the x-intercept and label it on the graph.

10. $y = -2x + 3$

11. $y = -\dfrac{6}{5}x + 5$

Are the two given lines parallel, perpendicular, or neither?

12. $4y = 8x - 11$

$2x - y = 44$

Find the equation of a line with the given slope and y-intercept.

13. Slope -6, y-intercept $(0, 5)$

Evaluate the given function.

14. $f(x) = 7 - 2x, f(-3)$

15. Find the slope–intercept form of the equation of a line with a slope of $-\frac{5}{2}$ that passes through the point $(-2, 9)$.

16. Find the slope–intercept form of the equation of a line with the given slope that passes through the points $(-6, 4)$ and $(-1, -6)$.

Graph. Label any intercepts.

17. $y = -4x + 6$

18. $-4x + 5y = 20$

19. $y = 2$

20. Graph the inequality on a plane. $x - 6y < -3$

Mathematicians in History
René Descartes

$\mathcal{R}$ené Descartes was an early 17th-century philosopher who made important contributions to mathematics. Cartesian geometry resulted from his application of algebra to geometry. The rectangular coordinate plane is also called the Cartesian plane in his honor. Descartes once said, "Mathematics is a more powerful instrument of knowledge than any other that has been bequeath to us by human agency."

Write a one-page summary (OR make a poster) of the life of René Descartes and his accomplishments. Also, look up Descartes's most famous quotes and list your favorite quote.

Interesting issues:

- Where and when was René Descartes born?
- How old was Descartes when he enrolled at the Jesuit College at La Fleche?
- Descartes's first major treatise on physics was *Le Monde, ou Traité de la Lumierè*. Why did he choose not to publish his results?
- Descartes's most famous quote, in Latin, is "Cogito, ergo sum." What is the English translation of this quote?
- How did a fly help Descartes come up with the idea for the rectangular coordinate system?
- Queen Christina of Sweden invited Descartes to Sweden in 1649, where he died shortly thereafter. Describe the circumstances that led to his death.

Step 1: Form groups of no more than 4 or 5 students. Write the member's names in the chart below. For the best results, try to form groups in which all members are of the same sex. Each group should have a meter stick, ruler, or some type of measuring device.

Step 2: Measure the length of each member's foot (without shoes). Then measure and record the height of each member. Record these values in the appropriate column of the following table:

Member's Name	Length of Foot	Height	Ordered Pair

Step 3: For each member, create an ordered pair in which the first coordinate is the length of the member's foot and the second coordinate is his or her height.

Step 4: Create an appropriate axis system with the horizontal axis being for the foot length and the vertical axis being for the heights.

Step 5: Plot your ordered pairs in the axis system your created. Try to be as accurate as possible. Starting from left to right, label your points A, B, C, D, etc. Visually, does it look like these points form a straight line? Would a straight line be an appropriate model for your data? Use a straightedge to visually check whether your group was correct.

Step 6: Find the slope of the line that would connect point A and point B. Find the slope of the line that would connect point B and point C. Continue until you have used all of your points. Are these slopes the same? Should the slopes be the same if the points don't line up in a straight line? Have a group discussion and write down the groups' thoughts on these questions.

Step 7: Using a straightedge, try to find a straight line that best fits your ordered pairs. Calculate the equation of this line and draw it on your plane.

Step 8: Along with your completed chart, you should turn in the linear equation you created which your group feels best fits your data.

4

SYSTEMS OF EQUATIONS

In this chapter, we will learn to set up and solve systems of equations. A system of equations is a set of two or more associated equations containing two or more variables. We will learn to solve systems of linear equations in two variables by three methods: graphing, the substitution method, and the addition method. We will also learn to set up systems of equations to solve applied problems. We will finish the chapter by learning to solve systems of linear inequalities in two variables.

Study Tip **DOING YOUR HOMEWORK**   *Doing a homework assignment should not be viewed as just some requirement. Homework exercises are assigned to help you to learn mathematics. In this chapter, we will discuss how to do your homework and get the most out of your effort.*

4.1

SYSTEMS
OF LINEAR
EQUATIONS;
SOLVING
SYSTEMS
GRAPHICALLY

Objectives

1 Determine whether an ordered pair is a solution of a system of equations.
2 Identify the solution of a system of linear equations from a graph.
3 Solve a system of linear equations graphically.
4 Identify systems of linear equations with no solution or infinitely many solutions.

Systems of Linear Equations and Their Solutions

A **system of linear equations** consists of two or more linear equations. Here are some examples:

$$5x + 4y = 20 \qquad\qquad y = \frac{3}{5}x \qquad\qquad x + y = 11$$
$$2x - y = 6 \qquad\qquad y = -4x + 3 \qquad\qquad y = 5$$

Objective 1 **Determine whether an ordered pair is a solution of a system of equations.** The equations in a system are examined together, and we are looking for the solution(s) that the equations have in common.

Solution of a System of Linear Equations

A **solution of a system of linear equations** is an ordered pair (x, y) that is a solution of each equation in the system.

EXAMPLE 1   Is the ordered pair $(2, -3)$ a solution of the given system of equations?

$$5x + y = 7$$
$$2x - 3y = 13$$

Solution

To determine whether the ordered pair $(2, -3)$ is a solution, we need to substitute 2 for x and -3 for y into each equation. If the resulting equations are both true, then the ordered pair is a solution of the system. Otherwise, it is not.

$$\begin{array}{cc} 5x + y = 7 & 2x - 3y = 13 \\ 5(2) + (-3) = 7 & 2(2) - 3(-3) = 13 \\ 10 - 3 = 7 & 4 + 9 = 13 \\ 7 = 7 & 13 = 13 \end{array}$$

Since $(2, -3)$ is a solution of each equation, it is a solution of the system of equations.

EXAMPLE 2   Is the ordered pair $(-5, -1)$ a solution of the given system of equations?

$$y = 2x + 9$$
$$y = 14 - 3x$$

Solution

Again, we begin by substituting -5 for x and -1 for y in each equation.

$$y = 2x + 9 \qquad\qquad y = 14 - 3x$$
$$(-1) = 2(-5) + 9 \qquad (-1) = 14 - 3(-5)$$
$$-1 = -10 + 9 \qquad\quad -1 = 14 + 15$$
$$-1 = -1 \qquad\qquad\quad -1 = 29$$

Although $(-5, -1)$ is a solution of the equation $y = 2x + 9$, it is not a solution of the equation $y = 14 - 3x$. Therefore $(-5, -1)$ is not a solution of the system of equations.

Quick Check 1

Is the ordered pair $(-6, 3)$ a solution of the given system of equations?

$$5x - 2y = -36$$
$$-x + 7y = 27$$

Objective 2 **Identify the solution of a system of linear equations from a graph.** Systems of two linear equations can be solved graphically. Since an ordered pair that is a solution of a system of two equations must be a solution of each equation, we know that this ordered pair is on the graph of each equation. To solve a system of two linear equations, we graph each line and look for points of intersection. Often, this will lead to a single solution. Here are some graphical examples of systems of equations that have the ordered pair $(3, 1)$ as a solution:

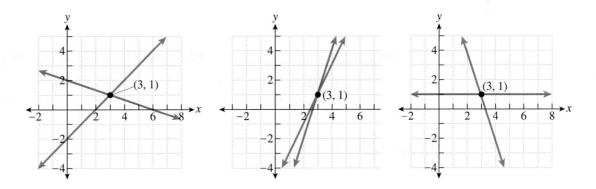

EXAMPLE 3 Find the solution of the system of equations plotted on the following graph:

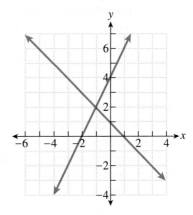

Solution

The two lines intersect at the point $(-1, 2)$, so the solution of the system of equations is $(-1, 2)$.

Quick Check   **2**   **Find the solution of the system of equations plotted on the graph.**

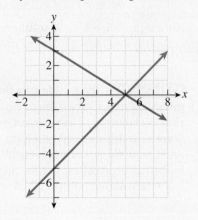

Solving Linear Systems of Equations by Graphing

Objective **3**   **Solve a system of linear equations graphically.**   Now we will solve a system of linear equations by graphing the lines and finding their point of intersection.

EXAMPLE **4**   Solve the following system by graphing:

$$y = 5x - 9$$
$$y = -\frac{7}{2}x + 8$$

Solution

Both equations are in slope–intercept form, so we will graph the lines by plotting the y-intercept and then using the slope of the line to find another point on the line.

The first line has a y-intercept at $(0, -9)$. From that point, we can find a second point by moving up five units and one unit to the right, since the slope is 5. The graph of the first line is shown at the top of the next page on the left.

The y-intercept of the second line is $(0, 8)$. Since the slope of this line is $-\frac{7}{2}$, we can move down seven units and two units to the right from this point to find a second point on this line. At the top of the next page on the right are the two lines on the same set of axes.

Examining the graph, we can see that the two lines intersect at the point $(2, 1)$. The solution of the system of equations is the ordered pair $(2, 1)$. We could check that this ordered pair is a solution by substituting its coordinates into both equations. The check is left to the reader.

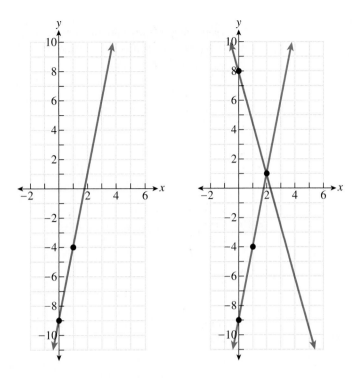

Quick Check **3**

Solve the following system by graphing:

$$y = 3x - 2$$
$$y = 5x - 6$$

Using Your Calculator   To solve a system of linear equations using the TI–83/84, we must graph each equation and then look for points of intersection. To graph the equations from Example 4, begin by pushing the $\boxed{Y=}$ key. Next to Y_1, type $5x - 9$, and type $-\frac{7}{2}x + 8$ next to Y_2. Press the $\boxed{\text{GRAPH}}$ key to display the graph.

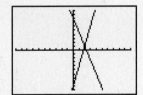

To find the points of intersection, push $\boxed{\text{2nd}}\boxed{\text{TRACE}}$ to access the CALC menu and select option **5: intersect.** You should see the following screen:

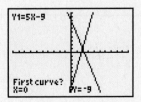

When prompted for the first curve, select Y_1 by pushing $\boxed{\text{ENTER}}$. When prompted for the second curve, select Y_2 by pushing $\boxed{\text{ENTER}}$.

We must now make an initial guess for one of the solutions. In the next screen shot, the TI–83/84 asks us if we would like $(0, 8)$ to be our guess. Accept this guess by pressing ENTER. The solution is the point $(2, 1)$.

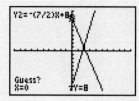 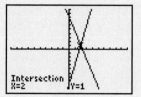

EXAMPLE 5   Solve the following system by graphing:

$$2x - 3y = 12$$
$$x + y = 1$$

Solution

Both equations are in general form, so we will graph them using their x- and y-intercepts.

2x − 3y = 12		x + y = 1	
x-intercept (y = 0)	**y-intercept (x = 0)**	**x-intercept (y = 0)**	**y-intercept (x = 0)**
$2x - 3(0) = 12$	$2(0) - 3y = 12$	$x + (0) = 1$	$(0) + y = 1$
$2x = 12$	$-3y = 12$	$x = 1$	$y = 1$
$x = 6$	$y = -4$		
$(6, 0)$	$(0, -4)$	$(1, 0)$	$(0, 1)$

Here are the graphs of each line on the same set of axes:

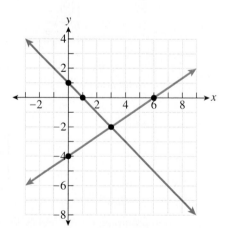

Quick Check **4**
Solve the following system by graphing:

$$x + 5y = 5$$
$$x - y = -7$$

The two lines intersect at the point $(3, -2)$, so the solution of this set of equations is the ordered pair $(3, -2)$.

Objective 4 Identify systems of linear equations with no solution or infinitely many solutions. A system of two linear equations that has a single solution is called an **independent system.** Not every system of linear equations has a single solution. Suppose that the graphs of the two equations are distinct parallel lines. Parallel lines do not intersect, so such a system has no solution. A system that has no solution is called an **inconsistent system.**

Occasionally, the two equations in a system will have the identical graphs. In this case, every ordered pair that lies on the line is a solution, so there are infinitely many solutions of the system. Such a system is called a **dependent system.**

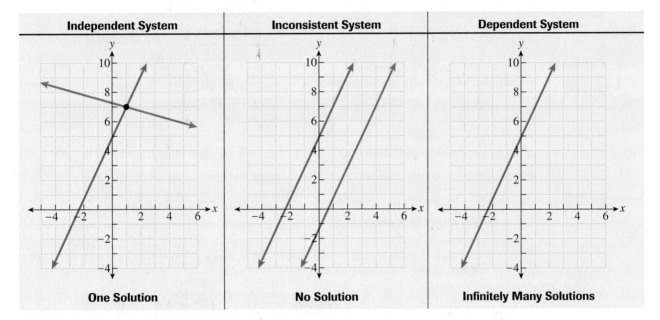

| Independent System | Inconsistent System | Dependent System |
| One Solution | No Solution | Infinitely Many Solutions |

EXAMPLE 6 Solve the following system by graphing:

$$y = -3x + 5$$
$$y = -3x - 3$$

Solution

Both equations are in slope–intercept form, so we can graph each line by plotting its y-intercept and then use the slope to find a second point on the line. The first line has a y-intercept at $(0, 5)$, and the slope is -3. The second line has a y-intercept at $(0, -3)$, and the slope is also -3. Here at the right are the graphs:

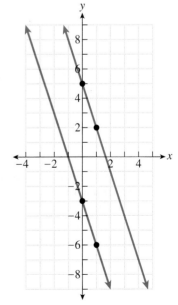

These two lines are parallel lines that do not intersect, so this system has no solution. The system is an inconsistent system. In general, if we know that the two lines have the same slope, but different y-intercepts, then we know that the system is an inconsistent system and has no solution. However, if the two lines have the same slope and the *same y*-intercept, then the system is a dependent system.

Quick Check **5**
Solve the following system by graphing:

$$y = -4x + 6$$
$$12x + 3y = 15$$

EXAMPLE ▶ 7   Solve the following system by graphing:

$$y = 2x - 4$$
$$4x - 2y = 8$$

Solution

The first line is in slope–intercept form, so we can graph it by first plotting its y-intercept at $(0, -4)$. Using its slope of 2, we see that the line also passes through the point $(1, -2)$ as shown in the graph below.

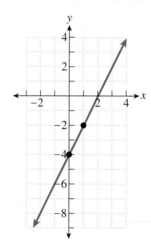

To graph the second line, we will find the x- and y-intercepts, as the equation is in general form.

x-intercept $(y = 0)$	y-intercept $(x = 0)$
$4x - 2(0) = 8$	$4(0) - 2y = 8$
$4x = 8$	$-2y = 8$
$x = 2$	$y = -4$
$(2, 0)$	$(0, -4)$

After plotting these intercepts on the graph, we see that the two lines are exactly the same. This system is a dependent system. We can display the solutions of a dependent system by showing the graph of either equation. Ordered pairs that are solutions of the system are of the form $(x, 2x - 4)$ for any real number x. To determine the form of these solutions, solve one of the equations for y and use that to replace y in the ordered pair (x, y). Notice that if we rewrite the equation $4x - 2y = 8$ in slope–intercept form, it is exactly the same as the second equation. Equations in a dependent linear system of equations are multiples of each other.

Quick Check **6**

Solve the following system by graphing:

$$4x + 6y = 12$$
$$y = -\tfrac{2}{3}x + 2$$

Solving systems of equations by graphing does have a major drawback. Often, we will not be able to accurately determine the coordinates of the solution of an independent system. Consider the graph of the system

$$y = 2x - 6$$
$$y = \frac{2}{5}x + 1$$

We can see that the *x*-coordinate of this solution is between 4 and 5 and that the *y*-coordinate of this solution is between 2 and 3. However, we cannot determine the exact coordinates of this solution from the graph. In the next two sections, we will develop algebraic techniques for finding solutions of linear systems of equations.

EXERCISES *4.1*

Is the ordered pair a solution of the given system of equations?

1. $(3, 1)$, $\begin{array}{l} 2x + y = 7 \\ 3x - 2y = 7 \end{array}$

2. $(0, 6)$, $\begin{array}{l} 7x - 2y = -12 \\ x + 5y = 30 \end{array}$

3. $(-4, 5)$, $\begin{array}{l} x + y = 1 \\ x - 3y = -11 \end{array}$

4. $(-2, -7)$, $\begin{array}{l} 4x + y = -15 \\ 6x - 2y = 2 \end{array}$

5. $(-1, -3)$, $\begin{array}{l} 4x - y = -1 \\ -2x + 4y = -10 \end{array}$

6. $(7, -8)$, $\begin{array}{l} 9x - 5y = 103 \\ x - y = -1 \end{array}$

Find the solution of the system of equations on each graph. If there is no solution, state this.

7.

8.

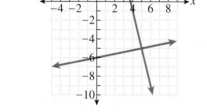

9.

10.

Solve the system by graphing. If the system is inconsistent and has no solution, state this. If the system is dependent, write the form of the solution for any real number x.

11. $y = 2x + 5$

$y = 4x + 1$

12. $y = -\dfrac{2}{3}x + 5$

$y = \dfrac{4}{3}x - 7$

13. $y = 5x - 3$

$x = 2$

14. $y = 3x$

$y = 3$

15. $x + 3y = 6$

$2x + y = -8$

16. $3x + 4y = -12$

$3x + 2y = 6$

17. $5x - 3y = 15$

$2x - y = 4$

18. $7x + 3y = 0$

$x + y = 4$

19. $4x - 5y = -20$

$-4x + 7y = 28$

20. $-2x + y = 10$

$2x + 3y = 6$

21. $y = 5x - 4$

$10x - 2y = 6$

22. $4x + 5y = 10$

$y = -\dfrac{3}{5}x + 3$

23. $9x - 2y = -18$

$y = \dfrac{1}{2}x - 7$

24. $y = -\dfrac{2}{7}x + 3$

$4x + 14y = 6$

25. Draw the graph of a linear equation that, along with the given line, forms an inconsistent system.

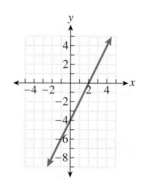

26. Draw the graph of a linear equation that, along with the given line, forms a dependent system.

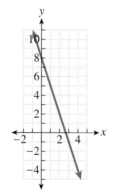

27. Draw the graph of a linear equation that, along with the given line, forms a system of equations whose single solution is the ordered pair $(5, 1)$.

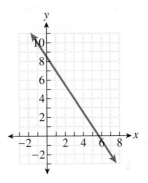

28. Draw the graphs of two linear equations that form an inconsistent system.

29. Draw the graphs of two linear equations that form a dependent system.

30. Draw the graphs of two linear equations that form a system of equations whose single solution is the ordered pair $(-3, 2)$.

Answer in complete sentences.

31. Explain why the solution of an independent system of equations is the ordered pair that is the point of intersection for the two lines.

32. When solving a system of equations by graphing, we may not be able to accurately determine the coordinates of the solution of an independent system. Explain why.

Study Tip **REVISITED**   Before beginning any homework assignment, it is a good idea to review first. Start by going over your notes from class to remind you of the types of problems covered by your instructor and of how your instructor solved them. Your notes may contain the steps for certain procedures and advice from your instructor about typical errors to avoid. After reviewing your notes, keep them handy for further reference as you proceed through the homework assignment.

You should then review the appropriate section in the text. Pay particular attention to the examples, as they show the solution to problems similar to the problems you are about to solve in the homework exercises. Look for the feature labeled A Word of Caution for advice on how to avoid making common errors. Finally, look for summaries of procedures introduced in the section. As you proceed through the homework assignment, refer back to the section as needed.

4.2

THE SUBSTITUTION METHOD FOR SOLVING SYSTEMS OF EQUATIONS

Objectives

1. Solve systems of linear equations by using the substitution method.
2. Use the substitution method to identify inconsistent systems of equations.
3. Use the substitution method to identify dependent systems of equations.
4. Solve systems of equations without variable terms with a coefficient of 1 or −1 by the substitution method.
5. Solve applied problems by using the substitution method.

Solving systems of equations by graphing each equation works well when the solution has integer coordinates. However, when the coordinates are not integers, or when the solution is outside of the region that was graphed, exact solutions are difficult to find. In the next two sections, we will examine algebraic methods for solving systems of linear equations. The goal of each method is to combine the two equations with two variables into a single equation with only one variable. After solving for this variable, we will substitute the value of that variable into one of the original equations and solve for the other variable.

The Substitution Method

Objective 1 Solve systems of linear equations by using the substitution method. Consider the following system of equations:

$$y = x - 5$$
$$2x + 3y = 20$$

Looking at the first equation, $y = x - 5$, we see that y is equal to the expression $x - 5$. We can use this fact to replace y by $x - 5$ in the second equation $2x + 3y = 20$. We are substituting the expression $x - 5$ for y, since the two expressions are equivalent. The first example will solve this system by using the substitution method.

The Substitution Method

> To solve a system of equations by using the substitution method, we solve one of the equations for one of the variables and then substitute that expression for the variable in the other equation.

EXAMPLE 1   Solve the following system by substitution:

$$y = x - 5$$
$$2x + 3y = 20$$

Solution

Since the first equation is solved for y in terms of x, we can substitute $x - 5$ for y in the second equation.

$$y = \boxed{x - 5}$$

$$2x + 3y = 20$$

This will produce a single equation with only one variable, x.

$$2x + 3y = 20$$

$2x + 3(x - 5) = 20$ Substitute $x - 5$ for y.

$2x + 3x - 15 = 20$ Distribute.

$5x - 15 = 20$ Combine like terms.

$5x = 35$ Add 15.

$x = 7$ Divide both sides by 5.

We are now halfway to our solution. We know that the x-coordinate of the solution is 7, and we now need to find the y-coordinate.

> ***A Word of Caution***   The solution of a system of equations in two variables is an *ordered pair*, not a single value.

We can find the y-coordinate by substituting 7 for x in either of the original equations. It is easier to substitute the value into the equation $y = x - 5$, since y is already isolated in this equation. We could have chosen the other equation; it would also produce the same solution. In general, it is a good idea to substitute the first value we find back into the equation that was used to make the original substitution.

$$y = x - 5$$

$y = (7) - 5$ Substitute 7 for x.

$y = 2$ Subtract.

The solution to this system is $(7, 2)$. This solution can be checked by substituting 7 for x and 2 for y in the equation $2x + 3y = 20$. The check is left to the reader.

Quick Check   1
Solve the following system by substitution:

$$y = x - 3$$
$$7x - 4y = 27$$

In the first example, one of the equations was already solved for y. To use the substitution method, we will often have to solve one of the equations for x or y. When choosing which equation to work with or which variable to solve for, look for an equation that has an x or y term with a coefficient of 1 or -1 (x, $-x$, y or $-y$). When we find one of those terms, solving for that variable can be done easily through addition or subtraction.

EXAMPLE 2   Solve the following system by substitution:

$$x + 2y = 7$$
$$4x - 3y = -16$$

Solution

The first equation can easily be solved for x by subtracting $2y$ from both sides of the equation. This produces the equation $x = 7 - 2y$. We can now substitute the expression $7 - 2y$ for x in the second equation $4x - 3y = -16$.

> ***A Word of Caution***   Do not substitute $7 - 2y$ into the equation that it came from, $x + 2y = 7$, as the resulting equation will be $7 = 7$. Substitute it into the other equation in the system.

$$4x - 3y = -16$$
$$4(7 - 2y) - 3y = -16 \qquad \text{Substitute } 7 - 2y \text{ for } x.$$
$$28 - 8y - 3y = -16 \qquad \text{Distribute 4.}$$
$$28 - 11y = -16 \qquad \text{Combine like terms.}$$
$$-11y = -44 \qquad \text{Subtract 28 from both sides.}$$
$$y = 4 \qquad \text{Divide both sides by } -11.$$

Since the y-coordinate of the solution is 4, we can substitute this value for y into the equation $x = 7 - 2y$ in order to find the x-coordinate.

$$x = 7 - 2(4) \qquad \text{Substitute 4 for } y.$$
$$x = -1 \qquad \text{Simplify.}$$

The solution of this system of equations is $(-1, 4)$. We could check this solution by substituting -1 for x and 4 for y in the equation $4x - 3y = -16$. This is left to the reader. Although the first coordinate we solved for was y, keep in mind that the ordered pair must be written in the order (x, y).

Quick Check **2**
Solve the following system by substitution:

$$x - y = -4$$
$$3x + 5y = 36$$

Inconsistent Systems

Objective **2** **Use the substitution method to identify inconsistent systems of equations.** Recall from the previous section that a linear system of equations with no solution is called an inconsistent system. The two lines associated with the equations of an inconsistent system are parallel lines. The next example shows how to determine that a system is an inconsistent system while using the substitution method.

EXAMPLE **3** Solve the following system using substitution:

$$x = 3y - 7$$
$$3x - 9y = 18$$

Solution

Since the first equation is solved for x, we will substitute $3y - 7$ for x in the second equation.

$$3x - 9y = 18$$
$$3(3y - 7) - 9y = 18 \qquad \text{Substitute } 3y - 7 \text{ for } x.$$
$$9y - 21 - 9y = 18 \qquad \text{Distribute.}$$
$$-21 = 18 \qquad \text{Combine like terms.}$$

The resulting equation is false, because -21 is not equal to 18. Since this equation can never be true, the system of equations has no solution ($\varnothing$) and is an inconsistent system.

Quick Check **3**
Solve the following system by substitution:

$$-2x + y = 3$$
$$8x - 4y = 10$$

Dependent Systems

Objective **3** **Use the substitution method to identify dependent systems of equations.** A dependent linear system of equations has two equations with identical graphs and is a system with infinitely many solutions. When using the substitution method to solve a dependent system, we will obtain an identity, such as $3 = 3$, after making the substitution.

EXAMPLE ▶ 4   Solve the following system by substitution:

$$4x - y = 3$$
$$-8x + 2y = -6$$

Solution

We will solve the first equation for y.

$$4x - y = 3$$

$\qquad -y = -4x + 3$ Subtract $4x$.

$\qquad y = 4x - 3$ Divide both sides by -1.

We now substitute $4x - 3$ for y in the second equation.

$$-8x + 2y = -6$$

$\qquad -8x + 2(4x - 3) = -6$ Substitute $4x - 3$ for y.

$\qquad -8x + 8x - 6 = -6$ Distribute 2.

$\qquad -6 = -6$ Combine like terms.

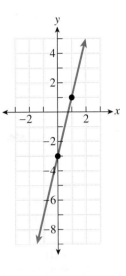

This equation is an identity that is true for all values of x. This system of equations is a dependent system with infinitely many solutions.

Since we have already shown that $y = 4x - 3$, the ordered pairs that are solutions have the form $(x, 4x - 3)$. The solutions are shown in the graph at the right, which is the graph of the line $y = 4x - 3$.

Quick Check ◀ 4
Solve the following system by substitution:

$$x - 6y = 4$$
$$3x - 18y = 12$$

Objective 4   Solve systems of equations without variable terms with a coefficient of 1 or −1 by the substitution method. All of the examples that have appeared so far in this section had at least one equation containing a variable term with a coefficient of 1 or −1. Such equations are easy to solve for one variable in terms of the other. In the next example, we will learn how to use the substitution method when this is not the case.

EXAMPLE ▶ 5   Solve the following system by substitution:

$$3x + 2y = -2$$
$$-2x + 7y = 43$$

Solution

There are no variable terms with a coefficient of 1 or −1 (x, $-x$, y, or $-y$). We will solve the first equation for y, although this selection is completely arbitrary. We could solve either equation for either variable and still find the same solution of the system.

$$3x + 2y = -2$$

$\qquad 2y = -3x - 2$ Subtract $3x$.

$\qquad y = -\dfrac{3}{2}x - 1$ Divide both sides by 2.

We now may substitute $-\frac{3}{2}x - 1$ for y in the second equation.

$$-2x + 7y = 43$$

$$-2x + 7\left(-\frac{3}{2}x - 1\right) = 43 \qquad \text{Substitute } -\frac{3}{2}x - 1 \text{ for } y.$$

$$-2x - \frac{21}{2}x - 7 = 43 \qquad \text{Distribute 7.}$$

$$2\left(-2x - \frac{21}{2}x - 7\right) = 2 \cdot 43 \qquad \text{Multiply both sides by 2 to clear fractions.}$$

$$2(-2x) - \overset{1}{\cancel{2}}\left(\frac{21}{\underset{1}{\cancel{2}}}x\right) - 2 \cdot 7 = 2 \cdot 43 \qquad \text{Distribute and divide out common factors.}$$

$$-4x - 21x - 14 = 86 \qquad \text{Multiply.}$$

$$-25x - 14 = 86 \qquad \text{Combine like terms.}$$

$$-25x = 100 \qquad \text{Add 14.}$$

$$x = -4 \qquad \text{Divide both sides by } -25.$$

We know that the x-coordinate of our solution is -4. We will substitute this value for x in the equation $y = -\frac{3}{2}x - 1$.

$$y = -\frac{3}{2}(-4) - 1 \qquad \text{Substitute } -4 \text{ for } x.$$

$$y = 5 \qquad \text{Simplify.}$$

Quick Check **5**

Solve the following system by substitution:

$$5x - 3y = 17$$
$$2x + 4y = -14$$

The solution to this system is $(-4, 5)$. The check is left to the reader.

Before turning our attention to an application of linear systems of equation, we present the following general strategy for solving these systems:

Solving a System of Equations by Using the Substitution Method

1. **Solve one of the equations for either variable.** If an equation has a variable term whose coefficient is 1 or -1, try to solve that equation for that variable.
2. **Substitute this expression for the variable in the other equation.** At this point, we will have an equation with only one variable.
3. **Solve this equation.** This gives us one coordinate of our ordered-pair solution.
4. **Substitute this value for the variable into the equation from step 1.** After simplifying, this will give us the other coordinate of our solution.
5. **Check your solution.**

Applications

Objective **5** Solve applied problems by using the substitution method.

EXAMPLE **6** A college performed a play. There were 75 people in the audience. If admission was $6 for adults and $2 for children and the total box office receipts were $310, how many adults and how many children attended the play?

Solution

There are two unknowns in this problem: the number of adults and the number of children. As in Chapter 2, we start with a table of unknowns.

> ***Unknowns***
>
> Number of Adults: A
> Number of Children: C

Since there are two variables, we need a system of two equations. The first equation comes from the total attendance of 75 people. Since all of these people are either adults or children, the number of adults (A) plus the number of children (C) must equal 75. The first equation is $A + C = 75$.

The second equation comes from the amount of money paid. The amount of money paid by adults ($\$6A$) plus the amount of money paid by children ($\$2C$) must equal the total amount of money paid. The second equation is $6A + 2C = 310$. Here is the system of equations that we need to solve:

$$A + C = 75$$
$$6A + 2C = 310$$

Before we solve this system of equations, let's look at a table that contains all of the pertinent information for this problem.

	Number of Attendees	**Cost per Ticket ($)**	**Money Paid ($)**
Adults	A	6	$6A$
Children	C	2	$2C$
Total	75		310

Notice that the first equation in the system ($A + C = 75$) can be found in the first column of the table (labeled Number of Attendees). The second equation in the system ($6A + 2C = 310$) can be found in the last column in the table (labeled Money Paid ($)).

The first equation in this system can be solved for either A or C; we will solve it for C. This produces the equation $C = 75 - A$, so the expression $75 - A$ can be substituted for C in the second equation.

$$
\begin{array}{ll}
6A + 2C = 310 & \\
6A + 2(75 - A) = 310 & \text{Substitute } 75 - A \text{ for } C. \\
6A + 150 - 2A = 310 & \text{Distribute.} \\
4A + 150 = 310 & \text{Combine like terms.} \\
4A = 160 & \text{Subtract 150.} \\
A = 40 & \text{Divide both sides by 4.}
\end{array}
$$

There were 40 adults. We now substitute this value for A in the equation $C = 75 - A$.

$$
\begin{array}{ll}
C = 75 - (40) & \text{Substitute 40 for } A. \\
C = 35 & \text{Subtract.}
\end{array}
$$

There were 40 adults and 35 children at the play.

We should check our solution for accuracy. If there were 40 adults and 35 children, then that is a total of 75 people. Also, the 40 adults each paid $6, which is $240 dollars. The 35 children each paid $2, which is another $70. The total paid is $310, so our solution checks.

Quick Check **6** A small-business owner bought 15 new computers for his company. He paid $700 for each desktop computer and $1000 for each laptop computer. If he paid $11,700 for the computers, how many desktop computers and how many laptop computers did he buy?

EXERCISES 4.2

Solve the system by substitution. If the system is inconsistent and has no solution, state this. If the system is dependent, write the form of the solution for any real number x.

1. $y = 3x - 5$
$x + 3y = 15$

2. $x = 4y + 3$
$2x + 5y = -7$

3. $x = 2 - 5y$
$7x + 10y = 89$

4. $x = 1 - 4y$
$4x + 2y = -17$

5. $y = 5x - 10$
$-10x + 2y = 20$

6. $y = 11 - 5x$
$-7x - 2y = 14$

7. $18x + 6y = -12$
$y = -3x - 2$

8. $x = 2y - 15$
$3x + 14y = 5$

9. $x + 6y = 5$
$10x - 4y = 2$

10. $2x + y = 3$
$12x - 2y = -12$

11. $x + y = 11$
$3x + 5y = 29$

12. $4x + 2y = 34$
$x - y = -8$

13. $3x + 8y = 20$
$2x + y = 9$

14. $x + 3y = 14$
$3x + 9y = 28$

15. $5x - y = 18$
$4x - 3y = 10$

16. $5x + 6y = 41$
$-x + 2y = 11$

17. $6x + y = 21$
$-18x - 3y = -63$

18. $10x - y = -23$
$-20x + 2y = -46$

19. $9x - 8y = 12$
$y = 3$

20. $1776x - 3y = 270$
$x = 0$

21. $4x + 2y = 10$
$5x - 4y = 6$

22. $9x + 3y = 36$
$7x - 6y = 78$

23. $3x + 5y = 14$
$6x - 5y = -32$

24. $9x + 4y = 23$
$5x - 2y = -2$

25. Give an equation that, along with the equation $y = 2x - 14$, forms an inconsistent system.

26. Give an equation that, along with the equation $y = 5x - 7$, forms a dependent system.

27. Give an equation that, along with the equation $y = 7x - 2$, forms a system whose only solution is $(-1, -9)$.

28. Make up two equations that form a system whose only solution is $(3, -2)$.

29. The college basketball team charges $5 admission to the general public for its games, while students at the college pay only $2. If the attendance for last night's game was 1200 people and the total receipts were $5250, how many of the people in attendance were students and how many were not students?

30. Kenny walks away from a blackjack table with a total of $62 in $1 and $5 chips. If Kenny has 26 chips, how many are $1 chips and how many are $5 chips?

31. A sorority set up a charity dinner to help raise money for a local children's support group. They sold chicken dinners for $8 and steak dinners for $10. If 142 people ate dinner at the fundraiser and the sorority raised $1326, how many people ordered chicken and how many people ordered steak?

32. Jeannie has some $10 bills and some $20 bills. If she has 273 bills worth a total of $4370, how many of the bills are $10 bills and how many are $20 bills?

33. To celebrate a productive week in his office, Devin stopped by a coffee shop on his way to work Friday.

He bought some regular coffees for $1.75 each and some café mochas for $3.25 each. If he bought 18 drinks and spent $39, how many regular coffees and how many café mochas did he buy?

34. Laurel scored 47 points in last night's basketball game. Eight of her points were scored on free throws, which count as 1 point each, and the rest were scored on 2-point shots and 3-point shots. If the number of 2-point shots that Laurel made was 2 more than twice the number of 3-point shots that she made, how many 2-point shots did she make?

35. The median age of women, y, at their first marriage in a particular year can be approximated by the equation $y = 0.12x + 23.2$, where x represents the number of years after 1970 and y is in years. The median age of men, y, at their first marriage in a particular year can be approximated by the equation $y = 0.14x + 20.8$, where x represents the number of years after 1970 and y is in years. In what year will the median age of women at their first marriage equal the median age of men at their first marriage? (*Source:* U.S. Census Bureau.)

36. The number of doctoral degrees conferred on women, y, in a particular year can be approximated by the equation $y = 473x + 17,811$, where x represents the number of years after 1995. The number of doctoral degrees conferred on men, y, in a particular year can be approximated by the equation $y = -423x + 26,841$, where x represents the number of years after 1995. In what year will the number of doctoral degrees conferred on women equal the number of doctoral degrees conferred on men? (*Source:* U.S. Department of Education, National Center for Educational Statistics)

Answer in complete sentences.

37. Explain, in your own words, how to solve a system of linear equations using the substitution method.

38. When using the substitution method, how can you tell that a system of equations is inconsistent and has no solutions? How can you tell that a system of equations is dependent and has infinitely many solutions? Explain your answer.

Study Tip **REVISITED**   Two important words to keep in mind when working on your homework exercises are *neat* and *complete.* When your homework is neat, it is easier to check your work, it helps an instructor spot an error if you need help with a particular problem, and it will be easier to understand when you review an old homework assignment prior to an exam.

Be as complete as you can when working on the homework exercises. By listing each step, you are increasing your chances of being able to remember all of the steps necessary to solve a similar problem on an exam or quiz. When you review an old homework assignment prior to an exam, you may have difficulty remembering how to solve a problem if you did not write down all of the necessary steps. If you make a mistake while working on a particular exercise, make note of the mistake that you made, to avoid making a similar mistake later.

**SOLVING
SYSTEMS
OF EQUATIONS
BY THE
ADDITION
METHOD**

Objectives

1 Solve systems of equations by using the addition method.
2 Use the addition method to identify inconsistent and dependent systems of equations.
3 Solve systems of equations with coefficients that are fractions by using the addition method.
4 Solve applied problems by using the addition method.

The Addition Method

Objective 1 **Solve systems of equations by using the addition method.** In this section, we will examine the **addition method** for solving systems of linear equations.

The Addition Method

> The **addition method** is an algebraic alternative to the substitution method. The two equations in a system of linear equations will be combined into a single equation with a single variable by adding them together.

In the last section, we saw that solving a system of equations by substitution is fairly easy when one of the equations contains a variable term with a coefficient of 1 or -1. The substitution method can become quite tedious for solving a system when this is not the case.

As with the substitution method, the goal of the addition method is to combine the two given equations into a single equation with only one variable. This method is based on the addition property of equality from Chapter 2, which tells us that when we add the same number to both sides of an equation, the equation remains true. (For any real numbers a, b, and c, if $a = b$, then $a + c = b + c$.) This property can be extended to cover adding equal expressions to both sides of an equation: if $a = b$ and $c = d$, then $a + c = b + d$.

Consider the system

$$3x + 2y = 20$$
$$5x - 2y = -4$$

If we add the left side of the first equation $(3x + 2y)$ to the left side of the second equation $(5x - 2y)$, this will be equal to the total of the two numbers on the right side of the equations. Here is the result of this addition:

$$
\begin{array}{r}
3x + 2y = 20 \\
5x - 2y = -4 \\
\hline
8x = 16
\end{array}
$$

Notice that when we added these two equations, the two terms containing y were opposites and this left an equation containing only the variable x. We next solve this equation for x and substitute this value for x in either of the two original equations to find y. The first example will walk through this entire process.

EXAMPLE ▸ 1   Solve the following system by addition:

$$3x + 2y = 20$$
$$5x - 2y = -4$$

Solution

Since the two terms containing y are opposites, we can use addition to eliminate y.

$$
\begin{array}{ll}
3x + 2y = 20 & \\
\underline{5x - 2y = -4} & \text{Add the two equations.} \\
8x \quad\;\; = 16 & \\
x = 2 & \text{Divide both sides by 8.}
\end{array}
$$

The x-coordinate of our solution is 2. We now substitute this value for x in the first original equation. (We could have chosen the other equation, since it will produce exactly the same solution.)

$$
\begin{array}{ll}
3x + 2y = 20 & \\
3(2) + 2y = 20 & \text{Substitute 2 for } x. \\
6 + 2y = 20 & \text{Multiply.} \\
2y = 14 & \text{Subtract 6.} \\
y = 7 & \text{Divide both sides by 2.}
\end{array}
$$

The solution of this system is the ordered pair $(2, 7)$. We could check this solution by substituting 2 for x and 7 for y in the equation $5x - 2y = -4$. This is left to the reader.

Quick Check 1
Solve the following system by addition:

$$4x + 3y = 31$$
$$-4x + 5y = -23$$

If the two equations in a system do not contain a pair of opposite variable terms, we can still use the addition method. We must first multiply both sides of one or both equations by a constant(s) in such a way that two of the variable terms become opposites. Then we proceed as in the first example.

EXAMPLE ▸ 2   Solve the following system by addition:

$$2x - 3y = -16$$
$$-6x + 5y = 32$$

Solution

Adding these two equations at this point would not be helpful, as the sum would still contain both variables. If we multiply the first equation by 3, the terms containing x will be opposites. This allows us to proceed with the addition method.

$$
\begin{array}{lcl}
2x - 3y = -16 & \xrightarrow{\text{Multiply by 3}} & 6x - 9y = -48 \\
-6x + 5y = 32 & & -6x + 5y = 32
\end{array}
$$

$$
\begin{array}{ll}
6x - \;\; 9y = -48 & \\
\underline{-6x + \;\; 5y = \;\; 32} & \text{Add.} \\
-4y = -16 & \\
y = 4 & \text{Divide both sides by } -4.
\end{array}
$$

A Word of Caution   Be sure to multiply *both* sides of the equation by the same number. Do not just multiply the side containing the variable terms.

We now substitute 4 for y in the equation $2x - 3y = -16$ and solve for x.

$$2x - 3(4) = -16 \qquad \text{Substitute 4 for } y.$$
$$2x - 12 = -16 \qquad \text{Multiply.}$$
$$2x = -4 \qquad \text{Add 12.}$$
$$x = -2 \qquad \text{Divide both sides by 2.}$$

The solution of this system is $(-2, 4)$. (Be sure to write the x-coordinate first in the ordered pair, even though we solved for y first.)

Quick Check 2
Solve the following system by addition:

$$2x - 9y = 33$$
$$4x + 3y = 3$$

Before continuing, let's outline the basic strategy for using the addition method to solve systems of linear equations.

Using the Addition Method

1. **Write each equation in standard form $(ax + by = c)$.** For each equation, gather all variable terms on the left side and all constants on the right side.
2. **Multiply one or both equations by the appropriate constant(s).** The goal of this step is to make either the terms containing x or the terms containing y opposites.
3. **Add the two equations together.** At this point, we should have one equation containing a single variable.
4. **Solve the resulting equation.** This will give us one of the coordinates of the solution.
5. **Substitute this value for the appropriate variable in either of the original equations, and solve for the other variable.** Choose the equation that you feel will be easier to solve. When we solve this equation, we will find the other coordinate of the solution.
6. **Write the solution as an ordered pair.**
7. **Check your solution.**

EXAMPLE 3   Solve the following system by addition:

$$5x + 8y + 11 = 0$$
$$7x = 6y + 19$$

Solution

To use the addition method, we must first write each equation in general form:

$$5x + 8y = -11$$
$$7x - 6y = 19$$

Now we must decide which variable to eliminate. A wise choice for this system is to eliminate y, as the two coefficients already have opposite signs. If we multiply the first equation by 3 and the second equation by 4, our terms containing y will be $24y$ and $-24y$.

$$5x + 8y = -11 \xrightarrow{\text{Multiply by 3}} 15x + 24y = -33$$
$$7x - 6y = 19 \xrightarrow{\text{Multiply by 4}} 28x - 24y = 76$$

$$
\begin{array}{rl}
15x + 24y = -33 & \\
\underline{28x - 24y = 76} & \quad \text{Add to eliminate } y. \\
43x = 43 & \\
x = 1 & \quad \text{Divide both sides by 43.}
\end{array}
$$

The x-coordinate of our solution is 1. Substituting 1 for x in the equation $5x + 8y = -11$ will allow us to find the y-coordinate.

$$
\begin{array}{rl}
5(1) + 8y = -11 & \quad \text{Substitute 1 for } x. \\
5 + 8y = -11 & \quad \text{Multiply.} \\
8y = -16 & \quad \text{Subtract 5.} \\
y = -2 & \quad \text{Divide both sides by 8.}
\end{array}
$$

Quick Check **3**
Solve the following system by addition.

$$3x + 2y = -26$$
$$5y = 2x + 11$$

The solution to this system is $(1, -2)$. The check is left to the reader.

Objective **2** **Use the addition method to identify inconsistent and dependent systems of equations.** Recall from the previous section that when solving an inconsistent system (no solution), we obtain an equation that is a contradiction such as $0 = -6$. When solving a dependent system (infinitely many solutions), we will end up with an equation that is an identity such as $7 = 7$. The same is true when applying the addition method.

EXAMPLE **4**   Solve the following system by addition:

$$3x + 5y = 15$$
$$6x + 10y = 30$$

Solution

Suppose we chose to eliminate the variable x. We can do so by multiplying the first equation by -2. This produces the following system:

$$3x + 5y = 15 \xrightarrow{\text{Multiply by } -2} -6x - 10y = -30$$
$$6x + 10y = 30 \phantom{\xrightarrow{\text{Multiply by } -2}} 6x + 10y = 30$$

$$
\begin{array}{rl}
-6x - 10y = -30 & \\
\underline{6x + 10y = 30} & \quad \text{Add to eliminate } x. \\
0 = 0 &
\end{array}
$$

Because the resulting equation is an identity, this is a dependent system with infinitely many solutions. To determine the form of our solutions, we can solve the equation $3x + 5y = 15$ for y.

$$
\begin{array}{rl}
3x + 5y = 15 & \\
5y = -3x + 15 & \quad \text{Subtract } 3x \text{ to isolate } 5y. \\
y = -\dfrac{3}{5}x + 3 & \quad \text{Divide both sides by 5 and simplify.}
\end{array}
$$

The ordered pairs that are solutions have the form $\left(x, -\frac{3}{5}x + 3\right)$. We could also display the solutions by graphing the line $3x + 5y = 15$, or $y = -\frac{3}{5}x + 3$.

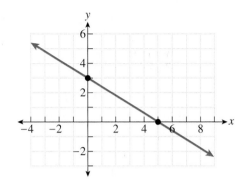

Quick Check 4
Solve the following system by addition:

$2x - 3y = 12$
$-4x + 6y = -24$

EXAMPLE 5   Solve the following system by addition:

$$6x - 3y = 7$$
$$-2x + y = -3$$

Solution

To eliminate the variable y, we can multiply the second equation by 3. The variable term containing y in the second equation would then be $3y$, which is the opposite of $-3y$ in the first equation.

$$
\begin{array}{ll}
6x - 3y = 7 & \\
-2x + \ y = -3 & \xrightarrow{\text{Multiply by 3}}
\end{array}
\qquad
\begin{array}{l}
6x - 3y = 7 \\
-6x + 3y = -9
\end{array}
$$

$$
\begin{array}{l}
6x - 3y = \ \ \ 7 \\
\underline{-6x + 3y = -9} \qquad \text{Add to eliminate } y. \\
 0 = -2
\end{array}
$$

Quick Check 5
Solve the following system by addition:

$3x = 4y + 11$
$9x - 12y = 30$

Since the resulting equation is false, this is an inconsistent system with no solution.

Systems of Equations with Coefficients That Are Fractions

Objective 3   **Solve systems of equations with coefficients that are fractions by using the addition method.** If one or both equations in a system of linear equations contain fractions, we can clear the fractions first and then solve the resulting system.

EXAMPLE 6   Solve the following system by addition:

$$\frac{5}{4}x + \frac{1}{3}y = 3$$
$$\frac{3}{8}x - \frac{5}{6}y = \frac{13}{2}$$

Solution

The LCM of the denominators in the first equation is 12, so if we multiply both sides of the first equation by 12, we can clear the equation of fractions.

$$12\left(\frac{5}{4}x + \frac{1}{3}y\right) = 12 \cdot 3 \qquad \text{Multiply both sides by 12.}$$

$$\overset{3}{\cancel{12}} \cdot \frac{5}{\underset{1}{\cancel{4}}}x + \overset{4}{\cancel{12}} \cdot \frac{1}{\underset{1}{\cancel{3}}}y = 12 \cdot 3 \qquad \text{Distribute and divide out common factors.}$$

$$15x + 4y = 36 \qquad \text{Simplify.}$$

Multiplying both sides of the second equation by 24 will clear the fractions from that equation.

$$24\left(\frac{3}{8}x - \frac{5}{6}y\right) = 24 \cdot \frac{13}{2} \qquad \text{Multiply both sides by 24.}$$

$$\overset{3}{\cancel{24}} \cdot \frac{3}{\underset{1}{\cancel{8}}}x - \overset{4}{\cancel{24}} \cdot \frac{5}{\underset{1}{\cancel{6}}}y = \overset{12}{\cancel{24}} \cdot \frac{13}{\underset{1}{\cancel{2}}} \qquad \text{Distribute and divide out common factors.}$$

$$9x - 20y = 156 \qquad \text{Simplify.}$$

Here is the resulting system of equations:

$$15x + 4y = 36$$
$$9x - 20y = 156$$

Since the coefficient of the term containing y in the second equation is a multiple of the coefficient containing y in the first equation, we will eliminate y. This can be done by multiplying the first equation by 5.

$$15x + 4y = 36 \xrightarrow{\text{Multiply by 5}} 75x + 20y = 180$$
$$9x - 20y = 156 \phantom{\xrightarrow{\text{Multiply by 5}}} 9x - 20y = 156$$

$$75x + 20y = 180$$
$$\underline{9x - 20y = 156} \qquad \text{Add to eliminate } y.$$
$$84x = 336$$

$$x = 4 \qquad \text{Divide both sides by 84.}$$

The x-coordinate of our solution is 4. To solve for y, we can substitute 4 for x in either of the original equations, or we can substitute into either of the equations that were created by clearing the fractions. We will use the equation $15x + 4y = 36$.

$$15(4) + 4y = 36 \qquad \text{Substitute 4 for } x.$$
$$60 + 4y = 36 \qquad \text{Multiply.}$$
$$4y = -24 \qquad \text{Subtract 60.}$$
$$y = -6 \qquad \text{Divide both sides by 4.}$$

The solution of this system is the ordered pair $(4, -6)$.

The substitution and addition methods for solving a system of equations will solve any linear system of equations. There are certain times when one method is easier to apply than the other, and knowing which method to choose for a particular system of equations can save time and prevent errors. In general, if one of the equations in the system is already solved for one of the variables, such as $y = 3x$ or $x = -4y + 5$, then using the substitution method is a good choice. The substitution method also works well when one

Quick Check 6
Solve the following system by addition:

$$-\frac{2}{3}x + \frac{5}{2}y = 14$$
$$\frac{5}{12}x + \frac{3}{4}y = \frac{1}{2}$$

of the equations contains a variable term with a coefficient of 1 or -1 (such as $x + 6y = 17$ or $7x - y = 14$). Otherwise, consider using the addition method.

Applications

Objective 4 **Solve applied problems by using the addition method.** We conclude this section with an application of solving systems of equations using the addition method:

EXAMPLE 7   Dylan has \$3.20 in change in his pocket. Each coin is either a dime or a quarter. If Dylan has 17 coins in his pocket, how many are dimes and how many are quarters?

Solution

There are two unknowns in this problem: the number of dimes and the number of quarters. We will start with a table of unknowns.

> ### *Unknowns*
> Number of dimes: d
> Number of quarters: q

The first equation in our system comes from the fact that Dylan has 17 coins. Since all of these coins are either dimes or quarters, the number of dimes (d) plus the number of quarters (q) must equal 17. The first equation is $d + q = 17$.

The second equation comes from the amount of money Dylan has in his pocket. The amount of money in dimes (\$0.10d) plus the amount of money in quarters (\$0.25q) must equal the total amount of money he has. The second equation is $0.10d + 0.25q = 3.20$.

	Number of Coins	**Value per Coin ($)**	**Money in Pocket ($)**
Dimes	d	0.10	0.10d
Quarters	q	0.25	0.25q
Total	17		3.20

Here is the system of equations that we will need to solve:

$$d + \quad q = 17$$
$$0.10d + 0.25q = 3.20$$

Multiplying both sides of the second equation by 100 will clear the equation of decimals, producing the following system:

$$d + \quad q = 17$$
$$10d + 25q = 320$$

We can now eliminate d by multiplying the first equation by -10.

$$d + q = 17 \quad \xrightarrow{\text{Multiply by } -10} \quad -10d - 10q = -170$$
$$10d + 25q = 320 \qquad\qquad\qquad 10d + 25q = 320$$

$$\begin{array}{r} -10d - 10q = -170 \\ \underline{10d + 25q = 320} \\ 15q = 150 \\ q = 10 \end{array}$$

Add to eliminate d.

Divide both sides by 15.

Quick Check 7

Erica has $5.35 in change in her pocket. Each coin is either a nickel or a dime. If Erica has 75 coins in her pocket, how many are nickels and how many are dimes?

There are 10 quarters in Dylan's pocket. To find the number of dimes that he has, we can substitute 10 for q in the equation $d + q = 17$ and solve for d.

$$d + (10) = 17 \qquad \text{Substitute 10 for } q.$$
$$d = 7 \qquad \text{Subtract 10.}$$

Dylan has 7 dimes and 10 quarters in his pocket. You can verify that the total amount of money in his pocket is $3.20.

EXERCISES 4.3

Solve the system by addition. If the system is inconsistent and has no solution, state this. If the system is dependent, write the form of the solution for any real number x.

1. $4x + 3y = -17$
$2x - 3y = 5$

2. $7x + 5y = 2$
$-7x + 4y = -11$

3. $-2x + y = -7$
$2x + 9y = 17$

4. $8x - 6y = -22$
$5x + 6y = -43$

5. $3x + 2y = 20$
$7x - 4y = 38$

6. $6x - 9y = -18$
$-3x + 8y = 23$

7. $2x - 4y = 11$
$-4x + 8y = -22$

8. $8x - 3y = 34$
$-2x + 9y = 8$

9. $x + 4y = 11$
$3x + 16y = 37$

10. $10x + 2y = 44$
$3x + 8y = 65$

11. $12x + 9y = -69$
$3x + 5y = -31$

12. $2x + 3y = 13$
$10x + 15y = 26$

13. $8x + 7y = -5$
$-x + 4y = -14$

14. $7x + 11y = -31$
$4x - y = 26$

15. $3x + 4y = 26$
$5x - 3y = 24$

16. $7x + 2y = -70$
$6x - 5y = -13$

17. $-8x + 5y = -28$
$3x + 8y = -29$

18. $5x - 5y = 45$
$-4x + 4y = -36$

19. $4x - 3y = -6$
$6x + 5y = 29$

20. $-7x + 4y = -25$
$5x + 14y = 60$

21. $-6x + 9y = 17$
$-4x + 6y = 12$

22. $3x - 11y = 43$
$2x - 7y = 27$

23. $5x + 7y = 16$
$4x + 8y = 14$

24. $4x + 6y = -20$
$6x + 13y = -48$

25. $4y = 3x - 13$
$6x + 5y = 52$

26. $7x = -18 - 4y$
$3x - 2y = -30$

27. $5x = 3y + 21$
$-5x + 4y = -23$

28. $5y = 2x + 13$
$4x + 3y = 65$

29. $\frac{1}{2}x + \frac{2}{5}y = \frac{7}{5}$
$\frac{1}{4}x - \frac{1}{6}y = \frac{1}{3}$

30. $\frac{1}{4}x + \frac{1}{2}y = \frac{11}{4}$
$-\frac{1}{9}x + \frac{1}{3}y = \frac{4}{9}$

31. $\frac{2}{3}x + \frac{1}{6}y = 4$
$\frac{5}{3}x - y = -7$

32. $\frac{1}{7}x + \frac{1}{28}y = \frac{1}{2}$
$\frac{1}{3}x - \frac{2}{9}y = -\frac{23}{18}$

33. A man bought roses and carnations for his wife on Valentine's Day. Each rose cost $6, and each carnation cost $4. If the man spent $68 for a total of 15 roses and carnations, how many of each type of flower did he buy?

34. During matinee times, a movie theater charges $7 per adult and $4 per child. If 84 people attended a matinee and paid a total of $432, how many adults and how many children were there?

35. To landscape their backyard, a family planted various trees that were in 5-gallon and 15-gallon containers. They paid $18 for each 5-gallon tree and $40 for each 15-gallon tree. If the family paid $514 for a total of 20 trees, how many trees came in 5-gallon containers and how many came in 15-gallon containers?

36. The Visalia Oaks minor league baseball team provides two seating options for their home games. Fans can purchase reserved box seats ($9) or general admission seats ($5). Last night, the team took in $12,955 with a total attendance of 2475 fans. How many of these fans paid for reserved box seats and how many paid for general admission seats?

37. Alycia opens her piggy bank to find only dimes and nickels inside. She counts the coins and finds that she has 81 coins. Her mom counts her money and tells her that she has $6.50. How many dimes were in the piggy bank, and how many nickels were in there?

38. Charlotte has 24 postage stamps. Some of the stamps are 15¢ stamps, and the rest are 37¢ stamps. If the total value of the stamps is $6.90, how many of each type of stamp does Charlotte have?

39. Jessica has a handful of nickels and quarters, worth a total of $6.95. If she has 7 more nickels than quarters, how many of each type of coin does she have?

40. The cash box at a bake sale contains nickels, dimes, and quarters. The number of quarters is 4 more than 3 times the number of dimes. The number of nickels is the same as the number of dimes. If the cash box contains $16.30 in coins, how many of each type of coin are in the box?

41. Suppose that your friend missed class on the day that your instructor covered the addition method. Using your own words, explain to your friend how to solve a system of linear equations using the addition method. Be sure to point out the differences between the addition method and the substitution method. Finally, explain to your friend how to determine which method to use for a particular system of equations.

Study Tip Revisited   Eventually, there will be a homework exercise that you cannot answer correctly. Rather than giving up, here are some options to consider:

- Review your class notes. There may be a similar problem that was discussed in class. If so, you can use the solution of this problem to help you figure out the homework exercise that you were unable to do.
- Review the related section in the text. You may be able to find a similar example, or there may be a Word of Caution warning you about typical errors on that type of problem.
- Call someone in your study group. A member of your study group may have already completed the problem and can help you figure out what to do.

If you still cannot solve the problem, move on and try the next problem. Be sure to ask your instructor the very next day about this problem.

4.4

APPLICATIONS OF SYSTEMS OF EQUATIONS

Objectives

1 Solve applied problems involving systems of equations by using the substitution method or the addition method.

2 Solve geometry problems by using a system of equations.

3 Solve interest problems by using a system of equations.

4 Solve mixture problems by using a system of equations.

5 Solve motion problems by using a system of equations.

Objective 1 Solve applied problems involving systems of equations by using the substitution method or the addition method. This section focuses on applications involving systems of linear equations with two variables. We will use both the substitution and addition methods to solve these systems. When trying to solve a system, use the method you feel will be more efficient.

We begin to solve an applied problem by identifying the two unknown quantities, and choosing a variable to represent each quantity. We then need to find two equations that relate these quantities.

EXAMPLE 1 Ross and Carolyn combined their compact disc collection when they got married. Ross had 42 more CDs than Carolyn had. Together, they have 148 CDs. How many CDs did each person have?

Solution

The two unknown quantities in this problem were the number of CDs that Ross had and the number of CDs that Carolyn had. We begin with a table of the unknown quantities.

> ***Unknowns***
> Number of Ross's CDs: r
> Number of Carolyn's CDs: c

We choose the variables r and c because it will remind us that they stand for **R**oss's CDs (r) and **C**arolyn's CDs (c). One sentence in the problem tells us that Ross had 42 more CDs than Carolyn had. This translates to the equation $r = c + 42$. We need a second equation, and this can be found from the sentence that tells us that together they have 148 CDs. This translates to the equation $r + c = 148$. Here is the system of equations that we need to solve.

$$r = c + 42$$
$$r + c = 148$$

We will solve this system of equations using the substitution method, since the first equation has already been solved for r. We will substitute the expression $c + 42$ for the variable r in the second equation.

$$r + c = 148$$
$$(c + 42) + c = 148 \qquad \text{Substitute } c + 42 \text{ for } r.$$
$$2c + 42 = 148 \qquad \text{Combine like terms.}$$
$$2c = 106 \qquad \text{Subtract 42.}$$
$$c = 53 \qquad \text{Divide both sides by 2.}$$

Carolyn had 53 CDs. To find out how many CDs Ross had, we can substitute 53 for c in the equation $r = c + 42$.

$$r = (53) + 42 \qquad \text{Substitute 53 for } c.$$
$$r = 95 \qquad\qquad \text{Add.}$$

Quick Check 1

At a certain community college, there are 68 instructors who teach math or English. The number of English instructors is 12 more than the number of math instructors. How many English instructors are there?

Ross had 95 CDs and Carolyn had 53 CDs. Combined, this is a total of 148 CDs.

The setup for the previous example can be used for any problem in which we know the sum and difference of two quantities. For example, the same system of equations would be used to find two numbers whose difference is 42 and whose sum is 148.

Geometry Problems

Objective 2 Solve geometry problems by using a system of equations. In Section 2.3, we solved problems involving the perimeter of a rectangle. We will now use a system of two equations in two variables to solve the same problems.

EXAMPLE 2   The length of a rectangle is 5 feet more than twice its width. If the perimeter of the rectangle is 124 feet, find the length and the width of the rectangle.

Solution

The two unknowns are the length and the width of the rectangle. We will use the variables l and w for the length and width, respectively.

Unknowns

Length: l
Width: w

We are told that the length is 5 feet more than twice the width of the rectangle. We express this relationship in an equation as $l = 2w + 5$. The second equation in the system comes from the fact that the perimeter is 124 feet. Since the perimeter of a rectangle is equal to twice the length plus twice the width, our second equation can be written as $2l + 2w = 124$. Here is the system we are solving:

$$l = 2w + 5$$
$$2l + 2w = 124$$

We will use the substitution method to solve this system, as the first equation is already solved for l. We begin by substituting the expression $2w + 5$ for l in the equation $2l + 2w = 124$.

$$2(2w + 5) + 2w = 124 \qquad \text{Substitute } 2w + 5 \text{ for } l.$$
$$4w + 10 + 2w = 124 \qquad \text{Distribute.}$$
$$6w + 10 = 124 \qquad \text{Combine like terms.}$$
$$6w = 114 \qquad \text{Subtract 10.}$$
$$w = 19 \qquad \text{Divide both sides by 6.}$$

The width of the rectangle is 19 feet. To solve for the length, we substitute 19 for w in the equation $l = 2w + 5$.

$$l = 2(19) + 5 \qquad \text{Substitute 19 for } w.$$
$$l = 43 \qquad \text{Simplify.}$$

Quick Check **2**   The length of the rectangle is 43 feet, and the width is 19 feet.

A rectangular concrete slab is being poured for a new house. The perimeter of the slab is 230 feet and the length of the slab is 35 feet longer than the width. Find the dimensions of the concrete slab.

Interest Problems

Objective **3**   **Solve interest problems by using a system of equations.** For certain problems, using a system of equations in two variables is easier to set up than trying to express both unknown quantities in terms of the same variable. For example, the mixture and interest problems covered in Section 2.4 are often easier to solve when working with a system of equations in two variables. We will begin with a problem involving interest.

EXAMPLE ▸**3**   Serena invested $8000 in two certificates of deposit (CDs). Some of the money was deposited in a CD that paid 7% annual interest, and the rest was deposited in a CD that paid 6% annual interest. If Serena earned $530 in interest in the first year, how much did she invest in each CD?

Solution

The two unknowns are the amount invested at 7% and the amount invested at 6%. We will let x represent the amount invested at 7% interest, and y represent the amount invested at 6% interest. To determine the interest earned, we multiply the principal by the interest rate by the time in years. If the first account earns 7% interest, then we can represent the interest earned in one year by $0.07x$. In a similar fashion, we can represent the interest earned in one year from the investment at 6% interest as $0.06y$. The following table summarizes this information:

Account	Principal	Interest Rate	Time (years)	Interest Earned
CD 1	x	0.07	1	$0.07x$
CD 2	y	0.06	1	$0.06y$
Total	8000			530

The first equation in our system comes from the fact that the amount invested at 7% interest plus the amount invested at 6% is equal to the total amount invested. As an equation, this can be represented as $x + y = 8000$. This equation can be found in our table in the column labeled Principal. The second equation comes from the amount of interest earned and can be found in the column labeled Interest Earned. We know that Serena earned $0.07x$ in interest from the first CD and she earned $0.06y$ in interest from the second CD. Since the total amount of interest earned comes from these two CDs, the second equation in our system is $0.07x + 0.06y = 530$. Here is the system we must solve:

$$x + y = 8000$$
$$0.07x + 0.06y = 530$$

We will clear the second equation of decimals by multiplying both sides of the equation by 100.

$$
\begin{aligned}
x + \quad y &= 8000 \\
0.07x + 0.06y &= 530
\end{aligned}
\quad\xrightarrow{\text{Multiply by 100}}\quad
\begin{aligned}
x + \quad y &= 8000 \\
7x + 6y &= 53000
\end{aligned}
$$

To solve this system of equations, we can use the addition method. (The substitution method would work just as well.) If we multiply the first equation by -6, then the variable terms containing y are opposites.

$$
\begin{aligned}
x + \quad y &= 8{,}000 \\
7x + 6y &= 53{,}000
\end{aligned}
\quad\xrightarrow{\text{Multiply by }-6}\quad
\begin{aligned}
-6x - 6y &= -48{,}000 \\
7x + 6y &= 53{,}000
\end{aligned}
$$

$$
\begin{array}{r}
-6x - 6y = -48{,}000 \\
7x + 6y = 53{,}000 \\
\hline
x = 5{,}000
\end{array}
\qquad \text{Add to eliminate } y.
$$

Quick Check 3
Annika invested $4200 in two certificates of deposit (CDs). Some of the money was deposited in a CD that paid 5% annual interest, and the rest was deposited in a CD that paid 4% annual interest. If Annika earned $200 in interest in the first year, how much did she invest in each CD?

We know that Serena invested $5000 at 7% interest. To find the amount invested at 6%, we substitute 5000 for x in the original equation $x + y = 8000$.

$$
\begin{aligned}
(5000) + y &= 8000 \qquad \text{Substitute 5000 for } x. \\
y &= 3000 \qquad \text{Subtract 5000.}
\end{aligned}
$$

Serena invested $5000 at 7% interest and $3000 at 6% interest.

Mixture Problems

Objective 4 Solve mixture problems by using a system of equations.

EXAMPLE 4 Gunther works at a coffee shop that sells Kona coffee beans for $40 per pound and Colombian coffee beans for $15 per pound. The coffee shop also sells a Kona blend, which is made up of Kona coffee and Colombian coffee, for $20 per pound. If Gunther's boss tells him to make 30 pounds of this Kona blend, how much Kona coffee and Colombian coffee must be mixed together?

Solution

The two unknowns are the number of pounds of Kona coffee (k) and the number of pounds of Colombian coffee (c) that need to be mixed together to form the Kona blend. Since Gunther needs to make 30 pounds of coffee with a value of $20 per pound, the total value of this mixture can be found by multiplying 30 by 20. The total cost of this Kona blend is $600. The following table summarizes this information:

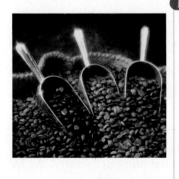

Coffee	Pounds	Cost per Pound	Total Cost
Kona	k	40	$40k$
Colombian	c	15	$15c$
Blend	30	20	600

The first equation in our system comes from the fact that the weight of the Kona coffee plus the weight of the Colombian coffee must equal 30 pounds. As an equation, this can be represented by $k + c = 30$. This equation can be found in our table in the column labeled Pounds. The second equation comes from the total cost of the coffee. We know that the total cost is $600, so the second equation in our system is $40k + 15c = 600$. This equation can be found in the column labeled Total Cost. Here is the system we must solve:

$$k + c = 30$$
$$40k + 15c = 600$$

Quick Check **4**

Hannibal is making 48 pounds of a mixture of almonds and cashews that his nut shop will sell for $6 per pound. If the nut shop sells almonds $5 per pound and cashews for $8 per pound, how many pounds of almonds and cashews should Hannibal mix together?

To solve this system of equations, we can use the addition method. If we multiply the first equation by -15, then the variable terms containing c are opposites.

$$\begin{array}{c} k + c = 30 \\ 40k + 15c = 600 \end{array} \xrightarrow{\text{Multiply by } -15} \begin{array}{c} -15k - 15c = -450 \\ 40k + 15c = 600 \end{array}$$

$$\begin{array}{r} -15k - 15c = -450 \\ \underline{40k + 15c = 600} \\ 25k = 150 \\ k = 6 \end{array}$$

Add to eliminate c.

Divide both sides by 25.

We know that Gunther must use 6 pounds of Kona coffee. To determine how much Colombian coffee Gunther must use, we substitute 6 for k in the equation $k + c = 30$.

$$(6) + c = 30 \qquad \text{Substitute 6 for } k.$$
$$c = 24 \qquad \text{Subtract 6.}$$

Gunther must use 6 pounds of Kona coffee and 24 pounds of Colombian coffee.

The previous example is a mixture problem. In this mixture problem, two items of different values or costs are combined into a mixture with a specific cost. In other mixture problems, two different solutions of different concentration levels are mixed together to form a single solution with a given strength, as in the next example.

EXAMPLE 5   Woody has one solution that is 28% alcohol and a second solution that is 46% alcohol. He wants to combine these solutions to make 45 liters of a solution that is 40% alcohol. How many liters of each original solution need to be used?

Solution

The two unknowns are the volume of the 28% alcohol solution and the volume of the 46% alcohol solution that need to be mixed together in order to make 45 liters of a solution that is 40% alcohol.

Solution	Volume of Solution	% Concentration of Alcohol	Volume of Alcohol
28%	x	0.28	$0.28x$
46%	y	0.46	$0.46y$
Mixture (40%)	45	0.40	18

The first equation in our system comes from the fact that the volume of the two solutions must equal 45 liters. As an equation, this can be represented by $x + y = 45$. This equation can be found in our table in the column labeled Volume of Solution. The second equation comes from the total volume of alcohol, under the column labeled Volume of Alcohol. We know that the total volume of alcohol is 18 liters (40% of the 45 liters in the mixture is alcohol), so the second equation in our system is $0.28x + 0.46y = 18$. Here is the system we must solve:

$$x + y = 45$$
$$0.28x + 0.46y = 18$$

We begin by clearing the second equation of decimals. This can be done by multiplying both sides of the second equation by 100.

$$\begin{array}{l} x + y = 45 \\ 0.28x + 0.46y = 18 \end{array} \xrightarrow{\text{Multiply by 100}} \begin{array}{l} x + y = 45 \\ 28x + 46y = 1800 \end{array}$$

To solve this system of equations, we can use the addition method. If we multiply the first equation by -28, then the variable terms containing x are opposites.

$$\begin{array}{l} x + y = 45 \\ 28x + 46y = 1800 \end{array} \xrightarrow{\text{Multiply by } -28} \begin{array}{l} -28x - 28y = -1260 \\ 28x + 46y = 1800 \end{array}$$

$$\begin{array}{r} -28x - 28y = -1260 \\ 28x + 46y = 1800 \\ \hline 18y = 540 \end{array} \qquad \text{Add to eliminate } x.$$

$$y = 30 \qquad \text{Divide both sides by 18.}$$

We know that Woody must use 30 liters of the solution that is 46% alcohol. To determine how much of the 28% alcohol solution that Woody must use, we substitute 30 for y in the equation $x + y = 45$.

$$x + (30) = 45 \qquad \text{Substitute 30 for } y.$$
$$x = 15 \qquad \text{Subtract 30.}$$

Woody must use 15 liters of the 28% alcohol solution and 30 liters of the 46% alcohol solution.

Motion Problems

Objective 5 **Solve motion problems by using a system of equations.** We will now turn our attention to motion problems, which were introduced in Section 2.3. Recall that the equation for motion problems is $d = r \cdot t$, where r is the rate of speed, t is the time and *d* is the distance traveled. For example, if a car is traveling at a speed of 50 miles per hour for 3 hours, then the distance traveled is 50 · 3 or 150 miles.

EXAMPLE 6   Efrain is finishing up his training for the Shaver Lake Triathlon. Yesterday he focused on running and biking, spending a total of 5 hours on the two activities. Efrain runs at a speed of 10 miles per hour and rides his bike at a speed of 30 miles per hour. If Efrain covered a total of 120 miles yesterday, how much time did he spend running and how much time did he spend riding?

Solution

The two unknowns in this problem are the amount of time Efrain ran and the amount of time he rode his bike. We will let *x* represent the time that Efrain ran and *y* represent the amount of time that he rode his bike. All of our information is displayed in the following table:

	Rate	Time	Distance ($d = r \cdot t$)
Running	10	*x*	10*x*
Biking	30	*y*	30*y*
Total		5	120

We know that Efrain spent a total of 5 hours on these two activities, so $x + y = 5$. We also know that the total distance traveled was 120 miles. This leads to the equation $10x + 30y = 120$. Here is the system of equations we must solve:

$$x + y = 5$$
$$10x + 30y = 120$$

We can solve the first equation for *x* ($x = 5 - y$), so we will use the substitution method to solve this system of equations. We substitute the expression $5 - y$ for the variable *x* in the equation $10x + 30y = 120$.

$10(5 - y) + 30y = 120$	Substitute $5 - y$ for *x*.
$50 - 10y + 30y = 120$	Distribute.
$50 + 20y = 120$	Combine like terms.
$20y = 70$	Subtract 50.
$y = \dfrac{7}{2}$	Divide both sides by 20 and simplify.

Writing the fraction $\frac{7}{2}$ as a mixed number, we find that Efrain spent $3\frac{1}{2}$ hours riding his bike. To find out how long he was running, we substitute $3\frac{1}{2}$ for *y* in the equation $x = 5 - y$ and solve for *x*.

$x = 5 - 3\dfrac{1}{2}$	Substitute $3\dfrac{1}{2}$ for *y*.
$x = 1\dfrac{1}{2}$	Subtract.

Efrain spent $1\frac{1}{2}$ hours running and $3\frac{1}{2}$ hours riding his bike.

***Quick Check* 6**
On a trip to Las Vegas, Dale drove at an average speed of 70 miles per hour, except for a period when he was driving through a construction zone. In the construction zone, Dale's average speed was 40 miles per hour. If it took Dale five hours to drive 335 miles, how long was he in the construction zone?

Suppose that a boat travels at a speed of 15 miles per hour in still water. The boat will travel faster than 15 miles per hour while heading downstream, because the speed of the current is pushing the boat forward as well. The boat will travel slower than 15 miles per hour while heading upstream, because the speed of the current is pushing the boat back. If the speed of the current is 3 miles per hour, this boat would travel 18 miles per hour downstream $(15 + 3)$ and 12 miles per hour upstream $(15 - 3)$. The speed of an airplane is affected in a similar fashion, depending on whether it is flying with or against the wind.

EXAMPLE 7   Vineta's motorboat can take her from her campsite downstream to the nearest store, which is 27 miles away, in 1 hour. Returning upstream from the store to the campsite takes 3 hours. How fast is Vineta's boat in still water, and what is the speed of the current?

Solution

The two unknowns in this problem are the speed of the boat in still water and the speed of the current. We will let b represent the speed of the boat in still water, and we will let c represent the speed of the current. The information for this problem can be summarized in the following table:

	Rate	Time	Distance ($d = r \cdot t$)
Downstream	$b + c$	1	$b + c$
Upstream	$b - c$	3	$3(b - c)$

Since the distance in each direction is 27 miles, we know that both $b + c$ and $3(b - c)$ are equal to 27. Here is the system we need to solve:

$$b + c = 27$$
$$3b - 3c = 27$$

We will solve this system of equations by using the addition method. Multiplying both sides of the first equation by 3 will help us eliminate the variable c.

$$
\begin{array}{ll}
b + c = 27 & \xrightarrow{\text{Multiply by 3}} \quad 3b + 3c = 81 \\
3b - 3c = 27 & \phantom{\xrightarrow{\text{Multiply by 3}}} \quad 3b - 3c = 27
\end{array}
$$

$$
\begin{array}{rl}
3b + 3c = & 81 \\
3b - 3c = & 27 \qquad \text{Add to eliminate } c. \\
\hline
6b = & 108 \\
b = & 18 \qquad \text{Divide both sides by 6.}
\end{array}
$$

The speed of the boat in still water is 18 miles per hour. To find the speed of the current, we substitute 18 for b in the original equation $b + c = 27$.

$$
\begin{array}{ll}
(18) + c = 27 & \text{Substitute 18 for } b. \\
c = 9 & \text{Subtract 18.}
\end{array}
$$

The speed of the current is 9 miles per hour and the speed of the boat is 18 miles per hour.

Quick Check 7

An airplane, traveling with a tailwind, can make a 2250-mile trip in $4\frac{1}{2}$ hours. However, traveling into the same wind would take 5 hours to fly 2250 miles. What is the speed of the plane in calm air, and what is the speed of the wind?

1. A mathematics instructor is teaching an elementary algebra class and a statistics class. She has 17 more students in her statistics class than she has in her elementary algebra class. The two classes have 93 students in total. How many students does she have in each class?

2. A hardcover math book is 289 pages longer than a paperback math book. If the total number of pages in these two books is 1691, how many pages are there in each book?

3. Andre's father is 28 years older than he is. If we added Andre's age to his father's age, the total would be 98. How old is Andre? How old is his father?

4. Maggie's ma tells everyone that she is 14 years younger than she actually is. If we add her actual age to the age that she tells people she is, the total would be 122. What is the actual age of Maggie's ma (as reported by Bob Dylan)?

5. Two baseball players hit 79 home runs combined last season. The first player hit 7 more home runs than twice the number of home runs hit by the second player. How many home runs did each player hit?

6. A piece of wire 240 centimeters long was cut into two pieces. The longer of the two pieces is 25 centimeters longer than four times the length of the shorter piece. How long is each piece?

7. On a history exam, each multiple-choice question is worth 3 points, and each true–false question is worth 2 points. Vanessa answered 28 questions correctly on the exam and earned a score of 77 points. How many multiple-choice questions did she answer correctly, and how many true–false questions did she answer correctly?

8. In a soccer league, teams are awarded 3 points for a win, 1 point for a tie and 0 points for a loss. Last season the Yellowjackets won 1 more game than they tied and finished the season with 31 points. How many games did they win and how many games did they tie?

9. In a recent game Shaquille O'Neal scored 37 points, making only free throws (1 point each) and field goals (2 points each). The number of field goals he made was one more than three times the number of free throws he made. How many field goals did

he make, and how many free throws did he make?

10. On a sociology exam, students earn 4 points for each correct response and lose 1 point for each incorrect response or unanswered question. The exam contained 25 questions, and Ron earned a score of 85 points. How many questions did Ron answer correctly?

11. Megan has 37 coins in her purse. Some are dimes and the rest are quarters. The value of the coins is $5.95. How many dimes does she have?

12. The cost to attend a college baseball game is $5 except for students, who are admitted for $1.50. Yesterday's game was attended by 310 fans and the college collected $1119.50. How many students attended the game?

13. Last week a softball team ordered three pizzas and two pitchers of soda after a game, and the bill was $41. This week they ordered five pizzas and three pitchers of soda, and the bill was $67. What is the price of a single pizza and what is the price of a pitcher of soda?

14. Two businessmen went shopping together. The first bought three shirts and one tie for a total of $85. The second bought two shirts and five ties for a total of $139. If the price for each shirt was the same and the price for each tie was the same, what is the cost of one shirt and what is the cost of one tie?

Recall from Chapter 2 that two positive angles are said to be complementary if their measures add up to be 90° and supplementary if their measures add up to be 180°.

15. Two angles are complementary. The measure of one angle is 22° more than the measure of the other angle. Find the measures of each angle.

16. Two angles are complementary. The measure of one angle is 5° more than four times the measure of the other angle. Find the measures of each angle.

17. Two angles are supplementary. The measure of one angle is 16° less than three times the measure of the other angle. Find the measures of each angle.

18. Two angles are supplementary. The measure of one angle is 24° more than 25 times the measure of the other angle. Find the measures of each angle.

19. The perimeter of a rectangle is 92 inches. The width of the rectangle is 8 inches less than the length of the rectangle. Find the dimensions of the rectangle.

20. An oil painting is in the shape of a rectangle. The perimeter of the painting is 127.4 inches, and the length of the painting is 1.6 times the width of the painting. Find the dimensions of the painting.

21. The Woodbury vegetable garden is in the shape of a rectangle, surrounded by 330 feet of fence. The length of the garden is 75 feet more than twice the width of the garden. Find the dimensions of this garden.

22. Tina needs 302 inches of fabric to use as binding around a rectangular quilt she is making. The width of the quilt is 10 inches less than the length of the quilt. Find the dimensions of the quilt.

23. A swimming pool is in the shape of a rectangle. The length of the pool is 15 feet more than the width. A concrete deck 6 feet wide is added around the pool, and the perimeter of the deck is 178 feet.

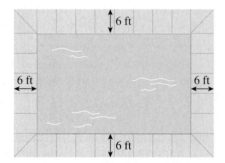

Find the dimensions of the swimming pool.

24. A farmer fences in a rectangular area as a pasture for her horses next to a building, using the existing building as the fourth side of the rectangle, as shown in the following diagram:

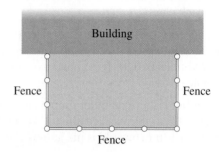

The side of the fence parallel to the building is 35 feet longer than either of the other two sides. If the farmer used 170 feet of fencing, find the dimensions of the pasture.

25. An investor invested a total of $2000 in two mutual funds. One fund earned a 6% profit while the other earned a 3% profit. If the investor's total profit was $84, how much was invested in each mutual fund?

26. Angela deposited a total of $5200 in two different certificates of deposit (CDs). One CD paid 3% annual interest and the other paid 4% annual interest. At the end of one year, she had earned $203 in interest from the two CDs. How much did Angela invest in each CD?

27. Mark deposited a total of $25,000 in two different CDs. One CD paid 4.25% annual interest and the other paid 3.5% annual interest. At the end of one year, he had earned $1025 in interest from the two CDs. How much did Mark invest in each CD?

28. Maria invested $5500 in two mutual funds. One fund earned an 8% profit, while the other earned a 1.5% profit. Between the two accounts, Maria made a profit of $180. How much was invested in each account?

29. Carter invested a total of $7500 in two mutual funds. While one fund earned a 4% profit, the other fund had a loss of 13%. Despite the loss, Carter made a profit of $45 from the two funds. How much was invested in each fund?

30. Colin has had a bad run of luck in the stock market. He invested a total of $40,000 in two mutual funds. The first fund had a loss of 11%, and the second fund had a loss of 30%. Between the two funds Colin lost $10,100. How much did he invest in each fund?

31. A chemist has one solution that is 80% acid and a second solution that is 44% acid. She wants to combine the solutions to make a new solution that is 53% acid. If she needs to make 320 milliliters of this new solution, how many milliliters of the two solutions should be mixed?

32. A metallurgist has one brass alloy that is 65% copper and a second brass alloy that is 41% copper. He needs to melt and combine these alloys in such a way to make 60 grams of an alloy that is 47% copper. How much of each alloy should he use?

33. A bartender has two liquors that are 20% and 50% alcohol respectively. She needs to mix these liquors together in such a way that she has 18 ounces of a mixture that is 30% alcohol. How much of each type of liquor does she need to mix together?

34. A farmer has two types of feed that are 12% protein and 18% protein respectively. How many pounds of each need to be combined to make a mixture of 36 pounds of feed that is 13.5% protein?

35. A bartender mixes a batch of vodka and tonic, which is made of vodka (40% alcohol) and tonic water (no alcohol). How much of each needs to be mixed to make 4 liters of vodka and tonic that is 22% alcohol?

36. An auto mechanic has an antifreeze solution that is 10% antifreeze. He needs to mix this solution with pure antifreeze to produce 5 liters of a solution that is 19% antifreeze. How much of each does he need to mix?

37. George and Tina drove from their house to Palm Springs for a vacation. During the first part of the trip, George drove at a speed of 70 miles per hour. They switched drivers and Tina drove the rest of the way at a speed of 85 miles per hour. It took them 6 hours to reach Palm Springs, which is 480 miles away. Find the length of time that George drove, as well as the length of time that Tina drove.

38. Matt was driving to attend a conference in Salt Lake City, Utah. During the first day of driving, Matt averaged 72 miles per hour. On the second day, he drove at an average speed of 80 miles per hour in order to arrive before sundown. If his total driving time for the two days was 15 hours and the distance traveled was 1148 miles, what was Matt's driving time for each day?

39. A racecar driver averaged 160 miles per hour over the first part of the Sequoia 500, but engine trouble dropped her average speed to 130 miles per hour for the rest of the race. If the 500-mile race took the driver 3.5 hours to complete, after what period did her engine trouble start?

40. A truck driver started for his destination at an average speed of 60 miles per hour. Once snow began to fall, he had to drop his speed to 40 miles per hour. It took the truck driver 9 hours to reach his destination, which was 425 miles away. How many hours after the driver started did the snow begin to fall?

41. A kayak can travel 18 miles downstream in 3 hours, while it would take 9 hours to make the same trip upstream. Find the speed of the kayak in still water, as well as the speed of the current.

42. A rowboat would take 3 hours to travel 6 miles upstream, while it would only take $\frac{3}{4}$ hours to travel 6 miles downstream. Find the speed of the rowboat in still water, as well as the speed of the current.

43. An airplane traveling with a tailwind can make a 3000-mile trip in 5 hours. However, traveling into the same wind would take 6 hours to fly 3000 miles. What is the speed of the plane in calm air, and what is the speed of the wind?

44. A small plane can fly 240 miles with a tailwind behind it in 3 hours. Flying the same distance into the same wind would take 4 hours. Find the speed of the plane with no wind.

45. Write a word problem whose solution is "The length of the rectangle is 15 m and the width is 8 m."

46. Write a word problem whose solution is "$5000 at 6% and $3000 at 5%."

47. Write a word problem whose solution is "80 ml of 40% acid solution and 10 ml of 58% acid solution."

48. Write a word problem whose solution is "Bill drove for 3 hours and Margie drove for 4 hours."

QUICK REVIEW EXERCISES

Section 4.4

Graph. Label any x- and y-intercepts.

1. $x - 4y = 6$

2. $y = 2x - 5$

3. $y = -\dfrac{1}{4}x - 2$

4. $y = 3$

> **Study Tip** REVISITED Once you finish the last exercise of a homework assignment, it is not necessarily time to stop. Try summarizing what you have just accomplished in a homework diary. In this diary, you can list the types of problems you have solved, the types of problems you struggled with, and important notes about these problems. When you are ready to begin preparing for a quiz or exam, this diary will help you to decide where to focus your attention. If you are using note cards as a study aide, this would be a good time to create a set of study cards for this section. You now know which problems are more difficult for you, as well as which procedures you will need to know in the future.
>
> Summarizing your homework efforts in these ways will help you to retain what you learned from the assignment.

SYSTEMS OF LINEAR INEQUALITIES

Objectives

1 Solve a system of linear inequalities by graphing.

2 Solve a system of linear inequalities with more than two inequalities.

Just as we had systems of linear equations, we can also have a **system of linear inequalities.** A system of linear inequalities is made up of two or more linear inequalities. As with systems of equations, an ordered pair is a solution to a system of linear inequalities if it is a solution to each linear inequality in the system.

Solving a System of Linear Inequalities by Graphing

Objective 1 **Solve a system of linear inequalities by graphing.** To find the solution set to a system of linear inequalities, we begin by graphing each inequality separately. If the solutions to the individual inequalities intersect, the region of intersection is the solution set to the system of inequalities.

EXAMPLE 1   Graph the following system of inequalities:

$$y \geq 2x - 5$$
$$y < 3x - 7$$

Solution

We begin by graphing the inequality $y \geq 2x - 5$. To do this, we graph the line $y = 2x - 5$ as a solid line. Recall that we use a solid line when the points on the line are solutions to the inequality. Since this equation is in slope–intercept form, an efficient way to graph it is by plotting the y-intercept at $(0, -5)$ and then using the slope of 2 to find additional points on the line.

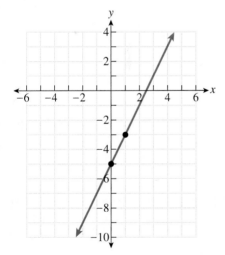

Since the origin is not on this line, we can use $(0, 0)$ as a test point. Substitute 0 for x and 0 for y in the inequality $y \geq 2x - 5$.

$$0 \geq 2(0) - 5$$
$$0 \geq 0 - 5$$
$$0 \geq -5$$

Substituting these coordinates into the inequality $y \geq 2x - 5$ produces a true statement, so we shade the half-plane containing $(0, 0)$ in the graph on the right.

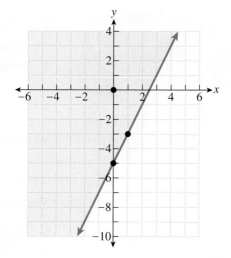

We now turn our attention to the second inequality in the system. To graph the inequality $y < 3x - 7$, we use a dashed line to graph the equation $y = 3x - 7$. Recall that the dashed line is used to signify that no point on the line is a solution to the inequality. Again, this equation is in slope–intercept form, so we graph the line by plotting the y-intercept at $(0, -7)$ and using the slope of 3 to find additional points in the graph on the right.

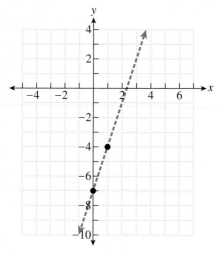

We can use the origin as a test point for this inequality as well. Substituting 0 for x and 0 for y produces a false statement.

$$0 < 3(0) - 7$$
$$0 < 0 - 7$$
$$0 < -7$$

We shade the half-plane on the opposite side of the line from the origin in the graph on the right.

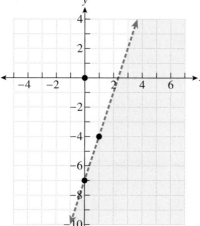

The solution of the system of linear inequalities is the region where the two solutions intersect.

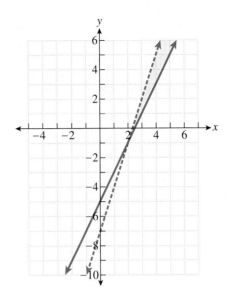

Quick Check 1

Graph the following system of inequalities:

$$y > x + 3$$
$$2x + 3y < 6$$

EXAMPLE 2   Graph the following system of inequalities:

$$y \leq -2x + 4$$
$$y \geq 3$$

Solution

We begin by graphing the line $y = -2x + 4$ using a solid line. This equation is in slope–intercept form, so we begin by plotting the y-intercept $(0, 4)$, and then we use the slope of -2 to find additional points on the line. Now we determine which half-plane to shade. Since the line does not pass through the origin, we can choose $(0, 0)$ as a test point.

$$y \leq -2x + 4$$
$$0 \leq -2(0) + 4$$
$$0 \leq 4$$

Substituting this ordered pair into the inequality produces a true statement. Thus, we shade the half-plane containing the origin in the graph on the right.

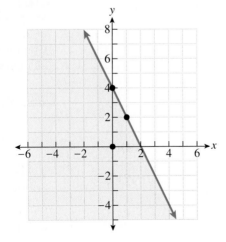

Now we turn our attention to the inequality $y \geq 3$. We begin by graphing the line $y = 3$ using a solid line. This line is a horizontal line with a y-intercept at $(0, 3)$. Again, we may choose the origin as a test point. Substituting the ordered pair $(0, 0)$ produces a false statement. We shade the half-plane above the line $y = 3$ that does not contain the origin. The graph is shown on the right.

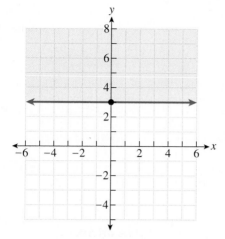

The solution of the system of equations is the region where the two shaded areas intersect in the graph on the right.

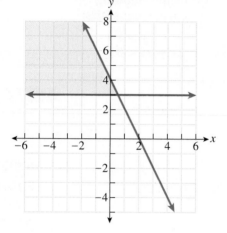

Quick Check 2

Graph the following system of inequalities:

$$-3x + 4y \leq 8$$
$$y \geq 0$$

Solving a System of Linear Inequalities with More than Two Inequalities

Objective 2 Solve a system of linear inequalities with more than two inequalities. The next example involves a system of three linear inequalities. Regardless of the number of inequalities in a system, we still find the solution in the same manner. Graph each inequality individually, and then find the region where the shaded regions all intersect.

EXAMPLE 3   Graph the following system of inequalities:

$$3x + 4y > 12$$
$$x < 6$$
$$y > 4$$

Solution

We begin by graphing the line $3x + 4y = 12$ with a dashed line. This is the line associated with the first inequality. Since the equation is in general form, we can graph it by finding its x- and y-intercepts. The x-intercept for this line can be found by substituting 0 for y in the equation.

3x + 4y = 12	
x-intercept ($y = 0$)	**y-intercept ($x = 0$)**
$3x + 4(0) = 12$	$3(0) + 4y = 12$
$3x = 12$	$4y = 12$
$x = 4$	$y = 3$
$(4, 0)$	$(0, 3)$

The x-intercept is at $(4, 0)$. The y-intercept of the line is at $(0, 3)$. Since the line does not pass through the origin, we can use $(0, 0)$ as a test point. When we substitute 0 for x and 0 for y in the inequality $3x + 4y > 12$, we obtain a false statement.

$$3(0) + 4(0) > 12$$
$$0 > 12$$

Since this statement is not true, we shade the half-plane on the side of the line that does not contain the point $(0, 0)$ in the graph on the right.

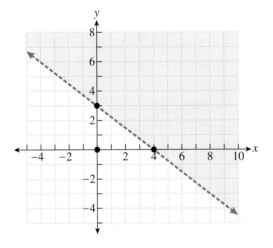

To graph the second inequality, we begin with the graph of the vertical line $x = 6$. This line is also dashed. The solutions of this inequality are to the left of this vertical line, as shown in the graph on the right.

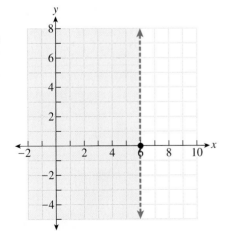

For the third inequality, we graph the dashed horizontal line $y = 4$. The solutions of this inequality are above the horizontal line $y = 4$, as shown in the graph on the right.

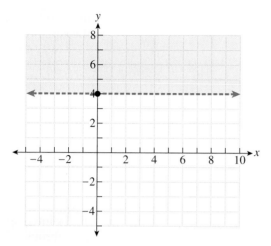

Quick Check 3

Graph the following system of inequalities:

$$x + y \leq 8$$
$$x \geq 1$$
$$y \geq -1$$

We now display the solutions to the system of inequalities by finding the region of intersection for these three inequalities shown in the graph on the right.

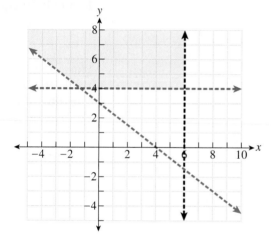

EXERCISES 4.5

Graph the system of inequalities.

1. $y \geq 3x - 4$
   $y \leq -x + 5$

2. $y \geq 2x + 7$
   $y < x + 3$

3. $y \le 5x - 4$
     $y > -2x - 6$

4. $y < 4x + 6$
     $y < 3$

9. $-7x + 3y > 21$
     $4x + 9y \le 36$

10. $8x + y \ge -8$
      $3x + 4y < -12$

5. $y < 5x + 5$
     $y > 5x - 5$

6. $y \ge \dfrac{1}{2}x + 3$
     $y < -4x$

11. $y \ge 5$
      $5x + 2y < 0$

12. $-4x + y < 6$
      $4x - y < 2$

7. $x + 4y < 8$
     $3x - 2y < 6$

8. $5x + 2y \le 10$
     $-3x + y > 5$

13. $y \geq 3x$
$y > 3$

18. $3x + 10y > 15$
$y \leq -\frac{5}{2}x + 7$

19. $y < 2x$
$y < 6$
$x > 2$

14. $x < -4$
$2x + 6y > 12$

20. $y > x - 4$
$y < -2x + 5$
$y > -2$

21. $3x - 2y \leq -10$
$y < -4x - 6$
$-x + 3y < 12$

15. $y > -\frac{3}{2}x + 6$
$4x - 3y \geq -6$

22. $x \geq -5$
$x \leq 4$
$y < \frac{1}{5}x + 2$

16. $x < 4$
$y \geq -7$

17. $x > 1$
$8x - 4y > 12$

23. Write a system of linear inequalities whose solution is the entire first quadrant, as shown in the following graph:

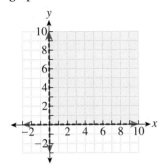

24. Write a system of linear inequalities whose solution is the entire third quadrant.

25. Write a system of linear inequalities that has no solution.

26. Is it possible to write a system of linear inequalities whose solution set is the entire rectangular coordinate plane? If so, write the system; if not, explain why in your own words.

27. Find the area of the region defined by the following system of linear inequalities:

$$x \geq 2$$
$$y \geq 1$$
$$y \leq -x + 9$$

28. Write a system of linear inequalities whose solution set is shown on the following graph:

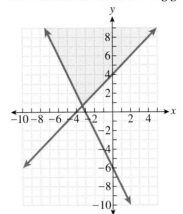

Study Tip REVISITED   Try to establish a regular schedule for doing your homework. Some students find it advantageous to work on their homework as soon as possible after class, while the material is still fresh in their minds. You may be able to find other students in your class that can work with you in the library. If you have to wait until the evening to do your homework, try to complete your assignment before you get tired. If you have more than one subject to work on, begin with math.

It is crucial that you be able to complete an assignment before the next class session. Since math is a sequential topic, if you do not understand the material from the previous day, then you will have difficulty understanding the next day's material.

Chapter 4 Summary

Section 4.1—Topic	Chapter Review Exercises
Solving Systems of Linear Equations by Graphing	1–8

Section 4.2—Topic	Chapter Review Exercises
Solving Systems of Linear Equations by Using the Substitution Method	9–12

Section 4.3—Topic	Chapter Review Exercises
Solving Systems of Linear Equations by Using the Addition Method	13–16
Solving Systems of Linear Equations by Any Method	17–26

Section 4.4—Topic	Chapter Review Exercises
Solving Applications Using Systems of Linear Equations	27–36

Section 4.5—Topic	Chapter Review Exercises
Solving Systems of Linear Inequalities	37–40

Summary of Chapter 4 Study Tips

A homework assignment should not be viewed as a chore to be completed, but rather as an opportunity to increase your understanding of mathematics. Here is a summary of the ideas presented in this chapter:

- Review your notes and the text before beginning a homework assignment.
- Be neat and complete when doing a homework assignment. *If your work is neat, it will be easier to spot errors. When reviewing for an exam, your homework will be easier to review if it is neat. Do not skip any steps, even if you feel that they are not necessary.*
- Strategies for homework problems you cannot answer:
  - *Look in your notes and the text for similar problems or suggestions about this type of problem.*
  - *Call another student in the class for help.*
  - *Get help from a tutor at the tutorial center on your campus.*
  - *Ask your instructor for help at the first opportunity.*
- Complete a homework session by summarizing your work. *Prepare notes on difficult problems or concepts. These hints will be helpful when you begin to review for an exam.*
- Keep up to date with your homework assignments. *Falling behind will make it difficult for you to learn the new material presented in class.*

Find the solution of the system of equations shown on each graph. If there is no solution, state this. [4.1]

1.

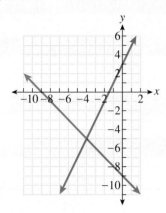

2.

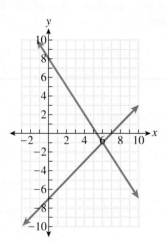

3.

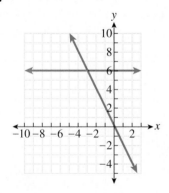

4.

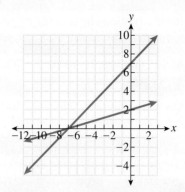

Solve the system by graphing. If the system is inconsistent and has no solution, state this. If the system is dependent, write the form of the solution for any real number x. [4.1]

5. $y = x - 5$
$y = -\dfrac{1}{2}x + 7$

6. $y = -2x + 8$
$y = \dfrac{1}{2}x + 3$

7. $y = \dfrac{1}{3}x - 4$
$y = -x$

8. $2x - y = -8$
$-3x + 2y = 10$

Solve the system by substitution. If the system is inconsistent and has no solution, state this. If the system is dependent, write the form of the solution for any real number x. [4.2]

9. $x = 2y + 1$
$2x - 5y = 4$

10. $y = -x - 3$
$3x + 4y = -5$

11. $4x + y = -5$
$5x - 3y = 15$

12. $-4x + 2y = 18$
$2x - y = 3$

Solve the system by addition. If the system is inconsistent and has no solution, state this. If the system is dependent, write the form of the solution for any real number x. [4.3]

13. $3x + 2y = 15$
$5x - 2y = 17$

14. $6x + 7y = -66$
$-3x + 4y = -27$

15. $2x - y = 4$
$-4x + 2y = -8$

16. $y = 4x + 5$
$5x - 2y = -4$

Solve the system by the method of your choice. If the system is inconsistent and has no solution, state this. If the system is dependent, write the form of the solution for any real number x. [4.2, 4.3]

17. $4x - 3y = 20$
$-x + 6y = -26$

18. $x + y = 4$
$-5x + 4y = 43$

19. $y = 2x - 1$
$3x - 7y = 29$

20. $8x - 9y = 67$
$x = 5$

21. $y = 3x - 2$
$-9x + 3y = -6$

22. $x + y = -4$
$4x - 6y = -1$

23. $3x - 2y = 12$
$y = \dfrac{4}{5}x - 6$

24. $x + 2y = -5$
$\dfrac{1}{3}x - \dfrac{1}{4}y = 13$

25. $-\dfrac{1}{2}x + 2y = 3$
$2x - 8y = -9$

26. $7x - 5y = -46$
$6x + 7y = 17$

27. The sum of two numbers is 87. One number is 18 more than twice the other number. Find the two numbers. [4.4]

28. Hal has 284 more baseball cards in his collection than his brother Mac does. Together, they have 1870 base-ball cards. How many cards does Hal have? [4.4]

29. A hockey team charges $10 for tickets to their games. On Wednesday nights, the team has a Ladies' Night promotion, and women of any age are admitted for half-price. If the total attendance on the last Ladies' Night was 1764 fans and the team collected $13,470, how many women were at the game? [4.4]

30. A movie theater charges $7.50 to see a movie, but senior citizens are admitted for only $4. If 206 people paid a total of $1384 to see a movie, how many were senior citizens? [4.4]

31. Jackie's home is on a rectangular lot. The fence that runs around the entire yard is 1320 feet long. If the length of the yard is 140 feet more than the width, find the dimensions of the yard. [4.4]

32. Sharon has a front door that lets cold drafts in. The height of the door is 4 inches more than twice its width. If she uses 224 inches of weather stripping around the entire door, how tall is her door? [4.4]

33. Greg deposits a total of $4000 in two different certificates of deposit (CDs). One CD pays 5% annual interest, while the other pays 4.25% annual interest. If Greg earned $179 interest during the first year, how much was put in each CD? [4.4]

34. Marcia bought shares in two mutual funds, investing a total of $2500. One fund's shares went up by 20%, while the other fund went up in value by 8%. This was a total profit of $320 for Marcia. How much was invested in each mutual fund? [4.4]

35. One type of cattle feed is 4% protein, while a second type is 13% protein. How many pounds of each type must be mixed together to make 45 pounds of a mixture that is 10% protein? [4.4]

36. A kayak can travel 12 miles upstream in 6 hours. It only takes $1\frac{1}{2}$ hours for the kayak to travel the same distance downstream. Find the speed of the kayak in still water. [4.4]

Solve the system of inequalities. [4.5]

37. $y \leq 2x + 7$
  $y > x - 5$

38. $3x + 4y < 24$
  $5x + y \geq 10$

39. $y \leq -4x + 7$
  $y \geq \dfrac{3}{2}x$

40. $y > \dfrac{1}{2}x - 3$
  $y > -2$

1. Find the solution of the system of equations shown on the graph. If there is no solution, state this.

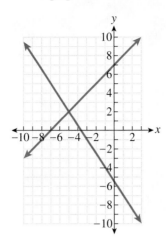

2. Solve the system that follows by graphing. If the system is inconsistent and has no solution, state this. If the system is dependent, write the form of the solution for any real number x.

$$y = -2x + 3$$
$$y = -3x + 7$$

3. Solve the system that follows by substitution. If the system is inconsistent and has no solution, state this. If the system is dependent, write the form of the solution for any real number x.

$$x + 4y = -9$$
$$2x - 5y = 21$$

4. Solve the system that follows by addition. If the system is inconsistent and has no solution, state this. If the system is dependent, write the form of the solution for any real number x.

$$5x + 2y = -22$$
$$2x - 5y = -3$$

Solve the given system by the method of your choice. If the system is inconsistent and has no solution, state this. If the system is dependent, write the form of the solution for any real number x.

5. $-x + 4y = 11$
$3x - 4y = -17$

6. $4x + 2y = 9$
$y = -2x + 18$

7. $x + y = 11$
$5x - 2y = 13$

8. $7x - 11y = -47$
$9x + 2y = 101$

9. $13x + 6y = 52$
$x = 7y + 4$

10. $4x - 6y = 18$
$-10x + 15y = -45$

11. Last season, a baseball player had 123 more walks than he had home runs. If we add the number of times he walked to the number of times he hit a home run, the total would be 237. How many walks did the player have?

12. Admission to an amusement park is $20, but children under eight years old are admitted for $5. If there were 13,900 people at the amusement park yesterday and they paid a total of $197,000, how many children under eight years old were at the park?

13. Steve's vegetable garden is in the shape of a rectangle, and it requires 66 feet of fencing to surround it. The width is 1 foot more than three times the length. Find the dimensions of Steve's garden.

14. Emma bought shares in two mutual funds, investing a total of $32,000. One fund's shares went up by 7%, while the other fund went up in value by 3%. This was a total profit of $1760 for Emma. How much was invested in each mutual fund?

15. Solve the system of inequalities.

$$y < \frac{1}{2}x + 4$$
$$5x + 2y > -16$$

Mathematicians in History
Hypatia of Alexandria

Hypatia of Alexandria was one of the first women known to make major contributions to the field of mathematics. She was also active in the fields of astronomy and philosophy. One quote that has been attributed to her is, "Reserve your right to think, for even to think wrongly is better than not to think at all."

Write a one-page summary (*or* make a poster) of the life of Hypatia of Alexandria and her accomplishments.

Interesting issues:

- Where and when was Hypatia born?
- Who was Hypatia's father, and what fields was he involved in?
- Hypatia lectured on the philosophy of Neoplatonism. What are the principle ideas of Neoplatonism?
- What was Hypatia's relationship to Cyril and Orestes?
- In the year A.D. 415, Hypatia was murdered. Describe the circumstances of her death.

Stretch Your Thinking ❯ Chapter 4

Have you ever wondered how math problems are created? Well, today you will be able to explore the answer to that question by creating your very own math application. You have studied various applications of systems of equations (Section 4.4), and now it's your turn to create an application problem which can be solved by using systems of equations.

The first step is to decide upon the context of your problem. You need to make sure and have two unknowns for which you will be solving. Will you be solving for the measure of two angles? Will you be looking for the quantity of two types of coffee beans being mixed together? Will you be looking for the dimensions of a rectangular object? Try and be as creative as possible when thinking about the context.

Once you have decided upon the context, you will need to decide on the values of your two unknowns. You will then need to create two distinct equations that relate these values to each other. Think about the equations you saw in Section 4.4 and about how you can make your own equations.

After you have your two equations, solve the system by using one of the methods you learned in this chapter. Did you arrive at the correct values of your unknowns? Have a friend solve your system, and see if he or she was able to correctly solve it. You may need to talk to your instructor if you are having difficulties in making sure that your equations correctly relate your unknowns to each other.

Once you have finalized your equations, you need a well-formed paragraph that describes the context of your problem and gives your equations in mathematical language. Again, refer to Section 4.4 for examples of wording and paragraph structure. After you feel that your paragraph correctly states the problem you intended, switch with someone in class and see if he or she can solve your application.

Turn in your paragraph, the two equations, and your solution to your instructor. Have fun!

Write the appropriate symbol, either < or >, between the following integers. [1.1]

1. -3 ____ -8

Simplify. [1.2]

2. $-9 - 25$

3. $15(-11)$

4. $-234 \div (-18)$

Write the prime factorization of the following numbers. (If the number is prime, state this.) [1.3]

5. 72

Simplify. Your answer should be in lowest terms. [1.4]

6. $\dfrac{6}{35} \cdot \dfrac{5}{21}$

7. $\dfrac{3}{4} + \dfrac{9}{10}$

Simplify the following decimal expressions. [1.5]

8. $9.8 - 3.72$

9. 4.7×3.8

10. A group of 32 baseball fans decided to attend a game together. If each ticket cost $14, how much did the group spend on tickets? [1.2]

11. If one recipe calls for $1\frac{1}{3}$ cups of flour and a second recipe calls for $2\frac{3}{4}$ cups of flour, how much flour is needed to make both recipes? [1.4]

Simplify the given expression. [1.7]

12. $9 + 3 \cdot 4 - 2^6$

13. $7 + 2(9 - 4 \cdot 5) - 6(-5)$

Evaluate the following algebraic expressions under the given conditions. [1.8]

14. $5x - 8$ for $x = -4$

Simplify. [1.8]

15. $8x - 11 - 3x + 8$

16. $3(3x - 8) - 5(4x + 9)$

Solve. [2.1, 2.2]

17. $3x - 4 = -31$

18. $-2x + 15 = 8$

19. $3x + 17 = 6x - 25$

20. $2(2x + 5) - 3(x - 8) = 20$

21. The width of a rectangle is 9 feet longer than its length, and the perimeter is 70 feet. Find the length and the width of the rectangle. [2.3]

22. Celia has $5 and $10 bills in her purse, worth a total of $205. She has 8 more $5 bills than $10 bills. How many $5 bills does Celia have? [2.3]

23. Twenty percent of 125 is what number? [2.4]

24. What percent of 275 is 66? [2.4]

25. Monisha bought a new house for $220,000. The house increased in value by 13% in one year. How much was the house worth after one year? [2.4]

Solve the proportion. [2.4]

26. $\dfrac{n}{8} = \dfrac{27}{36}$

27. At a certain community college, five out of every nine students plan to transfer to a four-year university. If the college has 8,307 students, how many plan to transfer to a four-year university? [2.4]

Solve. Graph your solution on a number line, and express it in interval notation. [2.5]

28. $2x - 5 \geq 3x - 8$

29. $-7 \leq 3x + 8 \leq 11$

Find the x- and y-intercepts, and use them to graph the equation. [3.2]

30. $2x - y = 6$

31. $-3x + 4y = -24$

Find the slope of the line that passes through the given points. [3.3]

32. $(-3, -7)$ and $(3, 5)$

Find the slope and the y-intercept of the given line. [3.3]

33. $2x + 3y = 21$

Graph using the slope and y-intercept. Find the x-intercept and label it on the graph. [3.3]

34. $y = 2x - 3$ **35.** $y = \dfrac{3}{4}x - 3$

Are the two given lines parallel, perpendicular, or neither? [3.5]

36. $5x + 3y = 9$
$3x - 5y = -5$

37. Bruce is saving his money to buy a new guitar. So far he has saved $220 and he plans to save $40 each month. [3.4]

   a) Create a function $f(x)$ for the amount of money that Bruce has saved after x months.

   b) Use the function from part a) to determine how much money Bruce will have saved after 8 months.

   c) Use the function from part a) to determine how many months it will take until Bruce has saved enough to buy the guitar, which will cost $800.

Evaluate the given function. [3.4]

38. $f(x) = 24x + 55, f(14)$

39. $g(x) = -7x + 905, g(-32)$

Find the slope–intercept form of the equation of a line with the given slope that passes through the given point. Graph the line and label any intercepts. [3.6]

40. Slope -2, through $(3, -4)$

Find the slope–intercept form of the equation of a line that passes through the two given points. Graph the line and label any intercepts. [3.6]

41. $(2, 1)$ and $(4, -5)$

Graph the inequality on a plane. [3.7]

42. $3x - 2y > 12$

Solve the system by the method of your choice. If the system is inconsistent and has no solution, state this. If the system is dependent, write the form of the solution for any real number x. [4.2, 4.3]

43. $4x + y = 22$
$-3x + 8y = 1$

44. $y = 3x - 2$
$2x - 3y = -15$

45. $9x + 8y = 25$
$5x - 6y = -7$

46. $4x + y = 6$
$\dfrac{3}{2}x + \dfrac{3}{5}y = \dfrac{9}{2}$

47. A minor league baseball team charges adults $5 and children $4 for admission. If 1050 people paid a total of $5000 to see last night's game, how many children were at the game? [4.4]

48. Bill's backyard is rectangular. The fence that runs around the backyard is 130 feet long. If the length of the backyard is 10 feet less than twice the width, find the dimensions of the backyard. [4.4]

49. Greg deposits a total of $15,000 in two different certificates of deposit (CDs). One CD pays 3% annual interest, while the other pays 2.5% annual interest. If Greg earned $430 interest during the first year, how much was put in each CD? [4.4]

Solve the system of inequalities. [4.5]

50. $y < x + 4$
$y \geq 4x - 2$

EXPONENTS AND POLYNOMIALS

The expressions and equations we have examined to this point in the text have been linear. In this chapter, we begin to examine expressions and functions that are nonlinear, including polynomials such as $-16t^2 + 32t + 128$ or $x^5 - 3x^4 + 8x^2 - 9x + 7$. After learning some basic definitions, we will learn to add, subtract, multiply, and divide polynomials. We will introduce properties for working with expressions containing exponents that will make using these operations easier. We will also cover scientific notation, which is used to represent very large or very small numbers and perform arithmetic calculations with them.

Study Tip TEST TAKING   *To be successful in a math class, understanding the material is important, but you must also be a good test taker. In this chapter, we will discuss the test-taking skills necessary for success in a math class.*

5.1
EXPONENTS

Objectives

1 Use the product rule for exponents.
2 Use the power rule for exponents.
3 Use the power of a product rule for exponents.
4 Use the quotient rule for exponents.
5 Use the zero exponent rule.
6 Use the power of a quotient rule for exponents.
7 Evaluate functions by using the rules for exponents.

We have used exponents to represent repeated multiplication of a particular factor. For example, the expression 3^7 is used to represent a product in which 3 is a factor 7 times.

$$3^7 = 3 \cdot 3 \cdot 3 \cdot 3 \cdot 3 \cdot 3 \cdot 3$$

Recall that in the expression 3^7, the number 3 is called the base and the number 7 is called the exponent.

In this section, we will introduce several properties of exponents and learn how to apply these properties to simplify expressions involving exponents.

Product Rule

Objective **1**   Use the product rule for exponents.

Product Rule

For any base x, $x^m \cdot x^n = x^{m+n}$

When multiplying two expressions with the same base, we keep the base and add the exponents. Consider the example $x^3 \cdot x^7$. We know that $x^3 = x \cdot x \cdot x$ and $x^7 = x \cdot x \cdot x \cdot x \cdot x \cdot x \cdot x$. So $x^3 \cdot x^7$ can be rewritten as $(x \cdot x \cdot x) \cdot (x \cdot x \cdot x \cdot x \cdot x \cdot x \cdot x)$.

Since x is being repeated as a factor 10 times, this expression can be rewritten as x^{10}.

$$x^3 \cdot x^7 = x^{3+7} = x^{10}$$

A Word of Caution   When multiplying two expressions with the same base, we keep the base and add the exponents. *Do not multiply the exponents.*

$$x^3 \cdot x^7 = x^{10}, \text{ not } x^{21}$$

EXAMPLE 1   Simplify $x^2 \cdot x^9$.

Solution

Since we are multiplying and the two bases are the same, we keep the base and add the exponents.

$$x^2 \cdot x^9 = x^{2+9} \qquad \text{Keep the base, add the exponents.}$$
$$= x^{11} \qquad \text{Add.}$$

EXAMPLE 2 Simplify $2 \cdot 2^5 \cdot 2^4$.

Solution

In this example, we are multiplying three expressions that have the same base. To do this, we keep the base and add the three exponents together. Keep in mind that if an exponent is not written for a factor, then the exponent is a 1.

$$\begin{aligned} 2 \cdot 2^5 \cdot 2^4 &= 2^{1+5+4} && \text{Keep the base, add the exponents.} \\ &= 2^{10} && \text{Add.} \\ &= 1024 && \text{Raise 2 to the 10}^{\text{th}} \text{ power.} \end{aligned}$$

Quick Check 1

Simplify.

a) $x^5 \cdot x^4$ b) $3^3 \cdot 3^2 \cdot 3$

EXAMPLE 3 Simplify $(x + y)^4 \cdot (x + y)^9$.

Solution

The base for these expressions is the sum $x + y$.

$$\begin{aligned} (x + y)^4 \cdot (x + y)^9 &= (x + y)^{4+9} && \text{Keep the base, add the exponents.} \\ &= (x + y)^{13} && \text{Add.} \end{aligned}$$

Quick Check 2

Simplify
$(a - 6)^{10} \cdot (a - 6)^{17}$.

EXAMPLE 4 Simplify $(x^2 y^7)(x^3 y)$.

Solution

When we have a product involving more than one base, the associative and commutative properties of multiplication can be used to simplify the product.

$$\begin{aligned} (x^2 y^7)(x^3 y) &= x^2 y^7 x^3 y && \text{Write without parentheses.} \\ &= x^2 x^3 y^7 y && \text{Group like bases together.} \\ &= x^5 y^8 && \text{For each base, add the exponents and keep the base.} \end{aligned}$$

For future problems, it will not be necessary to show the application of the associative and commutative properties of multiplication. Look for bases that are the same, add their exponents, and write each base to the calculated power.

Quick Check 3

Simplify $(a^6 b^5)(a^9 b^5)$.

Power Rule

Objective 2 **Use the power rule for exponents.** The second property of exponents involves raising an expression with an exponent to a power.

Power Rule

$$\text{For any base } x, (x^m)^n = x^{m \cdot n}$$

In essence, this property tells us that when we raise a power to a power, we keep the base and multiply the exponents. To show why this is true, consider the expression $(x^3)^7$.

Whenever we raise an expression to the seventh power, this means that the expression is repeated as a factor seven times. So $(x^3)^7 = x^3 \cdot x^3 \cdot x^3 \cdot x^3 \cdot x^3 \cdot x^3 \cdot x^3$. Applying the first property from this section, $x^3 \cdot x^3 \cdot x^3 \cdot x^3 \cdot x^3 \cdot x^3 \cdot x^3 = x^{3+3+3+3+3+3+3}$ or x^{21}. This exponent could be found by multiplying 3 by 7.

$$(x^3)^7 = x^{3 \cdot 7} = x^{21}$$

> **A Word of Caution** When raising an expression with an exponent to another power, we keep the base and multiply the exponents. *Do not add the exponents.*
>
> $$(x^3)^7 = x^{21}, \text{ not } x^{10}$$

EXAMPLE ▶ **5**   Simplify $(x^5)^4$.

Solution

In this example, we are raising a power to another power, so we keep the base and multiply the exponents.

$$
\begin{aligned}
(x^5)^4 &= x^{5 \cdot 4} &&\text{Keep the base, multiply the exponents.}\\
&= x^{20} &&\text{Multiply.}
\end{aligned}
$$

Quick Check ◀ **4**
Simplify $(x^9)^7$.

EXAMPLE ▶ **6**   Simplify $(x^3)^9(x^7)^8$.

Solution

To simplify this expression, we must use both properties that have been introduced in this section.

$$
\begin{aligned}
(x^3)^9(x^7)^8 &= x^{3 \cdot 9} x^{7 \cdot 8} &&\text{Apply the power rule.}\\
&= x^{27} x^{56} &&\text{Multiply.}\\
&= x^{83} &&\text{Keep the base, add the exponents.}
\end{aligned}
$$

Quick Check ◀ **5**
Simplify $(x^4)^6(x^5)^2$.

Power of a Product Rule

Objective **3** **Use the power of a product rule for exponents.** The third property introduced in this section involves raising a product to a power.

Power of a Product Rule

> For any bases x and y, $(xy)^n = x^n y^n$

This property tells us that when we raise a product to a power, we may raise each factor to that power. Consider the expression $(xy)^5$. We know that this can be rewritten as $xy \cdot xy \cdot xy \cdot xy \cdot xy$. Using the associative and commutative properties of multiplication, we can rewrite this expression as $x \cdot x \cdot x \cdot x \cdot x \cdot y \cdot y \cdot y \cdot y \cdot y$, which simplifies to $x^5 y^5$.

EXAMPLE ▶ **7**   Simplify $(x^2 y^3)^6$.

Solution

This example combines two properties. When we raise each factor to the sixth power, we are raising a power to a power and therefore multiply the exponents.

Quick Check ◀ **6**
Simplify $(a^5 b^3)^8$.

$$
\begin{aligned}
(x^2 y^3)^6 &= (x^2)^6 (y^3)^6 &&\text{Raise each factor to the 6}^{\text{th}} \text{ power.}\\
&= x^{12} y^{18} &&\text{Keep the base, multiply the exponents.}
\end{aligned}
$$

EXAMPLE 8   Simplify $(x^2y^3)^7(3x^5)^2$.

Solution

In this example, we begin by raising each factor to the appropriate power. Then we multiply expressions with the same base.

$$(x^2y^3)^7(3x^5)^2 = x^{14}y^{21} \cdot 3^2 x^{10} \qquad \text{Raise each base to the appropriate power.}$$
$$= 9x^{24}y^{21} \qquad \qquad \text{Simplify.}$$

It is a good idea to write the variables in an expression in alphabetical order, as shown in this example, although it is not necessary.

Quick Check 7
Simplify
$(4a^2b^3)^3(a^4b^2c^7)^5$.

A Word of Caution   This property applies only when we raise a *product* to a power, not a sum or a difference. Although $(xy)^2 = x^2y^2$, a similar approach does not work for $(x + y)^2$ or $(x - y)^2$.

$$(x + y)^2 \neq x^2 + y^2$$
$$(x - y)^2 \neq x^2 - y^2$$

For example, let $x = 12$ and $y = 5$. In this case, $(x + y)^2$ is equal to 17^2 or 289, but $x^2 + y^2$ is equal to $12^2 + 5^2$ or 169. This shows that $(x + y)^2$ is not equal to $x^2 + y^2$ in general.

Quotient Rule

Objective 4   Use the quotient rule for exponents.   The next property involves fractions containing the same base in their numerators and denominators. It also applies to dividing exponential expressions with the same base.

Quotient Rule

$$\text{For any base } x, \; \frac{x^m}{x^n} = x^{m-n} \; (x \neq 0)$$

Note that this property has a restriction; it does not apply when the base is 0. This is because division by 0 is undefined. This property tells us that when we are dividing two expressions with the same base, we subtract the exponent of the denominator from the exponent of the numerator. Consider the expression $\dfrac{x^7}{x^3}$. This can be rewritten as $\dfrac{x \cdot x \cdot x \cdot x \cdot x \cdot x \cdot x}{x \cdot x \cdot x}$. From our previous work, we know that this fraction can be simplified by dividing out common factors in the numerator and denominator.

$$\frac{\cancel{x} \cdot \cancel{x} \cdot \cancel{x} \cdot x \cdot x \cdot x \cdot x}{\cancel{x} \cdot \cancel{x} \cdot \cancel{x}} = x^4$$

One way to think of this is that the number of factors in the numerator has been reduced by three, which was the number of factors in the denominator. The property tells us that we can simplify this expression by subtracting the exponents, which produces the same result.

$$\frac{x^7}{x^3} = x^{7-3} = x^4$$

EXAMPLE ▶ **9**   Simplify $\dfrac{y^9}{y^2}$. (Assume $y \neq 0$.)

Solution

Since the bases are the same, we keep the base and subtract the exponents.

$$\dfrac{y^9}{y^2} = y^{9-2} \qquad \text{Keep the base, subtract the exponents.}$$
$$= y^7 \qquad \text{Subtract.}$$

EXAMPLE ▶ **10**   Simplify $a^{21} \div a^3$. (Assume $y \neq 0$.)

Solution

Although in a different form than the previous example, $a^{21} \div a^3$ is equivalent to $\dfrac{a^{21}}{a^3}$. We keep the base and subtract the exponents.

$$a^{21} \div a^3 = a^{21-3} \qquad \text{Keep the base, subtract the exponents.}$$
$$= a^{18} \qquad \text{Subtract.}$$

Quick Check **8**
Simplify. Assume $x \neq 0$.

a) $\dfrac{x^{20}}{x^4}$ b) $x^{24} \div x^8$

> **A Word of Caution**   When dividing two expressions with the same base, keep the base and subtract the exponents. *Do not divide the exponents.*
>
> $$a^{21} \div a^3 = a^{21-3}, \text{ not } a^7$$

EXAMPLE ▶ **11**   Simplify $\dfrac{16x^{10}y^9}{2x^7y^4}$. (Assume $x, y \neq 0$.)

Solution

Although we will subtract the exponents for the bases x and y, we will not subtract the numerical factors 16 and 2. We simplify numerical factors in the same way that we originally did in Chapter 1, so $\frac{16}{2} = 8$.

$$\dfrac{16x^{10}y^9}{2x^7y^4} = 8x^3y^5 \qquad \text{Simplify } \tfrac{16}{2}. \text{ Keep the variable bases, subtract the exponents.}$$

Quick Check **9**
Simplify $\dfrac{20a^{10}b^{12}c^{15}}{5a^2b^6c^9}$. (Assume $a, b, c \neq 0$.)

Zero Exponent Rule

Objective **5**   **Use the zero exponent rule.**   Consider the expression $\dfrac{2^7}{2^7}$. We know that any number (except 0) divided by itself is equal to 1. The previous property also tells us that we can subtract the exponents in this expression, so $\dfrac{2^7}{2^7} = 2^0$. Since we have shown that $\dfrac{2^7}{2^7}$ is equal to both 2^0 and 1, then the expression 2^0 must be equal to 1. The following property involves raising a base to the zero power.

Zero Exponent Rule

> For any base x, $x^0 = 1$ $(x \neq 0)$.

EXAMPLE 12   Simplify 7^0.

Solution

By applying the zero-exponent rule, we see that $7^0 = 1$.

> **A Word of Caution**   When raising a nonzero base to the zero power, the result is 1, not 0.
>
> $$7^0 = 1, \text{ not } 0$$

The expressions $(4x)^0$ and $4x^0$ are different $(x \neq 0)$. In the first expression, the base that is being raised to the zero power is $4x$. So $(4x)^0 = 1$, since any nonzero base raised to the zero power is equal to 1. In the second expression, the base that is being raised to the zero power is x, not $4x$. The expression $4x^0$ simplifies to be $4 \cdot 1$ or 4.

Quick Check   10

Simplify. Assume $x \neq 0$.

a) 124^0 b) $(3x^2)^0$
c) $9x^0$

Power of a Quotient Rule

Objective   6   Use the power of a quotient rule for exponents.   The concluding property in this section deals with raising a quotient to a power. It is similar to the property involving raising a product to a power.

Power of a Quotient Rule

> For any bases x and y, $\left(\dfrac{x}{y}\right)^n = \dfrac{x^n}{y^n}$ $(y \neq 0)$

To raise a quotient to a power, we raise both the numerator and the denominator to that power. The restriction $y \neq 0$ is again due to the fact that division by 0 is undefined. Consider the expression $\left(\frac{r}{s}\right)^3$, where $s \neq 0$. The property tells us that this is equivalent to $\dfrac{r^3}{s^3}$, and here's why:

$$\left(\frac{r}{s}\right)^3 = \frac{r}{s} \cdot \frac{r}{s} \cdot \frac{r}{s} \qquad \text{Repeat } \frac{r}{s} \text{ as a factor three times.}$$

$$= \frac{r \cdot r \cdot r}{s \cdot s \cdot s} \qquad \text{Use the definition of multiplication for fractions.}$$

$$= \frac{r^3}{s^3} \qquad \text{Rewrite the numerator and denominator using exponents.}$$

A Word of Caution   When raising a quotient to a power, raise both the numerator and the denominator to that power. Do not just raise the numerator to that power.

$$\left(\frac{r}{s}\right)^3 = \frac{r^3}{s^3}, \text{ not } \frac{r^3}{s} \qquad (\text{Assume } s \neq 0.)$$

EXAMPLE 13   Simplify $\left(\dfrac{x^2 y^3}{z^4}\right)^8$. (Assume $z \neq 0$.)

Solution

We begin by raising each factor in the numerator and denominator to the eighth power, and then we simplify the resulting expression.

$$\left(\frac{x^2 y^3}{z^4}\right)^8 = \frac{(x^2)^8 (y^3)^8}{(z^4)^8}$$ Raise each factor in the numerator and denominator to the 8th power.

$$= \frac{x^{16} y^{24}}{z^{32}}$$ Keep the bases, multiply the exponents.

Quick Check 11   Simplify $\left(\dfrac{ab^7}{c^3 d^4}\right)^5$. (Assume $c, d \neq 0$.)

Here is a brief summary of the properties introduced in this section:

Properties of Exponents

1. **Product Rule:** For any base x, $x^m \cdot x^n = x^{m+n}$
2. **Power Rule:** For any base x, $(x^m)^n = x^{m \cdot n}$
3. **Power of a Product Rule:** For any bases x and y, $(xy)^n = x^n y^n$
4. **Quotient Rule:** For any base x, $\dfrac{x^m}{x^n} = x^{m-n}$ $(x \neq 0)$
5. **Zero Exponent Rule:** For any base x, $x^0 = 1$ $(x \neq 0)$
6. **Power of a Quotient Rule:** For any bases x and y, $\left(\dfrac{x}{y}\right)^n = \dfrac{x^n}{y^n}$ $(y \neq 0)$

Objective 7   **Evaluate functions by using the rules for exponents.** We conclude this section by investigating functions containing exponents.

EXAMPLE 14   For the function $f(x) = x^4$, evaluate $f(3)$.

Solution

We substitute 3 for x and simplify the resulting expression.

$$f(3) = (3)^4$$ Substitute 3 for x.
$$= 81$$ Simplify.

Quick Check 12 For the function $f(x) = x^6$ evaluate $f(-4)$.

EXAMPLE 15 For the function $f(x) = x^3$, evaluate $f(a^5)$.

Solution

We substitute a^5 for x and simplify.

$$
\begin{aligned}
f(a^5) &= (a^5)^3 & &\text{Substitute } a^5 \text{ for } x. \\
&= a^{53} & &\text{Simplify using the power rule } ((x^m)^n = x^{m \cdot n}). \\
&= a^{15} & &\text{Multiply.}
\end{aligned}
$$

Quick Check 13 For the function $f(x) = x^7$, evaluate $f(a^8)$.

EXERCISES 5.1

Simplify.

1. $2^2 \cdot 2^3$ **2.** $5^5 \cdot 5$ **3.** $x^9 \cdot x^7$

4. $a^3 \cdot a^6$ **5.** $m^{13} \cdot m^{22}$ **6.** $n^{11} \cdot n^{40}$

7. $b^5 \cdot b^6 \cdot b^3$ **8.** $x^8 \cdot x^6 \cdot x$

9. $(2x - 3)^4 \cdot (2x - 3)^{10}$

10. $(4x + 1)^{12} \cdot (4x + 1)^9$

11. $(x^6 y^5)(x^2 y^8)$

12. $(5x^8 y^8)(2x^5 y^{11})$

Find the missing factor.

13. $x^4 \cdot ? = x^9$ **14.** $x^7 \cdot ? = x^{10}$

15. $? \cdot b^5 = b^6$ **16.** $? \cdot a^{11} = a^{20}$

Simplify.

17. $(x^2)^5$ **18.** $(x^3)^3$ **19.** $(a^7)^8$

20. $(b^{12})^5$ **21.** $(2^3)^2$ **22.** $(3^2)^5$

23. $(x^3)^7 (x^4)^2$ **24.** $(x^2)^8 (x^5)^3$

Find the missing exponent.

25. $(x^6)^? = x^{42}$ **26.** $(p^?)^8 = p^{112}$

27. $(x^3)^8 (x^?)^4 = x^{44}$ **28.** $(x^9)^? (x^{11})^{12} = x^{195}$

Simplify.

29. $(5x)^3$ **30.** $(3a)^6$

31. $(-4y^3)^3$

32. $(10x^5)^5$

33. $(x^3 y^4)^7$

34. $(x^8 y^3)^6$

35. $(3x^4 y)^4$

36. $(2x^5 y^6)^8$

37. $(x^2 y^5 z^3)^7 (y^4 z^5)^8$

38. $(a^{10} b^{15} c^{20})^7 (a^{11} b^7 c^3)^8$

Find the missing exponent(s).

39. $(x^5 y^4)^? = x^{35} y^{28}$

40. $(c^7 d^{11})^? = c^{21} d^{33}$

41. $(2s^? t^?)^4 = 16s^{24} t^{52}$

42. $(m^? n^?)^9 = m^{54} n^{216}$

Simplify. (Assume all variables are nonzero.)

43. $\dfrac{x^7}{x^5}$ **44.** $\dfrac{x^{11}}{x^3}$ **45.** $d^{35} \div d^5$

46. $z^{87} \div z^{86}$ **47.** $\dfrac{32x^8}{8x^3}$ **48.** $\dfrac{91y^{15}}{7y^3}$

49. $\dfrac{(a + 5b)^{13}}{(a + 5b)^{11}}$

50. $\dfrac{(2x - 17)^{11}}{(2x - 17)^5}$

51. $\dfrac{a^8 b^6}{a^3 b}$ **52.** $\dfrac{28r^6 s^7 t^9}{7r^3 s^7 t^6}$

MyMathLab
F MyMathLab
O
R *MathXL*
MathXL

E
X
T Interactmath.com
R
A MathXL
Tutorials on CD

H
E DVT CD
L Videotape
P
Tutor Center

Addison-Wesley
Math Tutor Center

Student Solutions
Manual

Find the missing exponent(s). (Assume all variables are nonzero.)

53. $\dfrac{x^{14}}{x^?} = x^2$

54. $\dfrac{y^?}{y^8} = y^9$

55. $m^? \div m^{11} = m^{19}$

56. $\dfrac{x^{15}y^?}{x^?y^{25}} = x^7y^6$

Simplify. (Assume all variables are nonzero.)

57. 9^0 **58.** -8^0 **59.** $(-16)^0$

60. 5^0 **61.** $\dfrac{3}{7^0}$ **62.** $\left(\dfrac{4}{13}\right)^0$

63. $(13x)^0$ **64.** $(22x^5y^{12}z^9)^0$

Simplify. (Assume all variables are nonzero.)

65. $\left(\dfrac{2}{5}\right)^4$ **66.** $\left(\dfrac{x}{3}\right)^5$ **67.** $\left(\dfrac{a}{b}\right)^7$

68. $\left(\dfrac{m}{n}\right)^9$ **69.** $\left(\dfrac{x^2}{y^3}\right)^7$ **70.** $\left(\dfrac{y^8}{x^4}\right)^8$

71. $\left(\dfrac{2a^6}{b^7}\right)^3$ **72.** $\left(\dfrac{x^{13}}{5y^9}\right)^5$

73. $\left(\dfrac{a^5b^4}{b^2c^7}\right)^9$ **74.** $\left(\dfrac{17x^5y^{19}z^{24}}{32a^{11}b^{70}c^{33}}\right)^0$

Find the missing exponent(s). (Assume all variables are nonzero.)

75. $\left(\dfrac{3}{4}\right)^? = \dfrac{81}{256}$ **76.** $\left(\dfrac{x^4}{y^7}\right)^? = \dfrac{x^{12}}{y^{21}}$

77. $\left(\dfrac{a^3b^6}{c^8}\right)^? = \dfrac{a^{36}b^{72}}{c^{96}}$ **78.** $\left(\dfrac{ab^8}{c^5d^9}\right)^? = \dfrac{a^{15}b^{120}}{c^{75}d^{135}}$

Evaluate the given function.

79. $f(x) = x^2$, $f(4)$ **80.** $g(x) = x^3$, $g(9)$

81. $g(x) = x^4$, $g(-2)$ **82.** $f(x) = x^{10}$, $f(-1)$

83. $f(x) = x^6$, $f(a^4)$ **84.** $f(x) = x^5$, $f(b^7)$

Simplify.

85. $(a^7)^6$ **86.** $(13a)^0$ **87.** $\left(\dfrac{5x^4}{2y^7}\right)^3$

88. $(x^6y^9z^2)^7$ **89.** $\dfrac{x^{13}}{x}$ **90.** $\dfrac{x^{100}}{x^{100}}$

91. $\left(\dfrac{a^8b^9}{2c^2}\right)^5$ **92.** $c^8 \cdot c^{12}$

93. $(b^{10})^2$ **94.** $\dfrac{x^8y^{13}}{xy^6}$

95. $(x^{15})^6$ **96.** $(2x^9yz^3)^4$

97. $\left(\dfrac{3a^2b^9}{7cd^4}\right)^4$ **98.** $(6x^7y^9z)^3$

Answer in complete sentences.

99. Which expression is equal to x^9: $(x^5)^4$ or $x^5 \cdot x^4$? Explain your answer.

100. Explain why the power of a product rule, $(xy)^n = x^ny^n$, does not apply to the expression $(x+y)^2$? In other words, why is $(x+y)^2$ not equal to $x^2 + y^2$? Explain your answer.

101. Explain, in your own words, why $8^0 = 1$.

102. Explain the difference between the expressions $\dfrac{a^3}{b}$ and $\left(\dfrac{a}{b}\right)^3$.

Objectives

1. Understand negative exponents.
2. Use the rules of exponents to simplify expressions containing negative exponents.
3. Convert numbers from scientific notation to decimal notation.
4. Convert numbers from decimal notation to scientific notation.
5. Perform arithmetic operations with numbers in scientific notation.
6. Use scientific notation to solve applied problems.

Negative Exponents

Objective 1   Understand negative exponents. In this section, we introduce the concept of a **negative exponent.**

Negative Exponents

$$\text{For any } \textit{nonzero} \text{ base } x, x^{-n} = \frac{1}{x^n}. \ (x \neq 0)$$

For example, $2^{-3} = \frac{1}{2^3}$ or $\frac{1}{8}$.

Let's examine this definition. Consider the expression $\frac{2^4}{2^7}$. If we apply the property $\frac{x^m}{x^n} = x^{m-n}$ from the previous section, we see that $\frac{2^4}{2^7}$ simplifies to equal 2^{-3}.

$$\frac{2^4}{2^7} = 2^{4-7}$$
$$= 2^{-3}$$

If we simplify $\frac{2^4}{2^7}$ by dividing out common factors, the result is $\frac{1}{2^3}$.

$$\frac{2^4}{2^7} = \frac{2 \cdot 2 \cdot 2 \cdot 2}{2 \cdot 2 \cdot 2 \cdot 2 \cdot 2 \cdot 2 \cdot 2}$$
$$= \frac{1}{2^3}$$

Since $\frac{2^4}{2^7}$ is equal to both 2^{-3} and $\frac{1}{2^3}$, $2^{-3} = \frac{1}{2^3}$.

A Word of Caution   Raising a positive base to a negative exponent is not the same as raising the opposite of that base to a positive power.

$$2^{-3} = \frac{1}{2^3}, \text{ not } -2^3$$

EXAMPLE 1   Rewrite the expression 7^{-2} without using negative exponents and simplify.

Solution

When raising a number to a negative exponent, we begin by rewriting the expression without negative exponents. We finish by actually raising the base to the appropriate positive power.

$$7^{-2} = \frac{1}{7^2} \qquad \text{Rewrite without negative exponents.}$$

$$= \frac{1}{49} \qquad \text{Raise 7 to the second power.}$$

Quick Check 1
Rewrite 4^{-3} without using negative exponents and simplify.

EXAMPLE 2   Rewrite the expression $9x^{-6}$ without using negative exponents. (Assume $x \neq 0$.)

Solution

The base in this example is x. The exponent does not apply to the 9, since there are no parentheses. When we rewrite the expression without a negative exponent, the number 9 is unaffected. Rather than writing x^{-6} as $\frac{1}{x^6}$ first, we can move the factor directly to the denominator of a fraction whose numerator is 9 by changing the sign of its exponent.

$$9x^{-6} = \frac{9}{x^6}$$

Quick Check 2
Rewrite the expression $a^5 b^{-4}$ without using negative exponents. (Assume $b \neq 0$.)

EXAMPLE 3   Rewrite the expression $\dfrac{x^8}{z^{-5}}$ without using negative exponents. (Assume $z \neq 0$.)

Solution

In this expression, we have a factor in the denominator with a negative exponent. To change the sign of its exponent to a positive number, we move it to the numerator to join the other factor.

$$\frac{x^8}{z^{-5}} = x^8 z^5$$

EXAMPLE 4   Rewrite the expression $\dfrac{a^7 b^{-3}}{c^{-5} d^{12}}$ without using negative exponents. (Assume all variables are nonzero.)

Solution

Of the four factors in this example, two of them (b^{-3} and c^{-5}) need to be rewritten without using negative exponents. This can be done by changing the sign of their exponents and rewriting the factors on the other side of the fraction bar.

$$\frac{a^7 b^{-3}}{c^{-5} d^{12}} = \frac{a^7 c^5}{b^3 d^{12}}$$

Quick Check **3** **Rewrite without using negative exponents. (Assume all variables are nonzero.)**

a) $\dfrac{x^{12}}{y^{-7}}$ b) $\dfrac{x^{-2}y^4}{z^{-1}w^{-9}}$

Using the Rules of Exponents with Negative Exponents

Objective **2** **Use the rules of exponents to simplify expressions containing negative exponents.** All of the properties of exponents developed in the last section hold true for negative exponents. When simplifying expressions involving negative exponents, there are two general routes we can take. We can choose to apply the appropriate property first, and then we can rewrite the expression without negative exponents. On the other hand, in some circumstances it will be more convenient to rewrite the expression without negative exponents before attempting to apply the appropriate property.

EXAMPLE **5** Simplify the expression $x^{11} \cdot x^{-6}$. Write the result without using negative exponents. (Assume $x \neq 0$.)

Solution

This example uses the product rule $x^m \cdot x^n = x^{m+n}$.

$$x^{11} \cdot x^{-6} = x^{11+(-6)} \qquad \text{Keep the base and add the exponents.}$$
$$= x^5 \qquad \text{Simplify.}$$

An alternative approach would be to rewrite the expression without negative exponents first.

Quick Check **4**

Simplify the expression $x^{-13} \cdot x^6$. Write the result without using negative exponents. (Assume $x \neq 0$.)

$$x^{11} \cdot x^{-6} = \dfrac{x^{11}}{x^6} \qquad \text{Rewrite without negative exponents.}$$
$$= x^5 \qquad \text{Keep the base and subtract the exponents.}$$

Use the approach that seems clearer to you.

EXAMPLE **6** Simplify the expression $(y^3)^{-2}$. Write the result without using negative exponents. (Assume $y \neq 0$.)

Solution

This example uses the power rule $(x^m)^n = x^{m \cdot n}$.

$$(y^3)^{-2} = y^{-6} \qquad \text{Keep the base and multiply the exponents.}$$
$$= \dfrac{1}{y^6} \qquad \text{Rewrite without using negative exponents.}$$

Quick Check **5** **Simplify the expression $(x^{-6})^{-7}$. Write the result without using negative exponents. (Assume $x \neq 0$.)**

EXAMPLE 7 Simplify the expression $(2a^{-7}b^9)^{-4}$. Write the result without using negative exponents. (Assume $a, b \neq 0$.)

Solution

This example uses the power of a product rule $(xy)^n = x^n y^n$.

$$(2a^{-7}b^9)^{-4} = 2^{-4}a^{-7(-4)}b^{9(-4)} \qquad \text{Raise each factor to the } -4 \text{ power.}$$
$$= 2^{-4}a^{28}b^{-36} \qquad \text{Simplify each exponent.}$$
$$= \frac{a^{28}}{2^4 b^{36}} \qquad \text{Rewrite without using negative exponents.}$$
$$= \frac{a^{28}}{16b^{36}} \qquad \text{Simplify } 2^4.$$

> **Quick Check 6**
> Simplify the expression $(9ab^{-3}c^2)^{-2}$. Write the result without using negative exponents. (Assume $a, b, c \neq 0$.)

EXAMPLE 8 Simplify the expression $\dfrac{x^3}{x^{15}}$. Write the result without using negative exponents. (Assume $x \neq 0$.)

Solution

This example uses the quotient rule $\dfrac{x^m}{x^n} = x^{m-n}$.

$$\frac{x^3}{x^{15}} = x^{-12} \qquad \text{Keep the base and subtract the exponents.}$$
$$= \frac{1}{x^{12}} \qquad \text{Rewrite without using negative exponents.}$$

> **Quick Check 7**
> Simplify the expression $\dfrac{x^8}{x^{14}}$. Write the result without using negative exponents. (Assume $x \neq 0$.)

EXAMPLE 9 Simplify the expression $\left(\dfrac{r^4}{s^2}\right)^{-3}$. Write the result without using negative exponents. (Assume $r, s \neq 0$.)

Solution

This example uses the power of a quotient rule $\left(\dfrac{x}{y}\right)^n = \dfrac{x^n}{y^n}$.

$$\left(\frac{r^4}{s^2}\right)^{-3} = \frac{r^{4(-3)}}{s^{2(-3)}} \qquad \text{Raise the numerator and the denominator to the } -3 \text{ power.}$$
$$= \frac{r^{-12}}{s^{-6}} \qquad \text{Multiply.}$$
$$= \frac{s^6}{r^{12}} \qquad \text{Rewrite without using negative exponents.}$$

> **Quick Check 8**
> Simplify the expression $\left(\dfrac{x^5}{y^{-4}}\right)^{-4}$. Write the result without using negative exponents. (Assume $x, y \neq 0$.)

Scientific Notation

Objective 3 Convert numbers from scientific notation to decimal notation.
Scientific notation is used to represent numbers that are very large, such as 93,000,000, or very small, such as 0.0000324.

Rewriting a Number in Scientific Notation

> To convert a number to scientific notation, we rewrite it in the form $a \times 10^b$, where $1 \leq a < 10$ and b is an integer.

First, consider the following table listing several powers of 10:

10^5	10^4	10^3	10^2	10^1	10^0	10^{-1}	10^{-2}	10^{-3}	10^{-4}	10^{-5}
100,000	10,000	1000	100	10	1	0.1	0.01	0.001	0.0001	0.00001

Notice that all of the positive powers of 10 are numbers that are 10 or higher. All of the negative powers of 10 are numbers between 0 and 1.

EXAMPLE ▶10 Convert 5.28×10^4 from scientific notation to decimal notation.

Solution

$$5.28 \times 10^4 = 5.28 \times 10,000 \qquad \text{Rewrite } 10^4 \text{ as } 10,000.$$
$$= 52,800 \qquad \text{Multiply.}$$

To multiply a decimal number by 10^4 or 10,000, we move the decimal point 4 places to the right. Note that the power of 10 is positive in this example.

EXAMPLE ▶11 Convert 2.0039×10^{-5} from scientific notation to decimal notation.

Solution

$$2.0039 \times 10^{-5} = 2.0039 \times 0.00001 \qquad \text{Rewrite } 10^{-5} \text{ as } 0.00001.$$
$$= 0.000020039 \qquad \text{Multiply.}$$

To multiply a decimal number by 10^{-5} or 0.00001, we move the decimal point five places to the left. Note that the power of 10 is negative in this example.

Quick Check 9
Convert from scientific notation to decimal notation.

a) 3.2×10^6
b) 7.21×10^{-4}

If you are unsure about which direction to move the decimal point, try thinking about whether we are making the number larger or smaller. Multiplying by positive powers of 10 makes the number larger, so move the decimal point to the right. Multiplying by negative powers of 10 makes the number smaller, so move the decimal point to the left.

Objective 4 Convert numbers from decimal notation to scientific notation.
To convert a number to scientific notation, we first move the decimal point so that it immediately follows the first nonzero digit in the number. Count the number of decimal places that the decimal point moves. This will give us the power of 10 when the number is written in scientific notation. If the original number was 10 or larger, then the exponent is positive, but if the original number was between 0 and 1, then the exponent is negative.

EXAMPLE ▶12 Convert 0.000321 to scientific notation.

Solution

Move the decimal point so that it follows the first nonzero digit, which is 3.

0.000321

To do this, we must move the decimal point four places to the right. Since 0.000321 is less than 1, this exponent must be negative.

$$0.000321 = 3.21 \times 10^{-4}$$

EXAMPLE 13   Convert 24,000,000,000 to scientific notation.

Solution

Move the decimal point so that it follows the digit 2. To do this, we must move the decimal point 10 places to the left. Since 24,000,000,000 is 10 or larger, the exponent will be a positive 10.

$$24,000,000,000 = 2.4 \times 10^{10}$$

Quick Check **10** **Convert to scientific notation.**

a) 0.0046

b) 3,570,000

Objective 5 **Perform arithmetic operations with numbers in scientific notation.** When performing calculations involving very large or very small numbers, using scientific notation can be very convenient.

EXAMPLE 14   Multiply $(2.2 \times 10^7)(2.8 \times 10^{13})$. Express your answer using scientific notation.

Solution

When multiplying two numbers that are in scientific notation, we may first multiply the two decimal numbers. We then multiply the powers of 10 using the product rule $x^m \cdot x^n = x^{m+n}$.

$$(2.2 \times 10^7)(2.8 \times 10^{13}) = (2.2)(2.8)(10^7)(10^{13})$$
$$= 6.16 \times 10^{20}$$

Reorder the factors. Multiply decimal numbers. Add the exponents for the base 10.

Quick Check **11** **Multiply** $(5.8 \times 10^4)(1.2 \times 10^9)$. **Express your answer using scientific notation.**

Using Your Calculator   The TI-83/84 can help us perform calculations with numbers in scientific notation. To enter a number in scientific notation, we use the second function **EE** above the key labeled ⊡ . For example, to enter the number 2.2×10^7, type 2.2 [2nd] ⊡ 7. Here is the screen shot showing how to multiply $(2.2 \times 10^7)(2.8 \times 10^{13})$.

```
2.2E7*2.8E13
              6.16E20
```

EXAMPLE 15   Divide $(1.2 \times 10^{-6}) \div (2.4 \times 10^{9})$. Express your answer using scientific notation.

Solution

To divide numbers that are in scientific notation, we begin by dividing the decimal numbers. We then divide the powers of 10 separately, using the property $\dfrac{x^m}{x^n} = x^{m-n}$.

$$(1.2 \times 10^{-6}) \div (2.4 \times 10^{9}) = \frac{1.2 \times 10^{-6}}{2.4 \times 10^{9}}$$ Rewrite as a fraction.

$$= \frac{1.2}{2.4} \times 10^{-6-9}$$ Divide the decimal numbers. Subtract the exponents for the powers of 10.

$$= 0.5 \times 10^{-15}$$ Divide 1.2 by 2.4. Simplify the exponent.

Although this is the correct quotient, our answer is not in scientific notation, as the number 0.5 does not have a nonzero digit to the left of the decimal point. In order to rewrite 0.5 as 5.0, we move the decimal point one place to the right. This will decrease the exponent by one from -15 to -16. $(1.2 \times 10^{-6}) \div (2.4 \times 10^{9}) = 5.0 \times 10^{-16}$.

Quick Check 12 Divide $(9.9 \times 10^{5}) \div (3.3 \times 10^{-7})$. Express your answer using scientific notation.

Applications

Objective 6   Use scientific notation to solve applied problems. We finish with an applied problem using scientific notation.

EXAMPLE 16   At its farthest point, the planet Mars is 155 million miles from the sun. How many seconds does it take for light from the sun to reach Mars if the speed of light is 1.86×10^{5} miles per second?

Solution

To determine the length of time that a trip takes, we divide the distance traveled by the rate of speed. To solve this problem, we need to divide the distance of 155 million miles by the speed of light. We will convert the distance to scientific notation, which is 1.55×10^{8}.

$$\frac{1.55 \times 10^{8}}{1.86 \times 10^{5}} \approx 0.833 \times 10^{3}$$ Divide the decimal numbers, approximate with a calculator. Subtract the exponents for the powers of 10.

To write this result in scientific notation, we move the decimal point one place to the right. This decreases the exponent by one. The light will reach Mars in approximately 8.33×10^{2} or 833 seconds, or 13 minutes and 53 seconds.

Quick Check 13 How many seconds does it take for light to travel 1,488,000,000 miles? (The speed of light is 1.86×10^{5} miles per second.)

Rewrite the expression without using negative exponents. (Assume all variables represent nonzero real numbers.)

1. 4^{-2} **2.** 3^{-2} **3.** 6^{-3} **4.** 5^{-3}

5. -7^{-3} **6.** -2^{-9} **7.** a^{-10}

8. a^{-8} **9.** $4x^{-3}$ **10.** $2x^{-9}$

11. $-5m^{-19}$ **12.** $-12m^{-6}$ **13.** $\dfrac{1}{x^{-5}}$

14. $\dfrac{1}{x^{-7}}$ **15.** $\dfrac{3}{y^{-4}}$ **16.** $\dfrac{x^7}{y^{-5}}$

17. $\dfrac{x^{-11}}{y^{-2}}$ **18.** $\dfrac{x^{-10}}{y^4}$ **19.** $\dfrac{-8a^{-6}b^5}{c^7}$

20. $\dfrac{-10x^3y^{-5}}{z^{-4}}$ **21.** $\dfrac{5a^{-6}b^{-9}}{c^{-4}d^5}$

22. $\dfrac{a^3b^{-3}}{c^{-4}d^{-9}}$

Simplify the expression. Write the result without using negative exponents. (Assume all variables represent nonzero real numbers.)

23. $x^{-7}\cdot x^{10}$ **24.** $x^8\cdot x^{-3}$ **25.** $a^{-5}\cdot a^{-11}$

26. $b^{-9}\cdot b^{-7}$ **27.** $m^4\cdot m^{-8}$

28. $n^{-11}\cdot n^9$ **29.** $\left(x^5\right)^{-4}$ **30.** $\left(x^{10}\right)^{-5}$

31. $\left(x^{-6}\right)^{-7}$ **32.** $\left(x^{-9}\right)^{-2}$

33. $\left(x^5y^4z^{-6}\right)^{-2}$ **34.** $\left(x^{-3}y^3z^{-2}\right)^{-7}$

35. $\left(4a^{-5}b^{-8}z^2\right)^{-3}$

36. $\left(-2x^4b^{-6}c^{-7}\right)^5$ **37.** $\dfrac{x^{-4}}{x^9}$

38. $\dfrac{x^5}{x^{-5}}$ **39.** $\dfrac{a^9}{a^{-3}}$ **40.** $\dfrac{a^{-8}}{a}$

41. $\dfrac{y^{-5}}{y^{-3}}$ **42.** $\dfrac{y^{-4}}{y^{-11}}$ **43.** $\dfrac{x^7}{x^{18}}$

44. $\dfrac{x^3}{x^{12}}$ **45.** $\dfrac{x^5\cdot x^{-11}}{x^{-6}}$ **46.** $\dfrac{x^{-13}}{x^{-17}\cdot x^4}$

47. $\left(\dfrac{x^3}{y^4}\right)^{-5}$ **48.** $\left(\dfrac{2x^8}{y^5}\right)^{-3}$

49. $\left(\dfrac{a^4b^7}{3c^5d^{-8}}\right)^{-2}$

50. $\left(\dfrac{x^{-8}y^5}{wz^{-6}}\right)^{-8}$

Convert the given number from scientific notation to decimal notation.

51. 3.07×10^{-7}

52. 2.3×10^5

53. 8.935×10^9

54. 6.001×10^{-8}

55. 9.021×10^4

56. 7.0×10^{10}

Convert the given number to scientific notation.

57. 0.0000045 **58.** $125{,}000$

59. $47{,}000{,}000$ **60.** 0.0031

61. 0.000000006

62. $5{,}000{,}000{,}000$

Perform the following calculations. Express your answer using scientific notation.

63. $(3.2\times 10^7)(2.9\times 10^{11})$

64. $(1.5\times 10^{-8})(6.1\times 10^{-9})$

65. $(2.0088\times 10^9)\div(5.4\times 10^{-6})$

66. $(2.6\times 10^{-13})\div(6.5\times 10^{10})$

67. $(5.32\times 10^{-15})(7.8\times 10^3)$

68. $(6.0\times 10^{13})\div(7.5\times 10^{-12})$

69. $(87{,}000{,}000{,}000)(0.000002)$

70. $0.0000000064\div 160{,}000{,}000{,}000$

71. If a computer can perform a calculation in 0.0000000008 seconds, how long would it take to perform $42{,}000{,}000{,}000{,}000$ calculations?

72. If a computer can perform a calculation in 0.0000000008 seconds, how many calculations can it perform in 2 minutes (120 seconds)?

73. The speed of light is 1.86×10^5 miles per second. How far can light from the sun travel in 10 minutes?

74. The speed of light is 1.86×10^5 miles per second. At its farthest point, the planet Pluto is 4,555,000,000 miles from the sun. How many seconds does it take for light from the sun to reach Pluto?

75. If the mass of a typical star is 2.2×10^{33} grams and there are approximately 500,000,000,000 stars in the Milky Way galaxy, what is the total mass of these stars?

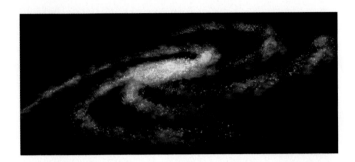

76. The speed of light is 1.86×10^5 miles per second. How far can light from the sun travel in 1 year? (This distance is often referred to as a "light-year.")

77. For the quarter ending in December 2003, Microsoft showed total revenues of $\$1.015 \times 10^{10}$ and Apple showed total revenues of $\$2.006 \times 10^9$. What was the combined revenue of the two companies for the quarter that ended in December 2003? (*Source:* Stargeek.com)

78. Burger King restaurants serve 7.7×10^6 customers daily. How many customers do Burger King restaurants serve in one year? (*Source:* Burger King Corporation)

79. The Intel Pentium 4 chip performs 4.2×10^7 calculations per second. How many calculations can it perform in one hour (3600 seconds)? (*Source:* Innovation Watch)

80. The Intel Pentium 4 chip performs 4.2×10^7 calculations per second. How many calculations can it perform in one year (31,536,000 seconds)? (*Source:* Innovation Watch)

81. The pie chart shows the percent of Burger King's business that comes from take-out sales and drive-thru sales, as well as from customers who eat in the restaurant. If Burger King's sales totaled to $\$1.11 \times 10^{10}$ in 2003, what were the total sales to drive-thru customers? (*Source:* Burger King Corporation)

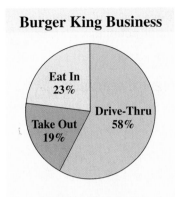

Burger King Business

Eat In 23%
Drive-Thru 58%
Take Out 19%

82. The pie chart shows the percent of U.S. college students who received some type of financial aid in 1999–2000. If there were approximately 1.67×10^7 U.S. college students in 1999–2000, how many of them received some type of financial aid? (*Source:* U.S. Department of Education, National Center for Educational Statistics, National Postsecondary Student Aid Study, 1999–2000)

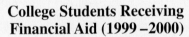

College Students Receiving Financial Aid (1999–2000)

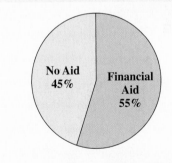

No Aid 45%
Financial Aid 55%

Answer in complete sentences.

83. Explain the difference between the expressions 4^{-2} and -4^2.

84. Describe three real-world examples of numbers using scientific notation.

85. Explain the statement "When performing calculations involving very large or very small numbers, using scientific notation can be very convenient."

QUICK REVIEW EXERCISES

Section 5.2

Simplify.

1. $8x + 3x$

2. $9a - 16a$

3. $-9x^2 + 4x + 3x^2 + 7x$

4. $10y^3 - y^2 + 5y + 13 + 2y^3 + 6y^2 - 18y - 50$

Study Tip **REVISITED**   As soon as you receive your test, write down any formulas, rules, or procedures that will help you during the exam. Once you have written these down, you can refer to them as you work through the test; for example, a table or formula that you use to solve a particular type of word problem. Write it down on the test while it is still fresh in your memory. That way, when you reach the appropriate problem on the test, you will not need to worry about not being able to remember the table or formula.

By writing all of this information on your test, you will eliminate memorization difficulties that arise when taking a test. However, if you do not understand the material or how to use what you have written down, then you will struggle on the test. There is no substitute for understanding what you are doing.

5.3

POLYNOMIALS; ADDITION AND SUBTRACTION OF POLYNOMIALS

Objectives

1. Identify polynomials and understand the vocabulary used to describe them.
2. Evaluate polynomials.
3. Add and subtract polynomials.
4. Understand polynomials in several variables.

Polynomials in a Single Variable

Objective 1 Identify polynomials and understand the vocabulary used to describe them. Many real-world phenomena cannot be described by linear expressions and linear functions. For example, if a car is traveling at x miles per hour on dry pavement, the distance in feet required for the car to come to a complete stop can be approximated by the expression $0.06x^2 + 1.1x + 0.02$. This nonlinear expression is an example of a **polynomial.**

Polynomials

A **polynomial in a single variable x** is a sum of **terms** of the form ax^n, where a is a real number and n is a whole number.

Here are some examples of a polynomial in a single variable x:

$$x^2 - 9x + 14, \qquad 4x^5 - 13, \qquad x^7 - 4x^5 + 8x^2 + 13x$$

An expression is *not* a polynomial if it contains a term with a variable that is raised to a power other than a whole number (such as $x^{3/4}$ or x^{-2}), or if it has a term with a variable in a denominator $\left(\text{such as } \dfrac{5}{x^3}\right)$.

Degree of a Term

For a polynomial in a single variable, the **degree of a term** is equal to the variable's exponent.

For example, the degree of the term $8x^9$ is 9. A **constant term** is a term that does not contain a variable. The degree of a constant term is 0, because a constant term such as -15 can be rewritten as $-15x^0$. The polynomial $4x^5 - 9x^4 + 2x^3 - 7x - 11$ has five terms, and their degrees are 5, 4, 3, 1, and 0 respectively.

Term	$4x^5$	$-9x^4$	$2x^3$	$-7x$	-11
Degree	5	4	3	1	0

A polynomial with only one term is called a **monomial.** A **binomial** is a polynomial that has two terms, while a **trinomial** is a polynomial that has three terms. (We do not have special names to describe polynomials with four or more terms.) Here are some examples:

Monomial	Binomial	Trinomial
$2x$	$x^2 - 25$	$x^2 - 15x - 76$
$9x^2$	$7x + 3$	$4x^2 + 12x + 9$
$-4x^3$	$5x^3 - 135$	$x^6 + 11x^5 - 26x^4$

EXAMPLE 1   Classify the polynomial as a monomial, binomial, or trinomial. List the degree of each term.
a) $16x^4 - 81$ **b)** $-13x^7$ **c)** $x^6 - 7x^3 - 18x$

Solution

a) $16x^4 - 81$ has two terms, so it is a binomial. The degrees of its two terms are 4 and 0.
b) $-13x^7$ has only one term, so it is a monomial. The degree of this term is 7.
c) $x^6 - 7x^3 - 18x$ is a trinomial as it has three terms. The degrees of those terms are 6, 3, and 1.

Quick Check 1
Classify as a monomial, binomial, or trinomial. List the degree of each term.

a) $x^2 - 9x + 20$
b) $x^2 - 81$
c) $-92x^7$

The **coefficient** of a term is the numerical part of a term. When determining the coefficient of a term, be sure to include its sign. For the polynomial $8x^3 - 6x^2 - 5x + 13$, the four terms have coefficients 8, -6, -5, and 13 respectively.

Term	$8x^3$	$-6x^2$	$-5x$	13
Coefficient	8	-6	-5	13

EXAMPLE 2   For the trinomial $x^2 - 9x + 18$, determine the coefficient of each term.

Solution

The first term of this trinomial, x^2, has a coefficient of 1. Even though we do not see a coefficient in front of the term, the coefficient is 1 because $x^2 = 1 \cdot x^2$. The second term, $-9x$, has a coefficient of -9. Finally, the coefficient of the constant term is 18.

Quick Check 2
For the polynomial $x^3 - 6x^2 - 11x + 32$, determine the coefficient of each term.

We will write polynomials in **descending order,** writing the term of highest degree first, followed by the term of next highest degree, and so on. For example, the polynomial $3x^2 - 9 - 5x^5 + 7x$ is written in descending order as $-5x^5 + 3x^2 + 7x - 9$. The term of highest degree is called the **leading term,** and its coefficient is called the **leading coefficient.**

Degree of a Polynomial

The degree of the leading term is also called the **degree of the polynomial.**

EXAMPLE ▸ 3   Rewrite the polynomial $7x^2 - 9 + 3x^5 - x$ in descending order and identify the leading term, the leading coefficient, and the degree of the polynomial.

Solution

Quick Check **3**
Rewrite the polynomial $x + 9 + 4x^2 - x^3$ in descending order and identify the leading term, the leading coefficient, and the degree of the polynomial.

To write a polynomial in descending order, we write the terms according to their degree from highest to lowest. In descending order, this polynomial is $3x^5 + 7x^2 - x - 9$. The leading term is $3x^5$, and the leading coefficient is 3. The leading term has degree 5, so this polynomial has degree 5.

Evaluating Polynomials

Objective **2**   **Evaluate polynomials.** To **evaluate a polynomial** for a particular value of a variable, we substitute the value for the variable in the polynomial and simplify the resulting expression.

EXAMPLE ▸ 4   Evaluate $4x^5 - 7x^3 + 12x^2 - 13x$ for $x = -2$.

Solution

$$\begin{aligned} & 4(-2)^5 - 7(-2)^3 + 12(-2)^2 - 13(-2) && \text{Substitute } -2 \text{ for } x. \\ & = 4(-32) - 7(-8) + 12(4) - 13(-2) && \text{Perform operations involving} \\ & && \text{exponents.} \\ & = -128 + 56 + 48 + 26 && \text{Multiply.} \\ & = 2 && \text{Simplify.} \end{aligned}$$

Quick Check **4**
Evaluate $x^4 - 5x^3 - 6x^2 + 10x - 21$ for $x = -3$.

A **polynomial function** is a function that is described by a polynomial, such as $f(x) = x^3 - 5x^2 - 5x + 7$. Linear functions of the form $f(x) = mx + b$ are first degree polynomial functions.

EXAMPLE ▸ 5   For the polynomial function $f(x) = x^2 - 9x + 16$, find $f(3)$.

Solution

Recall that the notation $f(3)$ tells us to substitute 3 for x in the function. After substituting we simplify the resulting expression.

$$\begin{aligned} f(3) & = (3)^2 - 9(3) + 16 && \text{Substitute 3 for } x. \\ & = 9 - 9(3) + 16 && \text{Perform operations involving exponents.} \\ & = 9 - 27 + 16 && \text{Multiply.} \\ & = -2 && \text{Simplify.} \end{aligned}$$

Quick Check **5**
For the polynomial function $f(x) = x^3 + 12x^2 - 21x$, find $f(-5)$.

Adding and Subtracting Polynomials

Objective **3**   **Add and subtract polynomials.** Just as we can perform arithmetic operations combining two numbers into one number, we can perform operations combining two polynomials into one polynomial, such as adding and subtracting.

Adding Polynomials

To add two polynomials together, we combine their like terms.

Recall that two terms are like terms if they have the same variables with the same exponents.

EXAMPLE 6   Add $(7x^2 - 5x - 8) + (4x^2 - 9x - 17)$.

Solution

To add these polynomials, we drop the parentheses and combine like terms.

$$(7x^2 - 5x - 8) + (4x^2 - 9x - 17) = 7x^2 - 5x - 8 + 4x^2 - 9x - 17$$

Rewrite without using parentheses.

$$= 11x^2 - 14x - 25 \qquad \text{Combine like terms. } 7x^2 + 4x^2 = 11x^2,$$
$$-5x - 9x = -14x, -8 - 17 = -25$$

Quick Check   6 Add $(x^3 - x^2 - 9x + 25) + (x^2 + 12x + 144)$.

A Word of Caution   The terms $11x^2$ and $-14x$ are not like terms, as the variables do not have the same exponents.

> To subtract one polynomial from another, such as $(x^2 + 5x - 9) - (3x^2 - 2x + 15)$, we change the sign of each term that is being subtracted and then combine like terms. Changing the sign of each term in the parentheses being subtracted is equivalent to applying the distributive property with -1.

EXAMPLE 7   Subtract $(x^2 + 5x - 9) - (3x^2 - 2x + 15)$.

Solution

We remove the parentheses by changing the sign of each term in the polynomial that is being subtracted. We then can combine like terms as follows:

$$(x^2 + 5x - 9) - (3x^2 - 2x + 15) = x^2 + 5x - 9 - 3x^2 + 2x - 15$$

Distribute to remove parentheses.

$$= -2x^2 + 7x - 24 \qquad \text{Combine like terms.}$$

Quick Check   7 Subtract $(5x^2 + 6x + 27) - (3x^3 + 6x^2 - 9x + 22)$.

A Word of Caution   When subtracting a polynomial from a second polynomial, be sure to change the sign of each term in the polynomial that is being subtracted.

EXAMPLE 8   Given the polynomial functions $f(x) = x^2 + 5x + 7$ and $g(x) = 2x^2 - 7x + 11$, find $f(x) - g(x)$.

Solution

We will substitute the appropriate expressions for $f(x)$ and $g(x)$, and simplify the resulting expression.

$$
\begin{aligned}
f(x) - g(x) &= (x^2 + 5x + 7) - (2x^2 - 7x + 11) && \text{Substitute.} \\
&= x^2 + 5x + 7 - 2x^2 + 7x - 11 && \text{Distribute to remove} \\
& && \text{parentheses.} \\
&= -x^2 + 12x - 4 && \text{Combine like terms.}
\end{aligned}
$$

Quick Check **8** Given the polynomial functions
$f(x) = 4x^2 + 30x - 45$ and $g(x) = -2x^2 + 17x + 52$, find $f(x) - g(x)$.

Polynomials in Several Variables

Objective **4** **Understand polynomials in several variables.** While the polynomials that we have examined to this point contained a single variable, some polynomials contain two or more variables. Here are some examples:

$$
x^2y^2 + 3xy - 10 \qquad 7a^3 - 8a^2b + 3ab^2 - 15b^3 \qquad x^2 + 2xh + h^2 + 5x + 5h
$$

We evaluate polynomials in several variables by substituting values for each variable and simplifying the resulting expression.

EXAMPLE **9** Evaluate the polynomial $x^2 - 8xy + 2y^2$ for $x = 5$ and $y = -4$.

Solution

We will substitute 5 for x and -4 for y. Be careful to substitute the right value for each variable.

$$
\begin{aligned}
x^2 &- 8xy + 2y^2 \\
(5)^2 &- 8(5)(-4) + 2(-4)^2 && \text{Substitute 5 for } x \text{ and } -4 \text{ for } y. \\
&= 25 - 8(5)(-4) + 2(16) && \text{Perform operations involving exponents.} \\
&= 25 + 160 + 32 && \text{Multiply.} \\
&= 217 && \text{Add.}
\end{aligned}
$$

Quick Check **9**
Evaluate the polynomial
$b^2 - 4ac$ for $a = 1$,
$b = -9$, and $c = -52$.

For a polynomial in several variables, the degree of a term is equal to the sum of the exponents of its variable factors. For example, the degree of the term $3x^2y^5$ is $2 + 5$ or 7.

EXAMPLE **10** Find the degree of each term in the polynomial $a^4b^3 - 3a^7b^5 + 9ab^6$. In addition, find the degree of the polynomial.

Quick Check **10** **Solution**

Find the degree of each
term in the polynomial
$x^5 - 3x^2y^6 + 8x^9y$. In
addition, find the degree
of the polynomial.

The first term, a^4b^3, has degree 7. The second term, $3a^7b^5$, has degree 12. The third term, $9ab^6$, has degree 7. The degree of a polynomial is equal to the highest degree of any of its terms, so the degree of this polynomial is 12.

To add or subtract polynomials in several variables, we need to combine like terms. Two terms are like terms if they have the same variables with the same exponents.

EXAMPLE 11   Add $4xy^2 + 7x^3y^5$ and $9x^2y - 3x^3y^5$.

Solution

Quick Check 11

Add $5x^4y^2 - 6x^3y^3$ and $8x^4y^2 + 2x^3y^3$.

$(4xy^2 + 7x^3y^5) + (9x^2y - 3x^3y^5)$ Write as a sum.

$= 4xy^2 + 7x^3y^5 + 9x^2y - 3x^3y^5$ Rewrite without parentheses.

$= 4xy^2 + 4x^3y^5 + 9x^2y$ Combine like terms ($7x^3y^5$ and $-3x^3y^5$).

EXERCISES **5.3**

List the degree of each term in the given polynomial.

1. $9x^4 - 7x^2 + 3x - 8$

2. $-4x^3 + 10x^2 + 5x + 4$

3. $13x - 8x^7 - 11x^4$

4. $5x^5 - 3x^4 + 2x^3 + x^2 - 15$

List the coefficient of each term in the given polynomial.

5. $7x^2 + x - 15$

6. $3x^3 + x^2 - 8x - 13$

7. $10x^5 - 17x^4 + 6x^3 - x^2 + 2$

8. $-5x^3 - 12x^2 - 9x - 4$

Identify the given polynomial as a monomial, binomial or trinomial.

9. $x^2 - 9x + 20$ **10.** $15x^9$

11. $4x - 7$ **12.** $2x^5 + 13x^4$

13. $-8x^3$ **14.** $5x^5 - 8x^3 + 17x$

Rewrite the polynomial in descending order. Identify the leading term, the leading coefficient, and the degree of the polynomial.

15. $8x - 7 + 3x^2$

16. $13 - 9x - x^2$

17. $6x^2 - 11x + 2x^4 + 10$

18. $2x^5 - x^6 + 24 - 3x^3 + x - 5x^2$

Evaluate the polynomial for the given value of the variable.

19. $x^2 + 5x + 8$ for $x = 2$

20. $2x^2 + 3x + 7$ for $x = 5$

21. $x^3 - 9x - 10$ for $x = -4$

22. $x^5 - 8x^2 + 15x$ for $x = -3$

23. $-3x^4 - 8x^2 + 21x$ for $x = 3$

24. $-x^4 + 3x^3 + 12$ for $x = 6$

Evaluate the given polynomial function.

25. $f(x) = x^2 - 7x - 10$, $f(8)$

26. $f(x) = x^2 + 12x - 13$, $f(-9)$

27. $g(x) = 3x^3 - 8x^2 + 5x + 9$, $g(-5)$

28. $g(x) = -2x^3 - 5x^2 + 10x + 35$, $g(6)$

Add or subtract.

29. $(5x^2 + 8x - 11) + (3x^2 - 14x + 14)$

30. $(2x^2 - 9x - 35) + (8x^2 - 6x + 17)$

31. $(4x^2 - 7x + 30) - (2x^2 + 10x - 50)$

32. $(x^2 + 12x - 42) - (6x^2 - 19x - 23)$

33. $(2x^4 + 7x^2 - 13x - 20) + (x^3 - 5x^2 - 14x)$

34. $(3x^5 - 5x^3 + 2x^2) - (8x^5 - 5x^3 - 2x^2 + 7)$

35. $(x^3 - 5x^2 + 17x) - (4x^2 - 3x + 16)$

36. $(3x^2 + 10x - 11) + (-4x^3 - 9x^2 + 32x)$

37. $(2x^9 - 5x^4 + 7x^2) - (-7x^6 + 4x^5 - 12)$

38. $(9x^3 + x^2 - x - 13) - (9x^3 + x^2 - x - 13)$

Find the missing polynomial.

39. $(3x^2 + 8x + 11) + ? = 7x^2 + 5x - 2$

40. $(2x^3 - 4x^2 - 7x + 19) + ? = x^2 + 3x + 24$

41. $(3x^4 - 5x^3 + 6x + 12) - ? =$
$x^4 - 2x^3 - x^2 - 5x + 1$

42. $? - (6x^3 - 11x^2 - 16x + 22) =$
$-x^3 + 19x^2 - 12x + 9$

For the given functions $f(x)$ and $g(x)$, find $f(x) + g(x)$ and $f(x) - g(x)$.

43. $f(x) = 7x^2 + 10x + 3$, $g(x) = 5x^2 - 9x + 6$

44. $f(x) = 2x^2 + 11x - 5$, $g(x) = -x^2 - 5x + 20$

45. $f(x) = x^3 - 8x - 31$, $g(x) = 4x^3 + x^2 + 3x + 25$

46. $f(x) = 2x^3 - 4x^2 + 8x - 16$, $g(x) = x^3 + 12x^2 - 15$

Evaluate the polynomial.

47. $x^2 - 7xy + 10y^2$ for $x = 2$ and $y = 5$

48. $3x^2 + 11xy + y^2$ for $x = 4$ and $y = 3$

49. $b^2 - 4ac$ for $a = -9$, $b = -2$ and $c = 4$

50. $b^2 - 4ac$ for $a = 5$, $b = -1$ and $c = 6$

51. $3x^2yz^3 - 4xy^2z - 6x^4y^2z$ for $x = 4$, $y = 5$ and $z = -2$

52. $7x^2yz + 9xy^2z - 10xyz^2$ for $x = 9$, $y = 3$ and $z = -2$

For each polynomial, list the degree of each term as well as the degree of the polynomial.

53. $6x^5y^5 - 7x^3y^3 + 9x^2y$

54. $-4x^8y^6 + 3x^3y^7 + x^5y^{11}$

55. $x^5y^2z - 5x^3y^3z^3 + 6x^2y^5z^4$

56. $a^{12}b^9c^{10} + 2a^{11}b^6c^{16} + 4a^{10}b^{10}c^3$

Add or subtract.

57. $(15x^2y + 8xy^2 - 7x^2y^3) + (-8x^2y + 3xy^2 + 4x^2y^3)$

58. $(x^4y - 5x^2y^3 - 14y^5) - (6x^4y + 11x^2y^3 - 7y^5)$

59. $(a^3b^2 + 11a^5b + 24a^2b^3) -$
$(19a^2b^3 - 2a^5b + 15a^3b^2)$

60. $(9m^4n^2 - 13m^3n^3 + 11m^2n^4) +$
$(-4m^2n^4 - 3m^4n^2 - m^3n^3)$

61. $(2x^3yz^2 - xy^4z^3 - 10x^2y^2z^5) -$
$(4x^2yz^3 + 14xy^4z^3 - 3x^2y^2z^5)$

62. $(22x^3y^9 - 21x^6y^4 - 7x^8y^2) -$
$(22x^3y^9 - 21x^6y^4 - 7x^8y^2)$

Answer in complete sentences.

63. Explain why it is a good idea to use parentheses when evaluating a polynomial for negative values of a variable. Do you feel that using parentheses is a good idea for evaluating any polynomial?

64. A classmate made the following error when subtracting two polynomials.

$(3x^2 + 2x - 7) - (x^2 - x + 9)$
$= 3x^2 + 2x - 7 - x^2 - x + 9$

Explain the error to your classmate, and give advice on how to avoid making this error.

Study Tip **REVISITED**   In the same way that you would begin to solve a word problem, you should begin to take a test by briefly reading through it. This will give you an idea of how many problems you have to solve, how many word problems there are, and roughly how much time you can devote to each problem. It is a good idea to establish a schedule, such as "I need to be done with 12 problems by the time half of the class period is over." This way you will know whether you need to speed up during the second half of the exam or if you have plenty of time.

Not all problems are assigned the same point value. Identify which problems are worth more points than others. It is important to not be in a position of having to rush through the problems that have higher point values because you didn't notice that they were worth more until you got to them at the end of the test.

5.4
MULTIPLYING POLYNOMIALS

Objectives

1 Multiply monomials.
2 Multiply a monomial by a polynomial.
3 Multiply polynomials.
4 Find special products.

Multiplying Monomials

Objective 1 **Multiply monomials.** Now that we have learned how to add and subtract polynomials, the next operation involving polynomials we will explore is multiplication. We begin by learning how to multiply monomials.

Multiplying Monomials

> When multiplying monomials, we begin by multiplying their coefficients. Then we multiply the variable factors, using the property of exponents which states that, for any real number x, $x^m \cdot x^n = x^{m+n}$.

EXAMPLE 1 Multiply $3x \cdot 5x$.

Solution

We multiply the coefficients first, and then we multiply variables.

$$
\begin{aligned}
3x \cdot 5x &= 3 \cdot 5 \cdot x \cdot x && \text{Multiply coefficients, then multiply variables.} \\
&= 15x^{1+1} && \text{Multiply variables using the property } x^m \cdot x^n = x^{m+n}. \\
&= 15x^2 && \text{Simplify the exponent.}
\end{aligned}
$$

Quick Check 1
Multiply $7x \cdot 8x$.

EXAMPLE 2 Multiply $2x^4y^3 \cdot 3x^2y^7z^4$.

Solution

The procedure for multiplying monomials containing more than one variable is the same as multiplying monomials containing a single variable. After multiplying the coefficients, we multiply the variables one variable at a time.

$$2x^4y^3 \cdot 3x^2y^7z^4 = 6x^6y^{10}z^4 \qquad \text{Multiply coefficients, then multiply variables.}$$

Notice that the variable z was only a factor of one of the monomials. The exponent of z was not changed.

Quick Check 2
Multiply $5x^6yz^7 \cdot 12x^8z^6$.

Multiplying a Monomial by a Polynomial

Objective 2 **Multiply a monomial by a polynomial.** We now advance to multiplying a monomial by a polynomial containing two or more terms, such as $3x(2x^2 - 5x + 2)$. To do this, we use the distributive property $a(b + c) = ab + ac$. To find the product

$3x(2x^2 - 5x + 2)$, we multiply the monomial $3x$ by each term of the polynomial $2x^2 - 5x + 2$.

Multiplying a Monomial by a Polynomial

> To multiply a monomial by a polynomial containing two or more terms, multiply the monomial by each term of the polynomial.

EXAMPLE 3   Multiply $3x(2x^2 - 5x + 2)$.

Solution

We begin by distributing the monomial $3x$ to each term of the polynomial.

$$3x(2x^2 - 5x + 2) = 3x \cdot 2x^2 - 3x \cdot 5x + 3x \cdot 2 \qquad \text{Distribute } 3x.$$
$$= 6x^3 - 15x^2 + 6x \qquad \text{Multiply.}$$

Quick Check 3

Multiply
$4x(7x^3 - 6x^2 + 5x - 8).$

Although we will continue to show the distribution of the monomial, your goal should be to perform this task mentally.

EXAMPLE 4   Multiply $-8x^3(x^5 - 4x^4 - 9x^2)$.

Solution

Notice that the coefficient of the monomial being distributed is negative. We must distribute $-8x^3$, and multiplying by this changes the sign of each term in the polynomial.

$$-8x^3(x^5 - 4x^4 - 9x^2) = (-8x^3)(x^5) - (-8x^3)(4x^4) - (-8x^3)(9x^2)$$
$$\text{Distribute } -8x^3.$$
$$= -8x^8 - (-32x^7) - (-72x^5) \qquad \text{Multiply.}$$
$$= -8x^8 + 32x^7 + 72x^5 \qquad \text{Simplify.}$$

Quick Check 4   Multiply $-3x^5(-2x^4 + x^3 - x^2 - 15x + 21).$

> *A Word of Caution*   When multiplying a polynomial by a term with a negative coefficient, be sure to change the sign of each term in the polynomial.

Multiplying Polynomials

Objective 3   **Multiply polynomials.**

Multiplying Two Polynomials

> To multiply two polynomials when each contains two or more terms, we multiply each term in the first polynomial by each term in the second polynomial.

Suppose that we wanted to multiply $(x + 9)(x + 7)$. We could distribute the factor $(x + 9)$ to each term in the second polynomial as follows:

$$(x + 9)(x + 7) = (x + 9) \cdot x + (x + 9) \cdot 7$$

We could then perform the two multiplications by first distributing x in the first product and then distributing 7 in the second product.

$$(x + 9)(x + 7) = (x + 9) \cdot x + (x + 9) \cdot 7$$
$$= x \cdot x + 9 \cdot x + x \cdot 7 + 9 \cdot 7$$
$$= x^2 + 9x + 7x + 63$$
$$= x^2 + 16x + 63$$

We end up with each term in the first polynomial being multiplied by each term in the second polynomial.

EXAMPLE 5   Multiply $(x + 6)(x + 4)$.

Solution

We begin by taking the first term, x, in the first polynomial and multiplying it by each term in the second polynomial. We then repeat this for the second term, 6, in the first polynomial.

$$(x + 6)(x + 4) = x \cdot x + x \cdot 4 + 6 \cdot x + 6 \cdot 4$$ Distribute the term x from the first polynomial, then distribute 6.
$$= x^2 + 4x + 6x + 24$$ Multiply.
$$= x^2 + 10x + 24$$ Combine like terms. ($4x$ and $6x$)

> **Quick Check 5**
> Multiply
> $(x + 11)(x + 9)$.

We often refer to the process of multiplying a binomial by another binomial as "FOIL." FOIL is an acronym for **F**irst, **O**uter, **I**nner, **L**ast, which describes the four multiplications that occur when we multiply two binomials. Here are the four multiplications that were performed in the previous example:

First	**O**uter	**I**nner	**L**ast
$x \cdot x$	$x \cdot 4$		
$(x + 6)(x + 4)$	$(x + 6)(x + 4)$	$(x + 6)(x + 4)$	$(x + 6)(x + 4)$
		$6 \cdot x$	$6 \cdot 4$

The term FOIL applies only when multiplying a binomial by another binomial. If we are multiplying a binomial by a trinomial, there are six multiplications that must be performed and the term FOIL cannot be used. Keep in mind that each term in the first polynomial must be multiplied by each term in the second polynomial.

EXAMPLE 6   Multiply $(x - 6)(4x - 5)$.

Solution

When multiplying polynomials we must be careful with our signs. When we distribute the second term of $x - 6$ to the second polynomial, we must distribute a negative 6.

$$(x - 6)(4x - 5) = x \cdot 4x - x \cdot 5 - 6 \cdot 4x + 6 \cdot 5$$ Distribute (FOIL). The product of two negative numbers is positive: $(-6)(-5) = 6 \cdot 5$.

> **Quick Check 6**
> Multiply
> $(5x - 2)(2x - 9)$.

$$= 4x^2 - 5x - 24x + 30$$ Multiply.
$$= 4x^2 - 29x + 30$$ Combine like terms.

EXAMPLE 7 Given the functions $f(x) = 3x - 2$ and $g(x) = x^2 + 4x - 7$, find $f(x) \cdot g(x)$.

Solution

We will substitute the appropriate expressions for $f(x)$ and $g(x)$ in parentheses, and simplify the resulting expression. We need to multiply each term in the first polynomial by each term in the second polynomial.

$$f(x) \cdot g(x) = (3x - 2)(x^2 + 4x - 7) \qquad \text{Substitute for } f(x) \text{ and } g(x).$$

$$= 3x \cdot x^2 + 3x \cdot 4x - 3x \cdot 7 - 2 \cdot x^2 - 2 \cdot 4x + 2 \cdot 7 \qquad \text{Distribute.}$$

$$= 3x^3 + 12x^2 - 21x - 2x^2 - 8x + 14 \qquad \text{Multiply.}$$

$$= 3x^3 + 10x^2 - 29x + 14 \qquad \text{Combine like terms.}$$

Quick Check 7
Given the functions
$f(x) = x + 8$ and
$g(x) = 3x^2 + 6x - 2$,
find $f(x) \cdot g(x)$.

Special Products

Objective 4 Find special products. We finish this section by examining some *special products*. The first special product is of the form $(a + b)(a - b)$. In words, this is the product of the sum and the difference of two terms. Here is the multiplication:

$$(a + b)(a - b) = a^2 - ab + ab - b^2 \qquad \text{Distribute.}$$

$$= a^2 - b^2 \qquad \text{Combine like terms.}$$

Notice that when we were combining like terms, two of the terms were opposites. This left us with only two terms. We can use this result to assist us anytime we multiply two binomials of the form $(a + b)(a - b)$.

$$(a + b)(a - b) = a^2 - b^2$$

EXAMPLE 8 Multiply $(x + 7)(x - 7)$.

Solution

$$(x + 7)(x - 7) = x^2 - 7^2 \qquad \text{Multiply using the pattern } (a + b)(a - b) = a^2 - b^2.$$

$$= x^2 - 49 \qquad \text{Simplify.}$$

Quick Check 8
Multiply
$(x + 10)(x - 10)$.

EXAMPLE 9 Multiply $(3x - 5)(3x + 5)$.

Solution

Although the difference is listed first, we can still multiply using the same pattern.

$$(3x - 5)(3x + 5) = (3x)^2 - 5^2 \qquad \text{Multiply using the pattern } (a + b)(a - b) = a^2 - b^2.$$

$$= 9x^2 - 25 \qquad \text{Simplify.}$$

Quick Check 9
Multiply
$(2x + 7)(2x - 7)$.

The other special product we will examine is the square of a binomial, such as $(x + 3)^2$ or $(9x + 4)^2$. Here are the patterns for squaring binomials of the form $a + b$ and $a - b$:

$$(a + b)^2 = a^2 + 2ab + b^2$$

$$(a - b)^2 = a^2 - 2ab + b^2$$

Let's derive the first of these patterns.

$$(a + b)^2 = (a + b)(a + b) \qquad \text{To square a binomial, we multiply it by itself.}$$
$$= a^2 + ab + ab + b^2 \qquad \text{Distribute (FOIL).}$$
$$= a^2 + 2ab + b^2 \qquad \text{Combine like terms.}$$

The second pattern can be derived in the same fashion.

EXAMPLE 10   Multiply $(x - 3)^2$.

Solution

We can use the pattern $(a - b)^2 = a^2 - 2ab + b^2$, substituting x for a and 3 for b.

Quick Check 10
Multiply $(5x - 8)^2$.

$$a^2 - 2ab + b^2$$
$$x^2 - 2 \cdot x \cdot 3 + 3^2 \qquad \text{Substitute } x \text{ for } a \text{ and 3 for } b.$$
$$= x^2 - 6x + 9 \qquad \text{Multiply.}$$

EXAMPLE 11   Multiply $(2x + 5)^2$.

Solution

We can use the pattern $(a + b)^2 = a^2 + 2ab + b^2$, substituting $2x$ for a and 5 for b.

$$a^2 + 2ab + b^2$$
$$(2x)^2 + 2 \cdot (2x) \cdot 5 + 5^2 \qquad \text{Substitute } 2x \text{ for } a \text{ and 5 for } b.$$

Quick Check 11
Multiply $(x + 6)^2$.

$$= 4x^2 + 20x + 25 \qquad \text{Multiply.}$$

Although the patterns developed for these three special products may help to save time when multiplying, keep in mind that we can find products of these types by multiplying as we did earlier in this section. When we square a binomial, we can start by rewriting the expression as the product of the binomial and itself. For example, we could rewrite $(8x - 7)^2$ as $(8x - 7)(8x - 7)$ and then multiply.

EXERCISES 5.4

Multiply.

1. $5x \cdot 7x^3$

2. $8x^2 \cdot 3x^2$

3. $-6x^5 \cdot 3x^4$

4. $-2x^6(-11x^9)$

5. $12x^3y^4 \cdot 8x^5y^2$

6. $7x^{10}y^7 \cdot 6x^{13}y^9$

7. $-5x^3y^2z^4 \cdot 14xz^5w^4$

8. $3a^5b^8c \cdot 9bc^7d^4$

9. $2x^5 \cdot 6x^3 \cdot 7x^2$

10. $-12x^8 \cdot 5x^{10} \cdot 6x^9$

Find the missing monomial.

11. $3x^4 \cdot ? = 15x^7$

12. $8x^6 \cdot ? = 56x^{11}$

13. $2x^9 \cdot ? = -24x^{18}$

14. $-6x^{10} \cdot ? = 84x^{23}$

Multiply.

15. $2(5x - 8)$

16. $7(3x + 2)$

17. $-4(2x - 9)$

18. $-3(4x + 7)$

19. $6x(3x + 5)$

20. $9x(4x - 13)$

21. $x^3(3x^2 - 4x + 7)$

22. $-x^3(2x^5 + 9x^4 - 6x)$

23. $2xy^2(3x^2 - 6xy + 7y^2)$

24. $11x^4y^3(2xy - 5x^4y^2 - 7x)$

Find the missing factor.

25. $?(2x^2 - 7x - 10) = 6x^5 - 21x^4 - 30x^3$

26. $?(3x^4 + 9x^2 + 16) = 18x^9 + 54x^7 + 96x^5$

27. $4x^4(?) = 12x^7 - 20x^6 - 48x^4$

28. $8x^9(?) = 56x^{17} + 104x^{14} - 96x^{11}$

Multiply.

29. $(x + 2)(x + 3)$

30. $(x + 6)(x - 2)$

31. $(x - 9)(x - 10)$

32. $(x - 3)(x - 8)$

33. $(2x + 1)(x - 5)$

34. $(5x - 4)(2x + 5)$

35. $(3x + 2)(x^2 - 5x - 9)$

36. $(7x - 4)(49x^2 + 28x + 16)$

37. $(x^2 - 7x + 10)(x^2 + 3x - 40)$

38. $(2x^2 - 5x - 8)(x^2 + 3x + 9)$

39. $(x + 2y)(x - 4y)$

40. $(x - 6y)(x - 7y)$

Find the missing term(s).

41. $(x + ?)(x + 5) = x^2 + 8x + 15$

42. $(x - 7)(x + ?) = x^2 + 3x - 70$

43. $(x + ?)(x + ?) = x^2 + 9x + 18$

44. $(? - 5)(? + 3) = 2x^2 + x - 15$

For the given functions $f(x)$ and $g(x)$, find $f(x) \cdot g(x)$.

45. $f(x) = x - 9, g(x) = x + 2$

46. $f(x) = 4x^3, g(x) = x^2 - 8x - 14$

47. $f(x) = 6x^5, g(x) = -3x^4$

48. $f(x) = 2x^7, g(x) = 11x^6$

49. $f(x) = x^3 - 5x^2 + 8x + 3, g(x) = -4x^5$

50. $f(x) = 7x + 4, g(x) = 6x - 13$

Find the special product using the appropriate formula.

51. $(x + 8)(x - 8)$

52. $(x - 13)(x + 13)$

53. $(2x + 5)(2x - 5)$

54. $(4x - 9)(4x + 9)$

55. $(x + 6)^2$

56. $(x - 1)^2$

57. $(2x - 3)^2$

58. $(4x + 7)^2$

Find the missing factor or term.

59. $(x + 9) \cdot ? = x^2 - 81$

60. $(5x + 7) \cdot ? = 25x^2 - 49$

61. $(?)^2 = x^2 - 12x + 36$

62. $(?)^2 = 4x^2 + 20x + 25$

Multiply.

63. $(x + 7)(x - 1)$

64. $2x^5 \cdot 3x^2$

65. $5x(x^2 - 8x - 9)$

66. $(x + 2)(x - 2)$

67. $-7x^2y^3 \cdot 4x^4y^6$

68. $(x - 7)^2$

69. $-5x^6(-4x^8)$

70. $(2x + 15)(3x - 7)$

71. $(5x + 3)(5x - 3)$

72. $-3x^2(x^2 - 10x - 13)$

73. $(x + 13)^2$

74. $(x + 8)(3x^2 + 7x - 6)$

75. $2x^2y(3x^2 - x^4y^3 - 7y)$

76. $9x^3(-6x^3)$

77. $(4x - 9)^2$

78. $(6 + 5x)^2$

79. $(x^2 + 3x + 4)(x^2 - 5x + 4)$

80. $-6x^5(-x^5 + 3x^3 - 11)$

Answer in complete sentences.

81. Explain the difference between simplifying the expression $8x^2 + 3x^2$ and simplifying the expression $(8x^2)(3x^2)$. Discuss how the coefficients and exponents are handled differently.

82. Explain why the term "FOIL" applies only when we multiply a binomial by another binomial and not when we multiply polynomials that have more than two terms. Explain how to multiply a binomial by a trinomial.

83. To simplify the expression $(x + 8)^2$, we can either multiply $(x + 8)(x + 8)$ or use the special product formula $(a + b)^2 = a^2 + 2ab + b^2$. Which method do you prefer? Which method do you feel you will be able to retain? Explain your answer.

Study Tip **REVISITED**   When you take a test, work on the easiest problems first, saving the more difficult problems for later. One benefit to this approach is that you will gain confidence as you progress through the test, so you are confident when you attempt to solve a difficult problem. Students who struggle on a problem on the test will lose confidence, and this may cause students to miss later problems that they know how to do. By completing the easier problems quickly first, you will save time to work on the few difficult problems.

One exception to this practice is to begin the test by first attempting the most difficult type of problem for you. For example, if a particular type of word problem has been difficult for you to solve throughout your preparation for the exam, you may want to focus on how to solve that type of problem just before arriving for the test. With the solution of this type of problem committed to your short-term memory, your best opportunity for success comes at the beginning of the test prior to solving any other problems.

5.5

DIVISION OF POLYNOMIALS

1. Divide a monomial by a monomial.
2. Divide a polynomial by a monomial.
3. Divide a polynomial by a polynomial by using long division.
4. Use placeholders when dividing a polynomial by a polynomial.

Dividing Monomials by Monomials

Objective 1 Divide a monomial by a monomial. In this section, we will learn to divide a polynomial by another polynomial. We will begin by reviewing how to divide a monomial by another monomial, such as $\dfrac{4x^5}{2x^2}$ or $\dfrac{30a^5b^4}{6ab^2}$.

Dividing a Monomial by a Monomial

To divide a monomial by another monomial, we divide the coefficients first. Then we divide the variables, using the quotient rule $\dfrac{x^m}{x^n} = x^{m-n}$, assuming that no variable in the denominator is equal to 0.

EXAMPLE 1 Divide $\dfrac{18x^5}{3x^3}$. (Assume $x \neq 0$.)

Solution

$$\frac{18x^5}{3x^3} = 6x^{5-3} \qquad \text{Divide coefficients. Subtract exponents for } x.$$
$$= 6x^2 \qquad \text{Simplify the exponent.}$$

We can check our quotients by using multiplication. If $\dfrac{18x^5}{3x^3} = 6x^2$, then we know that $3x^3 \cdot 6x^2$ should be equal to $18x^5$.

Quick Check 1
Divide $\dfrac{32x^9}{4x^4}$.
(Assume $x \neq 0$.)

Check:
$$3x^3 \cdot 6x^2 = 18x^{3+2} \qquad \text{Multiply coefficients. Keep the base } x, \text{ add the exponents.}$$
$$= 18x^5 \qquad \text{Simplify the exponent.}$$

Our quotient of $6x^2$ checks.

EXAMPLE 2 Divide $\dfrac{24x^3y^2}{3xy^2}$. (Assume $x, y \neq 0$.)

Solution

When there is more than one variable, as in this example, we divide the coefficients and then divide the variables one at a time.

$$\frac{24x^3y^2}{3xy^2} = 8x^2y^0 \qquad \text{Divide coefficients, subtract exponents.}$$
$$= 8x^2 \qquad \text{Rewrite without } y \text{ as a factor. } (y^0 = 1)$$

Quick Check 2

Divide $\dfrac{40x^3y^{11}}{8xy^2}$. (Assume $x, y \neq 0$.)

Notice that the powers of y in the numerator and denominator were equal. When the exponent of a variable in the numerator is equal to the exponent of the same variable in the denominator, that variable is not part of the quotient.

Dividing Polynomials by Monomials

Objective 2 **Divide a polynomial by a monomial.** Now we move on to dividing a polynomial by a monomial, such as $\dfrac{3x^5 - 9x^3 - 18x^2}{3x}$.

Dividing a Polynomial by a Monomial

> To divide a polynomial by a monomial, divide each term of the polynomial by the monomial.

EXAMPLE 3 Divide $\dfrac{15x^2 + 10x - 5}{5}$.

Solution

We will divide each term in the numerator by 5.

$$\frac{15x^2 + 10x - 5}{5} = \frac{15x^2}{5} + \frac{10x}{5} - \frac{5}{5} \qquad \text{Divide each term in the numerator by 5.}$$
$$= 3x^2 + 2x - 1 \qquad \text{Divide.}$$

If $\dfrac{15x^2 + 10x - 5}{5} = 3x^2 + 2x - 1$, then $5(3x^2 + 2x - 1)$ should equal $15x^2 + 10x - 5$. We can use this to check our work.

Quick Check 3

Divide $\dfrac{7x^2 - 21x - 49}{7}$.

Check:
$$5(3x^2 + 2x - 1) = 5 \cdot 3x^2 + 5 \cdot 2x - 5 \cdot 1 \qquad \text{Distribute.}$$
$$= 15x^2 + 10x - 5 \qquad \text{Multiply.}$$

Our quotient of $3x^2 + 2x - 1$ checks.

EXAMPLE 4 Divide $\dfrac{3x^5 - 9x^3 - 18x^2}{3x}$. (Assume $x \neq 0$.)

Solution

Quick Check 4

Divide $\dfrac{48x^{10} + 12x^7 + 30x^5}{6x^2}$.

(Assume $x \neq 0$.)

In this example, we will divide each term in the numerator by $3x$.

$$\frac{3x^5 - 9x^3 - 18x^2}{3x}$$
$$= \frac{3x^5}{3x} - \frac{9x^3}{3x} - \frac{18x^2}{3x} \qquad \text{Divide each term in the numerator by } 3x.$$
$$= x^4 - 3x^2 - 6x \qquad \text{Divide.}$$

Dividing a Polynomial by a Polynomial (Long Division)

Objective 3 **Divide a polynomial by a polynomial by using long division.** To divide a polynomial by another polynomial containing at least two terms, we use a procedure similar to long division. Before outlining this procedure, let's review some of the terms associated with long division.

$$3 \longleftarrow \text{Quotient}$$
$$\text{Divisor} \longrightarrow 2\overline{)6} \longleftarrow \text{Dividend}$$

Suppose that we were asked to divide $\dfrac{x^2 - 10x + 16}{x - 2}$. The polynomial in the numerator is the dividend and the polynomial in the denominator is the divisor. We may rewrite this division as $x - 2\overline{)x^2 - 10x + 16}$. We must be sure to write both the divisor and the dividend in descending order. We perform the division using the following steps:

Division by a Polynomial

1. Divide the term in the dividend with the highest degree by the term in the divisor with the highest degree. Add this result to the quotient.
2. Multiply the monomial obtained in step 1 by the divisor, writing the result underneath the dividend. (Align like terms vertically.)
3. Subtract the product obtained in step 2 from the dividend. (Recall that to subtract a polynomial from another polynomial, we change the signs of its terms and then combine like terms.)
4. Repeat steps 1–3 with the result of step 3 as the new dividend. Keep repeating this procedure until the degree of the new dividend is less than the degree of the divisor.

EXAMPLE 5 Divide $\dfrac{x^2 - 10x + 16}{x - 2}$.

Solution

We begin by writing $x - 2\overline{)x^2 - 10x + 16}$. We divide the term in the dividend with the highest degree (x^2) by the term in the divisor with the highest degree (x). Since $\dfrac{x^2}{x} = x$, we will write x in the quotient and multiply x by the divisor $x - 2$, writing this product underneath the dividend.

$$
\begin{array}{r}
x \\
x - 2\overline{)x^2 - 10x + 16} \\
x^2 - 2x
\end{array}
\qquad \text{Multiply } x(x - 2).
$$

To subtract, we may change the signs of the second polynomial and then combine like terms.

$$
\begin{array}{r}
x \\
x - 2\overline{)x^2 - 10x + 16} \\
\underline{\overset{-}{x^2} \overset{+}{\diagup} 2x} \downarrow \\
- 8x + 16
\end{array}
\qquad \text{Change the signs and combine like terms.}
$$

We now begin the process again by dividing $-8x$ by x, which equals -8. Multiply -8 by $x - 2$ and subtract.

$$
\begin{array}{r}
x - 8 \\
x - 2 \overline{)\, x^2 - 10x + 16} \\
\underline{x^2 - 2x} \quad \downarrow \\
-8x + 16 \\
\underline{-8x + 16} \\
0
\end{array}
$$

Multiply -8 by $x - 2$.
Subtract the product.

The remainder of 0 tells us that we are through, since its degree is less than the degree of the divisor $x - 2$. The expression written above the division box $(x - 8)$ is the quotient.

$$\frac{x^2 - 10x + 16}{x - 2} = x - 8.$$

We can check our work by multiplying the quotient $(x - 8)$ by the divisor $(x - 2)$, which should equal the dividend $(x^2 - 10x + 16)$.

Check:

$$
\begin{aligned}
(x - 8)(x - 2) &= x^2 - 2x - 8x + 16 \\
&= x^2 - 10x + 16
\end{aligned}
$$

Our division checks; the quotient is $x - 8$.

Quick Check **5** Divide $\dfrac{x^2 + 13x + 36}{x + 4}$.

In the previous example, the remainder of 0 also tells us that $x - 2$ divides into $x^2 - 10x + 16$ evenly, so $x - 2$ is a **factor** of $x^2 - 10x + 16$. The quotient, $x - 8$, is also a factor of $x^2 - 10x + 16$.

In the next example, we will learn how to write a quotient when there is a nonzero remainder.

EXAMPLE 6 Divide $x^2 - 10x + 9$ by $x - 3$.

Solution

Since the divisor and dividend are already written in descending order, we may begin to divide.

$$
\begin{array}{r}
x - 7 \\
x - 3 \overline{)\, x^2 - 10x + 9} \\
\underline{x^2 - 3x} \quad \downarrow \\
-7x + 9 \\
\underline{-7x + 21} \\
-12
\end{array}
$$

Multiply x by $x - 3$ and subtract.

Multiply -7 by $x - 3$ and subtract.

The remainder is -12. After the quotient, we write a fraction with the remainder in the numerator and the divisor in the denominator. Since the remainder is negative, we subtract this fraction from the quotient. If the remainder had been positive, we would add this fraction to the quotient.

$$\frac{x^2 - 10x + 9}{x - 3} = x - 7 - \frac{12}{x - 3}$$

Quick Check **6** Divide $\dfrac{x^2 - 3x - 8}{x + 7}$.

EXAMPLE **7**   Divide $6x^2 + 17x + 17$ by $2x + 3$.

Solution

Since the divisor and dividend are already written in descending order, we may begin to divide.

$$
\begin{array}{r}
3x + 4 \\
2x + 3\overline{)6x^2 + 17x + 17} \\
\underline{6x^2 + 9x} \downarrow \\
8x + 17 \\
\underline{8x + 12} \\
5
\end{array}
$$

Multiply $3x$ by $2x + 3$ and subtract.

Multiply 4 by $2x + 3$ and subtract.

The remainder is 5. We write the quotient, and add a fraction with the remainder in the numerator and the divisor in the denominator.

$$\frac{6x^2 + 17x + 17}{2x + 3} = 3x + 4 + \frac{5}{2x + 3}$$

Quick Check **7** Divide $\dfrac{12x^2 - 7x + 4}{3x - 4}$.

Using Placeholders when Dividing a Polynomial by a Polynomial

Objective **4** **Use placeholders when dividing a polynomial by a polynomial.**
Suppose that we wanted to divide $x^3 - 12x - 11$ by $x - 3$. Notice that the dividend is missing an x^2 term. When this is the case, we add the term $0x^2$ as a **placeholder.** We add placeholders to dividends that are missing terms of a particular degree.

EXAMPLE 8 Divide $(x^3 - 12x - 11) \div (x - 3)$.

Solution

The degree of the dividend is 3, and each degree lower than 3 must be represented in the dividend. We will add the term $0x^2$ as a placeholder. The divisor does not have any missing terms.

$$
\begin{array}{r}
x^2 + 3x - 3 \\
x - 3 \overline{) x^3 + 0x^2 - 12x - 11} \\
\underline{-x^3 + 3x^2} \quad\downarrow\quad\downarrow \\
3x^2 - 12x - 11 \\
\underline{-3x^2 + 9x}\quad\downarrow \\
-3x - 11 \\
\underline{+3x - 9} \\
-20
\end{array}
$$

Multiply x^2 by $x - 3$ and subtract.

Multiply $3x$ by $x - 3$ and subtract.

Multiply -3 by $x - 3$ and subtract.

$$(x^3 - 12x - 11) \div (x - 3) = x^2 + 3x - 3 - \frac{20}{x - 3}.$$

Quick Check 8

Divide $\dfrac{x^3 - 3x^2 - 9}{x - 2}$.

EXERCISES 5.5

Divide.

1. $\dfrac{30x^5}{6x^3}$ 2. $\dfrac{14x^8}{7x^2}$ 3. $\dfrac{-55x^{11}}{5x}$

4. $\dfrac{24x^9}{8x^8}$ 5. $\dfrac{34x^5y^4}{2xy^3}$ 6. $\dfrac{20x^9y^9}{4x^9y^3}$

7. $(15x^6) \div (3x^2)$ 8. $(70x^{13}) \div (10x^5)$

9. $\dfrac{12x^6}{8x^4}$ 10. $\dfrac{-35x^9}{25x^5}$ 11. $\dfrac{7x^{12}}{21x^4}$

12. $\dfrac{3x^{15}}{18x^7}$

Find the missing monomial.

13. $\dfrac{?}{5x^4} = 2x^7$ 14. $\dfrac{?}{3x^6} = -15x^8$

15. $\dfrac{24x^9}{?} = -6x^4$ 16. $\dfrac{20x^9}{?} = \dfrac{5x^4}{3}$

Divide.

17. $\dfrac{24x^2 + 28x - 16}{4}$

18. $\dfrac{3x^2 - 33x}{3}$

19. $\dfrac{5x^4 + 35x^3 - 45x^2 + 15x}{5x}$

20. $\dfrac{22x^5 + 26x^4 - 14x^2}{2x}$

21. $(6x^7 + 9x^6 + 15x^5) \div (3x)$

22. $(-14x^8 + 21x^6 + 7x^4) \div (7x)$

23. $\dfrac{20x^6 - 30x^4}{-10x^2}$

24. $\dfrac{-8x^5 + 2x^4 - 6x^3 - 2x^2}{-2x^2}$

25. $\dfrac{x^6y^6 - x^4y^5 + x^2y^4}{xy^2}$

26. $\dfrac{2a^{10}b^7 - 8a^3b^6 - 10a^5b}{2a^3b}$

Find the missing dividend or divisor.

27. $\dfrac{24x^3 - 48x^2 + 36x}{?} = 4x^2 - 8x + 6$

28. $\dfrac{10x^7 - 50x^5 + 35x^3}{?} = 2x^5 - 10x^3 + 7x$

29. $\dfrac{?}{2x^3} = 3x^4 - 4x^2 - 9$

30. $\dfrac{?}{7x^5} = -x^2 - 9x + 1$

Divide using long division.

31. $\dfrac{x^2 + 10x + 21}{x + 7}$

32. $\dfrac{x^2 + 11x + 18}{x + 2}$

33. $\dfrac{x^2 - 7x - 30}{x - 10}$

34. $\dfrac{x^2 + x - 30}{x - 5}$

35. $(x^2 - 15x + 56) \div (x - 7)$

36. $(x^2 - 22x + 119) \div (x - 13)$

37. $\dfrac{x^2 - 4x - 29}{x - 8}$

38. $\dfrac{x^2 + 7x - 7}{x + 5}$

39. $\dfrac{x^3 - 11x^2 - 37x + 14}{x + 1}$

40. $\dfrac{x^3 - x^2 - 24x - 19}{x + 3}$

41. $\dfrac{2x^2 + 3x - 32}{x + 5}$

42. $\dfrac{3x^2 + 26x + 24}{x + 9}$

43. $\dfrac{12x^2 - 11x + 15}{3x - 5}$

44. $\dfrac{14x^2 - 3x - 39}{7x + 2}$

45. $\dfrac{x^2 - 100}{x - 10}$

46. $(4x^2 - 121) \div (2x + 11)$

47. $\dfrac{x^4 + 2x^2 - 15x + 32}{x - 3}$

48. $\dfrac{x^5 - 9x^3 - 17x^2 + x - 15}{x + 8}$

49. $\dfrac{x^3 - 125}{x - 5}$

50. $\dfrac{x^3 + 343}{x + 7}$

Find the missing dividend or divisor.

51. $\dfrac{?}{x + 8} = x - 3$

52. $\dfrac{?}{x - 5} = x - 11$

53. $\dfrac{x^2 + 10x - 39}{?} = x - 3$

54. $\dfrac{6x^2 - 35x + 49}{?} = 3x - 7$

55. Is $x + 9$ a factor of $x^2 + 28x + 171$?

56. Is $x - 6$ a factor of $x^2 + 4x - 54$?

57. Is $2x - 5$ a factor of $4x^2 - 8x - 15$?

58. Is $3x + 4$ a factor of $9x^2 + 24x + 16$?

Divide.

59. $\dfrac{x^2 - 8x + 9}{x - 3}$

60. $\dfrac{x^2 + 3x - 15}{x + 7}$

61. $\dfrac{x^3 + 8x - 19}{x + 4}$

62. $\dfrac{8x^9}{2x^3}$

63. $\dfrac{12x^6 - 8x^4 + 20x^3 - 4x^2}{4x^2}$

64. $\dfrac{x^2 - 4x - 192}{x + 12}$

65. $\dfrac{8x^2 + 42x - 25}{x + 6}$

66. $\dfrac{4x^2 - 25x - 100}{x - 8}$

67. $\dfrac{20x^2 - 51x - 6}{4x - 3}$

68. $\dfrac{6x^2 - 19x + 30}{2x - 5}$

69. $\dfrac{27x^9 - 18x^7 - 6x^6 - 3x^3}{-3x^2}$

70. $\dfrac{6x^2 + x - 176}{3x - 16}$

71. $\dfrac{8x^3 - 86x - 63}{4x + 10}$

72. $\dfrac{16a^4b^8c^7}{4ab^7c^7}$

73. $\dfrac{6x^3 - 37x^2 + 11x + 153}{2x - 9}$

74. $\dfrac{6x^5 - 12x^4 + 18x^3}{4x^3}$

75. $\dfrac{21x^9y^6z^{17}}{-7x^7y^5z^{12}}$

76. $\dfrac{x^7 + 127}{x + 2}$

Answer in complete sentences.

77. Explain how to divide a polynomial by another poly-nomial. Use $\dfrac{6x^2 + x - 18}{2x - 3}$ as your example.

78. Explain the use of placeholders when dividing a poly-nomial by another polynomial.

Study Tip REVISITED   Try to leave yourself enough time to review the test at the end of the test period. Check for careless errors, which can cost you a fair amount of points. Also be sure that your answers make sense within the context of the problem. For example, if the question asks how tall a person is and your answer is 68 feet tall, then chances are that something has gone astray.

Check for problems, or parts of problems, that you may have skipped and left blank. There is no worse feeling in a math class than getting a test back and realizing that you simply forgot to do one or more of the problems.

Take all of the allotted time to review your test. There is no reward for turning in a test early, and the more that you work on the test, the more likely it is that you will find a mistake or a problem where you can gain points. Keep in mind that a majority of the students who turn in tests early do so because they are not prepared for the test and cannot do several of the problems.

Section 5.1—Topic	Chapter Review Exercises
Simplifying Expressions by Using the Rules of Exponents	1–10

Section 5.2—Topic	Chapter Review Exercises
Simplifying Expressions with Negative Exponents	11–24
Rewriting Numbers by Using Scientific Notation	25–27
Rewriting Numbers That Are in Scientific Notation	28–30
Performing Calculations Involving Scientific Notation	31–32
Solving Applied Problems by Using Scientific Notation	33–34

Section 5.3—Topic	Chapter Review Exercises
Evaluating Polynomials	35–40
Adding and Subtracting Polynomials	41–46
Evaluating Polynomial Functions	47–50
Adding and Subtracting Polynomial Functions	51–52

Section 5.4—Topic	Chapter Review Exercises
Multiplying Polynomials	53–68
Multiplying Polynomial Functions	69–70

Section 5.5—Topic	Chapter Review Exercises
Dividing Polynomials	71–80

Summary of Chapter 5 Study Tips

Attending class each day and doing all of your homework does not guarantee a good grade when taking an exam. Through careful preparation and the adoption of the test-taking strategies introduced in this chapter, you can maximize your grade on a math exam. Here is a summary of the points that have been introduced:

- Make the most of your time before the exam.
- Write down important facts as soon as you get your test.
- Briefly read through the test.
- Begin by solving the easier problems first.
- Review your test as thoroughly as possible before turning it in.

Simplify the expression. Write the result without using negative exponents. (Assume all variables represent nonzero real numbers.) [5.1–5.2]

1. $\dfrac{x^9}{x^3}$

2. $(4x^7)^3$

3. $7x^0$

4. $\left(\dfrac{5x^9}{2y^2}\right)^4$

5. $7x^7 \cdot 4x^{12}$

6. $3x^8y^5 \cdot 8x^8y^3$

7. $(-6a^5b^3c^6)^2$

8. $15^0 - 3x^0$

9. $(x^{10}y^{13}z^4)^4$

10. $\dfrac{x^{16}}{x^6}$

11. 5^{-2}

12. 10^{-3}

13. $7x^{-4}$

14. $-6x^{-6}$

15. $\dfrac{-8}{y^{-5}}$

16. $(3x^{-4})^3$

17. $(2x^{-5})^{-2}$

18. $\dfrac{x^{10}}{x^{-5}}$

19. $\left(\dfrac{a^{-4}}{b^{-7}}\right)^3$

20. $m^{13} \cdot m^{-19}$

21. $(4x^{-5}y^4z^{-7})^{-3}$

22. $\left(\dfrac{x^{-11}}{y^{-6}}\right)^6$

23. $a^{-15} \cdot a^{-5}$

24. $\dfrac{x^{-20}}{x^{-9}}$

Rewrite in scientific notation. [5.2]

25. 1,400,000,000

26. 0.0000000000021

27. 0.000005002

Convert from scientific to decimal notation. [5.2]

28. 1.23×10^{-4}

29. 4.075×10^7

30. 6.1275×10^{12}

Perform the following calculations. Express your answer using scientific notation. [5.2]

31. $(2.5 \times 10^7)(6.0 \times 10^6)$

32. $(3.0 \times 10^{-13}) \div (1.2 \times 10^5)$

33. The mass of a hydrogen atom is 1.66×10^{-24} grams. What is the mass of 1,500,000,000,000 hydrogen atoms? [5.2]

34. If a computer can perform a calculation in 0.000000000005 seconds, how long would it take to perform 3,000,000,000,000,000 calculations? [5.2]

Evaluate the polynomial for the given value of the variable. [5.3]

35. $x^2 - 8x + 15$ for $x = -5$

36. $x^2 - 11x - 29$ for $x = 2$

37. $-x^2 + 16x - 30$ for $x = 6$

38. $-3x^2 - 4x + 7$ for $x = -7$

39. $5x^2 + 10x + 12$ for $x = -3$

40. $7x^2 - 14x + 35$ for $x = 8$

Add or subtract. [5.3]

41. $(x^2 + 3x - 15) + (x^2 - 9x + 6)$

42. $(3x^2 - 31) + (5x^2 - 19x - 19)$

43. $(x^2 - 6x - 13) - (x^2 - 14x + 8)$

44. $(2x^2 + 8x - 7) - (x^2 - x + 15)$

45. $(3x^3 + x^2 - 10) - (6x^2 - 13x + 20)$

46. $(10x^2 - 21x - 35) - (4x^3 + 6x^2 - 15x + 80)$

Evaluate the given polynomial function. [5.3]

47. $f(x) = x^2 - 25, f(-4)$

48. $f(x) = 2x^2 - 6x + 17, f(5)$

49. $f(x) = x^2 - 7x + 10, f(a^4)$

50. $f(x) = x^2 + 3x - 30, f(2a^3)$

Given the functions $f(x)$ and $g(x)$, find $f(x) + g(x)$ and $f(x) - g(x)$. [5.3]

51. $f(x) = 3x^2 - 8x - 9, g(x) = 7x^2 + 6x + 40$

52. $f(x) = x^2 + 5x - 30, g(x) = -8x^2 - 15x + 6$

Multiply. [5.4]

53. $(x - 5)^2$

54. $(x - 4)(x + 13)$

55. $4x(3x^2 - 7x - 16)$

56. $(2x - 7)^2$

57. $5x^7 \cdot 3x^6$

58. $(x - 8)(x + 8)$

59. $(3x + 10)(2x - 13)$

60. $-9x^5 \cdot 2x$

61. $(6x + 5)(6x - 5)$

62. $(x + 10)^2$

63. $(4x + 7)^2$

64. $-8x^2(3x^5 - 7x^4 - 6x^3)$

65. $(x + 9)(x^2 - 6x + 12)$

66. $(x - 2)(3x^2 - 5x - 9)$

67. $(10x^5)(-3x^7)(4x^6)$

68. $-2x^4(-5x^2 - 11x + 12)$

Given the functions $f(x)$ and $g(x)$, find $f(x) \cdot g(x)$. [5.4]

69. $f(x) = x + 10, g(x) = x - 7$

70. $f(x) = -6x^6, g(x) = -4x^2 - 9x + 20$

Divide. [5.5]

71. $\dfrac{6x^4 - 8x^3 + 20x^2}{2x^2}$

72. $\dfrac{3x^2 - 8x - 25}{x - 4}$

73. $\dfrac{4x^5y^7}{-2x^5y}$

74. $\dfrac{x^3 + 3x^2 - 15}{x + 5}$

75. $\dfrac{x^4 + 81}{x + 3}$

76. $\dfrac{18x^3y^7z^6}{3x^3y^6z}$

77. $\dfrac{8x^2 + 87x + 171}{x + 9}$

78. $\dfrac{6x^2 - 30x - 84}{x - 7}$

79. $\dfrac{15x^2 - 43x - 238}{5x + 14}$

80. $\dfrac{16x^2 + 2x + 20}{8x - 3}$

Simplify the expression. Write the result without using negative exponents. (Assume all variables represent nonzero real numbers.)

1. $\dfrac{x^{10}}{x^2}$

2. $\left(\dfrac{x^5}{3y^4}\right)^6$

3. $(x^9 y^5 z^8)^8$

4. 4^{-4}

5. $(2x^{-7})^5$

6. $\dfrac{x^4}{x^{-13}}$

7. $x^{22} \cdot x^{-15}$

8. $(3xy^{-6}z^5)^{-2}$

Rewrite in scientific notation.

9. 23,500,000

Convert from scientific to decimal notation.

10. 4.7×10^{-8}

Perform the following calculations. Express your answer using scientific notation.

11. $(4.3 \times 10^{13})(1.8 \times 10^{-9})$

Evaluate the polynomial for the given value of the variable.

12. $x^2 + 12x - 38$ for $x = -8$

Add or subtract.

13. $(x^2 - 6x - 32) - (5x^2 + 8x - 33)$

14. $(4x^2 - 3x - 15) + (-2x^2 + 17x - 49)$

Evaluate the given polynomial function.

15. $f(x) = x^2 + 3x - 14$, $f(-7)$

Multiply.

16. $5x^3(6x^2 + 8x - 17)$

17. $(x + 6)(x - 6)$

18. $(5x - 9)(4x + 7)$

Divide.

19. $\dfrac{21x^7 + 33x^6 - 15x^5}{3x^2}$

20. $\dfrac{6x^2 - 11x + 38}{x + 3}$

Mathematicians in History
Carl Friedrich Gauss

Carl Friedrich Gauss was a German mathematician who lived in the 18th and 19th centuries and was one of the most prominent and prolific mathematicians of his day. Gauss made significant contributions to the fields of analysis, probability, statistics, number theory, geometry, and astronomy, among others. The Prussian mathematician Leopold Kronecker once said of him, "Almost everything, which the mathematics of our century has brought forth in the way of original scientific ideas, attaches to the name of Gauss." Gauss once said "It is not knowledge, but the act of learning, not possession but the act of getting there, which grants the greatest enjoyment."

Write a one-page summary (*or* make a poster) of the life of Carl Friedrich Gauss and his accomplishments. Also, look up Gauss's most famous quotes and list your favorite quote.

Interesting issues:

- Where and when was Carl Friedrich Gauss born?
- Gauss displayed incredible genius at an early age. Relay the story of how Gauss found his father's accounting error at only three years of age.
- At what age did Gauss find a shortcut for summing the integers from 1 to 100?
- Gauss's motto was "Few, but ripe." In your own words, explain the meaning of this motto.
- In 1798, Gauss showed how to construct a regular 17-gon by ruler and compass. This was the first major advance in the area of regular polygons in approximately 2000 years. What is a regular 17-gon?
- Gauss earned his doctorate in 1799 with a proof of what major mathematical theorem?
- Gauss gained notoriety in the field of astronomy by accurately predicting the orbit of the newly discovered asteroid Ceres by inventing the method of least squares. Describe the method of least squares.
- In 1821, Gauss built the first heliotrope. What is a heliotrope?
- Describe the circumstances that led to Gauss's death.

At a recent school carnival, George agreed to be launched out of a cannon to become the world's first flying mathematician. He landed after 3.4625 seconds. George's height, in feet, after t seconds is given by the function $h(t) = -16t^2 + 55.4t$.

a) Complete the following chart listing George's height at 0.5-second intervals:

Time (in seconds)	Height (in feet)
0	
0.5	
1	
1.5	
2	
2.5	
3	
3.5	

b) Create a coordinate system in which the horizontal axis represents time (in seconds) and the vertical axis represents height (in feet). Plot these eight points and connect them to see George's course of flight through the air. (Your graph should be a smooth curve.)

Use the table and graph to answer the following questions.

c) What is George's height when he was launched out of the cannon?

d) What is George's height after 0.5 seconds?

e) How long did it take George to reach a height of 39.4 feet?

f) What is George's height when he lands on the ground? How long did it take (from the time of being launched) for George to land on the ground?

g) When does it appear that George reaches his maximum height? What is his maximum height?

FACTORING AND QUADRATIC EQUATIONS

*I*n this chapter, we will begin to learn how to solve quadratic equations. A **quadratic equation** is an equation of the form $ax^2 + bx + c = 0$ $(a \neq 0)$, where a, b, and c are real numbers. To solve a quadratic equation, we will attempt to rewrite the quadratic expression $ax^2 + bx + c$ as a product of two expressions. In other words, we will factor the quadratic expression. The first five sections of this chapter focus on factoring techniques.

Once we have learned how to factor polynomials, we will learn how to solve quadratic equations as well as apply quadratic equations to real-world problems. We will also examine quadratic functions, which are functions of the form

$$f(x) = ax^2 + bx + c \ (a \neq 0).$$

Study Tip OVERCOMING MATH ANXIETY   *In this chapter, we will focus on how to overcome math anxiety, a condition shared by many students, and develop strategies to overcome it. Math anxiety can prevent a student from learning mathematics, and it can seriously impact a student's performance on quizzes and exams.*

6.1

AN INTRODUCTION TO FACTORING; THE GREATEST COMMON FACTOR; FACTORING BY GROUPING

Objectives

1 **Find the greatest common factor (GCF) of two or more integers.**
2 **Find the GCF of two or more variable terms.**
3 **Factor the GCF out of each term of a polynomial.**
4 **Factor out a common binomial factor.**
5 **Factor a polynomial by grouping.**

In the previous chapter, we learned about polynomials; in this section we will begin to learn how to **factor** a polynomial. A polynomial has been **factored** when it is represented as the product of two or more polynomials.

Greatest Common Factor

Objective 1 Find the greatest common factor (GCF) of two or more integers.
Before we begin to learn how to factor polynomials, we will go over the procedure for finding the **greatest common factor (GCF)** of two or more integers. The GCF of two or more integers is the largest whole number that is a factor of each integer.

Finding the GCF of Two or More Integers

> Find the prime factorization of each integer.
> Write the prime factors that are common to each integer; the GCF is the product of these prime factors.

EXAMPLE 1 Find the GCF of 30 and 42.

Solution

The prime factorization of 30 is $2 \cdot 3 \cdot 5$ and the prime factorization of 42 is $2 \cdot 3 \cdot 7$. (For a refresher on finding the prime factorization of a number, refer back to Section 1.3.) The prime factors that they share in common are 2 and 3, so the GCF is $2 \cdot 3$ or 6. (This means that 6 is the greatest number that divides evenly into both 30 and 42.)

Quick Check 1
Find the GCF of 16 and 20.

EXAMPLE 2 Find the GCF of 108, 504, and 720.

Solution

We begin with the prime factorizations.

$$108 = 2^2 \cdot 3^3 \qquad 504 = 2^3 \cdot 3^2 \cdot 7 \qquad 720 = 2^4 \cdot 3^2 \cdot 5$$

The only primes that are factors of all three numbers are 2 and 3. The smallest power of 2 that is a factor of any of the three numbers is 2^2, and the smallest power of 3 that is a factor is 3^2. The GCF is $2^2 \cdot 3^2$ or 36.

Quick Check 2
Find the GCF of 48, 120, and 156.

Note that if two or more integers do not have any common prime factors, then their GCF is 1. For example, the GCF of 8, 12, and 15 is 1, since there are not any prime numbers that are factors of all three numbers.

Objective 2 Find the GCF of two or more variable terms. We can also find the GCF of two or more variable terms. For a variable factor to be included in the GCF, it must be a factor of each term. As with prime factors, the exponent for a variable factor used in the GCF is the smallest exponent that can be found for that variable in any one term.

EXAMPLE 3 Find the GCF of $6x^2$ and $15x^4$.

Solution

The GCF of the two coefficients is 3. The variable x is a factor of each term, and its smallest exponent is 2. The GCF of these two terms is $3x^2$.

EXAMPLE 4 Find the GCF of $a^5b^4c^2$, a^7b^3, $a^6b^9c^5$, and $a^3b^8c^3$.

Solution

Only the variables a and b are factors of each term. (The variable c is not a factor of the second term.) The smallest power of a in any one term is 3, and this is the case for the variable b as well. The GCF is a^3b^3.

Quick Check **3**
Find the GCF.

a) $12x^5$ and $28x^3$
b) $x^3y^4z^9$, $x^6y^2z^{10}$, and x^7z^4

Factoring Out the Greatest Common Factor

Objective 3 Factor the GCF out of each term of a polynomial. The first step for factoring any polynomial is to factor out the GCF of all of the terms. This process uses the distributive property, and you may think of it as "undistributing" the GCF from each term. Consider the polynomial $4x^2 + 8x + 20$. The GCF of these three terms is 4, and the polynomial can be rewritten as the product $4(x^2 + 2x + 5)$. Notice that the GCF has been factored out of each term, and the polynomial inside the parentheses is the polynomial that we would multiply 4 by in order to equal $4x^2 + 8x + 20$.

$$4(x^2 + 2x + 5) = 4 \cdot x^2 + 4 \cdot 2x + 4 \cdot 5 \quad \text{Distribute.}$$
$$= 4x^2 + 8x + 20 \quad \text{Multiply.}$$

EXAMPLE 5 Factor $5x^3 - 30x^2 + 10x$ by factoring out the GCF.

Solution

We begin by finding the GCF, which is $5x$. The next task is to fill in the missing terms of the polynomial inside the parentheses so that the product of $5x$ and that polynomial is $5x^3 - 30x^2 + 10x$.

$$5x(? - ? + ?)$$

Quick Check **4**

Factor
$6x^4 - 42x^3 - 90x^2$ by factoring out the GCF.

To find the missing terms, we can divide each term of the original polynomial by $5x$.

$$5x^3 - 30x^2 + 10x = 5x(x^2 - 6x + 2)$$

We can check our answer by multiplying $5x(x^2 - 6x + 2)$, which should equal $5x^3 - 30x^2 + 10x$. The check is left to the reader.

EXAMPLE ▶ 6   Factor $6x^6 + 8x^4 + 2x^3$ by factoring out the GCF.

Solution

The GCF for these three terms is $2x^3$. Notice that this is also the third term.

$$6x^6 + 8x^4 + 2x^3 = 2x^3(3x^3 + 4x + 1) \qquad \text{Factor out the GCF.}$$

When a term in a polynomial is the GCF of the polynomial, factoring out the GCF leaves a 1 in that term's place. Why? Because we need to determine what we multiply $2x^3$ by in order to equal $2x^3$, and $2x^3 \cdot 1 = 2x^3$.

Quick Check 5
Factor
$15x^7 - 30x^5 + 3x^4$ by factoring out the GCF.

> **A Word of Caution**   Be sure to write a 1 in the place of a term that was the GCF when factoring out the GCF from a polynomial.

Objective 4 Factor out a common binomial factor. Occasionally, the GCF of two terms will be a binomial or some other polynomial with more than one term. Consider the expression $8x(x - 5) + 3(x - 5)$, which contains the two terms $8x(x - 5)$ and $3(x - 5)$. Each term has the binomial $x - 5$ as a factor. This common factor can be factored out of this expression. Working with this type of common factor is presented in the next example.

EXAMPLE ▶ 7   Factor $x(x + 3) + 7(x + 3)$ by factoring out the GCF.

Solution

The binomial $x + 3$ is a common factor for these two terms. We begin by factoring out this common factor.

$$x(x + 3) + 7(x + 3)$$

$$= (x + 3)(? + ?)$$

After factoring out $x + 3$, what factors remain in the first term? The only factor that remains is x. Using the same method, the only factor remaining in the second term is 7.

$$x(x + 3) + 7(x + 3) = (x + 3)(x + 7)$$

EXAMPLE ▶ 8   Factor $9x(2x + 3) - 8(2x + 3)$ by factoring out the GCF.

Quick Check 6
Factor by factoring out the GCF.

a) $5x(x - 9) + 14(x - 9)$
b) $7x(x - 8) - 6(x - 8)$

Solution

The GCF of these two terms is $2x + 3$, which can be factored out as follows:

$$9x(2x + 3) - 8(2x + 3) = (2x + 3)(9x - 8) \qquad \text{Factor out the common factor } 2x + 3.$$

Notice that the second term was negative, which led to $9x - 8$, rather than $9x + 8$, as the other factor.

Again, factoring out the GCF will be our first step in factoring any polynomial. Often this will make our factoring easier, and sometimes the expression cannot be factored without factoring out the GCF.

Factoring by Grouping

Objective 5 **Factor a polynomial by grouping.** We now turn our attention to a factoring technique known as **factoring by grouping.** Consider the polynomial $3x^3 + 33x^2 + 7x + 77$. The GCF for the four terms is 1, so we cannot factor the polynomial by factoring out the GCF. However, the first pair of terms has a common factor of $3x^2$ and the second pair of terms has a common factor of 7. If we factor $3x^2$ out of the first two terms and 7 out of the last two terms, we produce the following:

$$3x^2(x + 11) + 7(x + 11)$$

Notice that the two resulting terms have a common factor of $x + 11$. This common factor can then be factored out, and the polynomial $3x^3 + 33x^2 + 7x + 77$ is equal to the product of the two binomials $x + 11$ and $3x^2 + 7$.

Factoring a Polynomial with Four Terms by Grouping

> Factor a common factor out of the first two terms and another common factor out of the last two terms. If the two "groups" share a common binomial factor, this binomial can then be factored out to complete the factoring of the polynomial.

If the two "groups" do not share a common binomial factor, we can try rearranging the terms of the polynomial in a different order. If we cannot find two "groups" that share a common factor, then the polynomial cannot be factored by this method.

EXAMPLE 9   Factor $5x^3 - 30x^2 + 4x - 24$ by grouping.

Solution

We first check for a factor that is common to each of the four terms. Since there is no common factor other than 1, we proceed to factoring by grouping:

$$5x^3 - 30x^2 + 4x - 24 = 5x^2(x - 6) + 4x - 24 \quad \text{Factor } 5x^2 \text{ out of first two terms.}$$
$$= 5x^2(x - 6) + 4(x - 6) \quad \text{Factor 4 out of last two terms.}$$
$$= (x - 6)(5x^2 + 4) \quad \text{Factor out the common factor } x - 6.$$

Quick Check 7

Factor $x^3 + 4x^2 + 7x + 28$ **by grouping.**

As was the case earlier, we can check our factoring by multiplying the two factors together. The check is left to the reader.

EXAMPLE 10   Factor $8x^2 + 32x - 7x - 28$ by grouping.

Solution

Since the four terms have no common factors other than 1, we factor by grouping. After factoring $8x$ out of the first two terms, we must factor a *negative* 7 out of the last two terms. If we factor a positive 7 out of the last two terms, rather than a negative 7, the two binomial factors would be different.

$$8x^2 + 32x - 7x - 28 = 8x(x + 4) - 7(x + 4)$$

Factor the common factor $8x$ out of the first two terms and the common factor -7 out of the last two terms.

$$= (x + 4)(8x - 7)$$

Factor out the common factor $x + 4$.

When the third of the four terms is negative, this often indicates that a negative common factor will need to be factored from the last two terms.

Quick Check 8

Factor
$2x^2 - 10x - 9x + 45$
by grouping.

> **A Word of Caution**   Factoring a negative common factor from two terms when factoring by grouping is often necessary.

EXAMPLE 11   Factor $10x^3 + 50x^2 + x + 5$ by grouping.

Solution

Again, the four terms have no common factor other than 1, so we may proceed to factor this polynomial by grouping. Notice that the last two terms have no common factor other than 1. When this is the case, we will actually factor out the common factor of 1 from the two terms.

$$10x^3 + 50x^2 + x + 5 = 10x^2(x + 5) + 1(x + 5)$$

Factor the common factors of $10x^2$ and 1 from the first two terms and the last two terms, respectively.

$$= (x + 5)(10x^2 + 1)$$

Factor out the common factor $x + 5$.

Quick Check 9

Factor
$4x^3 - 36x^2 + x - 9$
by grouping.

In the next example, the four terms have a common factor other than 1. We will factor out the GCF from the polynomial before attempting to use factoring by grouping.

EXAMPLE 12   Factor $2x^2 - 6x - 20x + 60$ completely.

Solution

The four terms share a common factor of 2, and this will be factored out before proceeding with factoring by grouping.

$$2x^2 - 6x - 20x + 60 = 2(x^2 - 3x - 10x + 30)$$

Factor out the common factor 2.

$$= 2[x(x - 3) - 10(x - 3)]$$

Factor out the common factor x from the first two terms. Factor out the common factor -10 from the last two terms.

$$= 2(x - 3)(x - 10)$$

Factor out the common factor $x - 3$.

If we had not factored out the common factor 2 before factoring this polynomial by grouping, we would have factored $2x^2 - 6x - 20x + 60$ to be $(x - 3)(2x - 20)$. If we stop here, we have not factored this polynomial completely, since $2x - 20$ has a common factor of 2.

Quick Check 10

Factor
$3x^2 + 24x - 12x - 96$
by grouping.

Occasionally, despite our best efforts, a polynomial cannot be factored. Here is an example of just such a polynomial. Consider the polynomial $x^2 - 5x + 3x + 15$. The first two terms have a common factor of x and the last two terms have a common factor of 3.

$$x^2 - 5x + 3x + 15 = x(x - 5) + 3(x + 5)$$

Since the two binomials are not the same, we cannot factor a common factor out of the two terms. Reordering the terms $-5x$ and $3x$ leads to the same problem. The polynomial cannot be factored.

There are instances when factoring by grouping fails, yet the polynomial can be factored by other techniques. For example, the polynomial $x^3 + 2x^2 - 7x - 24$ cannot be factored by grouping. However, it can be shown that $x^3 + 2x^2 - 7x - 24 = (x - 3)(x^2 + 5x + 8)$.

EXERCISES 6.1

Find the GCF.

1. $6, 8$

2. $15, 21$

3. $30, 42$

4. $5, 50$

5. $16, 40, 60$

6. $8, 24, 27$

7. x^3, x^7

8. y^8, y^4

9. a^2b^3, a^5b^2

10. $a^8b^3c^4, a^5bc^2, a^3b^4c^5$

11. $4x^4, 6x^3$

12. $27x^5, 24x^{11}$

13. $15a^5b^2c, 25a^2c^3, 5a^3bc^2$

14. $30x^3y^4z^5, 12x^6y^{11}z^3, 24y^9z$

Factor the GCF out of the given expression.

15. $7x - 14$

16. $3a + 12$

17. $5x^2 + 4x$

18. $9x^2 - 20x$

19. $8x^3 + 20x$

20. $11x^5 - 33x^3$

21. $7x^2 + 56x + 70$

22. $9a^2 - 6a + 36$

23. $18x^6 - 14x^4 - 8x^3$

24. $18a^3 - 27a^5 + 9a^4$

25. $m^2n^3 - m^5n^2 + m^3n$

26. $x^5y^5 + x^3y^2 - x^2y^4$

27. $16a^9b^7 - 72a^5b^5 - 80a^3b^6$

28. $25r^3s^4 + 10r^2s^5 - 15r^4s$

29. $-8x^3 + 12x^2 - 16x$

30. $-18a^5 - 30a^3 + 24a^2$

31. $5x(2x - 7) + 8(2x - 7)$

32. $x(4x + 3) + 4(4x + 3)$

33. $x(3x + 11) - 7(3x + 11)$

34. $4x(x - 9) - 9(x - 9)$

35. $2x(x + 10) + (x + 10)$

36. $3x(x - 2) - (x - 2)$

Factor by grouping.

37. $x^2 + 8x + 7x + 56$

38. $x^2 + 3x + 9x + 27$

39. $x^2 - 5x + 4x - 20$

40. $x^2 - 6x + 9x - 54$

41. $x^2 + 2x - 9x - 18$

42. $x^2 - 8x - 5x + 40$

43. $3x^2 + 15x + 4x + 20$

44. $4x^2 - 24x + 7x - 42$

45. $7x^2 + 28x - 6x - 24$

46. $5x^2 - 25x - 12x + 60$

47. $3x^2 + 21x + x + 7$

48. $4x^2 - 48x + x - 12$

49. $2x^2 - 10x - x + 5$

50. $9x^2 - 81x - x + 9$

51. $x^3 + 8x^2 + 6x + 48$

52. $2x^3 - 12x^2 + 5x - 30$

53. $4x^3 + 12x + 3x^2 + 9$

54. $7x^3 + 49x + 4x^2 + 28$

55. $2x^2 + 10x + 6x + 30$

56. $3x^2 + 27x + 21x + 189$

Answer in complete sentences.

57. When factoring a polynomial, explain how to check your work.

58. Give an example of a four term polynomial that can be factored by grouping. Use this example to explain the process of factoring by grouping.

59. Explain why it is necessary to factor a negative common factor from the last two terms of the polynomial $x^3 - 5x^2 - 7x + 35$ in order to factor it by grouping.

Study Tip **REVISITED** If you have been avoiding taking a math class, if you panic when asked a mathematical question, or if you feel that you cannot learn mathematics, then you may have a condition known as math anxiety. Like many other anxiety-related conditions, math anxiety may be traced back to an event that first triggered the negative feelings towards mathematics. If you are going to overcome your anxiety, the first step is to understand the cause of your anxiety.

Many students can recall one embarrassing moment in their life that may be the initial cause of their anxiety. Were you ridiculed as a child by a teacher or another student when you couldn't solve a math problem? Were you expected to be a mathematical genius like a parent or older sibling but were not able to measure up to their reputation? One negative event can start math anxiety, and the journey to overcome math anxiety can begin with a single success. If you are able to complete a homework assignment, or show improvement on a quiz or exam, then celebrate your success. Let this be your vindication, and consider your slate to have been wiped clean.

6.2

FACTORING TRINOMIALS OF THE FORM $x^2 + bx + c$

Objectives

1 Factor a trinomial of the form $x^2 + bx + c$ when c is positive.

2 Factor a trinomial of the form $x^2 + bx + c$ when c is negative.

3 Factor a perfect square trinomial.

4 Determine that a trinomial is prime.

5 Factor a trinomial by first factoring out a common factor.

6 Factor a trinomial in several variables.

In this section, we will learn how to factor trinomials of degree 2 with a leading coefficient of 1. Some examples of this type of polynomial are $x^2 + 12x + 32$, $x^2 - 9x + 14$, $x^2 + 8x - 20$, and $x^2 - x - 12$.

Factoring Trinomials of the Form $x^2 + bx + c$

> If $x^2 + bx + c$ where b and c are integers, is factorable, it can be factored to be the product of two binomials of the form $(x + m)(x + n)$, where m and n are integers.

We will begin by multiplying two binomials of the form $(x + m)(x + n)$ and using the result to help us learn to factor trinomials of the form $x^2 + bx + c$.

Multiply the two binomials $x + 4$ and $x + 8$.

$$(x + 4)(x + 8) = x^2 + 8x + 4x + 32 \qquad \text{Distribute.}$$
$$= x^2 + 12x + 32 \qquad \text{Combine like terms.}$$

The two binomials have a product that is a trinomial of the form $x^2 + bx + c$ with $b = 12$ and $c = 32$. Notice that the two numbers 4 and 8 have a product of 32 and a sum of 12; in other words, their product is equal to c and their sum is equal to b.

$$4 \cdot 8 = 32$$
$$4 + 8 = 12$$

We will use this pattern to help us factor trinomials of the form $x^2 + bx + c$. We will look for two integers m and n with a product equal to c and a sum equal to b. If we can find two such integers, then the trinomial factors to be $(x + m)(x + n)$.

Factoring $x^2 + bx + c$ when c Is Positive

Objective 1 Factor a trinomial of the form $x^2 + bx + c$ when c is positive.

EXAMPLE 1   Factor $x^2 + 9x + 18$.

Solution

We begin, as always, by looking for common factors. The terms in this trinomial have no common factors other than 1. We are looking for two integers m and n whose product is 18 and whose sum is 9. Here are the factors of 18:

Factors $1 \cdot 18$ $2 \cdot 9$ $3 \cdot 6$

The pair of factors that have a sum of 9 are 3 and 6. The trinomial $x^2 + 9x + 18$ factors to be $(x + 3)(x + 6)$. Be aware that this could also be written as $(x + 6)(x + 3)$.

We can check our work by multiplying $x + 3$ by $x + 6$. If the product equals $x^2 + 9x + 18$, then we have factored correctly. The check is left to the reader.

EXAMPLE ▶ **2** Factor $x^2 - 11x + 24$.

Solution

The major difference between this trinomial and the previous one is that the x term has a negative coefficient. We are looking for two integers m and n whose product is 24 and whose sum is -11. Since the product of these two integers is positive, they must have the same sign. The sum of these two integers is negative, so each integer must be negative. Here are the negative factors of 24:

$$(-1)(-24) \quad (-2)(-12) \quad (-3)(-8) \quad (-4)(-6)$$

Quick Check 1

Factor.

a) $x^2 + 7x + 10$
b) $x^2 - 11x + 30$

The pair that has a sum of -11 is -3 and -8. $x^2 - 11x + 24 = (x - 3)(x - 8)$.

From the previous examples, we see that when the constant term c is positive, such as in $x^2 + 9x + 18$ and $x^2 - 11x + 24$, then the two numbers we are looking for (m and n) will both have the same sign. Both numbers will be positive if b is positive, and both numbers will be negative if b is negative.

Factoring $x^2 + bx + c$ when c Is Negative

Objective 2 **Factor a trinomial of the form $x^2 + bx + c$ when c is negative.** If the product of two numbers is negative, one of the numbers must be negative and the other number must be positive. So if the constant term c is negative, then m and n must have opposite signs.

EXAMPLE ▶ **3** Factor $x^2 + 8x - 33$.

Solution

These three terms do not have any common factors, so we look for two integers m and n such that their product is -33 and their sum is 8. Since the product is negative, we are looking for one negative number and one positive number. Here are the factors of -33, along with their sums:

Factors	$-1 \cdot 33$	$-3 \cdot 11$	$-11 \cdot 3$	$-33 \cdot 1$
Sum	32	8	-8	-32

The pair of integers we are looking for are -3 and 11. $x^2 + 8x - 33 = (x - 3)(x + 11)$.

EXAMPLE ▶ **4** Factor $x^2 - 3x - 40$.

Solution

Since there are no common factors to factor out, we are looking for two integers m and n that have a product of -40 and a sum of -3. Here are the factors of -40:

Factors	$-1 \cdot 40$	$-2 \cdot 20$	$-4 \cdot 10$	$-5 \cdot 8$	$-8 \cdot 5$	$-10 \cdot 4$	$-20 \cdot 2$	$-40 \cdot 1$
Sum	39	18	6	3	-3	-6	-18	-39

Our integers m and n are -8 and 5. $x^2 - 3x - 40 = (x - 8)(x + 5)$.

Quick Check **2** Factor.

a) $x^2 + 9x - 36$ b) $x^2 - x - 42$

It is not necessary to list each set of factors as we did in the previous examples. We can often find the two integers m and n quickly through trial and error.

EXAMPLE 5 Factor.

a) $x^2 - 10x - 24$

Solution

The two integers whose product is -24 and sum is -10 are -12 and 2.

$$x^2 - 10x - 24 = (x - 12)(x + 2)$$

b) $x^2 + 10x + 24$

Solution

This example is similar to the last example, but we are looking for two integers that have a product of *positive* 24 instead of -24. Also, the sum of these two integers is 10 instead of -10. The two integers whose product is 24 and whose sum is 10 are 4 and 6.

$$x^2 + 10x + 24 = (x + 4)(x + 6)$$

c) $x^2 + 10x - 24$

Quick Check **3**

Factor.

a) $x^2 + 5x - 6$
b) $x^2 + 5x + 6$
c) $x^2 - 5x + 6$
d) $x^2 - 5x - 6$

Solution

The two integers that have a product of -24 and a sum of 10 are -2 and 12.

$$x^2 + 10x - 24 = (x - 2)(x + 12)$$

d) $x^2 - 10x + 24$

Solution

The integers -4 and -6 have a product of 24 and a sum of -10.

$$x^2 - 10x + 24 = (x - 4)(x - 6)$$

Factoring a Perfect Square Trinomial

Objective **3** Factor a perfect square trinomial.

EXAMPLE 6 Factor $x^2 - 6x + 9$.

Solution

Quick Check **4**

Factor $x^2 + 10x + 25$.

We begin by looking for two integers that have a product of 9 and a sum of -6. Since the product of -3 and -3 is 9 and their sum is -6, $x^2 - 6x + 9 = (x - 3)(x - 3)$. Notice that the same factor is listed twice. We can rewrite this as $(x - 3)^2$.

When a trinomial factors to equal the square of a binomial, we call it a **perfect square trinomial.**

Trinomials That Are Prime

Objective 4 Determine that a trinomial is prime. Not every trinomial of the form $x^2 + bx + c$ can be factored. For example, there may not be a pair of integers that have a product of c and a sum of b. In this case, we say that the trinomial cannot be factored; it is **prime.** For example, $x^2 + 9x + 12$ is prime because we cannot find two numbers with a product of 12 and whose sum is 9.

Factoring a Trinomial Whose Terms Have a Common Factor

Objective 5 Factor a trinomial by first factoring out a common factor. We now turn our attention to factoring trinomials whose terms contain common factors other than 1. After we factor out a common factor, we will attempt to factor the remaining polynomial factor.

EXAMPLE 7 Factor $4x^2 + 48x + 80$.

Solution

The GCF of these three terms is 4.

> **A Word of Caution** When factoring a polynomial, begin by factoring out the GCF of all of the terms.

After factoring out the GCF, we have $4(x^2 + 12x + 20)$. We now try to factor the trinomial $x^2 + 12x + 20$ by finding two integers that have a product of 20 and a sum of 12. The integers that satisfy these conditions are 2 and 10.

$$
\begin{aligned}
4x^2 + 48x + 80 &= 4(x^2 + 12x + 20) &&\text{Factor out the GCF (4).} \\
&= 4(x + 2)(x + 10) &&\text{Factor } x^2 + 12x + 20.
\end{aligned}
$$

Quick Check 5
Factor $5x^2 + 40x + 60$.

> **A Word of Caution** When we factor out a common factor from a polynomial, that common factor must be written in all following stages of factoring. *Don't "lose" the common factor!*

EXAMPLE 8 Factor $-x^2 - 7x + 60$.

Solution

The leading coefficient for this trinomial is -1, but to factor the trinomial using our technique, the leading coefficient must be a *positive* 1. We begin by factoring out a -1 from each term, which will change the sign of each term.

$$-x^2 - 7x + 60 = -(x^2 + 7x - 60)$$ Factor out -1, so that the leading coefficient is 1.

$$= -(x - 5)(x + 12)$$ Factor $x^2 + 7x - 60$ by finding two integers whose product is -60 and whose sum is 7.

Quick Check 6

Factor $-x^2 + 14x - 48$.

If a trinomial's leading coefficient is not 1, or if the degree of the trinomial is not 2, then we should factor out a common factor so that the trinomial factor is of the form $x^2 + bx + c$. At this point in the text, this is the only type of trinomial we know how to factor. We will learn techniques to factor other trinomials in the next section.

Factoring Trinomials in Several Variables

Objective 6 **Factor a trinomial in several variables.** Trinomials in several variables can also be factored. In the next example, we will explore the similarities and the differences between factoring trinomials in two variables and trinomials in one variable.

EXAMPLE 9   Factor $x^2 + 13xy + 36y^2$.

Solution

These three terms have no common factors, and the leading coefficient is 1. Suppose that the variable y were not included; in other words, suppose that we were asked to factor $x^2 + 13x + 36$. After looking for two integers whose product is 36 and whose sum is 13, we would see that $x^2 + 13x + 36 = (x + 4)(x + 9)$. Now we can examine the changes that are necessary because of the second variable y. Note that $4 \cdot 9 = 36$, not $36y^2$. However, $4y \cdot 9y$ does equal $36y^2$, and $x \cdot 9y + x \cdot 4y$ does equal $13xy$. This trinomial factors to be $(x + 4y)(x + 9y)$.

Quick Check 7

Factor $x^2 - 4xy - 32y^2$.

We can factor a trinomial in two or more variables by first ignoring all of the variables except for the first variable. After determining how the trinomial factors for the first variable, we can then determine where the rest of the variables appear in the factored form of the trinomial. We can check that our factoring is correct by multiplying, making sure that the product is equal to the trinomial.

EXAMPLE 10   Factor $p^2r^2 - 9pr - 10$.

Solution

If the variable r did not appear in the trinomial, and we were factoring $p^2 - 9p - 10$, we would look for two integers that have a product of -10 and a sum of -9. The two integers are -10 and 1, so the polynomial $p^2 - 9p - 10 = (p - 10)(p + 1)$. Now we work on the variable r. Since the first term in the trinomial is p^2r^2, the first term in each binomial must be pr. $p^2r^2 - 9pr - 10 = (pr - 10)(pr + 1)$.

Quick Check 8

Factor
$x^2y^2 + 10xy - 24$.

Factor completely. If the polynomial cannot be factored, write "prime."

1. $x^2 - 8x - 20$

2. $x^2 - 2x - 15$

3. $x^2 + 5x - 36$

4. $x^2 - 10x - 39$

5. $x^2 + 12x + 24$

6. $x^2 - 12x + 36$

7. $x^2 + 14x + 48$

8. $x^2 + 14x - 33$

9. $x^2 + 13x - 30$

10. $x^2 - 13x + 30$

11. $x^2 - 5x + 36$

12. $x^2 - 15x + 54$

13. $x^2 - 17x - 60$

14. $x^2 - 17x + 60$

15. $x^2 + 11x - 12$

16. $x^2 + 15x + 56$

17. $x^2 + 14x + 49$

18. $x^2 - 4x - 12$

19. $x^2 + 20x + 91$

20. $x^2 + 14x + 13$

21. $x^2 - 24x + 144$

22. $x^2 + 5x - 14$

23. $x^2 - 13x + 40$

24. $x^2 - 5x - 24$

25. $x^2 - 13x - 30$

26. $x^2 + 18x + 81$

27. $x^2 - 9x + 20$

28. $x^2 - 11x - 28$

29. $x^2 - 16x + 60$

30. $x^2 + 16x + 48$

31. $5x^2 + 30x - 55$

32. $-x^2 - 7x + 18$

33. $21x^2 + 147x + 210$

34. $9x^2 - 45x - 450$

35. $-3x^2 + 48x - 192$

36. $8x^2 - 128x + 504$

37. $x^3 + 8x^2 + 16x$

38. $-x^4 + x^3 + 30x^2$

39. $-3x^5 - 6x^4 + 240x^3$

40. $2x^3 - 42x^2 + 196x$

41. $10x^8 + 20x^7 - 990x^6$

42. $35x^4 - 70x^3 + 35x^2$

43. $x^2 + xy - 42y^2$

44. $3x^2 - 36xy - 84y^2$

45. $x^2y^2 - 7xy + 12$

46. $x^2 - 20xy + 100y^2$

47. $x^2 - 10xy - 39y^2$

48. $x^2y^2 - 6xy - 7$

49. $5x^5 + 50x^4y + 105x^3y^2$

50. $x^2y^2 + 11xy - 60$

Factor completely, using the appropriate technique. (If the polynomial cannot be factored, write "prime.")

51. $x^4 + 9x^3 + 5x + 45$

52. $x^2 - 26x + 169$

53. $x^2 + x + 30$

54. $x^2 - 17x + 42$

55. $x^2 + 22x + 40$

56. $x^2 + 10x$

57. $x^2 + 40x + 400$

58. $x^2 + 4x - 60$

59. $4x^2 + 8x - 96$

60. $x^2 + 16x + 15$

61. $x^4 - 15x^3$

62. $x^2 + 2x - 120$

63. $x^2 - 16x + 63$

64. $x^2 - 7x - 78$

65. $x^2 + 10x - 25$

66. $x^3 - 12x^2 - 6x + 72$

67. $x^2 - 21x - 100$

68. $-x^2 + 17x - 60$

Answer in complete sentences.

69. Explain why it is a good idea to factor a common factor of -1 from the expression $-x^2 + 7x + 30$ before attempting to factor the trinomial.

70. Explain the process for factoring a trinomial of the form $x^2 + bx + c$. Use examples of various types to illustrate the process.

Study Tip REVISITED One effective tool for understanding your past difficulties in mathematics and how those past difficulties are hindering your ability to learn mathematics today is to write a mathematical autobiography. Write down your successes and your failures, as well as the reasons why you think you succeeded or failed. Go back into your mathematical past as far as you can remember. Write down how friends, relatives, and teachers affected you.

Let your autobiography sit for a few days and then read it over. Read it on an analytical level as if another person had written it and detach yourself from it personally. Look for patterns, and think about strategies to reverse the bad patterns and continue any good patterns that you can find.

6.3

FACTORING TRINOMIALS OF THE FORM $ax^2 + bx + c$ $(a \neq 1)$

Objectives

1 Factor a trinomial of the form $ax^2 + bx + c$ $(a \neq 1)$ by grouping.

2 Factor a trinomial of the form $ax^2 + bx + c$ $(a \neq 1)$ by trial and error.

In this section, we continue to learn how to factor second degree trinomials, focusing on trinomials whose leading coefficient is not 1. Some examples of this type of trinomial are $2x^2 + 13x + 20$, $8x^2 - 22x - 21$, and $3x^2 - 11x + 8$.

There are two methods for factoring this type of trinomial that we will examine: factoring by grouping and factoring by trial and error. Your instructor may prefer one of these methods to the other, or an altogether different method, and may ask you to use one method only. However, if you are allowed to use either method, use the method that you feel more comfortable using.

Factoring by Grouping

Objective 1 Factor a trinomial of the form $ax^2 + bx + c$ $(a \neq 1)$ by grouping.
We begin with factoring by grouping. Consider the product $(3x + 2)(2x + 5)$.

$$(3x + 2)(2x + 5) = 6x^2 + 15x + 4x + 10 \qquad \text{Distribute.}$$
$$= 6x^2 + 19x + 10 \qquad \text{Combine like terms.}$$

Since $(3x + 2)(2x + 5) = 6x^2 + 19x + 10$, we know that $6x^2 + 19x + 10$ can be factored as $(3x + 2)(2x + 5)$. By rewriting the middle term $19x$ as $15x + 4x$, we can factor by grouping. The important skill is being able to determine that $15x + 4x$ is the correct way to rewrite $19x$, instead of $16x + 3x$, $9x + 10x$, $22x - 3x$, or any other two terms whose sum is $19x$.

Factoring $ax^2 + bx + c$ $(a \neq 1)$ by Grouping

1. Multiply $a \cdot c$.
2. Find two integers whose product is $a \cdot c$ and whose sum is b.
3. Rewrite the term bx as two terms, using the two integers found in step 2.
4. Factor the resulting polynomial by grouping.

To factor the polynomial $6x^2 + 19x + 10$ by grouping, we begin by multiplying $6 \cdot 10$, which equals 60. We now look for two integers whose product is 60 and whose sum is 19, and those two integers are 15 and 4. We can then rewrite $19x$ as $15x + 4x$. (These terms could be written in the opposite order and would still lead to the correct factoring.) We then factor the polynomial $6x^2 + 15x + 4x + 10$ by grouping.

$$6x^2 + 19x + 10 = 6x^2 + 15x + 4x + 10 \qquad \text{Rewrite } 19x \text{ as } 15x + 4x.$$
$$= 3x(2x + 5) + 2(2x + 5) \qquad \text{Factor the common factor } 3x \text{ from the first two terms and the common factor 2 from the last two terms.}$$
$$= (2x + 5)(3x + 2) \qquad \text{Factor out the common factor } 2x + 5.$$

Now we will examine several examples using this technique.

EXAMPLE 1   Factor $2x^2 + 7x + 6$ by grouping.

Solution

We first check to see whether there are any common factors; and in this case, there are not. We proceed with factoring by grouping. Since $2 \cdot 6 = 12$, we look for two integers with a product of 12 and a sum of 7, which is the coefficient of the middle term. The integers 3 and 4 meet these criteria, so we rewrite $7x$ as $3x + 4x$.

$$2x^2 + 7x + 6 = 2x^2 + 3x + 4x + 6$$ Find two integers whose product is $2 \cdot 6 = 12$ and whose sum is 7. Rewrite $7x$ as $3x + 4x$.

$$= x(2x + 3) + 2(2x + 3)$$ Factor the common factor x from the first two terms and the common factor 2 from the last two terms.

Quick Check **1**

Factor $3x^2 + 13x + 12$ by grouping.

$$= (2x + 3)(x + 2)$$ Factor out the common factor $2x + 3$.

We check that this factoring is correct by multiplying $2x + 3$ by $x + 2$, which should equal $2x^2 + 7x + 6$. The check is left to the reader.

EXAMPLE 2   Factor $5x^2 - 7x + 2$ by grouping.

Solution

We begin by multiplying $5 \cdot 2$ as there are no common factors other than 1. We look for two integers whose product is 10 and whose sum is -7. The integers are -5 and -2, so we rewrite $-7x$ as $-5x - 2x$ and then factor by grouping.

$$5x^2 - 7x + 2 = 5x^2 - 5x - 2x + 2$$ Find two integers with a product of 10 and a sum of -7. Rewrite $-7x$ as $-5x - 2x$.

Quick Check **2**

Factor $4x^2 - 25x + 36$ by grouping.

$$= 5x(x - 1) - 2(x - 1)$$ Factor the common factor $5x$ from the first two terms and the common factor -2 from the last two terms.

$$= (x - 1)(5x - 2)$$ Factor out the common factor $x - 1$.

EXAMPLE 3   Factor $12x^2 + 11x - 15$ by grouping.

Solution

The terms have no common factor other than 1, so we begin by multiplying $12(-15)$, which equals -180. We need to find two integers with a product of -180 and a sum of 11. We can start by listing the different factors of 180. Since we know that one of the integers will be positive and one will be negative, we look for two factors in this list that have a difference of 11.

$$1 \cdot 180 \quad 2 \cdot 90 \quad 3 \cdot 60 \quad 4 \cdot 45 \quad 5 \cdot 36 \quad 6 \cdot 30 \quad 9 \cdot 20 \quad 10 \cdot 18 \quad 12 \cdot 15$$

Notice that the pair 9 and 20 has a difference of 11. The two integers are -9 and 20; their product is -180 and their sum is 11. We rewrite $11x$ as $-9x + 20x$ and then factor by grouping.

$$12x^2 + 11x - 15 = 12x^2 - 9x + 20x - 15$$

Find two integers whose product is -180 and whose sum is 11. Rewrite $11x$ as $-9x + 20x$.

$$= 3x(4x - 3) + 5(4x - 3)$$

Factor the common factor $3x$ out of the first two terms and the common factor 5 out of the last two terms.

$$= (4x - 3)(3x + 5)$$

Factor out the common factor $4x - 3$.

Quick Check **3**

Factor $16x^2 - 38x + 21$ by grouping.

Factoring by Trial and Error

Objective **2** **Factor a trinomial of the form $ax^2 + bx + c$ ($a \neq 1$) by trial and error.** We now turn our attention to the method of trial and error. We need to factor a trinomial $ax^2 + bx + c$, and, once factored, it will have the following form:

$$(\text{variable term } + \text{ constant})(\text{variable term } + \text{ constant})$$

The product of the variable terms will be ax^2, and the product of the constants will equal c. We will use these facts to get started by listing all of the factors of ax^2 and c.

Consider the trinomial $2x^2 + 7x + 6$, which has no common factors other than 1. We begin by listing all of the factors of $2x^2$ and 6.

Factors of $2x^2$: $x \cdot 2x$.   Factors of 6: $1 \cdot 6$, $2 \cdot 3$.

Since there is only one pair of factors whose product is $2x^2$, we know that if this trinomial is factorable, it will be of the form $(2x + ?)(x + ?)$. We will substitute the different factors of 6 in all possible orders in place of the question marks, satisfying the following products.

$$\overset{2x^2}{(2x + ?)(x + ?)}\\\underset{6}{}$$

We will keep substituting factors of 6 until the middle term of the product of the two binomials is $7x$. Rather than fully distributing for each trial, we need to look only at the two products shown in the following graphic.

$$(2x + ?)(x + ?)$$

When the sum of these two products is $7x$, we have found the correct factors.

We begin by using the factors 1 and 6, which leads to the binomial factors $(2x + 1)(x + 6)$. Since the middle term for these factors is $13x$, we have not found the correct factors. We then switch the positions of 1 and 6 and try again. However, when we multiply out $(2x + 6)(x + 1)$, the middle term equals $8x$, not $7x$. We need to try another pairing, so we try the other factors of 6, which are 2 and 3. When we multiply out $(2x + 3)(x + 2)$, the middle term is $7x$, so these are the correct factors:

$$\overset{4x}{(2x + 3)(x + 2)}\\\underset{3x}{}$$

$$2x^2 + 7x + 6 = (2x + 3)(x + 2).$$

EXAMPLE 4   Factor $10x^2 + 13x - 3$ by trial and error.

Solution

There are no common factors other than 1, so we begin by listing the factors of $10x^2$ and -3.

$$\text{Factors of } 10x^2: \quad x \cdot 10x, \quad 2x \cdot 5x$$
$$\text{Factors of } -3: \quad -1 \cdot 3, \quad -3 \cdot 1$$

We are looking for two factors that produce the middle term $13x$ when multiplied. We will begin by using x and $10x$ for the variable terms and -1 and 3 for the constants.

Factors	Middle Term
$(x - 1)(10x + 3)$	$-7x$
$(x + 3)(10x - 1)$	$29x$

Neither of these middle terms is equal to $13x$, so we have not found the correct factoring. Also, since neither middle term is the opposite of $13x$, switching the signs of the constants 1 and 3 will not lead to the correct factors either. So $(x + 1)(10x - 3)$ and $(x - 3)(10x + 1)$ are not correct. We switch to the pair $2x$ and $5x$.

Factors	Middle Term
$(2x - 1)(5x + 3)$	x
$(2x + 3)(5x - 1)$	$13x$

The second of these middle terms is the one that we are looking for, so $10x^2 + 13x - 3 = (2x + 3)(5x - 1)$.

Quick Check 4
Factor $2x^2 + x - 28$ by trial and error.

Now we will try to factor another trinomial by trial and error, using a trinomial whose terms have more factors.

EXAMPLE 5   Factor $12x^2 + 11x - 15$ by trial and error.

Solution

Since there are no common factors to factor out, we begin by listing the factors of $12x^2$ and -15.

$$\text{Factors of } 12x^2: \quad x \cdot 12x, \quad 2x \cdot 6x, \quad 3x \cdot 4x$$
$$\text{Factors of } -15: \quad -1 \cdot 15, \quad -15 \cdot 1, \quad -3 \cdot 5, \quad -5 \cdot 3$$

At first glance, this may seem like it will take a while, but there are some ways that we can shorten the process. For example, we may skip over any pairings resulting in a binomial factor in which both terms have a common factor other than 1, such as $3x - 3$. If the terms of a binomial factor had a common factor other than 1, then the original polynomial would also have a common factor, but we know that $12x^2 + 11x - 15$ does not have a common factor that can be factored out.

If the pair of constants -1 and 15 does not produce the desired middle term and does not produce the opposite of the desired middle term, then we do not need to use the pair of constants -15 and 1. Instead, we could move on to the next pair of constants. Eventually, through trial and error, we find that $12x^2 + 11x - 15 = (3x + 5)(4x - 3)$.

Quick Check 5

Factor $20x^2 + 17x - 24$ by trial and error.

> **A Word of Caution**   If a trinomial has a leading coefficient other than 1, be sure to check if the three terms have a common factor. If there is a common factor, we may be able to factor a trinomial using the techniques from Section 6.2.

EXAMPLE 6   Factor $8x^2 + 40x + 48$.

Solution

The three terms have a common factor of 8. After we factor out this common factor, the trinomial factor will be $x^2 + 5x + 6$, which is a trinomial with a leading coefficient of 1. We look for two integers whose product is 6 and whose sum is 5. Those two integers are 2 and 3.

$$8x^2 + 40x + 48 = 8(x^2 + 5x + 6)$$ Factor out the common factor 8.
$$= 8(x + 2)(x + 3)$$ To factor $x^2 + 5x + 6$ we need to find two integers whose product is 6 and whose sum is 5. The two integers are 2 and 3.

Quick Check 6

Factor $9x^2 + 27x - 162$.

Factoring out the common factor in the previous example makes factoring the trinomial much more manageable. If we were to try factoring by grouping without factoring out the common factor, we would begin by multiplying 8 by 48, which equals 384. We would look for two integers with a product of 384 and a sum of 40, which would be challenging. If we tried to factor by trial and error, $8x^2$ has two pairs of factors ($x \cdot 8x, 2x \cdot 4x$), but 48 has several pairs of factors ($1 \cdot 48, 2 \cdot 24, 3 \cdot 16, 4 \cdot 12, 6 \cdot 8$). Factoring by trial and error would be tedious at best.

EXERCISES 6.3

Factor completely.

1. $5x^2 - 18x - 8$
2. $3x^2 + 11x + 6$
3. $2x^2 + 9x + 4$
4. $7x^2 - 25x + 12$
5. $3x^2 - x - 4$
6. $4x^2 - 13x + 10$
7. $4x^2 + 4x - 3$
8. $21x^2 - 18x - 3$
9. $6x^2 + 7x + 2$
10. $8x^2 + 2x - 21$
11. $9x^2 - 3x - 20$

12. $10x^2 - 29x + 10$
13. $20x^2 + 161x + 8$
14. $18x^2 + 93x - 16$
15. $3x^2 + 13x - 10$
16. $20x^2 + 47x + 24$
17. $12x^2 + 11x - 56$
18. $24x^2 + 55x - 24$
19. $16x^2 - 24x + 9$
20. $16x^2 + 66x - 27$
21. $15x^2 - 24x - 12$
22. $15x^2 + 32x - 60$
23. $16x^2 + 24x - 40$

24. $15x^2 + 45x - 150$

25. $16x^2 - 64x + 64$

26. $40x^2 - 100x - 60$

27. $15x^2 - 25x - 560$

28. $-12x^2 - 36x + 120$

Factor completely, using the appropriate technique.

29. $x^2 + 14x + 13$

30. $x^2 - 10x - 11$

31. $18x - 9$

32. $3x^2 + 33x + 48$

33. $2x^2 - 7x + 6$

34. $x^2 + 30x + 225$

35. $x^2 + 19x - 20$

36. $32x^2 - 20x - 25$

37. $x^5 - 6x^4$

38. $x^2 - 8x - 48$

39. $4x^2 - 20x$

40. $x^2 + 17x - 60$

41. $-12x^2 - 23x - 10$

42. $4x^2 + 20x + 21$

43. $9x^2 - 30x + 13$

44. $3x^2 - 7x - 6$

45. $x^6 + 3x^4 + 8x^2 + 24$

46. $24x^3 + 168x^2 + 54x + 378$

47. $x^3 + 2x^2 - 28x - 56$

48. $x^2 + 25x + 150$

49. $x^2 - 24x + 144$

50. $x^2 - 3x + 2$

51. $9x^2 - 45x - 54$

52. $3x^3 - 30x^2 - 7x + 70$

53. $-5x^2 + 28x + 12$

54. $x^2 - 6x + 5$

55. $2x^2 - 15x - 27$

56. $x^2 + 8x + 16$

57. $x^2 + 12x - 45$

58. $3x^5 + 6x^4$

59. $-3x^2 - 12x + 96$

60. $x^2 - 17x + 60$

61. $15x^2 + 16x - 15$

62. $18x^2 + 51x + 8$

63. $2x^2 + 9x + 4$

64. $x^2 - 7x - 44$

Answer in complete sentences.

65. Compare the two different factoring methods for trinomials of the form $ax^2 + bx + c$ $(a \neq 1)$: factoring by grouping and factoring by trial and error. Which method do you prefer? Explain your reasoning.

66. Explain why it is a good idea to factor out a common factor from a trinomial of the form $ax^2 + bx + c$ before trying other factoring techniques.

> *Study Tip* **REVISITED**   Many students confuse test anxiety with math anxiety. If you feel that you can learn mathematics and you understand the material, but you "freeze up" on tests, then you may have test anxiety, especially if this is true in your classes besides your math class. Speak to your academic counselor about this. Many colleges offer seminars or short-term classes about how to reduce test anxiety; ask your counselor whether such a course would be appropriate for you.

6.4
FACTORING SPECIAL BINOMIALS

Objectives

1 **Factor a difference of squares.**
2 **Factor the factors of a difference of squares.**
3 **Factor a difference of cubes.**
4 **Factor a sum of cubes.**

Difference of Squares

Objective 1 **Factor a difference of squares.** Recall the special product $(a + b)(a - b) = a^2 - b^2$ from Section 5.4. The binomial $a^2 - b^2$ is a **difference of squares.** Based on this special product, a difference of squares factors in the following manner:

Difference of Squares

$$a^2 - b^2 = (a + b)(a - b)$$

To identify a binomial as a difference of squares, we must verify that each term is a perfect square. Variable factors must have exponents that are multiples of 2 such as x^2, y^4 and a^6. Any constant must also be a perfect square. Here are the first 10 squares:

$$1, 4, 9, 16, 25, 36, 49, 64, 81, 100$$

EXAMPLE 1 Factor $x^2 - 49$.

Solution

This binomial is a difference of squares, as it can be rewritten as $(x)^2 - (7)^2$. Rewriting the binomial in this form helps us to use the formula $a^2 - b^2 = (a + b)(a - b)$.

$$\begin{aligned} x^2 - 49 &= (x)^2 - (7)^2 \qquad && \text{Rewrite each term as a square.} \\ &= (x + 7)(x - 7) && \text{Factor using the formula for a difference of squares} \end{aligned}$$

We can check our work by multiplying $x + 7$ by $x - 7$, which should equal $x^2 - 49$. The check is left to the reader.

Quick Check 1
Factor $x^2 - 25$.

Although we will continue to write each term as a perfect square, if you can factor a difference of squares without writing each term as a perfect square, then do not feel that you *must* rewrite each term as a perfect square before proceeding.

EXAMPLE 2 Factor $64a^{10} - 25b^4$.

Solution

The first term of this binomial is a square, as 64 is a square and the exponent for the variable factor a is even. The first term can be rewritten as $(8a^5)^2$. In a similar fashion, $25b^4$ can be rewritten as $(5b^2)^2$. This binomial is a difference of squares.

Quick Check 2
Factor $81x^{12} - 100y^{16}$.

$$\begin{aligned} 64a^{10} - 25b^4 &= (8a^5)^2 - (5b^2)^2 \qquad && \text{Rewrite each term as a square.} \\ &= (8a^5 + 5b^2)(8a^5 - 5b^2) && \text{Factor as a difference of squares.} \end{aligned}$$

This binomial $8x^2 - 81$ is not a difference of squares because the coefficient of the first term (8) is not a square. Since there are no common factors other than 1, this binomial cannot be factored.

EXAMPLE 3   Factor $3x^2 - 108$.

Solution

Although this binomial is not a difference of squares (3 and 108 are not squares), the common factor of 3 can be factored out. After we factor out the common factor 3, we have $3x^2 - 108 = 3(x^2 - 36)$. The binomial in parentheses is a difference of squares and can be factored accordingly.

$$3x^2 - 108 = 3(x^2 - 36) \qquad \text{Factor out the common factor 3.}$$
$$= 3[(x)^2 - (6)^2] \qquad \text{Rewrite each term in the binomial factor as a square.}$$
$$= 3(x + 6)(x - 6) \qquad \text{Factor as a difference of squares.}$$

Quick Check **3**
Factor $7x^2 - 175$.

If a binomial is a **sum of squares,** then it cannot be factored unless the two terms have a common factor. Some examples of a sum of squares are $x^2 + 49$, $4a^2 + 9b^2$, and $64a^{10} + 25b^4$.

Sum of Squares

> A sum of squares, $a^2 + b^2$, cannot be factored.

Objective 2 **Factor the factors of a difference of squares.** Occasionally, after factoring a difference of squares, one or more of the factors may still be factorable. For example, consider the binomial $x^4 - 16$, which factors to be $(x^2 + 4)(x^2 - 4)$. The first factor, $x^2 + 4$, is a sum of squares and cannot be factored. However, the second factor, $x^2 - 4$, is a difference of squares and can be factored further.

EXAMPLE 4   Factor $x^4 - 81$ completely.

Solution

This is a difference of squares and we factor it accordingly. After we factor the binomial as a difference of squares, one of the binomial factors $(x^2 - 9)$ is a difference of squares and must be factored as well.

$$x^4 - 81 = (x^2)^2 - (9)^2 \qquad \text{Rewrite each term as a perfect square.}$$
$$= (x^2 + 9)(x^2 - 9) \qquad \text{Factor as a difference of squares.}$$
$$= (x^2 + 9)[(x)^2 - (3)^2] \qquad \text{Rewrite each term in the binomial } x^2 - 9 \text{ as a square.}$$
$$= (x^2 + 9)(x + 3)(x - 3) \qquad \text{Factor } x^2 - 9 \text{ as a difference of squares.}$$

Quick Check **4**
Factor $x^4 - y^4$.

A Word of Caution   When factoring a difference of squares, check the binomial factor containing a difference to see if it can be factored.

Difference of Cubes

Objective 3 **Factor a difference of cubes.** Another special binomial that can be factored is a **difference of cubes**. In this case, both terms are perfect cubes rather than squares. Here is the formula for factoring a difference of cubes.

Difference of Cubes

$$a^3 - b^3 = (a - b)(a^2 + ab + b^2)$$

Before proceeding, let's multiply out $(a - b)(a^2 + ab + b^2)$ to show that it actually equals $a^3 - b^3$.

$$
\begin{aligned}
(a - b)(a^2 + ab + b^2) &= a \cdot a^2 + a \cdot ab + a \cdot b^2 - b \cdot a^2 - b \cdot ab - b \cdot b^2 && \text{Distribute.}\\
&= a^3 + a^2b + ab^2 - a^2b - ab^2 - b^3 && \text{Multiply.}\\
&= a^3 - b^3 && \text{Combine like terms.}
\end{aligned}
$$

We see that the formula is correct.

For a term to be a cube, its variable factors must have exponents that are multiples of 3, such as x^3, y^6, and z^9. Also, each constant factor must be a cube. Here are the first 10 cubes:

$$1, 8, 27, 64, 125, 216, 343, 512, 729, 1000$$

We need to memorize the formula for factoring a difference of cubes. There are some patterns that can help us to remember the formula. A difference of cubes, $a^3 - b^3$, has two factors: a binomial $(a - b)$ and a trinomial $(a^2 + ab + b^2)$. The binomial looks just like the difference of cubes without the cubes, including the sign between the terms. The signs between the three terms in the trinomial are both addition signs. To find the actual terms in the trinomial factor, the following diagram may be helpful.

$$
\begin{array}{ccc}
\text{1st} & \text{2nd} & \text{1st} \cdot \text{2nd}\\
\end{array}
$$

$$a^3 - b^3 = (a - b)(a^2 + ab + b^2)$$

$$
\begin{array}{cc}
\text{1st} \cdot \text{1st} & \text{2nd} \cdot \text{2nd}
\end{array}
$$

EXAMPLE 5   Factor $x^3 - 27$.

Solution

We begin by looking for common factors other than 1 that the two terms share, but there are none. This binomial is not a difference of squares, as the exponent of the variable term is not a multiple of 2. Also, the constant 27 is not a square. The binomial is, however, a difference of cubes. We can rewrite the term x^3 as $(x)^3$ and we can rewrite 27 as $(3)^3$. Rewriting $x^3 - 27$ as $(x)^3 - (3)^3$ will help us to identify the terms that are in the binomial and trinomial factors.

$$x^3 - 27 = (x)^3 - (3)^3$$
$$= (x - 3)(x \cdot x + x \cdot 3 + 3 \cdot 3)$$

Rewrite each term as a perfect cube.

Factor as a difference of cubes. The binomial factor is the same as $(x)^3 - (3)^3$ without the cubes. Treating x as the 1st term in the binomial factor and 3 as the 2nd term, the terms in the trinomial are 1st · 1st + 1st · 2nd + 2nd · 2nd

$$= (x - 3)(x^2 + 3x + 9)$$

Simplify each term in the trinomial factor.

Quick Check 5

Factor $x^3 - 125$.

A Word of Caution A difference of cubes $x^3 - y^3$ cannot be factored as $(x - y)^3$.

EXAMPLE 6 Factor $y^{15} - 64$.

Solution

Although the number 64 is a square, this binomial is not a difference of squares because y^{15} is not a square. For a term to be a square, its variable factors must have exponents that are multiples of 2. This binomial is a difference of cubes. The exponent in the first term is a multiple of 3, and the number 64 is equal to 4^3.

$$y^{15} - 64 = (y^5)^3 - (4)^3$$

Rewrite each term as a perfect cube.

$$= (y^5 - 4)(y^5 \cdot y^5 + y^5 \cdot 4 + 4 \cdot 4)$$
$$= (y^5 - 4)(y^{10} + 4y^5 + 16)$$

Factor as a difference of cubes.

Simplify each term in the trinomial factor.

Quick Check 6

Factor $x^{12} - 8$.

A Word of Caution When factoring a difference of cubes $a^3 - b^3$, do not attempt to factor the trinomial factor $a^2 + ab + b^2$.

Sum of Cubes

Objective 4 Factor a sum of cubes. Unlike a sum of squares, a **sum of cubes** can be factored. Here is the formula:

Sum of Cubes

$$a^3 + b^3 = (a + b)(a^2 - ab + b^2)$$

Notice that the terms in the factors are the same as the factors in a *difference* of cubes, with the exception of some of their signs. When factoring a sum of cubes, the binomial factor is a sum rather than a difference. The sign of the middle term in the trinomial is negative rather than positive. The last term in the trinomial factor is positive, just as it was in the formula for a difference of cubes. The diagram shows the differences between the formula for a difference of cubes and a sum of cubes.

$$a^3 - b^3 = (a - b)(a^2 + ab + b^2)$$
$$a^3 + b^3 = (a + b)(a^2 - ab + b^2)$$

EXAMPLE 7   Factor $z^3 + 125$.

Solution

This binomial is a sum of cubes and can be rewritten as $(z)^3 + (5)^3$. We then can factor the sum of cubes using the formula.

$$
\begin{aligned}
z^3 + 125 &= (z)^3 + (5)^3 \\
&= (z + 5)(z \cdot z - z \cdot 5 + 5 \cdot 5) \\
&= (z + 5)(z^2 - 5z + 25)
\end{aligned}
$$

Rewrite each term as a perfect cube.
Factor as a sum of cubes.
Simplify each term in the trinomial factor.

Quick Check 7

Factor $x^3 + 216$.

A Word of Caution   A sum of cubes $x^3 + y^3$ cannot be factored as $(x + y)^3$.

We finish this section by summarizing the strategies for factoring binomials.

Factoring Binomials

- Factor out the GCF if there is a common factor other than 1.
- Determine whether both terms are perfect squares.
  If the binomial is a sum of squares, it cannot be factored.
  If the binomial is a difference of squares, we can factor it by using the formula
  $a^2 - b^2 = (a + b)(a - b)$. Keep in mind that some of the resulting factors can be differences of squares as well.
- If both terms are not perfect squares, determine whether they both are perfect cubes.
  If the binomial is a difference of cubes, we factor using the formula
  $a^3 - b^3 = (a - b)(a^2 + ab + b^2)$.
  If the binomial is a sum of cubes, we factor using the formula
  $a^3 + b^3 = (a + b)(a^2 - ab + b^2)$.

EXERCISES 6.4

Factor completely. If the polynomial cannot be factored, write "prime."

1. $x^2 - 4$

2. $x^2 - 9$

3. $x^2 - 36$

4. $a^2 - 64$

5. $100x^2 - 81$

6. $49x^2 - 64$

7. $x^2 - 2$

8. $x^2 - 18$

9. $a^2 - 16b^2$

10. $m^2 - 9n^2$

11. $25x^2 - 64y^2$

12. $36x^2 - 121y^2$

13. $16 - x^2$

14. $144 - x^2$

15. $3x^2 - 75$

16. $8x^2 - 392$

17. $x^2 + 25$

18. $x^2 + 1$

19. $5x^2 + 20$

20. $4x^2 + 100y^2$

21. $x^4 - 16$

22. $x^4 - 1$

23. $x^4 - 81y^2$

24. $x^8 - 64y^4$

25. $x^3 - 1$

26. $x^3 - y^3$

27. $x^3 - 8$

28. $a^3 - 216$

29. $1000 - y^3$

30. $343 - x^3$

31. $x^3 - 8y^3$

32. $y^3 - 27x^3$

33. $8a^3 - 729b^3$

34. $64r^3 - 125s^3$

35. $x^3 + y^3$

36. $x^3 + 27$

37. $x^3 + 8$

38. $b^3 + 64$

39. $x^6 + 27$

40. $y^{12} + 125$

41. $64m^3 + n^3$

42. $x^3 + 343y^3$

43. $6x^3 - 162$

44. $72a^3 + 243b^3$

45. $16x^4 + 250xy^3$

46. $64a^7b^{11} - a^4b^5$

Factor completely, using the appropriate technique. If the polynomial cannot be factored, write "prime."

47. $x^2 - 16x + 64$

48. $4x^3 + 16x^2 + x + 4$

49. $x^2 + 16x + 28$

50. $2x^2 + 16x - 18$

51. $14x - 6$

52. $x^2 - 8x + 33$

53. $6x^2 - 5x - 25$

54. $-8x^2 + 32x + 18$

55. $x^4 - 2x^3 - 63x^2$

56. $x^2 + 18x + 77$

57. $-8x^3 - 72$

58. $25 - x^2$

59. $-6x^2 + 60x - 144$

60. $x^2 - 36x$

61. $x^2 - 10x - 39$

62. $x^2 + 4x - 140$

63. $25x^2 - 36y^8$

64. $x^2 + 3x - 40$

65. $2x^2 + x - 13$

66. $x^3 - 216y^3$

67. $x^2 - 64y^6$

68. $x^3 - 64y^6$

69. $9x^5 - 24x^4 - 45x^2$

70. $20x^2 - 61x + 36$

71. $4x^2 + 196$

72. $36x^2 + 6x - 210$

73. $x^2 + 12x + 35$

74. $343x^3 - 8y^6z^9$

75. $3x^2 - 13x - 38$

76. $2x^2 + 17x + 26$

77. $x^3 - 10x^2 - 3x + 30$

78. $x^2 + 16y^2$

79. $x^3 + 125y^3$

80. $125x^6 - 64y^9$

81. $9x^7 + 30x^6 + 25x^5$

82. $x^3 - 5x^2 + 2x - 10$

Answer in complete sentences.

83. Explain how to identify a binomial as a difference of squares.

84. A student was asked to factor $4 - x^2$. The student's answer was $(x + 2)(x - 2)$. Was the student correct? If not, what was the student's error. Explain fully.

85. Explain how to identify a binomial as a difference of cubes or as a sum of cubes.

86. Explain, in your own words, how to remember the formula for factoring a difference of cubes.

Study Tip REVISITED   Many students think they have math anxiety because they constantly feel that they are in over their heads, when in fact they may not be prepared for their particular class. Taking a course that you do not have the prerequisite skills for is not a good idea, and success will be difficult to achieve. For example, if you are taking an elementary algebra course, but are struggling with performing arithmetic operations on fractions, order of operations problems, and operations with positive and negative numbers, then perhaps you would be better taking a prealgebra class.

If this situation applies to you, talk to your instructor. Your instructor will be able to advise you whether you should be taking another course or whether you have the skills to be successful in this course.

6.5

FACTORING POLYNOMIALS: A GENERAL STRATEGY

Objective

1 **Understand the strategy for factoring a general polynomial.**

Objective 1 **Understand the strategy for factoring a general polynomial.** In this section, we will review the different factoring techniques introduced in this chapter and develop a general strategy for factoring any polynomial. Keep in mind that a polynomial must be factored completely, meaning each of its polynomial factors cannot be factored further. For example, if we factored the trinomial $4x^2 + 24x + 32$ to be $(2x + 4)(2x + 8)$, this would not be factored completely because the two terms in each binomial factor have a common factor of 2 that must be factored out as follows:

$$\begin{aligned}
4x^2 + 24x + 32 &= (2x + 4)(2x + 8) \\
&= 2(x + 2) \cdot 2(x + 4) \\
&= 4(x + 2)(x + 4)
\end{aligned}$$

If we factor out all common factors as our first step when factoring a polynomial, we can avoid having to factor out common factors at the end of the process. We could have factored out the common factor of 4 from the polynomial $4x^2 + 24x + 32$ at the very beginning. In addition to ensuring that the polynomial has been factored completely, factoring out common factors will often make our work easier, and will occasionally allow us to factor polynomials that we would not be able to factor otherwise.

Here is a general strategy for factoring any polynomial:

Factoring Polynomials

1 Factor out any common factors.

2 Determine the number of terms in the polynomial.
   (a) If there are only **two terms,** check to see if the binomial is one of the special binomials discussed in Section 6.4.
   - Difference of Squares: $a^2 - b^2 = (a + b)(a - b)$
   - Sum of Squares: $a^2 + b^2$ is not factorable
   - Difference of Cubes: $a^3 - b^3 = (a - b)(a^2 + ab + b^2)$
   - Sum of Cubes: $a^3 + b^3 = (a + b)(a^2 - ab + b^2)$

   (b) If there are **three terms,** try to factor the trinomial using the techniques of Sections 6.2 and 6.3.
   - $x^2 + bx + c = (x + m)(x + n)$: find two integers m and n whose product is c and whose sum is b.
   - $ax^2 + bx + c\ (a \neq 1)$: we have two methods for factoring this type of trinomial, either by grouping or by trial and error.
     To use factoring by grouping, refer to objective 1 in Section 6.3.
     To factor by trial and error, refer to objective 2 in Section 6.3.

   (c) If there are **four terms,** try factoring by grouping, discussed in Section 6.1.

3 After the polynomial has been factored, be sure that any factor with two or more terms does not have any common factors other than 1. If there are common factors, factor them out.

4 Check your factoring through multiplication.

We will now factor several polynomials of various forms. Some of the examples will have twists that we did not see in the previous sections. The focus will be on identifying the best technique.

EXAMPLE 1 Factor $-8x^2 - 80x + 192$ completely.

Solution

These three terms have a common factor of -8, and after we factor it out, we have $-8(x^2 + 10x - 24)$. This is a quadratic trinomial with a leading coefficient of 1, so we factor it using the method introduced in Section 6.3.

$$-8x^2 - 80x + 192 = -8(x^2 + 10x - 24)$$
$$= -8(x + 12)(x - 2)$$

Factor out the GCF -8.
Find two integers whose product is   and whose sum is 10. The integers are 12 and -2.

Quick Check 1
Factor $-6x^2 + 54x + 60$ completely.

EXAMPLE 2 Factor $343r^9 - 64s^6t^{12}$ completely.

Solution

There are no common factors other that 1, so we begin by noticing that this is a binomial. This is not a difference of squares, as $343r^9$ cannot be rewritten as a perfect square. It is, however, a difference of cubes. The variable factors all have exponents that are multiples of 3, and the coefficients are perfect cubes. (Recall the list of the first 10 cubes given in Section 6.4, $343 = 7^3$ and $64 = 4^3$.) We begin by rewriting each term as a perfect cube.

$$343r^9 - 64s^6t^{12} = (7r^3)^3 - (4s^2t^4)^3$$

Rewrite each term as a perfect cube.

$$= (7r^3 - 4s^2t^4)(7r^3 \cdot 7r^3 + 7r^3 \cdot 4s^2t^4 + 4s^2t^4 \cdot 4s^2t^4)$$

Factor as a difference of cubes. (Recall the pattern for the trinomial factor: 1st · 1st + 1st · 2nd + 2nd · 2nd.)

$$= (7r^3 - 4s^2t^4)(49r^6 + 28r^3s^2t^4 + 16s^4t^8)$$

Simplify each term in the trinomial factor.

For a difference or sum of cubes, check to be sure that the binomial factor cannot be factored further as one of our special binomials.

Quick Check 2
Factor $216x^{12}y^{21} + 125$ completely.

EXAMPLE 3 Factor $2x^3 - 22x^2 - 8x + 88$ completely.

Solution

The four terms have common factors of 2, so we will first factor out the GCF. Then we will try to factor by grouping. If we have trouble factoring the polynomial as written, we can always rearrange the order of its terms.

$$2x^3 - 22x^2 - 8x + 88 = 2(x^3 - 11x^2 - 4x + 44)$$

Factor out the GCF.

$$= 2[x^2(x - 11) - 4(x - 11)]$$

Factor the common factor x^2 from the first two terms and factor the common factor -4 from the last two terms.

$$= 2(x - 11)(x^2 - 4)$$

Factor out the common factor $x - 11$.

Quick Check 3

Factor
$x^4 + 5x^3 - 8x - 40$
completely.

Notice that the factor $x^2 - 4$ is a difference of squares and must be factored further. Since $x^2 - 4 = (x + 2)(x - 2)$, the polynomial $2x^3 - 22x^2 - 8x + 88$ factors to be $2(x - 11)(x + 2)(x - 2)$.

If a binomial is both a difference of squares and a difference of cubes, such as $x^6 - 64$, to factor it completely we must start by factoring it as a difference of squares. The next example illustrates this.

EXAMPLE 4   Factor $x^6 - 64$ completely.

Solution

There are no common factors to factor out, so we begin by factoring this binomial as a difference of squares.

$$x^6 - 64 = (x^3)^2 - (8)^2$$ Rewrite each term as a perfect square.
$$= (x^3 + 8)(x^3 - 8)$$ Factor as a difference of squares.

Notice that each factor can be factored further. The binomial $x^3 + 8$ is a sum of cubes and the binomial $x^3 - 8$ is a difference of cubes. We use the fact that $x^3 + 8 = (x + 2)(x^2 - 2x + 4)$ and $x^3 - 8 = (x - 2)(x^2 + 2x + 4)$ to complete the factoring.

Quick Check 4

Factor $x^{12} - y^6$
completely.

$$x^6 - 64 = (x + 2)(x^2 - 2x + 4)(x - 2)(x^2 + 2x + 4).$$

EXAMPLE 5   Factor $x^{12} + 1$ completely.

Solution

The two terms do not have a common factor other than 1, so we need to determine whether this binomial matches one of our special forms. The binomial can be rewritten as a sum of cubes, $(x^4)^3 + (1)^3$.

Quick Check 5

Factor $x^6 + 64$
completely.

$$x^{12} + 1 = (x^4)^3 + (1)^3$$ Rewrite each term as a perfect cube.
$$= (x^4 + 1)(x^8 - x^4 + 1)$$ Factor as a sum of cubes.

EXAMPLE 6   Factor $15x^5 - 55x^4 + 40x^3$ completely.

Solution

The three terms have a common factor of $5x^3$, so we begin by factoring this out of the trinomial. After doing this, the trinomial factor $3x^2 - 11x + 8$ can be factored using either of the techniques in Section 6.3. We will factor it by grouping, although you may prefer to use trial and error.

$$15x^5 - 55x^4 + 40x^3 = 5x^3(3x^2 - 11x + 8)$$

Factor out the common factor $5x^3$.

$$= 5x^3(3x^2 - 3x - 8x + 8)$$

To factor by grouping we need to find two integers whose product is equal to $3 \cdot 8 = 24$ and whose sum is -11. The two integers are -3 and -8, so we rewrite the term $-11x$ as $-3x - 8x$.

Quick Check 6
Factor $8x^2 + 42x - 36$ completely.

$$= 5x^3[3x(x - 1) - 8(x - 1)]$$

Factor the common factor $3x$ from the first two terms and factor the common factor -8 from the last two terms.

$$= 5x^3(x - 1)(3x - 8)$$

Factor out the common factor $x - 1$.

EXERCISES 6.5

Factor completely. If the polynomial cannot be factored, write "prime."

1. $27x^3 - 8y^3$

2. $x^2 - 8x - 65$

3. $4x^3 - 28x^2 - x + 7$

4. $x^2 - 10x + 20$

5. $x^6 - 1$

6. $-10x^2 + 70x + 180$

7. $x^2 - 18x + 45$

8. $x^2 - 64$

9. $x^6 + 64y^{12}$

10. $x^2 + 17x - 30$

11. $4x^2 + 56x + 192$

12. $18x^2 - 60x + 42$

13. $3x^2 + 10x - 48$

14. $x^3 - 1000y^{18}$

15. $x^2 + 13x + 42$

16. $x^4 + 81y^2$

17. $x^4 + 16x^3 + x + 16$

18. $9x^2 + 54x + 72$

19. $9x^2 - 30x + 25$

20. $a^5b^7 - 3a^4b^6 + 8a^2b^9$

21. $x^2 - 14xy + 49y^2$

22. $x^2 + 7xy - 60y^2$

23. $x^2 + 4x - 165$

24. $x^7 + 18x^5 - 36x$

25. $12x + 20$

26. $x^3 - 9x^2 - 4x + 36$

27. $x^2 + 13x - 14$

28. $x^8 - 16y^6$

29. $6x^2 - 29x - 5$

30. $16x^{10} + 49y^6$

31. $84x^2 + 4x - 20$

32. $x^2y^2 + 17xy + 16$

33. $7x^2 + 29x + 4$

34. $1 + 25a^2b^8$

35. $9x^3 + 18x^2 - 4x - 8$

36. $7x^3 + 189$

37. $x^2 - 6x - 280$

38. $x^{16} - 1$

39. $4x^2 + 4x - 224$

40. $12x^2 - 41x + 22$

41. $x^6 - y^6$

42. $8 + x^{15}$

43. $-x^2 + 7x - 12$

44. $2x^2 - 3x - 12$

45. $6x^2 + 11x - 10$

46. $x^2 - 23x + 42$

47. $x^6 + 729$

48. $9x^2 - 29x + 6$

49. $27x^9 + 125y^{15}$

50. $x^4 - 64$

51. $x^2 - 3x - 154$

52. $8x^2 + 19x + 6$

53. $x^2 + 20x + 100$

54. $x^{21} - 1$

55. $6x^2 + 75x + 225$

56. $x^4 - 25x^2$

57. In this chapter, we have covered several strategies for factoring polynomials:

- Factoring out the GCF;
- Factoring by grouping;
- Factoring trinomials of the form $x^2 + bx + c$;
- Factoring polynomials of the form $ax^2 + bx + c$ ($a \neq 1$) by grouping;
- Factoring polynomials of the form $ax^2 + bx + c$ ($a \neq 1$) by trial and error;
- Factoring a difference of squares;
- Factoring a difference of cubes; and
- Factoring a sum of cubes.

Explain how to identify when to use each method of factoring. Describe each method, listing any pitfalls to avoid. Provide examples of each technique.

QUICK REVIEW EXERCISES

Section 6.5

Solve.

1. $x - 7 = 0$

2. $x + 3 = 0$

3. $5x - 12 = 0$

4. $2x + 29 = 0$

Study Tip REVISITED   Many students who perform poorly in their math class will attribute their performance to math anxiety, when, in reality, poor study skills are to blame. Be sure that you are giving your fullest possible effort, including

- working with a study group;
- reading the text before the material is covered in class;
- rereading the text after the material is covered in class;
- completing each homework assignment;
- making note cards for particularly difficult problems or procedures;
- getting help if there is a problem that you do not understand;
- asking your instructor questions when you do not understand; and
- seeing a tutor.

6.6

SOLVING QUADRATIC EQUATIONS BY FACTORING

Objectives

1. Solve an equation by using the zero-factor property of real numbers.
2. Solve a quadratic equation by factoring.
3. Solve a quadratic equation that is not in standard form.
4. Solve a quadratic equation with coefficients that are fractions.
5. Find a quadratic equation given its solutions.

Quadratic Equations

> A **quadratic equation** is an equation that can be written as $ax^2 + bx + c = 0$, where a, b, and c are real numbers and $a \neq 0$. This form is referred to as the **standard form of a quadratic equation.**

We have already learned to solve linear equations ($ax + b = 0$). The difference between these two types of equations is that a quadratic equation has a second-degree term, ax^2. Because of the second-degree term, the techniques used to solve linear equations will not work for quadratic equations.

The Zero-Factor Property of Real Numbers

Objective 1   **Solve an equation by using the zero-factor property of real numbers.** To solve a quadratic equation, we will use the **zero-factor property of real numbers.**

Zero-Factor Property of Real Numbers

> If $a \cdot b = 0$, then $a = 0$ or $b = 0$.

The principle behind this property is that if two or more unknown numbers have a product of zero, then at least one of the numbers must be zero. This property only holds true when the product is equal to 0, not for any other numbers.

EXAMPLE 1   Use the zero-factor property to solve the equation $(x + 3)(x - 7) = 0$.

Solution

In this example, the product of two unknown numbers, $x + 3$ and $x - 7$, is equal to 0. The zero-factor property tells us that either $x + 3 = 0$ or $x - 7 = 0$. Essentially, we have taken an equation that we did not know how to solve yet, $(x + 3)(x - 7) = 0$, and rewritten it as two linear equations that we do know how to solve.

$$(x + 3)(x - 7) = 0$$
$$x + 3 = 0 \quad \text{or} \quad x - 7 = 0 \qquad \text{Set each factor equal to 0.}$$
$$x = -3 \quad \text{or} \quad x = 7 \qquad \text{Solve each linear equation.}$$

There are two solutions to this equation, -3 and 7. We write these solutions in a solution set as $\{-3, 7\}$.

EXAMPLE 2   Use the zero-factor property to solve the equation $x(3x - 5) = 0$.

Solution

Applying the zero-factor property gives us the equations $x = 0$ and $3x - 5 = 0$. The first equation, $x = 0$, is already solved, so we only need to solve the equation $3x - 5 = 0$ to find the second solution.

$$x(3x - 5) = 0$$
$$x = 0 \quad \text{or} \quad 3x - 5 = 0 \qquad \text{Set each factor equal to 0.}$$
$$x = 0 \quad \text{or} \quad 3x = 5 \qquad \text{Solve each linear equation.}$$
$$x = 0 \quad \text{or} \quad x = \frac{5}{3}$$

Quick Check 1

Use the zero-factor property to solve the equation.

a) $(x - 2)(x + 8) = 0$
b) $x(4x + 9) = 0$

The solution set for this equation is $\{0, \frac{5}{3}\}$.

If an equation has the product of more than two factors equal to 0, such as $(x + 1)(x - 2)(x + 4) = 0$, we set each factor equal to 0 and solve. The solution set to this equation is $\{-1, 2, -4\}$.

Solving Quadratic Equations by Factoring

Objective 2 Solve a quadratic equation by factoring. To solve a quadratic equation, we will use the following procedure.

Solving Quadratic Equations by Factoring

1. Write the equation in standard form: $ax^2 + bx + c = 0$. *We need to collect all of the terms on one side of the equation. It helps to collect all of the terms so that the coefficient of the squared term is positive.*
2. Factor the expression $ax^2 + bx + c$ completely. *If you are struggling with factoring, refer back to Sections 6.1 through 6.5.*
3. Set each factor equal to 0 and solve the resulting equations. *Each of these equations should be a linear equation.*
4. Finish by checking the solutions. *This is an excellent opportunity to catch mistakes in factoring.*

EXAMPLE 3   Solve $x^2 + 7x - 30 = 0$.

Solution

This equation is already in standard form, so we begin by factoring the expression $x^2 + 7x - 30$.

$$x^2 + 7x - 30 = 0$$
$$(x + 10)(x - 3) = 0 \qquad \text{Factor.}$$
$$x + 10 = 0 \quad \text{or} \quad x - 3 = 0 \qquad \text{Set each factor equal to 0.}$$
$$x = -10 \quad \text{or} \quad x = 3 \qquad \text{Solve each equation.}$$

The solution set is $\{-10, 3\}$. Now we will check our solutions.

Check ($x = -10$)	Check ($x = 3$)
$x^2 + 7x - 30 = 0$	$x^2 + 7x - 30 = 0$
$(-10)^2 + 7(-10) - 30 = 0$	$(3)^2 + 7(3) - 30 = 0$
$100 + 7(-10) - 30 = 0$	$9 + 7(3) - 30 = 0$
$100 - 70 - 30 = 0$	$9 + 21 - 30 = 0$
$0 = 0$	$0 = 0$

Quick Check   2

Solve $x^2 - x - 20 = 0$.

The check shows that our solutions are correct.

A Word of Caution   Pay close attention to the directions of a problem. If you are asked to solve a quadratic equation, do not just factor the quadratic expression and stop. If you are asked to factor a quadratic expression, do not set each factor equal to 0 and solve the resulting equations.

EXAMPLE 4   Solve $3x^2 - 18x - 48 = 0$.

Solution

The equation is in standard form. To factor $3x^2 - 18x - 48$, we begin by factoring out the common factor 3.

$3x^2 - 18x - 48 = 0$	
$3(x^2 - 6x - 16) = 0$	Factor out the GCF.
$3(x - 8)(x + 2) = 0$	Factor $x^2 - 6x - 16$.
$x - 8 = 0$ or $x + 2 = 0$	Set each factor containing a variable equal to 0. We can ignore the numerical factor 3.
$x = 8$ or $x = -2$	Solve each equation.

The solution set is $\{8, -2\}$. The check is left to the reader.

We do not need to set a numerical factor equal to 0, because such an equation will not have a solution. For example, if we had set the common factor 3 equal to 0, the equation $3 = 0$ has no solution. The zero-factor property tells us that at least one of the three factors must equal 0; we just know that it cannot be the factor 3 that is equal to 0.

Quick Check   3

Solve
$4x^2 + 60x + 224 = 0$.

A Word of Caution   A common numerical factor does not affect the solutions of a quadratic equation.

EXAMPLE 5   Solve $2x^2 - 7x - 15 = 0$.

Solution

When factoring the expression $2x^2 - 7x - 15$, there is not a common factor other than 1 that can be factored out. So the leading coefficient of this trinomial is not 1. We can factor by trial and error or by grouping. We will use factoring by grouping. We look for two integers whose product is equal to $2 \cdot 15$ or 30 and whose sum is -7. The two integers are

-10 and 3, so we can rewrite the term $-7x$ as $-10x + 3x$ and then factor by grouping. (Refer to Section 6.3 to review this technique.)

$$2x^2 - 7x - 15 = 0$$
$$2x^2 - 10x + 3x - 15 = 0 \qquad \text{Rewrite } -7x \text{ as } -10x + 3x.$$
$$2x(x - 5) + 3(x - 5) = 0 \qquad \text{Factor the common factor } 2x \text{ from the}$$

Factor the common factor $2x$ from the first two terms and factor the common factor 3 from the last two terms.

$$(x - 5)(2x + 3) = 0 \qquad \text{Factor out the common factor } x - 5.$$
$$x - 5 = 0 \quad \text{or} \quad 2x + 3 = 0 \qquad \text{Set each factor equal to 0.}$$
$$x = 5 \quad \text{or} \quad 2x = -3 \qquad \text{Solve each equation.}$$
$$x = 5 \quad \text{or} \quad x = -\frac{3}{2}$$

Quick Check 4
Solve
$6x^2 - 23x + 7 = 0.$

The solution set is $\{5, -\frac{3}{2}\}$. The check is left to the reader.

Objective 3 **Solve a quadratic equation that is not in standard form.** We now turn our attention to equations that are not already in standard form. In each of the examples that follow, the check of the solutions is left to the reader.

EXAMPLE 6 Solve $x^2 = 49$.

Solution

We begin by rewriting this equation in standard form. This can be done by subtracting 49 from each side of the equation. Once this has been done, the expression to be factored in this example is a difference of squares.

$$x^2 = 49$$
$$x^2 - 49 = 0 \qquad \text{Subtract 49.}$$
$$(x + 7)(x - 7) = 0 \qquad \text{Factor.}$$
$$x + 7 = 0 \quad \text{or} \quad x - 7 = 0 \qquad \text{Set each factor equal to 0.}$$
$$x = -7 \quad \text{or} \quad x = 7 \qquad \text{Solve each equation.}$$

The solution set is $\{-7, 7\}$.

EXAMPLE 7 Solve $x^2 + 25 = 10x$.

Solution

To rewrite this equation in standard form, we need to subtract $10x$ so that all terms will be on the left side of the equation. When we subtract $10x$, we must be sure to write the terms in descending order.

$$x^2 + 25 = 10x$$
$$x^2 - 10x + 25 = 0 \qquad \text{Subtract } 10x.$$
$$(x - 5)(x - 5) = 0 \qquad \text{Factor.}$$
$$x - 5 = 0 \quad \text{or} \quad x - 5 = 0 \qquad \text{Set each factor equal to 0.}$$
$$x = 5 \quad \text{or} \quad x = 5 \qquad \text{Solve each equation.}$$

Quick Check 5
Solve $x^2 + 4x = 45$.

Notice that both solutions are identical. In this case, we only need to write the repeated solution once. The solution set is $\{5\}$.

Occasionally, we will need to simplify one or both sides of an equation in order to rewrite the equation in standard form. For instance, to solve the equation $x(x + 9) = 10$, we must first multiply x by $x + 9$. You may be wondering why we would want to multiply out the left side, since it is already factored. Although it is factored, the product is equal to 10, not 0.

EXAMPLE 8   Solve $x(x + 9) = 10$.

Solution

As mentioned, we must first multiply x by $x + 9$. Then we can rewrite the equation in standard form.

$$
\begin{aligned}
x(x + 9) &= 10 \\
x^2 + 9x &= 10 & &\text{Multiply.} \\
x^2 + 9x - 10 &= 0 & &\text{Subtract 10.} \\
(x + 10)(x - 1) &= 0 & &\text{Factor.} \\
x + 10 = 0 \quad &\text{or} \quad x - 1 = 0 & &\text{Set each factor equal to 0.} \\
x = -10 \quad &\text{or} \quad x = 1 & &\text{Solve each equation.}
\end{aligned}
$$

Quick Check 6

Solve $x(x - 3) = 70$.

The solution set is $\{-10, 1\}$.

> **A Word of Caution**   Be sure that the equation you are solving is written as a product equal to 0 before setting each factor equal to 0 and solving.

Objective 4   Solve a quadratic equation with coefficients that are fractions.
If an equation contains fractions, clearing those fractions makes it easier to factor the quadratic expression. We can clear the fractions by multiplying both sides of the equation by the LCM of the denominators.

EXAMPLE 9   Solve $\frac{1}{6}x^2 + x + \frac{4}{3} = 0$.

Solution

The LCM for these two denominators is 6, so we can clear the fractions by multiplying both sides of the equation by 6.

$$
\frac{1}{6}x^2 + x + \frac{4}{3} = 0
$$

$$
6 \cdot \left(\frac{1}{6}x^2 + x + \frac{4}{3} \right) = 6 \cdot 0 \qquad \text{Multiply both sides by 6, which is the LCM of the denominators.}
$$

$$
\overset{1}{\cancel{6}} \cdot \frac{1}{\cancel{6}} x^2 + 6 \cdot x + \overset{2}{\cancel{6}} \cdot \frac{4}{\cancel{3}} = 6 \cdot 0 \qquad \text{Distribute and divide out common factors.}
$$

$$
\begin{aligned}
x^2 + 6x + 8 &= 0 & &\text{Multiply.} \\
(x + 2)(x + 4) &= 0 & &\text{Factor.} \\
x + 2 = 0 \quad &\text{or} \quad x + 4 = 0 & &\text{Set each factor equal to 0.} \\
x = -2 \quad &\text{or} \quad x = -4 & &\text{Solve each equation.}
\end{aligned}
$$

Quick Check 7

Solve $\frac{2}{45}x^2 + \frac{4}{15}x - \frac{6}{5} = 0$.

The solution set is $\{-2, -4\}$.

Finding a Quadratic Equation, Given Its Solutions

Objective 5   Find a quadratic equation given its solutions.   If we know the two solutions to a quadratic equation, then we can determine an equation with these solutions. The next example illustrates this process.

EXAMPLE 10   Find a quadratic equation in standard form, with integer coefficients, that has the solution set $\{4, -9\}$.

Solution

We know that $x = 4$ is a solution to the equation. This tells us that $x - 4$ is a factor of the quadratic expression. Similarly, knowing that $x = -9$ is a solution tells us that $x + 9$ is a factor of the quadratic expression. Multiplying these two factors will tell us a quadratic equation with these two solutions.

$$x = 4 \qquad \text{or} \qquad x = -9 \qquad \text{Begin with the solutions.}$$
$$x - 4 = 0 \qquad \text{or} \qquad x + 9 = 0 \qquad \text{Rewrite each equation so the right side is equal to 0.}$$
$$(x - 4)(x + 9) = 0 \qquad \text{Write an equation that has these two expressions as factors.}$$
$$x^2 + 5x - 36 = 0 \qquad \text{Multiply.}$$

A quadratic equation that has the solution set $\{4, -9\}$ is $x^2 + 5x - 36 = 0$.

Notice that we say "*a*" quadratic equation rather than "*the*" quadratic equation. There are infinitely many quadratic equations with integer coefficients that have this solution set. For example, multiplying both sides of our equation by 2 gives us the equation $2x^2 + 10x - 72 = 0$, which has the same solution set.

Quick Check 8
Find a quadratic equation in standard form, with integer coefficients, that has the solution set $\{-6, 8\}$.

EXERCISES 6.6

Solve.

1. $(x + 7)(x - 100) = 0$

2. $(x - 2)(x - 10) = 0$

3. $x(x - 12) = 0$

4. $x(x + 5) = 0$

5. $9(x - 4)(x - 8) = 0$

6. $-2(x + 2)(x - 16) = 0$

7. $(2x + 7)(x - 5) = 0$

8. $(x + 1)(5x - 18) = 0$

9. $x^2 - 5x + 4 = 0$

10. $x^2 - 11x + 24 = 0$

11. $x^2 + 11x + 30 = 0$

12. $x^2 + 7x + 12 = 0$

13. $x^2 + 6x - 40 = 0$

14. $x^2 - 3x - 88 = 0$

15. $x^2 - 12x - 45 = 0$

16. $x^2 + 7x - 44 = 0$

17. $x^2 - 16x + 64 = 0$

18. $x^2 + 12x + 36 = 0$

19. $x^2 - 81 = 0$

20. $x^2 - 4 = 0$

21. $x^2 + 7x = 0$

22. $x^2 - 10x = 0$

23. $5x^2 - 22x = 0$

24. $14x^2 + 13x = 0$

25. $2x^2 + 8x - 154 = 0$

26. $7x^2 - 7x - 630 = 0$

27. $-x^2 + 14x - 13 = 0$

28. $-x^2 + 5x + 66 = 0$

29. $-4x^2 - 44x + 168 = 0$

30. $-9x^2 - 81x - 126 = 0$

31. $27x^2 - 3x - 2 = 0$

32. $6x^2 + 7x - 68 = 0$

33. $x^2 - 2x = 35$

34. $x^2 + 8x = -15$

35. $x^2 = 8x - 16$

36. $x^2 - 13x = 48$

37. $x^2 + 13x = 6x - 10$

38. $x^2 - x = 3x + 12$

39. $2x^2 - x - 13 = x^2 + 11x + 15$

40. $x^2 + 5x = 2x^2 + 7x - 8$

41. $x^2 = 9$

42. $x^2 = 25$

43. $x^2 + 11x = 11x + 36$

44. $x^2 - 3x = -3x + 100$

45. $x(x + 4) = 4(x + 16)$

46. $x(x + 7) = (4x + 3) + (3x + 13)$

47. $x(x + 13) = -40$

48. $x(x - 2) = 24$

49. $(x - 2)(x - 5) = 40$

50. $(x + 9)(x - 7) = -55$

51. $(x + 2)(x + 3) = (x + 7)(x - 4)$

52. $(x - 6)(x - 4) = (x + 12)(x - 10)$

53. $\dfrac{1}{6}x^2 - \dfrac{5}{4}x - \dfrac{9}{4} = 0$

54. $\dfrac{1}{2}x^2 + \dfrac{11}{12}x + \dfrac{1}{3} = 0$

55. $\dfrac{2}{15}x^2 - \dfrac{7}{10}x + \dfrac{2}{3} = 0$

56. $\dfrac{3}{4}x^2 - \dfrac{3}{4}x - \dfrac{5}{6} = 0$

Find a quadratic equation with integer coefficients that has the given solution set.

57. $\{4, 5\}$

58. $\{7, -10\}$

59. $\{0, 6\}$

60. $\{4\}$

61. $\{8, -8\}$

62. $\left\{-4, \dfrac{5}{4}\right\}$

63. $\left\{-\dfrac{3}{7}, \dfrac{7}{3}\right\}$

64. $\left\{-\dfrac{2}{5}, -\dfrac{9}{2}\right\}$

65. Use the fact that $x = \frac{7}{2}$ is a solution to the equation $6x^2 + 7x - 98 = 0$ to find the other solution to the equation.

66. Use the fact that $x = \frac{13}{3}$ is a solution to the equation $3x^2 - 28x + 65 = 0$ to find the other solution to the equation.

Answer in complete sentences.

67. Explain the procedure for solving a quadratic equation by factoring. Give an example to illustrate.

68. Explain the zero-product property of real numbers: if $a \cdot b = 0$, then $a = 0$ or $b = 0$. Describe how this property is used when solving quadratic equations by factoring.

69. When solving a quadratic equation by factoring, explain why we should not factor until one side of the quadratic equation is set equal to 0.

70. Explain why the common factor of 2 has no affect on the solutions of the equation $2(x - 7)(x + 5) = 0$.

Study Tip **REVISITED**   Learning to relax in stressful situations will help you to keep your anxiety under control. By being able to relax, you will be able to suppress the physical symptoms associated with math anxiety such as feeling tense and uncomfortable, sweating, shortness of breath, and an accelerated heartbeat.

Wanting to relax and being able to relax are often two different things. Many students use deep-breathing exercises. Close your eyes and take a deep breath. Try to clear your mind and relax your entire body. Slowly exhale, and then repeat the process until you feel relaxed.

There are other relaxation methods that may work for you, but you must make sure that you can practice this technique in class. For instance, primal scream therapy would be inappropriate during an exam.

6.7

QUADRATIC FUNCTIONS

Objectives

1. Evaluate quadratic functions.
2. Solve equations involving quadratic functions.
3. Solve applied problems involving quadratic functions.

We first investigated linear functions in Chapter 3. A linear function is a function of the form $f(x) = mx + b$. In this section we turn our attention to **quadratic functions.**

Quadratic Functions

> A quadratic function is a function of the form $f(x) = ax^2 + bx + c$ $(a \neq 0)$.

Evaluating Quadratic Functions

Objective 1 **Evaluate quadratic functions.** We begin our investigation of quadratic functions by learning to evaluate these functions for particular values of the variable. For example, if we were asked to find $f(3)$, we are being asked to evaluate the function $f(x)$ at $x = 3$. To do this, we substitute 3 for the variable x and simplify the resulting expression.

EXAMPLE 1 For the function $f(x) = x^2 - 5x - 8$, find $f(-4)$.

Solution

We will substitute -4 for the variable x and then simplify.

$$
\begin{aligned}
f(-4) &= (-4)^2 - 5(-4) - 8 && \text{Substitute } -4 \text{ for } x.\\
&= 16 - 5(-4) - 8 && \text{Square } -4.\ (-4)(-4) = 16\\
&= 16 + 20 - 8 && \text{Multiply.}\\
&= 28 && \text{Simplify.}
\end{aligned}
$$

Quick Check 1
For the function
$f(x) = x^2 - 9x + 405$,
find $f(-6)$.

Solving Equations Involving Quadratic Functions

Objective 2 **Solve equations involving quadratic functions.** Now that we have learned to solve quadratic equations, we can solve equations involving quadratic functions. Suppose that we were trying to find all values x for which some function $f(x)$ was equal to 0. We replace $f(x)$ by its formula and solve the resulting quadratic equation.

EXAMPLE 2 Let $f(x) = x^2 + 11x - 26$. Find all values x for which $f(x) = 0$.

Solution

We begin by replacing the function by its formula, and then we solve the resulting equation.

$$
\begin{aligned}
f(x) &= 0\\
x^2 + 11x - 26 &= 0 && \text{Replace } f(x) \text{ by its formula}\\
& && x^2 + 11x - 26.\\
(x + 13)(x - 2) &= 0 && \text{Factor.}\\
x + 13 = 0 \quad &\text{or} \quad x - 2 = 0 && \text{Set each factor equal to 0.}\\
x = -13 \quad &\text{or} \quad\quad x = 2 && \text{Solve each equation.}
\end{aligned}
$$

The two values x for which $f(x) = 0$ are -13 and 2.

Quick Check 2 Let $f(x) = x^2 - 17x + 72$. Find all values x for which $f(x) = 0$.

EXAMPLE 3 Let $f(x) = x^2 - 24$. Find all values x for which $f(x) = 25$.

Solution

After setting the function equal to 25, we need to rewrite the equation in standard form in order to solve it.

$$
\begin{aligned}
f(x) &= 25 \\
x^2 - 24 &= 25 &&\text{Replace } f(x) \text{ by its formula.} \\
x^2 - 49 &= 0 &&\text{Subtract 25.} \\
(x + 7)(x - 7) &= 0 &&\text{Factor.} \\
x + 7 = 0 \quad\text{or}\quad x - 7 &= 0 &&\text{Set each factor equal to 0.} \\
x = -7 \quad\text{or}\quad x &= 7 &&\text{Solve each equation.}
\end{aligned}
$$

The two values x for which $f(x) = 25$ are -7 and 7.

Quick Check 3 Let $f(x) = x^2 - 6x + 20$. Find all values x for which $f(x) = 92$.

Applications

Objective 3 Solve applied problems involving quadratic functions. We conclude the section with an applied problem that requires interpreting the graph of a quadratic function, a U-shaped curve called a **parabola.** Here are some examples. Note that these graphs are not linear.

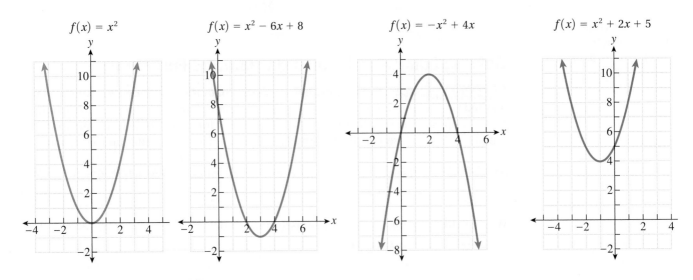

$f(x) = x^2$ $\qquad$ $f(x) = x^2 - 6x + 8$ $\qquad$ $f(x) = -x^2 + 4x$ $\qquad$ $f(x) = x^2 + 2x + 5$

EXAMPLE 4 A rocket is fired into the air from the ground at a speed of 96 feet per second. Its height, in feet, after t seconds is given by the function $h(t) = -16t^2 + 96t$, which is graphed as follows:

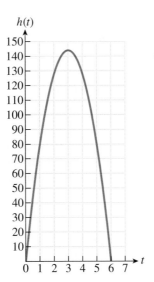

Use the graph to answer the following questions:

a) How high above the ground is the rocket after 5 seconds?

Solution

To answer this question, we need to find the function's value when $t = 5$. Looking at the graph, we see that it passes through the point $(5, 80)$. The height of the rocket is 80 feet after 5 seconds have passed.

b) How long will it take for the rocket to land on the ground?

Solution

When the rocket lands on the ground, its height is 0 feet. We need to find the values of t for which $h(t) = 0$. There are two such values, $t = 0$ and $t = 6$. The time $t = 0$ represents the instant the rocket was fired into the air, so the time that it takes to land on the ground is represented by $t = 6$. It takes the rocket 6 seconds to land on the ground.

c) After how many seconds does the rocket reach its greatest height?

Solution

On the graph, we are looking for the point at which the function reaches its highest point. (This point where the parabola changes from rising to falling is called the **vertex** of the parabola.) We see that this point corresponds to a t value of 3, so it reaches its greatest height after 3 seconds.

d) What is the greatest height that the rocket reaches?

Solution

From the graph, the best that we can say is that the maximum height is between 140 feet and 150 feet. However, since we know that the object reaches its greatest height after 3 seconds, we can evaluate the function $h(t) = -16t^2 + 96t$ at $t = 3$.

$$h(t) = -16t^2 + 96t$$
$$h(3) = -16(3)^2 + 96(3) \qquad \text{Replace } x \text{ by 3.}$$
$$= -16(9) + 96(3) \qquad \text{Square 3.}$$
$$= -144 + 288 \qquad \text{Multiply.}$$
$$= 144 \qquad \text{Simplify.}$$

The rocket reaches a height of 144 feet.

Quick Check   **4** A ball is thrown into the air from the top of a building 48 feet above the ground at a speed of 32 feet per second. Its height, in feet, after t seconds is given by the function $h(t) = -16t^2 + 32t + 48$, which is graphed as follows.

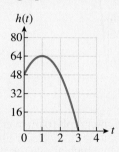

Use the graph to answer the following questions.

a) How high above the ground is the ball after 2 seconds?
b) How long will it take for the ball to land on the ground?
c) After how many seconds does the ball reach its greatest height?
d) What is the greatest height that the ball reaches?

EXERCISES 6.7

Evaluate the given function.

1. $f(x) = x^2 + 6x + 10, f(3)$

2. $f(x) = x^2 + 2x + 15, f(6)$

3. $f(x) = x^2 + 8x + 5, f(-2)$

4. $f(x) = x^2 + 4x + 20, f(-4)$

5. $f(x) = x^2 - 7x + 10, f(5)$

6. $f(x) = x^2 - 10x - 39, f(8)$

7. $g(x) = x^2 + 12x - 45, g(-15)$

8. $g(x) = x^2 - 11x + 22, g(-10)$

9. $g(x) = 3x^2 + 2x - 14, g(7)$

10. $g(x) = 5x^2 - 4x + 8, g(-9)$

11. $f(x) = -2x^2 - 9x + 8, f(-3)$

12. $f(x) = -4x^2 + 16x + 13, f(4)$

13. $h(t) = -16t^2 + 96t + 32, h(4)$

14. $h(t) = -4.9t^2 + 26t + 10, h(2)$

15. $f(x) = x^2 + 13x + 30, f(3a)$

16. $f(x) = x^2 - 5x + 17, f(4a)$

17. $f(x) = x^2 - 9x - 21, f(a + 2)$

18. $f(x) = x^2 + 14x - 36, f(a - 6)$

19. Let $f(x) = x^2 + 4x - 60$. Find all values x for which $f(x) = 0$.

20. Let $f(x) = x^2 - 11x + 30$. Find all values x for which $f(x) = 0$.

21. Let $f(x) = x^2 + 13x + 42$. Find all values x for which $f(x) = 0$.

22. Let $f(x) = x^2 - 7x - 78$. Find all values x for which $f(x) = 0$.

23. Let $f(x) = x^2 - 23x + 144$. Find all values x for which $f(x) = 18$.

24. Let $f(x) = x^2 + 14x + 14$. Find all values x for which $f(x) = -31$.

25. Let $f(x) = x^2 + 11$. Find all values x for which $f(x) = 60$.

26. Let $f(x) = x^2 + 27$. Find all values x for which $f(x) = 108$.

> A **fixed point** for a function $f(x)$ is a value a for which $f(a) = a$. For example, if $f(5) = 5$, then $x = 5$ is a fixed point for the function $f(x)$.

Find all fixed points for the following functions by setting the function equal to x and solving the resulting equation for x.

27. $f(x) = x^2 + 7x - 16$

28. $f(x) = x^2 - 5x + 8$

29. $f(x) = x^2 + 14x + 40$

30. $f(x) = x^2 - 9x + 25$

31. A ball is dropped from an airplane flying at an altitude of 400 feet. The ball's height above the ground, in feet, after t seconds is given by the function $h(t) = 400 - 16t^2$, whose graph is shown.

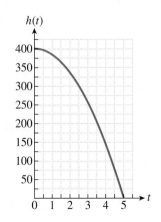

Use the function and its graph to answer the following questions.

a) How long will it take for the ball to land on the ground?

b) How high above the ground is the ball after 2 seconds?

32. A football is kicked up with an initial velocity of 64 feet/second. The football's height above the ground, in feet, after t seconds is given by the function $h(t) = -16t^2 + 64t$, whose graph is shown.

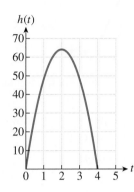

Use the function and its graph to answer the following questions:

a) Use the function to determine how high above the ground the football is after 1 second.

b) How long will it take for the football to land on the ground?

c) After how many seconds does the football reach its greatest height?

d) What is the greatest height that the football reaches?

33. A man is standing on a cliff 240 feet above a beach. He throws a rock up off the cliff with an initial velocity of 32 feet/second. The rock's height above the beach, in feet, after t seconds is given by the function $h(t) = -16t^2 + 32t + 240$, whose graph is shown.

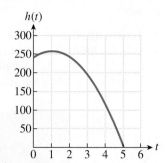

Use the function and its graph to answer the following questions:

a) Use the function to determine how high above the beach the rock is after 3 seconds.

b) How long will it take for the rock to land on the beach?

c) After how many seconds does the rock reach its greatest height above the beach?

d) What is the greatest height above the beach that the rock reaches?

34. The average price in dollars to produce x lawn chairs is given by the function
$f(x) = 0.00000016x^2 - 0.0024x + 14.55$,
whose graph is shown.

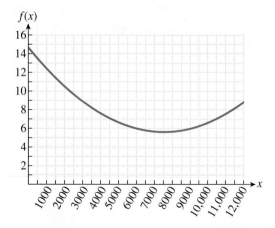

Use the function and its graph to answer the following questions:

a) Use the function to determine the average cost to produce 3000 lawn chairs.

b) What number of lawn chairs corresponds to the lowest average cost?

c) What is the lowest average cost that is possible?

35. A teenager starts a company selling personalized coffee mugs. The profit function, in dollars, for producing and selling x mugs is $f(x) = -0.4x^2 + 16x - 70$, which is graphed at the top of the next column.

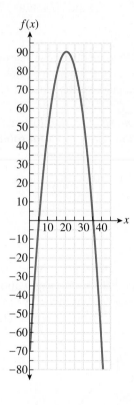

a) What are the start-up costs for the teenager's company?

b) How many mugs must the teenager sell before she breaks even?

c) How many mugs will give the maximum profit?

d) What will the profit be if 25 mugs are produced and sold?

36. A peach farmer must determine how many peaches to thin from his trees. By thinning the peaches, the trees will produce fewer peaches, but they will be larger and of better quality. The expected profit, in dollars, if x percent of the peaches are removed during thinning is given by the function
$$f(x) = -38x^2 + 2280x + 244625,$$
whose graph is shown.

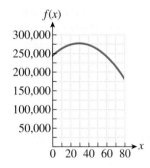

Use the function and its graph to answer the following questions:

a) What will the profit be if the farmer does not thin any peaches?

b) Use the function to determine the profit if 10% of the peaches are thinned.

c) What percent of the peaches must be thinned to produce the maximum profit?

d) What is the maximum potential profit?

Answer in complete sentences.

37. Suppose that the height, in feet, of a projectile after t seconds is given by a function $h(t)$. If we are asked to find the instant that the projectile lands on the ground, explain why we set $h(t)$ equal to 0 and solve for t.

38. Suppose that the height, in feet, of a projectile after t seconds is given by a function $h(t)$. If we are asked to find the height of the projectile after 8 seconds, explain why we evaluate $h(8)$.

6.8

APPLICATIONS
OF QUADRATIC
EQUATIONS
AND
QUADRATIC
FUNCTIONS

Objectives

1 Solve applied problems involving consecutive integers.

2 Solve applied problems involving the product of two unknown numbers.

3 Solve applied geometry problems.

4 Solve applied problems involving projectile motion.

5 Solve applied problems involving the sum of the first n natural numbers.

In this section, we learn to solve applied problems that involve quadratic equations. When problems become more complex, solving them by trial and error becomes difficult at best, so it is important to have a procedure to use. Even though some of the examples may be simple enough to solve by trial and error, they help us to learn the new procedure.

Consecutive Integer Problems

Objective 1 **Solve applied problems involving consecutive integers.** We will begin with problems involving consecutive integers. Recall that consecutive integers follow the pattern x, $x + 1$, $x + 2$, and so on. Consecutive even integers, as well as consecutive odd integers, follow the pattern x, $x + 2$, $x + 4$, and so on.

EXAMPLE 1   The product of two consecutive positive integers is 56. Find the two integers.

Solution

We begin by creating a table of unknowns. It is important that we can represent each unknown in terms of the same variable. In this problem, we are looking for two consecutive positive integers. We can let x represent the first integer. Since the integers are consecutive, we can represent the second integer by $x + 1$.

Unknowns
First: x
Second: $x + 1$

We know the product is 56, which leads to the equation $x(x + 1) = 56$.

$$
\begin{aligned}
x(x + 1) &= 56 & &\\
x^2 + x &= 56 & &\text{Multiply } x \text{ by } x + 1.\\
x^2 + x - 56 &= 0 & &\text{Subtract 56.}\\
(x + 8)(x - 7) &= 0 & &\text{Factor.}\\
x + 8 = 0 \quad &\text{or} \quad x - 7 = 0 & &\text{Set each factor equal to 0.}\\
x = -8 \quad &\text{or} \quad x = 7 & &\text{Solve each equation.}
\end{aligned}
$$

Since we know the integers are positive, we omit the solution $x = -8$. We use the table of our unknowns with the solution $x = 7$ to find the solution to our problem.

First: $x = 7$
Second: $x + 1 = 7 + 1 = 8$

Quick Check 1
The product of two consecutive positive integers is 132. Find the two integers.

The two integers are 7 and 8. The product of these two integers is 56.

A Word of Caution   Be sure to check the practicality of your answers to an applied problem. In the previous example, the integers that we were looking for were positive, so we omitted solutions that led to negative integers.

Problems Involving the Product of Two Unknown Numbers

Objective 2 **Solve applied problems involving the product of two unknown numbers.**

EXAMPLE 2   The sum of two numbers is 16 and their product is 55. Find the two numbers.

Solution

There are two unknowns in this problem: the two numbers. We let x represent the first number. Since the sum of the two numbers is 16, we represent the second number by $16 - x$.

Unknowns

First Number: x
Second Number: $16 - x$

Knowing that the product of these two numbers is 55 leads to the equation $x(16 - x) = 55$, which we now solve for x.

$$x(16 - x) = 55$$
$$16x - x^2 = 55 \qquad \text{Multiply.}$$
$$0 = x^2 - 16x + 55 \qquad \text{Collect all terms on the right side of the}$$
$$\text{equation. This makes the leading}$$
$$\text{coefficient positive.}$$
$$0 = (x - 5)(x - 11) \qquad \text{Factor.}$$
$$x - 5 = 0 \quad \text{or} \quad x - 11 = 0 \qquad \text{Set each factor equal to 0.}$$
$$x = 5 \quad \text{or} \quad x = 11 \qquad \text{Solve each equation.}$$

We return to the table of unknowns with the first solution, $x = 5$.

First Number: $x = 5$
Second Number: $16 - x = 16 - 5 = 11$

For this solution, the two numbers are 5 and 11. If we use the second solution, $x = 11$, we find the same two numbers.

> First Number: $x = 11$
> Second Number: $16 - x = 16 - 11 = 5$

Quick Check 2
The sum of two numbers is 21 and their product is 108. Find the two numbers.

So the two numbers are 5 and 11. Their product is indeed 55.

Geometry Problems

Objective 3 **Solve applied geometry problems.** The next example involves the area of a rectangle. Recall that the area of a rectangle is equal to the product of its length and width.

$$\text{Area} = \text{Length} \cdot \text{Width}$$

EXAMPLE 3   The area of a rectangle is 108 square meters. If the length of the rectangle is 3 meters more than its width, find the dimensions of the rectangle.

Solution

For this problem, the unknowns are the length and the width of the rectangle. Since the length is defined in terms of the width, we let w represent the width. The length can be represented by the expression $w + 3$ as it is 3 meters longer than the width.

> ***Unknowns***
> Length: $w + 3$
> Width: w

Since the product of the length and the width is equal to the area of the rectangle, the equation we need to solve is $(w + 3) \cdot w = 108$.

$$
\begin{array}{lll}
(w + 3) \cdot w = 108 & & \\
w^2 + 3w = 108 & & \text{Multiply.} \\
w^2 + 3w - 108 = 0 & & \text{Subtract 108.} \\
(w + 12)(w - 9) = 0 & & \text{Factor.} \\
w + 12 = 0 \quad \text{or} \quad w - 9 = 0 & & \text{Set each factor equal to 0.} \\
w = -12 \quad \text{or} \quad w = 9 & & \text{Solve each equation.}
\end{array}
$$

Quick Check 3
The area of a rectangle is 105 square feet. If the width of the rectangle is 8 feet more than its length, find the dimensions of the rectangle.

We omit the negative solution $w = -12$, as a rectangle cannot have a negative width or length. We can use the solution $w = 9$ and the table of unknowns to find the length and the width of the rectangle.

> Length: $w + 3 = 9 + 3 = 12$
> Width: $w = 9$

The length of the rectangle is 12 meters and the width is 9 meters. We can verify that the area of a rectangle with these dimensions is 108 square meters.

EXAMPLE 4   A homeowner has installed an in-ground spa in the backyard. The spa is rectangular in shape, with a length that is 4 feet more than its width. The homeowner put a one-foot-wide concrete border around the spa. If the area covered by the spa and the border is 96 square feet, find the dimensions of the spa.

Solution

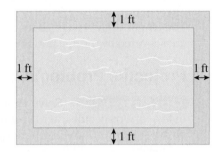

The unknowns are the length and width of the rectangular spa. We will let x represent the width of the spa.

> ***Unknowns***
>
> Length: $x + 4$
> Width: x

Since we know the area covered by the spa and the border, our equation must involve the outer rectangle in the picture. The width of the border can be represented by $x + 2$ because we need to add 2 feet to the width of the spa (x). The width of the spa is increased by 1 foot on 2 sides. In a similar fashion, the length of the border is $x + 6$ because we add 2 feet to the length of the spa.

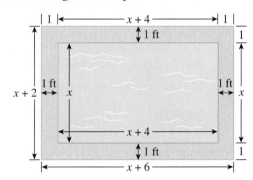

The equation we will solve is $(x + 6)(x + 2) = 96$, since the area covered is 96 square feet.

$$
\begin{aligned}
(x + 6)(x + 2) &= 96 \\
x^2 + 2x + 6x + 12 &= 96 & &\text{Multiply.} \\
x^2 + 8x + 12 &= 96 & &\text{Combine like terms.} \\
x^2 + 8x - 84 &= 0 & &\text{Subtract 96.} \\
(x + 14)(x - 6) &= 0 & &\text{Factor.} \\
x + 14 = 0 \quad &\text{or} \quad x - 6 = 0 & &\text{Set each factor equal to 0.} \\
x = -14 \quad &\text{or} \quad\quad x = 6 & &\text{Solve each equation.}
\end{aligned}
$$

Quick Check 4
A homeowner has poured a rectangular concrete slab in her backyard to use as a barbecue area. The length is 3 feet more than its width. There is a 2-foot–wide flower bed around the barbecue area. If the area covered by the barbecue area and the flower bed is 270 square feet, find the dimensions of the barbecue area.

We will omit the negative solution $x = -14$, as the dimensions of the spa cannot be negative. We use the solution $x = 6$ and the table of unknowns to find the length and the width of the rectangle.

$$\text{Length: } x + 4 = 6 + 4 = 10$$
$$\text{Width: } x = 6$$

The length of the spa is 10 feet and the width of the spa is 6 feet. We can verify that the area covered by the spa and border is indeed 96 square feet.

Projectile Problems

Objective 4 **Solve applied problems involving projectile motion.** The height, in feet, of a projectile after t seconds can be found using the function $h(t) = -16t^2 + v_0t + s$, where v_0 is the initial velocity of the projectile and s is the initial height.

$$h(t) = -16t^2 + v_0t + s$$
$$t: \text{Time (seconds)}$$
$$v_0: \text{Initial Velocity (feet/second)}$$
$$s: \text{Initial Height (feet)}$$

This formula does not cover projectiles that continue to propel themselves, such as a rocket with an engine.

EXAMPLE 5   A cannonball is fired from a platform that is 256 feet above the ground. The initial velocity of the object is 96 feet per second.

a) Find the function $h(t)$ that gives the height of the cannonball in feet after t seconds.

Solution

Since the initial velocity is 96 feet per second and the initial height is 256 feet, the function is $h(t) = -16t^2 + 96t + 256$.

b) How long will it take for the cannonball to land on the ground?

Solution

The cannonball's height when it lands on the ground is 0 feet, so we set the function equal to 0 and solve for the time t in seconds.

$-16t^2 + 96t + 256 = 0$	Set $h(t)$ equal to 0.
$0 = 16t^2 - 96t - 256$	Collect all terms on the right side of the equation so the leading coefficient is positive.
$0 = 16(t^2 - 6t - 16)$	Factor out the GCF (16).
$0 = 16(t - 8)(t + 2)$	Factor $t^2 - 6t - 16$.
$t - 8 = 0$ or $t + 2 = 0$	Set each variable factor equal to 0.
$t = 8$ or $t = -2$	Solve each equation.

We will omit the negative solution $t = -2$, as the time must be positive. It takes the cannonball 8 seconds to land on the ground.

c) What is the domain of the function $h(t)$?

Solution

Since the cannonball lands on the ground after 8 seconds, any length of time greater than 8 seconds does not apply. The domain of the function is $0 \leq t \leq 8$, which can be written in interval notation as $[0, 8]$.

Quick Check **5** A model rocket is launched from the ground with an initial velocity of 144 feet per second.

a) Find the function $h(t)$ that gives the height of the rocket in feet after t seconds.
b) How long will it take for the rocket to land on the ground?
c) What is the domain of the function $h(t)$?

Problems Involving the Sum of the First n Natural Numbers

Objective 5 **Solve applied problems involving the sum of the first n natural numbers.** The famous German mathematician Carl Gauss found a function for totaling the natural numbers from 1 to n, which is $f(n) = \frac{1}{2}n^2 + \frac{1}{2}n$. We use this function in the last example of the section.

EXAMPLE 6 **a)** Use Gauss's function to find the sum of the natural numbers from 1 to 46.

Solution

We evaluate Gauss's function at $n = 46$.

$$f(46) = \frac{1}{2}(46)^2 + \frac{1}{2}(46) \qquad \text{Substitute 46 for } n.$$

$$= \frac{1}{2}(2116) + \frac{1}{2}(46) \qquad \text{Square 46.}$$

$$= 1058 + 23 \qquad \text{Simplify each term.}$$

$$= 1081 \qquad \text{Add.}$$

The sum of the first 46 natural numbers is 1081.

b) Find a natural number n such that the sum of the first n natural numbers is 78.

Solution

We begin by setting the function $f(n)$ equal to 78 and solving for n.

$$\frac{1}{2}n^2 + \frac{1}{2}n = 78 \qquad \text{Set the function equal to 78.}$$

$$2\left(\frac{1}{2}n^2 + \frac{1}{2}n\right) = 2 \cdot 78 \qquad \text{Multiply both sides of the equation by 2 to clear the fractions.}$$

$$n^2 + n = 156 \qquad \text{Distribute and simplify each term.}$$

$$n^2 + n - 156 = 0 \qquad \text{Collect all terms on the left side of the equation.}$$

$$(n + 13)(n - 12) = 0 \qquad \text{Factor.}$$

$$n + 13 = 0 \qquad \text{or} \qquad n - 12 = 0 \qquad \text{Set each factor equal to 0.}$$

$$n = -13 \qquad \text{or} \qquad n = 12 \qquad \text{Solve each equation.}$$

We omit the solution $n = -13$ because -13 is not a natural number. The sum of the first 12 natural numbers is 78.

Quick Check **6** a) Use Gauss's function to find the sum of the natural numbers from 1 to 100.
b) Find a natural number n such that the sum of the first n natural numbers is 45.

EXERCISES 6.8

1. Two consecutive positive integers have a product of 182. Find the integers.

2. Two consecutive positive integers have a product of 380. Find the integers.

3. Two consecutive even positive integers have a product of 288. Find the integers.

4. Two consecutive even negative integers have a product of 80. Find the integers.

5. Two consecutive odd negative integers have a product of 99. Find the integers.

6. Two consecutive odd positive integers have a product of 399. Find the integers.

7. The product of two consecutive positive integers is 27 more than 5 times the larger integer. Find the integers.

8. The product of two consecutive odd positive integers is 14 more than 3 times the larger integer. Find the integers.

9. The product of two consecutive even positive integers is 398 more than the sum of the two integers. Find the integers.

10. The product of two consecutive positive integers is 131 more than the sum of the two integers. Find the integers.

11. One positive number is 5 more than a second number, and their product is 66. Find the two numbers.

12. One positive number is 9 more than a second number, and their product is 112. Find the two numbers.

13. One positive number is 4 less than 3 times a second number, and their product is 84. Find the two numbers.

14. One positive number is 3 more than twice a second number, and their product is 189. Find the two numbers.

15. The sum of two numbers is 32 and their product is 240. Find the two numbers.

16. The sum of two numbers is 24 and their product is 95. Find the two numbers.

17. The sum of two numbers is 17 and their product is 42. Find the two numbers.

18. The sum of two numbers is 20 and their product is 91. Find the two numbers.

19. The difference of two positive numbers is 3 and their product is 88. Find the two numbers.

20. The difference of two positive numbers is 9 and their product is 220. Find the two numbers.

21. Gabriela is 5 years older than her sister Graciela. If the product of their ages is 126, how old is Gabriela?

22. Marco is 8 years younger than his brother Paolo. If the product of their ages is 105, how old is Marco?

23. Clyde's age is 5 years more than twice Bonnie's age. If the product of their ages is 168, how old is Clyde?

24. Edith's age is 1 year less than 3 times Archie's age. If the product of their ages is 70, how old is Edith?

25. The length of a rectangular rug is 3 feet more than its width. If the area of the rug is 40 square feet, find the length and width of the rug.

26. The length of a rectangular swimming pool is twice its width, and the area covered by the pool is 800 square feet. Find the length and width of the swimming pool.

27. The width of a rectangular classroom is 3 times its length. If the area covered by the classroom is 432 square feet, find the length and width of the classroom.

28. The width of a rectangular photo frame is 8 centimeters less than its length. If the area of this rectangle is 240 square centimeters, find the length and width of the frame.

29. Mark has a rectangular garden in his back yard that covers 105 square meters. The length of the garden is 1 meter more than twice its width. Find the dimensions of the garden.

30. The height of a doorway is 7 feet less than 5 times its width. If the area of the doorway is 24 square feet, find the dimensions of the doorway.

31. A photo's width is 3 inches more than its length. A border of 1 inch is placed around the photo, and the area covered by the photo and its border is 70 square inches. Find the dimensions of the photo itself.

32. The length of a rectangular quilt is 4 inches more than its width. After a 2-inch border is placed around the quilt, its area is 320 square inches. Find the original dimensions of the quilt.

33. A rectangular flower garden has a length that is 1 foot less than twice its width. A 2-foot brick border is added around the garden, and the area of the garden and brick border is a total of 117 square feet. Find the dimensions of the garden without the brick border.

34. Larry has a rectangular pen for his pet goats that is adjacent to his barn, as shown in the following diagram:

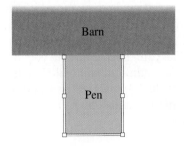

The side adjacent to the barn is 5 feet less than the side that projects out from the barn. He decides to expand the pen by 5 feet in all three directions, as shown in the following diagram:

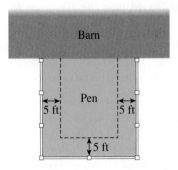

This expansion makes the area of the pen 625 square feet. Find the dimensions of the original pen.

For the following exercises, recall that the area of a square with side s is s^2 and that the area of a triangle with base b and height h is $\frac{1}{2}bh$.

35. The area of the floor in a square room is 64 square meters. Find the length of a wall of the room.

36. A dartboard is enclosed in a square that has an area of 1600 square centimeters. Find the length of a side of the square.

37. The height of a triangular sail is 5 feet more than twice its base. If the sail is made of 84 square feet of fabric, find the base and height of the triangle.

38. The three towns of Visalia, Hanford, and Fresno form a triangle. The base of the triangle extends from Visalia to Hanford, and the height of the triangle extends from Visalia to Fresno, as shown in the following diagram:

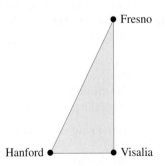

continued

The distance from Visalia to Fresno is twice the distance from Visalia to Hanford. If the area of the triangle formed by these three cities is 400 square miles, find the distance from Visalia to Fresno.

For problems 39–48, use the function $h(t) = -16t^2 + v_0t + s$.

39. A projectile was launched from a building 320 feet tall. If the initial velocity of the projectile was 128 feet per second, how long will it take the projectile to land on the ground?

40. Standing on a platform, Cory throws a softball upward from a height 48 feet above the ground at a speed of 32 feet per second. How long will it take for the softball to land on the ground?

41. Standing on the edge of a cliff, Randy throws a rock upward from a height 80 feet above the water at a speed of 64 feet per second. How long will it take for the rock to land in the water?

42. A projectile is launched from the top of a building 384 feet tall with an initial velocity of 160 feet per second. How long will it take for the projectile to land on the ground?

43. A projectile is launched from a platform 40 feet above the ground with an initial velocity of 64 feet per second. After how many seconds is the projectile at a height of 88 feet?

44. Standing atop a building, Angela throws a rock upward from a height 55 feet above the ground with an initial velocity of 32 feet per second. How long will it take before the rock is 7 feet above the ground?

45. A projectile is launched from the ground with an initial velocity of 160 feet per second. How long will it take for the projectile to land on the ground?

46. A projectile is launched from the ground with an initial velocity of 112 feet per second. How long will it take for the projectile to land on the ground?

47. A projectile is launched upward from the ground, and it lands on the ground after 8 seconds. What was the initial velocity of the projectile?

48. A projectile is launched upward from an initial height of 432 feet, and it lands on the ground after 9 seconds. What was the initial velocity of the projectile?

The height, in feet, of a free-falling object t seconds after being dropped from an initial height s can be found using the function $h(t) = -16t^2 + s$.

49. An object is dropped from a helicopter 144 feet above the ground. How long will it take for the object to land on the ground?

50. An object is dropped from an airplane that is 400 feet above the ground. How long will it take for the object to land on the ground?

Use Gauss's function for totaling the natural numbers from 1 to n, $f(n) = \frac{1}{2}n^2 + \frac{1}{2}n$, in the following exercises.

51. Find a natural number n such that the sum of the first n natural numbers is 36.

52. Find a natural number n such that the sum of the first n natural numbers is 55.

53. Find a natural number n such that the sum of the first n natural numbers is 210.

54. Find a natural number n such that the sum of the first n natural numbers is 325.

Answer in complete sentences.

55. Explain why the first three odd integers cannot be represented as x, $x + 1$, and $x + 3$.

56. Write a word problem involving a rectangle whose length is 24 feet and width is 16 feet. Your problem must lead to a quadratic equation.

57. Write a word problem involving a rectangle that leads to the equation $(x + 9)(x + 5) = 192$. Explain how you created your problem,

Study Tip REVISITED   Some students put off doing their math homework or studying due to negative feelings for mathematics. Procrastination is your enemy. Convince yourself that you can do it and get started. If you wait until you are tired, you will not be able to give your best effort. If you do not do the homework assignment, you will have trouble following the next day's material. Try scheduling a time to devote to mathematics each day, and stick to your schedule.

Chapter 6 Summary

Summary of Chapter 6 Study Tips

Math anxiety can hinder your success in a mathematics class, but it can be overcome.

- The first step is to understand what has caused your anxiety to begin with.
- Relaxation techniques can help you to overcome the physical symptoms of math anxiety, allowing you to give your best effort.
- Developing a positive attitude and confidence in your abilities will help as well.
- Avoid procrastinating.
- Many students believe that they have math anxiety, but poor performance can be caused by other factors as well. Do you completely understand the material, only to "freeze up" on the tests? Does this happen in other classes as well? If so, you may have test anxiety, not math anxiety.
- Some students do poorly in their class because they are taking a class that they are not prepared for. Discuss your correct placement with your instructor and your academic counselor if you feel that this may be the case for you.
- Finally, be sure that poor study skills are not the cause of your difficulties. If you are not giving your fullest effort, you cannot expect to learn and understand mathematics.

Factor completely. [6.1–6.5]

1. $3x - 21$

2. $x^2 - 20x$

3. $5x^7 + 15x^4 - 20x^3$

4. $6a^5b^2 - 10a^3b^3 + 15a^4b$

5. $x^3 + 6x^2 + 4x + 24$

6. $x^3 + 9x^2 - 7x - 63$

7. $x^3 - 3x^2 - 11x + 33$

8. $x^3 - 3x^2 - 9x + 27$

9. $x^2 - 8x + 16$

10. $x^2 + 3x - 28$

11. $x^2 - 11x + 24$

12. $-5x^2 - 20x + 160$

13. $x^2 + 14xy + 45y^2$

14. $x^2 + 9x - 23$

15. $x^2 + 17x + 30$

16. $x^2 - x - 42$

17. $10x^2 - 83x + 24$

18. $4x^2 - 3x - 10$

19. $18x^2 + 12x - 6$

20. $4x^2 - 12x + 9$

21. $3x^2 - 75$

22. $4x^2 - 25y^2$

23. $x^3 + 8y^3$

24. $x^2 + 121$

25. $x^3 - 216$

26. $7x^3 - 189$

Solve. [6.6]

27. $(x - 5)(x + 7) = 0$

28. $(3x - 8)(2x + 1) = 0$

29. $x^2 - 25 = 0$

30. $x^2 - 14x + 40 = 0$

31. $x^2 + 16x + 64 = 0$

32. $x^2 + 3x - 108 = 0$

33. $x^2 - 9x - 70 = 0$

34. $x^2 - 16x = 0$

35. $x^2 + 19x + 60 = 0$

36. $x^2 - 18x + 80 = 0$

37. $x^2 - 15x - 16 = 0$

38. $2x^2 - 7x - 60 = 0$

39. $7x^2 - 252 = 0$

40. $4x^2 - 60x + 144 = 0$

41. $x(x + 10) = -21$

42. $(x + 4)(x + 5) = 72$

Find a quadratic equation with integer coefficients that has the given solution set. [6.6]

43. $\{6, 7\}$

44. $\{-4, 4\}$

45. $\left\{2, -\dfrac{3}{5}\right\}$

46. $\left\{-\dfrac{5}{4}, \dfrac{2}{7}\right\}$

Evaluate the given quadratic function. [6.7]

47. $f(x) = x^2, f(-3)$

48. $f(x) = x^2 - 8x, f(4)$

49. $f(x) = x^2 + 10x + 24, f(-8)$

50. $f(x) = 3x^2 + 8x - 19, f(5)$

For the given function $f(x)$, find all values x for which $f(x) = 0$. [6.7]

51. $f(x) = x^2 - 36$

52. $f(x) = x^2 + 3x - 54$

53. $f(x) = x^2 + 12x + 20$

54. $f(x) = x^2 - 7x - 44$

55. A man is standing on a cliff above a beach. He throws a rock upward from a height 128 feet above the beach with an initial velocity of 32 feet/second. The rock's height above the beach, in feet, after t seconds is given by the function $h(t) = -16t^2 + 32t + 128$, whose graph is shown. [6.7]

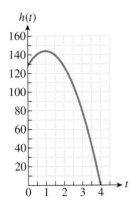

Use the function and its graph to answer the following questions:

a) Use the graph to determine how high above the beach the rock is after three seconds.

b) How long will it take for the rock to land on the beach?

c) After how many seconds does the rock reach its greatest height above the beach?

d) Use the function to determine the greatest height above the beach that the rock reaches.

56. The average price in dollars for a manufacturer to produce x toys is given by the function $f(x) = 0.0001x^2 - 0.08x + 20.85$, whose graph is shown. [6.7]

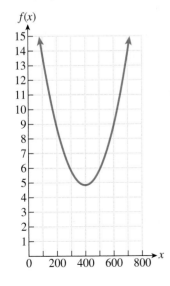

Use the function and its graph to answer the following questions:

a) Use the function to determine the average cost to produce 800 toys.

b) What number of toys corresponds to the lowest average cost?

c) Use the function to determine the lowest average cost that is possible.

57. Two consecutive even positive integers have a product of 288. Find the two integers. [6.8]

58. Rosa is 8 years younger than Dale. If the product of their ages is 105, how old is each person? [6.8]

59. The length of a rectangle is 3 meters less than twice the width. The area of the rectangle is 104 square meters. Find the length and the width of the rectangle. [6.8]

60. A ball is thrown upward with an initial velocity of 80 feet per second from the edge of a cliff that is 384 feet above a river. [6.8]

a) How long will it take for the ball to land in the river?

b) When is the ball 480 feet above the river?

Factor completely.

1. $x^2 - 25x$

2. $x^3 - 3x^2 - 5x + 15$

3. $x^2 - 9x + 20$

4. $x^2 - 14x - 72$

5. $3x^2 + 60x + 57$

6. $6x^2 - x - 5$

7. $x^2 - 25y^2$

8. $x^3 - 216y^3$

Solve.

9. $x^2 - 100 = 0$

10. $x^2 - 13x + 36 = 0$

11. $x^2 + 7x - 60 = 0$

12. $(x + 2)(x - 9) = 60$

Find a quadratic equation with integer coefficients that has the given solution set.

13. $\{-2, 2\}$

14. $\left\{-6, \dfrac{9}{5}\right\}$

Evaluate the given quadratic function.

15. $f(x) = x^2 - 12x + 35,\ f(-5)$

16. $f(x) = -x^2 + 2x + 17,\ f(7)$

For the given function $f(x)$, find all values x for which $f(x) = 0$.

17. $f(x) = x^2 + 14x + 48$

18. The average price in dollars for a manufacturer to produce x baseball hats is given by the function

$f(x) = 0.00000006x^2 - 0.00096x + 5.95$, whose graph is shown below.

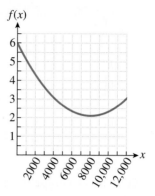

Use the function and its graph to answer the following questions:

a) Use the function to determine the average cost to produce 6500 baseball hats.

b) What number of baseball hats corresponds to the lowest average cost?

c) Use the function to determine the lowest average cost that is possible.

19. The length of a rectangle is 5 feet more than three times the width. The area of the rectangle is 100 square feet. Find the length and the width of the rectangle.

20. A ball is thrown upward with an initial velocity of 32 feet per second from the edge of a cliff 240 feet above a beach. Use the function $h(t) = -16t^2 + v_0 t + s$.

a) How long will it take for the ball to land on the beach?

b) When is the ball 192 feet above the beach?

Mathematicians in History
Sir Isaac Newton

$\mathscr{S}$ir Isaac Newton was an English mathematician and scientist who lived in the 17th and 18th centuries. His work with motion and gravity helped us to better understand our world, as well as our solar system. Alexander Pope once said "Nature and Nature's laws lay hid in the night; God said, Let Newton be! And all was light."

Write a one-page summary (OR make a poster) of the life of Sir Isaac Newton and his accomplishments.

Interesting issues:

- Where and when was Sir Isaac Newton born?
- Describe Newton's upbringing, and his relationship with his mother and stepfather.
- It has been said that an apple was the inspiration for Newton's ideas about the force of gravity. Explain how the apple is believed to have inspired Newton's ideas.
- In a letter to Robert Hooke, Newton wrote "If I have been able to see further, it was only because I stood on the shoulders of giants." Explain what this statement means.
- Newton is often referred to as the "Father of Calculus." What is calculus?
- What are Newton's three laws of motion? Explain what they mean in your own words.
- What did Newton invent for his pets?
- Where was Newton buried?

Stretch Your Thinking ❭ Chapter 6

In the year 1997, 23,100 Americans had weight-loss surgery. In the year 2002 there were 63,100 such surgeries, and in the year 2003 there were an estimated 103,200 surgeries.

(*Source:* American Society for Bariatric Surgery)

1. Do the data, as plotted here, suggest that the number of surgeries is increasing in a linear fashion? Explain why or why not in your own words.

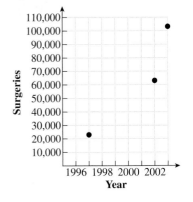

2. The number of weight-loss surgeries in a particular year can be modeled by the quadratic function $f(x) = 5350x^2 - 18,750x + 23,100$, where x represents the number of years after 1997.

   a) Use the function to predict the number of weight-loss surgeries in the year 2005.

   b) Use the function to predict the number of weight-loss surgeries that there will be in the year 2015. Does this number of surgeries seem reasonable? Why or why not?

7

RATIONAL EXPRESSIONS AND EQUATIONS

In this chapter, we will examine rational expressions, which are fractions whose numerator and denominator are polynomials. We will learn to simplify rational expressions, as well as learn how to add, subtract, multiply, and divide two rational expressions. Rational expressions are involved in many applied problems, such as determining the maximum load that a wooden beam can support, finding the illumination from a light source, and solving work-rate problems.

Study Tip PREPARING FOR A CUMULATIVE EXAM In this chapter, we will focus on how to prepare for a cumulative exam, such as a final exam. Although some of the strategies are similar to those used to prepare for a chapter test or quiz, there are some differences as well.

7.1

RATIONAL EXPRESSIONS AND FUNCTIONS

Objectives

1. Evaluate rational expressions.
2. Find the values for which a rational expression is undefined.
3. Simplify rational expressions to lowest terms.
4. Identify factors that are opposites of each other.
5. Evaluate rational functions.
6. Find the domain of rational functions.

A **rational expression** is a quotient of two polynomials, such as $\dfrac{x^2 + 15x + 44}{x^2 - 16}$. The denominator of a rational expression must not be zero, as division by zero is undefined. A major difference between rational expressions and linear or quadratic expressions is that a rational expression has one or more variables in the denominator.

Evaluating Rational Expressions

Objective 1 Evaluate rational expressions. We can evaluate a rational expression for a particular value of the variable just as we evaluated polynomials. We substitute the value for the variable in the expression and then simplify. When simplifying, we evaluate the numerator and denominator separately and then simplify the resulting fraction.

EXAMPLE 1   Evaluate the rational expression $\dfrac{x^2 + 2x - 24}{x^2 - 7x + 12}$ for $x = -5$.

Solution

We begin by substituting -5 for x.

$$\frac{(-5)^2 + 2(-5) - 24}{(-5)^2 - 7(-5) + 12} \qquad \text{Substitute} \quad \text{for } x.$$

$$= \frac{25 - 10 - 24}{25 + 35 + 12} \qquad \text{Simplify the numerator and denominator separately. Note: } (-5)^2 = 25.$$

$$= \frac{-9}{72} \qquad \text{Simplify the numerator and denominator.}$$

$$= -\frac{1}{8} \qquad \text{Simplify.}$$

Quick Check 1
Evaluate the rational expression $\dfrac{x^2 - 3x - 18}{x^2 - 5x - 6}$ for $x = -4$.

Finding Values for Which a Rational Expression Is Undefined

Objective 2 Find the values for which a rational expression is undefined. Rational expressions are undefined for values of the variable that cause the denominator to equal 0, as division by 0 is undefined. In general, to find the values for which a rational expression is undefined, we set the denominator equal to 0, ignoring the numerator, and solve the resulting equation.

EXAMPLE 2   Find the values for which the rational expression $\dfrac{3}{5x+4}$ is undefined.

Solution

We begin by setting the denominator, $5x + 4$, equal to 0 and then we solve for x.

$$5x + 4 = 0 \qquad \text{Set the denominator equal to 0.}$$
$$5x = -4 \qquad \text{Subtract 4.}$$
$$x = -\frac{4}{5} \qquad \text{Divide by 5.}$$

Quick Check 2
Find the values for which $\dfrac{3}{2x-7}$ is undefined.

The expression $\dfrac{3}{5x+4}$ is undefined for $x = -\dfrac{4}{5}$.

EXAMPLE 3   Find the values for which $\dfrac{x^2 - 7x + 12}{x^2 + 5x - 36}$ is undefined.

Solution

We begin by setting the denominator $x^2 + 5x - 36$ equal to 0, ignoring the numerator. Notice that the resulting equation is quadratic and can be solved by factoring.

$$x^2 + 5x - 36 = 0 \qquad \text{Set the denominator equal to 0.}$$
$$(x + 9)(x - 4) = 0 \qquad \text{Factor } x^2 + 5x - 36.$$
$$x + 9 = 0 \quad \text{or} \quad x - 4 = 0 \qquad \text{Set each factor equal to 0.}$$
$$x = -9 \quad \text{or} \quad x = 4 \qquad \text{Solve.}$$

Quick Check 3
Find the values for which $\dfrac{x-4}{x^2-8x-9}$ is undefined.

The expression $\dfrac{x^2 - 7x + 12}{x^2 + 5x - 36}$ is undefined when $x = -9$ or $x = 4$.

Simplifying Rational Expressions to Lowest Terms

Objective 3   **Simplify rational expressions to lowest terms.** Rational expressions are often referred to as *algebraic fractions*. As with numerical fractions, we will learn to simplify rational expressions to lowest terms. In later sections, we will learn to add, subtract, multiply, and divide rational expressions.

We simplified a numerical fraction to lowest terms by dividing out factors that were common to the numerator and denominator. For example, consider the fraction $\dfrac{30}{84}$. To simplify this fraction, we could begin by factoring the numerator and denominator.

$$\frac{30}{84} = \frac{2 \cdot 3 \cdot 5}{2 \cdot 2 \cdot 3 \cdot 7}$$

The numerator and denominator have common factors of 2 and 3, which are divided out to simplify the fraction to lowest terms.

$$\frac{\overset{1}{\cancel{2}} \cdot \overset{1}{\cancel{3}} \cdot 5}{\underset{1}{\cancel{2}} \cdot 2 \cdot \underset{1}{\cancel{3}} \cdot 7} = \frac{5}{2 \cdot 7} \text{ or } \frac{5}{14}$$

Simplifying Rational Expressions

> To simplify a rational expression to lowest terms, we first factor the numerator and denominator completely. Then we may divide out common factors in the numerator and denominator.
>
> If P, Q, and R are polynomials, $Q \neq 0$, and $R \neq 0$, then $\dfrac{PR}{QR} = \dfrac{P}{Q}$.

EXAMPLE 4   Simplify the rational expression $\dfrac{10x^3}{8x^5}$. (Assume $x \neq 0$.)

Solution

This rational expression has a numerator and denominator that are monomials. In this case, we can simplify the expression using the properties of exponents developed in Chapter 5.

$$\frac{10x^3}{8x^5} = \frac{\overset{5}{\cancel{10}}x^3}{\underset{4}{\cancel{8}}x^5} \qquad \text{Divide out the common factor 2.}$$

Quick Check 4
Simplify the rational expression $\dfrac{15x^4}{21x^3}$.
(Assume $x \neq 0$.)

$$= \frac{5x^3}{4x^5} \qquad \text{Simplify.}$$

$$= \frac{5}{4x^2} \qquad \text{Divide numerator and denominator by } x^3.$$

EXAMPLE 5   Simplify $\dfrac{(x-3)(x+7)}{(x+7)(x+2)(x-3)}$. (Assume the denominator is nonzero.)

Solution

In this example, the numerator and denominator have already been factored. Notice that they share the common factors $x - 3$ and $x + 7$.

$$\frac{(x-3)(x+7)}{(x+7)(x+2)(x-3)} = \frac{\overset{1}{\cancel{(x-3)}}\,\overset{1}{\cancel{(x+7)}}}{\underset{1}{\cancel{(x+7)}}(x+2)\underset{1}{\cancel{(x-3)}}} \qquad \text{Divide out common factors.}$$

$$= \frac{1}{x+2} \qquad \text{Simplify.}$$

Notice that both factors were divided out of the numerator. Be careful to note that the numerator is equal to 1 in such a situation.

Quick Check 5
Simplify
$\dfrac{(x+9)(x-2)(x-8)}{(x-2)(x-8)(x+3)}$.
(Assume the denominator is nonzero.)

A Word of Caution   If each factor in the numerator of a rational expression is divided out when simplifying the expression, we must be sure to write a 1 in the numerator.

EXAMPLE 6   Simplify $\dfrac{x^2 + 8x - 20}{x^2 - 7x + 10}$. (Assume the denominator is nonzero.)

Solution

The trinomials in the numerator and denominator must be factored before we can simplify this expression. (For a review of factoring techniques, you may refer to Sections 6.1 through 6.5.)

$$\frac{x^2 + 8x - 20}{x^2 - 7x + 10} = \frac{(x - 2)(x + 10)}{(x - 2)(x - 5)}$$ Factor numerator and denominator.

$$= \frac{\overset{1}{\cancel{(x - 2)}}(x + 10)}{\underset{1}{\cancel{(x - 2)}}(x - 5)}$$ Divide out common factors.

$$= \frac{x + 10}{x - 5}$$ Simplify.

Quick Check 6

Simplify $\dfrac{x^2 + 10x + 24}{x^2 - 2x - 48}$.
(Assume the denominator is nonzero.)

A Word of Caution   When simplifying a rational expression, we must be very careful that we divide out only expressions that are common factors of the numerator and denominator. We cannot *reduce* individual terms in the numerator and denominator as in the following examples:

$$\frac{x + 8}{x - 6} \neq \frac{x + \overset{4}{\cancel{8}}}{x - \underset{3}{\cancel{6}}}$$

$$\frac{x^2 - 25}{x^2 - 36} \neq \frac{\overset{1}{\cancel{x^2}} - 25}{\underset{1}{\cancel{x^2}} - 36}$$

To avoid this, keep in mind that we must factor the numerator and denominator completely before attempting to divide out common factors.

Identifying Factors in the Numerator and Denominator That Are Opposites

Objective 4   **Identify factors that are opposites of each other.**   Two expressions of the form $a - b$ and $b - a$ are **opposites**. Subtraction in the opposite order produces the opposite result. Consider the expressions $a - b$ and $b - a$ when $a = 10$ and $b = 4$. In this case, $a - b = 10 - 4$ or 6 and $b - a = 4 - 10$ or -6. We can also see that $a - b$ and $b - a$ are opposites by noting that their sum, $(a - b) + (b - a)$, is equal to 0.

This is a useful result when simplifying rational expressions. The rational expression $\dfrac{a - b}{b - a}$ simplifies to -1, as any fraction whose numerator is the opposite of its denominator is equal to -1. If a rational expression has a factor in the numerator that is the opposite of a factor in the denominator, these two factors can be divided out to equal -1, as is done in the next example. We write the -1 in the numerator.

EXAMPLE 7   Simplify $\dfrac{49 - x^2}{x^2 - 12x + 35}$. (Assume the denominator is nonzero.)

Solution

We begin by factoring the numerator and denominator completely.

$$\dfrac{49 - x^2}{x^2 - 12x + 35} = \dfrac{(7 + x)(7 - x)}{(x - 5)(x - 7)}$$ Factor the numerator and denominator.

$$= \dfrac{(7 + x)(7 - x)^{-1}}{(x - 5)(x - 7)_{1}}$$ Divide out the opposite factors.

$$= -\dfrac{7 + x}{x - 5}$$ Simplify, writing the negative sign in front of the fraction.

> **Quick Check 7**
> Simplify $\dfrac{x^2 + 8x - 9}{1 - x^2}$.
> (Assume the denominator is nonzero.)

A Word of Caution   Two expressions of the form $a + b$ and $b + a$ are not opposites but are equal to each other. Addition in the opposite order produces the same result. When we divide two expressions of the form $a + b$ and $b + a$, the result is 1, not -1. For example, $\dfrac{x + 2}{2 + x} = 1$.

Rational Functions

> A **rational function** $r(x)$ is a function of the form $r(x) = \dfrac{f(x)}{g(x)}$, where $f(x)$ and $g(x)$ are polynomials and $g(x) \neq 0$.

Objective 5   Evaluate rational functions. We begin our investigation of rational functions by learning to evaluate them.

EXAMPLE 8   For $r(x) = \dfrac{x^2 + 6x - 17}{x^2 - 5x - 22}$, find $r(8)$.

Solution

We begin by substituting 8 for x in the function.

$$r(8) = \dfrac{(8)^2 + 6(8) - 17}{(8)^2 - 5(8) - 22}$$ Substitute 8 for x.

$$= \dfrac{64 + 48 - 17}{64 - 40 - 22}$$ Simplify each term in the numerator and denominator.

$$= \dfrac{95}{2}$$ Simplify the numerator and denominator.

> **Quick Check 8**
> For
> $r(x) = \dfrac{x^2 - 3x - 12}{x^2 + 7x - 15}$,
> find $r(-6)$.

Finding the Domain of a Rational Function

Objective 6   Find the domain of rational functions. Rational functions differ from linear functions and quadratic functions in that the domain of a rational function is not always the set of real numbers. We have to exclude any value that causes the function to be undefined, namely any value for which the denominator is equal to 0. Suppose that the function $r(x)$ was undefined for $x = 6$. Then the domain of $r(x)$ is the set of all real numbers except 6. This can be expressed in interval notation as $(-\infty, 6) \cup (6, \infty)$, which is the union of the set of all real numbers that are less than 6 with the set of all real numbers that are greater than 6.

EXAMPLE 9   Find the domain of $r(x) = \dfrac{x^2 - 32x + 60}{x^2 - 9x}$.

Solution

We begin by setting the denominator equal to 0 and solving for x.

Quick Check 9
Find the domain of
$r(x) = \dfrac{x^2 + 11x + 24}{x^2 - 4x - 45}$.

$$\begin{array}{lll} x^2 - 9x = 0 & & \text{Set the denominator equal to 0.} \\ x(x - 9) = 0 & & \text{Factor } x^2 - 9x. \\ x = 0 \quad \text{or} \quad x - 9 = 0 & & \text{Set each factor equal to 0.} \\ x = 0 \quad \text{or} \quad x = 9 & & \text{Solve each equation.} \end{array}$$

The domain of the function is the set of all real numbers except 0 and 9. In interval notation, this can be written as $(-\infty, 0) \cup (0, 9) \cup (9, \infty)$.

EXERCISES 7.1 >

Evaluate the rational expression for the given value of the variable.

1. $\dfrac{6}{x + 4}$ for $x = 4$

2. $\dfrac{9}{x - 20}$ for $x = 5$

3. $\dfrac{x + 3}{x - 8}$ for $x = -25$

4. $\dfrac{x + 1}{x + 13}$ for $x = -7$

5. $\dfrac{x^2 - 13x - 48}{x^2 - 6x - 12}$ for $x = 3$

6. $\dfrac{x^2 + 5x - 23}{x^2 - 7x + 20}$ for $x = 2$

7. $\dfrac{x^2 - 6x + 15}{x^2 + 14x - 7}$ for $x = -9$

8. $\dfrac{x^2 + 10x - 56}{x - 4}$ for $x = -7$

Find all values of the variable for which the rational expression is undefined.

9. $\dfrac{9}{x - 3}$

10. $\dfrac{x - 5}{10x - 13}$

11. $\dfrac{x + 6}{x(x - 8)}$

12. $\dfrac{x + 2}{(x + 2)(x + 3)}$

13. $\dfrac{x - 10}{x^2 - 6x - 16}$

14. $\dfrac{x + 1}{x^2 + 11x + 24}$

15. $\dfrac{7x}{x^2 - 49}$

16. $\dfrac{x - 3}{x^2 - 9x}$

Simplify the given rational expression. (Assume all denominators are nonzero.)

17. $\dfrac{3x^5}{9x^8}$

18. $\dfrac{10x^7}{12x^{10}}$

19. $\dfrac{x + 5}{(x + 6)(x + 5)}$

20. $\dfrac{(x - 7)(x + 7)}{x - 7}$

21. $\dfrac{(x + 2)(x - 5)}{(x - 2)(x - 5)}$

22. $\dfrac{(x - 9)(x - 3)}{(x - 3)(x + 3)}$

23. $\dfrac{x^2 - 4x - 32}{x^2 - 15x + 56}$

24. $\dfrac{x^2 - 3x}{x^2 - 8x + 15}$

25. $\dfrac{x^2 - 4x - 45}{x^2 + 11x + 30}$

26. $\dfrac{x^2 + 6x - 40}{x^2 - 10x + 24}$

27. $\dfrac{x^2 + 9x + 8}{x^2 - 3x - 4}$

28. $\dfrac{x^2 - 36}{x^2 + 9x + 18}$

29. $\dfrac{x^2 - 11x + 10}{x^2 - 3x - 70}$

30. $\dfrac{x^2 + 7x + 12}{x^2 - 5x - 24}$

31. $\dfrac{3x^2 - 6x - 105}{x^2 - 11x + 28}$

32. $\dfrac{2x^2 + 14x + 20}{5x^2 - 25x - 70}$

Determine whether the two given binomials are or are not opposites.

33. $x + 5$ and $5 + x$

34. $x - 11$ and $11 - x$

35. $14 - x$ and $x - 14$

36. $3x - 2$ and $2x - 3$

37. $2x + 13$ and $-2x - 13$

38. $6x - 7$ and $6x + 7$

Simplify the given rational expression. (Assume all denominators are nonzero.)

39. $\dfrac{3x - 21}{(5 + x)(7 - x)}$

40. $\dfrac{(8 + x)(8 - x)}{x^2 - 2x - 48}$

41. $\dfrac{x^2 - 3x - 54}{81 - x^2}$

42. $\dfrac{5x - x^2}{x^2 + 2x - 35}$

43. $\dfrac{9 - x^2}{x^2 + 12x + 27}$

44. $\dfrac{x^2 - x - 72}{36 - 4x}$

Evaluate the given rational function.

45. $r(x) = \dfrac{20}{x^2 - 5x + 10}$, $r(-5)$

46. $r(x) = \dfrac{x}{x^2 - 13x - 20}$, $r(8)$

47. $r(x) = \dfrac{x^2 + 10x + 24}{x^2 - 5x - 66}$, $r(4)$

48. $r(x) = \dfrac{x^2 - 100}{x^2 + 13x + 30}$, $r(-3)$

49. $r(x) = \dfrac{x^3 - 7x^2 - 11x + 20}{x^2 + 8x - 20}$, $r(10)$

50. $r(x) = \dfrac{x^3 + 8x^2 + 17x + 10}{x^3 + 3x^2 - 18x - 40}$, $r(2)$

Find the domain of the given rational function.

51. $r(x) = \dfrac{x^2 + 18x + 77}{x^2 + 10x}$

52. $r(x) = \dfrac{x^2 - 16x + 60}{x^2 - 9x + 8}$

53. $r(x) = \dfrac{x^2 + 2x - 3}{x^2 - 2x - 15}$

54. $r(x) = \dfrac{x^2 + 7x - 8}{x^2 - 64}$

55. $r(x) = \dfrac{x^2 + 4x - 60}{x^2 + 3x - 18}$

56. $r(x) = \dfrac{x^2 - 13x + 36}{x^2 + 14x + 45}$

Identify the given function as a linear function, a quadratic function, or a rational function.

57. $f(x) = x^2 - 11x + 30$

58. $f(x) = 3x - 8$

59. $f(x) = \dfrac{x^2}{x^3 - 5x^2 + 11x - 35}$

60. $f(x) = \dfrac{x^2 + 3x + 8}{x^2 - 5x - 5}$

61. $f(x) = \dfrac{1}{5}x + \dfrac{4}{9}$

62. $f(x) = -\dfrac{1}{2}x^2 + 3x - \dfrac{1}{7}$

Use the given graph of a rational function $r(x)$ to solve the problems that follow.

63.

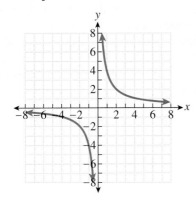

a) Find $r(1)$.

b) Find all values x such that $r(x) = -4$.

64.

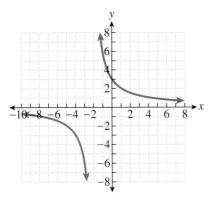

a) Find $r(0)$.

b) Find all values x such that $r(x) = -1$.

65.

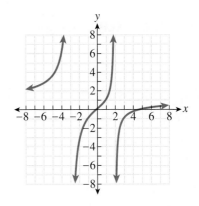

a) Find $r(-4)$.

b) Find all values x such that $r(x) = 0$.

66.

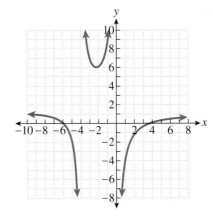

a) Find $r(1)$.

b) Find all values x such that $r(x) = 0$.

Answer in complete sentences.

67. Explain how to find the values for which a rational expression is undefined.

68. Is the rational expression $\dfrac{(x + 2)(x - 7)}{(x - 5)(x + 2)}$ undefined for the value $x = -2$? Explain your answer.

69. Explain how to simplify a rational expression to lowest terms. Use an example to illustrate the process.

70. Explain how to determine whether two factors are opposites. Use examples to illustrate your explanation.

QUICK REVIEW EXERCISES

Section 7.1

Simplify.

1. $\dfrac{5}{12} + \dfrac{4}{15}$

2. $\dfrac{17}{24} - \dfrac{3}{10}$

3. $\dfrac{14}{45} \cdot \dfrac{165}{308}$

4. $\dfrac{20}{99} \div \dfrac{35}{51}$

Study Tip REVISITED   To prepare for a cumulative exam effectively, develop a schedule and study plan. Ideally, you should begin studying approximately two weeks prior to the exam. Your schedule should include study time every day, and you should increase the time spent studying as you get closer to the exam date.

Before you begin to study, find out what material will and will not be covered on the exam by visiting your instructor during office hours. You can find out what topics or chapters will be emphasized and also what score you need to achieve on the exam in order to earn the grade you desire in the course.

Objectives

1 **Multiply two rational expressions.**
2 **Multiply two rational functions.**
3 **Divide a rational expression by another rational expression.**
4 **Divide a rational function by another rational function.**

7.2

**MULTIPLICATION
AND DIVISION
OF RATIONAL
EXPRESSIONS**

Objectives

1 **Multiply two rational expressions.**
2 **Multiply two rational functions.**
3 **Divide a rational expression by another rational expression.**
4 **Divide a rational function by another rational function.**

Multiplying Rational Expressions

Objective 1 Multiply two rational expressions. In this section, we will learn how to multiply and divide rational expressions. Multiplying rational expressions is similar to multiplying numerical fractions. Suppose that we needed to multiply $\frac{4}{9} \cdot \frac{21}{10}$. Before multiplying, we can divide out factors common to one of the numerators and one of the denominators. For example, the first numerator (4) and the second denominator (10) have a common factor of 2 that can be divided out of each. The second numerator (21) and the first denominator (9) have a common factor of 3 that can be divided out as well.

$$\frac{4}{9} \cdot \frac{21}{10} = \frac{2 \cdot 2}{3 \cdot 3} \cdot \frac{3 \cdot 7}{2 \cdot 5}$$ Factor each numerator and denominator.

$$= \frac{\overset{1}{\cancel{2}} \cdot 2}{\underset{1}{\cancel{3}} \cdot 3} \cdot \frac{\overset{1}{\cancel{3}} \cdot 7}{\underset{1}{\cancel{2}} \cdot 5}$$ Divide out common factors.

$$= \frac{2 \cdot 7}{3 \cdot 5}$$ Multiply remaining factors.

$$= \frac{14}{15}$$

Multiplying Rational Expressions

$$\frac{A}{B} \cdot \frac{C}{D} = \frac{AC}{BD} \quad B \neq 0 \text{ and } D \neq 0$$

To multiply two rational expressions, we will begin by factoring each numerator and denominator completely. After dividing out factors common to a numerator and a denominator, we will express the product of the two rational expressions as a single rational expression, leaving the numerator and denominator in factored form.

EXAMPLE 1 Multiply $\dfrac{x^2 - 11x + 30}{x^2 + 5x - 24} \cdot \dfrac{x^2 - 9}{x^2 - 2x - 15}$.

Solution

We begin by factoring each numerator and denominator. Then we proceed to divide out common factors.

$$\dfrac{x^2 - 11x + 30}{x^2 + 5x - 24} \cdot \dfrac{x^2 - 9}{x^2 - 2x - 15}$$

$$= \dfrac{(x - 5)(x - 6)}{(x + 8)(x - 3)} \cdot \dfrac{(x + 3)(x - 3)}{(x - 5)(x + 3)} \qquad \text{Factor numerators and denominators completely.}$$

$$= \dfrac{\overset{1}{\cancel{(x - 5)}}(x - 6)}{(x + 8)\cancel{(x - 3)}} \cdot \dfrac{\overset{1}{\cancel{(x + 3)}}\overset{1}{\cancel{(x - 3)}}}{\cancel{(x - 5)}\cancel{(x + 3)}} \qquad \text{Divide out common factors.}$$

$$= \dfrac{x - 6}{x + 8} \qquad \text{Simplify.}$$

Quick Check 1 Multiply $\dfrac{x^2 - 3x - 18}{x^2 - 9x + 20} \cdot \dfrac{x^2 - 3x - 10}{x^2 + 13x + 30}$.

EXAMPLE 2 Multiply $\dfrac{x^2 - 10x}{x^2 - 2x - 8} \cdot \dfrac{16 - x^2}{x^2 - 9x - 10}$.

Solution

Again, we begin by completely factoring both numerators and denominators.

$$\dfrac{x^2 - 10x}{x^2 - 2x - 8} \cdot \dfrac{16 - x^2}{x^2 - 9x - 10}$$

$$= \dfrac{x(x - 10)}{(x + 2)(x - 4)} \cdot \dfrac{(4 + x)(4 - x)}{(x + 1)(x - 10)} \qquad \text{Factor completely.}$$

$$= \dfrac{x\overset{1}{\cancel{(x - 10)}}}{(x + 2)\cancel{(x - 4)}} \cdot \dfrac{(4 + x)\overset{-1}{\cancel{(4 - x)}}}{(x + 1)\cancel{(x - 10)}} \qquad \begin{array}{l}\text{Divide out common} \\ \text{factors. Notice that the factors} \\ 4 - x \text{ and } x - 4 \text{ are opposites,} \\ \text{and divide out to equal } -1.\end{array}$$

$$= \dfrac{-x(4 + x)}{(x + 2)(x + 1)} \qquad \text{Multiply remaining factors.}$$

$$= -\dfrac{x(4 + x)}{(x + 2)(x + 1)} \qquad \begin{array}{l}\text{Write the negative sign in the} \\ \text{numerator in front of the fraction}\end{array}$$

Quick Check 2 Multiply $\dfrac{36 - x^2}{x^2 + 9x + 14} \cdot \dfrac{x^2 - 7x - 18}{x^2 - 15x + 54}$.

Here is a summary of the procedure for multiplying rational expressions:

Multiplying Rational Expressions

- Completely factor each numerator and each denominator.
- Divide out factors that are common to a numerator and a denominator, as well as factors in a numerator and denominator that are opposites.
- Multiply the remaining factors, leaving the numerator and denominator in factored form.

Objective 2 **Multiply two rational functions.**

EXAMPLE 3   For $f(x) = \dfrac{x^2 - 3x - 18}{x^2 + 8x + 16}$ and $g(x) = \dfrac{x^2 + 3x - 4}{9 - x^2}$, find $f(x) \cdot g(x)$.

Solution

We replace $f(x)$ and $g(x)$ by their formulas and proceed to multiply.

$$f(x) \cdot g(x) = \frac{x^2 - 3x - 18}{x^2 + 8x + 16} \cdot \frac{x^2 + 3x - 4}{9 - x^2}$$ Replace $f(x)$ and $g(x)$ by their formulas.

$$= \frac{(x - 6)(x + 3)}{(x + 4)(x + 4)} \cdot \frac{(x + 4)(x - 1)}{(3 + x)(3 - x)}$$ Factor completely.

$$= \frac{(x - 6)\overset{1}{\cancel{(x + 3)}}}{\cancel{(x + 4)}(x + 4)} \cdot \frac{\overset{1}{\cancel{(x + 4)}}(x - 1)}{\cancel{(3 + x)}(3 - x)}$$ Divide out common factors.

$$= \frac{(x - 6)(x - 1)}{(x + 4)(3 - x)}$$ Multiply remaining factors.

Quick Check 3 For $f(x) = \dfrac{x + 4}{49 - x^2}$ and $g(x) = \dfrac{x^2 - 10x + 21}{x^2 + x - 12}$, find $f(x) \cdot g(x)$.

Dividing a Rational Expression by Another Rational Expression

Objective 3 **Divide a rational expression by another rational expression.**
Dividing a rational expression by another rational expression is similar to dividing a numerical fraction by another numerical fraction. We replace the divisor, which is the rational expression by which we are dividing, by its reciprocal and then multiply.

Dividing Rational Expressions

$$\frac{A}{B} \div \frac{C}{D} = \frac{A}{B} \cdot \frac{D}{C} \qquad B \neq 0, \ C \neq 0, \text{ and } D \neq 0$$

EXAMPLE 4   Divide $\dfrac{x^2 + 13x + 42}{x^2 + x - 20} \div \dfrac{x^2 + 8x + 12}{x^2 - 4x}$.

Solution

We begin by inverting the divisor and multiplying. In this example, we must factor each numerator and denominator completely.

$$\dfrac{x^2 + 13x + 42}{x^2 + x - 20} \div \dfrac{x^2 + 8x + 12}{x^2 - 4x}$$

$$= \dfrac{x^2 + 13x + 42}{x^2 + x - 20} \cdot \dfrac{x^2 - 4x}{x^2 + 8x + 12}$$ Invert the divisor and multiply.

$$= \dfrac{(x + 6)(x + 7)}{(x + 5)(x - 4)} \cdot \dfrac{x(x - 4)}{(x + 2)(x + 6)}$$ Factor completely.

$$= \dfrac{\overset{1}{\cancel{(x + 6)}}(x + 7)}{(x + 5)\cancel{(x - 4)}} \cdot \dfrac{x\cancel{(x - 4)}}{(x + 2)\underset{1}{\cancel{(x + 6)}}}$$ Divide out common factors.

$$= \dfrac{x(x + 7)}{(x + 5)(x + 2)}$$ Multiply remaining factors.

Quick Check **4** Divide $\dfrac{x^2 - 4}{x^2 - 5x + 4} \div \dfrac{x^2 - 6x - 16}{x^2 + 3x - 28}$.

EXAMPLE 5   Divide $\dfrac{x^2 - 14x - 15}{x^2 + 2x - 35} \div \dfrac{x^2 - 1}{5 - x}$.

Solution

We rewrite the problem as a multiplication problem by inverting the divisor. We then factor each numerator and denominator completely before dividing out common factors. (You may want to factor at the same time you invert the divisor.)

$$\dfrac{x^2 - 14x - 15}{x^2 + 2x - 35} \div \dfrac{x^2 - 1}{5 - x}$$

$$= \dfrac{x^2 - 14x - 15}{x^2 + 2x - 35} \cdot \dfrac{5 - x}{x^2 - 1}$$ Invert the divisor and multiply.

$$= \dfrac{(x - 15)(x + 1)}{(x + 7)(x - 5)} \cdot \dfrac{5 - x}{(x + 1)(x - 1)}$$ Factor completely.

$$= \dfrac{(x - 15)\overset{1}{\cancel{(x + 1)}}}{(x + 7)\underset{1}{\cancel{(x - 5)}}} \cdot \dfrac{\overset{-1}{\cancel{5 - x}}}{\underset{1}{\cancel{(x + 1)}}(x - 1)}$$ Divide out common factors. Note that $5 - x$ and $x - 5$ are opposites.

$$= -\dfrac{x - 15}{(x + 7)(x - 1)}$$ Multiply remaining factors.

Quick Check   5 Divide $\dfrac{x^2 - 13x + 36}{x^2 - x - 12} \div \dfrac{9 - x}{x^2 + 8x + 15}$.

Here is a summary of the procedure for dividing rational expressions.

Dividing Rational Expressions

- Invert the divisor and change the operation from division to multiplication.
- Completely factor each numerator and each denominator.
- Divide out factors that are common to a numerator and a denominator, and factors in a numerator and denominator that are opposites.
- Multiply the remaining factors, leaving the numerator and denominator in factored form.

Objective 4   Divide a rational function by another rational function.

EXAMPLE 6   For $f(x) = \dfrac{x^2 - x - 12}{x + 7}$ and $g(x) = x^2 + 3x - 28$, find $f(x) \div g(x)$.

Solution

Replace $f(x)$ and $g(x)$ by their formulas and divide. Treat $g(x)$ as a rational function with a denominator of 1. We can see that its reciprocal is $\dfrac{1}{(x + 7)(x - 4)}$.

$$f(x) \div g(x)$$

$$= \frac{x^2 - x - 12}{x + 7} \div x^2 + 3x - 28$$
Replace $f(x)$ and $g(x)$ by their formulas.

$$= \frac{x^2 - x - 12}{x + 7} \cdot \frac{1}{x^2 + 3x - 28}$$
Invert the divisor and multiply.

$$= \frac{(x - 4)(x + 3)}{x + 7} \cdot \frac{1}{(x + 7)(x - 4)}$$
Factor completely.

$$= \frac{\overset{1}{\cancel{(x - 4)}}(x + 3)}{x + 7} \cdot \frac{1}{(x + 7)\underset{1}{\cancel{(x - 4)}}}$$
Divide out common factors.

$$= \frac{x + 3}{(x + 7)^2}$$
Multiply remaining factors.

Quick Check   6 For $f(x) = \dfrac{x^2 - 12x + 36}{x^2 - 11x + 10}$ and $g(x) = x^2 - 8x + 12$, find $f(x) \div g(x)$.

EXERCISES 7.2 >

Multiply.

1. $\dfrac{x+2}{(x-2)(x-4)} \cdot \dfrac{(x-4)(x+5)}{(x-3)(x+2)}$

2. $\dfrac{x+1}{x+10} \cdot \dfrac{(x+10)(x-6)}{(x-6)(x-1)}$

3. $\dfrac{x^2+18x+72}{x^2+2x-63} \cdot \dfrac{x^2-6x-7}{x^2+10x-24}$

4. $\dfrac{x^2+8x+16}{x^2+14x+48} \cdot \dfrac{x^2+4x-60}{x^2+x-12}$

5. $\dfrac{x^2-x}{x^2+2x-99} \cdot \dfrac{x^2+13x+22}{x^2+9x-10}$

6. $\dfrac{x^2-144}{x^2-16x+63} \cdot \dfrac{x^2-7x-18}{x^2+8x-48}$

7. $\dfrac{x^2-8x-33}{x^2-8x+16} \cdot \dfrac{x^2+2x-24}{x^2+5x+6}$

8. $\dfrac{x^2+12x+20}{x^2-13x+40} \cdot \dfrac{x^2-x-56}{x^2+18x+80}$

9. $\dfrac{x^2-11x}{x^2+5x+4} \cdot \dfrac{x^2+10x+9}{x^2+6x}$

10. $\dfrac{x^2-6x-55}{36-x^2} \cdot \dfrac{x^2-4x-12}{x^2+15x+50}$

11. $\dfrac{4x^2+48x+128}{x^2-6x+5} \cdot \dfrac{x^2+4x-45}{6x^2+36x+48}$

12. $\dfrac{x^2+22x+120}{x^2+11x+28} \cdot \dfrac{x^2+4x-21}{x^2+14x+24}$

13. $\dfrac{4-x^2}{x^2-x-30} \cdot \dfrac{x^2+9x+20}{x^2+7x-18}$

14. $\dfrac{x^2+3x-70}{x^2+11x+18} \cdot \dfrac{x^2+9x+14}{7x-x^2}$

15. $\dfrac{x^2-11x+24}{x^2-7x+12} \cdot \dfrac{x^2+4x-32}{x^2-64}$

16. $\dfrac{2x^2+24x+22}{8x^2+104x+320} \cdot \dfrac{x^2+3x-10}{x^2+7x+6}$

17. $\dfrac{x^2+15x+44}{x^2+x-90} \cdot \dfrac{9x-x^2}{x^2-2x-24}$

18. $\dfrac{x^2-3x-10}{x^2+2x} \cdot \dfrac{x^2+8x}{x^2+3x-40}$

For the given functions $f(x)$ and $g(x)$, find $f(x) \cdot g(x)$.

19. $f(x)=\dfrac{2}{x-3}, g(x)=\dfrac{x^2-12x+27}{x^2-5x-36}$

20. $f(x)=\dfrac{x^2-9x+20}{x^2+x-2}, g(x)=\dfrac{x^2+6x-16}{x^2-25}$

21. $f(x)=\dfrac{x^2-13x+22}{x^2+11x-12}, g(x)=\dfrac{x^2+4x-5}{x^2-6x+8}$

22. $f(x)=\dfrac{x^2+5x-84}{x^2-x-12}, g(x)=\dfrac{x^2+11x+24}{x^2-9x+14}$

23. $f(x) = \dfrac{x^2 + 6x - 16}{x^2 - x - 20}, g(x) = \dfrac{x^2 + 6x - 55}{x^2 + 17x + 72}$

24. $f(x) = x + 7, g(x) = \dfrac{x^2 + 3x - 28}{x^2 + 14x + 49}$

Find the missing numerator and denominator.

25. $\dfrac{(x + 5)(x - 2)}{(2x + 1)(x - 7)} \cdot \dfrac{?}{?} = \dfrac{(x + 5)}{(2x + 1)}$

26. $\dfrac{x^2 - 5x - 6}{x^2 + 15x + 54} \cdot \dfrac{?}{?} = \dfrac{(x + 1)(x + 5)}{(x + 6)^2}$

27. $\dfrac{x^2 - 10x + 16}{x^2 + 4x - 77} \cdot \dfrac{?}{?} = -\dfrac{x^2 - 11x + 24}{x^2 - 5x - 14}$

28. $\dfrac{x^2 + 4x}{x^2 - 81} \cdot \dfrac{?}{?} = \dfrac{x^2 - 6x}{x^2 - 7x - 18}$

Divide.

29. $\dfrac{(2x + 3)(x - 6)}{(3x + 1)(x - 8)} \div \dfrac{2x + 3}{x - 8}$

30. $\dfrac{(x + 3)(x - 4)}{x - 9} \div \dfrac{(x - 4)(x + 1)}{(x - 9)(x - 2)}$

31. $\dfrac{x^2 + 2x - 15}{x^2 - 49} \div \dfrac{x^2 + 7x - 30}{x^2 + 4x - 21}$

32. $\dfrac{x^2 + 19x + 88}{x^2 - 4x + 4} \div \dfrac{x^2 + 5x - 24}{x^2 - x - 6}$

33. $\dfrac{x^2 - 13x + 42}{x^2 + 8x - 20} \div \dfrac{x^2 - 3x - 28}{x^2 + 2x - 8}$

34. $\dfrac{x^2 - 2x - 80}{x^2 - 3x - 18} \div \dfrac{x^2 - 17x + 70}{x^2 + 4x + 3}$

35. $\dfrac{4x^2 + 4x - 80}{x^2 + 12x + 35} \div \dfrac{x^2 + 5x - 36}{3x^2 + 15x - 42}$

36. $\dfrac{x^2 - 13x + 30}{x^2 - 9x + 18} \div \dfrac{x^2 - 15x + 50}{x^2 - 11x + 30}$

37. $\dfrac{x^2 + 6x - 7}{x^2 + 16x + 55} \div \dfrac{x^2 + 13x + 42}{x^2 + 7x - 44}$

38. $\dfrac{x^2 + 13x + 36}{x^2 - 12x + 32} \div \dfrac{x^2 + 6x - 27}{16 - x^2}$

39. $\dfrac{x^2 - 4x - 5}{x^2 + x - 30} \div \dfrac{x^2 - x - 2}{x^2 + 4x - 12}$

40. $\dfrac{x^2 - 12x + 20}{x^2 + 6x + 5} \div \dfrac{x^2 - 11x + 18}{x^2 - 3x - 40}$

41. $\dfrac{x^2 - 2x - 63}{x^2 - 4x - 21} \div \dfrac{81 - x^2}{x^2 - 7x - 30}$

42. $\dfrac{x^2 - 6x - 27}{x^2 + 3x - 4} \div \dfrac{x^2 - 18x + 81}{x^2 - 8x + 7}$

43. $\dfrac{x^2 + 10x + 24}{x^2 - 7x + 10} \div \dfrac{x^2 + 16x + 60}{x^2 + 5x - 50}$

44. $\dfrac{x^2 - 15x + 54}{x^2 - 16x + 48} \div \dfrac{x^2 - 14x + 48}{x^2 - 2x - 8}$

45. $\dfrac{x^2 + 16x + 63}{2x - x^2} \div \dfrac{x^2 + x - 72}{x^2 - 5x + 6}$

46. $\dfrac{x^2 - 12x + 36}{x^2 + 8x + 15} \div \dfrac{6x^2 - x^3}{x^2 + 5x}$

For the given functions $f(x)$ and $g(x)$, find $f(x) \div g(x)$.

47. $f(x) = \dfrac{x^2 + 8x - 9}{x^2 - 10x + 25}, g(x) = x + 9$

48. $f(x) = 2x + 14, g(x) = \dfrac{x^2 + 15x + 56}{x^2 - 10x + 21}$

49. $f(x) = \dfrac{x^2 + 3x - 54}{x^2 - 2x - 3}, g(x) = \dfrac{x^2 + 9x}{x^2 - 7x - 8}$

50. $f(x) = \dfrac{x^2 - 14x + 24}{x^2 + x - 6}, g(x) = \dfrac{x^2 + 2x - 80}{x^2 + 13x + 30}$

51. $f(x) = \dfrac{2x^2 - 5x}{x^2 - 1}, g(x) = \dfrac{2x^2 + x - 15}{x^2 - 8x - 9}$

52. $f(x) = \dfrac{x^2 + 4x + 4}{x^2 + 14x + 40}, g(x) = \dfrac{3x^2 + 7x + 2}{5x^2 + 19x - 4}$

Find the missing numerator and denominator.

53. $\dfrac{(x + 9)(x - 6)}{(x - 3)(x - 7)} \div \dfrac{?}{?} = \dfrac{x + 9}{x - 7}$

54. $\dfrac{x^2 - 5x + 4}{x^2 - 14x + 45} \div \dfrac{?}{?} = \dfrac{(x - 1)(x - 6)}{(x - 9)^2}$

55. $\dfrac{x^2 - 10x + 9}{x^2 + 7x} \div \dfrac{?}{?} = \dfrac{x^2 - 13x + 12}{x^2 + 11x + 28}$

56. $\dfrac{x^2 + 12x + 36}{x^2 - 100} \div \dfrac{?}{?} = \dfrac{x^2 - 36}{x^2 + 10x}$

Answer in complete sentences.

57. Explain how to multiply rational expressions. Use an example to illustrate the process.

58. Explain how to divide rational expressions. Use an example to illustrate the process.

59. Explain the similarities between dividing numerical fractions and dividing rational expressions. Are there any differences?

60. Explain why the restrictions $B \neq 0$, $C \neq 0$, and $D \neq 0$ are necessary when dividing $\dfrac{A}{B} \div \dfrac{C}{D}$.

> **Study Tip** **Revisited**   When preparing for a cumulative exam, begin by reviewing old exams and quizzes. Begin by reviewing and understanding any mistakes that you made on the exam. By understanding what you did wrong, you are minimizing your chances of making the same mistakes on the cumulative exam.
>
> Once you understand the errors that you made, try reworking all of the problems on the exam or quiz. You should do this without referring to your notes or textbook, to help you to learn which topics you need to review and which topics you have under control. You need to focus your time on the topics you are struggling with and spend less time on the topics that you understand; retaking your tests will help you determine which topics you need to spend more time on.

7.3

ADDITION AND SUBTRACTION OF RATIONAL EXPRESSIONS WITH THE SAME DENOMINATOR

Objectives

1. **Add rational expressions with the same denominator.**
2. **Subtract rational expressions with the same denominator.**
3. **Add or subtract rational expressions with opposite denominators.**

Now that we have learned how to multiply and divide rational expressions, we move on to addition and subtraction. We know from our work with numerical fractions that two fractions must have the same denominator before we can add or subtract them. The same holds true for rational expressions. In this section, we will begin with rational expressions that already have the same denominator, and in the next section we will learn how to add and subtract rational expressions with different denominators.

Adding Rational Expressions with the Same Denominator

Objective 1 Add rational expressions with the same denominator. To add fractions that have the same denominator, we add the numerators and place the result over the common denominator. We will follow the same procedure when adding two rational expressions. Of course, we should check that our result is in simplest terms.

Adding Rational Expressions with the Same Denominator

$$\frac{A}{C} + \frac{B}{C} = \frac{A+B}{C} \qquad C \neq 0$$

EXAMPLE 1   Add $\dfrac{7}{x+3} + \dfrac{5}{x+3}$.

Solution

These two fractions have the same denominator, so we add the two numerators and place the result over the common denominator $x + 3$.

$$\frac{7}{x+3} + \frac{5}{x+3} = \frac{7+5}{x+3} \qquad \text{Add numerators, placing the sum over the common denominator.}$$

$$= \frac{12}{x+3} \qquad \text{Simplify the numerator.}$$

Quick Check 1

Add $\dfrac{9}{2x+7} + \dfrac{12}{2x+7}$.

The numerator and denominator do not have any common factors, so this is our final result. A common error is to attempt to divide a common factor out of 12 in the numerator and 3 in the denominator, but the number 3 is a term of the denominator and not a factor.

EXAMPLE 2   Add $\dfrac{6x-7}{x^2+5x+6} + \dfrac{2x+23}{x^2+5x+6}$.

Solution

The two denominators are the same, so we may add.

$$\frac{6x - 7}{x^2 + 5x + 6} + \frac{2x + 23}{x^2 + 5x + 6}$$

$$= \frac{(6x - 7) + (2x + 23)}{x^2 + 5x + 6} \qquad \text{Add numerators.}$$

$$= \frac{8x + 16}{x^2 + 5x + 6} \qquad \text{Combine like terms.}$$

$$= \frac{8\cancel{(x + 2)}^{\,1}}{\cancel{(x + 2)}_{\,1}(x + 3)} \qquad \begin{array}{l}\text{Factor numerator and denominator}\\ \text{and divide out the common factor.}\end{array}$$

$$= \frac{8}{x + 3} \qquad \text{Simplify.}$$

Quick Check 2 Add $\dfrac{2x + 54}{x^2 + 4x - 32} + \dfrac{3x - 14}{x^2 + 4x - 32}$.

EXAMPLE 3 Add $\dfrac{x^2 - 10x}{x^2 - 5x - 36} + \dfrac{7x - 54}{x^2 - 5x - 36}$.

Solution

The denominators are the same, so we add the numerators and then simplify.

$$\frac{x^2 - 10x}{x^2 - 5x - 36} + \frac{7x - 54}{x^2 - 5x - 36}$$

$$= \frac{(x^2 - 10x) + (7x - 54)}{x^2 - 5x - 36} \qquad \text{Add numerators.}$$

$$= \frac{x^2 - 3x - 54}{x^2 - 5x - 36} \qquad \text{Combine like terms.}$$

$$= \frac{(x + 6)\cancel{(x - 9)}^{\,1}}{(x + 4)\cancel{(x - 9)}_{\,1}} \qquad \begin{array}{l}\text{Factor numerator and denominator}\\ \text{and divide out the common factor.}\end{array}$$

$$= \frac{x + 6}{x + 4} \qquad \text{Simplify.}$$

Quick Check 3 Add $\dfrac{x^2 + 6x + 4}{x^2 + 4x - 21} + \dfrac{5x + 24}{x^2 + 4x - 21}$.

Subtracting Rational Expressions with the Same Denominator

Objective 2 Subtract rational expressions with the same denominator.
Subtracting two rational expressions with the same denominator is just like adding them, except that we subtract the two numerators rather than adding them.

Subtracting Rational Expressions with the Same Denominator

$$\frac{A}{C} - \frac{B}{C} = \frac{A - B}{C} \qquad C \neq 0$$

EXAMPLE ▶ **4**   Subtract $\dfrac{9}{2x + 8} - \dfrac{5}{2x + 8}$.

Solution

Since the two fractions have the same denominator, we subtract the numerators and place the result over the common denominator.

$$\frac{9}{2x + 8} - \frac{5}{2x + 8} = \frac{4}{2x + 8}$$ Subtract numerators, place difference over $2x + 8$.

$$= \frac{\overset{2}{\cancel{4}}}{\underset{1}{\cancel{2}}(x + 4)}$$ Factor denominator and divide out common factors.

$$= \frac{2}{x + 4}$$ Simplify.

Quick Check **4** ▶ Subtract $\dfrac{25}{6x + 36} - \dfrac{17}{6x + 36}$.

EXAMPLE ▶ **5**   Subtract $\dfrac{5x}{x - 2} - \dfrac{10}{x - 2}$.

Solution

The two denominators are the same, so we may subtract these two rational expressions.

$$\frac{5x}{x - 2} - \frac{10}{x - 2} = \frac{5x - 10}{x - 2}$$ Subtract numerators.

$$= \frac{5(\overset{1}{\cancel{x - 2}})}{\underset{1}{\cancel{x - 2}}}$$ Factor numerator and divide out the common factor.

$$= 5$$ Simplify.

Quick Check **5**

Subtract
$\dfrac{3x - 7}{x - 9} - \dfrac{20}{x - 9}$.

Notice that the denominator $x - 2$ was also a factor of the numerator, leaving a denominator of 1.

EXAMPLE ▶ **6**   Subtract $\dfrac{2x^2 - 3x - 9}{7x - x^2} - \dfrac{x^2 + 9x - 44}{7x - x^2}$.

Solution

The denominators are the same, so we can subtract these two rational expressions. When the numerator of the second fraction has more than one term, we must remember that we are subtracting the whole numerator and not just the first term. When we subtract the numerators and place the difference over the common denominator, it is a good idea to write each numerator in a set of parentheses. This will remind us to subtract each term in the second numerator.

$$\frac{2x^2 - 3x - 9}{7x - x^2} - \frac{x^2 + 9x - 44}{7x - x^2}$$

$$= \frac{(2x^2 - 3x - 9) - (x^2 + 9x - 44)}{7x - x^2} \qquad \text{Subtract numerators.}$$

$$= \frac{2x^2 - 3x - 9 - x^2 - 9x + 44}{7x - x^2} \qquad \text{Distribute.}$$

$$= \frac{x^2 - 12x + 35}{7x - x^2} \qquad \text{Combine like terms.}$$

$$= \frac{\overset{-1}{\cancel{(x - 7)}}(x - 5)}{x\underset{1}{\cancel{(7 - x)}}} \qquad \begin{array}{l}\text{Factor numerator and denominator} \\ \text{and divide out the common factor.} \\ \text{The } -1 \text{ results from the fact that} \\ x - 7 \text{ and } 7 - x \text{ are opposites.}\end{array}$$

$$= -\frac{x - 5}{x} \qquad \text{Simplify.}$$

Quick Check **6** Subtract $\dfrac{3x^2 + 6x - 37}{64 - x^2} - \dfrac{2x^2 + 17x - 61}{64 - x^2}$.

A Word of Caution   When subtracting a rational expression whose numerator contains more than one term, be sure to subtract the *entire* numerator and not just the first term. One way to remember this is by placing the numerators inside sets of parentheses.

Adding or Subtracting Rational Expressions with Opposite Denominators

Objective 3 **Add or subtract rational expressions with opposite denominators.** Consider the expression $\dfrac{10}{x - 2} + \dfrac{3}{2 - x}$. Are the two denominators the same? No, but they are opposites. We can rewrite the denominator $2 - x$ as its opposite $x - 2$ if we also rewrite the operation (addition) as its opposite (subtraction). In other words, we can rewrite the expression $\dfrac{10}{x - 2} + \dfrac{3}{2 - x}$ as $\dfrac{10}{x - 2} - \dfrac{3}{x - 2}$. Once the denominators are the same, we can subtract the numerators.

EXAMPLE 7   Add $\dfrac{7x}{2x - 16} + \dfrac{56}{16 - 2x}$.

Solution

The two denominators are opposites, so we may change the second denominator to $2x - 16$ by changing the operation from addition to subtraction.

Exercises 7.3　　**415**

$$\frac{7x}{2x-16} + \frac{56}{16-2x}$$

$$= \frac{7x}{2x-16} - \frac{56}{2x-16}$$

Rewrite the second denominator as $2x - 16$ by changing the operation from addition to subtraction.

$$= \frac{7x - 56}{2x - 16}$$

Subtract the numerators.

$$= \frac{7(\overset{1}{\cancel{x-8}})}{2(\underset{1}{\cancel{x-8}})}$$

Factor the numerator and denominator and divide out the common factor.

$$= \frac{7}{2}$$

Simplify.

> **Quick Check 7**
> Add $\dfrac{x}{5x-20} + \dfrac{4}{20-5x}$.

EXAMPLE 8 Subtract $\dfrac{x^2-4x}{x-3} - \dfrac{11x-30}{3-x}$.

Solution

These two denominators are opposites, so we begin by rewriting the second rational expression in such a way that the two rational expressions have the same denominator.

$$\frac{x^2-4x}{x-3} - \frac{11x-30}{3-x}$$

$$= \frac{x^2-4x}{x-3} + \frac{11x-30}{x-3}$$

Rewrite the second denominator as $x - 3$ by changing the operation from subtraction to addition.

$$= \frac{(x^2-4x)+(11x-30)}{x-3}$$

Add the numerators.

$$= \frac{x^2+7x-30}{x-3}$$

Combine like terms.

$$= \frac{(x+10)(\overset{1}{\cancel{x-3}})}{\underset{1}{\cancel{x-3}}}$$

Factor the numerator and divide out the common factor.

$$= x + 10$$

Simplify.

> **Quick Check 8**
> Subtract
> $\dfrac{x^2-2x-11}{x^2-16} - \dfrac{7x-25}{16-x^2}$.

EXERCISES 7.3

Add.

1. $\dfrac{5}{x+3} + \dfrac{8}{x+3}$

2. $\dfrac{9}{x-6} + \dfrac{7}{x-6}$

3. $\dfrac{x}{x-5} + \dfrac{10}{x-5}$

4. $\dfrac{3x}{x+14} + \dfrac{26}{x+14}$

5. $\dfrac{2x}{x+10} + \dfrac{20}{x+10}$

6. $\dfrac{3x}{4x+28} + \dfrac{21}{4x+28}$

7. $\dfrac{x}{x^2+7x-18} + \dfrac{9}{x^2+7x-18}$

8. $\dfrac{x}{x^2+12x+35} + \dfrac{7}{x^2+12x+35}$

9. $\dfrac{2x+4}{x^2+14x+33} + \dfrac{x+5}{x^2+14x+33}$

10. $\dfrac{x+14}{x^2+7x+6} + \dfrac{3x+10}{x^2+7x+6}$

11. $\dfrac{x^2+3x+11}{x^2-2x-48} + \dfrac{12x+43}{x^2-2x-48}$

F O R   E X T R A   H E L P

MyMathLab
MyMathLab

MathXL
MathXL

Interactmath.com

MathXL
Tutorials on CD

DVT CD
Videotape

Tutor
Center

Addison-Wesley
Math Tutor Center

Student Solutions
Manual

12. $\dfrac{x^2 + 5x + 20}{x^2 - 5x - 14} + \dfrac{5x - 4}{x^2 - 5x - 14}$

13. $\dfrac{x^2 - 16x - 45}{x^2 - 3x - 18} + \dfrac{x^2 - 2x - 27}{x^2 - 3x - 18}$

14. $\dfrac{x^2 - 11x - 34}{x^2 + 16x + 64} + \dfrac{x^2 + 17x - 46}{x^2 + 16x + 64}$

For the given rational functions $f(x)$ and $g(x)$, find $f(x) + g(x)$.

15. $f(x) = \dfrac{3}{x}, g(x) = \dfrac{5}{x}$

16. $f(x) = \dfrac{x}{5x + 30}, g(x) = \dfrac{6}{5x + 30}$

17. $f(x) = \dfrac{x^2 + 3x - 15}{x^2 - 4x - 45}, g(x) = \dfrac{4x + 25}{x^2 - 4x - 45}$

18. $f(x) = \dfrac{2x - 21}{x^2 + 20x + 99}, g(x) = \dfrac{x^2 + 2x - 56}{x^2 + 20x + 99}$

Subtract.

19. $\dfrac{13}{x + 1} - \dfrac{9}{x + 1}$

20. $\dfrac{6}{x - 4} - \dfrac{11}{x - 4}$

21. $\dfrac{x}{x + 7} - \dfrac{7}{x + 7}$

22. $\dfrac{3x}{2x + 5} - \dfrac{9}{2x + 5}$

23. $\dfrac{4x}{x - 3} - \dfrac{12}{x - 3}$ **24.** $\dfrac{2x}{3x - 15} - \dfrac{10}{3x - 15}$

25. $\dfrac{3x - 7}{x - 8} - \dfrac{x + 9}{x - 8}$ **26.** $\dfrac{5x - 23}{x - 9} - \dfrac{3x - 5}{x - 9}$

27. $\dfrac{x}{x^2 - 13x + 30} - \dfrac{3}{x^2 - 13x + 30}$

28. $\dfrac{x}{x^2 + 2x - 48} - \dfrac{6}{x^2 + 2x - 48}$

29. $\dfrac{4x - 7}{x^2 - 16} - \dfrac{2x - 15}{x^2 - 16}$

30. $\dfrac{6x + 1}{x^2 - 16x + 63} - \dfrac{3x + 22}{x^2 - 16x + 63}$

31. $\dfrac{x^2 - 7x + 10}{x^2 + 4x - 12} - \dfrac{5x - 10}{x^2 + 4x - 12}$

32. $\dfrac{x^2 + 3x - 18}{x^2 - 8x + 7} - \dfrac{6x + 10}{x^2 - 8x + 7}$

33. $\dfrac{2x^2 + x - 4}{x^2 + 8x} - \dfrac{x^2 + 9x - 4}{x^2 + 8x}$

34. $\dfrac{3x^2 - 7x - 19}{x^2 - 6x - 27} - \dfrac{2x^2 + 7x - 64}{x^2 - 6x - 27}$

35. $\dfrac{(x - 6)(x + 3)}{x^2 - 2x - 15} - \dfrac{9(x - 5)}{x^2 - 2x - 15}$

36. $\dfrac{(2x + 9)(x + 4)}{x^2 - 2x - 24} - \dfrac{(x + 6)(x - 6)}{x^2 - 2x - 24}$

For the given rational functions $f(x)$ and $g(x)$, find $f(x) - g(x)$.

37. $f(x) = \dfrac{12}{x + 8}, g(x) = \dfrac{23}{x + 8}$

38. $f(x) = \dfrac{21}{x - 7}, g(x) = \dfrac{3x}{x - 7}$

39. $f(x) = \dfrac{x^2 + 3x - 5}{x^2 - 11x + 18}, g(x) = \dfrac{8x + 9}{x^2 - 11x + 18}$

40. $f(x) = \dfrac{2x^2 + 10x - 13}{x^2 + 15x + 50}, g(x) = \dfrac{x^2 + 11x + 17}{x^2 + 15x + 50}$

Add or subtract.

41. $\dfrac{2x}{x - 5} + \dfrac{10}{5 - x}$ **42.** $\dfrac{5x}{2x - 14} + \dfrac{35}{14 - 2x}$

43. $\dfrac{x^2 + 8x}{x - 1} - \dfrac{2x - 11}{1 - x}$

44. $\dfrac{x^2 - 8x - 10}{x - 3} + \dfrac{x - 28}{3 - x}$

45. $\dfrac{x^2 - 3x - 9}{x - 9} + \dfrac{x^2 - 7x + 27}{9 - x}$

46. $\dfrac{3x^2 - 14x - 45}{x - 6} - \dfrac{-2x^2 + 14x + 9}{6 - x}$

Add or subtract.

47. $\dfrac{x^2 - 3x + 5}{2x - 20} + \dfrac{4x + 35}{20 - 2x}$

48. $\dfrac{x^2 - 7x + 7}{x^2 + x - 2} + \dfrac{2x - 3}{x^2 + x - 2}$

49. $\dfrac{2x^2 + x - 9}{x^2 - 8x + 15} - \dfrac{x^2 + 10x - 29}{x^2 - 8x + 15}$

50. $\dfrac{x^2 - 7x - 50}{x - 8} - \dfrac{x^2 + 3x - 46}{8 - x}$

51. $\dfrac{4x - 15}{x^2 - 3x} + \dfrac{x^2 + x - 9}{x^2 - 3x}$

52. $\dfrac{4x^2 - 6x + 3}{x^2 + 17x + 70} - \dfrac{3x^2 - 6x + 52}{x^2 + 17x + 70}$

53. $\dfrac{x^2 - 3x + 45}{x^2 - 81} + \dfrac{x^2 + 29x + 27}{x^2 - 81}$

54. $\dfrac{x^2 - 8x + 6}{x^2 + 10x + 21} + \dfrac{5x - 24}{x^2 + 10x + 21}$

55. $\dfrac{x^2 - 3x + 6}{x^2 - 12x + 32} - \dfrac{3x + 22}{x^2 - 12x + 32}$

56. $\dfrac{10}{5x - 25} + \dfrac{2x}{25 - 5x}$

57. $\dfrac{2x^2 - 15x + 39}{x^2 + 3x - 4} + \dfrac{x^2 - 18x - 9}{x^2 + 3x - 4}$

58. $\dfrac{2x^2 - 3x + 20}{x^2 + 6x - 55} - \dfrac{x^2 + 9x - 15}{x^2 + 6x - 55}$

59. $\dfrac{6x}{3x - 2} - \dfrac{4}{2 - 3x}$

60. $\dfrac{x^2 + 20x + 45}{x^2 + 15x + 56} - \dfrac{3x - 27}{x^2 + 15x + 56}$

61. $\dfrac{5x^2 + 3x - 13}{x^2 - 8x - 9} - \dfrac{4x^2 - 7x - 22}{x^2 - 8x - 9}$

62. $\dfrac{5x - 12}{x^2 - 8x + 12} + \dfrac{x^2 + 8x - 18}{x^2 - 8x + 12}$

63. $\dfrac{2x^2 + 5x + 20}{x^2 + 5x + 6} + \dfrac{3x^2 - 7x - 4}{x^2 + 5x + 6} - \dfrac{4x^2 - 2x + 25}{x^2 + 5x + 6}$

64. $\dfrac{x^2 + 3x - 7}{x^2 - 2x - 8} - \dfrac{x^2 + 5x - 9}{x^2 - 2x - 8} + \dfrac{x^2 - 6x + 14}{x^2 - 2x - 8}$

Find the missing numerator.

65. $\dfrac{?}{x + 6} + \dfrac{18}{x + 6} = 3$

66. $\dfrac{x^2 + 5x}{x - 2} - \dfrac{?}{x - 2} = x + 7$

67. $\dfrac{x^2 + 7x + 17}{(x + 8)(x + 5)} + \dfrac{?}{(x + 8)(x + 5)} = \dfrac{x + 3}{x + 5}$

68. $\dfrac{x^2 - 5x - 24}{(x + 2)(x - 4)} - \dfrac{?}{(x + 2)(x - 4)} = \dfrac{x + 2}{x - 4}$

Answer in complete sentences.

69. Explain how to add two rational expressions that have the same denominator. Use an example to illustrate the process.

70. Explain how to subtract two rational expressions that have the same denominator. Use an example to illustrate the process.

71. Explain how to determine whether two rational expressions have opposite denominators.

72. Explain how to add two rational expressions that have opposite denominators. Use an example to illustrate the process.

Study Tip REVISITED   When studying a particular topic, look back at your old homework for that topic. If you struggled with a particular type of problem, your homework will reflect that. Hopefully, your homework will contain some notes to yourself about how to do certain problems or mistakes to avoid.

Look over your class notes from that topic. Your notes should include examples of the problems in that section of the textbook, as well as pointers from your instructor. You can also review that section, paying close attention to the examples and any procedures developed in that section.

If you have been creating note cards during the semester, they will be most advantageous to you at this time. These note cards should focus on problems that you considered difficult at the time and contain strategies for solving these types of problems. A quick glance at these note cards will really speed up your review of a particular topic.

ADDITION AND SUBTRACTION OF RATIONAL EXPRESSIONS WITH UNLIKE DENOMINATORS

Objectives

1 **Find the least common denominator (LCD) of two or more rational expressions.**
2 **Add or subtract rational expressions with unlike denominators.**

The Least Common Denominator of Two or More Rational Expressions

Objective 1 Find the least common denominator (LCD) of two or more rational expressions. If two numerical fractions do not have the same denominator, we cannot add or subtract the fractions until we rewrite them as equivalent fractions with a common denominator. The same holds true for rational expressions with unlike denominators. We will begin this section by learning how to find the **least common denominator (LCD)** for two or more rational expressions. Then we will learn how to add or subtract rational expressions with unlike denominators.

Finding the LCD of Two Rational Expressions

> Begin by completely factoring each denominator, and then identify each expression that is a factor of one or both denominators. The LCD is equal to the product of these factors.

If an expression is a repeated factor of one or more of the denominators, then we repeat it as a factor in the LCD as well. The exponent used for this factor is equal to the greatest power that the factor is raised to in any one denominator.

EXAMPLE 1 Find the LCD of $\dfrac{7}{24a^3b}$ and $\dfrac{9}{16a}$.

Solution

Quick Check 1
Find the LCD of $\dfrac{10}{9r^2s^3}$ and $\dfrac{1}{12rs^6}$.

We begin with the coefficients 24 and 16. The smallest number that both divide into evenly is 48. Moving on to variable factors in the denominator, we see that the variables a and b are factors of one or both denominators. Note that the variable a is raised to the third power in the first denominator, so the LCD must contain a factor of a^3. The LCD is $48a^3b$.

EXAMPLE 2 Find the LCD of $\dfrac{8}{x^2 - 4x + 3}$ and $\dfrac{9}{x^2 - 9}$.

Quick Check 2
Find the LCD of $\dfrac{10}{x^2 - 13x + 40}$ and $\dfrac{3}{x^2 - 4x - 32}$.

Solution

We begin by factoring each denominator.

$$\frac{8}{x^2 - 4x + 3} = \frac{8}{(x - 1)(x - 3)} \qquad \frac{9}{x^2 - 9} = \frac{9}{(x + 3)(x - 3)}$$

The factors in the denominators are $x - 1$, $x - 3$, and $x + 3$. Since no expression is repeated as a factor in any one denominator, the LCD is $(x - 1)(x - 3)(x + 3)$.

EXAMPLE 3   Find the LCD of $\dfrac{2}{x^2 - 2x - 35}$ and $\dfrac{x}{x^2 + 10x + 25}$.

Solution

Again, we begin by factoring each denominator.

$$\frac{2}{x^2 - 2x - 35} = \frac{2}{(x - 7)(x + 5)} \qquad \frac{x}{x^2 + 10x + 25} = \frac{x}{(x + 5)(x + 5)}$$

The two expressions that are factors are $x - 7$ and $x + 5$; the factor $x + 5$ is repeated twice in the second denominator. So the LCD must have $x + 5$ as a factor twice as well. The LCD is $(x - 7)(x + 5)(x + 5)$, or $(x - 7)(x + 5)^2$.

Quick Check **3**
Find the LCD of $\dfrac{x + 8}{x^2 - 81}$
and $\dfrac{x - 7}{x^2 + 18x + 81}$.

Adding or Subtracting Rational Expressions with Unlike Denominators

Objective 2   **Add or subtract rational expressions with unlike denominators.**
To add or subtract two rational expressions that do not have the same denominator, we begin by finding the LCD. We then convert each rational expression to an equivalent rational expression that has the LCD as its denominator. We can then add or subtract as we did in the previous section. As always, we should attempt to simplify the resulting rational expression.

EXAMPLE 4   Add $\dfrac{2a}{3} + \dfrac{3a}{7}$.

Solution

The LCD of these two fractions is 21. We will rewrite each fraction as an equivalent fraction whose denominator is 21. Since $\frac{7}{7} = 1$, multiplying the first fraction by $\frac{7}{7}$ will produce an equivalent fraction whose denominator is 21. In a similar fashion, we multiply the second fraction by $\frac{3}{3}$. Once both fractions have the same denominator, we add the numerators and write the sum above the common denominator.

$$\frac{2a}{3} + \frac{3a}{7} = \frac{2a}{3} \cdot \frac{7}{7} + \frac{3a}{7} \cdot \frac{3}{3}$$ Multiply the first fraction by $\frac{7}{7}$ and the second fraction by $\frac{3}{3}$ to rewrite each fraction with a denominator of 21.

$$= \frac{14a}{21} + \frac{9a}{21}$$ Simplify each fraction.

$$= \frac{23a}{21}$$ Add the numerators and place the sum over the common denominator.

EXAMPLE 5   Add $\dfrac{8}{x + 6} + \dfrac{2}{x - 7}$.

Solution

The two denominators are not the same, so we begin by finding the LCD for these two rational expressions. Each denominator has a single factor, and the LCD is the product of these two denominators. The LCD is $(x + 6)(x - 7)$. We will multiply $\dfrac{8}{x + 6}$ by $\dfrac{x - 7}{x - 7}$ to write it as an equivalent fraction whose denominator is the LCD. We need to multiply $\dfrac{2}{x - 7}$ by $\dfrac{x + 6}{x + 6}$ to write it as an equivalent fraction whose denominator is the LCD.

$$\dfrac{8}{x + 6} + \dfrac{2}{x - 7}$$

$$= \dfrac{8}{x + 6} \cdot \dfrac{x - 7}{x - 7} + \dfrac{2}{x - 7} \cdot \dfrac{x + 6}{x + 6}$$
Multiply to rewrite each expression as an equivalent rational expression that has the LCD as its denominator.

$$= \dfrac{8x - 56}{(x + 6)(x - 7)} + \dfrac{2x + 12}{(x - 7)(x + 6)}$$
Distribute in each numerator, but do not distribute in the denominators.

$$= \dfrac{(8x - 56) + (2x + 12)}{(x + 6)(x - 7)}$$
Add the numerators, writing the sum over the common denominator.

$$= \dfrac{10x - 44}{(x + 6)(x - 7)}$$
Combine like terms.

$$= \dfrac{2(5x - 22)}{(x + 6)(x - 7)}$$
Factor the numerator.

Since the numerator and denominator do not have any common factors, this rational expression cannot be simplified any further.

Quick Check 4

Add $\dfrac{5}{x - 2} + \dfrac{6}{x + 6}$.

A Word of Caution   When adding two rational expressions, we cannot simply add the two numerators together and place their sum over the sum of the two denominators.

$$\dfrac{8}{x + 6} + \dfrac{2}{x - 7} \neq \dfrac{8 + 2}{(x + 6) + (x - 7)}$$

We must first find a common denominator and rewrite each rational expression as an equivalent expression whose denominator is equal to the common denominator.

When adding or subtracting rational expressions, we simplify the numerator, but leave the denominator in factored form. After we simplify the numerator, we factor if possible and check the denominators for common factors that can be divided out.

EXAMPLE 6   Subtract $\dfrac{x}{x^2 + 8x + 15} - \dfrac{5}{x^2 + 12x + 35}$.

Solution

In this example, we must factor each denominator to find the LCD.

$$\dfrac{x}{x^2 + 8x + 15} - \dfrac{5}{x^2 + 12x + 35}$$

$$= \dfrac{x}{(x + 3)(x + 5)} - \dfrac{5}{(x + 5)(x + 7)}$$
Factor each denominator. The LCD is $(x + 3)(x + 5)(x + 7)$.

$$= \frac{x}{(x+3)(x+5)} \cdot \frac{x+7}{x+7} - \frac{5}{(x+5)(x+7)} \cdot \frac{x+3}{x+3}$$

Multiply to rewrite each expression as an equivalent rational expression that has the LCD as its denominator.

$$= \frac{x^2+7x}{(x+3)(x+5)(x+7)} - \frac{5x+15}{(x+5)(x+7)(x+3)}$$

Distribute in each numerator.

$$= \frac{(x^2+7x)-(5x+15)}{(x+3)(x+5)(x+7)}$$

Subtract the numerators, writing the sum over the LCD.

$$= \frac{x^2+7x-5x-15}{(x+3)(x+5)(x+7)}$$

Distribute.

$$= \frac{x^2+2x-15}{(x+3)(x+5)(x+7)}$$

Combine like terms.

$$= \frac{\overset{1}{\cancel{(x+5)}}(x-3)}{(x+3)\underset{1}{\cancel{(x+5)}}(x+7)}$$

Factor the numerator and divide out the common factor.

$$= \frac{x-3}{(x+3)(x+7)}$$

Simplify.

We must be careful that we subtract the entire second numerator, not just the first term. In other words, the subtraction must change the sign of each term in the second numerator before we combine like terms. The use of parentheses when subtracting the two numerators will help with this.

Quick Check **5** Add $\dfrac{x}{x^2-6x+5} + \dfrac{1}{x^2+2x-3}$.

EXAMPLE 7 Subtract $\dfrac{x-6}{x^2-1} - \dfrac{5}{x^2-4x+3}$.

Solution

Notice that the first numerator contains a binomial. We must be careful when multiplying to create equivalent rational expressions with the same denominator. We begin by factoring each denominator to find the LCD.

$$\frac{x-6}{x^2-1} - \frac{5}{x^2-4x+3}$$

$$= \frac{x-6}{(x+1)(x-1)} - \frac{5}{(x-3)(x-1)}$$

Factor each denominator. The LCD is $(x+1)(x-1)(x-3)$.

$$= \frac{x-6}{(x+1)(x-1)} \cdot \frac{x-3}{x-3} - \frac{5}{(x-3)(x-1)} \cdot \frac{x+1}{x+1}$$

Multiply to rewrite each expression as an equivalent rational expression with the LCD as its denominator.

$$= \frac{x^2 - 9x + 18}{(x+1)(x-1)(x-3)} - \frac{5x+5}{(x-3)(x-1)(x+1)}$$

Distribute in each numerator.

$$= \frac{(x^2 - 9x + 18) - (5x+5)}{(x+1)(x-1)(x-3)}$$

Subtract the numerators, writing the sum over the LCD.

$$= \frac{x^2 - 9x + 18 - 5x - 5}{(x+1)(x-1)(x-3)}$$

Distribute.

$$= \frac{x^2 - 14x + 13}{(x+1)(x-1)(x-3)}$$

Combine like terms.

$$= \frac{(x-1)(x-13)}{(x+1)(x-1)(x-3)}$$

Factor the numerator and divide out the common factor.

$$= \frac{x-13}{(x+1)(x-3)}$$

Simplify.

Quick Check 6 Add $\dfrac{x+5}{x^2 + 8x + 12} + \dfrac{x+9}{x^2 - 36}$.

EXERCISES 7.4

Find the LCD of the given rational expressions.

1. $\dfrac{5}{2a}, \dfrac{1}{3a}$

2. $\dfrac{9}{8x^2}, \dfrac{7}{6x^3}$

3. $\dfrac{6}{x-9}, \dfrac{8}{x+5}$

4. $\dfrac{x}{x-7}, \dfrac{x+2}{2x+3}$

5. $\dfrac{2}{x^2-9}, \dfrac{x+7}{x^2-9x+18}$

6. $\dfrac{x-6}{x^2-3x-28}, \dfrac{x+1}{x^2-x-20}$

7. $\dfrac{3}{x^2+7x+10}, \dfrac{x-1}{x^2+4x+4}$

8. $\dfrac{x-10}{x^2-12x+36}, \dfrac{8}{x^2-36}$

Add or subtract.

9. $\dfrac{3x}{4} + \dfrac{x}{6}$

10. $\dfrac{7x}{12} - \dfrac{3x}{20}$

11. $\dfrac{8}{5a} - \dfrac{2}{3a}$

12. $\dfrac{1}{8b} + \dfrac{6}{7b}$

13. $\dfrac{4}{m^2 n^3} + \dfrac{5}{m^4 n}$

14. $\dfrac{6}{st^5} + \dfrac{7}{t^6}$

15. $\dfrac{5}{x+2} + \dfrac{3}{x+4}$

16. $\dfrac{7}{x+5} + \dfrac{2}{x-8}$

17. $\dfrac{10}{x+3} - \dfrac{4}{x-3}$

18. $\dfrac{6}{x} - \dfrac{9}{x - 4}$

19. $\dfrac{6}{x + 4} + \dfrac{3}{x - 2}$

20. $\dfrac{5}{x + 5} - \dfrac{5}{x - 6}$

21. $\dfrac{3}{(x + 5)(x + 2)} + \dfrac{4}{(x + 2)(x - 2)}$

22. $\dfrac{8}{(x - 6)(x - 10)} - \dfrac{6}{(x - 8)(x - 10)}$

23. $\dfrac{7}{(x + 1)(x - 6)} - \dfrac{3}{(x - 3)(x - 6)}$

24. $\dfrac{5}{(x - 8)(x + 2)} + \dfrac{1}{(x + 2)(x + 4)}$

25. $\dfrac{4}{x^2 - 6x - 27} - \dfrac{2}{x^2 - 12x + 27}$

26. $\dfrac{9}{x^2 - 11x + 28} + \dfrac{3}{x^2 - 7x + 12}$

27. $\dfrac{5}{x^2 - 15x + 50} + \dfrac{9}{x^2 - x - 20}$

28. $\dfrac{4}{x^2 + 12x} - \dfrac{2}{x^2 + 18x + 72}$

29. $\dfrac{3}{x^2 - 13x + 40} - \dfrac{4}{x^2 - 12x + 32}$

30. $\dfrac{2}{x^2 + x - 6} - \dfrac{4}{x^2 - 4x - 21}$

31. $\dfrac{5}{2x^2 + 3x - 2} + \dfrac{7}{2x^2 - 9x + 4}$

32. $\dfrac{7}{2x^2 - 15x - 27} - \dfrac{3}{2x^2 - 3x - 9}$

33. $\dfrac{x + 6}{x^2 - 9} - \dfrac{1}{x^2 + 7x + 12}$

34. $\dfrac{x - 2}{x^2 + x - 56} + \dfrac{4}{x^2 + 10x + 16}$

35. $\dfrac{x - 3}{x^2 - x - 20} + \dfrac{6}{x^2 - 2x - 24}$

36. $\dfrac{x - 10}{x^2 - 2x - 3} - \dfrac{3}{x^2 - 9x + 18}$

37. $\dfrac{x - 7}{x^2 + 5x - 24} + \dfrac{3}{x^2 - 9}$

38. $\dfrac{x + 5}{x^2 + 2x - 24} - \dfrac{4}{x^2 + 16x + 60}$

39. $\dfrac{x + 14}{x^2 + 5x - 50} - \dfrac{8}{x^2 + 15x + 50}$

40. $\dfrac{x - 6}{x^2 - 16} - \dfrac{2}{x^2 + 8x + 16}$

41. $\dfrac{x - 3}{x^2 + 17x + 70} + \dfrac{10}{x^2 + 10x}$

42. $\dfrac{x + 9}{x^2 - 8x} + \dfrac{8}{x^2 - 16x + 64}$

43. $\dfrac{x + 10}{x^2 + 5x - 66} - \dfrac{7}{x^2 + 20x + 99}$

44. $\dfrac{x - 4}{x^2 + 20x + 96} - \dfrac{14}{x^2 + 6x - 16}$

Answer in complete sentences.

45. Explain how to find the LCD of two rational expressions. Use an example to illustrate the process.

46. Explain how to add two rational expressions with unlike denominators. Use an example to illustrate the process.

47. Explain how to subtract two rational expressions with unlike denominators. Use an example to illustrate the process.

48. Here is a student's solution to a problem on an exam. Describe the student's error, and provide the correct solution. Assuming that the problem was worth 10 points, how many points would you give to the student for their solution? Explain your reasoning.

$$\frac{5}{(x+3)(x+2)} + \frac{7}{(x+3)(x+6)}$$

$$= \frac{5}{(x+3)(x+2)} \cdot \frac{x+6}{x+6} + \frac{7}{(x+3)(x+6)} \cdot \frac{x+2}{x+2}$$

$$= \frac{5(x+6) + 7(x+2)}{(x+3)(x+2)(x+6)}$$

$$= \frac{5(\overset{1}{\cancel{x+6}}) + 7(\overset{1}{\cancel{x+2}})}{(x+3)\underset{1}{(\cancel{x+2})}\underset{1}{(\cancel{x+6})}}$$

$$= \frac{12}{x+3}$$

Study Tip REVISITED   Some of the most valuable materials for preparing for a cumulative exam are the materials provided by your instructor, such as a review sheet for the final or a practice final, or a list of problems from the text to review. If your instructor feels that these problems are important enough for you to review, then they are definitely important enough to appear on the cumulative exam.

As you work through your study schedule, try to solve your instructor's problems when you review that section or chapter. Your performance on these problems will help you to know which topics you understand and which topics require further study.

One mistake that students make is to try to memorize the solution of each and every problem; this will not help you solve a similar problem on the exam. If you are having trouble with a certain problem, write down the steps to solve that problem on a note card. Review your note cards for a short period each day before the exam. This will help you understand how to solve a problem, and it increases your chance of solving a similar problem on the exam.

7.5

COMPLEX FRACTIONS

Objectives

1 Simplify complex numerical fractions.
2 Simplify complex fractions containing variables.

Complex Fractions

A **complex fraction** is a fraction or rational expression containing one or more fractions in its numerator or denominator. Here are some examples:

$$\frac{\dfrac{1}{2}+\dfrac{5}{3}}{\dfrac{10}{3}-\dfrac{7}{4}} \qquad \frac{\dfrac{x+5}{x-9}}{1-\dfrac{5}{x}} \qquad \frac{\dfrac{1}{2}+\dfrac{1}{x}}{\dfrac{1}{4}-\dfrac{1}{x^2}} \qquad \frac{1+\dfrac{3}{x}-\dfrac{28}{x^2}}{1+\dfrac{7}{x}}$$

Objective 1 Simplify complex numerical fractions. To simplify a complex fraction, we must rewrite it in such a way that its numerator and denominator do not contain fractions. This can be done by finding the LCD of all fractions within the complex fraction and then multiplying the numerator and denominator by this LCD. This will clear the fractions within the complex fraction. We finish by simplifying the resulting rational expression, if possible.

We begin with a complex fraction made up of numerical fractions.

EXAMPLE 1   Simplify the complex fraction $\dfrac{1+\dfrac{4}{3}}{\dfrac{2}{3}+\dfrac{11}{4}}$.

Solution

The LCD of the three denominators (3, 3, and 4) is 12, so we begin by multiplying the complex fraction by $\frac{12}{12}$. Notice that when we multiply by $\frac{12}{12}$, we are really multiplying by 1, which does not change the value of the original expression.

$$\frac{1+\dfrac{4}{3}}{\dfrac{2}{3}+\dfrac{11}{4}} = \frac{12}{12}\cdot\frac{1+\dfrac{4}{3}}{\dfrac{2}{3}+\dfrac{11}{4}}$$ Multiply the numerator and denominator by the LCD.

$$= \frac{12\cdot 1 + \overset{4}{\cancel{12}}\cdot\dfrac{4}{\cancel{3}}}{\underset{1}{\overset{4}{\cancel{12}}}\cdot\dfrac{2}{\cancel{3}} + \underset{1}{\overset{3}{\cancel{12}}}\cdot\dfrac{11}{\cancel{4}}}$$ Distribute and divide out common factors.

$$= \frac{12+16}{8+33}$$ Multiply.

$$= \frac{28}{41}$$ Simplify the numerator and denominator.

There is another method for simplifying complex fractions. We can rewrite the numerator as a single fraction by adding $1 + \frac{4}{3}$, which would equal $\frac{7}{3}$. We can also rewrite the denominator as a single fraction by adding $\frac{2}{3} + \frac{11}{4}$, which would equal $\frac{41}{12}$.

$$\frac{1 + \dfrac{4}{3}}{\dfrac{2}{3} + \dfrac{11}{4}} = \frac{\dfrac{7}{3}}{\dfrac{41}{12}}$$

Then we can rewrite $\dfrac{\dfrac{7}{3}}{\dfrac{41}{12}}$ as a division problem $\left(\dfrac{7}{3} \div \dfrac{41}{12} \right)$ and simplify from there. This method would also produce a result of $\dfrac{28}{41}$.

In most of the remaining examples, we will use the LCD method, as the technique is somewhat similar to the technique used to solve rational equations in the next section.

Quick Check 1
Simplify the complex fraction $\dfrac{\dfrac{2}{5} + \dfrac{3}{8}}{\dfrac{1}{4} + \dfrac{7}{10}}$.

Simplifying Complex Fractions Containing Variables

Objective 2 **Simplify complex fractions containing variables.**

EXAMPLE 2   Simplify $\dfrac{1 - \dfrac{9}{x^2}}{1 + \dfrac{3}{x}}$.

Solution

The LCD for the two simple fractions with denominators x and x^2 is x^2, so we will begin by multiplying the complex fraction by $\dfrac{x^2}{x^2}$.

$$\dfrac{1 - \dfrac{9}{x^2}}{1 + \dfrac{3}{x}} = \dfrac{x^2}{x^2} \cdot \dfrac{1 - \dfrac{9}{x^2}}{1 + \dfrac{3}{x}}$$

Multiply the numerator and denominator by the LCD.

$$= \dfrac{x^2 \cdot 1 - \overset{1}{\cancel{x^2}} \cdot \dfrac{9}{\underset{1}{\cancel{x^2}}}}{x^2 \cdot 1 + \overset{x}{\cancel{x^2}} \cdot \dfrac{3}{\underset{1}{\cancel{x}}}}$$

Distribute and divide out common factors, clearing the fractions.

$$= \dfrac{x^2 - 9}{x^2 + 3x}$$

Multiply.

$$= \dfrac{\overset{1}{\cancel{(x + 3)}}(x - 3)}{x\underset{1}{\cancel{(x + 3)}}}$$

Factor the numerator and denominator and divide out the common factor.

$$= \dfrac{x - 3}{x}$$

Simplify.

Once we have cleared the fractions, resulting in a rational expression such as $\dfrac{x^2 - 9}{x^2 + 3x}$, we simplify the rational expression by factoring and dividing out common factors.

Quick Check 2 Simplify $\dfrac{1 + \dfrac{3}{x} - \dfrac{10}{x^2}}{1 - \dfrac{2}{x}}$.

EXAMPLE 3   Simplify $\dfrac{\dfrac{1}{6} + \dfrac{1}{x}}{\dfrac{1}{36} - \dfrac{1}{x^2}}$.

Solution

We begin by multiplying the numerator and denominator by the LCD of all denominators. In this case, the LCD is $36x^2$.

$$\dfrac{\dfrac{1}{6} + \dfrac{1}{x}}{\dfrac{1}{36} - \dfrac{1}{x^2}} = \dfrac{36x^2}{36x^2} \cdot \dfrac{\dfrac{1}{6} + \dfrac{1}{x}}{\dfrac{1}{36} - \dfrac{1}{x^2}}$$

Multiply the numerator and denominator by the LCD.

$$= \dfrac{\overset{6}{\cancel{36}}x^2 \cdot \dfrac{1}{\cancel{6}} + 36\overset{x}{\cancel{x^2}} \cdot \dfrac{1}{\cancel{x}}}{\underset{1}{\cancel{36}}x^2 \cdot \dfrac{1}{\underset{1}{\cancel{36}}} - 36\overset{1}{\underset{1}{\cancel{x^2}}} \cdot \dfrac{1}{\underset{1}{\cancel{x^2}}}}$$

Distribute and divide out common factors.

$$= \dfrac{6x^2 + 36x}{x^2 - 36}$$

Multiply.

$$= \dfrac{6x\cancel{(x + 6)}}{\cancel{(x + 6)}(x - 6)}$$

Factor the numerator and denominator and divide out the common factor.

$$= \dfrac{6x}{x - 6}$$

Simplify.

Quick Check 3

Simplify $\dfrac{\dfrac{1}{64} - \dfrac{1}{x^2}}{\dfrac{1}{8} - \dfrac{1}{x}}$.

EXAMPLE 4   Simplify $\dfrac{\dfrac{3}{x + 1} - \dfrac{2}{x + 2}}{\dfrac{x}{x + 2} + \dfrac{6}{x + 1}}$.

Solution

The LCD for the four simple fractions is $(x + 1)(x + 2)$.

$$\dfrac{\dfrac{3}{x + 1} - \dfrac{2}{x + 2}}{\dfrac{x}{x + 2} + \dfrac{6}{x + 1}} = \dfrac{(x + 1)(x + 2)}{(x + 1)(x + 2)} \cdot \dfrac{\dfrac{3}{x + 1} - \dfrac{2}{x + 2}}{\dfrac{x}{x + 2} + \dfrac{6}{x + 1}}$$

Multiply the numerator and denominator by the LCD.

$$= \frac{(x+1)(x+2) \cdot \dfrac{3}{x+1} - (x+1)(x+2) \cdot \dfrac{2}{x+2}}{(x+1)(x+2) \cdot \dfrac{x}{x+2} + (x+1)(x+2) \cdot \dfrac{6}{x+1}}$$

Distribute and divide out common factors, clearing the fractions.

$$= \frac{3(x+2) - 2(x+1)}{x(x+1) + 6(x+2)}$$

Multiply.

$$= \frac{3x + 6 - 2x - 2}{x^2 + x + 6x + 12}$$

Distribute.

$$= \frac{x+4}{x^2 + 7x + 12}$$

Combine like terms.

$$= \frac{x+4}{(x+3)(x+4)}$$

Factor the denominator and divide out the common factor.

$$= \frac{1}{x+3}$$

Simplify.

Quick Check 4

Simplify $\dfrac{\dfrac{6}{x-4} + \dfrac{5}{x+7}}{\dfrac{x+2}{x-4}}$.

EXAMPLE 5   Simplify $\dfrac{\dfrac{x^2 + 5x - 14}{x^2 - 9}}{\dfrac{x^2 - 6x + 8}{x^2 + 8x + 15}}$.

Solution

In this case, it will be easier to rewrite the complex fraction as a division problem rather than multiplying the numerator and denominator by the LCD. This is a wise idea when we have a complex fraction with a single rational expression in its numerator and a single rational expression in its denominator. We have used this method when dividing rational expressions.

$$\frac{\dfrac{x^2 + 5x - 14}{x^2 - 9}}{\dfrac{x^2 - 6x + 8}{x^2 + 8x + 15}}$$

$$= \frac{x^2 + 5x - 14}{x^2 - 9} \div \frac{x^2 - 6x + 8}{x^2 + 8x + 15}$$

Rewrite as a division problem.

$$= \frac{x^2 + 5x - 14}{x^2 - 9} \cdot \frac{x^2 + 8x + 15}{x^2 - 6x + 8}$$

Invert the divisor and multiply.

$$= \frac{(x+7)(x-2)}{(x+3)(x-3)} \cdot \frac{(x+3)(x+5)}{(x-2)(x-4)}$$

Factor each numerator and denominator.

$$= \frac{(x+7)(x-2)}{(x+3)(x-3)} \cdot \frac{(x+3)(x+5)}{(x-2)(x-4)}$$

Divide out common factors.

$$= \frac{(x+5)(x+7)}{(x-3)(x-4)}$$

Simplify.

Quick Check 5

Simplify $\dfrac{\dfrac{x^2 - 2x - 48}{x^2 + 7x - 30}}{\dfrac{x^2 + 7x + 6}{x^2 - 5x + 6}}$.

Simplify the complex fraction.

1. $\dfrac{\dfrac{2}{5} - \dfrac{1}{4}}{\dfrac{9}{10} + \dfrac{5}{2}}$

2. $\dfrac{\dfrac{3}{7} + \dfrac{2}{3}}{\dfrac{16}{21} - \dfrac{2}{7}}$

3. $\dfrac{2 - \dfrac{3}{8}}{\dfrac{5}{4} + \dfrac{1}{3}}$

4. $\dfrac{\dfrac{5}{6} - 3}{\dfrac{3}{4} + \dfrac{11}{12}}$

5. $\dfrac{x + \dfrac{3}{5}}{x + \dfrac{4}{7}}$

6. $\dfrac{x - \dfrac{2}{9}}{x + \dfrac{5}{4}}$

7. $\dfrac{6 + \dfrac{15}{x}}{x + \dfrac{5}{2}}$

8. $\dfrac{x + \dfrac{3}{7}}{14 + \dfrac{6}{x}}$

9. $\dfrac{14 - \dfrac{4}{x}}{21 - \dfrac{6}{x}}$

10. $\dfrac{10 + \dfrac{8}{x}}{25 + \dfrac{20}{x}}$

11. $\dfrac{2 + \dfrac{8}{x}}{1 - \dfrac{16}{x^2}}$

12. $\dfrac{5 - \dfrac{20}{x^2}}{1 - \dfrac{2}{x}}$

13. $\dfrac{\dfrac{6}{x + 3} - \dfrac{2}{x + 5}}{\dfrac{x + 6}{x + 3}}$

14. $\dfrac{\dfrac{7}{x + 4} - \dfrac{1}{x - 2}}{\dfrac{5x - 15}{x - 2}}$

15. $\dfrac{\dfrac{4}{x} + \dfrac{8}{x - 6}}{1 + \dfrac{4}{x - 6}}$

16. $\dfrac{\dfrac{3}{x - 1} - \dfrac{2}{x + 2}}{1 + \dfrac{6}{x + 2}}$

17. $\dfrac{1 + \dfrac{4}{x} - \dfrac{32}{x^2}}{1 + \dfrac{13}{x} + \dfrac{40}{x^2}}$

18. $\dfrac{1 - \dfrac{9}{x} + \dfrac{18}{x^2}}{1 + \dfrac{2}{x} - \dfrac{15}{x^2}}$

19. $\dfrac{\dfrac{8}{x^2} - \dfrac{8}{x} + 2}{1 - \dfrac{4}{x^2}}$

20. $\dfrac{3 - \dfrac{18}{x}}{\dfrac{1}{x} - \dfrac{42}{x^2} + 1}$

21. $\dfrac{\dfrac{x^2 + 9x + 14}{x^2 - 2x - 63}}{\dfrac{x^2 - 3x - 10}{x^2 - 15x + 54}}$

22. $\dfrac{\dfrac{x^2 - x - 20}{x^2 - 9}}{\dfrac{x^2 + 12x + 32}{x^2 - 4x - 21}}$

23. $\dfrac{\dfrac{x^2 + 2x}{x^2 - 11x + 24}}{\dfrac{x^2 + 11x + 18}{x^2 - 7x + 12}}$

24. $\dfrac{\dfrac{x^2 + 19x + 88}{x^2 - 4x - 5}}{\dfrac{x^2 + 5x - 66}{x^2 + x - 30}}$

25. $\dfrac{\dfrac{1}{x^2 + 11x + 10}}{\dfrac{1}{x^2 - 6x - 7}}$

26. $\dfrac{\dfrac{1}{x^2 - 16x + 60}}{\dfrac{1}{x^2 - 5x - 50}}$

Simplify the given rational expression, using the techniques developed in Sections 7.1 through 7.5.

27. $\dfrac{x^2 + 4x}{x^2 - x - 42} + \dfrac{x - 6}{x^2 - x - 42}$

28. $\dfrac{x^2 - 14x + 40}{x^2 + 4x + 3} \div \dfrac{x^2 - 6x + 8}{x^2 + 10x + 9}$

29. $\dfrac{x + 3}{x^2 - 2x - 24} + \dfrac{5}{x^2 - 8x + 12}$

30. $\dfrac{6}{x^2 - 2x - 35} - \dfrac{4}{x^2 + 2x - 15}$

31. $\dfrac{x^2 + 7x}{x^2 - 11x + 30} \div \dfrac{x^2 + 12x + 35}{x^2 - 3x - 10}$

32. $\dfrac{x^2 - 36}{x^2 + 12x + 36}$

33. $\dfrac{x + 5}{x^2 - 49} - \dfrac{6}{x^2 - 7x}$

34. $\dfrac{x^2 + 3x - 40}{x^2 + x - 6} \cdot \dfrac{x^2 - 7x - 30}{x^2 + 12x + 32}$

35. $\dfrac{5x}{x - 2} + \dfrac{10}{2 - x}$

36. $\dfrac{1 - \dfrac{11}{x} + \dfrac{18}{x^2}}{1 - \dfrac{2}{x} - \dfrac{63}{x^2}}$

37. $\dfrac{\dfrac{7}{x + 2} - \dfrac{2}{x - 3}}{\dfrac{x - 5}{x + 2}}$

38. $\dfrac{3}{x^2 - x - 2} + \dfrac{5}{x^2 + 7x + 6}$

39. $\dfrac{3}{x^2 + 2x - 80} - \dfrac{1}{x^2 - 11x + 24}$

40. $\dfrac{x^2 + 6x - 16}{x^2 - 4x - 45} \cdot \dfrac{x^2 + 8x + 15}{x^2 - 4x + 4}$

41. $\dfrac{\dfrac{1}{8} + \dfrac{1}{x}}{\dfrac{1}{64} - \dfrac{1}{x^2}}$

42. $\dfrac{x^2 + 9x}{x - 5} - \dfrac{6x + 50}{5 - x}$

43. $\dfrac{8}{x^2 + 7x + 12} + \dfrac{4}{x^2 + 10x + 24}$

44. $\dfrac{8x}{2x - 5} + \dfrac{20}{5 - 2x}$

45. $\dfrac{x^2 + 18x + 77}{x^2 - 4x - 32} \cdot \dfrac{x^2 + 6x + 8}{x^2 + 9x + 14}$

46. $\dfrac{x + 3}{x^2 - 2x - 3} + \dfrac{3}{x^2 - 8x + 15}$

47. $\dfrac{x^2 + 7x}{x^2 - 2x - 3} - \dfrac{4x + 18}{x^2 - 2x - 3}$

48. $\dfrac{x + \dfrac{5}{8}}{x - \dfrac{7}{2}}$

49. $\dfrac{x^2 - 10x}{x^2 + 10x - 11} \div \dfrac{x^2 - 17x + 70}{x^2 + 4x - 5}$

50. $\dfrac{x + 5}{x^2 - 3x + 2} - \dfrac{8}{x^2 - 8x + 12}$

Answer in complete sentences.

51. Explain what a complex fraction is. Compare and contrast complex fractions and the rational expressions found in Section 7.1.

52. One method for simplifying complex fractions is to multiply the numerator and denominator by the LCD of the fractions within the complex fraction. Explain how to simplify a complex fraction using this method. Give an example to illustrate the process.

53. Refer to exercise 52. A second method is to rewrite the complex fraction as one rational expression divided by another rational expression. When do you feel that this is the most efficient way to simplify a complex fraction? Explain how to simplify a complex fraction using this method. Give an example to illustrate the process.

Study Tip **Revisited**   Use your cumulative review exercises to prepare for a cumulative exam. These exercise sets contain problems representative of the material covered up to that point in the text.

After you have been preparing for a while, try these exercises without referring to your notes, note cards, or the text. In this way, these problems can be used to determine which topics require more study. If you made a mistake while solving a problem, make note of the mistake and how to avoid it in the future. If there are problems that you do not recognize or know how to begin, then these topics will require additional studying. Ask your instructor or a tutor for help.

Objectives

1 **Solve rational equations.**

2 **Solve literal equations.**

Solving Rational Equations

Objective 1 **Solve rational equations.** In this section, we learn how to solve **rational equations,** which are equations containing at least one rational expression. The main goal is to rewrite the equation as an equivalent equation that does not contain a rational expression. We then solve the equation using methods developed in earlier chapters.

In Chapter 2, we learned how to solve an equation containing fractions such as the equation $\frac{1}{4}x - \frac{3}{5} = \frac{9}{10}$. We began by finding the LCD of all fractions and then multiplied both sides of the equation by that LCD to clear the equation of fractions. We will employ the same technique in this section. There is a major difference, though, when solving equations containing a variable in a denominator. Occasionally, we will find a solution that causes one of the rational expressions in the equation to be undefined. If a denominator of a rational expression is equal to 0 when the value of a solution is substituted for the variable, then the solution must be omitted and is called an **extraneous solution.** We must check each solution that we find to make sure that it is not an extraneous solution.

Solving Rational Equations

1. Find the LCD of all denominators in the equation.
2. Multiply both sides of the equation by the LCD to clear the equation of fractions.
3. Solve the resulting equation.
4. Check for extraneous solutions.

EXAMPLE ▶ 1   Solve $\dfrac{4}{x} + \dfrac{1}{3} = \dfrac{5}{6}$.

Solution

We begin by finding the LCD of these three fractions, which is $6x$. Now we multiply both sides of the equation by the LCD to clear the equation of fractions. Once this has been done, we can solve the resulting equation.

$$\frac{4}{x} + \frac{1}{3} = \frac{5}{6}$$

$$6x\left(\frac{4}{x} + \frac{1}{3}\right) = 6x \cdot \frac{5}{6}$$ Multiply both sides of the equation by the LCD.

$$6\overset{1}{\cancel{x}} \cdot \frac{4}{\underset{1}{\cancel{x}}} + \overset{2}{\cancel{6}}x \cdot \frac{1}{\underset{1}{\cancel{3}}} = \overset{1}{\cancel{6}}x \cdot \frac{5}{\underset{1}{\cancel{6}}}$$ Distribute and divide out common factors.

$$24 + 2x = 5x$$ Multiply. The resulting equation is linear.

$$24 = 3x$$ Subtract   to collect all variable terms on one side of the equation.

$$8 = x$$ Divide both sides by 3.

Check:

$$\frac{4}{(8)} + \frac{1}{3} = \frac{5}{6}$$ Substitute 8 for x.

$$\frac{1}{2} + \frac{1}{3} = \frac{5}{6}$$ Simplify the fraction $\frac{4}{8}$. The LCD of these fractions is 6.

$$\frac{3}{6} + \frac{2}{6} = \frac{5}{6}$$ Write each fraction with a common denominator of 6.

$$\frac{5}{6} = \frac{5}{6}$$ Add.

Since $x = 8$ does not make any rational expression in the original equation undefined, this value is a solution. The solution set is $\{8\}$.

Quick Check 1

Solve $\dfrac{6}{x} - \dfrac{1}{8} = \dfrac{7}{40}$.

When checking whether a solution is an extraneous solution, we need only determine whether the solution causes the LCD to equal 0. If the LCD is equal to 0 for this solution, then one or more rational expressions are undefined and the solution is an extraneous solution. Also, if the LCD is equal to 0, then we have multiplied both sides of the equation by 0. The multiplication property of equality says that we can multiply both sides of an equation by any *nonzero* number without affecting the equality of both sides. In the previous example, the only solution that could possibly be an extraneous solution is $x = 0$, because that is the only value of x for which the LCD is equal to 0.

EXAMPLE 2   Solve $x - 5 - \dfrac{36}{x} = 0$.

Solution

The LCD in this example is x. The LCD is equal to 0 only if $x = 0$. If we find that $x = 0$ is a solution, then we must omit that solution as an extraneous solution.

$$x - 5 - \frac{36}{x} = 0$$

$$x \cdot \left(x - 5 - \frac{36}{x} \right) = x \cdot 0$$ Multiply each side of the equation by the LCD, x.

$$x \cdot x - x \cdot 5 - \overset{1}{\cancel{x}} \cdot \frac{36}{\underset{1}{\cancel{x}}} = 0$$ Distribute and divide out common factors.

$$x^2 - 5x - 36 = 0$$ Multiply. The resulting equation is quadratic.

$$(x - 9)(x + 4) = 0$$ Factor.

$$x = 9 \quad \text{or} \quad x = -4$$ Set each factor equal to 0 and solve.

Quick Check 2

Solve $1 = \dfrac{5}{x} + \dfrac{24}{x^2}$.

You may verify that neither solution causes the LCD to equal 0. The solution set is $\{9, -4\}$.

A Word of Caution   When solving a rational equation, the use of the LCD is completely different than when we are adding or subtracting rational expressions. We use the LCD to clear the denominators of the rational expressions when solving a rational equation. When adding or subtracting rational expressions, we rewrite each expression as an equivalent expression whose denominator is the LCD.

EXAMPLE 3   Solve $\dfrac{x}{x+2} - 5 = \dfrac{3x+4}{x+2}$.

Solution

The LCD is $x + 2$, so we will begin to solve this equation by multiplying both sides of the equation by $x + 2$.

$$\frac{x}{x+2} - 5 = \frac{3x+4}{x+2}$$ The LCD is $x + 2$.

$$(x+2)\left(\frac{x}{x+2} - 5\right) = (x+2) \cdot \frac{3x+4}{x+2}$$ Multiply both sides by the LCD.

$$\cancel{(x+2)} \cdot \frac{x}{\cancel{x+2}} - (x+2) \cdot 5 = \cancel{(x+2)} \cdot \frac{3x+4}{\cancel{x+2}}$$ Distribute and divide out common factors.

$$x - 5x - 10 = 3x + 4$$ Multiply. The resulting equation is linear.

$$-4x - 10 = 3x + 4$$ Combine like terms.

$$-10 = 7x + 4$$ Add $4x$.

$$-14 = 7x$$ Subtract 4.

$$-2 = x$$ Divide both sides by 7.

The LCD is equal to 0 when $x = -2$, and two rational expressions in the original equation are undefined when $x = -2$. This solution is an extraneous solution, and since there are no other solutions, this equation has no solution. Recall that we write the solution set as $\varnothing$ when there is no solution.

> **Quick Check 3**
> **Solve**
> $\dfrac{7}{x-4} + 3 = \dfrac{2x-1}{x-4}$.

EXAMPLE 4   Solve $\dfrac{x+9}{x^2+9x+8} = \dfrac{2}{x^2+2x-48}$.

Solution

Again, we begin by factoring the denominators to find the LCD.

$$\frac{x+9}{x^2+9x+8} = \frac{2}{x^2+2x-48}$$

$$\frac{x+9}{(x+1)(x+8)} = \frac{2}{(x+8)(x-6)}$$ The LCD is $(x+1)(x+8)(x-6)$.

$$\cancel{(x+1)}\cancel{(x+8)}(x-6) \cdot \frac{x+9}{\cancel{(x+1)}\cancel{(x+8)}} = (x+1)\cancel{(x+8)}\cancel{(x-6)} \cdot \frac{2}{\cancel{(x+8)}\cancel{(x-6)}}$$

Multiply by the LCD. Divide out common factors.

$$(x-6)(x+9) = 2(x+1)$$ Multiply remaining factors.

$$x^2 + 3x - 54 = 2x + 2$$ Multiply.

$$x^2 + x - 56 = 0$$ Collect all terms on the left side. The resulting equation is quadratic.

$$(x+8)(x-7) = 0$$ Factor.

$$x = -8 \quad \text{or} \quad x = 7$$ Set each factor equal to 0 and solve.

The solution $x = -8$ is an extraneous solution, it makes the LCD equal to 0. It is left to the reader to verify that the solution $x = 7$ checks. The solution set for this equation is $\{7\}$.

Quick Check **4** Solve $\dfrac{x+2}{x^2 - 3x - 54} = \dfrac{2}{x^2 - 12x + 27}$.

EXAMPLE 5 Solve $\dfrac{x+10}{x^2 + 4x - 5} - \dfrac{1}{x - 3} = \dfrac{x-6}{x^2 - 4x + 3}$.

Solution

$\dfrac{x+10}{x^2 + 4x - 5} - \dfrac{1}{x - 3} = \dfrac{x-6}{x^2 - 4x + 3}$

$\dfrac{x+10}{(x+5)(x-1)} - \dfrac{1}{x - 3} = \dfrac{x-6}{(x-1)(x-3)}$ The LCD is $(x+5)(x-1)(x-3)$.

$(x+5)(x-1)(x-3)\left(\dfrac{x+10}{(x+5)(x-1)} - \dfrac{1}{x-3}\right) = (x+5)(x-1)(x-3) \cdot \dfrac{x-6}{(x-1)(x-3)}$

Multiply by the LCD.

$\cancel{(x+5)}^{1}\cancel{(x-1)}^{1}(x-3) \cdot \dfrac{x+10}{\cancel{(x+5)}_{1}\cancel{(x-1)}_{1}} - (x+5)(x-1)\cancel{(x-3)}^{1} \cdot \dfrac{1}{\cancel{(x-3)}_{1}}$

$= (x+5)\cancel{(x-1)}^{1}\cancel{(x-3)}^{1} \cdot \dfrac{x-6}{\cancel{(x-1)}_{1}\cancel{(x-3)}_{1}}$ Distribute and divide out common factors.

$(x-3)(x+10) - (x+5)(x-1) = (x+5)(x-6)$ Multiply remaining factors.

$(x^2 + 7x - 30) - (x^2 + 4x - 5) = x^2 - x - 30$ Multiply.

$x^2 + 7x - 30 - x^2 - 4x + 5 = x^2 - x - 30$ Distribute.

$3x - 25 = x^2 - x - 30$ Combine like terms.

$0 = x^2 - 4x - 5$ Collect all terms on the right side. The resulting equation is quadratic.

$0 = (x+1)(x-5)$ Factor.

$x = -1$ or $x = 5$ Set each factor equal to 0 and solve.

Check to verify that neither solution is extraneous. The solution set is $\{-1, 5\}$.

Quick Check **5** Solve $\dfrac{x+3}{x^2 + x - 12} - \dfrac{3}{x^2 - 2x - 3} = \dfrac{x+7}{x^2 + 5x + 4}$.

Literal Equations

Objective 2 Solve literal equations. Recall that a literal equation is an equation containing two or more variables, and we solve the equation for one of the variables by isolating that variable on one side of the equation. In this section, we will learn how to solve literal equations containing one or more rational expressions.

EXAMPLE ▸ 6   Solve the literal equation $\dfrac{1}{x} + \dfrac{1}{y} = \dfrac{2}{5}$ for x.

Solution

We begin by multiplying by the LCD $(5xy)$ to clear the equation of fractions.

$$\frac{1}{x} + \frac{1}{y} = \frac{2}{5}$$

$$5xy\left(\frac{1}{x} + \frac{1}{y}\right) = 5xy \cdot \frac{2}{5} \qquad \text{Multiply by the LCD, } 5xy.$$

$$5\overset{1}{\cancel{x}}y \cdot \frac{1}{\cancel{x}} + 5x\overset{1}{\cancel{y}} \cdot \frac{1}{\cancel{y}} = \overset{1}{\cancel{5}}xy \cdot \frac{2}{\cancel{5}} \qquad \text{Distribute and divide out common factors.}$$

$$5y + 5x = 2xy \qquad \text{Multiply remaining factors.}$$

Notice that there are two terms that contain the variable we are solving for. We need to collect both of these terms on the same side of the equation and then factor x out of those terms. This will allow us to divide and isolate x.

$$5y + 5x = 2xy$$

$$5y = 2xy - 5x \qquad \begin{array}{l}\text{Subtract } 5x \text{ to collect all terms with } x \\ \text{on the right side of the equation.}\end{array}$$

$$5y = x(2y - 5) \qquad \text{Factor out the common factor } x.$$

$$\frac{5y}{2y - 5} = \frac{x(\overset{1}{\cancel{2y - 5}})}{(\underset{1}{\cancel{2y - 5}})} \qquad \text{Divide both sides by } 2y - 5 \text{ to isolate } x.$$

$$\frac{5y}{2y - 5} = x \qquad \text{Simplify.}$$

$$x = \frac{5y}{2y - 5} \qquad \begin{array}{l}\text{Rewrite with the variable we are solving for} \\ \text{on the left side of the equation.}\end{array}$$

Quick Check **6**

Solve the literal equation $\dfrac{2}{x} + \dfrac{3}{y} = \dfrac{4}{z}$ **for** x**.**

As in Chapter 2, we will rewrite our solution so that the variable we are solving for is on the left side.

EXAMPLE ▸ 7   Solve the literal equation $\dfrac{1}{a + h} = \dfrac{1}{b} - 5$ for a.

Solution

We begin by multiplying both sides of the equation by the LCD, which is $(a + h)b$.

$$\frac{1}{a + h} = \frac{1}{b} - 5$$

$$(a + h)b \cdot \frac{1}{a + h} = (a + h)b\left(\frac{1}{b} - 5\right) \qquad \begin{array}{l}\text{Multiply both sides by the} \\ \text{LCD.}\end{array}$$

$$(\overset{1}{\cancel{a + h}})b \cdot \frac{1}{(\underset{1}{\cancel{a + h}})} = (a + h)\overset{1}{(\cancel{b})} \cdot \frac{1}{\underset{1}{\cancel{b}}} - (a + h)b \cdot 5 \qquad \begin{array}{l}\text{Distribute and divide out} \\ \text{common factors.}\end{array}$$

$$b = a + h - 5b(a + h) \qquad \text{Multiply remaining factors.}$$

$$b = a + h - 5ab - 5bh \qquad \text{Distribute.}$$

$$b - h + 5bh = a - 5ab \qquad \begin{array}{l}\text{Subtract } h \text{ and add } 5bh \text{ so} \\ \text{only the terms containing} \\ a \text{ are on the right side.}\end{array}$$

$b - h + 5bh = a(1 - 5b)$ Factor out the common factor a.

$\dfrac{b - h + 5bh}{1 - 5b} = \dfrac{a(\cancel{1 - 5b})}{\cancel{(1 - 5b)}}$ Divide both sides by $1 - 5b$ to isolate a.

$a = \dfrac{b - h + 5bh}{1 - 5b}$ Rewrite with a on the left side.

EXERCISES 7.6

Solve.

1. $\dfrac{x}{3} - \dfrac{7}{4} = \dfrac{9}{2}$

2. $\dfrac{x}{5} + \dfrac{1}{6} = \dfrac{8}{3}$

3. $\dfrac{3x}{4} + \dfrac{5}{6} = \dfrac{x}{12} + \dfrac{7}{8}$

4. $\dfrac{6x}{7} - \dfrac{9}{2} = \dfrac{x}{4} + \dfrac{3}{14}$

5. $\dfrac{9}{x} - \dfrac{3}{4} = \dfrac{7}{8}$

6. $\dfrac{9}{10} + \dfrac{6}{x} = \dfrac{17}{12}$

7. $\dfrac{1}{x} = \dfrac{21}{80}$

8. $\dfrac{1}{2x} + 7 = \dfrac{1}{3x} + \dfrac{1}{6x}$

9. $x - 5 + \dfrac{12}{x} = 2$

10. $x - \dfrac{25}{x} = 2 + \dfrac{23}{x}$

11. $\dfrac{x}{2} + \dfrac{6}{x} = 4$

12. $\dfrac{x}{3} + 2 - \dfrac{9}{x} = 0$

13. $\dfrac{8x - 3}{x + 7} = \dfrac{2x + 15}{x + 7}$

14. $\dfrac{6x + 13}{x - 1} = \dfrac{4x + 15}{x - 1}$

15. $\dfrac{7x - 11}{5x - 2} = \dfrac{2x - 9}{5x - 2}$

16. $\dfrac{x^2 + 3x}{x + 8} = \dfrac{3x + 64}{x + 8}$

17. $8 + \dfrac{6}{x - 4} = \dfrac{x - 12}{x - 4}$

18. $1 + \dfrac{2x - 7}{x + 3} = \dfrac{4x - 1}{x + 3}$

19. $\dfrac{3x + 4}{x + 2} - 7 = \dfrac{x}{x + 2}$

20. $\dfrac{6x + 7}{5x - 8} - \dfrac{2x - 13}{5x - 8} = 3$

21. $x + \dfrac{x + 11}{x - 6} = \dfrac{8x - 31}{x - 6}$

22. $x + \dfrac{3x - 10}{x + 1} = \dfrac{9x + 14}{x + 1}$

23. $\dfrac{3}{x - 7} = \dfrac{7}{x + 5}$

24. $\dfrac{6}{x + 4} = \dfrac{9}{2x - 3}$

25. $\dfrac{6}{x + 2} = \dfrac{25}{3x + 13}$

26. $\dfrac{2}{x} = \dfrac{11}{4x - 9}$

27. $\dfrac{x + 1}{x + 13} = \dfrac{3}{x + 4}$

28. $\dfrac{x - 3}{x + 9} = \dfrac{6}{x + 2}$

29. $\dfrac{x - 1}{3x - 7} = \dfrac{x + 1}{2x + 7}$

30. $\dfrac{2x}{x + 9} = \dfrac{x - 2}{x - 1}$

31. $\dfrac{3}{x + 2} - \dfrac{1}{x + 1} = \dfrac{x + 3}{x^2 + 3x + 2}$

32. $\dfrac{2}{x} + \dfrac{3}{x + 2} = \dfrac{7x - 8}{x^2 + 2x}$

33. $\dfrac{4}{x+4} + \dfrac{3}{x-4} = \dfrac{24}{x^2-16}$

34. $\dfrac{x}{x+5} + \dfrac{3}{x-7} = \dfrac{36}{x^2-2x-35}$

35. $\dfrac{2}{x^2-x-2} + \dfrac{10}{x^2-2x-3} = \dfrac{x+12}{x^2-x-2}$

36. $\dfrac{20}{x^2+12x+27} - \dfrac{4}{x^2+14x+45} = \dfrac{x+10}{x^2+8x+15}$

37. $\dfrac{x-8}{x-5} + \dfrac{x-9}{x-4} = \dfrac{x+7}{x^2-9x+20}$

38. $\dfrac{x}{x+3} + \dfrac{x-4}{x-3} = \dfrac{9x-5}{x^2-9}$

39. $\dfrac{4}{x^2+4x-5} + \dfrac{x+9}{x^2-1} = \dfrac{41}{x^2+6x+5}$

40. $\dfrac{x+4}{x^2+5x-14} + \dfrac{2}{x^2+3x-10} = \dfrac{38}{x^2+12x+35}$

41. $\dfrac{x+3}{x^2-4x-12} + \dfrac{x-11}{x^2-2x-24} = \dfrac{x+1}{x^2+6x+8}$

42. $\dfrac{x-2}{x^2+13x+40} + \dfrac{x+5}{x^2+7x-8} = \dfrac{x+3}{x^2+4x-5}$

43. $\dfrac{3x-4}{x^2-10x+21} - \dfrac{x-8}{x^2-18x+77} = \dfrac{x-5}{x^2-14x+33}$

44. $\dfrac{5x+4}{x^2+x-90} - \dfrac{1}{x-9} = \dfrac{3x-2}{x^2-100}$

Solve for the specified variable.

45. $L = \dfrac{A}{W}$ for W

46. $b = \dfrac{2A}{h}$ for A

47. $y = \dfrac{x}{2x+5}$ for x

48. $y = \dfrac{3x-7}{2x}$ for x

49. $\dfrac{x}{r} + \dfrac{y}{2r} = 1$ for r

50. $r = \dfrac{d}{t}$ for t

51. $\dfrac{1}{a} + \dfrac{1}{b} = \dfrac{1}{c}$ for b

52. $\dfrac{a}{b} + \dfrac{c}{d} = \dfrac{e}{f}$ for b

53. $m = \dfrac{y-y_1}{x-x_1}$ for x

54. $\dfrac{1}{R} = \dfrac{1}{R_1} + \dfrac{1}{R_2}$ for R

55. $n = \dfrac{n_1+n_2}{n_1 \cdot n_2}$ for n_1

56. $p = \dfrac{x_1+x_2}{n_1+n_2}$ for n_1

Answer in complete sentences.

57. Explain how to solve a rational equation. Use an example to illustrate the process.

58. Explain how to determine whether a solution is an extraneous solution.

59. We use the LCD when we add two rational expressions, as well as when we solve a rational equation. Explain how the LCD is used differently for these two types of problems.

60. A student was asked to solve the rational equation $\dfrac{1}{x} + \dfrac{1}{y} = \dfrac{3}{8}$ for x, and his answer was $x = \dfrac{3xy-8y}{8}$. Explain why this is incorrect, then explain how to solve the equation for x.

> ***Study Tip*** **REVISITED**   Many students have a difficult time with applied problems on a cumulative exam. Some students are unable to recognize what type of problem an applied problem is, and others cannot recall how to start to solve that type of problem. You will find the following strategy helpful:
>
> - Make a list of the different applied problems that you have covered this semester. Create a study sheet for each type of problem.
> - Write down an example or two of each type of problem.
> - List the steps necessary to solve each type of problem.
>
> Review these study sheets frequently as the exam approaches. This should help you identify the applied problems on the exam, as well as to remember how to solve the problems.

7.7

APPLICATIONS OF RATIONAL EQUATIONS

Objectives

1. Solve applied problems involving the reciprocal of a number.
2. Solve applied work-rate problems.
3. Solve applied uniform motion problems.
4. Solve variation problems.

In this section, we will look at applied problems requiring the use of rational equations to solve them. We begin with problems involving reciprocals.

Reciprocals

Objective 1 Solve applied problems involving the reciprocal of a number.

EXAMPLE 1   The sum of the reciprocal of a number and $\frac{1}{3}$ is $\frac{1}{2}$. Find the number.

Solution

There is only one unknown in this problem, and we will let x represent the unknown number.

Unknown
Number: x

The reciprocal of this number can be written as $\frac{1}{x}$. We are told that the sum of this reciprocal and $\frac{1}{3}$ is $\frac{1}{2}$, which leads to the equation $\frac{1}{x} + \frac{1}{3} = \frac{1}{2}$.

$$\frac{1}{x} + \frac{1}{3} = \frac{1}{2} \qquad \text{The LCD is } 6x.$$

$$6x \cdot \left(\frac{1}{x} + \frac{1}{3}\right) = 6x \cdot \frac{1}{2} \qquad \text{Multiply both sides by the LCD.}$$

$$\overset{1}{6\!\!\!/x} \cdot \frac{1}{\overset{}{\underset{1}{\cancel{x}}}} + \overset{2}{6\!\!\!/x} \cdot \frac{1}{\overset{}{\underset{1}{\cancel{3}}}} = \overset{3}{6\!\!\!/x} \cdot \frac{1}{\overset{}{\underset{1}{\cancel{2}}}} \qquad \text{Distribute and divide out common factors.}$$

$$6 + 2x = 3x \qquad \begin{array}{l}\text{Multiply remaining factors. The resulting}\\\text{equation is linear.}\end{array}$$

$$6 = x \qquad \text{Subtract } 2x.$$

Quick Check 1
The sum of the reciprocal of a number and $\frac{3}{8}$ is $\frac{19}{40}$. Find the number.

The unknown number is 6. The reader should verify that $\frac{1}{6} + \frac{1}{3} = \frac{1}{2}$.

EXAMPLE 2   One positive number is four larger than another positive number. If the reciprocal of the smaller number is added to six times the reciprocal of the larger number, the sum is equal to 1. Find the two numbers.

Solution

In this problem, there are two unknown numbers. If we let x represent the smaller number, then we can write the larger number as $x + 4$.

> **Unknowns**
>
> Smaller Number: x
> Larger Number: $x + 4$

The reciprocal of the smaller number is $\dfrac{1}{x}$ and six times the reciprocal of the larger number is $6 \cdot \dfrac{1}{x + 4}$ or $\dfrac{6}{x + 4}$. This leads to the equation $\dfrac{1}{x} + \dfrac{6}{x + 4} = 1$.

$$\frac{1}{x} + \frac{6}{x + 4} = 1 \qquad \text{The LCD is } x(x + 4).$$

$$x(x + 4)\left(\frac{1}{x} + \frac{6}{x + 4}\right) = x(x + 4) \cdot 1 \qquad \text{Multiply both sides by the LCD.}$$

$$\overset{1}{\cancel{(x)}}(x + 4) \cdot \frac{1}{\underset{1}{\cancel{(x)}}} + x\overset{1}{\cancel{(x + 4)}} \cdot \frac{6}{\underset{1}{\cancel{(x + 4)}}} = x(x + 4) \cdot 1 \qquad \begin{array}{l}\text{Distribute and divide out common factors.}\end{array}$$

$$x + 4 + 6x = x(x + 4) \qquad \text{Multiply remaining factors.}$$

$$x + 4 + 6x = x^2 + 4x \qquad \begin{array}{l}\text{Distribute. The resulting equation is quadratic.}\end{array}$$

$$7x + 4 = x^2 + 4x \qquad \text{Combine like terms.}$$

$$0 = x^2 - 3x - 4 \qquad \begin{array}{l}\text{Collect all terms on the right side of the equation.}\end{array}$$

$$0 = (x + 1)(x - 4) \qquad \text{Factor.}$$

$$x = -1 \quad \text{or} \quad x = 4 \qquad \begin{array}{l}\text{Set each factor equal to 0 and solve.}\end{array}$$

Since we were told that the numbers must be positive, we can omit the solution $x = -1$. We now return to the table of unknowns to find the two numbers.

> Smaller Number: $x = 4$
> Larger Number: $x + 4 = 4 + 4 = 8$

The two numbers are 4 and 8. The reader should verify that $\frac{1}{4} + 6 \cdot \frac{1}{8} = 1$.

Quick Check **2**

One positive number is 9 less than another positive number. If two times the reciprocal of the smaller number is added to four times the reciprocal of the larger number, the sum is equal to 1. Find the two numbers.

Work-Rate Problems

Objective **2** **Solve applied work-rate problems.** Now we turn our attention to problems known as **work-rate problems**. These problems usually involve two or more people or objects working together to perform a job, such as Jenny and Kim painting a room together or two copy machines processing an exam. In general, the equation we will be solving corresponds to the following:

| Portion of the job completed by person 1 | + | Portion of the job completed by person 2 | = | 1 (Completed job) |

To determine the portion of the job completed by each person, we must know the **rate** at which the person works. If it takes Kim four hours to paint a room, how much of the room could she paint in one hour? She could paint $\frac{1}{4}$ of the room in one hour, and this is her working rate. In general, the work rate for a person is equal to the reciprocal of the time it takes for that person to complete the whole job. If we multiply the work rate for a person by the time that person has been working, then this tells us the portion of the job that the person has completed.

EXAMPLE 3   Working alone, Kim can paint a room in 4 hours. Jenny can paint the same room in only 3 hours. How long would it take the two of them to paint the room if they work together?

Solution

A good approach to any work-rate problem is to start with the following table and fill in the information:

Person	Time to Complete the Job Alone	Work Rate	Time Working	Portion of the Job Completed
Person 1				
Person 2				

- Since we know it would take Kim 4 hours to paint the room, her work rate is $\frac{1}{4}$ room per hour. Similarly, Jenny's work rate is $\frac{1}{3}$ room per hour.
- The unknown in this problem is the amount of time they will be working together, which we will represent by the variable t.
- Finally, to determine the portion of the job completed by each person, we multiply the person's work rate by the time they have been working.

Person	Time to Complete the Job Alone	Work Rate	Time Working	Portion of the Job Completed
Kim	4 hours	$\frac{1}{4}$	t	$\frac{t}{4}$
Jenny	3 hours	$\frac{1}{3}$	t	$\frac{t}{3}$

To find the equation we need to solve, we add the portion of the room painted by Kim to the portion of the room painted by Jenny and set this sum equal to 1. The equation for this problem is $\frac{t}{4} + \frac{t}{3} = 1$.

$$\frac{t}{4} + \frac{t}{3} = 1 \qquad \text{The LCD is 12.}$$

$$12\left(\frac{t}{4} + \frac{t}{3}\right) = 12 \cdot 1 \qquad \text{Multiply both sides by the LCD.}$$

Quick Check 3

Jacob's new printer, working alone, can print a complete set of brochures in 20 minutes. His old printer can print the set of brochures in 35 minutes. How long would it take the two printers to print the set of brochures if they work together?

$$\overset{3}{\cancel{12}} \cdot \frac{t}{\underset{1}{\cancel{4}}} + \overset{4}{\cancel{12}} \cdot \frac{t}{\underset{1}{\cancel{3}}} = 12 \cdot 1 \qquad \text{Distribute and divide out common factors.}$$

$$3t + 4t = 12 \qquad \text{Multiply remaining factors.}$$
$$7t = 12 \qquad \text{Combine like terms.}$$
$$t = \frac{12}{7} \quad \text{or} \quad 1\frac{5}{7} \qquad \text{Divide both sides by 7.}$$

It would take them $1\frac{5}{7}$ hours to paint the room if they worked together. To convert the answer to hours and minutes we multiply $\frac{5}{7}$ of an hour by 60 minutes per hour, which is approximately 43 minutes. It would take them approximately one hour and 43 minutes. The check of this solution is left to the reader.

EXAMPLE 4   Two drainpipes working together can drain a tank in 6 hours. Working alone, the smaller pipe would take 9 hours longer than the larger pipe to drain the tank. How long would it take the smaller pipe alone to drain the tank?

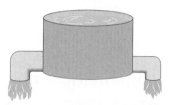

Solution

In this example, we know the amount of time it would take the pipes to drain the tank if they were working together, but we do not know how long it would take each pipe working alone. If we let t represent the time, in hours, that it takes the larger pipe to drain the tank, then the smaller pipe takes $t + 9$ hours to drain the tank. We will begin with the following table:

Pipe	Time to Complete the Job Alone	Work Rate	Time Working	Portion of the Job Completed
Smaller	$t + 9$ hours	$\frac{1}{t+9}$	6 hours	$\frac{6}{t+9}$
Larger	t hours	$\frac{1}{t}$	6 hours	$\frac{6}{t}$

Adding the portion of the tank drained by the smaller pipe in 6 hours to the portion of the tank drained by the larger pipe, we find that the sum equals 1. The equation is $\frac{6}{t+9} + \frac{6}{t} = 1$.

$$\frac{6}{t+9} + \frac{6}{t} = 1 \qquad \text{The LCD is } t(t+9).$$

$$t(t+9) \cdot \left(\frac{6}{t+9} + \frac{6}{t} \right) = t(t+9) \cdot 1 \qquad \text{Multiply both sides by the LCD.}$$

$$t(\overset{1}{\cancel{t+9}}) \cdot \frac{6}{\cancel{(t+9)}} + \overset{1}{\cancel{t}}(t+9) \cdot \frac{6}{\cancel{t}} = t(t+9) \cdot 1 \qquad \text{Distribute and divide out common factors.}$$

$$6t + 6(t+9) = t(t+9) \qquad \text{Multiply remaining factors.}$$

$$6t + 6t + 54 = t^2 + 9t \qquad \text{Multiply. The resulting equation is quadratic.}$$

$$12t + 54 = t^2 + 9t \qquad \text{Combine like terms.}$$

$$0 = t^2 - 3t - 54 \qquad \text{Collect all terms on the right side of the equation.}$$

$$0 = (t-9)(t+6) \qquad \text{Factor.}$$

$$t = 9 \quad \text{or} \quad t = -6 \qquad \text{Set each factor equal to 0 and solve.}$$

Since the time spent by each pipe must be positive, we may immediately omit the solution $t = -6$. The reader may check the solution $t = 9$ by verifying that $\dfrac{6}{9+9} + \dfrac{6}{9} = 1$. We now use $t = 9$ to find the amount of time it would take the smaller pipe to drain the tank. Since $t + 9$ represents the amount of time it would take for the smaller pipe to drain the tank, it would take $9 + 9$, or 18, hours to drain the tank.

Quick Check 4

Two drainpipes working together can drain a tank in 6 hours. Working alone, the smaller pipe would take 16 hours longer than the larger pipe to drain the tank. How long would it take the smaller pipe alone to drain the tank?

Uniform Motion Problems

Objective **3** **Solve applied uniform motion problems.** We now turn our attention to uniform motion problems. Recall that if an object is moving at a constant rate of speed for a certain amount of time, then the distance traveled by the object is equal to the product of its rate of speed and the length of time it traveled. The formula we used earlier in the text is rate $\cdot$ time = distance, or $r \cdot t = d$. If we solve this formula for the time traveled, we have

$$\text{time} = \frac{\text{distance}}{\text{rate}} \quad \text{or} \quad t = \frac{d}{r}$$

EXAMPLE **5** Nick drove 24 miles, one way, to deliver a package. On the way home he drove 20 miles per hour faster than he did on his way to deliver the package. If the total driving time was 1 hour for the entire trip, find Nick's driving speed on the way home.

Solution

- We know that the distance traveled in each direction is 24 miles.
- Nick's speed on the way home was 20 miles per hour faster than it was on the way to deliver the package. We will let r represent his rate of speed on the way to deliver the package, and we can represent his rate of speed on the way home as $r + 20$.
- To find an expression for the time spent on each part of the trip, we divide the distance by the rate.

We can summarize this information in a table:

	Distance (*d*)	Rate (*r*)	Time (*t*)
To Deliver Package	24 miles	r	$\dfrac{24}{r}$
Return Trip	24 miles	$r + 20$	$\dfrac{24}{r + 20}$

The equation we need to solve comes from the fact that the driving time on the way to deliver the package plus the driving time on the way home is equal to 1 hour. If we add the time spent on the way to deliver the package $\left(\dfrac{24}{r}\right)$ to the time spent on the way home $\left(\dfrac{24}{r + 20}\right)$, this will be equal to 1. The equation we need to solve is $\dfrac{24}{r} + \dfrac{24}{r + 20} = 1$.

$$\frac{24}{r} + \frac{24}{r + 20} = 1 \qquad \text{The LCD is } r(r + 20).$$

$$r(r + 20) \cdot \left(\frac{24}{r} + \frac{24}{r + 20}\right) = r(r + 20) \cdot 1 \qquad \text{Multiply both sides by the LCD.}$$

$$\overset{1}{\cancel{r}}(r + 20) \cdot \frac{24}{\cancel{r}} + r\overset{1}{\cancel{(r + 20)}} \cdot \frac{24}{\cancel{(r + 20)}} = r(r + 20) \cdot 1 \qquad \begin{array}{l}\text{Distribute and divide out} \\ \text{common factors.}\end{array}$$

$$24(r + 20) + 24r = r(r + 20) \cdot 1 \qquad \text{Multiply remaining factors.}$$
$$24r + 480 + 24r = r^2 + 20r \qquad \text{Multiply.}$$
$$48r + 480 = r^2 + 20r \qquad \text{Combine like terms.}$$
$$0 = r^2 - 28r - 480 \qquad \begin{array}{l}\text{Collect all terms on the right side} \\ \text{of the equation by subtracting} \\ 48r \text{ and } 480.\end{array}$$

$$0 = (r - 40)(r + 12) \qquad \text{Factor.}$$
$$r = 40 \quad \text{or} \quad r = -12 \qquad \begin{array}{l}\text{Set each factor equal to 0} \\ \text{and solve.}\end{array}$$

Quick Check **5**

Gail drove 300 miles to pick up a friend and then returned home. On the way home, she drove 15 miles per hour faster than she did on her way to pick up her friend. If the total driving time for the trip was 9 hours for the entire trip, find Gail's driving speed on the way home.

We omit the solution $r = -12$, as Nick's speed cannot be negative. The expression for Nick's driving speed on the way home is $r + 20$, so his speed on the way home was $40 + 20$, or 60, miles per hour. The reader can check the solution by verifying that $\dfrac{24}{40} + \dfrac{24}{60} = 1$.

EXAMPLE 6 Gretchen has taken her kayak to the Pawuxet River, which flows downstream at a rate of 3 kilometers per hour. If the time it takes Gretchen to paddle 12 kilometers upstream is 1 hour longer than the time it takes for her to paddle 12 kilometers downstream, find the speed that Gretchen can paddle in still water.

Solution

We will let r represent the speed that Gretchen can paddle in still water. Since the current of the river is 3 kilometers per hour, Gretchen's kayak travels at a speed of $r - 3$ kilometers per hour when she is paddling upstream. This is because the current is

pushing against the kayak. Gretchen travels at a speed of $r + 3$ kilometers per hour when she is paddling downstream, as the current is flowing in the same direction as the kayak. The equation we will solve involves the time spent paddling upstream and downstream. To find expressions in terms of r for the time spent in each direction, we divide the distance (12 km) by the rate of speed. Here is a table containing the relevant information:

	Distance	Rate	Time
Upstream	12 km	$r - 3$	$\frac{12}{r - 3}$
Downstream	12 km	$r + 3$	$\frac{12}{r + 3}$

We are told that the time needed to paddle 12 kilometers upstream is 1 hour more than the time needed to paddle 12 kilometers downstream. In other words, the time spent paddling upstream is equal to the time spent paddling downstream plus 1 hour, or $\frac{12}{r - 3} = \frac{12}{r + 3} + 1$.

$$\frac{12}{r - 3} = \frac{12}{r + 3} + 1 \qquad \text{The LCD is } (r - 3)(r + 3).$$

$$(r - 3)(r + 3) \cdot \frac{12}{r - 3} = (r - 3)(r + 3) \cdot \left(\frac{12}{r + 3} + 1 \right) \qquad \begin{array}{l}\text{Multiply both sides by} \\ \text{the LCD.}\end{array}$$

$$\cancel{(r - 3)}(r + 3) \cdot \frac{12}{\cancel{(r - 3)}} = (r - 3)\cancel{(r + 3)} \cdot \frac{12}{\cancel{(r + 3)}} + (r - 3)(r + 3) \cdot 1$$

Distribute and divide out common factors.

$$12(r + 3) = 12(r - 3) + (r - 3)(r + 3) \qquad \text{Multiply remaining factors.}$$
$$12r + 36 = 12r - 36 + r^2 - 9 \qquad \text{Multiply.}$$
$$12r + 36 = r^2 + 12r - 45 \qquad \begin{array}{l}\text{Combine like terms, writing} \\ \text{the right side in descending} \\ \text{order.}\end{array}$$

$$0 = r^2 - 81 \qquad \begin{array}{l}\text{Collect all terms on the right} \\ \text{side of the equation by} \\ \text{subtracting } 12r \text{ and } 36.\end{array}$$

$$0 = (r + 9)(r - 9) \qquad \text{Factor.}$$
$$r = -9 \quad \text{or} \quad r = 9 \qquad \begin{array}{l}\text{Set each factor equal to 0} \\ \text{and solve.}\end{array}$$

We omit the negative solution, as the speed of the kayak in still water must be positive. Gretchen's kayak travels at a speed of 9 kilometers per hour in still water. The reader can check this solution by verifying that $\frac{12}{9 - 3} = \frac{12}{9 + 3} + 1$.

Variation

Objective 4 **Solve variation problems.** In the remaining examples, we will investigate the concept of **variation** between two or more quantities. Two quantities are said to **vary directly** if an increase in one quantity produces a proportional increase in

Quick Check 6

Lucy takes her canoe to a river that flows downstream at a rate of 2 miles per hour. She paddled 8 miles downstream, and then returned back to the camp she started from. If the round trip took her 3 hours, find the speed that Lucy can paddle in still water.

the other quantity and a decrease in one quantity produces a proportional decrease in the other quantity. For example, suppose that you have a part-time job that pays by the hour. The hours you work in a week and the amount of money you earn (before taxes) vary directly. As the hours you work increase, the amount of money you earn increases by the same factor. If there is a decrease in the number of hours you work, the amount of money that you earn decreases by the same factor.

Direct Variation

If a quantity y varies directly as a quantity x, then the two quantities are related by the equation

$$y = kx$$

where k is called the **constant of variation.**

For example, if you are paid $12 per hour at your part-time job, then the amount of money you earn (y) and the number of hours you work (x) are related by the equation $y = 12x$. The value of k in this situation is 12, and it tells us that each time x increases by 1 hour, y increases by $12.

EXAMPLE ▶ 7 y varies directly as x. If $y = 42$ when $x = 7$, find y when $x = 13$.

Solution

Since the variation is direct, we will use the equation $y = kx$. We begin by finding k. Using $y = 42$ when $x = 7$, we can use the equation $42 = k \cdot 7$ to find k.

$$\begin{aligned} y &= kx \\ 42 &= k \cdot 7 \qquad \text{Substitute 42 for } y \text{ and 7 for } x \text{ into } y = kx. \\ 6 &= k \qquad \text{Divide both sides by 7.} \end{aligned}$$

Now we use this value of k to find y when $x = 13$.

$$\begin{aligned} y &= 6 \cdot 13 \qquad \text{Substitute 6 for } k \text{ and 13 for } x \text{ into } y = kx. \\ y &= 78 \qquad \text{Multiply.} \end{aligned}$$

EXAMPLE ▶ 8   The distance a train travels varies directly as the time it is traveling. If a train can travel 424 miles in 8 hours, how far can it travel in 15 hours?

Solution

The first step in a variation problem is to find k. We will let y represent the distance traveled and x represent the time. To find the variation constant, we will substitute the related information (424 miles in 8 hours) into $y = kx$, the equation for direct variation.

$$\begin{aligned} y &= kx \\ 424 &= k \cdot 8 \qquad \text{Substitute 424 for } y \text{ and 8 for } x \text{ into } y = kx. \\ 53 &= k \qquad \text{Divide both sides by 8.} \end{aligned}$$

Now we will use this variation constant to find the distance traveled in 15 hours.

$$\begin{aligned} y &= 53 \cdot 15 \qquad \text{Substitute 53 for } k \text{ and 15 for } x \text{ into } y = kx. \\ y &= 795 \qquad \text{Multiply.} \end{aligned}$$

The train travels 795 miles in 15 hours.

Quick Check ▶ 7
The tuition a college student pays varies directly as the number of units the student is taking. If a student pays $288 to take 12 units in a semester, how much would a student pay to take 15 units?

In some cases, an increase in one quantity produces a *decrease* in another quantity. In this case, we say that the quantities **vary inversely.** For example, suppose that you and some of your friends are going to buy a birthday gift for someone, splitting the cost equally. As the number of people who are contributing increases, the cost for each person decreases. The cost for each person varies inversely as the number of people contributing.

Inverse Variation

> If a quantity y varies inversely as a quantity x, then the two quantities are related by the equation
>
> $$y = \frac{k}{x}$$
>
> where k is the constant of variation.

EXAMPLE 9 y varies inversely as x. If $y = 10$ when $x = 8$, find y when $x = 5$.

Solution

Since the quantities vary inversely, we will use the equation $y = \dfrac{k}{x}$. We begin by finding k.

$$y = \frac{k}{x}$$

$$10 = \frac{k}{8} \qquad \text{Substitute 10 for } y \text{ and 8 for } x \text{ into } y = \tfrac{k}{x}.$$

$$80 = k \qquad \text{Multiply both sides by 8.}$$

Now we use this value of k to find y when $x = 5$.

$$y = \frac{80}{5} \qquad \text{Substitute 80 for } k \text{ and 5 for } x \text{ into } y = \tfrac{k}{x}.$$

$$y = 16 \qquad \text{Divide.}$$

EXAMPLE 10 The time required to drive from Visalia to San Francisco varies inversely as the average speed of the car. If it takes 3 hours to make the drive at 70 miles per hour, how long would it take to make the drive at 60 miles per hour?

Solution

Again, we begin by finding k. We will let y represent the time required to drive from Visalia to San Francisco and x represent the average speed of the car. To find k, we will substitute the related information (3 hours to make the trip at 70 miles per hour) into the equation for inverse variation.

$$y = \frac{k}{x}$$

$$3 = \frac{k}{70} \qquad \text{Substitute 3 for } y \text{ and 70 for } x \text{ into } y = \tfrac{k}{x}.$$

$$210 = k \qquad \text{Multiply both sides by 70.}$$

Now we will use this variation constant to find the time required to drive from Visalia to San Francisco at 60 miles per hour.

$$y = \frac{210}{60} \qquad \text{Substitute 210 for } k \text{ and 60 for } x \text{ into } y = \tfrac{k}{x}.$$

$$y = 3.5 \qquad \text{Divide.}$$

It would take 3.5 hours to drive from Visalia to San Francisco at 60 miles per hour.

> **Quick Check** **8** The time required to complete the Boston Marathon varies inversely as the average speed of the runner. If it takes 131 minutes to run the marathon at an average speed of 5 miles per hour, how long would it take to finish the marathon at an average speed of 8 miles per hour?

Often, a quantity varies depending on two or more variables. A quantity y is said to **vary jointly** as two quantities x and z if it varies directly as the product of these two quantities. The equation in such a case is $y = kxz$.

EXAMPLE ▶11 y varies jointly as x and the square of z. If $y = 2400$ when $x = 8$ and $z = 10$, find y when $x = 12$ and $z = 20$.

Solution

We begin by finding k, using the equation $y = kxz^2$.

$$y = kxz^2$$
$$2400 = k \cdot 8 \cdot 10^2 \qquad \text{Substitute 2400 for } y\text{, 8 for } x\text{, and 10 for } z \text{ into } y = kxz^2.$$
$$2400 = k \cdot 800 \qquad \text{Simplify.}$$
$$3 = k \qquad \text{Divide both sides by 800.}$$

> **Quick Check** **9**
> y varies jointly as x and the square root of z. If $y = 1000$ when $x = 50$ and $z = 25$, find y when $x = 72$ and $z = 9$.

Now we use this value of k to find y when $x = 12$ and $z = 20$.

$$y = 3 \cdot 12 \cdot 20^2 \qquad \text{Substitute 3 for } k\text{, 12 for } x\text{, and 20 for } z \text{ into } y = kxz^2.$$
$$y = 14{,}400 \qquad \text{Simplify.}$$

EXERCISES 7.7 ❯

1. The sum of the reciprocal of a number and $\frac{2}{3}$ is $\frac{13}{15}$. Find the number.

2. The sum of the reciprocal of a number and $\frac{4}{7}$ is $\frac{43}{63}$. Find the number.

3. The sum of four times the reciprocal of a number and $\frac{3}{4}$ is $\frac{49}{44}$. Find the number.

4. The sum of seven times the reciprocal of a number and $\frac{5}{8}$ is $\frac{3}{2}$. Find the number.

5. The difference of the reciprocal of a number and $\frac{5}{9}$ is $-\frac{41}{90}$. Find the number.

6. The difference of the reciprocal of a number and $\frac{3}{10}$ is $\frac{1}{5}$. Find the number.

7. One positive number is three larger than another positive number. If four times the reciprocal of the smaller number is added to three times the reciprocal of the larger number, the sum is equal to 1. Find the two numbers.

8. One positive number is six less than another positive number. If two times the reciprocal of the smaller number is added to five times the reciprocal of the larger number, the sum is equal to 1. Find the two numbers.

9. One positive number is 10 less than another positive number. If the reciprocal of the smaller number is added to three times the reciprocal of the larger number, the sum is equal to $\frac{1}{4}$. Find the two numbers.

10. One positive number is nine larger than another positive number. If the reciprocal of the smaller number is added to four times the reciprocal of the larger number, the sum is equal to $\frac{2}{3}$. Find the two numbers.

11. One copy machine can run off copies in 15 minutes. A newer machine can do the same job in 10 minutes. How long would it take the two machines, working together, to make all of the necessary copies?

12. Dylan can mow a lawn in 40 minutes, while Alycia takes 50 minutes to mow the same lawn. If Dylan and Alycia work together, using two lawn mowers, how long would it take them to mow the lawn?

13. Tina can completely weed her vegetable garden in 1 hour and 30 minutes. Her friend Marisa can do the same task in 2 hours. If they worked together, how long would it take Tina and Marisa to weed the vegetable garden?

14. One hose can fill a 40,000-gallon swimming pool in 18 hours. A hose from the neighbor's house can fill a swimming pool of that size in 24 hours. If the two hoses run at the same time, how long would they take to fill the swimming pool?

15. After a room has been prepared for painting, Andria can paint the entire room in 2 hours. Tonya can do the same job in 3 hours, while it would take Linda 5 hours to paint the entire room. If the three friends work together, how long would it take them to paint an entire room?

16. A company has printed out 500 surveys to be put into preaddressed envelopes and mailed out. Dusty can fill all of the envelopes in 5 hours, Felipe can do the job in 6 hours, and Grady can do it in nine hours. If all three work together, how long will it take them to stuff the 500 envelopes?

17. Two drainpipes working together can drain a pool in 12 hours. Working alone, the smaller pipe would take 7 hours longer than the larger pipe to drain the pool. How long would it take the smaller pipe alone to drain the pool?

18. Working together, Rosa and Dianne can plant 20 flats of pansies in 3 hours. If Rosa were working alone, it would take her 8 hours longer than it would take Dianne to plant all 20 flats. How long would it take Rosa to plant the pansies by herself?

19. Steve and his assistant Ross can wire a house in 6 hours. If they worked alone, Ross would take 9 hours longer than Steve to wire the house. How long would it take Ross to wire the house?

20. Earl and John can clean an entire building in 4 hours. Earl can clean the entire building by himself in 6 fewer hours than John can. How long would it take Earl to clean the building by himself?

21. Working together, Sarah and Jeff can feed all of the animals at the zoo in 2 hours. Jeff uses a motorized cart, whereas Sarah uses a pushcart to carry the food. If each person were to feed all of the animals without the help of the other, it would take Sarah twice as long as it would take Jeff. How long would it take Sarah to feed all of the animals?

22. A small pipe takes three times longer to fill a tank than a larger pipe takes. If both pipes are working at the same time, it takes 12 minutes to fill the tank. How long would it take the smaller pipe, working alone, to fill the tank?

23. When he works alone, Brent takes 2 hours longer to buff the gymnasium floor than Tracy takes when she works alone. After working for 3 hours, Brent quits and Tracy takes over. If it takes Tracy 2 hours to finish the rest of the floor, how long would it have taken Brent to buff the entire floor?

24. Vern can wallpaper a room in 3 fewer hours than Doug can. After they worked together for 2 hours, Vern had to leave. It took Doug an additional 4 hours to finish wallpapering the room. How long would it take Vern to wallpaper a room by himself?

25. Ruben is training for a triathlon. He rides his bicycle at a speed that is 15 miles per hour faster than his running speed. If Ruben can cycle 40 miles in the same amount of time that it takes him to run 16 miles, what is Ruben's running speed?

26. Ariel drives 15 miles per hour faster than Sharon does. Ariel can drive 100 miles in the same amount of time that it takes for Sharon to drive 80 miles. Find the speed that Ariel drives at.

27. Jared is training to run a marathon. Today he ran 14 miles in 2 hours. After running the first 9 miles at a certain speed, he increased his speed by 4 miles per hour for the remaining 5 miles. Find the speed at which Jared was running for the first 9 miles.

28. Ellen ran 7 miles this morning. After running the first 3 miles, she increased her speed by 2 miles per hour. If it took her exactly 1 hour to finish her run, find the speed at which she was running for the last 4 miles.

29. A salmon is swimming in a river that is flowing downstream at a speed of 10 kilometers per hour. The salmon can swim 3 kilometers upstream in the same time that it would take to swim 5 kilometers downstream. What is the speed of the salmon in still water?

30. Denae is swimming in a river that flows downstream at a speed of 0.5 meter per second. It takes her the same amount of time to swim 500 meters upstream as it does to swim 1000 meters downstream. Find Denae's swimming speed (in meters per second) in still water.

31. An airplane flies 72 miles with a 30-mile-per-hour tailwind and then flies back into the 30-mile-per-hour wind. If the time for the round trip was 1 hour, find the speed of the airplane in calm air.

32. Selma is kayaking in a river that flows downstream at a rate of 1 mile per hour. Selma paddles 5 miles downstream and then turns around and paddles 6 miles upstream, and the trip takes 3 hours.

   a) How fast can Selma paddle in still water?

   b) Selma is now 1 mile upstream of her starting point. How many minutes will it take her to paddle back to her starting point?

33. y varies directly as x. If $y = 30$ when $x = 3$, find y when $x = 8$.

34. y varies directly as x. If $y = 78$ when $x = 6$, find y when $x = 15$.

35. y varies directly as x. If $y = 60$ when $x = 16$, find y when $x = 24$.

36. y varies directly as x. If $y = 80$ when $x = 35$, find y when $x = 21$.

37. y varies inversely as x. If $y = 8$ when $x = 7$, find y when $x = 4$.

38. y varies inversely as x. If $y = 6$ when $x = 12$, find y when $x = 9$.

39. y varies inversely as x. If $y = 20$ when $x = 4$, find y when $x = 15$.

40. y varies inversely as x. If $y = 15$ when $x = 18$, find y when $x = 100$.

41. y varies directly as the square of x. If $y = 150$ when $x = 5$, find y when $x = 7$.

42. y varies inversely as the square of x. If $y = 4$ when $x = 6$, find y when $x = 3$.

43. y varies jointly as x and z. If $y = 270$ when $x = 10$ and $z = 6$, find y when $x = 3$ and $z = 8$.

44. y varies directly as x and inversely as z. If $y = 78$ when $x = 26$ and $z = 7$, find y when $x = 30$ and $z = 18$.

45. Sam's gross pay varies directly as the number of hours he works. In a week that he worked 28 hours, his gross pay was $238. What would Sam's gross pay be if he worked 32 hours?

46. The height that a ball bounces varies directly as the height from which it is dropped. If a ball dropped from a height of 54 inches bounces 24 inches, how high would the ball bounce if it were dropped from a height of 72 inches?

47. *Ohm's law.* In a circuit, the electric current (in amperes) varies directly as the voltage. If the current is 8 amperes when the voltage is 24 volts, find the current when the voltage is 9 volts.

48. *Hooke's law.* The distance that a hanging object stretches a spring varies directly as the mass of the object. If a 5-kilogram weight stretches a spring by 32 centimeters, how far would a 2-kilogram weight stretch the spring?

49. The amount of money that each person must contribute to buy a retirement gift for a coworker varies inversely as the number of people contributing. If 10 people would each have to contribute $24, how much would each person have to contribute if there were 16 people?

50. The maximum load that a wooden beam can support varies inversely as its length. If a beam that is 8 feet long can support 700 pounds, what is the maximum load that can be supported by a beam that is 5 feet long?

51. In a circuit, the electric current (in amperes) varies inversely as the resistance (in ohms). If the current is 30 amperes when the resistance is 3 ohms, find the current when the resistance is 5 ohms.

52. *Boyle's law.* The volume of a gas varies inversely as the pressure upon it. The volume of a gas is 128 cubic centimeters when it is under a pressure of 50 kilograms per square centimeter. Find the volume when the pressure is reduced to 40 kilograms per square centimeter.

53. The illumination of an object varies inversely as the square of its distance from the source of light. If a light source provides an illumination of 30 foot-candles at a distance of 10 feet, find the illumination at a distance of 20 feet. (One foot-candle is the amount of illumination produced by a standard candle at a distance of 1 foot.)

54. The illumination of an object varies inversely as the square of its distance from the source of light. If a light source provides an illumination of 80 foot-candles at a distance of 5 feet, find the illumination at a distance of 40 feet.

Answer in complete sentences.

55. Write a work-rate word problem that can be solved by the equation $\frac{t}{8} + \frac{t}{10} = 1$. Explain how you created your problem.

56. Write a work-rate word problem that can be solved by the equation $\frac{12}{t + 8} + \frac{2}{t} = 1$. Explain how you created your problem.

57. Write a uniform-motion word problem that can be solved by the equation $\frac{10}{r - 4} + \frac{10}{r + 4} = 6$. Explain how you created your problem.

58. Explain why the distance a car travels would vary directly as the time it is traveling but would vary inversely as its average speed.

Study Tip REVISITED   If you have participated in a study group throughout the semester, now is not the time to start studying exclusively on your own. Have each student in the group bring problems that he or she is struggling with or feels are important, and work through them as a group.

At the end of your group study session, try to write a practice exam as a group. In trying to determine which problems to include on the practice exam, you are really focusing on which types of problems are most likely to appear on you cumulative exam.

In some groups, members bring problems and challenge others in the group to solve them. Although this can be an effective way to study, it is not a good idea to do this the night before the exam. You may find that if you are unable to solve two or three problems in a row, your confidence may be shattered, and you need plenty of confidence when you sit down to take an exam. Spend the night before the exam alone, reviewing what you have done.

Chapter 7 Summary

Summary of Chapter 7 Study Tips

- When preparing for a cumulative exam, start approximately two weeks before the exam by determining what material will be on the exam.
- Begin your studying by going over your old exams and quizzes. Although you should rework each problem, pay particular attention to the problems you got wrong on the exam or quiz. Also, look over your old homework assignments for problems you struggled with and review these topics.
- Use review materials from your instructor and the cumulative review exercises in the text to check your progress. Try to work the exercises without referring to your notes, note cards, or text. This will give you a clear assessment as to which problems require further study.
- Create study sheets for each type of applied problem that may appear on the exam. These sheets should contain examples of these types of problems, the procedure for solving them, and the solutions of your examples. Such study sheets could also be created for other types of problems that you are struggling with.
- Continue to meet regularly with your study group. If your group has helped you to succeed on the previous exams, then they will be able to help you succeed on this exam as well.

Evaluate the rational expression for the given value of the variable. [7.1]

1. $\dfrac{3}{x-9}$ for $x = -6$

2. $\dfrac{x-7}{x+13}$ for $x = 19$

3. $\dfrac{x^2 + 9x - 11}{x^2 + 2x - 5}$ for $x = 3$

4. $\dfrac{x^2 - 5x - 15}{x^2 + 17x + 66}$ for $x = -4$

Find all values for which the rational expression is undefined. [7.1]

5. $\dfrac{-6}{x+9}$

6. $\dfrac{x^2 + 9x + 18}{x^2 - 3x - 54}$

Simplify the given rational expression. (Assume all denominators are nonzero.) [7.1]

7. $\dfrac{x+7}{x^2 + 4x - 21}$

8. $\dfrac{x^2 + 6x - 40}{x^2 - 9x + 20}$

9. $\dfrac{36 - x^2}{x^2 - 7x + 6}$

10. $\dfrac{x^2 + 6x + 9}{x^2 + 11x + 24}$

Evaluate the given rational function. [7.1]

11. $r(x) = \dfrac{x+7}{x^2 + 8x - 15}$, $r(3)$

12. $r(x) = \dfrac{x^2 - 15x}{x^2 - 11x + 18}$, $r(-2)$

Find the domain of the given rational function. [7.1]

13. $r(x) = \dfrac{x^2 + 3x - 40}{x^2 + 6x}$

14. $r(x) = \dfrac{x^2 + 7x - 8}{x^2 - 5x + 4}$

Multiply. [7.2]

15. $\dfrac{x+9}{x+5} \cdot \dfrac{x^2 - 7x}{x^2 + 2x - 63}$

16. $\dfrac{x^2 + 14x + 45}{x^2 - 6x - 7} \cdot \dfrac{x^2 + 9x + 8}{x^2 + 3x - 10}$

17. $\dfrac{x^2 - 4x}{x^2 - 15x + 54} \cdot \dfrac{x^2 - x - 72}{x^2 + 8x}$

18. $\dfrac{16 - x^2}{x + 2} \cdot \dfrac{x^2 + 5x + 6}{x^2 + 2x - 24}$

For the given functions $f(x)$ and $g(x)$, find $f(x) \cdot g(x)$. [7.2]

19. $f(x) = \dfrac{x-6}{x-1}$, $g(x) = \dfrac{x^2 + 6x + 5}{x^2 - 3x - 18}$

20. $f(x) = \dfrac{x^2 + 16x + 64}{x^2 + x - 42}$, $g(x) = \dfrac{49 - x^2}{x^2 + 11x + 24}$

Divide. [7.2]

21. $\dfrac{x^2 + 9x + 18}{x^2 - 7x + 12} \div \dfrac{x+3}{x-3}$

22. $\dfrac{x^2 - x - 6}{x^2 - 19x + 88} \div \dfrac{x^2 + 10x + 16}{x^2 + 4x - 96}$

23. $\dfrac{x^2 - 3x - 4}{x^2 + x} \div \dfrac{x^2 - 16x + 48}{x^2 - 1}$

24. $\dfrac{9 - x^2}{x - 7} \div \dfrac{x^2 - 8x + 15}{x^2 - 5x - 14}$

For the given functions $f(x)$ and $g(x)$, find $f(x) \div g(x)$. [7.2]

25. $f(x) = \dfrac{x^2 + 3x}{x^2 + 10x + 25}$, $g(x) = \dfrac{x^2 - 6x - 27}{x^2 - x - 30}$

26. $f(x) = \dfrac{x^2 + 8x + 12}{x^2 + 5x + 4}$, $g(x) = \dfrac{x^2 - 4x - 60}{x^2 + 7x + 12}$

Add or subtract. [7.3/7.4]

27. $\dfrac{5}{x+6} + \dfrac{7}{x+6}$

28. $\dfrac{x^2 + 10x}{x^2 + 15x + 56} - \dfrac{5x + 24}{x^2 + 15x + 56}$

29. $\dfrac{x^2 - 6x - 13}{x^2 - x - 20} + \dfrac{10x - 32}{x^2 - x - 20}$

30. $\dfrac{5x - 34}{x - 7} - \dfrac{x - 8}{7 - x}$

31. $\dfrac{2x^2 + 7x - 3}{x - 2} + \dfrac{x^2 + 3x + 9}{2 - x}$

32. $\dfrac{x^2 - 3x - 12}{x^2 - 25} - \dfrac{2x - 18}{25 - x^2}$

33. $\dfrac{5}{x^2 + 3x - 4} + \dfrac{2}{x^2 - 4x + 3}$

34. $\dfrac{6}{x^2 - 10x + 16} - \dfrac{1}{x^2 - 15x + 56}$

35. $\dfrac{x}{x^2 - 5x - 50} - \dfrac{4}{x^2 - 14x + 40}$

36. $\dfrac{x - 5}{x^2 + 14x + 33} + \dfrac{4}{x^2 + 20x + 99}$

37. $\dfrac{x + 1}{x^2 + 2x - 24} + \dfrac{x - 5}{x^2 - 6x + 8}$

For the given rational functions $f(x)$ and $g(x)$, find $f(x) + g(x)$. [7.4]

38. $f(x) = \dfrac{x + 3}{x^2 - x - 2}, g(x) = \dfrac{5}{x^2 - 7x + 10}$

For the given rational functions $f(x)$ and $g(x)$, find $f(x) + g(x)$. [7.4]

39. $f(x) = \dfrac{x + 2}{x^2 + 4x - 5}, g(x) = \dfrac{1}{x^2 + 12x + 35}$

Simplify the complex fraction. [7.5]

40. $\dfrac{1 - \dfrac{9}{x}}{1 - \dfrac{81}{x^2}}$

41. $\dfrac{\dfrac{x - 5}{x - 3} - \dfrac{3}{x - 7}}{\dfrac{x^2 + 7x - 44}{x^2 - 10x + 21}}$

42. $\dfrac{1 + \dfrac{2}{x} - \dfrac{35}{x^2}}{1 - \dfrac{7}{x} + \dfrac{10}{x^2}}$

43. $\dfrac{\dfrac{x^2 - 8x + 15}{x^2 - 11x + 18}}{\dfrac{x^2 + x - 30}{x^2 - 6x - 27}}$

Solve. [7.6]

44. $\dfrac{9}{x} - \dfrac{3}{4} = \dfrac{7}{8}$

45. $x - 5 + \dfrac{12}{x} = 2$

46. $\dfrac{3}{x - 7} = \dfrac{7}{x + 5}$

47. $\dfrac{2x - 11}{x - 1} = \dfrac{x - 7}{x + 3}$

48. $\dfrac{3}{x + 2} - \dfrac{1}{x + 1} = \dfrac{x + 3}{x^2 + 3x + 2}$

49. $\dfrac{x}{x + 5} + \dfrac{3}{x - 7} = \dfrac{36}{x^2 - 2x - 35}$

50. $\dfrac{2}{x^2 - x - 2} + \dfrac{10}{x^2 - 2x - 3} = \dfrac{x + 12}{x^2 - x - 2}$

51. $\dfrac{20}{x^2 + 12x + 27} - \dfrac{4}{x^2 + 14x + 45} = \dfrac{x + 10}{x^2 + 8x + 15}$

Solve for the specified variable. [7.6]

52. $\dfrac{1}{2} = \dfrac{A}{bh}$ for h

53. $y = \dfrac{3x}{4x - 7}$ for x

54. $\dfrac{x}{2r} - \dfrac{y}{3r} = \dfrac{1}{5}$ for r

55. The sum of the reciprocal of a number and $\frac{5}{6}$ is $\frac{23}{24}$. Find the number. [7.7]

56. One positive number is five more than another positive number. If three times the reciprocal of the smaller number is added to four times the reciprocal of the larger number, the sum is equal to 1. Find the two numbers. [7.7]

57. Rob can mow a lawn in 30 minutes, while Sean takes 60 minutes to mow the same lawn. If Rob and Sean work together, using two lawn mowers, how long would it take them to mow the lawn? [7.7]

58. Two pipes, working together, can fill a tank in 6 hours. Working alone, the smaller pipe takes 9 hours longer than the larger pipe to fill the tank. How long would it take the smaller pipe alone to fill the tank? [7.7]

59. Linda had to make a 275-mile drive to Omaha. After driving the first 100 miles, she increased her speed by 15 miles per hour. If the drive took her exactly 4 hours, find the speed at which she was driving for the first 100 miles. [7.7]

60. Dominique is kayaking in a river that is flowing downstream at a speed of 2 miles per hour. Dominique can paddle 10 miles downstream in the same amount of time that he can paddle 5 miles upstream. What is the speed that Dominique can paddle in still water? [7.7]

61. The number of calories in a glass of milk varies directly as the amount of milk. If a 12-ounce serving of milk has 195 calories, how many calories are there in an 8-ounce glass of milk? [7.7]

62. The distance required for a car to stop after applying the brakes varies directly as the square of the speed of the car. If it takes 100 feet for a car traveling at 40 miles per hour to stop, how far would it take for a car traveling 60 miles per hour to come to a stop? [7.7]

63. The maximum load that a wooden beam can support varies inversely as its length. If a beam that is 6 feet long can support 900 pounds, what is the maximum load that can be supported by a beam that is 8 feet long? [7.7]

64. The illumination of an object varies inversely as the square of its distance from the source of light. If a light source provides an illumination of 20 foot-candles at a distance of 12 feet, find the illumination at a distance of 6 feet. [7.7]

Evaluate the rational expression for the given value of the variable.

1. $\dfrac{10}{x-5}$ for $x = -7$

Find all values for which the rational expression is undefined.

2. $\dfrac{x^2 + 5x - 14}{x^2 + 10x + 21}$

Simplify the given rational expression. (Assume all denominators are nonzero.)

3. $\dfrac{x^2 - 8x + 15}{x^2 + 7x - 30}$

Evaluate the rational function.

4. $r(x) = \dfrac{x^2 + 3x - 13}{x^2 - 4x - 20},\ r(-3)$

Multiply.

5. $\dfrac{x^2 + x - 42}{x^2 + 6x - 7} \cdot \dfrac{x^2 - 2x + 1}{x^2 - 9x + 18}$

Divide.

6. $\dfrac{x^2 - 10x + 25}{x^2 + 11x + 24} \div \dfrac{25 - x^2}{x^2 + 3x - 40}$

Add or subtract.

7. $\dfrac{5x + 3}{x - 6} - \dfrac{2x + 21}{x - 6}$

8. $\dfrac{2x^2 + 6x - 11}{x - 4} + \dfrac{x^2 + 2x + 21}{4 - x}$

9. $\dfrac{6}{x^2 + 8x + 7} + \dfrac{5}{x^2 - 3x - 4}$

10. $\dfrac{x - 4}{x^2 - 13x + 40} - \dfrac{8}{x^2 - 10x + 16}$

Simplify the complex fraction.

11. $\dfrac{1 + \dfrac{4}{x} - \dfrac{45}{x^2}}{1 + \dfrac{2}{x} - \dfrac{35}{x^2}}$

Simplify.

12. $\dfrac{x + 5}{x^2 + 3x - 54} \cdot \dfrac{x^2 - 11x + 30}{x^2 + 16x + 55}$

13. $\dfrac{x^2 - 3x - 40}{x^2 + 7x + 6} \div \dfrac{x^2 - 8x}{x^2 + 10x + 9}$

14. $\dfrac{x^2 + 7x - 14}{x^2 + 6x + 8} + \dfrac{3x + 38}{x^2 + 6x + 8}$

15. $\dfrac{x - 7}{x^2 + 7x - 8} - \dfrac{2}{x^2 + x - 2}$

Solve.

16. $\dfrac{15}{x} + \dfrac{3}{8} = \dfrac{19}{24}$

17. $\dfrac{x + 2}{x^2 + 5x + 4} - \dfrac{4}{x^2 + x - 12} = \dfrac{2}{x^2 - 2x - 3}$

Solve for the specified variable.

18. $x = \dfrac{2y}{7y - 5}$ for y

19. Two pipes working together can fill a tank in 40 minutes. Working alone, the smaller pipe would take 18 minutes longer than the larger pipe to fill the tank. How long would it take the smaller pipe alone to fill the tank?

20. Preparing for a race, Greg went for a 25-mile bicycle ride. After the first 10 miles, he increased his speed by 10 miles per hour. If the ride took him exactly 1 hour, find the speed at which he was riding for the first 10 miles.

Mathematicians in History

John Nash

$\mathcal{J}$ohn Nash is an American mathematician whose research has greatly affected mathematics and economics, as well as many other fields. While applying to go to graduate school at Princeton, one of his math professors said quite simply, "This man is a genius."

Write a one-page summary (OR make a poster) of the life of John Nash and his accomplishments.

Interesting issues:

- Where and when was John Nash born?
- Describe Nash's childhood, as well as his life as a college student.
- What mental illness struck Nash in the late 1950s?
- Nash was influential in the field of game theory. What is game theory?
- In 1994, Nash won the Nobel Prize for Economics. Exactly what did Nash win the prize for?
- One of Nash's nicknames is "The Phantom of Fine Hall." Why was this nickname chosen for him?
- What color sneakers did Nash wear?
- Sylvia Nasar wrote a biography of Nash's life, which was made into an Academy Award–winning movie. What was the title of the book and movie?
- What actor played John Nash in the movie?

The weight W of an object varies inversely as the square of the distance d from the center of the Earth. At sea level (3978 miles from the center of the Earth), a person weighs 150 pounds. The formula used to compute the weight of this person at different distances from the center of the Earth is $W = \dfrac{2{,}373{,}672{,}600}{d^2}$.

a) Use the given formula to calculate the weight of this person at different distances from the center of the Earth. Round to the nearest tenth of a pound.

b) What do you notice about the weight of the person as the distance from the center of the Earth increases?

c) Why do you think your observations from part b) are happening?

d) As accurately as possible, plot these points on an axis system in which the horizontal axis represents the distance and the vertical axis represents the weight. Do the points support your observations from part (b)?

e) How far from the center of the Earth would this 150-pound person have to travel to weigh 100 pounds?

f) How far from the center of the Earth would this 150-pound person have to travel to weigh 75 pounds?

g) How far from the center of the Earth would this 150-pound person have to travel to weigh 0 pounds?

	Distance from center of Earth	Weight
On top of the tallest building Taipei 101 (Taipei, Taiwan)	3978.316 miles	
On top of the tallest structure KVLY-TV mast (Mayville, ND)	3978.391 miles	
In an airplane	3984.629 miles	
In the space station	4201.402 miles	
Halfway to the moon	123,406 miles	
Halfway to Mars	4,925,728 miles	

Simplify the expression. Write the result without using negative exponents. (Assume all variables represent nonzero real numbers.) [5.1/5.2]

1. $\dfrac{x^{11}}{x^4}$

2. $12m^{10}n^7 \cdot 9m^6n^{14}$

3. $\left(\dfrac{5a^6b^{11}}{c^7}\right)^4$

4. $\dfrac{t^{12}}{t^{-9}}$

5. $(a^{-6}b^7)^{-8}$

6. $x^{-16} \cdot x^{-19}$

Rewrite in scientific notation. [5.2]

7. 330,000,000,000

8. 0.000000000000714

9. If a computer can perform a calculation in 0.000000000005 second, how long would it take to perform 200,000,000 calculations? [5.2]

Evaluate the polynomial for the given value of the variable. [5.3]

10. $x^2 + 11x - 16$ for $x = -7$

Add or subtract. [5.3]

11. $(x^2 + 6x - 20) + (x^2 - 13x - 39)$

12. $(x^2 - 15x - 9) - (2x^2 - 3x - 31)$

Evaluate the given function. [5.3]

13. $f(x) = x^2 + 5x - 37, \ f(8)$

Multiply. [5.4]

14. $3x(5x^2 - 8x + 12)$

15. $(4x - 7)(3x - 8)$

Divide. [5.5]

16. $\dfrac{10x^{12} + 24x^9 - 18x^6}{2x^5}$

17. $\dfrac{x^2 + 13x + 39}{x + 6}$

Factor completely. [6.1–6.5]

18. $x^3 - 4x^2 + 6x - 24$

19. $2x^2 - 26x + 80$

20. $x^2 + 5x - 36$

21. $2x^2 - x - 10$

22. $x^3 - 125$

23. $x^2 - 81$

Solve. [6.6]

24. $x^2 - 49 = 0$

25. $x^2 - 11x + 24 = 0$

26. $x^2 + 14x + 49 = 0$

27. $x(x + 8) = 20$

Find a quadratic equation with integer coefficients that has the given solution set. [6.6]

28. $\{-2, 7\}$

For the given function $f(x)$, find all values x for which $f(x) = 0$. [6.7]

29. $f(x) = x^2 - 100$

30. $f(x) = x^2 + 2x - 35$

31. A man is standing on a cliff above a beach. He throws a rock upward from a height of 192 feet above the beach with an initial velocity of 64 feet/second. The rock's height above the beach, in feet, after t seconds is given by the function $h(t) = -16t^2 + 64t + 192$, which is graphed as follows:

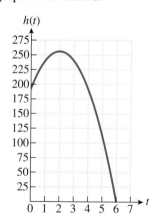

Use the function and its graph to answer the following questions: [6.7]

a) Use the function to determine how high above the beach the rock is after 3 seconds.

b) How long will it take for the rock to land on the beach?

c) After how many seconds does the rock reach its greatest height above the beach?

d) Use the function to determine the greatest height above the beach that the rock reaches.

32. Two consecutive even positive integers have a product of 168. Find the two integers. [6.8]

33. The length of a rectangle is 5 meters more than twice the width. The area of the rectangle is 133 square meters. Find the length and the width of the rectangle. [6.8]

Find all values for which the rational expression is undefined. [7.1]

34. $\dfrac{x^2 + 11x + 24}{x^2 - 14x + 45}$

Simplify the given rational expression. (Assume all denominators are nonzero.) [7.1]

35. $\dfrac{x^2 + 3x - 54}{x^2 + 16x + 63}$

Evaluate the given rational function. [7.1]

36. $r(x) = \dfrac{x + 12}{x^2 + 7x + 5}, \ r(3)$

Multiply. [7.2]

37. $\dfrac{x^2 + 3x - 18}{x^2 + 15x + 44} \cdot \dfrac{x^2 + 2x - 8}{x^2 + 8x + 12}$

Divide. [7.2]

38. $\dfrac{x^2 + x - 90}{x^2 + 14x + 49} \div \dfrac{x^2 - 8x - 9}{x^2 + 5x - 14}$

Add or subtract. [7.3/7.4]

39. $\dfrac{5x + 7}{x^2 - 2x - 8} - \dfrac{2x + 19}{x^2 - 2x - 8}$

40. $\dfrac{6}{x^2 + x - 2} + \dfrac{2}{x^2 - 6x + 5}$

41. $\dfrac{x - 5}{x^2 - 10x + 21} - \dfrac{3}{x^2 - 9}$

42. $\dfrac{x + 4}{x^2 + 13x + 42} + \dfrac{6}{x^2 + 12x + 35}$

Simplify the complex fraction. [7.5]

43. $\dfrac{\dfrac{x - 5}{x + 3} - \dfrac{1}{x - 6}}{\dfrac{x^2 + 5x - 24}{x^2 - 3x - 18}}$

44. $\dfrac{1 + \dfrac{1}{x} - \dfrac{20}{x^2}}{1 + \dfrac{13}{x} + \dfrac{40}{x^2}}$

Solve. [7.6]

45. $6 + \dfrac{4}{x - 3} = \dfrac{3x + 1}{x - 3}$

46. $\dfrac{2x - 3}{x + 4} = \dfrac{x + 3}{x - 4}$

47. $\dfrac{5}{x + 7} + \dfrac{4}{x - 3} = \dfrac{x - 11}{x^2 + 4x - 21}$

Solve for the specified variable. [7.6]

48. $\dfrac{1}{x} + \dfrac{1}{y} = 2$ for x

49. The sum of the reciprocal of a number and $\frac{4}{15}$ is $\frac{11}{30}$. Find the number. [7.7]

50. Two pipes working together can fill a tank in 3 hours. Working alone, the smaller pipe would take 8 hours longer than the larger pipe to fill the tank. How long would it take the smaller pipe alone to fill the tank? [7.7]

A TRANSITION

This chapter provides a transition between beginning algebra and intermediate algebra. We will review concepts covered in the first half of the text and extend these ideas to new topics.

Study Tip **SUMMARY OF PREVIOUS STUDY TIPS**   *In this chapter, we will summarize the study tips presented in the first half of the text.*

8.1
LINEAR EQUATIONS AND ABSOLUTE VALUE EQUATIONS

Objectives

1 **Solve linear equations.**
2 **Solve absolute value equations.**

Linear Equations

Objective 1   Solve linear equations. We begin this section by reviewing linear equations and their solutions. Recall the following guidelines we developed for solving linear equations in Chapter 2:

- Simplify each side of the equation. This includes performing any distributive multiplications, clearing all fractions by multiplying each side of the equation by the least common denominator (LCD), and combining like terms.
- Collect all variable terms on one side of the equation and all constant terms on the other side. It is a good idea to collect the variable terms in such a way that the coefficient of the variable term will be positive.
- Divide both sides of the equation by the coefficient of the variable term to find the solution.
- Check your solution and write it in solution set notation.

We will proceed with some examples that are intended as a review.

EXAMPLE 1   Solve $2x - 6 = 10$.

Solution   We begin by isolating the variable term on the left side of the equation.

$$
\begin{aligned}
2x - 6 &= 10 \\
2x &= 16 &&\text{Add 6 to both sides to isolate } 2x. \\
x &= 8 &&\text{Divide both sides by 2.}
\end{aligned}
$$

We will now check this solution.

Check:
$$
\begin{aligned}
2x - 6 &= 10 \\
2(8) - 6 &= 10 &&\text{Substitute 8 for } x. \\
16 - 6 &= 10 &&\text{Multiply.} \\
10 &= 10 &&\text{Subtract.}
\end{aligned}
$$

Quick Check 1
Solve $9x + 13 = -14$.

Since both sides simplify to be 10, our solution is valid. The solution set is $\{8\}$.

EXAMPLE 2   Solve $3(2x + 7) - 4x = 5x - 6$

Solution   We begin by performing the distributive multiplication on the left side of the equation.

$$
\begin{aligned}
3(2x + 7) - 4x &= 5x - 6 \\
6x + 21 - 4x &= 5x - 6 &&\text{Distribute 3.} \\
2x + 21 &= 5x - 6 &&\text{Combine like terms.} \\
27 &= 3x &&\text{Collect variable terms on right side by subtracting } 2x. \text{ Add 6 to collect constants on the left side.} \\
9 &= x &&\text{Divide both sides by 3.}
\end{aligned}
$$

The check is left to the reader. The solution set is $\{9\}$.

Quick Check 2 Solve $3x - 2(4x - 5) = 2x - 11$.

EXAMPLE 3 Solve $\frac{4}{3}x + \frac{5}{6} = \frac{9}{4}x - 1$

Solution

This equation includes fractions. We can clear these fractions by multiplying each side of the equation by the LCD, which is 12.

$$\frac{4}{3}x + \frac{5}{6} = \frac{9}{4}x - 1$$

$$12\left(\frac{4}{3}x + \frac{5}{6}\right) = 12\left(\frac{9}{4}x - 1\right) \qquad \text{Multiply both sides by 12.}$$

$$\overset{4}{\cancel{12}} \cdot \frac{4}{\underset{1}{\cancel{3}}}x + \overset{2}{\cancel{12}} \cdot \frac{5}{\underset{1}{\cancel{6}}} = \overset{3}{\cancel{12}} \cdot \frac{9}{\underset{1}{\cancel{4}}}x - 12 \cdot 1 \qquad \begin{array}{l}\text{Distribute and divide out common}\\\text{factors.}\end{array}$$

$$16x + 10 = 27x - 12 \qquad \text{Multiply.}$$

$$10 = 11x - 12 \qquad \text{Subtract } 16x.$$

$$22 = 11x \qquad \text{Add 12.}$$

$$2 = x \qquad \text{Divide both sides by 11.}$$

Quick Check 3
Solve $\frac{2}{3}x - \frac{1}{5} = \frac{1}{2}x + \frac{5}{6}$.

The check is left to the reader. The solution set is $\{2\}$.

Equations That Are Identities

In the first three examples of this section, the linear equations have had exactly one solution, but recall that this will not always be the case. An equation that is always true regardless of the value substituted for the variable x is called an identity. The solution set for an identity is the set of all real numbers, denoted by $\Re$.

EXAMPLE 4 Solve $2(2x - 3) + 1 = 4x - 5$.

Solution

$$2(2x - 3) + 1 = 4x - 5$$

$$4x - 6 + 1 = 4x - 5 \qquad \text{Distribute 2.}$$

$$4x - 5 = 4x - 5 \qquad \text{Combine like terms.}$$

$$-5 = -5 \qquad \text{Subtract } 4x \text{ from both sides.}$$

When we subtract $4x$ from each side in an attempt to collect all variable terms on one side of the equation, the variable x has been eliminated from the resulting equation. Is the equation $-5 = -5$ a true statement? Yes, and it tells us that our equation is an identity and that our solution set is the set of all real numbers $\Re$.

Quick Check 4 Solve $(2x + 1) - (5 - 3x) = 2(3x - 2) - x$.

Notice that at one point in the previous example, the equation was $4x - 5 = 4x - 5$. If you reach a line where both sides of the equation are identical, then the equation is an identity.

Equations That Are Contradictions

Just as there are equations that are always true regardless of the value we choose for the variable, there are equations that are never true for any value of the variable. These equations are called contradictions and have no solution. The solution set for these equations is the empty set, or null set, and is denoted $\varnothing$. We can tell that an equation is a contradiction when we are solving it because the variable terms on each side of the equation will be eliminated but in this case will leave an equation that is false, such as $1 = 2$.

EXAMPLE 5   Solve $4x + 3 = 2(2x + 3) - 1$.

Solution

$$4x + 3 = 2(2x + 3) - 1$$
$$4x + 3 = 4x + 6 - 1 \qquad \text{Distribute 2.}$$
$$4x + 3 = 4x + 5 \qquad \text{Combine like terms.}$$
$$3 = 5 \qquad \text{Subtract } 4x \text{ from both sides.}$$

After we subtract $4x$ from each side of the equation, the resulting equation $3 = 5$ is obviously false. This equation is a contradiction and its solution set is the empty set $\varnothing$.

Quick Check **5**

Solve
$5x - 7 = 6(x + 2) - x.$

Absolute Value Equations

Objective 2   **Solve absolute value equations.**   We now introduce equations involving absolute values of linear expressions.

EXAMPLE 6   Solve $|x| = 5$.

Solution

Recall that the absolute value of a number x, denoted $|x|$, is a measure of the distance between 0 and that number x on a real number line. In this equation, we are looking for a number x that is 5 units away from 0.

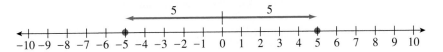

Quick Check **6**   There are two such numbers: 5 and -5. The solution set is $\{5, -5\}$.

Solve $|x| = 7$.

The equation in the previous example is an **absolute value equation.** An absolute value equation relates the absolute value of an expression to a constant, such as $|2x - 3| = 7$, or the absolute value of an expression to the absolute value of another expression, such as $|3x - 4| = |2x + 11|$.

Consider the absolute value equation $|2x - 3| = 7$. This equation involves a number, represented by $2x - 3$, whose absolute value is equal to 7. That number must either be 7 or -7. We begin to solve this equation by converting it into the following two equations.

$$2x - 3 = 7 \qquad \text{or} \qquad 2x - 3 = -7$$

In general, when we have an equation with the absolute value of an expression equal to a positive number, we will set the expression (without the absolute value bars) equal to that number and its opposite. Solving those two equations will give us our solutions.

Solving Absolute Value Equations

> For any expression X and any positive number a, the solutions to the equation
>
> $$|X| = a$$
>
> can be found by solving the two equations
>
> $$X = a \quad \text{and} \quad X = -a$$

EXAMPLE 7   Solve $|2x - 3| = 7$.

Solution

We begin by converting the absolute value equation into two equations that do not involve absolute values.

$$|2x - 3| = 7$$

$2x - 3 = 7$	or	$2x - 3 = -7$	Convert to two linear equations.
$2x = 10$	or	$2x = -4$	Add 3.
$x = 5$	or	$x = -2$	Divide by 2.

Check:

$$\begin{array}{cc}
\underline{x = 5} & \underline{x = -2} \\
|2(5) - 3| = 7 & |2(-2) - 3| = 7 \\
|10 - 3| = 7 & |-4 - 3| = 7 \\
|7| = 7 & |-7| = 7 \\
7 = 7 & 7 = 7
\end{array}$$

Quick Check 7
Solve $|3x + 8| = 5$.

Since both values check, our solution set is $\{5, -2\}$.

> **A Word of Caution**   When solving an absolute value equation such as $|2x - 3| = 7$, be sure to rewrite the equation as two equations. Do not simply drop the absolute value bars and solve the resulting equation.

EXAMPLE 8   Solve $3|2x + 7| - 4 = 20$.

Solution

We must isolate the absolute value before converting the equation to two linear equations.

$3|2x + 7| - 4 = 20$			
$3|2x + 7| = 24$			Add 4 to both sides.
$|2x + 7| = 8$			Divide both sides by 3.
$2x + 7 = 8$	or	$2x + 7 = -8$	Convert to two linear equations.
$2x = 1$	or	$2x = -15$	Subtract 7 from both sides of each equation.
$x = \dfrac{1}{2}$	or	$x = -\dfrac{15}{2}$	Divide both sides of each equation by 2.

Quick Check 8
Solve
$4\lvert 2x - 5 \rvert - 9 = 7.$

The solution set is $\left\{ \frac{1}{2}, -\frac{15}{2} \right\}$. The check of these solutions is left as an exercise.

> *A Word of Caution*   When solving an absolute value equation, we must isolate the absolute value before rewriting the equation as two equations without absolute values.

Since the absolute value of a number is a measure of its distance from 0 on the number line, the absolute value of a number cannot be negative. Therefore, an equation that has an absolute value equal to a negative number has no solution. For example, the equation $\lvert x \rvert = -3$ has no solution because there is no number whose absolute value is -3. We write $\varnothing$ for the solution set.

EXAMPLE 9   Solve $\lvert 3x - 4 \rvert + 8 = 6.$

Solution

We begin by isolating the absolute value.

$$\lvert 3x - 4 \rvert + 8 = 6$$
$$\lvert 3x - 4 \rvert = -2 \qquad \text{Subtract 8 from both sides.}$$

Since an absolute value cannot equal -2, this equation has no solution. The solution set is $\varnothing$.

Quick Check 9
Solve $\lvert 2x + 3 \rvert = -6.$

> *A Word of Caution*   An equation in which an absolute value is equal to a negative number has no solution. Do not try to rewrite the equation as two equivalent equations.

If two unknown numbers have the same absolute values, then either the two numbers are equal or they are opposites. We will use this idea to solve absolute value equations such as $\lvert 3x - 4 \rvert = \lvert 2x + 11 \rvert$. If these two absolute values are equal, then either the two expressions inside the absolute value bars are equal $(3x - 4 = 2x + 11)$, or the first expression is the opposite of the second expression $(3x - 4 = -(2x + 11))$. To find the solution of the original equation, we will solve these two resulting equations.

Solving Absolute Value Equations Involving Two Absolute Values

> For any expressions X and Y, the solutions of the equation
>
> $$\lvert X \rvert = \lvert Y \rvert$$
>
> can be found by solving the two equations
>
> $$X = Y \qquad \text{and} \qquad X = -Y$$

EXAMPLE 10   Solve $\lvert 3x - 4 \rvert = \lvert 2x + 11 \rvert.$

Solution

We begin by rewriting this equation as two equations that do not contain absolute values. For the first equation, we set $3x - 4$ equal to $2x + 11$; for the second equation, we set $3x - 4$ equal to the opposite of $2x + 11$. We finish by solving each equation.

$$3x - 4 = 2x + 11 \qquad \text{or} \qquad 3x - 4 = -(2x + 11)$$

We will now solve each equation separately.

$$3x - 4 = 2x + 11$$
$$x - 4 = 11 \quad \text{Subtract } 2x.$$
$$x = 15 \quad \text{Add 4.}$$

$$3x - 4 = -(2x + 11)$$
$$3x - 4 = -2x - 11 \quad \text{Distribute.}$$
$$5x - 4 = -11 \quad \text{Add } 2x.$$
$$5x = -7 \quad \text{Add 4.}$$
$$x = -\frac{7}{5} \quad \begin{array}{l}\text{Divide both sides}\\ \text{by 5.}\end{array}$$

Quick Check 10

Solve
$$|x - 9| = |2x + 13|.$$

The solution set is $\left\{15, -\frac{7}{5}\right\}$. The check of these solutions is left to the reader.

EXERCISES 8.1 ❯

Solve.

1. $x - 4 = 7$

2. $n + 2 = 6$

3. $m - 5 = -9$

4. $x + 7 = -3$

5. $2x = 10$

6. $-3x = 12$

7. $5t = -8$

8. $-2b = -28$

9. $2n - 5 = -11$

10. $3n + 4 = 19$

11. $-4x + 7 = -13$

12. $6x - 1 = 8$

13. $3x + 25 = -26$

14. $-5x + 33 = 48$

15. $2n - 5 = 5n + 11$

16. $8n - 27 = 5n + 12$

17. $4x + 18 = -x + 3$

18. $-6x + 29 = -2x + 5$

19. $-7b + 5 = 3b - 13$

20. $8t - 33 = 42 - 12t$

21. $2(2x - 4) - 7 = 6x - 11$

22. $5(3 - 2x) + 4x = 3(3x - 1) + 7$

23. $5x - 3(2x - 9) = 4(3x - 8) + 7x$

24. $10 - 6(2x + 1) = 13 - 9x$

25. $3(x + 4) - 2(x + 4) = 4(2x - 3) - 8(x - 2)$

26. $7x - 4(3x - 10) = 2(x - 8) + 6(3x + 1)$

27. $\frac{3}{4}t - 6 = \frac{1}{3}t - 1$

28. $\frac{1}{8}n - 7 = -\frac{1}{4}n - 13$

29. $\frac{1}{2}(2x - 7) - \frac{1}{3}x = \frac{4}{3}x - 5$

30. $\frac{2}{3}(x - 6) - \frac{1}{4}x = \frac{3}{8}(8 - x) + \frac{1}{2}x$

31. $2b - 9 = 2b - 9$

32. $4t + 11 = 4t - 11$

33. $3(2n - 1) + 5 = 2(3n + 2)$

34. $5(n - 3) - 3n = 2(n - 7) - 1$

35. $5x - 2(3 - 2x) = 3(3x - 2)$

36. $7x - (3x - 4) = 4x - 4$

37. $|x| = 2$

38. $|x| = 13$

39. $|x| = 929$

40. $|x| = 0$

41. $|x - 8| = 7$

42. $|2x - 11| = 5$

43. $|3x + 4| = 8$

44. $|4 - 5x| = 13$

45. $|x - 9| = -4$

46. $|x| - 9 = -4$

47. $|2x + 8 - 3x| = 6$

48. $|3(2 - x) + 7x| = 14$

49. $|x - 3| + 7 = 12$

50. $|x + 12| - 10 = 10$

51. $|4x - 11| - 19 = -8$

52. $|7x - 52| + 18 = 14$

53. $2|3x + 2| - 9 = 17$

54. $2|5x - 3| + 7 = 21$

55. $-|x + 3| - 10 = -16$

56. $-3|2x + 1| + 13 = -29$

57. $|4x - 5| = |3x - 23|$

58. $|x - 14| = |3x + 2|$

59. $|x + 12| = |2x - 9|$

60. $|2x + 1| = |x - 19|$

61. $|5x| = |2x - 21|$

62. $|7x - 11| = |4x + 6|$

63. Find an absolute value equation whose solution is $\{-2, 2\}$.

64. Find an absolute value equation whose solution is $\{-6, 10\}$.

65. Find an absolute value equation whose solution is $\{3, -5\}$.

66. Find an absolute value equation whose solution is $\{\frac{1}{2}, \frac{7}{2}\}$.

Answer in complete sentences.

67. Explain how to determine that a linear equation is an identity.

68. Explain how to determine that a linear equation is an contradiction.

69. Explain how to solve a linear equation. Use an example to illustrate the process.

70. Explain how to solve an absolute value equation in which an absolute value is equal to a positive number. Use an example to illustrate the process.

71. Explain why the equation $|2x - 7| + 6 = 4$ has no solution.

72. Here is a student's work for solving the equation $|x - 3| = -5$:

$$|x - 3| = -5$$
$$x - 3 = -5 \quad \text{or} \quad x - 3 = 5$$
$$x = -2 \quad \text{or} \quad x = 8$$
$$\{-2, 8\}$$

Explain the student's error and how to avoid making a similar error in the future.

QUICK REVIEW EXERCISES

Section 8.1

Evaluate.

1. $f(x) = 3x - 14, f(-6)$

2. $f(x) = x^2 + 17x - 60, f(-5)$

3. $f(x) = \dfrac{x^2 - 6x + 25}{3x + 7}, f(2)$

4. $f(x) = |x - 7| + 3, f(0)$

Study Tip **REVISITED**   **Study Groups** Creating a study group is one of the best suggestions to help you learn mathematics and improve your performance on quizzes and exams. When working with a group of students, as long as one student understands the material being covered, then there is a good chance that each student in the group will end up understanding the material. Sometimes, a concept will be easier for you to understand if it is explained to you by a peer.

When you explain a certain topic to a fellow student or show a student how to solve a particular problem, you increase your chances of retaining this knowledge. Because you are forced to choose your words in such a way that the other student will completely understand, you have demonstrated that you truly understand.

You will also find it helpful to have a support group: students who can lean on each other when times are bad. By being supportive of each other, you are increasing the chances that you will all learn and be successful in your class.

8.2

LINEAR INEQUALITIES AND ABSOLUTE VALUE INEQUALITIES

Objectives

1 Graph the solutions of a linear inequality on a number line and express the solutions by using interval notation.

2 Solve linear inequalities.

3 Solve compound linear inequalities.

4 Solve absolute value inequalities.

Linear Inequalities and Their Solutions

Objective 1 **Graph the solutions of a linear inequality on a number line and express the solutions by using interval notation.** In this section, we will examine solving inequalities. We begin by reviewing linear inequalities and the presentation of their solutions. The inequality $x < 4$ has any real number that is less than 4 as its solution. This can be represented on a number line as follows:

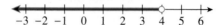

This can be expressed in interval notation as $(-\infty, 4)$.

The following table shows the number line and interval notation associated with several types of simple linear inequalities:

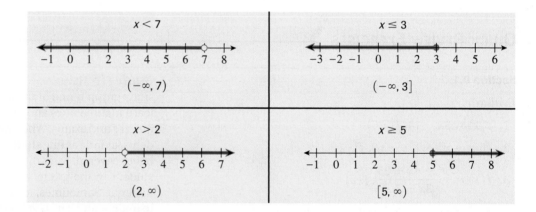

Recall that we use a closed circle whenever the endpoint is included in the interval ($\leq$ or $\geq$) and an open circle when the endpoint is not included ($<$ or $>$). When expressing the interval in interval notation, we use square brackets when the endpoint is included, and we use parentheses when the endpoint is not included.

Solving Linear Inequalities

Objective 2 **Solve linear inequalities.** Solving a linear inequality is much the same as solving a linear equation. The exception is if we multiply or divide both sides of the inequality by a negative number; then the direction of the inequality changes.

EXAMPLE 1 Solve $3x - 5 < -14$.

Solution

$$3x - 5 < -14$$
$$3x < -9 \qquad \text{Add 5 to both sides.}$$
$$x < -3 \qquad \text{Divide both sides by 3.}$$

Here is the number line showing our solution:

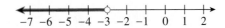

The solution expressed in interval notation is $(-\infty, -3)$.

Quick Check **1** Solve $2x + 5 \le 9$.

EXAMPLE 2 Solve $5x + 11 \le 2(2x + 4) - 3$.

Solution

$$5x + 11 \le 2(2x + 4) - 3$$
$$5x + 11 \le 4x + 8 - 3 \qquad \text{Distribute 2.}$$
$$5x + 11 \le 4x + 5 \qquad \text{Combine like terms.}$$
$$x + 11 \le 5 \qquad \text{Collect variables on the left side by subtracting } 4x \text{ from both sides.}$$
$$x \le -6 \qquad \text{Subtract 11 from both sides to collect constant terms on the right side.}$$

Here is the number line showing our solution:

The solution expressed in interval notation is $(-\infty, -6]$.

Quick Check **2** Solve $3(2x - 1) + 4 \ge 2x - 7$.

EXAMPLE 3 Solve $-2x + 3 > 13$.

Solution

$$-2x + 3 > 13$$
$$-2x > 10 \qquad \text{Subtract 3 from both sides.}$$
$$\frac{-2x}{-2} < \frac{10}{-2} \qquad \text{Divide both sides by } -2. \text{ This changes the direction of the inequality from } > \text{ to } <.$$
$$x < -5 \qquad \text{Simplify.}$$

Here is the number line showing our solution:

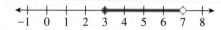

The solution expressed in interval notation is $(-\infty, -5)$.

Quick Check **3** Solve $5 - 2x > -5$.

> *A Word of Caution*   When we multiply or divide both sides of an inequality by a negative number, we must change the direction of the inequality.

Solving Compound Linear Inequalities

Objective 3 **Solve compound linear inequalities.** A compound inequality is made up of two simple inequalities. For instance, the compound inequality $3 \le x < 7$ represents numbers that are both greater than or equal to 3 $(x \ge 3)$ and less than 7 $(x < 7)$. This can be represented graphically by the following number line:

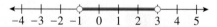

The interval notation for this interval is $[3, 7)$.

EXAMPLE 4   Solve $-3 < 2x - 1 < 5$.

Solution

The goal when solving a compound inequality such as this one is to isolate the variable x between two constants. We will work on all three parts of the inequality at the same time.

$$-3 < 2x - 1 < 5$$
$$-2 < 2x < 6 \qquad \text{Add 1 to each part of the inequality.}$$
$$-1 < x < 3 \qquad \text{Divide both sides by 2.}$$

Our solution consists of all real numbers greater than -1 and also less than 3. Here is the number line showing the solution:

The solution expressed in interval notation is $(-1, 3)$.

Quick Check **4** Solve $-8 \le 3x + 7 \le 1$.

Another type of compound inequality involves the word *or*. Consider the compound inequality $x < -2$ *or* $x > 1$. The solutions to this inequality are values that are solutions to either of the two inequalities. In other words, any number that is less than -2 $(x < -2)$ *or* is greater than 1 $(x > 1)$ is a solution of the compound inequality. Here are the solutions of this compound inequality, graphed on a number line:

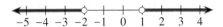

Notice that we have two different intervals that contain solutions. The solution expressed in interval notation is $(-\infty, -2) \cup (1, \infty)$. The symbol $\cup$ represents a union in set theory and combines the two intervals into one solution.

EXAMPLE 5   Solve $x + 4 \le 7$ or $2x - 1 > 11$.

Solution

To solve a compound inequality of this type, we first solve each inequality separately. We begin by solving $x + 4 \le 7$.

$$x + 4 \le 7$$
$$x \le 3 \qquad \text{Subtract 4 from both sides.}$$

Now we solve the other inequality.

$$2x - 1 > 11$$
$$2x > 12 \qquad \text{Add 1 to both sides.}$$
$$x > 6 \qquad \text{Divide both sides by 2.}$$

Here is a number line showing the solution:

The solution expressed in interval notation is $(-\infty, 3] \cup (6, \infty)$.

Quick Check   **5** Solve $x + 3 < -3$ or $2x + 5 > 14$.

Absolute Value Inequalities

Objective 4 **Solve absolute value inequalities.** As in the previous section, we will expand the topic to include absolute values. Consider the inequality $|x| < 4$. Any real number in the following interval has an absolute value less than 4:

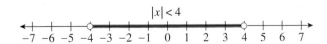

This can be expressed as the compound inequality $-4 < x < 4$. Using this idea, anytime we have an absolute value of an expression that is less than a positive number, we will begin by "trapping" the expression between that positive number and its opposite, such as $-4 < x < 4$. We then proceed to solve the resulting compound inequality.

Solving Absolute Value Inequalities of the Form $|X| < a$ or $|X| \le a$

> For any expression X and any positive number a, the solutions of the inequality $|X| < a$ can be found by solving the compound inequality
>
> $$-a < X < a$$
>
> Similarly, for any expression X and any positive number a, the solutions of the inequality $|X| \le a$ can be found by solving the compound inequality
>
> $$-a \le X \le a$$

EXAMPLE ⟩6 Solve $|x - 4| < 2$.

Solution

The first step is to "trap" $x - 4$ between -2 and 2.

$$
\begin{aligned}
|x - 4| &< 2 \\
-2 < x - 4 &< 2 \qquad \text{"Trap" } x - 4 \text{ between } -2 \text{ and } 2. \\
2 < x &< 6 \qquad \text{Add 4 to each part of the inequality.}
\end{aligned}
$$

Here is the number line showing our solution:

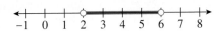

The solution expressed in interval notation is $(2, 6)$.

Quick Check 6 ⟩ Solve $|x + 3| \le 1$.

As with equations involving absolute values, we must first isolate the absolute value before rewriting the inequality as a compound inequality.

EXAMPLE ⟩7 Solve $|3x + 7| - 5 \le 4$.

Solution

$$
\begin{aligned}
|3x + 7| - 5 &\le 4 \\
|3x + 7| &\le 9 \qquad \text{Add 5 to both sides.} \\
-9 \le 3x + 7 &\le 9 \qquad \text{Rewrite as a compound inequality.} \\
-16 \le 3x &\le 2 \qquad \text{Subtract 7 from each part of the inequality.} \\
-\frac{16}{3} \le x &\le \frac{2}{3} \qquad \text{Divide each part of the inequality by 3.}
\end{aligned}
$$

Here is the number line showing our solution:

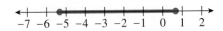

The solution expressed in interval notation is $\left[-\frac{16}{3}, \frac{2}{3}\right]$.

Quick Check   7 Solve $|2x - 1| + 4 < 9$.

A Word of Caution When solving an absolute value inequality, we must isolate the absolute value before rewriting the inequality as a compound inequality.

If we have an inequality in which an absolute value is less than a negative number, such as $|x| < -2$, or less than 0, such as $|x| < 0$, then this inequality has no solution. Since an absolute value is always 0 or greater, it can never be *less than* a negative number.

EXAMPLE 8   Solve $|4x + 3| + 7 < 5$.

Solution

$$|4x + 3| + 7 < 5$$
$$|4x + 3| < -2 \qquad \text{Subtract 7.}$$

Since an absolute value cannot be less than a negative number, this inequality has no solution: $\varnothing$.

Quick Check 8
Solve $|x + 1| - 3 < -8$.

A Word of Caution An inequality in which an absolute value is less than a negative number has no solution. Do not try to rewrite the inequality as a compound inequality.

Consider the inequality $|x| > 4$. Notice that we now have the absolute value of x *greater than* 4, rather than *less than* 4. Inequalities with an absolute value *greater* than a positive number must be approached in a different manner than absolute value inequalities with an absolute value *less* than a positive number. We can't "trap" the expression between two constants. Solutions of the inequality $|x| > 4$ are in one of two categories. First, any positive number greater than 4 will have an absolute value greater than 4 as well. Second, any negative number less than -4 will also have an absolute value greater than 4. The following number line shows where the solutions of this inequality can be found:

This can be expressed symbolically as $x < -4$ or $x > 4$.

Solving Absolute Value Inequalities of the Form $|X| > a$ or $|X| \geq a$

For any expression X and any positive number a, the solutions of the inequality $|X| > a$ can be found by solving the compound inequality

$$X < -a \qquad \text{or} \qquad X > a$$

Similarly, for any expression X and any positive number a, the solutions of the inequality $|X| \geq a$ can be found by solving the compound inequality

$$X \leq -a \qquad \text{or} \qquad X \geq a$$

EXAMPLE ▶9   Solve $|x + 3| > 2$.

Solution

We begin by converting this inequality into the compound inequality $x + 3 < -2$ or $x + 3 > 2$. We then solve each inequality.

$$|x + 3| > 2$$

$x + 3 < -2$	or	$x + 3 > 2$	Rewrite as a compound inequality.
$x < -5$	or	$x > -1$	Subtract 3 to solve each inequality.

Here is the number line showing our solution:

The solution expressed in interval notation is $(-\infty, -5) \cup (-1, \infty)$.

Quick Check   **9** ▶ Solve $|x - 2| > 4$.

Whether we are solving an absolute value equation or an absolute value inequality, the first step is always to isolate the absolute value. The following table shows the correct step to take when comparing an absolute value with a nonnegative number:

"Equal To"	$|2x + 9| = 5$	$2x + 9 = 5$ or $2x + 9 = -5$
"Less Than"	$|x - 4| < 3$	$-3 < x - 4 < 3$
or	or	or
"Less Than or Equal To"	$|x - 4| \leq 3$	$-3 \leq x - 4 \leq 3$
"Greater Than"	$|3x + 5| > 7$	$3x + 5 < -7$ or $3x + 5 > 7$
or	or	or
"Greater Than or Equal To"	$|3x + 5| \geq 7$	$3x + 5 \leq -7$ or $3x + 5 \geq 7$

EXAMPLE ▶10   Solve $|4x - 9| - 3 \geq 5$.

Solution

We begin by isolating the absolute value.

$$|4x - 9| - 3 \geq 5$$
$$|4x - 9| \geq 8 \qquad \text{Add 3 to both sides.}$$

Since this inequality involves an absolute value that is *greater than* or equal to 8, we will rewrite this absolute value inequality as two linear inequalities.

$4x - 9 \leq -8$	or	$4x - 9 \geq 8$	Convert to two linear inequalities.
$4x \leq 1$	or	$4x \geq 17$	Add 9 to both sides.
$x \leq \dfrac{1}{4}$	or	$x \geq \dfrac{17}{4}$	Divide both sides by 4.

Here is the number line showing our solution:

The solution expressed in interval notation is $\left(-\infty, \frac{1}{4}\right] \cup \left[\frac{17}{4}, \infty\right)$.

Quick Check   10 Solve $|2x + 7| + 5 > 10$.

Consider the inequality $|x| > -2$. Since the absolute value of any number must be 0 or greater, then $|x|$ must always be greater than -2 for any real number x. The solution set is the set of all real numbers $\Re$.

EXAMPLE 11   Solve $|3x + 5| - 3 > -4$.

Solution

We begin by isolating the absolute value.

$$|3x + 5| - 3 > -4$$
$$|3x + 5| > -1 \qquad \text{Add 3 to both sides to isolate the absolute value.}$$

This inequality must always be true, so our solution is the set of all real numbers $\Re$. Here is the number line showing our solution:

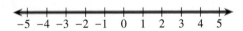

The interval notation associated with the set of real numbers is $(-\infty, \infty)$.

Quick Check   11 Solve $|7x - 13| \geq -1$.

If an equation or inequality compares an absolute value with a negative number, the following table summarizes the solutions:

Equation or Inequality	Solution
$|x + 3| = -2$	$\varnothing$
$|2x - 1| < -4$	$\varnothing$
$|3x + 4| > -3$	All real numbers: $\Re$

Solve the inequality. Graph your solution on a number line, and write your solution in interval notation.

1. $x - 4 < 3$

2. $x + 7 > 2$

3. $4x \leq -20$

4. $-3x \leq 24$

5. $3x + 2 \geq -10$

6. $2x + 5 > 9$

7. $\dfrac{2}{3}x - \dfrac{5}{6} > -2$

8. $\dfrac{3}{5}x - \dfrac{1}{3} < \dfrac{2}{5}$

9. $5(2x + 1) - 7x \leq 4x - 13$

10. $2x - 5(x - 3) < x + 9$

11. $8 - 3(6x - 4) > (2x + 15) - 7(3x + 4)$

12. $2(3x + 2) - (4x - 5) \geq -3(4 - 2x)$

13. $5 \leq x - 3 \leq 11$

14. $-2 < x + 5 < 4$

15. $-15 \leq 5x \leq 35$

16. $10 < 4x < 20$

17. $-9 < 2x + 5 < -4$

18. $-8 \leq 3x + 7 \leq 28$

19. $-6 < \dfrac{3}{5}x + \dfrac{2}{5} < -2$

20. $\dfrac{7}{4} \leq \dfrac{1}{2}x - \dfrac{3}{8} \leq \dfrac{5}{2}$

21. $x + 6 < 8 \text{ or } x - 3 > 5$

22. $x - 4 \leq -7 \text{ or } x - 8 \geq -1$

23. $3x + 4 \leq -17 \text{ or } 2x - 9 \geq 9$

24. $\dfrac{2}{3}x - 2 < -4 \text{ or } x + \dfrac{5}{3} > \dfrac{17}{2}$

25. $5x - 11 < 3x - 3 \text{ or } 6x + 7 > 2x + 31$

26. $4x - 3 \leq -21 \text{ or } 4x - 3 \geq 21$

27. $|x + 2| < 3$

28. $|x - 4| > 1$

29. $|2x + 6| > 4$

30. $|3x - 7| > 5$

31. $|3x| < 15$

32. $|2x - 9| < 7$

33. $|5x + 4| \geq -6$

34. $|4x - 6| < 10$

35. $|8x - 5| + 7 \geq 10$

36. $|x + 4| + 4 > 2$

37. $|4x + 9| - 7 \leq -3$

38. $|2x - 9| - 5 \geq 5$

39. $3|x - 7| \geq 9$

40. $4|3x - 5| \leq 16$

41. $|7x + 14| + 9 < 6$

42. $|3x - 1| + 5 \leq 9$

43. $|5x - 3| - 11 \leq 11$

44. $|6x - 11| - 5 \geq 8$

45. $|8 - x| \leq 5$

46. $|4 - 3x| \geq 10$

47. $2|3x + 2| + 7 > 19$

48. $3|x - 9| - 5 < 4$

49. $4|x + 5| - 11 > 17$

50. $2|2x + 10| + 19 \leq 11$

51. Find an absolute value inequality whose solution is $(-2, 2)$.

52. Find an absolute value inequality whose solution is $[-2, 6]$.

53. Find an absolute value inequality whose solution is $(-\infty, 1) \cup (9, \infty)$.

54. Find an absolute value inequality whose solution is $(-\infty, -4] \cup [-1, \infty)$.

Answer in complete sentences.

55. Explain how to solve an inequality of the form $|X| < a$. Use an example to illustrate the process.

56. Explain how to solve an inequality of the form $|X| > a$. Use an example to illustrate the process.

57. Explain why the inequality $|x + 5| < -3$ has no solutions, while every real number is a solution to the inequality $|x + 5| > -3$.

58. Explain why the inequality $|x| - 7 < -4$ has solutions, but the inequality $|x - 7| < -4$ does not.

***Study Tip* REVISITED Using Your Textbook and Other Resources** The textbook is a valuable resource, when used properly. Read a section the night before your instructor covers it to familiarize yourself with the material in the section. This way, you can have questions prepared for your instructor in class. After a section is covered in class, reread the section.

As you reread a section, you can use the textbook to create note cards to help you memorize new terms or procedures or to show you how to solve particular problems. After you read through an example, attempt the Quick Check exercise that follows it to reinforce the concept presented in the example. As you work through the homework exercises, be sure to refer back to the examples found in that section for assistance.

In each section, the feature titled A Word of Caution points out common student mistakes and offers advice for not making the same mistake yourself.

Your instructor is a valuable resource for answering questions and providing advice. Take advantage of your instructor's office hours in addition to the time in class. The tutorial center is another valuable resource. A quality tutor can help you to understand the material.

Finally, the online supplement to this textbook at MyMathLab.com can be quite helpful. At this site, you will find video clips, tutorial exercises, and much, much more.

GRAPHING
LINEAR
EQUATIONS
AND LINEAR
FUNCTIONS;
GRAPHING
ABSOLUTE
VALUE
FUNCTIONS

Objectives

1 Graph linear equations by using *x*- and *y*-intercepts.
2 Graph linear equations by using the slope and *y*-intercept.
3 Graph linear functions.
4 Graph absolute value functions.

In Chapter 3, we learned how to graph linear equations in two variables, such as $4x - y = 8$ or $y = -\frac{2}{3}x + 2$. We will begin this section with a brief review of how to graph this type of equation.

Graphing Linear Equations in Two Variables Using the *x*- and *y*-Intercepts

Objective 1 Graph linear equations by using *x*- and *y*-intercepts. We developed two techniques for graphing linear equations. The first involved finding the *x*- and *y*-intercepts of the line and then using these points to graph the line. Recall that the *x*-intercept of a line is the point where the line intersects the *x*-axis. To find the *x*-intercept, we substitute 0 for *y* in the equation and solve for *x*. The *y*-intercept is the point where the line crosses the *y*-axis, and can be found by substituting 0 for *x* in the equation and solving for *y*.

EXAMPLE 1 Find any intercepts and graph $4x - y = 8$.

Solution

We find the *x*-intercept by substituting 0 for *y* and solving for *x*.

$$
\begin{aligned}
4x - 0 &= 8 && \text{Substitute 0 for } y. \\
4x &= 8 && \text{Simplify.} \\
x &= 2 && \text{Divide both sides by 4.}
\end{aligned}
$$

The *x*-intercept is at $(2, 0)$. To find the *y*-intercept we substitute 0 for *x* and solve for *y*.

$$
\begin{aligned}
4(0) - y &= 8 && \text{Substitute 0 for } x. \\
-y &= 8 && \text{Simplify.} \\
y &= -8 && \text{Divide both sides by } -1.
\end{aligned}
$$

The *y*-intercept is at $(0, -8)$.

It is helpful to find a third point before graphing the line to serve as a check for our other two points. If the three points do not lie on a straight line, then at least one of the points is incorrect. To find a third point, we may select any value to substitute for *x* and solve for *y*. In this example, we will use 1 for *x*, but we could really use any value we choose.

$$
\begin{aligned}
4(1) - y &= 8 && \text{Substitute 1 for } x. \\
4 - y &= 8 && \text{Multiply.} \\
-y &= 4 && \text{Subtract 4.} \\
y &= -4 && \text{Divide both sides by } -1.
\end{aligned}
$$

This tells us that the point $(1, -4)$ lies on the line. Following is a graph showing the line, as well as the three points that we have found:

Quick Check 1
Find any intercepts and
graph $5x + 2y = -10$.

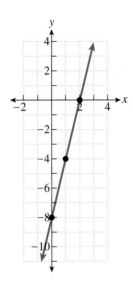

Graphing a Linear Equation Using Its Slope and *y*-Intercept

Objective 2   **Graph linear equations by using the slope and *y*-intercept.** A linear equation is in slope–intercept form if it is in the form $y = mx + b$. The constant m that is multiplied by x is the slope of the line, and the constant b represents the *y*-coordinate of the *y*-intercept. For example, for the equation $y = -\frac{2}{3}x + 2$, the slope of the line is $-\frac{2}{3}$ and the *y*-intercept is $(0, 2)$. Recall that slope is a measure of how quickly the line rises or falls as it moves to the right. A line whose slope is 4 moves up four units for every one unit it moves to the right, while a line whose slope is $-\frac{2}{3}$ moves down two units for every three units it moves to the right.

If an equation is in slope–intercept form, it provides us with the necessary information to graph the line. We begin by placing the *y*-intercept on the graph. From that point, we use the slope of the line to locate other points that are on the graph.

EXAMPLE 2   Graph $y = 2x + 4$.

Solution

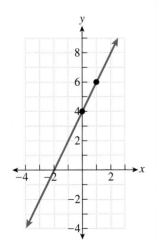

Since this equation is in slope–intercept form, we know that the *y*-intercept is $(0, 4)$, and we plot this point on the graph. The slope of this line is 2, so we can find another point on the line by starting at the *y*-intercept, $(0, 4)$, and moving up two units and one unit to the right. This tells us that the point $(1, 6)$ is also on the line. We could continue to find more points on the line using the same procedure (up two and to the right one). At the left is a graph of the line.

To check that our work is valid, we will find the *x*-intercept from the equation and see if it matches the *x*-intercept on the graph.

$$0 = 2x + 4 \qquad \text{Substitute 0 for } y.$$
$$-4 = 2x \qquad \text{Subtract 4.}$$
$$-2 = x \qquad \text{Divide both sides by 2.}$$

Quick Check 2

Graph $y = \frac{7}{2}x - 4$ and find any intercepts.

The x-intercept is $(-2, 0)$, which matches our graph.

Linear Functions and Their Graphs

Objective **3** **Graph linear functions.** A linear function is a function of the form $f(x) = mx + b$. Recall that the notation $f(x)$, read "f of x," stands for the function f evaluated at x. For example, consider the linear function $f(x) = 3x - 8$. If we wanted to evaluate $f(6)$, we would substitute 6 for x in the function and simplify.

$$
\begin{aligned}
f(6) &= 3(6) - 8 &&\text{Substitute 6 for } x.\\
&= 18 - 8 &&\text{Multiply.}\\
&= 10 &&\text{Subtract.}
\end{aligned}
$$

Graphing a linear function $f(x) = mx + b$ is identical to graphing a linear equation $y = mx + b$. We plot points of the form $(x, f(x))$, beginning with the y-intercept. We then use the slope of the line to find other points on the line.

EXAMPLE 3   Graph $f(x) = -\frac{3}{2}x + 6$.

Solution

The y-intercept of this graph is at $(0, 6)$, so we begin by placing that point on the graph. The slope of this line is $-\frac{3}{2}$, so, beginning at the y-intercept, we move down three units and two units to the right. This tells us that the point $(2, 3)$ is on the graph.

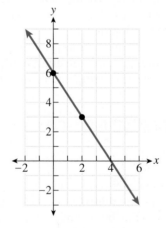

Quick Check 3

Graph the linear function $f(x) = -x + 6$ and find any intercepts.

It is left to the reader to verify that the x-intercept is at $(4, 0)$, which matches our graph.

Absolute Value Functions

Objective **4** **Graph absolute value functions.** We will now graph our first nonlinear function: the **absolute value function** $f(x) = |x|$. We will begin to graph this function by creating the following table of function values for various values of x:

x	$f(x) = |x|$	$(x, f(x))$
-2	$f(-2) = |-2| = 2$	$(-2, 2)$
-1	$f(-1) = |-1| = 1$	$(-1, 1)$
0	$f(0) = |0| = 0$	$(0, 0)$
1	$f(1) = |1| = 1$	$(1, 1)$
2	$f(2) = |2| = 2$	$(2, 2)$

Here is the graph. Notice that the graph is not a straight line but instead is V-shaped. The most important point for graphing an absolute value function is the point of the V, where the graph changes from falling to rising. The turning point is the origin for this particular function.

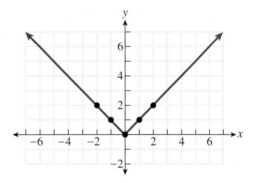

To graph an absolute value function, we find the x-coordinate of the turning point, select two values to the left and right of this value, and create a table of values. Choosing the values for x is crucial. Suppose that we had chosen only positive values for x when graphing $f(x) = |x|$. All of our points would have been on the straight line located to the right of the origin, and our graph would have been a straight line rather than being V-shaped.

The turning point occurs at the value of x that makes the expression inside the absolute value bars equal to 0. This is because the minimum output of an absolute value occurs when we take the absolute value of 0.

Graphing an Absolute Value Function

- Determine the value of x for which the expression inside the absolute value bars is equal to 0.
- In addition to this value, select two values that are less than this value and two that are greater.
- Create a table of function values for these values of x.
- Place the points $(x, f(x))$ on the graph and draw the V-shaped graph that passes through these points.

EXAMPLE 4 Graph $f(x) = |x + 3| - 2$.

Solution

For an absolute value function, we begin by finding the value of x that makes the expression inside the absolute value bars equal to 0.

$$x + 3 = 0 \qquad \text{Set } x + 3 \text{ equal to 0.}$$
$$x = -3 \qquad \text{Subtract 3.}$$

Now we select two values that are less than -3, such as -5 and -4, and two values that are greater than -3, such as -2 and -1. Next we create a table of values.

x	$f(x) = |x + 3| - 2$	$(x, f(x))$
-5	$f(-5) = |-5 + 3| - 2 = 2 - 2 = 0$	$(-5, 0)$
-4	$f(-4) = |-4 + 3| - 2 = 1 - 2 = -1$	$(-4, -1)$
-3	$f(-3) = |-3 + 3| - 2 = 0 - 2 = -2$	$(-3, -2)$
-2	$f(-2) = |-2 + 3| - 2 = 1 - 2 = -1$	$(-2, -1)$
-1	$f(-1) = |-1 + 3| - 2 = 2 - 2 = 0$	$(-1, 0)$

Here is the graph, based on the points from our table:

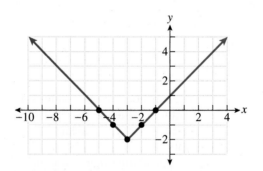

Quick Check 4

Graph the function
$f(x) = |x + 1| + 3$.

Using Your Calculator We can graph absolute value functions using the TI-83/84. (Recall that to access the absolute value function on the TI-83/84, press the [MATH] key; you will find the absolute value function under the **NUM** menu.) To enter the equation of the function from Example 4, $f(x) = |x + 3| - 2$, press the [Y=] key. Enter $|x + 3| - 2$ next to Y_1. Now press the key labeled [GRAPH] to graph the function.

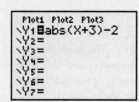
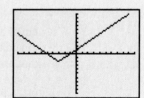

Recall that the domain of a function is the set of all possible input values for a function and the range of a function is the set of all possible output values for a function. From the graph of a function, the domain can be read from left to right and the range can be read from bottom to top. Let's look at the graph of $f(x) = |x|$ once again.

Notice that the graph extends out to infinity on the left and to infinity on the right. Its domain is the set of real numbers $\Re$, which is expressed in interval notation as $(-\infty, \infty)$. The lowest value of this function is 0, but it has no upper bound. The range of this function is $[0, \infty)$.

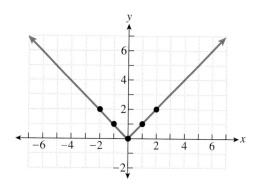

Let's reexamine the graph of $f(x) = |x + 3| - 2$. The domain of this function is also the set of real numbers. What is the range? The lowest function value for this function is -2, so the range is $[-2, \infty)$.

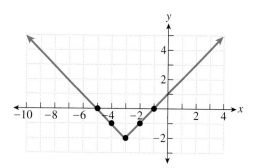

EXAMPLE 5   Graph the function $f(x) = |x - 1| + 4$, and find the domain and range of the function.

Solution

We begin by finding the x-coordinate of the turning point of the graph. To do this, we set the expression inside the absolute value bars equal to 0 and solve for x.

$$x - 1 = 0 \qquad \text{Set } x - 1 \text{ equal to 0.}$$
$$x = 1 \qquad \text{Add 1.}$$

Now we can create a table of values, using two values for x that are less than 1 and two that are greater than 1.

x	$f(x) = \lvert x - 1 \rvert + 4$	$(x, f(x))$
-1	$f(-1) = \lvert -1 - 1 \rvert + 4 = 6$	$(-1, 6)$
0	$f(0) = \lvert 0 - 1 \rvert + 4 = 5$	$(0, 5)$
1	$f(1) = \lvert 1 - 1 \rvert + 4 = 4$	$(1, 4)$
2	$f(2) = \lvert 2 - 1 \rvert + 4 = 5$	$(2, 5)$
3	$f(3) = \lvert 3 - 1 \rvert + 4 = 6$	$(3, 6)$

Below is the graph. From this graph, we can see that the domain is the set of all real numbers and the range is $[4, \infty)$.

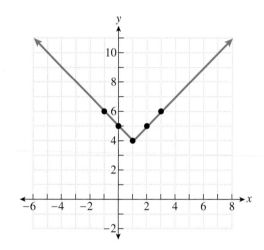

Quick Check 5

Graph the function $f(x) = \lvert x - 5 \rvert - 4$, and state the domain and range.

A Word of Caution   When graphing an absolute value function, choosing values for x wisely will result in a series of points that forms a "V." If all of our points appear to lie on a straight line, then either we have chosen values for x that are all on one side of the point where the graph changes from decreasing to increasing or we have made some mistakes when evaluating the function. In either case, we should go back and check our work.

EXAMPLE 6   Graph $f(x) = -\lvert x \rvert + 2$, and state the domain and range.

Solution

Notice that this absolute value function is slightly different than the first few we graphed. There is a negative sign in front of the absolute value bars, and this shall have a significant impact on how the graph looks. However, we will use the same approach to graphing this function as the previous absolute value functions. We begin by finding the x-coordinate of the turning point by setting the expression inside the absolute value bars equal to 0. In this case, $x = 0$ is the x-coordinate of the turning point. Now we can create a table of values, using two values for x that are less than 0 and two that are greater than 0.

x	$f(x) = -|x| + 2$	$(x, f(x))$
-2	$f(-2) = -|-2| + 2 = -2 + 2 = 0$	$(-2, 0)$
-1	$f(-1) = -|-1| + 2 = -1 + 2 = 1$	$(-1, 1)$
0	$f(0) = -|0| + 2 = 0 + 2 = 2$	$(0, 2)$
1	$f(1) = -|1| + 2 = -1 + 2 = 1$	$(1, 1)$
2	$f(2) = -|2| + 2 = -2 + 2 = 0$	$(2, 0)$

Quick Check 6

Graph the function
$f(x) = -|x - 4| - 5$,
and state the domain
and range.

Below is the graph. From this graph, we can see that the domain is the set of all real numbers. This time the function has a maximum value of 2 with no lower bound, so the range is $(-\infty, 2]$.

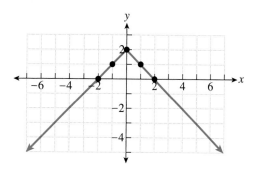

EXAMPLE 7   Find the absolute value function $f(x)$ on the graph.

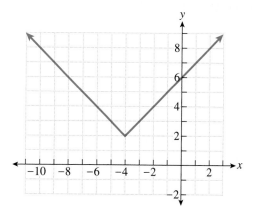

Solution

To find the function, let's focus on the turning point of this graph, which is at $(-4, 2)$. Since the x-coordinate is -4, the expression $x + 4$ must be inside the absolute value bars. The y-coordinate of the turning point is 2, which tells us that 2 is added to the absolute value. The function is of the form $f(x) = a|x + 4| + 2$, where a is a real number. To determine the value of a, we need to use the coordinates of another point on the graph, such as the y-intercept $(0, 6)$. This point tells us that $f(0) = 6$. We will use this fact to find a.

$$f(0) = 6$$

$a|0 + 4| + 2 = 6$ Substitute 0 for x in the function $f(x)$.

$4a + 2 = 6$ Simplify the absolute value.

$4a = 4$ Subtract 2.

$a = 1$ Divide both sides by 4.

Substituting 1 for a, we find that the function is $f(x) = |x + 4| + 2$.

Quick Check 7 Find the absolute value function on the graph.

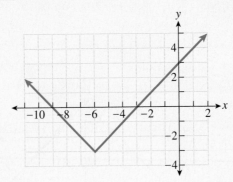

EXERCISES 8.3 ❯

Find the intercepts and then graph the line.

1. $x + 2y = -8$

3. $3x - 4y = -12$

2. $3x + y = 9$

4. $-5x + 7y = 35$

5. $\frac{2}{9}x + \frac{1}{3}y = 2$

6. $\frac{1}{6}x - \frac{1}{3}y = -\frac{2}{3}$

7. $-5x + 3y = 9$ **8.** $4x + y = 7$

9. $y = -3x + 4$ **10.** $y = -\frac{1}{5}x + 2$

Graph the line using the slope and y-intercept.

11. $y = x + 3$ **12.** $y = x - 4$

13. $y = 2x + 1$ **14.** $y = 3x - 3$

15. $y = -3x + 6$ **16.** $y = -2x - 5$

17. $y = \frac{2}{3}x - 6$

18. $y = -\dfrac{5}{2}x + 5$

19. $y = 4x$

24. $f(x) = 2x + 3$

25. $f(x) = -5x + 8$

20. $y = -\dfrac{1}{3}x$

26. $f(x) = -3x - 6$

27. $f(x) = \dfrac{7}{3}x - 4$

Graph the linear function.

21. $f(x) = x + 7$

28. $f(x) = \dfrac{4}{9}x$

22. $f(x) = x - 3$

23. $f(x) = 4x - 5$

29. $f(x) = -5$

30. $f(x) = 4$

35. $f(x) = |x| + 3$

Graph the absolute value function. State the domain and range of the function.

36. $f(x) = |x| - 4$

31. $f(x) = |x + 2|$

37. $f(x) = |x| - 6$

32. $f(x) = |x - 4|$

38. $f(x) = |x| + 1$

33. $f(x) = |x - 7|$

39. $f(x) = |x + 3| - 4$

34. $f(x) = |x + 6|$

40. $f(x) = |x + 1| + 2$

46. $f(x) = |8x + 21| + 3$ **47.** $f(x) = 2|x - 1| + 3$

41. $f(x) = |x - 6| + 3$

48. $f(x) = 3|x + 5| - 7$ **49.** $f(x) = -|x + 6|$

42. $f(x) = |x - 2| - 5$ **43.** $f(x) = |2x + 1|$

44. $f(x) = |3x - 8|$

50. $f(x) = -|x| + 6$

45. $f(x) = |4x - 5| - 8$

51. $f(x) = -|x - 3| - 4$

60. $f(x) = |x + 7| + 7$

Use the graph of the absolute value function $f(x)$ to solve the inequality.

61. $f(x) \le 0$

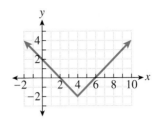

52. $f(x) = -|x + 8| + 5$

62. $f(x) < 0$

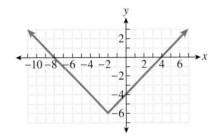

53. $f(x) = -3|x + 2| + 9$

63. $f(x) > 0$

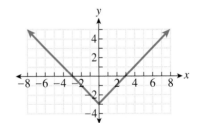

54. $f(x) = -4|x - 4| - 1$

64. $f(x) \ge 0$

Find the intercepts of the absolute value function.

55. $f(x) = |x + 5| - 4$

56. $f(x) = |x - 2| - 6$

57. $f(x) = |x - 8|$

58. $f(x) = |x| - 2$

59. $f(x) = |x - 3| + 1$

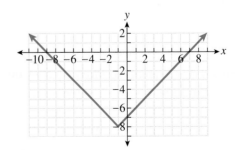

Find the absolute value function on the graph.

65.

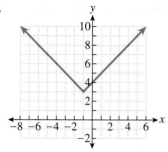

66.

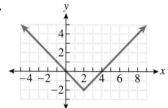

67.

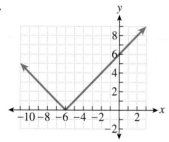

68.

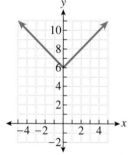

Answer in complete sentences.

69. Explain how to graph an absolute value function. Use an example to illustrate the process.

70. A student is trying to graph the function $f(x) = |x - 5| + 1$ and evaluates the function for the following values of x: $-2, -1, 0, 1, 2$. Will the student's graph be correct? If not, explain what the error is and how to avoid that error.

Study Tip **Revisited**   **Homework** Here are some tips for using homework assignments to increase your understanding of mathematics:

- Review your notes and the text before beginning a homework assignment.
- Be neat and complete when doing a homework assignment.
- Use strategies for homework problems you cannot do:
  - *Look in your notes and the text for similar problems or suggestions about this type of problem.*
  - *Ask members of your study group for help.*
  - *Call another student in the class for help.*
  - *Get help from a tutor at the tutorial center on your campus.*
  - *Ask your instructor for help at the first opportunity.*
- Complete a homework session by summarizing your work.
- Keep up to date with your homework assignments.

8.4

REVIEW OF
FACTORING;
QUADRATIC
EQUATIONS
AND RATIONAL
EQUATIONS

Objectives

1 Factor polynomials.
2 Solve quadratic equations.
3 Solve rational equations.

Factoring Polynomials

Objective 1 Factor polynomials. A polynomial is factored when it has been written as a product of two or more expressions. Every time we attempt to factor a polynomial, we should begin by attempting to factor out the greatest common factor (GCF) of all of the terms.

EXAMPLE 1   Factor $12x^5 - 3x^4 + 15x^2$.

Solution

The largest number that divides into these three coefficients is 3. The variable x is a factor of all three terms. For the GCF, we use the smallest power of x that appears in the terms, which is x^2. The GCF of these three terms is $3x^2$. We finish by factoring the GCF out of each term.

$$12x^5 - 3x^4 + 15x^2 = 3x^2(4x^3 - x^2 + 5)$$

Recall that we can check our factoring through multiplication. The reader can verify that $3x^2(4x^3 - x^2 + 5)$ equals $12x^5 - 3x^4 + 15x^2$.

Quick Check 1 Factor $36x^6 + 60x^5 - 12x^3$.

An important factoring technique is factoring by grouping. This technique is often used when a polynomial has four or more terms. To factor a polynomial by grouping, we split the terms of the polynomial into two groups. We then factor out a common factor of each group. At this point, if the two groups share a common factor, this common factor can be factored out of each group, leaving the polynomial in factored form.

EXAMPLE 2   Factor $x^3 - 6x^2 - 10x + 60$.

Solution

There are no factors (other than 1) that are common to all four terms, so we proceed to factor by grouping. We shall consider $x^3 - 6x^2$ to be the first group and $-10x + 60$ to be the second group.

$$x^3 - 6x^2 - 10x + 60 = x^2(x - 6) - 10(x - 6) \qquad \text{Factor } x^2 \text{ out of the first two terms and } -10 \text{ out of the last two terms.}$$

$$= (x - 6)(x^2 - 10) \qquad \text{Factor out the common factor } x - 6.$$

Recall that we can check our factoring by multiplying these two binomials. Their product should equal the original polynomial.

We will now discuss factoring quadratic trinomials of the form $x^2 + bx + c$, where b and c are integers. To factor such a polynomial, we look for two integers m and n whose product is c ($m \cdot n = c$) and whose sum is b ($m + n = b$). If we can find two integers that satisfy these conditions, then the polynomial factors to be $(x + m)(x + n)$.

When the constant term of the trinomial is positive, this tells us that the two integers we are looking for have the same sign. The sign of b tells us what the sign of each integer will be. When the constant term is negative, we are looking for one positive integer and one negative integer. The sign of the integer with the largest absolute value is the same as the sign of b.

EXAMPLE ▶ **3** Factor $x^2 + 8x + 15$.

Solution

Since the three terms do not contain a common factor other than 1, we begin by looking for two integers whose product is 15 and whose sum is 8. The integers that satisfy the conditions are 3 and 5, so $x^2 + 8x + 15$ can be factored as $(x + 3)(x + 5)$.

$$x^2 + 8x + 15 = (x + 3)(x + 5)$$

EXAMPLE ▶ **4** Factor $x^2 - 10x + 16$.

Solution

We cannot factor out a common factor, so we look for the integers with a product of 16 and a sum of -10. The two integers are -2 and -8.

$$x^2 - 10x + 16 = (x - 2)(x - 8)$$

EXAMPLE ▶ **5** Factor $3x^2 - 15x - 42$.

Solution

We begin by factoring out the common factor 3. This gives us $3(x^2 - 5x - 14)$. We then look to factor the trinomial factor by finding two integers with a product of -14 and a sum of -5. Again, since the product must be negative, we are looking for one negative integer and one positive integer. In this case the integers are -7 and 2.

$$3x^2 - 15x - 42 = 3(x^2 - 5x - 14)$$
$$= 3(x - 7)(x + 2)$$

If the leading coefficient of a quadratic trinomial is not equal to 1 and that coefficient cannot be factored out as a common factor, then we must use a different technique to factor the polynomial. We will use factoring by grouping to factor polynomials of the form $ax^2 + bx + c$ $(a \neq 1)$. (You could also use the method of trial and error, as presented in Section 6.3.)

We begin by multiplying a by c, finding two integers m and n whose product is equal to $a \cdot c$ and whose sum is b. Once we find these two integers, we rewrite the polynomial as $ax^2 + mx + nx + c$ and factor by grouping.

EXAMPLE 6   Factor $2x^2 - 3x - 20$.

Solution

We cannot factor out any common factors, so we begin by multiplying $2(-20)$, which equals -40. Next we look for two integers whose product is -40 and whose sum is -3. The two integers are -8 and 5, so we will replace the term $-3x$ in the original polynomial by $-8x + 5x$ and then factor by grouping.

$$
\begin{aligned}
2x^2 - 3x - 20 &= 2x^2 - 8x + 5x - 20 && \text{Rewrite } -3x \text{ as } -8x + 5x. \\
&= 2x(x - 4) + 5(x - 4) && \text{Factor the common factor } 2x \text{ from} \\
& && \text{the first two terms, and the common} \\
& && \text{factor } 5 \text{ from the last two terms.} \\
&= (x - 4)(2x + 5) && \text{Factor out the common factor} \\
& && x - 4.
\end{aligned}
$$

Quick Check 5
Factor $6x^2 - 7x - 20$.

A binomial of the form $a^2 - b^2$ is called a difference of squares, and can be factored using the following formula:

$$a^2 - b^2 = (a + b)(a - b)$$

EXAMPLE 7   Factor $4x^2 - 25$.

Solution

We cannot factor any common factors out of this binomial, so we check to see whether it is a difference of squares. The binomial can be rewritten as $(2x)^2 - (5)^2$, so this is indeed a difference of squares and can be factored using the formula.

$$
\begin{aligned}
4x^2 - 25 &= (2x)^2 - (5)^2 && \text{Rewrite as the difference of two squares.} \\
&= (2x + 5)(2x - 5) && \text{Factor using the formula} \\
& && a^2 - b^2 = (a + b)(a - b).
\end{aligned}
$$

Quick Check 6
Factor $9x^2 - 100$.

There are three other special binomials that should be mentioned. A binomial that is a sum of squares $(a^2 + b^2)$ cannot be factored. A binomial that is a difference of cubes $(a^3 - b^3)$ or a sum of cubes $(a^3 + b^3)$ can be factored using the following formulas:

$$
\begin{aligned}
\text{Difference of Cubes:} \quad & a^3 - b^3 = (a - b)(a^2 + ab + b^2) \\
\text{Sum of Cubes:} \quad & a^3 + b^3 = (a + b)(a^2 - ab + b^2)
\end{aligned}
$$

Solving Quadratic Equations

Objective 2 **Solve quadratic equations.** We now review the solution of quadratic equations by factoring. A quadratic equation is an equation that can be written in the form $ax^2 + bx + c = 0$, where a, b, and c are real numbers and $a \neq 0$. Recall the following guidelines for solving quadratic equations:

- Simplify each side of the equation. This includes performing any distributive multiplications, clearing all fractions by multiplying each side of the equation by the LCD, and combining like terms.
- Collect all terms on one side of the equation, leaving 0 on the other side. It is a good idea to move all terms to one side in such a way that the coefficient of the second-degree term is positive.
- Factor the quadratic expression.
- Set each factor equal to 0, and solve each equation.

The principle behind this procedure is the zero-factor property, which tells us that if two unknown numbers have a product of zero, then at least one of these factors must be equal to zero.

EXAMPLE 8 Solve $x^2 + 7x = 30 - 6x$.

Solution

We begin to solve this equation by collecting all terms on the left side of the equation.

$$x^2 + 7x = 30 - 6x$$
$$x^2 + 13x - 30 = 0 \qquad \text{Add } 6x, \text{ subtract } 30.$$
$$(x + 15)(x - 2) = 0 \qquad \text{Factor.}$$
$$x + 15 = 0 \qquad \text{or} \qquad x - 2 = 0 \qquad \text{Set each factor equal to 0.}$$
$$x = -15 \qquad \text{or} \qquad x = 2 \qquad \text{Solve.}$$

The solution set is $\{-15, 2\}$.

Quick Check 7

Solve
$x(x + 3) = 7x + 32$.

EXAMPLE 9 Solve $\frac{1}{2}x^2 - \frac{1}{4}x = 2x^2 - 3$.

Solution

We begin to solve this equation by clearing all fractions. This is done by multiplying each side of the equation by 4, which is the least common denominator (LCD).

$$\frac{1}{2}x^2 - \frac{1}{4}x = 2x^2 - 3$$

$$4 \cdot \left(\frac{1}{2}x^2 - \frac{1}{4}x\right) = 4 \cdot (2x^2 - 3) \qquad \text{Multiply by 4.}$$

$$\overset{2}{\cancel{4}} \cdot \frac{1}{\cancel{2}}x^2 - \overset{1}{\cancel{4}} \cdot \frac{1}{\cancel{4}}x = 4 \cdot 2x^2 - 4 \cdot 3 \qquad \begin{array}{l}\text{Distribute and divide out common} \\ \text{factors.}\end{array}$$

$$2x^2 - x = 8x^2 - 12 \qquad \text{Simplify.}$$

$$0 = 6x^2 + x - 12 \qquad \begin{array}{l}\text{Collect all terms on right side. Subtract} \\ 2x^2, \text{ add } x.\end{array}$$

$$0 = (3x - 4)(2x + 3) \qquad \text{Factor.}$$

$$3x - 4 = 0 \quad \text{or} \quad 2x + 3 = 0 \qquad \text{Set each factor equal to 0.}$$

$$x = \frac{4}{3} \quad \text{or} \quad x = -\frac{3}{2} \qquad \text{Solve.}$$

Quick Check 8

Solve $\frac{1}{5}x^2 - \frac{1}{10}x - 1 = 0$.

The solution set is $\{\frac{4}{3}, -\frac{3}{2}\}$.

Solving Rational Equations

Objective 3 **Solve rational equations.** We have also learned to solve rational equations, which are equations that contain one or more rational expressions. An example of such an equation is $\dfrac{2x}{x - 2} - \dfrac{5}{x + 1} = 3$. A first step in solving a rational equation is to multiply each side of the equation by the LCD. Then we proceed to solve the resulting equation. We must check that a solution does not cause a denominator to be 0. Such solutions are called extraneous solutions and are omitted from the solution set.

To find the LCD for two or more rational expressions:

- Completely factor each denominator.
- Identify each factor that is a factor of at least one denominator. The LCD is equal to the product of these factors.
- If an expression is a repeated factor of one or more of the denominators, then we repeat it as a factor in the LCD as well. The exponent used for this factor is equal to the greatest power that the factor is raised to in any one denominator.

EXAMPLE 10 Solve $\dfrac{1}{x + 3} + \dfrac{1}{x} = \dfrac{9}{x(x + 3)}$.

Solution

The LCD for these rational expressions is $x(x + 3)$. We see that the values 0 and -3 cause the LCD to be equal to 0 and therefore cannot be solutions of this equation.

$$\frac{1}{x + 3} + \frac{1}{x} = \frac{9}{x(x + 3)}$$

$$x(x + 3)\left(\frac{1}{x + 3} + \frac{1}{x}\right) = x(x + 3)\left(\frac{9}{x(x + 3)}\right) \qquad \text{Multiply by LCD.}$$

$$x(x+3) \cdot \frac{1}{x+3} + x(x + 3) \cdot \frac{1}{x} = x(x+3) \cdot \frac{9}{x(x+3)} \qquad \begin{array}{l}\text{Distribute and} \\ \text{divide out common} \\ \text{factors.}\end{array}$$

$$x + x + 3 = 9 \qquad \text{Simplify.}$$

$$2x + 3 = 9 \qquad \begin{array}{l}\text{Combine like} \\ \text{terms. The equa-} \\ \text{tion is linear.}\end{array}$$

Quick Check 9

Solve
$\dfrac{x}{x - 4} - \dfrac{8}{x - 3} = 1$.

$$2x = 6 \qquad \text{Subtract 3.}$$

$$x = 3 \qquad \text{Divide by 2.}$$

This value does not cause a denominator to equal 0. The solution set is $\{3\}$.

EXAMPLE 11   Solve $\dfrac{1}{x-5} + \dfrac{1}{x+5} = \dfrac{10}{x^2-25}$.

Solution

We must factor each denominator to find the LCD.

$$\frac{1}{x-5} + \frac{1}{x+5} = \frac{10}{x^2-25}$$

$$\frac{1}{x-5} + \frac{1}{x+5} = \frac{10}{(x+5)(x-5)} \qquad \text{Factor. The LCD is}$$
$$\qquad\qquad\qquad\qquad\qquad\qquad\qquad (x+5)(x-5).$$

$$(x+5)(x-5)\left(\frac{1}{x-5} + \frac{1}{x+5}\right) = \frac{10}{(x+5)(x-5)} \cdot (x+5)(x-5)$$

$$\text{Multiply by LCD.}$$

$$(x+5)\overset{1}{\cancel{(x-5)}} \cdot \frac{1}{\cancel{x-5}} + \overset{1}{\cancel{(x+5)}}(x-5) \cdot \frac{1}{\cancel{x+5}} = \frac{10}{\cancel{(x+5)}\cancel{(x-5)}} \cdot \overset{1}{\cancel{(x+5)}}\overset{1}{\cancel{(x-5)}}$$

Distribute and divide
out common factors.

$$x + 5 + x - 5 = 10 \qquad \text{Simplify.}$$
$$2x = 10 \qquad \text{Combine like terms.}$$
$$x = 5 \qquad \text{Divide both sides by 2.}$$

This solution must be omitted, as $x = 5$ causes a denominator to be equal to 0. Therefore, this equation has no solution, and its solution set is $\varnothing$.

Quick Check 10 Solve $\dfrac{x}{x+3} + \dfrac{1}{x+2} = \dfrac{7x}{x^2+5x+6}$.

Exercises 8.4

Factor completely.

1. $36x^8 - 16x^4 + 50x^3$

2. $9a^4 + 24a^3 - 3$

3. $20a^3b - 28a^2b^4 - 16a^5b^2$

4. $15x^3y^2z^4 - 5xy^6z^{10} + 40x^6y - 25x^4y^4z^4$

5. $n^3 + 4n^2 + 5n + 20$

6. $x^3 - 9x^2 + 2x - 18$

7. $x^3 + 7x^2 - 3x - 21$

8. $2t^3 - 16t^2 - 10t + 80$

9. $x^2 - 13x + 40$

10. $n^2 + 3n - 54$

11. $2x^2 - 8x - 64$

12. $x^2 - 6x - 16$

13. $a^2 + 13ab + 36b^2$

14. $m^2n^2 - 11mn + 28$

15. $3x^2 - 6x - 189$

16. $5x^2 + 5x - 10$

17. $5x^2 - 14x - 3$

18. $4x^2 - 15x + 14$

19. $6t^2 + 17t + 12$

20. $24x^2 + 18x - 15$

21. $x^2 - 100$

22. $x^2 - 49$

23. $3a^2 - 75$

24. $9b^2 - 64$

25. $x^2 + 16$

26. $x^2 + 81$

27. $x^3 - 8$

28. $x^3 + 64$

29. $27x^3 + 125$

30. $8x^3 - 343y^3$

Solve.

31. $x^2 + 3x - 4 = 0$

32. $x^2 - 7x - 30 = 0$

33. $t^2 + 17t + 72 = 0$

34. $3a^2 - 18a + 24 = 0$

35. $x^2 - 2x - 48 = 0$

36. $x^2 + 12x + 36 = 0$

37. $4x^2 - 16x - 20 = 0$

38. $x^2 + 20x + 91 = 0$

39. $2x^2 + 5x + 2 = 0$

40. $3x^2 - 4x + 1 = 0$

41. $x^2 - 16 = 0$

42. $x^2 - 81 = 0$

43. $9x^2 - 25 = 0$

44. $4x^2 - 49 = 0$

45. $x^2 - 64x = 0$

46. $x^2 + 21x = 0$

47. $x^2 + 4x = 21$

48. $x^2 = 11x - 18$

49. $x^2 + 11x = 4x - 6$

50. $x^2 + x + 5 = 9x - 7$

51. $\frac{4}{3}x^2 + \frac{14}{3}x + 2 = 0$

52. $\frac{1}{15}x^2 + \frac{3}{5}x + \frac{4}{3} = 0$

Solve.

53. $|x^2 - 13x| = 30$

54. $|x^2 + 5x| = 6$

55. $|x^2 + 10x| = 24$

56. $|x^2 + 8x + 6| = 6$

57. Find a quadratic equation whose solution set is $\{-3, 4\}$.

58. Find a quadratic equation whose solution set is $\{0, 7\}$.

59. Find a quadratic equation whose solution set is $\{3, \frac{1}{2}\}$.

60. Find a quadratic equation whose solution set is $\{\frac{2}{3}, -\frac{1}{4}\}$.

61. Find a quadratic equation whose solution set is $\{-8, 8\}$.

62. Find a quadratic equation whose solution set is $\{5\}$.

Solve.

63. $\frac{8}{x} = \frac{7}{x - 2}$

64. $\frac{5}{x - 6} = \frac{3}{x + 12}$

65. $x - 3 = \frac{10}{x}$

66. $x + \frac{4}{x} = 5$

67. $\frac{x^2 - 23}{x - 3} + 7 = 0$

68. $x + \frac{6x - 17}{x - 5} = -3$

69. $\frac{x + 1}{x + 2} = \frac{6}{x + 6}$

70. $\frac{x - 1}{x + 9} = \frac{4}{x - 3}$

71. $\frac{x}{x - 3} + \frac{4}{x + 5} = \frac{8x}{(x - 3)(x + 5)}$

72. $\dfrac{x + 2}{x + 4} + \dfrac{3}{x + 7} = \dfrac{6}{(x + 4)(x + 7)}$

73. $\dfrac{x - 4}{x - 1} + \dfrac{4}{x + 1} = \dfrac{x + 17}{x^2 - 1}$

74. $\dfrac{x}{x^2 + 13x + 40} + \dfrac{3}{x^2 + 2x - 15} = \dfrac{4}{x^2 + 5x - 24}$

75. $\dfrac{3}{x - 1} + \dfrac{7}{x + 2} = \dfrac{9}{x^2 + x - 2}$

76. $\dfrac{3}{x + 1} + \dfrac{1}{x - 5} = \dfrac{2x - 4}{x^2 - 4x - 5}$

Solve.

77. $\left| \dfrac{2x - 15}{x} \right| = 3$

78. $\left| \dfrac{3x + 1}{2x - 5} \right| = 4$

79. $\left| \dfrac{x^2 + 5x - 48}{x} \right| = 3$

80. $\left| \dfrac{x^2 + 9x - 20}{x} \right| = 10$

Answer in complete sentences.

81. When solving a quadratic equation, why do we set one side of the equation equal to 0 before factoring the quadratic expression? Explain your answer.

82. What is an extraneous solution of a rational equation? Why do we omit them from the solution set? Explain your answer.

> **Study Tip** REVISITED   **Test-Taking Strategies; Preparing for a Cumulative Exam** Through careful preparation and effective test-taking strategies, you can maximize your grade on a math exam. Here is a summary of some test-taking strategies:
>
> - Make the most of your time before the exam.
> - Write down important facts as soon as you get your test.
> - Briefly read through the test.
> - Begin by solving the easier problems first.
> - Review your test as thoroughly as possible before turning it in.
>
> When preparing for a cumulative exam,
>
> - Start approximately two weeks before the exam by determining what material will be on the exam.
> - Begin by going over your old exams and quizzes. Look over your old homework assignments for problems that you struggled with and review these topics.
> - Use review materials from your instructor and the cumulative review exercises in the text.
> - Create study sheets for each type of applied problem that may appear on the exam, as well as for other types of problems that you are struggling with.
> - Continue to meet regularly with your study group.

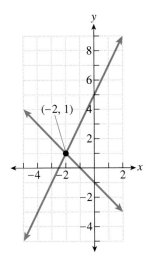

Objectives

1. Solve systems of two linear equations in two unknowns.
2. Determine whether an ordered triple is a solution of a system of three equations in three unknowns.
3. Solve systems of three equations in three unknowns.
4. Solve applied problems by using a system of three equations in three unknowns.

Systems of Two Linear Equations in Two Unknowns

Objective 1 Solve systems of two linear equations in two unknowns. We begin this section by reviewing how to solve a system of two linear equations in two variables. A solution of a system of two linear equations is an ordered pair that satisfies both equations simultaneously. In Chapter 4, we learned two techniques for solving a system of linear equations: by substitution and by addition.

The graph of each equation in a system of linear equations is a line, and if these two lines cross at exactly one point, then this ordered pair is the only solution of the system. In this case, the system is said to be independent.

The graph at the left shows the lines associated with $y = 2x + 5$ and $y = -x - 1$. These two lines intersect at the ordered pair $(-2, 1)$, and this point of intersection is the solution to this independent system of equations.

If the two lines associated with a system of equations are parallel lines, then the system has no solution and is said to be inconsistent.

If the two lines associated with a system of equations are actually the same line, then the system is said to be a dependent system. For example, consider the system
$$\begin{aligned} x + y &= 5 \\ 2x + 2y &= 10 \end{aligned}.$$
Here is the graph associated with this system. Each ordered pair that is on the line is a solution. We write the solution in the form $(x, 5 - x)$ by solving one of the equations for y in terms of x.

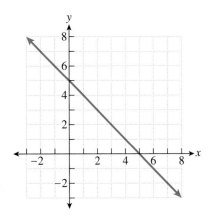

The Substitution Method

Now we turn our attention to solving a system of equations by substitution. This technique requires us to solve one of the equations for one of the variables in terms of the other variable. The expression that we find is then *substituted* for that variable in the other equation, giving us one equation with one variable. After solving that equation, we substitute this value into the expression we originally substituted to solve for the other variable.

EXAMPLE ‣1   Solve by substitution. $\begin{aligned} x + 3y &= 1 \\ 3x - 2y &= 14 \end{aligned}$

Solution

If there is a variable with a coefficient of 1, then it is a wise choice to solve the equation for that variable. (Using this strategy will help us avoid the use of fractions.) Solving the first equation for x gives us $x = 1 - 3y$. The expression $1 - 3y$ is then substituted into the equation $3x - 2y = 14$ for x.

$$
\begin{aligned}
3x - 2y &= 14 \\
3(1 - 3y) - 2y &= 14 \qquad & \text{Substitute } 1 - 3y \text{ for } x. \\
3 - 9y - 2y &= 14 \qquad & \text{Distribute.} \\
3 - 11y &= 14 \qquad & \text{Combine like terms.} \\
-11y &= 11 \qquad & \text{Subtract 3.} \\
y &= -1 \qquad & \text{Divide both sides by } -11.
\end{aligned}
$$

This value is then substituted for y into the equation $x = 1 - 3y$.

$$
\begin{aligned}
x &= 1 - 3(-1) \qquad & \text{Substitute } -1 \text{ for } y. \\
x &= 1 + 3 \qquad & \text{Multiply.} \\
x &= 4 \qquad & \text{Add.}
\end{aligned}
$$

The ordered pair solution is $(4, -1)$.

Quick Check 1
Solve by substitution.
$2x - 5y = -24$
$4x + y = -4$

The Addition Method

Another technique for solving a system of equations is solving by addition. We multiply one or both equations by a constant in such a way that one of the variables has opposite coefficients in the two equations. When we add the two equations together, this variable will be eliminated, resulting in an equation that only contains one variable. We finish finding the solution just as we did when solving by substitution.

EXAMPLE ‣2   Solve. $\begin{aligned} 4x + 5y &= 13 \\ 2x + 7y &= 11 \end{aligned}$

Solution

If we multiply the second equation by -2, the coefficient for the x term will be -4, which is the opposite of the coefficient of the x term in the first equation. (Recall that there will be times when you must multiply both equations by different constants to eliminate one variable.)

$$4x + 5y = 13$$
$$2x + 7y = 11 \xrightarrow{\text{Multiply by } -2.} \begin{array}{r} 4x + 5y = 13 \\ -4x - 14y = -22 \end{array}$$

$$\begin{array}{r} 4x + 5y = 13 \\ -4x - 14y = -22 \\ \hline -9y = -9 \end{array} \quad \text{Add.}$$

$$y = 1 \qquad \text{Divide both sides by } -9.$$

We now substitute 1 for y into either of the original equations to find x.

$$\begin{array}{rl} 4x + 5y = 13 & \\ 4x + 5(1) = 13 & \text{Substitute 1 for } y. \\ 4x + 5 = 13 & \text{Multiply.} \\ 4x = 8 & \text{Subtract 5.} \\ x = 2 & \text{Divide both sides by 4.} \end{array}$$

Quick Check 2
Solve by addition.
$3x + 2y = 13$
$5x - 3y = 9$

Our solution is $(2, 1)$.

Solutions of a System of Three Equations in Three Unknowns

Objective 3 **Determine whether an ordered triple is a solution of a system of three equations in three unknowns.** We now turn our attention towards **systems of three equations in three unknowns.** Each of the three equations will be of the form $ax + by + cz = d$, where a, b, c, and d are real numbers. If there is a unique solution of such a system, it will be an **ordered triple** (x, y, z).

EXAMPLE 3 Is $(3, 2, 4)$ a solution of the system of equations
$$\begin{array}{r} 2x - 3y + z = 4 \\ x + 2y - z = 3? \\ 2x - y - 2z = -4 \end{array}$$

Solution

To determine whether the ordered triple is a solution of the system of equations, we will substitute 3 for x, 2 for y, and 4 for z in each equation. If each equation is true for these three values, then the ordered pair is a solution of the system of equations.

Quick Check 3
Is $(4, -2, 5)$ a solution of the system of equations
$x - 2y - z = 3$
$3x + 2y + 2z = 26$?
$5x - 3y - 4z = 6$

$2x - 3y + z = 4$	$x + 2y - z = 3$	$2x - y - 2z = -4$
$2(3) - 3(2) + (4) = 4$	$(3) + 2(2) - (4) = 3$	$2(3) - (2) - 2(4) = -4$
$6 - 6 + 4 = 4$	$3 + 4 - 4 = 3$	$6 - 2 - 8 = -4$
$4 = 4$	$3 = 3$	$-4 = -4$

Since the ordered triple makes each equation in the system true, the ordered triple $(3, 2, 4)$ is a solution of the system of equations.

Solving a System of Three Equations in Three Unknowns

Objective 3 Solve systems of three equations in three unknowns. The graph of a linear equation with three variables is a two-dimensional **plane.**

For a system of three linear equations in three unknowns, if the planes associated with the equations intersect at a single point, this ordered triple (x, y, z) is the solution of the system of equations.

Some systems of three equations in three unknowns have no solution. Such a system is called an inconsistent system. While solving an inconsistent system, we will obtain an equation that is a contradiction, such as $0 = 1$. Graphically, the following systems are inconsistent, as the three planes do not have a point in common:

Some systems of three equations in three unknowns have infinitely many solutions. Such a system is called a dependent system of equations. In solving a dependent system of equations, we will obtain an equation that is an identity, such as $0 = 0$. Graphically, if at least two of the planes are identical, or if the intersection of the three planes is a line, then the system is a dependent system.

Since graphing equations in three dimensions is beyond the scope of this text, we will focus on solving systems of three linear equations in three unknowns algebraically.

Solving a System of Three Equations in Three Unknowns

- Select two of the equations and use the addition method to eliminate one of the variables.
- Select a different pair of equations and use the addition method to eliminate the same variable, leaving us with two equations in two unknowns.
- Solve the system of two equations for the two unknowns.
- Substitute these two values in any of the original three equations and solve for the unknown variable.

We will now apply this technique to the system of equations from the previous example.

EXAMPLE ▸ 4   Solve the system.

$$
\begin{aligned}
2x - 3y + z &= 4 \quad &\text{(Eq. 1)} \\
x + 2y - z &= 3 \quad &\text{(Eq. 2)} \\
2x - y - 2z &= -4 \quad &\text{(Eq. 3)}
\end{aligned}
$$

Solution

Notice that the three equations have been labeled. This will help to keep track of the equations as we proceed to solve the system. We must choose a variable to eliminate first, and in this example, we will eliminate z first. We may begin by adding Equations 1 and 2 together, as the coefficients of the z-terms are opposites.

$$
\begin{array}{rl}
2x - 3y + z = 4 & \text{(Eq. 1)} \\
\underline{x + 2y - z = 3} & \text{(Eq. 2)} \\
3x - y = 7 & \text{(Eq. 4)}
\end{array}
$$

We need to select a different pair of equations and eliminate z. By multiplying Equation 1 by 2 and adding the product to Equation 3, we can eliminate z once again.

$$
\begin{array}{ll}
4x - 6y + 2z = 8 & (2 \cdot \text{Eq. 1}) \\
\underline{2x - y - 2z = -4} & (\text{Eq. 3}) \\
6x - 7y = 4 & (\text{Eq. 5})
\end{array}
$$

We now work to solve the system of equations formed by Equations 4 and 5, which is a system of two equations in two unknowns. We will solve this system using the addition method as well, although we could also choose the substitution method if it was convenient. We will eliminate the variable y by multiplying Equation 4 by -7.

$$
\begin{array}{lll}
3x - y = 7 & \xrightarrow{\text{Multiply by } -7} & -21x + 7y = -49 \\
6x - 7y = 4 & & 6x - 7y = 4
\end{array}
$$

$$
\begin{array}{l}
-21x + 7y = -49 \\
\underline{6x - 7y = 4} \\
-15x = -45
\end{array}
$$

$$x = 3 \qquad\qquad \text{Divide both sides by } -15.$$

We now substitute 3 for x in either of the equations that contained only two variables. We will use Equation 4.

$$
\begin{array}{ll}
3x - y = 7 & \\
3(3) - y = 7 & \text{Substitute 3 for } x \text{ in Equation 4.} \\
9 - y = 7 & \text{Multiply.} \\
-y = -2 & \text{Subtract 9.} \\
y = 2 & \text{Divide both sides by } -1.
\end{array}
$$

To find z, we will substitute 3 for x and 2 for y in either of the original equations that contained three variables. We will use Equation 1.

$$
\begin{array}{ll}
2x - 3y + z = 4 & \\
2(3) - 3(2) + z = 4 & \text{Substitute 3 for } x \text{ and 2 for } y \text{ in Equation 1.} \\
6 - 6 + z = 4 & \text{Multiply.} \\
0 + z = 4 & \text{Combine like terms.} \\
z = 4 &
\end{array}
$$

The solution of this system is the ordered triple $(3, 2, 4)$.

Quick Check 4

Solve the system.

$$
\begin{array}{rcr}
x + y + z = & 0 \\
2x - y + 3z = & -9 \\
-3x + 2y - 4z = & 13
\end{array}
$$

EXAMPLE 5   Solve the system.

$$
\begin{array}{ll}
3x + 2y + 2z = 7 & (\text{Eq. 1}) \\
4x + 4y - 3z = -17 & (\text{Eq. 2}) \\
-x - 6y + 4z = 13 & (\text{Eq. 3})
\end{array}
$$

Solution

In this example, we will eliminate x first. It is a good idea to choose x because the coefficient of the x-term in Equation 3 can easily be multiplied to become the opposite of the coefficients of the x-terms in Equations 1 and 2. We begin by multiplying Equation 3 by 3 and adding it to Equation 1.

$$
\begin{array}{rll}
3x + 2y + 2z = 7 & (\text{Eq. 1}) \\
\underline{-3x - 18y + 12z = 39} & (3 \cdot \text{Eq. 3}) \\
-16y + 14z = 46 & (\text{Eq. 4})
\end{array}
$$

We need to select a different pair of equations and eliminate x. By multiplying Equation 3 by 4 and adding it to Equation 2, we can eliminate x once again.

$$
\begin{array}{rll}
4x + 4y - 3z = -17 & (\text{Eq. 2}) \\
\underline{-4x - 24y + 16z = 52} & (4 \cdot \text{Eq. 3}) \\
-20y + 13z = 35 & (\text{Eq. 5})
\end{array}
$$

We now solve the system of equations formed by Equations 4 and 5. We will use the addition method to eliminate the variable y. To determine what to multiply each equation by, we should find the LCM of 16 and 20, which is 80. If we multiply Equation 4 by 5, then the coefficient of the y-term will be -80. If we multiply Equation 5 by -4, then the coefficient of the y-term will be 80. This will make the coefficients of the y-terms opposites, allowing us to add the equations together and eliminate y.

$$
\begin{array}{l}
-16y + 14z = 46 \xrightarrow{\text{Multiply by 5}} -80y + 70z = 230 \\
-20y + 13z = 35 \xrightarrow{\text{Multiply by } -4} 80y - 52z = -140
\end{array}
$$

$$
\begin{array}{r}
-80y + 70z = 230 \\
\underline{80y - 52z = -140} \\
18z = 90
\end{array}
$$

$$z = 5 \qquad \text{Divide both sides by 18.}$$

We now substitute 5 for z in either of the equations that contained only two variables. We will use Equation 4.

$$
\begin{array}{ll}
-16y + 14z = 46 & \\
-16y + 14(5) = 46 & \text{Substitute 5 for } z \text{ in Equation 4.} \\
-16y + 70 = 46 & \text{Multiply.} \\
-16y = -24 & \text{Subtract 70.} \\
y = \dfrac{3}{2} & \text{Divide both sides by } -16 \text{ and simplify.}
\end{array}
$$

To find x, we will substitute 5 for z and $\frac{3}{2}$ for y in either of the original equations that contained three variables. We will use Equation 1.

Quick Check 5

Solve the system.

$$
\begin{array}{rrr}
2x + y - z = -8 \\
-4x - 3y + 3z = 25 \\
6x + 5y + 4z = 3
\end{array}
$$

$$
\begin{array}{ll}
3x + 2y + 2z = 7 & \\
3x + 2\left(\dfrac{3}{2}\right) + 2(5) = 7 & \text{Substitute } \frac{3}{2} \text{ for } y \text{ and 5 for } z \text{ in Equation 1.} \\
3x + 3 + 10 = 7 & \text{Multiply.} \\
3x + 13 = 7 & \text{Combine like terms.} \\
3x = -6 & \text{Subtract 13.} \\
x = -2 & \text{Divide both sides by 3.}
\end{array}
$$

The solution of this system is the ordered triple $\left(-2, \frac{3}{2}, 5\right)$.

Applications of Systems of Three Equations in Three Unknowns

Objective 4 **Solve applied problems by using a system of three equations in three unknowns.** We finish the section with an application involving a system of three equations in three unknowns.

EXAMPLE 6 A baseball team has three prices for admission to their games. Adults are charged $7, senior citizens are charged $5, and children are charged $4. At last night's game there were 1000 people in attendance, generating $5600 in receipts. If we know that there were 300 more children than senior citizens at the game, find how many adult tickets, senior citizen tickets, and children tickets were sold.

Solution

There are three unknowns in this problem: the number of adult tickets sold, the number of senior citizen tickets sold, and the number of children tickets sold. We shall let a, s, and c represent these three quantities, respectively. Here is a table of the unknowns:

> ***Unknowns***
> Number of adult tickets sold: a
> Number of senior citizen tickets sold: s
> Number of children tickets sold: c

Solving a problem with three unknowns requires that we establish a system of three equations. Since we know that there were 1000 people at last night's game, the first equation in our system is $a + s + c = 1000$.

We also know that the revenue from all of the tickets was $5600, which will lead to our second equation. The amount received from the adult tickets was $7a$ (a tickets at $7 each). In the same way, the amount received from senior citizen tickets was $5s$, and the amount received from children tickets was $4c$. The second equation in our system is $7a + 5s + 4c = 5600$.

Our third equation comes from knowing that there were 300 more children than senior citizens at the game. In other words, the number of children at the game was equal to the number of senior citizens plus 300. Our third equation is $c = s + 300$, which can be rewritten as $-s + c = 300$.

Here is the system that we will be working with:

$$
\begin{aligned}
a + s + c &= 1000 \quad &\text{(Eq. 1)} \\
7a + 5s + 4c &= 5600 \quad &\text{(Eq. 2)} \\
-s + c &= 300 \quad &\text{(Eq. 3)}
\end{aligned}
$$

Since Equation 3 has only two variables, we will use the addition method to combine Equations 1 and 2, eliminating the variable a. We begin by multiplying Equation 1 by -7 and adding it to Equation 2.

$$-7a - 7s - 7c = -7000 \quad (-7 \cdot \text{Eq. 1})$$
$$\underline{7a + 5s + 4c = 5600 \quad (\text{Eq. 2})}$$
$$-2s - 3c = -1400 \quad (\text{Eq. 4})$$

We now have two equations that contain only the two variables s and c. We can solve the system of equations formed by Equations 3 and 4. We will use the addition method to eliminate the variable c by multiplying Equation 3 by 3 and adding it to Equation 4.

$$-s + c = 300 \qquad \xrightarrow{\text{Multiply by 3}} \qquad -3s + 3c = 900$$
$$-2s - 3c = -1400 \qquad\qquad\qquad\qquad -2s - 3c = -1400$$

$$-3s + 3c = 900$$
$$\underline{-2s - 3c = -1400}$$
$$-5s \qquad\quad = -500$$

$$s = 100 \qquad\qquad \text{Divide both sides by } -5$$

There were 100 senior citizens at the game. To find the number of children that were at the game, we will substitute 100 for s in Equation 3.

$$-s + c = 300$$
$$-100 + c = 300 \qquad \text{Substitute 100 for } s \text{ in Equation 3.}$$
$$c = 400 \qquad \text{Add 100.}$$

There were 400 children at the game. (Since we knew that there were 300 more children than senior citizens at the game, we could have just added 300 to 100.) To find the number of adults that were at the game, we will substitute 100 for s and 400 for c in Equation 1.

$$a + s + c = 1000$$
$$a + (100) + (400) = 1000 \qquad \text{Substitute 100 for } s \text{ and 400 for } c \text{ in Equation 1.}$$
$$a = 500 \qquad \text{Solve for } a.$$

There were 500 adults, 100 senior citizens, and 400 children at last night's game. The reader can verify that the total revenue for 500 adults, 100 senior citizens, and 400 children is $5600.

Quick Check   6 Gilbert looks into his wallet and finds $1, $5, and $10 bills totaling $111. There are 30 bills in all, and Gilbert has 4 more $5 bills than he has $10 bills. How many bills of each type does Gilbert have in his wallet?

Solve by addition or substitution.

1. $x + 3y = 1$
$-x + 2y = -11$

2. $8x + y = 92$
$5x - y = 25$

3. $10x + 7y = -5$
$4x + 7y = -23$

4. $2x - 9y = 6$
$2x + 4y = 32$

5. $x + 3y = 17$
$-2x + 2y = 6$

6. $3x + 7y = 14$
$5x - 4y = -8$

7. $5x + 9y = 33$
$y = 2$

8. $-2x + 3y = -1$
$5x = 25$

9. $2x - 3y = 10$
$4x - 2y = 28$

10. $x + 2y = -12$
$3x - 4y = 26$

11. $x - 4y = 7$
$3x + 2y = 14$

12. $5x + y = -2$
$3x - 7y = -43$

13. $4x + 3y = -28$
$3x - 2y = -4$

14. $-5x + 4y = 22$
$2x + 5y = 11$

15. $6x - 4y = 11$
$9x - 6y = 17$

16. $4x + y = 13$
$12x + 3y = 39$

17. $\frac{3}{4}x + \frac{1}{3}y = 1$
$\frac{2}{5}x - \frac{1}{4}y = \frac{31}{10}$

18. $\frac{1}{6}x + \frac{1}{2}y = -3$
$\frac{2}{3}x - \frac{3}{4}y = \frac{7}{4}$

Is the ordered triple a solution to the given system of equations?

19. $(3, 2, 6)$

$x + y + z = 11$
$x - y + 2z = 17$
$4x - 3y - 2z = -6$

20. $(2, -4, -1)$

$x + y - z = -1$
$5x + 2y + 2z = 0$
$\frac{1}{2}x + 2y - 6z = -1$

21. $(-5, 2, 0)$

$x + 2y + 3z = -1$
$3x - y - 6z = -17$
$4x - 3y + z = -26$

22. $(1, -1, -6)$

$x - y - z = 8$
$5x + 7y + z = -8$
$-2x + 5y - 4z = 21$

Solve.

23. $x + 3y - 2z = -3$
$-2x - 2y + 3z = -2$
$-3x + 4y - 5z = -17$

24. $3x - y + 2z = -10$
$4x + 2y + 3z = 2$
$2x + 3y + 2z = 9$

25. $x + y + z = 1$
$3x + 2y + z = 7$
$4x + 4y - z = 19$

26. $x + y + z = 6$
$-2x - y + 3z = 43$
$3x + 2y + 2z = 3$

27. $x - 4y + 7z = -5$
$3x - 6y + 5z = 4$
$-5x + 10y + 4z = -19$

28.
$$x + y + 4z = -6$$
$$-2x + y - 4z = 6$$
$$3x + 2y - 8z = -2$$

29.
$$2x + 3y + z = 5$$
$$-2x - y + 4z = 60$$
$$x + 2y - 8z = -30$$

30.
$$x - y - z = 16$$
$$2x + 3y + 2z = -5$$
$$4x + 5y + 4z = -5$$

31.
$$x + 2y + z = 8$$
$$3x + 4y + 2z = 9$$
$$3x - 4y = -37$$

32.
$$2x + 3y - z = 10$$
$$-x + 4y + 2z = 200$$
$$5x + 2y = 88$$

33.
$$x + 2y + 3z = 14$$
$$3y - z = 3$$
$$z = 3$$

34.
$$3x + 8y - 2z = 3$$
$$5x + 4z = 49$$
$$3x = 15$$

35. Jonah opens his wallet and finds that all of the bills are either $1 bills, $5 bills, or $10 bills. There are 35 bills in his wallet. There are 4 more $5 bills than $1 bills. If the total amount of bills is $246, how many $10 bills does Jonah have?

36. A cash register has no $1 bills in it. All of the bills are either $5 bills, $10 bills, or $20 bills. Altogether, there are 37 bills in the cash register, and the total value of these bills is $310. If the number of $10 bills in the register is twice the number of $20 bills, how many $5 bills are in the register?

37. A campus club held a bake sale as a fundraiser, selling coffee, muffins, and bacon-and-egg sandwiches. They charged $1 for a cup of coffee, $2 for a muffin, and $3 for a bacon-and-egg sandwich. They sold a total of 50 items, raising $85 dollars. If they sold 5 more muffins than cups of coffee, how many bacon-and-egg sandwiches did they sell?

38. A grocery store was supporting a local charity by having shoppers donate $1, $5, or $20 at the checkout stand. A total of 110 shoppers donated $430. The amount of money raised from $5 donations was $50 higher than the amount of money raised from $20 donations. How many shoppers donated $20?

39. The measures of the three angles in a triangle must total 180°. The measure of angle A is 20° less than the measure of angle C. The measure of angle C is twice the measure of angle B. Find the measure of each angle.

40. The measures of the three angles in a triangle must total 180°. The measure of angle A is 15° more than the measure of angle B. The measure of angle B is 15° greater than four times the measure of angle C. Find the measure of each angle.

(Recall the introduction to problems about interest in Section 4.4.)

41. Mary invested a total of $40,000 in three different bank accounts. One account pays an annual interest rate of 3%, the second account pays 5% annual interest, and the third account pays 6% annual interest. In one year, Mary earned a total of $1960 in interest from these three accounts. If Mary invested $8000 more in the account that pays 5% interest than she did in the account that pays 6% interest, find the amount invested in each account.

42. Dennis invested a total of $100,000 in three different mutual funds. After one year, the first fund showed a profit of 8%, the second fund showed a loss of 5%, and the third fund showed a profit of 2%. Dennis invested $10,000 more in the fund that lost 5% than he invested in the fund that made a 2% profit. In the first year, Dennis made a profit of $2900 from the three funds. How much was initially invested in each fund?

Answer in complete sentences.

43. Write a word problem whose solution is "There were 40 children, 200 adults, and 60 senior citizens in attendance." Your problem must lead to a system of three equations in three unknowns. Explain how you created your problem.

44. Write a word problem for the following system of equations.

$$x + y + z = 350{,}000$$
$$0.05x + 0.07y - 0.04z = 17{,}000$$
$$y = 2x$$

Explain how you created your problem.

Study Tip **REVISITED** Math anxiety can hinder your success in a mathematics class, but it can be overcome. The first step is to understand what has caused your anxiety to begin with. Relaxation techniques can help you to overcome the physical symptoms of math anxiety. Developing a positive attitude and confidence in your abilities will help as well.

Poor performance on math exams can be caused by other factors, such as test anxiety, improper placement, and poor study skills. If you "freeze up" on exams in your other classes and do poorly on exams when you understand the material that you are being tested on, then you may have test anxiety; an academic counselor or learning resource specialist can help.

If you feel that you are taking a class for which you lack the prerequisite skills, discuss your correct placement with your instructor and your academic counselor.

Finally, be sure that poor study skills are not the cause of your difficulties. If you are not giving your best effort, you cannot expect to learn and understand mathematics.

Chapter 8 Summary

Section 8.1—Topic	Chapter Review Exercises
Solving Linear Equations	1–4
Solving Absolute Value Equations	5–10

Section 8.2—Topic	Chapter Review Exercises
Solving Linear Inequalities	11–16
Solving Absolute Value Inequalities	17–22

Section 8.3—Topic	Chapter Review Exercises
Graphing Linear Equations	23–26
Graphing Linear Functions	27–30
Graphing Absolute Value Functions	31–34

Section 8.4—Topic	Chapter Review Exercises
Factoring Polynomials	35–52
Solving Quadratic Equations	53–56
Solving Rational Equations	57–60

Section 8.5—Topic	Chapter Review Exercises
Solving Systems of Two Linear Equations in Two Unknowns	61–64
Solving Systems of Three Linear Equations in Three Unknowns	65–68
Solving Applied Problems Involving Systems of Equations	69–70

Summary of Chapter 8 Study Tips

The study tips in this chapter summarized the ideas presented in the first seven chapters of this text. For further information on any particular topic, you can refer back to the appropriate chapter.

Chapter 1—Study Groups
Chapter 2—Using Your Textbook Effectively
Chapter 3—Making the Best Use of Your Resources
Chapter 4—Doing Your Homework
Chapter 5—Test Taking
Chapter 6—Overcoming Math Anxiety
Chapter 7—Preparing for a Cumulative Exam
Chapter 8—Review

Solve. [8.1]

1. $2x - 17 = -5$

2. $3x - 9 = 17 - 5x$

3. $\dfrac{1}{3}x - \dfrac{1}{6} = \dfrac{2}{15}x + \dfrac{1}{3}$

4. $4x + 2(3x - 1) = 17x - 3(4x - 11)$

Solve. [8.1]

5. $|x| = 6$

6. $|x| = -6$

7. $|x + 7| - 14 = -6$

8. $|3x + 13| + 20 = 28$

9. $|6x + 25| = |x + 10|$

10. $|4x - 3| = |2x - 15|$

Solve the inequality. Graph your solution on a number line and write your solution in interval notation. [8.2]

11. $3x + 16 < 10$

12. $-5x + 19 \geq -16$

13. $2x - 13 \leq 4x - 31$

14. $13 \leq 2x - 21 \leq 18$

15. $-14 < 3x + 10 < 37$

16. $6x + 13 < 25$ or $2x - 19 \geq -7$

Solve the inequality. Graph your solution on a number line and write your solution in interval notation. [8.2]

17. $|x| \leq 10$

18. $|3x - 11| < 7$

19. $|2x + 1| - 9 < 6$

20. $|x| + 4 > 13$

21. $|6x + 21| - 10 \geq 5$

22. $|x - 8| + 13 \geq 8$

Find the intercepts and then graph the line. [8.3]

23. $5x + 6y = 30$

24. $8x - 3y = -12$

Graph the line using the slope and y-intercept. [8.3]

25. $y = -2x + 9$

26. $y = \dfrac{2}{3}x + 6$

Graph the absolute value function. State the domain and range of the function. [8.3]

31. $f(x) = |x| + 3$

Graph the linear function. [8.3]

27. $f(x) = -x + 5$

32. $f(x) = |x - 5|$

28. $f(x) = 3x$

33. $f(x) = |x + 2| - 4$

29. $f(x) = \dfrac{3}{5}x - 6$

34. $f(x) = -|x - 1| + 2$

30. $f(x) = -\dfrac{1}{6}x + 1$

Factor completely. [8.4]

35. $6x^9 - 4x^7 + 10x^3 - 18x^2 + 2x$

36. $5a^4b^3 - 10a^2b^5 - 25a^3b$

37. $x^3 - 8x^2 + 11x - 88$

38. $x^3 + 9x^2 - 4x - 36$

39. $x^2 - 5x - 14$

40. $3x^2 - 33x + 72$

41. $x^2 + 18xy + 80y^2$

42. $x^2 + 13x - 30$

43. $4x^2 + 20x - 24$

44. $x^2 - 3x - 54$

45. $6x^2 - 29x + 9$

46. $4x^2 - 8x - 21$

47. $x^2 - 36$

48. $9x^2 - 1$

49. $4x^2 - 49$

50. $25x^2 - 64y^2$

51. $x^3 + 27$

52. $x^3 - 729$

Solve. [8.4]

53. $x^2 - x - 6 = 0$

54. $x^2 + 10x + 24 = 0$

55. $x^2 + 13x + 25 = 6x + 43$

56. $x^2 = 144$

Solve. [8.4]

57. $\dfrac{9}{x + 2} = \dfrac{3}{x - 8}$

58. $x - 5 - \dfrac{24}{x} = 0$

59. $\dfrac{4}{x + 5} + \dfrac{1}{x - 4} = \dfrac{9}{x^2 + x - 20}$

60. $\dfrac{x - 2}{x + 3} - \dfrac{3}{x + 6} = \dfrac{9}{x^2 + 9x + 18}$

Solve. [8.5]

61. $2x + y = 7$
$3x + 4y = 8$

62. $x = 7 - 4y$
$6x - 5y = -45$

63. $2x + 3y = 23$
$-6x + 2y = 8$

64. $3x + 5y = 19$
$5x - 4y = -67$

Solve. [8.5]

65. $x - 4y + 2z = -15$
$2x + 3y = 22$
$y = 4$

66. $x + y - z = -9$
$3x + 5y + 6z = -7$
$4y + 7z = -4$

67. $x + y + z = 0$
$2x + 3y - z = 24$
$3x - 2y + 4z = -36$

68. $x + 3y - 4z = -13$
$-3x + 9y + 8z = -5$
$-5x + 6y + 6z = -19$

69. A golf course charges children under 13 years old $8 to play a round of golf. Senior citizens are charged $10 to play a round. Everyone else pays $15 per round. Yesterday, 200 people played a round of golf at the course, paying a total of $2540. If we know that there were 20 more senior citizens that played than there were children under 13 years old that played, find the number of senior citizens who played at the course. [8.5]

70. The measures of the three angles in a triangle must total 180°. The measure of angle *A* is 24° more than the measure of angle *B*. The sum of the measures of angles *A* and *B* is twice the measure of angle *C*. Find the measure of each angle. [8.5]

Solve.

1. $6(2x - 5) + 8 = 5(3x + 7) - 9$

Solve.

2. $|x + 9| = 7$

3. $|x - 5| - 11 = -3$

Solve the inequality. Graph your solution on a number line and write your solution in interval notation.

4. $-5x < 3x + 32$

5. $-47 \le 6x - 11 \le 19$

6. $2x + 15 < 21$ or $3x - 5 > 10$

Solve the inequality. Graph your solution on a number line and write your solution in interval notation.

7. $|x + 4| \le 5$

8. $|2x - 5| + 8 > 13$

Graph the line.

9. $4x - 3y = 6$

Graph the linear function.

10. $f(x) = x - 2$

Graph the absolute value function. State the domain and range of the function.

11. $f(x) = |x - 6| + 3$

Factor completely.

12. $x^2 + 3x - 40$

13. $4x^2 - 28x + 40$

14. $6x^2 + 19x + 10$

15. $9x^2 - 100$

Solve.

16. $x^2 + 6x - 27 = 0$

Solve.

17. $\dfrac{x + 7}{x^2 + 5x + 4} + \dfrac{6}{x^2 - 3x - 4} = \dfrac{3}{x^2 - 16}$

Solve.

18. $\begin{aligned} 5x + \ \ y &= -11 \\ 4x + 3y &= \ \ \ \ 0 \end{aligned}$

Solve.

19. $\begin{aligned} x + \ y + \ z &= \ \ 11 \\ 3x - 2y + 4z &= \ \ 24 \\ -2x + 3y - 5z &= -15 \end{aligned}$

20. The measures of the three angles in a triangle must total 180°. The measure of angle *B* is 25° more than the measure of angle *C*. The measure of angle *A* is 25° more than twice the measure of angle *B*. Find the measure of each angle.

Mathematicians in History
Pierre de Fermat

Pierre de Fermat was a French mathematician who lived in the 17th century. His work focused on number theory and probability. Although he did little publishing in his lifetime, he did pose many problems as challenges to the mathematical community of Europe.

Write a one-page summary (*or* make a poster) of the life of Pierre de Fermat and his accomplishments.

Interesting issues:

- Where and when was Pierre de Fermat born?
- What is number theory?
- Although Fermat lived until 1665, his health was so bad that it was believed by many that he had died in 1653. What was the cause of his near death?
- The Fronde greatly disrupted Fermat's communication with the math community in 1648. What was the Fronde?
- Fermat is best known today for Fermat's Last Theorem. What is Fermat's Last Theorem, and who finally proved this theorem in November 1994?
- Fermat had a public feud with which well-known mathematician? Describe what the feud was about and how it was resolved.

On a coordinate system, accurately plot the following points: A: $(-3, 1)$*, B:* $(-1, 3)$*, C:* $(1, 5)$*, and D:* $(2, 6)$*.*

a) Do these points appear to lie in a straight line? What kind of equations have graphs that are straight lines?

On the same coordinate system, now plot the following points: E: $(4, 4)$*, F:* $(6, 2)$*, G:* $(7, 1)$*.*

b) Does it still appear that the points lie in a straight line?

Connect the points in alphabetical order.

c) What shape is the graph? What kind of equation has graphs of this shape?

Extend your graph out past points A and G.

d) What are the x- and y-intercepts of this graph?

e) Find the equation that forms the graph.

RADICAL EXPRESSIONS AND EQUATIONS

In this chapter, we will investigate radical expressions and equations and their applications. Among the applications is the method for finding the distance between any two objects and a way to determine the speed a car was traveling by measuring the skid marks left by its tires.

Study Tip NOTE TAKING   *If you have poor note-taking skills, then you will not learn as much during the class period as you might be able to if your skills were better. In this chapter, we will focus on how to take notes in a math class. We will discuss how to be more efficient when taking notes, what material should be in your notes, and how to rework your notes.*

9.1
SQUARE ROOTS; RADICAL NOTATION

Objectives

1. **Find the square root of a number.**
2. **Simplify the square root of a variable expression.**
3. **Approximate the square root of a number by using a calculator.**
4. **Find nth roots.**
5. **Multiply radical expressions.**
6. **Divide radical expressions.**
7. **Evaluate radical functions.**
8. **Find the domain of a radical function.**

Square Roots

Consider the equation $x^2 = 36$. There are two solutions of this equation: $x = 6$ and $x = -6$.

Square Root

> A number a is a **square root** of a number b if $a^2 = b$.

The numbers 6 and -6 are square roots of 36, since $6^2 = 36$ and $(-6)^2 = 36$. The number 6 is the positive square root of 36, while the number -6 is the negative square root of 36.

Objective 1 Find the square root of a number.

Principal Square Root

> The **principal square root** of b, denoted $\sqrt{b}$, for $b > 0$, is the positive number a such that $a^2 = b$.

The expression $\sqrt{b}$ is called a **radical expression.** The sign $\sqrt{}$ is called a **radical sign,** while the expression contained inside the radical sign is called the **radicand.**

EXAMPLE 1   Simplify $\sqrt{25}$.

Solution

We are looking for a positive number a such that $a^2 = 25$. The number is 5, so $\sqrt{25} = 5$.

EXAMPLE 2   Simplify $\sqrt{\dfrac{1}{9}}$.

Solution

Since $\left(\dfrac{1}{3}\right)^2 = \dfrac{1}{9}$, $\sqrt{\dfrac{1}{9}} = \dfrac{1}{3}$.

EXAMPLE 3 Simplify $-\sqrt{49}$.

Solution

In this example, we are looking for the negative square root of 49, which is -7. So $-\sqrt{49} = -7$. We can first find the principal square root of 49 and then make it negative.

The principal square root of a negative number, such as $\sqrt{-49}$, is not a real number because there is no real number a such that $a^2 = -49$. We will learn in Section 9.7 that $\sqrt{-49}$ is called an *imaginary* number.

Quick Check 1

Simplify.

a) $\sqrt{49}$ b) $\sqrt{\dfrac{4}{25}}$

c) $-\sqrt{36}$

Square Roots of Variable Expressions

Objective 2 Simplify the square root of a variable expression. Now we turn our attention towards simplifying radical expressions containing variables, such as $\sqrt{x^8}$.

Simplifying $\sqrt{a^2}$

For any real number a, $\sqrt{a^2} = |a|$.

You may be wondering why the absolute value bars are necessary. For any nonnegative number a, $\sqrt{a^2} = a$. For example, $\sqrt{8^2} = 8$. The absolute value bars are necessary for negative values of a. Suppose that $a = -10$. Then $\sqrt{a^2} = \sqrt{(-10)^2}$, which is equal to $\sqrt{100}$, or 10. So $\sqrt{a^2}$ equals the opposite of a. In either case, the principal square root of a^2 will be a positive number.

$$\sqrt{a^2} = \begin{cases} a & \text{if } a \ge 0 \\ -a & \text{if } a < 0 \end{cases}$$

The absolute value bars address both cases. We must use absolute values when dealing with variables, because we do not know if the variable is negative.

EXAMPLE 4 Simplify $\sqrt{x^6}$.

Solution

We begin by rewriting the radicand as a square. Note that $x^6 = (x^3)^2$.

$$\sqrt{x^6} = \sqrt{(x^3)^2} \qquad \text{Rewrite radicand as a square.}$$
$$= |x^3| \qquad \text{Simplify.}$$

EXAMPLE 5 Simplify $\sqrt{64y^{10}}$.

Solution

Again, we rewrite the radicand as a square.

$$\sqrt{64y^{10}} = \sqrt{(8y^5)^2} \qquad \text{Rewrite radicand as a square.}$$
$$= |8y^5| \qquad \text{Simplify.}$$

Quick Check 2
Since we know that 8 is a positive number, it could be removed from the absolute value bars. This allows us to write the expression as $8|y^5|$.

Simplify.

a) $\sqrt{a^{14}}$ b) $\sqrt{25b^{22}}$

From this point on, we will assume all variable factors in a radicand represent nonnegative real numbers. This eliminates the need to use absolute value bars when simplifying radical expressions whose radicand contains variables.

Approximating Square Roots Using a Calculator

Objective 3 **Approximate the square root of a number by using a calculator.** Consider the expression $\sqrt{12}$. There is not a positive integer that is the square root of 12. In such a case, we can use a calculator to approximate the radical expression. All calculators have a function for calculating square roots. Rounding to the nearest thousandth, we see that $\sqrt{12} \approx 3.464$. The symbol "$\approx$" is read as *"is approximately equal to,"* so the principal square root of 12 is approximately equal to 3.464. If we square 3.464, it is equal to 11.999296, which is very close to 12.

EXAMPLE 6   Approximate $\sqrt{42}$ to the nearest thousandth, using a calculator.

Solution

Since $6^2 = 36$ and $7^2 = 49$, we know that $\sqrt{42}$ must be a number between 6 and 7.

$$\sqrt{42} \approx 6.481$$

Quick Check 3

Approximate $\sqrt{109}$ to the nearest thousandth, using a calculator.

Using Your Calculator   We can approximate square roots using the TI-83/84.

```
√(42)
        6.480740698
```

*n*th Roots

Objective 4 **Find *n*th roots.** Now we move on to discuss roots other than square roots.

Principal *n*th Root

> For any positive integer $n > 1$ and any number b, if $a^n = b$ and a and b both have the same sign, then a is the **principal *n*th root** of b, denoted $a = \sqrt[n]{b}$. The number n is called the **index** of the radical.

A square root has an index of 2, and the radical is written without the index. If the index is 3, this is called a **cube root**. If n is even, then the principal nth root is a nonnegative number, but if n is odd, the principal nth root is of the same sign as the radicand. As with square roots, an even root of a negative number is not a real number. For example, $\sqrt[6]{-64}$ is not a real number. However, an odd root of a negative number, such as $\sqrt[3]{-64}$, is a negative real number.

EXAMPLE 7 Simplify $\sqrt[3]{125}$.

Solution

We are looking for a number that, when cubed, is equal to 125. Since $125 = 5^3$, $\sqrt[3]{125} = 5$.

EXAMPLE 8 Simplify $\sqrt[4]{81}$.

Solution

In this example, we are looking for a number that, when raised to the fourth power, is equal to 81. Since $81 = 3^4$, $\sqrt[4]{81} = 3$.

Using Your Calculator The function for calculating nth roots on the TI-83/84 can be found by pressing the [MATH] key and selecting option 5 under the MATH menu.

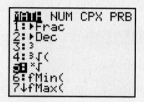

Press the index for the radical, then the nth root function, and then the radicand inside a set of parentheses. Here is the screen shot of the calculation of $\sqrt[4]{81}$:

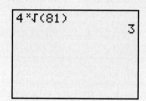

EXAMPLE 9 Simplify $\sqrt[5]{-32}$.

Solution

Quick Check 4
Simplify.
a) $\sqrt[3]{27}$ b) $\sqrt[4]{256}$
c) $\sqrt[3]{-343}$

Notice that the radicand is negative. So we are looking for a negative number that, when raised to the fifth power, is equal to -32. (If the index were even, then this expression would not be a real number.) Since $(-2)^5 = -32$, $\sqrt[5]{-32} = -2$.

For any nonnegative number x, $\sqrt[n]{x^n} = x$.

EXAMPLE 10 Simplify $\sqrt[4]{x^{20}}$. (Assume x is nonnegative.)

Solution

Quick Check 5
Simplify $\sqrt[6]{x^{42}}$. (Assume x is nonnegative.)

We begin by rewriting the radicand as an expression raised to the fourth power. We can rewrite x^{20} as $(x^5)^4$.

$$\sqrt[4]{x^{20}} = \sqrt[4]{(x^5)^4} \quad \text{Rewrite radicand as an expression raised to the fourth power.}$$
$$= x^5 \quad \text{Simplify.}$$

EXAMPLE 11 Simplify $\sqrt[3]{64a^3b^9c^{21}}$. (Assume a, b, and c are nonnegative.)

Solution

Quick Check 6
Simplify $\sqrt[5]{-32x^{35}y^{40}}$.
(Assume x and y are
nonnegative.)

We begin by rewriting the radicand as a cube. We can rewrite $64a^3b^9c^{21}$ as $(4ab^3c^7)^3$.

$$\sqrt[3]{64a^3b^9c^{21}} = \sqrt[3]{(4ab^3c^7)^3} \qquad \text{Rewrite radicand as a cube.}$$
$$= 4ab^3c^7 \qquad \text{Simplify.}$$

EXAMPLE 12 Simplify $\sqrt{x^2 - 8x + 16}$. (Assume $x \geq 4$.)

Solution

This is a square root, so we must begin by rewriting the radicand as a square. If we factor the radicand, we see that it can be expressed as $(x - 4)(x - 4)$ or $(x - 4)^2$.

$$\sqrt{x^2 - 8x + 16} = \sqrt{(x - 4)^2} \qquad \text{Rewrite the radicand as a square.}$$
$$= x - 4 \qquad \text{Simplify. (Since $x \geq 4$, $x - 4$ is nonnegative.)}$$

Quick Check 7
Simplify
$\sqrt{x^2 + 10x + 25}$.
(Assume $x \geq -5$.)

Multiplying Radical Expressions

Objective 5 Multiply radical expressions. We know that $\sqrt{9} \cdot \sqrt{100} = 3 \cdot 10$ or 30. We also know that $\sqrt{9 \cdot 100} = \sqrt{900}$ or 30. In this case, we see that $\sqrt{9} \cdot \sqrt{100} = \sqrt{9 \cdot 100}$. If two radical expressions with nonnegative radicands have the same index, then we can multiply the two expressions by multiplying the two radicands and writing the product inside the same radical.

Product Rule for Radicals

For any root n and any nonnegative numbers a and b, $\sqrt[n]{a} \cdot \sqrt[n]{b} = \sqrt[n]{ab}$.

For example, we can multiply $\sqrt{2}$ by $\sqrt{8}$. The product of these two square roots is $\sqrt{16}$, which simplifies to equal 4.

EXAMPLE 13 Multiply $\sqrt{45n} \cdot \sqrt{5n}$. (Assume n is nonnegative.)

Solution

Quick Check 8
Multiply
$\sqrt{6b^5} \cdot \sqrt{150b^3}$.
(Assume b is
nonnegative.)

Since both radicals are square roots, we can multiply the radicands.

$$\sqrt{45n} \cdot \sqrt{5n} = \sqrt{225n^2} \qquad \text{Multiply the radicands.}$$
$$= 15n \qquad \text{Simplify the square root.}$$

EXAMPLE 14 Multiply $7\sqrt{2} \cdot 9\sqrt{2}$.

Solution

We multiply the factors in front of the radicals by each other and multiply the radicands by each other.

Quick Check 9
Multiply $3\sqrt{6} \cdot 8\sqrt{6}$.

$$7\sqrt{2} \cdot 9\sqrt{2} = 63\sqrt{4} \qquad \text{Multiply the factors in front of the radicals } (7 \cdot 9).$$
$$\qquad\qquad\qquad \text{Multiply the radicands.}$$
$$= 63 \cdot 2 \qquad \text{Simplify the square root.}$$
$$= 126 \qquad \text{Multiply.}$$

Dividing Radical Expressions

Objective 6 **Divide radical expressions.** In a similar fashion, we can rewrite the quotient of two radical expressions that have the same index as the quotient of the two radicands inside the same radical.

Quotient Rule for Radicals

For any root n and any nonnegative numbers a and b, $b \neq 0$, $\dfrac{\sqrt[n]{a}}{\sqrt[n]{b}} = \sqrt[n]{\dfrac{a}{b}}$.

EXAMPLE 15 Simplify $\dfrac{\sqrt{108}}{\sqrt{3}}$.

Solution

Since both radicals are square roots, we begin by dividing the radicands. We write the quotient of the radicands under a single square root.

$$\frac{\sqrt{108}}{\sqrt{3}} = \sqrt{\frac{108}{3}} \qquad \text{Rewrite as the square root of the quotient of the radicands.}$$

$$= \sqrt{36} \qquad \text{Divide.}$$

$$= 6 \qquad \text{Simplify the square root.}$$

EXAMPLE 16 Simplify $\dfrac{\sqrt[3]{40b^7}}{\sqrt[3]{5b^4}}$. (Assume b is nonnegative.)

Solution

Since the index of both radicals are the same, we begin by dividing the radicands.

$$\frac{\sqrt[3]{40b^7}}{\sqrt[3]{5b^4}} = \sqrt[3]{\frac{40b^7}{5b^4}} \qquad \text{Divide the radicands.}$$

$$= \sqrt[3]{8b^3} \qquad \text{Simplify the radicand.}$$

$$= 2b \qquad \text{Simplify the radical.}$$

Quick Check **10** Simplify.

a) $\dfrac{\sqrt{350}}{\sqrt{14}}$ b) $\dfrac{\sqrt[5]{2916a^{18}}}{\sqrt[5]{12a^3}}$. (Assume a is nonnegative.)

Radical Functions

Objective 7 **Evaluate radical functions.** A **radical function** is a function that involves radicals, such as $f(x) = \sqrt{x - 4} + 3$. We will begin by learning to evaluate radical functions.

EXAMPLE 17   For the radical function $f(x) = \sqrt{x + 5} - 2$, find $f(-1)$.

Solution

To evaluate this function, we substitute -1 for x and simplify.

$$
\begin{aligned}
f(-1) &= \sqrt{(-1) + 5} - 2 & &\text{Substitute } -1 \text{ for } x.\\
&= \sqrt{4} - 2 & &\text{Simplify radicand.}\\
&= 2 - 2 & &\text{Take the square root of 4.}\\
&= 0 & &\text{Subtract.}
\end{aligned}
$$

Quick Check 11
For the radical function
$f(x) = \sqrt{3x + 13} + 33$,
find $f(-3)$.

Finding the Domain of Radical Functions

Objective 8 Find the domain of a radical function. Radical functions involving even roots are different than the functions we have seen to this point in that their domain is restricted. To find the domain of a radical function involving an even root, we need to find the values of the variable that make the radicand nonnegative. In other words, set the radicand greater than or equal to zero and solve. The domain of a radical function involving odd roots is the set of all real numbers. Remember that square roots are considered to be even roots, with an index of 2.

EXAMPLE 18   Find the domain of the radical function $f(x) = \sqrt{x - 9} + 7$. Express your answer in interval notation.

Solution

Since the radical has an even index, we begin by setting the radicand $(x - 9)$ greater than or equal to zero. We solve this inequality to find the domain.

Quick Check 12
Find the domain of the
radical function
$f(x) = \sqrt[6]{x + 18} - 30$.
Express your answer in
interval notation.

$$
\begin{aligned}
x - 9 &\geq 0 & &\text{Set the radicand greater than or equal to 0.}\\
x &\geq 9 & &\text{Add 9.}
\end{aligned}
$$

The domain of the function is $[9, \infty)$.

EXAMPLE 19   Find the domain of the radical function $f(x) = \sqrt[5]{14x - 9} + 21$. Express your answer in interval notation.

Solution

Since this radical function involves an odd root, its domain is the set of all real numbers $\Re$. This can be expressed using interval notation as $(-\infty, \infty)$.

Quick Check 13 Find the domain of the radical function $f(x) = \sqrt[3]{16x - 409} + 38$. Express your answer in interval notation.

Simplify the radical expression.

1. $\sqrt{36}$ **2.** $\sqrt{64}$ **3.** $\sqrt{4}$

4. $\sqrt{100}$ **5.** $\sqrt{\dfrac{1}{16}}$ **6.** $\sqrt{\dfrac{1}{25}}$

7. $\sqrt{\dfrac{36}{25}}$ **8.** $\sqrt{\dfrac{4}{81}}$ **9.** $-\sqrt{16}$

10. $-\sqrt{36}$ **11.** $\sqrt{-144}$

12. $\sqrt{-25}$

Simplify the radical expression. Where appropriate, include absolute values.

13. $\sqrt{a^{16}}$ **14.** $\sqrt{b^{12}}$ **15.** $\sqrt{x^{20}}$

16. $\sqrt{x^8}$ **17.** $\sqrt{9x^6}$ **18.** $\sqrt{16x^2}$

19. $\sqrt{\dfrac{1}{49}x^4}$ **20.** $\sqrt{\dfrac{1}{100}x^{24}}$

21. $\sqrt{m^{14}n^{10}}$ **22.** $\sqrt{a^{18}b^6}$

23. $\sqrt{x^4y^8z^{32}}$ **24.** $\sqrt{x^{20}y^2z^{10}}$

Find the missing number or expression. Assume all variables represent nonnegative real numbers.

25. $\sqrt{?} = 9$ **26.** $\sqrt{?} = 16$

27. $\sqrt{?} = 3x$ **28.** $\sqrt{?} = 5a^3$

Approximate to the nearest thousandth, using a calculator.

29. $\sqrt{55}$ **30.** $\sqrt{98}$

31. $\sqrt{326}$ **32.** $\sqrt{409}$

33. $\sqrt{0.53}$ **34.** $\sqrt{0.06}$

Simplify the radical expression. Assume all variables represent nonnegative real numbers.

35. $\sqrt[3]{8}$ **36.** $\sqrt[3]{-27}$

37. $\sqrt[4]{625}$ **38.** $\sqrt[4]{256}$

39. $\sqrt[3]{a^{18}}$ **40.** $\sqrt[4]{b^{44}}$

41. $\sqrt[3]{343m^{21}}$ **42.** $\sqrt[3]{216n^{42}}$

43. $\sqrt[4]{81x^{12}y^{20}}$ **44.** $\sqrt[6]{64s^{36}t^{54}}$

45. $\sqrt{x^2 + 6x + 9}$, $x \geq -3$

46. $\sqrt{x^2 + 12x + 36}$, $x \geq -6$

Simplify.

47. $\sqrt{27} \cdot \sqrt{3}$ **48.** $\sqrt{8} \cdot \sqrt{8}$

49. $\sqrt{10} \cdot \sqrt{90}$ **50.** $\sqrt{2} \cdot \sqrt{72}$

51. $\dfrac{\sqrt{180}}{\sqrt{5}}$ **52.** $\dfrac{\sqrt{63}}{\sqrt{7}}$

53. $\dfrac{\sqrt{800}}{\sqrt{8}}$ **54.** $\dfrac{\sqrt{1872}}{\sqrt{13}}$

55. $\sqrt[3]{6} \cdot \sqrt[3]{36}$ **56.** $\sqrt[3]{12} \cdot \sqrt[3]{18}$

57. $\sqrt[4]{48} \cdot \sqrt[4]{27}$ **58.** $\sqrt[5]{16} \cdot \sqrt[5]{64}$

59. $\dfrac{\sqrt[3]{297}}{\sqrt[3]{11}}$ **60.** $\dfrac{\sqrt[4]{144}}{\sqrt[4]{9}}$

61. $4\sqrt{5} \cdot 10\sqrt{5}$

62. $16\sqrt{3} \cdot 3\sqrt{3}$

63. $10\sqrt{80} \cdot \sqrt{45}$

64. $2\sqrt{49} \cdot 15\sqrt{121}$

Find the missing number.

65. $\sqrt{20} \cdot \sqrt{?} = 10$ **66.** $\sqrt{21} \cdot \sqrt{?} = 42$

67. $\dfrac{\sqrt{?}}{\sqrt{8}} = 7$ **68.** $\dfrac{\sqrt{?}}{\sqrt{6}} = 6$

Evaluate the radical function. Round to the nearest thousandth if necessary.

69. $f(x) = \sqrt{x - 5}$; find $f(21)$.

70. $f(x) = \sqrt{3x + 4}$; find $f(20)$.

71. $f(x) = \sqrt{2x + 19} - 2$; find $f(3)$.

72. $f(x) = \sqrt{4x - 11} - 10$; find $f(5)$.

73. $f(x) = \sqrt{x + 6}$; find $f(8)$.

74. $f(x) = \sqrt{3x + 1} + 5$; find $f(4)$.

75. $f(x) = \sqrt{x^2 - 8x + 16}$; find $f(9)$.

76. $f(x) = \sqrt{x^2 - 7x + 19}$; find $f(10)$.

Find the domain of the radical function. Express your answer in interval notation.

77. $f(x) = \sqrt{x - 8} + 10$

78. $f(x) = \sqrt{4x + 3} - 9$

79. $f(x) = \sqrt{2x - 15} - 8$

80. $f(x) = \sqrt{5x} - 95$

81. $f(x) = \sqrt[4]{4x + 10} + 2$

82. $f(x) = \sqrt[6]{8x - 50} - 11$

83. $f(x) = \sqrt[3]{15x - 42} + 6$

84. $f(x) = \sqrt[5]{10x + 115} - 45$

Answer in complete sentences.

85. Which of the following are real numbers, and which are not real numbers: $-\sqrt{64}$, $\sqrt{-64}$, $-\sqrt[3]{64}$, $\sqrt[3]{-64}$? Explain your reasoning.

86. If we do not know whether x is a nonnegative number, explain why $\sqrt{x^2} = |x|$ rather than x.

87. Explain how to find the domain of a radical function. Use examples to illustrate the process.

88. True or False: for any real number x, $\sqrt{x} \geq \sqrt[3]{x}$. Explain your reasoning.

Study Tip **REVISITED**   One important, yet frequently overlooked, aspect of effective note taking is choosing the appropriate seat in the classroom. You must sit in a location where you can clearly see the entire board, preferably in the center of the classroom. Students who sit in the front rows on the left or right side of the classroom may struggle with seeing material on the opposite side of the room. This problem is compounded in classrooms equipped with a whiteboard, as these types of boards can create a glare. Finally, try not to sit behind anyone who will obstruct your vision, such as an extremely tall person.

Location also affects your ability to hear your instructor clearly. You will find that your instructor speaks towards the middle of the classroom, so finding a seat towards the center of the classroom should help you to hear everything your instructor says. Try not to sit close to students who talk to each other during class; their discussions may distract you from what your instructor is saying.

In summary, try to choose your classroom seat in the same fashion that you choose a seat at a movie theater. Be sure that you can see and hear everything.

9.2
RATIONAL EXPONENTS

Objectives

1 **Simplify expressions containing exponents of the form 1/n.**
2 **Simplify expressions containing exponents of the form m/n.**
3 **Simplify expressions containing rational exponents.**
4 **Simplify expressions containing negative rational exponents.**
5 **Use rational exponents to simplify radical expressions.**

Rational Exponents of the Form 1/n

Objective 1 Simplify expressions containing exponents of the form 1/n. We have used exponents to represent repeated multiplication. For example, x^n tells us that the base x is a factor n times.

$$x^3 = x \cdot x \cdot x$$
$$x^6 = x \cdot x \cdot x \cdot x \cdot x \cdot x$$

We run into a problem with this definition when we encounter an expression with a fractional exponent, such as $x^{1/2}$. Saying that the base x is a factor $\frac{1}{2}$ times does not make any sense. Using the properties of exponents introduced in Chapter 5, we know that $x^{1/2} \cdot x^{1/2} = x^{1/2+1/2}$ or x. If we multiply $x^{1/2}$ by itself, the result is x. The same is true when we multiply $\sqrt{x}$ by itself, suggesting that $x^{1/2} = \sqrt{x}$.

For any integer $n > 1$, we define $a^{1/n}$ to be the nth root of a, or $\sqrt[n]{a}$.

EXAMPLE 1   Rewrite $81^{1/4}$ as a radical expression and simplify if possible.

Solution

$$81^{1/4} = \sqrt[4]{81}$$ Rewrite as a radical expression. An exponent of 1/4 is equivalent to a fourth root.
$$= \sqrt[4]{3^4}$$ Rewrite radicand as a number to the fourth power.
$$= 3$$ Simplify.

EXAMPLE 2   Rewrite $(-125x^6)^{1/3}$ as a radical expression and simplify if possible.

Solution

$$(-125x^6)^{1/3} = \sqrt[3]{(-5x^2)^3}$$ Rewrite as a radical expression. Rewrite the radicand as a cube.
$$= -5x^2$$ Simplify. Keep in mind that an odd root of a negative number is negative.

Quick Check   **1** Rewrite as a radical expression and simplify if possible.

a) $36^{1/2}$ b) $(-32x^{15})^{1/5}$

EXAMPLE 3   Rewrite $(a^{10}b^5c^{20})^{1/5}$ as a radical expression and simplify if possible.

Solution

$$(a^{10}b^5c^{20})^{1/5} = \sqrt[5]{a^{10}b^5c^{20}} \qquad \text{Rewrite as a radical expression}$$
$$= a^2bc^4 \qquad\qquad\qquad \text{Simplify.}$$

Quick Check **2** Rewrite $(x^{32}y^4z^{24})^{1/4}$ as a radical expression and simplify if possible. Assume all variables represent nonnegative values.

For any negative number x and even integer n, $x^{1/n}$ is not a real number. For example, $(-64)^{1/6}$ is not a real number, because $(-64)^{1/6} = \sqrt[6]{-64}$ and an even root of a negative number is not a real number.

Rational Exponents of the Form *m/n*

Objective 2   **Simplify expressions containing exponents of the form *m/n*.** We now turn our attention to rational exponents of the form m/n, for any integers m and $n > 1$. The expression $a^{m/n}$ can be rewritten as $(a^{1/n})^m$, which is equivalent to $(\sqrt[n]{a})^m$.

> For any integers m and n, $n > 1$, we define $a^{m/n}$ to be $(\sqrt[n]{a})^m$. This is also equivalent to $\sqrt[n]{a^m}$. The denominator in the exponent, n, is the root we are taking. The numerator in the exponent, m, is the power to which we raise this radical.

EXAMPLE 4   Rewrite $(243x^{10})^{3/5}$ as a radical expression and simplify if possible.

Solution

$$(243x^{10})^{3/5} = (\sqrt[5]{243x^{10}})^3 \qquad \text{Rewrite as a radical expression.}$$
$$= (3x^2)^3 \qquad\qquad\quad \text{Simplify the radical.}$$
$$= 27x^6 \qquad\qquad\quad\; \text{Raise } 3x^2 \text{ to the third power.}$$

Quick Check **3** Rewrite $(4096x^{18})^{5/6}$ as a radical expression and simplify if possible. Assume x is nonnegative.

Simplifying Expressions Containing Rational Exponents

Objective 3 **Simplify expressions containing rational exponents.** The properties of exponents developed in Chapter 5 for integer exponents are true for fractional exponents as well. Here is a summary of those properties:

Properties of Exponents

1. For any base x, $x^m \cdot x^n = x^{m+n}$
2. For any base x, $(x^m)^n = x^{m \cdot n}$
3. For any bases x and y, $(xy)^n = x^n y^n$
4. For any base x, $\dfrac{x^m}{x^n} = x^{m-n}$ $(x \neq 0)$
5. For any base x, $x^0 = 1$ $(x \neq 0)$
6. For any bases x and y, $\left(\dfrac{x}{y}\right)^n = \dfrac{x^n}{y^n}$ $(y \neq 0)$
7. For any *nonzero* base x, $x^{-n} = \dfrac{1}{x^n}$. $(x \neq 0)$

EXAMPLE 5   Simplify the expression $x^{4/3} \cdot x^{5/6}$. Assume x is nonnegative. Express your answer in radical notation.

Solution

When multiplying two expressions with the same base, we add the exponents and keep the base.

$$
\begin{aligned}
x^{4/3} \cdot x^{5/6} &= x^{\frac{4}{3}+\frac{5}{6}} & &\text{Add the exponents, keep the base.} \\
&= x^{\frac{8}{6}+\frac{5}{6}} & &\text{Rewrite the fractions with a common denominator.} \\
&= x^{13/6} & &\text{Add.} \\
&= \left(\sqrt[6]{x}\right)^{13} & &\text{Rewrite using radical notation.}
\end{aligned}
$$

Quick Check 4
Simplify the expression $x^{5/8} \cdot x^{7/12}$. Assume x is nonnegative. Express your answer in radical notation.

EXAMPLE 6   Simplify the expression $(x^{3/4})^{2/5}$. Assume x is nonnegative. Express your answer in radical notation.

Solution

When raising an exponential expression to another power, we multiply the exponents and keep the base.

$$
\begin{aligned}
(x^{3/4})^{2/5} &= x^{\frac{3}{4} \cdot \frac{2}{5}} & &\text{Multiply the exponents, keep the base.} \\
&= x^{3/10} & &\text{Multiply.} \\
&= \left(\sqrt[10]{x}\right)^3 & &\text{Rewrite using radical notation.}
\end{aligned}
$$

Quick Check 5
Simplify the expression $\left(x^{7/10}\right)^{4/9}$. Assume x is nonnegative. Express your answer in radical notation.

EXAMPLE 7   Simplify the expression $\left(\dfrac{b^4}{c^2 d^6}\right)^{3/2}$, $(c \neq 0, d \neq 0)$. Assume all variables represent nonnegative values.

Solution

We begin by raising each factor to the $\frac{3}{2}$ power.

$$\left(\frac{b^4}{c^2 d^6}\right)^{3/2} = \frac{b^{4\left(\frac{3}{2}\right)}}{c^{2\left(\frac{3}{2}\right)} d^{6\left(\frac{3}{2}\right)}} \qquad \text{Raise each factor to the } \frac{3}{2} \text{ power.}$$

$$= \frac{b^6}{c^3 d^9} \qquad \text{Multiply exponents.}$$

Quick Check 6
Simplify the expression
$\left(\dfrac{x^9}{y^3 z^{12}}\right)^{5/3}$,

$(y \neq 0, z \neq 0).$

Simplifying Expressions Containing Negative Rational Exponents

Objective 4 Simplify expressions containing negative rational exponents.

EXAMPLE 8   Simplify the expression $64^{-5/6}$.

Solution

Again, we begin by rewriting the expression with a positive exponent.

$$64^{-5/6} = \frac{1}{64^{5/6}} \qquad \text{Rewrite the expression with a positive exponent.}$$

$$= \frac{1}{\left(\sqrt[6]{64}\right)^5} \qquad \text{Rewrite in radical notation.}$$

$$= \frac{1}{2^5} \qquad \text{Simplify the radical.}$$

$$= \frac{1}{32} \qquad \text{Raise 2 to the fifth power.}$$

Quick Check 7
Simplify the expression
$81^{-3/2}$.

Using Rational Exponents to Simplify Radical Expressions

Objective 5 Use rational exponents to simplify radical expressions. In the previous section, we learned that we may multiply two radicals if they have the same index. In other words, $\sqrt[n]{a} \cdot \sqrt[n]{b} = \sqrt[n]{ab}$. If the two indices are not the same, we can use fractional exponents to multiply the radicals.

EXAMPLE 9   Simplify the expression $\sqrt{a} \cdot \sqrt[5]{a}$. Assume a is nonnegative. Express your answer in radical notation.

Solution

We will begin by rewriting the radicals using fractional exponents.

Quick Check 8
Simplify the expression
$\sqrt[3]{x} \cdot \sqrt[4]{x}$. Assume x is nonnegative. Express your answer in radical notation.

$$\sqrt{a} \cdot \sqrt[5]{a} = a^{1/2} \cdot a^{1/5} \qquad \text{Rewrite both radicals using fractional exponents.}$$
$$= a^{\frac{1}{2}+\frac{1}{5}} \qquad \text{Add the exponents, keeping the base.}$$
$$= a^{\frac{5}{10}+\frac{2}{10}} \qquad \text{Rewrite each fraction with a common denominator.}$$
$$= a^{7/10} \qquad \text{Add.}$$
$$= \sqrt[10]{a^7} \qquad \text{Rewrite in radical notation.}$$

Rewrite each radical expression using rational exponents.

1. $\sqrt[4]{x}$

2. $\sqrt[5]{a}$

3. $\sqrt{7}$

4. $\sqrt[3]{10}$

5. $9\sqrt[4]{d}$

6. $\sqrt[4]{9d}$

Rewrite as a radical expression and simplify if possible. Assume all variables represent nonnegative real numbers.

7. $64^{1/2}$

8. $25^{1/2}$

9. $64^{1/3}$

10. $1296^{1/4}$

11. $(-343)^{1/3}$

12. $(-243)^{1/5}$

13. $(x^{12})^{1/4}$

14. $(y^{15})^{1/3}$

15. $(x^{40}y^{35})^{1/5}$

16. $(x^{22}y^{36})^{1/2}$

17. $(16x^{32}y^{60})^{1/4}$

18. $(27x^{15}y^{30}z^{45})^{1/3}$

19. $(-216x^{18}y^{15}z^{21})^{1/3}$

20. $(-32x^{5}y^{25}z^{125})^{1/5}$

Rewrite each radical expression, using rational exponents. Assume all variables represent nonnegative real numbers.

21. $(\sqrt[5]{x})^3$

22. $(\sqrt[4]{x})^7$

23. $(\sqrt[9]{y})^8$

24. $(\sqrt[12]{a})^{17}$

25. $\sqrt[8]{x^5}$

26. $\sqrt[4]{a^{15}}$

27. $(\sqrt[3]{3x^2})^7$

28. $(\sqrt[9]{2x^4})^2$

29. $(\sqrt[8]{10x^4y^5})^3$

30. $(\sqrt[5]{a^6b^7c^8})^4$

Rewrite as a radical expression and simplify if possible. Assume all variables represent nonnegative real numbers.

31. $25^{3/2}$

32. $16^{3/4}$

33. $32^{2/5}$

34. $1000^{7/3}$

35. $(x^{12})^{2/3}$

36. $(x^{20})^{6/5}$

37. $(256a^8b^{24})^{3/4}$

38. $(4x^{16}y^{32}z^{64})^{5/2}$

39. $(-125x^9y^{15}z^3)^{4/3}$

40. $(256a^4b^8c^{20})^{7/4}$

Simplify the expression. Assume all variables represent nonnegative real numbers. Express your answer in radical notation.

41. $x^{7/10} \cdot x^{1/10}$

42. $x^{3/4} \cdot x^{11/4}$

43. $x^{2/5} \cdot x^{1/4}$

44. $x^{5/8} \cdot x^{1/6}$

45. $x^{4/3} \cdot x^{1/2}$

46. $x^{10/7} \cdot x^{1/14}$

47. $a^{5/3} \cdot a^{3/4} \cdot a^{1/6}$

48. $m^{4/5} \cdot m^{5/2} \cdot m^{7/10}$

49. $x^{1/8}y^{2/5} \cdot x^{3/8}y^{1/10}$

50. $x^{5/7}y^{2/9} \cdot x^{3/2}y^{2/3}$

51. $(x^{1/6})^{3/5}$

52. $(x^{5/4})^{2/3}$

53. $(b^{2/3})^{2/3}$

54. $(n^{3/7})^{5/9}$

55. $(x^4)^{7/8}$

56. $(x^{7/12})^3$

57. $\dfrac{x^{4/5}}{x^{3/10}} \ (x \neq 0)$

58. $\dfrac{x^{7/6}}{x^{11/12}} \ (x \neq 0)$

59. $\dfrac{x^{5/8}}{x^{1/3}} \ (x \neq 0)$

F O R **E X T R A** **H E L P**

MyMathLab
MyMathLab

MathXL
MathXL

Interactmath.com

MathXL
Tutorials on CD

DVT CD
Videotape

Tutor
Center
Addison-Wesley
Math Tutor Center

Student Solutions
Manual

60. $\dfrac{x^{9/10}}{x^{21/40}}\ (x \neq 0)$

61. $\dfrac{x^{13/14}}{x^{13/14}}\ (x \neq 0)$

62. $\dfrac{x^{9/8}}{x^{9/8}}\ (x \neq 0)$

63. $\left(x^{5/4}\right)^0\ (x \neq 0)$

64. $\left(x^{2/9}\right)^0\ (x \neq 0)$

65. $8^{-2/3}$

66. $81^{-1/4}$

67. $49^{-1/2}$

68. $25^{-3/2}$

69. $32^{-3/5} \cdot 32^{-4/5}$

70. $27^{-5/3} \cdot 27^{-4/3}$

71. $6^{-1/4} \cdot 6^{-3/4}$

72. $10^{-3/2} \cdot 10^{-5/2}$

73. $\dfrac{125^{2/3}}{125^{7/3}}$

74. $\dfrac{16^{3/4}}{16^{10/4}}$

75. $\dfrac{216^{7/3}}{216^{8/3}}$ **76.** $\dfrac{4^{3/2}}{4^5}$

Simplify the expression. Assume all variables represent nonnegative real numbers. Express your answer in radical notation.

77. $\sqrt[4]{x} \cdot \sqrt[5]{x}$

78. $\sqrt[9]{x} \cdot \sqrt[18]{x}$

79. $\sqrt[12]{a} \cdot \sqrt[4]{a}$

80. $\sqrt[3]{b} \cdot \sqrt[8]{b}$

81. $\dfrac{\sqrt[4]{m}}{\sqrt[12]{m}}\ (m \neq 0)$

82. $\dfrac{\sqrt{n}}{\sqrt[3]{n}}\ (n \neq 0)$

83. $\dfrac{\sqrt[30]{x}}{\sqrt[5]{x}}\ (x \neq 0)$

84. $\dfrac{\sqrt[10]{x}}{\sqrt[5]{x}}\ (x \neq 0)$

Answer in complete sentences.

85. Is $-16^{1/2}$ a real number? Explain your answer.

86. Is $16^{-1/2}$ a real number? Explain your answer.

87. Explain how to rewrite an expression with a fractional exponent as a radical expression. Use an example to illustrate the process.

88. Explain how to rewrite a radical expression using fractional exponents. Use an example to illustrate the process.

Study Tip **Revisited**   When your instructor solves a problem during class, copy down every step your instructor writes. Many students feel that they completely understand each step of a solution when it is written on the board, only to forget when they start to do their homework. Writing down each step prevents this.

If you become confused about a problem, writing down each step will help you when you try to make sense of the material after class. Leave yourself some room for adding explanations later. Place a large question mark in the margin of your notes next to steps you do not understand, and ask your instructor, study group, or tutor for clarification after class.

If you ask your instructor for clarification about a particular step, write down the explanation in your notes right next to the step in question.

9.3

**SIMPLIFYING
RADICAL
EXPRESSIONS;
ADDING AND
SUBTRACTING
RADICAL
EXPRESSIONS**

Objectives

1 Simplify radical expressions by using the product property.
2 Add or subtract radical expressions containing like radicals.
3 Simplifying radical expressions before adding or subtracting.

Simplifying Radical Expressions Using the Product Property

Objective 1 **Simplify radical expressions by using the product property.** A radical expression is considered simplified if the radicand contains no factors with exponents greater than or equal to the index of the radical. For example, $\sqrt[3]{x^5}$ is not simplified because there is a power inside the radical that is greater than the index of the radical. The goal for simplifying radical expressions is to remove as many factors as possible from the radicand. We will use the product property for radical expressions to help us with this.

We could rewrite $\sqrt[3]{x^5}$ as $\sqrt[3]{x^3 \cdot x^2}$, and then rewrite this radical expression as the product of two radicals. Using the product property for radicals, $\sqrt[n]{a} \cdot \sqrt[n]{b} = \sqrt[n]{ab}$, we know that $\sqrt[3]{x^3 \cdot x^2} = \sqrt[3]{x^3} \cdot \sqrt[3]{x^2}$. The reason for rewriting $\sqrt[3]{x^5}$ as $\sqrt[3]{x^3} \cdot \sqrt[3]{x^2}$ is that we can simplify the radical $\sqrt[3]{x^3}$ as x.

$$\sqrt[3]{x^5} = \sqrt[3]{x^3 \cdot x^2}$$
$$= \sqrt[3]{x^3} \cdot \sqrt[3]{x^2}$$
$$= x \cdot \sqrt[3]{x^2}$$

Now the radicand contains no factors with an exponent that is greater than or equal to the index 3 and is simplified.

Simplifying Radical Expressions

- Completely factor any numerical factors in the radicand.
- Rewrite each factor as a product of two factors. The exponent for the first factor should be the largest multiple of the radical's index that is less than or equal to the factor's original exponent.
- Use the product property to remove factors from the radicand.

EXAMPLE 1   Simplify $\sqrt[4]{a^{23}}$. Assume a is nonnegative.

Solution

We begin by rewriting a^{23} as a product of two factors. The largest multiple of the index (4) that is less than or equal to the exponent for this factor (23) is 20, so we will rewrite a^{23} as $a^{20} \cdot a^3$.

Quick Check **1**

Simplify $\sqrt[5]{x^{17}}$. Assume x is nonnegative.

$$\sqrt[4]{a^{23}} = \sqrt[4]{a^{20} \cdot a^3} \qquad \text{Rewrite } a^{23} \text{ as the product of two factors.}$$
$$= \sqrt[4]{a^{20}} \cdot \sqrt[4]{a^3} \qquad \text{Use the product property of radicals to rewrite the radical as the product of two radicals.}$$
$$= a^5 \sqrt[4]{a^3} \qquad \text{Simplify the radical.}$$

EXAMPLE 2 Simplify $\sqrt{x^{11}y^{10}z^5}$. Assume all variables represent nonnegative values.

Solution

Again, we begin by rewriting factors as a product of two factors where necessary. In this example, the exponent of the factor y is a multiple of the index 2. We do not need to rewrite this factor as the product of two factors.

$$\begin{aligned} \sqrt{x^{11}y^{10}z^5} &= \sqrt{(x^{10} \cdot x)y^{10}(z^4 \cdot z)} & \text{Rewrite factors.} \\ &= \sqrt{x^{10}y^{10}z^4} \cdot \sqrt{xz} & \text{Rewrite as the product of two radicals.} \\ &= x^5y^5z^2\sqrt{xz} & \text{Simplify the radical.} \end{aligned}$$

Quick Check 2
Simplify $\sqrt{a^8b^{15}c^7d}$. Assume all variables represent nonnegative values.

EXAMPLE 3 Simplify $\sqrt{24}$.

Solution

We begin by rewriting 24 using its prime factorization $(2^3 \cdot 3)$.

$$\begin{aligned} \sqrt{24} &= \sqrt{2^3 \cdot 3} & \text{Factor 24.} \\ &= \sqrt{(2^2 \cdot 2) \cdot 3} & \text{Rewrite } 2^3 \text{ as } 2^2 \cdot 2. \\ &= \sqrt{2^2} \cdot \sqrt{2 \cdot 3} & \text{Rewrite as the product of two radicals.} \\ &= 2\sqrt{6} & \text{Simplify.} \end{aligned}$$

We could have used a different tactic to simplify this square root. What is the largest factor of 24 that is a perfect square? The largest perfect square that is a factor of 24 is 4, so we could begin by rewriting 24 as $4 \cdot 6$. Since we know that the square root of 4 is 2, we can factor 4 out of the radicand and write it as 2 in front of the radical.

$$\begin{aligned} \sqrt{24} &= \sqrt{4 \cdot 6} \\ &= 2\sqrt{6} \end{aligned}$$

Quick Check 3
Simplify $\sqrt{90}$.

EXAMPLE 4 Simplify $\sqrt[3]{324}$.

Solution

We begin by factoring 324 to be $2^2 \cdot 3^4$.

$$\begin{aligned} \sqrt[3]{324} &= \sqrt[3]{2^2 \cdot 3^4} & \text{Factor 324.} \\ &= \sqrt[3]{2^2 \cdot (3^3 \cdot 3)} & \text{Rewrite } 3^4 \text{ as } 3^3 \cdot 3. \\ &= \sqrt[3]{3^3} \cdot \sqrt[3]{2^2 \cdot 3} & \text{Rewrite as the product of two radicals.} \\ &= 3\sqrt[3]{12} & \text{Simplify.} \end{aligned}$$

Quick Check 4
Simplify $\sqrt[3]{280}$.

There is an alternative approach for simplifying radical expressions. Suppose we were trying to simplify $\sqrt[5]{a^{48}}$. Using the previous method, we find that this simplifies to $a^9\sqrt[5]{a^3}$.

$$\begin{aligned} \sqrt[5]{a^{48}} &= \sqrt[5]{a^{45} \cdot a^3} & \text{Rewrite } a^{48} \text{ as } a^{45} \cdot a^3, \text{ since 45 is the highest multiple} \\ & & \text{of 5 that is less than or equal to 48.} \\ &= \sqrt[5]{a^{45}} \cdot \sqrt[5]{a^3} & \text{Rewrite as the product of two radicals.} \\ &= a^9\sqrt[5]{a^3} & \text{Simplify.} \end{aligned}$$

We know that for every five times that a is repeated as a factor in the radicand, we can take a^5 out of the radicand and write it as a in front of the radical. We need to determine how many groups of five can be removed from the radicand, using division. Notice that if we divide the exponent 48 by the index 5, the quotient is 9 with a remainder of 3. When we divide the exponent of a factor in the radicand by the index of the radical, the quotient tells us the exponent of the factor removed from the radicand and the remainder tells us the exponent of the factor remaining in the radicand.

An Alternative Approach for Simplifying $\sqrt[n]{x^p}$

- Divide p by n: $\frac{p}{n} = q + \frac{r}{n}$
- The quotient q tells us how many times that x will be a factor in front of the radical.
- The remainder r tells us how many times that x will remain as a factor in the radicand.

$$\sqrt[n]{x^p} = x^q \sqrt[n]{x^r}$$

EXAMPLE 5   Simplify $\sqrt[6]{a^{31}b^{18}c^5d^{53}}$. Assume all variables represent nonnegative values.

Solution

We will work with one factor at a time, beginning with a. The index, 6, divides into the exponent, 31, five times with a remainder of one. This tells us we can write a^5 as a factor in front of the radical and that a^1 or a can be written in the radicand.

$$\sqrt[6]{a^{31}b^{18}c^5d^{53}} = a^5\sqrt[6]{ab^{18}c^5d^{53}}$$

For the factor b, $18 \div 6 = 3$ with a remainder of 0. We will write b^3 as a factor in front of the radical, and since the remainder is 0, we will not write b as a factor in the radicand. For the factor c, the index does not divide into 5, so c^5 remains as a factor in the radicand. Finally, for the factor d, $53 \div 6 = 8$ with a remainder of 5. We will write d^8 as a factor in front of the radical and d^5 as a factor in the radicand.

$$\sqrt[6]{a^{31}b^{18}c^5d^{53}} = a^5b^3d^8\sqrt[6]{ac^5d^5}$$

Quick Check 5
Simplify $\sqrt[5]{x^{33}y^6z^{50}w^{18}}$.

Adding and Subtracting Radical Expressions Containing Like Radicals

Objective 2 Add or subtract radical expressions containing like radicals.

Like Radicals

Two radical expressions are called **like radicals** if they have the same index and the same radicand.

The radical expressions $5\sqrt[3]{4x}$ and $9\sqrt[3]{4x}$ are like radicals because they have the same index (3) and the same radicand ($4x$). Here are some examples of radical expressions that are not like radicals:

$$\sqrt{5} \quad \text{and} \quad \sqrt[3]{5} \qquad \text{The two radicals have different indices.}$$
$$\sqrt[4]{7x^2y^3} \quad \text{and} \quad \sqrt[4]{7x^3y^2} \qquad \text{The two radicands are different.}$$

We can add and subtract radical expressions by combining like radicals in a fashion similar to the way we combine like terms. We add or subtract the factors in front of the like radicals.

$$6\sqrt{2} + 3\sqrt{2} = 9\sqrt{2}$$

EXAMPLE 6   Simplify $5\sqrt[3]{4x} - 15\sqrt[3]{4x}$.

Solution

Notice that the radicals are like radicals. We can combine these two expressions by subtracting the coefficients.

Quick Check 6

Simplify
$7\sqrt[5]{2x^2} + 14\sqrt[5]{2x^2}$.

$$5\sqrt[3]{4x} - 15\sqrt[3]{4x} = -10\sqrt[3]{4x}$$ Combine the like radicals by subtracting the coefficients and keeping the radical.

EXAMPLE 7   Simplify $12\sqrt{13} + \sqrt{3} + \sqrt{3} - 6\sqrt{13}$.

Solution

There are two pairs of like radicals in this example. There are two radical expressions containing $\sqrt{13}$ and two radical expressions containing $\sqrt{3}$.

$$12\sqrt{13} + \sqrt{3} + \sqrt{3} - 6\sqrt{13}$$
$$= 6\sqrt{13} + \sqrt{3} + \sqrt{3}$$ Subtract $12\sqrt{13} - 6\sqrt{13}$.
$$= 6\sqrt{13} + 2\sqrt{3}$$ Add $\sqrt{3} + \sqrt{3}$.

Quick Check 7 Simplify $9\sqrt{10} - 13\sqrt{5} + 6\sqrt{10} + 8\sqrt{5}$.

EXAMPLE 8   Simplify $18\sqrt[5]{x^3y^2} - 2\sqrt[5]{x^2y^3} - 8\sqrt[5]{x^3y^2}$.

Solution

Of the three terms, only the first and third contain like radicals.

$$18\sqrt[5]{x^3y^2} - 2\sqrt[5]{x^2y^3} - 8\sqrt[5]{x^3y^2} = 10\sqrt[5]{x^3y^2} - 2\sqrt[5]{x^2y^3}$$

Subtract $18\sqrt[5]{x^3y^2} - 8\sqrt[5]{x^3y^2}$.

Quick Check 8 Simplify $16\sqrt[5]{a^4b^3} - 11\sqrt[7]{a^4b^3} - 5\sqrt[5]{a^4b^3}$.

Objective 3   Simplifying radical expressions before adding or subtracting. Are the expressions $\sqrt{24}$ and $\sqrt{54}$ like radicals? Although at first our answer might be no, we must simplify each radical completely before we can determine whether the two expressions are like radicals. In this case, $\sqrt{24} = 2\sqrt{6}$ and $\sqrt{54} = 3\sqrt{6}$, so $\sqrt{24}$ and $\sqrt{54}$ are like radicals.

EXAMPLE 9 Simplify $\sqrt{12} + \sqrt{3}$.

Solution

We begin by simplifying each radical completely. Since 12 can be written as $4 \cdot 3$, $\sqrt{12}$ can be simplified to be $2\sqrt{3}$. We could also use the prime factorization of 12 ($2^2 \cdot 3$) to simplify $\sqrt{12}$.

$$\begin{aligned} \sqrt{12} + \sqrt{3} &= \sqrt{4 \cdot 3} + \sqrt{3} &&\text{Factor 12.} \\ &= 2\sqrt{3} + \sqrt{3} &&\text{Simplify } \sqrt{4 \cdot 3}. \\ &= 3\sqrt{3} &&\text{Add.} \end{aligned}$$

Quick Check 9
Simplify $\sqrt{63} + \sqrt{7}$.

EXAMPLE 10 Simplify $\sqrt{45} - \sqrt{80} - \sqrt{20}$.

Solution

Quick Check 10
Simplify
$\sqrt{18} - \sqrt{32} + \sqrt{98}$.

In this example, we must simplify all three radicals before proceeding.

$$\begin{aligned} \sqrt{45} - \sqrt{80} - \sqrt{20} &= \sqrt{9 \cdot 5} - \sqrt{16 \cdot 5} - \sqrt{4 \cdot 5} &&\text{Factor each radicand.} \\ &= 3\sqrt{5} - 4\sqrt{5} - 2\sqrt{5} &&\text{Simplify each radical.} \\ &= -3\sqrt{5} &&\text{Combine like radicals.} \end{aligned}$$

EXERCISES 9.3

Simplify.

1. $\sqrt{12}$

2. $\sqrt{18}$

3. $\sqrt{80}$

4. $\sqrt{175}$

5. $\sqrt{363}$

6. $\sqrt{288}$

7. $\sqrt[3]{81}$

8. $\sqrt[3]{48}$

9. $\sqrt[3]{750}$

10. $\sqrt[3]{1029}$

11. $\sqrt[4]{1200}$

12. $\sqrt[5]{448}$

Simplify the radical expression. Assume all variables represent nonnegative real numbers.

13. $\sqrt{x^9}$

14. $\sqrt{a^{15}}$

15. $\sqrt[3]{m^{13}}$

16. $\sqrt[3]{x^{40}}$

17. $\sqrt[5]{x^{74}}$

18. $\sqrt[4]{b^{82}}$

19. $\sqrt{x^{15}y^{12}}$

20. $\sqrt{a^{23}b^3}$

21. $\sqrt{xy^7}$

22. $\sqrt{x^{20}y^{11}}$

23. $\sqrt{x^{33}y^{17}z^{16}}$

24. $\sqrt{r^{19}s^{18}t^{28}}$

25. $\sqrt[3]{a^{12}b^2c^{19}}$

26. $\sqrt[3]{x^{30}y^{21}z^{25}}$

27. $\sqrt[6]{a^{69}b^{35}c^5}$

28. $\sqrt[5]{x^{80}y^{40}z^{11}}$

29. $\sqrt{8a^9b^8}$

30. $\sqrt{180x^{25}y^{21}}$

31. $\sqrt[4]{112a^3b^{13}c^{23}}$

32. $\sqrt[5]{64x^{72}y^{54}z^{104}}$

Add or subtract. Assume all variables represent non-negative real numbers.

33. $10\sqrt{5} + 17\sqrt{5}$

34. $8\sqrt{11} - 19\sqrt{11}$

35. $9\sqrt[3]{4} - 15\sqrt[3]{4}$

36. $14\sqrt[3]{20} + 6\sqrt[3]{20}$

37. $40\sqrt{21} - 12\sqrt{21} - 33\sqrt{21}$

38. $6\sqrt{7} - 44\sqrt{7} + 16\sqrt{7}$

39. $7\sqrt{10} + 6\sqrt{15} - 12\sqrt{15} + 9\sqrt{10}$

40. $\sqrt{19} + 18\sqrt{6} - 13\sqrt{19} + 24\sqrt{6}$

41. $(9\sqrt{2} + 5\sqrt{3}) - (6\sqrt{2} - 5\sqrt{3})$

42. $(4\sqrt{5} - 3\sqrt{14}) - (8\sqrt{14} - 16\sqrt{5})$

43. $5\sqrt{x} - 7\sqrt{x}$

44. $13\sqrt{y} + 12\sqrt{y}$

45. $3\sqrt[5]{a} + 8\sqrt[5]{a}$

46. $11\sqrt[4]{x} - 2\sqrt[4]{x}$

47. $16\sqrt{x} - 9\sqrt{x} + 15\sqrt{x}$

48. $-24\sqrt{x} - 17\sqrt{x} + 3\sqrt{x}$

49. $4\sqrt{x} - 11\sqrt{y} - 6\sqrt{y} + \sqrt{x}$

50. $-2\sqrt{a} + 9\sqrt{b} - 18\sqrt{a} - 5\sqrt{b}$

51. $8\sqrt[3]{x} + 7\sqrt[4]{x} - 10\sqrt[4]{x} + 13\sqrt[3]{x}$

52. $\sqrt[4]{ab^3} + 12\sqrt[4]{a^2b^3} + 23\sqrt[4]{ab^3} - 25\sqrt[4]{a^2b^3}$

53. $8\sqrt{2} + \sqrt{50}$

54. $2\sqrt{48} + 5\sqrt{3}$

55. $7\sqrt{28} - 3\sqrt{63} + 16\sqrt{7}$

56. $6\sqrt{8} - 13\sqrt{18} + 9\sqrt{200}$

57. $2\sqrt[3]{81} - 5\sqrt[3]{192} - 10\sqrt[3]{3}$

58. $\sqrt[4]{2} + 6\sqrt[4]{32} + 9\sqrt[4]{512}$

59. $10\sqrt{75} - 3\sqrt{216} + 6\sqrt{96} + 14\sqrt{192}$

60. $13\sqrt{128} + 6\sqrt{169} - 7\sqrt{121} - 11\sqrt{162}$

Find the missing radical expression.

61. $(8\sqrt{3} + 7\sqrt{2}) + (?) = 15\sqrt{3} - 4\sqrt{2}$

62. $(3\sqrt{200} - 6\sqrt{108}) + (?) = 19\sqrt{2} - 50\sqrt{3}$

63. $(3\sqrt{50} - \sqrt{405}) - (?) = 3\sqrt{98} - 13\sqrt{20}$

64. $(4\sqrt{224} - 2\sqrt{360}) - (?) = \sqrt{640} + 4\sqrt{350}$

Answer in complete sentences.

65. Explain how to simplify $\sqrt{360}$.

66. Explain how to simplify radical expressions. Use an example to illustrate the process.

67. Explain how to determine whether radical expressions are like radicals.

68. Explain how to add and subtract radical expressions. Use examples to illustrate the process.

> **Study Tip** REVISITED   The belief that "I'm a slow writer; there is no way that I can copy down all of the notes in time" is a common concern for students. Try to focus on writing down enough information so that you will understand the material later, rather than feeling you must copy *every single word* your instructor has written or spoken. Then rewrite your notes, filling in any blanks or adding explanations, as soon as possible after class.
>
> Your handwriting should be legible, but don't feel that it must be as neat as possible. Try to use abbreviations whenever possible. Write your notes using phrases rather than complete sentences. These ideas should help you improve your speed.
>
> With your instructor's approval, you may want to try using a tape recorder to record a lecture. After class you can use the tape to help fill in holes in your notes. If you are falling behind in your note taking, leave space in your notes. You can always borrow a classmate's notes to get the information you missed.

9.4

MULTIPLICATION AND DIVISION OF RADICAL EXPRESSIONS

Objectives

1 **Multiply radical expressions.**
2 **Use the distributive property to multiply radical expressions.**
3 **Multiply radical expressions with two or more terms.**
4 **Multiply radical expressions that are conjugates.**
5 **Rationalize a denominator with one term.**
6 **Rationalize a denominator with two terms.**

Multiplying Two Radical Expressions

Objective 1 **Multiply radical expressions.** In Section 9.1, we learned how to multiply one radical by another, as well as how to divide one radical by another.

Multiplying Radicals with the Same Index

For any nonnegative expressions x and y and any positive integer $n > 1$,

$$\sqrt[n]{x} \cdot \sqrt[n]{y} = \sqrt[n]{xy} \quad \text{and} \quad \frac{\sqrt[n]{x}}{\sqrt[n]{y}} = \sqrt[n]{\frac{x}{y}}$$

In this section, we will build on that knowledge and learn how to multiply and divide expressions containing two or more radicals. We begin with a review of multiplying radicals.

EXAMPLE 1 Multiply $\sqrt{18} \cdot \sqrt{8}$.

Solution

Since the index of each radical is the same and neither radicand is a perfect square, we begin by multiplying the two radicands.

$$\sqrt{18} \cdot \sqrt{8} = \sqrt{144} \qquad \text{Multiply the radicands.}$$
$$= 12 \qquad \text{Simplify the radical.}$$

EXAMPLE 2 Multiply $\sqrt[3]{a^{13}b^7c^2} \cdot \sqrt[3]{a^{10}b^8c^{17}}$.

Solution

Since there are powers in each radical greater than the index of that radical, we could simplify each radical first. However, we would then have to multiply and simplify the radical again. A more efficient approach is to multiply first and then simplify only once.

$$\sqrt[3]{a^{13}b^7c^2} \cdot \sqrt[3]{a^{10}b^8c^{17}} = \sqrt[3]{a^{23}b^{15}c^{19}} \qquad \text{Multiply the radicands by adding the exponents for each factor.}$$
$$= a^7b^5c^6\sqrt[3]{a^2c} \qquad \text{Simplify the radical. For each factor we take out as many groups of 3 as possible.}$$

EXAMPLE 3   Multiply $9\sqrt{6} \cdot 7\sqrt{10}$.

Solution

Since the index is the same for each radical, we can multiply the radicands together. The factors in front of each radical, 9 and 7, will be multiplied by each other as well. After multiplying, we finish by simplifying the radical completely.

$$9\sqrt{6} \cdot 7\sqrt{10} = 63\sqrt{60}$$ Multiply factors in front of the radicals and multiply the radicands.

$$= 63\sqrt{2^2 \cdot 3 \cdot 5}$$ Factor the radicand.

$$= 63 \cdot 2\sqrt{3 \cdot 5}$$ Simplify the radical.

$$= 126\sqrt{15}$$ Multiply.

Quick Check 1

Multiply.

a) $\sqrt{45} \cdot \sqrt{80}$
b) $\sqrt[3]{x^4y^2z} \cdot \sqrt[3]{x^8yz^4}$
c) $4\sqrt{8} \cdot 9\sqrt{6}$

It is important to note that whenever we multiply the square root of an expression by the square root of the same expression, the product is equal to the expression itself as long as the expression is nonnegative.

Multiplying a Square Root by Itself

For any nonnegative x, $\sqrt{x} \cdot \sqrt{x} = x$.

Using the Distributive Property with Radical Expressions

Objective 2 Use the distributive property to multiply radical expressions.
Now we will use the distributive property to multiply radical expressions.

EXAMPLE 4   Multiply $\sqrt{5}(\sqrt{10} - \sqrt{5})$.

Solution

We begin by distributing $\sqrt{5}$ to each term in the parentheses, and then we multiply the radicals as in the previous examples.

$$\sqrt{5}(\sqrt{10} - \sqrt{5}) = \sqrt{5} \cdot \sqrt{10} - \sqrt{5} \cdot \sqrt{5}$$ Distribute $\sqrt{5}$.

$$= \sqrt{50} - 5$$ Multiply. Recall that $\sqrt{5} \cdot \sqrt{5} = 5$.

$$= 5\sqrt{2} - 5$$ Simplify the radical.

Quick Check 2

Multiply
$\sqrt{12}(\sqrt{3} + \sqrt{15})$.

EXAMPLE 5   Multiply $\sqrt[4]{x^{11}y^6}(\sqrt[4]{x^5y^{19}} + \sqrt[4]{x^{11}y^2})$. Assume x and y are nonnegative.

Solution

We begin by using the distributive property. Since each radical has an index of 4, we can then multiply the radicals.

$$\sqrt[4]{x^{11}y^6}\left(\sqrt[4]{x^5y^{19}} + \sqrt[4]{x^{11}y^2}\right) = \sqrt[4]{x^{11}y^6} \cdot \sqrt[4]{x^5y^{19}} + \sqrt[4]{x^{11}y^6} \cdot \sqrt[4]{x^{11}y^2}$$

Distribute $\sqrt[4]{x^{11}y^6}$.

$$= \sqrt[4]{x^{16}y^{25}} + \sqrt[4]{x^{22}y^8}$$

Multiply by adding exponents for each factor.

$$= x^4y^6\sqrt[4]{y} + x^5y^2\sqrt[4]{x^2}$$

Simplify each radical.

Since the radicals are not like radicals, we cannot simplify this expression any further.

> **Quick Check 3**
> Multiply
> $\sqrt{x^9y^6}\left(\sqrt{x^3y^6} - \sqrt{x^8y^7}\right)$.
> Assume x and y are nonnegative.

Multiplying Radical Expressions with at Least Two Terms

Objective 3 Multiply radical expressions with two or more terms.

EXAMPLE 6   Multiply $\left(8\sqrt{6} + \sqrt{2}\right)\left(3\sqrt{12} - 4\sqrt{3}\right)$.

Solution

Using the distributive property, we begin by multiplying each term in the first set of parentheses by each term in the second set of parentheses. Since there are two terms in each set of parentheses, we can use the FOIL technique. Multiply factors outside of a radical by factors outside of a radical, and multiply radicands by radicands.

$$\left(8\sqrt{6} + \sqrt{2}\right)\left(3\sqrt{12} - 4\sqrt{3}\right)$$
$$= 8 \cdot 3\sqrt{6 \cdot 12} - 8 \cdot 4\sqrt{6 \cdot 3} + 3\sqrt{2 \cdot 12} - 4\sqrt{2 \cdot 3}$$

Distribute.

$$= 24\sqrt{72} - 32\sqrt{18} + 3\sqrt{24} - 4\sqrt{6}$$

Multiply.

$$= 24 \cdot 6\sqrt{2} - 32 \cdot 3\sqrt{2} + 3 \cdot 2\sqrt{6} - 4\sqrt{6}$$

Simplify each radical.

$$= 144\sqrt{2} - 96\sqrt{2} + 6\sqrt{6} - 4\sqrt{6}$$

Multiply.

$$= 48\sqrt{2} + 2\sqrt{6}$$

Combine like radicals.

EXAMPLE 7   Multiply $\left(\sqrt{5} + \sqrt{6}\right)^2$.

Solution

To square any binomial, we multiply it by itself.

$$\left(\sqrt{5} + \sqrt{6}\right)^2 = \left(\sqrt{5} + \sqrt{6}\right)\left(\sqrt{5} + \sqrt{6}\right)$$

Square the binomial $\sqrt{5} + \sqrt{6}$ by multiplying it by itself.

$$= \sqrt{5} \cdot \sqrt{5} + \sqrt{5} \cdot \sqrt{6} + \sqrt{6} \cdot \sqrt{5} + \sqrt{6} \cdot \sqrt{6}$$

Distribute.

$$= 5 + \sqrt{30} + \sqrt{30} + 6$$

Multiply.

$$= 11 + 2\sqrt{30}$$

Combine like terms.

Quick Check **4** Multiply.

a) $(5\sqrt{3} + 4\sqrt{2})(7\sqrt{3} - 6\sqrt{2})$ b) $(\sqrt{7} - \sqrt{10})^2$

A Word of Caution   Whenever squaring a binomial, such as $(\sqrt{5} + \sqrt{6})^2$, we must multiply the binomial by itself. We cannot simply square each term.

$$(a + b)^2 \neq a^2 + b^2$$

Multiplying Conjugates

Objective **4**   **Multiply radical expressions that are conjugates.** The binomials $\sqrt{13} + \sqrt{5}$ and $\sqrt{13} - \sqrt{5}$ are called **conjugates.** Two binomials are conjugates if they are of the form $x + y$ and $x - y$. Notice that the two terms are the same, with the exception of the sign of the second term.

The multiplication of two conjugates follows a pattern. Let's look at the product $(\sqrt{x} + \sqrt{y})(\sqrt{x} - \sqrt{y})$.

$$(\sqrt{x} + \sqrt{y})(\sqrt{x} - \sqrt{y}) = x - \sqrt{xy} + \sqrt{xy} - y$$

Distribute. Note that $\sqrt{x} \cdot \sqrt{x} = x$ and $\sqrt{y} \cdot \sqrt{y} = y$.

$$= x - y$$

Combine the two opposite terms $-\sqrt{xy}$ and $\sqrt{xy}$.

Whenever we multiply conjugates, the two middle terms will be opposites of each other and therefore will add to be 0. We can multiply the first term in the first set of parentheses by the first term in the second set of parentheses, then multiply the second term in the first set of parentheses by the second term in the second set of parentheses, and, finally, place a minus sign between two products.

Multiplication of Two Conjugates

$$(a + b)(a - b) = a^2 - b^2$$

with a^2 labeled over $a \cdot a$ and b^2 labeled under $b \cdot b$

EXAMPLE **8**   Multiply $(\sqrt{17} + \sqrt{23})(\sqrt{17} - \sqrt{23})$.

Solution

These two binomials are conjugates and we multiply them accordingly.

$$(\sqrt{17} + \sqrt{23})(\sqrt{17} - \sqrt{23}) = \sqrt{17} \cdot \sqrt{17} - \sqrt{23} \cdot \sqrt{23}$$

Multiply first term by first term, second term by second term.

$$= 17 - 23$$ Multiply.

$$= -6$$ Subtract.

EXAMPLE 9   Multiply $(2\sqrt{6} - 8\sqrt{5})(2\sqrt{6} + 8\sqrt{5})$.

Solution

When we multiply two conjugates, we must remember to multiply the factors in front of the radicals by each other and to multiply the radicands by each other.

$$(2\sqrt{6} - 8\sqrt{5})(2\sqrt{6} + 8\sqrt{5}) = 2\sqrt{6} \cdot 2\sqrt{6} - 8\sqrt{5} \cdot 8\sqrt{5}$$

Multiply first term by first term, second term by second term.

$$= 4 \cdot 6 - 64 \cdot 5 \qquad \text{Multiply.}$$
$$= 24 - 320 \qquad \text{Multiply.}$$
$$= -296 \qquad \text{Subtract.}$$

Quick Check **5** Multiply.

a) $(\sqrt{38} - \sqrt{29})(\sqrt{38} + \sqrt{29})$ b) $(8\sqrt{11} - 5\sqrt{7})(8\sqrt{11} + 5\sqrt{7})$

Rationalizing the Denominator

Objective 5 **Rationalize a denominator with one term.** Earlier in this chapter, we introduced a criterion for determining whether a radical was simplified. We stated that for a radical to be simplified, its index must be greater than any power within the radical. There are two other rules that we now add.

- There can be no fractions in a radicand.
- There can be no radicals in the denominator of a fraction.

For example, we would not consider the following expressions simplified: $\sqrt{\dfrac{3}{10}}$, $\dfrac{9}{\sqrt{2}}$, $\dfrac{6}{\sqrt{4} - \sqrt{3}}$, and $\dfrac{\sqrt{6} + \sqrt{12}}{\sqrt{3} - 8}$. The process of rewriting an expression without a radical in its denominator is called **rationalizing the denominator.**

The rational expression $\dfrac{\sqrt{16}}{\sqrt{49}}$ is not simplified, as there is a radical in the denominator. However, we know that $\sqrt{49} = 7$, so we can simplify the denominator in such a way that it no longer contains a radical.

$$\frac{\sqrt{16}}{\sqrt{49}} = \frac{4}{7} \qquad \text{Simplify the numerator and denominator.}$$

The radical expression $\sqrt{\dfrac{75}{3}}$ is not simplified, as there is a fraction inside the radical. We can simplify $\frac{75}{3}$ to be 25, rewriting the radical without a fraction inside.

$$\sqrt{\frac{75}{3}} = \sqrt{25} = 5 \qquad \text{Simplify the fraction first, then } \sqrt{25}.$$

Suppose that we needed to simplify $\dfrac{\sqrt{15}}{\sqrt{2}}$. We cannot simplify $\sqrt{2}$, and the fraction itself cannot be simplified either. In such a case, we will multiply both the numerator and denominator by an expression that will allow us to rewrite the denominator without a radical. Then we simplify.

EXAMPLE 10 Rationalize the denominator: $\dfrac{\sqrt{15}}{\sqrt{2}}$.

Solution

If we multiply the denominator by $\sqrt{2}$, the denominator would equal 2, and the denominator will be rationalized.

$$\frac{\sqrt{15}}{\sqrt{2}} = \frac{\sqrt{15}}{\sqrt{2}} \cdot \frac{\sqrt{2}}{\sqrt{2}} \qquad \text{Multiply by } \frac{\sqrt{2}}{\sqrt{2}}, \text{ which makes the denominator equal}$$

to 2. Multiplying by $\dfrac{\sqrt{2}}{\sqrt{2}}$ is equivalent to multiplying by 1.

$$= \frac{\sqrt{30}}{2} \qquad \text{Multiply.}$$

Since $\sqrt{30}$ cannot be simplified, this expression cannot be simplified further.

> **Quick Check 6**
> Rationalize the denominator: $\dfrac{\sqrt{70}}{\sqrt{3}}$

EXAMPLE 11 Rationalize the denominator: $\dfrac{11}{\sqrt{12}}$.

Solution

At first glance, we might think that multiplying the numerator and denominator by $\sqrt{12}$ is the correct way to proceed. However, if we multiply the numerator and denominator by $\sqrt{3}$, the radicand in the denominator will be 36, which is a perfect square.

$$\frac{11}{\sqrt{12}} = \frac{11}{\sqrt{12}} \cdot \frac{\sqrt{3}}{\sqrt{3}} \qquad \text{Multiply by } \frac{\sqrt{3}}{\sqrt{3}} \text{ to make the radicand in the}$$

denominator a perfect square.

$$= \frac{11\sqrt{3}}{\sqrt{36}} \qquad \text{Multiply.}$$

$$= \frac{11\sqrt{3}}{6} \qquad \text{Simplify the radical in the denominator.}$$

Multiplying by $\dfrac{\sqrt{12}}{\sqrt{12}}$ would also be valid, but there would be a great deal of simplifying from there. One way to determine the best expression to multiply by is to completely factor the radicand in the denominator. In this example, $12 = 2^2 \cdot 3$. The factor 2 is already a perfect square, but the factor 3 is not. Multiplying by $\sqrt{3}$ makes the factor 3 a perfect square as well.

> **Quick Check 7**
> Rationalize the denominator: $\dfrac{2}{\sqrt{20}}$.

EXAMPLE 12 Rationalize the denominator: $\sqrt{\dfrac{5a^3 b^{14} c^6}{80 a^9 b^7 c^{11}}}$. Assume all variables represent nonnegative values.

Solution

Notice that the numerator and denominator have common factors. We will begin by simplifying the fraction to lowest terms.

$$\sqrt{\frac{5a^3b^{14}c^6}{80a^9b^7c^{11}}} = \sqrt{\frac{b^7}{16a^6c^5}}$$ Divide out common factors and simplify.

$$= \frac{\sqrt{b^7}}{\sqrt{16a^6c^5}}$$ Rewrite as the quotient of two square roots. Notice that the factors 16 and a^6 are already perfect squares, but c^5 is not.

$$= \frac{\sqrt{b^7}}{\sqrt{16a^6c^5}} \cdot \frac{\sqrt{c}}{\sqrt{c}}$$ Multiply by $\dfrac{\sqrt{c}}{\sqrt{c}}$.

$$= \frac{\sqrt{b^7c}}{\sqrt{16a^6c^6}}$$ Multiply.

$$= \frac{b^3\sqrt{bc}}{4a^3c^3}$$ Simplify both radicals.

Quick Check **8**
Rationalize the denominator: $\sqrt{\dfrac{12x^2y^8z^5}{75x^5y^3z^{15}}}$.
Assume all variables represent nonnegative values.

Rationalizing a Denominator with Two Terms

Objective 6 **Rationalize a denominator with two terms.** In the previous examples, each denominator had only one term. If a denominator is a binomial that contains one or two square roots, then we rationalize the denominator by multiplying the numerator and denominator by the conjugate of the denominator. For example, consider the expression $\dfrac{6}{\sqrt{11} + \sqrt{7}}$. We know from earlier in the section that multiplying $\sqrt{11} + \sqrt{7}$ by its conjugate $\sqrt{11} - \sqrt{7}$ will produce a product that does not contain a radical.

EXAMPLE 13   Rationalize the denominator: $\dfrac{6}{\sqrt{11} + \sqrt{7}}$.

Solution

Since this denominator is a binomial, we will multiply the numerator and denominator by the conjugate of the denominator.

$$\frac{6}{\sqrt{11} + \sqrt{7}} = \frac{6}{\sqrt{11} + \sqrt{7}} \cdot \frac{\sqrt{11} - \sqrt{7}}{\sqrt{11} - \sqrt{7}}$$ Multiply the numerator and denominator by the conjugate of the denominator $(\sqrt{11} - \sqrt{7})$.

$$= \frac{6\sqrt{11} - 6\sqrt{7}}{\sqrt{11} \cdot \sqrt{11} - \sqrt{7} \cdot \sqrt{7}}$$ Multiply. Since $\sqrt{11} + \sqrt{7}$ and $\sqrt{11} - \sqrt{7}$ are conjugates, we only need to multiply $\sqrt{11}$ by $\sqrt{11}$ and $\sqrt{7}$ by $\sqrt{7}$.

$$= \frac{6\sqrt{11} - 6\sqrt{7}}{11 - 7}$$ Simplify the products in the denominator.

$$= \frac{6\sqrt{11} - 6\sqrt{7}}{4} \qquad \text{Subtract.}$$

$$= \frac{6(\sqrt{11} - \sqrt{7})}{4} \qquad \text{Factor the numerator.}$$

$$= \frac{\overset{3}{6}(\sqrt{11} - \sqrt{7})}{\underset{2}{4}} \qquad \text{Divide out the common factor 2.}$$

$$= \frac{3(\sqrt{11} - \sqrt{7})}{2} \qquad \text{Simplify.}$$

Quick Check 9
Rationalize the denominator: $\dfrac{\sqrt{15}}{\sqrt{5} + \sqrt{3}}$.

EXAMPLE 14   Rationalize the denominator: $\dfrac{4\sqrt{3} - 3\sqrt{5}}{2\sqrt{3} - \sqrt{5}}$.

Solution

Since the denominator is a binomial, we begin by multiplying the numerator and denominator by the conjugate of the denominator, which is $2\sqrt{3} + \sqrt{5}$.

$$\frac{4\sqrt{3} - 3\sqrt{5}}{2\sqrt{3} - \sqrt{5}} = \frac{4\sqrt{3} - 3\sqrt{5}}{2\sqrt{3} - \sqrt{5}} \cdot \frac{2\sqrt{3} + \sqrt{5}}{2\sqrt{3} + \sqrt{5}}$$

Multiply the numerator and denominator by the conjugate of the denominator.

$$= \frac{4\sqrt{3} \cdot 2\sqrt{3} + 4\sqrt{3} \cdot \sqrt{5} - 3\sqrt{5} \cdot 2\sqrt{3} - 3\sqrt{5} \cdot \sqrt{5}}{2\sqrt{3} \cdot 2\sqrt{3} - \sqrt{5} \cdot \sqrt{5}}$$

Multiply. We must multiply each term in the first numerator by each term in the second numerator. To multiply the denominators we can take advantage of the fact that the two denominators are conjugates.

$$= \frac{8 \cdot 3 + 4\sqrt{15} - 6\sqrt{15} - 3 \cdot 5}{4 \cdot 3 - 5} \qquad \text{Simplify each product.}$$

$$= \frac{24 + 4\sqrt{15} - 6\sqrt{15} - 15}{12 - 5} \qquad \text{Multiply.}$$

$$= \frac{9 - 2\sqrt{15}}{7} \qquad \text{Combine like terms and like radicals.}$$

Quick Check 10
Rationalize the denominator: $\dfrac{2\sqrt{6} - 7\sqrt{2}}{2\sqrt{6} + 3\sqrt{2}}$.

Multiply. Assume all variables represent nonnegative real numbers.

1. $\sqrt{4} \cdot \sqrt{9}$

2. $\sqrt{25} \cdot \sqrt{81}$

3. $\sqrt{2} \cdot \sqrt{18}$

4. $\sqrt{3} \cdot \sqrt{48}$

5. $\sqrt[3]{5} \cdot \sqrt[3]{25}$

6. $\sqrt[3]{12} \cdot \sqrt[3]{42}$

7. $5\sqrt{8} \cdot 3\sqrt{2}$

8. $6\sqrt{12} \cdot 3\sqrt{75}$

9. $\sqrt{x^{11}} \cdot \sqrt{x^5}$

10. $\sqrt{x^{21}} \cdot \sqrt{x}$

11. $\sqrt{m^{10}n^7} \cdot \sqrt{m^9n^9}$

12. $\sqrt{s^3t} \cdot \sqrt{s^5t^{19}}$

13. $\sqrt[4]{x^8y^{13}} \cdot \sqrt[4]{x^{33}y^{15}}$

14. $\sqrt[7]{a^{22}b^{20}} \cdot \sqrt[7]{a^{34}b^{19}}$

Multiply. Assume all variables represent nonnegative real numbers.

15. $\sqrt{2}(\sqrt{8} - \sqrt{6})$

16. $\sqrt{6}(\sqrt{2} + \sqrt{3})$

17. $\sqrt{98}(\sqrt{18} + \sqrt{10})$

18. $\sqrt{15}(\sqrt{30} - \sqrt{35})$

19. $\sqrt{10}(3\sqrt{6} - 8\sqrt{15})$

20. $12\sqrt{15}(7\sqrt{3} - 11\sqrt{5})$

21. $\sqrt{m^5n^9}(\sqrt{m^3n^4} + \sqrt{m^8n^7})$

22. $\sqrt{xy^{11}}(\sqrt{x^7y^3} - \sqrt{x^{39}y^{15}})$

23. $\sqrt[3]{a^7b^{10}c^{13}}(\sqrt[3]{a^{11}b^{35}} + \sqrt[3]{a^{23}c^8})$

24. $\sqrt[3]{x^4y^7z^{10}}(\sqrt[3]{x^8y^{16}z^{24}} - \sqrt[3]{x^{16}y^{12}z^2})$

Multiply.

25. $(\sqrt{2} + \sqrt{5})(\sqrt{2} - \sqrt{3})$

26. $(\sqrt{3} + \sqrt{8})(\sqrt{6} - \sqrt{2})$

27. $(\sqrt{10} - \sqrt{5})(\sqrt{10} + \sqrt{5})$

28. $(\sqrt{14} + \sqrt{6})(\sqrt{7} - \sqrt{54})$

29. $(5\sqrt{3} + \sqrt{2})(6\sqrt{2} - 11\sqrt{3})$

30. $(5\sqrt{12} + \sqrt{5})(3\sqrt{20} + 2\sqrt{21})$

31. $(9 + 3\sqrt{8})(\sqrt{6} - 7\sqrt{50})$

32. $(3 + 4\sqrt{7})(5\sqrt{14} - 8)$

33. $(4\sqrt{5} - 3\sqrt{2})^2$

34. $(6 + 7\sqrt{3})^2$

Multiply the conjugates.

35. $(\sqrt{7} - \sqrt{10})(\sqrt{7} + \sqrt{10})$

36. $(\sqrt{26} + \sqrt{3})(\sqrt{26} - \sqrt{3})$

37. $(9\sqrt{6} + 3\sqrt{8})(9\sqrt{6} - 3\sqrt{8})$

38. $(5\sqrt{18} - 2\sqrt{24})(5\sqrt{18} + 2\sqrt{24})$

39. $(8 - 11\sqrt{2})(8 + 11\sqrt{2})$

40. $(4\sqrt{3} + 15)(4\sqrt{3} - 15)$

Simplify. Assume all variables represent nonnegative real numbers.

41. $\dfrac{\sqrt{36}}{\sqrt{25}}$

42. $\dfrac{\sqrt{49}}{\sqrt{100}}$

43. $\sqrt{\dfrac{12}{x^4y^6}}$

44. $\sqrt{\dfrac{b^7}{a^{10}c^2}}$

45. $\sqrt{\dfrac{126}{7}}$

46. $\sqrt{\dfrac{100}{5}}$

47. $\sqrt{\dfrac{135}{20}}$

48. $\sqrt{\dfrac{14}{18}}$

Rationalize the denominator and simplify. Assume all variables represent nonnegative real numbers.

49. $\dfrac{\sqrt{8}}{\sqrt{3}}$

50. $\dfrac{\sqrt{24}}{\sqrt{7}}$

51. $\dfrac{1}{\sqrt{2}}$

52. $\dfrac{5}{\sqrt{5}}$

53. $\sqrt{\dfrac{27}{11a^7}}$

54. $\sqrt{\dfrac{80}{3n^{10}}}$

55. $\sqrt[3]{\dfrac{s^8}{2r^2t}}$

56. $\sqrt[3]{\dfrac{2x^{10}}{7y^{16}z^{22}}}$

57. $\dfrac{15}{\sqrt{18}}$

58. $\dfrac{2\sqrt{3}}{\sqrt{32}}$

59. $\dfrac{ab^6}{\sqrt{a^5b^4c^7}}$

60. $\dfrac{2x^8}{\sqrt{10xy^3}}$

Rationalize the denominator.

61. $\dfrac{9}{\sqrt{13} - \sqrt{7}}$

62. $\dfrac{10}{\sqrt{20} + \sqrt{2}}$

63. $\dfrac{4\sqrt{3}}{\sqrt{3} + \sqrt{11}}$

64. $\dfrac{5\sqrt{5}}{\sqrt{15} - \sqrt{5}}$

65. $\dfrac{6\sqrt{3}}{\sqrt{13} - 3}$

66. $\dfrac{8\sqrt{2}}{4 - \sqrt{6}}$

67. $\dfrac{\sqrt{3} + \sqrt{8}}{\sqrt{6} - \sqrt{2}}$

68. $\dfrac{\sqrt{5} + \sqrt{2}}{\sqrt{10} - \sqrt{2}}$

69. $\dfrac{\sqrt{8} + \sqrt{32}}{\sqrt{8} - \sqrt{32}}$

70. $\dfrac{\sqrt{13} - \sqrt{11}}{\sqrt{11} - \sqrt{13}}$

71. $\dfrac{4\sqrt{2} - \sqrt{3}}{2\sqrt{2} - 3\sqrt{3}}$

72. $\dfrac{\sqrt{12} - 9\sqrt{10}}{6\sqrt{3} - 4\sqrt{5}}$

Simplify. Assume all variables represent nonnegative real numbers.

73. $\sqrt{\dfrac{90}{98}}$

74. $\sqrt{10x^5} \cdot \sqrt{18x^9}$

75. $\left(8\sqrt{7} + 5\sqrt{5}\right)^2$

76. $\dfrac{28}{\sqrt{32}}$

77. $\sqrt[3]{25a^8bc^7} \cdot \sqrt[3]{25ab^5c^7}$

78. $\sqrt[3]{\dfrac{2x^4y^{13}}{54x^{10}y^2}}$

79. $\dfrac{10\sqrt{5} - 7\sqrt{7}}{\sqrt{5} + \sqrt{7}}$

80. $\sqrt{60}\left(7\sqrt{3} - 2\sqrt{15}\right)$

81. $\left(3\sqrt{10} - 7\sqrt{2}\right)\left(4\sqrt{10} + 9\sqrt{5}\right)$

82. $\dfrac{8\sqrt{2}}{\sqrt{10}}$

83. $-\sqrt{7x^9}\left(2\sqrt{14x} - \sqrt{7x^{13}}\right)$

84. $\left(9\sqrt{7} - 8\sqrt{8}\right)\left(9\sqrt{7} + 8\sqrt{8}\right)$

85. $\sqrt{\dfrac{a^5 b^8}{12 c^9}}$

86. $\dfrac{13\sqrt{5} + 6}{10 - 7\sqrt{5}}$

87. $\dfrac{10x^2}{\sqrt[3]{36x^8}}$

88. $(20\sqrt{3} - 7\sqrt{10})(8\sqrt{3} + 15\sqrt{2})$

89. $(18\sqrt{7} - \sqrt{11})(18\sqrt{7} + \sqrt{11})$

90. $(4\sqrt{14} - 3\sqrt{7})^2$

91. Develop a general formula for the product $(\sqrt{a} + \sqrt{b})^2$.

92. Develop a general formula for the product $(a\sqrt{b} + c\sqrt{d})^2$.

Answer in complete sentences.

93. Explain how to determine that two radical expressions are conjugates.

94. Explain how to multiply conjugates. Use an example to illustrate the process.

95. Explain how to rationalize the denominator of a radical expression whose denominator has only one term. Use an example to illustrate the process.

96. Explain how to rationalize the denominator of a radical expression whose denominator has two terms, including at least one square root. Use an example to illustrate the process.

QUICK REVIEW EXERCISES

Section 9.4

Find the prime factorization of the given number.

1. 144

2. 1400

3. 1024

4. 29,106

> ***Study Tip*** **REVISITED** What belongs in your notes? Begin by including anything that your instructor writes on the board. If your instructor solves several problems of the same type on the board, make a note of this fact as well; this should alert you that this type of problem is important.
>
> Instructors also give verbal cues when they feel that something should be taken down in your notes. Your instructor may pause to give you enough time to finish writing in your notebook; or may repeat the same phrase to make sure that you accurately write down the statement in your notes.
>
> Some instructors will warn the class that a particular topic, problem, or step is difficult. When you hear this, your instructor is telling you to take the best notes that you can because you will need them later.
>
> Finally, keep an open ear for what may be the most important six-word phrase in a math class: "this will be on the test." When you hear this, your instructor is trying to make sure that you are prepared for the exam. Try writing the words "ON TEST," circled, next to the notes you take.

9.5

RADICAL EQUATIONS

Objectives

1 Solve radical equations.
2 Solve equations containing radical functions.
3 Solve equations containing rational exponents.
4 Solve equations in which a radical is equal to a variable expression.
5 Solve equations containing two radicals.

Solving Radical Equations

Objective 1 Solve radical equations. A **radical equation** is an equation containing one or more radicals. Here are some examples of radical equations:

$$\sqrt{x} = 9 \qquad \sqrt[3]{2x - 5} = 3 \qquad x + \sqrt{x} = 20$$
$$\sqrt[4]{3x - 8} = \sqrt[4]{2x + 11} \qquad \sqrt{x - 4} + \sqrt{x + 8} = 6$$

In this section, we will learn how to solve radical equations. We will find a way to convert a radical equation to an equivalent equation that we already know how to solve.

Raising Equal Numbers to the Same Power

> If two numbers a and b are equal, then for any n, $a^n = b^n$.

If we raise two equal numbers to the same power, they remain equal to each other. We will use this idea to solve radical equations.

Solving Radical Equations

> - Isolate one radical containing the variable on one side of the equation.
> - Raise both sides of the equation to the nth power, where n is the index of the radical. *For any nonnegative number x and any integer n > 1, $(\sqrt[n]{x})^n = x$.*
> - If the resulting equation does not contain a radical, solve the equation. If the resulting equation does contain a radical, begin the process again by isolating the radical on one side of the equation.
> - Check the solution(s).

It is crucial that we check all solutions when solving a radical equation, as raising both sides of an equation to a power can introduce **extraneous solutions.** An extraneous solution is a solution of the resulting equation when we raise both sides to a certain power but is not a solution of the original equation. A value may be a solution of the equation we obtain after raising both sides of the equation to the nth power without being a solution of the original equation.

EXAMPLE 1   Solve $\sqrt{x - 5} = 3$.

Solution

Since the radical $\sqrt{x - 5}$ is isolated on the left side of the equation, we may begin by squaring both sides of the equation.

$$\sqrt{x-5} = 3$$
$$(\sqrt{x-5})^2 = 3^2 \qquad \text{Square both sides.}$$
$$x - 5 = 9 \qquad \text{Simplify.}$$
$$x = 14 \qquad \text{Add 5.}$$

We need to check this solution, using the original equation.

$$\sqrt{(14) - 5} = 3 \qquad \text{Substitute 14 for } x.$$
$$\sqrt{9} = 3 \qquad \text{Subtract.}$$
$$3 = 3 \qquad \text{Simplify the square root.}$$

▶ The solution $x = 14$ checks, so the solution set is $\{14\}$.

EXAMPLE 2   Solve $\sqrt[3]{6x + 4} + 7 = 11$.

Solution

In this example, we begin by isolating the radical.

$$\sqrt[3]{6x + 4} + 7 = 11$$
$$\sqrt[3]{6x + 4} = 4 \qquad \text{Subtract 7 to isolate the radical.}$$
$$(\sqrt[3]{6x + 4})^3 = 4^3 \qquad \text{Raise both sides of the equation to the third power.}$$
$$6x + 4 = 64 \qquad \text{Simplify.}$$
$$6x = 60 \qquad \text{Subtract 4.}$$
$$x = 10 \qquad \text{Divide both sides by 6.}$$

Now we check the solution, using the original equation.

$$\sqrt[3]{6(10) + 4} + 7 = 11 \qquad \text{Substitute 10 for } x.$$
$$\sqrt[3]{64} + 7 = 11 \qquad \text{Simplify the radicand.}$$
$$4 + 7 = 11 \qquad \text{Simplify the cube root.}$$
$$11 = 11 \qquad \text{Add. The solution checks.}$$

Quick Check 1

Solve.

a) $\sqrt{x + 2} = 7$
b) $\sqrt[3]{x + 9} + 10 = 6$

The solution set is $\{10\}$. Note that raising both sides of an equation to an odd power will not introduce extraneous solutions.

EXAMPLE 3   Solve $\sqrt{x} + 8 = 5$.

Solution

We begin by isolating $\sqrt{x}$ on the left side of the equation.

$$\sqrt{x} + 8 = 5$$
$$\sqrt{x} = -3 \qquad \text{Subtract 8 to isolate the square root.}$$
$$(\sqrt{x})^2 = (-3)^2 \qquad \text{Square both sides of the equation.}$$
$$x = 9 \qquad \text{Simplify.}$$

Now we check this solution, using the original equation.

$$\sqrt{9} + 8 = 5 \qquad \text{Substitute 9 for } x.$$
$$3 + 8 = 5 \qquad \text{Simplify the square root. The principal square root of 9 is 3, not } -3.$$
$$11 = 5 \qquad \text{Add.}$$

This solution does not check, so it is an extraneous solution. The equation has no solution: the solution set is $\varnothing$.

If we obtain an equation in which an even root is equal to a negative number, such as $\sqrt{x} = -3$, this equation will not have any solutions. This is because the principal even root of a number, if it exists, must be nonnegative.

Quick Check 2
Solve
$\sqrt{2x - 9} - 8 = -11.$

Solving Equations Involving Radical Functions

Objective 2 Solve equations containing radical functions.

EXAMPLE 4 For $f(x) = \sqrt{2x + 11} - 6$, find all values x for which $f(x) = 3$.

Solution

We begin by setting the function equal to 3.

$$f(x) = 3$$
$$\sqrt{2x + 11} - 6 = 3 \qquad \text{Replace } f(x) \text{ by its formula.}$$
$$\sqrt{2x + 11} = 9 \qquad \text{Add 6 to isolate the radical.}$$
$$(\sqrt{2x + 11})^2 = 9^2 \qquad \text{Square both sides.}$$
$$2x + 11 = 81 \qquad \text{Simplify.}$$
$$2x = 70 \qquad \text{Subtract 11.}$$
$$x = 35 \qquad \text{Divide both sides by 2.}$$

It is left to the reader to verify that the solution is not an extraneous solution. $f(35) = 3$.

Quick Check 3
For
$f(x) = \sqrt{3x - 8} + 4$,
find all values x for which $f(x) = 9$.

Solving Equations with Rational Exponents

Objective 3 Solve equations containing rational exponents. The next example involves equations containing fractional exponents. Recall that $x^{1/n} = \sqrt[n]{x}$.

EXAMPLE 5 Solve $x^{1/3} + 7 = -3$.

Solution

We begin by rewriting $x^{1/3}$ as $\sqrt[3]{x}$.

$$x^{1/3} + 7 = -3$$
$$\sqrt[3]{x} + 7 = -3 \qquad \text{Rewrite } x^{1/3} \text{ using radical notation as } \sqrt[3]{x}.$$
$$\sqrt[3]{x} = -10 \qquad \text{Subtract 7 to isolate the radical.}$$
$$(\sqrt[3]{x})^3 = (-10)^3 \qquad \text{Raise both sides to the third power.}$$
$$x = -1000 \qquad \text{Simplify.}$$

It is left to the reader to verify that the solution is not an extraneous solution. The solution set is $\{-1000\}$.

EXAMPLE 6 Solve $(1 - 5x)^{1/2} - 3 = 3$.

Solution

We begin by rewriting the equation using a radical.

$$(1 - 5x)^{1/2} - 3 = 3$$

$$\sqrt{1 - 5x} - 3 = 3 \qquad \text{Rewrite using radical notation.}$$

$$\sqrt{1 - 5x} = 6 \qquad \text{Add 3 to isolate the radical.}$$

$$(\sqrt{1 - 5x})^2 = 6^2 \qquad \text{Square both sides.}$$

$$1 - 5x = 36 \qquad \text{Simplify.}$$

$$-5x = 35 \qquad \text{Subtract 1.}$$

$$x = -7 \qquad \text{Divide both sides by } -5.$$

It is left to the reader to verify that the solution is not an extraneous solution. The solution set is $\{-7\}$.

Quick Check 4

Solve.

a) $x^{1/2} - 10 = -7$
b) $(x + 4)^{1/3} + 8 = 2$

Solving Equations in Which a Radical Is Equal to a Variable Expression

Objective 4 **Solve equations in which a radical is equal to a variable expression.** After isolating the radical in all of the previous examples, the resulting equation had a radical expression equal to a number. In the next example, we will learn how to solve equations that result in a radical being equal to a variable expression.

EXAMPLE 7 Solve $\sqrt{6x + 16} = x$.

Solution

Since the radical is already isolated, we begin by squaring both sides. This will result in a quadratic equation, which we solve by collecting all terms on one side of the equation and factoring.

$$\sqrt{6x + 16} = x$$

$$(\sqrt{6x + 16})^2 = x^2 \qquad \text{Square both sides.}$$

$$6x + 16 = x^2 \qquad \text{Simplify.}$$

$$0 = x^2 - 6x - 16 \qquad \text{Collect all terms on the right side of the equation by subtracting } 6x \text{ and 16.}$$

$$0 = (x - 8)(x + 2) \qquad \text{Factor.}$$

$$x = 8 \quad \text{or} \quad x = -2 \qquad \text{Set each factor equal to 0 and solve.}$$

Now we check both solutions.

$x = 8$	$x = -2$
$\sqrt{6(8) + 16} = 8$	$\sqrt{6(-2) + 16} = -2$
$\sqrt{48 + 16} = 8$	$\sqrt{-12 + 16} = -2$
$\sqrt{64} = 8$	$\sqrt{4} = -2$
$8 = 8$	$2 = -2$

Quick Check 5

Solve $\sqrt{12x - 20} = x$.

The solution $x = -2$ is an extraneous solution. The solution set is $\{8\}$.

EXAMPLE 8   Solve $\sqrt{x} + 6 = x$.

Solution

We begin by isolating the radical.

$$\sqrt{x} + 6 = x$$
$$\sqrt{x} = x - 6 \qquad \text{Subtract 6 to isolate the radical.}$$
$$(\sqrt{x})^2 = (x - 6)^2 \qquad \text{Square both sides.}$$
$$x = (x - 6)(x - 6) \qquad \text{Square the binomial by multiplying it by itself.}$$
$$x = x^2 - 12x + 36 \qquad \text{Multiply. The resulting equation is quadratic.}$$
$$0 = x^2 - 13x + 36 \qquad \text{Subtract } x \text{ to collect all terms on the right side of}$$
$$\text{the equation.}$$
$$0 = (x - 4)(x - 9) \qquad \text{Factor.}$$
$$x = 4 \quad \text{or} \quad x = 9 \qquad \text{Set each factor equal to 0 and solve.}$$

Now we check both solutions.

$x = 4$	$x = 9$
$\sqrt{4} + 6 = 4$	$\sqrt{9} + 6 = 9$
$2 + 6 = 4$	$3 + 6 = 9$
$8 = 4$	$9 = 9$

Quick Check 6

Solve $\sqrt{2x + 4} = x$.

The solution $x = 4$ is an extraneous solution. The solution set is $\{9\}$.

Solving Radical Equations Containing Two Radicals

Objective 5 Solve equations containing two radicals.

EXAMPLE 9   Solve $\sqrt[5]{6x + 5} = \sqrt[5]{4x - 3}$.

Solution

We have a radical isolated on the left side of the equation, so we will raise both sides to the fifth power. Since both radicals have the same index, this will result in an equation that does not contain a radical.

$$\sqrt[5]{6x + 5} = \sqrt[5]{4x - 3}$$
$$(\sqrt[5]{6x + 5})^5 = (\sqrt[5]{4x - 3})^5 \qquad \text{Raise both sides to the 5}^{\text{th}} \text{ power.}$$
$$6x + 5 = 4x - 3 \qquad \text{Simplify.}$$
$$2x + 5 = -3 \qquad \text{Subtract } 4x \text{ from both sides.}$$
$$2x = -8 \qquad \text{Subtract 5.}$$
$$x = -4 \qquad \text{Divide both sides by 2.}$$

Quick Check 7

Solve
$\sqrt[3]{5x - 11} = \sqrt[3]{7x + 33}$.

It is left to the reader to verify that the solution is not an extraneous solution. This solution checks, and the solution set is $\{-4\}$.

Occasionally, equations containing two square roots will still contain a square root after squaring both sides. This will require us to square both sides a second time.

EXAMPLE 10   Solve $\sqrt{x+6} - \sqrt{x-1} = 1$.

Solution

We must begin by isolating one of the two radicals on the left side of the equation. We will isolate $\sqrt{x+6}$, as it is positive.

$$\sqrt{x+6} - \sqrt{x-1} = 1$$
$$\sqrt{x+6} = 1 + \sqrt{x-1}$$ Add $\sqrt{x-1}$ to isolate the radical $\sqrt{x+6}$ on the left side.

$$(\sqrt{x+6})^2 = (1 + \sqrt{x-1})^2$$ Square both sides.

$$x+6 = (1 + \sqrt{x-1})(1 + \sqrt{x-1})$$ Square the binomial on the right side by multiplying it by itself.

$$x+6 = 1 \cdot 1 + 1 \cdot \sqrt{x-1} + 1 \cdot \sqrt{x-1} + \sqrt{x-1} \cdot \sqrt{x-1}$$
    Distribute.

$$x+6 = 1 + 2\sqrt{x-1} + x - 1$$ Simplify.
$$x+6 = 2\sqrt{x-1} + x$$ Combine like terms.
$$6 = 2\sqrt{x-1}$$ Subtract x to isolate the radical.

$$3 = \sqrt{x-1}$$ Divide both sides by 2.
$$3^2 = (\sqrt{x-1})^2$$ Square both sides.
$$9 = x - 1$$ Simplify.
$$10 = x$$ Add 1.

Quick Check **8**

Solve.
$\sqrt{x+3} - \sqrt{x-2} = 1$.

It is left to the reader to verify that the solution is not an extraneous solution. The solution set is $\{10\}$.

EXERCISES 9.5 ❯

Solve. Check for extraneous solutions.

1. $\sqrt{x+2} = 5$

2. $\sqrt{x-4} = 4$

3. $\sqrt[3]{2x-5} = 3$

4. $\sqrt[4]{3x-8} = 2$

5. $\sqrt{5x-19} = -6$

6. $\sqrt[3]{2x+15} = -1$

7. $\sqrt{4x-3} - 9 = -4$

8. $\sqrt{7x+13} + 19 = 11$

9. $\sqrt[6]{x-6} - 9 = -10$

10. $\sqrt[3]{x+7} - 5 = -3$

11. $\sqrt{x^2 + 3x - 3} = 5$

12. $\sqrt{x^2 - 9} = 4$

13. For the function $f(x) = \sqrt{3x+9}$, find all values x for which $f(x) = 12$.

14. For the function $f(x) = \sqrt{x-8} + 7$, find all values x for which $f(x) = 13$.

15. For the function $f(x) = \sqrt[3]{5x-1} + 5$, find all values x for which $f(x) = 9$.

16. For the function $f(x) = \sqrt[4]{2x-3} - 7$, find all values x for which $f(x) = -4$.

17. For the function $f(x) = \sqrt{x^2 - 5x + 2} + 10$, find all values x for which $f(x) = 14$.

18. For the function $f(x) = \sqrt{x^2 + 8x + 40} - 3$, find all values x for which $f(x) = 2$.

Solve. Check for extraneous solutions.

19. $x^{1/2} + 8 = 10$

20. $x^{1/3} - 11 = -20$

21. $(x + 10)^{1/2} - 4 = 3$

22. $(2x - 35)^{1/2} + 6 = 17$

23. $(5x + 6)^{1/3} - 8 = -2$

24. $(x^2 - 10x + 49)^{1/2} + 2 = 7$

25. $x = \sqrt{2x + 48}$

26. $\sqrt{3x + 10} = x$

27. $\sqrt{4x + 13} = x - 2$

28. $x + 9 = \sqrt{6x + 46}$

29. $\sqrt{2x - 5} + 4 = x$

30. $\sqrt{3x + 13} - 3 = x$

31. $x = \sqrt{49 - 8x} + 7$

32. $x = \sqrt{2x + 9} - 5$

33. $2x - 3 = \sqrt{30 - 7x}$

34. $3x + 5 = \sqrt{27x + 27}$

35. $3x = 1 + \sqrt{4x^2 + x + 7}$

36. $x = \sqrt{54 + 5x - x^2} - 3$

37. $\sqrt{4x - 15} = \sqrt{3x + 11}$

38. $\sqrt[3]{6x + 7} = \sqrt[3]{x - 5}$

39. $\sqrt[4]{x^2 - 8x + 4} = \sqrt[4]{3x - 14}$

40. $\sqrt{5x^2 + 3x - 11} = \sqrt{4x^2 - 6x - 25}$

41. $\sqrt{3x^2 + 6x + 10} = \sqrt{x^2 + 5x + 55}$

42. $\sqrt{8x^2 - 3x - 15} = \sqrt{2x^2 - 8x - 11}$

43. $\sqrt{x + 4} = \sqrt{x - 1} + 1$

44. $\sqrt{x + 14} - \sqrt{x - 10} = 2$

45. $\sqrt{x - 9} + \sqrt{5x - 14} = 7$

46. $\sqrt{4x - 15} + 3 = \sqrt{6x}$

47. $\sqrt{3x - 5} = 5 + \sqrt{2x + 3}$

48. $\sqrt{5x + 1} = 2 - \sqrt{5x - 1}$

Answer in complete sentences.

49. What is an extraneous solution to an equation? Explain how to determine that a solution to a radical equation is actually an extraneous solution.

50. Explain how to solve a radical equation. Use an example to illustrate the process.

> *Study Tip* **Revisited**   Some students fall into the trap of mechanically taking notes without thinking about the material at all. In a math class, there can be no learning without thinking. Try to be an active learner while taking notes. Do your best to understand each statement that you write down in your notes.
>
> When your instructor writes a problem on the board, try to solve it yourself in your notes. When you have finished solving the problem, compare your solution with your instructor's solution. If you make a mistake, or if your solution varies from your instructor's solution, you can make editor's notes on your solution.

Objectives

1 Solve applied problems involving a pendulum and its period.
2 Solve other applied problems involving radicals.

A Pendulum and its Period

Objective 1 Solve applied problems involving a pendulum and its period.
The **period** of a pendulum is the amount of time it takes to swing from one extreme to the other and then back again. The period T of a pendulum in seconds can be found using the formula $T = 2\pi\sqrt{\dfrac{L}{32}}$, where L is the length of the pendulum in feet.

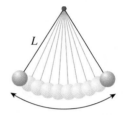

EXAMPLE 1   A pendulum has a length of 3 feet. Find its period, rounded to the nearest hundredth of a second.

Solution

We substitute 3 for L in the formula and simplify to find the period T.

$$T = 2\pi\sqrt{\frac{L}{32}}$$

$$T = 2\pi\sqrt{\frac{3}{32}} \qquad \text{Substitute 3 for } L.$$

$$T \approx 1.92 \qquad \text{Approximate using a calculator.}$$

The period of a pendulum that is 3 feet long is approximately 1.92 seconds.

Quick Check 1
A pendulum has a length of 6 feet. Find its period, rounded to the nearest hundredth of a second.

EXAMPLE 2   If a pendulum has a period of 1 second, find its length in feet. Round to the nearest hundredth of a foot.

Solution

In this example, we substitute 1 for T and solve for L. To solve this equation for L, we isolate the radical and then square both sides.

$$T = 2\pi\sqrt{\frac{L}{32}}$$

$$1 = 2\pi\sqrt{\frac{L}{32}} \qquad \text{Substitute 1 for } T.$$

$$\frac{1}{2\pi} = \frac{2\pi\sqrt{\dfrac{L}{32}}}{2\pi}$$ Divide by 2π to isolate the radical.

$$\frac{1}{2\pi} = \sqrt{\frac{L}{32}}$$ Simplify.

$$\left(\frac{1}{2\pi}\right)^2 = \left(\sqrt{\frac{L}{32}}\right)^2$$ Square both sides.

$$\frac{1}{4\pi^2} = \frac{L}{32}$$ Simplify.

$$32 \cdot \frac{1}{4\pi^2} = 32 \cdot \frac{L}{32}$$ Multiply by 32 to isolate L.

$$\overset{8}{\cancel{32}} \cdot \frac{1}{\underset{1}{\cancel{4}}\pi^2} = \overset{1}{\cancel{32}} \cdot \frac{L}{\underset{1}{\cancel{32}} \cdot}$$ Divide out common factors.

$$\frac{8}{\pi^2} = L$$ Simplify.

$$L \approx 0.81$$ Approximate using a calculator.

Quick Check 2
If a pendulum has a period of 3 seconds, find its length in feet. Round to the nearest hundredth of a foot.

The length of the pendulum is approximately 0.81 feet.

Other Applications Involving Radicals

Objective 2 **Solve other applied problems involving radicals.**

EXAMPLE 3   A vehicle involved in an accident made 150 feet of skid marks on the asphalt before crashing. The speed, s, that the vehicle was traveling in miles per hour can be approximated by the formula $s = \sqrt{30df}$, where d represents the length of the skid marks in feet and f represents the drag factor of the road. If the drag factor for asphalt is 0.75, find the speed the car was traveling. Round to the nearest mile per hour.

Solution

We begin by substituting 150 for d and 0.75 for f.

$$s = \sqrt{30df}$$
$$s = \sqrt{30(150)(0.75)}$$ Substitute 150 for d and 0.75 for f.
$$s = \sqrt{3375}$$ Simplify the radicand.
$$s \approx 58$$ Approximate, using a calculator.

The car was traveling at approximately 58 miles per hour.

Quick Check 3 A vehicle involved in an accident made 215 feet of skid marks on the asphalt before crashing. The speed, s, that the vehicle was traveling in miles per hour can be approximated by the formula $s = \sqrt{30df}$, where d represents the length of the skid marks in feet and f represents the drag factor of the road. If the drag factor for asphalt is 0.75, find the speed that the car was traveling. Round to the nearest mile per hour.

For Exercises 1–6, use the formula $T = 2\pi\sqrt{\dfrac{L}{32}}$.

1. A pendulum has a length of 4 feet. Find its period, rounded to the nearest hundredth of a second.

2. A pendulum has a length of 6 feet. Find its period, rounded to the nearest hundredth of a second.

3. A pendulum has a length of 2.5 feet. Find its period, rounded to the nearest hundredth of a second.

4. A pendulum has a length of 1.4 feet. Find its period, rounded to the nearest hundredth of a second.

5. If a pendulum has a period of 0.6 second, find its length in feet. Round to the nearest tenth of a foot.

6. If a pendulum has a period of 1.3 seconds, find its length in feet. Round to the nearest tenth of a foot.

If a pendulum has a length of L inches, then its period in seconds, T, can be found using the formula $T = 2\pi\sqrt{\dfrac{L}{384}}$.

7. A pendulum has a length of 17 inches. Find its period, rounded to the nearest hundredth of a second.

8. If a pendulum has a period of 0.45 second, find its length in inches. Round to the nearest tenth of an inch.

If a pendulum has a length of L meters, then its period in seconds, T, can be found using the formula $T = 2\pi\sqrt{\dfrac{L}{9.8}}$.

9. A pendulum has a length of 3 meters. Find its period, rounded to the nearest hundredth of a second.

10. If a pendulum has a period of 1.2 seconds, find its length in meters. Round to the nearest tenth of a meter.

Skid-mark analysis is one way to estimate the speed a car was traveling prior to an accident. The speed, s, that the vehicle was traveling in miles per hour can be approximated by the formula $s = \sqrt{30df}$, *where d represents the length of the skid marks in feet and f represents the drag factor of the road.*

11. A vehicle that was involved in an accident made 70 feet of skid marks on the asphalt before crashing. If the drag factor for asphalt is 0.75, find the speed the car was traveling. Round to the nearest mile per hour.

12. A vehicle that was involved in an accident made 240 feet of skid marks on the asphalt before crashing. If the drag factor for asphalt is 0.75, find the speed the car was traveling. Round to the nearest mile per hour.

13. A vehicle that was involved in an accident made 185 feet of skid marks on a concrete road before crashing. If the drag factor for concrete is 0.95, find the speed the car was traveling. Round to the nearest mile per hour.

14. A vehicle that was involved in an accident made 60 feet of skid marks on a concrete road before crashing. If the drag factor for concrete is 0.95, find the speed the car was traveling. Round to the nearest mile per hour.

A water tank has a hole at the bottom, and the rate r at which water flows out of the hole in gallons per minute can be found using the formula $r = 19.8\sqrt{d}$, *where d represents the depth of the water in the tank in feet.*

15. Find the rate of water flow if the depth of water in the tank is 49 feet.

16. Find the rate of water flow if the depth of water in the tank is 9 feet.

17. Find the rate of water flow if the depth of water in the tank is 13 feet. Round to the nearest tenth of a gallon per minute.

18. Find the rate of water flow if the depth of water in the tank is 22 feet. Round to the nearest tenth of a gallon per minute.

19. If water is flowing out of the tank at a rate of 30 gallons per minute, find the depth of water in the tank. Round to the nearest tenth of a foot.

20. If water is flowing out of the tank at a rate of 100 gallons per minute, find the depth of water in the tank. Round to the nearest tenth of a foot.

The hull speed of a sailboat is the maximum speed that the hull can attain from wind power. The hull speed, h, in knots can be calculated using the formula $h = 1.34\sqrt{L}$, where L is the length of the water line in feet.

21. Find the hull speed of a sailboat with a waterline of 36 feet.

22. Find the hull speed of a sailboat with a waterline of 28 feet. Round to the nearest hundredth of a knot.

23. If the hull speed of a sailboat is 9.5 knots, find the length of its waterline. Round to the nearest tenth of a foot.

24. If the hull speed of a sailboat is 6.4 knots, find the length of its waterline. Round to the nearest tenth of a foot.

A tsunami is a great sea wave caused by natural phenomena such as earthquakes or volcanic activity. The speed of a tsunami, s, in miles per hour at any particular point, can be found from the formula $s = 308.29\sqrt{d}$, where d is the depth of the ocean in miles at that point.

25. Find the speed of a tsunami if the depth of the water is 3 miles. Round to the nearest mile per hour.

26. Find the speed of a tsunami if the depth of the water is 0.4 mile. Round to the nearest mile per hour.

27. If a tsunami is traveling at 600 miles per hour, what is the depth of the ocean at that location? Round to the nearest tenth of a mile.

28. If a tsunami is traveling at 450 miles per hour, what is the depth of the ocean at that location? Round to the nearest tenth of a mile.

Answer in complete sentences.

29. Write a real-world word problem that involves finding the period of a pendulum. Solve your problem, explaining each step of the process.

30. Write a real-world word problem that involves estimating the speed a car was travelling based on the length of its skid marks. Solve your problem, explaining each step of the process.

Study Tip REVISITED   How should you format your notes? This is really a matter of personal preference, but it is important to come up with a system that works for you that you consistently use.

A good note-taking system takes advantage of the margins for specific tasks. You can use the left margin for writing down key words or as a place to denote important material. You can then use the right margin to clarify steps in problems or to take down advice from your instructor.

9.7

**THE COMPLEX
NUMBERS**

1 Rewrite square roots of negative numbers as imaginary numbers.

2 Add or subtract complex numbers.

3 Multiply imaginary numbers.

4 Multiply complex numbers.

5 Divide by a complex number.

6 Divide by an imaginary number.

7 Simplify expressions containing powers of *i*.

Imaginary Numbers

Objective 1 **Rewrite square roots of negative numbers as imaginary numbers.** In Section 9.1, we stated that the square root of a negative number, such as $\sqrt{-1}$ or $\sqrt{-25}$, was not a real number. This is because there is not a real number that equals a negative number when it is squared. The square root of a negative number is an **imaginary number.**

We define the **imaginary unit** *i* to be a number that is equal to $\sqrt{-1}$. The number *i* has the property that $i^2 = -1$.

Imaginary Unit *i*

$$i = \sqrt{-1}$$
$$i^2 = -1$$

All imaginary numbers can be expressed in terms of *i* because $\sqrt{-1}$ is a factor of every imaginary number.

EXAMPLE 1 Express $\sqrt{-25}$ in terms of *i*.

Solution

Whenever we have a square root with a negative radicand, we begin by factoring out *i*. Then we simplify the resulting square root.

$$\sqrt{-25} = \sqrt{25(-1)} \qquad \text{Rewrite } -25 \text{ as } 25(-1).$$
$$= \sqrt{25} \cdot \sqrt{-1} \qquad \text{Rewrite as the product of two square roots.}$$
$$= \sqrt{25}\, i \qquad \text{Rewrite } \sqrt{-1} \text{ as } i.$$
$$= 5i \qquad \text{Simplify the square root.}$$

As you become more experienced at working with imaginary numbers, you may wish to combine a few of the previous steps into one step.

EXAMPLE 2 Express $\sqrt{-40}$ in terms of *i*.

Solution

$$\sqrt{-40} = \sqrt{40} \cdot \sqrt{-1} \qquad \text{Rewrite as the product of two square roots.}$$
$$= 2\sqrt{10}\, i \qquad \text{Simplify the square root. Rewrite } \sqrt{-1} \text{ as } i.$$

A Word of Caution   After rewriting the square root of a negative number, such as $\sqrt{-40}$, as an imaginary number, be sure that i does not appear in the radicand. In other words, *be sure to write $2\sqrt{10}\,i$ and not $2\sqrt{10i}$.*

EXAMPLE 3   Express $-\sqrt{-12}$ in terms of i.

Solution

Notice that in this example there is a negative sign in front of the square root as well as in the radicand. We simplify the square root first, making the result negative.

$$-\sqrt{-12} = -\sqrt{12} \cdot \sqrt{-1} \qquad \text{Rewrite as the product of two square roots.}$$
$$= -2\sqrt{3}\,i \qquad\qquad \text{Simplify the square root. Rewrite } \sqrt{-1} \text{ as } i.$$

Quick Check 1

Express in terms of i.

a) $\sqrt{-36}$
b) $\sqrt{-63}$
c) $-\sqrt{-54}$

Complex Numbers

The set of imaginary numbers and the set of real numbers are subsets of the set of **complex numbers.**

Complex Numbers

> A **complex number** is a number of the form $a + bi$, where a and b are real numbers.

Real numbers are complex numbers for which $b = 0$, while imaginary numbers are complex numbers for which $a = 0$ but $b \neq 0$. We often associate the word complex with something that is difficult, but here *complex* refers to the fact that these numbers are made up of two parts.

Real and Imaginary Parts of a Complex Number

> For the complex number $a + bi$, the number a is the **real part** and the number b is the **imaginary part.**

Addition and Subtraction of Complex Numbers

Objective 2   **Add or subtract complex numbers.** We now focus on operations involving complex numbers. We add two complex numbers by adding the two real parts and adding the two imaginary parts. The same holds true for subtraction.

EXAMPLE 4   Simplify $(6 + 5i) + (7 - 2i)$.

Solution

We begin by removing the parentheses, and then we combine the two real parts and the two imaginary parts of these complex numbers.

$$(6 + 5i) + (7 - 2i) = 6 + 5i + 7 - 2i \qquad \text{Remove parentheses.}$$
$$= 13 + 3i \qquad\qquad\;\; \text{Combine the two real parts.}$$
$$\text{Combine the two imaginary parts.}$$

Notice that the process of adding these two complex numbers is identical to simplifying the expression $(6 + 5x) + (7 - 2x)$.

EXAMPLE 5   Simplify $(-2 + 9i) - (8 - 5i)$.

Solution

As with subtracting variable expressions, we must distribute the negative sign to both parts of the second complex number.

$$(-2 + 9i) - (8 - 5i) = -2 + 9i - 8 + 5i \qquad \text{Distribute.}$$
$$= -10 + 14i \qquad\qquad\qquad \text{Combine the two real parts and combine the two imaginary parts.}$$

Quick Check 2 Simplify.

a) $(3 + 12i) + (-8 + 15i)$
b) $(14 - 6i) - (3 + 22i)$

Multiplying Imaginary Numbers

Objective 3   Multiply imaginary numbers. Before learning to multiply two complex numbers, we will discuss the multiplication of two imaginary numbers. Suppose that we wanted to multiply $6i$ by $9i$. Just as $6x \cdot 9x = 54x^2$, the product $6i \cdot 9i$ is equal to $54i^2$. However, recall that $i^2 = -1$. So this product is $54(-1)$ or -54. When multiplying two imaginary numbers, we substitute -1 for i^2.

EXAMPLE 6   Multiply $8i \cdot 13i$.

Solution

$$8i \cdot 13i = 104i^2 \qquad \text{Multiply.}$$
$$= 104(-1) \qquad \text{Rewrite } i^2 \text{ as } -1.$$
$$= -104 \qquad\quad \text{Multiply.}$$

Quick Check 3
Multiply $4i \cdot 15i$.

EXAMPLE 7   Multiply $\sqrt{-24} \cdot \sqrt{-45}$.

Solution

Although we may think that we can multiply -24 by -45 and combine the two square roots into one, we cannot do this. The property $\sqrt{x} \cdot \sqrt{y} = \sqrt{xy}$ holds true only if either x or y is nonnegative. We must rewrite each square root as an imaginary number before multiplying.

$$\sqrt{-24} \cdot \sqrt{-45} = \sqrt{24}\,i \cdot \sqrt{45}\,i \qquad \text{Rewrite each radical as an imaginary number.}$$
$$= \sqrt{2^3 \cdot 3}\,i \cdot \sqrt{3^2 \cdot 5}\,i \qquad \text{Factor each radicand.}$$

$$= \sqrt{2^3 \cdot 3^3 \cdot 5}\, i^2 \qquad \text{Multiply the two radicands.}$$
$$= 2 \cdot 3\sqrt{2 \cdot 3 \cdot 5}\, i^2 \qquad \text{Simplify the square root.}$$
$$= -6\sqrt{30} \qquad \text{Rewrite } i^2 \text{ as } -1 \text{ and simplify.}$$

When multiplying $\sqrt{24}$ by $\sqrt{45}$, our work will be easier if we factor 24 and 45 before multiplying, rather than trying to simplify $\sqrt{1080}$.

Quick Check 4
Multiply $\sqrt{-5} \cdot \sqrt{-120}$.

Multiplying Complex Numbers

Objective 4 Multiply complex numbers. We multiply two complex numbers by using the distributive property. Often, a product of two complex numbers will contain a term with i^2, and we will rewrite i^2 as -1.

EXAMPLE 8   Multiply $3i(4 - 5i)$.

Solution

We begin by multiplying $3i$ by both terms in the parentheses.

$$
\begin{aligned}
3i(4 - 5i) &= 3i \cdot 4 - 3i \cdot 5i & &\text{Distribute.}\\
&= 12i - 15i^2 & &\text{Multiply.}\\
&= 12i + 15 & &\text{Rewrite } i^2 \text{ as } -1 \text{ and simplify.}\\
&= 15 + 12i & &\text{Rewrite in the form } a + bi.
\end{aligned}
$$

Quick Check 5
Multiply $-6i(7 + 8i)$.

EXAMPLE 9   Multiply $(9 + i)(2 + 7i)$.

Solution

In this example, we must multiply each term in the first set of parentheses by each term in the second set.

$$
\begin{aligned}
(9 + i)(2 + 7i) &= 9 \cdot 2 + 9 \cdot 7i + i \cdot 2 + i \cdot 7i & &\text{Distribute (FOIL).}\\
&= 18 + 63i + 2i + 7i^2 & &\text{Multiply.}\\
&= 18 + 65i - 7 & &\text{Add } 63i + 2i. \text{ Rewrite } i^2 \text{ as}\\
& & &-1 \text{ and simplify}\\
&= 11 + 65i & &\text{Combine like terms.}
\end{aligned}
$$

Quick Check 6
Multiply $(3 - 2i)(8 + i)$.

EXAMPLE 10   Multiply $(7 + 2i)^2$.

Solution

Recall that we square a binomial by multiplying it by itself.

$$
\begin{aligned}
(7 + 2i)^2 &= (7 + 2i)(7 + 2i) & &\text{Multiply } 7 + 2i \text{ by itself.}\\
&= 7 \cdot 7 + 7 \cdot 2i + 2i \cdot 7 + 2i \cdot 2i & &\text{Distribute.}\\
&= 49 + 14i + 14i + 4i^2 & &\text{Multiply.}\\
&= 49 + 28i - 4 & &\text{Add } 14i + 14i. \text{ Rewrite } i^2 \text{ as } -1\\
& & &\text{and simplify.}\\
&= 45 + 28i & &\text{Combine like terms.}
\end{aligned}
$$

Quick Check 7
Multiply $(9 - 4i)^2$.

EXAMPLE 11 Multiply $(5 + 6i)(5 - 6i)$.

Solution

$$
\begin{aligned}
(5 + 6i)(5 - 6i) &= 5 \cdot 5 - 5 \cdot 6i + 6i \cdot 5 - 6i \cdot 6i && \text{Distribute.}\\
&= 25 - 30i + 30i - 36i^2 && \text{Multiply.}\\
&= 25 - 36i^2 && \text{Combine like terms.}\\
&= 25 + 36 && \text{Rewrite } i^2 \text{ as } -1 \text{ and}\\
&&& \text{simplify.}\\
&= 61 && \text{Add.}
\end{aligned}
$$

Quick Check 8
Multiply
$(7 + 4i)(7 - 4i)$.

In the previous example, the two complex numbers that were multiplied were conjugates. Their product was a real number and did not have an imaginary part. Two complex numbers of the form $a + bi$ and $a - bi$ are conjugates, and their product will always be equal to $a^2 + b^2$.

$$
\begin{aligned}
(a + bi)(a - bi) &= a^2 - abi + abi - b^2 i^2 && \text{Distribute.}\\
&= a^2 - b^2 i^2 && \text{Combine like terms.}\\
&= a^2 + b^2 && \text{Rewrite } i^2 \text{ as } -1 \text{ and simplify.}
\end{aligned}
$$

EXAMPLE 12 Multiply $(10 - 3i)(10 + 3i)$.

Solution

We will use the fact that $(a + bi)(a - bi) = a^2 + b^2$ for two complex numbers that are conjugates.

$$
\begin{aligned}
(10 - 3i)(10 + 3i) &= 10^2 + 3^2 && \text{The product equals } a^2 + b^2.\\
&= 100 + 9 && \text{Square 10 and 3.}\\
&= 109 && \text{Add.}
\end{aligned}
$$

Quick Check 9
Multiply
$(11 - 4i)(11 + 4i)$.

Dividing by a Complex Number

Objective 5 **Divide by a complex number.** Since the imaginary number i is a square root $(\sqrt{-1})$, a simplified expression cannot contain i in its denominator. We will use conjugates to rewrite the expression without i in the denominator, in a procedure similar to rationalizing a denominator in Section 9.4.

EXAMPLE 13 Simplify $\dfrac{2}{3 + i}$.

Solution

We begin by multiplying the numerator and denominator by the conjugate of the denominator, which is $3 - i$.

$$\frac{2}{3+i} = \frac{2}{3+i} \cdot \frac{3-i}{3-i}$$

Multiply the numerator and denominator by the conjugate of the denominator.

$$= \frac{6-2i}{3^2+1^2}$$

Multiply the numerators by distributing 2 to both terms in the second numerator. Multiply the denominators by using the fact that $(a+bi)(a-bi) = a^2+b^2$.

$$= \frac{6-2i}{10}$$

Simplify the denominator.

$$= \frac{\overset{1}{\cancel{2}}(3-i)}{\underset{5}{\cancel{10}}}$$

Divide out factors common to the numerator and denominator.

$$= \frac{3-i}{5}$$

Simplify.

$$= \frac{3}{5} - \frac{1}{5}i$$

Rewrite in the form $a+bi$.

Quick Check 10

Simplify $\dfrac{15}{8+6i}$.

Answers in the back of the text will be written in both forms: as a single fraction and in $a+bi$ form. Your instructor will let you know which form is preferred for your class.

EXAMPLE 14   Simplify $\dfrac{1+6i}{2+5i}$.

Solution

In this example, the numerator has two terms, and we must multiply by using the distributive property.

$$\frac{1+6i}{2+5i} = \frac{1+6i}{2+5i} \cdot \frac{2-5i}{2-5i}$$

Multiply the numerator and denominator by the conjugate of the denominator.

$$= \frac{2-5i+12i-30i^2}{2^2+5^2}$$

Multiply.

$$= \frac{2+7i-30i^2}{4+25}$$

Combine like terms in the numerator. Square 2 and 5 in the denominator.

$$= \frac{2+7i+30}{29}$$

Rewrite i^2 as -1 and simplify. Simplify the denominator.

$$= \frac{32+7i}{29}$$

Combine like terms.

$$= \frac{32}{29} + \frac{7}{29}i$$

Rewrite in the form $a+bi$.

Quick Check 11

Simplify $\dfrac{9+2i}{7-3i}$.

Note that the expression $\dfrac{1+6i}{2+5i}$ is equivalent to $(1+6i) \div (2+5i)$. If asked to divide a complex number by another complex number, we begin by rewriting the expression as a fraction and then proceed as in the previous example.

Dividing by an Imaginary Number

Objective 6 Divide by an imaginary number. When a denominator is an imaginary number, we multiply the numerator and denominator by i to rewrite the fraction without i in the denominator.

EXAMPLE 15   Simplify $\dfrac{5 + 2i}{3i}$.

Solution

Since the denominator is an imaginary number, we will begin by multiplying the fraction by $\frac{i}{i}$.

$$\frac{5 + 2i}{3i} = \frac{5 + 2i}{3i} \cdot \frac{i}{i} \qquad \text{Multiply the numerator and denominator by } i.$$

$$= \frac{5i + 2i^2}{3i^2} \qquad \text{Multiply.}$$

$$= \frac{5i - 2}{-3} \qquad \text{Rewrite } i^2 \text{ as } -1 \text{ and simplify.}$$

$$= -\frac{5i - 2}{3} \qquad \text{Factor } -1 \text{ out of the denominator.}$$

$$= \frac{-5i + 2}{3} \qquad \text{Distribute.}$$

$$= \frac{2}{3} - \frac{5}{3}i \qquad \text{Rewrite in } a + bi \text{ form.}$$

Quick Check 12
Simplify $\dfrac{15 - 8i}{12i}$.

Powers of i

Objective 7 Simplify expressions containing powers of i. Occasionally, i may be raised to a power greater than 2. We finish this section by learning to simplify such expressions. The expression i^3 can be rewritten as $i^2 \cdot i$, which is equivalent to $-1 \cdot i$ or $-i$. The expression i^4 can be rewritten as $i^2 \cdot i^2$, which is equivalent to $-1(-1)$ or 1. The following table shows the first four positive powers of i.

$$
\begin{array}{cccc}
i & i^2 & i^3 & i^4 \\
\downarrow & \downarrow & \downarrow & \downarrow \\
i & -1 & -i & 1
\end{array}
$$

We can use the fact that $i^4 = 1$ to simplify greater powers of i.

$$i^5 = i^4 \cdot i = 1 \cdot i = i \qquad\qquad i^6 = i^4 \cdot i^2 = 1(-1) = -1$$
$$i^7 = i^4 \cdot i^3 = 1 \cdot i^3 = 1(-i) = -i \qquad i^8 = i^4 \cdot i^4 = 1 \cdot 1 = 1$$

We can now see that there is a pattern.

$$
\begin{array}{cccccccc}
i & i^2 & i^3 & i^4 & i^5 & i^6 & i^7 & i^8 \\
\downarrow & \downarrow & \downarrow & \downarrow & \downarrow & \downarrow & \downarrow & \downarrow \\
i & -1 & -i & 1 & i & -1 & -i & 1
\end{array}
$$

In general, to simplify i^n, where n is a whole number, we divide n by 4. If the remainder is equal to r, then $i^n = i^r$.

Remainder	0	1	2	3
i^n	1	i	-1	$-i$

EXAMPLE 16   Simplify i^{27}.

Solution

Quick Check 13   If we divide the exponent 27 by 4, the remainder is 3. So, $i^{27} = i^3 = -i$.

Simplify i^{30}.

EXERCISES 9.7

Express in terms of i.

1. $\sqrt{-4}$

2. $\sqrt{-16}$

3. $\sqrt{-81}$

4. $\sqrt{-100}$

5. $-\sqrt{-121}$

6. $-\sqrt{-196}$

7. $\sqrt{-27}$

8. $\sqrt{-80}$

9. $\sqrt{-32}$

10. $-\sqrt{-735}$

Add or subtract the complex numbers.

11. $(8 + 9i) + (3 + 5i)$

12. $(13 + 11i) - (6 + 2i)$

13. $(6 - 7i) - (2 + 10i)$

14. $(1 + 6i) + (5 - 12i)$

15. $(-3 + 8i) + (3 - 4i)$

16. $(5 - 9i) + (5 + 9i)$

17. $(12 + 13i) - (12 - 13i)$

18. $(14 - i) - (21 + 11i)$

19. $(1 - 12i) - (8 + 8i) + (5 - 4i)$

20. $(-9 + 2i) - (5 + 6i) - (10 - 20i)$

Find the missing complex number.

21. $(4 - 7i) + ? = 10 + 2i$

22. $(11 + 10i) + ? = 20 - 3i$

23. $(3 + 4i) - ? = 1 - 5i$

24. $? - (-9 + 7i) = -15 - 3i$

Multiply.

25. $7i \cdot 14i$

26. $3i \cdot 8i$

27. $-4i \cdot 5i$

28. $i(-9i)$

29. $\sqrt{-49} \cdot \sqrt{-64}$

30. $\sqrt{-25} \cdot \sqrt{-25}$

31. $\sqrt{-12} \cdot \sqrt{-18}$

32. $\sqrt{-28} \cdot \sqrt{-7}$

33. $\sqrt{-20} \cdot \sqrt{-45}$

34. $\sqrt{-27} \cdot \sqrt{-50}$

Find two imaginary numbers with the given product. (Answers may vary.)

35. -65

36. -14

37. 40

38. 12

Multiply.

39. $6i(8 - 3i)$

40. $4i(2 + 5i)$

41. $-7i(4 - 7i)$

42. $-3i(9 + 8i)$

43. $(8 + i)(3 - 5i)$

44. $(7 + 6i)(7 - 6i)$

45. $(6 - 4i)^2$

46. $(5 - 2i)(8 - 10i)$

47. $(9 - i)(9 + i)$

48. $(10 + 3i)^2$

49. $(1 + 7i)^2$

50. $(5i + 4)(4i - 5)$

Multiply the conjugates using the fact that
$(a + bi)(a - bi) = a^2 + b^2.$

51. $(5 + 2i)(5 - 2i)$

52. $(7 + 4i)(7 - 4i)$

53. $(12 - 8i)(12 + 8i)$

54. $(6 - i)(6 + i)$

Rationalize the denominator.

55. $\dfrac{5}{2 + i}$

56. $\dfrac{4}{5 - 3i}$

57. $\dfrac{15i}{7 - 4i}$

58. $\dfrac{6i}{9 + 5i}$

59. $\dfrac{4 - 9i}{3 + 2i}$

60. $\dfrac{1 + 3i}{6 - i}$

61. $\dfrac{5 + 6i}{5 - 6i}$

62. $\dfrac{7 + 5i}{5 + 7i}$

Rationalize the denominator.

63. $\dfrac{8}{i}$

64. $\dfrac{5}{6i}$

65. $\dfrac{5 - 7i}{3i}$

66. $\dfrac{3 + 10i}{5i}$

67. $\dfrac{6 + i}{-4i}$

68. $\dfrac{2 - 9i}{-8i}$

Divide.

69. $4i \div (7 + 3i)$

70. $(9 + 6i) \div (3i)$

71. $(6 - i) \div (3 + 4i)$

72. $(5 - 10i) \div (12i - 5)$

Find the missing complex number.

73. $(4 + i)(?) = 4 + 35i$

74. $(7 - 5i)(?) = 31 - i$

75. $\dfrac{?}{1 + 9i} = 4 - 3i$

76. $\dfrac{?}{10 - 3i} = 6 - 4i$

Simplify.

77. i^{15}

78. i^{34}

79. i^{200}

80. i^{65}

81. $i^{13} \cdot i^{17}$

82. $\dfrac{i^{51}}{i^{20}}$

83. $i^{39} + i^{29}$

84. $i^{103} - i^{161}$

Simplify.

85. $(7 + 2i)(6 - 3i)$

86. $(4 - 6i)(4 + 6i)$

87. $\dfrac{9 + 4i}{5i}$

88. $(18 - 13i) - (8 - 5i)$

89. $12i \div (4 - 2i)$

90. $\dfrac{10 - 9i}{5 - i}$

91. $(2 - 5i)(2 + 5i)$

92. $(14 - 11i) + (-23 + 6i)$

93. $\sqrt{-24} \cdot \sqrt{-27}$

94. $(6 - 4i) \div (8i)$

95. $\dfrac{15i}{9 - 2i}$

96. $(16 - 9i)^2$

97. $\sqrt{-252}$

98. $(11 + 4i)(6 - 13i)$

99. $(9 - 3i)^2$

100. $7i \cdot 12i$

101. $(8 - 11i) + (15 + 21i)$

102. i^{207}

103. $-14i \cdot 9i$

104. $\sqrt{-20} \cdot \sqrt{-15}$

105. i^{323}

106. $\dfrac{7 - 6i}{-2i}$

107. $(30 - 17i) - (54 - 33i)$

108. $\sqrt{-2673}$

Answer in complete sentences.

109. Explain how to rewrite the square root of a negative number in terms of i. Use an example to illustrate the process.

110. Explain how adding two complex numbers is similar to adding two variable expressions.

111. Explain how to multiply two complex numbers. Use an example to illustrate the process.

112. Explain how to divide a number by a complex number that has both a real part and an imaginary part. Use an example to illustrate the process.

Study Tip REVISITED   The best note-taking tip is to rewrite your notes as soon as possible after class. The simple task of rewriting your notes serves as a review of the material that was covered in class. If you rewrite your notes while the material is still fresh, then you have a much better chance of being able to actually read what you have written.

As you rework your notes, take the time to supplement your notes. Replace a brief definition with the full definition from the text. Add notes to clarify what you wrote down in class, including notes to yourself about what has been done in each step of the solution to a problem. This includes any calculator steps that are necessary. Each time a new term or procedure is introduced in your notes, add them to your set of note cards.

Finally, consider creating a page in your notes that contains the problems (without solutions) your instructor solved during class. When you begin to review for the exam, this list of problems will be a good review sheet.

Summary of Chapter 9 Study Tips

- A good set of class notes is one of the most valuable study resources you can create. To ensure that you create the best set of notes that you can, begin by choosing your seat wisely. Be sure that you can see the entire board and that you can clearly hear the instructor from your location.
- Write down each step of the solution to a problem, whether you feel that the step is necessary or not. Often, solutions seem clear while sitting in class, but they will not be so clear when you sit down to work the homework exercises. If you find that you are completely lost, do your best to write down every step so that you can try to make sense of the steps after class when reworking your notes.
- If you find that you are having trouble keeping up with your instructor, it is important that you write down enough information so that you will be able to understand the material after class has ended. Try using abbreviations when appropriate, and write down phrases rather than complete sentences. The use of a tape recorder, with your instructor's permission, can help you to fill in missing sections in your notes. A classmate can help you fill in missing notes as well. If your instructor gives you a cue that material is important, be sure that you write that material in your notes regardless of where you are in your notes. You can catch up on missing material after class.
- Reworking your notes after class serves as an excellent review of what was covered during the class period, as well as giving you an opportunity to make your notes more legible and user friendly. During this time, you can fill in any holes in your notes and supplement your notes with extra material.

Simplify the radical expression. Assume all variables represent nonnegative real numbers. [9.1]

1. $\sqrt{25}$

2. $\sqrt{169}$

3. $\sqrt[3]{-64}$

4. $\sqrt[3]{x^9}$

5. $\sqrt{81x^{12}}$

6. $\sqrt[5]{243x^{15}y^{10}z^{25}}$

Approximate to the nearest thousandth, using a calculator. [9.1]

7. $\sqrt{17}$

8. $\sqrt{41}$

Evaluate the radical function. Round to the nearest thousandth if necessary. [9.1]

9. $f(x) = \sqrt{x-10}, f(91)$

10. $f(x) = \sqrt{x^2 + 6x - 19}, f(3)$

Find the domain of the radical function. Express your answer in interval notation. [9.1]

11. $f(x) = \sqrt{2x+6}$

12. $f(x) = \sqrt[4]{3x+25}$

Rewrite each radical expression using rational exponents. Assume all variables represent nonnegative real numbers. [9.2]

13. $\sqrt[5]{a}$

14. $(\sqrt[4]{x})^3$

15. $(\sqrt{x})^{13}$

16. $\sqrt[5]{n^8}$

Rewrite as a radical expression and simplify if possible. Assume all variables represent nonnegative real numbers. [9.2]

17. $4^{1/2}$

18. $(343x^{21})^{2/3}$

Simplify the expression. Assume all variables represent nonnegative real numbers. Express your answer in radical notation. [9.2]

19. $x^{3/2} \cdot x^{1/6}$

20. $(x^{3/8})^{7/6}$

21. $\dfrac{x^{2/3}}{x^{5/12}}$

22. $49^{-1/2}$

23. $64^{-5/3}$

24. $\dfrac{27^{4/3}}{27^{8/3}}$

Simplify the radical expression. Assume all variables represent nonnegative real numbers. [9.3]

25. $\sqrt{28}$

26. $\sqrt[3]{x^{10}}$

27. $\sqrt[5]{x^{11}y^8z^{15}}$

28. $\sqrt[3]{405a^{25}b^{12}c^{16}}$

29. $\sqrt{180r^3s^{11}t^{12}}$

30. $\sqrt{175r^6s^8t^4}$

Add or subtract. Assume all variables represent nonnegative real numbers. [9.3]

31. $4\sqrt{2} + 8\sqrt{2}$

32. $7\sqrt[3]{x} - \sqrt[3]{x}$

33. $5\sqrt{20} + 3\sqrt{125}$

34. $6\sqrt{12} - 8\sqrt{50} + 9\sqrt{75}$

Multiply. Assume all variables represent nonnegative real numbers. [9.4]

35. $\sqrt[3]{4x^2} \cdot \sqrt[3]{18x^4}$

36. $5\sqrt{3}(2\sqrt{6} - 4\sqrt{3})$

37. $(\sqrt{3} + \sqrt{8})(\sqrt{6} - \sqrt{2})$

38. $(10\sqrt{5} - 7\sqrt{11})(9\sqrt{5} + 2\sqrt{11})$

39. $(2\sqrt{15} - 9\sqrt{3})^2$

40. $(2\sqrt{13} - 10\sqrt{7})(2\sqrt{13} + 10\sqrt{7})$

Simplify. Assume all variables represent nonnegative real numbers. [9.4]

41. $\dfrac{\sqrt{224x^5}}{\sqrt{2x}}$

42. $\sqrt{\dfrac{50a^9b^8c^3}{8a^3bc^{13}}}$

Rationalize the denominator and simplify. Assume all variables represent nonnegative real numbers. **[9.4]**

43. $\sqrt[3]{\dfrac{16}{5}}$

44. $\dfrac{8}{\sqrt{14}}$

45. $\sqrt{\dfrac{11}{18}}$

46. $\dfrac{a^5bc^3}{\sqrt{a^4b^5c}}$

47. $\dfrac{8}{\sqrt{11} - \sqrt{5}}$

48. $\dfrac{2\sqrt{3}}{10\sqrt{15} - 3\sqrt{12}}$

49. $\dfrac{\sqrt{7} + 5\sqrt{2}}{2\sqrt{7} - 3\sqrt{2}}$

50. $\dfrac{3 - \sqrt{5}}{6 + 7\sqrt{5}}$

Solve. **[9.5]**

51. $\sqrt{4x + 1} = 7$

52. $\sqrt[3]{7x - 15} = 5$

53. $\sqrt{3x + 22} + 2 = x$

54. $\sqrt{17 - 4x} - x = 1$

55. $\sqrt{x + 5} = \sqrt{3x - 13}$

56. $\sqrt{x + 8} = 2 - \sqrt{x - 12}$

57. For the function $f(x) = \sqrt{3x + 4}$, find all values of x for which $f(x) = 4$. **[9.5]**

58. For the function $f(x) = \sqrt[3]{x^2 - 6x + 11} + 8$, find all values of x for which $f(x) = 11$. **[9.5]**

59. A pendulum has a length of 5 feet. Find its period, rounded to the nearest hundredth of a second. Use the formula $T = 2\pi\sqrt{\dfrac{L}{32}}$. **[9.6]**

60. A vehicle involved in an accident made 200 feet of skid marks on the asphalt. The speed, s, that the vehicle was traveling in miles per hour can be approximated by the formula $s = \sqrt{30df}$, where d

represents the length of the skid marks in feet and f represents the drag factor of the road. If the drag factor for asphalt is 0.75, find the speed the car was traveling. Round to the nearest mile per hour. **[9.6]**

Express in terms of i. **[9.7]**

61. $\sqrt{-49}$

62. $\sqrt{-40}$

63. $\sqrt{-252}$

64. $-\sqrt{-675}$

Add or subtract the complex numbers. **[9.7]**

65. $(4 + 2i) + (7 + 3i)$

66. $(9 - 4i) + (7 + 2i)$

67. $(11 - 8i) - (10 - 13i)$

68. $(6 + 12i) - (5 + 5i)$

Multiply. **[9.7]**

69. $2i \cdot 8i$

70. $\sqrt{-6} \cdot \sqrt{-50}$

71. $7i(3 - 4i)$

72. $(10 + 2i)(13 + 8i)$

73. $(9 - 3i)^2$

74. $(3 + 14i)(3 - 14i)$

Rationalize the denominator. **[9.7]**

75. $\dfrac{6}{7 - 3i}$

76. $\dfrac{15i}{6 + 8i}$

77. $\dfrac{2 + i}{9 - 5i}$

78. $\dfrac{16}{i}$

Simplify. **[9.7]**

79. i^{53}

80. i^{22}

Simplify the radical expression. Assume all variables represent nonnegative real numbers.

1. $\sqrt{25x^{10}y^{14}}$

2. For $f(x) = \sqrt{x^2 - 9x + 6}$, evaluate $f(-5)$. Round to the nearest thousandth.

Simplify the expression. Assume all variables represent nonnegative real numbers. Express your answer in radical notation.

3. $a^{3/10} \cdot a^{1/5}$

4. $\dfrac{b^{7/8}}{b^{5/12}}$

5. $125^{-2/3}$

Simplify the radical expression. Assume all variables represent nonnegative real numbers.

6. $\sqrt{72}$

7. $\sqrt[4]{a^{21}b^{42}c^4}$

Add or subtract. Assume all variables represent nonnegative real numbers.

8. $8\sqrt{18} + 11\sqrt{2} - 3\sqrt{200}$

Multiply. Assume all variables represent nonnegative real numbers.

9. $4\sqrt{5}(7\sqrt{10} - 2\sqrt{35})$

10. $(3\sqrt{6} - 4\sqrt{8})(9\sqrt{6} - 2\sqrt{8})$

Rationalize the denominator and simplify. Assume all variables represent nonnegative real numbers.

11. $\dfrac{x^3y^2}{\sqrt{x^3y^{11}}}$

12. $\dfrac{15\sqrt{7} - 8\sqrt{2}}{4\sqrt{7} + 9\sqrt{2}}$

Solve.

13. $\sqrt[3]{6x + 5} = 5$

14. $\sqrt{x + 28} + 2 = x$

15. If a pendulum has a period of 4 seconds, find its length, rounded to the nearest tenth of a foot. Use the formula $T = 2\pi\sqrt{\dfrac{L}{32}}$.

16. Express $\sqrt{-99}$ in terms of i.

Add or subtract the complex numbers.

17. $(2 - 3i) - (20 + 8i)$

Multiply.

18. $(6 + 7i)(6 - 7i)$

Rationalize the denominator.

19. $\dfrac{4 + 9i}{3 - 5i}$

Mathematicians in History
Pythagoras of Samos

Pythagoras of Samos, often referred to simply as Pythagoras, was the leader of a society known as the Pythagoreans. This society was partially religious and partially scientific in nature. One of their mathematical achievements is the first proof of a theorem that related the lengths of the sides of a right triangle. This theorem is known as the Pythagorean theorem.

Write a one-page summary (*or* make a poster) of the life of Pythagoras, the mathematical achievements of Pythagoras and his society, and the beliefs of the Pythagoreans.

Interesting issues:

- What was the "semicircle?"
- Who were the mathematikoi, and by what rules did they lead their life?
- What number did the Pythagoreans consider to be the "best" number, and why?
- Which society was the first to know of the Pythagorean theorem?
- The Pythagoreans believed that all relationships could be expressed numerically as a ratio of two numbers. They discovered, however, that the diagonal of a square whose side has length 1 was an irrational number. This number is $\sqrt{2}$. This discovery rocked the foundation of their system of beliefs, and they swore each other to secrecy regarding this discovery. A Pythagorean named Hippasus told others outside of the society of this irrational number. What was the fate of Hippasus?

Have you ever watched a child swing back and forth on a playground swing? Do you remember seeing in the movies a person swinging a pocket-watch back and forth in order to put another person into a trance? This motion of the swing and pocket-watch is called a pendulum. Today, you will be creating your own pendulum and using the data you collect to discover a mathematical model.

Step 1: Form groups of no more than four or five students. Each group should have a yard stick or ruler, a piece of string (approximately 2 yards long), a weight tied to the end of the string (hardware washers work well), and a stop watch (or some method for counting seconds).

Step 2: Measure a length for your pendulum (in inches) and fasten your string to the ceiling at this length. For example, you might want to start with a 36-inch pendulum. Write the length of your pendulum in the following table:

Length of Pendulum (Inches)	Number of Seconds for 10 Swings	Average Time

Step 3: Pull the pendulum back and release. Calculate the number of seconds it takes for your pendulum to make 10 swings. (Note that a swing is one complete back-and-forth motion.) Record the time in the chart. Repeat this process three to five times for the same length. Calculate the average time needed for this given length.

Step 4: Change the length of your pendulum and record the length in the chart. Record the time needed for 10 swings. Repeat this process three to five times for the same length. Calculate the average time needed for this given length.

Step 5: Repeat Step 4 a few more times for different lengths of the pendulum.

Step 6: What did you notice about the length of the pendulum in relation to the length of time needed for 10 swings? Did the time increase as the length increased? Or did the time decrease as the length increased?

Step 7: Fill in the following table:

Length of Pendulum	Square Root Length of Pendulum	Average Time for 10 Swings	Average Time for 1 Swing	Average Time for 1 Swing Divided by Square Root Length of Pendulum

Step 8: What do you notice about the last column? Are all the values relatively the same? As it turns out, these values should be relatively the same and should be around 0.32. There is actually a formula which uses square roots which relates the time it takes for a pendulum to make one swing and the length of the pendulum. This formula is $T = 2\pi\sqrt{\dfrac{L}{384}}$, where T is the time for one swing and L is length of the pendulum in inches.

Step 9: See how close your times were throughout the experiment to what they should be according to the formula. Were your actual times close to the predicted times?

QUADRATIC EQUATIONS

In this chapter, we will learn how to apply techniques other than factoring to solve quadratic equations. One important goal of this chapter is determining which technique will provide the most efficient solution to a particular equation. We will also apply these techniques to new application problems that previously could not be solved because they produce equations which are not factorable. Later in the chapter, we will examine the graph of a quadratic equation. The graph of a quadratic equation is called a parabola and is U shaped. The techniques used to graph linear equations will also be used in graphing quadratic equations, along with some new techniques. The chapter ends by examining inequalities that involve quadratic and rational expressions.

Study Tip TIME MANAGEMENT   *When asked why they are having difficulties in a particular class, many students claim that they simply do not have enough time. However, a great number of these students have enough time, but lack the time-management skills to make the most of their time. It seems as if they fritter and waste the hours in an offhand way. In this chapter, we will discuss effective time-management strategies, focusing on how to make more efficient use of available time.*

10.1

SOLVING QUADRATIC EQUATIONS BY EXTRACTING SQUARE ROOTS; COMPLETING THE SQUARE

Objectives

1 Solve quadratic equations by factoring.
2 Solve quadratic equations by extracting square roots.
3 Solve quadratic equations by extracting square roots involving a linear expression that is squared.
4 Solve quadratic equations by completing the square.

Solving Quadratic Equations by Factoring

Objective 1 **Solve quadratic equations by factoring.** In Chapter 6, we learned how to solve quadratic equations through the use of factoring. We begin this section with a review of this technique. We start by simplifying both sides of the equation completely. This includes any distributing that must be done and combining like terms when possible. After that, we need to set the equation equal to 0 by collecting all of the terms on one side of the equation. Recall that it is a good idea to move the terms to the side of the equation that will produce a positive second-degree term. After the equation contains an expression equal to 0, we need to factor the expression. Finally, set each factor equal to 0 and solve the resulting equation.

EXAMPLE 1 Solve $x^2 - 7x = -10$.

Solution

$$x^2 - 7x = -10$$
$$x^2 - 7x + 10 = 0 \qquad \text{Add 10 to collect all terms on the left side.}$$
$$(x - 2)(x - 5) = 0 \qquad \text{Factor.}$$
$$x - 2 = 0 \quad \text{or} \quad x - 5 = 0 \qquad \text{Set each factor equal to 0.}$$
$$x = 2 \quad \text{or} \quad x = 5 \qquad \text{Solve the resulting equations.}$$

The solution set is $\{2, 5\}$.

EXAMPLE 2 Solve $x^2 = 49$.

Solution

$$x^2 = 49$$
$$x^2 - 49 = 0 \qquad \text{Set the equation equal to 0.}$$
$$(x + 7)(x - 7) = 0 \qquad \text{Factor. (Difference of Squares)}$$
$$x + 7 = 0 \quad \text{or} \quad x - 7 = 0 \qquad \text{Set each factor equal to 0.}$$
$$x = -7 \quad \text{or} \quad x = 7 \qquad \text{Solve the resulting equations.}$$

Quick Check 1

The solution set is $\{-7, 7\}$.

Solve.

a) $x^2 = 2x + 15$
b) $x^2 = 121$

Solving Quadratic Equations by Extracting Square Roots

Objective 2 **Solve quadratic equations by extracting square roots.** In the preceding example, we were trying to find a number x that, when squared, equals 49. The principal square root of 49 is 7, which gives us one of the two solutions. In an equation where we have a squared term equal to a constant, we can take the square root of both sides of the equation to find the solution, as long as we take both the positive and negative square root of the constant. The symbol " $\pm$ " is used to represent both the positive and negative square root and is read as "plus or minus." For example, $x = \pm 7$ means $x = 7$ or $x = -7$. This technique for solving quadratic equations is called **extracting square roots.**

Extracting Square Roots

1. Isolate the squared term.
2. Take the square root of each side. (Remember to take both the *positive* and *negative* ($\pm$) square root of the constant.)
3. Simplify the square root.
4. Solve by isolating the variable.

Now we will use extracting square roots to solve an equation that we would not have been able to solve by factoring.

EXAMPLE ▶ 3   Solve $x^2 - 50 = 0$.

Solution

The expression $x^2 - 50$ cannot be factored because 50 is not a perfect square. We proceed to solve this equation by extracting square roots.

$$x^2 - 50 = 0$$
$$x^2 = 50 \qquad \text{Add 50.}$$
$$\sqrt{x^2} = \pm\sqrt{50} \qquad \text{Take the square root of each side.}$$
$$x = \pm 5\sqrt{2} \qquad \text{Simplify the square root. } \sqrt{50} = \sqrt{25 \cdot 2} = 5\sqrt{2}$$

The solution set is $\{5\sqrt{2}, -5\sqrt{2}\}$. Using a calculator, we find that the solutions are approximately ± 7.07.

A Word of Caution   When taking the square root of both sides of an equation, do not forget to use the symbol $\pm$.

This technique can also be used to find complex solutions to equations, as well as real solutions.

EXAMPLE ▶ 4   Solve $x^2 = -16$.

Solution

If we added the 16 to the left side of the equation, then the resulting equation would have been $x^2 + 16 = 0$. The expression $x^2 + 16$ is not factorable (recall that the sum of two squares is not factorable). The only way for us to find a solution would be to take the square roots of both sides of the original equation. We proceed to solve this equation by extracting square roots.

Quick Check 2

Solve.

a) $x^2 = 28$
b) $x^2 + 32 = 0$

$$x^2 = -16$$
$$\sqrt{x^2} = \pm\sqrt{-16} \qquad \text{Take the square root of each side.}$$
$$x = \pm 4i \qquad \text{Simplify the square root.}$$

The solution set is $\{4i, -4i\}$.

Objective 3 **Solve quadratic equations by extracting square roots involving a linear expression that is squared.** Extracting square roots is an excellent technique to solve quadratic equations whenever our equation is made up of a squared term and a constant term. Consider the equation $(2x - 7)^2 = 25$. To use factoring to solve this equation, we would first have to square $2x - 7$, then collect all terms on the left side of the equation by subtracting 25, and hope that the resulting expression can be factored. Extracting square roots is a more efficient way to solve this equation.

EXAMPLE 5   Solve $(2x - 7)^2 = 25$.

Solution

$$(2x - 7)^2 = 25$$

$$\sqrt{(2x - 7)^2} = \pm\sqrt{25} \qquad \text{Take the square root of each side.}$$

$$2x - 7 = \pm 5 \qquad \text{Simplify the square root. Note that the square root of } (2x - 7)^2 \text{ is } 2x - 7.$$

$$2x = 7 \pm 5 \qquad \text{Add 7.}$$

$$x = \frac{7 \pm 5}{2} \qquad \text{Divide by 2.}$$

$\dfrac{7 + 5}{2} = 6$ and $\dfrac{7 - 5}{2} = 1$, so the solution set is $\{6, 1\}$.

EXAMPLE 6   Solve $(3x + 4)^2 + 35 = 15$.

Solution

$$(3x + 4)^2 + 35 = 15$$

$$(3x + 4)^2 = -20 \qquad \text{Subtract 35 to isolate the squared term.}$$

$$\sqrt{(3x + 4)^2} = \pm\sqrt{-20} \qquad \text{Take the square root of each side.}$$

$$3x + 4 = \pm 2\sqrt{5}\, i \qquad \text{Simplify the square root.}$$

$$3x = -4 \pm 2\sqrt{5}\, i \qquad \text{Subtract 4.}$$

$$x = \frac{-4 \pm 2\sqrt{5}\, i}{3} \qquad \text{Divide both sides by 3.}$$

Since we cannot simplify the numerator in this case, these are our solutions. The solution set is $\left\{ \dfrac{-4 + 2\sqrt{5}\, i}{3}, \dfrac{-4 - 2\sqrt{5}\, i}{3} \right\}$.

***Quick Check* 3** Solve.

a) $(3x + 1)^2 = 100$
b) $2(2x - 3)^2 + 17 = -1$

Solving Quadratic Eqxuations by Completing the Square

Objective 4 **Solve quadratic equations by completing the square.** We can solve any quadratic equation by converting it to an equation with a squared term equal to a constant. Rewriting the equation in this form allows us to solve it by extracting square roots. We will now examine a procedure for doing this called **completing the square**. This procedure is used for an equation such as $x^2 - 6x - 16 = 0$, which has both a second-degree term (x^2) and a first-degree term $(-6x)$. The first step is to isolate variable terms on one side of the equation with the constant term on the other side. The next step is to add a number to both sides of the equation that makes the side of the equation containing the variable terms into a perfect-square trinomial. This number is found by taking half of the coefficient of the first-degree term and then squaring it. Half of -6 is -3, which equals 9 when squared: $\frac{1}{2}(-6) = -3$, $(-3)^2 = 9$. Add 9 to both sides of the equation.

$$x^2 - 6x - 16 = 0$$
$$x^2 - 6x = 16 \qquad \text{Add 16, to isolate the constant term.}$$
$$x^2 - 6x + 9 = 16 + 9 \qquad \text{Add 9, to make the expression on the left side of}$$
$$\text{the equation a perfect square trinomial.}$$
$$x^2 - 6x + 9 = 25 \qquad \text{Simplify the right side of the equation.}$$

The resulting expression on the left side $(x^2 - 6x + 9)$ is now a perfect square trinomial that factors to be $(x - 3)^2$. The resulting equation is now in the correct form for extracting square roots: $(x - 3)^2 = 25$.

Here is the procedure for solving a quadratic equation by completing the square, provided that the coefficient of the squared term is 1.

Completing the Square

1. Isolate all variable terms on one side of the equation, with the constant term on the other side of the equation.
2. Identify the coefficient of the first-degree term. Take half of that number, square it, and add that to both sides of the equation.
3. Factor the resulting perfect square trinomial.
4. Take the square root of each side of the equation. Be sure to include $\pm$ on the side where the constant is.
5. Solve the resulting equation.

Here is the example $x^2 - 6x - 16 = 0$ worked out completely:

EXAMPLE 7   Solve $x^2 - 6x - 16 = 0$ by completing the square.

Solution

$$x^2 - 6x - 16 = 0$$
$$x^2 - 6x = 16 \qquad \text{Add 16.}$$
$$x^2 - 6x + 9 = 16 + 9 \qquad \left(\frac{-6}{2}\right)^2 = (-3)^2 = 9.$$
$$\text{Add 9 to complete the square.}$$
$$(x - 3)^2 = 25 \qquad \text{Simplify. Factor the left side.}$$
$$\sqrt{(x - 3)^2} = \pm\sqrt{25} \qquad \text{Take the square root of each side.}$$
$$x - 3 = \pm 5 \qquad \text{Simplify the square root.}$$
$$x = 3 \pm 5 \qquad \text{Add 3.}$$

$3 + 5 = 8$ and $3 - 5 = -2$, so the solution set is $\{8, -2\}$.

Note that the equation in the previous example could have been solved with less work by factoring $x^2 - 6x - 16$. Always use factoring whenever possible, because it often leads to the quickest and most direct solutions.

EXAMPLE 8   Solve $x^2 + 8x + 20 = 0$ by completing the square.

Solution

The expression $x^2 + 8x + 20$ does not factor, so we proceed with completing the square.

$$x^2 + 8x + 20 = 0$$
$$x^2 + 8x = -20 \qquad \text{Subtract 20.}$$
$$x^2 + 8x + 16 = -20 + 16 \qquad \text{Half of 8 is 4, which equals 16 when squared:}$$
$$\left(\tfrac{8}{2}\right)^2 = (4)^2 = 16. \text{ Add 16.}$$
$$x^2 + 8x + 16 = -4 \qquad \text{Simplify.}$$
$$(x + 4)^2 = -4 \qquad \text{Factor the left side.}$$
$$\sqrt{(x + 4)^2} = \pm\sqrt{-4} \qquad \text{Take the square root of each side.}$$
$$x + 4 = \pm 2i \qquad \text{Simplify the square root.}$$
$$x = -4 \pm 2i \qquad \text{Subtract 4.}$$

The solution set is $\{-4 + 2i, -4 - 2i\}$.

Quick Check 4

Solve by completing the square.

a) $x^2 + 8x + 12 = 0$
b) $x^2 - 6x + 10 = 0$

In the first two examples of completing the square, the coefficient of the first-degree term was an even integer. If this is not the case, we must use fractions to complete the square.

EXAMPLE 9   Solve $x^2 - 5x - 5 = 0$ by completing the square.

Solution

The expression $x^2 - 5x - 5$ does not factor, so we proceed with completing the square.

$$x^2 - 5x - 5 = 0$$
$$x^2 - 5x = 5 \qquad \text{Add 5.}$$
$$x^2 - 5x + \frac{25}{4} = 5 + \frac{25}{4} \qquad \text{Half of } -5 \text{ is } -\tfrac{5}{2}, \text{ which equals } \tfrac{25}{4} \text{ when}$$
$$\text{squared: } \left(-\tfrac{5}{2}\right)^2 = \tfrac{25}{4}. \text{ Add } \tfrac{25}{4}.$$
$$x^2 - 5x + \frac{25}{4} = \frac{45}{4} \qquad \text{Add 5 and } \tfrac{25}{4} \text{ by rewriting 5 as a fraction whose}$$
$$\text{denominator is 4: } 5 + \tfrac{25}{4} = \tfrac{20}{4} + \tfrac{25}{4}.$$
$$\left(x - \frac{5}{2}\right)^2 = \frac{45}{4} \qquad \text{Factor.}$$
$$\sqrt{\left(x - \frac{5}{2}\right)^2} = \pm\sqrt{\frac{45}{4}} \qquad \text{Take the square root of each side.}$$
$$x - \frac{5}{2} = \pm\frac{3\sqrt{5}}{2} \qquad \text{Simplify the square root.}$$
$$x = \frac{5}{2} \pm \frac{3\sqrt{5}}{2} \qquad \text{Add } \tfrac{5}{2}.$$

Quick Check 5

Solve $x^2 + 7x - 18 = 0$ by completing the square.

The solution set is $\left\{\dfrac{5 + 3\sqrt{5}}{2}, \dfrac{5 - 3\sqrt{5}}{2}\right\}$.

A Word of Caution   To solve a quadratic equation by completing the square, we must be sure that the leading coefficient is positive 1. If the leading coefficient is not equal to 1, then divide both sides of the equation by the leading coefficient.

EXERCISES 10.1

Solve by factoring.

1. $x^2 - 3x - 10 = 0$

2. $x^2 + 9x + 14 = 0$

3. $x^2 + 11x + 24 = 0$

4. $x^2 + 7x - 18 = 0$

5. $x^2 + 6x + 8 = 0$

6. $x^2 + 15x + 56 = 0$

7. $x^2 - 2x = 24$

8. $x^2 - 45 = -4x$

9. $x^2 - 64 = 0$

10. $x^2 - 25 = 0$

11. $x^2 - 4x = 7x + 26$

12. $x^2 + 8x + 13 = 5x + 41$

Solve by extracting square roots.

13. $x^2 = 36$

14. $x^2 = 4$

15. $x^2 - 98 = 0$

16. $x^2 - 48 = 0$

17. $x^2 = -25$

18. $x^2 = -12$

19. $x^2 + 14 = 42$

20. $x^2 + 75 = 30$

21. $(x - 5)^2 = 36$

22. $(x - 9)^2 = 54$

23. $(x + 3)^2 = -27$

24. $(x + 14)^2 = -144$

25. $(x + 5)^2 + 33 = 15$

26. $(x + 3)^2 - 11 = 38$

27. $2(x - 9)^2 - 17 = 83$

28. $3(x + 1)^2 + 11 = 83$

29. $\left(x - \dfrac{2}{5}\right)^2 = \dfrac{49}{25}$

30. $\left(x + \dfrac{3}{4}\right)^2 = -\dfrac{25}{16}$

Fill in the missing term that makes the expression a perfect square trinomial. Factor the resulting expression.

31. $x^2 + 12x +$ ___

32. $x^2 - 2x +$ ___

33. $x^2 - 5x +$ ___

34. $x^2 + 13x +$ ___

35. $x^2 - 8x +$ ___

36. $x^2 - 16x +$ ___

37. $x^2 + \dfrac{1}{4}x +$ ___

38. $x^2 + x +$ ___

Solve by completing the square.

39. $x^2 - 8x - 33 = 0$

40. $x^2 + 6x - 7 = 0$

41. $x^2 + 4x + 7 = 0$

42. $x^2 - 10x + 18 = 0$

43. $x^2 + 8x = 9$

44. $x^2 + 12x = -27$

45. $x^2 - 16 = -6x$

46. $x^2 - 48 = 2x$

47. $x^2 + 7x + 6 = 0$

48. $x^2 + 13x - 30 = 0$

49. $x^2 + 3x = 5$

50. $x^2 - x = 7$

51. $x^2 + \dfrac{5}{2}x + 1 = 0$

52. $x^2 - \dfrac{23}{6}x + \dfrac{7}{2} = 0$

Find a quadratic equation with integer coefficients that has the following solution set.

53. $\{2, -5\}$

54. $\{-3, -4\}$

55. $\{6, 8\}$

56. $\{0, 5\}$

57. $\left\{-2, \dfrac{3}{4}\right\}$

58. $\left\{\dfrac{1}{2}, \dfrac{11}{4}\right\}$

59. $\{3, -3\}$

60. $\{6, -6\}$

61. $\{5i, -5i\}$

62. $\{i, -i\}$

Solve by any method (factoring, extracting square roots, or completing the square).

63. $x^2 + 13x + 30 = 0$

64. $x^2 + 12 = 0$

65. $x^2 + 6x - 17 = 0$

66. $(x - 5)^2 = 1$

67. $x^2 - 6x - 7 = 0$

68. $x^2 - 6x = -10$

69. $(x - 4)^2 - 11 = 16$

70. $x^2 - 5x - 36 = 0$

71. $x^2 - 4x = 20$

72. $x^2 + 85 = 18x$

73. $x^2 - 5x - 6 = 0$

74. $(2x + 8)^2 - 5 = 11$

75. $x^2 - 8x + 19 = 0$

76. $x^2 - 20x + 91 = 0$

77. $3(x + 3)^2 + 13 = -11$

78. $x^2 - 9 = 0$

79. $x^2 + 3x + 9 = 0$

80. $x^2 - 2x + 50 = 0$

81. $x(x + 5) - 7(x + 5) = 0$

82. $(4x - 3)(4x - 3) = 25$

83. $-16x^2 + 64x + 80 = 0$

84. $\left(x + \dfrac{b}{2a}\right)^2 = \dfrac{b^2 - 4ac}{4a^2}$, where a, b, and c are con-

stants and $a \neq 0$.

85. $\dfrac{2}{3}x^2 + \dfrac{8}{3}x - \dfrac{10}{3} = 0$

86. $x^2 - 10x + 11 = 0$

For the following rational expressions, list the values that are excluded from the domain. (Recall that a real number is excluded from the domain of a rational expression if it causes the denominator to equal 0 when it is substituted into the expression.)

87. $\dfrac{7}{x^2 - 7x + 12}$

88. $\dfrac{5}{x^2 - 3x}$

89. $\dfrac{x + 2}{x^2 - 40}$

90. $\dfrac{x - 6}{x^2 + 25}$

91. $\dfrac{x^2 - 11x + 28}{(x - 5)^2 - 16}$

92. $\dfrac{x^2 + 49}{3(4x - 3)^2 - 24}$

Answer in complete sentences.

93. Explain how to solve an equation by extracting square roots. Use an example to illustrate the process.

94. Explain why we use the symbol $\pm$ when taking the square root of each side of an equation.

95. Explain how to solve an equation by completing the square. Use an example to illustrate the process.

96. If the coefficient of the second-degree term is not 1, we divide both sides of the equation by that coefficient before attempting to complete the square. Explain why dividing both sides of the equation by this coefficient does not affect the solutions of the equation.

Study Tip **REVISITED**   The first step in achieving effective time management is keeping track of your time over a one-week period. Mark down the times that you spend in class, working, cooking, watching TV, eating, sleeping, spending time with family, socializing, getting tutorial help, getting ready for school, and most importantly, studying. This will give you a good idea of how much extra time you have to devote to studying, as well as how much time you devote to other activities. Most students do not realize how much time they "waste" in a day.

Once you have made any changes to your current schedule, try to schedule more time for studying math. You should be spending between two and four hours studying per week for each hour that you spend in the classroom. Try to schedule some time each day, rather than cramming it all into the weekend. Leave some blank periods for flexibility; these periods can be used for emergencies for any of your classes.

10.2 THE QUADRATIC FORMULA

Objectives

1. Derive the quadratic formula.
2. Identify the coefficients a, b, and c of a quadratic equation.
3. Solve quadratic equations by using the quadratic formula.
4. Use the discriminant to determine the number and types of solutions of a quadratic equation.
5. Use the discriminant to determine whether a quadratic expression is factorable.

The Quadratic Formula

Objective 1 Derive the quadratic formula. Completing the square to solve a quadratic equation can be tedious. As an alternative, in this section we will develop and use the **quadratic formula**, which is a numerical formula that gives the solutions of *any* quadratic equation.

We derive the quadratic formula by solving the general equation $ax^2 + bx + c = 0$ (where a, b, and c are real numbers and $a \neq 0$) by completing the square. Recall that the coefficient of the second-degree term must be equal to 1, so the first step is to divide both sides of the equation $ax^2 + bx + c = 0$ by a.

$$ax^2 + bx + c = 0$$

$$x^2 + \frac{b}{a}x + \frac{c}{a} = 0 \qquad \text{Divide both sides by a.}$$

$$x^2 + \frac{b}{a}x = -\frac{c}{a} \qquad \text{Subtract the constant term, } \tfrac{c}{a}.$$

$$x^2 + \frac{b}{a}x + \frac{b^2}{4a^2} = -\frac{c}{a} + \frac{b^2}{4a^2} \qquad \text{Half of } \tfrac{b}{a} \text{ is } \tfrac{b}{2a}. \left(\tfrac{b}{2a}\right)^2 = \tfrac{b^2}{4a^2}. \text{ Add } \tfrac{b^2}{4a^2} \text{ to both sides of the equation.}$$

$$x^2 + \frac{b}{a}x + \frac{b^2}{4a^2} = \frac{b^2 - 4ac}{4a^2} \qquad \text{Simplify the right side of the equation as a single fraction by rewriting } -\tfrac{c}{a} \text{ as } -\tfrac{4ac}{4a^2}.$$

$$\left(x + \frac{b}{2a}\right)^2 = \frac{b^2 - 4ac}{4a^2} \qquad \text{Factor the left side of the equation.}$$

$$\sqrt{\left(x + \frac{b}{2a}\right)^2} = \pm\sqrt{\frac{b^2 - 4ac}{4a^2}} \qquad \text{Take the square root of both sides.}$$

$$x + \frac{b}{2a} = \pm\frac{\sqrt{b^2 - 4ac}}{\sqrt{4a^2}} \qquad \text{Simplify the square root on the left side. Rewrite the right side as the quotient of two square roots.}$$

$$x + \frac{b}{2a} = \pm\frac{\sqrt{b^2 - 4ac}}{2a} \qquad \text{Simplify the square root in the denominator.}$$

$$x = -\frac{b}{2a} \pm \frac{\sqrt{b^2 - 4ac}}{2a} \qquad \text{Subtract } \tfrac{b}{2a}.$$

$$x = \frac{-b \pm \sqrt{b^2 - 4ac}}{2a} \qquad \text{Rewrite as a single fraction.}$$

The Quadratic Formula

> Given a general quadratic equation $ax^2 + bx + c = 0$ (where a, b, and c are real numbers and $a \neq 0$), the quadratic formula tells us that the solutions of this equation are given by
>
> $$x = \frac{-b \pm \sqrt{b^2 - 4ac}}{2a}$$

If we can identify the coefficients a, b, and c in a quadratic equation, then we can find the solutions of the equation by substituting these values for a, b, and c in the quadratic formula.

Identifying the Coefficients to Be Used in the Quadratic Formula

Objective 2   Identify the coefficients a, b, and c of a quadratic equation. The first step in using the quadratic formula is to identify the coefficients a, b, and c. We can do this only after our equation is in standard form: $ax^2 + bx + c = 0$.

EXAMPLE 1   For the quadratic equation, identify a, b, and c.

a) $x^2 - 6x + 8 = 0$

Solution

$a = 1$, $b = -6$, and $c = 8$. Be sure to include the negative sign when identifying coefficients that are negative.

b) $3x^2 - 5x = 7$

Solution

Before identifying our coefficients, we must convert this equation to standard form: $3x^2 - 5x - 7 = 0$. So $a = 3$, $b = -5$, and $c = -7$.

c) $x^2 + 8 = 0$

Solution

In this example, we do not see a first-degree term. In a case like this, $b = 0$. The coefficients that we do see tell us that $a = 1$ and $c = 8$.

> **Quick Check 1**
>
> For the given quadratic equation, identify a, b, and c.
>
> a) $x^2 + 11x - 13 = 0$
> b) $5x^2 - 11 = 9x$
> c) $x^2 = 30$

Solving Quadratic Equations by Using the Quadratic Formula

Objective 3   Solve quadratic equations by using the quadratic formula. Now we will solve several equations using the quadratic formula.

EXAMPLE 2   Solve $x^2 - 6x + 8 = 0$.

Solution

We can use the quadratic formula with $a = 1$, $b = -6$, and $c = 8$.

$$x = \frac{-b \pm \sqrt{b^2 - 4ac}}{2a}$$

$$x = \frac{6 \pm \sqrt{(-6)^2 - 4(1)(8)}}{2(1)}$$

Substitute 1 for a, -6 for b, and 8 for c. When using the formula, it is a good idea to think of the $-b$ in the numerator as the "opposite of b." The opposite of -6 is 6.

$$x = \frac{6 \pm \sqrt{36 - 32}}{2}$$

Simplify each term in the radicand.

$$x = \frac{6 \pm \sqrt{4}}{2}$$

Subtract.

$$x = \frac{6 \pm 2}{2}$$

Simplify the square root.

$\dfrac{6 + 2}{2} = 4$ and $\dfrac{6 - 2}{2} = 2$, so the solution set is $\{4, 2\}$.

Note that the equation in the previous example could have been solved much quicker by factoring $x^2 - 6x + 8$ to be $(x - 4)(x - 2)$ and then solving. Use factoring whenever you can, treating the quadratic formula as an alternative.

EXAMPLE 3   Solve $3x^2 + 8x = -3$.

Solution

To solve this equation, we must rewrite the equation in standard form by collecting all terms on the left side of the equation. In other words, we will rewrite the equation as $3x^2 + 8x + 3 = 0$.

If an equation does not quickly appear to factor, it is wise to proceed directly to the quadratic formula rather than trying to factor an expression that may not be factorable. Here $a = 3$, $b = 8$, and $c = 3$.

$$x = \frac{-8 \pm \sqrt{(8)^2 - 4(3)(3)}}{2(3)}$$

Substitute 3 for a, 8 for b, and 3 for c in the quadratic formula.

$$x = \frac{-8 \pm \sqrt{28}}{6}$$

Simplify the radicand.

$$x = \frac{-8 \pm 2\sqrt{7}}{6}$$

Simplify the square root.

$$x = \frac{\overset{1}{2}(-4 \pm \sqrt{7})}{\underset{3}{6}}$$

Factor the numerator and divide out the common factor.

$$x = \frac{-4 \pm \sqrt{7}}{3}$$

Simplify.

The solution set is $\left\{ \dfrac{-4 + \sqrt{7}}{3}, \dfrac{-4 - \sqrt{7}}{3} \right\}$. These solutions are approximately -0.45 and -2.22, respectively.

Using Your Calculator   Here is the screen that shows how to approximate $\dfrac{-4 + \sqrt{7}}{3}$ and $\dfrac{-4 - \sqrt{7}}{3}$:

```
(-4+√(7))/3
            -.4514162296
(-4-√(7))/3
            -2.215250437
```

Quick Check   2 Solve by using the quadratic formula.

a) $x^2 + 7x - 30 = 0$

b) $7x^2 + 21x = -9$

EXAMPLE 4   Solve $x^2 - 4x = -5$.

Solution

We begin by rewriting the equation in standard form: $x^2 - 4x + 5 = 0$. The quadratic expression in this equation does not factor, so we use the quadratic formula with $a = 1$, $b = -4$, and $c = 5$.

$$x = \dfrac{4 \pm \sqrt{(-4)^2 - 4(1)(5)}}{2(1)}$$ Substitute 1 for a, -4 for b, and 5 for c in the quadratic formula.

$$x = \dfrac{4 \pm \sqrt{-4}}{2}$$ Simplify the radicand.

$$x = \dfrac{4 \pm 2i}{2}$$ Simplify the square root. Be sure to include "i" as we are taking the square root of a negative number.

$$x = \dfrac{\overset{1}{\cancel{2}}(2 \pm i)}{\underset{1}{\cancel{2}}}$$ Factor the numerator and divide out the common factor.

$$x = 2 \pm i$$ Simplify.

Quick Check   3   The solution set is $\{2 + i, 2 - i\}$.

Solve $x^2 - 7x = -19$.

If an equation has coefficients that are fractions, multiply each side of the equation by the LCD to clear the equation of fractions. If we can solve the equation by factoring, it will be easier to factor without the fractions involved. If we cannot solve by factoring, the quadratic formula will be easier to simplify using integers rather than fractions.

EXAMPLE 5 Solve $\frac{1}{2}x^2 - x + \frac{1}{3} = 0$.

Solution

The first step is to clear the fractions by multiplying each side of the equation by the LCD, which in this example is 6.

$$6 \cdot \left(\frac{1}{2}x^2 - x + \frac{1}{3} \right) = 6 \cdot 0 \qquad \text{Multiply both sides of the equation by the LCD.}$$

$$3x^2 - 6x + 2 = 0 \qquad \text{Distribute and simplify.}$$

Now we can use the quadratic formula with $a = 3$, $b = -6$, and $c = 2$.

$$x = \frac{6 \pm \sqrt{(-6)^2 - 4(3)(2)}}{2(3)} \qquad \begin{array}{l}\text{Substitute 3 for } a, -6 \text{ for } b, \text{ and 2 for } c \text{ in the}\\ \text{quadratic formula.}\end{array}$$

$$x = \frac{6 \pm \sqrt{12}}{6} \qquad \text{Simplify the radicand.}$$

$$x = \frac{6 \pm 2\sqrt{3}}{6} \qquad \text{Simplify the square root.}$$

$$x = \frac{\overset{1}{2}(3 \pm \sqrt{3})}{\underset{3}{6}} \qquad \begin{array}{l}\text{Factor the numerator and divide out common}\\ \text{factors.}\end{array}$$

$$x = \frac{3 \pm \sqrt{3}}{3} \qquad \text{Simplify.}$$

The solution set is $\left\{ \dfrac{3 + \sqrt{3}}{3}, \dfrac{3 - \sqrt{3}}{3} \right\}$. These solutions are approximately 1.58 and 0.42, respectively.

> **Quick Check 4**
> Solve $\dfrac{1}{4}x^2 + \dfrac{1}{3}x + \dfrac{1}{2} = 0$.

EXAMPLE 6

In addition to clearing any fractions, make sure that the coefficient (a) of the second-degree term is positive. If this term is negative, you can collect all terms on the other side of the equation or multiply each side of the equation by negative 1.

$$\text{Solve } -2x^2 + 11x + 6 = 0.$$

Solution

We begin by rewriting the equation in such a way that a is positive.

$$0 = 2x^2 - 11x - 6 \qquad \text{Collect all terms on the right side of the equation.}$$

Often, when the leading coefficient is not 1, factoring can be difficult or time consuming if the expression is factorable at all. In such a situation, go right to the quadratic formula. In this example, we can use the quadratic formula with $a = 2$, $b = -11$, and $c = -6$.

$$x = \frac{11 \pm \sqrt{(-11)^2 - 4(2)(-6)}}{2(2)} \qquad \begin{array}{l}\text{Substitute 2 for } a, -11 \text{ for } b, \text{ and } -6 \text{ for } c\\ \text{in the quadratic formula.}\end{array}$$

$$x = \frac{11 \pm \sqrt{169}}{4} \qquad \text{Simplify the radicand.}$$

$$x = \frac{11 \pm 13}{4} \qquad \text{Simplify the square root.}$$

> **Quick Check 5**
> Solve
> $-6x^2 - 7x + 20 = 0$.

$\frac{11 + 13}{4} = 6$ and $\frac{11 - 13}{4} = -\frac{1}{2}$, so the solution set is $\left\{ 6, -\frac{1}{2} \right\}$.

Although the quadratic formula can be used to solve *any* quadratic equation, it does not always provide the most efficient way to solve a particular equation. We should always check to see whether factoring or extracting square roots can be used before using the quadratic formula.

Using the Discriminant to Determine the Number and Type of Solutions of a Quadratic Equation

Objective 4 **Use the discriminant to determine the number and types of solutions of a quadratic equation.** In the quadratic formula, the expression $b^2 - 4ac$ is called the **discriminant**. The discriminant can be used to give us some information about our solutions. If the discriminant is negative ($b^2 - 4ac < 0$), then the equation has two nonreal complex solutions. This is because we take the square root of a negative number in the quadratic formula. It is important to know that an equation has no solutions that are real numbers when working on an applied problem. (It will also be important when graphing quadratic equations, which is covered later in this chapter.) The following chart summarizes what the discriminant tells us about the number and type of solutions of an equation:

Solutions of a Quadratic Equation Based on the Discriminant

$b^2 - 4ac$	Number and Type of Solutions
Negative	Two Nonreal Complex Solutions
Zero	One Real Solution
Positive	Two Real Solutions

EXAMPLE 7 For the quadratic equation, use the discriminant to determine the number and type of solutions.

a) $x^2 - 6x - 16 = 0$

Solution

$$(-6)^2 - 4(1)(-16) = 100 \qquad \text{Substitute 1 for } a, -6 \text{ for } b, \text{ and } -16 \text{ for } c.$$

Since the discriminant is positive, this equation has two real solutions.

b) $x^2 + 36 = 0$

Solution

$$(0)^2 - 4(1)(36) = -144 \qquad \text{Substitute 1 for } a, 0 \text{ for } b, \text{ and } 36 \text{ for } c.$$

The discriminant is negative. This equation has two nonreal complex solutions.

c) $x^2 + 10x + 25 = 0$

Solution

$$(10)^2 - 4(1)(25) = 0 \qquad \text{Substitute 1 for } a, 10 \text{ for } b, \text{ and } 25 \text{ for } c.$$

Since the discriminant equals 0, this equation has one real solution.

Quick Check 6
For the given quadratic equation, use the discriminant to determine the number and type of solutions.

a) $x^2 + 18x - 63 = 0$
b) $x^2 + 5x + 42 = 0$
c) $x^2 - 20x + 100 = 0$

Using the Discriminant to Determine Whether a Quadratic Expression Is Factorable

Objective 5 **Use the discriminant to determine whether a quadratic expression is factorable.** The discriminant can also be used to tell us whether a quadratic expression is factorable. If the discriminant is equal to 0 or a positive number that is a perfect square (1, 4, 9, etc.), then the expression is factorable.

EXAMPLE ▶ 8   For the quadratic expression, use the discriminant to determine whether the expression can be factored.

a) $x^2 - 6x - 27$

Solution

$$(-6)^2 - 4(1)(-27) = 144 \qquad \text{Substitute 1 for } a, -6 \text{ for } b, \text{ and } -27 \text{ for } c.$$

The discriminant is a perfect square ($\sqrt{144} = 12$), so the expression can be factored.

$$x^2 - 6x - 27 = (x - 9)(x + 3)$$

b) $2x^2 + 7x + 4$

Solution

$$(7)^2 - 4(2)(4) = 17 \qquad \text{Substitute 2 for } a, 7 \text{ for } b, \text{ and } 4 \text{ for } c.$$

The discriminant is not a perfect square, so the expression is not factorable.

Quick Check 7
For the given quadratic expression, use the discriminant to determine whether the expression can be factored.

a) $x^2 + 14x + 12$
b) $5x^2 - 36x - 32$

EXERCISES *10.2* ❯

Solve by using the quadratic formula.

1. $x^2 - 5x - 36 = 0$

2. $x^2 + 4x - 45 = 0$

3. $x^2 - 4x + 2 = 0$

4. $x^2 + 10x + 13 = 0$

5. $x^2 + x + 7 = 0$

6. $x^2 - 3x + 18 = 0$

7. $x^2 + 6x = 0$

8. $x^2 + 8x + 10 = 0$

9. $x^2 + 3x - 10 = 0$

10. $x^2 + 5x + 12 = 0$

11. $x^2 + 5x + 6 = 0$

12. $x^2 + 11x + 28 = 0$

13. $x^2 - 4 = 0$

14. $2x^2 + 3x - 35 = 0$

15. $x^2 - 15 = 2x$

16. $x^2 + 8x = 12$

17. $x^2 - 3x = 9$

18. $x^2 = 7x$

19. $-4 = 19x - 5x^2$

20. $4x - x^2 = 3$

21. $x^2 - 6x + 9 = 0$

22. $x^2 + 10x + 25 = 0$

23. $x^2 - 24 = 0$

F
O
R

E
X
T
R
A

H
E
L
P

MyMathLab
MyMathLab

Math*XP*
MathXL

Interactmath.com

MathXL
Tutorials on CD

DVT CD
Videotape

Tutor
Center

Addison-Wesley
Math Tutor Center

Student Solutions
Manual

24. $x^2 + 49 = 0$

25. $x(x - 4) + 3x = 20$

26. $(2x + 1)(x - 3) = -9$

27. $x^2 - \dfrac{1}{5}x + \dfrac{3}{4} = 0$

28. $\dfrac{2}{3}x^2 - \dfrac{3}{5}x + \dfrac{1}{4} = 0$

29. $-x^2 + 7x - 12 = 0$

30. $-2x^2 + 15x = 8$

For the following quadratic equations, use the discriminant to determine the number and type of solutions.

31. $x^2 + 12x - 30 = 0$

32. $x^2 - 9x + 21 = 0$

33. $2x^2 - 3x + 5 = 0$

34. $25x^2 - 20x + 4 = 0$

35. $x^2 + \dfrac{2}{5}x + \dfrac{5}{6} = 0$

36. $x^2 - 5x - 9 = 0$

37. $9x^2 - 12x + 4 = 0$

38. $x^2 - \dfrac{2}{3}x + \dfrac{1}{9} = 0$

Use the discriminant to determine whether the following quadratic expressions are factorable. If the expression can be factored, write "factorable." Otherwise, write "prime."

39. $x^2 - 8x + 19$

40. $x^2 + 6x - 40$

41. $x^2 + 15x + 54$

42. $x^2 - 11x + 27$

43. $3x^2 - 5x - 8$

44. $2x^2 - 15x - 21$

45. $5x^2 + 29x + 20$

46. $10x^2 + 11x + 3$

Solve the following quadratic equations using the most efficient technique (factoring, extracting square roots, completing the square, or quadratic formula).

47. $x^2 - 5x - 15 = 0$

48. $x^2 - 68 = 0$

49. $3x^2 + 2x - 1 = 0$

50. $x^2 - 20x + 91 = 0$

51. $(5x - 4)^2 = 36$

52. $7x(8x - 3) + 4(8x - 3) = 0$

53. $2x^2 + 8x = -9$

54. $x^2 + 179x = 0$

55. $x^2 + 324 = 0$

56. $x^2 + \dfrac{3}{5}x - \dfrac{1}{12} = 0$

57. $6x^2 - 29x + 28 = 0$

58. $x^2 + 8x - 9 = 0$

59. $3(2x + 1)^2 - 7 = 23$

60. $5x^2 + 11x - 9 = 0$

61. $x^2 - 9x - 21 = 0$

62. $4x^2 - 25 = 0$

63. $16x^2 - 24x + 9 = 0$

64. $x^2 - 15x + 50 = 0$

65. $x^2 + x + 20 = 0$

66. $x^2 + 13x + 36 = 0$

67. $x^2 - 4x - 2 = 0$

68. $3x^2 - 2x - 16 = 0$

69. $x^2 - 6x + 10 = 0$

70. $(x - 8)^2 + 13 = 134$

71. $\dfrac{3}{4}x^2 + \dfrac{2}{3}x - \dfrac{1}{2} = 0$

72. $x^2 + 3x - 18 = 0$

73. $2x(3x + 7) - 5(3x + 7) = 0$

74. $(2x + 3)^2 - 10 = 71$

75. $2(2x - 9)^2 + 13 = 77$

76. $x^2 + x - 72 = 0$

77. $x^2 + 3x - 4 = 0$

78. $x^2 + 10x + 21 = 0$

79. $x^2 - 10x + 18 = 0$

80. $x^2 + 2x + 4 = 0$

81. $x^2 - 16x + 63 = 0$

82. $4x^2 - 12x - 11 = 0$

83. Derive a formula that would provide the general solution of the linear equation $ax + b = 0$, where a and b are real numbers and $a \neq 0$.

Answer in complete sentences.

84. Explain how to use the quadratic formula to solve a quadratic equation. Use an example to illustrate the process.

85. When both methods are possible, which method of solving quadratic equations do you prefer: solving by factoring or the quadratic formula? Explain your answer.

86. Explain how the discriminant tells us whether an equation has two real solutions, one real solution, or two complex solutions.

Study Tip **REVISITED**   The best time to study new material is as soon as possible after the class period. After you have made a schedule of your current commitments, look for a block of time that is as close to your class period as possible. You can study the new material on campus if necessary. Try to study each day at the same time. It is ideal to begin each study session by reworking your notes; then move on to attempting the homework exercises.

If possible, establish a second study period during the day that will be used for review purposes. This second study period should take place later in the day and can be used to review homework or notes or even to read ahead for the next class period.

10.3
APPLICATIONS USING QUADRATIC EQUATIONS

Objectives

1 **Solve applied geometric problems.**
2 **Solve problems by using the Pythagorean theorem.**
3 **Solve applied problems by using the Pythagorean theorem.**
4 **Solve projectile motion problems.**

In this section, we will learn how to solve applied problems resulting in quadratic equations.

Solving Applied Geometric Problems

Objective 1 **Solve applied geometric problems.** Problems involving the area of a geometric figure often lead to quadratic equations, as area is measured in square units. Here are some useful area formulas:

Figure	Square	Rectangle	Triangle	Circle
Dimensions	Side s	Length l, Width w	Base b, Height h	Radius r
Area	$A = s^2$	$A = l \cdot w$	$A = \dfrac{1}{2}bh$	$A = \pi r^2$

EXAMPLE 1   The length of a rectangle is 5 cm less than its width. The area is 36 cm². Find the length and the width of the rectangle.

Solution

In this problem, the unknown quantities are the length and the width, while we know that the area is 36 cm². Since the length is given in terms of the width, a wise choice is to represent the width of the rectangle by x. Since the length is 5 cm less than the width, it can be represented by $x - 5$. This information is summarized in the following table:

Unknowns	***Known***	
Length: $x - 5$ Width: x	Area: 36 cm²	

Since the area of a rectangle is equal to its length times its width, the equation we need to solve is $(x - 5)x = 36$.

$$(x - 5)x = 36$$
$$x^2 - 5x = 36 \qquad \text{Distribute.}$$
$$x^2 - 5x - 36 = 0 \qquad \text{Rewrite in standard form.}$$
$$(x - 9)(x + 4) = 0 \qquad \text{Factor.}$$
$$x - 9 = 0 \quad \text{or} \quad x + 4 = 0 \qquad \text{Set each factor equal to 0.}$$
$$x = 9 \quad \text{or} \quad x = -4 \qquad \text{Solve.}$$

Look back to the table of unknowns. If $x = -4$, then the length is -9 cm and the width is -4 cm, which is not possible. The solutions derived from $x = -4$ are omitted. If $x = 9$, then the length is $9 - 5$ or 4 cm, while the width is 9 cm.

Length: $x - 5 = 9 - 5 = 4$
Width: $x = 9$

Now put the answer in a complete sentence with the proper units. The length of the rectangle is 4 cm and the width is 9 cm.

Quick Check 1

The base of a triangle is 2 inches longer than 3 times its height. If the area of the triangle is 60 square inches, find the base and height of the triangle.

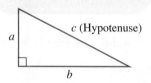

Solving Problems Using the Pythagorean Theorem

Objective 2   Solve problems by using the Pythagorean theorem. Other applications of geometry that lead to quadratic equations involve the Pythagorean theorem, which is an equation that relates the length of the three sides of a right triangle.

The side opposing the right angle is called the **hypotenuse** and is labeled c in the figure at the left. The other two sides that form the right angle are called the **legs** of the triangle and are labeled a and b. (It makes no difference which is a and which is b.)

The Pythagorean theorem states that, for any right triangle whose hypotenuse has length c and whose other two legs have lengths a and b, respectively, $a^2 + b^2 = c^2$.

EXAMPLE 2   A right triangle has a hypotenuse that measures 18 inches, and one of its legs is 6 inches long. Find the length of the other leg, to the nearest tenth of an inch.

Solution

In this problem, the length of one of the legs is unknown. We can label the unknown leg as either a or b.

Unknowns	Known
a	b: 6 in.
	Hypotenuse (c): 18 in.

$$a^2 + 6^2 = 18^2 \qquad \text{Substitute 6 for } b \text{ and 18 for } c \text{ in } a^2 + b^2 = c^2.$$
$$a^2 + 36 = 324 \qquad \text{Square 6 and 18.}$$
$$a^2 = 288 \qquad \text{Subtract 36.}$$
$$\sqrt{a^2} = \pm\sqrt{288} \qquad \text{Solve by extracting square roots.}$$
$$a = \pm 12\sqrt{2} \qquad \text{Simplify the square root.}$$

Since the length of a leg must be a positive number, we are only concerned about $12\sqrt{2}$, which rounds to be 17.0 inches. The length of the other leg is approximately 17.0 inches.

Quick Check **2**

One leg of a right triangle measures 5 inches, while the hypotenuse measures 11 inches. Find, to the nearest hundredth of an inch, the length of the other leg of the triangle.

Applications of the Pythagorean Theorem

Objective **3**   **Solve applied problems by using the Pythagorean theorem.**
Now we turn our attention to solving applied problems using the Pythagorean theorem. In these problems, we begin by drawing a picture of the situation. We must be able to identify a right triangle in our figure in order to apply the Pythagorean theorem.

EXAMPLE **3**   A 5-foot ladder is leaning against a wall. If the bottom of the ladder is 3 feet from the base of the wall, how high up the wall is the top of the ladder?

Solution

The ladder, the wall, and the ground form a right triangle, with the ladder being the hypotenuse. In this problem the height of the wall, which is the length of one of the legs in the right triangle, is unknown.

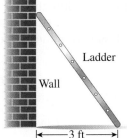

Unknowns	*Known*
a	b: 3 ft
	Hypotenuse (c): 5 ft

Quick Check **3**

An 8-foot ladder is leaning against a wall. If the bottom of the ladder is 2 feet from the base of the wall, how high up the wall is the top of the ladder? (Round to the nearest tenth of a foot.)

$$a^2 + 3^2 = 5^2 \qquad \text{Substitute 3 for } b \text{ and 5 for } c \text{ in } a^2 + b^2 = c^2.$$
$$a^2 + 9 = 25 \qquad \text{Square 3 and 5.}$$
$$a^2 = 16 \qquad \text{Subtract 9.}$$
$$\sqrt{a^2} = \pm\sqrt{16} \qquad \text{Solve by extracting square roots.}$$
$$a = \pm 4 \qquad \text{Simplify the square root.}$$

Again, the negative solution does not make sense in this problem. The ladder is resting at a point on the wall that is 4 feet above the ground.

EXAMPLE **4**   The Modesto airport is located 120 miles north and 50 miles west of the Visalia airport. If a plane flies directly from Visalia to Modesto, how many miles is the flight?

Solution

The directions of north and west form a 90-degree angle, so our picture shows us that we have a right triangle whose hypotenuse (the direct distance from Visalia to Modesto) is unknown.

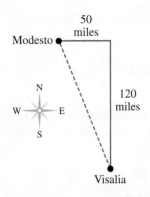

Modesto ●
50 miles

120 miles

N
W — E
S

Visalia

Unknowns	*Known*
Hypotenuse: c	a: 120 miles
	b: 50 miles

$$120^2 + 50^2 = c^2$$ Substitute 120 for a and 50 for b in $a^2 + b^2 = c^2$.
$$14{,}400 + 2500 = c^2$$ Square 120 and 50.
$$16{,}900 = c^2$$ Simplify.
$$\pm\sqrt{16{,}900} = \sqrt{c^2}$$ Solve by extracting square roots.
$$\pm 130 = c$$ Simplify the square root.

The negative solution does not make sense in this problem. The direct distance from the Visalia airport to the Modesto airport is 130 miles.

Projectile Motion Problems

Objective 4   Solve projectile motion problems. Another application that leads to a quadratic equation involves the height in feet of an object propelled into the air after t seconds. Recall that the height, in feet, of a projectile after t seconds can be found using the function $h(t) = -16t^2 + v_0t + s$, where v_0 is the initial velocity of the projectile and s is the initial height.

Height of a Projectile

$$h(t) = -16t^2 + v_0t + s$$
t: Time (seconds)
v_0: Initial Velocity (feet/second)
s: Initial Height (feet)

EXAMPLE 5   A rock is thrown from ground level at a speed of 48 feet/second.

a) How long will it take until the rock lands on the ground?

Solution

The initial velocity of the rock is 48 feet per second, so $v_0 = 48$. Since the rock is thrown from ground level, the initial height is 0 feet. The function for the height of the rock after t seconds is $h(t) = -16t^2 + 48t$.

   The rock's height when it lands on the ground is 0 feet, so we set the function equal to 0 and solve for the time t in seconds.

$$-16t^2 + 48t = 0$$ Set the function equal to 0.
$$0 = 16t^2 - 48t$$ Collect all terms on the right side of the equation so the leading coefficient is positive.

$$0 = 16t(t - 3)$$ Factor out the GCF ($16t$).
$$16t = 0 \quad \text{or} \quad t - 3 = 0$$ Set each variable factor equal to 0.
$$t = 0 \quad \text{or} \quad t = 3$$ Solve each equation.

Quick Check 4

Cassie's backyard is in the shape of a rectangle whose dimensions are 70 feet by 240 feet. She needs a hose that will extend from one corner of her yard to the corner that is diagonally opposite to it. How long does the hose have to be?

The time of 0 seconds corresponds to the precise moment that the rock was thrown and does not represent the time required to land on the ground. Our solution is 3 seconds.

b) When will the rock be at a height of 32 feet?

Solution

This problem is similar to part (a), except that we will set the function $h(t)$ equal to 32 rather than 0.

Quick Check **5**

A rock is thrown upward from ground level at an initial velocity of 80 feet per second.

a) After how many seconds will the rock land on the ground?
b) When will the rock be at a height of 96 feet?

$$-16t^2 + 48t = 32 \qquad \text{Set the function equal to 32.}$$
$$0 = 16t^2 - 48t + 32 \qquad \text{Collect all terms on the right side of the equation.}$$
$$0 = 16(t^2 - 3t + 2) \qquad \text{Factor out the GCF (16).}$$
$$0 = 16(t - 1)(t - 2) \qquad \text{Factor the trinomial.}$$
$$t - 1 = 0 \quad \text{or} \quad t - 2 = 0 \qquad \text{Set each variable factor equal to 0.}$$
$$t = 1 \quad \text{or} \quad t = 2 \qquad \text{Solve each equation.}$$

Do both of these answers make sense? The rock first passes a height of 32 feet after 1 second while it is on its way up. It is at a height of 32 feet after two seconds while it is on its way down. The rock is 32 feet above the ground after 1 second and again after 2 seconds.

EXAMPLE 6 A golf ball is launched by a slingshot from a platform that is 95 feet high at an initial velocity of 88 feet per second.

a) When will the golf ball be at a height of 175 feet? (Round to the nearest hundredth of a second.)

Solution

The initial velocity of the golf ball is 88 feet per second, so $v_0 = 88$. Since the golf ball was launched from a platform 95 feet high, the initial height $s = 95$. The function for the height of the golf ball after t seconds is $h(t) = -16t^2 + 88t + 95$.

To find when the golf ball is at a height of 175 feet, we set the function equal to 175 and solve for the time t in seconds.

$$-16t^2 + 88t + 95 = 175 \qquad \text{Set the function equal to 175.}$$
$$0 = 16t^2 - 88t + 80 \qquad \text{Collect all terms on the right side of the equation.}$$
$$0 = 8(2t^2 - 11t + 10) \qquad \text{Factor out the GCF (8).}$$
$$0 = 2t^2 - 11t + 10 \qquad \text{Divide both sides by 8.}$$

This trinomial does not factor, so we will use the quadratic formula with $a = 2$, $b = -11$, and $c = 10$ to solve this equation.

$$t = \frac{11 \pm \sqrt{(-11)^2 - 4(2)(10)}}{2(2)} \qquad \text{Substitute 2 for } a, -11 \text{ for } b, \text{ and } 10 \text{ for } c.$$

$$t = \frac{11 \pm \sqrt{41}}{4} \qquad \text{Simplify the radicand and the denominator.}$$

We use a calculator to approximate these solutions.

$$\frac{11 + \sqrt{41}}{4} \approx 4.35 \qquad \frac{11 - \sqrt{41}}{4} \approx 1.15$$

The golf ball is at a height of 175 feet after approximately 1.15 seconds and again after approximately 4.35 seconds.

b) Will the golf ball ever reach a height of 250 feet? If so, when will it be at this height?

Solution

We begin by setting the function equal to 250 and solving for t.

$$-16t^2 + 88t + 95 = 250 \qquad \text{Set the function equal to 250.}$$
$$0 = 16t^2 - 88t + 155 \qquad \text{Collect all terms on the right side of the equation.}$$

This trinomial does not have any common factors other than 1, so we will use the quadratic formula with $a = 16$, $b = -88$, and $c = 155$ to solve this equation.

$$t = \frac{88 \pm \sqrt{(-88)^2 - 4(16)(155)}}{2(16)} \qquad \text{Substitute 16 for } a, -88 \text{ for } b, \text{ and 155 for } c.$$

$$t = \frac{88 \pm \sqrt{-2176}}{32} \qquad \text{Simplify the radicand and the denominator.}$$

Since the radicand is negative, this equation has no real number solutions. The golf ball does not reach a height of 250 feet.

> **A Word of Caution**   When using the quadratic formula to solve an applied problem, a negative radicand indicates that there are no real solutions to this problem.

Quick Check 6

A projectile is launched from the top of a building 40 feet high at an initial velocity of 36 feet per second.

a) When will the projectile be at a height of 50 feet? (Round to the nearest hundredth of a second.)

b) Will the projectile ever reach a height of 80 feet? If so, when will it be at this height?

EXERCISES *10.3*

For all problems, approximate to the nearest tenth when necessary.

1. The area of a square is 81 square meters. Find the length of a side of the square.

2. The area of a square is 128 square inches. Find the length of a side of the square.

3. The area of a circle is 32π square feet. Find the radius of the circle.

4. The area of a circle is 49π square meters. Find the radius of the circle.

5. The area of a circle is 100 square inches. Find the radius of the circle.

6. The area of a circle is 60 square centimeters. Find the radius of this circle.

7. The length of a rectangle is 2 inches less than its width. If the area of the rectangle is 80 square inches, find the length and width of the rectangle.

8. The width of a rectangle is 7 inches less than three times the length. If the area of the rectangle is 40 square inches, find the length and width of the rectangle.

9. The width of a rectangle is twice its length. If the area of the rectangle is 98 square inches, find the length and width of the rectangle.

10. The length of a rectangle is 1 foot more that five times its width. If the area of the rectangle is 328 square feet, find the length and width of the rectangle.

11. The length of a rectangle is 1 inch less than seven times its width. If the area of the rectangle is 50 square inches, find the dimensions of the rectangle. Round to the nearest tenth of an inch.

12. The length of a rectangle is half of its width. If the area of the rectangle is 300 square inches, find the dimensions of the rectangle. Round to the nearest tenth of an inch.

13. The base of a triangle is 5 inches more than its height. If the area of the triangle is 42 square inches, find the base and height of the triangle.

14. The height of a triangle is 1 foot less than three times its base. If the area of the triangle is 22 square feet, find the base and height of the triangle.

15. The height of a triangle is 1 inch more than twice its base. If the area of the triangle is 15 square inches, find the base and height of the triangle. Round to the nearest tenth of an inch.

16. The height of a triangle is 7 inches less than its base. If the area of the triangle is 32 square inches, find the base and height of the triangle. Round to the nearest tenth of an inch.

17. A rectangular photograph has an area of 40 square inches. If the length of the photograph is 3 inches more than its height, find the dimensions of the photograph.

18. The area of a rectangular poster is 1400 square centimeters. If the length of the poster is 5 centimeters less than its height, find the dimensions of the poster.

19. The length of a rectangular quilt is twice its width, and the area of the quilt is 12.5 square feet. Find the dimensions of the quilt.

20. The length of a rectangular room is 3 feet more than twice its width. If the area of the room is 160 square feet, find the dimensions of the room. Round to the nearest tenth of a foot.

21. The length of a rectangular table is 16 inches less than its width. If the area of the table is 540 square inches, find the dimensions of the table. Round to the nearest tenth of an inch.

22. Steve has a rectangular lawn, and the width of the lawn is 25 feet more than the length of the lawn. If the area of the lawn is 7000 square feet, find the dimensions of the lawn. Round to the nearest tenth of a foot.

23. A kite is in the shape of a triangle. The base of the kite is twice its height. If the area of the kite is 256 square inches, find the base and height of the kite.

24. A hang glider is triangular in shape. If the base of the hang glider is 2 feet more than twice its height, and the area of the hang glider is 30 square feet, find the hang glider's base and height.

25. The sail on a sailboat is shaped like a triangle, with an area of 46 square feet. The height of the sail is 8 feet more than the base of the sail. Find the base and height of the sail. Round to the nearest tenth of a foot.

26. A kite is in the shape of a triangle and is made with 250 square inches of material. The base of the kite is 20 inches longer than the height of the kite. Find the base and the height of the kite. Round to the nearest tenth of an inch.

27. The two legs of a right triangle are 7 inches and 24 inches, respectively. Find the hypotenuse of the triangle.

28. The two legs of a right triangle are 6 inches and 11 inches, respectively. Find the hypotenuse of the triangle, rounded to the nearest tenth of an inch.

29. A right triangle with a hypotenuse of 10 centimeters has a leg that measures 5 centimeters. Find the length of the other leg, rounded to the nearest tenth of a centimeter.

30. A right triangle has a leg that measures 15 inches and a hypotenuse that measures 30 inches. Find the length of the other leg, rounded to the nearest tenth of an inch.

31. Patti drove 80 miles to the west, and then drove 60 miles south. How far is she from her starting location?

32. Victor flew to a city that was 700 miles north and 2400 miles east of his starting point. How far did he fly to reach this city?

33. George is casting a shadow on the ground. If George is 2 feet shorter than the length of the shadow on the ground and the tip of the shadow is 10 feet from the top of George's head, how tall is George?

34. The base of a 15-foot ladder is leaning against the wall. If the distance between the base of the ladder and the wall is 3 feet less than the height of the top of the ladder on the wall, how high up the wall does the ladder reach?

35. A guy wire 40 feet long runs from the top of a pole to a spot on the ground. If the height of the pole is 5 feet more than the distance from the base of the pole to the spot where the guy wire is anchored, how tall is the pole? Round to the nearest tenth of a foot.

40 ft

36. A 12-foot ramp leads to a doorway. The height of the doorway is 8 feet less than the horizontal distance covered by the ramp. How high is the doorway? Round to the nearest tenth of a foot.

37. A rectangular computer screen is 13 inches wide and 10 inches high. Find the length of its diagonal. Round to the nearest tenth of an inch.

38. The bases on a baseball diamond form a square whose side is 90 feet. How far is it from home plate to second base? Round to the nearest tenth of a foot.

39. The length of a rectangular quilt is 1 foot more than its width. If the diagonal of the quilt is 6.5 feet, find the length and width of the quilt. Round to the nearest tenth of a foot.

40. The diagonal of a rectangular table is 9 feet, and the length of the table is 5 feet more than its width. Find the length and width of the table. Round to the nearest tenth of a foot.

Use the following fact about right triangles for Exercises 41–44: the legs of a right triangle represent the base and height of that triangle.

41. The height of a right triangle is 3 feet less than the base of the triangle. The area of the triangle is 54 square feet.

a) Use this information to find the base and height of this triangle.

b) Find the hypotenuse of the triangle.

42. The base of a right triangle is 4 feet less than twice the height of the triangle. The area of the triangle is 24 square feet.

a) Use this information to find the base and height of this triangle.

b) Then use that information to find the hypotenuse of the triangle.

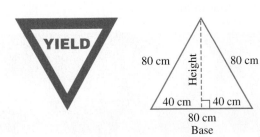

43. A yield sign is in the shape of an equilateral triangle, with each side measuring 80 centimeters.

a) Use the Pythagorean theorem to find the height of the triangle. Round to the nearest tenth of a centimeter.

b) Find the area of the sign.

44. A farmer fenced in a corral in the shape of an equilateral triangle, with each side measuring 30 feet.

30 ft   Height   30 ft

15 ft   15 ft

30 ft
Base

  a) Use the Pythagorean theorem to find the height of the triangle. Round to the nearest tenth of a foot.

  b) Find the area of the corral.

For Exercises 45 through 54, use the function
$$h(t) = -16t^2 + v_0 t + s.$$

45. An object is launched from the ground with an initial speed of 128 feet per second. How long will it take until the object lands on the ground?

46. An object is launched from a platform 160 feet above the ground with an initial speed of 48 feet per second. How long will it take until it lands on the ground?

47. Nick is standing on a cliff above a beach. He throws a rock upward from a height 90 feet above the beach at a speed of 70 feet per second. How long will it take until the rock lands on the beach? Round to the nearest tenth of a second.

48. Jan is standing on the roof of a building. She launches a water balloon upwards from a height of 20 feet at an initial speed of 44 feet per second. How long will it take until it lands on the ground? Round to the nearest tenth of a second.

49. An object is launched from ground level with an initial velocity of 64 feet per second. At what time(s) is the object 48 feet above the ground?

50. An object is launched from ground level with an initial speed of 75 feet per second. At what time(s) is the object 70 feet above the ground? Round to the nearest tenth of a second.

51. An object is launched from the top of a building 80 feet high, with an initial velocity of 100 feet per second. At what time(s) is the object 200 feet above the ground? Round to the nearest tenth of a second.

52. An object is launched from a cliff 240 feet above a beach, with an initial velocity of 80 feet per second. At what time(s) is the object 300 feet above the ground? Round to the nearest tenth of a second.

53. A boy throws a rock with an initial velocity of 16 feet per second at a streetlight that is 25 feet above the ground. Does the rock ever reach the height of the streetlight? Explain why in your own words.

54. A football player kicks a football to start the game. If it took 4 seconds for the ball to land on the ground, find the initial velocity of the football.

Answer in complete sentences.

55. Write a word problem whose solution is "The length of the rectangle is 14 feet, and the width is 9 feet." The problem must lead to a quadratic equation.

56. Write a word problem associated with the equation $60^2 + b^2 = 90^2$. Explain how you created the problem. Solve your problem, explaining each step.

57. Write a word problem associated with the equation $-16t^2 + 80t + 42 = 138$. Explain how you created the problem. Solve your problem, explaining each step.

58. Write a word problem whose solution is "The object will land on the ground after 10 seconds." Explain how you created the problem.

Study Tip REVISITED   When should you study math? Try to schedule your math study sessions for the time of day when you feel that you are at your sharpest. If you feel that you are most alert in the mornings, then you should reserve as much time as possible in the mornings to study math. Other students will find that it is better for them to study math in the middle of the day, in the evening, or at night. If you constantly get tired while studying late at night, change your study schedule so that you can study math before you get tired.

10.4
EQUATIONS THAT ARE QUADRATIC IN FORM

Objectives

1 **Solve equations by making a *u*-substitution.**
2 **Solve radical equations.**
3 **Solve rational equations.**
4 **Solve work-rate problems.**

In this section, we will learn how to solve several types of equations that are **quadratic in form**. For example, $x^4 - 13x^2 + 36 = 0$ is not a quadratic equation, but if we rewrite it as $(x^2)^2 - 13(x^2) + 36 = 0$, then we can see that it looks like a quadratic equation.

Solving Equations by Making a u-Substitution

Objective 1 Solve equations by making a *u*-substitution. One approach to solving equations that are quadratic in form is to use a ***u*-substitution**. We substitute the variable *u* for an expression, such as x^2, so that the resulting equation is a quadratic equation in *u*. In other words, the equation can be rewritten in the form $au^2 + bu + c = 0$. We can then solve this quadratic equation using the methods of Sections 10.1 and 10.2 (factoring, extracting square roots, completing the square, and quadratic formula). After solving this equation for *u*, we replace *u* by the expression it previously substituted for and then solve the resulting equations for the original variable.

EXAMPLE 1 Solve $x^4 - 13x^2 + 36 = 0$.

Solution

Let $u = x^2$. We can then replace x^2 in the original equation by *u*, and we can replace x^4 by u^2. The resulting equation will be $u^2 - 13u + 36 = 0$, which is quadratic.

$$x^4 - 13x^2 + 36 = 0$$
$$u^2 - 13u + 36 = 0 \qquad \text{Substitute } u \text{ for } x^2.$$
$$(u - 4)(u - 9) = 0 \qquad \text{Factor.}$$
$$u - 4 = 0 \quad \text{or} \quad u - 9 = 0 \qquad \text{Set each factor equal to 0.}$$
$$u = 4 \quad \text{or} \quad u = 9 \qquad \text{Solve.}$$

Now we replace *u* by x^2 and solve the resulting equations for *x*.

$$u = 4 \quad \text{or} \quad u = 9$$
$$x^2 = 4 \quad \text{or} \quad x^2 = 9 \qquad \text{Substitute } x^2 \text{ for } u.$$
$$\sqrt{x^2} = \pm\sqrt{4} \quad \text{or} \quad \sqrt{x^2} = \pm\sqrt{9} \qquad \text{Solve by taking the square root of both sides of the equation.}$$
$$x = \pm 2 \quad \text{or} \quad x = \pm 3 \qquad \text{Simplify the square root.}$$

Quick Check 1 The solution set is $\{2, -2, 3, -3\}$.
Solve $x^4 - x^2 - 12 = 0$.

> ***A Word of Caution*** When solving an equation by using a *u*-substitution, *do not stop after solving for u.* You must solve for the variable in the original equation.

The challenge is determining when a *u*-substitution will be helpful and determining what to let *u* represent. Look for an equation in which the variable part of the first term is the square of the variable part of a second term; in other words, its exponent is twice the exponent of the second term. We can then let u represent the variable part with the smaller exponent.

EXAMPLE 2   Find the u-substitution that will convert the equation to a quadratic equation.

a) $x - 7\sqrt{x} - 30 = 0$

Solution

Let $u = \sqrt{x}$. This will allow us to replace $\sqrt{x}$ by u and x by u^2, since $u^2 = (\sqrt{x})^2 = x$. The resulting equation will be $u^2 - 7u - 30 = 0$, which is quadratic.

b) $x^{2/3} + 9x^{1/3} + 8 = 0$

Solution

Let $u = x^{1/3}$. We can then replace $x^{2/3}$ by u^2, since $(x^{1/3})^2 = x^{2/3}$. The resulting equation will be $u^2 + 9u + 8 = 0$.

c) $(x^2 + 6x)^2 + 13(x^2 + 6x) + 40 = 0$

Solution

Let $u = x^2 + 6x$ as this is the expression that is being squared. The resulting equation will be $u^2 + 13u + 40 = 0$.

Quick Check   2   **Find the u-substitution that will convert the given equation to a quadratic equation.**

a) $x + 11\sqrt{x} - 26 = 0$
b) $2x^{2/3} - 17x^{1/3} + 8 = 0$
c) $(x^2 - 4x)^2 - 9(x^2 - 4x) - 36 = 0$

EXAMPLE 3   Solve $x + 3\sqrt{x} - 10 = 0$.

Solution

Let $u = \sqrt{x}$. The resulting equation is $u^2 + 3u - 10 = 0$, which is a quadratic equation solvable by factoring. If we could not use factoring, then we would have to use the quadratic formula to solve for u.

$$x + 3\sqrt{x} - 10 = 0$$

$u^2 + 3u - 10 = 0$			Replace $\sqrt{x}$ by u.
$(u - 2)(u + 5) = 0$			Factor.
$u - 2 = 0$	or	$u + 5 = 0$	Set each factor equal to 0.
$u = 2$	or	$u = -5$	Solve for u.
$\sqrt{x} = 2$	or	$\sqrt{x} = -5$	Replace u by
$(\sqrt{x})^2 = 2^2$	or	$(\sqrt{x})^2 = (-5)^2$	Square both sides of the equation.
$x = 4$	or	$x = 25$	Simplify.

Recall that any time we square each side of an equation, we must check for extraneous roots.

Check $(x = 25)$:

$(25) + 3\sqrt{(25)} - 10 = 0$ Substitute 25 for x in the original equation.

$25 + 3 \cdot 5 - 10 = 0$ Simplify the square root.

$30 = 0$ Simplify.

So $x = 25$ is not a solution of the equation. We could have seen this before squaring each side of the equation $\sqrt{x} = -5$. The square root of x cannot be negative, so the equation $\sqrt{x} = -5$ cannot have a solution. The check that $x = 4$ is actually a solution is left to the reader. The solution set is $\{4\}$.

> **A Word of Caution**   Whenever we square both sides of an equation, such as in the previous example, we must check our solutions for extraneous roots.

EXAMPLE 4   Solve $x^{2/3} - 6x^{1/3} - 7 = 0$.

Solution

Let $u = x^{1/3}$. The resulting equation is $u^2 - 6u - 7 = 0$, which can be solved by factoring.

$$x^{2/3} - 6x^{1/3} - 7 = 0$$

$$u^2 - 6u - 7 = 0 \qquad \text{Replace } x^{1/3} \text{ by } u.$$

$$(u - 7)(u + 1) = 0 \qquad \text{Factor.}$$

$$u - 7 = 0 \quad \text{or} \quad u + 1 = 0 \qquad \text{Set each factor equal to 0.}$$

$$u = 7 \quad \text{or} \quad u = -1 \qquad \text{Solve for } u.$$

$$x^{1/3} = 7 \quad \text{or} \quad x^{1/3} = -1 \qquad \text{Replace } u \text{ by } x^{1/3}.$$

$$(x^{1/3})^3 = 7^3 \quad \text{or} \quad (x^{1/3})^3 = (-1)^3 \qquad \begin{array}{l}\text{Raise each side to the}\\ \text{third power.}\end{array}$$

$$x = 343 \quad \text{or} \quad x = -1 \qquad \text{Simplify.}$$

Quick Check 3   The solution set is $\{343, -1\}$.

Solve.

a) $x - 6x^{1/2} + 5 = 0$
b) $x^{2/3} - 4x^{1/3} + 3 = 0$.

Solving Radical Equations

Objective 2   Solve radical equations. Some equations that contain square roots cannot be solved by using a u-substitution. When this happens, we will use the techniques developed in Section 9.5. We begin by isolating the radical, and then we proceed to square both sides of the equation. This can lead to an equation that is quadratic.

EXAMPLE 5   Solve $\sqrt{x + 7} + 5 = x$.

Solution

We begin by isolating the radical so that we may square each side of the equation.

$$\sqrt{x + 7} + 5 = x$$

$$\sqrt{x + 7} = x - 5 \qquad \text{Subtract 5 to isolate the radical.}$$

$$(\sqrt{x + 7})^2 = (x - 5)^2 \qquad \text{Square both sides.}$$

$$x + 7 = (x - 5)(x - 5) \qquad \begin{array}{l}\text{Square the binomial by multiplying}\\ \text{it by itself.}\end{array}$$

$$x + 7 = x^2 - 10x + 25 \qquad \text{Multiply.}$$

$$0 = x^2 - 11x + 18 \qquad \begin{array}{l}\text{Subtract } x \text{ and 7 to collect all terms}\\ \text{on the right side of the equation.}\end{array}$$

$$0 = (x - 2)(x - 9) \qquad \text{Factor.}$$
$$x - 2 = 0 \quad \text{or} \quad x - 9 = 0 \qquad \text{Set each factor equal to 0.}$$
$$x = 2 \quad \text{or} \quad x = 9 \qquad \text{Solve.}$$

Since we have squared each side of the equation, we must check for extraneous roots.

x = 2	**x = 9**
$\sqrt{(2) + 7} + 5 = (2)$	$\sqrt{(9) + 7} + 5 = (9)$
$\sqrt{9} + 5 = 2$	$\sqrt{16} + 5 = 9$
$3 + 5 = 2$	$4 + 5 = 9$
$8 = 2$	$9 = 9$

Quick Check 4

Solve $x + 7 = \sqrt{x + 9}$.

The solution $x = 2$ is an extraneous root and must be omitted. The solution set is $\{9\}$.

Solving Rational Equations

Objective 3   Solve rational equations. Solving rational equations, which were covered in Chapter 7, often requires that we solve a quadratic equation. We begin to solve a rational equation by finding the LCD and multiplying each side of the equation by it to clear the equation of fractions. The resulting equation could be quadratic, as demonstrated in the following example. Once we solve the resulting equation, any solution that causes a denominator in the original equation to be equal to 0 must be omitted.

EXAMPLE 6   Solve $\dfrac{x}{x - 4} + \dfrac{2}{x + 3} = \dfrac{6}{x^2 - x - 12}$.

Solution

We begin by factoring the denominators to find the LCD. The LCD is $(x - 4)(x + 3)$, and solutions of $x = 4$ and $x = -3$ must be omitted, since either would result in a denominator of 0.

$$\frac{x}{x - 4} + \frac{2}{x + 3} = \frac{6}{x^2 - x - 12}$$

$$\frac{x}{x - 4} + \frac{2}{x + 3} = \frac{6}{(x - 4)(x + 3)} \qquad \text{The LCD is}$$

$$(x - 4)(x + 3)\left(\frac{x}{x - 4} + \frac{2}{x + 3}\right) = (x - 4)(x + 3) \cdot \frac{6}{(x - 4)(x + 3)}$$

Multiply by the LCD.

$$\overset{1}{\cancel{(x - 4)}}(x + 3) \cdot \frac{x}{\underset{1}{\cancel{(x - 4)}}} + (x - 4)\overset{1}{\cancel{(x + 3)}} \cdot \frac{2}{\underset{1}{\cancel{(x + 3)}}} = \overset{1}{\cancel{(x - 4)}}\overset{1}{\cancel{(x + 3)}} \cdot \frac{6}{\underset{1}{\cancel{(x - 4)}}\underset{1}{\cancel{(x + 3)}}}$$

Distribute and divide out common factors.

$$x(x + 3) + 2(x - 4) = 6$$

Multiply remaining factors.

$$x^2 + 3x + 2x - 8 = 6$$

Multiply.

$$x^2 + 5x - 8 = 6$$

Combine like terms.

$$x^2 + 5x - 14 = 0$$

Collect all terms on the left side by subtracting 6.

$$(x + 7)(x - 2) = 0$$

Factor.

$$x + 7 = 0 \quad \text{or} \quad x - 2 = 0$$

Set each factor equal to 0.

$$x = -7 \quad \text{or} \quad x = 2$$

Solve.

Since neither solution causes a denominator to equal 0, we do not need to omit either solution. The solution set is $\{-7, 2\}$.

Quick Check 5

Solve $\dfrac{2}{x + 1} + \dfrac{1}{x - 1} = 1$.

Solving Work-Rate Problems

Objective 4   Solve work-rate problems. The last example of the section is a work-rate problem. Work-rate problems involve rational equations and were introduced in Chapter 7.

EXAMPLE 7   A water tower has two drainpipes attached to it. Working alone, the smaller pipe would take 15 minutes longer than the larger pipe to empty the tower. If both drainpipes work together, the tower can be drained in 30 minutes. How long would it take the small pipe, working alone, to drain the tower? (Round your answer to the nearest tenth of a minute.)

Solution

If we let t represent the amount of time that it takes for the larger pipe to drain the tower, then the time required for the small pipe to drain the tower can be represented by $t + 15$. Recall that the work rate is the reciprocal of the time required to complete the entire job. So the work rate for the smaller pipe is $\frac{1}{t + 15}$ and the work-rate for the large pipe is $\frac{1}{t}$. To determine the portion of the job completed by each pipe when they work together, we multiply the work-rate for each pipe by the amount of time that it takes for the two pipes to drain the tower while working together. Here is a table showing the important information:

Pipe	Time to Complete the Job Alone	Work Rate	Time Working	Portion of the Job Completed
Smaller	$t + 15$ minutes	$\dfrac{1}{t + 15}$	30	$\dfrac{30}{t + 15}$
Larger	t minutes	$\dfrac{1}{t}$	30	$\dfrac{30}{t}$

After adding the portion of the tower drained by the smaller pipe in 30 minutes to the portion of the tower drained by the larger pipe, the sum will equal 1, which represents finishing the entire job. The equation is $\frac{30}{t + 15} + \frac{30}{t} = 1$.

$$\frac{30}{t + 15} + \frac{30}{t} = 1$$

The LCD is $t(t + 15)$.

$$t(t + 15)\left(\frac{30}{t + 15} + \frac{30}{t}\right) = t(t + 15) \cdot 1$$

Multiply both sides by the LCD.

$$t(\cancel{t + 15}) \frac{30}{\cancel{(t + 15)}} + \cancel{t}(t + 15) \cdot \frac{30}{\cancel{t}} = t(t + 15) \cdot 1$$

Distribute and divide out common factors.

$$30t + 30(t + 15) = t(t + 15)$$ Multiply remaining factors.

$$30t + 30t + 450 = t^2 + 15t$$ Multiply. The resulting equation is quadratic.

$$60t + 450 = t^2 + 15t$$ Combine like terms.

$$0 = t^2 - 45t - 450$$ Collect all terms on the right side of the equation.

The quadratic expression does not factor, so we will use the quadratic formula.

$$t = \frac{45 \pm \sqrt{(-45)^2 - 4(1)(-450)}}{2(1)}$$ Substitute 1 for a, -45 for b, and -450 for c.

$$t = \frac{45 \pm \sqrt{3825}}{2}$$ Simplify the radicand.

At this point, we must use a calculator to approximate the solutions for t.

$$\frac{45 + \sqrt{3825}}{2} \approx 53.4 \qquad \frac{45 - \sqrt{3825}}{2} \approx -8.4$$

We omit the negative solution, so $t \approx 53.4$. The amount of time required by the small pipe is represented by $t + 15$, so it would take the small pipe approximately $53.4 + 15$, or 68.4, minutes to drain the tank.

Quick Check 6

Working alone, Gabe can clean the gymnasium floor in 50 minutes less time than it takes Rob. If both janitors work together, it takes them 45 minutes to clean the gymnasium floor. How long would it take Gabe, working alone, to clean the gymnasium floor? (Round your answer to the nearest tenth of a minute.)

EXERCISES *10.4*

Solve by making a u-substitution.

1. $x^4 - 5x^2 + 4 = 0$

2. $x^4 + 5x^2 - 36 = 0$

3. $x^4 - 6x^2 + 9 = 0$

4. $x^4 + 12x^2 + 32 = 0$

5. $x^4 + 7x^2 - 18 = 0$

6. $x^4 + x^2 - 2 = 0$

7. $x^4 - 13x^2 + 36 = 0$

8. $x^4 + 8x^2 + 16 = 0$

9. $x - 9\sqrt{x} + 8 = 0$

10. $x + 3\sqrt{x} - 40 = 0$

11. $x - 20\sqrt{x} + 64 = 0$

12. $x + 13\sqrt{x} - 30 = 0$

13. $x - 2x^{1/2} - 3 = 0$

14. $x + 11x^{1/2} + 18 = 0$

15. $x - 7x^{1/2} + 10 = 0$

16. $x + 5x^{1/2} - 14 = 0$

17. $x^6 - 28x^3 + 27 = 0$

18. $x^6 - 19x^3 - 216 = 0$

19. $x^6 + 16x^3 + 64 = 0$

20. $x^6 - 17x^3 + 16 = 0$

21. $x^6 + 1001x^3 + 1000 = 0$

22. $x^6 - 2x^3 + 1 = 0$

23. $x^{2/3} + 5x^{1/3} - 6 = 0$

24. $x^{2/3} - 12x^{1/3} + 20 = 0$

25. $x^{2/3} + 9x^{1/3} + 20 = 0$

26. $x^{2/3} - 4x^{1/3} - 45 = 0$

27. $(x - 3)^2 + 5(x - 3) + 4 = 0$

28. $(x + 7)^2 + 2(x + 7) - 24 = 0$

29. $(2x - 9)^2 - 6(2x - 9) - 27 = 0$

F O R   E X T R A   H E L P

MyMathLab

MathXL
MathXL

Interactmath.com

MathXL Tutorials on CD

DVT CD
Videotape

Addison-Wesley Math Tutor Center

Student Solutions Manual

30. $(4x + 2)^2 - 4(4x + 2) - 60 = 0$

Solve.

31. $\sqrt{x^2 + 6x} = 4$

32. $\sqrt{x^2 - 8x} = 3$

33. $\sqrt{3x - 6} = x - 2$

34. $\sqrt{4x + 52} = x + 5$

35. $\sqrt{x + 15} - x = 3$

36. $\sqrt{3x^2 + 8x + 5} - 5 = 2x$

37. $\sqrt{x - 1} + 2 = \sqrt{2x + 5}$

38. $\sqrt{2x + 6} = 5 - \sqrt{x - 4}$

Solve.

39. $x = \dfrac{10}{x - 3}$

40. $x + 13 = \dfrac{30}{x}$

41. $1 + \dfrac{7}{x} + \dfrac{6}{x^2} = 0$

42. $1 + \dfrac{5}{x} - \dfrac{24}{x^2} = 0$

43. $\dfrac{1}{x} + \dfrac{7}{x + 2} = \dfrac{10}{x(x + 2)}$

44. $\dfrac{4}{x + 3} + \dfrac{1}{x - 5} = \dfrac{3}{x^2 - 2x - 15}$

45. $\dfrac{2}{x + 3} + \dfrac{x + 7}{x + 1} = \dfrac{9}{4}$

46. $\dfrac{x}{x + 7} - \dfrac{4}{x + 2} = \dfrac{7}{x^2 + 9x + 14}$

47. One small pipe takes twice as long to fill a tank as a larger pipe does. If it takes the two pipes 40 minutes to fill the tank when working together, how long will it take each pipe individually to fill the tank?

48. Rob takes 5 hours longer than Genevieve to paint a room. If they work together, they can paint a room in 6 hours. How long does it take Rob to paint a room by himself?

49. Daniel takes 1 hour more than Bryan to rake the leaves from their yard in the fall. If the two work together, they can rake the leaves in 2 hours. How long does it take Daniel, working alone, to rake the leaves? Round to the nearest tenth of an hour.

50. A water tank has two drainpipes attached to it. The larger pipe can drain the tank in 2 hours less than the smaller pipe. If both pipes are being used, they can drain the tank in 4 hours. How long would it take the smaller pipe to drain the tank working alone? Round to the nearest tenth of an hour.

51. A new copy machine can print a set of brochures in 20 minutes less than an older machine. If both machines work simultaneously, they can print the set of brochures in 30 minutes. How long would it take the older machine to print the set of brochures if it works alone? Round to the nearest tenth of a minute.

52. It takes Leo 10 minutes more than Joyce to plant a flat of marigolds. If the two can plant a flat of marigolds in 16 minutes, how long does it take Leo to plant a flat of marigolds? Round to the nearest tenth of a minute.

Solve using the technique of your choice.

53. $x^2 - 8x - 13 = 0$

54. $(7x - 11)^2 - 6(7x - 11) = 0$

55. $x - 2\sqrt{x} - 48 = 0$

56. $x^2 + 15x + 54 = 0$

57. $(3x - 2)^2 = 32$

58. $x^{2/3} + 3x^{1/3} - 4 = 0$

59. $x^2 - 5x + 14 = 0$

60. $x^4 - 5x^2 - 36 = 0$

61. $x^2 - 6x - 91 = 0$

62. $(2x + 15)^2 = 18$

63. $x^6 - 7x^3 - 8 = 0$

64. $3x^2 + x - 8 = 0$

65. $\sqrt{3x - 2} = x - 2$

66. $x + \sqrt{x} - 12 = 0$

67. $(5x + 3)^2 + 2(5x + 3) - 15 = 0$

68. $x = \sqrt{6x - 27} + 3$

69. $(x + 8)^2 = -144$

70. $\dfrac{x + 3}{x - 4} + \dfrac{9}{x + 2} = \dfrac{2}{x^2 - 2x - 8}$

71. $x - 3\sqrt{x} - 4 = 0$

72. $x^2 + 13x + 50 = 0$

73. $x^{2/3} + 12x^{1/3} + 35 = 0$

74. $x^2 - 15x + 56 = 0$

75. $x^2 + 10x + 25 = 0$

76. $(x + 4)^2 = -108$

Answer in complete sentences.

77. Suppose that you are solving an equation that is quadratic in form by using the substitution $u = x^2$. Once you have solved the equation for u, explain how you would solve for x.

78. Suppose that you are solving an equation that is quadratic in form by using the substitution $u = \sqrt{x}$. Once you have solved the equation for u, explain how you would solve for x.

79. Give an example of an equation containing a square root that would be best solved by using a u-substitution. Explain why you feel that using a u-substitution is a more efficient method for solving this equation than isolating the square root and squaring both sides of the equation.

80. Give an example of an equation containing a square root that would be best solved by isolating the square root and squaring both sides of the equation. Explain why you feel that isolating the square root and squaring both sides of the equation is a more efficient method for solving this equation than using a u-substitution.

QUICK REVIEW EXERCISES

Solve.

1. $(x - 3)^2 = 49$

2. $x^2 - 9x - 36 = 0$

3. $x^2 - 6x - 13 = 0$

4. $x^2 + 11x + 40 = 0$

Study Tip **REVISITED** One way to get the most from a study session is to establish a set of goals to accomplish for each session. For example, during a 90-minute session, you may set a goal of completing the homework assignment, in addition to creating a set of note cards for the most recent section in the textbook. Setting goals will encourage you to work quickly and efficiently. Many students set a goal of studying for a certain amount of time, but time alone is not a worthy goal. Create a "to do" list each time you start a study session, and you will find that you will have a much greater chance of reaching your goals.

10.5

GRAPHING QUADRATIC EQUATIONS AND QUADRATIC FUNCTIONS

Objectives

1 **Graph quadratic equations.**
2 **Graph parabolas that open downward.**
3 **Find the vertex of a parabola by completing the square.**
4 **Graph quadratic equations of the form $y = a(x - h)^2 + k$.**

Graphing Quadratic Equations

Objective 1 **Graph quadratic equations.** The graphs of quadratic equations are not lines like the graphs of linear equations, or V-shaped like the graphs of absolute value equations. The graphs of quadratic equations are U-shaped and called parabolas. Let's consider the graph of the most basic quadratic equation: $y = x^2$. We will first create a table of ordered pairs that will represent points on our graph.

x	$y = x^2$		x	$y = x^2$
-2			-2	4
-1			-1	1
0			0	0
1			1	1
2			2	4

The figure below shows these points and the graph of $y = x^2$. Notice that the shape is not a straight line but instead is U-shaped.

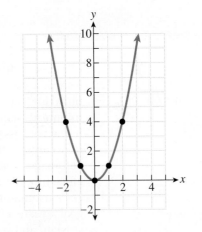 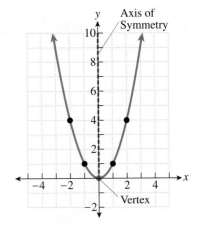

The point where the graph changes from decreasing to increasing is called the vertex. In the previous example, the vertex is located at the bottom of the parabola. Notice that if we drew a vertical line through the vertex of the parabola, the left and right sides are mirror images. The graph of an equation is said to be **symmetric** if we can fold the graph along a line and the two sides of the graph coincide; in other words, a graph is symmetric if one side of the graph is a mirror image of the other side. A parabola is always symmetric, and we call the vertical line through the vertex the **axis of symmetry**.

We graph parabolas by plotting points, and the choice of our points is very important. We look for the y-intercept, the x-intercept(s) if there are any, and the vertex. We also use the axis of symmetry to help us find "mirror" points that are symmetric to points we have already graphed.

As before, we find the y-intercept by substituting 0 for x and solving for y. We will notice that the y-intercept of a quadratic equation in standard form ($y = ax^2 + bx + c$) is always the point $(0, c)$. We find the x-intercepts, if there are any, by substituting 0 for y and solving for x. This equation will be quadratic, and we solve it by previous techniques.

A parabola opens upward if $a > 0$, and the vertex will be at the lowest point of the parabola. To learn how to find the coordinates of the vertex, we begin by completing the square for the equation $y = ax^2 + bx + c$.

$$y = ax^2 + bx + c$$

$$y - c = a\left(x^2 + \frac{b}{a}x\right)$$

Factor a from the two terms containing x. Subtract c from both sides.

$$y - c + \frac{b^2}{4a} = a\left(x^2 + \frac{b}{a}x + \frac{b^2}{4a^2}\right)$$

Half of $\dfrac{b}{a}$ is $\dfrac{b}{2a}$. Add $\left(\dfrac{b}{2a}\right)^2$ or $\dfrac{b^2}{4a^2}$ to the terms inside the parentheses. Since there is a factor in front of the parentheses, we add $a \cdot \dfrac{b^2}{4a^2}$ or $\dfrac{b^2}{4a}$ to the left side.

$$y = a\left(x + \frac{b}{2a}\right)^2 + \frac{4ac - b^2}{4a}$$

Factor the trinomial inside the parentheses. Collect all terms on the right side of the equation. $c - \dfrac{b^2}{4a} = \dfrac{4ac - b^2}{4a}$.

Since a squared expression cannot be negative, the minimum value of y occurs when $x + \dfrac{b}{2a} = 0$, or, in other words, when $x = \dfrac{-b}{2a}$.

If the equation is in standard form, $y = ax^2 + bx + c$, then we can find the x-coordinate of the vertex using the formula $x = \dfrac{-b}{2a}$. We then find the y-coordinate of the vertex by substituting this value for x in the original equation.

EXAMPLE 1   Graph $y = x^2 + 6x + 8$.

Solution

We begin by setting $x = 0$ and solving for y to find the y-intercept.

$$y = (0)^2 + 6(0) + 8 = 8 \qquad \text{Substitute 0 for } x \text{ in the original equation.}$$

The y-intercept is $(0, 8)$. To find the x-intercepts, we substitute 0 for y and attempt to solve for x.

$$0 = x^2 + 6x + 8 \qquad \text{Substitute 0 for } y \text{ in the original equation.}$$
$$0 = (x + 2)(x + 4) \qquad \text{Factor the trinomial.}$$
$$x = -2 \quad \text{or} \quad x = -4 \qquad \text{Set each factor equal to 0 and solve.}$$

The x-intercepts are $(-2, 0)$ and $(-4, 0)$. Finally, we find the vertex. We begin by finding the x-coordinate of the vertex.

$$x = \frac{-6}{2(1)} = -3 \qquad \text{Substitute 1 for } a \text{ and 6 for } b \text{ into } x = \frac{-b}{2a}.$$

Now we substitute this value for x into the original equation and solve for y.

$$y = (-3)^2 + 6(-3) + 8 \qquad \text{Substitute } -3 \text{ for } x.$$
$$y = -1 \qquad\qquad\qquad\quad \text{Simplify.}$$

The vertex is at $(-3, -1)$.

Below left is a sketch showing the locations of the four points we have found. If we add the axis of symmetry, we see that the y-intercept $(0, 8)$ is three units to the right of the axis of symmetry. There is a mirror point with the same y-coordinate located three units to the left of axis of symmetry.

This produces the graph shown at the right:

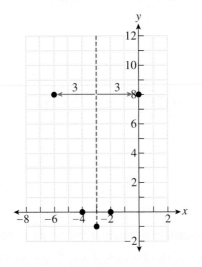 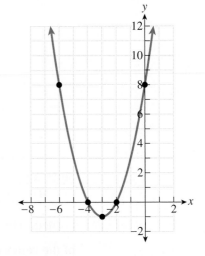

Quick Check 1

Graph $y = x^2 - 8x + 7$.

Using Your Calculator   We can use the TI–83/84 to graph parabolas. To graph $y = x^2 + 6x + 8$, begin by pushing the $\boxed{\text{Y=}}$ key and typing $x^2 + 6x + 8$ next to Y_1 as shown in the screen shot below left.

To graph the parabola, press the $\boxed{\text{GRAPH}}$ key. Below right is the screen that you should see in the standard viewing window:

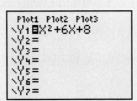 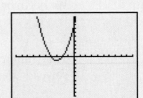

EXAMPLE 2   Graph $y = x^2 - 4x - 7$.

Solution

We begin by substituting 0 for x in the original equation and solving for y to find the y-intercept.

$$y = (0)^2 - 4(0) - 7 = -7 \qquad \text{Substitute 0 for } x.$$

The y-intercept is $(0, -7)$. Again, the y-intercept is the point $(0, c)$. To find the x-intercepts, we substitute 0 for y in the original equation and attempt to solve for x.

$$0 = x^2 - 4x - 7 \qquad \text{Substitute 0 for } y.$$

Since we cannot factor this expression, we must use the quadratic formula to find the x-intercepts.

$$x = \frac{4 \pm \sqrt{(-4)^2 - 4(1)(-7)}}{2(1)} \qquad \text{Substitute 1 for } a, -4 \text{ for } b, \text{ and } -7 \text{ for } c.$$

$$x = \frac{4 \pm \sqrt{44}}{2} \qquad \text{Simplify the radicand and denominator.}$$

$$x = \frac{4 \pm 2\sqrt{11}}{2} \qquad \text{Simplify the square root.}$$

$$x = \frac{\overset{1}{\cancel{2}}(2 \pm \sqrt{11})}{\underset{1}{\cancel{2}}} \qquad \text{Divide out common factors.}$$

$$x = 2 \pm \sqrt{11} \qquad \text{Simplify.}$$

Since $2 + \sqrt{11} \approx 5.3$ and $2 - \sqrt{11} \approx -1.3$, the x-intercepts are approximately $(5.3, 0)$ and $(-1.3, 0)$. Now we find the vertex, using the formula $x = \frac{-b}{2a}$.

Quick Check 2

Graph $y = x^2 + 6x - 9$.

$$x = \frac{4}{2(1)} = 2 \qquad \text{Substitute 1 for } a \text{ and } -4 \text{ for } b.$$

We substitute 2 for x in the original equation and solve for y to find the y-coordinate of the vertex.

$$y = (2)^2 - 4(2) - 7 \qquad \text{Substitute 2 for } x.$$
$$y = -11 \qquad \text{Simplify.}$$

The vertex is at $(2, -11)$.

At the right is a sketch of the parabola showing the vertex, x-intercepts, y-intercept, and the axis of symmetry $(x = 2)$. Also shown is the point $(4, -7)$, which is symmetric to the y-intercept.

Occasionally, a parabola will not have any x-intercepts. In this case, we will have a negative discriminant $(b^2 - 4ac)$ when using the quadratic formula. However, we can determine that a parabola does not have any x-intercepts by altering the order we have been using to find our intercepts and vertex.

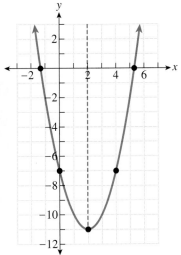

EXAMPLE 3 Graph $y = x^2 - 2x + 2$.

Solution

In this example, we begin by finding the vertex rather than the intercepts. The x-coordinate of the vertex is found using the formula $x = \frac{-b}{2a}$.

$$x = \frac{2}{2(1)} = 1 \qquad \text{Substitute 1 for } a \text{ and } -2 \text{ for } b.$$

Now we substitute 1 for x in the original equation and solve for y.

$$y = (1)^2 - 2(1) + 2 \qquad \text{Substitute 1 for } x.$$
$$y = 1 \qquad \text{Simplify.}$$

The vertex is at $(1, 1)$. To find the y-intercept, we substitute 0 for x in the original equation.

$$y = (0)^2 - 2(0) + 2 = 2 \qquad \text{Substitute 0 for } x.$$

Quick Check 3

Graph $y = x^2 + 6x + 12$.

The y-intercept is $(0, 2)$.

If we plot the vertex and the y-intercept on the graph, we see there are no x-intercepts. The vertex is above the x-axis, and the parabola only moves in an upward direction from there. Therefore, this parabola does not have any x-intercepts. If we chose to use the quadratic formula to find the x-intercepts, we would have ended up with $x = \frac{2 \pm \sqrt{-4}}{2}$. Since the discriminant is negative, the equation $0 = x^2 - 2x + 2$ does not have any real solutions and the parabola does not have any x-intercepts.

At the right is the graph, showing the axis of symmetry $(x = 1)$ and a third point, $(2, 2)$, that is symmetric to the y-intercept.

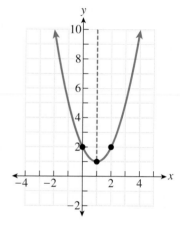

A Word of Caution If a parabola that opens upward has its vertex above the x-axis, then there are no x-intercepts.

Graphing Parabolas That Open Downward

Objective 2 **Graph parabolas that open downward.** Some parabolas open downward rather than upward. The way to determine which way a parabola will open is by writing the equation in standard form: $y = ax^2 + bx + c$. If a is positive, as it was in our previous examples, then the parabola will open upward. If a is negative, then the parabola will open downward. For instance, the graph of $y = -3x^2 + 5x - 7$ would open downward because the coefficient of the second-degree term is negative. The following example shows how to graph a parabola that opens downward.

EXAMPLE 4 Graph $y = -x^2 + 4x + 12$.

Solution

We begin by finding the x-coordinate of the vertex.

$$x = \frac{-4}{2(-1)} = 2 \qquad \text{Substitute } -1 \text{ for } a \text{ and } 4 \text{ for } b.$$

Now we substitute 2 for x in the original equation and solve for y.

$$y = -(2)^2 + 4(2) + 12 \qquad \text{Substitute 2 for } x.$$
$$y = 16 \qquad\qquad\qquad\quad \text{Simplify.}$$

The vertex is at $(2, 16)$. To find the y-intercept, we substitute 0 for x and solve for y.

$$y = -(0)^2 + 4(0) + 12 = 12 \qquad \text{Substitute 0 for } x.$$

The y-intercept is $(0, 12)$. If we plot the vertex and the y-intercept on the graph, we will see that there must be two x-intercepts. The vertex is above the x-axis, and the parabola opens downward, so the graph must cross the x-axis. To find the x-intercepts, we substitute 0 for y in the original equation and solve for x.

$$0 = -x^2 + 4x + 12 \qquad \text{Substitute 0 for } y.$$
$$x^2 - 4x - 12 = 0 \qquad\;\; \text{Collect all terms on the left side of}$$
$$\qquad\qquad\qquad\qquad\qquad \text{the equation, so that the coefficient}$$
$$\qquad\qquad\qquad\qquad\qquad \text{of the squared term is positive.}$$
$$(x - 6)(x + 2) = 0 \qquad\;\; \text{Factor.}$$
$$x = 6 \quad \text{or} \quad x = -2 \qquad \text{Set each factor equal to 0 and solve.}$$

The x-intercepts are $(6, 0)$ and $(-2, 0)$.

Quick Check 4

Graph
$y = -x^2 + 9x + 10.$

Below is the graph, showing the axis of symmetry ($x = 2$) and the point, $(4, 12)$, that is symmetric to the y-intercept.

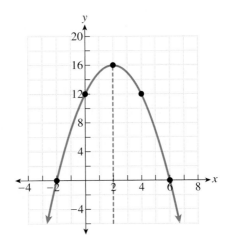

EXAMPLE 5   Graph $y = -\dfrac{1}{2}x^2 - x + 4.$

Solution

This parabola will open downward because the second-degree term has a negative coefficient. We begin by finding the x-coordinate of the vertex.

$$x = \frac{1}{2\left(-\dfrac{1}{2}\right)} = -1 \qquad \text{Substitute } -\tfrac{1}{2} \text{ for } a \text{ and } -1 \text{ for } b.$$

Now we substitute -1 for x in the original equation and solve for y.

$$y = -\frac{1}{2}(-1)^2 - (-1) + 4 \qquad \text{Substitute } -1 \text{ for } x.$$

$$y = -\frac{1}{2} + 1 + 4 \qquad \text{Simplify each term.}$$

$$y = -\frac{1}{2} + \frac{2}{2} + \frac{8}{2} \qquad \text{Rewrite each fraction with } a \text{ common denominator of 2.}$$

$$y = \frac{9}{2} \qquad \text{Simplify.}$$

The vertex is at $\left(-1, \frac{9}{2}\right)$. For an equation of the form $y = ax^2 + bx + c$, the y-intercept is at the point $(0, c)$. The y-intercept of this parabola is $(0, 4)$.

If we put the vertex and the y-intercept on the graph, we will be able to see that there must be two x-intercepts. The vertex is above the x-axis, and the parabola opens downward, so the graph must cross the x-axis. To find the x-intercepts, we substitute 0 for y in the original equation and attempt to solve for x.

$$0 = -\frac{1}{2}x^2 - x + 4 \qquad \text{Substitute 0 for } y.$$

$$\frac{1}{2}x^2 + x - 4 = 0 \qquad \text{Collect all terms on the left side of the equation so that the coefficient of the squared term is positive.}$$

$$2\left(\frac{1}{2}x^2 + x - 4\right) = 2 \cdot 0 \qquad \text{Multiply both sides of the equation by the LCD (2) to clear the equation of fractions.}$$

$$x^2 + 2x - 8 = 0 \qquad \text{Distribute and simplify.}$$
$$(x - 2)(x + 4) = 0 \qquad \text{Factor.}$$
$$x = 2 \qquad \text{or} \qquad x = -4 \qquad \text{Set each factor equal to 0 and solve.}$$

The x-intercepts are $(2, 0)$ and $(-4, 0)$.

Below is the graph, using the axis of symmetry to find a fifth point that is symmetric to the y-intercept.

Quick Check **5**

Graph

$$y = -\frac{1}{4}x^2 - \frac{5}{2}x - 4.$$

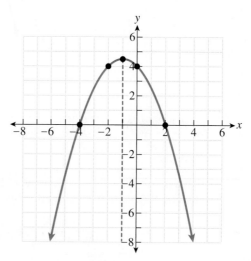

Here is a summary for graphing parabolas.

Graphing Parabolas

- Determine whether the parabola opens upward or downward. For an equation in standard form ($y = ax^2 + bx + c$), the parabola will open upward if a is positive. If a is negative, the parabola will open downward.
- Find the vertex of the parabola. Use the formula $x = \dfrac{-b}{2a}$ to find the x-coordinate of the vertex. Substitute this result for x in the original equation to find the y-coordinate.
- Find the y-intercept of the parabola. We find the y-intercept by letting $x = 0$.
- After plotting the vertex and the y-intercept, determine whether there are any x-intercepts. If there are any x-intercepts, we find them by letting $y = 0$ and solving the resulting quadratic equation for x. If we use the quadratic formula to solve this equation and the discriminant, $b^2 - 4ac$, is negative, then there are no x-intercepts.

Finding the Vertex of a Parabola by Completing the Square

Objective 3 **Find the vertex of a parabola by completing the square.** As an alternative, we could find the coordinates of the vertex by completing the square.

Finding the Vertex of $y = a(x - h)^2 + k$

The graph of the quadratic equation $y = a(x - h)^2 + k$ is a parabola with axis of symmetry $x = h$ and vertex (h, k). The parabola opens upward if a is positive and opens downward if a is negative.

EXAMPLE 6   Find the vertex and axis of symmetry for the parabola.

a) $y = (x - 4)^2 + 3$

Solution

Quick Check **6**
Find the vertex and axis of symmetry for the parabola.

a) $y = (x + 2)^2 - 8$
b) $y = -2(x - 8)^2 + 7.$

This equation is in the form $y = a(x - h)^2 + k$. The axis of symmetry is $x = 4$ and the vertex is $(4, 3)$.

b) $y = -(x + 1)^2 - 4$

Solution

This parabola opens downward, but this does not affect how we find the axis of symmetry or the vertex. The axis of symmetry is $x = -1$ and the vertex is $(-1, -4)$.

EXAMPLE 7   Find the vertex of the parabola $y = x^2 - 14x + 33$ by completing the square.

Solution

We will begin by isolating the terms containing x on the right side of the equation. Then we will complete the square.

$$y = x^2 - 14x + 33$$
$$y - 33 = x^2 - 14x$$
$$y - 33 + 49 = x^2 - 14x + 49$$

Subtract 33 to isolate the terms containing x.
Take half of the coefficient of the first-degree term, square it, and add it to both sides of the equation. $\left(\frac{-14}{2}\right)^2 = (-7)^2 = 49$

$$y + 16 = (x - 7)^2$$
$$y = (x - 7)^2 - 16$$

Factor the trinomial as a perfect square.
Subtract 16 to isolate y.

The vertex of the parabola is $(7, -16)$.

EXAMPLE 8   Find the vertex of the parabola $y = -2x^2 - 12x + 21$ by completing the square.

Solution

After isolating the terms containing x, we must factor out -2 so that the coefficient of the squared term is 1.

$$y = -2x^2 - 12x + 21$$
$$y - 21 = -2x^2 - 12x$$

Subtract 21 to isolate the terms containing x.

$$y - 21 = -2(x^2 + 6x)$$

Factor out the common factor -2.

$$y - 21 - 18 = -2(x^2 + 6x + 9)$$

$\left(\frac{6}{2}\right)^2 = 3^2 = 9$

Add 9 in the parentheses. This is equivalent to subtracting 18 on the right side because of the factor -2 that is multiplied by 9. Subtract 18 from the left side.

$$y - 39 = -2(x + 3)^2$$
$$y = -2(x + 3)^2 + 39$$

Factor the trinomial as a perfect square.
Add 39 to isolate y.

The vertex is $(-3, 39)$.

> **Quick Check 7**
>
> Find the vertex of the parabola by completing the square.
>
> a) $y = x^2 + 6x - 40$
> b) $y = -x^2 + 12x + 45$

Graphing Quadratic Equations of the Form $y = a(x - h)^2 + k$

Objective 4   Graph quadratic equations of the form $y = a(x - h)^2 + k$.

EXAMPLE 9   Graph $y = (x + 4)^2 - 9$.

Solution

This parabola opens upward, and the vertex is $(-4, -9)$. Next we find the y-intercept.

$$y = (0 + 4)^2 - 9 \qquad \text{Substitute 0 for } x.$$
$$y = 7 \qquad \text{Simplify.}$$

The y-intercept is $(0, 7)$. Since the parabola opens upward and the vertex is below the x-axis, the parabola has two x-intercepts. We find the coordinates of the x-intercepts by substituting 0 for y and solving for x by extracting square roots.

$$0 = (x + 4)^2 - 9 \qquad \text{Substitute 0 for } y.$$
$$9 = (x + 4)^2 \qquad \text{Add 9 to isolate } (x + 4)^2.$$
$$\pm\sqrt{9} = \sqrt{(x + 4)^2} \qquad \text{Take the square root of each side.}$$
$$\pm 3 = x + 4 \qquad \text{Simplify each square root.}$$
$$-4 \pm 3 = x \qquad \text{Subtract 4 to isolate } x.$$

The x-coordinates of the x-intercepts are $-4 + 3 = -1$ and $-4 - 3 = -7$. The x-intercepts are $(-1, 0)$ and $(-7, 0)$. The axis of symmetry is $x = -4$, and the point $(-8, 7)$ is symmetric to the y-intercept.

Quick Check **8**

Graph
$y = -(x - 2)^2 + 1.$

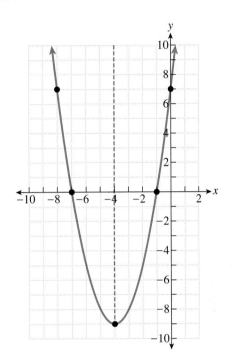

EXERCISES *10.5*

Find the vertex of the parabola associated with the following quadratic equations, as well as the equation of the axis of symmetry.

1. $y = x^2 - 6x - 21$

2. $y = x^2 + 8x - 15$

3. $y = x^2 + 14x + 40$

4. $y = x^2 - 4x - 45$

5. $y = -x^2 - 10x + 32$

6. $y = -x^2 + 4x + 13$

7. $y = x^2 - 9x + 14$

8. $y = x^2 + 7x + 7$

9. $y = 3x^2 + 12x - 20$

10. $y = -2x^2 - 6x + 17$

11. $y = x^2 - 7$

12. $y = x^2 - 8x$

13. $y = (x - 3)^2 - 4$

14. $y = (x + 5)^2 + 1$

15. $y = -(x + 4)^2 + 4$

16. $y = -(x - 10)^2 + 15$

Find the x- and y-intercepts of the parabola associated with the following quadratic equations. If necessary, round to the nearest tenth. If the parabola does not have any x-intercepts, state "no x-intercepts."

17. $y = x^2 + 6x + 5$

18. $y = x^2 - 8x - 6$

19. $y = -x^2 + 4x - 7$

20. $y = x^2 - 8x + 7$

21. $y = 2x^2 - 7x - 13$

22. $y = -x^2 + x + 72$

23. $y = x^2 + 4x$

24. $y = -x^2 + 12x - 43$

25. $y = x^2 - 6x + 9$

26. $y = -x^2 + 16x - 64$

27. $y = (x + 2)^2 - 1$

28. $y = (x - 7)^2 - 8$

29. $y = -(x - 5)^2 - 16$

30. $y = -(x + 6)^2 + 9$

Graph the given parabolas. Label the vertex and all intercepts.

31. $y = x^2 - 2x - 3$ **32.** $y = x^2 - 7x + 5$

33. $y = -x^2 + 4x$ **34.** $y = x^2 - 8x - 20$

35. $y = -x^2 + 3x - 4$ **36.** $y = -x^2 + 4$

37. $y = 2x^2 + 8x - 25$

41. $y = -x^2 + 6x - 6$

42. $y = x^2 - 5x$

43. $y = x^2 + 8x + 16$

44. $y = x^2 + x - 6$

38. $y = x^2 - 2x - 1$

45. $y = x^2 - 12$

46. $y = -x^2 + 4x - 4$

39. $y = x^2 + 6x + 10$

40. $y = x^2 + 8x + 18$

47. $y = \dfrac{1}{2}x^2 - 2x + \dfrac{3}{2}$

48. $y = \frac{1}{3}x^2 + 2x - 9$ **49.** $y = (x - 2)^2 + 3$

54. Explain how to determine that a parabola does not have any x-intercepts.

55. Explain how to find the vertex of a parabola by completing the square. Use an example as an illustration of the process.

56. Explain how to graph a parabola whose equation has the form $y = a(x - h)^2 + k$. Include instructions for how to find the vertex, the y-intercept, and the x-intercepts, if there are any.

50. $y = (x - 5)^2 + 1$ **51.** $y = -(x + 1)^2 + 3$

> **Study Tip** REVISITED   If you have more than one subject to study, as most students do, then the order in which you study the subjects is important. A wise idea is to arrange your study schedule so that you will be studying the most difficult subject first, while you are most alert. If possible, avoid studying two difficult subjects back to back. If you cannot arrange a study break between these two topics, try to study a less demanding subject between two difficult subjects. Another suggestion to keep your mental energy at its highest while studying is to take short 10-minute study breaks. One study break per hour will help to keep you from feeling fatigued.

52. $y = -(x + 3)^2 + 8$

Answer in complete sentences.

53. Explain how to graph a parabola whose equation has the form $y = ax^2 + bx + c$. Include instructions for how to find the vertex, the y-intercept, and the x-intercepts, if there are any.

Objectives

1. **Solve quadratic inequalities.**
2. **Solve quadratic inequalities graphically.**
3. **Solve rational inequalities.**
4. **Solve inequalities involving functions.**
5. **Solve applied problems involving inequalities.**

Quadratic Inequalities

Objective 1   Solve quadratic inequalities. In this section, we will expand our knowledge of inequalities to include two different types of inequalities: quadratic inequalities and rational inequalities.

Quadratic Inequalities

A **quadratic inequality** is an inequality that can be rewritten as $ax^2 + bx + c < 0$, $ax^2 + bx + c \leq 0$, $ax^2 + bx + c > 0$, or $ax^2 + bx + c \geq 0$.

The first step to solving a quadratic inequality is to find its **zeros**, which are values of x for which $ax^2 + bx + c = 0$. For example, to find the zeros for the inequality $x^2 + 9x - 22 > 0$, we set $x^2 + 9x - 22$ equal to 0 and solve for x.

$$x^2 + 9x - 22 = 0 \qquad \text{Set the quadratic expression equal to 0.}$$
$$(x + 11)(x - 2) = 0 \qquad \text{Factor.}$$
$$x = -11 \quad \text{or} \quad x = 2 \qquad \text{Set each factor equal to 0 and solve.}$$

The zeros for this inequality are -11 and 2.

The zeros of an inequality divide the number line into intervals and often act as boundary points. In any particular interval created by the zeros of an inequality, either each real number in the interval is a solution of the inequality, or each real number in the interval is not a solution of the inequality. By picking one value in each interval and testing it, we can determine which intervals are solutions and which are not.

When solving a strict inequality involving the symbols $<$ or $>$, the zeros are not included in the solutions of the inequality, and we place an open circle on each zero on the number line. When solving a weak inequality involving the symbols $\leq$ or $\geq$, the zeros are included in the solutions of the inequality, and we place a closed circle on each zero.

EXAMPLE 1   Solve $x^2 + x - 20 \leq 0$.

Solution

We begin by finding the zeros.

$$x^2 + x - 20 = 0 \qquad \text{Set the quadratic expression equal to 0.}$$
$$(x + 5)(x - 4) = 0 \qquad \text{Factor.}$$
$$x = -5 \quad \text{or} \quad x = 4 \qquad \text{Set each factor equal to 0 and solve.}$$

The zeros for this inequality are -5 and 4. We plot these zeros on a number line and create boundaries between the three intervals, which have been labeled as I, II, and III in the figure that follows. The zeros are included as solutions and we place closed circles on the number line at $x = -5$ and $x = 4$.

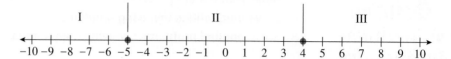

Now we choose a value from each of the three intervals we have created to be our test points. A **test point** is a value of the variable x that we use to evaluate the expression in the inequality, allowing us to determine which intervals are solutions of the inequality. We will use $x = -6$, $x = 0$, and $x = 5$. Because we are looking for intervals where $x^2 + x - 20 \le 0$, our solution will be made up of the intervals whose test points produce a negative result when substituted into the expression $x^2 + x - 20$.

Test Point (x)	-6	0	5
$x^2 + x - 20$	$(-6)^2 + (-6) - 20$ $= 36 - 6 - 20$ $= 10$	$(0)^2 + (0) - 20$ $= 0 + 0 - 20$ $= -20$	$(5)^2 + (5) - 20$ $= 25 + 5 - 20$ $= 10$

The only test point for which the expression $x^2 + x - 20$ is negative is $x = 0$, so the interval $[-5, 4]$ is the solution of this inequality.

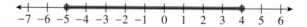

When the expression we are working with can be factored, we can use a sign chart to determine which intervals are solutions to the inequality. A **sign chart** is a chart that can be used to determine whether an expression is positive or negative for certain intervals of real numbers. A sign chart focuses on whether a factor is positive or negative for each interval, using a test point. If we know the sign of each factor, then we can easily find the sign of the product, allowing us to solve our inequality. The initial sign chart should look like the following.

Test Point	$x + 5$	$x - 4$	Product
-6			
0			
5			

We fill in the first row by determining the sign of each factor when $x = -6$. Since $-6 + 5$ is negative, we put a negative sign in the first row underneath $x + 5$. In the same way, $x - 4$ is negative when $x = -6$, so we put another negative sign in the first row underneath $x - 4$. The product of two negative factors is positive, so we put a positive sign in the first row under Product. Here is the sign chart after it has been completed.

Test Point	$x + 5$	$x - 4$	Product
-6	$-$	$-$	$+$
0	$+$	$-$	$-$
5	$+$	$+$	$+$

Notice that the product is negative in the interval containing the test point $x = 0$. Therefore our solution is the interval $[-5, 4]$.

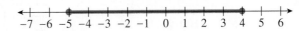

Quick Check   1 Solve $x^2 - 14x + 48 \leq 0$.

If the quadratic expression does not factor, then we cannot use a sign chart. In such a case, we will find the zeros of the expression by using the quadratic formula. We will then substitute the values for our test points directly into the quadratic expression.

EXAMPLE 2   Solve $x^2 + 6x + 4 > 0$.

Solution

We find the zeros by solving the equation $x^2 + 6x + 4 = 0$. Since $x^2 + 6x + 4$ does not factor, we will use the quadratic formula to solve the equation.

$$x = \frac{-6 \pm \sqrt{6^2 - 4(1)(4)}}{2(1)}$$ Substitute 1 for a, 6 for b, and 4 for c.

$$x = \frac{-6 \pm \sqrt{20}}{2}$$ Simplify the radicand.

$$x = \frac{-6 \pm 2\sqrt{5}}{2}$$ Simplify the square root.

$$x = \frac{\overset{1}{\cancel{2}}(-3 \pm \sqrt{5})}{\underset{1}{\cancel{2}}}$$ Divide out common factors.

$$x = -3 \pm \sqrt{5}$$ Simplify.

The zeros for this inequality are $-3 + \sqrt{5}$ and $-3 - \sqrt{5}$. We need an approximate value for each zero to determine where each zero belongs on the number line. These two zeros are approximately equal to -0.8 and -5.2. The zeros are not included as soutions, so we place open circles at the zeros.

We will use $x = -6$, $x = -1$, and $x = 0$ as test points. We are looking for intervals where the expression $x^2 + 6x + 4$ is greater than 0, and our solution will be made up of the intervals whose test points produce a positive result when substituted into $x^2 + 6x + 4$.

Test Point	$x^2 + 6x + 4$	Sign
-6	$(-6)^2 + 6(-6) + 4 = 4$	$+$
-1	$(-1)^2 + 6(-1) + 4 = -1$	$-$
0	$(0)^2 + 6(0) + 4 = 4$	$+$

Notice that the product is positive in the intervals containing the test points $x = -6$ and $x = 0$. Here is our solution:

$$-3 - \sqrt{5} \qquad -3 + \sqrt{5}$$

In interval notation, we can express this solution as $(-\infty, -3 - \sqrt{5}) \cup (-3 + \sqrt{5}, \infty)$. (Notice that we used the exact values in our solution, not the approximate values.)

Quick Check **2** Solve $x^2 + 3x - 15 \geq 0$.

If the expression in our inequality has no zeros, then either every real number is a solution of the inequality or it has no solutions at all.

EXAMPLE 3 Solve $x^2 - 5x + 7 \leq 0$.

Solution

We begin by looking for the zeros of $x^2 - 5x + 7$. Since this expression does not factor, we will use the quadratic formula.

$$x = \frac{5 \pm \sqrt{(-5)^2 - 4(1)(7)}}{2(1)} \qquad \text{Substitute 1 for } a, -5 \text{ for } b, \text{ and 7 for } c.$$

$$x = \frac{5 \pm \sqrt{-3}}{2} \qquad \text{Simplify the radicand and denominator.}$$

The discriminant is negative, so this expression has no real zeros. In this case, we can pick any real number and use it as our only test point. When $x = 0$, we can see that $x^2 - 5x + 7$ is equal to 7, which is not less than or equal to 0. This inequality has no solution.

Quick Check **3** Solve $x^2 - 5x + 20 > 0$.

Solving Quadratic Inequalities from the Graph of a Quadratic Equation

Objective 2 **Solve quadratic inequalities graphically.** Consider the graph of $y = x^2 - 6x + 8$. The graph of the equation drops below the x-axis in the interval $(2, 4)$. This tells us that the interval $(2, 4)$ is the solution of the inequality $x^2 - 6x + 8 < 0$. For any value of x that is less than 2 or greater than 4, the expression $x^2 - 6x + 8$ is greater than 0.

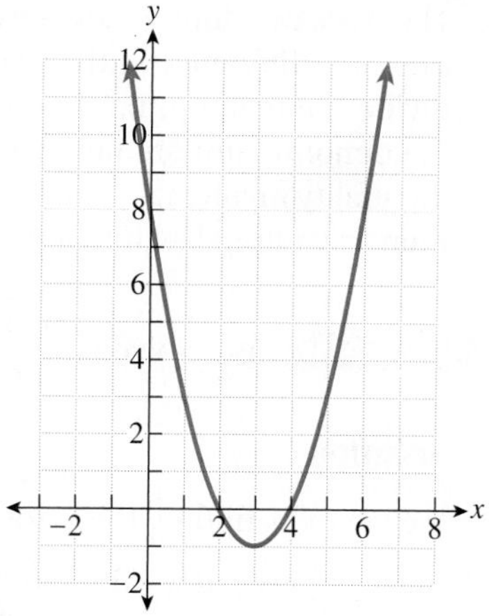

EXAMPLE 4 Use the graph of $y = -x^2 + 3x + 10$ to solve the inequality $-x^2 + 3x + 10 > 0$.

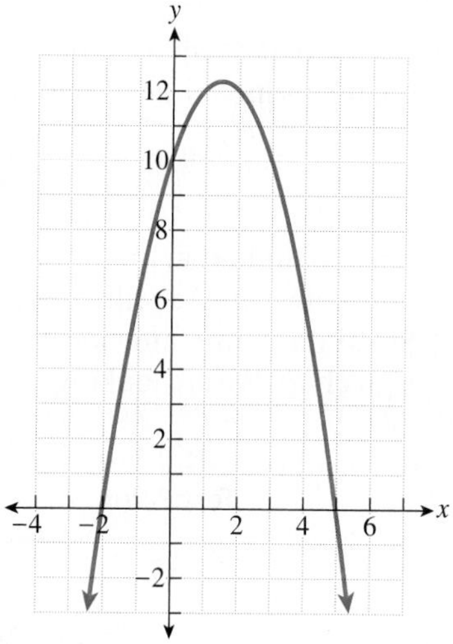

Solution

We are looking for intervals for which the graph is above the x-axis. Our solution in this case is the interval $(-2, 5)$.

***Quick Check* 4**

Use the graph of $y = x^2 - 5x - 14$ to solve the inequality $x^2 - 5x - 14 > 0$.

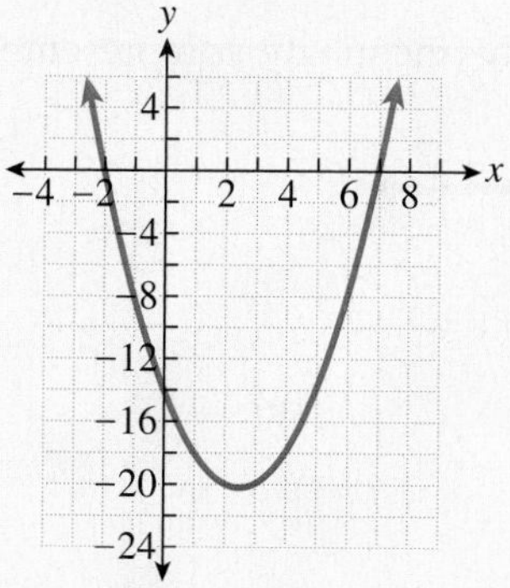

Rational Inequalities

Objective 3 **Solve rational inequalities.**

Rational Inequalities

A **rational inequality** is an inequality that involves a rational expression, such as $\dfrac{x+2}{x-4} < 0$.

There are two differences between solving a rational inequality and a quadratic inequality. The first difference is that we use the zeros of the numerator and the denominator to divide the real number line into intervals. The second difference is that the zeros from the denominator are never included in the solution, even when the inequality is a weak inequality involving $\leq$ or $\geq$. This is because if a value causes a denominator to equal 0, then the rational expression is undefined for that value.

EXAMPLE 5 Solve $\dfrac{(x + 3)(x + 5)}{x - 1} \geq 0$.

Solution

We begin by finding the zeros of the numerator.

$$(x + 3)(x + 5) = 0 \qquad \text{Set the numerator equal to 0.}$$
$$x = -3 \quad \text{or} \quad x = -5 \qquad \text{Set each factor equal to 0 and solve.}$$

The zeros of the numerator are -3 and -5. Now we find the zeros of the denominator.

$$x - 1 = 0 \qquad \text{Set the denominator equal to 0.}$$
$$x = 1 \qquad \text{Solve.}$$

The zero of the denominator is 1. Here are all of the zeros on a single number line, dividing the number line into four intervals.

Keep in mind that the value $x = 1$ cannot be included in any solution, as it is a zero of the denominator. We will use the values $x = -6$, $x = -4$, $x = 0$, and $x = 2$ as test points. Since the numerator factors and the denominator is a linear expression, we can use a sign chart. We are looking for intervals where the expression is positive.

Test Point	$x + 3$	$x + 5$	$x - 1$	$\dfrac{(x + 3)(x + 5)}{x - 1}$
-6	$-$	$-$	$-$	$-$
-4	$-$	$+$	$-$	$+$
0	$+$	$+$	$-$	$-$
2	$+$	$+$	$+$	$+$

The intervals containing $x = -4$ and $x = 2$ are solutions of this inequality, as represented on the following number line.

This can be represented in interval notation by $[-5, -3] \cup (1, \infty)$.

Quick Check 5 Solve $\dfrac{(x + 2)(x - 6)}{(x - 8)(x + 7)} \leq 0$.

A Word of Caution   When solving a rational inequality, the zeros of the denominator are always excluded as solutions because the rational expression is undefined for those values.

EXAMPLE 6   Solve $\dfrac{x^2 - 4x - 12}{x^2 - 9} < 0$.

Solution

We begin by finding the zeros of the numerator.

$x^2 - 4x - 12 = 0$	Set the numerator equal to 0.
$(x - 6)(x + 2) = 0$	Factor.
$x - 6 = 0$ or $x + 2 = 0$	Set each factor equal to 0.
$x = 6$ or $x = -2$	Solve.

The zeros of the numerator are 6 and -2. Now we find the zeros of the denominator.

$x^2 - 9 = 0$	Set the denominator equal to 0.
$(x + 3)(x - 3) = 0$	Factor.
$x + 3 = 0$ or $x - 3 = 0$	Set each factor equal to 0.
$x = -3$ or $x = 3$	Solve.

The zeros of the denominator are -3 and 3. Here are all of the zeros on a single number line, dividing the number line into five intervals:

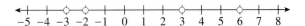

In this example, none of the zeros can be included in any solution, as the inequality was strictly less than 0 and not less than or equal to 0. We will use the values $x = -4$, $x = -2.5$, $x = 0$, $x = 4$, and $x = 7$ as test points. Here is the sign chart; keep in mind that we are looking for intervals in which this expression is negative.

Test Point	$x - 6$	$x + 2$	$x + 3$	$x - 3$	$\dfrac{(x - 6)(x + 2)}{(x + 3)(x - 3)}$
-4	$-$	$-$	$-$	$-$	$+$
-2.5	$-$	$-$	$+$	$-$	$-$
0	$-$	$+$	$+$	$-$	$+$
4	$-$	$+$	$+$	$+$	$-$
7	$+$	$+$	$+$	$+$	$+$

The intervals containing $x = -2.5$ and $x = 4$ are solutions of this inequality, as represented on the following number line:

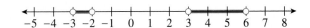

This can be represented in interval notation by $(-3, -2) \cup (3, 6)$.

Quick Check   6 Solve $\dfrac{x^2 - 6x}{x^2 + 5x + 4} > 0.$

Solving Inequalities Involving Functions

Objective **4** **Solve inequalities involving functions.** We finish this section by examining inequalities involving functions.

EXAMPLE   7   Given $f(x) = x^2 - 2x - 16$, find all values x for which $f(x) \le 8$.

Solution

We begin by setting the function less than or equal to 8.

$$f(x) \le 8$$
$$x^2 - 2x - 16 \le 8 \qquad \text{Set the function less than or equal to 8.}$$
$$x^2 - 2x - 24 \le 0 \qquad \text{Subtract 8 to write the inequality in general form.}$$

We now find the zeros for this inequality.

$$x^2 - 2x - 24 = 0 \qquad\qquad \text{Set the quadratic expression equal to 0.}$$
$$(x - 6)(x + 4) = 0 \qquad\qquad \text{Factor.}$$
$$x - 6 = 0 \quad \text{or} \quad x + 4 = 0 \qquad \text{Set each factor equal to 0.}$$
$$x = 6 \quad \text{or} \quad x = -4 \qquad \text{Solve.}$$

The zeros for this inequality are 6 and -4.

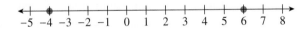

We will use $x = -5$, $x = 0$, and $x = 7$ as our test points. Our solution will be made up of the intervals whose test points produce a negative result when substituted into the expression $x^2 - 2x - 24$. Here is the sign chart after it has been completed:

Test Point	$x - 6$	$x + 4$	Product
-5	$-$	$-$	$+$
0	$-$	$+$	$-$
7	$+$	$+$	$+$

Quick Check   7

Given
$f(x) = x^2 + 10x + 35$,
find all values x for
which $f(x) < 14$.

The product is negative in the interval containing the test point $x = 0$. Here is our solution represented on a number line:

In interval notation, the solution is expressed as $[-4, 6]$.

Solving Applied Problems Involving Inequalities

Objective **5** Solve applied problems involving inequalities.

EXAMPLE 8   A projectile is fired from the roof of a building 72 feet tall, with an initial velocity of 96 feet per second. The height of the projectile, in feet, after t seconds is given by the function $h(t) = -16t^2 + 96t + 72$. For what length of time is the projectile at least 200 feet above the ground?

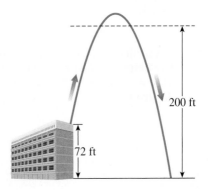

Solution

To find the time interval that the projectile is at least 200 feet above the ground, we must solve the inequality $h(t) \geq 200$.

$$h(t) \geq 200$$
$$-16t^2 + 96t + 72 \geq 200 \qquad \text{Replace } h(t) \text{ by } -16t^2 + 96t + 72.$$
$$-16t^2 + 96t - 128 \geq 0 \qquad \text{Subtract 200 to rewrite the inequality in standard form.}$$
$$16t^2 - 96t + 128 \leq 0 \qquad \text{Multiply both sides of the inequality by } -1. \text{ Change the direction of the inequality.}$$

Now we find the zeros of the inequality.

$$16t^2 - 96t + 128 = 0$$
$$16(t^2 - 6t + 8) = 0 \qquad \text{Factor out the common factor 16.}$$
$$16(t - 2)(t - 4) = 0 \qquad \text{Factor the trinomial.}$$
$$t = 2 \quad \text{or} \quad t = 4 \qquad \text{Set each variable factor equal to 0 and solve.}$$

The zeros of the inequality are $t = 2$ and $t = 4$.

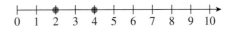

We will use $t = 1$, $t = 3$, and $t = 5$ as our test points. (Note that t must be greater than or equal to 0, since it represents the amount of time since the projectile was fired.) In this case, we will substitute the values for our test points into the function $h(t)$. If the function's output for one of the test points is 200 or higher, then each point in the interval containing the test point is a solution to the inequality $h(t) \geq 200$.

$t = 1$	$t = 3$	$t = 5$
$h(1) = -16(1)^2 + 96(1) + 72$	$h(3) = -16(3)^2 + 96(3) + 72$	$h(5) = -16(5)^2 + 96(5) + 72$
$= -16(1) + 96 + 72$	$= -16(9) + 288 + 72$	$= -16(25) + 480 + 72$
$= -16 + 96 + 72$	$= -144 + 288 + 72$	$= -400 + 480 + 72$
$= 152$	$= 216$	$= 152$

The inequality $h(t) \geq 200$ is true only in the interval containing the test point $t = 3$, so the projectile's height is at least 200 feet from 2 seconds after launch until 4 seconds after launch.

Quick Check **8** A projectile is fired from ground level with an initial velocity of 272 feet per second. The height of the projectile, in feet, after t seconds is given by the function $h(t) = -16t^2 + 272t$. For what length of time is the projectile at least 960 feet above the ground?

EXERCISES 10.6 ›

Solve the quadratic inequality. Express your solution on a number line using interval notation.

1. $(x - 5)(x - 1) < 0$

2. $(x + 3)(x - 2) > 0$

3. $-2(x - 6)(x + 3) \leq 0$

4. $x(x - 4) > 0$

5. $x^2 - 13x + 30 \leq 0$

6. $x^2 + 2x - 63 < 0$

7. $x^2 + 8x + 15 > 0$

8. $x^2 - 7x + 10 \leq 0$

9. $x^2 - 7x < 0$

10. $x^2 - 16 > 0$

11. $x^2 + 6x + 3 > 0$

12. $x^2 - 5x - 11 < 0$

13. $3x^2 - 5x - 1 \geq 0$

14. $2x^2 + 7x + 4 \leq 0$

15. $x^2 + 3x + 5 > 0$

16. $x^2 + x + 8 < 0$

17. $-x^2 + 10x - 5 > 0$

18. $x^2 + 6x \leq 27$

19. $x^2 - 4x \geq 45$

20. $-x^2 + 6x + 1 < 0$

Solve the polynomial inequality. Express your solution on a number line using interval notation.

21. $(x - 1)(x + 2)(x - 4) \geq 0$

22. $(x + 8)(x - 5)(x - 2) \leq 0$

23. $(x + 6)(x + 2)(x + 3) < 0$

24. $x(x - 1)(x + 8)(x + 2) > 0$

25. $(x - 7)^2(x + 5)(x + 9) \leq 0$

26. $(x - 10)^3(x + 4)^2(x - 4) \geq 0$

Solve the rational inequality. Express your solution on a number line using interval notation.

27. $\dfrac{x - 3}{x + 4} < 0$

28. $\dfrac{x + 8}{x - 6} > 0$

29. $\dfrac{(x + 2)(x - 2)}{x + 1} \leq 0$

30. $\dfrac{(x + 7)(x - 4)}{(x - 5)(x + 10)} < 0$

31. $\dfrac{x^2 - 4x - 5}{x + 2} < 0$

32. $\dfrac{x + 7}{x^2 - 14x + 48} \leq 0$

33. $\dfrac{x^2 + 7x - 8}{x^2 - 49} \geq 0$

34. $\dfrac{x^2 - 13x + 36}{x^2 + 12x + 36} > 0$

35. $\dfrac{x^2 - 12x + 35}{x^2 + 6x - 16} \leq 0$

36. $\dfrac{x^2 + 15x + 44}{x^2 - 7x - 60} \geq 0$

37. $\dfrac{x^2 - 4x + 3}{x^2 + 4x - 32} < 0$

38. $\dfrac{x^2 - 81}{x^2 - 15x + 54} \leq 0$

39. A projectile is fired from the roof of a building 27 feet tall with an initial velocity of 64 feet per second. The height of the projectile, in feet, after t seconds is given by the function $h(t) = -16t^2 + 64t + 27$. For what length of time is the projectile at least 75 feet above the ground?

40. A projectile is fired from the roof of a building 98 feet tall with an initial velocity of 208 feet per second. The height of the projectile, in feet, after t seconds is given by the function $h(t) = -16t^2 + 208t + 98$. For what length of time is the projectile at least 450 feet above the ground?

41. A projectile is fired upwards from ground level with an initial velocity of 32 feet per second. The height of the projectile, in feet, after t seconds is given by the function $h(t) = -16t^2 + 32t$. For what length of time is the projectile at least 12 feet above the ground?

42. A projectile is fired upwards from ground level with an initial velocity of 48 feet per second. The height of the projectile, in feet, after t seconds is given by the function $h(t) = -16t^2 + 48t$. For what length of time is the projectile at least 20 feet above the ground?

43. A projectile is fired from the roof of a building 276 feet tall with an initial velocity of 144 feet per second. The height of the projectile, in feet, after t seconds is given by the function $h(t) = -16t^2 + 144t + 276$. For how long is the projectile at least 500 feet above the ground?

44. A projectile is fired from the roof of a building 20 feet tall with an initial velocity of 84 feet per second. The height of the projectile, in feet, after t seconds is given by the function $h(t) = -16t^2 + 84t + 20$. For how long is the projectile at least 100 feet above the ground?

45. Given $f(x) = x^2 - 8x + 12$, find all values x for which $f(x) \le 0$.

46. Given $f(x) = x^2 + 13x - 5$, find all values x for which $f(x) \ge 0$.

47. Given $f(x) = x^2 - 5x + 13$, find all values x for which $f(x) \ge 7$.

48. Given $f(x) = x^2 + 4x - 31$, find all values x for which $f(x) < 14$.

49. Given $f(x) = x^2 - 8x + 24$, find all values x for which $f(x) \le -9$.

50. Given $f(x) = x^2 + 6x - 9$, find all values x for which $f(x) < -49$.

51. Given $f(x) = \dfrac{x + 9}{x - 6}$, find all values x for which $f(x) \le 0$.

52. Given $f(x) = \dfrac{x^2 - 17x + 70}{x^2 + 2x - 48}$, find all values x for which $f(x) < 0$.

53. Given $f(x) = \dfrac{x^2 - 8x - 33}{x^2 + 13x + 42}$, find all values x for which $f(x) > 0$.

54. Given $f(x) = \dfrac{x^2 + 14x + 24}{x^2 + 6x + 5}$, find all values x for which $f(x) \ge 0$.

Find a quadratic inequality whose solution is given.

55.

56.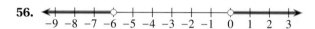

57.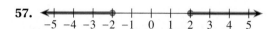

58.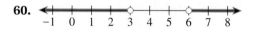

Find a rational inequality whose solution is given.

59.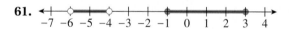

60.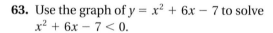

61.

62.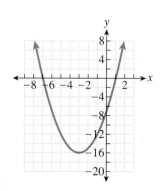

63. Use the graph of $y = x^2 + 6x - 7$ to solve $x^2 + 6x - 7 < 0$.

64. Use the graph of $y = -x^2 + 3x + 4$ to solve $-x^2 + 3x + 4 \geq 0$.

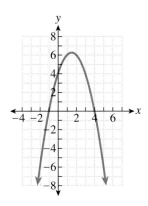

65. Use the graph of $y = -x^2 + 8x - 19$ to solve $-x^2 + 8x - 19 > 0$.

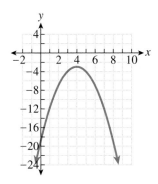

66. Use the graph of $y = x^3 + x^2 - 10x + 8$ to solve $x^3 + x^2 - 10x + 8 \geq 0$.

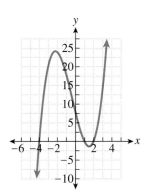

Answer in complete sentences.

67. Explain how to solve a quadratic inequality. Use an example to illustrate the process. Be sure to explain how to determine whether to use open or closed circles when graphing the solution on the number line.

68. Explain how to solve a rational inequality. Use an example to illustrate the process. Be sure to explain how to determine whether to use open or closed circles when graphing the solution on the number line.

Study Tip REVISITED   Many students have a job, and this commitment takes a great deal of time away from their studies. If possible, find a job that allows you opportunities to study while you work. For example, if you are working at the check-out counter of the library, your boss may allow you to study while there are no students waiting to check out books. If you are unable to study at work, be sure to take full advantage of any breaks that you receive. A 10-minute work break may not seem like a lot of time, but you can read through a series of note cards or review your class notes during this time.

Chapter 10 Summary

Review of Chapter 10 Study Tips

Poor time management can cause a student to do poorly in a math class. To maximize the time you have available to study mathematics, and to make your use of this time effective, consider the following:

- Record how you spend your time over the period of one week. Look for how you might be wasting time. Create an efficient schedule to follow.
- Study new material as soon as possible after class. Save review work for later in the day.
- Arrange your schedule to allow you to study mathematics at the time of day when your mental energy is at its highest.
- Set goals for each study session, and record these goals in a "to do" list.
- Study your most difficult subject first, before you become fatigued.
- If you must work, try to find a job that will allow you to do some studying while you work. At the very least, make efficient use of your work breaks to study.

Solve by factoring. [10.1]

1. $x^2 + 11x + 28 = 0$

2. $x^2 + 7x - 30 = 0$

3. $x^2 + 4x = 0$

4. $2x^2 - 13x + 15 = 0$

5. $x^2 + 7x - 12 = 3x - 7$

6. $x^2 - 9x + 23 = 9(x - 6)$

Solve by extracting square roots. [10.1]

7. $x^2 = 49$

8. $x^2 - 72 = 0$

9. $x^2 + 4 = 0$

10. $(x - 6)^2 = -36$

11. $(x - 8)^2 + 30 = 21$

12. $(2x + 7)^2 + 20 = 141$

Solve by completing the square. [10.1]

13. $x^2 - 2x - 8 = 0$

14. $x^2 - 6x + 3 = 0$

15. $x^2 + 14x + 47 = 0$

16. $x^2 + 5x + 10 = 0$

Solve by using the quadratic formula. [10.2]

17. $x^2 + 7x + 1 = 0$

18. $x^2 + 4x - 2 = 0$

19. $x^2 - 3x + 9 = 0$

20. $4x^2 - 20x + 29 = 0$

21. $2x^2 - 5x + 9 = x^2 - 9x + 30$

22. $x(x + 4) = 16x - 37$

Solve by using a u-substitution. [10.4]

23. $x^4 - 5x^2 + 4 = 0$

24. $x - 4\sqrt{x} - 32 = 0$

25. $x^6 + 7x^3 - 8 = 0$

26. $x^{2/3} - 4x^{1/3} - 21 = 0$

Solve. [10.4]

27. $\sqrt{x^2 + x - 2} = 2$

28. $\sqrt{x + 15} - x = 9$

Solve. [10.4]

29. $1 + \dfrac{11}{x} + \dfrac{30}{x^2} = 0$

30. $\dfrac{x}{x - 3} - \dfrac{7}{x - 2} = \dfrac{3}{x^2 - 5x + 6}$

Solve using the most effecient method. [10.1, 10.2, 10.4]

31. $x^2 - 19x + 90 = 0$

32. $x^4 + 5x^2 - 36 = 0$

33. $(x + 6)^2 + 11 = 3$

34. $x^2 + 4x + 16 = 0$

35. $x^2 + 10x = 0$

36. $x^2 - 6x - 20 = 0$

37. $\sqrt{3x + 49} = x + 3$

38. $\sqrt{2x + 9} = \sqrt{4x + 1} - 2$

39. $x - 1 - \dfrac{20}{x} = 0$

40. $3x^2 - 8x + 2 = 0$

41. $x + 5x^{1/2} - 24 = 0$

42. $(x + 9)^2 + 64 = 0$

For the given function $f(x)$, *find all values x for which* $f(x) = 0$. [10.1, 10.2, 10.4]

43. $f(x) = (x + 12)^2 - 16$

44. $f(x) = x^2 - 7x - 20$

Find a quadratic equation with integer coefficients that has the given solutions. [10.1, 10.2, 10.4]

45. $\{4, -9\}$

46. $\{4\}$

47. $\{\frac{4}{3}, \frac{13}{2}\}$

48. $\{2\sqrt{2}, -2\sqrt{2}\}$

49. $\{5i, -5i\}$

50. $\{8i, -8i\}$

Find the vertex and axis of symmetry of the parabola associated with the given quadratic equation. [10.5]

51. $y = x^2 - 8x - 20$

52. $y = x^2 + 5x - 16$

53. $y = -x^2 - 2x + 35$

54. $y = (x - 3)^2 - 4$

Find the x- and y-intercepts of the parabola associated with the given equation. If the parabola does not have any x-intercepts, state "no x-intercepts." [10.5]

55. $y = x^2 + 6x + 11$

56. $y = x^2 - 4x - 3$

57. $y = -x^2 + 10x - 25$

58. $y = (x + 4)^2 - 8$

Graph the quadratic equation. Label the vertex, the y-intercept, and any x-intercepts. [10.5]

59. $y = x^2 + 8x + 15$

60. $y = -x^2 + 2x + 3$

61. $y = x^2 + 4x - 2$

62. $y = x^2 + 4x + 8$

63. $y = -(x + 3)^2 + 4$

Solve the rational inequality. Express your solution on a number line using interval notation. [10.6]

69. $\dfrac{x^2 - 15x + 54}{x - 7} \le 0$

70. $\dfrac{x^2 - 4}{x^2 + 2x - 80} \ge 0$

Given the function $f(x)$, solve the inequality $f(x) \ge 0$. Express your solution on a number line using interval notation. [10.6]

71. $f(x) = x^2 + 15x + 50$

64. $y = -(x - 6)^2 + 1$

72. $f(x) = \dfrac{x^2 - 16}{x^2 - 10x + 9}$

73. A rectangular photograph has an area of 288 square centimeters. If the width of the photo is 2 centimeters longer that its length, find the dimensions of the photo. [10.3]

74. A rectangular swimming pool covers an area of 500 square feet. If the length of the pool is 13 feet more than its width, find the dimensions of the pool. Round to the nearest tenth of a foot. [10.3]

75. Ann needs to fly to a city that is 1000 miles north and 800 miles east of where she lives. If the airplane flies a direct route, how far will the flight be? Round to the nearest tenth of a mile. [10.3]

76. The diagonal of a rectangular workbench is 12 feet long. If the width of the workbench is 3 feet less than the length of the workbench, find the dimensions of the workbench. Round to the nearest tenth of a foot. [10.3]

Solve the quadratic inequality. Express your solution on a number line using interval notation. [10.6]

65. $x^2 - 9x + 14 \le 0$

66. $x^2 + 9x + 8 > 0$

67. $x^2 + 4x - 11 > 0$

For Exercises 77 and 78, use the function
$h(t) = -16t^2 + v_0 t + s.$ [10.3]

77. A projectile is launched upwards from the ground at a speed of 104 feet per second. How long will it take until the projectile lands on the ground?

78. A projectile is launched upward from the roof of a building 30 feet high with an initial velocity of 66 feet per second. How long will it take until the projectile lands on the ground? Round to the nearest hundredth of a second.

68. $x^2 + 3x + 40 \ge 0$

79. A water tank is connected to two pipes. When both pipes are open, they can fill the tank in 4 hours. The larger of the pipes can fill the tank in 2 hours less time than the smaller pipe. How long would it take the smaller pipe, working alone, to fill the tank? Round to the nearest tenth of an hour. [10.4]

80. A tree-trimming company has two crews. If both crews work together, they can trim all of the trees on campus in 32 hours of work. The faster crew could trim all of the trees in 6 hours less than the slower crew. How long would it take the faster crew, working alone, to trim all of the trees on campus? Round to the nearest tenth of an hour. [10.4]

1. Solve by factoring. $x^2 + 9x + 18 = 0$

2. Solve by extracting square roots. $(3x - 8)^2 - 6 = 58$

3. Solve by completing the square. $x^2 - 14x - 3 = 0$

4. Solve by using the quadratic formula.
$x^2 - 6x + 12 = 0$

5. Solve by using a u-substitution. $x^4 - 8x^2 - 48 = 0$

6. Solve by using a u-substitution. $x + 8x^{1/2} - 9 = 0$

Solve.

7. $(4x - 1)^2 - 13 = 12$

8. $2x^2 - 15x - 27 = 0$

9. $x^2 - 14x + 40 = 0$

10. $\dfrac{x}{x + 8} - \dfrac{6}{x - 4} = \dfrac{8}{x^2 + 4x - 32}$

11. $x^2 - 10x + 34 = 0$

Find a quadratic equation with integer coefficients that has the given solutions.

12. $\left\{5, \frac{4}{7}\right\}$

13. $\{4i, -4i\}$

14. Find the x- and y-intercepts of the parabola associated with $y = x^2 - 10x + 20$. Round to the nearest tenth if necessary.

15. Graph $y = -x^2 - 10x + 11$. Label the vertex, the y-intercept, and any x-intercepts.

16. Graph $y = (x - 5)^2 - 12$. Label the vertex, the y-intercept, and any x-intercepts.

17. Solve. $x^2 + 3x - 28 > 0$

18. Solve. $\dfrac{x^2 + 11x + 24}{x^2 - 5x - 14} \leq 0$

19. The area of a rectangular playing field is 3600 square yards. If the width of the field is 60 yards less than the length of the field, find the dimensions of the field. Round to the nearest tenth of a yard.

20. Daniel flies to a resort town 1500 miles south and 800 miles west of where he lives. How far away is the resort town from where he lives?

Mathematicians in History
Evariste Galois

Evariste Galois was a brilliant mathematician who was interested in the algebraic solutions of equations. The amount and depth of work accomplished by Galois, who died before his 21st birthday, is legendary. Galois showed that there is no general solution of equations that are fifth degree or higher:

$$ax^5 + bx^4 + cx^3 + dx^2 + ex + f = 0.$$

Write a one-page summary (OR make a poster) of the mathematical achievements of Galois, his fascinating (but short) life, and the details surrounding his untimely death.

Interesting issues:

- When and where was Galois born?
- What was the fate of Galois's father?
- Galois twice failed the admittance exam to École Polytechnique. What did he do to one of the examiners on his second attempt?
- Galois was expelled from École Normale in December 1830. Why was he expelled?
- Galois was arrested in 1831 after raising a toast to King Louis-Philippe. What was the toast, and how did he make it?
- In 1831, Galois was arrested on Bastille Day. What was the charge?
- On the night before his death, Galois wrote a letter to his friend Auguste Chevalier. What were the contents of the letter?
- What were the circumstances that led to Galois's death? How old was Galois at the time?
- Which mathematician published Galois's papers in 1846?

The pilot of a small plane wished to fly due east. However, there is a powerful wind blowing due north which is causing the plane to fly 7° off of its eastern course. The wind has also increased the speed of the plane so that it is flying at 287 miles per hour. The speed of the plane in still air is 250 miles per hour faster than the speed of the wind. Find the speed of the plane in still air and of the wind.

11

FUNCTIONS

In this chapter, we will continue our work with functions. In addition to linear and quadratic functions, we will examine new functions such as the square root function and the cubic function. We will also cover the algebra of functions, learning ways to create a new function from two or more existing functions. We finish the chapter by discussing inverse functions.

Study Tip STUDY ENVIRONMENT   *It is very important to study in the proper environment. In this chapter, we will focus on how to create an environment that allows you to get the most out of your study sessions. We will discuss where to study and what the conditions should be like in that location.*

11.1

**REVIEW
OF FUNCTIONS**

Objectives

1 Review of functions, domain, and range.
2 Represent a function as a set of ordered pairs.
3 Use the vertical-line test to determine whether a graph represents a function.
4 Use function notation.
5 Evaluate functions.
6 Interpret graphs of functions.

Functions; Domain and Range

Objective 1   **Review of functions, domain, and range.** A college charges its students a $50 registration fee in addition to an enrollment fee of $18 per unit. A student who is taking 12 units this semester would be charged $216 for their classes (12 units times $18 per unit) plus the $50 registration fee, or a total of $266. The amount a student pays in fees depends on the number of units in which a student is enrolled. We say that the amount a student pays is a function of the number of units in which a student is enrolled. In general, if a student is taking x units, then the student's fees can be found from the formula $50 + 18x$.

Recall from Chapter 3 that a **function** is a rule that takes an input value and assigns a particular output value to it. In the previous example, the input value would be the number of units the student was taking, and the output value would be the total fees for that student. For a rule to define a function, it is necessary that each input value be assigned one and only one output value. In terms of the previous example, this means that students who are taking the same number of units pay the same amount of fees. It is not possible for two students to each take 12 units, but pay a different amount of fees.

As we first saw in Chapter 3, the set of input values for a function is called the **domain** of the function. The domain for the previous example is the set of units in which a student can be enrolled. The possible values in the domain would be the set $\{1, 2, 3, \ldots\}$. The set of output values for a function is called the **range** of the function. Here is a table showing some of the possible values in the range of this function:

Domain (Units)	1	2	3	. . .	x
Range (Fees, $)	$50 + 18(1) = 68$	$50 + 18(2) = 86$	$50 + 18(3) = 104$	. . .	$50 + 18x$

If there is no maximum number of units that a student can take, then this table would continue indefinitely.

Although many of our functions will be given by a formula, there are many functions that will not. For instance, consider the set of students enrolled in your math class. If we asked each student in your class to tell us the month of his or her birthday, then a function would exist in which the input value is a student and the output value is the month of the student's birthday. The domain of this function is the set of students in your math class, and the range is the set of months from January through December.

EXAMPLE 1   The World Cup is a soccer tournament that is held every four years. Here are the winners of the World Cup for the years 1930–2002 (the tournament was not held in 1942 or 1946 due to World War II):

Year	Winner	Year	Winner
1930	Uruguay	1974	West Germany
1934	Italy	1978	Argentina
1938	Italy	1982	Italy
1950	Uruguay	1986	Argentina
1954	West Germany	1990	West Germany
1958	Brazil	1994	Brazil
1962	Brazil	1998	France
1966	England	2002	Brazil
1970	Brazil		

a) Does a function exist for which the input value is the year of the tournament and the output value is the winner of that tournament?

Solution

Since each year listed has only one winner, this correspondence is a function. This is true even though some teams, such as Italy, have won more than once. By the definition of a function, each input value must be associated with one and only one output value. It is possible that more than one input value can be associated with the same output value.

   The domain of this function is the set of years in which the World Cup has been held, and the range of this function is the set of teams that have won the World Cup.

b) Does a function exist for which the input value is the winner of a tournament and the output value is the year that the country won the tournament?

Solution

No, this correspondence is not a function, as one input value could be associated with more than one output value. For instance, the input value Italy is associated with the years 1934, 1938, and 1982.

Quick Check   **1**   **Here are the names and ages of the five members of a study group:**

Name	Tiffany	Tyler	Teresa	Tonya	Tina
Age	19	22	19	20	24

a) Does a function exist for which the input value is the name of a member of the study group and the output value is the age of that person?
b) Does a function exist for which the input value is the age of a member of the study group and the output value is the name of the person?

Ordered-Pair Notation

Objective 2   **Represent a function as a set of ordered pairs.**   Functions can be expressed as a set of ordered pairs. The first coordinate of the ordered pairs will be the input values of the function, and the second coordinate will be the corresponding output value. A set of ordered pairs represents a function if none of the first coordinates are repeated.

EXAMPLE 2   Determine whether the set of ordered pairs is a function. If it is a function, state its domain and range.
$$\{(-3, 9), (-2, 4), (-1, 1), (0, 0), (1, 1), (2, 4), (3, 9)\}$$

Solution

This set of ordered pairs is a function because no first coordinate is repeated. Notice that three pairs of ordered pairs share the same second coordinate. This does not violate the definition of a function. Although each possible input value of a function can be associated with only one output value, a particular output value of a function can be associated with several different input values.

The domain of this function is the set $\{-3, -2, -1, 0, 1, 2, 3\}$. The range is $\{0, 1, 4, 9\}$.

Quick Check **2** Determine whether the set of ordered pairs is a function. If it is a function, state its domain and range.
$$\{(16, -3), (13, -2), (10, -1), (7, 0), (10, 1)\}$$

Vertical-Line Test

Objective **3** **Use the vertical-line test to determine whether a graph represents a function.** Although we have only examined sets of ordered pairs that have a **finite,** or limited, number of ordered pairs, the graph of an equation may represent a function as well. The graph of an equation represents an **infinite,** or unlimited, number of ordered pairs. If there are two ordered pairs on the graph of an equation that share the same x-coordinate, but have different y-coordinates, then the graph does not represent a function.

EXAMPLE 3   Determine whether the graph below represents a function.

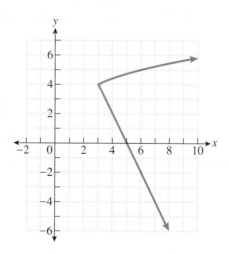

Solution

This graph does not represent a function, as there are points on the graph that have the same x-coordinates, but different y-coordinates. For example, the points $(6, 5)$ and $(6, -2)$ have the same x-coordinate, but different y-coordinates. This means that one input value is associated with more than one output value, violating the definition of a function.

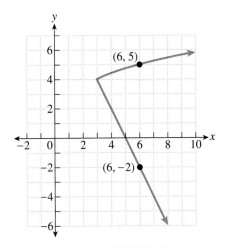

To determine whether a graph represents a function, we will apply the **vertical-line test.**

The Vertical-Line Test

> If a vertical line can be drawn that intersects a graph at more than one point, then the graph does not represent a function.

The graph from the previous example fails the vertical-line test and therefore does not represent a function.

EXAMPLE 4   Use the vertical-line test to determine whether the graph below represents a function.

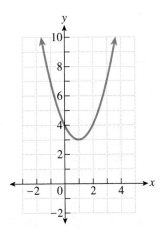

Solution

We cannot draw a vertical line that intersects this graph at more than one point. This graph passes the vertical-line test, so it does represent a function.

Quick Check **3** Use the vertical-line test to determine whether the graph at the right represents a function.

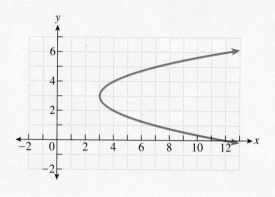

Function Notation

Objective 4 Use function notation. Consider the following graph of the equation $y = 2x + 5$:

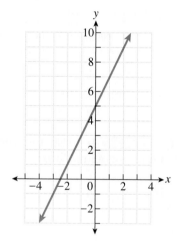

Since the graph passes the vertical-line test, the graph represents a function. In this case, the equation is solved for y in terms of x and we say that y is a function of x. The input of this function is x and the output is y. The value of y depends on the value of x, and we say that y is the **dependent variable** in the equation. Typically, the equation will be solved for the dependent variable in terms of the other variable. The other variable in an equation is called the **independent variable.** In this equation, the variable x is the independent variable.

If we say that y is a function of x, this can be expressed symbolically as $y = f(x)$. We read $f(x)$ as "f of x." The letter f is the name of the function, and x represents the input value of the function. The input variable is the independent variable. The notation $f(x)$ represents the output value of a function f for the input value x.

> **A Word of Caution**   The notation $f(x)$ represents the *output* of function f for the input value x. It does not indicate the product of f and x. We often use the letter f for the name of a function, but we may choose any letter that we desire, such as g, h, F, or even a Greek letter such as ϕ (phi).

Since $y = 2x + 5$ is a function, we can also write it equivalently as $f(x) = 2x + 5$. When an equation contains $f(x)$ instead of y, the equation is expressed in **function notation.**

$f(x) = 2x + 5$	$f(x) = 2x + 5$	$f(x) = 2x + 5$
⇑	⇑	⇑
The variable inside the parentheses is the input variable for the function.	$f(x)$ is the output value of the function f when the input value is x.	The expression on the right side of the equation is the formula for this function.

Evaluating Functions

Objective 5   **Evaluate functions.**   When we find the output value of a function $f(x)$ for a particular value of x, this is called **evaluating** the function. To evaluate a function for a particular value of the variable, we substitute that value for the variable in the function's formula, and then we simplify the resulting expression.

EXAMPLE 5   Let $h(x) = x^2 + 7x - 30$. Find $h(-8)$.

Solution

In this example, we need to square the input value. Keep in mind that when -8 is squared the result is positive. The use of parentheses should help.

$$
\begin{aligned}
h(-8) &= (-8)^2 + 7(-8) - 30 && \text{Substitute } -8 \text{ for } x. \\
&= -22 && \text{Simplify.}
\end{aligned}
$$

Quick Check 4

Let
$f(x) = x^2 - 3x - 20.$
Find $f(-5).$

EXAMPLE 6   Let $f(x) = 7x - 5$. Find $f(3a + 2)$.

Solution

In this example, we are substituting a variable expression, $3a + 2$, for the input variable x rather than substituting a number as in previous examples.

Quick Check 5

Let $f(x) = -4x + 21.$
Find $f(3m - 9).$

$$
\begin{aligned}
f(3a + 2) &= 7(3a + 2) - 5 && \text{Substitute } 3a + 2 \text{ for } x. \\
&= 21a + 14 - 5 && \text{Multiply.} \\
&= 21a + 9 && \text{Combine like terms.}
\end{aligned}
$$

Interpreting Graphs

Objective 6 **Interpret graphs of functions.** The ability to read and interpret a graph is a valuable skill, as the graph of a function reveals important information about the function itself. To begin, we can determine the domain and range of a function from its graph. Recall that the domain of a function is the set of all possible input values for that function. This can be read from left to right along the x-axis on a graph. Since the range of a function is the set of all possible output values for the function, this can be read vertically from the bottom of the graph to the top along the y-axis. The following graph of a function $f(x)$ will help us to summarize these ideas:

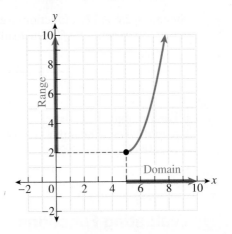

The graph begins at the point $(5, 2)$ and extends upward to the right. We can express the domain of the function in interval notation as $[5, \infty)$, as the set of possible input values begins at $x = 5$ and continues without bound. The range of the function is $[2, \infty)$, as the lowest function value is 2, and the values of the function increase without bound from there.

Consider the graph of $f(x)$ shown below. This graph extends indefinitely to the left and to the right, so the domain of the function $f(x)$ is the set of all real numbers $\Re$. However, this function does have a minimum value of -6, as the point $(-2, -6)$ is the lowest point on the graph. The function increases without limit, so the range of the function $f(x)$ is $[-6, \infty)$.

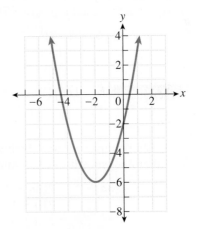

We can use the graph of a function to evaluate the function for a particular input value, even if we do not know the formula for the function.

Suppose we wanted to use the graph of a function $f(x)$ to evaluate the function when $x = 6$, or, in other words, find $f(6)$. The point on the graph that has an x-coordinate of 6 is the point $(6, 5)$. Since the y-coordinate is 5, we know that $f(6) = 5$.

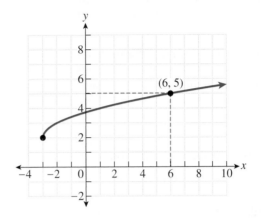

Finally, we can use the graph of a function to determine which input value(s) produce a particular output value.

Suppose that we wanted to use the graph of a function $f(x)$ to determine the values of x for which the function $f(x) = -7$. We are looking for a point or points on the graph whose y-coordinate is -7. There are two such points, $(-4, -7)$ and $(-6, -7)$, on this graph with a y-coordinate of -7. $f(x) = -7$ when $x = -4$ and when $x = -6$.

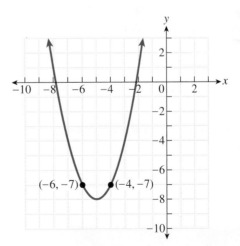

1. Does a function exist for which the input value is a college graduate and the output value is the starting salary for that graduate's first job? Explain why or why not.

2. Does a function exist for which the input value is a woman and the output value is the number of children that the woman has? Explain why or why not.

3. Does a function exist for which the input value is a person and the output value is the number of jobs that the person has held in their lifetime? Explain why or why not.

4. Does a function exist for which the input value is an orchestra and the output value is the number of players in that orchestra? Explain why or why not.

5. Does a function exist for which the input value is a temperature in degrees Fahrenheit and the output value is a city that had that temperature as its high temperature last Tuesday? Explain why or why not.

6. Does a function exist for which the input value is the price per share of a stock and the output value is the name of the stock? Explain why or why not.

Determine whether the set of ordered pairs is a function. If it is a function, state its domain and range.

7. {(Bruce Springsteen, "Born to Run"), (Bob Dylan, "Tangled Up in Blue"), (Madonna, "Material Girl"), (Nirvana, "Smells Like Teen Spirit"), (Tone Loc, "Wild Thing"), (TLC, "Waterfalls")}

8. {(1999, New York Yankees), (2000, New York Yankees), (2001, Arizona Diamondbacks), (2002, Anaheim Angels)}

9. $\{(5, -5), (3, -3), (1, -1), (-1, 1), (-3, 3)(-5, 5)\}$

10. $\{(-2, 8), (-1, 6), (0, 4), (1, 2), (2, 0), (3, -2), (4, -4), (5, -6), (6, -8)\}$

11. $\{(-6, 5), (-3, 5), (0, 5), (3, 5), (6, 5)\}$

12. $\{(9, -3), (4, -2), (1, -1), (0, 0), (1, 1), (4, 2), (9, 3)\}$

Use the vertical-line test to determine whether the graph represents a function.

13.

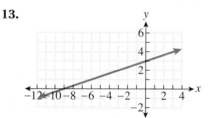

14.

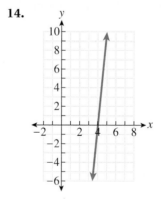

15.

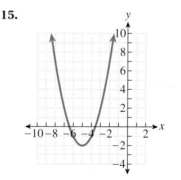

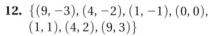

16.

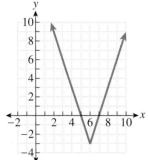

17.

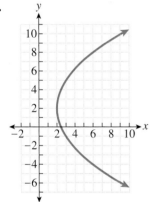

18.

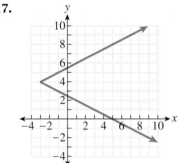

19.

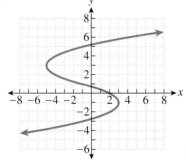

20.

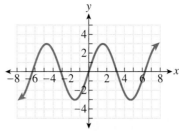

Evaluate the given function.

21. $f(x) = x - 6, f(-2)$

22. $f(x) = x + 19, f(-26)$

23. $f(x) = -2x + 20, f(7)$

24. $f(x) = 8x - 43, f(-9)$

25. $f(x) = \dfrac{3}{4}x - 9, f(16)$

26. $f(x) = \dfrac{5}{8}x + \dfrac{13}{6}, f\left(\dfrac{5}{3}\right)$

27. $f(x) = 3x - 15, f(a - 5)$

28. $f(x) = 2x + 21, f(2a + 10)$

29. $f(x) = -7x + 19, f(4n - 3)$

30. $f(x) = -5x + 32, f(-3m - 24)$

31. $f(x) = x^2 + 5x + 12, f(8)$

32. $f(x) = x^2 + 3x + 20, f(5)$

33. $f(x) = x^2 + 13x + 4, f(-6)$

34. $f(x) = x^2 + 8x + 11, f(-9)$

35. $f(x) = (x - 4)^2 + 17, f(11)$

36. $f(x) = (x + 12)^2 + 35, f(-6)$

37. $f(x) = |x - 4| + 11, f(16)$

38. $f(x) = |2x + 3| - 13, f(8)$

Determine the domain and range of the function $f(x)$ on the graph.

39.

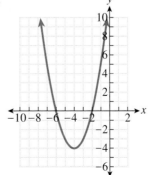

40.

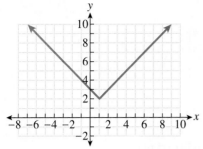

41.

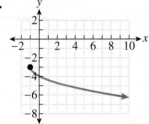

42.

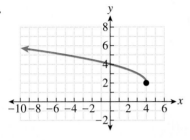

Determine the x- and y-intercepts, if they exist, of the function f(x) on the graph.

43.

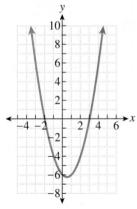

44.

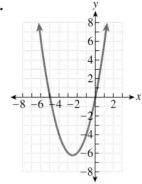

Use the graph of the function f(x) to find the indicated function value.

45. $f(-3)$

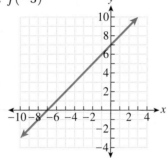

46. $f(8)$

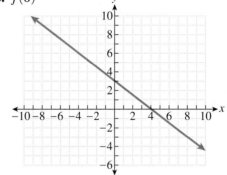

47. $f(-2)$

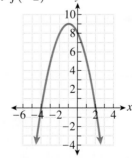

48. $f(4)$

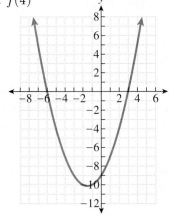

Use the graph of the function $f(x)$ to determine which values of x satisfy the given equation.

49. $f(x) = -4$

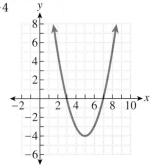

50. $f(x) = -2$

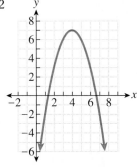

51. $f(x) = -9$

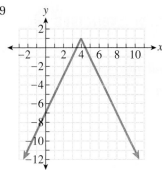

52. $f(x) = -4$

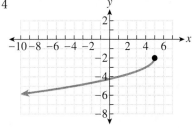

Answer in complete sentences.

53. If a vertical line can be drawn that intersects a graph at two points, explain why this graph does not represent a function.

54. Explain how to find the domain and range of a function from its graph.

55. Given the graph of a function $f(x)$, explain how to find all values x for which $f(x) = 2$. Use an example to illustrate the process.

56. Given the graph of a function $f(x)$, explain how to find $f(2)$. Use an example to illustrate the process.

QUICK REVIEW EXERCISES

Section 11.1

Graph. Label all intercepts.

1. $y = 3x - 9$

2. $4x - 3y = -24$

3. $y = x^2 - 8x + 7$

4. $y = -(x - 4)^2 + 1$

Study Tip **REVISITED**   When choosing a location to study, it is important to select a location that is free from distractions. For example, if you study in a room that has a television on, you will find yourself repeatedly glancing up at the television. This steals time and attention from the task at hand. Working in a room that constantly has people entering and exiting can be distracting as well.

Establish locations that will only be used for studying, not for other purposes. When you sit down to study, you will feel as if you have entered your workplace. This will help you to focus on your studies. As an exercise, keep a journal of things that distract your attention while studying. In the future, choose a location that will minimize your exposure to these distractions.

11.2
LINEAR FUNCTIONS

Objectives

1 Graph linear functions.
2 Determine a linear function from its graph.
3 Determine linear functions from data.
4 Solve applied problems involving linear profit functions.

As discussed in Chapter 3, a linear function is a function that can be written in the form $f(x) = mx + b$, where m and b are real numbers.

Graphing Linear Functions

Objective 1 Graph linear functions. Evaluating a function $f(x)$ produces ordered pairs of the form $(x, f(x))$. If $f(x)$ is a linear function, then these ordered pairs will be points on a line. Recall from Chapter 3 that the graph of a linear function is a line, and the two characteristics that determine the graph of a line are its y-intercept and its slope. To graph a linear function, $f(x) = mx + b$, we begin by putting the y-intercept, the point $(0, b)$, on the graph.

Once the y-intercept is on the graph, we use the slope m of the line to find additional points on the graph. The slope of a line is a measure of how quickly the function increases or decreases as x increases. In other words, the slope of a line is the rate of change of the function with respect to changes in x. A line with positive slope rises as it moves to the right. A line with negative slope falls as it moves to the right. In general, the numerator of the slope is the amount of increase or decrease in y as x increases by the amount in the denominator.

By using the slope m, we are able to find a second point on the line in addition to the y-intercept. We only need to find two points in order to graph a line, but we will add the x-intercept to our graph as a third point. In addition to serving as a check of our work with the y-intercept and the point found by using the slope, the x-intercept of a function is an important point in its own right. The x-intercept of a function tells us where the function is equal to 0. In an applied problem, the x-intercept can represent the break-even point in a business problem or the instant that a projectile lands on the ground. To find the x-intercept, we set the function equal to 0 and solve the resulting equation.

EXAMPLE 1   Graph $f(x) = -\frac{2}{3}x + 4$.

Solution

We begin by plotting the y-intercept $(0, 4)$ on the graph.

The slope of this line is $m = -\frac{2}{3}$. Beginning at the y-intercept, we can find a second point on the line by moving down two units and to the right three units, which is at the point $(3, 2)$. We can find a third point on the graph by determining the coordinates of the x-intercept.

$$-\frac{2}{3}x + 4 = 0 \qquad \text{Set } f(x) \text{ equal to } 0.$$

$$3\left(-\frac{2}{3}x + 4\right) = 3 \cdot 0 \qquad \text{Multiply both sides of the equation by 3 to clear the equation of fractions.}$$

$$\overset{1}{\cancel{3}} \cdot \left(-\frac{2}{\underset{1}{\cancel{3}}}x\right) + 3 \cdot 4 = 3 \cdot 0 \qquad \text{Distribute and divide out common factors.}$$

$$-2x + 12 = 0 \qquad \text{Simplify.}$$

$$12 = 2x \qquad \text{Add } 2x \text{ to both sides to isolate the term containing } x.$$

$$6 = x \qquad \text{Divide both sides by 2.}$$

The x-intercept is at $(6, 0)$. We finish by graphing a line that passes through these three points.

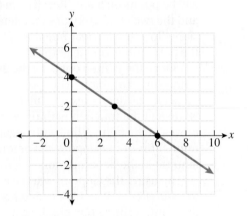

Quick Check **1** Graph $f(x) = \frac{3}{4}x - 6$.

Some functions do not have an x-intercept. One example is a **constant function** of the form $f(x) = b$. The graph of this function is a horizontal line that has a y-intercept at the point $(0, b)$. The reason that this type of function is called a constant function is that its output value remains the same, or stays constant, regardless of the input value x.

EXAMPLE **2**   Graph $f(x) = 2$.

Solution

The graph of this constant function is a horizontal line with a y-intercept at $(0, 2)$.

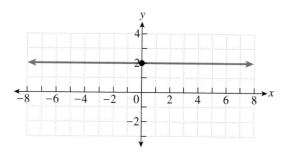

Quick Check **2** Graph $f(x) = -5$.

Finding a Linear Function for a Given Situation

Objective 2 **Determine a linear function from its graph.** We can determine the linear function $f(x) = mx + b$ that is graphed if we can find the coordinates of at least two points that lie on its graph. We begin by finding m. If the line passes through the points (x_1, y_1) and (x_2, y_2), then we can find the slope of the line using the formula $m = \dfrac{y_2 - y_1}{x_2 - x_1}$.

Consider the line passing through the points $(1, 7)$ and $(3, 4)$, as shown in the graph at the right. We can substitute 1 for x_1, 7 for y_1, 3 for x_2 and 4 for y_2 into the formula $m = \dfrac{y_2 - y_1}{x_2 - x_1}$ to find m.

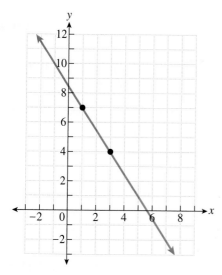

$$m = \frac{4 - 7}{3 - 1} \qquad \text{Substitute into } m = \frac{y_2 - y_1}{x_2 - x_1}.$$

$$m = -\frac{3}{2} \qquad \text{Simplify the numerator and denominator.}$$

Now that we have found the slope, we know that the function is of the form $f(x) = -\frac{3}{2}x + b$. We turn our attention to finding b. We choose one of the points on the line, substitute its coordinates in the function, and solve for b. Since the point $(1, 7)$ is on the graph of the function, we know that $f(1) = 7$.

$$f(1) = 7$$

$$-\frac{3}{2}(1) + b = 7 \qquad \text{Substitute 1 for } x \text{ in the function } f(x).$$

$$-\frac{3}{2} + b = 7 \qquad \text{Multiply.}$$

$$2 \cdot \left(-\frac{3}{2} + b\right) = 2 \cdot 7 \qquad \begin{array}{l}\text{Multiply both sides of the equation by 2 to clear} \\ \text{the equation of fractions.}\end{array}$$

$$\overset{1}{\cancel{2}} \cdot \left(-\frac{3}{\underset{1}{\cancel{2}}}\right) + 2 \cdot b = 2 \cdot 7 \qquad \text{Distribute and divide out common factors.}$$

$$-3 + 2b = 14 \qquad \text{Simplify.}$$

$$2b = 17 \qquad \text{Add 3 to isolate the term containing } b.$$

$$b = \frac{17}{2} \qquad \text{Divide both sides by 2.}$$

Replacing b by $\frac{17}{2}$, we find that the function is $f(x) = -\frac{3}{2}x + \frac{17}{2}$.

To Determine a Linear Function from Its Graph

1. Determine the coordinates of two points (x_1, y_1) and (x_2, y_2) on the line.
2. Find the slope m of the line using the formula $m = \dfrac{y_2 - y_1}{x_2 - x_1}$.
3. Substitute the value found for m in the formula for a linear function, $f(x) = mx + b$. Then set $f(x_1) = y_1$, or $f(x_2) = y_2$, and solve for b.
4. Substitute the value for b into the formula $f(x) = mx + b$.

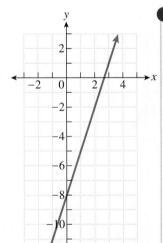

EXAMPLE 3 Use the given graph to determine the linear function $f(x)$ on the graph.

Solution

We begin by determining the coordinates of any two points on the line. This line passes through the points $(2, -2)$ and $(3, 1)$. Now we find the slope m of the line.

$$m = \frac{1 - (-2)}{3 - 2} \qquad \text{Substitute into } m = \frac{y_2 - y_1}{x_2 - x_1}.$$

$$m = 3 \qquad \text{Simplify.}$$

The function is of the form $f(x) = 3x + b$. Since the point $(3, 1)$ is on the graph of the function, we know that $f(3) = 1$. We can use this fact to find b.

$$f(3) = 1$$

$$3(3) + b = 1 \qquad \text{Substitute 3 for } x \text{ in the function } f(x).$$

$$9 + b = 1 \qquad \text{Multiply.}$$

$$b = -8 \qquad \text{Subtract 9 to isolate } b.$$

Replacing b by -8, we find that the function is $f(x) = 3x - 8$.

Quick Check **3** Use the given graph to determine the linear function $f(x)$ on the graph.

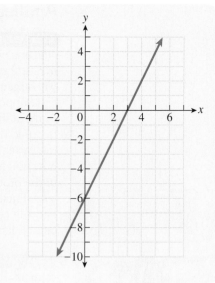

Finding a Linear Function from Data

Objective **3** Determine linear functions from data.

EXAMPLE **4** The freezing point of water on the Celsius scale is 0° C, while on the Fahrenheit scale it is 32° F. The boiling point of water on the Celsius scale is 100° C, while on the Fahrenheit scale it is 212° F. Find a linear function $F(x) = mx + b$ whose input is a Celsius temperature and whose output is the corresponding Fahrenheit temperature.

Solution

Since a Celsius temperature of 0° C corresponds to a Fahrenheit temperature of 32° F, we know that $F(0) = 32$. In a similar fashion, we know that $F(100) = 212$. We can find the slope of this linear function by using $(0, 32)$ for (x_1, y_1) and $(100, 212)$ for (x_2, y_2).

$$m = \frac{212 - 32}{100 - 0} \qquad \text{Substitute.}$$

$$m = \frac{9}{5} \qquad \text{Simplify the fraction to lowest terms.}$$

The function is of the form $F(x) = \frac{9}{5}x + b$. We will now use the fact that $F(0) = 32$ to find b.

$$F(0) = 32$$

$$\frac{9}{5}(0) + b = 32 \qquad \text{Substitute 0 for } x \text{ in the function } F(x).$$

$$b = 32 \qquad \text{Simplify.}$$

The function that converts a Celsius temperature x to its corresponding Fahrenheit temperature is $F(x) = \frac{9}{5}x + 32$.

Quick Check **4**
Mary held her wedding reception at a country club. There were 150 guests at the reception, and Mary was charged $3500. Juliet held her wedding reception at the same country club and was charged $2100 for 80 guests. Find a linear function $f(x) = mx + b$ whose input is the number of guests at a reception and whose output is the corresponding charge by the country club.

Applications of Linear Functions

Objective **4** **Solve applied problems involving linear profit functions.** A specific business application that can involve linear functions is determining the cost to manufacture x units of a product. Functions can also be used to determine the revenue that will be generated by selling those x units, as well as the profit that will result from their sale.

   Suppose that a company is manufacturing items. Such a company has fixed costs (rent, utilities, etc.), as well as variable costs that depend on the number of items manufactured (materials, payroll, etc.). The **cost function** $C(x)$ is equal to the fixed costs plus the variable costs associated with producing x items. The **revenue function** $R(x)$ tells us how much money will be taken in by selling these x items. The revenue function is equal to the product of the selling price of each item and the number of those particular items sold. Finally, the **profit function** $P(x)$ is equal to the difference of the revenue function and the cost function; that is, $P(x) = R(x) - C(x)$. Recall that profit is how much revenue is left over after paying costs.

EXAMPLE 5   A student club decides to sell burgers on campus to raise money for a scholarship fund. The club must pay the college a $25 fee to reserve space for their booth. In addition, they must spend another $40 for condiments and paper goods. They determine that it will cost $0.55 for ingredients to make each burger, and they plan to sell their burgers for $3 each.

a) Find the cost function $C(x)$, the revenue function $R(x)$, and the profit function $P(x)$.

Solution

The cost function is equal to the fixed costs plus the cost to make x burgers. There are $65 in fixed costs ($25 for booth space and $40 for condiments and paper goods). Each burger costs $0.55 for ingredients, so it costs $0.55x$ for the ingredients in x burgers. The cost function is $C(x) = 65 + 0.55x$.

   Since each burger is sold for $3, the revenue function for selling x burgers is $R(x) = 3x$. The profit function is equal to the difference of the revenue function and the cost function, $P(x) = R(x) - C(x)$. In this example, $P(x) = 3x - (65 + 0.55x)$, which simplifies to $P(x) = 2.45x - 65$.

b) How much profit will be generated if the club makes and sells 50 burgers?

Solution

To determine the profit, we evaluate the profit function $P(x) = 2.45x - 65$ when $x = 50$.

$$P(50) = 2.45(50) - 65 \qquad \text{Substitute 50 for } x.$$
$$= 57.5 \qquad \text{Simplify.}$$

Making and selling 50 burgers will generate a profit of $57.50.

c) How many burgers do they need to make and sell in order to break even?

Solution

The club breaks even when the profit is 0.

$$P(x) = 0 \qquad \text{Set } P(x) \text{ equal to 0.}$$
$$2.45x - 65 = 0 \qquad \text{Replace } P(x) \text{ by its formula.}$$
$$2.45x = 65 \qquad \text{Add 65.}$$
$$x \approx 26.5 \qquad \text{Divide by 2.45.}$$

Since the club cannot break even until they sell 26.5 burgers, they will need to make and sell 27 burgers before they can break even.

d) If the goal of the club is to raise $500, how many burgers do they need to make and sell?

Solution

We begin by setting the profit function equal to $500 and solving for x.

$$P(x) = 500 \qquad \text{Set } P(x) \text{ equal to 500.}$$
$$2.45x - 65 = 500 \qquad \text{Replace } P(x) \text{ by } 2.45x - 65.$$
$$2.45x = 565 \qquad \text{Add 65.}$$
$$x \approx 230.6 \qquad \text{Divide by 2.45.}$$

This solution needs to be a whole number, as the club cannot make 230.6 burgers. Making and selling only 230 burgers would result in a profit of only $2.45(230) - 65$, or $498.50. The club will need to make and sell 231 burgers in order to raise at least $500.

Quick Check 5 Ross decides to start selling a calendar that features the pictures of 12 renowned mathematicians. In order to do this, he had to pay $50 for a business license. Each calendar costs Ross $1.25 to produce, and he plans to sell them for $9.95.

a) Find the cost function $C(x)$, the revenue function $R(x)$, and the profit function $P(x)$.

b) How many calendars will Ross have to produce and sell to make a $1000 profit?

EXERCISES *11.2*

Graph each function. Label the x- and y-intercepts.

1. $f(x) = x - 7$

2. $f(x) = x + 4$ **3.** $f(x) = -x + 3$

4. $f(x) = -x - 5$

5. $f(x) = 2x + 4$

6. $f(x) = 3x - 9$ **7.** $f(x) = -3x + 3$

12. $h(x) = -\dfrac{1}{5}x - 2$

13. $f(x) = 3x - 30$ **14.** $f(x) = -4x + 36$

8. $f(x) = -4x - 16$ **9.** $g(x) = 4x - 6$

15. $f(x) = 5x$ **16.** $f(x) = -3x$

10. $g(x) = -6x + 10$ **11.** $F(x) = \dfrac{2}{3}x - 4$

17. $f(x) = 7$

21.

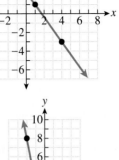

18. $f(x) = -2$

22.

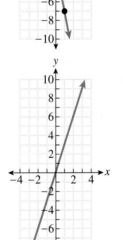

Use the given graph to determine the linear function $f(x)$ on the graph.

19.

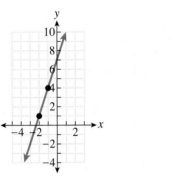

23.

20.

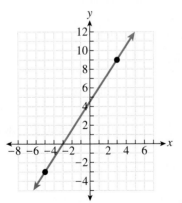

24.

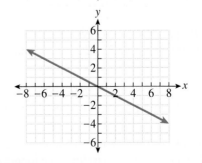

25.

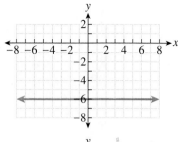

26.

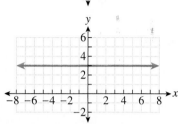

27. John's telephone plan charges him a connection fee for each call in addition to charging him per minute. A 10-minute phone call costs a total of $0.69, while a 30-minute phone call costs $1.29.

a) Find a linear function $f(x) = mx + b$ whose input is the length of a call in minutes and whose output is the corresponding charge for that call.

b) Explain, in your own words, what the slope and y-intercept of this function represent.

c) Use the function from part (a) to determine the cost of a phone call that lasts 44 minutes.

28. Janie and Julie went to the same carnival. In addition to paying the admission charge, Janie bought 20 ride tickets and Julie bought 36 ride tickets. Janie paid a total of $15, while Julie paid $23.

a) Find a linear function $f(x) = mx + b$ whose input is the number of ride tickets purchased and whose output is the total amount of money paid by a person attending the carnival.

b) Explain, in your own words, what the slope and y-intercept of this function represent.

c) Use the function from part (a) to determine the total amount paid by a person who purchased 12 ride tickets.

29. A car rental company charges a base fee for each car rented, in addition to charging a fee for each mile

driven. If a person drives a car for 100 miles, then the total charge is $35. If a person drives a car for 150 miles, then the total charge is $42.50.

a) Find a linear function $f(x) = mx + b$ whose input is the number of miles driven and whose output is the corresponding total charge for renting the car.

b) Explain, in your own words, what the slope and y-intercept of this function represent.

c) Use the function from part (a) to determine the cost of renting a car and driving it 260 miles.

30. A certain county charges a flat fine for speeding, in addition to adding on a fee for each mile per hour a driver goes over the speed limit. A driver who was going 86 mph in a 55 mph zone was fined $410, while another driver who was traveling 50 mph in a 35 mph zone was fined $250.

a) Find a linear function $f(x) = mx + b$ whose input is the number of miles per hour over the speed limit that a person was driving and whose output is the fine that must be paid.

b) Use the function from part (a) to determine the fine for a person who was driving 62 mph in a 40 mph zone.

31. An elementary school booster club is wrapping gifts at a local mall as a fundraiser. In addition to paying $50 to rent space for the day, the booster club also had to spend $75 on materials. They decide to charge $4 to wrap each gift.

a) Find the cost function $C(x)$, the revenue function $R(x)$, and the profit function $P(x)$.

b) How much profit will be generated if the booster club wraps 75 gifts?

32. Members of a sorority have set up a car wash to raise money for a scholarship fund. They paid a gas station $20 to use the parking lot and water. They also paid $10 to buy soap and buckets. They are charging $5 per car.

a) Find the cost function $C(x)$, the revenue function $R(x)$, and the profit function $P(x)$.

b) How much profit will be generated if they wash 185 cars?

33. The art club at a community college is selling burgers at lunchtime as a fundraiser. The club spent $12 for condiments and supplies. The cost for the meat and bun for each burger is $0.40, and the club is charging $3 for each burger.

    a) Find the cost function $C(x)$, the revenue function $R(x)$, and the profit function $P(x)$.

    b) How much profit will be generated if the art club sells 60 burgers?

34. A student has started her own business by making and selling calendars featuring the professors at her college. She paid each of the 12 professors in the calendar $50, and each calendar costs her $2 to make. She sells the calendars for $9.95.

    a) Find the cost function $C(x)$, the revenue function $R(x)$, and the profit function $P(x)$.

    b) How much profit will be generated if she sells 280 calendars?

35. Bill has decided to go into business manufacturing and selling computers out of the garage of his home. He spent $2000 to remodel his garage and to buy tools. It costs him $350 for the parts to make a computer, which he sells for $600.

    a) Find the cost function $C(x)$, the revenue function $R(x)$ and the profit function $P(x)$.

    b) How many computers will he have to sell in order to make a profit of $10,000?

36. Tina has started a business in which she sells a book of her favorite recipes online. She spent $1500 on her office and technology, and each book costs her $0.50 to produce. She sells each book for $8.

    a) Find the cost function $C(x)$, the revenue function $R(x)$, and the profit function $P(x)$.

    b) How many books will she have to sell in order to break even?

37. If $f(x) = mx + 7$ and $f(5) = 22$, find m.

38. If $f(x) = mx - 6$ and $f(8) = 34$, find m.

39. If $f(x) = mx - 13$ and $f(-6) = -25$, find m.

40. If $f(x) = mx + 24$ and $f(-9) = 51$, find m.

Answer in complete sentences.

41. Explain how to graph a linear function. Use an example to illustrate the process.

42. Explain how to determine a linear function $f(x)$ from its graph, provided that we know the coordinates of two points on the graph. Use an example to illustrate the process.

43. Write a word problem whose solution is "The company will break even if they make and sell 2385 items."

44. Explain why the break-even point for a profit function $P(x)$ corresponds to the function's x-intercept.

Study Tip **Revisited**   Noise can be a major distraction while studying. Some students need total silence while they are working; if you are one of those students, then be sure to select a location that will be free of noise. This is not easy to do in a public location such as the library, but you may be able to find a room or cubicle that is devoted to silent studying.

    Other students study better when there is some background noise. For instance, some students prefer to have a radio playing soft music in the background while they study.

    The amount of noise that you can handle may depend on the type of studying that you are doing. Music and other noises may provide a pleasant background while reviewing but can be distracting while one is learning new material.

Objectives

1. Graph quadratic functions of the form $f(x) = ax^2 + bx + c$.
2. Graph quadratic functions of the form $f(x) = a(x - h)^2 + k$.
3. Find the maximum or minimum value of a quadratic function.
4. Solve applied maximum–minimum problems.
5. Find the range of a quadratic function without graphing.
6. Simplify difference quotients for quadratic functions.

Recall that a **quadratic function** is a function that can be written in the form $f(x) = ax^2 + bx + c$, where a, b, and c are real numbers and $a \neq 0$.

Graphing Quadratic Functions of the Form $f(x) = ax^2 + bx + c$

Objective 1 Graph quadratic functions of the form $f(x) = ax^2 + bx + c$. We begin by learning how to graph quadratic functions of the form $f(x) = ax^2 + bx + c$, $a \neq 0$. The graphs of quadratic equations were covered in Section 10.5.

The graph of a quadratic function $f(x) = ax^2 + bx + c$ is U-shaped and is called a parabola. If a is positive, then the parabola opens upward, and if a is negative, it opens downward. Important points include the vertex of the parabola as well as the y-intercept and the x-intercept(s).

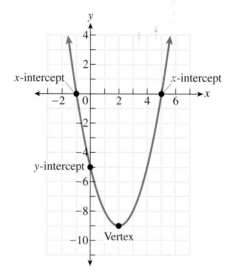

The vertex of a parabola is the "turning point" of the parabola, the point where the graph changes from decreasing to increasing or vice versa. The x-coordinate of the vertex of the graph of a quadratic function $f(x) = ax^2 + bx + c$ can be found using the formula $x = \frac{-b}{2a}$. Once we know the x-coordinate of the vertex, we evaluate the function for that value to find the y-coordinate of the vertex. Thus, the vertex of the parabola is at the point $\left(\frac{-b}{2a}, f\left(\frac{-b}{2a}\right)\right)$.

A parabola is symmetric about the vertical line that runs through the vertex. Recall that this line is called the axis of symmetry, and its equation is $x = \frac{-b}{2a}$. The graph to the left of this line is a mirror image of the graph to the right of this line. We can use the axis of symmetry to find additional points on the parabola.

To find the y-intercept of a parabola, determine $f(0)$. If the function is of the form $f(x) = ax^2 + bx + c$, then the y-intercept will be located at the point $(0, c)$.

To find the x-intercepts of a parabola, we set the function $f(x)$ equal to 0 and solve for x. The resulting equation is quadratic, and we have several methods for solving quadratic equations. In general, attempt to solve them by factoring if possible. If the quadratic expression cannot be factored, we can use the quadratic formula to solve the equation.

Parabolas can have as many as two x-intercepts. If the vertex of the parabola is on the x-axis, then the vertex is the only x-intercept.

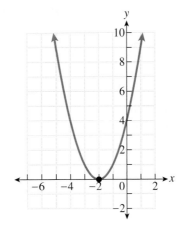

Some parabolas do not have any x-intercepts, and in this case we will obtain a negative discriminant when using the quadratic formula to solve the equation $f(x) = 0$. We can also tell whether a parabola has any x-intercepts by considering the location of its vertex.

Cases in Which a Parabola Has No x-intercepts

Parabola opens upward; vertex is above the x-axis

Parabola opens downward; vertex is below the x-axis

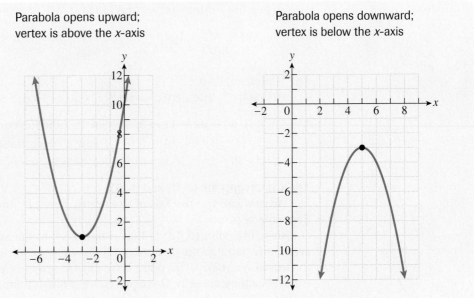

Graphing Quadratic Functions of the Form $f(x) = ax^2 + bx + c$

- Determine whether the parabola opens upward or downward.
- Find the vertex of the parabola.
- Find the y-intercept of the parabola.
- After plotting the vertex and the y-intercept, determine whether there are any x-intercepts. If there are x-intercepts, find them.
- Use the axis of symmetry, $x = \frac{-b}{2a}$, to find points that are symmetric to points that have already been plotted.
- Draw a smooth, U-shaped curve through the points that have been plotted.

EXAMPLE 1   Graph $f(x) = x^2 - 8x + 12$.

Solution

In this quadratic function, $a = 1$, $b = -8$, and $c = 12$. Since a is positive, this parabola opens upward. We begin by finding the coordinates of the vertex. We find the x-coordinate of the vertex using the formula $x = \dfrac{-b}{2a}$.

$$x = \frac{-(-8)}{2(1)} = 4 \qquad \text{Substitute 1 for } a \text{ and } -8 \text{ for } b \text{ and simplify.}$$

To find the y-coordinate of the vertex, we evaluate the function when $x = 4$.

$$f(4) = (4)^2 - 8(4) + 12 \qquad \text{Substitute 4 for } x.$$
$$= -4 \qquad \text{Simplify.}$$

The vertex is $(4, -4)$. Because the vertex is below the x-axis and the parabola opens upward, the graph must have two x-intercepts.

To find the y-intercept, we find $f(0)$.

$$f(0) = (0)^2 - 8(0) + 12 \qquad \text{Substitute 0 for } x.$$
$$= 12 \qquad \text{Simplify.}$$

The y-intercept is $(0, 12)$.

To find the x-intercepts, we set $f(x)$ equal to 0 and solve for x.

$$x^2 - 8x + 12 = 0 \qquad \text{Set the function equal to 0.}$$
$$(x - 2)(x - 6) = 0 \qquad \text{Factor the quadratic expression.}$$
$$x = 2 \quad \text{or} \quad x = 6 \qquad \text{Set each factor equal to 0 and solve.}$$

The x-intercepts are $(2, 0)$ and $(6, 0)$.

We now can use the axis of symmetry, $x = 4$, to find the point that is symmetric to the y-intercept.

Since the point $(0, 12)$ is 4 units to the left of the axis of symmetry, there must be a symmetric point on the parabola with the same y-coordinate that is 4 units to the right of the axis of symmetry. This point is $(8, 12)$. Once all five points have been plotted, we finish by drawing a smooth, U-shaped curve through them.

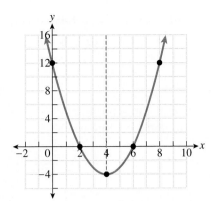

Quick Check **1** Graph $f(x) = x^2 + 2x - 24$.

In the next example, we will graph a parabola that opens downward. In addition, we will have to use the quadratic formula to find the x-intercepts.

EXAMPLE **2**   Graph $f(x) = -x^2 + 6x + 5$.

Solution

In this quadratic function, $a = -1$. Since a is negative, this parabola opens downward. We begin by finding the coordinates of the vertex. We find the x-coordinate of the vertex using the formula $x = \dfrac{-b}{2a}$.

$$x = \frac{-6}{2(-1)} = 3 \qquad \text{Substitute } -1 \text{ for } a \text{ and } 6 \text{ for } b \text{ and simplify.}$$

To find the y-coordinate of the vertex, we evaluate the function when $x = 3$.

$$\begin{aligned} f(3) &= -(3)^2 + 6(3) + 5 \qquad &\text{Substitute 3 for } x.\\ &= 14 &\text{Simplify.} \end{aligned}$$

The vertex is $(3, 14)$. Since the vertex is above the x-axis and the parabola opens downward, the graph must have two x-intercepts.

The y-intercept is $(0, 5)$, since the y-intercept of $f(x) = ax^2 + bx + c$ is the point $(0, c)$.

To find the x-intercepts, we set $f(x)$ equal to 0 and solve for x.

$$-x^2 + 6x + 5 = 0$$ Set the function equal to 0.

$$0 = x^2 - 6x - 5$$ Collect all terms on the right side of the equation, so that the coefficient of the second-degree term is positive.

Since the expression $x^2 - 6x - 5$ cannot be factored, we will use the quadratic formula to find the x-intercepts.

$$x = \frac{6 \pm \sqrt{(-6)^2 - 4(1)(-5)}}{2(1)}$$ Substitute 1 for a, -6 for b, and -5 for c in the quadratic formula.

$$x = \frac{6 \pm \sqrt{56}}{2}$$ Simplify the radicand and denominator.

$$x = \frac{6 \pm 2\sqrt{14}}{2}$$ Simplify the radical.

$$x = \frac{\overset{1}{\cancel{2}}(3 \pm \sqrt{14})}{\underset{1}{\cancel{2}}}$$ Divide out common factors.

$$x = 3 \pm \sqrt{14}$$ Simplify.

Using a calculator, we find that $3 + \sqrt{14} \approx 6.7$ and $3 - \sqrt{14} \approx -0.7$, so the x-intercepts are located at approximately $(6.7, 0)$ and $(-0.7, 0)$.

The graph is shown at the right. The point $(6, 5)$ can be found using the axis of symmetry $(x = 3)$ and the y-intercept. Once all five points have been plotted, we draw an inverted U-shaped curve through them.

Quick Check **2**
Graph
$f(x) = -x^2 + 4x + 3.$

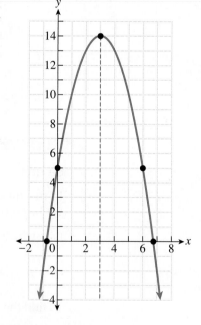

Graphing Quadratic Functions of the Form $f(x) = a(x - h)^2 + k$

Objective 2 Graph quadratic functions of the form $f(x) = a(x - h)^2 + k$.
We will now learn how to graph quadratic functions of the form $f(x) = a(x - h)^2 + k$. The graph of the quadratic function $f(x) = a(x - h)^2 + k$ is a parabola with axis of

symmetry $x = h$ and vertex (h, k). The parabola opens upward if a is positive and opens downward if a is negative. We still find the y-intercept of the function by finding $f(0)$. We also find the x-intercepts, if there are any, by setting the function $f(x)$ equal to 0 and solving for x. However, we will often solve this equation by extracting square roots, rather than solving by factoring or using the quadratic formula.

EXAMPLE 3   Graph $f(x) = -(x - 3)^2 + 16$.

Solution

For this quadratic function, $a = -1$. Since a is negative, this parabola opens downward and its vertex is $(3, 16)$. Since the vertex is above the x-axis and the parabola opens downward, this parabola has two x-intercepts.

We next find the y-intercept by evaluating the function at $x = 0$.

$$f(0) = -(0 - 3)^2 + 16 \qquad \text{Substitute 0 for } x.$$
$$= 7 \qquad \text{Simplify.}$$

The y-intercept is $(0, 7)$.

To find the x-intercepts, we set the function equal to 0 and solve for x. We will solve the resulting equation by extracting square roots.

$$-(x - 3)^2 + 16 = 0 \qquad \text{Set the function equal to 0.}$$
$$16 = (x - 3)^2 \qquad \text{Add } (x - 3)^2 \text{ to isolate the squared expression.}$$
$$\pm\sqrt{16} = \sqrt{(x - 3)^2} \qquad \text{Take the square root of both sides.}$$
$$\pm 4 = x - 3 \qquad \text{Simplify each square root.}$$
$$3 \pm 4 = x \qquad \text{Add 3 to isolate } x.$$

$3 + 4 = 7$ and $3 - 4 = -1$, so the x-intercepts are $(7, 0)$ and $(-1, 0)$.

The graph of the function is shown at the right. The point $(6, 7)$ can be found using the axis of symmetry, which is $x = 3$.

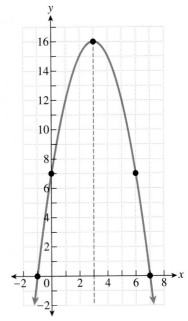

Quick Check 3

Graph
$f(x) = -(x + 1)^2 + 8.$

Minimum and Maximum Values of Quadratic Functions

Objective **3** **Find the maximum or minimum value of a quadratic function.**

Maximum and Minimum Values of a Function

> The least possible output of a function is called the **minimum value** of the function. The greatest possible output of a function is called the **maximum value** of the function.

Each parabola that opens upward has a minimum value at its vertex.

We can see that this parabola does not go below the line $y = -4$, which is the y-coordinate of the vertex. The minimum value of this parabola is -4.

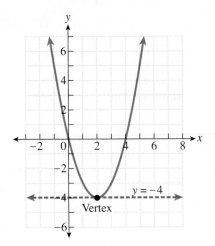

To find the minimum value of a quadratic function whose graph is a parabola that opens upward, we need to find the y-coordinate of its vertex.

EXAMPLE ▶ 4   Find the minimum value of the function $f(x) = 3x^2 - 12x + 17$.

Solution

Since the graph of this function is a parabola that opens upward ($a > 0$), the function has a minimum value at its vertex. To find the vertex, we begin by finding the x-coordinate by using the formula $x = \dfrac{-b}{2a}$.

$$x = \frac{-(-12)}{2(3)} = 2 \qquad \text{Substitute 3 for } a \text{ and } -12 \text{ for } b \text{ and simplify.}$$

The minimum value of this function occurs when $x = 2$. Now, to find the minimum value, we evaluate the function when $x = 2$.

$$\begin{aligned} f(2) &= 3(2)^2 - 12(2) + 17 &&\text{Substitute 2 for } x. \\ &= 5 &&\text{Simplify.} \end{aligned}$$

▶ The minimum value for this function is 5.

Do all quadratic functions have a minimum value? No, if the graph of a quadratic function is a parabola that opens downward, then the function does not have a minimum value. Such a function does have a maximum value, and this maximum value can be found at the vertex.

EXAMPLE ▶ 5   Find the maximum or minimum value of the function
$$f(x) = -2x^2 - 8x - 13.$$

Solution

Since this parabola will open downward, the function has a maximum value at the vertex.

$$x = \frac{-(-8)}{2(-2)} = -2 \qquad \text{Substitute } -2 \text{ for } a \text{ and } -8 \text{ for } b \text{ and simplify.}$$

The maximum value occurs when $x = -2$. Now we evaluate the function for this value of x to find the maximum value of the function.

$$\begin{aligned}
f(-2) &= -2(-2)^2 - 8(-2) - 13 & &\text{Substitute } -2 \text{ for } x. \\
&= -2(4) - 8(-2) - 13 & &\text{Square } -2. \\
&= -8 + 16 - 13 & &\text{Multiply.} \\
&= -5 & &\text{Simplify.}
\end{aligned}$$

The maximum value of this function is -5 and occurs when $x = -2$.

Quick Check   4 ▶ Find the maximum or minimum value of the quadratic function.

a) $f(x) = -x^2 + 10x - 35$
b) $f(x) = 5x^2 - 20x + 67$

Applied Maximum–Minimum Problems

Objective   4   Solve applied maximum–minimum problems.

EXAMPLE ▶ 6   What is the maximum product of two numbers whose sum is 40?

Solution

There are two unknown numbers in this problem. If we let x represent the first number, then we can represent the second number by $40 - x$.

Unknowns
1: x
2: $40 - x$

The product of these two numbers is given by the function $f(x) = x(40 - x)$. This function simplifies to be $f(x) = -x^2 + 40x$. Since this function is quadratic and its graph is a parabola that opens downward $(a < 0)$, its maximum value can be found at the vertex. The maximum product occurs when $x = \dfrac{-b}{2a}$.

$$x = \frac{-40}{2(-1)} = 20 \qquad \text{Substitute } -1 \text{ for } a \text{ and } 40 \text{ for } b \text{ and simplify.}$$

The maximum product occurs when $x = 20$.

1: $x = 20$

2: $40 - x = 40 - 20 = 20$

Quick Check 5

What is the maximum product of two numbers whose sum is 62?

The two numbers whose sum is 40 that have the greatest product are 20 and 20. The maximum product is 400.

EXAMPLE 7   A farmer wants to fence off a rectangular pen for some pigs. If he has 120 feet of fencing, what dimensions would give the pen the largest possible area?

Solution

There are two unknowns in this problem: the length and the width of the rectangle.

Suppose that we let x represent the length of the rectangle. Since the perimeter is equal to 120 feet, the total length of the remaining two sides is $120 - 2x$.

Therefore, each remaining side is $\dfrac{120 - 2x}{2}$ or $60 - x$.

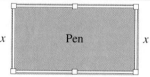

In general, the width of a rectangle is equal to the difference of half the perimeter and the length, $W = \frac{P}{2} - L$. To verify this, solve the formula for the perimeter of a rectangle, $P = 2L + 2W$, for W.

Unknowns

Length: x

Width: $60 - x$

The area of a rectangle is equal to its length times its width. The area of this pen is given by the function $A(x) = x(60 - x)$. This function simplifies to be $A(x) = -x^2 + 60x$. Since this function is quadratic and its graph is a parabola that opens downward $(a < 0)$, its maximum value can be found at the vertex. The maximum area occurs when $x = \dfrac{-b}{2a}$.

$$x = \frac{-60}{2(-1)} = 30 \qquad \text{Substitute } -1 \text{ for } a \text{ and } 60 \text{ for } b \text{ and simplify.}$$

The maximum area occurs when $x = 30$.

Length: $x = 30$
Width: $60 - x = 60 - 30 = 30$

The maximum area occurs when the length and the width are both 30 feet. The maximum area that can be enclosed is 900 square feet.

Quick Check 6
What is the largest possible rectangular area that can be fenced in with 264 meters of fencing?

Finding the Range of a Quadratic Function Without Graphing

Objective 5   Find the range of a quadratic function without graphing. If we are able to find the maximum or minimum value of a quadratic function, then we are able to determine the range of the function without actually having to graph it.

EXAMPLE 8   Determine the range of the function $f(x) = x^2 - 10x + 19$.

Solution

The graph of this function is a parabola that opens upward, so the function has a minimum value, but does not have a maximum value. We begin by finding the minimum value, which occurs at the vertex.

$$x = \frac{-(-10)}{2(1)} = 5 \qquad \text{Substitute 1 for } a \text{ and } -10 \text{ for } b \text{ and simplify.}$$

The minimum value occurs when $x = 5$. To find the minimum value, we evaluate the function when $x = 5$.

$$f(5) = (5)^2 - 10(5) + 19 \qquad \text{Substitute 5 for } x.$$
$$= -6 \qquad \text{Simplify.}$$

The minimum value of this function is -6. Since the function does not have a maximum value, the range of the function is $[-6, \infty)$.

Quick Check 7
Determine the range of the quadratic function.
$f(x) = 5x^2 + 100x + 216$

Difference Quotients

Objective 6   Simplify difference quotients for quadratic functions.

The **difference quotient** $\dfrac{f(x + h) - f(x)}{h}$ is used to determine the rate of change for a function $f(x)$ at a particular value of x.

We will now learn how to simplify this difference quotient for quadratic functions.

EXAMPLE 9   For the function $f(x) = x^2 - 8x + 14$, simplify the difference quotient $\dfrac{f(x + h) - f(x)}{h}$.

Solution

We begin to simplify the difference quotient by first simplifying $f(x + h)$.

$$f(x + h) = (x + h)^2 - 8(x + h) + 14 \qquad \text{Substitute } x + h \text{ for } x.$$
$$= (x + h)(x + h) - 8(x + h) + 14 \qquad \text{Square } x + h \text{ by multiplying it by itself.}$$
$$= x^2 + xh + xh + h^2 - 8(x + h) + 14 \qquad \text{Multiply using the distributive property.}$$
$$= x^2 + 2xh + h^2 - 8(x + h) + 14 \qquad \text{Combine like terms.}$$
$$= x^2 + 2xh + h^2 - 8x - 8h + 14 \qquad \text{Distribute.}$$

There are no like terms to combine, so we can now substitute into the difference quotient.

$$\frac{f(x + h) - f(x)}{h} = \frac{(x^2 + 2xh + h^2 - 8x - 8h + 14) - (x^2 - 8x + 14)}{h}$$

Substitute for $f(x + h)$ and $f(x)$.

$$= \frac{x^2 + 2xh + h^2 - 8x - 8h + 14 - x^2 + 8x - 14}{h}$$

Simplify the numerator by removing parentheses.

$$= \frac{2xh + h^2 - 8h}{h} \qquad \text{Combine like terms: } x^2 \text{ and } -x^2, -8x \text{ and } 8x, 14 \text{ and } -14.$$

$$= \frac{\overset{1}{\cancel{h}}(2x + h - 8)}{\underset{1}{\cancel{h}}} \qquad \text{Divide out common factors.}$$

$$= 2x + h - 8 \qquad \text{Simplify.}$$

$$\frac{f(x + h) - f(x)}{h} = 2x + h - 8 \qquad \text{for} \qquad f(x) = x^2 - 8x + 14.$$

> **Quick Check 9**
> For the function
> $f(x) = x^2 + 3x + 317$,
> simplify the difference
> quotient
> $\dfrac{f(x + h) - f(x)}{h}$.

EXERCISES 11.3 ▶

Graph the function. Label the vertex, y-intercept, and any x-intercepts.

1. $f(x) = x^2 - 4x + 3$

2. $f(x) = x^2 - 2x - 8$

3. $f(x) = x^2 + 4x - 21$ **4.** $f(x) = x^2 + 10x + 16$ **7.** $f(x) = x^2 - 16$ **8.** $f(x) = x^2 - 9$

9. $f(x) = x^2 + 6x - 5$ **10.** $f(x) = x^2 - 3x - 8$

5. $f(x) = -x^2 + 6x + 7$ **6.** $f(x) = -x^2 + 2x + 15$

11. $f(x) = x^2 - 8$

12. $f(x) = x^2 - 12$

17. $f(x) = -x^2 - 4x - 6$ **18.** $f(x) = -x^2 - 2x - 7$

19. $f(x) = (x + 4)^2 - 9$ **20.** $f(x) = (x + 1)^2 - 4$

13. $f(x) = 4x^2 - 8x - 21$ **14.** $f(x) = -5x^2 + 10x + 3$

15. $f(x) = x^2 - 6x + 10$ **16.** $f(x) = x^2 + 4x + 8$

21. $f(x) = -(x + 1)^2 + 16$ **22.** $f(x) = -(x - 3)^2 + 9$

23. $f(x) = (x + 6)^2$ **24.** $f(x) = -(x + 2)^2$ **27.** $f(x) = -(x + 4)^2 + 10$

28. $f(x) = -(x - 6)^2 + 2$

25. $f(x) = (x - 4)^2 - 5$

26. $f(x) = (x + 2)^2 - 3$ **29.** $f(x) = (x + 2)^2 + 10$ **30.** $f(x) = -(x + 3)^2 - 16$

31. $f(x) = 2(x + 2)^2 - 18$ **32.** $f(x) = -2(x - 6)^2 + 8$

Determine whether the given quadratic function has a maximum value or a minimum value. Then find that maximum or minimum value.

33. $f(x) = x^2 - 8x + 20$

34. $f(x) = x^2 - 6x - 35$

35. $f(x) = x^2 + 12x - 42$

36. $f(x) = x^2 + 9x + 60$

37. $f(x) = -x^2 + 4x - 30$

38. $f(x) = -x^2 + 14x - 76$

39. $f(x) = -x^2 - 20x - 57$

40. $f(x) = -x^2 - 10x + 15$

41. $f(x) = (x - 7)^2 + 18$

42. $f(x) = (x + 3)^2 - 19$

43. $f(x) = -(x + 16)^2 - 33$

44. $f(x) = -(x - 14)^2 + 47$

45. $f(x) = 7(x - 4)^2 + 46$

46. $f(x) = -8(x + 11)^2 - 50$

47. $f(x) = 3x^2 - 20x + 121$

48. $f(x) = -2x^2 - 14x + 207$

49. $f(x) = -x^2 + 224x + 54160$

50. $f(x) = x^2 - 320x + 27500$

51. A girl throws a rock upward with an initial velocity of 32 feet/second. The height of the rock (in feet) after t seconds is given by the function $h(t) = -16t^2 + 32t + 5$. What is the maximum height that the rock reaches?

52. A pilot flying at a height of 5000 feet determines that she must eject from her plane. The ejection seat launches with an initial velocity of 600 feet/second. The height of the pilot (in feet) t seconds after ejection is given by the function $h(t) = -16t^2 + 600t + 5000$. What is the maximum height that the pilot reaches?

53. At the start of a college football game, a referee tosses a coin to determine which team will receive the opening kickoff. The height of the coin (in meters) after t seconds is given by the function $h(t) = -4.9t^2 + 2.1t + 1.5$. What is the maximum height that the coin reaches?

54. If an astronaut on the moon throws a rock upward with an initial velocity of 78 feet per second, the height of the rock (in feet) after t seconds is given by the function $h(t) = -2.6t^2 + 78t + 5$. What is the maximum height that the rock reaches?

55. What is the maximum product of two numbers whose sum is 24?

56. What is the maximum product of two numbers whose sum is 100?

57. A farmer has 300 feet of fencing to make a rectangular corral. What dimensions will make a corral that has the maximum area? What is the maximum area possible?

58. Emeril wants to fence off a rectangular herb garden adjacent to his house, using the house to form the fourth side of the rectangle as shown.

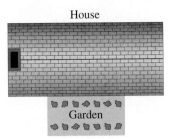

If Emeril has 80 feet of fencing, what dimensions of the herb garden will produce the maximum area? What is the maximum possible area?

59. A company produces small frying pans. The average cost per pan in dollars is given by the function $f(x) = 0.000625x^2 - 0.0875x + 6.5$, where x is the number of pans produced (in thousands). For what number of pans is the average cost per pan minimized? What is the minimum average cost per pan that is possible?

60. A company produces mini CD players/stereos. The average cost per radio in dollars is given by the function $f(x) = 0.009x^2 - 0.432x + 9.3$, where x is the number of radios produced (in thousands). For what number of radios is the average cost per radio minimized? What is the minimum cost per radio that is possible?

Determine the range of the quadratic function $f(x)$.

61. $f(x) = x^2 - 10x + 32$

62. $f(x) = x^2 + 11x - 33$

63. $f(x) = -x^2 - 8x - 2$

64. $f(x) = -x^2 + 14x - 99$

65. $f(x) = -(x - 8)^2 - 42$

66. $f(x) = (x + 23)^2 + 1729$

For the function $f(x)$, simplify the difference quotient $\dfrac{f(x + h) - f(x)}{h}$.

67. $f(x) = x^2 - 4$

68. $f(x) = x^2 - 10x$

69. $f(x) = x^2 - 9x + 17$

70. $f(x) = x^2 + 8x + 45$

71. $f(x) = 3x^2 - 11x + 49$

72. $f(x) = -x^2 - 16x + 105$

Answer in complete sentences.

73. Explain how to graph a quadratic function. Use an example to illustrate the process.

74. Explain how to determine whether a quadratic function has a minimum value or has a maximum value. Explain how to find the minimum or maximum value of a quadratic function. Use an example to illustrate the process.

75. Write a word problem whose solution is "The company will maximize their profit if they make and sell 400 items." Explain how you created your problem.

76. Write a word problem whose solution is "The projectile reaches its maximum height of 180 feet after 3 seconds." Explain how you created your problem.

Study Tip **Revisited**　When you sit down to study, you need a space large enough to accommodate all of your materials. Be sure that you can easily fit your textbook, notes, homework notebook, study cards, calculator, ruler, etc., on the desk or table that you are working on. Anything that you might use should be easily within your reach. If your workspace is unorganized and cluttered, you will waste time searching for pages and shuffling through your materials.

Objectives

1 **Graph square-root functions.**

2 **Graph cubic functions.**

3 **Determine a function from its graph.**

In this section, we will examine the graphs of two functions: the square-root function and the cubic function.

Graphing Square-Root Functions

Objective 1 **Graph square-root functions.** We will begin by learning to graph the **square-root function.** The basic square-root function is the function $f(x) = \sqrt{x}$. Recall that the square root of a negative number is an imaginary number, so the radicand must be nonnegative. We begin to graph this function by selecting values for x and evaluating the function for these values. It is a good idea to choose values of x that are perfect squares, such as 0, 1, 4, and 9.

x	$f(x) = \sqrt{x}$	$(x, f(x))$
0	$f(0) = \sqrt{0} = 0$	$(0, 0)$
1	$f(1) = \sqrt{1} = 1$	$(1, 1)$
4	$f(4) = \sqrt{4} = 2$	$(4, 2)$
9	$f(9) = \sqrt{9} = 3$	$(9, 3)$

The domain of this function, which is read from left to right on this graph, is $[0, \infty)$. The range of this function, which is read from the bottom to the top, is also $[0, \infty)$.

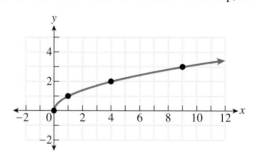

We now turn our attention to graphing square-root functions of the form $f(x) = a\sqrt{x - h} + k$.

EXAMPLE 1 Graph $f(x) = \sqrt{x + 5} + 4$ and state the domain and range.

Solution

We begin by choosing values for x. To find the value of x where the graph starts, we set the radicand equal to 0 and solve for x.

$$x + 5 = 0 \qquad \text{Set the radicand equal to 0.}$$
$$x = -5 \qquad \text{Subtract 5.}$$

To find other values of x to use, we can set $x + 5$ equal to the perfect squares 1, 4, and 9 and solve for x. The numbers 1, 4, and 9 are the first three positive integers that are perfect squares.

$$x + 5 = 1 \qquad x + 5 = 4 \qquad x + 5 = 9$$
$$x = -4 \qquad\quad x = -1 \qquad\quad x = 4$$

Now we evaluate the function for these values of x and then draw the graph.

x	$f(x) = \sqrt{x + 5} + 4$	$(x, f(x))$
-5	$f(-5) = \sqrt{-5 + 5} + 4 = 4$	$(-5, 4)$
-4	$f(-4) = \sqrt{-4 + 5} + 4 = 5$	$(-4, 5)$
-1	$f(-1) = \sqrt{-1 + 5} + 4 = 6$	$(-1, 6)$
4	$f(4) = \sqrt{4 + 5} + 4 = 7$	$(4, 7)$

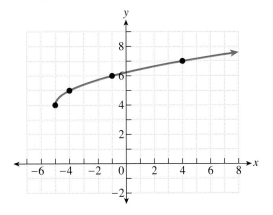

To find the y-intercept of this graph, we need to find $f(0)$.

$$\begin{aligned} f(0) &= \sqrt{(0) + 5} + 4 && \text{Substitute 0 for } x. \\ &= \sqrt{5} + 4 && \text{Simplify the radicand.} \\ &\approx 6.2 && \text{Approximate with a calculator.} \end{aligned}$$

The y-intercept is approximately at $(0, 6.2)$. We can see from the graph that this function has no x-intercepts. The domain of this function is $[-5, \infty)$, and the range is $[4, \infty)$.

Using Your Calculator   We can graph square-root functions using the TI-83/84. To enter the function from Example 1, $f(x) = \sqrt{x + 5} + 4$, press the Y= key. Enter $\sqrt{x + 5} + 4$ next to Y_1. Use parentheses to separate the radicand from the rest of the expression. Press the key labeled GRAPH to graph the function in the standard window.

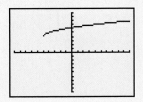

Graphing a Square-Root Function

- Determine the values of x for which the radicand is equal to 0, 1, 4, and 9.
- Create a table of function values for these values of x.
- Plot the points in the table on a coordinate system and draw a smooth curve that passes through these points, beginning at the point whose x-coordinate causes the radicand to be equal to 0.

EXAMPLE 2   Graph $f(x) = \sqrt{x + 8} - 4$ and state the domain and range.

Solution

We begin by choosing values for x. Again, set the radicand equal to 0, 1, 4, and 9, and solve for x.

$$x + 8 = 0 \qquad x + 8 = 1 \qquad x + 8 = 4 \qquad x + 8 = 9$$
$$x = -8 \qquad\quad x = -7 \qquad\quad x = -4 \qquad\quad x = 1$$

Now we evaluate the function for these values of x and then draw the graph.

x	$f(x) = \sqrt{x + 8} - 4$	$(x, f(x))$
-8	$f(-8) = \sqrt{-8 + 8} - 4 = -4$	$(-8, -4)$
-7	$f(-7) = \sqrt{-7 + 8} - 4 = -3$	$(-7, -3)$
-4	$f(-4) = \sqrt{-4 + 8} - 4 = -2$	$(-4, -2)$
1	$f(1) = \sqrt{1 + 8} - 4 = -1$	$(1, -1)$

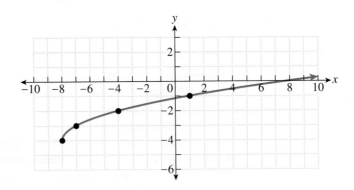

To find the y-intercept of this function, we evaluate $f(0)$.

$$f(0) = \sqrt{0 + 8} - 4 \qquad \text{Substitute 0 for } x.$$
$$= \sqrt{8} - 4 \qquad\qquad \text{Simplify the radicand.}$$
$$\approx -1.2 \qquad\qquad\quad \text{Approximate using a calculator.}$$

The y-intercept is located approximately at $(0, -1.2)$. To find the x-intercept, we set the function equal to 0 and solve for x.

$$\sqrt{x + 8} - 4 = 0 \qquad \text{Set the function equal to 0.}$$
$$\sqrt{x + 8} = 4 \qquad \text{Add 4 to isolate the square root.}$$
$$(\sqrt{x + 8})^2 = 4^2 \qquad \text{Square both sides.}$$
$$x + 8 = 16 \qquad \text{Simplify each side.}$$
$$x = 8 \qquad \text{Subtract 8.}$$

The x-intercept is at $(8, 0)$. The domain of this function is $[-8, \infty)$, and the range is $[-4, \infty)$.

Quick Check **1** Graph $f(x) = \sqrt{x + 6} + 5$ and state its domain and range.

EXAMPLE **3**   Graph $f(x) = \sqrt{-x} + 3$ and state the domain and range.

Solution

Notice that the coefficient of x is negative. In this case, the graph will move to the left rather than to the right. We begin by setting the radicand equal to 0, 1, 4, and 9, and solve for x.

$$-x = 0 \qquad -x = 1 \qquad -x = 4 \qquad -x = 9$$
$$x = 0 \qquad x = -1 \qquad x = -4 \qquad x = -9$$

Now we evaluate the function for these values of x and then draw the graph.

x	$f(x) = \sqrt{-x} + 3$	$(x, f(x))$
0	$f(0) = \sqrt{-0} + 3 = 3$	$(0, 3)$
-1	$f(-1) = \sqrt{-(-1)} + 3 = 4$	$(-1, 4)$
-4	$f(-4) = \sqrt{-(-4)} + 3 = 5$	$(-4, 5)$
-9	$f(-9) = \sqrt{-(-9)} + 3 = 6$	$(-9, 6)$

The y-intercept is $(0, 3)$ and is already on the graph. This function does not have an x-intercept. The domain of this function is $(-\infty, 0]$, and the range is $[3, \infty)$.

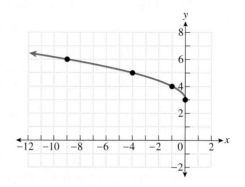

Graphing Cubic Functions

Objective **2**   **Graph cubic functions.** The other function we will investigate in this section is the **cubic function.** The basic cubic function is the function $f(x) = x^3$. Here is a table of function values for values of x ranging from -2 to 2:

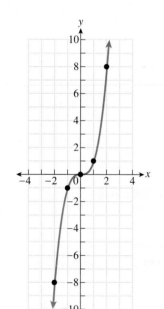

x	$f(x) = x^3$	$(x, f(x))$
-2	$f(-2) = (-2)^3 = -8$	$(-2, -8)$
-1	$f(-1) = (-1)^3 = -1$	$(-1, -1)$
0	$f(0) = 0^3 = 0$	$(0, 0)$
1	$f(1) = 1^3 = 1$	$(1, 1)$
2	$f(2) = 2^3 = 8$	$(2, 8)$

We plot these points and draw a smooth curve through them. The graph at the left extends forever to the left and to the right, so the domain of this function is the set of all real numbers $\Re$. The graph also extends forever upwards and downwards, so the range of this function is also the set of all real numbers $\Re$. This is true for all cubic functions.

EXAMPLE **4**   Graph $f(x) = (x + 2)^3 - 1$ and state the domain and range.

Solution

To choose values for x, begin by setting the expression that is cubed equal to 0 and solving for x. Then choose two values for x that are less than this value and two others that are greater than this value.

$$x + 2 = 0 \qquad \text{Set the expression cubed equal to 0.}$$
$$x = -2 \qquad \text{Subtract 2.}$$

We will create a table of function values for values of x ranging from -4 to 0 and then draw the graph.

x	$f(x) = (x + 2)^3 - 1$	$(x, f(x))$
-4	$f(-4) = (-4 + 2)^3 - 1 = -9$	$(-4, -9)$
-3	$f(-3) = (-3 + 2)^3 - 1 = -2$	$(-3, -2)$
-2	$f(-2) = (-2 + 2)^3 - 1 = -1$	$(-2, -1)$
-1	$f(-1) = (-1 + 2)^3 - 1 = 0$	$(-1, 0)$
0	$f(0) = (0 + 2)^3 - 1 = 7$	$(0, 7)$

From the points that we have plotted in the graph below, we can see that the y-intercept is $(0, 7)$ and the x-intercept is $(-1, 0)$. The domain of this function is the set of all real numbers $\Re$, and the range is also the set of all real numbers $\Re$.

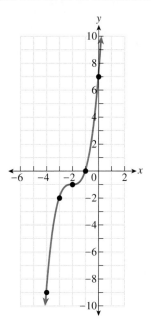

Graphing a Cubic Function

- Determine the values of x for which the cubed expression is equal to 0.
- Select two additional values that are less than this value of x, as well as two values that are greater than this value of x.
- Create a table of function values using these five values of x.
- Plot the points in the table on a coordinate system and draw a smooth curve that passes through these points.

EXAMPLE 5   Graph $f(x) = -(x - 4)^3 + 2$ and state the domain and range.

Solution

We begin by setting the expression that is cubed equal to 0 and solving for x.

$$x - 4 = 0 \qquad \text{Set the expression cubed equal to 0.}$$
$$x = 4 \qquad \text{Add 4.}$$

We will create a table of function values for values of *x* ranging from 2 to 6 and then draw the graph.

x	f(x) = -(x - 4)³ + 2	(x, f(x))
2	$f(2) = -(2 - 4)^3 + 2 = 10$	$(2, 10)$
3	$f(3) = -(3 - 4)^3 + 2 = 3$	$(3, 3)$
4	$f(4) = -(4 - 4)^3 + 2 = 2$	$(4, 2)$
5	$f(5) = -(5 - 4)^3 + 2 = 1$	$(5, 1)$
6	$f(6) = -(6 - 4)^3 + 2 = -6$	$(6, -6)$

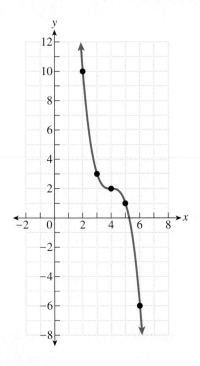

To find the *y*-intercept, we need to evaluate $f(0)$.

$$f(0) = -(0 - 4)^3 + 2 \qquad \text{Substitute 0 for } x.$$
$$= 66 \qquad\qquad\qquad \text{Simplify.}$$

The *y*-intercept is at $(0, 66)$. To find the *x*-intercept, we set $f(x)$ equal to 0 and solve for *x*. When dealing with a cubic function, this will involve using a cube root.

$$-(x - 4)^3 + 2 = 0 \qquad\qquad \text{Set the function equal to 0.}$$
$$2 = (x - 4)^3 \qquad\qquad\quad \text{Add } (x - 4)^3 \text{ to isolate the cubed expression.}$$
$$\sqrt[3]{2} = \sqrt[3]{(x - 4)^3} \qquad \text{Take the cube root of both sides.}$$
$$\sqrt[3]{2} = x - 4 \qquad\qquad\quad \text{Simplify.}$$
$$4 + \sqrt[3]{2} = x \qquad\qquad\quad \text{Add 4 to isolate } x.$$
$$x \approx 5.3 \qquad\qquad\qquad\; \text{Approximate using a calculator.}$$

To approximate the x-coordinate of the x-intercept, we have to approximate $\sqrt[3]{2}$ using a calculator or software package. To do this on a calculator, we can raise 2 to the 1/3 power. (Recall that $\sqrt[3]{x} = x^{1/3}$.) The x-intercept is at approximately $(5.3, 0)$.

Next we present two graphs showing the function and its y-intercept. In the first graph, the scale on the horizontal axis is not equal to the scale on the vertical axis. This helps us to see all of the points that we had plotted. In the second graph, the horizontal axis and the vertical axis use equal scales.

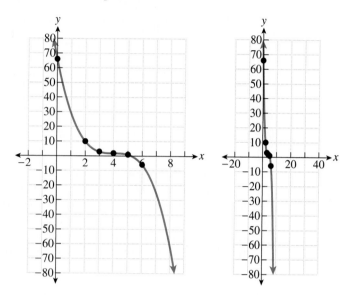

The domain of this function is the set of all real numbers $\Re$, and the range is also the set of all real numbers $\Re$.

Quick Check **3** Graph the function and state its domain and range.

a) $f(x) = (x - 1)^3 + 1$ b) $f(x) = -(x + 3)^3$

Determining a Function from Its Graph

Objective 3 Determine a function from its graph.

EXAMPLE 6   Determine the function $f(x)$ that has been graphed.

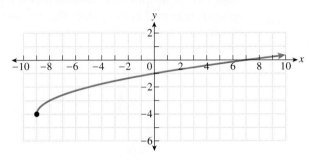

Solution

The graph of this function tells us that $f(x)$ is a square-root function of the form $f(x) = a\sqrt{x - h} + k$. To find the function, let's focus on the point where this graph begins, which is at $(-9, -4)$. Since the x-coordinate is -9, the expression $x + 9$ must be inside the square root. The y-coordinate is -4, which tells us that 4 is subtracted from the square root. The function is of the form $f(x) = a\sqrt{x + 9} - 4$. To find a, we will again use the coordinates of another point on the graph, such as $(-8, -3)$.

$$f(-8) = -3 \qquad \text{Set } f(-8) \text{ equal to } -3.$$
$$a\sqrt{-8 + 9} - 4 = -3 \qquad \text{Substitute } -8 \text{ for } x \text{ in the function } f(x).$$
$$a - 4 = -3 \qquad \text{Simplify the square root.}$$
$$a = 1 \qquad \text{Add 4.}$$

Since $a = 1$, the function is $f(x) = \sqrt{x + 9} - 4$.

EXAMPLE 7   Determine the function $f(x)$ that has been graphed.

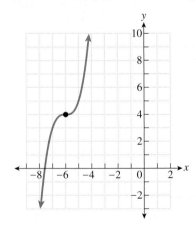

Solution

In this example, $f(x)$ is a cubic function of the form $f(x) = a(x - h)^3 + k$. To find the function, let's focus on the point where this graph flattens out, which is at $(-6, 4)$.

Since the x-coordinate is -6, the expression $x + 6$ must be the cubed expression. The y-coordinate is 4, which tells us that 4 is added to the cubed expression. The function is of the form $f(x) = a(x + 6)^3 + 4$. It is left to the reader to show that $a = 1$ for this function. The function is $f(x) = (x + 6)^3 + 4$.

Quick Check **4** Determine the function $f(x)$ that has been graphed.

a)

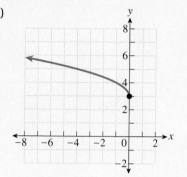

b)

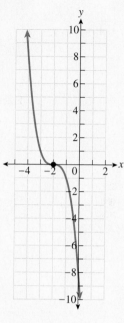

EXERCISES 11.4 ❭

Graph the given square-root function, and state its domain and range.

1. $f(x) = \sqrt{x} + 2$

2. $f(x) = \sqrt{x} - 3$

3. $f(x) = \sqrt{x + 2}$

4. $f(x) = \sqrt{x} - 3$

8. $f(x) = \sqrt{x - 9} - 10$

5. $f(x) = \sqrt{x + 4} + 6$

9. $f(x) = -\sqrt{x + 4}$

6. $f(x) = \sqrt{x + 1} - 7$

10. $f(x) = -\sqrt{x} + 8$

7. $f(x) = \sqrt{x - 5} - 2$

11. $f(x) = -\sqrt{x - 2} + 3$

12. $f(x) = -\sqrt{x+5} - 4$

16. $f(x) = x^3 + 8$

17. $f(x) = (x-3)^3$

13. $f(x) = \sqrt{3-x} + 1$

18. $f(x) = (x+1)^3$

14. $f(x) = \sqrt{-2-x} + 4$

Graph the given cubic function, and state its domain and range.

19. $f(x) = (x+4)^3 + 1$

15. $f(x) = x^3 - 1$

20. $f(x) = (x - 2)^3 - 4$ **21.** $f(x) = (x + 2)^3 + 7$ **24.** $f(x) = -(x + 3)^3$ **25.** $f(x) = 2x^3$

22. $(x + 1)^3 - 10$ **26.** $f(x) = -\dfrac{1}{2}x^3$

23. $f(x) = -x^3 - 27$

Determine the function $f(x)$ that has been graphed. The function will be of the form $f(x) = a\sqrt{x - h} + k$ or $f(x) = a(x - h)^3 + k$. You may assume $a = 1$ or $a = -1$.

27.

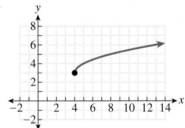

28.

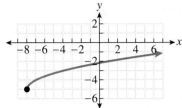

29.

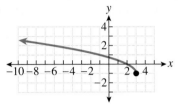

30.

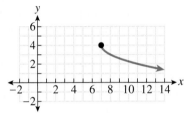

31.

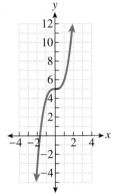

32.

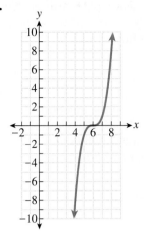

33. a)   On the same set of axes, graph the functions
$f(x) = \sqrt{x}$, $g(x) = \sqrt{x + 2}$, and $h(x) = \sqrt{x - 6}$.

b)   Using the results from part a), explain how the graph of $f(x) = \sqrt{x + 19}$ would differ from the graph of the function $f(x) = \sqrt{x}$. Also, explain how the graph of $f(x) = \sqrt{x - 27}$ would differ from the graph of the function $f(x) = \sqrt{x}$.

34. a)   On the same set of axes, graph the functions
$f(x) = \sqrt{x}$, $g(x) = \sqrt{x} + 4$, and $h(x) = \sqrt{x} - 7$.

b)   Using the results from part a), explain how the graph of $f(x) = \sqrt{x} + 35$ would differ from the graph of the function $f(x) = \sqrt{x}$. Also, explain how the graph of $f(x) = \sqrt{x} - 31$ would differ from the graph of the function $f(x) = \sqrt{x}$.

35. On the same set of axes, graph the functions $f(x) = \sqrt{x}$ and $g(x) = -\sqrt{x}$. Explain, in general, how the graph of any square-root function is affected by placing a negative sign in front of the square root.

36. Using the concepts developed in Exercises 33–35, explain how the graph of $g(x) = -\sqrt{x+8} - 1$ differs from the graph of the function $f(x) = \sqrt{x}$. Check your explanation by actually graphing the functions $f(x)$ and $g(x)$ on the same set of axes.

Answer in complete sentences.

37. Explain how to graph a square-root function and how to determine the domain and range of a square-root function. Give an example to illustrate the process.

38. Explain how to graph a cubic function. Give an example to illustrate the process.

39. Explain the similarities and differences between graphing a quadratic function and a square-root function.

40. Explain why both the domain and the range of a cubic function is the set of real numbers.

Study Tip REVISITED   Lighting is an important consideration when selecting a study area. Some students prefer a bright, well-lit room, while this type of lighting makes other students fidgety and uncomfortable. Some students prefer to work in a room with soft, warm light, while this type of lighting makes other students drowsy. Some students prefer to work in a room with ample natural light. Whatever your preference, be sure that there is enough light to see clearly without straining your eyes. Working with insufficient light will make your eyes tired, which will make you tired as well.

11.5

THE ALGEBRA OF FUNCTIONS

Objectives

1 Find the sum, difference, product, and quotient functions for two functions $f(x)$ and $g(x)$.

2 Find the graph of the sum function or difference function from the graphs of two functions $f(x)$ and $g(x)$.

3 Solve applications involving the sum function or the difference function.

4 Find the composite function of two functions $f(x)$ and $g(x)$.

5 Find the domain of a composite function $(f \circ g)(x)$.

In this section, we will examine several ways to combine two or more functions into a single function. Just as we use addition, subtraction, multiplication, and division to combine two numbers, we can use these operations to combine functions as well.

The Sum Function and the Difference Function

Objective 1 **Find the sum, difference, product, and quotient functions for two functions $f(x)$ and $g(x)$.** The function that results when two functions $f(x)$ and $g(x)$ are added together is called the **sum function,** and is denoted by $(f + g)(x)$. The function $(f - g)(x)$ is the difference of the two functions $f(x)$ and $g(x)$.

The Sum Function

For any two functions $f(x)$ and $g(x)$, $(f + g)(x) = f(x) + g(x)$.

The Difference Function

For any two functions $f(x)$ and $g(x)$, $(f - g)(x) = f(x) - g(x)$.

EXAMPLE 1 If $f(x) = x^2 + 3x - 10$ and $g(x) = 3x^2 - 5x - 9$, find the following.

a) $(f + g)(x)$

Solution

We will begin by rewriting $(f + g)(x)$ as $f(x) + g(x)$.

$$\begin{aligned}(f + g)(x) &= f(x) + g(x) &&\text{Rewrite as the sum of the} \\ &&&\text{two functions.}\\ &= (x^2 + 3x - 10) + (3x^2 - 5x - 9) &&\text{Substitute for each} \\ &&&\text{function.}\\ &= x^2 + 3x - 10 + 3x^2 - 5x - 9 &&\text{Remove parentheses.}\\ &= 4x^2 - 2x - 19 &&\text{Combine like terms.}\end{aligned}$$

b) $(f - g)(x)$

Solution

We will begin by rewriting $(f - g)(x)$ as $f(x) - g(x)$.

$$(f - g)(x) = f(x) - g(x)$$ Rewrite as the difference of the two functions.

$$= (x^2 + 3x - 10) - (3x^2 - 5x - 9)$$ Substitute for each function.

$$= x^2 + 3x - 10 - 3x^2 + 5x + 9$$ Distribute to remove parentheses.

$$= -2x^2 + 8x - 1$$ Combine like terms.

c) $(f + g)(-3)$

Solution

Since we have already found that $(f + g)(x) = 4x^2 - 2x - 19$, we can substitute -3 for x in this function.

$$(f + g)(x) = 4x^2 - 2x - 19$$
$$(f + g)(-3) = 4(-3)^2 - 2(-3) - 19$$ Substitute -3 for x.
$$= 23$$ Simplify.

$(f + g)(-3) = 23$. An alternative method to find $(f + g)(-3)$ would be to substitute -3 for x in the functions $f(x)$ and $g(x)$ and then add the results.

Quick Check 1
If $f(x) = x^2 - 9x + 20$ and $g(x) = 3x - 44$, find the given function.

a) $(f + g)(x)$
b) $(f - g)(x)$
c) $(f + g)(7)$
d) $(f - g)(-2)$

The Product Function and the Quotient Function

The function $(f \cdot g)(x)$ is the product of the two functions $f(x)$ and $g(x)$. The function $\left(\dfrac{f}{g}\right)(x)$ is the quotient of the two functions $f(x)$ and $g(x)$.

The Product Function

    For any two functions $f(x)$ and $g(x)$, $(f \cdot g)(x) = f(x) \cdot g(x)$.

The Quotient Function

    For any two functions $f(x)$ and $g(x)$, $\left(\dfrac{f}{g}\right)(x) = \dfrac{f(x)}{g(x)}$, $g(x) \neq 0$.

EXAMPLE 2   If $f(x) = x - 5$ and $g(x) = x^2 + 6x + 8$, find $(f \cdot g)(x)$.

Solution

We will begin by writing $(f \cdot g)(x)$ as $f(x) \cdot g(x)$.

$$(f \cdot g)(x) = f(x) \cdot g(x)$$ Rewrite as the product of the two functions.

$$= (x - 5)(x^2 + 6x + 8)$$ Substitute for each function.

$$= x^3 + 6x^2 + 8x - 5x^2 - 30x - 40$$ Multiply using the distributive property.

$$= x^3 + x^2 - 22x - 40$$ Combine like terms.

Quick Check 2
If $f(x) = 3x + 4$ and $g(x) = 4x - 7$, find $(f \cdot g)(x)$.

EXAMPLE › 3   If $f(x) = x^2 - 3x - 40$ and $g(x) = x^2 - 25$, find $\left(\dfrac{f}{g}\right)(x)$.

Solution

We will begin by writing $\left(\dfrac{f}{g}\right)(x)$ as $\dfrac{f(x)}{g(x)}$. We then simplify the resulting rational expression, if possible, using the techniques of Section 7.1.

$$\left(\frac{f}{g}\right)(x) = \frac{f(x)}{g(x)}$$ Rewrite as the quotient of the two functions.

$$= \frac{x^2 - 3x - 40}{x^2 - 25}$$ Substitute for each function.

$$= \frac{(x - 8)(x + 5)}{(x + 5)(x - 5)}$$ Factor the numerator and denominator.

$$= \frac{(x - 8)\overset{1}{\cancel{(x + 5)}}}{\underset{1}{\cancel{(x + 5)}}(x - 5)}$$ Divide out common factors.

$$= \frac{x - 8}{x - 5}$$ Simplify.

> *Quick Check* **3**
>
> If $f(x) = x^2 + 3x - 18$ and
> $g(x) = x^2 + 13x + 42$,
> find $\left(\dfrac{f}{g}\right)(x)$.

Using the Graphs of Two Functions to Graph Their Sum or Difference Functions

Objective **2** Find the graph of the sum function or difference function from the graphs of two functions $f(x)$ and $g(x)$.

EXAMPLE › 4   Use the graphs of the two functions $f(x)$ and $g(x)$, shown below on the same set of axes, to graph the function $(f + g)(x)$.

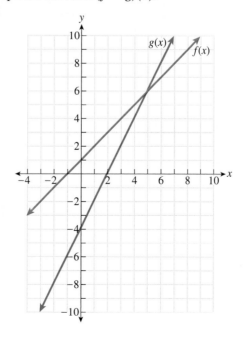

Solution

We will choose certain values for x and then determine the function values of $f(x)$ and $g(x)$ for these values. Adding these function values will give us the function values of $(f + g)(x)$ for the chosen values of x. These ordered pairs can then be plotted on a graph. We will begin by choosing $-2, -1, 0, 1,$ and 2 for x. The function values for $f(x)$ and $g(x)$ can be read from the graph. The sum of these function values is listed in the column on the right.

x	$f(x)$	$g(x)$	$(f + g)(x)$
-2	-1	-8	$(-1) + (-8) = -9$
-1	0	-6	$0 + (-6) = -6$
0	1	-4	$1 + (-4) = -3$
1	2	-2	$2 + (-2) = 0$
2	3	0	$3 + 0 = 3$

The ordered pairs $(x, (f + g)(x))$ that we will plot on the graph are $(-2, -9)$, $(-1, -6)$, $(0, -3)$, $(1, 0)$, and $(2, 3)$. The points appear to line up, so we draw a straight line through them. In general, the sum or difference of two linear functions will also be a linear function.

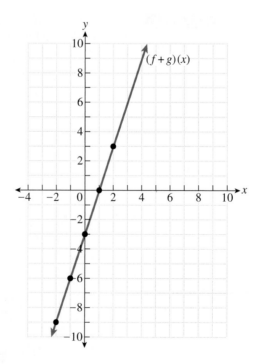

Quick Check **4** Given the graphs of the functions $f(x)$ and $g(x)$, graph the function $(f + g)(x)$.

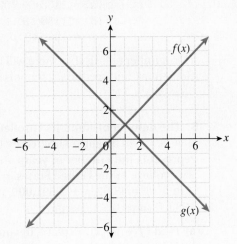

Applications

Objective 3 **Solve applications involving the sum function or the difference function.**

EXAMPLE 5 The number of public elementary and secondary schools in the United States in a particular year can be approximated by the function $f(x) = 950x + 82{,}977$, where x represents the number of years after 1990. The number of private elementary and secondary schools in the United States in a particular year can be approximated by the function $g(x) = 188x + 25{,}941$, where again x represents the number of years after 1990. (*Source:* U.S. Department of Education, National Center for Education Statistics)

a) Find $(f + g)(x)$. Explain, in your own words, what this function represents.

Solution

$$(f + g)(x) = f(x) + g(x)$$ Rewrite as the sum of the two functions.

$$= (950x + 82{,}977) + (188x + 25{,}941)$$ Substitute for each function.

$$= 950x + 82{,}977 + 188x + 25{,}941$$ Remove parentheses.
$$= 1138x + 108{,}918$$ Combine like terms.

$(f + g)(x) = 1138x + 108{,}918$. This function tells the combined number of public and private elementary and secondary schools there are x years after 1990.

b) Find $(f + g)(19)$. Explain, in your own words, what this number represents.

Solution

$$(f + g)(x) = 1138x + 108{,}918$$
$$(f + g)(19) = 1138(19) + 108{,}918 \qquad \text{Substitute 19 for } x.$$
$$= 130{,}540 \qquad \text{Simplify.}$$

This number tells us that there will be a total of approximately 130,540 public and private elementary and secondary schools in the year 2009, which is 19 years after 1990.

Quick Check **5** The number of students enrolled at a public college in the United States in a particular year can be approximated by the function $f(x) = 111{,}916x + 9{,}528{,}000$, where x represents the number of years after 1990. The number of students enrolled at a private college in the United States in a particular year can be approximated by the function $g(x) = 43{,}973x + 2{,}591{,}000$, where again x represents the number of years after 1990. (*Source:* U.S. Department of Education, National Center for Education Statistics)

a) Find $(f + g)(x)$. Explain, in your own words, what this function represents.
b) Find $(f - g)(x)$. Explain, in your own words, what this function represents.

Composition of Functions

Objective 4 **Find the composite function of two functions $f(x)$ and $g(x)$.**
Another way to combine two functions is to use the output of one function as the input for the other function. When this is done, it is called the **composition** of the two functions. For example, consider the functions $f(x) = x + 5$ and $g(x) = 2x + 1$. If we evaluated the function $g(x)$ when $x = 7$, the output would be 15.

$$g(7) = 2(7) + 1 \qquad \text{Substitute 7 for } x.$$
$$= 15 \qquad \text{Simplify.}$$

We can visualize this with the following picture of a function "machine." The machine takes an input of 7 and creates an output of 15.

$$7 \longrightarrow \boxed{g(x) = 2x + 1} \longrightarrow 15$$

Now, if we evaluate the function $f(x)$ when $x = 15$, this is a composition of the two functions. The output of function $g(x)$ is the input of the function $f(x)$.

$$f(15) = 15 + 5 \qquad \text{Substitute 15 for } x.$$
$$= 20 \qquad \text{Add.}$$

The function $f(x)$ takes an input of 15 and creates an output of 20.

$$7 \longrightarrow \boxed{g(x) = 2x + 1} \longrightarrow 15 \longrightarrow \boxed{f(x) = x + 5} \longrightarrow 20$$

Symbolically, this can be represented as $f(g(7)) = 20$.

Composite Function

For any two functions $f(x)$ and $g(x)$, the **composite function** $(f \circ g)(x)$ is defined as $(f \circ g)(x) = f(g(x))$ and read "f of g of x."
The symbol "$\circ$" is used to denote the **composition of two functions.**

Suppose that $f(x) = 3x - 5$ and $g(x) = 4x + 9$, and we wanted to find $(f \circ g)(5)$. We will rewrite $(f \circ g)(5)$ as $f(g(5))$ and begin by evaluating $g(5)$.

$$g(5) = 4(5) + 9 \qquad \text{Substitute 5 for } x.$$
$$= 29 \qquad \text{Simplify.}$$

Now we evaluate $f(29)$.

$$(f \circ g)(5) = f(g(5))$$
$$= f(29) \qquad \text{Substitute 29 for } g(5).$$
$$= 3(29) - 5 \qquad \text{Substitute 29 for } x \text{ in the function } f(x).$$
$$= 82 \qquad \text{Simplify.}$$

$$5 \longrightarrow \boxed{g(x) = 4x + 9} \longrightarrow 29 \longrightarrow \boxed{f(x) = 3x - 5} \longrightarrow 82$$

Suppose again that $f(x) = 3x - 5$ and $g(x) = 4x + 9$ and that we wanted to find the composite function $(f \circ g)(x)$. We begin by rewriting $(f \circ g)(x)$ as $f(g(x))$; then we replace $g(x)$ by the expression $4x + 9$.

$$(f \circ g)(x) = f(g(x))$$
$$= f(4x + 9) \qquad \text{Replace } g(x) \text{ by } 4x + 9.$$
$$= 3(4x + 9) - 5 \qquad \text{Substitute } 4x + 9 \text{ for } x \text{ in the function } f(x).$$
$$= 12x + 27 - 5 \qquad \text{Distribute.}$$
$$= 12x + 22 \qquad \text{Combine like terms.}$$

Notice that if we evaluate this composite function for $x = 5$, the output would be $12(5) + 22$ or 82. This is the same result that we obtained by first evaluating $g(5)$ and then evaluating $f(x)$ for this value.

EXAMPLE 6 Given that $f(x) = 2x + 11$ and $g(x) = 3x - 10$, find $(f \circ g)(x)$.

Solution

When finding a composite function, it is crucial to substitute the correct expression into the correct function. A safe way to ensure this is by rewriting $(f \circ g)(x)$ as $f(g(x))$, and then replacing the "inner" function by the appropriate expression. In this example we will replace $g(x)$ by $3x - 10$. Then we evaluate the "outer" function for this expression.

$$(f \circ g)(x) = f(g(x))$$
$$= f(3x - 10) \qquad \text{Replace } g(x) \text{ by } 3x - 10.$$
$$= 2(3x - 10) + 11 \qquad \text{Substitute } 3x - 10 \text{ for } x \text{ in the function } f(x).$$
$$= 6x - 20 + 11 \qquad \text{Distribute.}$$
$$= 6x - 9 \qquad \text{Combine like terms.}$$

Quick Check 6

Given that
$f(x) = 4x - 9$ and
$g(x) = 2x + 7$, find
$(f \circ g)(x)$

EXAMPLE ▸ 7   Given that $f(x) = 5x - 6$ and $g(x) = x^2 + 2x - 8$, find the given composite functions.

a) $(f \circ g)(x)$

Solution

$$
\begin{aligned}
(f \circ g)(x) &= f(g(x)) \\
&= f(x^2 + 2x - 8) && \text{Replace } g(x) \text{ by } x^2 + 2x - 8. \\
&= 5(x^2 + 2x - 8) - 6 && \text{Substitute } x^2 + 2x - 8 \text{ for } x \text{ in} \\
& && \text{the function } f(x). \\
&= 5x^2 + 10x - 40 - 6 && \text{Distribute.} \\
&= 5x^2 + 10x - 46 && \text{Combine like terms.}
\end{aligned}
$$

b) $(g \circ f)(x)$

Solution

In this example, the outer function $g(x)$ is a quadratic function.

$$
\begin{aligned}
(g \circ f)(x) &= g(f(x)) \\
&= g(5x - 6) && \text{Replace } f(x) \text{ by} \\
& && 5x - 6. \\
&= (5x - 6)^2 + 2(5x - 6) - 8 && \text{Substitute } 5x - 6 \text{ for} \\
& && x \text{ in the function} \\
& && g(x). \\
&= (5x - 6)(5x - 6) + 2(5x - 6) - 8 && \text{Square the binomial} \\
& && 5x - 6 \text{ by multiplying} \\
& && \text{it by itself.} \\
&= 25x^2 - 30x - 30x + 36 + 10x - 12 - 8 && \text{Multiply} \\
& && (5x - 6)(5x - 6) \\
& && \text{and } 2(5x - 6). \\
&= 25x^2 - 50x + 16 && \text{Combine like terms.}
\end{aligned}
$$

> **Quick Check 7**
>
> Given that $f(x) = x + 3$ and
> $g(x) = x^2 - 3x - 28$,
> find the given composite functions.
>
> **a)** $(f \circ g)(x)$
> **b)** $(g \circ f)(x)$

The Domain of a Composite Function

Objective 5   **Find the domain of a composite function $(f \circ g)(x)$.** The domain of a composite function $(f \circ g)(x)$ is the set of all input values x whose output from $g(x)$ is in the domain of $f(x)$. In all of the previous examples, the domain has been the set of all real numbers $\Re$, as none of the functions had any restrictions on their domains.

EXAMPLE ▸ 8   Given that $f(x) = \dfrac{x + 9}{2x - 1}$ and $g(x) = 6x + 15$, find $(f \circ g)(x)$ and state its domain.

Solution

Since $g(x)$ is a linear function, there are no restrictions on its domain. Once we find the composite function $(f \circ g)(x)$, we will find the domain of that function.

$$
\begin{aligned}
(f \circ g)(x) &= f(g(x)) \\
&= f(6x + 15) && \text{Replace } g(x) \text{ by } 6x + 15.
\end{aligned}
$$

$$= \frac{(6x + 15) + 9}{2(6x + 15) - 1}$$ Substitute $6x + 15$ for x in the function $f(x)$.

$$= \frac{6x + 15 + 9}{12x + 30 - 1}$$ Distribute.

$$= \frac{6x + 24}{12x + 29}$$ Combine like terms.

Quick Check 8

Given that

$f(x) = \dfrac{2x + 5}{x + 7}$ and

$g(x) = x - 4$, find

$(f \circ g)(x)$ and state its domain.

So $(f \circ g)(x) = \frac{6x + 24}{12x + 29}$. The composite function is a rational function, and we find the restrictions on the domain by setting the denominator equal to 0 and solving.

$$12x + 29 = 0$$ Set the denominator equal to 0.

$$12x = -29$$ Subtract 29.

$$x = -\frac{29}{12}$$ Divide by 12.

The domain of this composite function $(f \circ g)(x)$ is the set of all real numbers except $-\frac{29}{12}$.

EXERCISES *11.5*

For the given functions $f(x)$ and $g(x)$, find

a) $(f + g)(x)$

b) $(f + g)(8)$

c) $(f + g)(-3)$

1. $f(x) = 4x + 11, g(x) = 3x - 17$

2. $f(x) = 2x - 13, g(x) = -7x + 6$

3. $f(x) = 5x + 8, g(x) = x^2 - x - 19$

4. $f(x) = x^2 - 7x + 12, g(x) = x^2 + 4x - 32$
$2x^2 - 3x - 20$

For the given functions $f(x)$ and $g(x)$, find

a) $(f - g)(x)$

b) $(f - g)(6)$

c) $(f - g)(-5)$

5. $f(x) = 6x - 11, g(x) = -2x - 5$

6. $f(x) = 8x - 13, g(x) = 8x + 13$

7. $f(x) = 10x + 37, g(x) = x^2 - 5x - 12$

8. $f(x) = x^2 + 7x + 100, g(x) = -x^2 + 12x - 25$

For the given functions $f(x)$ and $g(x)$, find

a) $(f \cdot g)(x)$

b) $(f \cdot g)(4)$

c) $(f \cdot g)(-10)$

9. $f(x) = x + 7, g(x) = 2x - 10$

10. $f(x) = 3x, g(x) = x - 9$

11. $f(x) = 4x + 15, g(x) = x^2 + 5x - 36$

12. $f(x) = x^2 - 3x - 54,$
$g(x) = x^2 + 9x + 14$

For the given functions $f(x)$ and $g(x)$, find

a) $\left(\dfrac{f}{g}\right)(x)$

b) $\left(\dfrac{f}{g}\right)(7)$

c) $\left(\dfrac{f}{g}\right)(-2)$

13. $f(x) = 3x + 1, g(x) = x + 7$

14. $f(x) = 6x - 22$, $g(x) = x^2 + 11x + 10$

15. $f(x) = x^2 + 8x + 15$, $g(x) = x^2 - x - 12$

16. $f(x) = x^2 - 17x + 72$, $g(x) = x^2 - 7x - 8$

Let $f(x) = 3x - 8$ *and* $g(x) = 2x + 15$. *Find the following.*

17. $(f + g)(6)$

18. $(f + g)(-4)$

19. $(f - g)(-2)$

20. $(f - g)(5)$

21. $(f \cdot g)(7)$

22. $(f \cdot g)(10)$

23. $\left(\dfrac{f}{g}\right)(1)$

24. $\left(\dfrac{f}{g}\right)(-4)$

Let $f(x) = 5x - 9$ *and* $g(x) = 2x - 17$. *Find the following, and simplify completely.*

25. $(f + g)(x)$

26. $(f - g)(x)$

27. $(f \cdot g)(x)$

28. $\left(\dfrac{f}{g}\right)(x)$

Let $f(x) = x - 10$ *and* $g(x) = x^2 - 8x - 20$. *Find the following, and simplify completely.*

29. $(f - g)(x)$

30. $(f + g)(x)$

31. $\left(\dfrac{f}{g}\right)(x)$

32. $(f \cdot g)(x)$

Find the unknown function $g(x)$ *that satisfies the given conditions.*

33. $f(x) = 3x - 20$, $(f + g)(x) = 7x - 11$

34. $f(x) = 9x + 13$, $(f + g)(x) = 5x + 22$

35. $f(x) = x + 8$, $(f - g)(x) = 5x + 2$

36. $f(x) = -2x + 37$, $(f - g)(x) = -9x + 60$

Given the graphs of the functions $f(x)$ *and* $g(x)$, *graph the function* $(f + g)(x)$.

37.

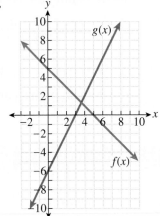

38.

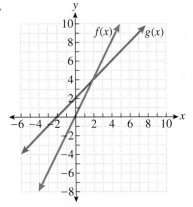

Given the graphs of the functions f(x) and g(x), graph the function (f − g)(x).

39.

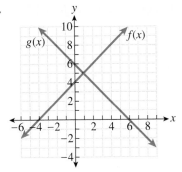

40.

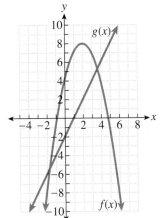

41. The number of male doctors, in thousands, in the United States in a particular year can be approximated by the function $f(x) = 10.4x + 398$, where x represents the number of years after 1980. The number of female doctors, in thousands, in the United States in a particular year can be approximated by the function $g(x) = 9.1x + 49$, where again x represents the number of years after 1980. (*Source: American Medical Association*)

a) Find $(f + g)(x)$. Explain, in your own words, what this function represents.

b) Find $(f + g)(30)$. Explain, in your own words, what this number represents.

42. The number of male inmates who are HIV positive in the United States in a particular year can be approximated by the function $f(x) = 223x + 20{,}969$, where x represents the number of years after 1995. The number of female inmates who are HIV positive in the United States in a particular year can be approximated by the function $g(x) = 83x + 2102$, where again x represents the number of years after 1995. (*Source:* U.S. Department of Justice, Bureau of Justice Statistics)

a) Find $(f + g)(x)$. Explain, in your own words, what this function represents.

b) Find $(f + g)(22)$. Explain, in your own words, what this number represents.

43. The number of master's degrees, in thousands, earned by females in the United States in a particular year can be approximated by the function $f(x) = 9.2x + 181.8$, where x represents the number of years after 1990. The number of master's degrees, in thousands, earned by males in the United States in a particular year can be approximated by the function $g(x) = 3.4x + 160.9$, where again x represents the number of years after 1990. (*Source:* U.S. Department of Education, National Center for Education Statistics)

a) Find $(f - g)(x)$. Explain, in your own words, what this function represents.

b) Find $(f - g)(30)$. Explain, in your own words, what this number represents.

44. The amount of money, in billions of dollars, spent on health care that was covered by insurance in the United States in a particular year can be approximated by the function $f(x) = x^2 + 11x + 244$, where x represents the number of years after 1990. The amount of money, in billions of dollars, spent on health care that was paid out of pocket in the United States in a particular year can be approximated by the function $g(x) = 6x + 129$, where again x represents the number of years after 1990. (*Source:* U.S. Centers for Medicare and Medicaid Services)

a) Find $(f - g)(x)$. Explain, in your own words, what this function represents.

b) Find $(f - g)(40)$. Explain, in your own words, what this number represents.

Let $f(x) = 4x + 7$ and $g(x) = x - 3$. Find the following.

45. $(f \circ g)(5)$

46. $(f \circ g)(-2)$

47. $(g \circ f)(-1)$

48. $(g \circ f)(10)$

Let $f(x) = 2x - 9$ and $g(x) = x^2 - 9x + 18$. Find the following.

49. $(f \circ g)(1)$

50. $(f \circ g)(4)$

51. $(g \circ f)(-5)$

52. $(g \circ f)(10)$

For the given functions $f(x)$ and $g(x)$, find
a) $(f \circ g)(x)$ b) $(g \circ f)(x)$
c) $(f \circ g)(3)$ d) $(g \circ f)(-4)$

53. $f(x) = 3x + 5, g(x) = 2x + 4$

54. $f(x) = 5x - 9, g(x) = -x + 6$

55. $f(x) = x + 4, g(x) = x^2 + 3x - 40$

56. $f(x) = x^2 - 7x - 18, g(x) = x - 9$

For the given functions $f(x)$ and $g(x)$, find $(f \circ g)(x)$ and state its domain.

57. $f(x) = \dfrac{5x}{x + 7}, g(x) = 2x + 13$

58. $f(x) = \dfrac{2x - 3}{3x + 5}, g(x) = x + 8$

59. $f(x) = \sqrt{2x - 10}, g(x) = x + 11$

60. $f(x) = \sqrt{x - 2}, g(x) = 3x + 14$

For the given function $f(x)$, find $(f \circ f)(x)$.

61. $f(x) = 3x - 10$

62. $f(x) = -5x + 18$

63. $f(x) = x^2 + 6x + 12$

64. $f(x) = x^2 - 2x - 15$

For the given functions $f(x)$ and $g(x)$, find a value of x for which $(f \circ g)(x) = (g \circ f)(x)$.

65. $f(x) = x^2 + 4x + 3, g(x) = x + 5$

66. $f(x) = x^2 + 6x - 9, g(x) = x - 3$

67. $f(x) = x + 7, g(x) = x^2 - 25$

68. $f(x) = x - 4, g(x) = 2x^2 + 15x + 18$

69. If $g(x) = x + 3$ and $(f \circ g)(x) = x + 11$, find $f(x)$.

70. If $g(x) = 3x - 4$ and $(f \circ g)(x) = 6x - 8$, find $f(x)$.

71. If $g(x) = 2x + 7$ and $(f \circ g)(x) = 10x + 19$, find $f(x)$.

72. If $g(x) = 5x - 2$ and $(f \circ g)(x) = -15x + 10$, find $f(x)$.

For the given function $g(x)$, find a function $f(x)$ such that $(f \circ g)(x) = x$.

73. $g(x) = x + 8$

74. $g(x) = -5x$

75. $g(x) = 3x + 8$

76. $g(x) = 4x - 17$

77. In general, is $(f + g)(x)$ equal to $(g + f)(x)$? If your answer is yes, explain why. If your answer is no, give an example that shows why they are not equal in general.

78. In general, is $(f - g)(x)$ equal to $(g - f)(x)$? If your answer is yes, explain why. If your answer is no, give an example that shows why they are not equal in general.

Study Tip **REVISITED**   It is important to be comfortable while studying, without being too comfortable. If the room is too warm, you may get tired. If the room is too cold, you may be distracted by thinking about how cold you are. If your chair is not comfortable, you will fidget and lose concentration. If your chair is too comfortable, you may get sleepy. You may have the same problem if you eat a large meal before sitting down to study. In sum, choose a situation where you will be comfortable, but not too comfortable.

11.6

INVERSE FUNCTIONS

Objectives

1 Determine whether a function is one to one.
2 Use the horizontal-line test to determine whether a function is one to one.
3 Understand inverse functions.
4 Determine whether two functions are inverse functions.
5 Find the inverse of a one-to-one function.
6 Find the inverse of a function from its graph.

One-to-One Functions

Objective 1 Determine whether a function is one to one. Recall that for any function $f(x)$, each value in the domain corresponds to only one value in the range.

One-to-One Function

> If each value in the range of a function $f(x)$ corresponds to only one value in the domain, then the function $f(x)$ is called a **one-to-one function.**

In other words, if $f(x)$ is different for each input value x in the domain, then $f(x)$ is a one-to-one function.

Suppose that for some function $f(x)$, both $f(-1)$ and $f(3)$ are equal to 2. This function is not a one-to-one function, because one output value is associated two different input values. However, the function shown in the following table is a one-to-one function, since each function value corresponds to only one input value:

x	0	1	4	9	16
$f(x)$	0	1	2	3	4

A function whose input is a person's name and whose output is that person's birthday is not a one-to-one function because several people (input) can have the same birthday (output). A function whose input is a person's name and whose output is that person's Social Security number is a one-to-one function because no two people can have the same Social Security number.

EXAMPLE 1   Determine whether the function represented by the set of ordered pairs $\{(-7, -5), (-5, -2), (-2, 4), (3, 3), (6, -7)\}$ is a one-to-one function.

Solution

Since each x-value corresponds to only one y-value and each y-value corresponds to only one x-value, the function is one to one.

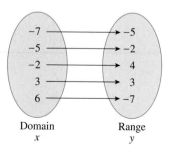

Domain   Range
x y

Quick Check 1
Determine whether the function represented by the set of ordered pairs $\{(-4, -9), (-2, -7), (1, -6), (2, -7), (6, -11)\}$ is a one-to-one function.

Horizontal-Line Test

Objective 2   **Use the horizontal-line test to determine whether a function is one to one.** We can determine whether a function is one to one by applying the **horizontal-line test** to its graph.

Horizontal-Line Test

> If a horizontal line can intersect the graph of a function at more than one point, then the function is not one to one.

Consider the graph of a quadratic function shown below. We can draw a horizontal line that intersects the function at two points, as indicated. The coordinates of these two points are $(-2, 6)$ and $(2, 6)$. This shows that two different input values, -2 and 2, correspond to the same output value (6). Therefore, the function is not one to one.

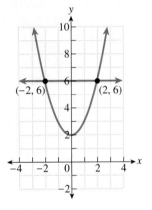

The linear function graphed below is a one-to-one function. No horizontal line crosses the graph of this function at more than one point, so the function passes the horizontal-line test.

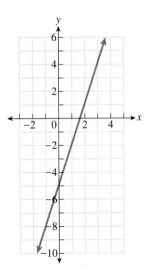

EXAMPLE 2 Use the horizontal-line test to determine whether the function is one to one.

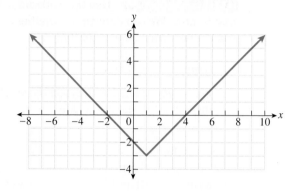

Solution

We can draw a horizontal line that crosses the graph of this absolute value function at more than one point, as shown below. This function fails the horizontal-line test and is not one to one.

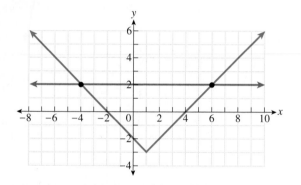

Quick Check 2 Determine whether the function is one to one.

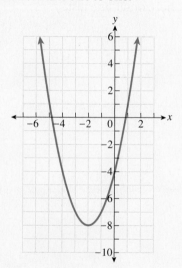

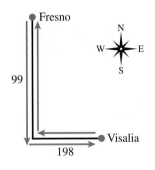

Fresno

99

Visalia

198

Inverse Functions

Objective 3   Understand inverse functions. Anne commutes to work from Visalia to Fresno each day. She starts by taking Highway 198 west and then takes Highway 99 north. How does she return home each day? She begins by taking Highway 99 south and then takes Highway 198 east. Her trips are depicted in the map on the left.

The second route takes Anne back to the starting point. The two routes have an **inverse** relationship. In this section, we will be examining **inverse functions.**

The inverse of a one-to-one function "undoes" what the function does. For example, if a function $f(x)$ takes an input value of 3 and produces an output value of 7, its inverse function takes an input value of 7 and produces an output value of 3.

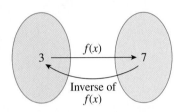

We see that the ordered pair $(3, 7)$ is in the function $f(x)$ and that the ordered pair $(7, 3)$ is in the inverse function of $f(x)$. The inputs of a function are the outputs of its inverse function, and the outputs of a function are the inputs of its inverse function. In general, if (a, b) is in a function, then (b, a) is in the inverse of that function.

Verifying That Two Functions Are Inverses

Objective 4   Determine whether two functions are inverse functions. We can determine whether two functions are inverse functions algebraically by finding their composite functions.

Inverse Functions

> Two one-to-one functions $f(x)$ and $g(x)$ are inverse functions if $(f \circ g)(x) = x$ and $(g \circ f)(x) = x.$

If $f(x)$ and $g(x)$ are inverse functions, then each function "undoes" the actions of the other function.

The function g takes an input value x and produces an output value $g(x)$.

The function f takes this value and produces an output value of x.

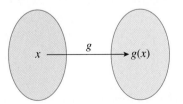

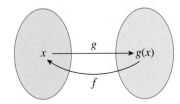

So $f(g(x)) = x$, or, in other words, $(f \circ g)(x) = x$. In a similar fashion, we could show that $(g \circ f)(x) = x$ as well.

EXAMPLE 3   Determine whether the following two functions are inverse functions:
$$f(x) = 5x - 2, g(x) = \frac{x + 2}{5}$$

Solution

Both of these functions are linear functions and are one to one. We will determine whether $(f \circ g)(x) = x$ and $(g \circ f)(x) = x$. (Replace the inner function by its formula, substitute that expression for x in the outer function, and simplify.)

$(f \circ g)(x) = x$	$(g \circ f)(x) = x$
$(f \circ g)(x) = f(g(x))$	$(g \circ f)(x) = g(f(x))$
$= f\left(\dfrac{x + 2}{5}\right)$	$= g(5x - 2)$
$= 5\left(\dfrac{x + 2}{5}\right) - 2$	$= \dfrac{(5x - 2) + 2}{5}$
$= \overset{1}{\cancel{5}}\left(\dfrac{x + 2}{\underset{1}{\cancel{5}}}\right) - 2$	$= \dfrac{5x - 2 + 2}{5}$
$= x + 2 - 2$	$= \dfrac{5x}{5}$
$= x$	$= \dfrac{\overset{1}{\cancel{5}}x}{\underset{1}{\cancel{5}}}$
	$= x$

Quick Check 3
Determine whether the following two functions are inverse functions:
$$f(x) = \frac{x - 7}{4},$$
$$g(x) = 4x + 7$$

Since $(f \circ g)(x) = x$ and $(g \circ f)(x) = x$, these two functions are inverses.

EXAMPLE 4   Determine whether the following two functions are inverse functions:
$$f(x) = \frac{1}{4}x - \frac{3}{4}, \qquad g(x) = 3x + 4$$

Solution

These two functions are linear and one to one. We will begin by determining whether $(f \circ g)(x) = x$.

$$
\begin{aligned}
(f \circ g)(x) &= f(g(x)) \\
&= f(3x + 4) &&\text{Replace } g(x) \text{ by } 3x + 4. \\
&= \frac{1}{4}(3x + 4) - \frac{3}{4} &&\text{Substitute } 3x + 4 \text{ for } x \text{ in the function } f(x). \\
&= \frac{3}{4}x + 1 - \frac{3}{4} &&\text{Distribute.} \\
&= \frac{3}{4}x + \frac{1}{4} &&\text{Combine like terms.}
\end{aligned}
$$

Quick Check 4
Determine whether the following two functions are inverse functions:
$$f(x) = \frac{1}{3}x + 4,$$
$$g(x) = 3x - 12$$

$(f \circ g)(x) \neq x$. These two functions are not inverses.

Note that it is not necessary to check $(g \circ f)(x)$, since both composite functions must simplify to x in order for the two functions to be inverse functions.

Finding an Inverse Function

Objective 5 **Find the inverse of a one-to-one function.** We use the notation $f^{-1}(x)$ to denote the inverse of a function $f(x)$. We read $f^{-1}(x)$ as "f *inverse of x.*"

> **A Word of Caution** When a superscript of -1 is written after the name of a function, the -1 is not an exponent; it is used to denote the inverse of a function. In other words, $f^{-1}(x)$ is not the same as $\dfrac{1}{f(x)}$.

Here is a procedure that can be used to find the inverse of a one-to-one function $f(x)$:

Finding $f^{-1}(x)$

1. **Determine whether $f(x)$ is one to one.** If $f(x)$ is not one to one, then it does not have an inverse function. We can use the horizontal-line test to determine this.
2. **Replace $f(x)$ by y.**
3. **Interchange x and y.** We interchange these two variables because we know that the input of $f(x)$ must be the output of $f^{-1}(x)$, and the output of $f(x)$ must be the input of $f^{-1}(x)$.
4. **Solve the resulting equation for y.** This will rewrite the inverse function as a function of x.
5. **Replace y by $f^{-1}(x)$.**

EXAMPLE 5   Find $f^{-1}(x)$ for the function $f(x) = 5x - 3$.

Solution

We begin by determining whether the function is one to one. Since the function is a linear function, it is one to one. Shown at the left is the graph of $f(x)$. The function clearly passes the horizontal-line test.

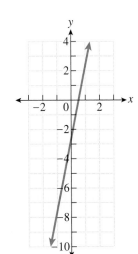

We begin to find $f^{-1}(x)$ by replacing $f(x)$ by y.

$$f(x) = 5x - 3$$
$$y = 5x - 3 \qquad \text{Replace } f(x) \text{ by } y.$$
$$x = 5y - 3 \qquad \text{Interchange } x \text{ and } y.$$
$$x + 3 = 5y \qquad \text{Add 3 to isolate the term containing } y.$$
$$\frac{x + 3}{5} = y \qquad \text{Divide both sides by 5 to isolate } y.$$
$$f^{-1}(x) = \frac{x + 3}{5} \qquad \begin{array}{l}\text{Replace } y \text{ by } f^{-1}(x). \text{ It is customary to write } f^{-1}(x) \text{ on the} \\ \text{left side.}\end{array}$$

The inverse of $f(x) = 5x - 3$ is $f^{-1}(x) = \dfrac{x + 3}{5}$, which could be written as $f^{-1}(x) = \dfrac{1}{5}x + \dfrac{3}{5}$.

Quick Check **5** Find $f^{-1}(x)$ for the function $f(x) = \frac{1}{2}x + 10$.

EXAMPLE 6   Find $f^{-1}(x)$ for the function $f(x) = \dfrac{8}{x+1}$.

Solution

In this example, $f(x)$ is a rational function. We have not yet learned to graph rational functions, but the graph of $f(x)$, which has been generated by technology, is shown below. (We could use a graphing calculator or software to generate this graph.) The function passes the horizontal-line test.

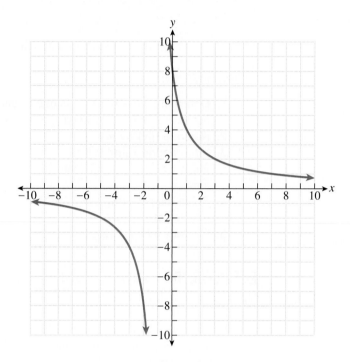

We begin to find $f^{-1}(x)$ by replacing $f(x)$ by y.

$$f(x) = \frac{8}{x+1}$$

$$y = \frac{8}{x+1} \qquad \text{Replace } f(x) \text{ by } y.$$

$$x = \frac{8}{y+1} \qquad \text{Interchange } x \text{ and } y.$$

$$x(y+1) = \frac{8}{\cancel{y+1}} \cdot \frac{\cancel{y+1}}{1} \qquad \begin{array}{l}\text{Multiply both sides by } y+1 \text{ to clear the equation}\\ \text{of fractions.}\end{array}$$

$$x(y+1) = 8 \qquad \text{Simplify.}$$
$$xy + x = 8 \qquad \text{Multiply.}$$
$$xy = 8 - x \qquad \text{Subtract } x \text{ to isolate the term containing } y.$$

$$y = \frac{8-x}{x} \qquad \text{Divide by } x \text{ to isolate } y.$$

$$f^{-1}(x) = \frac{8-x}{x} \qquad \text{Replace } y \text{ by } f^{-1}(x).$$

The inverse of $f(x) = \dfrac{8}{x+1}$ is $f^{-1}(x) = \dfrac{8-x}{x}$.

Note that the domain of this inverse function is the set of all real numbers except 0, as the denominator of the function is equal to 0 when $x = 0$. This makes the inverse function undefined when $x = 0$.

Quick Check **6**

Find $f^{-1}(x)$ for the function $f(x) = \dfrac{3 + 5x}{x}$.

Recall that the output values of a one-to-one function are the input values of its inverse function. So the range of a one-to-one function $f(x)$ is the domain of its inverse function $f^{-1}(x)$. Occasionally, if the range of a one-to-one function $f(x)$ is not the set of real numbers, we will have to restrict the domain of $f^{-1}(x)$ accordingly. This is illustrated in the next example.

EXAMPLE 7   Find $f^{-1}(x)$ for the function $f(x) = \sqrt{x + 4} - 2$. State the domain of $f^{-1}(x)$.

Solution

We begin by showing that the function is one to one. Below is the graph of $f(x)$. (For help on graphing square-root functions, refer back to Section 11.4.) The function passes the horizontal-line test, so it is a one-to-one function that has an inverse function. The range of $f(x)$ is $[-2, \infty)$, and this is the domain of its inverse function $f^{-1}(x)$.

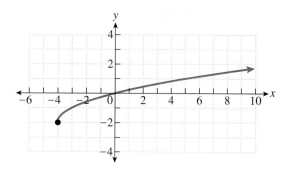

We begin to find $f^{-1}(x)$ by replacing $f(x)$ by y.

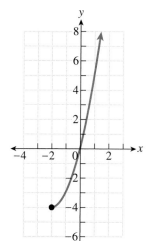

$$f(x) = \sqrt{x+4} - 2$$

$$y = \sqrt{x+4} - 2 \qquad \text{Replace } f(x) \text{ by } y.$$

$$x = \sqrt{y+4} - 2 \qquad \text{Interchange } x \text{ and } y.$$

$$x + 2 = \sqrt{y+4} \qquad \text{Add 2 to isolate the radical.}$$

$$(x+2)^2 = (\sqrt{y+4})^2 \qquad \text{Square both sides.}$$

$$(x+2)(x+2) = y + 4 \qquad \text{Square } x + 2 \text{ by multiplying it by itself.}$$

$$x^2 + 4x + 4 = y + 4 \qquad \text{Multiply } (x+2)(x+2).$$

$$x^2 + 4x = y \qquad \text{Subtract 4 to isolate } y.$$

$$f^{-1}(x) = x^2 + 4x \qquad \text{Replace } y \text{ by } f^{-1}(x).$$

The inverse of $f(x) = \sqrt{x+4} - 2$ is $f^{-1}(x) = x^2 + 4x$. The domain of $f^{-1}(x)$ is $[-2, \infty)$.

At the left is the graph of $f^{-1}(x)$. Notice that the graph does pass the horizontal-line test with this restricted domain. If we did not restrict this domain, the graph of $f^{-1}(x)$ would be a parabola and the function would not be one to one.

Quick Check **7** Find $f^{-1}(x)$ for the function $f(x) = \sqrt{x - 9} + 8$. State the domain of $f^{-1}(x)$.

Finding an Inverse Function from a Graph

Objective **6** **Find the inverse of a function from its graph.** We know that if $f(x)$ is a one-to-one function and $f(a) = b$, then $f^{-1}(b) = a$. This tells us that if the point (a, b) is on the graph of $f(x)$, then the point (b, a) is on the graph of $f^{-1}(x)$.

EXAMPLE **8** If the function is one to one, find its inverse.

$$\{(-2, -3), (-1, -2), (0, 1), (1, 5), (2, 13)\}$$

Solution

Since each value of x corresponds to only one value of y, and each value of y corresponds to only one value of x, this is a one-to-one function.

To find the inverse function, we interchange each x-coordinate and y-coordinate. The inverse function is $\{(-3, -2), (-2, -1), (1, 0), (5, 1), (13, 2)\}$.

EXAMPLE **9** For the given graph of a one-to-one function $f(x)$, graph its inverse function $f^{-1}(x)$.

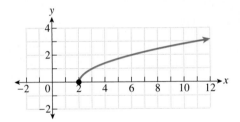

Solution

We begin by identifying the coordinates of some points that are on the graph of this function. We will use the points $(2, 0)$, $(3, 1)$, and $(6, 2)$.

By interchanging each x-coordinate with its corresponding y-coordinate, we know that the points $(0, 2)$, $(1, 3)$, and $(2, 6)$ are on the graph of $f^{-1}(x)$. We finish by drawing a graph that passes through these points. Here on the right is the graph of $f^{-1}(x)$.

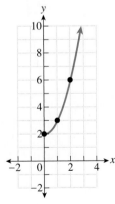

Quick Check **8** For the given graph of a one-to-one function $f(x)$, graph its inverse function $f^{-1}(x)$.

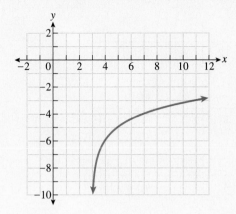

Here are the graphs of $f(x)$ and $f^{-1}(x)$ from the previous example on the same set of axes.

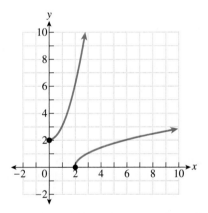

Notice that the two functions look somewhat similar; in fact, they are mirror images of each other. For any one-to-one function $f(x)$ and its inverse function $f^{-1}(x)$, their graphs are symmetric to each other about the line $y = x$. If we fold our graph along the line $y = x$, the graphs of $f(x)$ and $f^{-1}(x)$ will lie on top of each other.

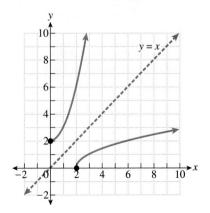

EXERCISES 11.6

F O R   E X T R A   H E L P

MyMathLab
MyMathLab

MathXL
MathXL

Interactmath.com

MathXL
Tutorials on CD

DVT CD
Videotape

Tutor
Center

Addison-Wesley
Math Tutor Center

Student Solutions
Manual

Determine whether the function $f(x)$ is a one-to-one function.

1.

x	−3	0	3	6	9
$f(x)$	−8	−2	4	10	16

2.

x	−5	−1	1	4	10
$f(x)$	9	5	3	0	−6

3.

x	−4	−2	0	2	4
$f(x)$	16	4	0	4	16

4.

x	−3	−2	−1	0	1
$f(x)$	3	2	1	0	1

Determine whether the function represented by the set of ordered pairs is a one-to-one function.

5. $\{(-9, -3), (-4, 0), (1, 1), (5, -2), (10, -3)\}$

6. $\{(1, 4), (3, 8), (5, 12), (7, 8), (9, 4)\}$

7. $\{(-8, -4), (-5, 3), (-1, 0), (4, 3), (9, 10)\}$

8. $\{(0, 1), (1, 2), (2, 4), (3, 8), (4, 16)\}$

9. Is the function whose input is a person and output is that person's mother a one-to-one function? Explain your answer in your own words.

10. Is the function whose input is a student and output is that student's favorite math teacher a one-to-one function? Explain your answer in your own words.

11. Is the function whose input is a state and output is that state's governor a one-to-one function? Explain your answer in your own words.

12. Is the function whose input is a licensed driver and output is that driver's license number a one-to-one function? Explain your answer in your own words.

Use the horizontal-line test to determine whether the function is one to one.

13.

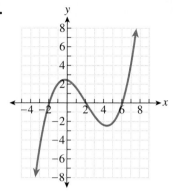

14.

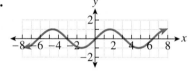

15.

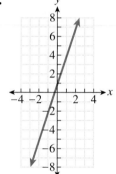

16.

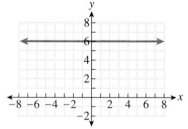

17.

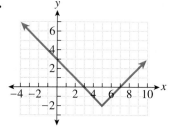

18.

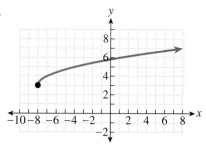

Determine whether the functions f(x) and g(x) are inverse functions by showing that (f ∘ g)(x) = x and (g ∘ f)(x) = x.

19. $f(x) = x - 7$, $g(x) = x + 7$

20. $f(x) = x + 12$, $g(x) = x - 12$

21. $f(x) = 3x$, $g(x) = -3x$

22. $f(x) = 2x$, $g(x) = -\frac{1}{2}x$

23. $f(x) = 4x + 3$, $g(x) = \frac{1}{4}x - 3$

24. $f(x) = 3x - 18$, $g(x) = \frac{1}{3}x + 6$

25. $f(x) = -x + 9$, $g(x) = x - 9$

26. $f(x) = 7x + 28$, $g(x) = \frac{1}{7}(x - 28)$

For the given function f(x), find f⁻¹(x).

27. $f(x) = x + 5$

28. $f(x) = x - 8$

29. $f(x) = 4x$

30. $f(x) = -\frac{1}{5}x$

31. $f(x) = -x + 9$

32. $f(x) = -x - 13$

33. $f(x) = 2x + 17$

34. $f(x) = -5x - 11$

35. $f(x) = -\frac{2}{3}x - 8$

36. $f(x) = \frac{5}{4}x + \frac{3}{8}$

37. $f(x) = mx$

38. $f(x) = mx + b$

39. $f(x) = \frac{1}{x} + 3$

40. $f(x) = \frac{1}{x} - 5$

41. $f(x) = \frac{1}{x + 3}$

42. $f(x) = \frac{1}{x - 5}$

43. $f(x) = \frac{5 - 6x}{x}$

44. $f(x) = \frac{1 + 10x}{x}$

45. $f(x) = \frac{8 + 4x}{3x}$

46. $f(x) = \frac{2x}{7x - 10}$

For the given function f(x), find f⁻¹(x). State the domain of f⁻¹(x).

47. $f(x) = \sqrt{x}$

48. $f(x) = \sqrt{x - 4}$

49. $f(x) = \sqrt{x} + 5$

50. $f(x) = \sqrt{x + 1} - 3$

51. $f(x) = \sqrt{x + 8} + 4$

52. $f(x) = \sqrt{x + 9} + 3$

53. $f(x) = x^2 - 9 \ (x \geq 0)$

54. $f(x) = (x + 6)^2 + 5 \ (x \geq -6)$

55. $f(x) = x^2 - 6x + 8$ $(x \geq 3)$ (*Hint:* try completing the square)

56. $f(x) = x^2 + 4x - 12$ $(x \geq -2)$ (*Hint:* try completing the square)

If the function represented by the set of ordered pairs is one to one, find its inverse.

57. $\{(-5, -17), (-1, -9), (1, -5), (4, 1), (7, 7)\}$

58. $\{(-2, -1), (0, 5), (1, 8), (4, 17), (7, 26)\}$

59. $\{(-2, 21), (-1, 17), (0, 13), (1, 9), (2, 13)\}$

60. $\{(-10, 4), (-9, 5), (-6, 6), (-1, 7), (6, 8)\}$

For the given graph of a one-to-one function $f(x)$, graph its inverse function $f^{-1}(x)$.

61.

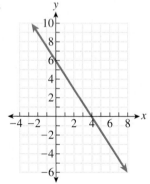

62.

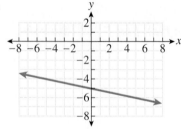

63.

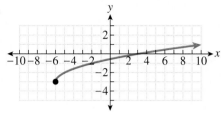

64.

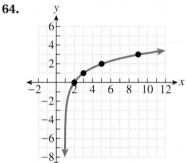

Graph a one-to-one function $f(x)$ that meets the given criteria.

65. $f(x)$ is a linear function, $f(4) = 9$, and $f^{-1}(5) = -2$.

66. $f(x)$ is a linear function, $f^{-1}(5) = -7$, and $f^{-1}(-6) = 8$.

67. $f(x)$ is a quadratic function, the domain of $f(x)$ is restricted to $[3, \infty)$, $f(3) = 1$, $f(4) = 0$, $f^{-1}(3) = 5$ and $f^{-1}(8) = 6$.

68. $f(x)$ is a quadratic function, the domain of $f(x)$ is restricted to $[1, \infty)$, $f^{-1}(-8) = 1$, $f^{-1}(-2) = 3$ and $f^{-1}(8) = 5$.

Answer in complete sentences.

69. True or False: if a linear function has negative slope, then its inverse function has positive slope.

70. Explain how the horizontal-line test shows whether a function is or is not one to one.

71. Describe a real-world function that is a one-to-one function, and explain why the function is one to one. Describe a real-world function that is not a one-to-one function, and explain why it is not a one-to-one function.

72. Explain how to find the inverse function of a one-to-one function $f(x)$. Use an example to illustrate the process.

Study Tip REVISITED   Under ideal circumstances, some of your studying should be accomplished on campus, and some of your studying should be done at home. The best time for studying new material is as soon as possible after class. That means that you should look for a location on campus. Since studying new material takes more concentration, find a location with as few distractions as possible. Most campus libraries have an area designated as a silent study area. Other libraries have cubicles that help to isolate you while studying.

While studying at home, stay away from the computer, phone, or television. These devices will steal your time away from you, and you may not even realize it. Keep all temptations as far away as possible.

Chapter 11 Summary

Summary of Chapter 11 Study Tip s

This chapter's study tips focused on creating the proper study environment.

- Find a location as free from distractions as possible.
- Determine what level of background noise you can tolerate without being distracted.
- Select a location that allows you to spread out your materials.
- Select a location that provides sufficient lighting.
- Select a location that is comfortable, with proper temperature control.
- Search out multiple locations, at home and on campus, that are conducive to studying.

1. Does a function exist for which the input value is a mathematics instructor and the output value is the state that the instructor was born in? [11.1]

2. Does a function exist for which the input value is the political party of a state's governor and the output value is the state? [11.1]

Use the vertical-line test to determine whether the graph represents a function. [11.1]

3.

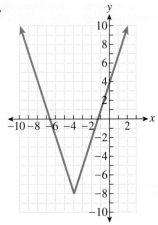

4.

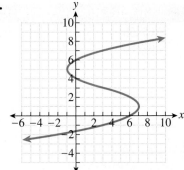

Evaluate the given function. [11.1]

5. $f(x) = 2x, f(-5)$

6. $f(x) = 3x - 7, f(4)$

7. $f(x) = 7x - 11, f(3b - 8)$

8. $f(x) = x^2 - 7x + 32, f(7)$

9. $f(x) = x^2 - 9x - 36, f(-3)$

10. $f(x) = x^2 + 10x - 25, f(2n + 5)$

11. $f(x) = |x + 2| - 17, f(-9)$

12. $f(x) = \sqrt{2x + 16} + 27, f(10)$

Determine the domain and range of the function $f(x)$ that has been graphed. [11.1]

13.

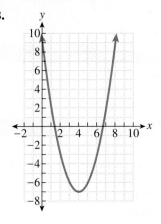

14.

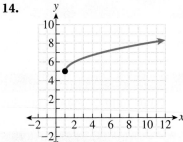

Use the graph of the function $f(x)$ to find the indicated function value. [11.1]

15. $f(4)$

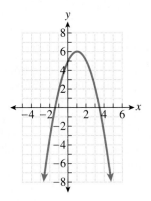

16. $f(2)$

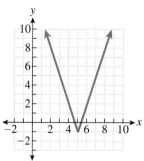

Use the graph of the function $f(x)$ to determine which values of x satisfy the given equation. [11.1]

17. $f(x) = 3$

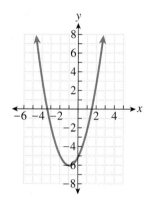

18. $f(x) = -7$

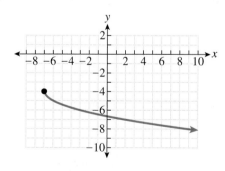

Graph each function. Label the x- and y-intercepts. [11.2]

19. $f(x) = x - 7$

20. $f(x) = -3x - 6$

21. $f(x) = \dfrac{2}{5}x - 5$

22. $f(x) = 2x + 4$

Use the given graph to determine the linear function f(x) on the graph. **[11.2]**

23.

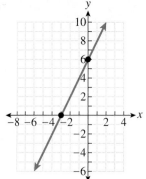

24.

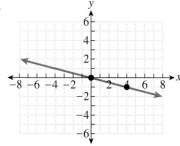

25. A rental car company charges a fixed fee for renting a car in addition to charging per mile driven. If a person rents a car and drives it 250 miles, the charge is $42.50. If a person rents a car and drives it 400 miles, the charge is $50. **[11.2]**

a) Find a linear function $f(x) = mx + b$ whose input is the number of miles driven and whose output is the corresponding charge for renting the car.

b) Use the function from part (a) to determine the cost of renting a car and driving it 75 miles.

26. A college club is selling snow cones to raise money for its scholarship fund. In addition to paying $40 to rent space for the day, the club also had to spend $60 on a snow cone machine. The supplies to make each snow cone costs $0.10, and the club is charging $1.50 for each snow cone. **[11.2]**

a) Find the cost function $C(x)$.

b) Find the revenue function $R(x)$.

c) Find the profit function $P(x)$.

d) How much profit will be generated if the club sells 400 snow cones?

Graph the function. Label the vertex, y-intercept, and any x-intercepts. **[11.3]**

27. $f(x) = x^2 - 4x - 12$

28. $f(x) = x^2 + 6x + 1$

29. $f(x) = x^2 - 8x + 19$

30. $f(x) = -x^2 - 4x + 3$

given by the function $h(t) = -16t^2 + 176t + 42$. What is the maximum height that the projectile reaches? [11.3]

38. A farmer has 144 feet of fencing to make a rectangular corral. What dimensions will make a corral with the maximum area? What is the maximum area possible? [11.3]

Determine the range of the quadratic function $f(x)$. [11.3]

39. $f(x) = x^2 + 9x + 56$

31. $f(x) = (x - 3)^2 - 7$

40. $f(x) = -x^2 + 14x + 32$

For the given function $f(x)$, simplify the difference quotient $\dfrac{f(x + h) - f(x)}{h}$. [11.3]

41. $f(x) = x^2 + 13$

42. $f(x) = x^2 - 6x - 33$

Graph the given square-root function, and state its domain and range. [11.4]

43. $f(x) = \sqrt{x + 4}$

32. $f(x) = -(x + 2)^2 - 1$

44. $f(x) = \sqrt{x - 2} + 7$

Determine whether the given quadratic function has a maximum value or a minimum value. Then find that maximum or minimum value. [11.3]

33. $f(x) = x^2 - 12x + 60$

34. $f(x) = x^2 + 7x - 18$

35. $f(x) = -x^2 + 16x - 70$

36. $f(x) = -(x + 8)^2 - 39$

37. A projectile is launched upward from the roof of a building with an initial velocity of 176 feet/second. The height of the projectile (in feet) after t seconds is

45. $f(x) = -\sqrt{x+1} + 2$

46. $f(x) = \sqrt{x+5} - 3$

Graph the given cubic function, and state its domain and range. [11.4]

47. $f(x) = (x-2)^3$

48. $f(x) = x^3 - 1$

Determine the function $f(x)$ that has been graphed. [11.4]

49.

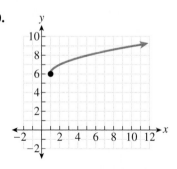

50.

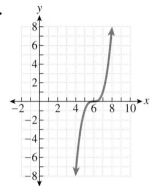

Let $f(x) = x^2 + 9x - 22$ and $g(x) = x + 11$. Find the following. [11.5]

51. $(f + g)(-5)$ **52.** $(f - g)(15)$

53. $(f \cdot g)(8)$ **54.** $\left(\dfrac{f}{g}\right)(-1)$

For the given functions $f(x)$ and $g(x)$, find $(f + g)(x)$. [11.5]

55. $f(x) = 6x - 11, g(x) = 4x + 35$

For the given functions $f(x)$ and $g(x)$, find $(f - g)(x)$. [11.5]

56. $f(x) = 8x - 33, g(x) = -8x + 55$

For the given functions $f(x)$ and $g(x)$, find $(f \cdot g)(x)$. [11.5]

57. $f(x) = x - 9, g(x) = x + 4$

For the given functions $f(x)$ and $g(x)$, find $\left(\frac{f}{g}\right)(x)$. [11.5]

58. $f(x) = x^2 - x - 6, g(x) = x^2 - 10x + 21$

Let $f(x) = x + 4$ and $g(x) = x^2 - 14x + 48$. Find the following. [11.5]

59. $(f \circ g)(8)$ **60.** $(g \circ f)(9)$

For the given functions $f(x)$ and $g(x)$, find $(f \circ g)(x)$ and $(g \circ f)(x)$. [11.5]

61. $f(x) = 6x - 17, g(x) = 3x + 20$

62. $f(x) = 2x + 15, g(x) = -4x + 9$

63. $f(x) = x - 8, g(x) = x^2 + 7x - 56$

64. $f(x) = 3x + 7, g(x) = x^2 - 9x + 41$

65. Is the function whose input is a person and output is that person's favorite fast-food restaurant a one-to-one function? Explain your answer in your own words. [11.6]

66. Is the function whose input is a player on your school's basketball team and output is that player's uniform number a one-to-one function? Explain your answer in your own words. [11.6]

Use the horizontal-line test to determine whether the function is one to one. [11.6]

67.

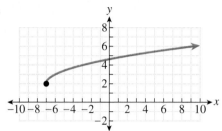

68.

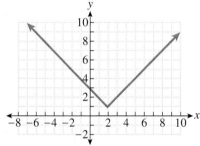

Determine whether the functions $f(x)$ and $g(x)$ are inverse functions by showing that $(f \circ g)(x) = x$ and $(g \circ f)(x) = x$. [11.6]

69. $f(x) = 3x + 4$, $g(x) = 4x - 3$

70. $f(x) = 2x - 10$, $g(x) = \dfrac{x + 10}{2}$

71. $f(x) = 4x + 18$, $g(x) = \dfrac{1}{4}x + \dfrac{9}{2}$

72. $f(x) = -5x + 9$, $g(x) = \dfrac{9 - x}{5}$

For the given function $f(x)$, find $f^{-1}(x)$. State the domain of $f^{-1}(x)$. [11.6]

73. $f(x) = x - 10$

74. $f(x) = 2x - 15$

75. $f(x) = -x + 19$

76. $f(x) = -8x + 27$

77. $f(x) = \dfrac{9x + 2}{3x}$

78. $f(x) = \dfrac{7x}{4x + 21}$

For the given graph of a one-to-one function $f(x)$, graph its inverse function $f^{-1}(x)$. [11.6]

79.

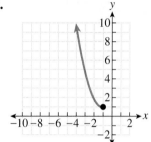

80.

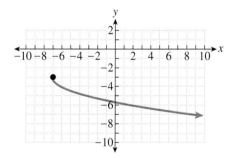

1. Does a function exist for which the input value is a state and the output value is the state's capital city?

2. Use the vertical-line test to determine whether the graph represents a function.

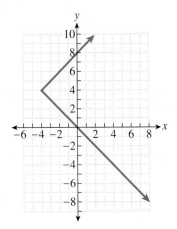

Evaluate the given function.

3. $f(x) = -2x + 35, f(-8)$

4. $f(x) = 5x + 16, f(6a - 5)$

5. $f(x) = x^2 - 15x + 54, f(-9)$

6. Determine the domain and range of the function $f(x)$ that has been graphed.

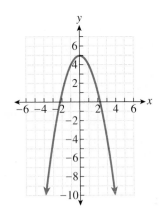

Graph each function. Label the x- and y-intercepts.

7. $f(x) = -4x + 6$

8. $f(x) = \dfrac{2}{3}x - 4$

9. A ceramics shop holds birthday parties for children. They charge a party fee in addition to charging a fee for each guest attending. A party with 10 guests costs $105, while a party with 25 guests costs $210. Find a linear function $f(x) = mx + b$ whose input is the number of guests and whose output is the cost of the party.

Graph the function. Label the vertex, y-intercept, and any x-intercepts.

10. $f(x) = x^2 + 8x + 10$

11. $f(x) = -(x + 3)^2 + 4$

12. A young child throws a ball upward with an initial velocity of 20 feet/second. The height of the ball (in feet) after t seconds is given by the function $h(t) = -16t^2 + 20t + 3$. What is the maximum height that the ball reaches?

Graph the given function, and state its domain and range.

13. $f(x) = \sqrt{x + 1} - 3$

14. $f(x) = (x + 1)^3 + 8$

15. Determine the function $f(x)$ on the graph.

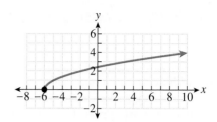

16. If $f(x) = 5x - 2$ and $g(x) = x^2 - 4x - 13$, find $(f + g)(-2)$.

17. If $f(x) = x - 6$ and $g(x) = x^2 + 6x + 36$, find $(f \cdot g)(x)$.

Let $f(x) = x + 7$ and $g(x) = 2x - 8$. Find the following.

18. $(f \circ g)(6)$ **19.** $(g \circ f)(6)$

For the given functions $f(x)$ and $g(x)$, find $(f \circ g)(x)$.

20. $f(x) = 4x + 13$, $g(x) = -2x + 23$

21. $f(x) = x^2 - 9x - 30$, $g(x) = x + 12$

22. Use the horizontal-line test to determine whether the function is one to one.

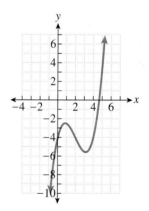

23. Determine whether $f(x) = 2x - 16$ and $g(x) = \frac{1}{2}x + 8$ are inverse functions. Recall that $f(x)$ and $g(x)$ are inverse functions if $(f \circ g)(x) = x$ and $(g \circ f)(x) = x$.

For the given function $f(x)$, find $f^{-1}(x)$.

24. $f(x) = -4x + 20$

25. $f(x) = \dfrac{x - 9}{4x}$

Mathematicians in History
Leonhard Euler

Leonhard Euler (pronounced "oiler") is believed by many to have been the greatest mathematician of the 18th century. He was clearly one of the most prolific, authoring over 800 papers and books. He is credited with introducing the function notation $f(x)$ that we still use today.

Write a one-page summary (*or* make a poster) of the life of Euler and his mathematical achievements.

Interesting issues:

- Where and when was Leonhard Euler born?
- What currency did Euler's face appear on?
- What are the Seven Bridges of Königsberg?
- Who invited Euler to Berlin in 1741?
- On the day that he died, Euler was lecturing to his grandchildren; what was the subject? What were his last words?
- Besides function notation, what other notation did Euler introduce?
- What physical condition did Euler have for the last 20 years of his life?

Every morning at 8 A.M., a couple of the members of a math department meet for coffee. Sharon orders a large cup which contains 240 mg of caffeine and Geoff orders a small cup which contains 100 mg of caffeine.

Caffeine is eliminated from the body by the kidneys. The kidneys eliminate a constant fraction of the caffeine from the bloodstream each hour. Each person's body eliminates the caffeine at a different rate. In one hour,

Sharon's kidneys remove 10% of the caffeine that was present at the start of the hour. This means that the fraction of caffeine remaining after an hour is 90%. In one hour, Geoff's kidneys remove 14% of the caffeine that was present at the start of the hour.

a) Fill in the following charts to show how much caffeine is present in Sharon and Geoff:

Time	Amount of Caffeine in Sharon's Bloodstream	Amount of Caffeine in Sharon's Bloodstream (Expanded Form)	Amount of Caffeine in Sharon's Bloodstream (Exponent Form)
8 am	240 mg	240 mg	$240(.90)^0$ mg
9 am			
10 am			
11 am			
12 noon		747	

Time	Amount of Caffeine in Geoff's Bloodstream	Amount of Caffeine in Geoff's Bloodstream (Expanded Form)	Amount of Caffeine in Geoff's Bloodstream (Exponent Form)
8 am	100 mg	100 mg	$100(.86)^0$
9 am			
10 am			
11 am			
12 noon			

b) Let the fraction of caffeine remaining after one hour for any person be represented by a. Write a polynomial function $C(a)$ for the amount of caffeine remaining in the bloodstream after four hours for any person who drinks a large cup at 8 A.M.

c) If a person drinks another cup of coffee at 10 A.M., write a two-term polynomial function for the

amount of caffeine remaining in his or her bloodstream at noon.

d) If a person drinks a third cup of coffee at noon, write a three-term polynomial function for the amount of caffeine remaining in his/her bloodstream at 5 P.M.

Solve. **[8.1]**

1. $|x + 3| - 8 = -2$

Solve the inequality. Graph your solution on a number line, and write your solution in interval notation. **[8.2]**

2. $-2x - 9 \geq -17$

3. $-11 \leq 4x - 13 \leq 39$

Solve the inequality. Graph your solution on a number line, and write your solution in interval notation. **[8.2]**

4. $|2x + 1| > 5$

5. $|x| - 8 < -3$

Factor completely. **[8.4]**

6. $x^2 - 3x - 54$

7. $4x^2 - 36x + 32$

8. $x^2 + 11x + 30$

9. $49x^2 - 64$

Solve. **[8.5]**

10. $3x - 2y = 22$
    $7x + 4y = 34$

Solve. **[8.5]**

11. $\quad x + y + z = 6$
    $\quad x + 3y - z = 20$
    $\quad 2x - 4y + z = -16$

Evaluate the radical function. (Round to the nearest thousandth if necessary.) **[9.1]**

12. $f(x) = \sqrt{6x - 58},\ f(19)$

Rewrite as a radical expression and simplify if possible. Assume all variables represent nonnegative values. **[9.2]**

13. $(16x^{10})^{3/2}$

Simplify the expression. Assume all variables represent nonnegative values. Express your answer in radical notation. **[9.2]**

14. $(x^{6/7})^{3/4}$

Simplify the radical expression. Assume all variables represent nonnegative values. **[9.3]**

15. $\sqrt{75r^{12}s^{10}t^9}$

Add or subtract. Assume all variables represent nonnegative values. **[9.3]**

16. $7\sqrt{108} - 2\sqrt{48}$

Multiply. Assume all variables represent nonnegative values. **[9.4]**

17. $(4\sqrt{3} - 2\sqrt{5})(5\sqrt{3} + 4\sqrt{5})$

Rationalize the denominator and simplify. Assume all variables represent nonnegative values. **[9.4]**

18. $\dfrac{ab^3c^2}{\sqrt{a^3b^8c}}$

19. $\dfrac{5\sqrt{5} + 6\sqrt{2}}{4\sqrt{5} - 3\sqrt{2}}$

20. For the function $f(x) = \sqrt{2x - 9}$, find all values x for which $f(x) = 5$. **[9.5]**

Express in terms of i. **[9.7]**

21. $\sqrt{-360}$

Multiply. **[9.7]**

22. $(2 + 5i)(9 - 2i)$

Rationalize the denominator. **[9.7]**

23. $\dfrac{5i}{3 + 4i}$

Solve by extracting square roots. **[10.1]**

24. $(x + 2)^2 + 41 = 23$

Solve by using the quadratic formula. **[10.2]**

25. $3x^2 - 10x - 21 = 0$

Solve by using a u-substitution. **[10.4]**

26. $x^4 - 11x^2 + 18 = 0$

27. $x + 3\sqrt{x} - 10 = 0$

Solve. **[10.4]**

28. $\sqrt{x + 23} - x = 3$

Solve. [10.1–10.2]

29. $x^2 - 12x + 27 = 0$

Graph the quadratic equation. Label the vertex, the y-intercept, and any x-intercepts. [10.5]

30. $y = -x^2 + 8x + 6$

31. $y = (x - 2)^2 + 3$

Solve the quadratic inequality. Express your solution on a number line using interval notation. [10.6]

32. $x^2 - 10x + 24 \le 0$

Solve the rational inequality. Express your solution on a number line using interval notation. [10.6]

33. $\dfrac{x^2 - 2x - 48}{x + 5} \le 0$

34. The width of a rectangular lawn is 5 feet more than its length. If the area of the lawn is 980 square feet, find the dimensions of the lawn. Round to the nearest tenth of a foot. [10.6]

35. A projectile is launched upward from the roof of a building 12 feet high with an initial velocity of 80 feet per second. How long will it take until the projectile lands on the ground? Round to the nearest hundredth of a second. (Use the function $h(t) = -16t^2 + v_0t + s$.) [10.3]

36. A water tank is connected to two pipes. When both pipes are open, they can fill the tank in 5 hours. The larger of the pipes can fill the tank in 3 hours less time than the smaller pipe. How long would it take the smaller pipe, working alone, to fill the tank? Round to the nearest tenth of an hour. [10.4]

Evaluate the given function. [11.1]

37. $f(x) = 3x - 10, f(4n - 11)$

38. $f(x) = x^2 - 10x + 51, f(-5)$

Graph the function. Label the x- and y-intercepts. [11.2]

39. $f(x) = -\dfrac{2}{3}x - 4$

40. A company charges a fixed fee for renting a moving van in addition to charging per mile driven. If a person rents a moving van and drives it 100 miles, the charge is $35. If a person rents a moving van and drives it 160 miles, the charge is $44. [11.2]

a) Find a linear function $f(x) = mx + b$ whose input is the number of miles driven and whose output is the corresponding charge for renting the moving van.

b) Use the function from part a) to determine the cost of renting a moving van and driving it 300 miles.

Determine whether the given quadratic function has a maximum value or a minimum value. Then find that maximum or minimum value. [11.3]

41. $f(x) = x^2 - 30x + 129$

42. A projectile is launched upward from the roof of a building with an initial velocity of 144 feet/second. The height of the projectile (in feet) after t seconds is given by the function $h(t) = -16t^2 + 144t + 50$. What is the maximum height that the projectile reaches? [11.3]

Graph the given function, and state its domain and range. [8.3, 11.4]

43. $f(x) = |x - 2| - 5$

44. $f(x) = \sqrt{x + 4} - 1$

45. $f(x) = (x + 1)^3$

46. For the functions $f(x) = 7x + 11$ and $g(x) = 3x - 42$, find $(f + g)(x)$. [11.5]

For the given functions $f(x)$ and $g(x)$, find $(f \circ g)(x)$ and $(g \circ f)(x)$. [11.5]

47. $f(x) = x + 6$, $g(x) = x^2 - 2x + 12$

Determine whether the functions $f(x)$ and $g(x)$ are inverse functions. Recall that $f(x)$ and $g(x)$ are inverse functions if $(f \circ g)(x) = x$ and $(g \circ f)(x) = x$. [11.6]

48. $f(x) = 3x + 8$, $g(x) = \dfrac{x - 8}{3}$

For the given function $f(x)$, find $f^{-1}(x)$. [11.6]

49. $f(x) = 3x - 19$

50. $f(x) = \dfrac{3}{x - 9}$

CHAPTER *12*

LOGARITHMIC AND EXPONENTIAL FUNCTIONS

In this chapter, we will investigate exponential functions and logarithmic functions. These types of functions have many practical applications, including calculating compound interest and population growth, radiocarbon dating, and determining the intensity of an earthquake or the pH of a substance.

Study Tip PRACTICE QUIZZES   *Creating practice quizzes is an excellent way to prepare for exams. It is important to put yourself in a test environment before taking a test, without facing the consequences of a test. Practice quizzes will tell you which topics you understand and which topics you need further work on. Throughout this chapter, we will revisit this study tip and help you to incorporate it into your study habits.*

12.1

EXPONENTIAL FUNCTIONS

Objectives

1. Define exponential functions.
2. Evaluate exponential functions.
3. Graph exponential functions.
4. Define the natural exponential function.
5. Solve exponential equations.
6. Use exponential functions in applications.

Definition of Exponential Functions

Objective 1 Define exponential functions. Would you rather have $1 million, or $1 that gets doubled every day for 30 days? You may be surprised by just how much that $1 is worth at the end of 30 days. The following table tells you:

Day	Amount ($)	Day	Amount ($)	Day	Amount ($)
1	$1	11	$1024	21	$1,048,576
2	$2	12	$2048	22	$2,097,152
3	$4	13	$4096	23	$4,194,304
4	$8	14	$8192	24	$8,388,608
5	$16	15	$16,384	25	$16,777,216
6	$32	16	$32,768	26	$33,554,432
7	$64	17	$65,536	27	$67,108,864
8	$128	18	$131,072	28	$134,217,728
9	$256	19	$262,144	29	$268,435,456
10	$512	20	$524,288	30	$536,870,912

After 30 days, the $1 would turn into over $500 million. Notice that, although the total grew quite slowly at first, it increased quickly towards the end of the 30 days. This is an example of exponential growth, which is based on exponential functions.

Exponential Function

> A function that can be written in the form $f(x) = b^x$, where $b > 0$ and $b \neq 1$, is called an **exponential function** with base b.
>
> (If $b = 1$, then $f(x)$ is simply the constant function $f(x) = 1$.)

This type of function is called an exponential function because the variable is in an exponent. Some examples of exponential functions are

$$f(x) = 2^x \qquad f(x) = 10^x \qquad f(x) = \left(\frac{1}{3}\right)^x \qquad f(x) = 4^{-x}$$

Evaluating Exponential Functions

Objective 2 **Evaluate exponential functions.** We evaluate exponential functions in the same manner that we evaluated other functions: We substitute the specified value for the function's variable and then simplify the resulting expression.

EXAMPLE 1   Let $f(x) = 3^x$. Find the following:

a) $f(4)$

Solution

We begin by substituting 4 for x and then simplify the resulting expression.

$$f(4) = 3^4 \qquad \text{Substitute 4 for } x.$$
$$= 81 \qquad \text{Simplify.}$$

b) $f(0)$

Solution

We substitute 0 for x.

$$f(0) = 3^0 \qquad \text{Substitute 0 for } x.$$
$$= 1 \qquad \text{Simplify. Recall that raising a nonzero base to the exponent 0 is equal to 1.}$$

c) $f(-2)$

Solution

We substitute -2 for x.

$$f(-2) = 3^{-2} \qquad \text{Substitute } -2 \text{ for } x.$$

$$= \frac{1}{3^2} \qquad \text{Rewrite the expression without a negative exponent. Recall that } b^{-n} = \frac{1}{b^n}.$$

$$= \frac{1}{9} \qquad \text{Simplify.}$$

Quick Check 1
Let $f(x) = 4^x$. Find $f(5)$ and $f(-3)$.

EXAMPLE 2   Let $f(x) = 5 \cdot 3^{x-4} + 16$. Find $f(6)$.

Solution

This function, although more complex than those in previous examples, is also an exponential function. We will substitute 6 for x and then simplify.

Quick Check 2
Let $f(x) = 2 \cdot 7^{x+3} - 19$. Find $f(-1)$.

$$f(6) = 5 \cdot 3^{6-4} + 16 \qquad \text{Substitute 6 for } x.$$
$$= 5 \cdot 3^2 + 16 \qquad \text{Simplify the exponent.}$$
$$= 61 \qquad \text{Simplify.}$$

Using Your Calculator   To simplify the expression $5 \cdot 3^2 + 16$ with the TI-83/84, we will use the button labeled ☐. Here is how the calculator screen should look.

```
5*3^2+16
              61
```

Graphs of Exponential Functions

Objective **3**   **Graph exponential functions.** To understand the behavior of exponential functions, we must examine the graphs of these functions. In this section, we will learn to create a quick sketch of the graph of an exponential function. A more thorough presentation of graphing exponential functions, including finding x-intercepts, will be given in Section 12.6. For now, we will plot points for selected values of x.

Consider the function $f(x) = 2^x$. Here is a table of values of $f(x)$ for values of x ranging from -3 to 3:

x	-3	-2	-1	0	1	2	3
$f(x)$	$\dfrac{1}{8}$	$\dfrac{1}{4}$	$\dfrac{1}{2}$	1	2	4	8

We plot the ordered pairs and then draw a smooth curve passing through those points. The domain of this function is the set of all real numbers $(-\infty, \infty)$. As we look to the left of the graph, the function gets closer and closer to the x-axis without actually touching it.

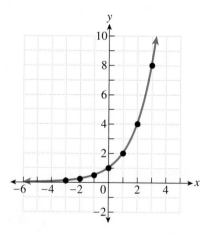

Here are some additional function values to illustrate this trend:

x	-4	-5	-6	-7	-8	-9	-10
$f(x)$	$\dfrac{1}{16}$	$\dfrac{1}{32}$	$\dfrac{1}{64}$	$\dfrac{1}{128}$	$\dfrac{1}{256}$	$\dfrac{1}{512}$	$\dfrac{1}{1024}$

As shown in the following graph, these function values are approaching 0 and the graph is approaching the *x*-axis:

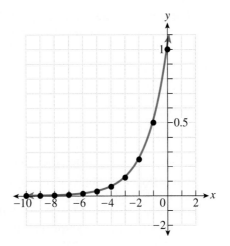

In this case, we say that the *x*-axis is a horizontal asymptote for the graph of the function $f(x) = 2^x$.

A **horizontal asymptote** is a horizontal line that the graph of a function approaches as it either moves to the left or to the right. The horizontal asymptote is placed on the graph using a dashed line, as the points on the asymptote are not part of the graph of the exponential function.

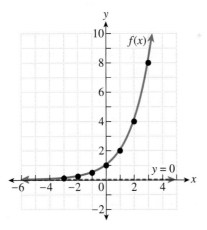

The horizontal asymptote also helps us to determine the range of an exponential function. As the values of *x* decrease, the values of $f(x)$ get closer and closer to 0, without ever reaching it. As the values of *x* increase, the function values increase without limit. The range of this function is $(0, \infty)$.

An exponential function of the form $f(x) = b^x$ ($b > 1$) is an **increasing function.** This means that as *x* increases, so does $f(x)$. The graph of an increasing function moves upward as it moves from left to right. Notice that the function increases at a very slow rate for values of *x* that are less than 0 but increases more rapidly as the values of *x* increase.

EXAMPLE 3   Graph $f(x) = 3^x$, and state the domain and range of $f(x)$.

Solution

For this example, we will use the values -2, -1, 0, 1, and 2 for x. This will be the case whenever the exponent is simply x.

x	-2	-1	0	1	2
$f(x) = 3^x$	$3^{-2} = \dfrac{1}{3^2} = \dfrac{1}{9}$	$3^{-1} = \dfrac{1}{3^1} = \dfrac{1}{3}$	$3^0 = 1$	$3^1 = 3$	$3^2 = 9$

These ordered pairs have been plotted and a smooth curve drawn through them. Notice that as the values of x get smaller, the points are getting closer and closer to the x-axis. The line $y = 0$ is the horizontal asymptote for the graph of this function and has been graphed as a dashed line.

The domain of this function is the set of all real numbers $(-\infty, \infty)$, while the range is $(0, \infty)$.

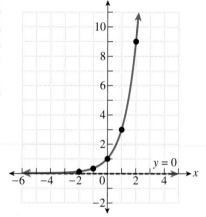

Quick Check 3

Graph $f(x) = 5^x$, and state the domain and range of $f(x)$.

Here are the graphs of $f(x) = 2^x$ and $g(x) = 3^x$ on the same set of axes. Notice that for positive values of x, the function $g(x) = 3^x$ increases much more quickly than the function $f(x) = 2^x$ does. For exponential functions of the form $f(x) = b^x$, $b > 1$, the function increases more quickly for larger values of the base b.

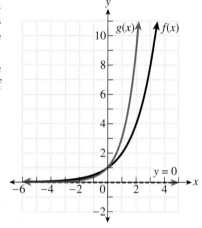

EXAMPLE 4   Graph $f(x) = \left(\frac{1}{2}\right)^x$, and state the domain and range of $f(x)$.

Solution

We will begin by creating a table of values for this function, using the same values of x that we used in the previous example.

x	-2	-1	0	1	2
$f(x) = \left(\dfrac{1}{2}\right)^{x}$	$\left(\dfrac{1}{2}\right)^{-2} = 2^2 = 4$	$\left(\dfrac{1}{2}\right)^{-1} = 2^1 = 2$	$\left(\dfrac{1}{2}\right)^{0} = 1$	$\left(\dfrac{1}{2}\right)^{1} = \dfrac{1}{2}$	$\left(\dfrac{1}{2}\right)^{2} = \dfrac{1}{4}$

Quick Check 4**

Graph $f(x) = \left(\frac{1}{3}\right)^{x}$, and state the domain and range of $f(x)$.

Notice that this time as the values of x get larger, the function values are getting closer and closer to 0. The horizontal asymptote is the line $y = 0$. The domain of this function is the set of all real numbers $(-\infty, \infty)$, while the range is $(0, \infty)$.

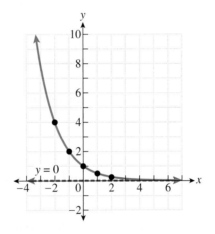

EXAMPLE 5   Graph $f(x) = 2^{x+1} - 4$, and state the domain and range of $f(x)$.

Solution

Quick Check 5**

Graph $f(x) = 6^{x-1} + 3$, and state the domain and range of $f(x)$.

The exponent is equal to 0 when $x = -1$. In addition to using this value for x, we will use two values that are less than -1 and two values that are greater than -1.

x	-3	-2	-1	0	1
$f(x) = 2^{x+1} - 4$	$2^{-2} - 4 = -3\dfrac{3}{4}$	$2^{-1} - 4 = -3\dfrac{1}{2}$	$2^{0} - 4 = -3$	$2^{1} - 4 = -2$	$2^{2} - 4 = 0$

Notice that the x-axis is not the horizontal asymptote for the graph of this function. As the values of x get smaller and smaller, the function values approach -4. The line $y = -4$ is the horizontal asymptote for the graph of this function. The domain of this function is the set of all real numbers $(-\infty, \infty)$, while the range is $(-4, \infty)$.

In general, the value that is added to or subtracted from the exponential expression will tell us the location of the horizontal asymptote.

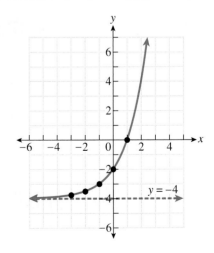

Using Your Calculator   To graph $f(x) = 2^{x+1} - 4$ and its horizontal asymptote using the TI-83/84, we begin by pushing the Y= key. Next to Y_1, type in the function, using parentheses around the exponent. The horizontal asymptote is the line $y = -4$, so type -4 next to Y_2. Press the GRAPH key to display the graph. Here is the screen showing how to enter the function and its corresponding graph:

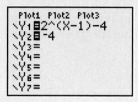
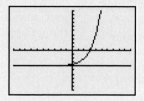

Here are some guidelines for graphing an exponential function with a base that is greater than 1:

Graphing an Exponential Function of the Form $f(x) = b^{x-h} + k$, $b > 1$

1. **Graph the horizontal asymptote.** Using a dashed line, graph the line $y = k$.
2. **Create a table of values.** In addition to using the value $x = h$, use two values of x that are less than h and two values that are greater than h. Evaluate the function for all five values of x to determine the coordinates of ordered pairs on the graph.
3. **Draw the graph.** Plot the five points and draw the graph that passes through these points. Be sure that the graph you draw approaches the horizontal asymptote. The graph should increase from left to right.

We graph a function of the form $f(x) = b^{x-h} + k$, where $0 < b < 1$, in the same fashion that we graph an exponential function whose base b is greater than 1. The major exception is that if $0 < b < 1$, then the function is a decreasing function, rather than an increasing function.

EXAMPLE ▶ 6   Graph $f(x) = 5^{-x} + 2$, and state the domain and range of $f(x)$.

Solution

We begin by rewriting this function as $f(x) = \left(\frac{1}{5}\right)^x + 2$. Notice that this base is less than 1. The line $y = 2$ is the horizontal asymptote for the graph of this function. Now we can create a table of values.

x	-2	-1	0	1	2
$f(x) = \left(\dfrac{1}{5}\right)^x + 2$	$\left(\dfrac{1}{5}\right)^{-2} + 2 = 27$	$\left(\dfrac{1}{5}\right)^{-1} + 2 = 7$	$\left(\dfrac{1}{5}\right)^{0} + 2 = 3$	$\left(\dfrac{1}{5}\right)^{1} + 2 = 2\dfrac{1}{5}$	$\left(\dfrac{1}{5}\right)^{2} + 2 = 2\dfrac{1}{25}$

Note that the scale on the *y*-axis is not the same as the scale on the *x*-axis. The domain of this function is the set of all real numbers $(-\infty, \infty)$, while the range is $(2, \infty)$.

Quick Check **6**

Graph
$f(x) = \left(\frac{1}{2}\right)^{x-3} - 2$,
and state the domain
and range of $f(x)$.

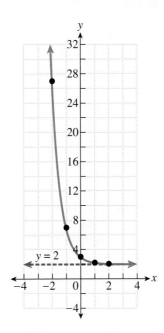

One last fact about exponential functions is that they are one-to-one functions. We can determine that exponential functions are one-to-one functions by applying the horizontal line test.

The Natural Exponential Function

Objective **4** **Define the natural exponential function.** One number that will be used as the base of an exponential function is an irrational number that is denoted by the letter *e* and is called the natural base.

The Natural Base e

$$e \approx 2.7182818284 \ldots$$

The use of the letter *e* to denote this number is credited to Leonhard Euler. You may think that he chose *e* because it was the first letter of his last name, but many math historians claim that he had been using the letter *a* in his writings and that *e* was simply the next vowel available.

The Natural Exponential Function

The function $f(x) = e^x$ is called the **natural exponential function.**

Calculators have a built-in function for evaluating the natural exponential function, and it usually is listed above the key labeled as ln.

EXAMPLE 7   Let $f(x) = e^x$. Find the following and round to the nearest thousandth.

a) $f(5)$

Solution

$$f(5) = e^5 \qquad \text{Substitute 5 for } x.$$
$$\approx 148.413 \qquad \text{Approximate using a calculator.}$$

b) $f(-1.7)$

Solution

$$f(-1.7) = e^{-1.7} \qquad \text{Substitute } -1.7 \text{ for } x.$$
$$\approx 0.183 \qquad \text{Approximate using a calculator.}$$

Quick Check 7

Let $f(x) = e^x$. Find $f(-1)$ and $f(4.3)$.

Using Your Calculator   The TI-83/84 has a built-in function e^x, which is a 2nd function above the key labeled ⃞LN. To calculate e^5, we begin by pushing the ⃞2nd key, followed by the ⃞LN key. Notice that the calculator gives you a left-hand parenthesis. After typing the exponent 5, we close the parentheses by pushing ⃞). After pressing ⃞ENTER, your calculator screen should look as follows:

```
e^(5)
        148.4131591
```

EXAMPLE 8   Graph $f(x) = e^x$, and state the domain and range of $f(x)$.

Solution

Quick Check 8

Graph $f(x) = e^{x+2} + 5$, and state the domain and range of $f(x)$.

We will begin by creating a table of values for this function.

x	-2	-1	0	1	2
$f(x) = e^x$	$e^{-2} \approx 0.14$	$e^{-1} \approx 0.37$	$e^0 = 1$	$e^1 \approx 2.72$	$e^2 \approx 7.39$

The horizontal asymptote is the line $y = 0$. The domain of this function is the set of all real numbers $(-\infty, \infty)$, while the range is $(0, \infty)$.

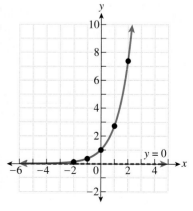

Solving Exponential Equations

Objective 5 **Solve exponential equations.** An **exponential equation** is an equation containing a variable in an exponent, such as $2^x = 16$. One technique for solving exponential equations is to write both sides of the equation in terms of the same base. If this is possible, we can then apply the one-to-one property of exponential functions to solve the equation.

One-to-One Property of Exponential Functions

For any positive real number b $(b \neq 1)$ and any real numbers r and s, if $b^r = b^s$, then $r = s$.

EXAMPLE 9 Solve $2^x = 16$.

Solution

We will begin by expressing 16 as a power of 2.

$$2^x = 16$$
$$2^x = 2^4 \qquad \text{Rewrite 16 as } 2^4.$$
$$x = 4 \qquad \text{Use the one-to-one property of exponential functions.}$$

The solution set is $\{4\}$.

EXAMPLE 10 Solve $3^{x+2} = \frac{1}{27}$.

Solution

We can rewrite $\frac{1}{27}$ as 3^{-3}.

$$3^{x+2} = \frac{1}{27}$$
$$3^{x+2} = 3^{-3} \qquad \text{Rewrite } \frac{1}{27} \text{ as } 3^{-3}.$$
$$x + 2 = -3 \qquad \text{If } b^r = b^s, \text{ then } r = s.$$
$$x = -5 \qquad \text{Subtract 2.}$$

The solution set is $\{-5\}$.

Quick Check **9** Solve.

a) $4^x = 256$ b) $9^{2x-17} = 9$

Applications

Objective 6 **Use exponential functions in applications.** We conclude this section with an application of exponential functions.

EXAMPLE 11   The population of a particular small town can be approximated by the function $f(x) = 75{,}000 \cdot e^{0.018x}$, where x represents the number of years after 1980. What will be the population of the town in the year 2010?

Solution

To predict the town's population in the year 2010, we will evaluate the function for $x = 30$, which is $2010 - 1980$.

$$\begin{aligned} f(30) &= 75{,}000 \cdot e^{0.018(30)} && \text{Substitute 30 for } x. \\ &= 75{,}000 \cdot e^{0.54} && \text{Simplify the exponent.} \\ &\approx 128{,}701 && \text{Approximate using a calculator. Round to the} \\ &&& \text{nearest person.} \end{aligned}$$

The town's population in the year 2010 will be approximately 128,701.

Quick Check 10   The total revenues generated in a given year by McDonald's (in millions of dollars) can be approximated by the function $f(x) = 7337 \cdot 1.08^x$, where x represents the number of years after 1992. Use this function to predict the total revenues of McDonald's in the year 2012. (*Source:* McDonald's Corporation)

EXERCISES 12.1

Let $f(x) = 2^x$. Find the following:

1. $f(4)$ 2. $f(7)$

3. $f(0)$ 4. $f(-1)$

Let $f(x) = \left(\frac{2}{5}\right)^x$. Find the following:

5. $f(2)$ 6. $f(3)$

7. $f(-4)$ 8. $f(-2)$

Evaluate the given function.

9. $f(x) = 3^x + 2, f(2)$

10. $f(x) = 3^{x+2}, f(2)$

11. $f(x) = 2^{x+5} - 5, f(3)$

12. $f(x) = 2^{x-3} + 27, f(6)$

13. $f(x) = \left(\frac{1}{4}\right)^{x-4} + 6, f(3)$

14. $f(x) = \left(\frac{1}{2}\right)^{x+1} + 5, f(-3)$

15. $f(x) = 4^{-x}, f(3)$

16. $f(x) = 4^{-x}, f(-2)$

17. $f(x) = 3^{2x-1} - 100, f(4)$

18. $f(x) = 5^{4x+3} + 175, f(0)$

Complete the table for $f(x)$.

19. $f(x) = 10^x$

x	-2	-1	0	1	2
f(x)					

20. $f(x) = \left(\frac{1}{4}\right)^x$

x	-2	-1	0	1	2
f(x)					

21. $f(x) = 4^{x-3} - 9$

x	1	2	3	4	5
f(x)					

22. $f(x) = \left(\frac{1}{2}\right)^{x+4} + 12$

x	-6	-5	-4	-3	-2
f(x)					

Graph $f(x)$. Label the horizontal asymptote. State the domain and range of the function.

23. $f(x) = 4^x$

24. $f(x) = 10^x$

25. $f(x) = \left(\frac{1}{4}\right)^x$

26. $f(x) = \left(\frac{3}{2}\right)^x$

27. $f(x) = 5^{x+1}$

28. $f(x) = 5^x + 1$

29. $f(x) = 3^x - 1$

30. $f(x) = 3^{x-1}$

31. $f(x) = 2^{x+2} - 8$

32. $f(x) = 2^{x-3} + 5$

33. $f(x) = \left(\frac{1}{2}\right)^{x+4} + 2$ **34.** $f(x) = \left(\frac{1}{3}\right)^{x-5} - 9$ **47.** $f(x) = e^x + 2$ **48.** $f(x) = e^x + 7$

Let $f(x) = e^x$. Find the following and round to the nearest thousandth.

35. $f(2)$ **36.** $f(3)$

49. $f(x) = e^{x+4} + 4$ **50.** $f(x) = e^{x-1} + 10$

37. $f(-1)$ **38.** $f(0)$

39. $f(-5.2)$ **40.** $f(4.13)$

Evaluate the given function.

41. $f(x) = e^{x+4}, f(-1)$

42. $f(x) = e^{x+2}, f(3)$

43. $f(x) = e^{3x-1}, f(-2)$

44. $f(x) = e^{2x-5}, f(2.7)$

Graph $f(x)$. Label the horizontal asymptote. State the domain and range of the function.

45. $f(x) = e^{x-2}$ **46.** $f(x) = e^{x+5}$

Solve.

51. $3^x = 81$ **52.** $2^x = 32$

53. $5^x = 125$ **54.** $10^x = 100{,}000$

55. $6^x = \frac{1}{6}$ **56.** $4^x = \frac{1}{64}$

57. $2^{x-5} = 128$ **58.** $5^{x+2} = 25$

59. $9^{x-3} = \frac{1}{81}$ **60.** $3^{x+4} = \frac{1}{81}$

61. $\left(\frac{1}{3}\right)^{2x-1} = \frac{1}{243}$ **62.** $\left(\frac{1}{4}\right)^{x-7} = \frac{1}{64}$

63. $4^{3x-5} = 256$ **64.** $7^{5x+12} = 49$

65. $3^{8x-27} = \frac{1}{27}$ **66.** $2^{4x+14} = \frac{1}{1024}$

67. $8^{17x-85} = 1$ **68.** $5^{-3x+9} = 1$

69. Wendy deposited $5000 in an account that pays 13% annual interest, compounded annually. The balance after t years is given by the function $f(t) = 5000(1.13)^t$. What will the balance of this account be after 40 years?

70. Mary deposited $350 in an account that pays 5% annual interest, compounded annually. The balance after t years is given by the function $f(t) = 350(1.05)^t$. What will the balance of this account be after 6 years?

71. Dylan deposited $10,000 in an account that pays 9% annual interest, compounded quarterly. The balance after t years is given by the function $f(t) = 10,000(1.0225)^{4t}$. What will the balance of this account be after 5 years?

72. Adam deposited $50,000 in an account that pays 6% annual interest, compounded quarterly. The balance after t years is given by the function $f(t) = 50,000(1.015)^{4t}$. What will the balance of this account be after 10 years?

73. In 1995, a particular Cal Ripken baseball card was valued at $8. By 2003, the card had increased in value to $20. The value of the card can be approximated by the function $f(t) = 8e^{0.1145t}$, where t represents the number of years after 1995. Assuming that the value of the card continues to grow exponentially, predict the value of the card in 2017.

74. The average ticket price for attending a movie has been increasing exponentially. The average ticket price in a particular year can be approximated by the function $f(t) = 4.69 \cdot e^{0.0627t}$, where t represents the number of years after 1998. (This average price includes both first run and subsequent runs, as well as all special pricing.) Use this function to predict the average ticket price in the year 2011. (*Source:* Motion Picture Association of America (MPAA))

75. The net revenues of Amazon.com in a particular year, in millions of dollars, can be approximated by the function $f(t) = 2762 \cdot e^{0.1767t}$, where t represents the number of years after 2000. If net revenues continue to grow at this rate, predict the net revenues of Amazon.com in the year 2009. (*Source:* Amazon.com, Inc.)

76. The number of Wal-Mart stores in a particular year can be approximated by the function $f(t) = 2440 \cdot e^{0.0726t}$, where t represents the number of years after 1993. Assuming that this rate of growth continues, use this function to predict the number of Wal-Mart stores in the year 2013. (*Source:* Wal-Mart Stores, Inc.)

77. If a container of water whose temperature is 80° C is placed in a refrigerator whose temperature remains a constant 4° C, then the temperature of the water, in degrees Celsius, after t minutes is given by the function $f(t) = 4 + 76e^{-0.014t}$. What will the temperature of the water be after 60 minutes?

78. If a person dies in a room that is 70° F, then the body's temperature in degrees Fahrenheit after t hours is given by the function $f(t) = 70 + 28.6e^{-0.44t}$. What will the body's temperature be after 3 hours?

79. The average cost of a loaf of white bread in a particular year can be approximated by the function $f(t) = 0.74 \cdot 1.033^t$, where t represents the number of years after 1993. Use this function to approximate the cost of a loaf of white bread in the year 2008. (*Source:* U.S. Bureau of Labor Statistics)

80. The number of U.S. wineries in a particular year can be approximated by the function $f(t) = 669 \cdot 1.055^t$, where t represents the number of years after 1975. Assuming that this rate of growth continues, use this function to predict the number of U.S. wineries in the year 2010. (*Source:* Alcohol and Tobacco Tax and Trade Bureau)

Plot the given ordered pairs $(x, f(x))$. Based on the graph, do you feel that $f(x)$ is an exponential function? Explain your reasoning.

81. $(-3, -17), (0, -8), (5, 7), (6, 10), (10, 22)$

82. $(-5, -8), (-4, -11), (-3, -12), (-1, -8), (0, -3), (1, 4)$

83. $(0, 4.125), (1, 4.25), (3, 5), (4, 6), (6, 12), (7, 20)$

84. $(-3, 91), (-2, 37), (-1, 19), (0, 13), (1, 11)$

Answer in complete sentences.

85. Explain how to graph an exponential function. Use an example to illustrate the process.

86. Explain the difference between the graph of $f(x) = b^x$ for $b > 1$ and $0 < b < 1$.

87. Explain the difference between the graph of $f(x) = 2^x$ and $g(x) = x^2$.

88. Create an exponential equation whose solution is $x = 4$, and explain how to solve the equation.

Study Tip **REVISITED**   Following is a practice quiz containing two problems for each objective from objective 2 through objective 6. After you take the quiz and check your work, you will have an idea about which objectives will require extra study and which ones will not.

Objective 2

1. Let $f(x) = 3^{x+4} - 29$. Find $f(-2)$.

2. Let $f(x) = \left(\frac{1}{2}\right)^{x+3}$. Find $f(-5)$.

Objective 3

3. Graph $f(x) = 2^{x+1}$, and state the domain and range of the function. Label the horizontal asymptote.

4. Graph $f(x) = \left(\frac{1}{4}\right)^x + 5$, and state the domain and range of the function. Label the horizontal asymptote.

Objective 4

5. Let $f(x) = e^{x+6} + 409$. Find $f(3)$. Round to the nearest thousandth.

6. Graph $f(x) = e^{x+2} - 7$, and state the domain and range of the function. Label the horizontal asymptote.

Objective 5

7. Solve $2^{x-3} = 64$.   **8.** Solve $3^{x^2-5x-12} = 9$.

Objective 6

9. David deposited $2000 in a bank account that pays 4.5% annual interest, compounded monthly. The balance after t years is given by the function $f(t) = 2000(1.00375)^{12t}$. How much money will be in the account after 3 years?

10. The value of a car, in dollars, after x years can be approximated by the function $f(x) = 20{,}000 \cdot 0.82^x$. Use this function to determine the value of the car 5 years after its purchase.

12.2
LOGARITHMIC FUNCTIONS

1 Define logarithms and logarithmic functions.
2 Evaluate logarithms and logarithmic functions.
3 Define the common logarithm and natural logarithm.
4 Convert back and forth between exponential form and logarithmic form.
5 Solve logarithmic equations.
6 Graph logarithmic functions.
7 Use logarithmic functions in applications.

Logarithms

Objective 1   **Define logarithms and logarithmic functions.** In this section, we will examine numbers called **logarithms.** Prior to the development of modern technology such as computers and handheld calculators, logarithms were used to help perform difficult numeric calculations. Although this is no longer a practice, logarithms are still quite important in today's society and are used in a wide variety of applications such as the study of earthquakes and the pH of a chemical substance.

Suppose that we were trying to determine what power of 2 is equal to 10; in other words, trying to solve the equation $2^a = 10$ for a. This unknown exponent a is a logarithm, and can be expressed as $\log_2 10$. We read $\log_2 10$ as *"the logarithm, base 2, of 10."*

Logarithm

> For any positive real number b $(b \neq 1)$ and any positive real number x, we define $\log_b x$ to be the exponent that b must be raised to in order to equal x. The number x is called the **argument** of the logarithm. The argument must be a positive number, as it is the result obtained from raising a positive number to a power.

So $\log_5 25$ is the exponent that 5 must be raised to in order to equal 25. In other words, $\log_5 25$ is the value of y that is a solution to the equation $5^y = 25$. Since we know that $5^2 = 25$, we know that $\log_5 25 = 2$.

Logarithmic Functions

> For any positive real number b such that $b \neq 1$, a function of the form $f(x) = \log_b x$ is called a **logarithmic function.** This function is defined for values of x that are greater than 0, so the domain of this function is the set of positive real numbers, $(0, \infty)$.

We will see later in this section that logarithmic functions are the inverses of exponential functions.

Evaluating Logarithms and Logarithmic Functions

Objective 2   **Evaluate logarithms and logarithmic functions.** Now we shift our focus to evaluating logarithms.

EXAMPLE 1 Evaluate $\log_2 8$.

Solution

To evaluate this logarithm, we need to determine what power of 2 is equal to 8. Since $2^3 = 8$, $\log_2 8 = 3$.

EXAMPLE 2 Evaluate $\log_4 \frac{1}{16}$.

Solution

We need to determine what power of 4 is equal to the fraction $\frac{1}{16}$. The only way that a number greater than 1 can be raised to a power and produce a result less than 1 is if the exponent is negative. Since $4^{-2} = \frac{1}{4^2}$ or $\frac{1}{16}$, $\log_4 \frac{1}{16} = -2$.

Quick Check **1**

Evaluate.

a) $\log_5 25$ b) $\log_3 \frac{1}{27}$
c) $\log_{12} 12$ d) $\log_6 1$

One important property of logarithms is that, for any positive number b $(b \neq 1)$, $\log_b b = 1$. For example, $\log_7 7 = 1$. Another important property of logarithms is that, for any positive number b $(b \neq 1)$, $\log_b 1 = 0$. For example, $\log_9 1 = 0$.

EXAMPLE 3 Given the function $f(x) = \log_2 x$, find $f(16)$.

Solution

We evaluate a logarithmic function in the same fashion that we evaluate any other function. We will substitute 16 for x and then simplify the resulting expression.

$$f(16) = \log_2 16 \qquad \text{Substitute 16 for } x.$$
$$= 4 \qquad \text{Since } 2^4 = 16, \log_2 16 = 4.$$

Quick Check **2**

Given the function $f(x) = \log_4 x$, find $f(\frac{1}{256})$.

Common Logarithms, Natural Logarithms

Objective 3 Define the common logarithm and natural logarithm. There are two logarithms that we will use more often than others. The first is called the common logarithm.

Common Logarithm

The **common logarithm** is a logarithm whose base is 10.

In this case, we will write $\log x$ rather than $\log_{10} x$, and the base is understood to be 10.

A Word of Caution When working with a logarithm whose base is not 10, we must be sure to write the base because $\log x$ is understood to have a base of 10.

Scientific calculators have built-in functions to calculate common logarithms. This function is typically denoted by a button labeled as log.

EXAMPLE ▶ 4   Evaluate the given common logarithm. Round to the nearest thousandth if necessary.

a) log 100

Solution

To evaluate log 100, we must determine what power of the base 10 is equal to 100. Since $10^2 = 100$, then log 100 = 2. A calculator would produce the same result.

b) log 321

Solution

To evaluate log 321, we will use a calculator, since we do not know what power of 10 is 321. Rounding to the nearest thousandth, we find that log 321 ≈ 2.507. To check this result, evaluate $10^{2.507}$ on a calculator, which should be approximately equal to 321.

Using Your Calculator   The TI-83/84 has a built-in function for calculating the common logarithm of a number. To evaluate log 321, press the button labeled LOG. Notice that the calculator gives you a left-hand parenthesis. The argument of the logarithm needs to be enclosed in parentheses. Then type 321 and close the parentheses by pushing ⑴. After you press ENTER, your calculator screen should look as follows:

```
log(321)
        2.506505032
```

Quick Check **3**
Evaluate the following. Round to the nearest thousandth if necessary.

a) log 10,000
b) log 75

The other logarithm that we will use frequently is called the natural logarithm.

Natural Logarithm

The natural logarithm, denoted ln *x*, is the logarithm whose base is the natural base *e*.

Recall from the previous section that *e* is a real number that is approximately equal to 2.718281828. Scientific calculators have built-in functions for calculating natural logarithms, using a button labeled as ln.

EXAMPLE ▶ 5   Evaluate the given natural logarithm. Round to the nearest thousandth if necessary.

a) ln e^7

Solution

To evaluate ln e^7, we must find an exponent *a* such that $e^a = e^7$. In this case, $a = 7$, so ln $e^7 = 7$.

This example illustrates another important property of logarithms.

$\log_b b^r$

> For any positive base b $(b \neq 1)$, $\log_b b^r = r$.

b) $\ln 45$

Solution

This logarithm must be evaluated by a calculator. Rounded to the nearest thousandth, $\ln 45 \approx 3.807$.

> **Using Your Calculator**   The TI-83/84 also has a built-in function for calculating the natural logarithm of a number. To evaluate ln 45, we begin by pushing the [LN] key. After typing the argument 45, we close the parentheses by pushing [)] and then press [ENTER]. Your calculator screen should look as follows:
>
> ```
> ln(45)
>           3.80666249
> ```

Quick Check 4
Evaluate the following and round to the nearest thousandth if necessary.

a) $\ln e^{-3}$
b) $\ln 605$

Exponential Form, Logarithmic Form

Objective 4   Convert back and forth between exponential form and logarithmic form.   Consider the equation $5^3 = 125$. This equation is said to be in **exponential form,** as it has a base (5) being raised to a power (3). Any equation in exponential form can be rewritten as an equivalent equation in **logarithmic form.** The equation $5^3 = 125$ can also be written in logarithmic form as $\log_5 125 = 3$.

Exponential Form	Logarithmic Form
$b^y = n$	$\log_b n = y$

The following images will help us when rewriting equations from one form to the other:

$b^y = n$ $\log_b n = y$ Exponent (y) $b^y = n$ $\log_b n = y$

Base (b) $b^y = n$ $\log_b n = y$ Number (n)

EXAMPLE 6   Write the exponential equation $2^{-8} = \frac{1}{256}$ in logarithmic form.

Solution

This equation can be written in logarithmic form as $\log_2\left(\frac{1}{256}\right) = -8$.

Quick Check 5 Write the exponential equation $7^x = 20$ in logarithmic form.

EXAMPLE ▸7   Write the logarithmic equation $\ln x = 7$ in exponential form.

Solution

Quick Check **6**
Write the logarithmic
equation $\log_3 x = 6$ in
exponential form.

Recall that the base is e. In exponential form, this equation can be written as $e^7 = x$.

Consider the function $f(x) = b^x$, where $b > 0$ and $b \neq 1$. To find the inverse of this function, we would replace $f(x)$ by y and then interchange x and y, which would produce the equation $x = b^y$. If we rewrite this equation in logarithmic form, we would obtain the equation $y = \log_b x$. Since this equation is solved for y, we have shown that $f^{-1}(x) = \log_b x$. The inverse of an exponential function is a logarithmic function.

Solving Logarithmic Equations

Objective 5   Solve logarithmic equations. When solving an equation that is in logarithmic form, it will often be helpful to rewrite the equation in its equivalent exponential form. We can then attempt to solve the resulting equation. For example, suppose that we were trying to solve the equation $\log_3 x = 2$. The exponential form of this equation is $3^2 = x$, which simplifies to $x = 9$.

EXAMPLE ▸8   Solve $\log_5 x = 4$.

Solution

We begin by rewriting the equation in exponential form.

$$\log_5 x = 4$$
$$5^4 = x \qquad \text{Rewrite in exponential form.}$$
$$625 = x \qquad \text{Simplify.}$$

Quick Check **7**
Solve $\log_4 x = 5$.

The solution set is $\{625\}$.

EXAMPLE ▸9   Let $f(x) = \ln(x - 8)$. Solve $f(x) = 3$. Round to the nearest thousandth.

Solution

We begin by rewriting the equation in exponential form, keeping in mind that the base of this logarithm is e.

$$\ln(x - 8) = 3 \qquad \text{Set the function equal to 3.}$$
$$e^3 = x - 8 \qquad \text{Rewrite in exponential form.}$$
$$e^3 + 8 = x \qquad \text{Add 8.}$$

The exact solution to the equation is $x = e^3 + 8$, so the solution set is $\{e^3 + 8\}$. Approximating this solution to the nearest thousandth by calculator, $x \approx 28.086$.

Quick Check **8**
Let $f(x) = \log_3(x + 6)$.
Solve $f(x) = 2$.

Graphing Logarithmic Functions

Objective 6   Graph logarithmic functions. We will graph logarithmic functions by plotting points. As with many other functions, the choice of which points to plot is crucial. This choice is made easier by rewriting a logarithmic function in exponential form. Suppose that we wanted to graph the function $f(x) = \log_2 x$. Replacing $f(x)$ by y, we can rewrite the equation $y = \log_2 x$ in exponential form as $2^y = x$. We can now choose values for y and find the corresponding values of x. We will use the values -2,

-1, 0, 1, and 2 for y. At this point, we notice that this is similar to the way we graphed exponential functions in the last section, with the exception that we are substituting values for y rather than for x.

$x = 2^y$	$2^{-2} = \dfrac{1}{4}$	$2^{-1} = \dfrac{1}{2}$	$2^0 = 1$	$2^1 = 2$	$2^2 = 4$
y	-2	-1	0	1	2

Below is the graph of $f(x) = \log_2 x$, which passes through the preceding points. Notice that the graph includes only positive values of x. Recall that the domain of a logarithmic function of the form $f(x) = \log_b x$ is the set of all positive real numbers, written in interval notation as $(0, \infty)$.

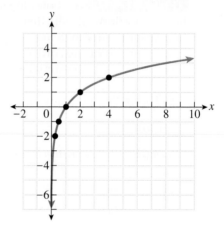

Also notice that as the values of x get closer and closer to 0, the graph of the function decreases without bound. The y-axis, whose equation is $x = 0$, is the **vertical asymptote** for the graph of this function.

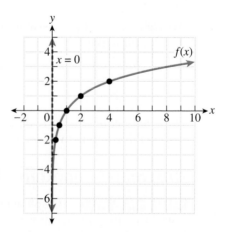

In general, the y-axis will be a vertical asymptote for a function of the form $f(x) = \log_b x$, since the domain of this function is the set of positive real numbers.

EXAMPLE 10   Graph $f(x) = \log_4 x$.

Solution

Replacing $f(x)$ by y, we see that the equation $y = \log_4 x$ can be rewritten in exponential form as $4^y = x$. Here is a table of values for this function:

$x = 4^y$	$4^{-2} = \dfrac{1}{16}$	$4^{-1} = \dfrac{1}{4}$	$4^0 = 1$	$4^1 = 4$	$4^2 = 16$
y	-2	-1	0	1	2

Below is the graph of $f(x) = \log_4 x$, which passes through those points, including the vertical asymptote at $x = 0$.

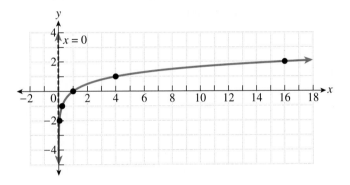

Note that as the values of x increase, the function $f(x) = \log_4 x$ does not increase nearly as quickly as the function $f(x) = \log_2 x$ does. What would we expect of the graph of the function $f(x) = \log x$ as x increases? It increases at a much slower rate than either of these two previous functions. In general, the larger the base of a logarithmic function, the slower the function increases as x increases. Here is a graph showing all three functions:

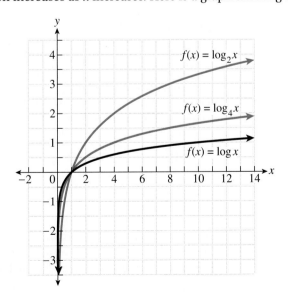

EXAMPLE ▶11   Graph $f(x) = \ln x$.

Solution

Replacing $f(x)$ by y, we see that the equation $y = \ln x$ can be rewritten in exponential form as $e^y = x$. Here is a table of values for this function, with values of x rounded to the nearest tenth:

$x = e^y$	$e^{-2} \approx 0.1$	$e^{-1} \approx 0.4$	$e^0 = 1$	$e^1 \approx 2.7$	$e^2 \approx 7.4$
y	-2	-1	0	1	2

Below is the graph of $f(x) = \ln x$, which passes through those points, including the vertical asymptote at $x = 0$.

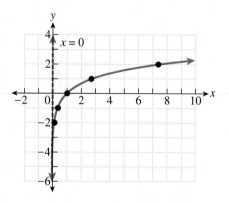

EXAMPLE ▶12   Graph $f(x) = \log_{1/3} x$.

Solution

In this case, the base of the logarithmic function is less than 1. This graph will have a different look than the logarithmic functions that were graphed in the previous examples. Replacing $f(x)$ by y, we find that the equation $y = \log_{1/3} x$ can be rewritten in exponential form as $\left(\frac{1}{3}\right)^y = x$. Here is a table of values for this function.

$x = \left(\dfrac{1}{3}\right)^y$	$\left(\dfrac{1}{3}\right)^{-2} = 9$	$\left(\dfrac{1}{3}\right)^{-1} = 3$	$\left(\dfrac{1}{3}\right)^0 = 1$	$\left(\dfrac{1}{3}\right)^1 = \dfrac{1}{3}$	$\left(\dfrac{1}{3}\right)^2 = \dfrac{1}{9}$
y	-2	-1	0	1	2

On the next page is the graph of $f(x) = \log_{1/3} x$, which passes through those points, including the vertical asymptote at $x = 0$. Notice that the graph of this function decreases as x increases, because the base is less than 1.

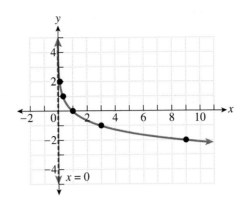

Quick Check **9** Graph.

a) $f(x) = \log_5 x$

b) $f(x) = \log_{1/4} x$

Consider the graphs of the functions $f(x) = 2^x$ and $g(x) = \log_2 x$, which have been graphed together on the same set of axes.

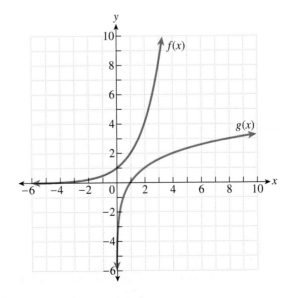

Adding the graph of the line $y = x$ allows us to see that the two functions are symmetric to each other about the line $y = x$ and therefore are inverses of each other.

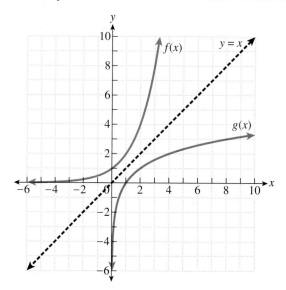

In general, for any positive real number b $(b \neq 1)$, the functions $f(x) = b^x$ and $g(x) = \log_b x$ are inverse functions. We will explore this idea more thoroughly in Section 12.4.

EXAMPLE 13   Use the given graph of $f(x) = 3^x$ to graph the function $g(x) = \log_3 x$.

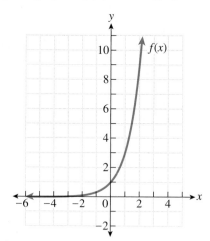

Solution

Since $f(x) = 3^x$ and $g(x) = \log_3 x$ are inverses of each other, we can sketch the graph of $g(x)$ by finding ordered pairs that are on the graph of $f(x)$ and interchanging their x- and y-coordinates.

From the graph, we can see that the points $(0, 1)$, $(1, 3)$, and $(2, 9)$ are on the graph of $f(x)$. This tells us that the points $(1, 0)$, $(3, 1)$, and $(9, 2)$ are on the graph of $g(x)$.

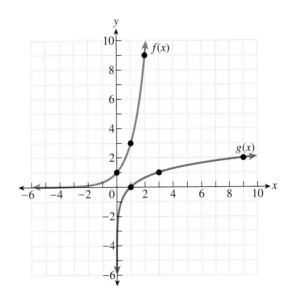

Above are those points, along with the graph of $f(x)$. We can now sketch the graph of $g(x)$, making sure that its graph is symmetric around the line $y = x$ with the graph of the function $f(x)$.

Applications of Logarithmic Functions

Objective 7 Use logarithmic functions in applications.

EXAMPLE 14 The average height of boys, in inches, during their first year of life can be approximated using the function $f(x) = 1.95 \ln x + 23.53$, where x is the age of the boy in months. Use this function to estimate the average height of boys who are 6 months old. (*Source:* Centers for Disease Control (CDC))

Solution

To estimate the average height of boys who are 6 months old, we need to evaluate the function when $x = 6$.

$$f(6) = 1.95 \ln 6 + 23.53 \qquad \text{Substitute 6 for } x.$$
$$\approx 27.02 \qquad \text{Approximate by calculator.}$$

The average height of 6-month-old boys is approximately 27.02 inches.

Quick Check **10** The annual sales of a company, in billions of dollars, can be approximated by the function $f(x) = 7.95 \ln x + 20.95$, where x represents the number of years after 1991. Use the function to predict the annual sales of this company in the year 2008.

The **Richter scale** is used to measure the magnitude or strength of an earthquake. The magnitude of an earthquake is a function of the size of the shock wave that it creates on a seismograph. If an earthquake creates a shock wave that is I times larger than the smallest measurable shock wave that is recordable on a seismograph, then its magnitude R on the Richter scale is given by the formula $R = \log I$.

> **EXAMPLE 15**   If an earthquake creates a shock wave that is 500 times the smallest measurable shock wave recordable by a seismograph, find the magnitude of the earthquake on the Richter scale.

Quick Check 11

If an earthquake creates a shockwave that is 2500 times the smallest measurable shock wave that is recordable by a seismograph, find the magnitude of the earthquake on the Richter scale.

Solution

In this problem, we are given the fact that $I = 500$, so we substitute 500 for I in the formula $R = \log I$ and solve for R.

$$R = \log I$$
$$R = \log 500 \quad \text{Substitute 500 for } I.$$
$$R \approx 2.7 \quad \text{Approximate on a calculator.}$$

The magnitude of this earthquake on the Richter scale is 2.7.

EXERCISES 12.2

Evaluate.

1. $\log_2 4$
2. $\log_2 32$
3. $\log_3 27$
4. $\log_3 81$
5. $\log_4 64$
6. $\log_5 625$
7. $\log_8 1$
8. $\log_5 1$
9. $\log_3\left(\frac{1}{9}\right)$
10. $\log_2\left(\frac{1}{16}\right)$
11. $\log_8\left(\frac{1}{512}\right)$
12. $\log_6\left(\frac{1}{6}\right)$

Evaluate the given function.

13. $f(x) = \log_3 x$ **a)** $f(9)$ **b)** $f(243)$
14. $f(x) = \log_5 x$ **a)** $f(25)$ **b)** $f(1)$
15. $f(x) = \log_2(x - 7)$ **a)** $f(8)$ **b)** $f(135)$
16. $f(x) = \log_3(2x + 11)$ **a)** $f(35)$ **b)** $f(116)$

Evaluate. Round to the nearest thousandth if necessary.

17. $\log 1000$
18. $\log 10$
19. $\log 0.01$
20. $\log\left(\frac{1}{10,000}\right)$
21. $\log 112$
22. $\log 47$
23. $\log 7.42$
24. $\log 0.0018$
25. $\ln e$
26. $\ln(e^3)$
27. $\ln\left(\frac{1}{e^6}\right)$
28. $\ln(e^{-2})$
29. $\ln 32$
30. $\ln 105$
31. $\ln 16.23$
32. $\ln 0.5$

Rewrite in logarithmic form.

33. $2^3 = 8$
34. $5^4 = 625$
35. $9^3 = 729$
36. $4^{-2} = \frac{1}{16}$
37. $6^x = 30$
38. $3^x = 17$
39. $10^x = 500$
40. $e^x = 7$

Rewrite in exponential form.

41. $\log_4 1024 = 5$
42. $\log_3 19{,}683 = 9$
43. $\log_{16} 8 = \frac{3}{4}$
44. $\log_{49} 7 = \frac{1}{2}$
45. $\log x = 3$
46. $\ln x = 5$
47. $\ln x - 3 = 4$
48. $\log(x + 2) - 10 = -5$

Solve. Round to the nearest thousandth if necessary.

49. $\log_2 x = 5$

50. $\log_2 x = -2$

51. $\log_3 x = -4$

52. $\log_3 x = 5$

53. $\log x = 4$

54. $\log x = 7$

55. $\ln x = 2$

56. $\ln x = -1$

57. $\log_3 x + 4 = 6$

58. $\log_3(x + 4) = 6$

59. $\ln(x - 5) + 13 = 11$

60. $\log(2x - 8) - 21 = -18$

61. Let $f(x) = \log_6 x$. Solve $f(x) = 3$.

62. Let $f(x) = \log_5(x + 12)$. Solve $f(x) = 2$.

63. Let $f(x) = \log(x - 10) - 3$. Solve $f(x) = -1$.

64. Let $f(x) = \ln(x - 5) + 17$. Solve $f(x) = 17$.

Solve.

65. $\log_5 125 = x$

66. $\log_2 256 = x$

67. $\log 100{,}000 = x$

68. $\log 0.001 = x$

69. $\log_9 27 = x$

70. $\log_4 32 = x$

Graph the given function $f(x)$. Label the vertical asymptote.

71. $f(x) = \log_3 x$

72. $f(x) = \log_6 x$

73. $f(x) = \log x$

74. $f(x) = \log_8 x$

75. $f(x) = \log_{1/5} x$

76. $f(x) = \log_{1/2} x$

Use the graph of the given exponential function $f(x)$ to graph its inverse logarithmic function $f^{-1}(x)$.

77.

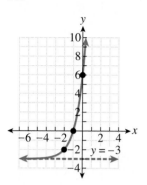

78.

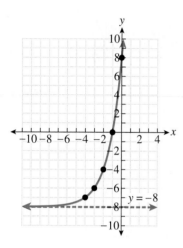

79.

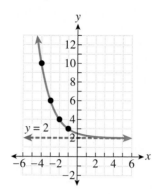

80.

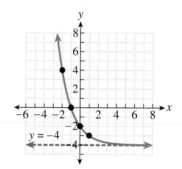

Use the table of values to determine the formula for the logarithmic function $f(x)$.

81.

x	$\dfrac{1}{4}$	$\dfrac{1}{2}$	1	2	4
$f(x)$	−2	−1	0	1	2

82.

x	$\dfrac{1}{36}$	$\dfrac{1}{6}$	1	6	36
$f(x)$	−2	−1	0	1	2

83.

x	9	3	1	$\dfrac{1}{3}$	$\dfrac{1}{9}$
$f(x)$	−2	−1	0	1	2

84.

x	$\dfrac{4}{49}$	$\dfrac{2}{7}$	1	$\dfrac{7}{2}$	$\dfrac{49}{4}$
$f(x)$	−2	−1	0	1	2

85. The number of U.S. airports in a particular year can be approximated by the function $f(x) = 14650 + 1344 \ln x$, where x represents the number of years after 1979. Use this function to predict the number of U.S. airports in the year 2030. (*Source:* Bureau of Transportation Statistics)

86. The percent of U.S. workers who drive themselves to work can be approximated by the function $f(x) = 72.5 + 2.01 \ln x$, where x represents the number of years after 1984. Use this function to predict what percent of U.S. workers will drive themselves to work in the year 2010. (*Source:* Bureau of Transportation Statistics)

87. The average height, in inches, of girls who are 3 years old or younger can be approximated by the function $f(x) = 18.6 + 4.78 \ln x$, where x represents the age in months. Use the function to determine the average height of girls who are 24 months old. (*Source:* CDC)

88. The average weight, in pounds, of boys who are 3 years old or younger can be approximated by the function $f(x) = 7.6 + 6.36 \ln x$, where x represents the age in months. Use the function to determine the average weight of boys who are 18 months old. (*Source:* CDC)

For Exercises 89–92, use the formula $R = \log I$, where R is the magnitude of an earthquake on the Richter scale whose shock wave is I times larger than the smallest measurable shock wave that is recordable on a seismograph.

89. If an earthquake creates a shock wave that is 16,000 times the smallest measurable shock wave that is recordable by a seismograph, find the magnitude of the earthquake on the Richter scale.

90. A 1950 earthquake along the India–China border created a shock wave that was 400,000,000 times the smallest measurable shock wave that is recordable by a seismograph. Find the magnitude of this earthquake on the Richter scale.

91. The 1906 earthquake off the coast of Ecuador had a magnitude of 8.8 on the Richter scale. How many times larger was the shock wave created by this earthquake than the smallest measurable shock wave?

92. The 1992 earthquake centered in Landers, California, had a magnitude of 7.3 on the Richter scale. How many times larger was the shock wave created by this earthquake than the smallest measurable shock wave?

Answer in complete sentences.

93. What is a logarithm? Explain clearly, using your own words.

94. Explain how to convert an expression in exponential form to an expression in logarithmic form.

95. Write a logarithmic equation whose solution is $x = 7$. Explain how to solve the equation.

96. Write a logarithmic equation whose solution is $x = \frac{3}{2}$. Explain how to solve the equation.

97. Explain how to graph a logarithmic function. Use an example to illustrate the process.

98. Explain why the graph of a logarithmic function has a vertical asymptote.

Quick Review Exercises

Section 12.2

Simplify. Assume all variables are nonzero real numbers.

1. $x^7 \cdot x^9$
2. $\dfrac{b^{16}}{b^4}$
3. $(n^8)^7$
4. z^0

Study Tip **Revisited** Make your own study quiz for this section by selecting odd-numbered problems from the exercise set. Here is a suggested outline:

Number of Problems	Topic
4	Evaluate logarithms and logarithmic functions (including common logarithms and natural logarithms)
2	Rewrite an equation in exponential form
2	Rewrite an equation in logarithmic form
4	Solve logarithmic equations
2	Graph logarithmic functions
1	Applied problem

Be sure to choose problems of varying levels of difficulty. By choosing odd problems from the exercise set, you will have an answer key for your practice quiz in the back of the text.

12.3
PROPERTIES OF LOGARITHMS

Objectives

1 Use the product rule for logarithms.
2 Use the quotient rule for logarithms.
3 Use the power rule for logarithms.
4 Use additional properties of logarithms.
5 Use properties to rewrite two or more logarithmic expressions as a single logarithmic expression.
6 Use properties to rewrite a single logarithmic expression as a sum or difference of logarithmic expressions whose arguments have an exponent of 1.
7 Use the change-of-base formula for logarithms.

In this section, we will learn several properties of logarithms. These properties, in addition to being useful for simplifying logarithmic expressions, will also be helpful when solving logarithmic equations.

Product Rule for Logarithms

Objective 1 **Use the product rule for logarithms.** We know that $\log_2 8 = 3$, because $2^3 = 8$. We also know that $\log_2 16 = 4$, since $2^4 = 16$. So $\log_2 8 + \log_2 16 = 3 + 4$ or 7. To find this sum, we had to evaluate two logarithms. Consider $\log_2(8 \cdot 16)$, which is equal to $\log_2 128$. This logarithm is also equal to 7, since $2^7 = 128$. We see that the sum $\log_2 8 + \log_2 16$ is equal to the logarithm whose argument is the product of their two arguments, $\log_2(8 \cdot 16)$. This leads to the **product rule for logarithms.**

Product Rule for Logarithms

> For any positive real number b ($b \neq 1$) and any positive real numbers x and y,
> $\log_b x + \log_b y = \log_b(xy)$.

This rule tells us that we can rewrite the sum of two logarithms as a single logarithm, provided that the two logarithms have the same base. It also tells us that the logarithm of a product can be rewritten as a sum of logarithms. Here is a proof of the product rule for logarithms:

> Let $M = \log_b x$ and $N = \log_b y$.
> Rewriting $M = \log_b x$ in exponential form, we see that $b^M = x$. Similarly, $b^N = y$.
> So $xy = b^M \cdot b^N = b^{M+N}$.
> Rewriting $xy = b^{M+N}$ in logarithmic form, we see that $\log_b(xy) = M + N$.
> But $M + N = \log_b x + \log_b y$, so $\log_b(xy) = \log_b x + \log_b y$.

EXAMPLE 1   Rewrite $\ln 7 + \ln x$ as a single logarithm. Assume x is a positive real number.

Solution

Both logarithms have the same base, so we will use the product rule to simplify this expression.

$$\ln 7 + \ln x = \ln(7x) \qquad \text{Rewrite as the logarithm of a product.}$$

Quick Check **1** Rewrite $\log 5 + \log 7 + \log x$ as a single logarithm. Assume x is a positive real number.

EXAMPLE 2 Rewrite $\log(10xy)$ as the sum of two or more logarithms. Simplify if possible. Assume all variables are positive real numbers.

Solution

This expression is the logarithm of three factors, and we may use the product rule to rewrite it as the sum of three logarithms.

$$\log(10xy) = \log 10 + \log x + \log y \qquad \text{Rewrite as the sum of three logarithms.}$$
$$= 1 + \log x + \log y \qquad \text{Simplify } \log 10.$$

Quick Check **2** Rewrite $\log_2(5b)$ as the sum of two or more logarithms. Assume b is a positive real number.

A Word of Caution The product rule for logarithms only applies to expressions of the form $\log_b x + \log_b y$, not to expressions such as $\log_b(x + y)$. In general,

$$\log_b(x + y) \neq \log_b x + \log_b y$$

Quotient Rule for Logarithms

Objective 2 Use the quotient rule for logarithms. Just as the sum of two logarithms can be expressed as a single logarithm, so can the difference of two logarithms.

Quotient Rule for Logarithms

For any positive real number b ($b \neq 1$) and any positive real numbers x and y,
$$\log_b x - \log_b y = \log_b\left(\tfrac{x}{y}\right).$$

This rule tells us that we can rewrite the difference of two logarithms as a single logarithm of a quotient, provided that the two logarithms have the same base. It also tells us that the logarithm of a quotient can be rewritten as a difference of logarithms. The proof of the quotient rule for logarithms is similar to the proof of the product rule for logarithms, using the fact that $\dfrac{b^M}{b^N} = b^{M-N}$. The proof is left as an exercise for the reader.

EXAMPLE 3 Rewrite $\log_3 36 - \log_3 4$ as a single logarithm and simplify if possible.

Solution

Both logarithms have the same base, so we will use the quotient rule to simplify this expression.

$$\log_3 36 - \log_3 4 = \log_3\left(\frac{36}{4}\right) \qquad \text{Rewrite as the logarithm of a quotient.}$$

$$= \log_3 9 \qquad \text{Divide.}$$

$$= 2 \qquad \text{Simplify the logarithm.}$$

Quick Check 3
Rewrite $\log_2 60 - \log_2 15$ as a single logarithm. Simplify if possible.

EXAMPLE 4 Rewrite $\ln\left(\frac{12x}{y}\right)$ in terms of two or more logarithms. Assume all variables are positive real numbers.

Solution

We begin by applying the quotient rule.

$$\ln\left(\frac{12x}{y}\right) = \ln(12x) - \ln y \qquad \begin{array}{l}\text{Rewrite as the logarithm of the numerator} \\ \text{minus the logarithm of the denominator.}\end{array}$$

$$= \ln 12 + \ln x - \ln y \qquad \text{Apply the product rule to } \ln(12x).$$

Quick Check 4
Rewrite $\log_b\left(\frac{x}{b}\right)$ in terms of two logarithms. Simplify, if possible. Assume x and b are positive real numbers.

A Word of Caution   The quotient rule for logarithms only applies to expressions of the form $\log_b x - \log_b y$, not to expressions such as $\log_b(x - y)$. In general,

$$\log_b(x - y) \neq \log_b x - \log_b y$$

Power Rule for Logarithms

Objective 3   Use the power rule for logarithms. The properties of logarithms discussed in this section will help us to solve exponential equations, and one of the most important properties is the power rule for logarithms.

Power Rule for Logarithms

For any positive real number b ($b \neq 1$), any positive real number x, and any real number r,

$$\log_b x^r = r \cdot \log_b x$$

This rule tells us that we can rewrite the logarithm of a positive number that is raised to a power by multiplying the exponent by the logarithm of the number. Here is a proof of the power rule, which uses the product rule for logarithms.

$$\log_b x^r = \log_b\left(\underbrace{x \cdot x \cdot \cdots \cdot x}_{r\,\text{times}}\right) \qquad \begin{array}{l}\text{Rewrite } x^r \text{ by listing } x \text{ as a factor } r \\ \text{times.}\end{array}$$

$$= \underbrace{\log_b x + \log_b x + \cdots + \log_b x}_{r\,\text{times}} \qquad \begin{array}{l}\text{Use the product rule for} \\ \text{logarithms.}\end{array}$$

$$= r \cdot \log_b x \qquad \begin{array}{l}\text{Rewrite the repeated addition of} \\ \log_b x \text{ as } r \cdot \log_b x.\end{array}$$

EXAMPLE 5   Rewrite $\log_2 x^7$ as a logarithm of a single factor without exponents. Assume x is a positive real number.

Quick Check 5

Rewrite $\log_3 (9x^2 y^4)$ using logarithms of single factors that do not have exponents. Simplify, if possible. Assume x and y are positive real numbers.

Solution

The power rule tells us that we may rewrite this expression by removing 7 as an exponent and multiplying it by $\log_2 x$. In other words, we are moving the number 7 from the exponent of x and placing it as a factor in front of $\log_2 x$.

$$\overset{\curvearrowleft}{\log_2 x^{\text{⑦}}}$$

$$\log_2 x^7 = 7 \log_2 x \qquad \text{Use the power rule.}$$

Recall that $\sqrt[n]{x} = x^{1/n}$. So, we can apply the power rule when the argument of a logarithm is an expression involving radicals. For example, $\log_9 \sqrt[4]{x} = \log_9 x^{1/4} = \frac{1}{4} \log_9 x$.

Quick Check 6   Rewrite $\ln \sqrt[n]{x}$ as a logarithm of a single factor without any radicals. Assume x is a positive real number.

EXAMPLE 6   Rewrite $15 \ln x$ in such a way that the logarithm is not being multiplied by a number. Assume x is a positive real number.

Solution

We will use the power rule to rewrite 15 as an exponent of x.

Quick Check 7

Rewrite $8 \log_8 x$ in such a way that the logarithm is not being multiplied by a number. Assume x is a positive real number.

$$15 \ln x = \ln x^{15} \qquad \text{Use the power rule.}$$

A Word of Caution   Be careful when working with logarithmic expressions involving exponents. In the expression $\log_b x^n$, only the number x is being raised to the nth power. However, in the expression $(\log_b x)^n$, it is the logarithm $\log_b x$ that is being raised to the nth power. We cannot apply the power rule for logarithms to the expression $(\log_b x)^n$. In general,

$$\log_b x^n \neq (\log_b x)^n$$

Objective 4   Use additional properties of logarithms.   The power rule leads us to another property of logarithms. Consider the expression $\log_5 5^7$. The power rule tells us that $\log_5 5^7 = 7 \log_5 5$. Since $\log_5 5$ is equal to 1, this expression simplifies to equal $7 \cdot 1$ or 7. This result can be generalized as follows:

$\log_b b^r$

For any positive real number b ($b \neq 1$) and any real number r, $\log_b b^r = r$.

This property will be particularly helpful when taking the natural logarithm of e raised to a power.

EXAMPLE ▶ 7   Simplify $\ln e^x$.

Solution

The base of this logarithm is e, so the property $\log_b b^r = r$ may be applied.

$$\ln e^x = x$$

Quick Check 8
Simplify $\ln e^{-4}$.

Another property of logarithms involves raising a positive number to a power containing a logarithm whose base is that number. This property follows from the fact that exponential and logarithmic functions are inverses.

$$b^{\log_b x}$$

For any positive real number b $(b \neq 1)$ and any positive real number x,

$$b^{\log_b x} = x$$

EXAMPLE ▶ 8   Simplify $8^{\log_8 15}$.

Solution

Quick Check 9
Simplify $e^{\ln 16}$.

By the previous property, $8^{\log_8 15} = 15$.

Rewriting Logarithmic Expressions as a Single Logarithm

Objective 5 **Use properties to rewrite two or more logarithmic expressions as a single logarithmic expression.** We will often need to rewrite an expression containing two or more logarithms in terms of a single logarithm. To do this, we will be using the power rule first if possible, and then we will use the quotient rule and the product rule.

EXAMPLE ▶ 9   Rewrite $\log a + \log b - \log c - \log d$ as a single logarithm. Assume a, b, c, and d are positive real numbers.

Solution

The first two logarithms can be combined by using the product rule, as can the last two logarithms. We can then combine these logarithms by using the quotient rule.

$$\log a + \log b - \log c - \log d$$
$$= \log a + \log b - (\log c + \log d) \quad \text{Factor } -1 \text{ from the logarithms being subtracted.}$$
$$= \log(ab) - \log(cd) \quad \text{Use the product rule.}$$
$$= \log\left(\frac{ab}{cd}\right) \quad \text{Use the quotient rule.}$$

Quick Check 10 Rewrite $\log_2 a - \log_2 b + \log_2 c + \log_2 d$ as a single logarithm. Assume a, b, c, and d are positive real numbers.

EXAMPLE 10   Rewrite $\frac{1}{4} \ln a - 10 \ln b - 2 \ln c$ as a single logarithm. Assume a, b, and c are positive real numbers.

Solution

We will begin by using the power rule to rewrite each logarithm. We can then use the quotient rule.

$$\frac{1}{4} \ln a - 10 \ln b - 2 \ln c = \ln a^{1/4} - \ln b^{10} - \ln c^2 \qquad \text{Use the power rule.}$$

$$= \ln a^{1/4} - (\ln b^{10} + \ln c^2) \qquad \begin{array}{l}\text{Factor } -1 \text{ from the last} \\ \text{two terms.}\end{array}$$

$$= \ln\left(\frac{a^{1/4}}{b^{10}c^2}\right) \qquad \begin{array}{l}\text{Use the quotient rule.} \\ \text{Since } \ln b^{10} \text{ and } \ln c^2 \\ \text{are both being sub-} \\ \text{tracted, } b^{10} \text{ and } c^2 \\ \text{are written in the} \\ \text{denominator.}\end{array}$$

$$= \ln\left(\frac{\sqrt[4]{a}}{b^{10}c^2}\right) \qquad \begin{array}{l}\text{Rewrite the numerator} \\ \text{using radical notation.}\end{array}$$

Quick Check   11 Rewrite $6 \log a + \frac{1}{5} \log b - 9 \log c - 12 \log d$ as a single logarithm. Assume a, b, c, and d are positive real numbers.

Rewriting a Logarithmic Expression as a Sum or Difference of Logarithms Whose Arguments Have an Exponent of 1

Objective 6 Use properties to rewrite a single logarithmic expression as a sum or difference of logarithmic expressions whose arguments have an exponent of 1. We now turn our attention to expanding a single logarithm into the sum or difference of separate logarithms. This will be done in such a way that each logarithm is of a single factor that does not have an exponent. To do this, we will be using the product rule, the quotient rule, and the power rule.

EXAMPLE 11   Rewrite $\log\left(\frac{a^5 b^8}{c^7}\right)$ in terms of two or more logarithms. Each argument should contain a single factor and each exponent should be written as a factor. Assume a, b, and c represent positive real numbers.

Solution

We will begin by applying the product and quotient rules to rewrite this logarithm in terms of three logarithms. The power rule can then be applied to each of these logarithms.

$$\log\left(\frac{a^5 b^8}{c^7}\right) = \log a^5 + \log b^8 - \log c^7 \qquad \begin{array}{l}\text{Use the product and quotient} \\ \text{rules.}\end{array}$$

$$= 5 \log a + 8 \log b - 7 \log c \qquad \text{Use the power rule.}$$

Change-of-Base Formula

Objective **7**   **Use the change-of-base formula for logarithms.** The final property of logarithms in this section is the change-of-base formula.

Change-of-Base Formula

For any positive numbers a, b, and x $(a, b \neq 1)$, $\log_b x = \dfrac{\log_a x}{\log_a b}$.

This formula allows us to rewrite any logarithm in terms of two logarithms that have a different base than the original logarithm. The advantage of this is that we can rewrite logarithms in terms of either common logarithms or natural logarithms, allowing us to approximate these logarithms using a calculator. For instance, suppose that we wanted to evaluate $\log_2 7$. Since we do not know what power of 2 is equal to 7, we would not be able to evaluate this logarithm. We can rewrite this logarithm as $\dfrac{\log 7}{\log 2}$ or as $\dfrac{\ln 7}{\ln 2}$ using the change-of-base formula. Either of these two expressions can be evaluated on a calculator. Rounding to four decimal places, we see that $\log_2 7 \approx 2.8074$.

EXAMPLE **12**   Evaluate $\log_5 13$ using the change-of-base formula.

Solution

We can rewrite this logarithm using common logarithms as $\dfrac{\log 13}{\log 5}$.

$$\log_5 13 = \frac{\log 13}{\log 5} \qquad \text{Use the change-of-base formula.}$$
$$\approx 1.5937 \qquad \text{Approximate by calculator.}$$

We can check this result by raising 5 to the 1.5937 power. Notice that the result is approximately equal to 13. Also notice that if we had used natural logarithms instead of common logarithms, $\dfrac{\ln 13}{\ln 5} \approx 1.5937$ as well.

Using Your Calculator We can evaluate logarithms such as $\log_5 13$ on the TI-83/84 using the change-of-base formula, which tells us that $\log_5 13 = \dfrac{\log 13}{\log 5}$. This can be entered as follows:

```
log(13)/log(5)
       1.593692641
```

EXERCISES 12.3 ▶

Rewrite as a single logarithm using the product rule for logarithms. Simplify, if possible. Assume all variables represent positive real numbers.

1. $\log_2 5 + \log_2 x$

2. $\log_3 8 + \log_3 a$

3. $\log_4 b + \log_4 c$

4. $\log 9 + \log x^3$

5. $\log_4 8 + \log_4 32$

6. $\log_{21} 9 + \log_{21} 49$

Rewrite as the sum of two or more logarithms using the product rule for logarithms. Simplify, if possible. Assume all variables represent positive real numbers.

7. $\log_3(10x)$

8. $\log_6(8b)$

9. $\ln(3e)$

10. $\log_5(25z)$

11. $\log_4(abc)$

12. $\log_2(16mn)$

Rewrite as a single logarithm using the quotient rule for logarithms. Simplify, if possible. Assume all variables represent positive real numbers.

13. $\log_5 42 - \log_5 7$

14. $\log_4 30 - \log_4 150$

15. $\log 39 - \log x$

16. $\log_7 a - \log_7 9$

17. $\log_4 448 - \log_4 7$

18. $\log_3 10 - \log_3 90$

Rewrite in terms of two or more logarithms using the quotient and product rules for logarithms. Simplify, if possible. Assume all variables represent positive real numbers.

19. $\log_8\left(\dfrac{10a}{b}\right)$

20. $\log_5\left(\dfrac{7x}{3y}\right)$

21. $\log_4\left(\dfrac{6xyz}{w}\right)$

22. $\log_7\left(\dfrac{1}{30}\right)$

23. $\log_5\left(\dfrac{xy}{25z}\right)$

24. $\log_2\left(\dfrac{8a}{bc}\right)$

Rewrite using the power and product rules. Simplify if possible. Assume all variables represent positive real numbers.

25. $\log a^{10}$

26. $\log_2 b^3$

27. $\log_7 x^2 y^3$

28. $\ln a^9 b^3$

29. $\log_3 \sqrt{x}$

30. $\log_4 \sqrt[6]{b}$

31. $\log_2 16 x^5 y^7$

32. $\log_5 25 a^6 \sqrt[3]{b}$

Rewrite using the power rule. Assume all variables represent positive real numbers.

33. $8 \log_5 x$ **34.** $3 \log_4 b$

35. $\dfrac{1}{3} \ln x$ **36.** $\dfrac{1}{2} \log_2 x$

37. $5 \log_3 x^2$ **38.** $4 \log b^6$

Simplify.

39. $\log_5 5^{12}$ **40.** $\log_2 2^{21}$ **41.** $\ln e^9$

42. $\log 10^{25}$ **43.** $2^{\log_2 7}$ **44.** $5^{\log_5 36}$

45. $3^{\log_3 b}$ **46.** $e^{\ln 0.026}$

Rewrite as a single logarithm. Assume all variables represent positive real numbers.

47. $\log_3 2 + \log_3 5 - \log_3 7$

48. $\log_2 7 - \log_2 10 - \log_2 9$

49. $3 \log_5 2 + 6 \log_5 3$

50. $3 \log 9 - 2 \log 11$

51. $\dfrac{1}{2} \ln 16 - \dfrac{1}{3} \ln 125$

52. $\dfrac{1}{2} \log_6 144 + 4 \log_6 5$

53. $\log_2 x - \log_2 y - \log_2 z - \log_2 w$

54. $5 \log_3 a - 7 \log_3 b + 2 \log_3 c$

55. $4 \ln x + 9 \ln y - 13 \ln z$

56. $10 \log_5 x - 3 \log_5 y - 6 \log_5 z$

57. $\log(x + 2) + \log(x - 6)$

58. $\log_4(x + 4) + \log_4(x - 4)$

59. $\log_3(2x + 7) + \log_3(x + 5)$

60. $\log_7(x^2 + 5x - 24) - \log_7(x + 8)$

61. $3 \ln a + \dfrac{1}{3} \ln b - 10 \ln c$

62. $\dfrac{1}{5} \ln x - 7 \ln y - 6 \ln z$

Suppose for some base $b > 0$ $(b \neq 1)$ that $\log_b 2 = A$, $\log_b 3 = B$, $\log_b 5 = C$, and $\log_b 7 = D$. Express the following logarithms in terms of A, B, C, or D.

(Hint: rewrite the logarithm in terms of $\log_b 2$, $\log_b 3$, $\log_b 5$, or $\log_b 7$.)

63. $\log_b 6$ **64.** $\log_b 15$

65. $\log_b 35$ **66.** $\log_b 14$

67. $\log_b 8$ **68.** $\log_b 9$ **69.** $\log_b 125$

70. $\log_b\left(\dfrac{1}{49}\right)$

Expand. Simplify, if possible. Assume all variables represent positive real numbers.

71. $\log_8\left(\dfrac{a}{bcd}\right)$

72. $\log_9\left(\dfrac{81xy}{zw}\right)$

73. $\log_2 32 x^4 y^8 z^5$

74. $\log_3\left(\dfrac{1}{27 x^9 y^{20}}\right)$

75. $\ln\left(\dfrac{a^6 \sqrt{b}}{c^7}\right)$

76. $\ln\left(\dfrac{\sqrt[7]{x}}{y^{10} z^3}\right)$

77. $-\ln\left(\dfrac{a^3 d^2}{bc^7}\right)$

78. $-2 \log\left(\dfrac{xy^2}{z^3}\right)$

Evaluate using the change-of-base formula. Round to the nearest thousandth.

79. $\log_2 7$

80. $\log_{11} 3000$

81. $\log_{12} 47$

82. $\log_9 250$

83. $\log_7 300{,}000$

84. $\log_{15} 650{,}000$

85. $\log_8 3$

86. $\log_{20} 5$

87. $\log_9 0.0354$

88. $\log_6 0.195$

Find the missing number. Round to the nearest thousandth. (Hint: rewrite in logarithmic form and use the change-of-base formula.)

89. $5^? = 19$

90. $3^? = 30$

91. $15^? = 100$

92. $2^? = 725$

Answer in complete sentences.

93. Explain how and when to use the change-of-base formula.

94. Give examples showing that the following statements are false:

- $\log_b(x + y) = \log_b x + \log_b y$
- $\log_b(x - y) = \log_b x - \log_b y$
- $\log_b x^n = (\log_b x)^n$

Study Tip REVISITED   Make your own practice quiz for the first half of this chapter by selecting odd-numbered problems from the exercise sets of the first three sections. Select five odd problems from each section that you feel are representative of the types of problems that you would be expected to solve on an exam. Be sure to choose problems of varying levels of difficulty. By choosing odd problems from the exercise set, you will have an answer key for your practice quiz in the back of the text.

12.4

EXPONENTIAL AND LOGARITHMIC EQUATIONS

Objectives

1. Solve an exponential equation in which both sides have the same base.
2. Solve an exponential equation by using logarithms.
3. Solve a logarithmic equation in which both sides have logarithms with the same base.
4. Solve a logarithmic equation by converting it to exponential form.
5. Use properties of logarithms to solve a logarithmic equation.
6. Find the inverse function for exponential and logarithmic functions.

In this section, we will be solving exponential and logarithmic equations. An exponential equation is an equation containing a variable in an exponent. A logarithmic equation is an equation involving logarithms of variable expressions.

Solving Exponential Equations

Objective 1 Solve an exponential equation in which both sides have the same base. Given an exponential equation to solve, we will first try to express both sides of the equation in terms of the same base. If we can do this, we can then solve the equation using the techniques developed in Section 12.1. Here are two examples reviewing this process.

EXAMPLE 1 Solve $2^{x+5} = 8$.

Solution

We will begin by rewriting 8 as 2^3, as both sides of the equation will then have the same base. We can then apply the one-to-one property of exponential functions.

$$2^{x+5} = 8$$
$$2^{x+5} = 2^3 \qquad \text{Rewrite 8 as } 2^3.$$
$$x + 5 = 3 \qquad \text{Apply the one-to-one property of exponential functions. If } b^r = b^s, \text{ then } r = s.$$
$$x = -2 \qquad \text{Subtract 5.}$$

The solution of the equation is $x = -2$. We could check this solution by substituting -2 for x in the original equation. This check is left to the reader. The solution set is $\{-2\}$.

EXAMPLE 2 Solve $9^{4x-1} = 27^{2x}$.

Solution

Although we cannot write 27 as a power of 9, we can rewrite both bases as powers of 3.

$$9^{4x-1} = 27^{2x}$$
$$(3^2)^{4x-1} = (3^3)^{2x} \qquad \text{Rewrite 9 as } 3^2 \text{ and 27 as } 3^3.$$
$$3^{8x-2} = 3^{6x} \qquad \text{Simplify the exponents.}$$
$$8x - 2 = 6x \qquad \text{If } b^x = b^y, \text{ then } x = y.$$
$$-2 = -2x \qquad \text{Subtract } 8x.$$
$$1 = x \qquad \text{Divide both sides by } -2.$$

The solution set is $\{1\}$.

Quick Check 1

Solve.

a) $3^{3x-2} = 81$
b) $4^{3x-1} = 32^{x+2}$

Solving Exponential Equations by Using Logarithms

Objective 2 Solve an exponential equation by using logarithms. Often, it is not possible to rewrite both sides of an exponential equation in terms of the same base, such as in the equation $3^x = 12$. We cannot rewrite 12 as a power of 3. In such a case we will apply the one-to-one property of logarithmic functions. We can see that logarithmic functions are one-to-one functions by applying the horizontal line test to the graph of any logarithmic function.

One-to-One Property of Logarithmic Functions

For any positive real number b $(b \neq 1)$ and any real numbers r and s, if $r = s$, then $\log_b r = \log_b s$. In addition, if $\log_b r = \log_b s$, then $r = s$.

This property tells us that if two positive numbers are equal, then the logarithms of those numbers are equal as well, as long as both logarithms have the same base. When solving an exponential equation like $3^x = 12$, we will first take the common logarithm (base 10) of both sides of the equation. Then we will use the power rule for logarithms to help us solve the resulting equation. We will use the common logarithm because of its availability on our calculators, although using the natural logarithm (base e) would be a good choice as well.

EXAMPLE 3 Solve $3^x = 12$. Find the exact solution and then approximate the solution to four decimal places.

Solution

We begin by taking the common logarithm of each side of the equation.

$$3^x = 12$$
$$\log 3^x = \log 12 \qquad \text{Take the common logarithm of both sides.}$$
$$x \cdot \log 3 = \log 12 \qquad \text{Use the power rule for logarithms. This moves the variable from the exponent.}$$
$$\frac{x \cdot \cancel{\log 3}}{\cancel{\log 3}} = \frac{\log 12}{\log 3} \qquad \text{Divide both sides by } \log 3.$$
$$x = \frac{\log 12}{\log 3} \qquad \text{This is the exact solution.}$$
$$x \approx 2.2619 \qquad \text{Approximate using a calculator.}$$

The solution set is $\left\{ \dfrac{\log 12}{\log 3} \right\}$. The exact solution is $x = \dfrac{\log 12}{\log 3}$, which is approximately equal to 2.2619. Substituting this value back into the original equation, we see that $3^{2.2619} \approx 12.0005$, which is approximately equal to 12.

Quick Check 2
Solve $4^{x-5} - 9 = 3$.

EXAMPLE 4 Let $f(x) = 13^{x+8} - 22$. Solve the equation $f(x) = 108$. Find the exact solution and then approximate the solution to the nearest thousandth.

Solution

We begin by setting the function equal to 108.

$$f(x) = 108 \qquad \text{Set } f(x) \text{ equal to 108.}$$
$$13^{x+8} - 22 = 108 \qquad \text{Replace } f(x) \text{ by } 13^{x+8} - 22.$$
$$13^{x+8} = 130 \qquad \text{Add 22.}$$
$$\log 13^{x+8} = \log 130 \qquad \text{Take the common logarithm of both sides.}$$
$$(x+8) \cdot \log 13 = \log 130 \qquad \text{Use the power rule for logarithms.}$$
$$\frac{(x+8) \cdot \overset{1}{\cancel{\log 13}}}{\underset{1}{\cancel{\log 13}}} = \frac{\log 130}{\log 13} \qquad \text{Divide both sides by } \log 13.$$
$$x + 8 = \frac{\log 130}{\log 13} \qquad \text{Simplify.}$$
$$x = \frac{\log 130}{\log 13} - 8 \qquad \text{Subtract 8. This is the exact solution.}$$
$$x \approx -6.102 \qquad \text{Approximate using a calculator.}$$

Quick Check 3
Let $f(x) = 9^{x+3} - 17$. Solve the equation $f(x) = -6$. Find the exact solution and then approximate the solution to the nearest thousandth.

The solution set is $\left\{ \dfrac{\log 130}{\log 13} - 8 \right\}$. The exact solution is $x = \dfrac{\log 130}{\log 13} - 8$, which is approximately equal to -6.102.

If the exponential expression in an equation has e as its base, then it is a good idea to use the natural logarithm rather than the common logarithm. This is because $\ln e^x = x$ for any real number x.

EXAMPLE 5 Solve $2 \cdot e^{x+4} + 19 = 107$. Find the exact solution and then approximate the solution to four decimal places.

Solution

Since the base is e, we will use the natural logarithm rather than the common logarithm. We begin by isolating the exponential expression.

$$2 \cdot e^{x+4} + 19 = 107$$
$$2 \cdot e^{x+4} = 88 \qquad \text{Subtract 19.}$$
$$e^{x+4} = 44 \qquad \text{Divide both sides by 2, isolating the exponential expression.}$$
$$\ln e^{x+4} = \ln 44 \qquad \text{Take the natural logarithm of both sides.}$$
$$x + 4 = \ln 44 \qquad \text{Simplify the left side of the equation.}$$
$$x = \ln 44 - 4 \qquad \text{Subtract 4. This is the exact solution. Do not subtract 4 from 44, as we must take the natural logarithm of 44 before subtracting.}$$
$$x \approx -0.2158 \qquad \text{Approximate using a calculator.}$$

Quick Check 4
Solve $e^{x+2} - 10 = 18$. Round to the nearest thousandth.

The solution set is $\{\ln 44 - 4\}$. The exact solution is $x = \ln 44 - 4$, which is approximately equal to -0.2158.

Solving Logarithmic Equations

Objective 3 Solve a logarithmic equation in which both sides have logarithms with the same base. Suppose that we have a logarithmic equation in which two logarithms with the same base are equal to each other, such as $\log_5 x = \log_5 2$. The one-to-one property of logarithmic functions can be applied to this situation, which states that for any positive real number b ($b \neq 1$) and any real numbers r and s, if $\log_b r = \log_b s$, then $r = s$. This means that for the equation $\log_5 x = \log_5 2$, x must be equal to 2.

When solving logarithmic equations, we must check our solutions for extraneous solutions. Recall that the domain for a logarithmic function is the set of positive real numbers. Any tentative solution that would require us to take the logarithm of a number that is not positive must be omitted from the solution set.

EXAMPLE ▶ 6   Solve $\ln(x^2 - 2x - 19) = \ln 5$.

Solution

Since both logarithms have the same base, we may use the one-to-one property of logarithmic functions to solve the equation.

$$\ln(x^2 - 2x - 19) = \ln 5$$

$$x^2 - 2x - 19 = 5 \qquad \text{Use the one-to-one property of logarithmic functions. The resulting equation is quadratic.}$$

$$x^2 - 2x - 24 = 0 \qquad \text{Subtract 5.}$$

$$(x - 6)(x + 4) = 0 \qquad \text{Factor.}$$

$$x = 6 \quad \text{or} \quad x = -4 \qquad \text{Set each factor equal to 0 and solve.}$$

We must check each solution in the original equation. Although one of these solutions is negative, it does not mean that it is an extraneous solution. We must determine whether a solution requires us to take a logarithm of a nonpositive number in the original equation.

Check

$x = 6$	$x = -4$
$\ln((6)^2 - 2(6) - 19) = \ln 5$	$\ln((-4)^2 - 2(-4) - 19) = \ln 5$
$\ln 5 = \ln 5$	$\ln 5 = \ln 5$

Both solutions check. The solution set is $\{6, -4\}$.

Quick Check   5 Solve $\log(x^2 + 15x - 30) = \log 2x$.

EXAMPLE ▶ 7   Solve $\log(4x - 20) = \log(x - 14)$.

Solution

Since both logarithms have the same base, we may use the one-to-one property of logarithmic functions to solve the equation.

$$\log(4x - 20) = \log(x - 14)$$

$$4x - 20 = x - 14 \qquad \text{Use the one-to-one property of logarithmic functions.}$$

$$3x - 20 = -14 \qquad \text{Subtract } x.$$

$$3x = 6 \qquad \text{Add 20.}$$

$$x = 2 \qquad \text{Divide both sides by 3.}$$

We must check this tentative solution in the original equation.

Check $(x = 2)$

$$\log(4(2) - 20) = \log((2) - 14) \qquad \text{Substitute 2 for } x.$$
$$\log(-12) = \log(-12) \qquad \text{Simplify.}$$

Since we cannot take the logarithm of a negative number, we must omit the solution $x = 2$. Since this was the only tentative solution, the equation has no solutions. The solution set is $\varnothing$.

Quick Check **6** Solve $\log_4(2x - 7) = \log_4(3x + 2)$.

Solving Logarithmic Equations by Rewriting the Equation in Exponential Form

Objective 4 **Solve a logarithmic equation by converting it to exponential form.** If we are solving an equation in which the logarithm of a variable expression is equal to a number, such as $\log_2(x - 3) = 4$, we begin by rewriting the equation in exponential form and then solve the resulting equation. As with the previous logarithmic equations, we must check our solutions.

EXAMPLE 8 Solve $\log_2(x - 3) = 4$.

Solution

We begin by rewriting the equation $\log_2(x - 3) = 4$ in exponential form, which is $2^4 = x - 3$. We then solve the resulting equation.

$$\log_2(x - 3) = 4$$
$$2^4 = x - 3 \qquad \text{Rewrite in exponential form.}$$
$$16 = x - 3 \qquad \text{Raise 2 to the fourth power.}$$
$$19 = x \qquad \text{Add 3.}$$

Quick Check **7**

Solve
$\log_7(x^2 - 3x - 3) = 1$.

The solution set is $\{19\}$. The check is left to the reader.

When solving an equation involving a natural logarithm, keep in mind that the base of this logarithm is e when rewriting the equation in exponential form.

EXAMPLE 9 Let $f(x) = \ln(x - 3) + 15$. Solve the equation $f(x) = 21$. Find the exact solution and then approximate the solution to the nearest thousandth.

Solution

We will begin by setting the function equal to 21. We then will rewrite the equation in exponential form and solve the resulting equation.

$$f(x) = 21 \qquad \text{Set } f(x) \text{ equal to 21.}$$
$$\ln(x - 3) + 15 = 21 \qquad \text{Replace } f(x) \text{ by } \ln(x - 3) + 15.$$
$$\ln(x - 3) = 6 \qquad \text{Subtract 15.}$$
$$e^6 = x - 3 \qquad \text{Convert to exponential form.}$$
$$e^6 + 3 = x \qquad \text{Add 3.}$$
$$x \approx 406.429 \qquad \text{Approximate the solution using a calculator.}$$

We now must check the tentative solution in the original equation. We will use the approximate solution.

$$f(x) = \ln(x - 3) + 15$$
$$f(406.429) = \ln(406.429 - 3) + 15 \qquad \text{Substitute 406.429 for } x.$$
$$= \ln 403.429 + 15 \qquad \text{Subtract.}$$
$$\approx 21 \qquad \text{Approximate using a calculator.}$$

This solution checks, and the solution set is $\{e^6 + 3\}$.

Quick Check **8** Let $f(x) = \log(3x - 14) + 9$. Solve the equation $f(x) = 15$.

Using the Properties of Logarithms when Solving Logarithmic Equations

Objective 5 **Use properties of logarithms to solve a logarithmic equation.**
We now know how to solve a logarithmic equation in which either two logarithms are equal to each other, $\log_b x = \log_b y$, or a logarithm is equal to a number, $\log_b x = a$. However, often we will solve a logarithmic equation that is not in one of these forms, such as $\log_3 x + \log_3(x - 6) = 3$. In such a case, we will use the properties of logarithms to rewrite the equation in the form of an equation that we already know how to solve. In other words, we need to rewrite the equation as an equation that has one logarithm equal to another logarithm with the same base, or as an equation that has a logarithm equal to a number. The product and quotient rules will be particularly helpful.

EXAMPLE 10   Solve $\log_3 x + \log_3(x - 6) = 3$.

Solution

We begin to solve this equation by using the product rule for logarithms to rewrite the left-hand side of this equation as a single logarithm. We can then rewrite the resulting equation in exponential form and solve it.

$$\log_3 x + \log_3(x - 6) = 3$$
$$\log_3(x(x - 6)) = 3 \qquad \text{Use the product rule for logarithms.}$$
$$\log_3(x^2 - 6x) = 3 \qquad \text{Multiply.}$$
$$3^3 = x^2 - 6x \qquad \text{Convert to exponential form.}$$
$$27 = x^2 - 6x \qquad \text{Raise 3 to the third power.}$$
$$0 = x^2 - 6x - 27 \qquad \text{Subtract 27.}$$
$$0 = (x - 9)(x + 3) \qquad \text{Factor.}$$
$$x = 9 \quad \text{or} \quad x = -3 \qquad \text{Set each factor equal to 0 and solve.}$$

We must check each tentative solution in the original equation. We must determine whether a solution requires us to take a logarithm of a nonpositive number.

Check

$$x = 9 \qquad\qquad\qquad x = -3$$
$$\log_3(9) + \log_3((9) - 6) = 3 \qquad \log_3(-3) + \log_3((-3) - 6) = 3$$
$$\log_3 9 + \log_3 3 = 3 \qquad\qquad \log_3(-3) + \log_3(-9) = 3$$
$$2 + 1 = 3$$
$$3 = 3$$

The solution $x = 9$ checks. Since the tentative solution $x = -3$ produces logarithms of negative numbers in the original equation, it must be omitted as a solution. The solution set is $\{9\}$.

EXAMPLE 11   Solve $\log(9x + 2) - \log(x - 7) = 1$.

Solution

We begin to solve this equation by using the quotient rule for logarithms to rewrite the left side of this equation as a single logarithm. We can then rewrite the resulting equation in exponential form and solve it.

$$\log(9x + 2) - \log(x - 7) = 1$$

$$\log\left(\frac{9x + 2}{x - 7}\right) = 1 \qquad \text{Use the quotient rule for logarithms.}$$

$$10^1 = \frac{9x + 2}{x - 7} \qquad \text{Rewrite in exponential form.}$$

$$10(x - 7) = \frac{9x + 2}{\cancel{x - 7}_1} \cdot \cancel{(x - 7)}^1 \qquad \text{Multiply both sides by } x - 7.$$

$$10x - 70 = 9x + 2 \qquad \text{Multiply.}$$

$$x - 70 = 2 \qquad \text{Subtract } 9x.$$

$$x = 72 \qquad \text{Add 70.}$$

The check of this solution is left to the reader. The solution set is $\{72\}$.

Quick Check **9** Solve.

a) $\log_2 x + \log_2(x + 4) = 5$
b) $\log_3(41x - 42) - \log_3(x - 2) = 4$

EXAMPLE 12   Solve $\ln(x - 1) + \ln(x + 5) = \ln(3x + 7)$.

Solution

We begin to solve this equation by using the product rule for logarithms to rewrite the left-hand side of this equation as a single logarithm. We can then solve the resulting equation by using the one-to-one property of logarithms.

$$\ln(x - 1) + \ln(x + 5) = \ln(3x + 7)$$

$$\ln((x - 1)(x + 5)) = \ln(3x + 7) \qquad \text{Use the product rule for logarithms.}$$

$$\ln(x^2 + 4x - 5) = \ln(3x + 7) \qquad \text{Multiply.}$$

$$x^2 + 4x - 5 = 3x + 7 \qquad \text{Use the one-to-one property of logarithmic functions.}$$

$$x^2 + x - 12 = 0 \qquad \text{Subtract } 3x \text{ and } 7.$$

$$(x + 4)(x - 3) = 0 \qquad \text{Factor.}$$

$$x = -4 \quad \text{or} \quad x = 3 \qquad \text{Set each factor equal to 0 and solve.}$$

When we substitute the tentative solution -4 for x in the original equation, it produces the following equation:

$$\ln(-5) + \ln(1) = \ln(-5)$$

Since we can only find the logarithm of positive numbers, the tentative solution $x = -4$ must be omitted. When checking the solution $x = 3$, we obtain the equation $\ln(2) + \ln(8) = \ln(16)$. This solution checks, and so the solution set is $\{3\}$.

Quick Check   **10**   Solve $\log_2(x - 3) + \log_2(x - 7) = \log_2(3x - 1)$.

Finding the Inverse Function of Exponential and Logarithmic Functions

Objective 6   **Find the inverse function for exponential and logarithmic functions.** One important characteristic of the exponential function $f(x) = b^x$ and the logarithmic function $f(x) = \log_b x$ $(b > 0, b \neq 1)$ is that they are inverse functions. Consider the function $f(x) = e^x$. We will find its inverse using the technique developed in Chapter 11.

$$f(x) = e^x$$
$$y = e^x \qquad \text{Replace } f(x) \text{ by } y.$$
$$x = e^y \qquad \text{Exchange } x \text{ and } y.$$
$$\ln x = \ln e^y \qquad \text{To solve for } y, \text{ take the natural logarithm of each side.}$$
$$\ln x = y \qquad \text{Simplify } \ln e^y.$$
$$f^{-1}(x) = \ln x \qquad \text{Replace } y \text{ by } f^{-1}(x).$$

The inverse function of $f(x) = e^x$ is $f^{-1}(x) = \ln x$.

In the examples that follow, we will learn to find the inverse function of exponential functions, as well as logarithmic functions.

EXAMPLE 13   Find the inverse function of $f(x) = e^{x-2} - 5$.

Solution

$$f(x) = e^{x-2} - 5$$
$$y = e^{x-2} - 5 \qquad \text{Replace } f(x) \text{ by } y.$$
$$x = e^{y-2} - 5 \qquad \text{Exchange } x \text{ and } y.$$
$$x + 5 = e^{y-2} \qquad \text{Add 5.}$$
$$\ln(x + 5) = \ln e^{y-2} \qquad \text{Take the natural logarithm of each side.}$$
$$\ln(x + 5) = y - 2 \qquad \text{Simplify } \ln e^{y-2}.$$
$$\ln(x + 5) + 2 = y \qquad \text{Add 2.}$$
$$f^{-1}(x) = \ln(x + 5) + 2 \qquad \text{Replace } y \text{ by } f^{-1}(x).$$

Quick Check 11
Find the inverse function of $f(x) = e^{x+6} + 9$.

The process for finding the inverse function of a logarithmic function is quite similar to finding the inverse function of an exponential function. However, rather than taking the logarithm of both sides of an equation, we will rewrite the equation in exponential form in order to solve for y.

EXAMPLE 14 Find the inverse function of $f(x) = \ln(x - 8) + 3$.

Solution

$$f(x) = \ln(x - 8) + 3$$

$y = \ln(x - 8) + 3$	Replace $f(x)$ by y.
$x = \ln(y - 8) + 3$	Exchange x and y.
$x - 3 = \ln(y - 8)$	Subtract 3.
$e^{x-3} = y - 8$	Rewrite in exponential form.
$e^{x-3} + 8 = y$	Add 8.
$f^{-1}(x) = e^{x-3} + 8$	Replace y by $f^{-1}(x)$.

Quick Check 12
Find the inverse
function of
$f(x) = \ln(x + 10) - 18$.

EXERCISES 12.4

Solve.

1. $3^x = 27$

2. $8^x = 64$

3. $5^{x-3} = 125$

4. $10^{x+7} = 1000$

5. $3^{5x-19} = 729$

6. $4^{-2x+11} = 64$

Solve. (Rewrite both sides of the equation in terms of the same base.)

7. $9^x = 27$

8. $4^x = 128$

9. $36^{x-3} = 216$

10. $16^{x+4} = 8$

11. $100^{x-6} = 1000^x$

12. $4^{x+9} = 512^x$

Solve. Round to the nearest thousandth.

13. $4^x = 17$

14. $9^x = 40$

15. $10^x = 129$

16. $e^x = 65$

17. $2^{x-5} = 27$

18. $2^{x+10} = 4295$

19. $12^{x+3} = 225$

20. $e^{x-6} = 290$

21. $3^{2x} = 68.3$

22. $5^{3x} = 4$

23. $e^{2x-3} = 37$

24. $10^{5x+17} = 29{,}345$

25. $3^{3x+8} = 600$

26. $3^{4x-5} = 52$

Solve.

27. $\ln(3x + 1) = \ln(5x - 7)$

28. $\log_3 10x = \log_3(4x + 12)$

29. $\log_8(3x + 13) = \log_8(7x - 23)$

30. $\ln(2x + 5) = \ln(6x - 1)$

31. $\log_4(4x - 1) = \log_4(5x + 3)$

32. $\log_{14}(x - 6) = \log_{14}(3x - 8)$

33. $\log_2(x^2 - 14x + 39) = \log_2 15$

34. $\log_7(x^2 + 6x - 20) = \log_7 20$

35. $\log(x^2 + 18x + 75) = \log(2x + 15)$

36. $\ln(x^2 + 8x - 30) = \ln(3x - 6)$

Solve. Round to the nearest thousandth.

37. $\log_3 x = 5$

38. $\ln x = 2$

39. $\log_5 x = -4$

40. $\log_3 x = 0$

41. $\ln(x - 4) = 6$

42. $\log(x + 11) = 4$

43. $\log_2(x + 8) - 7 = -2$

44. $\ln(x - 3) + 5 = 9$

45. $\log(x^2 + 4x + 4) = 2$

46. $\log_2(x^2 + 11x + 60) = 5$

47. $\log_{12}(x^2 - 3x - 39) = 0$

48. $\log_7(x^2 - 15) = 2$

49. $\log_4\left(\dfrac{3x + 8}{x - 6}\right) = 2$

50. $\log_2\left(\dfrac{x - 1}{9x - 4}\right) = -3$

Solve.

51. $\log(2x + 7) - \log(x - 9) = \log 3$

52. $\log_3 x + \log_3(x - 3) = \log_3 10$

53. $\log_5(x + 1) + \log_5(x + 6) = \log_5 24$

54. $\log_2(3x + 89) - \log_2(x + 8) = \log_2(x + 3)$

55. $\log_3(5x + 3) - \log_3(x - 1) = 2$

56. $\log_8(x - 3) + \log_8(x + 9) = 2$

57. $\log_2 x + \log_2(x - 4) = 5$

58. $\log(x + 15) + \log x = 3$

59. $\log_4(3x - 1) + \log_4(x + 5) = 3$

60. $\log_3 3x + \log_3(x + 6) = 4$

61. $\log_6(7x + 11) - \log_6(2x - 9) = 1$

62. $\log_2(9x + 50) - \log_2(2x + 5) = 5$

63. Let $f(x) = 3^{2x-15} + 4$. Solve $f(x) = 31$.

64. Let $f(x) = \ln(x - 6) + 8$. Solve $f(x) = 13$.
Round to the nearest thousandth.

65. Let $f(x) = \log_5(x^2 + 3x - 45) + 6$. Solve $f(x) = 8$.

66. Let $f(x) = 5^{x+2} - 7$. Solve $f(x) = 29$.
Round to the nearest thousandth.

Solve. Round to the nearest thousandth.

67. $\log_8(x^2 - 10x - 20) = \log_8(3x + 10)$

68. $4^{6x-15} = \dfrac{1}{64}$

69. $5^{x+9} = 407$

70. $\log_6(7x + 8) = 2$

71. $9^{x-4} = 32$

72. $9^{x+6} = 27$

73. $2^{7x+30} = \dfrac{1}{32}$

74. $\log_5(x^2 - 4x - 20) = 2$

75. $\log(3x - 41) = 3$

76. $\log_3(x^2 + 8x - 3) = 4$

77. $2^{6x+25} = 128$

78. $3^{x^2-x-15} = 243$

79. $7^{x-8} = 56$

80. $\log(7x - 185) = 4$

81. $\log_{12}(9x - 11) = \log_{12} 4$

82. $4^{5x-2} = 32^{x+3}$

83. $\log_2 x + \log_2(x + 4) = 5$

84. $\log_6\left(\dfrac{5x + 13}{x - 8}\right) = 1$

85. $\log_7(4x - 19) - \log_7(x + 6) = \log_7 3$

86. $\log_3(x + 1) + \log_3(x - 7) = 2$

87. $\log_8(5x - 43) + 4 = 7$

88. $5^{5x-3} = 25^{2x+11}$

89. $\log_4(x^2 + 4x + 16) = \log_4(x^2 + x - 5)$

90. $\ln(7x - 13) - 8 = -3$

91. $\log_5(x - 7) + \log_5(x + 3) = \log_5(5x + 15)$

92. $3\log_2(x - 10) - 4 = 14$

93. $\ln(2x + 5) = -2$

94. $3^{2x-11} = 49$

95. $2^{x^2+12x+42} = 64$

96. $\log(15x - 27) = \log(11x - 17)$

For the given function $f(x)$, find its inverse function $f^{-1}(x)$.

97. $f(x) = e^{x+3}$

98. $f(x) = e^x + 3$

99. $f(x) = e^{x-5} + 4$

100. $f(x) = 3^{x+4} + 6$

101. $f(x) = \ln(x - 6)$

102. $f(x) = \ln x - 6$

103. $f(x) = \ln(x + 2) + 5$

104. $f(x) = \ln(x - 8) + 3$

Answer in complete sentences.

105. Explain why we must check logarithmic equations, and not exponential equations, for extraneous roots.

106. Explain why we take the natural logarithm of both sides of an equation involving the natural base e.

107. Solve the exponential equation $25^{3x-7} = 125^{2x+9}$ in two ways: by rewriting each base in terms of 5 and by taking the common logarithm of both sides. Which method do you prefer? Clearly explain your reasoning.

108. Explain how to find the inverse of an exponential function. Explain how to find the inverse of a logarithmic function. Use examples to illustrate the process.

Study Tip **Revisited**   Practice quizzes can be effective when working with classmates. Make a practice quiz for a classmate by selecting odd-numbered problems associated with each objective in this section. Select problems for each objective that you feel are representative of the types of problems you would be expected to solve on an exam. Be sure that you understand how to do each of the problems that you have selected. Have this classmate prepare a practice quiz for you as well.

After two students trade practice quizzes, the quiz should be returned to the student who created it, and that student should grade the quiz. Do not simply mark the problems as correct or incorrect, but provide feedback on each problem. If a solution is incorrect, write down where the solution went wrong and how to find the correct solution. After you receive your practice quiz back, go over any mistakes that you made. When you understand what went wrong, select a similar problem from the exercise set and try it.

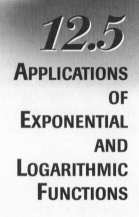

12.5

APPLICATIONS OF EXPONENTIAL AND LOGARITHMIC FUNCTIONS

Objectives

1. Solve compound interest problems.
2. Solve exponential growth problems.
3. Solve exponential decay problems.
4. Solve applications involving exponential functions.
5. Solve applications involving logarithmic functions.

In this section, we shall be solving applied problems that involve exponential and logarithmic equations and functions.

Compound Interest

Objective 1   Solve compound interest problems. Suppose that you deposit $1000 in a savings account at a bank that pays interest monthly. At the end of the first month, the bank calculates the interest earned and places it into your account. During the second month, not only does the original $1000 earn interest, but the interest earned during the first month earns interest as well. This type of interest is called **compound interest.**

Compound Interest Formula

If P dollars are deposited in an account that pays an annual interest rate r that is compounded n times per year, then the amount A in the account after t years is given by the formula

$$A = P\left(1 + \frac{r}{n}\right)^{nt}$$

The amount P is referred to as the principal. The annual interest rate r is often reported as a percent but must be converted to a decimal number for this formula.

Interest that is compounded monthly is compounded 12 times per year. Here are some other common types of compound interest:

	Quarterly	Semiannually	Annually
n	4	2	1

EXAMPLE 1   Mona is planning on taking a trip in 4 years and deposits $3000 in an account that pays 6% annual interest, compounded monthly. If she does not withdraw any money from this account, how much money will be in the account after 4 years?

Solution

It is often helpful to collect pertinent information in a table before attempting to use the equation.

Amount in the Account	A	Unknown
Principal	P	$3000
Interest Rate	r	0.06 (6%)
Number of Times Compounded/Year	n	12 (monthly)
Time (Years)	t	4

Now we can use the equation for compound interest.

Quick Check **1**

Bob deposited $400 in an account that pays 3% annual interest, compounded monthly. What will the balance of this account be in five years?

$$A = P\left(1 + \frac{r}{n}\right)^{nt}$$

$$A = 3000\left(1 + \frac{0.06}{12}\right)^{12 \cdot 4}$$ Substitute for P, r, n, and t in the equation.

$$A = 3000(1.005)^{48}$$ Simplify the expression inside parentheses.

$$A = 3811.47$$ Approximate using a calculator. Round to the nearest cent.

After 4 years, Mona's account balance will be $3811.47.

EXAMPLE 2   Albert deposited $20,000 in an account that pays 8% annual interest, compounded quarterly. How long will it take his account to reach $40,000?

Solution

Here is a table containing the pertinent information:

A	P	r	n	t
$40,000	$20,000	0.08 (8%)	4 (quarterly)	Unknown

Now we can use the equation for compound interest. In this problem, the unknown is the variable t. Since this variable is an exponent, we will need to use logarithms to solve this equation.

$$40,000 = 20,000\left(1 + \frac{0.08}{4}\right)^{4t}$$ Substitute for A, P, r, and n in the equation.

$$40,000 = 20,000(1.02)^{4t}$$ Simplify the expression inside parentheses.

$$2 = 1.02^{4t}$$ Divide both sides by 20,000.

$$\log 2 = \log 1.02^{4t}$$ Take the common logarithm of both sides.

$$\log 2 = 4t \log 1.02$$ Use the power rule for logarithms.

$$\frac{\log 2}{4 \log 1.02} = \frac{4t \log 1.02}{4 \log 1.02}$$ Divide by 4 log 1.02 to isolate t.

$$\frac{\log 2}{4 \log 1.02} = t$$ Simplify.

$$t \approx 8.75$$ Approximate using a calculator. Enter the denominator inside a set of parentheses on your calculator.

Quick Check **2**

If $25,000 is deposited in an account that pays 10% annual interest, compounded quarterly, how long will it take until there is $100,000 in the account?

Albert's account will grow to $40,000 in approximately 8.75 years.

Using Your Calculator   To approximate the expression $\dfrac{\log 2}{4 \log 1.02}$ by using the TI-83/84, it is necessary to enter the denominator in a pair of parentheses, as shown in the following screen shot:

```
log(2)/(4*log(1.
02))
        8.750697195
```

In the previous example, we were trying to determine the length of time for the balance of a bank account to double. The length of time for any quantity to double in size is called the **doubling time.**

Some banks use a method for calculating compound interest known as **continuous compounding.**

Continuous Compound Interest Formula

If *P* dollars are deposited in an account that pays an annual interest rate *r* that is compounded continuously, then the amount *A* in the account after *t* years is given by the formula

$$A = Pe^{rt}$$

EXAMPLE 3   If an account pays 12% annual interest, compounded continuously, how long will it take a deposit of $50,000 to produce an account balance of $1,000,000?

Solution

Here is a table containing the pertinent information:

Amount in the Account	*A*	$1,000,000
Principal	*P*	$50,000
Interest Rate	*r*	0.12 (12%)
Time (Years)	*t*	Unknown

Now we can use the equation for continuous compound interest, $A = Pe^{rt}$.

$$1,000,000 = 50,000e^{0.12t}$$ Substitute for *A*, *P*, and *r* in the equation.
$$20 = e^{0.12t}$$ Divide both sides by 50,000.
$$\ln 20 = \ln e^{0.12t}$$ Take the natural logarithm of both sides.
$$\ln 20 = 0.12t$$ Simplify the right side of the equation.
$$\frac{\ln 20}{0.12} = t$$ Divide both sides by 0.12 to isolate *t*.
$$t \approx 24.96$$ Approximate using a calculator.

The account balance will reach $1 million in approximately 25 years.

Quick Check 3
If $50,000 is invested in an account that pays 15% annual interest, compounded continuously, how long will it take until the balance of the account is $500,000?

Exponential Growth

Objective 2 Solve exponential growth problems. The previous example involving interest that is compounded continuously is an example of a quantity that grows exponentially.

Exponential Growth Formula

If a quantity increases at an exponential rate, then its size P at time t is given by the formula

$$P = P_0 e^{kt}$$

P_0 represents the initial size of the quantity, while k is the exponential growth rate.

Often, we will not know the exponential growth rate k, but we can determine what it is by knowing the size of the quantity at two different times.

EXAMPLE 4   The average annual tuition and fees paid to attend a public two-year college increased from \$641 in 1985 to \$1735 in 2002. If tuition and fees are growing exponentially, what will it cost to attend a public two-year college in 2012? (*Source:* The College Board)

Solution

We begin by determining the exponential growth rate k. We will treat \$641 as the initial cost P_0. Since it took 17 years for the costs to reach \$1735, we can substitute these values for t and P in the exponential growth formula. Here is a table containing the pertinent information:

P	P_0	k	t
1735	641	Unknown	17

We will now substitute into the equation $P = P_0 e^{kt}$ and solve for k.

$$1735 = 641 e^{k \cdot 17} \qquad \text{Substitute for } P, P_0, \text{ and } t.$$

$$\frac{1735}{641} = e^{k \cdot 17} \qquad \text{Divide both sides by } 641.$$

$$\ln\left(\frac{1735}{641}\right) = \ln e^{k \cdot 17} \qquad \text{Take the natural logarithm of both sides.}$$

$$\ln\left(\frac{1735}{641}\right) = k \cdot 17 \qquad \text{Simplify the right side of the equation.}$$

$$\frac{\ln\left(\dfrac{1735}{641}\right)}{17} = k \qquad \text{Divide both sides by } 17.$$

$$k \approx 0.058573 \qquad \text{Approximate using a calculator.}$$

Now that we have found the exponential growth rate k, we can turn our attention to finding the cost to attend a public two-year college in 2012. If we continue to let 641 be the initial cost (P_0), then $t = 2012 - 1985$ or 27. Here is a table containing information that is pertinent to this part of the problem:

	P	**P₀**	**k**	**t**
	Unknown	641	0.058573	27

Now we substitute the appropriate values into the formula for exponential growth to determine the cost to attend a public two-year college in 2012.

$$P = P_0 e^{kt}$$
$$P = 641 e^{0.058573 \cdot 27} \qquad \text{Substitute for } P_0, k, \text{ and } t.$$
$$P = 641 e^{1.581471} \qquad \text{Simplify the exponent.}$$
$$P \approx 3116.61 \qquad \text{Approximate using a calculator.}$$

Quick Check **4**

In 1974, a first-class U.S. postage stamp cost 10 cents. By 1994, the cost of a stamp had risen to 32 cents. If the cost of a first-class U.S. postage stamp continues to rise exponentially, how much would it cost to buy a stamp in 2020?

The cost to attend a public two-year college in 2012 will be approximately $3116.61.

Exponential Decay

Objective **3** **Solve exponential decay problems.** Just as some quantities grow exponentially, other quantities decay exponentially. The equation for exponential decay is the same as the formula for exponential growth, with the exception that k is the exponential *decay* rate. The exponential decay rate will be a negative number.

Exponential Decay Formula

If a quantity decreases at an exponential rate, then its size P at time t is given by the formula

$$P = P_0 e^{kt}$$

P_0 represents the initial size of the quantity, while k is the exponential decay rate.

The **half-life** of a radioactive element is the amount of time that it takes for half of the substance to decay into another element. For example, the half-life of carbon-14 is 5730 years. This means that after 5730 years, 50%, or half, of the carbon-14 that was present in a sample will have decayed into another element.

EXAMPLE **5** All living things contain the radioactive element carbon-14. When a living thing dies, the radioactive carbon-14 begins to change to the stable nitrogen-14. By measuring the percentage of carbon-14 that remains, scientists can determine how much time has passed since death.

If an old table is made of wood that still contains 95% of its original carbon-14, how much time has passed since the tree it was made out of died? (The half-life of carbon-14 is 5730 years.)

Solution

We will use the half-life of carbon-14 to determine the value of the exponential decay rate k. We know that after 5730 years the wood from the tree will contain 50% of its original carbon-14.

P	**P₀**	**k**	**t**
0.5 (50%)	1 (100%)	Unknown	5730

We will now substitute into the equation and solve for k.

$$0.5 = 1e^{k \cdot 5730}$$ Substitute for P, P_0, and t.

$$\ln 0.5 = \ln e^{k \cdot 5730}$$ Take the natural logarithm of both sides.

$$\ln 0.5 = k \cdot 5730$$ Simplify the right side of the equation.

$$\frac{\ln 0.5}{5730} = k$$ Divide both sides by 5730.

$$k \approx -0.000121$$ Approximate using a calculator.

Now we are able to determine how long it has been since the death of the tree. Here is a table containing information that is pertinent to this part of the problem:

P	P_0	k	t
0.95 (95%)	1 (100%)	-0.000121	Unknown

Quick Check 5

The radioactive element strontium-89 has a half-life of 50.6 days. How long would it take for 50 grams of strontium-89 to decay to 10 grams?

Now we substitute the appropriate values into the formula for exponential growth.

$$0.95 = 1e^{-0.000121t}$$ Substitute for P, P_0, and k.

$$\ln 0.95 = \ln e^{-0.000121t}$$ Take the natural logarithm of both sides.

$$\ln 0.95 = -0.000121t$$ Simplify the right side of the equation.

$$\frac{\ln 0.95}{-0.000121} = t$$ Divide both sides by -0.000121.

$$t \approx 423.9$$ Approximate using a calculator.

The tree that the table was made out of died approximately 423.9 years ago.

Applications of Other Exponential Functions

Objective 4   Solve applications involving exponential functions.

EXAMPLE 6   The number of Subway restaurants in business in a particular year can be described by the function $f(x) = 1828.65 \cdot 1.06^x$, where x represents the number of years after 1964. If this pattern of exponential growth continues, when will there be 50,000 Subway restaurants? (*Source:* Doctor's Associates, Inc.)

Solution

We will begin by setting the function equal to 50,000 and solving for x.

$$1828.65 \cdot 1.06^x = 50,000$$ Set the function equal to 50,000.

$$1.06^x = \frac{50,000}{1828.65}$$ Divide both sides by 1828.65.

$$\log 1.06^x = \log\left(\frac{50,000}{1828.65}\right)$$ Take the common logarithm of both sides.

$$x \cdot \log 1.06 = \log\left(\frac{50,000}{1828.65}\right)$$ Apply the power rule for logarithms.

$$x = \frac{\log\left(\frac{50,000}{1828.65}\right)}{\log 1.06}$$ Divide both sides by $\log 1.06$.

$$x \approx 57$$ Approximate using a calculator. Round to the nearest year.

Recall that x represents the number of years after 1964, so Subway will have 50,000 restaurants in 2021 (1964 + 57).

Quick Check **6** The average annual income of a person with a bachelor's degree can be approximated by the function $f(x) = 13400 \cdot 1.055^x$, where x represents the number of years after 1975. If this exponential growth in income continues, in what year will the average annual income of a person with a bachelor's degree be $100,000? (*Source:* U.S. Census Bureau)

Applications of Logarithmic Functions

Objective **5** Solve applications involving logarithmic functions.

pH

Every liquid can be classified as either acid, base, or neutral. Acids contain an excess of hydrogen (H^+) ions, while bases contain an excess of hydroxide (OH^-) ions. In a neutral liquid, such as pure water, the concentration of these two ions is equal. The pH scale, based on the concentration of hydrogen ions in the solution, is used to determine how acidic or basic a solution is.

pH of a Liquid

The pH of a liquid is calculated using the formula $pH = -\log[H^+]$, where $[H^+]$ is the concentration of hydrogen ions in moles per liter.

Solutions that are neutral have a pH of 7. The pH of an acid is less than 7 and the pH of a base is greater than 7. The pH of a liquid is generally between 0 and 14. The stronger an acid is, the closer its pH will be to 0. The stronger a base is, the closer its pH will be to 14.

EXAMPLE **7**   Determine the pH of ammonia, whose concentration of hydrogen ions is 3.16×10^{-12} moles/liter. Determine whether ammonia is an acid or a base.

Solution

Quick Check **7**
The concentration of hydrogen ions $[H^+]$ in vinegar is 1.58×10^{-3} moles/liter. Find the pH of vinegar. Is vinegar an acid or a base?

We substitute 3.16×10^{-12} for $[H^+]$ in the formula $pH = -\log[H^+]$ to determine the pH of ammonia.

$$pH = -\log(3.16 \times 10^{-12}) \qquad \text{Substitute } 3.16 \times 10^{-12} \text{ for } [H^+].$$
$$= -(-11.5) \qquad \text{Approximate } \log(3.16 \times 10^{-12})$$
$$\text{using a calculator.}$$
$$= 11.5 \qquad \text{Simplify.}$$

The pH of ammonia is 11.5. Ammonia is a base, as it has a pH greater than 7.

EXAMPLE **8**   The pH of the water in a swimming pool is 7.8. Find the concentration of hydrogen ions in moles per liter.

Solution

We will substitute 7.8 for pH in the formula for pH and solve for the concentration of hydrogen ions, $[H^+]$.

$$7.8 = -\log[H^+]$$ Substitute 7.8 for pH.
$$-7.8 = \log[H^+]$$ Divide by -1.
$$10^{-7.8} = [H^+]$$ Convert to exponential form.
$$[H^+] \approx 1.58 \times 10^{-8}$$ Approximate using a calculator.

The concentration of hydrogen ions in the water of the swimming pool is approximately 1.58×10^{-8} moles per liter.

Quick Check **8**
The pH of battery acid is 0.3. Find the concentration of hydrogen ions in moles per liter.

Volume of a Sound

The volume of a sound is measured in decibels (db) and depends on the intensity of the sound which is measured in watts per square meter. 85 db is considered the threshold level for safety, and prolonged exposure to noise at 85 db can cause hearing loss.

Volume of a Sound

The volume (L) of a sound in decibels is given by the formula $L = 10 \log\left(\dfrac{I}{10^{-12}}\right)$, where I is the intensity of the sound in watts per square meter. The constant 10^{-12} is the intensity of the softest sound that can be heard by humans.

EXAMPLE 9   The intensity of a sound generated by a personal stereo's headphones is 10^{-2} watts per square meter. Find the volume of this sound.

Solution

We begin by substituting 10^{-2} for I in the formula for the volume of a sound, $L = 10 \log\left(\dfrac{I}{10^{-12}}\right)$, and then we will find L.

$$L = 10 \log\left(\frac{10^{-2}}{10^{-12}}\right)$$ Substitute 10^{-2} for I.

$$L = 10 \log(10^{-2-(-12)})$$ Simplify the fraction using the rule $\dfrac{x^m}{x^n} = x^{m-n}$.

Quick Check **9**
The intensity of a sound generated by normal conversation is 10^{-6} watts per square meter. Find the volume of this sound.

$$L = 10 \log(10^{10})$$ Simplify the exponent.
$$L = 10 \cdot 10$$ Simplify the logarithm.
$$L = 100$$ Multiply.

The volume generated by a personal stereo's headphones is 100 db.

EXAMPLE 10   The threshold for pain is defined to be at a volume of 130 db. Find the intensity of a sound with this volume.

Solution

We begin by substituting 130 for L in the formula for the volume of a sound, and then we solve for I.

$$130 = 10 \log\left(\frac{I}{10^{-12}}\right) \qquad \text{Substitute 130 for } L.$$

$$13 = \log\left(\frac{I}{10^{-12}}\right) \qquad \text{Divide both sides by 10.}$$

$$\frac{I}{10^{-12}} = 10^{13} \qquad \text{Convert to exponential form.}$$

$$10^{-12} \cdot \frac{I}{10^{-12}} = 10^{-12} \cdot 10^{13} \qquad \text{Multiply by } 10^{-12} \text{ to isolate } I.$$

$$I = 10^1 \qquad \text{Simplify.}$$

Quick Check 10
The volume of a vacuum cleaner is 80 db. Find the intensity of a sound with this volume.

A sound that meets the threshold for pain has an intensity of 10 watts per square meter.

EXERCISES 12.5

FOR EXTRA HELP

MyMathLab MyMathLab

MathXL MathXL

Interactmath.com

MathXL Tutorials on CD

DVT CD Videotape

Tutor Center Addison-Wesley Math Tutor Center

Student Solutions Manual

1. Tori has deposited $5000 in a certificate of deposit (CD) account that pays 4% annual interest, compounded quarterly. How much will the balance of the account be in 3 years?

2. Nils deposited $750 in an account that pays 6% annual interest, compounded monthly. What will the balance of the account be in 2 years?

3. If Tina deposits $12,413 in an account that pays 8% annual interest, compounded semiannually, how much money will be in the account in 10 years?

4. Greg sold his baseball card collection for $3325 and deposited the money in a bank account that pays 7% interest, compounded annually. How much money will be in the account in 20 years?

5. Mark needs $5000 to take a family vacation to Florida. If he deposits $2000 in an account that pays 9% annual interest, compounded monthly, how long will it take until there is enough money in the account to pay for the vacation?

6. Kika deposited $4250 in an account that pays 5% annual interest, compounded quarterly. How long will it take until she has $7500 in the account?

7. Nomar started a bank account with an initial deposit of $240. If the bank account pays 3% annual interest, compounded monthly, how long will it take until the account's balance doubles?

8. If a bank account pays 10% annual interest, compounded annually, find the amount of time that it would take a deposit to double in value.

9. The interest on a bank account is compounded continuously, with an annual interest rate of 5.9%. If a deposit of $1000 is made, what will the balance of the account be in 5 years?

10. Anne has a bank account that pays 8% annual interest, compounded continuously. If Anne made an initial deposit of $23,000, what will the balance of the account be in 9 years?

11. How long would it take an investment of $4500 to grow to $6750 if it is invested at 7% annual interest, compounded continuously?

12. An investor has deposited $80,000 in an account that pays 5.3% annual interest, compounded continually. How long will it take until the account's balance is $100,000?

13. In the year 2000, Home Depot had 1134 stores nationwide. By 2002, this total had grown to 1532. If the number of stores continues to grow exponentially at the same rate, how many stores will there be in the year 2007? (*Source:* The Home Depot, Inc.)

14. In 1990, 7-Eleven® stores in Japan had annual sales of 9.3 billion yen. In 1999, they had annual sales of 19.6 billion yen. If sales continue to increase exponentially at this rate, what will the annual sales for 7-Eleven stores in Japan be in 2012? (*Source:* 7-Eleven, Inc.)

15. In 1974, the total world population reached 4 billion people. The total world population reached 6 billion people in 1999. If the total world population continues to grow exponentially at this rate, what will the population be in the year 2050? (*Source:* U.S. Census Bureau)

16. During its exponential growth phase, a colony of the bacteria *Bacillus megaterium* grew from 20,000 cells to 40,000 cells in 25 minutes. How many cells will be present after 60 minutes?

17. In 1999, a total of 1 billion spam email messages were sent daily worldwide. By 2002, this total had reached 5.6 billion per day. If the number of spam email messages continues to grow exponentially, when will the total reach 15 billion per day? (*Source:* IDC)

18. In 1997, the average value per acre of U.S. farm cropland was $1270. The average value had increased to $1650 by 2002. If the average value continues to grow exponentially at the same rate, when will the average value be $3000 per acre? (*Source:* National Agricultural Statistics Service)

19. The average monthly bill for U.S. cell-phone users rose from $39.43 in 1998 to $48.40 in 2002. If the average monthly bill continues to increase exponentially at this rate, when will it reach $75? (*Source:* Cellular Telecommunications and Internet Associates)

20. The average tuition and fees paid by students at private four-year universities grew from $9340 in 1990 to $16,233 in 2000. If the average tuition and fees continue to grow exponentially at this rate, when will the average tuition and fees reach $30,000? (*Source:* The College Board)

21. During its exponential growth phase, a colony of the bacterium *E. coli* grew from 5000 cells to 40,000 cells in 60 minutes. How long would it take the colony of 5000 cells to grow to 100,000 cells?

22. During its exponential growth phase, a colony of the bacterium *E. coli* grew from 15,000 cells to 60,000 cells in 40 minutes. How long would it take the colony of 15,000 cells to grow to 250,000 cells?

23. In 1993, there were 593 drive-in theaters in the United States. By the year 2001, the number of drive-in theaters had dropped to 474. If the number of drive-in theaters continues to decline exponentially, how many drive-in theaters will there be in the year 2011? (*Source:* MPAA)

24. The circulation of all daily U.S. newspapers in 1992 was 60.1 million. Total circulation had dropped to 55.1 million by the year 2002. If circulation continues to decline exponentially, what will the total circulation of all daily U.S. newspapers be in the year 2025? (*Source: USA Today*)

25. Fluorine-18 is a radioactive version of glucose, and is used in PET scanning. The half-life of fluorine-18 is 112 minutes. If 100 milligrams of fluorine-18 is injected into a patient, how much will remain after 240 minutes?

26. The half-life of bismuth-210 is 5 days. If 10 grams of bismuth-210 are present initially, how much will remain after 2 weeks?

27. If a bone contains 98% of its original carbon-14, how old is the bone? (Carbon-14 has a half-life of 5730 years.)

28. If a wooden bowl contains 85% of its original carbon-14, how old is the bowl? (Carbon-14 has a half-life of 5730 years.)

29. Calcium-41 is being used in studies testing the effectiveness of drugs for preventing osteoporosis. The half-life of calcium-41 is 100,000 years. If 5 grams of calcium-14 are present initially, how long would it take until only 1 gram remains?

30. Radon-222 has a half-life of 3.8 days. If 200 milligrams of radon-222 are present initially, how long would it take until only 10 milligrams remain?

31. In 1940, there were 1878 daily newspapers in the United States. In 2002, this number had dropped to 1457 daily newspapers. If the number of daily newspapers in the United States continues to decline exponentially at this rate, when will there be only 1000 daily newspapers in the United States? (*Source:* USA Today)

32. The number of drive-in screens in the U.S. dropped from 3561 in 1980 to only 717 in the year 2000. If the number of screens continues to decline exponentially at this rate, when will there be only 100 drive-in screens in the United States? (*Source:* MPAA)

33. The value of a car purchased in 2001 for $30,000 had dropped to $24,000 in 2002. If the value of the car continues to decrease exponentially at this rate, when will it be worth $5000?

34. The value of a car purchased in 2001 for $12,000 had dropped in value to $9720 in 2003. If the value of the car continues to decrease exponentially at this rate, when will it be worth $2000?

35. The number of Starbucks stores in business in a particular year can be described by the function $f(x) = 4.88 \cdot 1.524^x$, where x represents the number of years after 1984. Use this function to predict the number of Starbucks stores in the year 2008. (*Source:* Starbucks Corporation)

36. The number of McDonald's restaurants in business in a particular year can be described by the function $f(x) = 12436 \cdot 1.095^x$, where x represents the number of years after 1991. Use this function to predict

the number of McDonald's restaurants in the year 2011. (*Source:* McDonald's Corporation)

37. The number of deaths each year caused by AIDS in the United States has been decreasing exponentially since the mid-1990's. The number of deaths caused by AIDS in a particular year can be described by the function $f(x) = 48,460 \cdot 0.719^x$, where x represents the number of years after 1995. Use this function to predict the number of deaths caused by AIDS in the United States in the year 2012. (*Source:* CDC)

38. The number of American children younger than 13 who died of AIDS in a particular year can be described by the function $f(x) = 768 \cdot 0.645^x$, where x represents the number of years after 1994. Use this function to predict the number of deaths for children younger than 13 caused by AIDS in the year 2015. (*Source:* CDC)

39. The average premium, in dollars, of homeowner's insurance in a particular year can be described by the function $f(x) = 410.57 \cdot 1.035^x$, where x represents the number of years after 1994. Use this function to determine when the average homeowner's premium is $1000. (*Source:* National Association of Insurance Commissioners and Insurance Information Institute)

40. The amount spent, in millions of dollars, by Americans to download songs can be described by the function $f(x) = 1.4 \cdot 2.85^x$, where x represents the number of years after 2001. Use this function to determine when the amount spent will be $900 million. (*Source:* PricewaterhouseCoopers)

41. The average annual income of a person without a high school diploma can be approximated by the function $f(x) = 6830 \cdot 1.038^x$, where x represents the number of years after 1975. Use this function to predict when the average annual income of a person without a high school diploma will be $100,000. (*Source:* U.S. Census Bureau)

42. The average annual income of a person with an advanced college degree (master's degree or higher) can be approximated by the function $f(x) = 17,775 \cdot 1.059^x$, where x represents the number of years after 1975. Use this function to predict when the average annual income of a person with an advanced college degree will be $100,000. (*Source:* U.S. Census Bureau)

43. The value of a new car, in dollars, x years after its purchase can be described by the function $f(x) = 24{,}000 \cdot 0.8^x$. Use this function to determine when the value of the car will be $5000.

44. The value of a new copy machine, in dollars, x years after its purchase can be described by the function $f(x) = 3500 \cdot 0.85^x$. Use this function to determine when the value of the copier will be $2000.

45. Body temperature is often used by forensic investigators to determine the time of death for a person. For a person who was found dead in a walk-in refrigerator that maintained a constant temperature of 35° F, the body's temperature, in degrees Fahrenheit, can be approximated by the function $f(t) = 35 + 61e^{-0.0165t}$, where $t = 0$ corresponds to the time that the body was found.

a) What will the body's temperature be 6 hours after it was found?

b) How many hours before being found did the person die? Assume the person had a normal temperature of 98.6° F when alive.

46. The body temperature, in degrees Fahrenheit, of a person who was found dead in an apartment that maintained a constant temperature of 70° F can be approximated by the function $f(t) = 70 + 12e^{-0.182t}$, where $t = 0$ corresponds to the time that the body was found.

a) What will the body's temperature be 12 hours after it was found?

b) How many hours before being found did the person die?

47. The average annual salary, in dollars, of teachers in public elementary and secondary schools in a particular year can be described by the function $f(x) = 32178 \cdot 1.027^x$, where x represents the number of years after 1989. Use this function to determine when the average annual salary will reach $70,000. (*Source:* National Education Association)

48. The expenditure, in dollars, per pupil in public elementary and secondary schools in a particular year can be described by the function $f(x) = 4454 \cdot 1.038^x$, where x represents the number of years after 1988. Use this function to determine

when the expenditure per pupil will reach $10,000. (*Source:* U.S. Department of Education)

49. The concentration of hydrogen ions $[\text{H}^+]$ in bleach is 2.51×10^{-13} moles/liter. Find the pH of bleach. Is bleach an acid or a base?

50. The concentration of hydrogen ions $[\text{H}^+]$ in orange juice is 6.31×10^{-5} moles/liter. Find the pH of orange juice. Is orange juice an acid or a base?

51. The concentration of hydrogen ions $[\text{H}^+]$ in lemon juice is 5.01×10^{-3} moles/liter. Find the pH of lemon juice.

52. The concentration of hydrogen ions $[\text{H}^+]$ in seawater is 1×10^{-8} moles/liter. Find the pH of seawater.

53. The pH of boric acid is 5.0. Find the concentration of hydrogen ions in moles per liter.

54. The pH of borax is 9.3. Find the concentration of hydrogen ions in moles per liter.

55. The pH of milk of magnesia is 10.2. Find the concentration of hydrogen ions in moles per liter.

56. The pH of corn is 6.2. Find the concentration of hydrogen ions in moles per liter.

57. The intensity of a sound generated by a vacuum cleaner is 10^{-4} watts per square meter. Find the volume of this sound.

58. The intensity of a sound generated by an alarm clock is 10^{-5} watts per square meter. Find the volume of this sound.

59. The intensity of a sound generated by a leaf blower is 10^{-1} watts per square meter. Find the volume of this sound.

60. The intensity of a sound generated by an airplane taking off is 10^2 watts per square meter. Find the volume of this sound.

61. The intensity of a sound generated in a noisy restaurant is $10^{-3.5}$ watts per square meter. Find the volume of this sound.

62. The intensity of a sound generated by a firecracker is 10^3 watts per square meter. Find the volume of this sound.

63. The volume of a crying baby is 110 db. Find the intensity of a sound with this volume.

64. The volume of a refrigerator is 50 db. Find the intensity of a sound with this volume.

65. The volume of an electric drill is 95 db. Find the intensity of a sound with this volume.

66. The volume of a chainsaw is 125 db. Find the intensity of a sound with this volume.

67. The life expectancy at birth for Americans of both sexes in a particular year can be described by the function $f(x) = 11.03 + 14.29 \ln x$, where x represents the number of years after 1900. Use this function to determine the life expectancy for Americans born in the year 2025. (*Source:* National Center for Health Statistics)

68. The number of U.S. cell-phone subscribers (in millions) in a particular year can be described by the function $f(x) = 63.1 + 45.43 \ln x$, where x represents the number of years after 1997. Use this function to predict how many subscribers that there will be in the year 2009. (*Source:* Cellular Telecommunications & Internet Associates)

69. The number of public elementary and secondary teachers (in millions) in a particular year can be described by the function $f(x) = 2.648 + 0.181 \ln x$, where x represents the number of years after 1995. Use this function to predict when there will be 3.2 million teachers. (*Source:* U.S. Department of Education)

70. The percent of U.S. full-time college students that receive some financial aid in a particular year can be described by the function $f(x) = 58.8 + 6.7 \ln x$, where x represents the number of years after 1991. Use this function to find when 75% of all students received some sort of financial aid. (*Source:* U.S. Department of Education)

Answer in complete sentences.

71. Write a compound interest word problem whose solution is "Gretchen's balance will reach the correct amount in 17 years."

72. Write a compound interest word problem whose solution is "Lucy needs to invest $10,000."

73. Explain what the half-life of a radioactive element is.

74. Write a word problem whose solution is "The pH of the liquid is 4.3."

Study Tip **REVISITED** Practice quizzes can be effective when working with a group of students. Here is a plan for incorporating a practice quiz into a group study session. Have each student bring 1 question to the session. (Although using odd exercises is a good idea, feel free to use even exercises or even make up a problem of your own based on problems you find in the book.) Once the session begins, randomly select a student to do each problem for the group. The student should explain each step of the solution as he or she works through it. The remainder of the group should act as coaches, using resources such as their notes or text as needed. If you are not the student solving the problem, ask any questions that you may have. At the end of the session, the group should create a practice quiz based on these problems. Each student should take this practice quiz at home and check back with the other group members if he or she has difficulty solving the problems.

12.6

**GRAPHING
EXPONENTIAL
AND
LOGARITHMIC
FUNCTIONS**

Objectives

1 **Graph exponential functions.**
2 **Graph logarithmic functions.**

In this section, we will learn how to graph logarithmic and exponential functions in greater detail than we did in Sections 12.1 and 12.2. In particular, now that we know how to solve exponential and logarithmic equations, we will be able to find the coordinates of the x-intercept if the intercept exists.

Graphing Exponential Functions

Objective **1** **Graph exponential functions.** We will focus on graphing exponential functions of the form $f(x) = b^{x-h} + k$, where b is a positive real number ($b \neq 1$) and h and k are real numbers. For a function of this form, the horizontal asymptote is the line $y = k$. For example, the graph of $f(x) = 2^{x+4} + 3$ has a horizontal asymptote of $y = 3$, while the graph of $f(x) = e^{x-7} - 1$ has a horizontal asymptote of $y = -1$. It helps to place the horizontal asymptote on the graph before plotting any points that are on the graph of the function, as it shows us how the function behaves.

The choice of values for x is important. We will begin by using $x = h$, which is the value of x for which the exponent is equal to 0. We will also choose two values that are less than this value and two other values that are greater. We will evaluate the function for these values of x to determine their corresponding y-coordinates.

We will finish by finding the coordinates of the y- and x-intercepts. To find the y-intercept we will evaluate $f(0)$. The coordinates of the y-intercept are $(0, f(0))$. To find the x-intercept, if one exists, we set $f(x)$ equal to 0 and solve this equation for x. This equation will be an exponential equation, and we will need to use logarithms to solve it.

Graphing Exponential Functions of the Form $f(x) = b^{x-h} + k$

1. **Graph the horizontal asymptote.** Using a dashed line, graph the line $y = k$.
2. **Create a table of values.** In addition to using the value $x = h$, use two values of x that are less than h and two values that are greater than h. Evaluate the function for all five values of x to determine the coordinates of ordered pairs on the graph.
3. **Find the y-intercept.** Evaluate $f(0)$. This is the y-coordinate of the y-intercept.
4. **Find the x-intercept if one exists.** Set the function $f(x)$ equal to 0 and solve for x. Solving this equation for x will often require the use of logarithms. Recall that we can only take the logarithms of positive numbers. If there is a solution, then this value is the x-coordinate of the x-intercept.

EXAMPLE 1   Graph $f(x) = 3^{x+2} - 7$. Label any intercepts and state the domain and range of the function.

Solution

We begin by finding the horizontal asymptote, which is the line $y = -7$. Notice that $x = -2$ is the value for which the exponent is equal to 0. In addition to using this value for x, we will use two values that are less than -2 (-3 and -4) and two values that are greater than -2 (-1 and 0). Here is a table of values. Again, you should verify these function values.

x	-4	-3	-2	-1	0
$f(x) = 3^{x+2} - 7$	$3^{-2} - 7 = -6\frac{8}{9}$	$3^{-1} - 7 = -6\frac{2}{3}$	$3^0 - 7 = -6$	$3^1 - 7 = -4$	$3^2 - 7 = 2$

The points $\left(-4, -6\frac{8}{9}\right)$, $\left(-3, -6\frac{2}{3}\right)$, $(-2, -6)$, $(-1, -4)$, and $(0, 2)$ are on the graph of $f(x)$. Notice that we have already found the y-intercept, which is at $(0, 2)$. Also notice that the graph of this function must have an x-intercept, since the horizontal asymptote is below the x-axis. To find the x-coordinate of the x-intercept, we set the function $f(x)$ equal to 0 and solve for x.

$$3^{x+2} - 7 = 0 \qquad \text{Set the function equal to 0.}$$
$$3^{x+2} = 7 \qquad \text{Add 7.}$$
$$\log 3^{x+2} = \log 7 \qquad \text{Take the common logarithm of each side.}$$
$$(x + 2) \cdot \log 3 = \log 7 \qquad \text{Apply the power rule for logarithms.}$$
$$\frac{(x + 2) \cdot \overset{1}{\cancel{\log 3}}}{\underset{1}{\cancel{\log 3}}} = \frac{\log 7}{\log 3} \qquad \text{Divide both sides by } \log 3.$$
$$x = \frac{\log 7}{\log 3} - 2 \qquad \text{Subtract 2.}$$
$$x \approx -0.2 \qquad \text{Approximate using a calculator.}$$

The x-intercept is approximately at $(-0.2, 0)$. Below is the graph of the function, including the x-intercept. The domain of this function is the set of all real numbers $(-\infty, \infty)$. The range is $(-7, \infty)$.

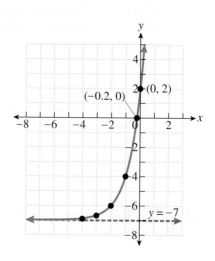

Using Your Calculator   We can use the TI-83/84 to graph $f(x) = 3^{x+2} - 7$, as well as its horizontal asymptote, $y = -7$. Begin by pushing the $\boxed{Y=}$ key. Next to Y_1, type $3^{x+2} - 7$. Type the number -7 next to Y_2. Press the $\boxed{\text{GRAPH}}$ key to display the graph.

<table>
<tr><td align="center">**Enter Expressions**</td><td align="center">**Graph**</td></tr>
<tr><td></td><td>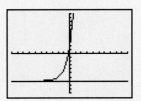</td></tr>
</table>

Quick Check   **1** Graph $f(x) = 4^{x-2} - 6$. Label any intercepts and state the domain and range of the function.

EXAMPLE **2**   Graph $f(x) = e^{x+1} - 15$. Label any intercepts and state the domain and range of the function.

Solution

We begin by finding the horizontal asymptote, which is the line $y = -15$.

Notice that $x = -1$ is the value for which the exponent is equal to 0. In addition to using this value for x, we will use two values that are less than -1 (-2 and -3) and two values that are greater than -1 (0 and 1). Here is a table of values:

x	-3	-2	-1	0	1
$f(x) = e^{x+1} - 15$	$e^{-2} - 15 \approx -14.9$	$e^{-1} - 15 \approx -14.6$	$e^{0} - 15 = -14$	$e^{1} - 15 \approx -12.3$	$e^{2} - 15 \approx -7.6$

Using approximate values, the points $(-3, -14.9)$, $(-2, -14.6)$, $(-1, -14)$, $(0, -12.3)$, and $(1, -7.6)$ are on the graph of $f(x)$. Notice that we have already found the y-intercept, which is approximately at $(0, -12.3)$. To find the x-coordinate of the x-intercept, we set the function $f(x)$ equal to 0 and solve for x.

$$e^{x+1} - 15 = 0 \qquad \text{Set the function equal to 0.}$$
$$e^{x+1} = 15 \qquad \text{Add 15.}$$
$$\ln e^{x+1} = \ln 15 \qquad \text{Take the natural logarithm of both sides.}$$
$$x + 1 = \ln 15 \qquad \text{Simplify } \ln e^{x+1}.$$
$$x = \ln 15 - 1 \qquad \text{Subtract 1.}$$
$$x \approx 1.7 \qquad \text{Approximate using a calculator.}$$

The x-intercept is approximately at $(1.7, 0)$.

Below is the graph of the function, including the x-intercept. The domain of this function is the set of all real numbers $(-\infty, \infty)$. The range is $(-15, \infty)$.

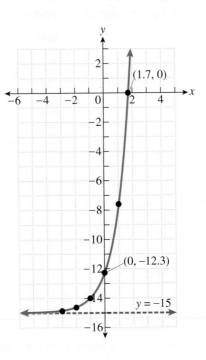

Quick Check 2 Graph $f(x) = e^{x-3} - 9$. Label any intercepts and state the domain and range of the function.

Graphing Logarithmic Functions

Objective 2 Graph logarithmic functions. We now turn our attention to graphing logarithmic functions of the form $f(x) = \log_b(x - h) + k$, where b is a positive real number $(b \neq 1)$ and h and k are real numbers. Recall that the graph of a logarithmic function has a vertical asymptote and that for a function of this form, the vertical asymptote is the line $x = h$. This is the value for which the argument of the logarithm is equal to 0. For example, the graph of $f(x) = \log_5(x - 8) + 12$ has a vertical asymptote of $x = 8$, while the graph of $f(x) = \ln(x + 2) - 9$ has a vertical asymptote of $x = -2$. Just as with exponential functions, it helps to place the asymptote on the graph before plotting any points that are on the graph of the function, as it shows us how the function behaves.

The domain of a logarithmic function, unlike an exponential function, is not the set of real numbers $(-\infty, \infty)$. To find the domain of a logarithmic function, we need to find the values of x for which the argument of the logarithm is positive. In other words, we can set the argument of the logarithm greater than 0 and solve the inequality for x. The range of a logarithmic function is the set of real numbers $(-\infty, \infty)$.

We will plot points to help us graph logarithmic functions. However, we will not choose values for x. We will first replace $f(x)$ by y, and then rewrite this equation in exponential form. In general, the equation $y = \log_b(x - h) + k$ will be rewritten as $b^{y-k} + h = x$. We will then choose values for y and find the corresponding values of x. The choice of values for y is important. We will begin by using $y = k$, which is the value of y for which the exponent is equal to 0. We will also choose two values of y that are less than this value and two other values of y that are greater.

We will finish by finding the coordinates of the y- and x-intercepts. To find the y-intercept, if it exists, we will evaluate the original function $f(x)$ at $x = 0$. Note that $x = 0$ may not be in the domain of the function. In such a case, the graph will not have a y-intercept. If a y-intercept does exist, the coordinates of the y-intercept are $(0, f(0))$.

To find the x-intercept, we will substitute 0 for y in the equation that is in exponential form and solve this equation for x.

Graphing Logarithmic Functions of the Form $f(x) = \log_b(x - h) + k$

1. **Graph the vertical asymptote.** Using a dashed line, graph the line $x = h$.
2. **Convert the function to exponential form as $b^{y-k} + h = x$.**
3. **Create a table of values.** We will choose values for y in order to find ordered pairs that are on the graph of the function. In addition to using the value $y = k$, use two values of y that are less than k and two values that are greater than k. Substitute all five values for y in the equation to determine the corresponding x-coordinates of the ordered pairs.
4. **Find the y-intercept if one exists.** Evaluate the original function for $x = 0$, if possible. Zero may not be in the domain of the function, and in this case there is no y-intercept. Otherwise, $f(0)$ is the y-coordinate of the y-intercept.
5. **Find the x-intercept.** Substitute 0 for y in the equation that is in exponential form and simplify for x. This is the x-coordinate of the x-intercept.

EXAMPLE 3   Graph $f(x) = \log_2(x - 4) + 1$. Label any intercepts and state the domain and range of the function.

Solution

We begin by finding the vertical asymptote, which is the line $x = 4$. This is the value of x for which the argument of the logarithm is equal to 0. Next we rewrite this function as an equation that is in exponential form.

$$
\begin{array}{ll}
y = \log_2(x - 4) + 1 & \text{Replace } f(x) \text{ by } y. \\
y - 1 = \log_2(x - 4) & \text{Subtract 1 to isolate the logarithm.} \\
2^{y-1} = x - 4 & \text{Rewrite in exponential form.} \\
2^{y-1} + 4 = x & \text{Add 4 to isolate } x.
\end{array}
$$

Notice that $y = 1$ is the value for which the exponent is equal to 0. In addition to using this value for y, we will use two values that are less than 1 (-1 and 0) and two values that are greater than 1 (2 and 3). Here is a table of values. It is crucial that we remember that the values we have chosen are values of y, and this is different than the way we usually graph functions by choosing values of x.

$x = 2^{y-1} + 4$	$2^{-2} + 4 = 4\frac{1}{4}$	$2^{-1} + 4 = 4\frac{1}{2}$	$2^{0} + 4 = 5$	$2^{1} + 4 = 6$	$2^{2} + 4 = 8$
y	-1	0	1	2	3

The points $\left(4\frac{1}{4}, -1\right)$, $\left(4\frac{1}{2}, 0\right)$, $(5, 1)$, $(6, 2)$, and $(8, 3)$ are on the graph of $f(x)$. Notice that we have already found the x-intercept, which is at $(4.5, 0)$. Notice also that the graph of this function cannot have a y-intercept, since the vertical asymptote is to the right of the y-axis. At right is the graph of the function, as well as the vertical asymptote $x = 4$. The domain of this function is $(4, \infty)$. The range is the set of all real numbers $(-\infty, \infty)$.

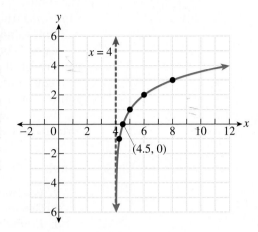

Using Your Calculator   We can use the TI-83/84 to graph the logarithmic function $f(x) = \log_2(x - 4) + 1$. Begin by pushing the $\boxed{\text{Y=}}$ key. Next to Y_1, we will enter the function.

We must use the change-of-base formula to rewrite $\log_2(x - 4) + 1$ as $\dfrac{\log(x - 4)}{\log 2} + 1$

in order to graph this function on our calculator. Press the $\boxed{\text{GRAPH}}$ key to display the graph. Using the standard window, your calculator screen should look like this (note that this function has a vertical asymptote at $x = 4$):

Enter Expressions

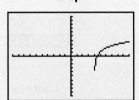

Graph

Quick Check   **3** Graph $f(x) = \log_3(x + 2) - 1$. Label any intercepts and state the domain and range of the function.

EXAMPLE ▶ 4   Graph $f(x) = \ln(x + 2) + 3$. Label any intercepts and state the domain and range of the function.

Solution

We begin by finding the vertical asymptote, which is the line $x = -2$. Next we convert this function to an equation that is in exponential form.

$$
\begin{array}{ll}
y = \ln(x + 2) + 3 & \text{Replace } f(x) \text{ by } y. \\
y - 3 = \ln(x + 2) & \text{Subtract 3 to isolate the logarithm.} \\
e^{y-3} = x + 2 & \text{Rewrite in exponential form.} \\
e^{y-3} - 2 = x & \text{Subtract 2 to isolate } x.
\end{array}
$$

The exponent is equal to 0 when $y = 3$. In addition to using this value for y, we will use 1, 2, 4, and 5. Here is a table of values:

$x = e^{y-3} - 2$	$e^{-2} - 2 \approx -1.9$	$e^{-1} - 2 \approx -1.6$	$e^{0} - 2 = -1$	$e^{1} - 2 \approx 0.7$	$e^{2} - 2 \approx 5.4$
y	1	2	3	4	5

The points that we will plot, $(-1.9, 1)$, $(-1.6, 2)$, $(-1, 3)$, $(0.7, 4)$, and $(5.4, 5)$ are on the graph of $f(x)$. The graph of this function does have x- and y-intercepts that must be found. We will begin with the y-intercept, which can be found by evaluating the original function $f(x) = \ln(x + 2) + 3$ at $x = 0$.

$$
\begin{array}{ll}
f(0) = \ln(0 + 2) + 3 & \text{Substitute 0 for } x. \\
\approx 3.7 & \text{Approximate using a calculator.}
\end{array}
$$

The y-intercept is approximately at $(0, 3.7)$. To find the x-intercept, we substitute 0 for y in the equation $e^{y-3} - 2 = x$.

$$
\begin{array}{ll}
x = e^{y-3} - 2 & \\
x = e^{0-3} - 2 & \text{Substitute 0 for } y. \\
\approx -1.95 & \text{Approximate using a calculator.}
\end{array}
$$

The x-intercept is at approximately $(-1.95, 0)$.

Below is the graph of the function. The domain of this function is $(-2, \infty)$. The range is the set of all real numbers $(-\infty, \infty)$.

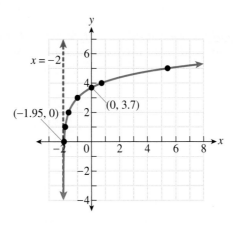

Quick Check **4** Graph $f(x) = \ln(x - 6) - 2$. Label any intercepts and state the domain and range of the function.

EXERCISES 12.6 >

Find all intercepts for the given function. Find exact values, and then round to the nearest tenth if necessary.

1. $f(x) = 2^{x+4} + 6$

2. $f(x) = 3^{x+4} - 9$

3. $f(x) = 6^{x+2} - 21$

4. $f(x) = e^{x+4} + 1$

5. $f(x) = e^{x+3} - 10$

6. $f(x) = e^{x-2} - 76$

7. $f(x) = \log_2(x - 5) - 3$

8. $f(x) = \log_3(x + 1) - 3$

9. $f(x) = \ln(x + 3) - 4$

10. $f(x) = \ln(x + 8) - 10$

Determine the equation of the horizontal asymptote for the graph of this function and state the domain and range of this function.

11. $f(x) = 3^{x-5} + 8$

12. $f(x) = 6^{x+2} - 10$

13. $f(x) = 4^{x-12} - 9$

14. $f(x) = 5^{x-6}$

Determine the equation of the vertical asymptote for the graph of this function and state the domain and range of this function.

15. $f(x) = \log_7(x - 5) - 9$

16. $f(x) = \log_8(x + 4) + 4$

17. $f(x) = \log(2x - 1) + 16$

18. $f(x) = \log_6 x - 12$

Graph. Label any intercepts and asymptotes. State the domain and range of the function.

19. $f(x) = 2^{x+2} + 4$

20. $f(x) = 3^{x+3} - 3$ **21.** $f(x) = 3^{x-1} - 15$ **24.** $f(x) = e^{x+2} - 8$

25. $f(x) = \log(x - 2) - 1$

22. $f(x) = 7^x - 30$ **23.** $f(x) = e^{x+4} + 6$

26. $f(x) = \log_3(x - 1) + 1$

27. $f(x) = \log_5(x + 2) - 2$

28. $f(x) = \log_3(x + 8) + 2$

32. $f(x) = \ln(x - 2) - 1$

$f(x) = \log_3(x + 4) - 3$

29. $f(x) = \ln(x - 1) - 3$

33.

34. $f(x) = 3^{x-3} - 6$

30. $f(x) = \ln(x + 4) + 5$ **31.** $f(x) = e^{x+3} - 2$

35. $f(x) = \ln(x + 6)$

36. $f(x) = 4^{x+2} + 4$

37. $f(x) = 2^{x-1} - 8$ **38.** $f(x) = \log_3(x + 10) + 1$

42.

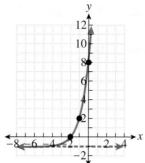

43.

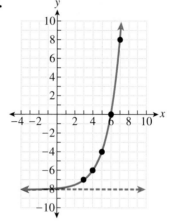

39. $f(x) = \left(\dfrac{1}{3}\right)^x - 3$ **40.** $f(x) = e^{x-4} + 13$

Given the graph of a function $f(x)$, graph the function $f^{-1}(x)$.

41.

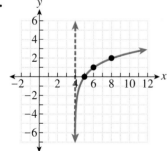

44.

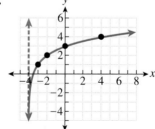

For the given function $f(x)$, find and graph $f^{-1}(x)$. Label all intercepts for $f^{-1}(x)$, as well as its asymptote.

45. $f(x) = \log_2 x$

46. $f(x) = \log_2(x - 4) + 1$

47. $f(x) = 2^{x-4} - 8$

48. $f(x) = e^{x-3} + 2$

Answer in complete sentences.

49. Explain how to graph an exponential function, including how to find any intercepts and the horizontal asymptote. Use an example to illustrate the process.

50. Explain how to graph a logarithmic function, including how to find any intercepts and the vertical asymptote. Use an example to illustrate the process.

Study Tip **REVISITED**   Make your own practice quiz for the second half of this chapter by selecting odd-numbered problems from the exercise sets of the last three sections. Select five odd problems from each section that you feel are representative of the types of problems that you would be expected to solve on an exam. Be sure to choose problems of varying levels of difficulty. By choosing odd problems from the exercise set, you will have an answer key for your practice quiz in the back of the text.

Chapter 12 Summary

Section 12.1—Topic	Chapter Review Exercises
Evaluating Exponential Functions	1–2

Section 12.2—Topic	Chapter Review Exercises
Simplifying Logarithmic Expressions	3–6
Evaluating Logarithmic Functions	7–8
Converting Exponential Equations to Logarithmic Form	9–10
Converting Logarithmic Equations to Exponential Form	11–12

Section 12.3—Topic	Chapter Review Exercises
Rewriting Logarithmic Expressions Using a Single Logarithm	13–16
Rewriting a Logarithm In Terms of Two Or More Logarithms	17–20
Applying the Change-of-Base Formula	21–24

Section 12.4—Topic	Chapter Review Exercises
Solving Exponential Equations	25–32
Solving Logarithmic Equations	33–40
Solving Equations Involving Functions	41–42
Finding Inverse Functions of Exponential and Logarithmic Functions	43–46

Section 12.5—Topic	Chapter Review Exercises
Solving Applied Problems	47–54

Section 12.6—Topic	Chapter Review Exercises
Graphing Exponential Functions	55–57
Graphing Logarithmic Functions	58–60

Summary of Chapter 12 Study Tips

The chapter test at the end of this chapter can be treated as a practice quiz. Take this chapter test as if it were an actual exam, without using your notes, homework, or text. Do this at least two days prior to the actual exam, if possible. After you check your answers, you will have a good idea as to which types of problems you understand and which types of problems you will need to practice more. At this point, review your notes and homework regarding these problems, and read the appropriate sections of the text again. Now you can create focused practice quizzes on these topics by choosing problems from the chapter review. If you are still having trouble at this point, review the material again and visit your instructor or a tutorial center to have your questions answered. When you feel that you do understand these topics, create another practice quiz by choosing problems from the appropriate sections to confirm that you are ready to go for the test.

Evaluate the given function. Round to the nearest thousandth. [12.1]

1. $f(x) = 4^x - 9$, $f(5)$

2. $f(x) = e^{x-8} - 13$, $f(11)$

Simplify. [12.2]

3. $\log_6 \dfrac{1}{36}$ 4. $\log_5 3125$

5. $\log_9 1$ 6. $\ln e^{12}$

Evaluate the given function. Round to the nearest thousandth. [12.2]

7. $f(x) = \ln(x - 9) + 20$, $f(52)$

8. $f(x) = \log(x + 23) - 6$, $f(1729)$

Rewrite in logarithmic form. [12.2]

9. $4^x = 1024$

10. $e^x = 20$

Rewrite in exponential form. [12.2]

11. $\log x = 3.2$

12. $\ln x - 3 = 8$

Rewrite as a single logarithmic expression. Simplify if possible. Assume all variables represent positive real numbers. [12.3]

13. $\log_2 13 + \log_2 8$

14. $2 \log a - 5 \log b$

15. $4 \log_2 x - 3 \log_2 y + 9 \log_2 z$

16. $-5 \log_b 3 - 2 \log_b 2$

Rewrite as the sum or difference of logarithmic expressions whose arguments have an exponent of 1. Simplify, if possible. Assume all variables represent positive real numbers. [12.3]

17. $\log_8 64x$

18. $\log_3 \left(\dfrac{n}{27} \right)$

19. $\log_b \left(\dfrac{a^6 c^3}{b^2} \right)$

20. $\log_2 \left(\dfrac{8x^9}{yz^7} \right)$

Evaluate using the change-of-base formula. Round to the nearest thousandth. [12.3]

21. $\log_4 25$

22. $\log_{1/2} 295$

23. $\log_2 1597$

24. $\log_{37} 6{,}403{,}200$

Solve. Round to the nearest thousandth. [12.4]

25. $10^x = 1000$

26. $8^{x-5} = 64$

27. $16^x = 32$

28. $9^{x+6} = 27^x$

29. $7^x = 11$

30. $6^{x+4} = 3000$

31. $e^{x-8} = 62$

32. $5^{2x-9} = 67$

Solve. Round to the nearest thousandth. [12.4]

33. $\log_4(x + 3) = \log_4 17$

34. $\log(x^2 + 12x + 6) = \log(3x + 16)$

35. $\ln x = 5$

36. $\log_3(x^2 - 13x + 57) - 1 = 2$

37. $\log_9 x + \log_9(x + 8) = \log_9 20$

38. $\ln(x + 10) - \ln(x - 6) = \ln 5$

39. $\log_3(x + 5) + \log_3(x - 1) = 3$

40. $\log x + \log(x + 3) + 5 = 6$

41. Let $f(x) = 2^{3x-7} + 8$. Solve $f(x) = 13$. Round to the nearest thousandth.

42. Let $f(x) = \ln x + 4$. Solve $f(x) = 8$. Round to the nearest thousandth.

For the given function $f(x)$, find its inverse function $f^{-1}(x)$. [12.4]

43. $f(x) = e^x + 4$

44. $f(x) = 2^{x-5} - 1$

45. $f(x) = \ln(x + 8)$

46. $f(x) = \log_3(x + 6) - 10$

47. Grady has deposited $3500 in an account that pays 8% annual interest, compounded quarterly. How

long will it take until the balance of the account is $10,000? [12.5]

48. The population of a city increased from 200,000 in 1992 to 250,000 in 2002. If the population continues to increase exponentially at the same rate, what will the population be in 2015? [12.5]

49. The half-life of bismuth-210 is 5 days. If 25 grams of bismuth-210 are present initially, how much will remain after 10 days? [12.5]

50. If researchers find a wooden arrow that contains 98% of its original carbon-14, how old is the arrow? (Carbon-14 has a half-life of 5730 years.) [12.5]

51. During the 1990s, the percentage of American adults who were obese increased exponentially. For any particular year, the percentage of adult Americans who were obese can be approximated by the function $f(x) = 11.5 \cdot 1.057^x$, where x represents the number of years after 1990. Use this function to predict when the percentage of Americans who will be obese reaches 30%. (*Source:* CDC)

52. The body temperature, in degrees Fahrenheit, of a person who was found dead in an apartment that maintained a constant temperature of 75° F can be approximated by the function $f(t) = 75 + 12e^{-0.206t}$, where $t = 0$ corresponds to the time that the body was found. What was the body's temperature 6 hours after it was found? [12.5]

53. If an earthquake creates a shock wave that is 20,000 times the smallest measurable shock wave that is recordable by a seismograph, find the magnitude of the earthquake on the Richter scale. [12.5]

54. The pH of a particular bottle of wine is 3.5. Find the concentration of hydrogen ions in moles per liter. [12.5]

Graph. Label any intercepts and asymptotes. State the domain and range of the function. [12.6]

55. $f(x) = 3^x + 2$

$f(x) = 2^{x+2} - 8$

56.

57. $f(x) = e^{x+1} - 6$

58. $f(x) = \log_2(x - 5)$

59. $f(x) = \log_3(x + 9) - 2$

60. $f(x) = \ln(x + 8) + 1$

Evaluate the given function. Round to the nearest thousandth.

1. $f(x) = e^{x-4} - 10, f(6)$

2. $f(x) = \ln(x + 7) + 4, f(11)$

Simplify.

3. $\log_{12} 144$

4. $\ln e^{-3}$

Rewrite as a single logarithmic expression. Simplify if possible. Assume all variables represent positive real numbers.

5. $\log_4 3136 - \log_4 49$

6. $2 \log_b x + 5 \log_b y - 3 \log_b z$

Rewrite as the sum or difference of logarithmic expressions whose arguments have an exponent of 1. Simplify, if possible. Assume all variables represent positive real numbers.

7. $\ln\left(\dfrac{a^4 b^6}{c^8 d}\right)$

Evaluate using the change-of-base formula. Round to the nearest thousandth.

8. $\log_9 62$

Solve. Round to the nearest thousandth.

9. $6^{x+4} = \dfrac{1}{36}$

10. $4^{x-9} = 76$

11. $e^{x+2} - 6 = 490$

Solve. Round to the nearest thousandth.

12. $\log_3(2x - 11) = \log_3 5$

13. $\log_5(x + 65) + 5 = 8$

14. $\log_2 x + \log_2(x - 4) = 5$

15. For the function $f(x) = e^{x+9} - 17$, find its inverse function $f^{-1}(x)$.

16. Sonia started a bank account with an initial deposit of $4200. The bank account pays 6% interest, compounded monthly. How long will it take for the deposit to grow to $6000?

17. A painting that was purchased for $15,000 in 1992 was valued at $25,000 in 2002. If the value of the painting keeps rising exponentially at the same rate, how much will it be worth in 2025?

18. If researchers find a wooden chalice that contains 90% of its original carbon-14, how old is the chalice? (Carbon-14 has a half-life of 5730 years.)

Graph. Label any intercepts and asymptotes. State the domain and range of the function.

19. $f(x) = e^{x-2} + 4$

20. $f(x) = \log_3(x - 5) + 3$

Mathematicians in History
Albert Einstein

*A*lbert Einstein was named as *Time* magazine's "Man of the Century" for the 20th century. That is quite an accomplishment for a man who once failed an exam that would have allowed him to study to be an electrical engineer. In 1901, working as a temporary high school math teacher, Einstein had given up the ambition to go to a university. However, while working in Switzerland's patent office, Einstein earned a doctorate in physics from the University of Zurich in 1908. Einstein introduced the theory of relativity, and became one of the most popular scientists ever. Einstein once said that compound interest was the most powerful force in the universe and that it is the greatest mathematical discovery of all time.

Write a one-page summary (*or* make a poster) of the life of Albert Einstein and his accomplishments. Also, look up Einstein's most famous quotes and list your five favorite quotes.

Interesting issues:

- Where and when was Albert Einstein born?
- What famous formula is credited to Einstein?
- Einstein married Mileva Maric in 1903. What became of their sons Hans Albert and Eduard?
- After divorcing Mileva in 1919, Einstein married his second wife Elsa. How did he know Elsa?
- When did Einstein win the Nobel prize, and what did he win it for?
- Einstein came to Princeton University in 1932, planning to teach part of the year at Princeton and the remainder of the year in Berlin. What event prohibited Einstein from returning to Berlin in 1933?
- In 1939, Einstein sent a letter to Franklin Roosevelt. What was the subject of that letter?
- What job was Einstein offered in 1952?
- When did Einstein die, and what were the circumstances of his death?

For this activity, you will need either a bag of plain M&M's or a bag of Skittles.

Step 1: Open the bag of candy and pour it out on some flat surface. Remove any broken candies. Count the total number of candies in the bag. Record the number of candies in the table provided. Pour all the candies back into the bag.

Step 2: Pour out all of the candies on a flat surface. Separate the candies which are letter up from those which are letter down. Record the number of each in the table.

Step 3: Put only those candies that were letter down back into the bag.

Step 4: Repeat Steps 2 and 3 until only at most one candy is letter down.

Total number of candies: _____

Pours	Letter up	Letter down
1		
2		
3		
4		
5		
6		
7		

Step 5: Create an axis system with the horizontal axis representing the pour number and the vertical axis representing the number of letter up candies counted. Plot the number of letter up candies for each pour, and connect the dots using a smooth graph.

a) Is the graph increasing or decreasing? Why do you think this is the case?

b) Is the graph linear? Why or why not?

c) If the graph is not linear, what kind of function would best represent the graph you drew?

d) What kind of function is $f(n) = 2^{6-n}$? If n represents the pour number, find the values for pour 1, pour 2, etc.

e) Do your values of letter up candies for each pour come close to the number found by evaluating the function at those different n-values? If so, why do you think a base value of 2 was appropriate to use?

Step 6: Using the same bag of M&M's or Skittles, try this experiment again. Did your results vary much?

CONIC SECTIONS

*In this chapter, we will learn to graph **conic sections:** parabolas, circles, ellipses, and hyperbolas. The equations of the different conic sections are all nonlinear equations. These graphs are called conic sections, because they can be constructed by intersecting a plane with a cone, as shown in the following figure:*

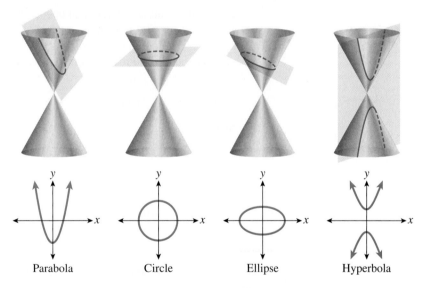

Parabola   Circle   Ellipse   Hyperbola

The chapter concludes with a section on systems of nonlinear equations.

Study Tip USING THE PREVIOUS STUDY TIPS TO PREPARE FOR A FINAL EXAM (PART 1)
The study tips in this chapter will focus on how to incorporate the study tips in previous chapters into your preparation for a final exam. These tips will be continued in Chapter 14.

13.1
PARABOLAS

1. Graph equations of the form $y = ax^2 + bx + c$.
2. Graph equations of the form $y = a(x - h)^2 + k$.
3. Graph equations of the form $x = ay^2 + by + c$.
4. Graph equations of the form $x = a(y - k)^2 + h$.
5. Find a vertex by completing the square.
6. Find the equation of a parabola that meets the given conditions.

In Chapters 10 and 11, we graphed parabolas that opened either upward or downward. These parabolas were associated with equations that could be expressed as functions of x. After reviewing these graphs, we will learn to graph parabolas that open to the right or to the left.

Graphing Equations of the Form $y = ax^2 + bx + c$

Objective 1   Graph equations of the form $y = ax^2 + bx + c$. Here is a brief summary for graphing equations of the form $y = ax^2 + bx + c$:

Graphing an Equation of the Form $y = ax^2 + bx + c$

- Determine whether the parabola opens upward or downward.
- Find the vertex of the parabola.
- Find the y-intercept of the parabola.
- After plotting the vertex and the y-intercept, determine whether there are any x-intercepts, and if there are any, find them.
- Find the axis of symmetry $\left(x = \dfrac{-b}{2a} \right)$ and use it to find the point that is symmetric to the y-intercept.

EXAMPLE 1   Graph $y = x^2 + 6x - 2$. Label the vertex, y-intercept, x-intercept(s) (if any), and the axis of symmetry.

Solution

Since a is positive, this parabola will open upward. We begin by finding the coordinates of the vertex.

$$x = \frac{-6}{2(1)} \qquad \text{Substitute 1 for } a \text{ and 6 for } b \text{ into } x = \frac{-b}{2a}.$$
$$x = -3 \qquad \text{Simplify.}$$

To find the y-coordinate, we substitute -3 for x.

$$y = (-3)^2 + 6(-3) - 2 \qquad \text{Substitute } -3 \text{ for } x.$$
$$y = -11 \qquad \text{Simplify.}$$

The vertex is $(-3, -11)$.

Now we turn our attention to the y-intercept. We substitute 0 for x in the original equation.

$$y = (0)^2 + 6(0) - 2 \qquad \text{Substitute 0 for } x.$$
$$y = -2 \qquad \text{Simplify.}$$

The y-intercept is $(0, -2)$.

Since the vertex is located below the x-axis and the parabola opens upward, the graph must cross the x-axis. To find the x-intercepts, we substitute 0 for y and solve for x.

$$0 = x^2 + 6x - 2 \qquad \text{Substitute 0 for } y. \text{ Since } \qquad \text{does not factor, we can use the quadratic formula.}$$

$$x = \frac{-6 \pm \sqrt{(6)^2 - 4(1)(-2)}}{2(1)} \qquad \text{Substitute 1 for } a, \text{ 6 for } b, \text{ and } -2 \text{ for } c.$$

$$x = \frac{-6 \pm \sqrt{44}}{2} \qquad \text{Simplify radicand.}$$

$$x = \frac{-6 \pm 2\sqrt{11}}{2} \qquad \text{Simplify the radical.}$$

$$x = \frac{\overset{1}{2}(-3 \pm \sqrt{11})}{\underset{1}{2}} \qquad \text{Factor the numerator and divide out common factors.}$$

$$x = -3 \pm \sqrt{11} \qquad \text{Simplify.}$$

Using a calculator, we find that $-3 + \sqrt{11} \approx 0.3$ and $-3 - \sqrt{11} \approx -6.3$. The x-intercepts are approximately at $(0.3, 0)$ and $(-6.3, 0)$.

Here is the graph of the given equation. The axis of symmetry, $x = -3$, has been used to find the coordinates of the point symmetric to the y-intercept, $(-6, -2)$.

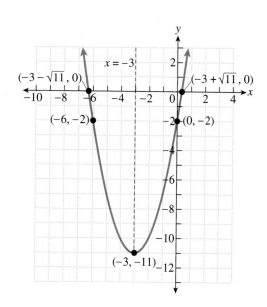

Quick Check   **1** Graph $y = -x^2 + 2x + 15$. Label the vertex, y-intercept, x-intercept(s) (if any), and the axis of symmetry.

Graphing Equations of the Form $y = a(x - h)^2 + k$

Objective 2 **Graph equations of the form $y = a(x - h)^2 + k$.** Now we turn our attention to reviewing the graphs of equations of the form $y = a(x - h)^2 + k$.

Graphing an Equation of the Form $y = a(x - h)^2 + k$

- Determine whether the parabola opens upward or downward.
- Find the vertex of the parabola, which is the point (h, k).
- Find the y-intercept of the parabola.
- After plotting the vertex and the y-intercept, find the x-intercepts, if there are any.
- Use the axis of symmetry $(x = h)$ to find the point that is symmetric to the y-intercept.

EXAMPLE 2   Graph $y = -(x - 1)^2 + 9$. Label the vertex, y-intercept, x-intercept(s) (if any), and the axis of symmetry.

Solution

Since a is negative, this parabola opens downward. The vertex of this parabola is at $(1, 9)$. We next find the y-intercept by substituting 0 for x.

$$y = -(0 - 1)^2 + 9 \qquad \text{Substitute 0 for } x.$$
$$y = 8 \qquad \text{Simplify.}$$

The y-intercept is $(0, 8)$.

Since the vertex is above the x-axis and the parabola opens downward, this parabola has two x-intercepts. To find the x-intercepts, we substitute 0 for y and solve for x. We will solve the resulting equation by extracting square roots.

$$0 = -(x - 1)^2 + 9 \qquad \text{Substitute 0 for } y.$$
$$(x - 1)^2 = 9 \qquad \text{Add } (x - 1)^2 \text{ to isolate the squared expression.}$$
$$\sqrt{(x - 1)^2} = \pm\sqrt{9} \qquad \text{Take the square root of both sides.}$$
$$x - 1 = \pm 3 \qquad \text{Simplify each square root.}$$
$$x = 1 \pm 3 \qquad \text{Add 1 to both sides to isolate } x.$$

$1 + 3 = 4$ and $1 - 3 = -2$, so the x-intercepts are $(4, 0)$ and $(-2, 0)$.

Here is the graph. The point $(2, 8)$ is symmetric to the y-intercept. It can be found using the axis of symmetry, $x = 1$.

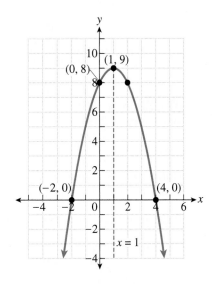

Quick Check   2   Graph $y = (x + 4)^2 + 3$. Label the vertex, y-intercept, x-intercept(s) (if any), and the axis of symmetry.

Graphing Equations of the Form $x = ay^2 + by + c$

Objective 3   Graph equations of the form $x = ay^2 + by + c$. If the variables x and y are interchanged in the equation $y = ax^2 + bx + c$, then the graph of the equation will be a parabola that opens to the right or to the left. The process for graphing this type of parabola is quite similar to graphing a parabola that opens up or down.

Graphing an Equation of the Form $x = ay^2 + by + c$

- Determine whether the parabola opens to the right or to the left. The parabola will open to the right if a is positive; to the left if a is negative.
- Find the vertex of the parabola by using the formula $y = \dfrac{-b}{2a}$. Substitute this result in the original equation for y to find the x-coordinate.

continued

- Find the *x*-intercept of the parabola by substituting 0 for *y* in the original equation.
- After plotting the vertex and the *x*-intercept, find the *y*-intercepts, if there are any, by letting *x* = 0 in the original equation and solving the resulting quadratic equation for *y*.
- Find the axis of symmetry $\left(y = \dfrac{-b}{2a} \right)$ and use it to find the point on the parabola that is symmetric to the *x*-intercept.

EXAMPLE 3 Graph $x = y^2 + 2y - 3$. Label the vertex, any intercepts, and the axis of symmetry.

Solution

Notice that the squared variable in the equation is *y*, not *x*, so the parabola will open either to the right or to the left. This parabola opens to the right since *a* is positive. We begin by finding the *y*-coordinate of the vertex, using the formula $y = \dfrac{-b}{2a}$.

$$y = \frac{-2}{2(1)} \qquad \text{Substitute 1 for } a \text{ and 2 for } b \text{ into } y = \frac{-b}{2a}.$$
$$y = -1 \qquad \text{Simplify.}$$

To find the *x*-coordinate of the vertex, we substitute -1 for *y* in the original equation.

$$x = (-1)^2 + 2(-1) - 3 \qquad \text{Substitute } -1 \text{ for } y.$$
$$x = -4 \qquad \text{Simplify.}$$

The vertex of the parabola is $(-4, -1)$.

> *A Word of Caution* Keep in mind that even though we find the *y*-coordinate first and the *x*-coordinate second, we must write the ordered pair as (x, y).

Next we find the *x*-intercept by substituting 0 for *y* in the equation.

$$x = (0)^2 + 2(0) - 3 \qquad \text{Substitute 0 for } y.$$
$$x = -3 \qquad \text{Simplify.}$$

The *x*-intercept is $(-3, 0)$.

Since the vertex is located to the left of the *y*-axis and the parabola opens to the right, the parabola has two *y*-intercepts. We can find these intercepts by substituting 0 for *x* and solve the resulting equation for *y*.

$$0 = y^2 + 2y - 3 \qquad \text{Substitute 0 for } x.$$
$$0 = (y + 3)(y - 1) \qquad \text{Factor.}$$
$$y = -3 \quad \text{or} \quad y = 1 \qquad \text{Set each factor equal to 0 and solve.}$$

The *y*-intercepts are $(0, -3)$ and $(0, 1)$.

We can find another point on the graph by finding the point on the parabola that is symmetrical to the *x*-intercept $(-3, 0)$. We can do this using the axis of symmetry, which in this case is the horizontal line $y = -1$.

Since the x-intercept is one unit above the axis of symmetry, the point symmetric to it will be one unit below the axis of symmetry at $(-3, -2)$. Here is the graph.

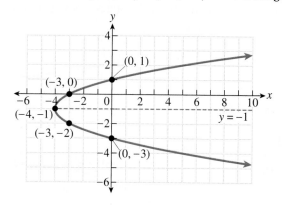

Quick Check 3 Graph $x = y^2 + 10y - 1$. Label the vertex, any intercepts, and the axis of symmetry.

If a parabola opens to the right and the vertex is to the right of the y-axis, then it does not have any y-intercepts. The same holds true for parabolas that open to the left whose vertex is to the left of the y-axis.

EXAMPLE 4   Graph $x = -y^2 + 6y - 10$. Label the vertex, any intercepts, and the axis of symmetry.

Solution

Since a is negative, the parabola opens to the left. We begin by finding the vertex.

$$y = \frac{-b}{2a}$$ $$x = -y^2 + 6y - 10$$

$$y = \frac{-6}{2(-1)}$$ Substitute for a and b. $$x = -(3)^2 + 6(3) - 10$$ Substitute 3 for y.

$$y = 3$$ Simplify. $$x = -1$$ Simplify.

The vertex of the parabola is $(-1, 3)$. Next we find the x-intercept.

$$x = -(0)^2 + 6(0) - 10$$ Substitute 0 for y.

$$x = -10$$ Simplify.

The x-intercept is $(-10, 0)$.

The vertex is located to the left of the y-axis. Since the parabola opens to the left, the parabola does not have any y-intercepts. The point $(-10, 6)$ shown on the graph is symmetric to the x-intercept and can be found by using the axis of symmetry $y = 3$.

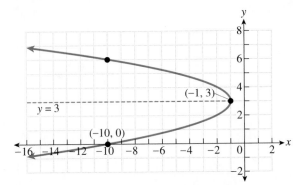

Quick Check 4

Graph
$x = -y^2 + 8y - 12$.
Label the vertex, any intercepts, and the axis of symmetry.

Graphing Equations of the Form $x = a(y - k)^2 + h$

Objective 4 **Graph equations of the form $x = a(y - k)^2 + h$.** Now we turn our attention to the graphs of equations of the form $x = a(y - k)^2 + h$. The graph of an equation in this form will be a parabola that opens to the right or to the left with a vertex at (h, k).

Graphing an Equation of the Form $x = a(y - k)^2 + h$

- Determine whether the parabola opens to the right or to the left.
- Find the vertex of the parabola, which is the point (h, k).
- Find the x-intercept of the parabola by substituting 0 for y in the original equation.
- After plotting the vertex and the x-intercept, find the y-intercepts, if there are any, by letting $x = 0$ in the original equation and solving for y. This equation can be solved by extracting square roots.
- Find the axis of symmetry $(y = k)$ and use it to find the point on the parabola that is symmetric to the x-intercept.

EXAMPLE 5 Graph $x = -(y - 1)^2 + 4$. Label the vertex, any intercepts, and the axis of symmetry.

Solution

Since a is negative, this parabola opens to the left. The vertex of this parabola is $(4, 1)$. We next find the x-intercept by substituting 0 for y.

$$x = -(0 - 1)^2 + 4 \qquad \text{Substitute 0 for } y.$$
$$x = 3 \qquad \text{Simplify.}$$

The x-intercept is $(3, 0)$.
Since the vertex is to the right of the y-axis and the parabola opens to the left, this parabola has two y-intercepts. To find the y-intercepts, we substitute 0 for x and solve for y. We will solve the resulting equation by extracting square roots.

$$0 = -(y - 1)^2 + 4 \qquad \text{Substitute 0 for } x.$$
$$(y - 1)^2 = 4 \qquad \text{Add } (y - 1)^2 \text{ to isolate the squared expression.}$$
$$\sqrt{(y - 1)^2} = \pm\sqrt{4} \qquad \text{Take the square root of both sides.}$$
$$y - 1 = \pm 2 \qquad \text{Simplify.}$$
$$y = 1 \pm 2 \qquad \text{Add 1 to both sides to isolate } y.$$

$1 + 2 = 3$ and $1 - 2 = -1$, so the y-intercepts are $(0, 3)$ and $(0, -1)$. Below is the graph. The point $(3, 2)$ can be found using the y-intercept and the axis of symmetry, which is $y = 1$.

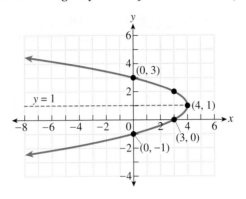

Quick Check 5 Graph $x = -(y - 4)^2 - 3$. Label the vertex, any intercepts, and the axis of symmetry.

EXAMPLE 6 Graph $x = (y - 2)^2$. Label the vertex, any intercepts, and the axis of symmetry.

Solution

Since a is positive, this parabola opens to the right. The vertex of this parabola is $(0, 2)$. To see this, we could rewrite the equation as $x = (y - 2)^2 + 0$. To find the x-intercept, we substitute 0 for y in the original equation and solve for x.

$$x = (0 - 2)^2 \qquad \text{Substitute 0 for } y.$$
$$x = 4 \qquad \text{Simplify.}$$

The x-intercept is $(4, 0)$.

Quick Check 6

Graph $x = (y - 3)^2 - 1$. Label the vertex, any intercepts, and the axis of symmetry.

Since the vertex $(0, 2)$ is located on the y-axis, the vertex is the only y-intercept. The point $(4, 4)$ can be found using the y-intercept and the axis of symmetry, which is $y = 2$.

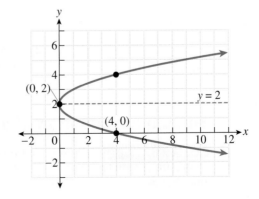

Finding the Vertex by Completing the Square

Objective 5 Find a vertex by completing the square. Consider the equation $y = x^2 - 8x + 33$. Although we could find the vertex by using the formula $x = \dfrac{-b}{2a}$, we can also find the vertex by completing the square to write the equation in the form $y = a(x - h)^2 + k$, where (h, k) is the vertex.

EXAMPLE 7 Find the vertex of the parabola defined by the equation $y = x^2 - 8x + 33$ by completing the square.

Solution:

$$y = x^2 - 8x + 33$$
$$y - 33 = x^2 - 8x \qquad \text{Subtract 33 from both sides.}$$
$$y - 33 + 16 = x^2 - 8x + 16 \qquad \left(\tfrac{-8}{2}\right)^2 = 16.\ \text{Add 16 to both sides.}$$
$$y - 17 = (x - 4)^2 \qquad \text{Combine like terms on the left side. Factor the quadratic expression on the right side.}$$
$$y = (x - 4)^2 + 17 \qquad \text{Add 17 to both sides.}$$

The vertex is $(4, 17)$.

Although it may seem easier to have used the formula $x = \dfrac{-b}{2a}$ to find the vertex in this case, the ability to complete the square will be an important skill in the sections that follow.

EXAMPLE 8 Find the vertex of the parabola defined by the equation $x = -y^2 + 4y - 50$ by completing the square.

Solution:

$$x = -y^2 + 4y - 50$$
$$x + 50 = -y^2 + 4y \qquad \text{Add 50 to both sides.}$$
$$x + 50 = -(y^2 - 4y) \qquad \text{Factor out a negative 1 on the right side.}$$
$$x + 50 - 4 = -(y^2 - 4y + 4) \qquad \left(\tfrac{-4}{2}\right)^2 = 4.\ \text{Add 4 to the expression in parentheses on the right side. Subtract 4 from the left side, because there is a factor of } -1 \text{ in front of the parentheses.}$$

$$x + 46 = -(y - 2)^2 \qquad \text{Combine like terms on the left side. Factor the quadratic expression on the right side.}$$

$$x = -(y - 2)^2 - 46 \qquad \text{Subtract 46 from both sides.}$$

The vertex is $(-46, 2)$.

> **Quick Check 7**
> Find the vertex of the parabola defined by the given equation by completing the square.
>
> a) $y = x^2 + 4x + 20$
> b) $x = -y^2 - 6y + 17$

Finding the Equation of a Parabola

Objective 6 Find the equation of a parabola that meets the given conditions. There are several techniques that can be used to find the equation of a parabola. Here is one that uses the intercepts of the parabola.

EXAMPLE ▶ 9   Find the equation of the parabola whose graph is shown at the accompanying figure.

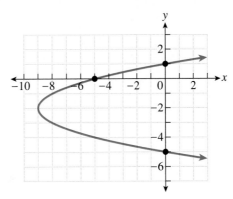

Solution:

We begin by noting that the parabola opens to the right, so its equation is of the form $x = ay^2 + by + c$, where $a > 0$. We can observe the y-intercepts are $(0, 1)$ and $(0, -5)$, and the x-intercept is $(-5, 0)$.

Since the y-coordinates of the y-intercept are $y = 1$ and $y = -5$, we know that $y - 1$ and $y + 5$ are factors of the expression $ay^2 + by + c$. Multiplying $(y - 1)(y + 5)$ yields the expression $y^2 + 4y - 5$, so the equation is of the form $x = a(y^2 + 4y - 5)$. To find the value of a, we use the coordinates of the x-intercept $(-5, 0)$. We could also use the coordinates of any other point on the graph of the parabola, including the vertex. After substituting -5 for x and 0 for y into the equation $x = a(y^2 + 4y - 5)$, we solve for a.

$$
\begin{aligned}
x &= a(y^2 + 4y - 5) \\
-5 &= a((0)^2 + 4(0) - 5) \qquad &&\text{Substitute } -5 \text{ for } x \text{ and } 0 \text{ for } y. \\
-5 &= a(-5) \qquad &&\text{Simplify.} \\
1 &= a \qquad &&\text{Divide both sides by } -5.
\end{aligned}
$$

Since $a = 1$, the equation of the parabola is $x = 1(y^2 + 4y - 5)$, or simply $x = y^2 + 4y - 5$.

Quick Check   **8**   **Find the equation of the parabola whose graph is shown at the accompanying figure.**

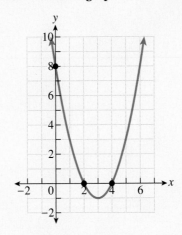

Graph the parabola. Label the vertex and any intercepts.

1. $y = x^2 + 8x + 11$ **2.** $y = x^2 - 9x$

8. $y = -x^2 + 11x - 20$

9. $y = (x + 5)^2 + 5$

3. $y = x^2 - 4x + 7$ **4.** $y = x^2 + 6x + 9$

10. $y = (x - 1)^2 - 15$ **11.** $y = -(x + 2)^2 + 9$

5. $y = -x^2 + x + 20$ **6.** $y = -x^2 - 9x - 14$

12. $y = -(x - 3)^2 - 1$ **13.** $x = y^2 + 2y - 24$

7. $y = -x^2 + 5$

14. $x = y^2 + 12y + 27$

18. $x = y^2 - 8y + 20$

19. $x = -y^2 + 8y - 15$

20. $x = -y^2 + 4y + 32$

15. $x = y^2 + 6y + 2$

16. $x = y^2 + 2y - 4$

21. $x = -y^2 - 9y - 6$

17. $x = y^2 + 3y + 10$

22. $x = -y^2 + 2y - 6$

23. $x = (y + 5)^2 - 1$

28. $x = -(y - 3)^2 + 7$

29. $x = -y^2 - 6y - 15$

24. $x = (y - 2)^2 - 5$

30. $x = -y^2 + 10y - 13$

25. $x = (y + 2)^2 + 2$

26. $x = (y - 7)^2 + 10$

31. $x = (y - 1)^2 - 7$

27. $x = -(y - 2)^2 + 4$

32. $x = y^2 - 8y - 20$

33. $y = x^2 + 7x + 4$

34. $x = -y^2 + 2y + 15$

35. $y = -(x - 1)^2 + 9$

36. $x = (y + 5)^2 - 4$

37. $y = x^2 + 5x - 24$

38. $x = -(y + 5)^2 - 3$

39. $y = (x - 4)^2 + 3$

40. $x = y^2 - 5y + 15$

Find the vertex of the parabola by completing the square.

41. $y = x^2 - 2x + 15$

42. $y = x^2 + 10x + 62$

43. $y = -x^2 + 14x + 76$

44. $y = -x^2 - 8x + 21$

45. $x = y^2 + 6y - 40$

46. $x = y^2 - 12y - 33$

47. $x = -y^2 + 4y - 51$

48. $x = -y^2 - 16y + 109$

Find the equation of the parabola that has been graphed by using the intercepts.

49.

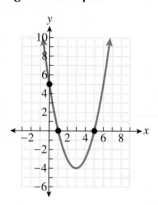

50.

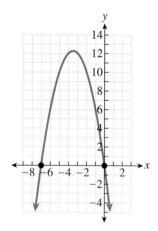

51.

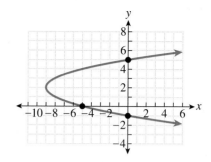

52.

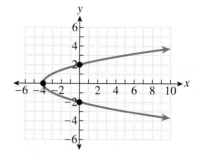

53. A footbridge in the Sequoia National Park is in the shape of a parabola. It spans a distance of 100 feet, and at its lowest point it drops 5 feet in elevation.

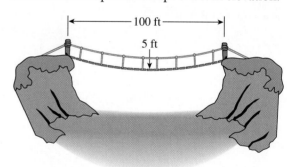

The bridge is displayed below on a rectangular coordinate plane.

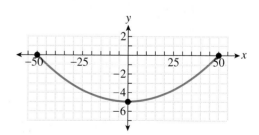

a) Find the equation for the parabola.

b) After a person walks 30 feet from one side of the bridge, how far has the person dropped in elevation?

54. A freeway overpass is built atop a parabolic arch. The width of the arch at the base is 120 feet, and its height at its peak is 30 feet.

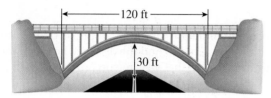

The overpass is displayed below on a rectangular coordinate plane.

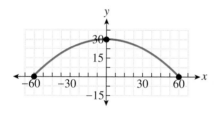

a) Find the equation for the parabola.

b) How far from the center of the arch does the height drop to below 16 feet?

Answer in complete sentences.

55. Explain how to determine whether a parabola opens to the left or to the right, as opposed to opening up or opening down.

56. Suppose that a parabola opens to the right. Explain how you would determine that the parabola has no y-intercepts.

57. Explain how to graph a parabola of the form $x = ay^2 + by + c$. Use an example to illustrate the process.

58. Explain how to complete the square to find the vertex of a parabola of the form $x = ay^2 + by + c$. Use an example to illustrate the process.

Study Tip **Revisited** To do well on a cumulative exam, such as a final exam, takes planning. Begin by determining what material will be covered on the exam and how many problems to expect. Also, you should know what part of your grade the exam represents, as well as how well you need to do in order to pass the class. If your instructor does not give you this information in class, make an appointment to visit his or her office. Preferably, the best time to begin this process is at least two weeks before the exam.

For further information on preparing for a cumulative exam, see the study tips in Chapter 7.

13.2

CIRCLES

Objectives

1 Use the distance and midpoint formulas.
2 Graph circles centered at the origin.
3 Graph circles centered at a point (h, k).
4 Find the center of a circle and its radius by completing the square.
5 Find the equation of a circle that meets the given conditions.

The Distance Formula

Objective 1 **Use the distance and midpoint formulas.** Suppose that we are given two points, (x_1, y_1) and (x_2, y_2), and are asked to find the distance between these two points. If we plot the points on a graph, we see that we can construct a right triangle whose hypotenuse has these points as its endpoints.

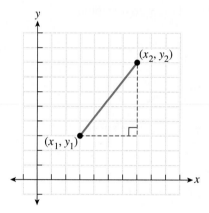

The distance between the two points, d, is the length of the hypotenuse. The length of the leg on the bottom of the triangle is $x_2 - x_1$, and the length of the leg on the right side of the triangle is $y_2 - y_1$.

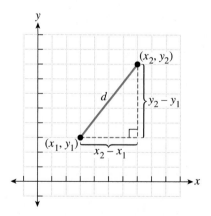

The Pythagorean theorem tells us that $d^2 = (x_2 - x_1)^2 + (y_2 - y_1)^2$. If we take the square root of both sides of this equation, we obtain a formula for the distance in terms of the coordinates of the two points, $d = \sqrt{(x_2 - x_1)^2 + (y_2 - y_1)^2}$. (We omit the negative square root because a distance is a nonnegative number.)

The Distance Formula

The distance, d, between two points (x_1, y_1) and (x_2, y_2) is given by the formula
$d = \sqrt{(x_2 - x_1)^2 + (y_2 - y_1)^2}$.

EXAMPLE 1   Find the distance between the points $(3, 9)$ and $(6, 5)$.

Solution

We will let $(3, 9)$ be the point (x_1, y_1) and $(6, 5)$ be the point (x_2, y_2). Now we can use the distance formula.

$$d = \sqrt{(x_2 - x_1)^2 + (y_2 - y_1)^2}$$
$$d = \sqrt{(6 - 3)^2 + (5 - 9)^2} \quad \text{Substitute for } x_2, x_1, y_2, \text{ and } y_1.$$
$$d = \sqrt{(3)^2 + (-4)^2} \quad \text{Simplify the expressions in the radicand.}$$
$$d = \sqrt{25} \quad \text{Simplify the radicand.}$$
$$d = 5 \quad \text{Simplify the radical.}$$

The distance between the two points is five units.

The choice of which point will be (x_1, y_1) and which point will be (x_2, y_2) is completely arbitrary. Verify that if we let $(6, 5)$ be the point (x_1, y_1) and $(3, 9)$ be the point (x_2, y_2), then d would still have been equal to 5.

EXAMPLE 2   Find the distance between the points $(2, -6)$ and $(-5, -11)$. Round to the nearest tenth.

Solution

We will let $(2, -6)$ be the point (x_1, y_1) and $(-5, -11)$ be the point (x_2, y_2).

$$d = \sqrt{(x_2 - x_1)^2 + (y_2 - y_1)^2}$$
$$d = \sqrt{(-5 - 2)^2 + (-11 - (-6))^2} \quad \text{Substitute for } x_2, x_1, y_2, \text{ and } y_1.$$
$$d = \sqrt{74} \quad \text{Simplify the radicand.}$$
$$d \approx 8.6 \quad \text{Approximate using a calculator.}$$

Quick Check 1   The distance between the two points is approximately 8.6 units.

Find the distance between the given points. Round to the nearest tenth if necessary.

a) $(1, 2)$ and $(9, 8)$
b) $(3, -6)$ and $(-1, 6)$

The Midpoint Formula

Suppose that we are given two points, (x_1, y_1) and (x_2, y_2), and are asked to find their **midpoint,** which is the point on the line segment connecting the two points that is located exactly halfway between the two points. The x-coordinate of the midpoint must be equal to the average of the x-coordinates of the two points, and the y-coordinate must be equal to the average of the y-coordinates of the two points.

The Midpoint Formula

The midpoint of a line segment connecting two points (x_1, y_1) and (x_2, y_2) is the point whose coordinates are $\left(\dfrac{x_1 + x_2}{2}, \dfrac{y_1 + y_2}{2} \right)$.

EXAMPLE 3 Find the midpoint of the line segment that connects the points $(2, -9)$ and $(6, 3)$.

Solution

We will use the midpoint formula, treating $(2, -9)$ as (x_1, y_1) and $(6, 3)$ as (x_2, y_2). This choice is arbitrary.

$$x\text{-coordinate} \qquad\qquad y\text{-coordinate}$$
$$\frac{x_1 + x_2}{2} = \frac{2 + 6}{2} \qquad\qquad \frac{y_1 + y_2}{2} = \frac{-9 + 3}{2}$$
$$= 4 \qquad\qquad\qquad\qquad = -3$$

The midpoint is $(4, -3)$.

Quick Check 2
Find the midpoint of the line segment that connects the points $(-4, -3)$ and $(5, 7)$.

Circles Centered at the Origin

Objective 2 Graph circles centered at the origin.

A **circle** is defined as the collection of all points (x, y) in a plane that are a fixed distance from a point called its **center**. The distance from the center to each point on the circle is called the **radius** of the circle.

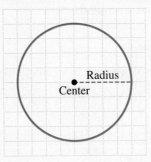

We will begin by learning to graph circles centered at the origin.

EXAMPLE 4 Graph the circle centered at the origin whose radius is 5.

Solution

We begin by plotting the center at the origin $(0, 0)$. From there we will move five units to the right of the origin and plot a point there. We repeat this process for points that are five units to the left, above, and below the center.

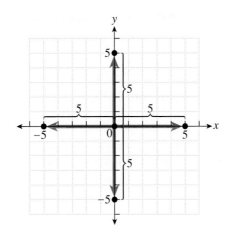

We finish by drawing the circle that passes through these points.

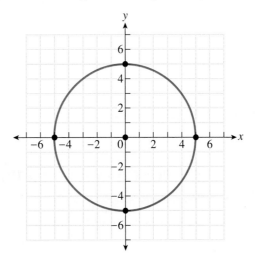

The circle in the previous example is the collection of all points (x, y) that are a distance of five units from the origin $(0, 0)$. We can use this fact and the distance formula to find the equation for this circle.

$$d = \sqrt{(x_2 - x_1)^2 + (y_2 - y_1)^2}$$
$$5 = \sqrt{(x - 0)^2 + (y - 0)^2}$$
$$5 = \sqrt{x^2 + y^2}$$
$$25 = x^2 + y^2$$

Substitute 5 for d, x for x_2, 0 for x_1, y for y_2, and 0 for y_1.
Simplify.
Square both sides.

The equation of the circle centered at the origin with a radius of 5 can be written as $x^2 + y^2 = 25$.

Equation of a Circle Centered at the Origin (Standard Form)

The equation for a circle with radius r centered at the origin is $x^2 + y^2 = r^2$.

EXAMPLE 5   Graph the circle $x^2 + y^2 = 10$. State the center and radius.

Solution

This circle also has its center at the origin, and since $r^2 = 10$, the radius r is equal to $\sqrt{10}$. Note that $\sqrt{10} \approx 3.2$.

We begin to graph the circle by plotting the center $(0, 0)$. Next, we plot the points that are approximately 3.2 units to the right and left of the center, as well as the points that are approximately 3.2 units above and below the center. Once these points have been plotted, draw the circle that passes through them.

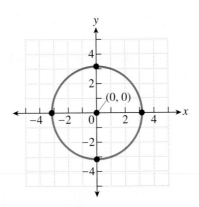

Quick Check 3

Graph the circle $x^2 + y^2 = 12$. State the center and radius.

Circles Centered at a Point Other than the Origin

Objective 3 Graph circles centered at a point (h, k). The equation for a circle centered at a point (h, k) can be derived by using the distance formula, just as the equation for a circle centered at the origin was derived.

Equation of a Circle (Standard Form)

> The equation for a circle with radius r centered at the point (h, k) is
> $(x - h)^2 + (y - k)^2 = r^2$.

EXAMPLE 6   Graph the circle $(x - 3)^2 + (y - 4)^2 = 4$. State the center and radius.

Solution

Since the equation of the circle is of the form $(x - h)^2 + (y - k)^2 = r^2$, the center of this circle is the point $(3, 4)$.

Since $r^2 = 4$, the radius r is equal to $\sqrt{4}$, or 2.

We begin by plotting the center $(3, 4)$. Next we plot the points that are 2 units to the right and left of the center, as well as the points that are 2 units above and

Quick Check **4**

Graph the circle
$(x - 6)^2 + (y - 5)^2 = 9$.
State the center and
radius.

below the center. Once these points have been plotted, draw the circle that passes through them.

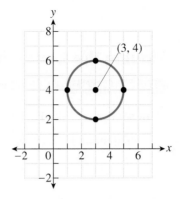

EXAMPLE 7 Graph the circle $(x + 3)^2 + (y - 6)^2 = 25$. State the center and radius.

Solution

The center of this circle is the point $(-3, 6)$. To find the x-coordinate of the center, we can rewrite $x + 3$ as $x - (-3)$. This shows us that h is -3. Alternatively, we could have set the expression containing x that was being squared, $x + 3$, equal to 0 and solved for x. A similar approach will yield the y-coordinate of the center.

Since $r^2 = 25$, the radius r is equal to $\sqrt{25}$, or 5.

We begin to graph the circle by plotting the center $(-3, 6)$. Next, we plot the points that are five units to the right and left of the center, as well as the points that are five units above and below the center. Once these points have been plotted, draw the circle that passes through them.

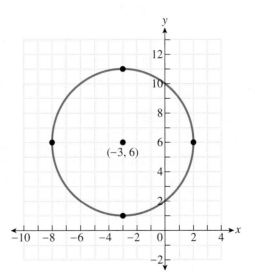

Quick Check **5** Graph the circle. State the center and radius.

a) $(x - 1)^2 + (y + 6)^2 = 16$ b) $(x + 2)^2 + (y - 3)^2 = 64$

Finding the Center and Radius by Completing the Square

Objective **4** **Find the center of a circle and its radius by completing the square.**

General Form of the Equation of a Circle

The **general form** of the equation of a circle is $Ax^2 + Ay^2 + Bx + Cy + D = 0$, $A \neq 0$.

To graph a circle in this form, we must rewrite the equation in the standard form, which will allow us to determine the center and radius of the circle. This is done by completing the square, which must be done for both x and y.

EXAMPLE 8   Graph the circle $x^2 + y^2 - 8x + 10y + 40 = 0$. State the center and radius, and find any intercepts.

Solution

We will rewrite the equation in standard form by completing the square for x and y.

$$x^2 + y^2 - 8x + 10y + 40 = 0$$
$$(x^2 - 8x) + (y^2 + 10y) = -40$$

Collect terms containing x. Collect terms containing y. Subtract the constant to the right side.

$$(x^2 - 8x + 16) + (y^2 + 10y + 25) = -40 + 16 + 25$$

Add 16 to both sides to complete the square for x. Add 25 to both sides to complete the square for y.

$$(x - 4)^2 + (y + 5)^2 = 1$$

Factor the two quadratic expressions on the left side of the equation.

Since the equation of this circle has been written in standard form, the center of this circle is the point $(4, -5)$. Since $r^2 = 1$, the radius r is equal to $\sqrt{1}$, or 1. We begin to graph the circle by plotting the center $(4, -5)$. Next we plot the points that are one unit to the right and left of the center, as well as the points that are one unit above and below the center. We finish by drawing the circle that passes through these four points.

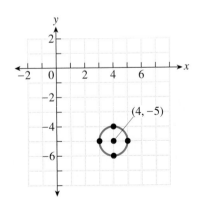

Quick Check 6 Graph the circle $x^2 + y^2 + 14x - 6y + 57 = 0$. State the center and radius.

Finding the Equation of a Circle Whose Center and Radius Are Known

Objective 5 Find the equation of a circle that meets the given conditions.

EXAMPLE 9   Find the equation of the circle.

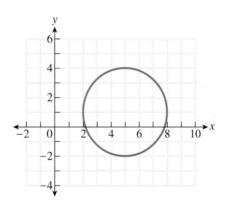

Solution

We begin by finding the center of the circle, which is the point $(5, 1)$. Next, we measure the distance from the center of the circle to a point on the circle. This distance is the radius. In this case the radius is 3.

We can now write the equation in standard form.

$$(x - h)^2 + (y - k)^2 = r^2$$
$$(x - 5)^2 + (y - 1)^2 = 3^2 \qquad \text{Substitute 5 for } h, 1 \text{ for } k, \text{ and 3 for } r.$$
$$(x - 5)^2 + (y - 1)^2 = 9 \qquad \text{Simplify.}$$

The equation of this circle is $(x - 5)^2 + (y - 1)^2 = 9$.

Quick Check   7 Find the equation of the circle.

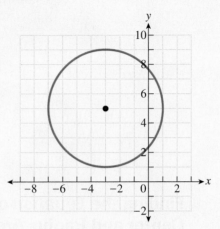

The **diameter** of a circle is a line segment that has both endpoints on the circle and passes through the center of the circle. The length of the diameter of a circle is equal to twice the length of the radius of the circle.

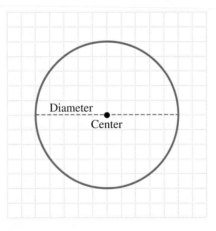

EXAMPLE 10   Find the equation in standard form of the circle that has a diameter with endpoints $(-1, -5)$ and $(7, 1)$.

Solution

We begin by finding the center of the circle, which is the midpoint of the given endpoints of a diameter.

$$x = \frac{x_1 + x_2}{2} \qquad y = \frac{y_1 + y_2}{2}$$
$$x = \frac{-1 + 7}{2} \qquad y = \frac{-5 + 1}{2}$$
$$= 3 \qquad\qquad = -2$$

The center of the circle is $(3, -2)$. Next, we find the radius by calculating the distance from the center of the circle to a point on the circle. We will use the point $(7, 1)$.

$$d = \sqrt{(x_2 - x_1)^2 + (y_2 - y_1)^2}$$
$$d = \sqrt{(7 - 3)^2 + (1 - (-2))^2} \qquad \text{Substitute for } x_2, x_1, y_2, \text{ and } y_1.$$
$$d = \sqrt{25} \qquad\qquad\qquad \text{Simplify the radicand.}$$
$$d = 5 \qquad\qquad\qquad\quad \text{Simplify.}$$

Quick Check **8**

Find the equation in general form of the circle that has a diameter with endpoints $(4, -7)$ and $(-2, 3)$.

The radius is 5. We can now write the equation in standard form.

$$(x - h)^2 + (y - k)^2 = r^2$$
$$(x - 3)^2 + (y - (-2))^2 = 5^2 \qquad \text{Substitute 3 for } h, -2 \text{ for } k, \text{ and 5 for } r.$$
$$(x - 3)^2 + (y + 2)^2 = 25$$

The equation of this circle is $(x - 3)^2 + (y + 2)^2 = 25$.

EXERCISES *13.2*

Find the distance between the given points. Round to the nearest tenth if necessary.

1. $(7, 6)$ and $(-1, -9)$

2. $(-5, -2)$ and $(4, 10)$

3. $(15, -53)$ and $(54, 27)$

4. $(-32, -15)$ and $(-77, 13)$

5. $(5, 2)$ and $(-8, 2)$

6. $(-4, -9)$ and $(-4, 9)$

7. $(7, 4)$ and $(10, 1)$

8. $(1, 2)$ and $(3, 8)$

9. $(-6, -5)$ and $(-1, -9)$

10. $(-2, 12)$ and $(4, -3)$

Find the midpoint of the line segment that connects the given points.

11. $(0, 0)$ and $(6, 4)$

12. $(3, 15)$ and $(7, 1)$

13. $(6, -9)$ and $(8, -25)$

14. $(-5, 10)$ and $(-12, 32)$

Graph the circle. State the center and radius of the circle.

15. $x^2 + y^2 = 1$

16. $x^2 + y^2 = 4$

17. $x^2 + y^2 = 16$

18. $x^2 + y^2 = 9$

19. $x^2 + y^2 = 100$

20. $x^2 + y^2 = 144$

21. $x^2 + y^2 = 6$

22. $x^2 + y^2 = 18$

23. $(x - 5)^2 + (y - 9)^2 = 16$

24. $(x - 7)^2 + (y - 4)^2 = 9$

25. $(x - 3)^2 + (y - 2)^2 = 4$

26. $(x - 1)^2 + (y - 6)^2 = 1$

27. $(x + 4)^2 + (y - 5)^2 = 9$

28. $(x - 7)^2 + (y + 3)^2 = 4$

29. $(x + 2)^2 + (y + 1)^2 = 49$

30. $(x + 6)^2 + (y + 4)^2 = 25$

31. $(x + 5)^2 + y^2 = 25$

32. $x^2 + (y - 9)^2 = 36$

33. $(x - 2)^2 + (y + 6)^2 = 18$

34. $(x + 8)^2 + (y + 3)^2 = 8$

35. $(x + 3)^2 + (y - 7)^2 = 45$

36. $(x - 1)^2 + (y - 10)^2 = 48$

Find the center and radius of the circle by completing the square.

37. $x^2 + y^2 + 4x + 12y - 41 = 0$

38. $x^2 + y^2 + 6x + 18y + 65 = 0$

39. $x^2 + y^2 - 10x + 2y - 23 = 0$

40. $x^2 + y^2 - 8x - 40y + 127 = 0$

41. $x^2 + y^2 + 16x - 57 = 0$

42. $x^2 + y^2 - 10y - 11 = 0$

43. $x^2 + y^2 + 4x + 14y - 27 = 0$

44. $x^2 + y^2 - 8x + 6y + 14 = 0$

Find the equation of the circle with the given center and radius.

45. Center $(0, 0)$, radius 6

46. Center $(0, 0)$, radius 9

47. Center $(4, 2)$, radius 7

48. Center $(1, 8)$, radius 4

49. Center $(-9, -10)$, radius 5

50. Center $(-6, 0)$, radius 2

51. Center $(0, -8)$, radius $\sqrt{15}$

52. Center $(-2, -13)$, radius $\sqrt{37}$

Find the equation of the circle.

53.

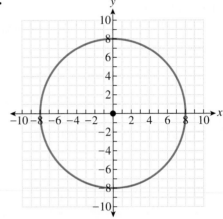

54.

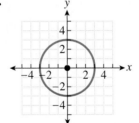

55.

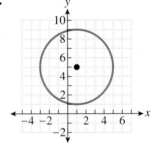

56.

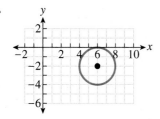

57.

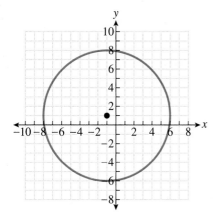

58.

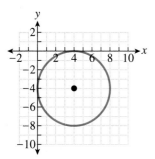

59.

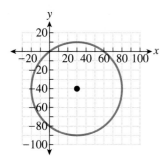

60.

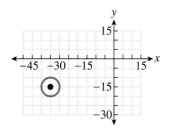

Find the equation of the circle that has a diameter with the given endpoints.

61. $(1, 7)$ and $(9, 7)$

62. $(-6, 3)$ and $(4, 3)$

63. $(-2, 10)$ and $(6, 4)$

64. $(8, -17)$ and $(-8, 13)$

65. $(4, 1)$ and $(12, -3)$

66. $(-3, 9)$ and $(-15, -11)$

Graph. For graphs that are parabolas, label the vertex and any intercepts. For graphs that are circles, label the center.

67. $x = -(y - 3)^2 + 8$

68. $(x + 4)^2 + (y - 6)^2 = 25$

69. $x^2 + y^2 + 18x - 6y - 10 = 0$

70. $y = -(x - 5)^2 + 9$

74. $x = -y^2 + 8y - 19$

75. $y = (x + 1)^2 + 6$

71. $x = y^2 - 4y - 4$

72. $y = x^2 + 6x - 27$

76. $x = (y + 3)^2 - 4$

77. $x^2 = -(y - 7)^2 + 25$

73. $x^2 + y^2 = 20$

78. $y = -x^2 + 4x + 1$

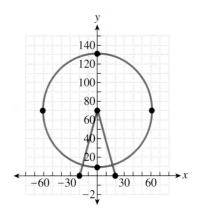

79. Bart has built a rectangular barbecue area that measures 20 feet by 50 feet. He plans to draw a circle around the barbecue area, as shown below, and fill in this area with grass.

50 ft

20 ft

Barbecue Area

a) Find the equation of the circle if we treat the point in the center of the barbecue area as the origin.

b) How many square feet of sod need to be ordered to cover the grass area?

80. The London Eye is a gigantic wheel built to celebrate the millennium. The diameter of the wheel is 122 meters, and its height is 131 meters.

The London Eye can be represented on a rectangular coordinate plane as shown at top right.

a) Find an equation for the circle.

b) How far to the left or right of center is a car that is 100 feet above the ground?

Answer in complete sentences.

81. Explain how to determine whether the graph of an equation will be a parabola that opens up, a parabola that opens down, a parabola that opens to the right, a parabola that opens to the left, or a circle. Give an example of each type of equation.

82. Explain how to determine the center and radius of a circle from its equation if it is in standard form.

83. Explain how to find the center and radius of a circle by completing the square. Use an example to illustrate the process.

84. Explain how to graph a circle. Use an example to illustrate the process.

> *Study Tip* **Revisited**   The end of the semester is a hectic time for students. The same can be said for employees of your college as well. Talk to your instructor and find out when he or she will be available on campus for questions in the period of time leading up to the final exam. Also, check with the tutorial center or math lab to find out if there will be any change in either of their hours of operation. If you ask early enough, but there are changes, then you will be able to make adjustments to your schedule to have access to help if you need it.
>
> For more advice on making effective use of all your resources, consult the study tips at the end of Chapter 3.

Objectives

1. Graph ellipses centered at the origin.
2. Graph ellipses centered at a point (h, k).
3. Find the center of an ellipse and the lengths of its axes by completing the square.
4. Find the equation of an ellipse that meets the given conditions.

In this section, we will investigate a conic section called an ellipse. An **ellipse** is the collection of all points (x, y) in the plane for which the sum of the distances d_1 and d_2 between the point (x, y) and two fixed points, F_1 and F_2, called **foci,** is a positive constant. Each fixed point is called a **focus.** The graph is similar to a circle but more oval in its shape. For any point (x, y) on the ellipse, the sum of the distances d_1 and d_2 remains the same.

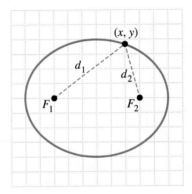

In this text, the foci will be either on a horizontal line or on a vertical line. An ellipse has a **center** like a circle, and it is the point located midway between the two foci.

A horizontal line passing through the center of an ellipse intersects the ellipse at two points, as does a vertical line passing through the center. The horizontal and vertical line segments connecting these points are called the **axes** of the ellipse. The longer line segment is called the **major axis,** and its endpoints are the **vertices** of the ellipse. The other line segment is called the **minor axis,** and its endpoints are the **co-vertices** of the ellipse.

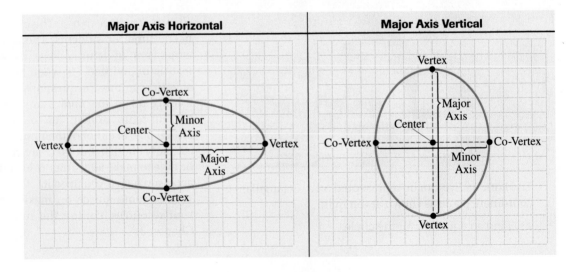

Ellipses Centered at the Origin

Objective **1** **Graph ellipses centered at the origin.**

Standard Form of the Equation of an Ellipse Centered at the Origin

> The equation of an ellipse that is centered at the origin is of the form $\dfrac{x^2}{a^2} + \dfrac{y^2}{b^2} = 1$.
>
> This is the **standard form** of the equation of an ellipse.

The ellipse will have x-intercepts at $(a, 0)$ and $(-a, 0)$ and y-intercepts at $(0, b)$ and $(0, -b)$. If $a > b$, then the major axis is horizontal and the minor axis is vertical. If $b > a$, then the major axis is vertical and the minor axis is horizontal.

To graph an ellipse centered at the origin, we begin by plotting a point at the origin. We then plot points that are a units to the left and right of the origin, as well as points that are b units above and below the origin.

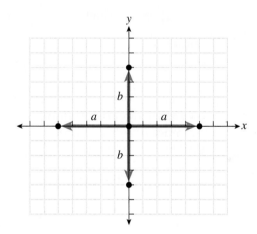

We finish by drawing the ellipse that passes through these four points.

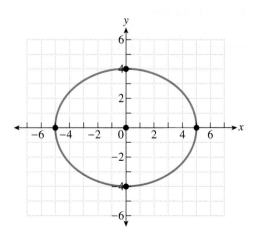

A Word of Caution   When drawing the ellipse that passes through the four points, be sure that your graph looks like an oval and not like a diamond.

EXAMPLE 1   Graph the ellipse $\dfrac{x^2}{9} + \dfrac{y^2}{4} = 1$.

Solution

This ellipse is centered at the origin, so we plot a point there.

Since $a^2 = 9$, we know that $a = \sqrt{9}$ or 3. Moving three units to the left and right of the origin, we see that the x-intercepts are at $(-3, 0)$ and $(3, 0)$.

Since $b^2 = 4$, we know that $b = \sqrt{4}$ or 2. Moving two units above and below the origin, we see that the y-intercepts are at $(0, 2)$ and $(0, -2)$.

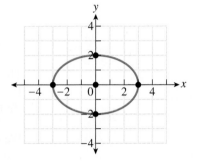

EXAMPLE 2   Graph $\dfrac{x^2}{4} + \dfrac{y^2}{36} = 1$.

Solution

This ellipse is centered at the origin, so we plot a point there.

Since $a^2 = 4$, we know that $a = \sqrt{4}$ or 2. Thus, the x-intercepts are at $(2, 0)$ and $(-2, 0)$.

Since $b^2 = 36$, we know that $b = \sqrt{36}$, or 6. It follows that the y-intercepts are at $(0, 6)$ and $(0, -6)$.

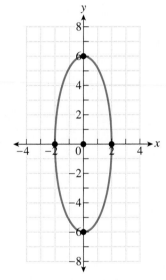

Quick Check **1** Graph.

a) $\dfrac{x^2}{25} + \dfrac{y^2}{9} = 1$

b) $\dfrac{x^2}{9} + \dfrac{y^2}{16} = 1$

In the next example, we will need to rewrite the equation in standard form before graphing the ellipse.

EXAMPLE 3   Graph $25x^2 + 16y^2 = 400$.

Solution

For the equation of an ellipse to be in standard form, the sum of the variable terms must be equal to 1. To rewrite this equation in standard form, we divide both sides of the equation by 400 and simplify.

$$25x^2 + 16y^2 = 400$$

$$\frac{25x^2 + 16y^2}{400} = \frac{400}{400}$$ Divide both sides by 400.

$$\frac{\overset{1}{25}x^2}{\underset{16}{400}} + \frac{\overset{1}{16}y^2}{\underset{25}{400}} = 1$$ Rewrite the left side of the equation as the sum of two fractions and simplify.

$$\frac{x^2}{16} + \frac{y^2}{25} = 1$$ Simplify.

The center of this ellipse is the origin. Since $a^2 = 16$, we know that $a = 4$. The x-intercepts are at $(4, 0)$ and $(-4, 0)$. Since $b^2 = 25$, we know that $b = 5$ and that the y-intercepts are at $(0, 5)$ and $(0, -5)$.

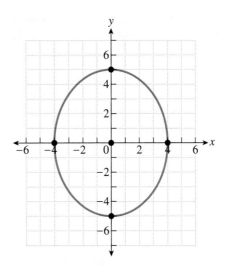

Quick Check **2** Graph $4x^2 + 36y^2 = 144$.

Ellipses Centered at a Point Other than the Origin

Objective 2 Graph ellipses centered at a point (h, k).

Standard Form of the Equation of an Ellipse with Center (h, k)

> The equation for an ellipse centered at the point (h, k) whose horizontal axis has length $2a$ and whose vertical axis has length $2b$ is $\dfrac{(x - h)^2}{a^2} + \dfrac{(y - k)^2}{b^2} = 1.$

To graph an ellipse centered at the point (h, k), we begin by plotting a point at the center. The endpoints of the horizontal axis can be found by moving a units to the left and right of the center. The endpoints of the vertical axis can be found by moving b units above and below the center. Once these four endpoints have been plotted, we draw the ellipse that passes through them.

EXAMPLE 4   Graph $\dfrac{(x - 3)^2}{4} + \dfrac{(y - 5)^2}{9} = 1.$

Solution

Since the equation has the form $\dfrac{(x - h)^2}{a^2} + \dfrac{(y - k)^2}{b^2} = 1$, the center of this ellipse is $(3, 5)$.

Since $a^2 = 4$, we know that $a = 2$. The endpoints of the horizontal axis are two units to the left and right of the center. Since $b^2 = 9$, we know that $b = 3$. The endpoints of the vertical axis are three units above and below the center. We finish by drawing the ellipse that passes through these four endpoints.

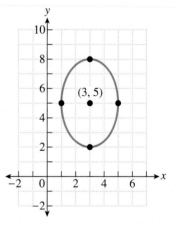

Quick Check 3 Graph $\dfrac{(x - 8)^2}{49} + \dfrac{(y - 7)^2}{25} = 1.$

EXAMPLE 5   Graph $(x + 3)^2 + \dfrac{(y + 2)^2}{10} = 1.$

Solution

The center of this ellipse is $(-3, -2)$.

When we do not see a denominator under the squared term containing x, $a^2 = 1$; so $a = 1$, and we plot the endpoints of the horizontal axis 1 unit to the left and right of the center. Since $b^2 = 10$, we know that $b = \sqrt{10}$. Since $\sqrt{10} \approx 3.2$, we plot the endpoints of the vertical axis approximately 3.2 units above and below the center. We finish by drawing the ellipse that passes through the four endpoints.

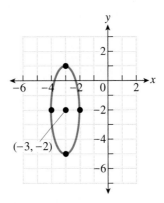

$(-3, -2)$

Quick Check **4**

Graph
$$\frac{(x - 4)^2}{20} + (y + 5)^2 = 1.$$
Find any intercepts.

Finding the Center of an Ellipse and the Lengths of Its Axes by Completing the Square

Objective **3**   Find the center of an ellipse and the lengths of its axes by completing the square.

General Form of the Equation of an Ellipse

The **general form** of the equation of an ellipse is $Ax^2 + By^2 + Cx + Dy + E = 0$, $A \neq 0$, $B \neq 0$, and $A \neq B$.

Notice that the coefficient of the x^2 term is not equal to the coefficient of the y^2 term. If those two coefficients are equal, then the graph of the equation is more specifically defined as a circle than as an ellipse. To graph an ellipse in general form, we must rewrite the equation in standard form. This will allow us to determine the center, as well as a and b. This is done by completing the square, which must be done for both x and y.

EXAMPLE **6**   Graph the ellipse $25x^2 + 4y^2 + 150x - 16y + 141 = 0$. Give the center.

Solution

We will convert the equation to standard form by completing the square for x and y.

$$25x^2 + 4y^2 + 150x - 16y + 141 = 0$$
$$(25x^2 + 150x) + (4y^2 - 16y) = -141$$

Collect terms containing x.
Collect terms containing y.
Subtract the constant to the right side.

$$25(x^2 + 6x) + 4(y^2 - 4y) = -141$$

Factor 25 from the terms containing x so that the coefficient of the x^2 term is 1. Factor 4 from the terms containing y.

$$25(x^2 + 6x + 9) + 4(y^2 - 4y + 4) = -141 + 225 + 16$$

Add 9 inside the parentheses containing x to complete the square for x. Add $25 \cdot 9$ or 225 to the right side of the equation. Add 4 inside the parentheses containing y to complete the square for y. Add $4 \cdot 4$ or 16 to the right side of the equation.

$$25(x + 3)^2 + 4(y - 2)^2 = 100$$

Factor the two quadratic expressions on the left side of the equation.

$$\frac{\overset{1}{\cancel{25}}(x + 3)^2}{\underset{4}{\cancel{100}}} + \frac{\overset{1}{\cancel{4}}(y - 2)^2}{\underset{25}{\cancel{100}}} = 1$$

Divide both sides of the equation by 100. Rewrite the left side of the equation as the sum of two fractions. Divide out common factors.

$$\frac{(x + 3)^2}{4} + \frac{(y - 2)^2}{25} = 1$$

Simplify.

Since the equation is now in standard form, the center of this ellipse is $(-3, 2)$. Since $a^2 = 4$, $a = 2$ and we plot the endpoints of the horizontal axis two units to the left and right of the center. Since $b^2 = 25$, $b = 5$ and we plot the endpoints of the vertical axis five units above and below the center. Draw the ellipse that passes through the four endpoints.

(h, k),

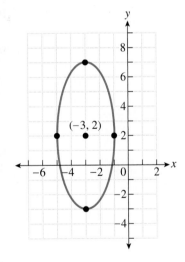

Quick Check **5** Graph the ellipse $9x^2 + 16y^2 - 36x + 160y + 292 = 0$. Give the center.

Finding the Equation of an Ellipse

Objective **4** **Find the equation of an ellipse that meets the given conditions.**
To find the equation of an ellipse, we begin by determining the center a, and b.

Then we write the equation of the ellipse in standard form $\dfrac{(x - h)^2}{a^2} + \dfrac{(y - k)^2}{b^2} = 1$.

EXAMPLE 7   Find the standard form equation of the ellipse.

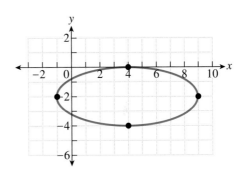

Solution

The center of the ellipse is $(4, -2)$.

We measure the distance from the center of the ellipse to an endpoint of its horizontal axis. This distance is a. In this case, $a = 5$.

Next, we measure the distance from the center of the ellipse to an endpoint of its vertical axis. This distance is b. In this case, $b = 2$.

We can now write the equation in standard form.

$$\frac{(x - 4)^2}{5^2} + \frac{(y - (-2))^2}{2^2} = 1$$ Substitute 4 for h, -2 for k, 5 for a, and 2 for b into the standard form of the equation of an ellipse.

$$\frac{(x - 4)^2}{25} + \frac{(y + 2)^2}{4} = 1$$ Simplify.

The equation of this ellipse is $\dfrac{(x - 4)^2}{25} + \dfrac{(y + 2)^2}{4} = 1$.

Quick Check   6   Find the standard-form equation of the ellipse.

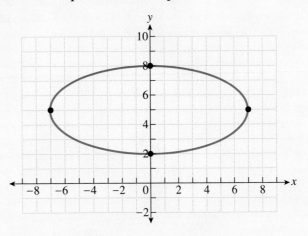

Graph the ellipse. Give the coordinates of the center, as well as the values of a and b.

1. $\dfrac{x^2}{4} + \dfrac{y^2}{9} = 1$

2. $\dfrac{x^2}{16} + \dfrac{y^2}{25} = 1$

3. $\dfrac{x^2}{25} + \dfrac{y^2}{4} = 1$

4. $\dfrac{x^2}{49} + \dfrac{y^2}{9} = 1$

5. $x^2 + \dfrac{y^2}{25} = 1$

6. $\dfrac{x^2}{4} + y^2 = 1$

7. $\dfrac{x^2}{36} + \dfrac{y^2}{12} = 1$

8. $\dfrac{x^2}{25} + \dfrac{y^2}{32} = 1$

9. $25x^2 + 9y^2 = 225$

10. $36x^2 + 16y^2 = 576$

15. $\dfrac{(x-9)^2}{81} + \dfrac{(y+4)^2}{9} = 1$

16. $\dfrac{(x+8)^2}{36} + \dfrac{(y-3)^2}{4} = 1$

11. $4x^2 + 49y^2 = 196$

12. $9x^2 + 16y^2 = 144$ **13.** $\dfrac{(x-6)^2}{25} + \dfrac{(y-8)^2}{36} = 1$ **17.** $\dfrac{(x+5)^2}{9} + \dfrac{(y+3)^2}{64} = 1$

14. $\dfrac{(x-4)^2}{4} + \dfrac{(y-7)^2}{16} = 1$

18. $\dfrac{(x+1)^2}{16} + \dfrac{(y+9)^2}{25} = 1$

19. $\dfrac{(x+5)^2}{4} + \dfrac{y^2}{36} = 1$ **20.** $\dfrac{x^2}{36} + \dfrac{(y+3)^2}{25} = 1$

24. $\dfrac{(x-3)^2}{8} + \dfrac{(y+8)^2}{25} = 1$

25. $4x^2 + 9y^2 - 24x - 90y + 225 = 0$

21. $\dfrac{(x+2)^2}{49} + (y-4)^2 = 1$

$(x+7)^2 + \dfrac{(y-1)^2}{25} = 1$

26. $25x^2 + 4y^2 - 300x + 56y + 996 = 0$

22.

27. $2x^2 + 8y^2 + 16x - 64y + 88 = 0$

23. $\dfrac{(x-6)^2}{20} + \dfrac{(y+6)^2}{4} = 1$

28. $49x^2 + 9y^2 + 98x + 36y - 356 = 0$

Find the standard-form equation of the ellipse.

35.

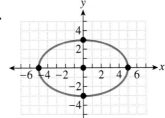

36.

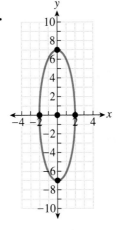

Find the standard-form equation of the ellipse that meets the given conditions.

	Center	Major Axis	Length of Major Axis	Length of Minor Axis
29.	$(-2, 8)$	vertical	20	4
30.	$(4, 7)$	vertical	8	2
31.	$(6, -3)$	horizontal	16	14
32.	$(-5, -9)$	horizontal	18	10
33.	$(0, -6)$	vertical	10	6
34.	$(8, -1)$	horizontal	32	24

37.

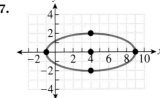

38.

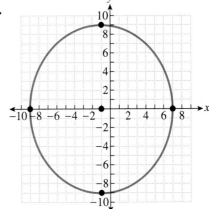

39.

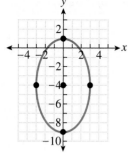

40.

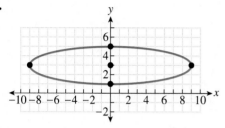

41.

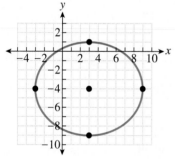

42.

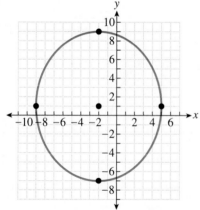

Graph each equation. For graphs that are parabolas, label the vertex and any intercepts. For graphs that are circles or ellipses, label the center.

43. $(x + 4)^2 + (y - 6)^2 = 9$ **44.** $y = -(x - 2)^2 + 5$

45. $x = y^2 + 6y + 10$

46. $\dfrac{x^2}{36} + \dfrac{y^2}{4} = 1$

47. $3x^2 + 5y^2 = 45$

48. $x = -y^2 + 2y + 7$

53. $x^2 + y^2 - 12x - 14y - 59 = 0$ **54.** $y = x^2 + 6x + 2$

49. $y = -x^2 + 2x + 63$

55. $x^2 + y^2 = 32$

50. $\dfrac{(x-1)^2}{9} + \dfrac{(y+5)^2}{16} = 1$

56. $x^2 + 49y^2 - 12x + 196y + 183 = 0$

57. $x = -(y+4)^2 - 9$ **58.** $x^2 + y^2 = 81$

51. $9x^2 + 25y^2 + 54x - 100y - 44 = 0$

52. $x = (y-4)^2 - 4$

59. $\dfrac{(x + 6)^2}{25} + \dfrac{(y + 1)^2}{4} = 1$ **60.** $y = (x - 3)^2 + 4$

61. An elliptical track can be described by the equation $1089x^2 + 2500y^2 = 2,722,500$, where x and y are in feet.

   a) Find the width of the track from west to east.

   b) Find the width of the track from north to south.

62. A tunnel through a hillside is in the shape of a semi-ellipse. The base of the tunnel is 80 feet across, and at its highest point the tunnel is 30 feet high. The tunnel is displayed below on a rectangular coordinate plane (not to scale).

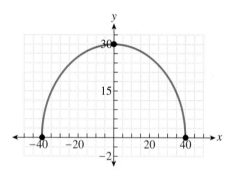

   a) Find the equation of the ellipse.

   b) How far from the center of the tunnel is the height 16 feet?

63. The area of an ellipse given by the equation $\dfrac{(x - h)^2}{a^2} + \dfrac{(y - k)^2}{b^2} = 1$ is πab. A homeowner is

building an elliptical swimming pool that is 30 feet from end to end at its widest point, and 20 feet from side to side at its widest point. Find the area covered by the swimming pool.

64. A small park in the shape of an ellipse covers an area of approximately 1372.88 square feet. If the park measures 46 feet from west to east, as shown below, what is the distance from south to north?

Answer in complete sentences.

65. Explain the difference between an ellipse and a circle. Are there any similarities between the two?

66. Explain how to find the center of an ellipse from its equation. Include the case in which we must complete the square to find the center.

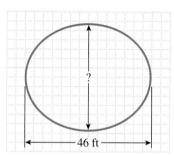

46 ft

Study Tip REVISITED   As you start to prepare for a final exam, realize that there is only a finite amount of time left before the exam, so you must use it wisely. Arrange your schedule so that there is an expanded allotment of time for studying math. It is important that you prepare for the final exam every day; do not take any days off from your preparation.

    If you have a job, try to work fewer hours. Ask for days off if at all possible.

    For further suggestions on effective time management, read the study tips in Chapter 10.

13.4

HYPERBOLAS

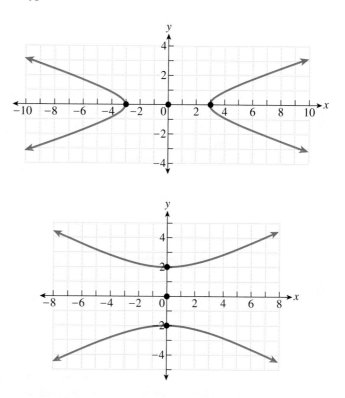

Objectives

1. Graph hyperbolas centered at the origin.
2. Graph hyperbolas centered at a point (h, k).
3. Find the center of a hyperbola and the lengths of its axes by completing the square.
4. Find the equation of a hyperbola that meets the given conditions.

In this section, we will investigate a conic section called a hyperbola. Here are two examples of graphs of hyperbolas.

Notice that the graph of a hyperbola is different from other graphs we have drawn in that it has two parts, called **branches.**

Hyperbola

> A **hyperbola** is the collection of all points (x, y) in the plane for which the *difference* of the distances, d_1 and d_2, between the point and two fixed points, F_1 and F_2, called foci is a constant.

This definition is similar to that of an ellipse, but for an ellipse the *sum* of the distances remains constant, not the difference. For any point (x, y) on the hyperbola, $|d_1 - d_2|$ remains the same.

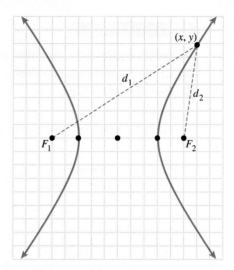

Like a circle and an ellipse, a hyperbola has a center. The **center** of a hyperbola is the point that is located midway between the two foci.

The line that passes through the two foci intersects the hyperbola at two points. These points are the **vertices** of the hyperbola, and the line segment between them is called the **transverse axis** of the hyperbola.

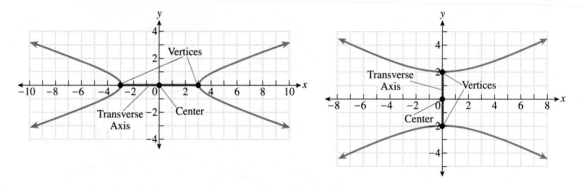

Each hyperbola has a pair of asymptotes. Each **asymptote** is a line that passes through the center of the hyperbola, showing us the behavior of the branches of the hyperbola as we move further out from the center of the hyperbola.

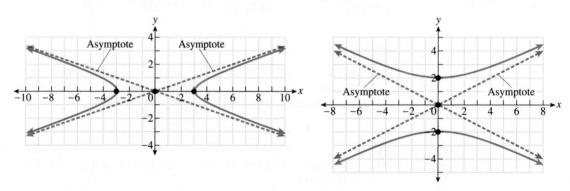

Hyperbolas Centered at the Origin

Objective **1** **Graph hyperbolas centered at the origin.** We will begin by investigating hyperbolas that have a horizontal transverse axis, whose branches open to the left and to the right.

Standard Form of the Equation of a Hyperbola with a Horizontal Transverse Axis

The standard-form equation of a hyperbola with a horizontal transverse axis centered at the origin has the form $\dfrac{x^2}{a^2} - \dfrac{y^2}{b^2} = 1$.

The hyperbola will have vertices at $(a, 0)$ and $(-a, 0)$. The asymptotes of the hyperbola will be the lines $y = \frac{b}{a}x$ and $y = -\frac{b}{a}x$.

To graph a hyperbola, we begin by plotting the center and each vertex. We then graph the asymptotes, using dashed lines to indicate that they are not part of the hyperbola. We finish the graph by drawing a branch through each vertex so that the branch approaches the asymptotes.

To graph the asymptotes, construct a rectangle around the center of the hyperbola that extends a units to the left and right of the center and b units above and below the center. Once this rectangle has been constructed, the asymptotes are the lines that pass through the diagonals of the rectangle.

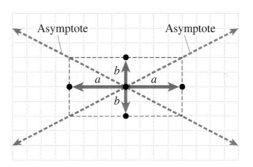

Notice that there are four points labeled on the rectangle. Two of those points are vertices and the endpoints of the transverse axis. The two points that are not vertices are the endpoints of what is known as the **conjugate axis.** If the branches of the hyperbola open to the left and the right, then the conjugate axis will be vertical, and its length is $2b$.

EXAMPLE **1** Graph the hyperbola $\dfrac{x^2}{4} - \dfrac{y^2}{25} = 1$. Give the center and the vertices, and the equations of the asymptotes.

Solution

Since the equation is of the form $\dfrac{x^2}{a^2} - \dfrac{y^2}{b^2} = 1$, the hyperbola is centered at the origin and its branches open to the left and to the right. We begin by plotting a point at the origin and graphing the asymptotes.

Since $a^2 = 4$ and $b^2 = 25$, we know that $a = 2$ and $b = 5$. The rectangle used to graph the asymptotes extends two units to the left and to the right of the center and five units above and below the center. The equations for these asymptotes are $y = \frac{5}{2}x$ and $y = -\frac{5}{2}x$.

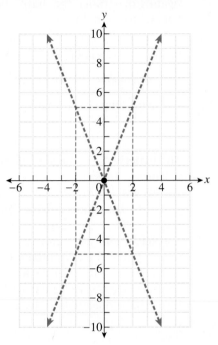

The vertices are located at the points $(-2, 0)$ and $(2, 0)$, on the rectangle that we drew to graph the asymptotes. Next, we draw the branches of the hyperbola opening to the left and to the right in such a way that they approach the asymptotes.

Quick Check   **1**

Graph the hyperbola $\dfrac{x^2}{16} - \dfrac{y^2}{9} = 1$. Give the center and the vertices, and the equations of the asymptotes.

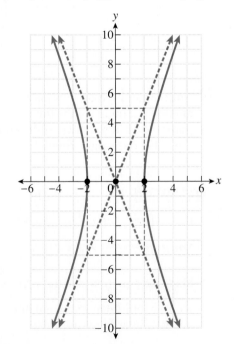

We now turn to hyperbolas that have a vertical transverse axis. For this type of hyperbola, the branches open upwards and downwards.

Standard Form of the Equation of a Hyperbola with a Vertical Transverse Axis

The equation of a hyperbola with a vertical transverse axis that is centered at the origin has the form $\dfrac{y^2}{b^2} - \dfrac{x^2}{a^2} = 1$.

The hyperbola will have vertices at $(0, b)$ and $(0, -b)$. The asymptotes of the hyperbola will be the lines $y = \dfrac{b}{a}x$ and $y = -\dfrac{b}{a}x$. The conjugate axis will be horizontal, and its length is $2a$.

EXAMPLE 2   Graph $\dfrac{y^2}{9} - \dfrac{x^2}{16} = 1$. Give the center, the vertices, and the equations of the asymptotes.

Solution

This hyperbola is centered at the origin and its branches open upward and downward. We begin by plotting a point at the origin and graphing the asymptotes.

Since $a^2 = 16$ and $b^2 = 9$, we know that $a = 4$ and $b = 3$. The rectangle used to graph the asymptotes extends four units to the left and to the right of the center and three units above and below the origin. The equations for these asymptotes are $y = \dfrac{3}{4}x$ and $y = -\dfrac{3}{4}x$.

The vertices are located at the points $(0, 3)$ and $(0, -3)$. After plotting the vertices, we draw the branches of the hyperbola opening upward and downward in such a way that they approach the asymptotes.

Quick Check **2**

Graph $\dfrac{y^2}{4} - \dfrac{x^2}{36} = 1$. Give the coordinates of the center and the vertices and the equations of the asymptotes.

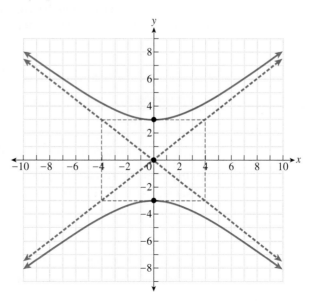

Here is a brief summary of hyperbolas centered at the origin:

Transverse Axis	Horizontal	Vertical
Equation	$\dfrac{x^2}{a^2} - \dfrac{y^2}{b^2} = 1$	$\dfrac{y^2}{b^2} - \dfrac{x^2}{a^2} = 1$
Sample Graph		
Vertices	$(a, 0), (-a, 0)$	$(0, b), (0, -b)$
Asymptotes	$y = \dfrac{b}{a}x, \; y = -\dfrac{b}{a}x$	$y = \dfrac{b}{a}x, \; y = -\dfrac{b}{a}x$

It is crucial to be able to tell what type of hyperbola we are graphing from the equation. For a hyperbola whose equation is of the form $\dfrac{x^2}{a^2} - \dfrac{y^2}{b^2} = 1$, the branches of the hyperbola open to the left and to the right. For a hyperbola whose equation is of the form $\dfrac{y^2}{b^2} - \dfrac{x^2}{a^2} = 1$, the branches of the hyperbola open upward and downward.

It is also important to be able to differentiate between the equation of a hyperbola and the equation of an ellipse. The major difference between the two equations is that the equation of a hyperbola involves the difference of the terms containing x and y, while the equation of an ellipse involves their sum.

Hyperbolas Centered at a Point Other than the Origin

Objective 2 **Graph hyperbolas centered at a point (h, k).** We will now learn how to graph hyperbolas centered at any point (h, k).

Standard-Form Equation of a Hyperbola with Center (h, k)

- The equation of a hyperbola centered at the point (h, k) with a horizontal transverse axis has the standard form

$$\frac{(x - h)^2}{a^2} - \frac{(y - k)^2}{b^2} = 1.$$

The vertices are the points $(h + a, k)$ and $(h - a, k)$. The equations for the asymptotes are $y = \frac{b}{a}(x - h) + k$ and $y = -\frac{b}{a}(x - h) + k$.

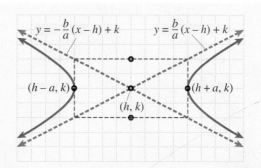

- The equation for a hyperbola centered at the point (h, k) with a vertical transverse axis has the standard form

$$\frac{(y - k)^2}{b^2} - \frac{(x - h)^2}{a^2} = 1.$$

The vertices are the points $(h, k + b)$ and $(h, k - b)$. The equations for the asymptotes are $y = \frac{b}{a}(x - h) + k$ and $y = -\frac{b}{a}(x - h) + k$.

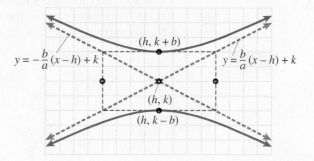

To graph a hyperbola centered at a point (h, k), we begin by plotting the center of the hyperbola and at its vertices. If the branches of the hyperbola open to the left and to the right, then the vertices can be found by moving a units to the left and right of the center. If the branches of the hyperbola open upward and downward, then the vertices can be found by moving b units above and below the center.

Once the center and vertices have been plotted, we graph the asymptotes of the hyperbola using dashed lines. Although we can graph these asymptotes using their equations, we can continue to use a rectangle in the same way as for hyperbolas centered at the origin.

We then draw the two branches of the hyperbola in such a way that each branch passes through a vertex and approaches the asymptotes.

EXAMPLE 3 Graph the hyperbola $\dfrac{(x - 4)^2}{9} - \dfrac{(y - 2)^2}{4} = 1$. Give the center, the vertices, and the equations of the asymptotes.

Solution

Since the equation is of the form $\dfrac{(x - h)^2}{a^2} - \dfrac{(y - k)^2}{b^2} = 1$, the hyperbola is centered at $(4, 2)$ and its branches open to the left and to the right. We begin by plotting the center and graphing the asymptotes.

Since $a^2 = 9$ and $b^2 = 4$, we know that $a = 3$ and $b = 2$. The rectangle used to graph the asymptotes extends three units to the left and to the right of the center and two units above and below the center. The equations for these asymptotes are $y = \frac{2}{3}(x - 4) + 2$ and $y = -\frac{2}{3}(x - 4) + 2$.

Graph the hyperbola
$$\frac{(x - 3)^2}{4} - \frac{(y - 5)^2}{16} = 1.$$
Give the center, the vertices, and the equations of the asymptotes.

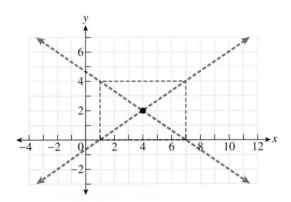

The vertices are located three units to the left and to the right of the center $(4, 2)$ at the points $(1, 2)$ and $(7, 2)$. After plotting the vertices, we draw the branches of the hyperbola opening to the left and to the right in such a way that they approach the asymptotes.

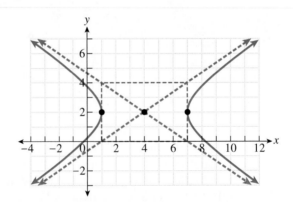

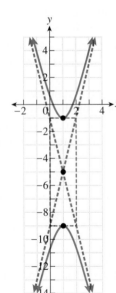

EXAMPLE 4   Graph the hyperbola $\dfrac{(y + 5)^2}{16} - (x - 1)^2 = 1$. Give the center, the vertices, and the equations of the asymptotes.

Solution

Since the equation is of the form $\dfrac{(y - k)^2}{b^2} - \dfrac{(x - h)^2}{a^2} = 1$, the hyperbola is centered at $(1, -5)$ and its branches open upward and downward. We begin by plotting a point at the center and graphing the asymptotes.

There is no denominator under the squared term containing x, so $a^2 = 1$. Therefore, $a = 1$. Since $b^2 = 16$, $b = 4$. The rectangle used to graph the asymptotes extends one unit to the left and to the right of the center and four units above and below the center. The equations for these asymptotes are $y = 4(x - 1) - 5$ and $y = -4(x - 1) - 5$.

The vertices are located 4 units above and below the center at the points $(1, -1)$ and $(1, -9)$. After plotting the vertices, we draw the branches of the hyperbola opening upward and downward in such a way that they approach the asymptotes.

Quick Check **4**

Graph the hyperbola
$$\frac{(y - 2)^2}{49} - \frac{(x + 3)^2}{9} = 1.$$
Give the center, the vertices, and the equations of the asymptotes.

Finding the Center of a Hyperbola and the Lengths of Its Axes by Completing the Square

Objective **3** Find the center of a hyperbola and the lengths of its axes by completing the square.

General Form of the Equation of a Hyperbola

The **general form** of the equation of a hyperbola is $Ax^2 + By^2 + Cx + Dy + E = 0$, where $A \neq 0$, $B \neq 0$, and A and B have opposite signs.

If both A and B have the same sign, then the graph will be a circle or an ellipse, not a hyperbola. To graph a hyperbola in general form, we must rewrite the equation in standard form. This will allow us to determine the center, as well as a and b. This is done by completing the square, which must be done for both x and y.

EXAMPLE **5** Graph the hyperbola $-4x^2 + 25y^2 + 24x + 50y - 111 = 0$. Give the center, the vertices, and the equations of the asymptotes.

Solution

We will convert the equation to standard form by completing the square for x and y.

$$-4x^2 + 25y^2 + 24x + 50y - 111 = 0$$

$(-4x^2 + 24x) + (25y^2 + 50y) = 111$ Collect terms containing x. Collect terms containing y. Add 111 to both sides.

$-4(x^2 - 6x) + 25(y^2 + 2y) = 111$ Factor -4 from the terms containing x so that the coefficient of the x^2 term is 1. In the same fashion, factor 25 from the terms containing y.

$-4(x^2 - 6x + 9) + 25(y^2 + 2y + 1) = 111 - 36 + 25$

Add 9 inside the parentheses containing x to complete the square for x. Since $-4 \cdot 9 = -36$, subtract 36 on the right side of the equation. Add 1 inside the parentheses containing y to complete the square for y. Add $25 \cdot 1$ or 25 to the right side of the equation.

$$-4(x - 3)^2 + 25(y + 1)^2 = 100$$

Factor the two quadratic expressions on the left side of the equation.

$$25(y + 1)^2 - 4(x - 3)^2 = 100$$

Rewrite so that the term containing x is being subtracted from the term containing y.

$$\frac{(y + 1)^2}{4} - \frac{(x - 3)^2}{25} = 1$$

Divide both sides by 100 and simplify.

We begin by plotting the center, $(3, -1)$, and graphing the asymptotes.

For this hyperbola, $a = 5$ and $b = 2$. The equations for the asymptotes are $y = \frac{2}{5}(x - 3) - 1$ and $y = -\frac{2}{5}(x - 3) - 1$.

Since the equation has the form $\frac{(y - k)^2}{b^2} - \frac{(x - h)^2}{a^2} = 1$, the branches of this hyperbola open upward and downward.

The vertices are $(3, 1)$ and $(3, -3)$. After plotting the vertices, draw the branches of the hyperbola opening upward and downward in such a way that they approach the asymptotes.

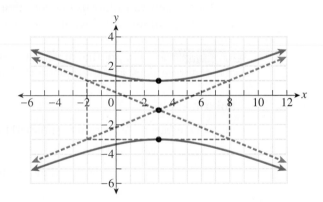

Quick Check   **5** Graph the hyperbola $16x^2 - 9y^2 - 64x - 54y - 161 = 0$. Give the center, the vertices, and the equations of the asymptotes.

Finding the Standard-Form Equation of a Hyperbola

Objective   4 **Find the equation of a hyperbola that meets the given conditions.** To find the equation of a hyperbola in standard form, we must find the center (h, k), a, and b.

EXAMPLE 6   Find the equation of the hyperbola whose graph is shown on the next page.

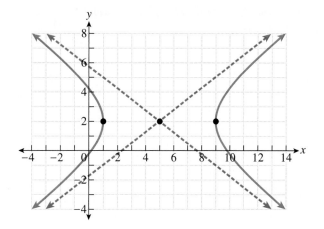

Solution:

Since the branches of the hyperbola open to the left and to the right, its equation is of the form $\dfrac{(x - h)^2}{a^2} - \dfrac{(y - k)^2}{b^2} = 1.$

We begin by finding the center of the hyperbola, which is at the point $(5, 2)$.

Then we measure the distance a from the center of the hyperbola to one of its vertices, and find that $a = 4$.

Next, we measure the vertical distance b from a vertex of the hyperbola to one of its asymptotes, and find that $b = 3$.

We can now write the equation in standard form.

$$\frac{(x - h)^2}{a^2} - \frac{(y - k)^2}{b^2} = 1$$

$$\frac{(x - 5)^2}{4^2} - \frac{(y - 2)^2}{3^2} = 1 \qquad \text{Substitute 5 for } h, \text{2 for } k, \text{4 for } a, \text{and 3 for } b.$$

$$\frac{(x - 5)^2}{16} - \frac{(y - 2)^2}{9} = 1 \qquad \text{Simplify.}$$

The equation of this hyperbola is $\dfrac{(x - 5)^2}{16} - \dfrac{(y - 2)^2}{9} = 1.$

Quick Check   **6**   **Find the equation of the hyperbola whose graph is shown.**

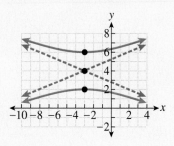

Graph the hyperbola. Give the coordinates of the center, as well as the values of a and b.

1. $\dfrac{x^2}{9} - \dfrac{y^2}{4} = 1$

2. $\dfrac{x^2}{16} - \dfrac{y^2}{25} = 1$

3. $\dfrac{y^2}{49} - \dfrac{x^2}{9} = 1$

4. $\dfrac{y^2}{16} - \dfrac{x^2}{36} = 1$

5. $\dfrac{x^2}{9} - \dfrac{y^2}{9} = 1$

6. $\dfrac{y^2}{16} - \dfrac{x^2}{16} = 1$

7. $\dfrac{y^2}{25} - x^2 = 1$

10. $\dfrac{y^2}{6} - \dfrac{x^2}{25} = 1$

11. $49x^2 - 4y^2 = 196$

8. $\dfrac{x^2}{16} - y^2 = 1$

9. $\dfrac{x^2}{12} - \dfrac{y^2}{9} = 1$

12. $9x^2 - 36y^2 = 324$

13. $y^2 - 9x^2 = 9$

16. $\dfrac{(y - 6)^2}{25} - \dfrac{(x - 3)^2}{9} = 1$

14. $y^2 - 4x^2 = 4$

17. $\dfrac{(x + 1)^2}{9} - \dfrac{(y + 4)^2}{4} = 1$

15. $\dfrac{(y - 5)^2}{16} - \dfrac{(x - 2)^2}{9} = 1$

18. $\dfrac{(x + 3)^2}{4} - \dfrac{(y - 3)^2}{25} = 1$

19. $\dfrac{(y-1)^2}{4} - \dfrac{x^2}{4} = 1$

22. $\dfrac{(y-2)^2}{9} - \dfrac{(x+4)^2}{49} = 1$

23. $(y-5)^2 - \dfrac{(x+3)^2}{9} = 1$

20. $\dfrac{(x+1)^2}{9} - \dfrac{(y+5)^2}{9} = 1$

24. $\dfrac{(y+1)^2}{4} - (x-6)^2 = 1$

21. $\dfrac{(x-4)^2}{49} - \dfrac{(y+2)^2}{16} = 1$

25. $\dfrac{(x+2)^2}{25} - \dfrac{(y+3)^2}{28} = 1$

28. $-16x^2 + 9y^2 - 160x + 72y - 400 = 0$

29. $-25x^2 + 16y^2 - 100x - 224y + 284 = 0$

26. $\dfrac{(x+1)^2}{18} - \dfrac{(y-2)^2}{4} = 1$

27. $9x^2 - 4y^2 + 18x - 24y - 63 = 0$

30. $x^2 - 64y^2 - 10x + 768y - 2343 = 0$

Find the equation of the hyperbola that meets the given conditions.

31. Center: $(4, 2)$
   Transverse axis: vertical
   Length of transverse axis: 6
   Length of conjugate axis: 10

32. Center: $(3, 7)$
   Transverse axis: horizontal
   Length of transverse axis: 10
   Length of conjugate axis: 12

33. Center: $(6, -2)$
   Transverse axis: horizontal
   Length of transverse axis: 8
   Length of conjugate axis: 8

34. Center: $(-3, 9)$
   Transverse axis: vertical
   Length of transverse axis: 2
   Length of conjugate axis: 16

Find the standard-form equation of the hyperbola whose graph is shown.

35.

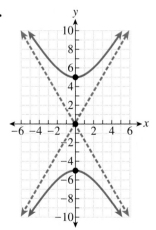

36.

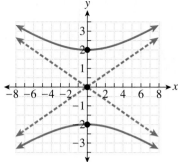

37.

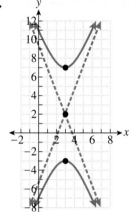

38.

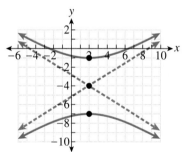

39.

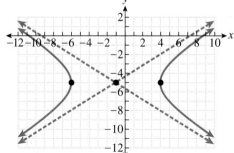

40.

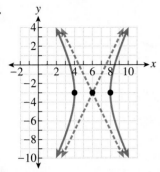

43. $y = x^2 + 8x + 19$

Graph each equation. For graphs that are parabolas, label the vertex and any intercepts. For graphs that are circles, ellipses, or hyperbolas, label the center.

41. $x = -y^2 + 4y + 9$

44. $\dfrac{y^2}{25} - x^2 = 1$

42. $y = -x^2 + 6x - 17$

45. $\dfrac{(x-6)^2}{9} + \dfrac{y^2}{49} = 1$

46. $x = (y - 1)^2 - 12$

50. $\dfrac{(y - 3)^2}{25} - \dfrac{(x + 7)^2}{36} = 1$

47. $y = (x - 6)^2 - 4$

51. $x^2 + y^2 = 64$

48. $(x + 5)^2 + (y + 1)^2 = 20$

52. $\dfrac{x^2}{36} + \dfrac{y^2}{4} = 1$

49. $\dfrac{(x + 2)^2}{25} + \dfrac{(y - 5)^2}{9} = 1$

53. $4x^2 - 9y^2 + 40x + 54y - 17 = 0$

54. $(x - 4)^2 + (y + 7)^2 = 9$

57. $\dfrac{(x + 6)^2}{9} - \dfrac{(y + 2)^2}{16} = 1$

55. $x^2 + y^2 - 8x + 18y + 48 = 0$

58. $x = y^2 + 10y + 16$

56. $16x^2 + 9y^2 - 256x + 90y + 1105 = 0$

Answer in complete sentences.

59. Explain how to determine whether the graph of an equation is a hyperbola whose branches open to the left and right. Also, explain how to determine that the graph of an equation is a hyperbola whose branches open up and down.

60. If the equations of the asymptotes of a hyperbola are known, is that enough information to determine the equation of the hyperbola? Explain why or why not.

Section 13.4

Solve.

1. $x + 8y = 19$
   $7x - 5y = 11$

2. $y = 5x + 6$
   $-3x + 4y = -27$

3. $2x - 3y = 11$
   $5x + 6y = 14$

4. $4x - 5y = 15$
   $-2x + 15y = -30$

Study Tip **REVISITED**   One of the most important decisions regarding your preparation for the final exam is choosing which problems you will work on. Consider the following options:

- **Textbook.** Each chapter in the textbook contains a chapter review assignment and a chapter test. Also, there are cumulative review assignments at the end of Chapters 4, 7, 11, and 14.
- **Old Exams/Quizzes.** Study your old exams and quizzes. Your instructor considered these problems important enough to test you on them.
- **Your Notes/Homework Journal.** If you have been keeping track of problems and concepts that you find difficult, include these problems in your preparation.
- **MyMathLab.com.** Search for supplementary problems on MyMathLab.com, including practice tests and tutorial exercises.

13.5

NONLINEAR SYSTEMS OF EQUATIONS

Objectives

1. Solve nonlinear systems by using the substitution method.
2. Solve nonlinear systems by using the addition method.
3. Solve applications of nonlinear systems.

A **nonlinear system of equations** is a system of equations in which the graph of at least one of the equations is not a line. For example, if the graph of one equation is a circle and the graph of the other equation is a line, then the system is a nonlinear system of equations. For a system of two equations in two unknowns, an ordered pair (x, y) is a solution of the system if it is a solution of each equation. Graphically, a solution of a system of two equations in two unknowns is a point of intersection of the graphs of the two equations. Some nonlinear systems can have one or more solutions. The following are examples of nonlinear systems of equations and their solutions, which are points of intersection for the two graphs:

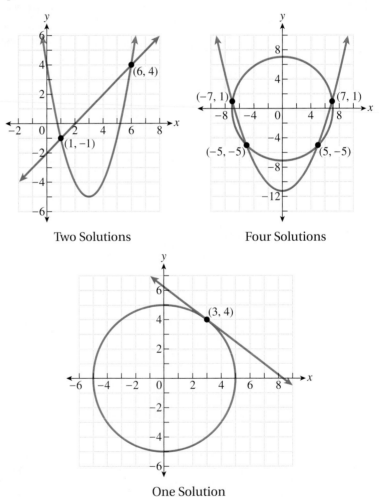

Two Solutions   Four Solutions

One Solution

If the graphs of the two equations do not intersect, then the system of equations has no solution.

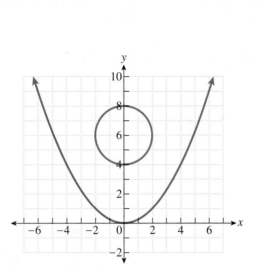

Circle and Parabola (No Intersection)

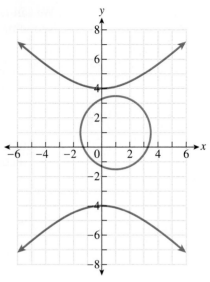

Circle and Hyperbola (No Intersection)

The Substitution Method

Objective 1 **Solve nonlinear systems by using the substitution method.** One effective technique for solving nonlinear systems of equations is the substitution method. As in Section 4.2, we will solve one of the equations for one of the variables in terms of the other variable and then substitute this expression for that variable in the other equation. This will result in an equation with only one variable, which we can then solve.

Solving a System of Equations by Using the Substitution Method

1. Solve one of the equations for either variable.
2. Substitute this expression for the variable in the other equation.
3. Solve this equation.
4. Substitute each value for the variable into the equation from step 1.

EXAMPLE 1 Solve the nonlinear system.
$$x^2 + y^2 = 25$$
$$x + 2y = 10$$

Solution

We will begin by solving the second equation for x, since the coefficient of that term is 1. If we subtract $2y$ from both sides of the equation, we find that $x = 10 - 2y$. This expression can then be substituted for x in the equation $x^2 + y^2 = 25$.

$$x^2 + y^2 = 25$$
$$(10 - 2y)^2 + y^2 = 25 \qquad \text{Substitute } 10 - 2y \text{ for } x.$$
$$5y^2 - 40y + 100 = 25 \qquad \text{Square } 10 - 2y \text{ and combine like terms.}$$
$$5y^2 - 40y + 75 = 0 \qquad \text{Subtract 25 from both sides.}$$
$$5(y - 3)(y - 5) = 0 \qquad \text{Factor completely.}$$
$$y = 3 \quad \text{or} \quad y = 5 \qquad \text{Set each variable factor equal to 0 and solve.}$$

We can now substitute these values for y in the equation $x = 10 - 2y$ to find the corresponding x-coordinates of the solutions.

$y = 3$ $y = 5$
$x = 10 - 2(3)$ $x = 10 - 2(5)$ Substitute for y.
$x = 4$ $x = 0$ Simplify.

The two solutions are $(4, 3)$ and $(0, 5)$.

The graph of the first equation in the system is a circle, and the graph of the second equation is a line. Below are the graphs of the two equations, showing their points of intersection.

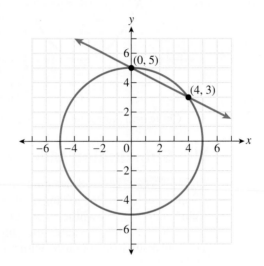

Quick Check 1

Solve the nonlinear system.
$4x^2 + y^2 = 41$
$x + 2y = -8$

Although it is not necessary, it is a good idea to graph both equations in the system. This will give us an idea of how many solutions we are looking for, as well as the approximate coordinates of the solutions.

EXAMPLE 2   Solve the nonlinear system. $y = x^2 - 3$
                                                     $x^2 + y^2 = 9$

Solution

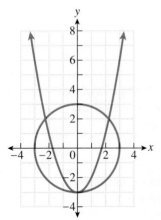

The graph on the left of the first equation is a parabola that opens upward, while the graph of the second equation is a circle. Based on the graph, there appear to be three solutions.

Solve the first equation for x^2, giving $x^2 = y + 3$, and substitute that expression for x^2 in the second equation. This is a useful technique when the graph of neither equation is a line.

$x^2 + y^2 = 9$
$(y + 3) + y^2 = 9$ Substitute $y + 3$ for x^2.
$y^2 + y - 6 = 0$ Collect all terms on the left side.
$(y + 3)(y - 2) = 0$ Factor.
$y = -3$ or $y = 2$ Set each factor equal to 0 and solve.

We can now substitute these values for y in the equation $x^2 = y + 3$ to find the corresponding x-coordinates of the solutions.

$$y = -3 \qquad\qquad y = 2$$
$$x^2 = (-3) + 3 \qquad x^2 = (2) + 3 \qquad \text{Substitute for } y.$$
$$x = 0 \qquad\qquad x = \pm\sqrt{5} \qquad \text{Simplify and take the square root of each side.}$$

Quick Check 2

Solve the nonlinear system.
$$y = x^2$$
$$5x^2 + y^2 = 24$$

The solutions are $(0, -3)$, $(\sqrt{5}, 2)$, and $(-\sqrt{5}, 2)$.

The Addition Method

Objective 2 Solve nonlinear systems by using the addition method. Another effective technique for solving nonlinear systems of equations is the addition method, which was introduced in Section 4.3. By rewriting the two equations in such a way that the coefficients of like terms are opposites, we can add the two equations to obtain a single equation with only one variable, which we can then solve.

EXAMPLE 3 Solve the nonlinear system. $16x^2 + 25y^2 = 400$
$x^2 + y^2 = 16$

Solution

If we multiply both sides of the second equation by -16, then the coefficients of the two terms containing x^2 will be opposites. We will then be able to eliminate terms containing x^2, leaving an equation whose only variable is y.

$$16x^2 + 25y^2 = 400 \qquad\qquad\qquad\qquad 16x^2 + 25y^2 = 400$$
$$x^2 + y^2 = 16 \qquad \xrightarrow{\text{Multiply by } -16} \qquad -16x^2 - 16y^2 = -256$$

$$\begin{array}{r} 16x^2 + 25y^2 = 400 \\ -16x^2 - 16y^2 = -256 \qquad \text{Add.} \\ \hline 9y^2 = 144 \end{array}$$

We can now solve the equation $9y^2 = 144$ for y.

$$9y^2 = 144$$
$$y^2 = 16 \qquad \text{Divide both sides by 9.}$$
$$y = \pm 4 \qquad \text{Take the square root of each side.}$$

Quick Check 3

Solve the nonlinear system.
$$4x^2 + 7y^2 = 211$$
$$x^2 + y^2 = 34$$

We can now substitute 4 and -4 for y in either equation to find the corresponding x-coordinates of the solutions. We will use the equation of the circle $x^2 + y^2 = 16$.

$$y = 4 \qquad\qquad y = -4$$
$$x^2 + (4)^2 = 16 \qquad x^2 + (-4)^2 = 16 \qquad \text{Substitute for } y.$$
$$x = 0 \qquad\qquad x = 0 \qquad\qquad \text{Solve for } x.$$

The solutions are $(0, 4)$ and $(0, -4)$.

We can attempt to solve nonlinear systems by graphing, but our solutions will depend upon the accuracy of our graph, not to mention our estimate of solutions that do not have integer coordinates. However, if we use technology such as a graphing calculator or computer software, these tools have built-in functions for finding the points of intersection of two graphs.

Applications of Nonlinear Systems of Equations

Objective **3** **Solve applications of nonlinear systems.** We now turn our attention to applied problems requiring solving a nonlinear system of equations to solve the problem.

EXAMPLE **4** A rectangular dorm room has a perimeter of 40 feet and an area of 91 square feet. Find its dimensions.

Solution

The unknowns in this problem are the length and width of the room. We will let x represent the length of the room and we will let y represent the width of the room.

> ***Unknowns***
>
> Length: x
> Width: y

Since the perimeter of the rectangle is 40 feet, we know that $2x + 2y = 40$. Also, since the area of the rectangle is 91 square feet, we know that $xy = 91$. If we solve the perimeter equation for either x or y, we can then substitute the expression for that variable into the area equation.

$$2x + 2y = 40$$
$$2x = 40 - 2y \qquad \text{Subtract } 2y.$$
$$x = 20 - y \qquad \text{Divide both sides by 2.}$$

We can now substitute $20 - y$ for x in the equation $xy = 91$.

$$xy = 91$$
$$(20 - y)y = 91 \qquad \text{Substitute } 20 - y \text{ for } x.$$
$$0 = y^2 - 20y + 91 \qquad \text{Simplify and collect all terms on the right side of the equation.}$$
$$0 = (y - 7)(y - 13) \qquad \text{Factor the quadratic expression.}$$
$$y = 7 \quad \text{or} \quad y = 13 \qquad \text{Set each factor equal to 0 and solve.}$$

We can find the corresponding x-coordinates of the solutions by substituting these values for y in one of the two equations and solving for x. We will use the perimeter equation.

$y = 7$	$y = 13$	
$2x + 2(7) = 40$	$2x + 2(13) = 40$	Substitute for y.
$x = 13$	$x = 7$	Solve for x.

Quick Check **4**

A rectangle has a perimeter of 30 feet and an area of 36 square feet. Find its dimensions.

We see that the length is 13 feet if the width is 7 feet, and the length is 7 feet if the width is 13 feet. In either case, the dimensions of the room are 7 feet by 13 feet.

EXAMPLE ▸ 5   A rectangular computer monitor has a perimeter of 46 inches and a diagonal that is 17 inches long. Find the dimensions of the monitor.

Solution

The unknowns in this problem are the length and width of the monitor. We will let x represent the length of the monitor and we will let y represent the width of the monitor.

Unknowns
Length: x
Width: y

Since the perimeter of the monitor is 46 inches, we know that $2x + 2y = 46$. Also, since the diagonal of the monitor is 17 inches, we know from the Pythagorean theorem that $x^2 + y^2 = 17^2$, or $x^2 + y^2 = 289$.

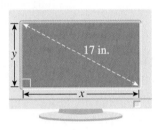

Solve the perimeter equation for x, giving $x = 23 - y$, and then substitute that expression for x into the equation $x^2 + y^2 = 289$.

$$x^2 + y^2 = 289$$
$$(23 - y)^2 + y^2 = 289 \qquad \text{Substitute } 23 - y \text{ for } x.$$
$$2y^2 - 46y + 240 = 0 \qquad \text{Square } 23 - y \text{ and collect all terms on the left side.}$$
$$2(y - 8)(y - 15) = 0 \qquad \text{Factor completely.}$$
$$y = 8 \quad \text{or} \quad y = 15 \qquad \text{Set each variable factor equal to 0 and solve.}$$

We can find the corresponding x-coordinates of the solutions by substituting these values for y in one of the two equations and solving for x. We will use the perimeter equation.

Quick Check 5

A rectangle has a perimeter of 42 inches and a diagonal 15 inches long. Find the dimensions of the rectangle.

$y = 8$	$y = 15$	
$2x + 2(8) = 46$	$2x + 2(15) = 46$	Substitute for y.
$x = 15$	$x = 8$	Solve for x.

We see that the length is 15 inches if the width is 8 inches and that the length is 8 inches if the width is 15 inches. In either case, the dimensions of the rectangle are 8 inches by 15 inches.

We conclude with an example involving the height of a projectile. Recall that if an object is launched with an initial velocity of v_0 feet per second from an initial height of s feet, then its height in feet after t seconds is given by the function $h(t) = -16t^2 + v_0 t + s$.

EXAMPLE ▸ 6   A rock is thrown upward from ground level with an initial velocity of 100 feet per second. At the same instant, a water balloon is dropped from a helicopter that is hovering 300 feet above the ground. At what time will the two objects be the same height above the ground? What is the height?

Solution

We will begin by finding the time that it takes for the two objects to be at the same height. The function describing the height of the rock is $h(t) = -16t^2 + 100t$, while the function describing the height of the water balloon dropped from the helicopter is $h(t) = -16t^2 + 300$. If we let the variable y represent the height of each object, we get the following nonlinear system of equations:

$$y = -16t^2 + 100t$$
$$y = -16t^2 + 300$$

Since each equation is already solved for y, we can set $-16t^2 + 100t$ equal to $-16t^2 + 300$ and solve for t.

$$-16t^2 + 100t = -16t^2 + 300$$
$$100t = 300 \qquad \text{Add } 16t^2 \text{ to both sides of the equation. The resulting equation is linear.}$$
$$t = 3 \qquad \text{Divide both sides by 100.}$$

The two objects will be at the same height after 3 seconds. We can substitute 3 for t in either function to find the height.

$$h(t) = -16t^2 + 100t$$
$$h(3) = -16(3)^2 + 100(3) \qquad \text{Substitute 3 for } t.$$
$$= 156 \qquad \text{Simplify.}$$

After 3 seconds, each object will be at a height of 156 feet.

> **Quick Check ▸ 6**
> A boy throws a ball upward from the ground with an initial velocity of 40 feet per second. At the same instant, a boy standing on a platform 10 feet above the ground throws another ball upward with an initial velocity of 35 feet per second. At what time will the two balls be the same height above the ground?

EXERCISES *13.5* ▸

Solve by the substitution method.

1. $x^2 + y^2 = 100$
$x - 7y = 50$

2. $x^2 + y^2 = 25$
$x - y = -1$

3. $x^2 + y^2 = 85$
$3x + y = -29$

4. $x^2 + y^2 = 50$
$-x + 2y = 15$

5. $4x^2 + 25y^2 = 100$
$2x + 5y = 10$

6. $9x^2 + y^2 = 9$
$-3x + y = 3$

7. $4x^2 + y^2 = 16$
$y = 2x - 4$

8. $2x^2 + 3y^2 = 14$
$x = -3y + 7$

9. $y = x^2 - 3x + 8$
$-2x + y = 4$

10. $y = x^2 + 6x$
$-4x + y = 3$

11. $y = -x^2 + 8x - 13$
$2x - y = 5$

12. $y = -x^2 - 9x + 25$
$-x + y = 25$

13. $x^2 - y^2 = 8$
$y = x - 2$

14. $y^2 - x^2 = 7$
$y = x + 7$

15. $x^2 + y^2 = 25$
$y = x^2 - 5$

16. $x^2 + y^2 = 16$
$y = x^2 - 4$

17. $x^2 + y^2 = 30$
$x = y^2$

18. $x^2 + y^2 = 42$
$x = -y^2$

19. $x^2 + 4y^2 = 4$
$y = x^2 + 1$

20. $x^2 + 9y^2 = 36$
$y = x^2 + 2$

21. $9x^2 + 4y^2 = 9$
$x = y^2 + 1$

22. $25x^2 + 9y^2 = 225$
$x = y^2 + 3$

23. $x^2 - y^2 = 1$
$x = y^2 + 11$

24. $x^2 - y^2 = 15$
$x = \dfrac{1}{3}y^2 - 1$

25. $x^2 - 4y^2 = 16$
$x = y^2 - 11$

26. $y^2 - 9x^2 = 9$
$y = x^2 + 1$

27. $y = x^2 - 7$
$y = -x^2 + 11$

28. $y = x^2 - 8$
$y = -x^2$

29. $x = y^2 - 5y$
$x = y^2 + 10$

30. $x = 3y^2 - 15y + 16$
$x = 2y^2 - 9y + 7$

Solve by the addition method.

31. $x^2 + y^2 = 5$
$x^2 - y^2 = 3$

32. $x^2 + y^2 = 20$
$y^2 - x^2 = 12$

33. $3x^2 + y^2 = 14$
$x^2 - y^2 = 2$

34. $5x^2 + y^2 = 45$
$x^2 - y^2 = 9$

35. $x^2 + 2y^2 = 18$
$y^2 - x^2 = 6$

36. $x^2 + 9y^2 = 9$
$y^2 - x^2 = 1$

37. $9x^2 + 4y^2 = 87$
$x^2 - 2y^2 = 6$

38. $8x^2 + 6y^2 = 44$
$4x^2 - 2y^2 = 12$

39. $x^2 + y^2 = 24$
$x^2 + 5y^2 = 60$

40. $x^2 + y^2 = 19$
$10x^2 + y^2 = 100$

41. $x^2 + y^2 = 6$
$4x^2 + 9y^2 = 39$

42. $x^2 + y^2 = 13$
$3x^2 + 18y^2 = 99$

43. $x^2 + 12y^2 = 117$
$x^2 + 4y^2 = 45$

44. $3x^2 + y^2 = 21$
$10x^2 + y^2 = 49$

45. $3x^2 + 2y^2 = 21$
$15x^2 + 4y^2 = 51$

46. $3x^2 + 4y^2 = 84$
$9x^2 + 21y^2 = 333$

47. A rectangular flower garden has a perimeter of 44 feet and an area of 105 square feet. Find the dimensions of the garden.

48. A rectangular classroom has a perimeter of 78 feet and an area of 360 square feet. Find the dimensions of the classroom.

49. The cover for a rectangular swimming pool measures 400 square feet. If the perimeter of the pool is 82 feet, find the dimensions of the pool.

50. The perimeter of a soccer field is 320 yards, and the area of the field is 6000 square yards. Find the dimensions of the soccer field.

51. John is remodeling his home office by replacing the molding around the floor of the room as well as the carpet. The door to the office is 4 feet wide, and no molding is required there. If John used 60 feet of molding and 192 square feet of carpet, find the dimensions of the office.

52. Paige is painting a rectangular wall in her home. The perimeter of the wall is 60 feet. The wall has two windows in it, each with an area of 15 square feet. If the painting area is 170 square feet, find the dimensions of the wall.

53. George's vegetable garden is in the shape of a rectangle, and has 210 feet of fencing around it. Diagonally, the garden measures 75 feet from corner to corner. Find the dimensions of the garden.

54. A rectangular lawn has a perimeter of 100 feet and a diagonal of $10\sqrt{13}$ feet. Find the dimensions of the lawn.

55. A rectangular sports court has an area of 800 square feet. If the diagonal of the court measures $20\sqrt{5}$ feet, find the dimensions of the court.

56. Laura's horses live in a rectangular pasture that has an area of 10,000 square feet. The diagonal of the pasture measures $50\sqrt{17}$ feet. Find the dimensions of the pasture.

57. An arrow is fired upward with an initial velocity of 120 feet per second. At the same instant, a ball is dropped from a helicopter that is hovering 540 feet above the ground. At what time will the arrow and ball be the same height above the ground?

58. A ball is thrown upward from a beach with an initial velocity of 32 feet per second. At the same instant, another ball is dropped from a cliff 48 feet above the beach. At what time will the two balls be the same height above the beach?

59. A model rocket is fired upward from the ground with an initial velocity of 60 feet per second. At the same instant, another model rocket is fired upward from a roof 84 feet above the ground with an initial velocity of 32 feet per second.

a) At what time will the two rockets be the same height above the ground?

b) How high above the ground are the two rockets at that time?

60. A slingshot fires a rock upward from the ground with an initial velocity of 96 feet per second. At the same instant, a cannonball is fired upward from a roof 120 feet above the ground with an initial velocity of 66 feet per second.

a) At what time will the rock and cannonball be the same height above the ground?

b) How high above the ground are the two objects at that time?

Answer in complete sentences,

61. Write a word problem leading to a system of nonlinear equation whose solution is "The rectangle is 50 feet by 30 feet."

62. Give an example of a system of nonlinear equations for which you feel it would be best to solve by the substitution method; give an example of another system for which you feel it would be best to solve by the addition method. Explain why the method is the best choice for each example.

> ***Study Tip*** **Revisited**   Choosing the location where you will be studying for the final exam is not a trivial matter. Be sure to study in a distraction-free zone; you cannot afford to waste time and effort at this point in the course. Study in a well-lit area that allows you to have easy access to all of your materials. See the study tips in Chapter 11 for further advice on creating a positive study environment.

Section 13.1—Topic	Chapter Review Exercises
Graphing Parabolas of the Form $y = ax^2 + bx + c$	1–2
Graphing Parabolas of the Form $y = a(x - h)^2 + k$	3–4
Graphing Parabolas of the Form $x = ay^2 + by + c$	5–8
Graphing Parabolas of the Form $x = a(y - k)^2 + h$	9–12

Section 13.2—Topic	Chapter Review Exercises
Distance Formula	13–14
Graphing a Circle	15–20
Finding the Center and Radius By Completing the Square	21–23
Finding the Equation of a Circle From Its Graph	24–25

Section 13.3—Topic	Chapter Review Exercises
Graphing an Ellipse	26–30
Finding the Center, a, and b By Completing the Square	31–33
Finding the Equation of an Ellipse From Its Graph	34–35

Section 13.4—Topic	Chapter Review Exercises
Graphing a Hyperbola	36–40
Finding the Center, a, and b By Completing the Square	41–43
Finding the Equation of a Hyperbola From Its Graph	44–45

Section 13.5—Topic	Chapter Review Exercises
Solving Nonlinear Systems of Equations By Substitution	46–53
Solving Nonlinear Systems of Equations By Addition	54–57
Applications of Nonlinear Systems of Equations	58–60

Summary of Chapter 13 Study Tips

The study tips in this chapter focused on preparing to take a final exam and serve as a summary of the various study tips presented throughout the textbook. After determining the types of problems that will be on the exam, you can look for problems to practice in various reviews throughout this textbook, in your notes, on your old exams, and on MyMath-Lab.com. Your instructor may also give you a review assignment for the final exam.

Try to free up as much time as possible in your daily schedule in order to maximize your studying, and be sure to start studying for the final exam at least two weeks prior to the actual exam.

Study tips for preparing for a final exam will continue in Chapter 14.

Graph. Label the vertex and any intercepts. [13.1]

1. $y = x^2 - 8x - 9$ **2.** $y = -x^2 + 6x - 3$

6. $x = y^2 + 3y - 28$ **7.** $x = -y^2 + y + 20$

8. $x = -y^2 - 6y - 14$

3. $y = (x - 4)^2 - 1$ **4.** $y = -(x + 3)^2 - 4$

9. $x = (y + 5)^2 - 4$

5. $x = y^2 - 10y + 15$

10. $x = -(y + 1)^2 + 10$

11. $x = -(y - 4)^2 + 9$

17. $(x + 3)^2 + (y - 4)^2 = 4$

18. $(x - 8)^2 + (y + 3)^2 = 9$

12. $x = (y - 3)^2 + 6$

19. $(x + 4)^2 + (y - 5)^2 = 25$

20. $(x - 2)^2 + y^2 = 16$

Find the distance between the given points. Round to the nearest tenth if necessary. [13.2]

13. $(-7, -8)$ and $(1, 7)$

14. $(3, -5)$ and $(-6, -2)$

Graph the circle. Label the center of the circle. Give its radius as well. [13.2]

15. $x^2 + y^2 = 36$ **16.** $x^2 + y^2 = 20$

21. $x^2 + y^2 + 12x - 2y - 12 = 0$

Find the center and radius of the circle by completing the square. [13.2]

22. $x^2 + y^2 + 6x - 18y - 10 = 0$

23. $x^2 + y^2 - 8x - 20y + 99 = 0$

Find the equation of the circle. [13.2]

24.

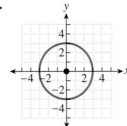

25.

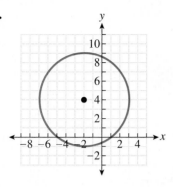

Graph the ellipse. Label the center. Give the values of a and b. [13.3]

26. $\dfrac{x^2}{25} + \dfrac{y^2}{9} = 1$

27. $x^2 + \dfrac{y^2}{16} = 1$

28. $\dfrac{(x+3)^2}{9} + \dfrac{(y-4)^2}{16} = 1$

29. $\dfrac{(x-8)^2}{25} + \dfrac{(y-1)^2}{4} = 1$

30. $\dfrac{(x+6)^2}{9} + \dfrac{y^2}{36} = 1$

31. $9x^2 + 16y^2 + 90x - 64y + 145 = 0$

Find the center of the ellipse by completing the square. Find the values of a and b as well. [13.3]

32. $4x^2 + 49y^2 - 48x - 98y - 3 = 0$

33. $81x^2 + 16y^2 + 324x + 128y - 716 = 0$

Find the equation of the ellipse. [13.3]

34.

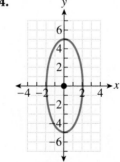

35.

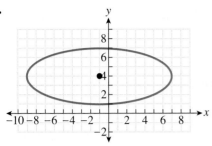

38. $\dfrac{(x-1)^2}{4} - \dfrac{(y-3)^2}{9} = 1$

Graph the hyperbola. Label the center. Give the values of a and b. [13.4]

36. $\dfrac{x^2}{9} - \dfrac{y^2}{25} = 1$

39. $\dfrac{(x+4)^2}{25} - \dfrac{(y-2)^2}{36} = 1$

37. $\dfrac{y^2}{16} - \dfrac{x^2}{16} = 1$

40. $\dfrac{(y+3)^2}{4} - \dfrac{(x+2)^2}{81} = 1$

41. $-4x^2 + 25y^2 - 24x + 300y + 764 = 0$

47. $4x^2 + 3y^2 = 48$
$y = 2x - 4$

48. $y = x^2 - 8x + 23$
$2x - y = 1$

49. $4y^2 - x^2 = 96$
$6y - x = 32$

50. $x^2 + y^2 = 17$
$y = x^2 + 3$

51. $3x^2 + 5y^2 = 72$
$y = x^2 - 6$

Find the center of the hyperbola by completing the square. Find the lengths of its transverse and conjugate axes as well. **[13.4]**

42. $-9x^2 + 4y^2 + 90x - 56y - 65 = 0$

52. $9x^2 - y^2 = 9$
$y = x^2 - 1$

53. $y = -x^2 + 2x + 12$
$y = x^2$

43. $49x^2 - 36y^2 + 784x + 288y + 796 = 0$

Solve by the addition method. **[13.5]**

54. $x^2 + y^2 = 100$
$y^2 - x^2 = 62$

Find the equation of the hyperbola. **[13.4]**

44.

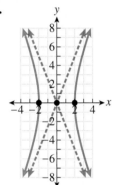

55. $3x^2 + 4y^2 = 31$
$x^2 - 2y^2 = 7$

56. $x^2 + y^2 = 8$
$2x^2 + 15y^2 = 68$

57. $6x^2 + y^2 = 159$
$2x^2 + 9y^2 = 131$

58. A rectangular flower garden has a perimeter of 70 feet and an area of 300 square feet. Find the dimensions of the garden. **[13.5]**

59. A rectangle has a perimeter of 42 inches and a diagonal of 15 inches. Find the dimensions of the rectangle.

60. A projectile is fired upward from the ground with an initial velocity of 76 feet per second. At the same instant, another projectile is fired upward from a roof 140 feet above the ground with an initial velocity of 41 feet per second. **[13.5]**

a) After how long will the two projectiles be the same height above the ground?

b) How high above the ground are the two projectiles at that time?

45.

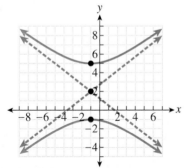

Solve by substitution. **[13.5]**

46. $x^2 + y^2 = 50$
$x + 2y = 15$

Graph. Label the vertex and any intercepts.

1. $y = x^2 - 4x + 7$

2. $y = -(x + 3)^2 + 9$

3. $x = -y^2 + 9y - 14$

4. $x = (y + 1)^2 - 8$

Find the distance between the given points. Round to the nearest tenth if necessary.

5. $(7, -3)$ and $(3, 5)$

Graph the circle. Identify the center and radius of the circle.

6. $(x + 4)^2 + (y - 2)^2 = 25$

Find the center and radius of the circle by completing the square.

7. $x^2 + y^2 - 12x - 6y + 13 = 0$

Find the equation of the circle with the given center and radius.

8. Center $(3, -8)$, radius 16

Graph the ellipse. Label the center. Give the values of a and b.

9. $\dfrac{(x - 4)^2}{16} + \dfrac{(y + 7)^2}{9} = 1$

Find the center of the ellipse by completing the square. Give the values of a and b.

10. $4x^2 + 25y^2 - 64x + 100y + 256 = 0$

Graph the hyperbola. Label the center. Give the values of a and b.

11. $\dfrac{x^2}{16} - y^2 = 1$

Find the equation of the conic section that has been graphed.

12.

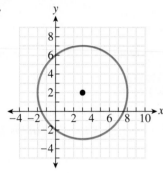

13.

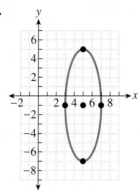

Solve the nonlinear system.

14. $x^2 + y^2 = 100$
$x + 3y = 10$

15. $x^2 + 6y^2 = 49$
$x - 2y = 1$

16. $y^2 - x^2 = 10$
$y = x^2 - 32$

17. $x^2 + 4y^2 = 85$
$3x^2 + 2y^2 = 165$

18. A rectangular corral has a perimeter of 340 feet and an area of 6000 square feet. Find the dimensions of the corral.

Mathematicians in History
Sofia Kovalevskaya

The story of 19th-century Russian mathematician Sofia Kovalevskaya is one of genius and determination. At a time when women were discouraged from studying mathematics, she developed into one of the most respected mathematicians of her day.

Write a one-page summary (*or* make a poster) of the life of Sofia Kovalevskaya and her accomplishments.

Interesting issues:

- Where and when was Sofia Kovalevskaya born?
- What were the walls of her nursery papered with when she was 11 years old?
- Who convinced Kovalevskaya's parents to allow her to study mathematics?
- Whom did Kovalevskaya marry, and what was his fate?
- By the spring of 1874, Kovalevskaya had completed three papers while studying with which prominent mathematician in Berlin?
- What job did Kovalevskaya hold after earning her doctorate?
- In what European city did Kovalevskaya finally obtain a position?
- When did Kovalevskaya die, and what were the circumstances of her death?
- Kovalevskaya once said, "It is impossible to be a mathematician without being a poet in soul." Explain what you think Kovalevskaya meant by that.

Given the following data, find the equation that describes Mercury's revolution around the sun.

(Hint: The sun is not at the center of Mercury's path; rather, it is at one of the foci.)

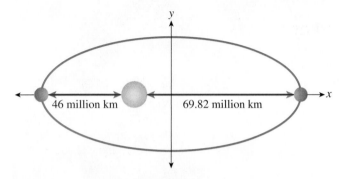

Radius of Sun = 0.696 *million km*

Radius of Mercury = *negligible*

SEQUENCES, SERIES AND THE BINOMIAL THEOREM

*In this chapter, we will learn about **sequences**, which are ordered lists of numbers. We will also investigate **series**, which are the sums of the numbers in a sequence. In particular, we will discuss two specific types of sequences and series: the arithmetic sequence and series and the geometric sequence and series.*

The chapter concludes with a section on the binomial theorem, which provides us with a method for raising a binomial to a power that is a natural number such as $(2x + 3y)^9$.

Study Tip USING THE PREVIOUS STUDY TIPS TO PREPARE FOR A FINAL EXAM (PART 2)
The study tips in this chapter continue to focus on how to incorporate the study tips in previous chapters into your preparation for a final exam.

14.1

SEQUENCES AND SERIES

1. Find the terms of a sequence given its general term.
2. Find the general term of a sequence.
3. Find partial sums of a sequence.
4. Use summation notation to evaluate a series.
5. Use sequences to solve applied problems.

Sequences

Suppose that you deposited $100 in a bank account that pays 6% interest, compounded monthly. Here is the amount of money in the account after each of the first five years if no further deposits or withdrawals are made:

At the end of year . . .	Principal Balance
1	$106.17
2	$112.72
3	$119.67
4	$127.05
5	$134.89

Using this information, we can create a function whose input is a natural number from 1 to 5 and whose output is the balance of the account after that number of years. This type of function is called a sequence.

Sequence

> A **sequence** is a function whose domain is a set of consecutive natural numbers beginning with 1. If the domain is the entire set of natural numbers $\{1, 2, 3, \ldots\}$, then the sequence is an **infinite sequence.** If the domain is the set $\{1, 2, 3, \ldots, n\}$, for some natural number n, then the sequence is a **finite sequence.**

The input for the function tells the order of the term in the sequence, while the output of the function is the actual term. We denote the first number in a sequence as a_1, the second number in a sequence as a_2, and so on. In the example of the bank account, $a_1 = \$106.17$, $a_2 = \$112.72$, and so on.

Finding the Terms of a Sequence

Objective 1 **Find the terms of a sequence given its general term.** A sequence is often denoted by its **general term** a_n. The general term of a sequence provides a formula for finding any term of a sequence, provided that we know its position in the sequence.

EXAMPLE 1   Find the first five terms of the sequence whose general term is $a_n = n^2 - 10$.

Solution

We will substitute the first five natural numbers for n in the general term.

$$a_1 = (1)^2 - 10 = -9 \qquad a_2 = (2)^2 - 10 = -6 \qquad a_3 = (3)^2 - 10 = -1$$
$$a_4 = (4)^2 - 10 = 6 \qquad a_5 = (5)^2 - 10 = 15$$

The first five terms of the sequence are $-9, -6, -1, 6$, and 15. The entire sequence could be represented as $-9, -6, -1, 6, 15, \ldots$.

Quick Check **1**
Find the first four terms of the sequence whose general term is
$$a_n = \frac{1}{3n + 4}.$$

A sequence whose terms alternate between being positive and negative is called an **alternating sequence.** An alternating sequence has a factor of $(-1)^n$ or $(-1)^{n+1}$ in its general term. Each time the value of n increases by 1, both $(-1)^n$ and $(-1)^{n+1}$ alternate between -1 and 1.

EXAMPLE 2   Find the first four terms of the sequence whose general term is
$$a_n = \frac{(-1)^n}{3^{n+2}}.$$

Solution

We will substitute the first four natural numbers for n in the general term.

$$a_1 = \frac{(-1)^1}{3^{1+2}} = \frac{-1}{3^3} = -\frac{1}{27} \qquad a_2 = \frac{(-1)^2}{3^{2+2}} = \frac{1}{3^4} = \frac{1}{81}$$
$$a_3 = \frac{(-1)^3}{3^{3+2}} = \frac{-1}{3^5} = -\frac{1}{243} \qquad a_4 = \frac{(-1)^4}{3^{4+2}} = \frac{1}{3^6} = \frac{1}{729}$$

Quick Check **2**
Find the first four terms of the sequence whose general term is
$$a_n = \frac{(-1)^{n+1}}{2n}.$$

The first four terms of the sequence are $-\frac{1}{27}, \frac{1}{81}, -\frac{1}{243}$, and $\frac{1}{729}$.

Finding the General Term of a Sequence

Objective 2   Find the general term of a sequence. To find the general term of a sequence, we must be able to detect the pattern of the terms of the sequence; consequently, we must know enough terms of the sequence. Suppose that we were trying to find the general term of the sequence $2, 4, \ldots$. There are not enough terms to determine the pattern. It could be the sequence whose general term is $a_n = 2^n$ $(2, 4, 8, 16, 32, \ldots)$, or it could be the sequence whose general term is $a_n = 2n$ $(2, 4, 6, 8, 10, \ldots)$. There are simply not enough terms to be sure.

EXAMPLE 3   Find the next three terms of the sequence $3, 6, 9, 12, 15, \ldots$, and find its general term a_n.

Solution

The terms are all multiples of 3, so continuing the pattern gives us the terms 18, 21, and 24. The general term is $a_n = 3n$.

When examining the terms of a sequence, determine whether the successive terms increase by the same amount each time. If they do, then the general term a_n will contain a multiple of n.

EXAMPLE 4 Find the next three terms of the sequence 9, 13, 17, 21, 25, . . . , and find its general term a_n.

Solution

First, we should notice that each successive term increases by 4. So the next three terms are 29, 33, and 37. When each successive term increases by 4, this tells us that the general term will contain the expression $4n$. However, $4n$ is not the general term of this sequence, as its terms would be 4, 8, 12, 16, 20, ... Notice that each term is five less than the corresponding term in the sequence 9, 13, 17, 21, 25, ...The general term can be found by adding 5 to $4n$, so the general term is $a_n = 4n + 5$.

We now turn to finding the general term of an alternating sequence.

EXAMPLE 5 Find the next three terms of the sequence $-1, 4, -9, 16, -25, \ldots$, and find its general term a_n.

Solution

Notice that if we disregard the signs, the first five terms are the squares of 1, 2, 3, 4, and 5. Since the signs of the terms are alternating, the next term will be positive. The next three terms are 36, -49, and 64.

When an alternating sequence begins with a negative term, the general term will contain the factor $(-1)^n$. (If the first term had been positive, the general term would have contained the factor $(-1)^{n+1}$.) The general term for this sequence is $a_n = (-1)^n \cdot n^2$.

On occasion we represent a sequence by giving its first term and explaining how to find any term after that from the preceding term. In this case, we are defining the general term **recursively.** For example, recall the sequence 9, 13, 17, 21, 25, . . . , from Example 4. We found its general term to be $a_n = 4n + 5$. Since each term in the sequence is four more than the previous term, we could define the general term recursively as $a_n = a_{n-1} + 4$, where a_{n-1} is the term that precedes the term a_n. A recursive formula for a general term a_n defines it in terms of the preceding term, a_{n-1}.

Quick Check **3**
Find the next three terms of the sequence $7, -13, 19, -25, 31, \ldots$, and find its general term a_n.

EXAMPLE 6 Find the first five terms of the sequence whose first term is $a_1 = 19$ and whose general term is $a_n = a_{n-1} - 7$.

Solution

We know that the first term is $a_1 = 19$. We can now use the recursive formula $a_n = a_{n-1} - 7$ to find the next four terms.

$$a_2 = a_{2-1} - 7 = a_1 - 7 = 19 - 7 = 12$$
$$a_3 = a_{3-1} - 7 = a_2 - 7 = 12 - 7 = 5$$
$$a_4 = a_{4-1} - 7 = a_3 - 7 = 5 - 7 = -2$$
$$a_5 = a_{5-1} - 7 = a_4 - 7 = -2 - 7 = -9$$

The first five terms of the sequence are 19, 12, 5, -2, and -9.

Quick Check **4**
Find the first five terms of the sequence whose first term is $a_1 = 2$ and whose general term is $a_n = (a_{n-1})^2 + 1$.

Series

We now turn our attention to series, which are sums of the terms in a sequence.

Infinite Series, Finite Series

> An **infinite series,** s, is the sum of the terms in an infinite sequence.
>
> $$s = a_1 + a_2 + a_3 + \cdots + a_n + \cdots$$
>
> A **finite series,** s_n, is the sum of the first n terms of a sequence.
>
> $$s_n = a_1 + a_2 + a_3 + \cdots + a_n$$
>
> The finite series s_n is also called the **nth partial sum** of the sequence.

Finding Partial Sums

Objective 3 Find partial sums of a sequence.

EXAMPLE 7   For the sequence whose general term is $a_n = (-1)^n \cdot (2n + 1)$, find s_6.

Solution

The first six terms of the sequence are $a_1 = -3$, $a_2 = 5$, $a_3 = -7$, $a_4 = 9$, $a_5 = -11$, and $a_6 = 13$. We can now find the partial sum.

$$s_6 = (-3) + 5 + (-7) + 9 + (-11) + 13 = 6$$

> **Quick Check 5**
> For the sequence whose general term is $a_n = n^2 + 3n$, find s_4.

EXAMPLE 8   For the sequence whose first term is $a_1 = -4$ and whose general term is $a_n = \dfrac{a_{n-1}}{2}$, find s_5.

Solution

We know that $a_1 = -4$. Since each term in the sequence is half of the preceding term, $a_2 = -2$, $a_3 = -1$, $a_4 = -\frac{1}{2}$, and $a_5 = -\frac{1}{4}$. We can now find the partial sum.

$$s_5 = (-4) + (-2) + (-1) + \left(-\frac{1}{2}\right) + \left(-\frac{1}{4}\right) = -7\frac{3}{4}$$

> **Quick Check 6**
> For the sequence whose first term is $a_1 = 1$ and whose general term is $a_n = 3a_{n-1} + 4$, find s_4.

Summation Notation

Objective 4 Use summation notation to evaluate a series. The nth partial sum of a sequence, $s_n = a_1 + a_2 + a_3 + \cdots + a_n$, can be represented as $\displaystyle\sum_{i=1}^{n} a_i$. This notation for a partial sum is called **summation notation.** The number n is the **upper limit of summation,** while the number 1 is the **lower limit of summation.** The variable i is called the **index of summation,** or simply the **index.** The symbol Σ is the capital

Greek letter sigma, and is often used in mathematics to represent a sum. This notation is also called **sigma notation.**

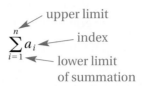

$$\sum_{i=1}^{n} a_i$$

upper limit

index

lower limit
of summation

EXAMPLE 9   Find the sum $\sum_{i=1}^{8}(3i + 5)$.

Solution

In this example, we need to find the sum of the first eight terms of this sequence. The first term of the sequence is 8, and each successive term increases by 3.

$$\sum_{i=1}^{8}(3i + 5) = 8 + 11 + 14 + 17 + 20 + 23 + 26 + 29$$ Rewrite as the sum of the first eight terms of the sequence.

$$= 148$$ Add.

Quick Check 7

Find the sum $\sum_{i=1}^{4} 6i.$

Applications

Objective 5   Use sequences to solve applied problems. We conclude this section with an applied problem involving sequences.

EXAMPLE 10   Erin found a new job with a starting salary of $40,000 per year. Each year, she will be given a 5% raise. Write a sequence showing Erin's annual salary for the first five years, and find the total amount she will earn during those five years.

Solution

Since Erin earned $40,000 in the first year, $a_1 = $40,000$. To find her salary for the second year, we calculate her 5% raise and add it to her salary. Five percent of $40,000 can be calculated by multiplying 0.05 by $40,000.

$$0.05(\$40,000) = \$2000$$

With her $2000 raise, Erin's salary for the second year is $42,000. So $a_2 = $42,000$. Continuing in the same fashion, we find that the sequence is $40,000, $42,000, $44,100, $46,305, $48,620.25. During the first five years, Erin will earn $221,025.25.

Quick Check 8 A copy machine was purchased for $8000. Its value decreases by 20% each year. Write a sequence showing the value of the copy machine at the end of each of the first four years.

1. Find the fifth term of the sequence whose general term is $a_n = 3n - 4$.

2. Find the third term of the sequence whose general term is $a_n = (n + 3)^2 - 5n$.

3. Find the sixth term of the sequence whose general term is $a_n = (-1)^n \cdot 3^{n+1}$.

4. Find the ninth term of the sequence whose general term is $a_n = (-1)^n \cdot \dfrac{1}{2n - 1}$.

Find the first five terms of the sequence with the given general term.

5. $a_n = -n$

6. $a_n = 2n$

7. $a_n = 4n - 3$

8. $a_n = 3n + 7$

9. $a_n = (-1)^n \cdot (2n + 3)$

10. $a_n = \dfrac{(-1)^{n+1}}{n^2}$

11. $a_n = n^2 - 3n$

12. $a_n = 4 \cdot 3^{n-1}$

Find the next three terms of the given sequence, and find its general term a_n.

13. $5, 10, 15, 20, 25, \ldots$

14. $3, 10, 17, 24, 31, \ldots$

15. $1, 9, 81, 729, 6561, \ldots$

16. $9, 16, 25, 36, 49, \ldots$

17. $\dfrac{1}{5}, \dfrac{1}{8}, \dfrac{1}{11}, \dfrac{1}{14}, \dfrac{1}{17}, \ldots$

18. $\dfrac{1}{2}, \dfrac{1}{4}, \dfrac{1}{6}, \dfrac{1}{8}, \dfrac{1}{10}, \ldots$

19. $1, 8, 27, 64, 125, \ldots$

20. $-1, 2, -4, 8, -16, \ldots$

Find the first five terms of the sequence with the given first term and general term.

21. $a_1 = 6, a_n = a_{n-1} + 3$

22. $a_1 = 10, a_n = 2a_{n-1}$

23. $a_1 = -5, a_n = 3a_{n-1} + 8$

24. $a_1 = -9, a_n = a_{n-1} - 10$

25. $a_1 = 16, a_n = -3a_{n-1}$

26. $a_1 = 8, a_n = \dfrac{-a_{n-1}}{2}$

27. $a_1 = -35, a_n = -a_{n-1} - 17$

28. $a_1 = 20, a_n = \dfrac{1}{2a_{n-1}}$

Find the indicated partial sum for the given sequence.

29. $s_8, 13, 20, 27, 34, \ldots$

30. $s_9, 25, 16, 7, -2, \ldots$

31. $s_7, -18, -31, -44, -57 \ldots$

32. $s_6, -19, -2, 15, 32, \ldots$

33. $s_9, 6, 12, 24, 48, \ldots$

34. $s_5, -3, 27, -243, 2187, \ldots$

35. $s_6, 25, -36, 49, -64, \ldots$

36. $s_{40}, 1, -1, 1, -1, \ldots$

Find the indicated partial sum for the sequence with the given general term.

37. $s_5, a_n = 3n + 10$

38. $s_4, a_n = 2n - 19$

39. $s_6, a_n = -7n$

40. $s_7, a_n = -n + 8$

41. $s_5, a_n = n^2$

42. $s_5, a_n = 2n - 1$

43. $s_{10}, a_n = (-1)^n \cdot (5n)$

44. $s_9, a_n = (-1)^n \cdot n^2$

Find the indicated partial sum for the sequence with the given first term and general term.

45. $s_6, a_1 = 2, a_n = a_{n-1} + 7$

46. $s_7, a_1 = 9, a_n = a_{n-1} + 4$

47. $s_8, a_1 = 5, a_n = a_{n-1} - 9$

48. $s_5, a_1 = -11, a_n = a_{n-1} - 14$

49. $s_4, a_1 = 3, a_n = 20a_{n-1}$

50. $s_5, a_1 = -4, a_n = -3a_{n-1} + 13$

51. $s_5, a_1 = 3, a_n = (a_{n-1})^2$

52. $s_5, a_1 = 2, a_n = (a_{n-1})^2 + 1$

Find the sum.

53. $\sum_{i=1}^{15} i$ **54.** $\sum_{i=1}^{7} (i - 3)$

55. $\sum_{i=1}^{8} (2i - 1)$ **56.** $\sum_{i=1}^{5} (21i + 35)$

57. $\sum_{i=1}^{10} i^2$ **58.** $\sum_{i=1}^{6} i^3$

59. $\sum_{i=1}^{15} \left(\dfrac{1}{i} - \dfrac{1}{i+1} \right)$ **60.** $\sum_{i=1}^{8} \dfrac{1}{2^i}$

Rewrite the sum using summation notation.

61. $7 + 14 + 21 + 28$

62. $2 + 4 + 6 + 8 + 10$

63. $5 + (-10) + 15 + (-20) + 25 + (-30)$

64. $(-3) + 6 + (-9) + 12 + (-15)$

65. $1 + 4 + 9 + 16 + 25 + 36 + 49$

66. $1 + \dfrac{1}{2} + \dfrac{1}{3} + \dfrac{1}{4} + \dfrac{1}{5}$

67. Theo purchased a new computer, valued at $2500, for his accounting firm. If the value of the computer decreases by 40% each year, construct a sequence showing the value of the computer after each of the first four years.

68. A quilt shop purchased a quilting machine for $10,000. If the resale value of the machine drops by 20% each year, construct a sequence showing the value of the quilting machine after each of the first six years.

69. Tina started a new job with an annual salary of $32,000. Each year, she will be given an $800 raise.

   a) Write a sequence showing Tina's annual salary for the first eight years.

   b) Find the total amount earned by Tina during those eight years.

70. Bart started a new job with an annual salary of $27,500. Each year, he will be given a 2% raise.

   a) Write a sequence showing Bart's annual salary for the first five years.

   b) Find the total amount earned by Bart during those five years.

Answer in complete sentences.

71. Describe a real-world application involving a sequence.

72. Describe a real-world application involving a series.

> **Study Tip** REVISITED   If you have been working with a study group while taking the course, try to incorporate working with your group into your preparation for the final. A group study session is a great place to bring difficult problems. Keep your focus on the job at hand while working with your group. Often, a group will have a member who would prefer to socialize rather than study, but you cannot afford to waste any time at this point in the course. For further information on working with a study group, see the study tips in Chapter 1.

14.2

ARITHMETIC SEQUENCES AND SERIES

Objectives

1. Find the general term of an arithmetic sequence.
2. Find partial sums of an arithmetic sequence.
3. Use an arithmetic sequence to solve an applied problem.

Arithmetic Sequences

In this section, we will examine a particular type of sequence called an arithmetic sequence.

Arithmetic Sequence

> An **arithmetic sequence** is a sequence in which each term, after the first term, differs from the preceding term by a constant amount d. This means that the general term a_n can be defined recursively by $a_n = a_{n-1} + d$, for $n \neq 1$. The number d is called the **common difference** of the arithmetic sequence.

If a sequence is an arithmetic sequence, successive terms must increase or decrease by the same amount for each term in the sequence after the first term. For example, the sequence 4, 13, 22, 31, 40, . . . is an arithmetic sequence with $d = 9$ because each term is nine greater than the preceding term. The sequence $\frac{5}{4}, \frac{3}{4}, \frac{1}{4}, -\frac{1}{4}, -\frac{3}{4}, \ldots$ is also an arithmetic sequence with $d = -\frac{1}{2}$, as each term is $\frac{1}{2}$ less than the preceding term.

The sequence 2, 4, 8, 16, 32, . . . is not an arithmetic sequence. The second term is two greater than the first term, but the third term is four greater than the second term. Since the terms are not increasing by the same amount for the entire sequence, the sequence is not an arithmetic sequence.

EXAMPLE 1   Find the first five terms of the arithmetic sequence whose first term is 5 and whose common difference is $d = 4$.

Solution

Since the first term of this sequence is 5 and each term is four greater than the previous term, the first five terms of the sequence are 5, 9, 13, 17, and 21.

Quick Check **1**

Find the first six terms of the arithmetic sequence whose first term is −8 and whose common difference is $d = 3$.

Finding the General Term of an Arithmetic Sequence

Objective 1 **Find the general term of an arithmetic sequence.** For any arithmetic sequence whose first term is a_1 and whose common difference is d, we know that the second term a_2 is equal to $a_1 + d$. Since we could add d to the second term to find the third term of the sequence, we know that $a_3 = a_1 + 2d$.

$$
\begin{aligned}
a_3 &= a_2 + d \\
&= (a_1 + d) + d \\
&= a_1 + 2d
\end{aligned}
$$

Continuing this pattern, we have $a_4 = a_1 + 3d$, $a_5 = a_1 + 4d$, and so on. This leads to the following result for the general term a_n of an arithmetic sequence:

General Term of an Arithmetic Sequence

> If an arithmetic sequence whose first term is a_1 has a common difference of d, then the general term of this sequence is $a_n = a_1 + (n - 1)d$.

EXAMPLE 2   Find the general term, a_n, of the arithmetic sequence 5, 11, 17, 23, 29, ...

Solution

The first term, a_1, of this sequence is 5. We will subtract the first term of the sequence from the second term to find the common difference d.

$$d = 11 - 5 = 6$$

We now can find the general term of this sequence using the formula $a_n = a_1 + (n - 1)d$.

$$a_n = 5 + (n - 1)6 \qquad \text{Substitute 5 for } a_1 \text{ and 6 for } d.$$
$$= 6n - 1 \qquad \text{Simplify.}$$

Quick Check **2**
Find the general term, a_n, of the arithmetic sequence 2, 14, 26, 38, 50, ...

The general term for this arithmetic sequence is $a_n = 6n - 1$.

After we find the general term, a_n, of an arithmetic sequence, we can go on to find any particular term of the sequence.

EXAMPLE 3   Find the 45th term, a_{45}, of the arithmetic sequence 3, 7, 11, 15, 19, ...

Solution

We begin by finding the general term, a_n, of this arithmetic sequence. We can then find a_{45}.
    The first term, a_1, of this sequence is 3, and the common difference d is 4. We now can find the general term of this sequence using the formula $a_n = a_1 + (n - 1)d$.

$$a_n = 3 + (n - 1)4 \qquad \text{Substitute 3 for } a_1 \text{ and 4 for } d.$$
$$= 4n - 1 \qquad \text{Simplify.}$$

The general term for this arithmetic sequence is $a_n = 4n - 1$. We can now find the 45th term.

$$a_{45} = 4(45) - 1 \qquad \text{Substitute 45 for } n \text{ into } a_n = 4n - 1.$$
$$= 179 \qquad \text{Simplify.}$$

Quick Check **3**
Find the 61st term, a_{61}, of the arithmetic sequence 57, 59, 61, 63, 65, ...

The 45th term of this arithmetic sequence is 179.

Arithmetic Series

Objective 3 Find partial sums of an arithmetic sequence. An **arithmetic series** is the sum of the terms of an arithmetic sequence. A finite arithmetic series consisting of only the first n terms of an arithmetic sequence is also known as the **nth partial sum** of the arithmetic sequence, and is denoted as s_n.

EXAMPLE 4   Find s_7 for the arithmetic sequence 4, 8, 12, 16, 20, ...

Solution

Since only the first five terms are given, we must begin by finding the sixth and seventh terms. Since each term in this arithmetic sequence is four greater than the preceding term, the next two terms of the sequence are 24 and 28. We now can find the seventh partial sum of this arithmetic sequence.

$$s_7 = 4 + 8 + 12 + 16 + 20 + 24 + 28$$
$$= 112 \qquad \text{Add.}$$

Quick Check **4**

Find s_9 for the arithmetic sequence 20, 26, 32, 38, 44, ...

We now present a formula for finding partial sums of an arithmetic sequence. This formula will be quite helpful, as it depends on only the first and nth terms of the arithmetic sequence (a_1 and a_n), as well as the number of terms being added.

nth Partial Sum of an Arithmetic Sequence

The sum of the first n terms of an arithmetic sequence is given by

$$s_n = \frac{n}{2}(a_1 + a_n)$$

where a_1 is the first term and a_n is the nth term of the sequence.

Before we apply this formula, let's investigate where the formula comes from. We will let a_1 be the first term of the arithmetic sequence, a_n be its nth term, and d be the common difference. Since s_n is the sum of the first n terms of the sequence, it can be expressed as

$$s_n = a_1 + (a_1 + d) + (a_1 + 2d) + (a_1 + 3d) + \cdots + (a_1 + (n-1)d)$$

The first n terms of an arithmetic sequence can also be generated by starting with the nth term a_n and subtracting d to find the preceding term: $a_n, a_n - d, a_n - 2d, \ldots$ $a_n - (n-1)d$. So the nth partial sum can also be expressed as

$$s_n = a_n + (a_n - d) + (a_n - 2d) + (a_n - 3d) + \cdots + (a_n - (n-1)d)$$

We now have two different expressions for s_n. If we add the two expressions, term by term, we obtain the following result:

$$
\begin{aligned}
s_n &= a_1 &&+ (a_1 + d) + (a_1 + 2d) + (a_1 + 3d) + \cdots + (a_1 + (n-1)d) \\
s_n &= a_n &&+ (a_n - d) + (a_n - 2d) + (a_n - 3d) + \cdots + (a_n - (n-1)d) \\
\hline
2s_n &= (a_1 + a_n) &&+ (a_1 + a_n) + (a_1 + a_n) + (a_1 + a_n) + \cdots + (a_1 + a_n)
\end{aligned}
$$

So $2s_n = n(a_1 + a_n)$, since the expression $(a_1 + a_n)$ appears n times on the right side of the equation. Dividing both sides of the equation $2s_n = n(a_1 + a_n)$ by 2 produces the result $s_n = \frac{n}{2}(a_1 + a_n)$.

The seventh term of the arithmetic sequence 4, 8, 12, 16, 20, . . . is 28, so s_7 can be found as follows:

$$s_7 = \frac{7}{2}(a_1 + a_7)$$

$$s_7 = \frac{7}{2}(4 + 28) \qquad \text{Substitute 7 for } n, \text{ 4 for } a_1, \text{ and 28 for } a_7.$$

$$= 112 \qquad \text{Simplify.}$$

To find the nth partial sum of an arithmetic sequence using the formula $s_n = \frac{n}{2}(a_1 + a_n)$, we must know the value of the first term and the last term that are being added. In the next example, we will find the general term of the arithmetic sequence,

$a_n = a_1 + (n - 1)d$, to help us find the last term in the sum. This becomes particularly useful as n becomes larger.

EXAMPLE 5   Find s_{20} for the arithmetic sequence 3, 8, 13, 18, 23, ...

Solution

We already know that $n = 20$ and $a_1 = 3$. To find the value of a_{20}, we will begin by finding the general term of this arithmetic sequence using the formula $a_n = a_1 + (n - 1)d$. The common difference for this sequence is $d = 5$.

$$a_n = 3 + (n - 1)5 \qquad \text{Substitute 3 for } a_1 \text{ and 5 for } d.$$
$$= 5n - 2 \qquad \text{Simplify.}$$

We can now use the fact that $a_n = 5n - 2$ to find the value of a_{20}.

$$a_{20} = 5(20) - 2 \qquad \text{Substitute 20 for } n.$$
$$= 98 \qquad \text{Simplify.}$$

We can now apply the formula $s_n = \frac{n}{2}(a_1 + a_n)$ to find s_{20}.

$$s_{20} = \frac{20}{2}(3 + 98) \qquad \text{Substitute 20 for } n, \text{ 3 for } a_1, \text{ and 98 for } a_{20}.$$
$$= 1010 \qquad \text{Simplify.}$$

Quick Check   5   Find s_{36} for the arithmetic sequence 25, 44, 63, 82, 101, ...

EXAMPLE 6   Find the sum of the first 147 odd positive integers.

Solution

The first 147 odd positive integers comprise the sequence that begins 1, 3, 5, 7, 9, ... This is an arithmetic sequence with $a_1 = 1$ and $d = 2$. We will find the general term of this sequence using the formula $a_n = a_1 + (n - 1)d$ to help us find a_{147}.

$$a_n = 1 + (n - 1)2 \qquad \text{Substitute 1 for } a_1 \text{ and 2 for } d.$$
$$= 2n - 1 \qquad \text{Simplify.}$$

We can now use $a_n = 2n - 1$ to find the value of a_{147}.

$$a_{147} = 2(147) - 1 \qquad \text{Substitute 147 for } n.$$
$$= 293 \qquad \text{Simplify.}$$

We can now apply the formula $s_n = \frac{n}{2}(a_1 + a_n)$ to find s_{147}.

$$s_{147} = \frac{147}{2}(1 + 293) \qquad \text{Substitute 147 for } n, \text{ 1 for } a_1, \text{ and 293 for } a_{147}.$$
$$= 21,609 \qquad \text{Simplify.}$$

The sum of the first 147 odd positive integers is 21,609.

Quick Check   6   Find the sum of the first 62 even positive integers.

Applications

Objective **4** **Use an arithmetic sequence to solve an applied problem.**

EXAMPLE **7**   Joyce found a new job with a starting salary of $40,000 per year. Each year, she will be given a $2500 raise.

a) Write a sequence showing Joyce's annual salary for each of the first six years.
b) Find the general term a_n for the sequence from part a).
c) What will Joyce's total earnings be if she stays with the company for 30 years?

Solution

a) Since the starting salary is $40,000 and increases by $2500 each year, the sequence is $40,000, $42,500, $45,000, $47,500, $50,000, $52,500.
b) The first term, a_1, of this sequence is $40,000, and the common difference d is $2500. We now can find the general term of this sequence using the formula $a_n = a_1 + (n - 1)d$.

$$a_n = 40,000 + (n - 1)2500 \qquad \text{Substitute 40,000 for } a_1 \text{ and 2500 for } d.$$
$$= 2500n + 37,500 \qquad \text{Simplify.}$$

The general term for this arithmetic sequence is $a_n = 2500n + 37,500$.
c) To find Joyce's total earnings over 30 years, we need to find s_{30}. We already know that $n = 30$ and $a_1 = 40,000$. To apply the formula $s_n = \frac{n}{2}(a_1 + a_n)$, we must find the value of a_{30}.

$$a_{30} = 2500(30) + 37,500 \qquad \text{Substitute 30 for } n \text{ in } a_n = 2500n + 37,500.$$
$$= 112,500 \qquad \text{Simplify.}$$

We can now apply the formula $s_n = \frac{n}{2}(a_1 + a_n)$.

$$s_{30} = \frac{30}{2}(40,000 + 112,500) \qquad \text{Substitute 30 for } n, \text{ 40,000 for } a_1, \text{ and 112,500 for } a_{30}.$$
$$= 2,287,500 \qquad \text{Simplify.}$$

During the first 30 years, Joyce will earn $2,287,500.

Find the common difference, d, of the given arithmetic sequence.

1. 3, 12, 21, 30, 39, . . .

2. 1, 22, 43, 64, 85, . . .

3. $5, \dfrac{15}{2}, 10, \dfrac{25}{2}, 15, \ldots$

4. $12, \dfrac{44}{3}, \dfrac{52}{3}, 20, \dfrac{68}{3}, \ldots$

5. $-13, -10, -7, -4, -1, \ldots$

6. $-4, -8, -12, -16, -20, \ldots$

7. $\dfrac{9}{4}, \dfrac{41}{4}, \dfrac{73}{4}, \dfrac{105}{4}, \dfrac{137}{4}, \ldots$

8. $\dfrac{31}{6}, \dfrac{17}{2}, \dfrac{71}{6}, \dfrac{91}{6}, \dfrac{37}{2}, \ldots$

Find the first five terms of the arithmetic sequence with the given first term and common difference.

9. $a_1 = 1, d = 5$

10. $a_1 = 1, d = -2$

11. $a_1 = -9, d = 12$

12. $a_1 = -13, d = -8$

13. $a_1 = -10, d = -21$

14. $a_1 = 9, d = \dfrac{5}{2}$

Find the general term, a_n, of the given arithmetic sequence.

15. 9, 22, 35, 48, 61, . . .

16. 7, 27, 47, 67, 87, . . .

17. $-8, -3, 2, 7, 12, \ldots$

18. $-9, -16, -23, -30, -37, \ldots$

19. $87, 52, 17, -18, -53, \ldots$

20. 62, 63, 64, 65, 66, . . .

21. $-\dfrac{3}{2}, \dfrac{5}{2}, \dfrac{13}{2}, \dfrac{21}{2}, \dfrac{29}{2}, \ldots$

22. $8, \dfrac{16}{3}, \dfrac{8}{3}, 0, -\dfrac{8}{3}, \ldots$

23. Find the 19th term, a_{19}, of the arithmetic sequence 16, 21, 26, 31, 36, ...

24. Find the 14th term, a_{14}, of the arithmetic sequence -9, 29, 67, 105, 143, ...

25. Which term of the arithmetic sequence 23, 29, 35, 41, 47, . . . , is equal to 389?

26. Which term of the arithmetic sequence $-99, -94, -89, -84, -79, \ldots$, is equal to 336?

27. Which term of the arithmetic sequence 611, 603, 595, 587, 579, . . . , is equal to -365?

28. Which term of the arithmetic sequence $-327, -336, -345, -354, -363, \ldots$, is equal to -3981?

Find the partial sum of the arithmetic sequence with the given first term and common difference.

29. $s_{13}, a_1 = 5, d = 6$

30. $s_9, a_1 = 11, d = 12$

31. $s_{16}, a_1 = -9, d = 8$

32. $s_{37}, a_1 = -10, d = -10$

33. $s_{40}, a_1 = 89, d = -7$

34. $s_{21}, a_1 = 810, d = 62$

Find the partial sum for the given arithmetic sequence.

35. $s_{10}, 1, 21, 41, 61, \ldots$

36. $s_{15}, 9, 22, 35, 48, \ldots$

37. $s_{17}, -17, -1, 15, 31, \ldots$

38. $s_9, -38, -31, -24, -17, \ldots$

39. $s_{20}, 86, 47, 8, -31, \ldots$

40. $s_{27}, -3, -16, -29, -42, \ldots$

41. $s_{32}, 109, 208, 307, 406, \ldots$

42. $s_{19}, 62, 17, -28, -73, \ldots$

43. Find the sum of the first 1000 positive integers.

44. Find the sum of the first 418 positive integers.

45. Find the sum of the first 67 odd positive integers.

46. Find the sum of the first 109 odd positive integers.

47. Find the sum of the first 200 even positive integers.

48. Find the sum of the first 82 even positive integers.

49. Andrea finds a job that pays $10 per hour. Each year, her pay rate will be increased by $1 per hour.

  a) Write a sequence showing Andrea's hourly pay rate for the first five years.

  b) Find the general term a_n for the sequence from part a).

50. During its first year of existence, a minor league baseball team sold 2800 season tickets. In each year after that, the number of season tickets sold decreased by 120.

  a) Write a sequence showing the number of season tickets sold in the first four years.

  b) Find the general term a_n for the number of season tickets sold in the team's nth year.

51. Jon starts a job with an annual salary of $38,500. Each year, he will be given a raise of $750.

  a) Write a sequence showing Jon's annual salary for the first six years.

  b) Find the general term a_n for Jon's annual salary in his nth year in this job.

  c) How much money will Jon earn from this job over the first 15 years?

52. A teacher has a retirement savings plan. In the first year, the teacher contributes $100 per month. Each year, the teacher will increase her contributions by $10 per month, so that she contributes $110 per month in the second year, $120 per month in the third year, and so on.

  a) Find the general term a_n of an arithmetic sequence showing the annual contribution of the teacher in the nth year.

  b) How much money will the teacher contribute over a period of 30 years?

Answer in complete sentences.

53. Describe a real-world application involving an arithmetic sequence.

54. Describe a real-world application involving an arithmetic series.

55. Explain how to find the general term of an arithmetic sequence. Use an example to illustrate the process.

56. Explain how to find the partial sum of an arithmetic sequence. Use an example to illustrate the process.

Study Tip **REVISITED**   Creating and taking practice tests is an effective way to prepare for a final exam. Start by creating a practice test for an entire chapter. If you determine that you are struggling with certain types of problems, review those problems and then try another practice test. Once you feel comfortable with that chapter's material, move on to create a practice test for another chapter. Repeat this until you have covered all of the chapters. You could arrange to exchange practice tests with a classmate. For further information on practice tests, see the study tips in Chapter 12.

14.3
GEOMETRIC SEQUENCES AND SERIES

Objectives

1. Find the common ratio of a geometric sequence.
2. Find the general term of a geometric sequence.
3. Find partial sums of a geometric sequence.
4. Find the sum of an infinite geometric series.
5. Use a geometric sequence to solve an applied problem.

Geometric Sequences

Consider the sequence 4, 12, 36, 108, 324, ... Each term in this sequence is equal to three times the preceding term. This is an example of a geometric sequence, and this type of sequence is the focus of this section.

Geometric Sequence

A **geometric sequence** is a sequence in which each term after the first term is a constant multiple of the preceding term. This means that the general term a_n can be defined recursively by $a_n = r \cdot a_{n-1}$, for $n \neq 1$. The number r is called the **common ratio** of the geometric sequence.

A sequence is a geometric sequence if the quotient of any term after the first term and its preceding term, $\dfrac{a_n}{a_{n-1}}$, is a constant. For example, the sequence 1, 5, 25, 125, 625, . . . , is a geometric sequence because $\dfrac{a_n}{a_{n-1}} = 5$ for each term after the first term in the sequence. In other words, each term is five times greater than the preceding term. The value 5 is the common ratio r of this geometric sequence.

The sequence 2, 6, 10, 14, 18, . . . , is not a geometric sequence because there is no common ratio. The second term is three times greater than the first term, but the third term is not three times greater than the second term.

Finding the Common Ratio of a Geometric Sequence

Objective 1 Find the common ratio of a geometric sequence. To find the common ratio, r, of a geometric sequence, take any term of the sequence after the first term and divide it by the preceding term.

EXAMPLE 1 Find the common ratio, r, of the geometric sequence 2, 8, 32, 128, 512, ...

Solution

To find r, we can divide the second term of the sequence by the first term.

$$r = \frac{8}{2} = 4$$

Quick Check 1
Find the common ratio, r, of the geometric sequence 256, 64, 16, 4, 1, ...

A geometric sequence can be an alternating sequence. In this case, the common ratio r will be a negative number.

EXAMPLE 2   Find the common ratio, r, of the geometric sequence $5, -10, 20, -40, 80, \ldots$.

Solution

Quick Check 2
Find the common ratio, r, of the geometric sequence $-120, 60, -30, 15, -\dfrac{15}{2}, \ldots$.

To find r, we can divide the second term of the sequence by the first term.

$$r = \frac{-10}{5} = -2$$

The common ratio for this geometric sequence is $r = -2$.

EXAMPLE 3   Find the first five terms of the geometric sequence whose first term is 7 and whose common ratio is $r = 2$.

Solution

Quick Check 3
Find the first five terms of the geometric sequence whose first term is 5 and whose common ratio is $r = -\frac{3}{4}$.

Since the first term of this sequence is 7, and each term is two times greater than the previous term, the first five terms of the sequence are 7, 14, 28, 56, and 112.

Finding the General Term of a Geometric Sequence

Objective 2   Find the general term of a geometric sequence. The general term of a geometric sequence can be determined from the first term of the sequence, a_1, and the sequence's common ratio r. We know that the second term a_2 is equal to $a_1 \cdot r$. Continuing this pattern, we obtain

$$a_3 = a_2 \cdot r = (a_1 \cdot r) \cdot r = a_1 \cdot r^2$$
$$a_4 = a_3 \cdot r = (a_1 \cdot r^2) \cdot r = a_1 \cdot r^3$$
$$a_5 = a_4 \cdot r = (a_1 \cdot r^3) \cdot r = a_1 \cdot r^4$$

This leads to the following result about the general term a_n of a geometric sequence:

General Term of a Geometric Sequence

If a geometric sequence whose first term is a_1 has a common ratio of r, then the general term of the sequence is $a_n = a_1 \cdot r^{n-1}$.

EXAMPLE 4   Find the general term, a_n, of the geometric sequence $3, 30, 300, 3000, \ldots$.

Solution

The first term, a_1, of this sequence is 3. We will divide the second term of the sequence by the first term to find the common ratio r.

$$r = \frac{30}{3} = 10$$

Quick Check 4

Find the general term, a_n, of the geometric sequence 9, 18, 36, 72, ...

We can now find the general term of this sequence, using the formula $a_n = a_1 \cdot r^{n-1}$.

$$a_n = 3 \cdot 10^{n-1}$$ Substitute 3 for a_1 and 10 for r.

The general term for this geometric sequence is $a_n = 3 \cdot 10^{n-1}$.

EXAMPLE 5   Find the general term, a_n, of the geometric sequence $-1, 4, -16, 64, \ldots$.

Solution

The first term, a_1, of this sequence is -1. We will divide the second term of the sequence by the first term to find the common ratio r.

$$r = \frac{4}{-1} = -4$$

We can now find the general term of this sequence, using the formula $a_n = a_1 \cdot r^{n-1}$.

$$a_n = -1 \cdot (-4)^{n-1}$$ Substitute -1 for a_1 and -4 for r.

Quick Check 5

Find the general term, a_n, of the geometric sequence 11, -55, 275, -1375, ...

The general term for this geometric sequence is $a_n = -1 \cdot (-4)^{n-1}$.

Geometric Series

Objective 3 Find partial sums of a geometric sequence. A **geometric series** is the sum of the terms of a geometric sequence. We begin by examining finite geometric series consisting of only the first n terms of a geometric sequence. Again, the sum of the first n terms of a sequence is also known as the nth partial sum of the sequence and is denoted by s_n. In the case of a finite geometric series,

$$s_n = a_1 + a_1 r + a_1 r^2 + a_1 r^3 + \cdots + a_1 r^{n-1}$$

EXAMPLE 6   Find s_7 for the geometric sequence 1, 2, 4, 8, 16, ...

Solution

Since only the first five terms are given, we begin by finding the sixth and seventh terms. Since each term in this geometric sequence is equal to two times the preceding term, the next two terms of the sequence are 32 and 64. We can now find the seventh partial sum of this geometric sequence.

$$s_7 = 1 + 2 + 4 + 8 + 16 + 32 + 64$$
$$= 127$$ Add.

Quick Check 6

Find s_8 for the geometric sequence 12, 36, 108, 324, ...

We now present a formula for finding partial sums of a geometric sequence. This formula will be quite helpful, as it depends only on the first term of the geometric sequence (a_1) and the common ratio r of the sequence.

nth Partial Sum of a Geometric Sequence

The sum of the first n terms of a geometric sequence is given by

$$s_n = a_1 \cdot \frac{1 - r^n}{1 - r}$$

where a_1 is the first term and r is the common ratio of the sequence.

Before we apply this formula, let's investigate where the formula comes from. Suppose we have a geometric sequence whose first term is a_1 and whose common ratio is r. Since s_n is the sum of the first n terms of the sequence, it can be expressed as

$$s_n = a_1 + a_1 r + a_1 r^2 + a_1 r^3 + \cdots + a_1 r^{n-1}$$

If we multiply both sides of this equation by r, the resulting equation is

$$r \cdot s_n = a_1 r + a_1 r^2 + a_1 r^3 + a_1 r^4 + \cdots + a_1 r^n$$

Now we may subtract corresponding terms of these two equations as follows:

$$s_n = a_1 + a_1 r + a_1 r^2 + a_1 r^3 + \cdots + a_1 r^{n-1}$$
$$r \cdot s_n = a_1 r + a_1 r^2 + a_1 r^3 + \cdots + a_1 r^{n-1} + a_1 r^n$$

$$s_n - r \cdot s_n = a_1 - a_1 r^n$$
$$s_n - r s_n = a_1 - a_1 r^n$$
$$s_n(1 - r) = a_1(1 - r^n)$$

Factor out the common factor s_n on the left side of the equation and the common factor a_1 on the right side of the equation.

$$\frac{s_n(1 - r)}{1 - r} = \frac{a_1(1 - r^n)}{1 - r}$$

Divide both sides by $1 - r$ to isolate s_n.

$$s_n = a_1 \cdot \frac{1 - r^n}{1 - r}$$

Simplify.

For the geometric sequence 1, 2, 4, 8, 16, . . . , in the previous example, $a_1 = 1$ and $r = 2$.

Now we can apply the formula $s_n = a_1 \cdot \dfrac{1 - r^n}{1 - r}$ to find s_7.

$$s_7 = 1 \cdot \frac{1 - 2^7}{1 - 2}$$

Substitute 1 for a_1, 2 for r, and 7 for n.

$$= 127$$

Simplify.

Note that this matches the result found in the previous example.

In the next example, we will find the partial sum of a geometric sequence whose common ratio r is negative.

EXAMPLE 7   Find s_9 for the geometric sequence 6, −42, 294, −2058, ...

Solution

For this sequence, $a_1 = 6$, and $r = \dfrac{-42}{6} = -7$. Now we can apply the formula

$$s_n = a_1 \cdot \frac{1 - r^n}{1 - r} \text{ to find } s_9.$$

Quick Check 7

Find s_{10} for the geometric sequence 1, −3, 9, −27, ...

$$s_9 = 6 \cdot \frac{1 - (-7)^9}{1 - (-7)}$$

Substitute 6 for a_1, −7 for r, and 9 for n.

$$= 30{,}265{,}206$$

Simplify.

Infinite Geometric Series

Objective 4 Find the sum of an infinite geometric series. We now turn our attention to **infinite geometric series,** $s = a_1 + a_1 r + a_1 r^2 + \cdots$. Consider the geometric sequence $\frac{1}{2}, \frac{1}{4}, \frac{1}{8}, \frac{1}{16}, \frac{1}{32}, \ldots$, where $a_1 = \frac{1}{2}$ and $r = \frac{1}{2}$. The first partial sum of this series is $s_1 = \frac{1}{2}$. Here are the next four partial sums:

$$s_2 = \frac{1}{2} + \frac{1}{4} = \frac{2}{4} + \frac{1}{4} = \frac{3}{4}$$

$$s_3 = \frac{1}{2} + \frac{1}{4} + \frac{1}{8} = \frac{4}{8} + \frac{2}{8} + \frac{1}{8} = \frac{7}{8}$$

$$s_4 = \frac{1}{2} + \frac{1}{4} + \frac{1}{8} + \frac{1}{16} = \frac{8}{16} + \frac{4}{16} + \frac{2}{16} + \frac{1}{16} = \frac{15}{16}$$

$$s_5 = \frac{1}{2} + \frac{1}{4} + \frac{1}{8} + \frac{1}{16} + \frac{1}{32} = \frac{16}{32} + \frac{8}{32} + \frac{4}{32} + \frac{2}{32} + \frac{1}{32} = \frac{31}{32}$$

As n increases, the partial sums of this sequence are getting closer and closer to 1. In such a case, we say that the **limit** of the infinite geometric series $\frac{1}{2} + \frac{1}{4} + \frac{1}{8} + \frac{1}{16} + \frac{1}{32} + \cdots$ is equal to 1.

Limit of an Infinite Geometric Series

> Suppose that a geometric sequence has a common ratio r. If $|r| < 1$, then the corresponding infinite geometric series has a limit, and this limit is given by the formula
> $$s = \frac{a_1}{1 - r}.$$
> If $|r| \geq 1$, as it is for the series $\frac{1}{2} + \frac{3}{4} + \frac{9}{8} + \cdots (r = \frac{3}{2})$, then no limit exists.

EXAMPLE 8   For the geometric sequence $16, 4, 1, \frac{1}{4}, \ldots$, does the infinite series have a limit? If so, what is that limit?

Solution

We begin by determining whether $|r| < 1$ for this sequence.

$$r = \frac{a_2}{a_1} = \frac{4}{16} = \frac{1}{4}$$

Since the absolute value of r is less than 1, the infinite series does have a limit. To find the limit we will use the formula $s = \dfrac{a_1}{1 - r}$.

$$s = \frac{16}{1 - \dfrac{1}{4}} = \frac{16}{3/4} = \frac{64}{3} \qquad \text{Substitute 16 for } a_1 \text{ and } \tfrac{1}{4} \text{ for } r. \text{ Simplify.}$$

Quick Check 8

For the geometric sequence $3, 2, \frac{4}{3}, \frac{8}{9}, \ldots$, does the infinite series have a limit? If so, what is that limit?

The limit of this infinite geometric series is $\frac{64}{3}$ or $21\frac{1}{3}$.

EXAMPLE 9   For the geometric sequence $4, 6, 9, \frac{27}{2}, \ldots$, does the infinite series have a limit? If so, what is that limit?

Solution

First determine whether $|r| < 1$.

$$r = \frac{a_2}{a_1} = \frac{6}{4} = \frac{3}{2}$$

Quick Check 9

For the geometric sequence $-2, 8, -32, 128, \ldots$, does the infinite series have a limit? If so, what is that limit?

Since the absolute value of r is greater than 1, the infinite series does not have a limit. In this case, the partial sums increase without limit as n continues to increase.

In the next example, we will find the limit of an infinite geometric series given the general term of the sequence.

EXAMPLE 10   For the geometric sequence whose general term is $a_n = 8 \cdot \left(\frac{1}{3}\right)^{n-1}$, does the infinite series have a limit? If so, what is that limit?

Solution

For this sequence, $r = \frac{1}{3}$. Since the absolute value of r is less than 1, the infinite series does have a limit. The first term of this series is $a_1 = 8$, and the limit can be found using $s = \frac{a_1}{1 - r}$.

$$s = \frac{8}{1 - \frac{1}{3}} \qquad \text{Substitute 8 for } a_1 \text{ and } \tfrac{1}{3} \text{ for } r.$$

$$= 12 \qquad \text{Simplify.}$$

The limit of this infinite geometric series is 12.

Applications

Objective 5 **Use a geometric sequence to solve an applied problem.** We finish this section with an application involving a geometric sequence.

EXAMPLE 11   Donna's parents have established a trust fund for their daughter. On her 21st birthday, Donna receives a payment of $10,000. On each successive birthday, she receives a payment that is 80% of the previous payment.

a) Write a sequence showing Donna's payments for the first four years.
b) Find the total amount of Donna's first 25 payments.

Solution

a) Since the first payment is $10,000 and the second payment is 80% of that amount, she will receive 0.8($10,000) or $8000 on her 22nd birthday. Repeating this process, we find that the sequence is $10,000, $8000, $6400, $5120.

b) To find the total of Donna's payments for the first 25 years, we need to find s_{25} for this geometric sequence. We already know that $n = 25$ and $a_1 = 10,000$. To apply the formula $s_n = a_1 \cdot \dfrac{1 - r^n}{1 - r}$, we must find the common ratio r.

$$r = \frac{a_2}{a_1} = \frac{8000}{10,000} = 0.8$$

It should be no surprise that $r = 0.8$ for this sequence, as each term is 80% of the preceding term. We can now apply the formula $s_n = a_1 \cdot \dfrac{1 - r^n}{1 - r}$.

$$s_{25} = 10,000 \cdot \frac{1 - (0.8)^{25}}{1 - 0.8} \qquad \text{Substitute 25 for } n, \text{ 10,000 for } a_1 \text{ and 0.8 for } r.$$
$$\approx 49,811.11 \qquad \text{Evaluate using a calculator.}$$

During the first 25 years, Donna will receive $49,811.11.

Quick Check 10 A car was purchased for $25,000. Its value decreases by 30% each year.

a) Write a sequence showing the value of the car at the end of each of the first six years.

b) Find the value of the car at the end of year 10.

EXERCISES *14.3*

Find the common ratio, r, of the given geometric sequence.

1. 1, 3, 9, 27, 81, . . .

2. 1, 6, 36, 216, 1296, . . .

3. $125, 25, 5, 1, \dfrac{1}{5}, \ldots$

4. $343, 49, 7, 1, \dfrac{1}{7}, \ldots$

5. 3, −6, 12, −24, 48, . . .

6. −4, 12, −36, 108, −324, . . .

7. $-6, 8, -\dfrac{32}{3}, \dfrac{128}{9}, -\dfrac{512}{27}, \ldots$

8. $32, -12, \dfrac{9}{2}, -\dfrac{27}{16}, \dfrac{81}{128}, \ldots$

Find the first five terms of the geometric sequence with the given first term and common ratio.

9. $a_1 = 1, r = 10$

10. $a_1 = 1, r = -3$

11. $a_1 = 4, r = -2$

12. $a_1 = 2, r = 5$

13. $a_1 = 8, r = \dfrac{3}{4}$

14. $a_1 = 12, r = \dfrac{1}{2}$

15. $a_1 = -20, r = -\dfrac{7}{4}$

16. $a_1 = \dfrac{1}{3}, r = -\dfrac{4}{3}$

Find the general term, a_n, of the given geometric sequence.

17. $1, 8, 64, 512, \ldots$

18. $1, \dfrac{5}{3}, \dfrac{25}{9}, \dfrac{125}{27}, \ldots$

19. $-1, -6, -36, -216, \ldots$

20. $-1, 10, -100, 1000, \ldots$

21. $9, 18, 36, 72, \ldots$

22. $5, 20, 80, 320, \ldots$

23. $16, -12, 9, -\dfrac{27}{4}, \ldots$

24. $\dfrac{3}{10}, -\dfrac{9}{100}, \dfrac{27}{1000}, -\dfrac{81}{10,000}, \ldots$

Find the partial sum of the geometric sequence with the given first term and common ratio.

25. $s_7, a_1 = 5, r = 4$

26. $s_8, a_1 = 15, r = 3$

27. $s_{20}, a_1 = 48, r = -2$

28. $s_5, a_1 = -6, r = -10$

29. $s_{10}, a_1 = 256, r = \dfrac{1}{2}$

30. $s_{12}, a_1 = 625, r = 0.2$

Find the indicated partial sum for the given geometric sequence.

31. $s_6, 1, -8, 64, -512, \ldots$

32. $s_8, 1, 10, 100, 1000, \ldots$

33. $s_7, 4, 12, 36, 108 \ldots$

34. $s_9, 50, 100, 200, 400, \ldots$

35. $s_6, 13, 156, 1872, 22,464, \ldots$

36. $s_7, 9, -27, 81, -243, \ldots$

37. $s_{14}, 1024, 512, 256, 128, \ldots$

38. $s_8, -24, 6, -\dfrac{3}{2}, \dfrac{3}{8}, \ldots$

For the given geometric sequence, does the infinite series have a limit? If so, find that limit.

39. $30, 10, \dfrac{10}{3}, \dfrac{10}{9}, \ldots$

40. $432, 72, 12, 2, \ldots$

41. $1, -\dfrac{4}{5}, \dfrac{16}{25}, -\dfrac{64}{125}, \ldots$

42. $\dfrac{1}{9}, -\dfrac{5}{27}, \dfrac{25}{81}, -\dfrac{125}{243}, \ldots$

43. $\dfrac{81}{100}, \dfrac{243}{200}, \dfrac{729}{400}, \dfrac{2187}{800}, \ldots$

44. $108, 90, 75, \dfrac{125}{2}, \ldots$

45. $\dfrac{1}{2}, -\dfrac{1}{4}, \dfrac{1}{8}, -\dfrac{1}{16}, \ldots$

46. $1, -1, 1, -1, \ldots$

For the geometric sequence whose general term is given, does the infinite series have a limit? If so, what is that limit?

47. $a_n = 5 \cdot \left(\dfrac{5}{6}\right)^{n-1}$

48. $a_n = 200 \cdot \left(\dfrac{1}{4}\right)^{n-1}$

49. $a_n = -8 \cdot \left(\dfrac{3}{10}\right)^{n-1}$

50. $a_n = 6 \cdot \left(-\dfrac{5}{3}\right)^{n-1}$

51. $a_n = \dfrac{5}{7} \cdot \left(\dfrac{11}{8}\right)^{n-1}$

52. $a_n = \dfrac{4}{25} \cdot \left(-\dfrac{5}{12}\right)^{n-1}$

53. A copy machine purchased for $8000 decreases in value by 20% each year.

    a) Write a geometric sequence showing the value of the copy machine at the end of each of the first three years after it was purchased.

    b) Find the value of the copy machine at the end of the eighth year.

54. A farmer purchased a new tractor for $30,000. The tractor's value decreases by 10% each year.

    a) Write a geometric sequence showing the value of the tractor at the end of each of the first six years after it was purchased.

b) Find the value of the tractor at the end of year 10.

55. A motor home valued at $40,000 decreases in value by 50% each year. Find the general term a_n of a geometric sequence showing the value of the motor home n years after it was purchased.

56. A diamond ring valued at $4000 increases in value by 5% each year. Find the general term a_n of a geometric sequence showing the value of the diamond ring n years after it was purchased.

57. A professional baseball player signed a 10-year contract. The salary for the first year was $10,000,000, with a 10% raise for each of the remaining 9 years. At the end of the contract, what is the total amount of money earned by the player?

58. Cathy takes a job as a high school teacher. The starting salary is $32,000 per year, with a 2% raise each year. If Cathy teaches for 30 years, what is the total amount that she will be paid?

Answer in complete sentences.

59. Describe a real-world application involving a geometric sequence.

60. Describe a real-world application involving a geometric series.

61. Explain the difference between an arithmetic sequence and a geometric sequence.

62. Explain how to find the general term of a geometric sequence. Use an example to illustrate the process.

63. Explain how to find the partial sum of a geometric sequence. Use an example to illustrate the process.

64. Explain how to determine if a geometric sequence has a limit. If it does have a limit, explain how to find that limit. Use an example to illustrate the process.

QUICK REVIEW EXERCISES

Section 14.3

Simplify.

1. $(x + 5)^2$

2. $(x - y)^2$

3. $(2x + 9y)^2$

4. $(5a^3 - 4b^4)^2$

Study Tip **REVISITED**   When you arrive to take your final exam, be sure to bring a positive attitude with you. Be confident. If you have prepared as well as possible, then trust your ability. If you sit down to take the final exam and you are convinced that something is going to go wrong, then you will probably be correct. For further information on dealing with the anxiety that accompanies test taking, see the study tips in Chapter 6.

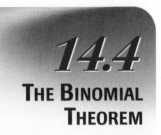

14.4
THE BINOMIAL THEOREM

Objectives

1 Evaluate factorials.
2 Calculate binomial coefficients.
3 Use the binomial theorem to expand a binomial raised to a power.
4 Use Pascal's triangle and the binomial theorem to expand a binomial raised to a power.

Recall that a binomial is a polynomial with two terms, such as $2x - 3$ or $5x + 8y$. In this section, we will learn a method for raising a binomial to a power, such as $(x + y)^9$.

Factorials

Objective 1   Evaluate factorials.

Factorials

> The product of all the positive integers from 1 through some positive integer n,
> $1 \cdot 2 \cdot 3 \cdot \; \cdots \; \cdot (n - 1) \cdot n$, is called **$n$ factorial,** and is denoted $n!$.

For example, $5! = 1 \cdot 2 \cdot 3 \cdot 4 \cdot 5$ or 120. By definition, $0! = 1$.

EXAMPLE 1   Evaluate $7!$.

Solution

To evaluate $7!$, we need to multiply the successive integers from 1 through 7.

$$7! = 1 \cdot 2 \cdot 3 \cdot 4 \cdot 5 \cdot 6 \cdot 7 \qquad \text{Multiply the positive integers from 1 through 7.}$$
$$= 5040 \qquad\qquad\qquad\quad \text{Simplify.}$$

Quick Check 1
Evaluate $9!$.

Most calculators have a built-in function for calculating factorials; consult your calculator manual for directions. Most nongraphing calculators will have a button labeled as $\boxed{n!}$ or $\boxed{x!}$, while many graphing calculators store the factorial function in the $\boxed{\text{MATH}}$ menu.

Binomial Coefficients

Objective 2   Calculate binomial coefficients.   We now will use factorials to calculate numbers called binomial coefficients.

Binomial Coefficients

> For any two positive integers n and r, $n > r$, the number $_nC_r$ is a **binomial coefficient** and is given by the formula $_nC_r = \dfrac{n!}{r! \cdot (n - r)!}$.

EXAMPLE 2   Evaluate the binomial coefficient $_9C_4$.

Solution

We begin by substituting 9 for n and 4 for r in the formula $_nC_r = \dfrac{n!}{r! \cdot (n-r)!}$.

$$_9C_4 = \frac{9!}{4! \cdot (9-4)!} \qquad \text{Substitute 9 for } n \text{ and 4 for } r.$$

$$= \frac{9!}{4! \cdot 5!} \qquad \text{Subtract } 9 - 4.$$

$$= \frac{362{,}880}{24 \cdot 120} \qquad \text{Evaluate each factorial.}$$

$$= \frac{362{,}880}{2880} \qquad \text{Simplify the denominator.}$$

$$= 126 \qquad \text{Simplify the fraction.}$$

$$_9C_4 = 126.$$

Quick Check 2
Evaluate the binomial coefficient $_6C_2$.

Using Your Calculator   To find binomial coefficients on the TI–83/84, we will be accessing the PRB menu. To evaluate the binomial coefficient $_9C_4$, begin by typing the first number, 9. Then press [MATH] to access the MATH menu. Use the right arrow to scroll over to the PRB menu. Select option **3: nCr** . When you return to the main screen, type the second number, 4, and press [ENTER]. Here is the calculator screen that you should see:

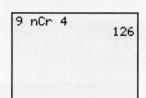

EXAMPLE 3   Evaluate the binomial coefficient $_8C_0$.

Solution

We will begin by substituting 8 for n and 0 for r in the formula $_nC_r = \dfrac{n!}{r! \cdot (n-r)!}$.

$$_8C_0 = \frac{8!}{0! \cdot 8!} \qquad \text{Substitute 8 for } n \text{ and 0 for } r.$$

$$= \frac{8!}{8!} \qquad 0! = 1$$

$$= 1 \qquad \text{Simplify the fraction.}$$

Quick Check 3
Evaluate the binomial coefficient $_{20}C_{20}$.

For any positive integer n, the binomial coefficient $_nC_0 = 1$.

The binomial coefficient $_nC_n = 1$ as well, since $_nC_n = \dfrac{n!}{n! \cdot 0!}$.

We finish our exploration of binomial coefficients with an example that will save time when calculating several binomial coefficients. It also demonstrates an interesting property of binomial coefficients.

EXAMPLE ▶ 4   Evaluate the binomial coefficients $_7C_0, _7C_1, _7C_2, \ldots, _7C_7$.

Solution

Verify the following results by using your calculator's built-in function for calculating binomial coefficients or by applying the formula.

$$_7C_0 = 1 \quad _7C_1 = 7 \quad _7C_2 = 21 \quad _7C_3 = 35 \quad _7C_4 = 35 \quad _7C_5 = 21 \quad _7C_6 = 7 \quad _7C_7 = 1$$

Notice the following properties:

- The first and last coefficients are both equal to 1.
- The second coefficient is equal to 7. For any positive integer n, $_nC_1 = n$.
- The coefficients follow a symmetric pattern. For any positive integer n and any non-negative integer $a \leq n$, $_nC_a = {_nC_{n-a}}$. This means that once we calculate $_7C_2$, we know that $_7C_5$ is equal to the same value.

These properties will be helpful when expanding binomials using the binomial theorem.

Quick Check   4   Complete the following table using the properties of binomial coefficients that were introduced in Example 4. You should not have to actually calculate any of these binomial coefficients.

$$_9C_0 = \qquad _9C_1 = \qquad _9C_2 = 36 \qquad _9C_3 = \qquad _9C_4 = 126$$
$$_9C_5 = \qquad _9C_6 = 84 \qquad _9C_7 = \qquad _9C_8 = \qquad _9C_9 =$$

The Binomial Theorem

Objective 3   Use the binomial theorem to expand a binomial raised to a power. The **binomial theorem** is used to raise a binomial expression to a power that is a positive integer without having to repeatedly multiply the binomial expression by itself. For instance, we can use the binomial theorem to expand $(x + y)^5$ rather than multiplying the expression $x + y$ by itself five times.

The Binomial Theorem

For any positive integer n,

$$(x + y)^n = {_nC_0}x^n y^0 + {_nC_1}x^{n-1}y^1 + {_nC_2}x^{n-2}y^2 + \cdots + {_nC_{n-1}}x^1 y^{n-1} + {_nC_n}x^0 y^n$$

This can be expressed using summation notation as

$$(x + y)^n = \sum_{r=0}^{n} {_nC_r} \cdot x^{n-r}y^r$$

EXAMPLE ▶ 5   Expand $(x + y)^5$ using the binomial theorem.

Solution

We will apply the binomial theorem with $n = 5$.

$$(x + y)^5 = {_5C_0} \cdot x^5 y^0 + {_5C_1} \cdot x^4 y^1 + {_5C_2} \cdot x^3 y^2 + {_5C_3} \cdot x^2 y^3 + {_5C_4} \cdot x^1 y^4 + {_5C_5} \cdot x^0 y^5$$

Quick Check 5

Expand $(x + y)^3$ using the binomial theorem.

After evaluating the binomial coefficients and simplifying factors whose exponent is 0 or 1, we obtain the following:

$$(x + y)^5 = x^5 + 5x^4y + 10x^3y^2 + 10x^2y^3 + 5xy^4 + y^5$$

If the binomial that is being raised to a power is a difference rather than a sum, such as $(x - y)^4$, then the signs of the terms in the binomial expansion alternate between positive and negative.

EXAMPLE 6   Expand $(x - y)^4$ using the binomial theorem.

Solution

We will apply the binomial theorem with $n = 4$, alternating the signs of the terms.

$$(x - y)^4 = {}_4C_0 \cdot x^4y^0 - {}_4C_1 \cdot x^3y^1 + {}_4C_2 \cdot x^2y^2 - {}_4C_3 \cdot x^1y^3 + {}_4C_4 \cdot x^0y^4$$
$$= x^4y^0 - 4x^3y^1 + 6x^2y^2 - 4x^1y^3 + x^0y^4 \qquad \text{Evaluate the binomial coefficients.}$$
$$= x^4 - 4x^3y + 6x^2y^2 - 4xy^3 + y^4 \qquad \text{Simplify factors whose exponents are 0 or 1.}$$

Quick Check 6

Expand $(x - 4)^4$ using the binomial theorem.

Pascal's Triangle

Objective 4 **Use Pascal's triangle and the binomial theorem to expand a binomial raised to a power.** One convenient way for finding binomial coefficients is through the use of a triangular array of numbers called **Pascal's triangle,** which is named in honor of the 17th-century French mathematician Blaise Pascal. Here is the portion of Pascal's triangle showing the binomial coefficients for values of n through 7:

$n = 0$							1						
$n = 1$						1		1					
$n = 2$					1		2		1				
$n = 3$				1		3		3		1			
$n = 4$			1		4		6		4		1		
$n = 5$		1		5		10		10		5		1	
$n = 6$	1		6		15		20		15		6		1
$n = 7$	1	7		21		35		35		21		7	1

Consider the row labeled $n = 5$, which contains the values 1, 5, 10, 10, 5, and 1. These are the binomial coefficients ${}_5C_0, {}_5C_1, {}_5C_2, {}_5C_3, {}_5C_4,$ and ${}_5C_5$, respectively. Each value in Pascal's triangle is the sum of the two entries diagonally above it.

$$n = 4 \qquad 1 \quad 4 \quad 6 \quad 4 \quad 1$$

$$n = 5 \qquad 1 \quad 5 \quad 10 \quad 10 \quad 5 \quad 1$$

To construct Pascal's triangle, begin with a single 1 in the first row $(n = 0)$, with two 1's written diagonally underneath it in the next row $(n = 1)$.

$$n = 0 \qquad 1$$
$$n = 1 \qquad 1 \qquad 1$$

The next row will have one more value than the previous row, and each number in the row is the sum of the values diagonally above it.

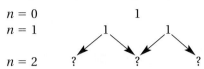

$n = 0$ 1

$n = 1$ 1   1

$n = 2$?   ?   ?

The values for this next row will be 1, 2, and 1. We then proceed to the next row, repeating this process until we have reached the desired row.

EXAMPLE 7   Expand $(x + y)^6$ using the binomial theorem and Pascal's triangle.

Solution

We will apply the binomial theorem with $n = 6$. Here is the row of Pascal's triangle associated with $n = 6$.

$n = 6$ 1   6   15   20   15   6   1

This tells us that the coefficients of the seven terms will be 1, 6, 15, 20, 15, 6, and 1, respectively. We know that the exponent of x in the first term will be 6, and will decrease by 1 in each successive term. We also know that the exponent of y in the first term will be 0, and will increase by 1 in each successive term.

Quick Check 7
Expand $(x + y)^7$ using the binomial theorem and Pascal's triangle.

$$(x + y)^6 = 1x^6y^0 + 6x^5y^1 + 15x^4y^2 + 20x^3y^3 + 15x^2y^4 + 6x^1y^5 + 1x^0y^6$$

Apply the binomial theorem, obtaining the binomial coefficients from the $n = 6$ row of Pascal's triangle.

$$= x^6 + 6x^5y + 15x^4y^2 + 20x^3y^3 + 15x^2y^4 + 6xy^5 + y^6$$

Simplify each term.

EXERCISES *14.4*

Evaluate.

1. 6!

2. 4!

3. 8!

4. 10!

5. 3!

6. 2!

7. 1!

8. 0!

Rewrite the given product as a factorial.

9. $1 \cdot 2 \cdot 3 \cdots \cdot 15$

10. $1 \cdot 2 \cdot 3 \cdots \cdot 24$

11. $1 \cdot 2 \cdot 3 \cdots \cdot 45$

12. $1 \cdot 2 \cdot 3 \cdots \cdot 32$

Simplify.

13. $\dfrac{11!}{4!}$

14. $\dfrac{10!}{8!}$

15. $\dfrac{9!}{3! \cdot 6!}$

16. $\dfrac{11!}{7! \cdot 4!}$

17. $8! + 7!$

18. $(5!)^2$

Evaluate the given binomial coefficient.

19. $_6C_3$

20. $_{13}C_7$

21. $_9C_2$

22. $_{10}C_6$

23. $_4C_2$

24. $_5C_4$

25. $_5C_1$

26. $_6C_0$

27. $_{20}C_{20}$

28. $_{30}C_{29}$

Complete the table that follows by using the properties of binomial coefficients. Do not actually calculate any of these binomial coefficients.

29.

$_{11}C_0 =$	$_{11}C_1 =$	$_{11}C_2 =$
$_{11}C_3 = 165$	$_{11}C_4 =$	$_{11}C_5 = 462$
$_{11}C_6 =$	$_{11}C_7 = 330$	$_{11}C_8 =$
$_{11}C_9 = 55$	$_{11}C_{10} =$	$_{11}C_{11} =$

30.

$_{10}C_0 =$	$_{10}C_1 =$	$_{10}C_2 = 45$
$_{10}C_3 = 120$	$_{10}C_4 =$	$_{10}C_5 = 252$
$_{10}C_6 = 210$	$_{10}C_7 =$	$_{10}C_8 =$
$_{10}C_9 =$	$_{10}C_{10} =$	

Expand using the binomial theorem.

31. $(x + y)^3$

32. $(x + a)^5$

33. $(x + 2)^3$

34. $(x + y)^4$

35. $(x + 6)^4$

36. $(x + 1)^3$

37. $(x - y)^5$

38. $(x - y)^2$

39. $(x - 3)^4$

40. $(x - 2)^3$

41. $(x + 3y)^3$

42. $(4x + y)^4$

43. $(4x - y)^3$

44. $(x - 2y)^6$

45. Construct the rows of Pascal's triangle for $n = 8$ through $n = 12$.

46. Find the sum of the binomial coefficients in each of the first six rows of Pascal's triangle. Do you notice a pattern? Find a formula for the sum of the binomial coefficients in the row associated with $n = k$. Predict the sum of the coefficients in the row associated with $n = 22$.

Expand using the binomial theorem and Pascal's triangle.

47. $(x + y)^5$

48. $(x + y)^3$

49. $(x + 2)^7$

50. $(x + 4)^6$

51. $(x - y)^3$

52. $(x - y)^8$

53. $(x - 3)^4$

54. $(x - 5)^5$

Answer in complete sentences.

55. If $a + b = n$, explain why $_nC_a = {_nC_b}$.

56. Explain how to construct Pascal's triangle.

Section 14.1—Topic	Chapter Review Exercises
Terms of a Sequence	1–8
Partial Sum of a Series	9–14
Summation Notation	15–16

Section 14.2—Topic	Chapter Review Exercises
Common Difference and Terms of an Arithmetic Sequence	17–28
Partial Sum of an Arithmetic Series	29–33
Applications of Arithmetic Sequences and Series	34

Section 14.3—Topic	Chapter Review Exercises
Common Ratio and Terms of a Geometric Sequence	35–41
Partial Sum of an Geometric Series	42–47
Limit of an Infinite Geometric Series	48–51
Applications of Geometric Sequences and Series	52

Section 14.4—Topic	Chapter Review Exercises
Factorials	53–54
Binomial Coefficients	55–56
Binomial Theorem	57–59
Pascal's Triangle	60

Summary of Chapter 14 Study Tips

The study tips in this chapter focused on preparing for, and doing well on, a final exam. One effective way to prepare for a final exam is by studying with a small group of students. In a small group, you can help each other to ensure that everyone is as prepared for the exam as possible. It is a great way to get answers to any questions that you may have.

Create a series of practice tests to help measure how ready you are for an exam. Start with practice tests that cover a wide range of material, then move on to more specialized practice tests for problems and concepts you find difficult.

Bring all the confidence and positive attitude that you can carry to your final exam. Students who arrive fearing for the worst often realize their fears.

Finally, even though a final exam may seem much more important than a chapter exam, do not abandon the test-taking strategies that have gotten you to this point in the course.

Find the first five terms of the sequence with the given general term. [**14.1**]

1. $a_n = n + 7$

2. $a_n = 5n - 19$

Find the next three terms of the given sequence, and find its general term a_n. [**14.1**]

3. 13, 26, 39, 52, 65, . . .

4. 12, 19, 26, 33, 40, . . .

Find the first five terms of the sequence with the given first term and general term. [**14.1**]

5. $a_1 = -9, a_n = (a_{n-1}) + 17$

6. $a_1 = -4, a_n = 2(a_{n-1}) - 23$

Find a recursive formula for the general term a_n of the given sequence. [**14.1**]

7. 3, 17, 31, 45, 59, . . .

8. 5, −10, 20, −40, 80, . . .

Find the indicated partial sum for the given sequence. [**14.1**]

9. s_6, 4, 14, 24, 34, . . .

10. s_9, 2, 12, 72, 432, . . .

Find the indicated partial sum for the sequence with the given general term. [**14.1**]

11. $s_7, a_n = 2n - 5$

12. $s_8, a_n = 5n + 6$

Find the indicated partial sum for the sequence with the given general term. [**14.1**]

13. $s_7, a_1 = 50, a_n = a_{n-1} - 8$

14. $s_5, a_1 = 20, a_n = 4a_{n-1} + 5$

Find the sum. [**14.1**]

15. $\displaystyle\sum_{i=1}^{12} i$

16. $\displaystyle\sum_{i=1}^{11} (5i + 6)$

Find the common difference, d, of the given arithmetic sequence. [**14.2**]

17. 7, 31, 55, 79, 103, . . .

18. −3, −9, −15, −21, −27, . . .

Find the first five terms of the arithmetic sequence with the given first term and common difference. [**14.2**]

19. $a_1 = 16, d = -9$

20. $a_1 = -18, d = 11$

Find the general term, a_n, of the given arithmetic sequence. [**14.2**]

21. 4, 11, 18, 25, 32, . . .

22. 8, 1, −6, −13, −20, . . .

23. −15, −13, −11, −9, −7, . . .

24. $\dfrac{5}{6}, \dfrac{19}{12}, \dfrac{7}{3}, \dfrac{37}{12}, \dfrac{23}{6}, . . .$

25. Find the 39th term, a_{39}, of the arithmetic sequence 25, 46, 67, 88, 109, ... [**14.2**]

26. Find the 27th term, a_{27}, of the arithmetic sequence −9, −32, −55, −78, −101, ... [**14.2**]

27. Which term of the arithmetic sequence −12, −5, 2, 9, 16, . . . , is equal to 1633? [**14.2**]

28. Which term of the arithmetic sequence 40, 21, 2, −17, −36, . . . , is equal to −9346? [**14.2**]

Find the indicated partial sum of the given arithmetic sequence. [**14.2**]

29. s_8, 10, 17, 24, 31, . . .

30. s_{15}, −102, −84, −66, −48, . . .

31. s_{24}, 16, 13, 10, 7 . . .

32. s_{30}, −10, −3, 4, 11, . . .

33. Find the sum of the first 340 positive integers.

34. The new CEO of a company started the job with an annual salary of $20,000,000. Her contract states that she will be given a raise of $2,000,000 each year. [**14.2**]

   a) Write a sequence showing the CEO's annual salary for the first five years.

   b) Find the general term a_n for the CEO's annual salary in her nth year in this job.

c) How much money will the CEO earn from this job over the first 10 years?

Find the common ratio, r, of the given geometric sequence. **[14.3]**

35. 1, 10, 100, 1000, 10,000, . . .

36. 5, 30, 180, 1080, 6480, . . .

Find the general term, a_n, of the given geometric sequence. **[14.3]**

37. 1, 12, 144, 1728, . . .

38. $8, 6, \dfrac{9}{2}, \dfrac{27}{8}, \ldots$

39. $-4, -12, -36, -108, \ldots$

40. $-24, 8, -\dfrac{8}{3}, \dfrac{8}{9}, \ldots$

41. Find the eighth term, a_8, of the geometric sequence 9, 18, 36, 72, ... [14.3]

Find the partial sum of the geometric sequence with the given first term and common ratio. **[14.3]**

42. $s_9, a_1 = 1, r = 3$

43. $s_{11}, a_1 = 10, r = -2$

Find the partial sum for the given geometric sequence. **[14.3]**

44. $s_7, 1, 6, 36, 216, \ldots$

45. $s_8, \dfrac{3}{2}, 12, 96, 768, \ldots$

46. $s_{10}, 320, 160, 80, 40 \ldots$

47. $s_9, 35, -140, 560, -2240, \ldots$

For the given geometric sequence, does the infinite series have a limit? If so, find that limit. **[14.3]**

48. 64, 80, 100, 125, . . .

49. $\dfrac{36}{7}, \dfrac{24}{7}, \dfrac{16}{7}, \dfrac{32}{21}, \ldots$

For the geometric sequence whose general term is given, does the infinite series have a limit? If so, find that limit. **[14.3]**

50. $a_n = -21 \cdot \left(\dfrac{2}{7}\right)^{n-1}$

51. $a_n = 45 \cdot \left(-\dfrac{7}{8}\right)^{n-1}$

52. A community service organization has started a homeless shelter with a budget of $150,000 for the first year. Their plans call for an annual increase in the budget of 10% over the previous year. How much will be budgeted for the shelter over the first eight years of operation? [14.3]

Evaluate. **[14.4]**

53. 9!

54. 7!

Evaluate the given binomial coefficient. **[14.4]**

55. $_8C_3$

56. $_{11}C_8$

Expand using the binomial theorem. **[14.4]**

57. $(x + y)^6$

58. $(x + 6)^4$

59. $(x - 9)^3$

Expand using the binomial theorem and Pascal's triangle. **[14.4]**

60. $(x + y)^9$

1. Find the next three terms of the sequence 11, 26, 41, 56, 71, . . . , and find its general term a_n.

2. Find the sum $\sum\limits_{i=1}^{9}(4i + 11)$.

3. Find the first five terms of the arithmetic sequence whose first term is $a_1 = 9$ and common difference is $d = 8$.

4. Find the general term, a_n, of the given arithmetic sequence 20, 13, 6, −1, −8, ...

5. Find the 15th partial sum, s_{15}, of the arithmetic sequence $14, \frac{35}{2}, 21, \frac{49}{2}, \ldots$.

6. Find the sum of the first 200 positive even integers.

7. Find the first five terms of the geometric sequence whose first term is $a_1 = 5$ and common ratio is $r = 8$.

8. Find the general term, a_n, of the geometric sequence 4, 28, 196, 1372, ...

9. Find the tenth partial sum, s_{10}, of the geometric sequence 3, 15, 75, 375, ...

10. For the geometric sequence $10, -6, \frac{18}{5}, -\frac{54}{25}, \ldots$, does the infinite series have a limit?

    If so, find that limit.

11. Jonah's parents have decided to start a college fund for him. On Jonah's first birthday his parents deposit $2500 in the fund. On each birthday after that, they increase the amount that they deposit by $500.

    a) Write a sequence showing the amounts that Jonah's parents deposit on his first six birthdays.

    b) Find the general term a_n for the amount deposited on Jonah's nth birthday.

    c) What is the total amount deposited in the fund for Jonah's first 18 birthdays?

12. A new car, valued at $18,000, decreases in value by 25% each year.

    a) Write a geometric sequence showing the value of the car at the end of each of the first 4 years after it was purchased.

    b) Find the general term, a_n, for the value of the car n years after it was purchased.

13. Evaluate the binomial coefficient $_{12}C_5$.

Expand using the binomial theorem.

14. $(x + 1)^4$

Expand using the binomial theorem and Pascal's triangle.

15. $(4x + y)^7$

Mathematicians in History
Fibonacci

Fibonacci was a mathematician in the 12th and 13th centuries who introduced the world to an important sequence of numbers known as the Fibonacci sequence.

Write a one-page summary (*or* make a poster) of the life of Fibonacci and his accomplishments.

Interesting issues:

- Where and when was Fibonacci born?
- Where was Fibonacci educated?
- What was Fibonacci's father's occupation?
- What was Fibonacci's real name?
- What is the Fibonacci sequence, and how is it created?
- Which problem in Fibonacci's book *Liber abaci* introduced the sequence that came to be known as the Fibonacci sequence?
- List at least 3 occurrences of the Fibonacci sequence in nature.
- Fibonacci also wrote problems involving perfect numbers. What is a perfect number?

*The Fibonacci sequence is one of the most famous sequences in history. In fact, it has an entire magazine dedicated to it called **The Fibonacci Quarterly**. The sequence begins with 1 and 1, and each following number is formed by adding the previous two numbers. This is what is known as a recursive sequence. Mathematically, this recursive sequence is formed using the notation $f_n = f_{n-1} + f_{n-2}$, where f_{n-1} and f_{n-2} are the predecessors of f_n and $n > 2$.*

a) List the first 20 Fibonacci numbers. Start with 1 and 1 as the first two Fibonacci numbers.

The Fibonacci sequence has many interesting properties in mathematics alone. These properties have been proven by many mathematicians, but prove to yourself that they are true by finding at least one example to back up each property.

b) The sum of 10 Fibonacci numbers is always divisible by 11. Do you notice anything interesting about the number that results in dividing the sum by 11?

c) Twice any Fibonacci number minus the next Fibonacci number equals the number that is two numbers before the original number in the sequence. $(2 \cdot f_n - f_{n+1} = f_{n-2})$

d) For any three consecutive Fibonacci numbers, the product of the first and third numbers $(f_n \cdot f_{n+2})$ differs from the square of the middle number $((f_{n+1})^2)$ by one.

e) Now that you have seen a few of the fascinating properties, try to come up with one of your own.

Evaluate the given function. Round to the nearest thousandth. [12.1–12.2]

1. $f(x) = e^{x+3} - 21, f(2)$

2. $f(x) = \ln(x + 7) + 32, f(25)$

Simplify. [12.2]

3. $\log_3 81$

4. $\ln e^8$

Rewrite as a single logarithmic expression. Simplify if possible. Assume all variables represent positive real numbers. [12.3]

5. $2 \ln x - 4 \ln y + 5 \ln z$

Rewrite as the sum or difference of logarithmic expressions whose arguments have an exponent of 1. Simplify, if possible. Assume all variables represent positive real numbers. [12.3]

6. $\log\left(\dfrac{b^4 c^5}{a}\right)$

Evaluate using the change-of-base formula. Round to the nearest thousandth. [12.3]

7. $\log_{13} 200$

8. $\log_9 3224$

Solve. Round to the nearest thousandth. [12.4]

9. $5^{x-6} = 25$

10. $7^x = 219$

11. $e^{x+7} = 6205$

Solve. Round to the nearest thousandth. [12.4]

12. $\log(x^2 + 7x - 9) = \log(3x + 23)$

13. $\log_3 (x - 9) = 4$

14. $\log_2 (x + 3) + \log_2 (x + 9) = 4$

For the given function $f(x)$, find its inverse function $f^{-1}(x)$. [12.4]

15. $f(x) = e^{x-2} + 3$

16. $f(x) = \ln(x + 9) - 16$

17. Jean has deposited $2000 in an account that pays 6% annual interest, compounded quarterly. How long will it take until the balance of the account is $10,000?

18. The population of a city increased from 120,000 in 1992 to 150,000 in 2002. If the population continues to increase exponentially at the same rate, what will the population be in 2022?

Graph. Label any intercepts and asymptotes. State the domain and range of the function. [12.6]

19. $f(x) = 2^x - 1$

20. $f(x) = \ln(x + 5) - 3$

Graph the parabola. Label the vertex and any intercepts. [13.1]

21. $y = x^2 + 6x - 8$

22. $y = (x + 3)^2 - 4$

23. $x = -y^2 + y + 6$

24. $x = (y - 2)^2 + 3$

Graph the hyperbola. Label the center. Give the values of a and b. [13.4]

29. $\dfrac{x^2}{25} - \dfrac{y^2}{4} = 1$

Graph the circle. Label the center of the circle. Identify the radius of the circle. [13.2]

25. $x^2 + y^2 = 64$

26. $(x + 3)^2 + (y - 2)^2 = 25$

30. $(y - 2)^2 - \dfrac{(x + 4)^2}{9} = 1$

Graph the ellipse. Label the center. Give the values of a and b. [13.3]

27. $\dfrac{x^2}{9} + \dfrac{y^2}{49} = 1$

Find the equation of the conic section. [13.2–13.4]

31.

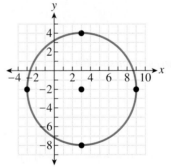

28. $36x^2 + 25y^2 - 288x - 300y + 576 = 0$

32.

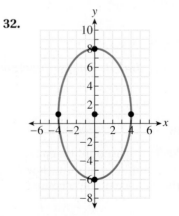

Solve by substitution. [13.5]

33. $x^2 + y^2 = 40$
$2x + y = 10$

34. $x^2 + 2y^2 = 18$
$y = x - 3$

Solve by the addition method. [13.5]

35. $x^2 + y^2 = 100$
$x^2 - y^2 = 28$

36. $2x^2 + 5y^2 = 38$
$4x^2 - 3y^2 = 24$

37. A rectangular lawn has a perimeter of 100 feet and an area of 600 square feet. Find the dimensions of the lawn.

Find the next three terms of the given sequence, and find its general term a_n. [14.1]

38. $-4, 9, 22, 35, 48, \ldots$

Find the first five terms of the sequence of the given first term and general term. [14.1]

39. $a_1 = 13, a_n = 2(a_{n-1}) + 5$

Find the general term, a_n, of the given arithmetic sequence. [14.2]

40. $3, 12, 21, 30, 39, \ldots$

Find the partial sum for the given arithmetic sequence. [14.2]

41. $s_{12}, -6, -1, 4, 9, \ldots$

42. Find the sum of the first 200 positive integers.

Find the general term, a_n, of the given geometric sequence. [14.3]

43. $12, 9, \dfrac{27}{4}, \dfrac{81}{16}, \ldots$

Find the partial sum of the geometric sequence of the given first term and common ratio. [14.3]

44. $s_8, a_1 = 1, r = 4$

Find the partial sum for the given geometric sequence. [14.3]

45. $s_7, 2, 14, 98, 686, \ldots$

For the geometric sequence whose general term is given, does the infinite series have a limit? If so, what is that limit? [14.3]

46. $a_n = 6 \cdot \left(\dfrac{3}{8}\right)^{n-1}$

Evaluate the given binomial coefficient. [14.4]

47. $_{10}C_4$

Expand using the binomial theorem. [14.4]

48. $(x + y)^4$

49. $(x - y)^6$

Expand using the binomial theorem and Pascal's triangle. [14.4]

50. $(x + y)^8$

Appendix A
USING MATRICES
TO SOLVE SYSTEMS
OF EQUATIONS

Matrix

A **matrix** is a rectangular array of numbers, such as $\begin{bmatrix} 1 & 5 & 7 & 20 \\ 0 & 4 & -3 & 5 \\ 0 & 0 & 6 & 6 \end{bmatrix}$. The numbers in a matrix are called **elements.** A matrix is often described by the number of **rows** and number of **columns** it contains. The matrix in this example is a three by four, or 3×4, matrix. (The plural form of matrix is **matrices.**) We refer to the first row of a matrix as R_1, the second row as R_2, and so on.

We can use matrices to solve systems of equations in a manner that is similar to using the addition method. For example, to solve the system of equations $\begin{array}{l} 2x - 5y = -9 \\ 4x + 3y = -5 \end{array}$, we begin by creating the **augmented matrix** $\left[\begin{array}{cc|c} 2 & -5 & -9 \\ 4 & 3 & -5 \end{array}\right]$. The first column of the matrix contains the coefficients of the variable x from the two equations. The second column contains the coefficients of the variable y. The dashed vertical line separates these coefficients from the constants in each equation.

We will use **row operations** to convert a matrix into a form in which we can easily solve the system of equations.

Row Operations

1. Any two rows of a matrix can be interchanged.
2. The elements in any row can be multiplied by any nonzero number.
3. The elements in any row can be changed by adding a nonzero multiple of another row's elements to them.

To solve a system of two equations in two unknowns, our goal is to transform the augmented matrix associated with that system to one of the form $\left[\begin{array}{cc|c} a & b & c \\ 0 & d & e \end{array}\right]$. This matrix corresponds to the system of equations $\begin{array}{l} ax + by = c \\ dy = e \end{array}$. We can then easily solve the second equation for y and substitute that value in the first equation to solve for x. So, the goal is to transform the matrix into one whose first element in the second row is 0.

EXAMPLE 1 Solve $\begin{array}{l} 2x - 5y = -9 \\ 4x + 3y = -5 \end{array}$.

Solution

The augmented matrix associated with this system is $\begin{bmatrix} 2 & -5 & -9 \\ 4 & 3 & -5 \end{bmatrix}$. We can change the first element in row 2 to 0 by adding -2 times the first row to the second row.

$$\begin{bmatrix} 2 & -5 & -9 \\ 4 & 3 & -5 \end{bmatrix}$$

$$\begin{bmatrix} 2 & -5 & -9 \\ 0 & 13 & 13 \end{bmatrix} \quad \text{Multiply } -2 \text{ by } R_1 \text{ and add to } R_2.$$

The second row of this matrix is equivalent to the equation $13y = 13$, which can be solved by dividing both sides of the equation by 13. The solution of this equation is $y = 1$.

The first row of this matrix is equivalent to the equation $2x - 5y = -9$. If we substitute 1 for y, then we can solve this equation for x.

$$2x - 5(1) = -9$$
$$2x - 5 = -9$$
$$2x = -4$$
$$x = -2$$

The solution of the system of equations is $(-2, 1)$.

EXAMPLE 2   Solve $\begin{aligned} 3x + 4y &= 6 \\ 2x - y &= -7 \end{aligned}$.

Solution

The augmented matrix associated with this system is $\begin{bmatrix} 3 & 4 & 6 \\ 2 & -1 & -7 \end{bmatrix}$. We begin by multiplying 3 by row 2, so that the first element in row 2 is a multiple of the first element in row 1.

$$\begin{bmatrix} 3 & 4 & 6 \\ 2 & -1 & -7 \end{bmatrix}$$

$$\begin{bmatrix} 3 & 4 & 6 \\ 6 & -3 & -21 \end{bmatrix} \quad \text{Multiply 3 by } R_2.$$

$$\begin{bmatrix} 3 & 4 & 6 \\ 0 & -11 & -33 \end{bmatrix} \quad \text{Multiply } -2 \text{ by } R_1 \text{ and add to } R_2.$$

This matrix is equivalent to the system of equations $\begin{aligned} 3x + 4y &= 6 \\ -11y &= -33 \end{aligned}$. We will solve the second equation for y.

$$-11y = -33$$
$$y = 3 \quad \text{Divide both sides by } -11.$$

Substitute 3 for y in the equation $3x + 4y = 6$ and solve for x.

$$3x + 4(3) = 6$$
$$3x + 12 = 6$$
$$3x = -6$$
$$x = -2$$

The solution of the system of equations is $(-2, 3)$.

To solve a system of three equations in three unknowns, we begin by setting up a 3×4 augmented matrix. Using row operations, our goal is to transform the matrix to one of the form $\begin{bmatrix} a & b & c & | & d \\ 0 & e & f & | & g \\ 0 & 0 & h & | & i \end{bmatrix}$.

EXAMPLE 3   Solve the system
$$x + y + z = 1$$
$$2x - 3y + 4z = 29.$$
$$-x - 2y + 3z = 8$$

Solution

The augmented matrix associated with this system is $\begin{bmatrix} 1 & 1 & 1 & | & 1 \\ 2 & -3 & 4 & | & 29 \\ -1 & -2 & 3 & | & 8 \end{bmatrix}$. We will begin by changing the first element in row 2 and row 3 to 0.

$$\begin{bmatrix} 1 & 1 & 1 & | & 1 \\ 2 & -3 & 4 & | & 29 \\ -1 & -2 & 3 & | & 8 \end{bmatrix}$$

$$\begin{bmatrix} 1 & 1 & 1 & | & 1 \\ 0 & -5 & 2 & | & 27 \\ 0 & -1 & 4 & | & 9 \end{bmatrix}$$ Multiply -2 by R_1 and add to R_2. Add R_1 to R_3.

Now we need to change the second element of row 3 to a 0. We begin to do that by changing that element to a multiple of the second element of row 2.

$$\begin{bmatrix} 1 & 1 & 1 & | & 1 \\ 0 & -5 & 2 & | & 27 \\ 0 & 5 & -20 & | & -45 \end{bmatrix}$$ Multiply -5 by R_3.

$$\begin{bmatrix} 1 & 1 & 1 & | & 1 \\ 0 & -5 & 2 & | & 27 \\ 0 & 0 & -18 & | & -18 \end{bmatrix}$$ Add R_2 to R_3.

The last row is equivalent to the equation $-18z = -18$, whose solution is $z = 1$. We substitute this value for z in the equation that is equivalent to row 2, $-5y + 2z = 27$.

$$-5y + 2(1) = 27$$
$$-5y + 2 = 27$$
$$-5y = 25$$
$$y = -5$$

Now we substitute -5 for y and 1 for z in the equation that is equivalent to row 1, $x + y + z = 1$.

$$x + (-5) + (1) = 1$$
$$x - 4 = 1$$
$$x = 5$$

The solution of this system is $(5, -5, 1)$.

EXERCISES

Solve the system of equations by using matrices.

1. $2x + 3y = 13$
    $5x - 4y = -2$

2. $3x - 2y = 16$
    $7x + 5y = 18$

3. $-4x + 9y = 75$
    $3x - 7y = -58$

4. $x + 2y = -23$
    $6x + y = -39$

5. $11x + 7y = 42$
    $19x - 13y = -78$

6. $15x + 8y = 9$
    $-10x + 9y = -6$

7. $x + 3y + 2z = 11$
    $3x - 2y - z = 4$
    $-2x + 4y + 3z = 5$

8. $x - y + z = 1$
    $4x + 3y + 2z = 10$
    $4x - 2y - 5z = 31$

9. $x + y + z = 3$
    $2x + 3y + 4z = 16$
    $3x - y + 4z = 23$

10. $x + y + z = 2$
    $-2x - 3y + 8z = -26$
    $5x - 4y - 9z = 20$

11. $3x + 7y - 8z = 60$
    $5x - 4y + 7z = -44$
    $-4x + y + 10z = -36$

12. $6x + y + 3z = -40$
    $-5x + \dfrac{3}{2}y - z = 20$
    $\dfrac{9}{2}x + \dfrac{1}{5}y + \dfrac{3}{4}z = 19$

FOR EXTRA HELP

 MyMathLab

 MathXL
MathXL

Interactmath.com

MathXL
Tutorials on CD

DVT CD
Videotape

Tutor Center
Addison-Wesley
Math Tutor Center

Student Solutions
Manual

Synthetic division is a shorthand alternative to polynomial long division when the divisor is of the form $x - b$. When using synthetic division, we do not write any variables, dealing only with the coefficients.

Suppose that we wanted to divide $\dfrac{2x^2 - x - 25}{x - 4}$. Note that $b = 4$ in this example. Before we begin, we check the dividend to be sure that it is written in descending order and that there are no missing terms. If there are missing terms, we will add placeholders. On the first line, we write down the value for b as well as the coefficients for the dividend.

$$\underline{4}\,\rvert \quad 2 \qquad -1 \qquad -25$$

Notice that we set the value of b apart from the coefficients of the divisor. After we use synthetic division, we will find the quotient and the remainder in the last of our three rows. To begin the actual division, we bring the first coefficient down to the bottom row as follows:

$$
\begin{array}{r|rrr}
4 & 2 & -1 & -25 \\
  & \downarrow & & \\
\hline
  & 2 & &
\end{array}
$$

This number will be the leading coefficient of our quotient. In the next step, we multiply 4 by this first coefficient and write it underneath the next coefficient in the dividend (-1). Totaling this column will give us the next coefficient in the quotient.

$$
\begin{array}{r|rrr}
4 & 2 & -1 & -25 \\
  & \downarrow & 8 & \\
\hline
  & 2 & 7 &
\end{array}
$$

We repeat this step until we total the last column. Multiply 4 by 7 and write the product underneath -25. Total this column as follows:

$$
\begin{array}{r|rrr}
4 & 2 & -1 & -25 \\
  & \downarrow & 8 & 28 \\
\hline
  & 2 & 7 & \boxed{3}
\end{array}
$$

We are now ready to read our quotient and remainder from the bottom row. The remainder is the last number in the bottom row, which in our example is 3. The numbers in front of the remainder are the coefficients of our quotient, written in descending order. In this example, the quotient is $2x + 7$. So, $\dfrac{2x^2 - x - 25}{x - 4} = 2x + 7 + \dfrac{3}{x - 4}$.

EXAMPLE 1   Use synthetic division to divide $\dfrac{x^3 - 11x^2 - 13x + 6}{x + 2}$.

Solution

The divisor is of the correct form, and $b = -2$. This is because $x + 2$ can be rewritten as $x - (-2)$. The dividend is in descending form with no missing terms, so we may begin.

$$
\begin{array}{r|rrrr}
-2 & 1 & -11 & -13 & 6 \\
& \downarrow & -2 & 26 & -26 \\
\hline
& 1 & -13 & 13 & \boxed{-20}
\end{array}
$$

So, $\dfrac{x^3 - 11x^2 - 13x + 6}{x + 2} = x^2 - 13x + 13 - \dfrac{20}{x + 2}$.

Now we turn to an example that has a dividend with missing terms, requiring the use of placeholders.

EXAMPLE 2   Use synthetic division to divide $\dfrac{3x^4 - 8x - 5}{x - 3}$.

Solution

The dividend, written in descending order with placeholders, is $3x^4 + 0x^3 + 0x^2 - 8x - 5$. The value for b is 3.

$$
\begin{array}{r|rrrrr}
3 & 3 & 0 & 0 & -8 & -5 \\
& \downarrow & 9 & 27 & 81 & 219 \\
\hline
& 3 & 9 & 27 & 73 & \boxed{214}
\end{array}
$$

So, $\dfrac{3x^4 - 8x - 5}{x - 3} = 3x^3 + 9x^2 + 27x + 73 + \dfrac{214}{x - 3}$.

EXERCISES

Divide by using synthetic division.

1. $\dfrac{x^2 - 18x + 77}{x - 7}$

2. $\dfrac{x^2 + 4x - 45}{x - 5}$

3. $(x^2 + 13x - 48) \div (x - 3)$

4. $(2x^2 - 43x + 221) \div (x - 13)$

5. $\dfrac{7x^2 - 11x - 12}{x + 2}$

6. $\dfrac{3x^2 + 50x - 30}{x + 17}$

7. $\dfrac{12x^2 - 179x - 154}{x - 14}$

8. $\dfrac{x^2 + 14x - 13}{x + 1}$

9. $\dfrac{x^3 - 34x + 58}{x - 5}$

10. $\dfrac{x^3 - 59x - 63}{x + 7}$

11. $\dfrac{x^4 + 8x^2 - 33}{x + 8}$

12. $\dfrac{5x^5 - 43x^3 - 8x^2 + 9x - 9}{x - 3}$

F O R   E X T R A   H E L P

MyMathLab
MyMathLab

MathXL
MathXL

Interactmath.com

MathXL
Tutorials on CD

DVT CD
Videotape

Tutor
Center

Addison-Wesley
Math Tutor Center

Student Solutions
Manual

Chapter 1

Quick Check 1.1 **1.** [number line 0–10 with point at 4] **2. a)** $>$ **b)** $<$ **3. a)** 13 **b)** -8 **4. a)** $<$ **b)** $>$ **5.** 9

Section 1.1 **1.** [number line 0–10 point at 7] **3.** [number line 0–10 point at 6] **5.** [number line 0–10 point at 1] **7.** $<$ **9.** $>$ **11.** $>$
13. [number line -10 to 0 point at -5] **15.** [number line 0–10 point at 3] **17.** [number line -15 to 0 point at -10] **19.** 7 **21.** -14 **23.** 0 **25.** $>$ **27.** $<$
29. $<$ **31.** $<$ **33.** 15 **35.** 0 **37.** 25 **39.** $-7, 7$ **41.** 0 **43.** $-6, 6$ **45.** No **47.** Negative

Quick Check 1.2 **1.** 8 **2.** -13 **3.** -11 **4.** 18 **5.** -4 **6.** -60 **7.** 63 **8.** -400 **9.** -9

Section 1.2 **1.** -2 **3.** -18 **5.** 1 **7.** 8 **9.** -11 **11.** 2 **13.** -6 **15.** -8 **17.** -21 **19.** 27 **21.** 15 **23.** -6 **25.** 27 **27.** 1
29. \$8 **31.** $4°C$ **33.** 2150 ft **35.** -48 **37.** -36 **39.** 143 **41.** -82 **43.** 0 **45.** 90 **47.** 180 **49.** -6 **51.** -9 **53.** 7
55. -14 **57.** 0 **59.** Undefined **61.** 132 **63.** -8 **65.** -21 **67.** -144 **69.** -216 **71.** 0 **73.** \$96 **75.** $-\$200$ **77.** True
79. True **81.** Answers will vary. **83.** Answers will vary. **85.** 26 **87.** -84

Quick Check 1.3 **1.** $\{1, 2, 3, 4, 6, 9, 12, 18, 36\}$ **2. a)** Composite **b)** Prime **c)** Composite **3.** $3 \cdot 3 \cdot 7$ **4. a)** $\frac{3}{14}$ **b)** $\frac{1}{16}$ **5.** $9\frac{4}{13}$ **6.** $\frac{49}{6}$

Section 1.3 **1.** No **3.** Yes **5.** Yes **7.** $\{1, 3, 9, 27\}$ **9.** $\{1, 2, 4, 5, 10, 20\}$ **11.** $\{1, 31\}$ **13.** $\{1, 2, 3, 4, 5, 6, 10, 12, 15, 20, 30, 60\}$
15. $\{1, 7, 49\}$ **17.** $\{1, 3, 23, 69\}$ **19.** $\{1, 13, 17, 221\}$ **21.** $2 \cdot 3 \cdot 3$ **23.** $2 \cdot 3 \cdot 7$ **25.** $3 \cdot 11$ **27.** $3 \cdot 3 \cdot 3$ **29.** $5 \cdot 5 \cdot 5$ **31.** Prime
33. $3 \cdot 3 \cdot 11$ **35.** Prime **37.** $7 \cdot 17$ **39.** $2 \cdot 2 \cdot 2 \cdot 3 \cdot 3 \cdot 5$ **41.** $2 \cdot 2 \cdot 2 \cdot 13$ **43.** $\frac{5}{8}$ **45.** $\frac{1}{5}$ **47.** $\frac{3}{7}$ **49.** $\frac{27}{64}$ **51.** $\frac{10}{11}$ **53.** $\frac{19}{5}$ **55.** $\frac{50}{17}$
57. $\frac{431}{19}$ **59.** $7\frac{4}{5}$ **61.** $14\frac{3}{7}$ **63.** $5\frac{6}{11}$ **65.** Answers will vary.

Quick Check 1.4 **1.** $\frac{5}{56}$ **2.** 23 **3.** $\frac{8}{105}$ **4.** $\frac{5}{8}$ **5.** $\frac{3}{5}$ **6.** 126 **7.** $\frac{26}{45}$ **8.** $\frac{5}{6}$

Section 1.4 **1.** $\frac{3}{20}$ **3.** $\frac{21}{25}$ **5.** $2\frac{2}{5}$ **7.** $31\frac{1}{2}$ **9.** $\frac{16}{27}$ **11.** $\frac{4}{3}$ **13.** $\frac{1}{2}$ **15.** $\frac{17}{20}$ **17.** $\frac{5}{44}$ **19.** $\frac{21}{50}$ **21.** $\frac{11}{15}$ **23.** 1 **25.** $\frac{2}{5}$ **27.** $\frac{3}{8}$ **29.** 36
31. 70 **33.** 36 **35.** 36 **37.** $\frac{5}{6}$ **39.** $\frac{31}{20}$ **41.** $\frac{53}{40}$ **43.** $\frac{8}{15}$ **45.** $11\frac{1}{5}$ **47.** $11\frac{5}{6}$ **49.** $\frac{1}{2}$ **51.** $\frac{1}{5}$ **53.** $\frac{13}{28}$ **55.** $\frac{2}{15}$ **57.** $4\frac{1}{4}$ **59.** $3\frac{3}{10}$
61. $\frac{8}{15}$ **63.** $\frac{4}{3}$ **65.** $\frac{8}{25}$ **67.** $-\frac{17}{24}$ **69.** $\frac{56}{45}$ **71.** 104 **73.** $2\frac{6}{35}$ **75.** 8 **77.** 14 **79.** $1\frac{7}{12}$ cups **81.** $\frac{9}{20}$ fluid ounces **83.** $\frac{7}{10}$ feet
85. No, the craftsman needs 3 feet of wood. **87.** $\frac{2}{3}$ cups **89.** Answers will vary.

Quick Check 1.5 **1.** 0.75 **2.** $0.2\overline{7}$ **3.** $\frac{1}{25}$ **4.** $\frac{17}{40}$ **5.** 70% **6.** $52\frac{1}{2}\%$ **7.** 42% **8.** $\frac{7}{20}$ **9.** $\frac{7}{60}$ **10.** 0.08 **11.** 2.4

Section 1.5 **1.** 7.85 **3.** 36.4 **5.** 4.7 **7.** 33.92 **9.** -8.03 **11.** 35.032 **13.** 55.2825 **15.** 94 **17.** 0.6 **19.** 0.875 **21.** 0.64
23. 24.58 **25.** $\frac{1}{5}$ **27.** $\frac{11}{25}$ **29.** $\frac{3}{4}$ **31.** $\frac{3}{8}$ **33.** $101.4°F$ **35.** \$245.95 **37.** 257.64 **39.** 50% **41.** 75% **43.** 80% **45.** $87\frac{1}{2}\%$
47. 675% **49.** 40% **51.** 15% **53.** 9% **55.** 320% **57.** $\frac{53}{100}$ **59.** $\frac{21}{25}$ **61.** $\frac{7}{100}$ **63.** $\frac{1}{9}$ **65.** $5\frac{1}{5}$ **67.** 0.32 **69.** 0.16 **71.** 0.04
73. 0.003 **75.** 4 **77.** Answers will vary.

Quick Check 1.6

1. Is There Adequate Parking at Your Campus?

2.

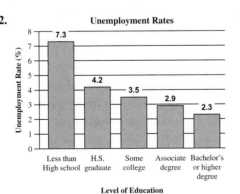

3. Collisions between Wildlife and Civilian Aircraft

Section 1.6 **1.**

Teachers

Men 25%
Women 75%

3. Proficient Level

Above 31%
Below 69%

5. Holiday Gift Shopping Online

Larger Selection 6%
Speed 4%
Free Shipping 34%
Comparison Shopping 23%
Convenience 33%

7. Do Companies Operate Fairly and Honestly

No 66%
A Few 34%

9. When Freshmen Planned on Attending College

Unsure 7%
Never 17%
After Working at Least 2 Years 12%
Right after Graduation 64%

11. Women That Shop Online   Men That Shop Online

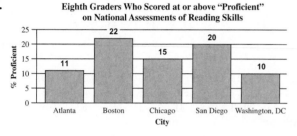

Don't Shop Online 62%
Shop Online 38%

Don't Shop Online 56%
Shop Online 44%

13.

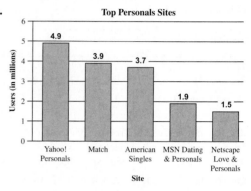

Top Personals Sites

Users (in millions)

Yahoo! Personals 4.9
Match 3.9
American Singles 3.7
MSN Dating & Personals 1.9
Netscape Love & Personals 1.5

Site

15.

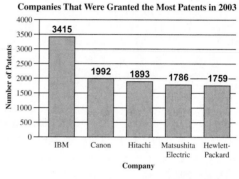

Eighth Graders Who Scored at or above "Proficient" on National Assessments of Reading Skills

% Proficient

Atlanta 11
Boston 22
Chicago 15
San Diego 20
Washington, DC 10

City

17.

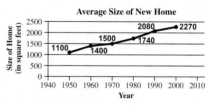

Companies That Were Granted the Most Patents in 2003

Number of Patents

IBM 3415
Canon 1992
Hitachi 1893
Matsushita Electric 1786
Hewlett-Packard 1759

Company

19.

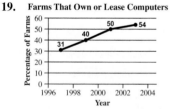

Farms That Own or Lease Computers

Percentage of Farms

31
40
50
54

1996 1998 2000 2002 2004

Year

21.

Worldwide Spam Messages Sent Daily

Spam Messages (in billions)

1
2.3
4
5.6
7.3

1998 1999 2000 2001 2002 2003 2004

Year

23.

Average Size of New Home

Size of Home (in square feet)

1100
1400
1500
1740
2080
2270

1940 1950 1960 1970 1980 1990 2000 2010

Year

25.

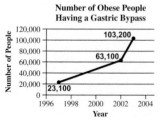

Number of Obese People Having a Gastric Bypass

Number of People

23,100
63,100
103,200

1996 1998 2000 2002 2004

Year

27. Answers will vary.

Quick Review   1. -7 **2.** -120 **3.** -20 **4.** $-\frac{110}{21}$

Quick Check 1.7   1. 64   **2.** $\frac{1}{4096}$ **3.** -84 **4.** 1   **5.** 360   **6.** -84

Section 1.7   1. 2^3 **3.** $\left(\frac{2}{9}\right)^6$ **5.** $(-2)^4$ **7.** -3^5 **9.** 5^3 **11.** 32   **13.** 2401   **15.** 1000   **17.** 1   **19.** $\frac{27}{64}$ **21.** 0.00032   **23.** 81
25. -128 **27.** 72   **29.** 576   **31.** 7   **33.** 10   **35.** 16   **37.** -71 **39.** 49   **41.** 51   **43.** $\frac{11}{14}$ **45.** $\frac{11}{125}$ **47.** $\frac{1}{21}$ **49.** $-\frac{1}{4}$ **51.** $\frac{10}{3}$ **53.** 14
55. $3 \cdot 5 - 9 + 8 = 14$ **57.** $(3 + 7 \cdot 9) \div 2 = 33$ **59.** Answers will vary.   **61.** Answers will vary.   **63.** 1024   **65.** 1,073,741,824
67. 6   **69.** 6   **71.** 5   **73.** Answers will vary. $(-2)^6 = 64, -2^6 = -64$

Quick Check 1.8   1. $x + 9$ **2.** $x - 25$ **3.** $2x$ **4.** $\frac{x}{20}$ **5.** 57   **6.** 128   **7.** 97   **8.** $24a$ **9.** $9x + 34$ **10.** $35x - 28$
11. $12x - 24y + 36z$ **12.** $-24x - 66$ **13.** Four terms: $x^3, -x^2, 23x$, and -59, Coefficients: $1, -1, 23, -59$ **14.** $-2x + 8y$
15. $19x + 7$ **16.** $-2x + 51$

Section 1.8   1. $x + 11$ **3.** $5x$ **5.** $13 - x$ **7.** $3x - 4$ **9.** $x + y$ **11.** $7(x - y)$ **13.** $200u$ **15.** $25{,}000 + 22a$
17. -1 **19.** 20   **21.** 6   **23.** 88   **25.** -20 **27.** 0   **29.** -754 **31.** 49   **33.** 44   **35.** -31.95 **37.** 8.91   **39. a)** Answers
will vary.   **b)** Yes, answers will vary.   **41.** $3x - 27$ **43.** $15 - 35x$ **45.** $16x$ **47.** $-5x$ **49.** $7x + 4$ **51.** $38a - 63$
53. $4a + 5b$ **55.** $-2x - 22y$ **57.** $10x - 8$ **59.** $-8x + 7$ **61.** $-9y + 85$ **63.** $30z + 6$ **65.** $-10a - 5b + 26c$
67. $2(3x - 9) + 5(3x - 9) = 6x - 18 + 15x - 45 = 21x - 63, 7(3x - 9) = 21x - 63$ **69.** $4(2a + 11) = 8a + 44$
71. a) 4   **b)** $5x^3, 3x^2, -7x, -15$ **c)** $5, 3, -7, -15$ **73. a)** 2   **b)** $3x, -17$ **c)** $3, -17$ **75. a)** 3   **b)** $15x^2, -41x, 55$ **c)** $15, -41, 55$
77. a) 4   **b)** $-35a, -33b, 52c, 69$ **c)** $-35, -33, 52, 69$ **79.** Answers will vary.

Chapter 1 Review   1. $<$ **2.** $>$ **3.** 8   **4.** 13   **5.** 6   **6.** -12 **7.** -7 **8.** -3 **9.** -41 **10.** 27   **11.** -37 **12.** 20   **13.** -54
14. 16   **15.** $\{1, 2, 3, 6, 7, 14, 21, 42\}$ **16.** $\{1, 2, 3, 4, 6, 9, 12, 18, 27, 36, 54, 108\}$ **17.** 2^5 **18.** $2^2 \cdot 3 \cdot 5$ **19.** $\frac{4}{7}$ **20.** $\frac{1}{8}$ **21.** $\frac{20}{117}$
22. 14   **23.** $\frac{17}{25}$ **24.** $\frac{307}{25}$ **25.** $3\frac{4}{5}$ **26.** $9\frac{1}{6}$ **27.** $\frac{21}{100}$ **28.** $2\frac{1}{3}$ **29.** $\frac{14}{25}$ **30.** $\frac{32}{75}$ **31.** $\frac{29}{24}$ **32.** $\frac{89}{60}$ **33.** $\frac{1}{6}$ **34.** $-\frac{29}{252}$ **35.** $\frac{25}{77}$ **36.** $\frac{5}{24}$
37. $\frac{21}{104}$ **38.** $\frac{5}{12}$ **39.** 12.62   **40.** 8.818   **41.** 30.24   **42.** 8.8   **43.** 0.32   **44.** 0.9375   **45.** $\frac{3}{4}$ **46.** $\frac{7}{25}$ **47.** \$17   **48.** $-\$47$
49. \$12,600   **50.** $4\frac{1}{6}$ cups   **51.** 64   **52.** $\frac{8}{125}$ **53.** -64 **54.** 6075   **55.** 37   **56.** 22   **57.** 48   **58.** 60   **59.** 51   **60.** $\frac{87}{100}$
61. $n + 14$ **62.** $n - 20$ **63.** $2n - 8$ **64.** $6n + 9$ **65.** $3.55c$ **66.** $20 + 0.15m$ **67.** 44   **68.** 25   **69.** 56   **70.** -69 **71.** 0
72. 84   **73.** $5x + 35$ **74.** $27x$ **75.** $5x - 8$ **76.** $-16y + 126$ **77.** $5k - 48$ **78.** $3x - 303$

	79.	**80.**
# of terms	4	3
Terms	$x^3, -4x^2, -10x, 41$	$-x^2, 5x, -30$
Coefficients	$1, -4, -10, 41$	$-1, 5, -30$

Chapter 1 Test   1. $>$ [1.1]   **2.** 17 [1.1]   **3.** -6 [1.2]   **4.** 63 [1.2]   **5.** $\{1, 3, 5, 9, 15, 45\}$ [1.3]   **6.** $2^2 \cdot 3^3$ [1.3]   **7.** $\frac{5}{7}$ [1.3]
8. $3\frac{13}{18}$ [1.3]   **9.** $\frac{5}{84}$ [1.4]   **10.** $\frac{7}{12}$ [1.4]   **11.** $\frac{91}{60}$ [1.4]   **12.** $-\frac{17}{72}$ [1.4]   **13.** 18.2735 [1.5]   **14.** $\frac{9}{25}$ [1.5]   **15.** \$6487 [1.2]
16. \$795.51 [1.5]   **17.** -24 [1.7]   **18.** 25 [1.7]   **19.** $\frac{5}{11}$ [1.7]   **20.** $4n - 7$ [1.8]   **21.** $50 + 20h$ [1.8]   **22.** 61 [1.8]   **23.** -1 [1.8]
24. $10x - 65$ [1.8]   **25.** $-9y + 240$ [1.8]

Chapter 2

Quick Check 2.1   1. Yes   **2.** $\{-16\}$ **3.** $\{14\}$ **4.** $\{-16\}$ **5.** $\left\{\frac{14}{3}\right\}$ **6.** $\{-30\}$ **7.** $\{15\}$ **8.** $\left\{\frac{31}{12}\right\}$ **9.** \$65

Section 2.1   1. No, there is a term containing x^2.   **3.** Yes   **5.** Yes   **7.** Yes   **9.** No, there is a variable in a denominator.   **11.** Yes
13. No   **15.** Yes   **17.** $\{18\}$ **19.** $\{-10\}$ **21.** $\left\{\frac{5}{3}\right\}$ **23.** $\{0\}$ **25.** $\{-7\}$ **27.** $\{14\}$ **29.** $\{-45\}$ **31.** $\{21\}$ **33.** $\{-96\}$ **35.** $\{8\}$
37. $\{-10\}$ **39.** $\{3\}$ **41.** $\{-4\}$ **43.** $\{17\}$ **45.** $\{2.5\}$ **47.** $\{9\}$ **49.** $\{20\}$ **51.** $\{3\}$ **53.** $\{23\}$ **55.** $\{-9\}$ **57.** $\left\{\frac{5}{36}\right\}$ **59.** $\{-4\}$
61. $\{12\}$ **63.** $\{16\}$ **65.** $\{30\}$ **67.** $\{-41\}$ **69.** $\{-77\}$ **71.** $\left\{-\frac{29}{30}\right\}$ **73.** $\{-56\}$ **75.** $\left\{\frac{3}{8}\right\}$ **77.** $\left\{-\frac{21}{10}\right\}$ **79.** $\{0\}$

Answers will vary for 81–87. The following are examples of correct answers.
81. $2x = 14$ **83.** $2n = 5$ **85.** $x + 2 = -4$ **87.** $b + \frac{5}{6} = 1$ **89.** 27 nickels   **91.** 159 people   **93.** 166 employees   **95.** $56°$ F
97.–99. Answers will vary.

Quick Check 2.2   1. $\{7\}$ **2.** $\{-8\}$ **3.** $\left\{\frac{35}{12}\right\}$ **4.** $\{-9\}$ **5.** $\{2\}$ **6.** $\{-3\}$ **7.** $\{4\}$ **8.** $\{-6\}$ **9.** $\varnothing$ **10.** $\Re$ **11.** $y = \frac{5 - x}{2}$ **12.** $x = \frac{5z}{y}$

Section 2.2   1. $\{-3\}$ **3.** $\left\{\frac{1}{2}\right\}$ **5.** $\left\{-\frac{5}{2}\right\}$ **7.** $\{0\}$ **9.** $\{3\}$ **11.** $\{37\}$ **13.** $\{-10\}$ **15.** $\left\{-\frac{15}{2}\right\}$ **17.** $\left\{\frac{17}{3}\right\}$ **19.** $\left\{\frac{104}{25}\right\}$ **21.** $\Re$ **23.** $\varnothing$
25. $\left\{\frac{5}{3}\right\}$ **27.** $\{-5\}$

Answers will vary for 29–37. The following are examples of correct answers.
29. $x - 5 = 4$ **31.** $9a - 1 = 0$ **33.** $5x - 6 = 5x - 7$ **35.** $2(x - 4) - 3 = 2x - 11$ **37.** $x + 3 = -8$ **39.** $y = -5x - 2$
41. $y = \frac{-8x + 4}{2}$ **43.** $y = \frac{-6x + 9}{-3}$ **45.** $b = P - a - c$ **47.** $t = \frac{d}{r}$ **49.** $r = \frac{C}{2\pi}$ **51.** $20°$ C   **53.** 354   **55. a)** 80 kites
b) \$6.50 **c)** \$12.75   **57.** 4   **59.** 15   **61.** Answers will vary.

Quick Check 2.3   **1.** 8   **2.** Length: 24 ft, Width: 16 ft   **3.** 14 in., 48 in., 50 in.   **4.** 70, 71, 72   **5.** 7 hours   **6.** 17 and 86
7. 22 nickels and 9 quarters

Section 2.3   **1.** 27, 44   **3.** 44, 52   **5.** 13, 18   **7.** Length: 4 m, Width: 9 m   **9.** Length: 17 ft, Width: 10 ft   **11.** Length: 13 cm, Width:
52 cm   **13.** 55 ft   **15.** 32 cm   **17.** 5 in., 9 in., 12 in.   **19.** 11 in.   **21.** A: $60°$, B: $30°$ **23.** $20°, 70°$ **25.** $42°, 138°$ **27.** $50°, 60°, 70°$
29. 667 sq in.   **31.** 52 in.   **33.** 152, 153   **35.** 157   **37.** 73, 75, 77   **39.** 93   **41.** 138, 139   **43.** 15, 16, 17   **45.** 500 miles   **47.** $5\frac{1}{2}$ hr
49. $1\frac{11}{14}$ m/sec   **51. a)** 15 hours **b)** 560 miles   **53.** 55   **55.** 18   **57.** 90   **59.** Answers will vary.   **61.** Answers will vary.

Quick Check 2.4   **1.** 11   **2.** 65   **3.** 70%   **4.** 140   **5.** 54%   **6.** \$60.90   **7.** \$5000 at 3%, \$7000 at 5%   **8.** 20 ml 30% and 60 ml 42%
9. $\{33\}$ **10.** 68

Section 2.4   **1.** 7   **3.** 140   **5.** 70%   **7.** 39   **9.** 163.2   **11.** 44   **13.** 84   **15.** 1025   **17.** 300 ml   **19.** \$18.40   **21.** \$3.25
23. \$21.21   **25.** 7.5%   **27.** 25%   **29.** \$700 at 2%, \$1400 at 5%   **31.** \$2500 at 4%, \$4000 at 5%   **33.** 24 gal of 70%, 36 gal of 40%
35. 400 gal of 5%, 600 gal of 2%   **37.** 240 ml   **39.** $\{32\}$ **41.** $\{36\}$ **43.** $\{50.4\}$ **45.** $\{87.5\}$ **47.** 5520   **49.** 22.5   **51.** 144
53. 144   **55.** Answers will vary.

Quick Review   **1.** $\{6\}$ **2.** $\{-16\}$ **3.** $\{13\}$ **4.** $\{16\}$

Quick Check 2.5   **1.** [number line]   **2.** $x < 7$, [number line] , $(-\infty, 7)$ **3.** $x \geq 5$, [number line] , $[5, \infty)$
4. $x < -3$, [number line] , $(-\infty, -3)$ **5.** $x \geq -6$, [number line] , $[-6, \infty)$ **6.** $x \leq 4$ or $x \geq 7$,
[number line] , $(-\infty, 4] \cup [7, \infty)$ **7.** $-3 < x < \frac{3}{2}$, [number line] , $(-3, \frac{3}{2})$ **8.** $x < 130$ **9.** 48 or lower

Section 2.5   **1.** [number line] , $(-\infty, 3)$ **3.** [number line] , $[-1, \infty)$ **5.** [number line] , $(-2, 8)$
7. [number line] , $(\frac{9}{2}, \infty)$ **9.** [number line] , $(-\infty, 2] \cup (8, \infty)$ **11.** [number line] , $(-\infty, \frac{3}{2}) \cup (\frac{13}{4}, \infty)$
13. $x > 5$ **15.** $-4 \leq x \leq 7$ **17.** $x < -4$ **19.** $x < -4$, [number line] , $(-\infty, -4)$ **21.** $x > 6$, [number line] , $(6, \infty)$
23. $x < -3$, [number line] , $(-\infty, -3)$ **25.** $x < 4$, [number line] , $(-\infty, 4)$ **27.** $x \leq 2$, [number line] , $(-\infty, 2]$
29. $x \geq -10$, [number line] , $[-10, \infty)$ **31.** $x > 7$, [number line] , $(7, \infty)$ **33.** $x > \frac{5}{3}$, [number line] , $(\frac{5}{3}, \infty)$
35. $x \leq -7$, [number line] , $(-\infty, -7]$ **37.** $x < 7$, [number line] , $(-\infty, 7)$ **39.** $x > -8$, [number line] ,
$(-8, \infty)$ **41.** $x > -\frac{11}{2}$, [number line] , $(-\frac{11}{2}, \infty)$ **43.** $x < -4$ or $x > -2$, [number line] , $(-\infty, -4) \cup (-2, \infty)$
45. $x < -2$ or $x > \frac{14}{3}$, [number line] , $(-\infty, -2) \cup (\frac{14}{3}, \infty)$ **47.** $x > 6$ or $x < 3$, [number line] ,
$(-\infty, 3) \cup (6, \infty)$ **49.** $x \leq \frac{3}{2}$ or $x \geq 3$, [number line] , $(-\infty, \frac{3}{2}] \cup [3, \infty)$ **51.** $-1 < x < 4$, [number line] , $(-1, 4)$
53. $-\frac{3}{2} \leq x \leq \frac{9}{4}$, [number line] , $[-\frac{3}{2}, \frac{9}{4}]$ **55.** $1 < x < 7$, [number line] , $(1, 7)$ **57.** $-3 < x \leq 6$, [number line] ,
$(-3, 6]$ **59.** $\frac{16}{15} \leq x \leq \frac{25}{6}$, [number line] , $[\frac{16}{15}, \frac{25}{6}]$ **61.** Answers will vary.   **63.** $x \geq 2$ **65.** $x < 30$ **67.** $x \geq 12{,}500$ **69.** At least
300   **71.** At most 929   **73.** At least \$5625   **75.** 34 to 100   **77.** Answers will vary.

Chapter 2 Review   **1.** No   **2.** Yes   **3.** $\{16\}$ **4.** $\{-6\}$ **5.** $\{-6\}$ **6.** $\{-6\}$ **7.** 118   **8.** 212   **9.** $\{7\}$ **10.** $\{-9\}$ **11.** $\{2\}$
12. $\{36\}$ **13.** $\{10\}$ **14.** $\{\frac{35}{4}\}$ **15.** $\{-33\}$ **16.** $\{-\frac{3}{5}\}$ **17.** $\{16\}$ **18.** $\{-7\}$ **19.** $y = 8 - 7x$ **20.** $y = \frac{30 - 15x}{10}$ **21.** $d = \frac{C}{\pi}$
22. $a = \frac{P - b - 2c}{3}$ **23.** 36, 62   **24.** Length: 9 ft, Width: 16 ft   **25.** 66, 68, 70   **26.** $4\frac{3}{4}$ hr   **27.** 41   **28.** 40%   **29.** 28%   **30.** 90%
31. 45%   **32.** $\frac{3}{10}$ **33.** $\frac{11}{20}$ **34.** $\frac{6}{25}$ **35.** $\frac{3}{50}$ **36.** 0.15   **37.** 0.9   **38.** 0.04   **39.** 3.4   **40.** 19   **41.** $37\frac{1}{2}\%$ **42.** 4250   **43.** 5786
44. $63\frac{3}{4}\%$ **45.** \$12,300   **46.** \$47.96   **47.** \$1700 at 6%, \$300 at 5%   **48.** 4   **49.** 32   **50.** 78   **51.** 50   **52.** 6408   **53.** 15
54. $(-\infty, -3)$ [number line]   **55.** $(-5, \infty)$ [number line]   **56.** $[5, \infty)$ [number line]   **57.** $(-\infty, -4]$
[number line]   **58.** $(-\infty, -6) \cup (7, \infty)$ [number line]   **59.** $(-\infty, -8] \cup [-3, \infty)$
[number line]   **60.** $[-5, 10]$ [number line]   **61.** $(-1, 10)$ [number line]   **62.** $x \geq 7$
63. At least 67   **64.** $79 \leq x \leq 100$

Chapter 2 Test   **1.** $\{-7\}$ [2.1]   **2.** $\{-39\}$ [2.1]   **3.** $\{8\}$ [2.2]   **4.** $\{-\frac{21}{4}\}$ [2.2]   **5.** $\{\frac{55}{2}\}$ [2.2]   **6.** $\{-3\}$ [2.2]   **7.** $\{12\}$ [2.2]
8. $\{55\}$ [2.4]   **9.** $y = (28 - 4x)/2$ [2.2]   **10.** $\frac{18}{25}$ [2.4]   **11.** 0.06 [2.4]   **12.** 28 [2.4]   **13.** 15% [2.4]   **14.** $\left(-\infty, -\frac{5}{2}\right]$ [2.5]

15. $\left(-\infty, -9\right] \cup \left[-\frac{3}{2}, \infty\right)$ [2.5]   **16.** $\left(-10, \frac{16}{3}\right)$ [2.5]

17. Length: 13 ft, Width: 23 ft [2.3]   **18.** 17 [2.3]   **19.** 88 [2.4]   **20.** At least 34 [2.5]

Chapter 3

Quick Check 3.1   **1.** Yes   **2.**

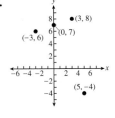

3. $A\left(-6, 2\right), B\left(-2, 6\right), C\left(0, -5\right), D\left(-4, -2\right), E\left(8, 1\right)$ **4.** $y = -4$

5.

x	4	9	−1
y	−2	3	−7

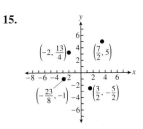

6.

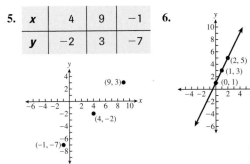

Section 3.1   **1.** Yes   **3.** No   **5.** Yes   **7.** Yes   **9.** Yes   **11.**

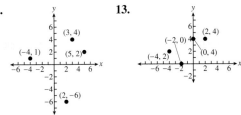

13.

15.

21. IV   **23.** I   **25.** I   **27.** IV   **29.** −1   **31.** −7   **33.** −2
35. 2   **37.** −2

	17.	**19.**
A	$(-3, 1)$	$(30, -15)$
B	$(2, -5)$	$(-40, 0)$
C	$(-6, -4)$	$(-10, -30)$
D	$(0, 3)$	$(-15, 15)$

39.

x	y
1	5
3	1
0	7

41.

x	y
−3	−1
0	4
3	9

43.

x	y
0	−9
1	−5
4	7

45.

x	y
−3	−4
0	−2
6	2

For Exercises 47–57, individual points will vary. The graph of the line is shown.

47.  **49.** 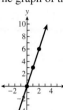 **51.** 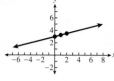 **53.**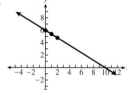

55.   **57.** 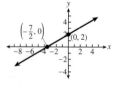 **59.** Answers will vary.

Quick Check 3.2 **1.** $(-4, 0), (0, 5)$ **2.** $(4, 0), (0, -3)$ **3.** $\left(\frac{9}{2}, 0\right), (0, -9)$

4. 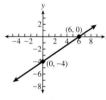 **5.** 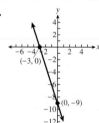 **6.**  **7.** 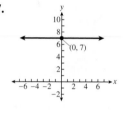 **8.**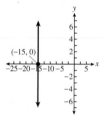

9. a) Nancy was initially \$10,000 in debt. **b)** It will take 5 months until Nancy breaks even.

Section 3.2 **1.** $(-3, 0), (0, 6)$ **3.** $(1, 0), (0, 5)$ **5.** No x-int., $(0, 2)$ **7.** $(6, 0), (0, 6)$ **9.** $(-8, 0), (0, 8)$ **11.** $(3, 0), (0, 9)$
13. $(-15, 0), (0, 9)$ **15.** $\left(-\frac{7}{2}, 0\right), \left(0, \frac{7}{3}\right)$ **17.** $\left(\frac{25}{3}, 0\right), \left(0, \frac{100}{3}\right)$ **19.** $(5, 0), (0, 6)$ **21.** $(8, 0), (0, 8)$ **23.** $(4, 0), (0, -12)$
25. $(0, 0), (0, 0)$ **27.** $\left(\frac{10}{3}, 0\right), (0, -2)$ **29.** No x-int., $(0, 2)$
31. $(4, 0), (0, 4)$ **33.** $(5, 0), (0, -2)$ **35.** $(8, 0), (0, -6)$ **37.** $(3, 0), (0, 9)$ **39.** $\left(-\frac{7}{2}, 0\right), (0, 2)$

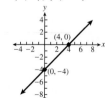  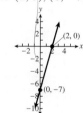

41. $(4, 0), (0, -4)$ **43.** $(0, 0), (0, 0)$ **45.** $\left(\frac{7}{2}, 0\right), (0, 7)$ **47.** $(2, 0), (0, -7)$

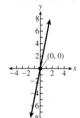 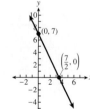

49. No x-int., $(0, 4)$ **51.** $(3, 0)$, No y-int. **53. a)** $(0, 12,000)$ The original value of the copy machine is $12,000. **b)** $(8, 0)$ The copy machine has no value after eight years. **c)** $7500 **55. a)** $(0, 5000)$ The cost of membership without playing any rounds of golf is $5000. **b)** Since the minimum fee is $5000, the costs to a member can never be $0. **c)** $8900

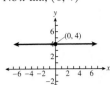

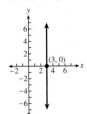

Quick Check 3.3 **1.** -2 **2.** 3 **3.** $-\frac{5}{2}$ **4.** Slope is undefined. **5.** $m = 0$ **6.** $m = \frac{5}{2}, (0, -6)$ **7.** $m = -3, (0, 5)$ **8.** $y = \frac{1}{2}x - \frac{4}{7}$
9. **10.** **11.** Slope is 218: Number of women accepted to medical school increases by 218 per year. y-intercept is (0, 7845): Approximately 7845 women were accepted to medical school in 1997.

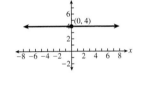

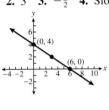

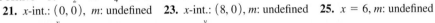

Section 3.3 **1.** Negative **3.** Positive **5.** $-\frac{1}{2}$ **7.** $-\frac{3}{4}$ **9.** 0 **11.** 2 **13.** 3 **15.** 0 **17.** Undefined
19. y-int.: $(0, 4), m: 0$ **21.** x-int.: $(0, 0), m:$ undefined **23.** x-int.: $(8, 0), m:$ undefined **25.** $x = 6, m:$ undefined

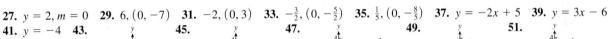

27. $y = 2, m = 0$ **29.** $6, (0, -7)$ **31.** $-2, (0, 3)$ **33.** $-\frac{3}{2}, (0, -\frac{5}{2})$ **35.** $\frac{1}{5}, (0, -\frac{8}{5})$ **37.** $y = -2x + 5$ **39.** $y = 3x - 6$
41. $y = -4$ **43.** **45.** **47.** **49.** **51.**

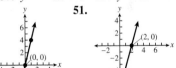

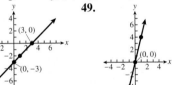

53. **55.** **57.** **59.** **61.**

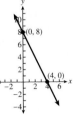

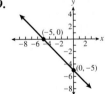

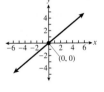

63. 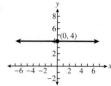   **65.** 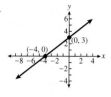   **67.** 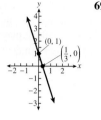   **69.**    **71.** $(1, -3)$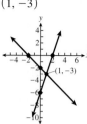

73. a) 4, The value increased by \$4/year. **b)** $(0, 50)$, The original value of the card was \$50. **c)** \$150   **75.** $\frac{1}{20}$ **77.** Negative, The heart rate (y) decreases as the animal's weight (x) increases.   **79.** Answers will vary.

Quick Check 3.4   **1.** $19.95 + 0.15x$, Domain: Set of all possible miles, Range: Set of all possible costs   **2.** -3 **3.** -23 **4.** $7a + 44$
5. 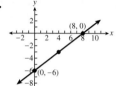   **6.** 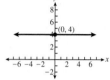   **7. a)** $(4, 0), (0, -8)$ **b)** 4 **c)** 7

Section 3.4   **1.** Yes, each player is listed with only one team.   **3.** Yes, each person has only one mother.   **5. a)** Yes **b)** No
7. a) Yes **b)** No   **9.** Yes   **11.** No, some x-coordinates are associated with more than one y-coordinate.   **13.** No, some x-coordinates are associated with more than one y-coordinate.   **15. a)** $F(x) = 1.8x + 32$ **b)** $32°F, 212°F, 86°F, 14°F, -40°F$
17. a) $f(x) = 7x + 36$ **b)** \$120   **19.** $f(x) = 4x + 3$ **21.** $f(x) = -3x - 4$ **23.** $f(x) = 6$ **25.** -3 **27.** 7
29. 32   **31.** -25 **33.** 1   **35.** $3a + 4$ **37.** $7a + 19$ **39.** $-6a + 31$ **41.** $6x + 6h + 4$
43. 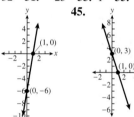   **45.**   **47.** 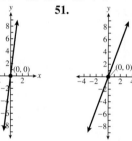   **49.** 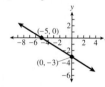   **51.**   **53.**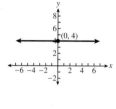

55. a) $(3, 0)$ **b)** $(0, -6)$ **c)** 4 **d)** -1 **61.** Answers will vary.

	57.	**59.**
Domain	$(-\infty, \infty)$	$(-\infty, \infty)$
Range	$(-\infty, \infty)$	$\{5\}$

Quick Check 3.5   **1.** No   **2.** Yes   **3.** $\frac{2}{9}$ **4.** Yes   **5.** Yes   **6.** $-\frac{3}{7}$ **7.** Neither   **8.** Parallel   **9.** Perpendicular   **10.** $y = -\frac{3}{2}x - 7$
Section 3.5   **1.** Yes   **3.** No   **5.** Yes   **7.** No   **9.** Yes   **11.** No   **13.** Neither   **15.** Perpendicular   **17.** Parallel   **19.** Parallel   **21.** Neither
23. Perpendicular   **25.** Parallel   **27.** Neither   **29.** Parallel   **31.** -6 **33.** 5   **35.** -3 **37.** -4 **39.** $-\frac{5}{3}$ **41.** $\frac{8}{5}$ **43.** $-\frac{A}{B}$ **45.** Yes
47. Yes   **49.** $y = 2x - 6$ **51.** $y = -4$ **53.** $y = -\frac{3}{5}x - 3$ **55.** Answers will vary.   **57.** Answers will vary.

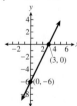 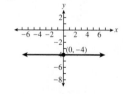

Quick Check 3.6   **1.** $y = 2x + 3$ **2.** $y = -3x + 6$ **3.** $f(x) = -4x + 5$ **4.** $y = -\frac{3}{2}x + 4$ **5.** $y = 4$ **6.** $y = 3.9x + 84.6$
7. $y = \frac{2}{5}x + 2$ **8.** $y = -\frac{4}{3}x + 6$

Section 3.6   **1.** $y = -2x + 4$ **3.** $y = \frac{1}{5}x - 2$ **5.** $y = 7x - 3$ **7.** $y = -3x + 5$ **9.** $y = \frac{2}{3}x - 3$ **11.** $y = 5x + 8$
13. $y = x - 1$ **15.** $y = -3x + 3$ **17.** $y = \frac{3}{2}x$ **19.** $y = 5$ **21.** $f(x) = -2x + 15$

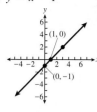 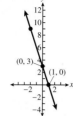 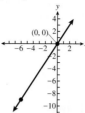 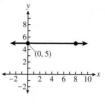

23. $f(x) = \frac{3}{5}x - 8$ **25.** $y = x - 5$ **27.** $y = 3x + 6$ **29.** $y = 2x + 8$ **31.** $x = -2$ **33.** $y = -\frac{1}{3}x + 2$

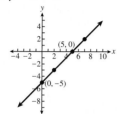 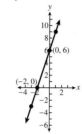 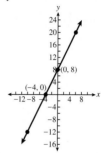 

35. $y = x - 4$ **37.** $y = x - 6$ **39. a)** $y = 10x + 36$ **b)** \$36   **41. a)** $y = 70x + 4200$ **b)** \$4200 **c)** \$70   **43.** $y = -2x - 15$
45. $y = 3x - 18$ **47.** $y = 7$ **49.** $y = x + 3$ **51.** $y = -\frac{1}{3}x + 7$ **53.** $y = \frac{5}{4}x + 7$ **55.** $x = -5$ **57.** $y = -\frac{1}{4}x + 2$ **59.** Answers will vary.

Quick Review   **1.**   **2.**   **3.**   **4.**

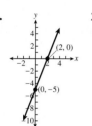 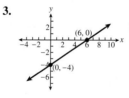 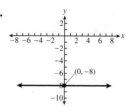

Quick Check 3.7   **1.**   **2.**   **3.**   **4.** $x + y \geq 60$

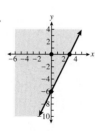 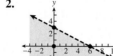 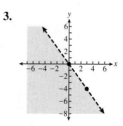 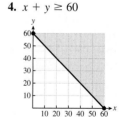

Section 3.7 **1. a)** Yes **b)** No **c)** Yes **d)** Yes **3. a)** Yes **b)** No **c)** No **d)** No **5. a)** Yes **b)** Yes **c)** No **d)** No

7. 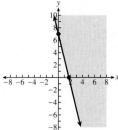 **9.** 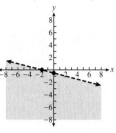 **11.** A **13.** A **15.** ≤ **17.** < **19.**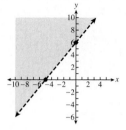

21. 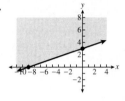 **23.** 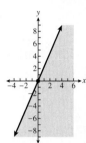 **25.** 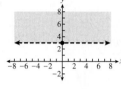 **27.**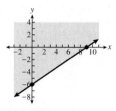

29. 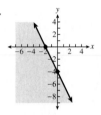 **31.** 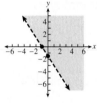 **33.** 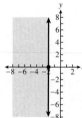 **35.** 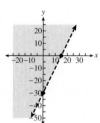 **37.** $y \le -x - 5$

39. $y > 8x$ **41.** $x + y > 50$ **43.** 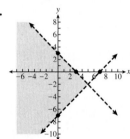 **45.** False. For some ordered pairs $3x - 8y = 24$.

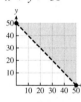

Chapter 3 Review **1.** A $(7, -2)$ B $(-5, -1)$ C$(-1, -5)$ D$(0, 8)$ **2.** III **3.** II **4.** No **5.** Yes

6. Yes **7.** $(-20, 0), (0, 5)$ **8.** $(-3, 0), (0, 15)$ **9.** $(0, 0)$ **10.** $\left(\frac{7}{2}, 0\right), (0, -2)$

11. x-int.: $(-4, 0)$ **12.** x-int.: $(-8, 0)$ **13.** x-int.: $(0, 0)$ **14.** x-int.: $\left(-\frac{3}{2}, 0\right)$ **15.** x-int.: $(4, 0)$ **16.** x-int.: $(2, 0)$
  y-int.: $(0, 6)$ y-int.: $(0, 2)$ y-int.: $(0, 0)$ y-int.: $(0, -2)$ y-int.: $(0, 6)$ y-int.: $(0, -4)$

  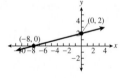 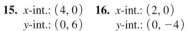

17. 3   **18.** -4 **19.** 1   **20.** 0   **21.** Slope -2, y-int.: $(0, 7)$ **22.** Slope 4, y-int.: $(0, -6)$ **23.** Slope $-\frac{2}{3}$, y-int.: $(0, 0)$
24. Slope 2, y-int.: $\left(0, -\frac{9}{2}\right)$ **25.** Slope $-\frac{5}{3}$, y-int.: $(0, 6)$ **26.** Slope 0, y-int.: $(0, -7)$

27.    **28.** 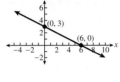   **29.** 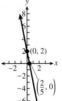   **30.** 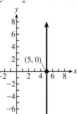   **31.** 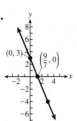   **32.**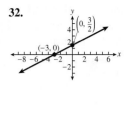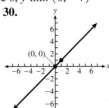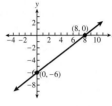

33. Neither   **34.** Perpendicular   **35.** Parallel   **36.** $y = -4x - 2$ **37.** $y = \frac{2}{5}x - 6$ **38. a)** $f(x) = 720 + 50x$ **b)** \$1270
c) 41 months   **39.** -11 **40.** 43   **41.** $15a + 4$ **42. a)** -7 **b)** 6 **c)** $(-\infty, \infty)$ **d)** $(-\infty, \infty)$ **43.** $y = x - 6$ **44.** $y = -5x + 8$
45. $y = -\frac{3}{2}x + 3$ **46.** $y = -2x - 3$ **47.** $y = \frac{3}{2}x + 4$ **48.** $y = 3$ **49.** $y = \frac{3}{2}x + 6$ **50.** $y = 3x$

51. 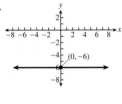   **52.** 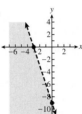   **53.** 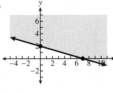   **54.** 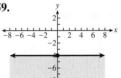   **55.**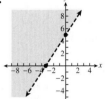

56.    **57.** 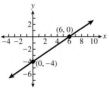 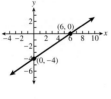   **58.**   **59.**   **60.**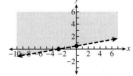

Chapter 3 Test   **1.** II [3.1]   **2.** Yes [3.1]   **3.** $(-3, 0), \left(0, \frac{5}{2}\right)$ [3.2]   **4.** $(6, 0), (0, 21)$ [3.2]
5. x-int.: $\left(\frac{3}{2}, 0\right)$ [3.2]   **6.** x-int.: $(6, 0)$ [3.2]   **7.** $-\frac{1}{5}$ [3.3]   **8.** $7, (0, -8)$ [3.3]   **9.** $-\frac{3}{5}, (0, 9)$ [3.3]   **10.** [3.3]   **11.** [3.3]
   y-int.: $(0, -6)$ y-int.: $(0, -4)$

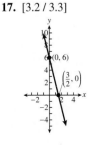

12. Parallel [3.5]   **13.** $y = -6x + 5$ [3.3]   **14.** 13 [3.4]   **15.** $y = -\frac{5}{2}x + 4$ [3.6]   **16.** $y = -2x - 8$ [3.6]
17. [3.2 / 3.3]   **18.** [3.2 / 3.3]   **19.** [3.2 / 3.3]   **20.** [3.7]

Chapter 4

Quick Check 4.1 **1.** Yes **2.** $(5, 0)$ **3.** $(2, 4)$ **4.** $(-5, 2)$ **5.** $\varnothing$ **6.** $(x, -\frac{2}{3}x + 2)$

Section 4.1 **1.** Yes **3.** No **5.** Yes **7.** $(2, 4)$ **9.** $(-3, 6)$ **11.** $(2, 9)$ **13.** $(2, 7)$ **15.** $(-6, 4)$ **17.** $(-3, -10)$ **19.** $(0, 4)$
21. Inconsistent, $\varnothing$ **23.** $(-4, -9)$

Answers to 25–29 will vary. Examples of correct solutions are given.

25. 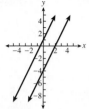 **27.** 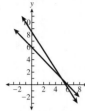 **29.** 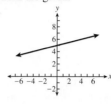 **31.** Answers will vary.

Quick Check 4.2 **1.** $(5, 2)$ **2.** $(2, 6)$ **3.** $\varnothing$ **4.** $(x, \frac{1}{6}x - \frac{2}{3})$ **5.** $(1, -4)$ **6.** 11 desktops and 4 laptops

Section 4.2 **1.** $(3, 4)$ **3.** $(17, -3)$ **5.** Inconsistent, $\varnothing$ **7.** Dependent, $(x, -3x - 2)$ **9.** $(\frac{1}{2}, \frac{3}{4})$ **11.** $(13, -2)$ **13.** $(4, 1)$
15. $(4, 2)$ **17.** Dependent, $(x, -6x + 21)$ **19.** $(4, 3)$ **21.** $(2, 1)$ **23.** $(-2, 4)$

Answers to 25–27 will vary. Examples of correct solutions are given.
25. $y = 2x - 6$ **27.** $x + y = -10$ **29.** 250 students, 950 nonstudents **31.** 47 chicken, 95 steak **33.** 13 regular coffees, 5 café mochas
35. 2090 **37.** Answers will vary.

Quick Check 4.3 **1.** $(7, 1)$ **2.** $(3, -3)$ **3.** $(-8, -1)$ **4.** $(x, \frac{2}{3}x - 4)$ **5.** $\varnothing$ **6.** $(-6, 4)$ **7.** 43 nickels, 32 dimes

Section 4.3 **1.** $(-2, -3)$ **3.** $(4, 1)$ **5.** $(6, 1)$ **7.** Dependent, $(x, \frac{1}{2}x - \frac{11}{4})$ **9.** $(7, 1)$ **11.** $(-2, -5)$ **13.** $(2, -3)$ **15.** $(6, 2)$
17. $(1, -4)$ **19.** $(\frac{3}{2}, 4)$ **21.** Inconsistent, $\varnothing$ **23.** $(\frac{5}{2}, \frac{1}{2})$ **25.** $(7, 2)$ **27.** $(3, -2)$ **29.** $(2, 1)$ **31.** $(3, 12)$ **33.** 4 roses,
11 carnations **35.** 13 5-gallon trees, 7 15-gallon trees **37.** 49 dimes, 32 nickels **39.** 29 nickels, 22 quarters **41.** Answers will vary.

Quick Check 4.4 **1.** 40 **2.** 75 feet by 40 feet **3.** $3200 at 5%, $1000 at 4% **4.** 32 pounds of almonds and 16 pounds of cashews
5. 160 ml of 60% solution and 240 ml of 50% solution **6.** $\frac{1}{2}$ hour **7.** Plane 475 mph, Wind 25 mph

Section 4.4 **1.** algebra: 38, statistics: 55 **3.** Andre: 35, father: 63 **5.** 24, 55 **7.** 21 multiple choice, 7 true–false **9.** 5 free throws,
16 field goals **11.** 22 **13.** pizza: $11, soda: $4 **15.** 34°, 56° **17.** 49°, 131° **19.** length: 27 in., width: 19 in. **21.** length: 135 ft,
width: 30 ft **23.** length: 40 ft, width: 25 ft **25.** $800 at 6%, $1200 at 3% **27.** $20,000 at 4.25%, $5000 at 3.5% **29.** $6000 at 4% profit,
$1500 at 13% loss **31.** 80 ml of 80%, 240 ml of 44% **33.** 12 oz of 20%, 6 oz of 50% **35.** 2.2 L of vodka, 1.8 L of tonic water
37. George: 2 hours, Tina: 4 hours **39.** after 1.5 hours **41.** kayak: 4 mph, current: 2 mph **43.** plane: 550 mph, wind: 50 mph
45. Answers will vary. **47.** Answers will vary.

Quick Review **1.** 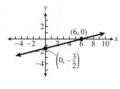 **2.** 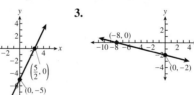 **3.** 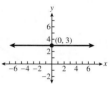 **4.**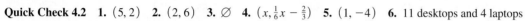

Quick Check 4.5 **1.** 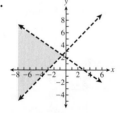 **2.** 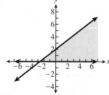 **3.**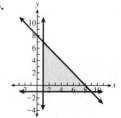

Section 4.5 **1.** 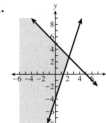 **3.** 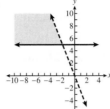 **5.** 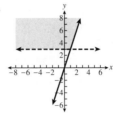 **7.**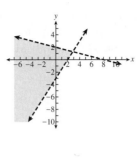

9. 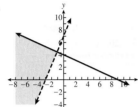 **11.** 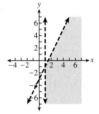 **13.** 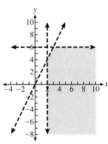 **15.**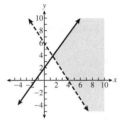

17. 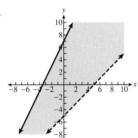 **19.** 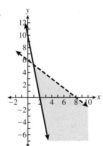 **21.**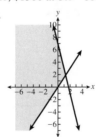

23. $x > 0$ **25.** $x + y > 4$ (Answers will vary.)
      $y > 0$ $x + y < -2$

27. 18 square units

Chapter 4 Review   **1.** $(-4, -5)$ **2.** $(6, -1)$ **3.** $(-3, 6)$ **4.** $(-7, 0)$ **5.** $(8, 3)$ **6.** $(2, 4)$ **7.** $(3, -3)$
8. $(-6, -4)$ **9.** $(-3, -2)$ **10.** $(-7, 4)$ **11.** $(0, -5)$ **12.** $\varnothing$ **13.** $(4, \frac{3}{2})$ **14.** $(-\frac{5}{3}, -8)$ **15.** $(x, 2x - 4)$
16. $(-2, -3)$ **17.** $(2, -4)$ **18.** $(-3, 7)$ **19.** $(-2, -5)$ **20.** $(5, -3)$ **21.** $(x, 3x - 2)$ **22.** $(-\frac{5}{2}, -\frac{3}{2})$
23. $(0, -6)$ **24.** $(27, -16)$ **25.** $\varnothing$ **26.** $(-3, 5)$ **27.** $23, 64$ **28.** 1077 **29.** 834 **30.** 46 **31.** 400 ft by 260 ft
32. 76 in.   **33.** $1200 at 5%, $2800 at 4.25%   **34.** $1000 at 20%, $1500 at 8%   **35.** 15 lb 4%, 30 lb 13%   **36.** 5 mph

37. 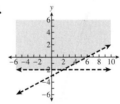 **38.** **39.** **40.**

Chapter 4 Test **1.** $(-5, 2)$ [4.1] **2.** $(4, -5)$ [4.1] **3.** $(3, -3)$ [4.2] **4.** $(-4, -1)$ [4.3] **5.** $(-3, 2)$ [4.2/4.3] **6.** $\varnothing$ [4.2/4.3]
7. $(5, 6)$ [4.2/4.3] **8.** $(9, 10)$ [4.2/4.3] **9.** $(4, 0)$ [4.2/4.3] **10.** $\left(x, \frac{2}{3}x - 3\right)$ [4.2/4.3] **11.** 180 [4.4] **12.** 5400 [4.4] **13.** Length:
8 ft, Width: 25 ft [4.4] **14.** \$20,000 at 7%, \$12,000 at 3% [4.4] **15.** [4.5]

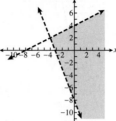

Cumulative Review 1–4 Answers **1.** $>$ **2.** -34 **3.** -165 **4.** 13 **5.** $2^3 \cdot 3^2$ **6.** $\frac{2}{49}$ **7.** $\frac{33}{20}$ **8.** 6.08 **9.** 17.86 **10.** \$448
11. $4\frac{1}{12}$ cups **12.** -43 **13.** 15 **14.** -28 **15.** $5x - 3$ **16.** $-11x - 69$ **17.** $\{-9\}$ **18.** $\left\{\frac{7}{2}\right\}$ **19.** $\{14\}$ **20.** $\{-14\}$
21. Length: 13 ft, Width: 22 ft **22.** 19 **23.** 25 **24.** 24% **25.** \$248,600 **26.** $\{6\}$ **27.** 4615 **28.** 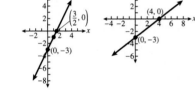 $(-\infty, 3]$
29. 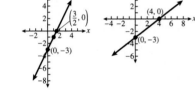 $[-5, 1]$ **30.** $(3, 0), (0, -6)$ **31.** $(8, 0), (0, -6)$ **32.** 2 **33.** $m = -\frac{2}{3}, (0, 7)$

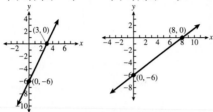

34. **35.** **36.** Perpendicular **37. a)** $f(x) = 220 + 40x$ **b)** \$540 **c)** 15 months **38.** 391 **39.** 1129

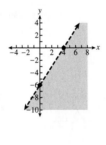

40. $y = -2x + 2$ **41.** $y = -3x + 7$ **42.** **43.** $(5, 2)$ **44.** $(3, 7)$ **45.** $(1, 2)$ **46.** $(-1, 10)$

47. 250 children **48.** Length: 40 ft, Width: 25 ft **49.** \$11,000 at 3%, \$4000 at 2.5% **50.**

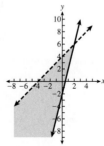

Chapter 5

Quick Check 5.1 **1. a)** x^9 **b)** 729 **2.** $(a-6)^{27}$ **3.** $a^{15}b^{10}$ **4.** x^{63} **5.** x^{34} **6.** $a^{40}b^{24}$ **7.** $64a^{26}b^{19}c^{35}$ **8. a)** x^{16} **b)** x^{16} **9.** $4a^8b^6c^6$

10. a) 1 **b)** 1 **c)** 9 **11.** $\dfrac{a^5b^{35}}{c^{15}d^{20}}$ **12.** 4096 **13.** a^{56}

Section 5.1 **1.** 32 **3.** x^{16} **5.** m^{35} **7.** b^{14} **9.** $(2x-3)^{14}$ **11.** x^8y^{13} **13.** x^5 **15.** b **17.** x^{10} **19.** a^{56} **21.** 64 **23.** x^{29} **25.** 7
27. 5 **29.** $125x^3$ **31.** $-64y^9$ **33.** $x^{21}y^{28}$ **35.** $81x^{16}y^4$ **37.** $x^{14}y^{67}z^{61}$ **39.** 7 **41.** 6, 13 **43.** x^2 **45.** d^{30} **47.** $4x^5$ **49.** $(a+5b)^2$

51. a^5b^5 **53.** 12 **55.** 30 **57.** 1 **59.** 1 **61.** 3 **63.** 1 **65.** $\dfrac{16}{625}$ **67.** $\dfrac{a^7}{b^7}$ **69.** $\dfrac{x^{14}}{y^{21}}$ **71.** $\dfrac{8a^{18}}{b^{21}}$ **73.** $\dfrac{a^{45}b^{18}}{c^{63}}$ **75.** 4 **77.** 12 **79.** 16

81. 16 **83.** a^{24} **85.** a^{42} **87.** $\dfrac{125x^{12}}{8y^{21}}$ **89.** x^{12} **91.** $\dfrac{a^{40}b^{45}}{32c^{10}}$ **93.** b^{20} **95.** x^{90} **97.** $\dfrac{81a^8b^{36}}{2401c^4d^{16}}$ **99.** $x^5 \cdot x^4$, explanations will vary.

101. Answers will vary.

Quick Check 5.2 **1.** $\dfrac{1}{64}$ **2.** $\dfrac{a^5}{b^4}$ **3. a)** $x^{12}y^7$ **b)** $\dfrac{y^4zw^9}{x^2}$ **4.** $\dfrac{1}{x^7}$ **5.** x^{42} **6.** $\dfrac{b^6}{81a^2c^4}$ **7.** $\dfrac{1}{x^6}$ **8.** $\dfrac{1}{x^{20}y^{16}}$ **9. a)** 3,200,000 **b)** 0.000721

10. a) 4.6×10^{-3} **b)** 3.57×10^6 **11.** 6.96×10^{13} **12.** 3.0×10^{12} **13.** 8000 seconds

Section 5.2 **1.** $\dfrac{1}{16}$ **3.** $\dfrac{1}{216}$ **5.** $-\dfrac{1}{343}$ **7.** $\dfrac{1}{a^{10}}$ **9.** $\dfrac{4}{x^3}$ **11.** $-\dfrac{5}{m^{19}}$ **13.** x^5 **15.** $3y^4$ **17.** $\dfrac{y^2}{x^{11}}$ **19.** $-\dfrac{8b^5}{a^6c^7}$ **21.** $\dfrac{5c^4}{a^6b^9d^5}$ **23.** x^3

25. $\dfrac{1}{a^{16}}$ **27.** $\dfrac{1}{m^4}$ **29.** $\dfrac{1}{x^{20}}$ **31.** x^{42} **33.** $\dfrac{z^{12}}{x^{10}y^8}$ **35.** $\dfrac{a^{15}b^{24}}{64z^6}$ **37.** $\dfrac{1}{x^{13}}$ **39.** a^{12} **41.** $\dfrac{1}{y^2}$ **43.** $\dfrac{1}{x^{11}}$ **45.** 1 **47.** $\dfrac{y^{20}}{x^{15}}$ **49.** $\dfrac{9c^{10}}{a^8b^{14}d^{16}}$
51. 0.000000307 **53.** 8,935,000,000 **55.** 90,210 **57.** 4.5×10^{-6} **59.** 4.7×10^7 **61.** 6.0×10^{-9} **63.** 9.28×10^{18}
65. 3.72×10^{14} **67.** 4.1496×10^{-11} **69.** 1.74×10^5 **71.** 3.36×10^4 seconds (33,600) **73.** 1.116×10^8 miles **75.** 1.1×10^{45} grams
77. \$12,156,000,000 **79.** 151,200,000,000 calculations **81.** \$6,438,000,000 **83.** Answers will vary. **85.** Answers will vary.

Quick Review **1.** $11x$ **2.** $-7a$ **3.** $-6x^2 + 11x$ **4.** $12y^3 + 5y^2 - 13y - 37$

Quick Check 5.3 **1. a)** Trinomial, 2, 1, 0 **b)** Binomial, 2, 0 **c)** Monomial, 7 **2.** 1, -6, -11, 32 **3.** $-x^3 + 4x^2 + x + 9$, Leading
Term: $-x^3$, Leading Coefficient: -1, Degree: 3 **4.** 111 **5.** 280 **6.** $x^3 + 3x + 169$ **7.** $-3x^3 - x^2 + 15x + 5$ **8.** $6x^2 + 13x - 97$
9. 289 **10.** Each Term: 5, 8, 10, Polynomial: 10 **11.** $13x^4y^2 - 4x^3y^3$

Section 5.3 **1.** 4, 2, 1, 0 **3.** 1, 7, 4 **5.** 7, 1, -15 **7.** 10, -17, 6, -1, 2 **9.** trinomial **11.** binomial **13.** monomial

	Descending Order	Leading Term	Leading Coefficient	Degree of Polynomial
15.	$3x^2 + 8x - 7$	$3x^2$	3	2
17.	$2x^4 + 6x^2 - 11x + 10$	$2x^4$	2	4

19. 22 **21.** -38 **23.** -252 **25.** -2 **27.** -591 **29.** $8x^2 - 6x + 3$ **31.** $2x^2 - 17x + 80$ **33.** $2x^4 + x^3 + 2x^2 - 27x - 20$
35. $x^3 - 9x^2 + 20x - 16$ **37.** $2x^9 + 7x^6 - 4x^5 - 5x^4 + 7x^2 + 12$ **39.** $4x^2 - 3x - 13$ **41.** $2x^4 - 3x^3 + x^2 + 11x + 11$
43. $12x^2 + x + 9, 2x^2 + 19x - 3$ **45.** $5x^3 + x^2 - 5x - 6, -3x^3 - x^2 - 11x - 56$ **47.** 184 **49.** 148 **51.** 75,680

	First Term	Second Term	Third Term	Polynomial
53.	10	6	3	10
55.	8	9	11	11

57. $7x^2y + 11xy^2 - 3x^2y^3$ **59.** $-14a^3b^2 + 13a^5b + 5a^2b^3$ **61.** $2x^3yz^2 - 15xy^4z^3 - 7x^2y^2z^5 - 4x^2yz^3$ **63.** Answers will vary.

Quick Check 5.4 **1.** $56x^2$ **2.** $60x^{14}yz^{13}$ **3.** $28x^4 - 24x^3 + 20x^2 - 32x$ **4.** $6x^9 - 3x^8 + 3x^7 + 45x^6 - 63x^5$ **5.** $x^2 + 20x + 99$
6. $10x^2 - 49x + 18$ **7.** $3x^3 + 30x^2 + 46x - 16$ **8.** $x^2 - 100$ **9.** $4x^2 - 49$ **10.** $25x^2 - 80x + 64$ **11.** $x^2 + 12x + 36$

Section 5.4 **1.** $35x^4$ **3.** $-18x^9$ **5.** $96x^8y^6$ **7.** $-70x^4y^2z^9w^4$ **9.** $84x^{10}$ **11.** $5x^3$ **13.** $-12x^9$ **15.** $10x - 16$ **17.** $-8x + 36$
19. $18x^2 + 30x$ **21.** $3x^5 - 4x^4 + 7x^3$ **23.** $6x^3y^2 - 12x^2y^3 + 14xy^4$ **25.** $3x^3$ **27.** $3x^3 - 5x^2 - 12$ **29.** $x^2 + 5x + 6$
31. $x^2 - 19x + 90$ **33.** $2x^2 - 9x - 5$ **35.** $3x^3 - 13x^2 - 37x - 18$ **37.** $x^4 - 4x^3 - 51x^2 + 310x - 400$ **39.** $x^2 - 2xy - 8y^2$

41. 3 **43.** 3, 6 **45.** $x^2 - 7x - 18$ **47.** $-18x^9$ **49.** $-4x^8 + 20x^7 - 32x^6 - 12x^5$ **51.** $x^2 - 64$ **53.** $4x^2 - 25$
55. $x^2 + 12x + 36$ **57.** $4x^2 - 12x + 9$ **59.** $x - 9$ **61.** $x - 6$ **63.** $x^2 + 6x - 7$ **65.** $5x^3 - 40x^2 - 45x$ **67.** $-28x^6y^9$
69. $20x^{14}$ **71.** $25x^2 - 9$ **73.** $x^2 + 26x + 169$ **75.** $6x^4y - 2x^6y^4 - 14x^2y^2$ **77.** $16x^2 - 72x + 81$ **79.** $x^4 - 2x^3 - 7x^2 - 8x + 16$
81. Answers will vary. **83.** Answers will vary.

Quick Check 5.5 **1.** $8x^5$ **2.** $5x^2y^9$ **3.** $x^2 - 3x - 7$ **4.** $8x^8 + 2x^5 + 5x^3$ **5.** $x + 9$ **6.** $x - 10 + \dfrac{62}{x + 7}$ **7.** $4x + 3 + \dfrac{16}{3x - 4}$
8. $x^2 - x - 2 - \dfrac{13}{x - 2}$

Section 5.5 **1.** $5x^2$ **3.** $-11x^{10}$ **5.** $17x^4y$ **7.** $5x^4$ **9.** $\dfrac{3x^2}{2}$ **11.** $\dfrac{x^8}{3}$ **13.** $10x^{11}$ **15.** $-4x^5$ **17.** $6x^2 + 7x - 4$ **19.** $x^3 + 7x^2 - 9x + 3$
21. $2x^6 + 3x^5 + 5x^4$ **23.** $-2x^4 + 3x^2$ **25.** $x^5y^4 - x^3y^3 + xy^2$ **27.** $6x$ **29.** $6x^7 - 8x^5 - 18x^3$ **31.** $x + 3$ **33.** $x + 3$
35. $x - 8$ **37.** $x + 4 + \dfrac{3}{x - 8}$ **39.** $x^2 - 12x - 25 + \dfrac{39}{x + 1}$ **41.** $2x - 7 + \dfrac{3}{x + 5}$ **43.** $4x + 3 + \dfrac{30}{3x - 5}$ **45.** $x + 10$
47. $x^3 + 3x^2 + 11x + 18 + \dfrac{86}{x - 3}$ **49.** $x^2 + 5x + 25$ **51.** $x^2 + 5x - 24$ **53.** $x + 13$ **55.** Yes **57.** No **59.** $x - 5 - \dfrac{6}{x - 3}$
61. $x^2 - 4x + 24 - \dfrac{115}{x + 4}$ **63.** $3x^4 - 2x^2 + 5x - 1$ **65.** $8x - 6 + \dfrac{11}{x + 6}$ **67.** $5x - 9 - \dfrac{33}{4x - 3}$ **69.** $-9x^7 + 6x^5 + 2x^4 + x$
71. $2x^2 - 5x - 9 + \dfrac{27}{4x + 10}$ **73.** $3x^2 - 5x - 17$ **75.** $-3x^2yz^5$ **77.** Answers will vary.

Chapter 5 Review **1.** x^6 **2.** $64x^{21}$ **3.** 7 **4.** $\dfrac{625x^{36}}{16y^8}$ **5.** $28x^{19}$ **6.** $24x^{16}y^8$ **7.** $36a^{10}b^6c^{12}$ **8.** -2 **9.** $x^{40}y^{52}z^{16}$ **10.** x^{10} **11.** $\dfrac{1}{25}$
12. $\dfrac{1}{1000}$ **13.** $\dfrac{7}{x^4}$ **14.** $-\dfrac{6}{x^6}$ **15.** $-8y^5$ **16.** $\dfrac{27}{x^{12}}$ **17.** $\dfrac{x^{10}}{4}$ **18.** x^{15} **19.** $\dfrac{b^{21}}{a^{12}}$ **20.** $\dfrac{1}{m^6}$ **21.** $\dfrac{x^{15}z^{21}}{64y^{12}}$ **22.** $\dfrac{y^{36}}{x^{66}}$ **23.** $\dfrac{1}{a^{20}}$ **24.** $\dfrac{1}{x^{11}}$
25. 1.4×10^9 **26.** 2.1×10^{-12} **27.** 5.002×10^{-6} **28.** 0.000123 **29.** 40,750,000 **30.** 6,127,500,000,000 **31.** 1.5×10^{14}
32. 2.5×10^{-18} **33.** 2.49×10^{-12} grams **34.** 15,000 seconds **35.** 80 **36.** -47 **37.** 30 **38.** -112 **39.** 27 **40.** 371
41. $2x^2 - 6x - 9$ **42.** $8x^2 - 19x - 50$ **43.** $8x - 21$ **44.** $x^2 + 9x - 22$ **45.** $3x^3 - 5x^2 + 13x - 30$ **46.** $-4x^3 + 4x^2 - 6x - 115$
47. -9 **48.** 37 **49.** $a^8 - 7a^4 + 10$ **50.** $4a^6 + 6a^3 - 30$ **51.** $(f + g)(x) = 10x^2 - 2x + 31, (f - g)(x) = -4x^2 - 14x - 49$
52. $(f + g)(x) = -7x^2 - 10x - 24, (f - g)(x) = 9x^2 + 20x - 36$ **53.** $x^2 - 10x + 25$ **54.** $x^2 + 9x - 52$ **55.** $12x^3 - 28x^2 - 64x$
56. $4x^2 - 28x + 49$ **57.** $15x^{13}$ **58.** $x^2 - 64$ **59.** $6x^2 - 19x - 130$ **60.** $-18x^6$ **61.** $36x^2 - 25$ **62.** $x^2 + 20x + 100$
63. $16x^2 + 56x + 49$ **64.** $-24x^7 + 56x^6 + 48x^5$ **65.** $x^3 + 3x^2 - 42x + 108$ **66.** $3x^3 - 11x^2 + x + 18$ **67.** $-120x^{18}$
68. $10x^6 + 22x^5 - 24x^4$ **69.** $(f \cdot g)(x) = x^2 + 3x - 70$ **70.** $(f \cdot g)(x) = 24x^8 + 54x^7 - 120x^6$ **71.** $3x^2 - 4x + 10$
72. $3x + 4 - \dfrac{9}{x - 4}$ **73.** $-2y^6$ **74.** $x^2 - 2x + 10 - \dfrac{65}{x + 5}$ **75.** $x^3 - 3x^2 + 9x - 27 + \dfrac{162}{x + 3}$ **76.** $6yz^5$ **77.** $8x + 15 + \dfrac{36}{x + 9}$
78. $6x + 12$ **79.** $3x - 17$ **80.** $2x + 1 + \dfrac{23}{8x - 3}$

Chapter 5 Test **1.** x^8 [5.1] **2.** $\dfrac{x^{30}}{729y^{24}}$ [5.1] **3.** $x^{72}y^{40}z^{64}$ [5.1] **4.** $\dfrac{1}{256}$ [5.2] **5.** $\dfrac{32}{x^{35}}$ [5.2] **6.** x^{17} [5.2] **7.** x^7 [5.2]
8. $\dfrac{y^{12}}{9x^2z^{10}}$ [5.2] **9.** 2.35×10^7 [5.2] **10.** 0.000000047 [5.2] **11.** 7.74×10^4 [5.2] **12.** -70 [5.3] **13.** $-4x^2 - 14x + 1$ [5.3]
14. $2x^2 + 14x - 64$ [5.3] **15.** 14 [5.3] **16.** $30x^5 + 40x^4 - 85x^3$ [5.4] **17.** $x^2 - 36$ [5.4] **18.** $20x^2 - x - 63$ [5.4]
19. $7x^5 + 11x^4 - 5x^3$ [5.5] **20.** $6x - 29 + \dfrac{125}{x + 3}$ [5.5]

Chapter 6

Quick Check 6.1 **1.** 4 **2.** 12 **3. a)** $4x^3$ **b)** x^3z^4 **4.** $6x^2(x^2 - 7x - 15)$ **5.** $3x^4(5x^3 - 10x + 1)$ **6. a)** $(x - 9)(5x + 14)$
b) $(x - 8)(7x - 6)$ **7.** $(x + 4)(x^2 + 7)$ **8.** $(x - 5)(2x - 9)$ **9.** $(x - 9)(4x^2 + 1)$ **10.** $3(x + 8)(x - 4)$
Section 6.1 **1.** 2 **3.** 6 **5.** 4 **7.** x^3 **9.** a^2b^2 **11.** $2x^3$ **13.** $5a^2c$ **15.** $7(x - 2)$ **17.** $x(5x + 4)$ **19.** $4x(2x^2 + 5)$
21. $7(x^2 + 8x + 10)$ **23.** $2x^3(9x^3 - 7x - 4)$ **25.** $m^2n(n^2 - m^3n + m)$ **27.** $8a^3b^5(2a^6b^2 - 9a^2 - 10b)$ **29.** $4x(-2x^2 + 3x - 4)$

31. $(2x - 7)(5x + 8)$ **33.** $(3x + 11)(x - 7)$ **35.** $(x + 10)(2x + 1)$ **37.** $(x + 8)(x + 7)$ **39.** $(x - 5)(x + 4)$
41. $(x + 2)(x - 9)$ **43.** $(x + 5)(3x + 4)$ **45.** $(x + 4)(7x - 6)$ **47.** $(x + 7)(3x + 1)$ **49.** $(x - 5)(2x - 1)$
51. $(x + 8)(x^2 + 6)$ **53.** $(x^2 + 3)(4x + 3)$ **55.** $2(x + 5)(x + 3)$ **57.** Answers will vary. **59.** Answers will vary.

Quick Check 6.2 **1. a)** $(x + 2)(x + 5)$ **b)** $(x - 5)(x - 6)$ **2. a)** $(x + 12)(x - 3)$ **b)** $(x + 6)(x - 7)$ **3. a)** $(x + 6)(x - 1)$
b) $(x + 2)(x + 3)$ **c)** $(x - 2)(x - 3)$ **d)** $(x - 6)(x + 1)$ **4.** $(x + 5)^2$ **5.** $5(x + 2)(x + 6)$ **6.** $-(x - 6)(x - 8)$
7. $(x - 8y)(x + 4y)$ **8.** $(xy + 12)(xy - 2)$

Section 6.2 **1.** $(x - 10)(x + 2)$ **3.** $(x + 9)(x - 4)$ **5.** Prime **7.** $(x + 6)(x + 8)$ **9.** $(x + 15)(x - 2)$ **11.** Prime
13. $(x - 20)(x + 3)$ **15.** $(x + 12)(x - 1)$ **17.** $(x + 7)^2$ **19.** $(x + 7)(x + 13)$ **21.** $(x - 12)^2$ **23.** $(x - 5)(x - 8)$
25. $(x - 15)(x + 2)$ **27.** $(x - 4)(x - 5)$ **29.** $(x - 6)(x - 10)$ **31.** $5(x^2 + 6x - 11)$ **33.** $21(x + 2)(x + 5)$ **35.** $-3(x - 8)^2$
37. $x(x + 4)^2$ **39.** $-3x^3(x + 10)(x - 8)$ **41.** $10x^6(x + 11)(x - 9)$ **43.** $(x + 7y)(x - 6y)$ **45.** $(xy - 3)(xy - 4)$
47. $(x - 13y)(x + 3y)$ **49.** $5x^3(x + 3y)(x + 7y)$ **51.** $(x + 9)(x^3 + 5)$ **53.** Prime **55.** $(x + 2)(x + 20)$ **57.** $(x + 20)^2$
59. $4(x + 6)(x - 4)$ **61.** $x^3(x - 15)$ **63.** $(x - 7)(x - 9)$ **65.** Prime **67.** $(x - 25)(x + 4)$ **69.** Answers will vary.

Quick Check 6.3 **1.** $(3x + 4)(x + 3)$ **2.** $(4x - 9)(x - 4)$ **3.** $(2x - 3)(8x - 7)$ **4.** $(x + 4)(2x - 7)$ **5.** $(4x - 3)(5x + 8)$
6. $9(x + 6)(x - 3)$

Section 6.3 **1.** $(5x + 2)(x - 4)$ **3.** $(2x + 1)(x + 4)$ **5.** $(3x - 4)(x + 1)$ **7.** $(2x + 3)(2x - 1)$ **9.** $(2x + 1)(3x + 2)$
11. $(3x - 5)(3x + 4)$ **13.** $(x + 8)(20x + 1)$ **15.** $(3x - 2)(x + 5)$ **17.** $(3x + 8)(4x - 7)$ **19.** $(4x - 3)^2$
21. $3(5x + 2)(x - 2)$ **23.** $8(2x + 5)(x - 1)$ **25.** $16(x - 2)^2$ **27.** $5(3x + 16)(x - 7)$ **29.** $(x + 1)(x + 13)$ **31.** $9(2x - 1)$
33. $(2x - 3)(x - 2)$ **35.** $(x + 20)(x - 1)$ **37.** $x^4(x - 6)$ **39.** $4x(x - 5)$ **41.** $-(3x + 2)(4x + 5)$ **43.** Prime
45. $(x^2 + 3)(x^4 + 8)$ **47.** $(x + 2)(x^2 - 28)$ **49.** $(x - 12)^2$ **51.** $9(x - 6)(x + 1)$ **53.** $-(5x + 2)(x - 6)$ **55.** $(x - 9)(2x + 3)$
57. $(x + 15)(x - 3)$ **59.** $-3(x + 8)(x - 4)$ **61.** $(5x - 3)(3x + 5)$ **63.** $(2x + 1)(x + 4)$ **65.** Answers will vary.

Quick Check 6.4 **1.** $(x + 5)(x - 5)$ **2.** $(9x^6 + 10y^8)(9x^6 - 10y^8)$ **3.** $7(x + 5)(x - 5)$ **4.** $(x^2 + y^2)(x + y)(x - y)$
5. $(x - 5)(x^2 + 5x + 25)$ **6.** $(x^4 - 2)(x^8 + 2x^4 + 4)$ **7.** $(x + 6)(x^2 - 6x + 36)$

Section 6.4 **1.** $(x + 2)(x - 2)$ **3.** $(x + 6)(x - 6)$ **5.** $(10x + 9)(10x - 9)$ **7.** Prime **9.** $(a + 4b)(a - 4b)$
11. $(5x + 8y)(5x - 8y)$ **13.** $(4 + x)(4 - x)$ **15.** $3(x + 5)(x - 5)$ **17.** Prime **19.** $5(x^2 + 4)$ **21.** $(x^2 + 4)(x + 2)(x - 2)$
23. $(x^2 + 9y)(x^2 - 9y)$ **25.** $(x - 1)(x^2 + x + 1)$ **27.** $(x - 2)(x^2 + 2x + 4)$ **29.** $(10 - y)(100 + 10y + y^2)$
31. $(x - 2y)(x^2 + 2xy + 4y^2)$ **33.** $(2a - 9b)(4a^2 + 18ab + 81b^2)$ **35.** $(x + y)(x^2 - xy + y^2)$ **37.** $(x + 2)(x^2 - 2x + 4)$
39. $(x^2 + 3)(x^4 - 3x^2 + 9)$ **41.** $(4m + n)(16m^2 - 4mn + n^2)$ **43.** $6(x - 3)(x^2 + 3x + 9)$ **45.** $2x(2x + 5y)(4x^2 - 10xy + 25y^2)$
47. $(x - 8)^2$ **49.** $(x + 2)(x + 14)$ **51.** $2(7x - 3)$ **53.** $(2x - 5)(3x + 5)$ **55.** $x^2(x - 9)(x + 7)$ **57.** $-8(x^3 + 9)$
59. $-6(x - 4)(x - 6)$ **61.** $(x - 13)(x + 3)$ **63.** $(5x + 6y^4)(5x - 6y^4)$ **65.** Prime **67.** $(x + 8y^3)(x - 8y^3)$
69. $3x^2(3x^3 - 8x^2 - 15)$ **71.** $4(x^2 + 49)$ **73.** $(x + 5)(x + 7)$ **75.** $(x + 2)(3x - 19)$ **77.** $(x - 10)(x^2 - 3)$
79. $(x + 5y)(x^2 - 5xy + 25y^2)$ **81.** $x^5(3x + 5)^2$ **83.** Answers will vary. **85.** Answers will vary.

Quick Check 6.5 **1.** $-6(x - 10)(x + 1)$ **2.** $(6x^4y^7 + 5)(36x^8y^{14} - 30x^4y^7 + 25)$ **3.** $(x + 5)(x - 2)(x^2 + 2x + 4)$
4. $(x^2 + y)(x^4 - x^2y + y^2)(x^2 - y)(x^4 + x^2y + y^2)$ **5.** $(x^2 + 4)(x^4 - 4x^2 + 16)$ **6.** $2(4x - 3)(x + 6)$

Section 6.5 **1.** $(3x - 2y)(9x^2 + 6xy + 4y^2)$ **3.** $(x - 7)(2x + 1)(2x - 1)$ **5.** $(x + 1)(x^2 - x + 1)(x - 1)(x^2 + x + 1)$
7. $(x - 3)(x - 15)$ **9.** $(x^2 + 4y^4)(x^4 - 4x^2y^4 + 16y^8)$ **11.** $4(x + 6)(x + 8)$ **13.** $(3x - 8)(x + 6)$ **15.** $(x + 6)(x + 7)$
17. $(x + 16)(x + 1)(x^2 - x + 1)$ **19.** $(3x - 5)^2$ **21.** $(x - 7y)^2$ **23.** $(x + 15)(x - 11)$ **25.** $4(3x + 5)$ **27.** $(x + 14)(x - 1)$
29. $(x - 5)(6x + 1)$ **31.** $4(21x^2 + x - 5)$ **33.** $(x + 4)(7x + 1)$ **35.** $(x + 2)(3x + 2)(3x - 2)$ **37.** $(x - 20)(x + 14)$
39. $4(x + 8)(x - 7)$ **41.** $(x + y)(x^2 - xy + y^2)(x - y)(x^2 + xy + y^2)$ **43.** $-(x - 3)(x - 4)$ **45.** $(3x - 2)(2x + 5)$
47. $(x^2 + 9)(x^4 - 9x^2 + 81)$ **49.** $(3x^3 + 5y^5)(9x^6 - 15x^3y^5 + 25y^{10})$ **51.** $(x - 14)(x + 11)$ **53.** $(x + 10)^2$
55. $3(x + 5)(2x + 15)$ **57.** Answers will vary.

Quick Review **1.** $\{7\}$ **2.** $\{-3\}$ **3.** $\left\{\frac{12}{5}\right\}$ **4.** $\left\{-\frac{29}{2}\right\}$

Quick Check 6.6 **1. a)** $\{2, -8\}$ **b)** $\left\{0, -\frac{9}{4}\right\}$ **2.** $\{5, -4\}$ **3.** $\{-7, -8\}$ **4.** $\left\{\frac{7}{2}, \frac{1}{3}\right\}$ **5.** $\{-9, 5\}$ **6.** $\{-7, 10\}$ **7.** $\{3, -9\}$
8. $x^2 - 2x - 48 = 0$

Section 6.6 **1.** $\{-7, 100\}$ **3.** $\{0, 12\}$ **5.** $\{4, 8\}$ **7.** $\left\{-\frac{7}{2}, 5\right\}$ **9.** $\{1, 4\}$ **11.** $\{-5, -6\}$ **13.** $\{-10, 4\}$ **15.** $\{15, -3\}$ **17.** $\{8\}$
19. $\{9, -9\}$ **21.** $\{0, -7\}$ **23.** $\left\{0, \frac{22}{5}\right\}$ **25.** $\{-11, 7\}$ **27.** $\{1, 13\}$ **29.** $\{-14, 3\}$ **31.** $\left\{-\frac{2}{9}, \frac{1}{3}\right\}$ **33.** $\{7, -5\}$ **35.** $\{4\}$ **37.** $\{-2, -5\}$
39. $\{14, -2\}$ **41.** $\{3, -3\}$ **43.** $\{6, -6\}$ **45.** $\{8, -8\}$ **47.** $\{-5, -8\}$ **49.** $\{10, -3\}$ **51.** $\{-17\}$ **53.** $\left\{-\frac{3}{2}, 9\right\}$ **55.** $\left\{\frac{5}{4}, 4\right\}$
57. $x^2 - 9x + 20 = 0$ **59.** $x^2 - 6x = 0$ **61.** $x^2 - 64 = 0$ **63.** $21x^2 - 40x - 21 = 0$ **65.** $x = -\frac{14}{3}$ **67.** Answers will vary.
69. Answers will vary.

Quick Check 6.7 **1.** 495 **2.** 8, 9 **3.** $-6, 12$ **4. a)** 48 feet **b)** 3 seconds **c)** 1 second **d)** 64 feet

Section 6.7   **1.** 37   **3.** -7 **5.** 0   **7.** 0   **9.** 147   **11.** 17   **13.** 160   **15.** $9a^2 + 39a + 30$ **17.** $a^2 - 5a - 35$ **19.** $-10, 6$
21. $-6, -7$ **23.** $9, 14$ **25.** $7, -7$ **27.** $-8, 2$ **29.** $-5, -8$ **31. a)** 5 seconds **b)** 336 feet   **33. a)** 192 feet **b)** 5 seconds
c) 1 second **d)** 256 feet   **35. a)** \$70 **b)** 5 mugs **c)** 20 mugs **d)** \$80   **37.** Answers will vary.

Quick Check 6.8   **1.** 11 and 12   **2.** 9 and 12   **3.** Length: 7 ft, Width: 15 ft   **4.** 11 ft by 14 ft   **5. a)** $h(t) = -16t^2 + 144t$
b) 9 seconds **c)** $[0, 9]$ **6. a)** 5050 **b)** 9

Section 6.8   **1.** 13, 14   **3.** 16, 18   **5.** $-9, -11$ **7.** 8, 9   **9.** 20, 22   **11.** 6, 11   **13.** 6, 14   **15.** 12, 20   **17.** 3, 14   **19.** 8, 11   **21.** 14
23. 21   **25.** Length: 8 feet, Width: 5 feet   **27.** Length: 12 feet, Width: 36 feet   **29.** Length: 15 m, Width: 7 m   **31.** Length: 5 inches,
Width: 8 inches   **33.** Length: 9 feet, Width: 5 feet   **35.** 8 m   **37.** Base: 8 feet, Height: 21 feet   **39.** 10 seconds   **41.** 5 seconds
43. 1 second, 3 seconds   **45.** 10 seconds   **47.** 128 feet/second   **49.** 3 seconds   **51.** 8   **53.** 20   **55.** Answers will vary.
57. Answers will vary.

Chapter 6 Review   **1.** $3(x - 7)$ **2.** $x(x - 20)$ **3.** $5x^3(x^4 + 3x - 4)$ **4.** $a^3b(6a^2b - 10b^2 + 15a)$ **5.** $(x + 6)(x^2 + 4)$
6. $(x + 9)(x^2 - 7)$ **7.** $(x - 3)(x^2 - 11)$ **8.** $(x - 3)^2(x + 3)$ **9.** $(x - 4)^2$ **10.** $(x + 7)(x - 4)$ **11.** $(x - 3)(x - 8)$
12. $-5(x + 8)(x - 4)$ **13.** $(x + 5y)(x + 9y)$ **14.** Prime   **15.** $(x + 2)(x + 15)$ **16.** $(x - 7)(x + 6)$ **17.** $(x - 8)(10x - 3)$
18. $(4x + 5)(x - 2)$ **19.** $6(3x - 1)(x + 1)$ **20.** $(2x - 3)^2$ **21.** $3(x + 5)(x - 5)$ **22.** $(2x + 5y)(2x - 5y)$
23. $(x + 2y)(x^2 - 2xy + 4y^2)$ **24.** Prime   **25.** $(x - 6)(x^2 + 6x + 36)$ **26.** $7(x - 3)(x^2 + 3x + 9)$ **27.** $\{5, -7\}$
28. $\{\frac{8}{3}, -\frac{1}{2}\}$ **29.** $\{5, -5\}$ **30.** $\{4, 10\}$ **31.** $\{-8\}$ **32.** $\{-12, 9\}$ **33.** $\{14, -5\}$ **34.** $\{0, 16\}$ **35.** $\{-4, -15\}$ **36.** $\{8, 10\}$
37. $\{16, -1\}$ **38.** $\{\frac{15}{2}, -4\}$ **39.** $\{6, -6\}$ **40.** $\{3, 12\}$ **41.** $\{-3, -7\}$ **42.** $\{-13, 4\}$ **43.** $x^2 - 13x + 42 = 0$ **44.** $x^2 - 16 = 0$
45. $5x^2 - 7x - 6 = 0$ **46.** $28x^2 + 27x - 10 = 0$ **47.** 9   **48.** -16 **49.** 8   **50.** 96   **51.** $6, -6$ **52.** $-9, 6$ **53.** $-2, -10$
54. $11, -4$ **55. a)** 80 feet **b)** 4 seconds **c)** 1 second **d)** 144 feet   **56. a)** \$20.85 **b)** 400 **c)** \$4.85   **57.** 16, 18   **58.** Rosa: 7, Dale: 15
59. Length: 13 m, Width: 8 m   **60. a)** 8 seconds **b)** 2 seconds, 3 seconds

Chapter 6 Test   **1.** $x(x - 25)$ [6.1]   **2.** $(x - 3)(x^2 - 5)$ [6.1]   **3.** $(x - 4)(x - 5)$ [6.2]   **4.** $(x - 18)(x + 4)$ [6.2]
5. $3(x + 1)(x + 19)$ [6.2]   **6.** $(6x + 5)(x - 1)$ [6.3]   **7.** $(x + 5y)(x - 5y)$ [6.4]   **8.** $(x - 6y)(x^2 + 6xy + 36y^2)$ [6.4]
9. $\{10, -10\}$ [6.6]   **10.** $\{4, 9\}$ [6.6]   **11.** $\{-12, 5\}$ [6.6]   **12.** $\{-6, 13\}$ [6.6]   **13.** $x^2 - 4 = 0$ [6.6]   **14.** $5x^2 + 21x - 54 = 0$ [6.6]
15. 120 [6.7]   **16.** -18 [6.7]   **17.** $-6, -8$ [6.7]   **18. a)** \$2.245 **b)** 8000 **c)** \$2.11 [6.7]   **19.** Length: 20 feet, Width: 5 feet [6.8]
20. a) 5 seconds **b)** 3 seconds [6.8]

Chapter 7

Quick Check 7.1   **1.** $\dfrac{1}{3}$ **2.** $\dfrac{7}{2}$ **3.** $-1, 9$ **4.** $\dfrac{5x}{7}$ **5.** $\dfrac{x + 9}{x + 3}$ **6.** $\dfrac{x + 4}{x - 8}$ **7.** $-\dfrac{x + 9}{1 + x}$ **8.** -2 **9.** All real numbers except 9 and -5

Section 7.1   **1.** $\dfrac{3}{4}$ **3.** $\dfrac{2}{3}$ **5.** $\dfrac{26}{7}$ **7.** $-\dfrac{75}{26}$ **9.** 3   **11.** $0, 8$ **13.** $-2, 8$ **15.** $-7, 7$ **17.** $\dfrac{1}{3x^3}$ **19.** $\dfrac{1}{x + 6}$ **21.** $\dfrac{x + 2}{x - 2}$ **23.** $\dfrac{x + 4}{x - 7}$
25. $\dfrac{x - 9}{x + 6}$ **27.** $\dfrac{x + 8}{x - 4}$ **29.** $\dfrac{x - 1}{x + 7}$ **31.** $\dfrac{3(x + 5)}{x - 4}$ **33.** Not opposites   **35.** Opposites   **37.** Opposites   **39.** $-\dfrac{3}{5 + x}$ **41.** $-\dfrac{x + 6}{9 + x}$
43. $\dfrac{3 - x}{x + 9}$ **45.** $\dfrac{1}{3}$ **47.** $-\dfrac{8}{7}$ **49.** $\dfrac{21}{16}$ **51.** All real numbers except 0 and -10.   **53.** All real numbers except 5 and -3.   **55.** All real
numbers except 3 and -6.   **57.** Quadratic   **59.** Rational   **61.** Linear   **63. a)** 4 **b)** -1 **65. a)** 6 **b)** $0, 5$ **67.** Answers will vary.
69. Answers will vary.

Quick Review   **1.** $\dfrac{41}{60}$ **2.** $\dfrac{49}{120}$ **3.** $\dfrac{1}{6}$ **4.** $\dfrac{68}{231}$

Quick Check 7.2   **1.** $\dfrac{(x - 6)(x + 2)}{(x - 4)(x + 10)}$ **2.** $-\dfrac{6 + x}{x + 7}$ **3.** $-\dfrac{1}{7 + x}$ **4.** $\dfrac{(x - 2)(x + 7)}{(x - 1)(x - 8)}$ **5.** $-(x + 5)$ **6.** $\dfrac{x - 6}{(x - 1)(x - 10)(x - 2)}$

Section 7.2   **1.** $\dfrac{x + 5}{(x - 2)(x - 3)}$ **3.** $\dfrac{(x + 6)(x + 1)}{(x + 9)(x - 2)}$ **5.** $\dfrac{x(x + 2)}{(x - 9)(x + 10)}$ **7.** $\dfrac{(x - 11)(x + 6)}{(x - 4)(x + 2)}$ **9.** $\dfrac{(x - 11)(x + 9)}{(x + 4)(x + 6)}$
11. $\dfrac{2(x + 8)(x + 9)}{3(x - 1)(x + 2)}$ **13.** $-\dfrac{(2 + x)(x + 4)}{(x - 6)(x + 9)}$ **15.** 1   **17.** $-\dfrac{x(x + 11)}{(x + 10)(x - 6)}$ **19.** $\dfrac{2}{x + 4}$ **21.** $\dfrac{(x - 11)(x + 5)}{(x + 12)(x - 4)}$

23. $\dfrac{(x-2)(x+11)}{(x+4)(x+9)}$ **25.** $\dfrac{x-7}{x-2}$ **27.** $\dfrac{(x-3)(x+11)}{(2-x)(x+2)}$ **29.** $\dfrac{x-6}{3x+1}$ **31.** $\dfrac{(x+5)(x-3)}{(x-7)(x+10)}$ **33.** $\dfrac{x-6}{x+10}$ **35.** $\dfrac{12(x-2)}{x+9}$

37. $\dfrac{(x-1)(x-4)}{(x+5)(x+6)}$ **39.** 1 **41.** $-\dfrac{(x+7)(x-10)}{(x-7)(9+x)}$ **43.** $\dfrac{x+4}{x-2}$ **45.** $-\dfrac{(x+7)(x-3)}{x(x-8)}$ **47.** $\dfrac{x-1}{(x-5)^2}$ **49.** $\dfrac{(x-6)(x-8)}{x(x-3)}$

51. $\dfrac{x(x-9)}{(x-1)(x+3)}$ **53.** $\dfrac{x-6}{x-3}$ **55.** $\dfrac{(x-9)(x+4)}{x(x-12)}$ **57.** Answers will vary. **59.** Answers will vary.

Quick Check 7.3 **1.** $\dfrac{21}{2x+7}$ **2.** $\dfrac{5}{x-4}$ **3.** $\dfrac{x+4}{x-3}$ **4.** $\dfrac{4}{3(x+6)}$ **5.** 3 **6.** $-\dfrac{x-3}{8+x}$ **7.** $\dfrac{1}{5}$ **8.** $\dfrac{x+9}{x+4}$

Section 7.3 **1.** $\dfrac{13}{x+3}$ **3.** $\dfrac{x+10}{x-5}$ **5.** 2 **7.** $\dfrac{1}{x-2}$ **9.** $\dfrac{3}{x+11}$ **11.** $\dfrac{x+9}{x-8}$ **13.** $\dfrac{2(x-12)}{x-6}$ **15.** $\dfrac{8}{x}$ **17.** $\dfrac{x+2}{x-9}$ **19.** $\dfrac{4}{x+1}$

21. $\dfrac{x-7}{x+7}$ **23.** 4 **25.** 2 **27.** $\dfrac{1}{x-10}$ **29.** $\dfrac{2}{x-4}$ **31.** $\dfrac{x-10}{x+6}$ **33.** $\dfrac{x-8}{x+8}$ **35.** $\dfrac{(x-3)(x-9)}{(x-5)(x+3)}$ **37.** $-\dfrac{11}{x+8}$

39. $\dfrac{(x-7)(x+2)}{(x-2)(x-9)}$ **41.** 2 **43.** $x+11$ **45.** 4 **47.** $\dfrac{x+3}{2}$ **49.** $\dfrac{x-4}{x-3}$ **51.** $\dfrac{x+8}{x}$ **53.** $\dfrac{2(x+4)}{x-9}$ **55.** $\dfrac{x+2}{x-4}$

57. $\dfrac{3(x-10)}{x+4}$ **59.** $\dfrac{2(3x+2)}{3x-2}$ **61.** $\dfrac{x+9}{x-9}$ **63.** $\dfrac{x-3}{x+2}$ **65.** $3x$ **67.** $4x+7$ **69.** Answers will vary. **71.** Answers will vary.

Quick Check 7.4 **1.** $36r^2s^6$ **2.** $(x-8)(x-5)(x+4)$ **3.** $(x+9)^2(x-9)$ **4.** $\dfrac{11x+18}{(x-2)(x+6)}$ **5.** $\dfrac{x+5}{(x-5)(x+3)}$

6. $\dfrac{2(x-1)}{(x+2)(x-6)}$

Section 7.4 **1.** $6a$ **3.** $(x-9)(x+5)$ **5.** $(x+3)(x-3)(x-6)$ **7.** $(x+2)^2(x+5)$ **9.** $\dfrac{11x}{12}$ **11.** $\dfrac{14}{15a}$ **13.** $\dfrac{4m^2+5n^2}{m^4n^3}$

15. $\dfrac{2(4x+13)}{(x+2)(x+4)}$ **17.** $\dfrac{6(x-7)}{(x+3)(x-3)}$ **19.** $\dfrac{9x}{(x+4)(x-2)}$ **21.** $\dfrac{7}{(x+5)(x-2)}$ **23.** $\dfrac{4}{(x+1)(x-3)}$

25. $\dfrac{2}{(x+3)(x-3)}$ **27.** $\dfrac{14}{(x-10)(x+4)}$ **29.** $-\dfrac{1}{(x-5)(x-4)}$ **31.** $\dfrac{6}{(x+2)(x-4)}$ **33.** $\dfrac{x^2+9x+27}{(x+3)(x-3)(x+4)}$

35. $\dfrac{x^2-3x-12}{(x-5)(x+4)(x-6)}$ **37.** $\dfrac{x^2-x+3}{(x+8)(x-3)(x+3)}$ **39.** $\dfrac{x^2+11x+110}{(x+10)(x-5)(x+5)}$ **41.** $\dfrac{x^2+7x+70}{x(x+7)(x+10)}$

43. $\dfrac{x^2+12x+132}{(x+11)(x-6)(x+9)}$ **45.** Answers will vary. **47.** Answers will vary.

Quick Check 7.5 **1.** $\dfrac{31}{38}$ **2.** $\dfrac{x+5}{x}$ **3.** $\dfrac{x+8}{8x}$ **4.** $\dfrac{11}{x+7}$ **5.** $\dfrac{(x-8)(x-2)}{(x+10)(x+1)}$

Section 7.5 **1.** $\dfrac{3}{68}$ **3.** $\dfrac{39}{38}$ **5.** $\dfrac{7(5x+3)}{5(7x+4)}$ **7.** $\dfrac{6}{x}$ **9.** $\dfrac{2}{3}$ **11.** $\dfrac{2x}{x-4}$ **13.** $\dfrac{4}{x+5}$ **15.** $\dfrac{12}{x}$ **17.** $\dfrac{x-4}{x+5}$ **19.** $\dfrac{2(x-2)}{x+2}$ **21.** $\dfrac{x-6}{x-5}$

23. $\dfrac{x(x-4)}{(x-8)(x+9)}$ **25.** $\dfrac{x-7}{x+10}$ **27.** $\dfrac{x-1}{x-7}$ **29.** $\dfrac{x^2+6x+14}{(x-6)(x+4)(x-2)}$ **31.** $\dfrac{x(x+2)}{(x-6)(x+5)}$ **33.** $\dfrac{x+6}{x(x+7)}$ **35.** 5

37. $\dfrac{5}{x-3}$ **39.** $\dfrac{2x-19}{(x+10)(x-8)(x-3)}$ **41.** $\dfrac{8x}{x-8}$ **43.** $\dfrac{12(x+5)}{(x+3)(x+4)(x+6)}$ **45.** $\dfrac{x+11}{x-8}$ **47.** $\dfrac{x+6}{x+1}$ **49.** $\dfrac{x(x+5)}{(x+11)(x-7)}$

51. Answers will vary. **53.** Answers will vary.

Quick Check 7.6 **1.** $\{20\}$ **2.** $\{-3,8\}$ **3.** $\varnothing$ **4.** $\{-3,6\}$ **5.** $\{4\}$ **6.** $x=\dfrac{2yz}{4y-3z}$ **7.** $x=\dfrac{16(y-1)}{3y-4}$

Section 7.6   **1.** $\left\{\dfrac{75}{4}\right\}$ **3.** $\left\{\dfrac{1}{16}\right\}$ **5.** $\left\{\dfrac{72}{13}\right\}$ **7.** $\left\{\dfrac{80}{21}\right\}$ **9.** $\{3, 4\}$ **11.** $\{2, 6\}$ **13.** $\{3\}$ **15.** $\varnothing$ **17.** $\{2\}$ **19.** $\varnothing$ **21.** $\{7\}$

23. $\{16\}$ **25.** $\{4\}$ **27.** $\{-7, 5\}$ **29.** $\{0, 9\}$ **31.** $\{2\}$ **33.** $\varnothing$ **35.** $\{-2, 5\}$ **37.** $\left\{\dfrac{7}{2}, 10\right\}$ **39.** $\{5, 18\}$ **41.** $\{1\}$ **43.** $\{-1, 15\}$

45. $W = \dfrac{A}{L}$ **47.** $x = -\dfrac{5y}{2y - 1}$ **49.** $r = \dfrac{2x + y}{2}$ **51.** $b = \dfrac{ac}{a - c}$ **53.** $x = \dfrac{y - y_1 + mx_1}{m}$ **55.** $n_1 = \dfrac{n_2}{nn_2 - 1}$ **57.** Answers will

vary.   **59.** Answers will vary.

Quick Check 7.7   **1.** 10   **2.** 3 and 12   **3.** $12\dfrac{8}{11}$ minutes   **4.** 24 hours   **5.** 75 mph   **6.** 6 mph   **7.** \$360   **8.** $81\dfrac{7}{8}$ minutes   **9.** 864

Section 7.7   **1.** 5   **3.** 11   **5.** 10   **7.** 6, 9   **9.** 10, 20   **11.** 6 minutes   **13.** $51\dfrac{3}{7}$ minutes   **15.** $\dfrac{30}{31}$ hours   **17.** 28 hours   **19.** 18 hours

21. 6 hours   **23.** 6 hours   **25.** 10 mph   **27.** 6 mph   **29.** 40 km/hr   **31.** 150 mph   **33.** 80   **35.** 90   **37.** 14   **39.** $\dfrac{16}{3}$ **41.** 294

43. 108   **45.** \$272   **47.** 3 amperes   **49.** \$15   **51.** 18 amperes   **53.** $7\dfrac{1}{2}$ foot-candles   **55.** Answers will vary.   **57.** Answers will vary.

Chapter 7 Review   **1.** $-\dfrac{1}{5}$ **2.** $\dfrac{3}{8}$ **3.** $\dfrac{5}{2}$ **4.** $\dfrac{3}{2}$ **5.** -9 **6.** $-6, 9$ **7.** $\dfrac{1}{x - 3}$ **8.** $\dfrac{x + 10}{x - 5}$ **9.** $-\dfrac{6 + x}{x - 1}$ **10.** $\dfrac{x + 3}{x + 8}$ **11.** $\dfrac{5}{9}$ **12.** $\dfrac{17}{22}$

13. All real numbers except -6 and 0.   **14.** All real numbers except 1 and 4.   **15.** $\dfrac{x}{x + 5}$ **16.** $\dfrac{(x + 9)(x + 8)}{(x - 7)(x - 2)}$ **17.** $\dfrac{x - 4}{x - 6}$

18. $-\dfrac{(4 + x)(x + 3)}{x + 6}$ **19.** $\dfrac{(x + 1)(x + 5)}{(x - 1)(x + 3)}$ **20.** $\dfrac{(x + 8)(7 - x)}{(x - 6)(x + 3)}$ **21.** $\dfrac{x + 6}{x - 4}$ **22.** $\dfrac{(x - 3)(x + 12)}{(x - 11)(x + 8)}$ **23.** $\dfrac{(x + 1)(x - 1)}{x(x - 12)}$

24. $-\dfrac{(x + 2)(x + 3)}{x - 5}$ **25.** $\dfrac{x(x - 6)}{(x - 9)(x + 5)}$ **26.** $\dfrac{(x + 2)(x + 3)}{(x - 10)(x + 1)}$ **27.** $\dfrac{12}{x + 6}$ **28.** $\dfrac{x - 3}{x + 7}$ **29.** $\dfrac{x + 9}{x + 4}$ **30.** 6   **31.** $x + 6$

32. $\dfrac{x - 6}{x - 5}$ **33.** $\dfrac{7}{(x - 3)(x + 4)}$ **34.** $\dfrac{5}{(x - 7)(x - 2)}$ **35.** $\dfrac{x + 2}{(x - 4)(x + 5)}$ **36.** $\dfrac{x - 3}{(x + 3)(x + 9)}$ **37.** $\dfrac{2(x + 4)}{(x - 2)(x + 6)}$

38. $\dfrac{x + 5}{(x - 5)(x + 1)}$ **39.** $\dfrac{x + 3}{(x - 1)(x + 7)}$ **40.** $\dfrac{x}{x + 9}$ **41.** $\dfrac{x - 11}{x + 11}$ **42.** $\dfrac{x + 7}{x - 2}$ **43.** $\dfrac{(x + 3)(x - 3)}{(x - 2)(x + 6)}$ **44.** $\left\{\dfrac{72}{13}\right\}$ **45.** $\{3, 4\}$

46. $\{16\}$ **47.** $\{-8, 5\}$ **48.** $\{2\}$ **49.** $\{-3\}$ **50.** $\{-2, 5\}$ **51.** $\{-1, -2\}$ **52.** $h = \dfrac{2A}{b}$ **53.** $x = \dfrac{7y}{4y - 3}$ **54.** $r = \dfrac{15x - 10y}{6}$

55. 8   **56.** 5, 10   **57.** 20 min   **58.** 18 hr   **59.** 60 mph   **60.** 6 mph   **61.** 130 calories   **62.** 225 ft   **63.** 675 lb   **64.** 80 foot-candles

Chapter 7 Test   **1.** $-\dfrac{5}{6}$ [7.1]   **2.** $-3, -7$ [7.1]   **3.** $\dfrac{x - 5}{x + 10}$ [7.1]   **4.** -13 [7.1]   **5.** $\dfrac{x - 1}{x - 3}$ [7.2]   **6.** $-\dfrac{(x - 5)(x - 5)}{(x + 3)(x + 5)}$ [7.2]

7. 3 [7.3]   **8.** $x + 8$ [7.3]   **9.** $\dfrac{11}{(x - 4)(x + 7)}$ [7.4]   **10.** $\dfrac{x - 6}{(x - 5)(x - 2)}$ [7.4]   **11.** $\dfrac{x + 9}{x + 7}$ [7.5]   **12.** $\dfrac{x - 5}{(x + 9)(x + 11)}$ [7.2]

13. $\dfrac{(x + 5)(x + 9)}{x(x + 6)}$ [7.2]   **14.** $\dfrac{x + 6}{x + 2}$ [7.3]   **15.** $\dfrac{(x - 10)(x + 3)}{(x - 1)(x + 2)(x + 8)}$ [7.4]   **16.** $\{36\}$ [7.6]   **17.** $\{-2, 9\}$ [7.6]

18. $y = \dfrac{5x}{7x - 2}$ [7.6]   **19.** 90 min [7.7]   **20.** 20 mph [7.7]

Cumulative Review 5–7 Answers   **1.** x^7 **2.** $108m^{16}n^{21}$ **3.** $\dfrac{625a^{24}b^{44}}{c^{28}}$ **4.** t^{21} **5.** $\dfrac{a^{48}}{b^{56}}$ **6.** $\dfrac{1}{x^{35}}$ **7.** 3.3×10^{11} **8.** 7.14×10^{-13}

9. 0.001 sec   **10.** -44 **11.** $2x^2 - 7x - 59$ **12.** $-x^2 - 12x + 22$ **13.** 67   **14.** $15x^3 - 24x^2 + 36x$ **15.** $12x^2 - 53x + 56$

16. $5x^7 + 12x^4 - 9x$ **17.** $x + 7 - \dfrac{3}{x+6}$ **18.** $(x-4)(x^2+6)$ **19.** $2(x-5)(x-8)$ **20.** $(x+9)(x-4)$ **21.** $(2x-5)(x+2)$

22. $(x-5)(x^2+5x+25)$ **23.** $(x+9)(x-9)$ **24.** $\{-7, 7\}$ **25.** $\{3, 8\}$ **26.** $\{-7\}$ **27.** $\{-10, 2\}$ **28.** $x^2 - 5x - 14 = 0$

29. $-10, 10$ **30.** $-7, 5$ **31. a)** 240 ft **b)** 6 sec **c)** 2 sec **d)** 256 ft **32.** $12, 14$ **33.** Length: 19 m, Width: 7 m **34.** $5, 9$ **35.** $\dfrac{x-6}{x+7}$

36. $\dfrac{3}{7}$ **37.** $\dfrac{(x-3)(x-2)}{(x+11)(x+2)}$ **38.** $\dfrac{(x+10)(x-2)}{(x+7)(x+1)}$ **39.** $\dfrac{3}{x+2}$ **40.** $\dfrac{2(4x-13)}{(x+2)(x-1)(x-5)}$ **41.** $\dfrac{x-2}{(x-7)(x+3)}$

42. $\dfrac{x+8}{(x+6)(x+5)}$ **43.** $\dfrac{x-9}{x+8}$ **44.** $\dfrac{x-4}{x+8}$ **45.** $\{5\}$ **46.** $\{0, 18\}$ **47.** $\{-3\}$ **48.** $x = \dfrac{y}{2y-1}$ **49.** 10 **50.** 12 hours

Chapter 8

Quick Check 8.1 **1.** $\{-3\}$ **2.** $\{3\}$ **3.** $\left\{\frac{31}{5}\right\}$ **4.** $\Re$ **5.** $\varnothing$ **6.** $\{7, -7\}$ **7.** $\left\{-1, -\frac{13}{3}\right\}$ **8.** $\left\{\frac{9}{2}, \frac{1}{2}\right\}$ **9.** $\varnothing$ **10.** $\left\{-22, -\frac{4}{3}\right\}$

Section 8.1 **1.** $\{11\}$ **3.** $\{-4\}$ **5.** $\{5\}$ **7.** $\left\{-\frac{8}{5}\right\}$ **9.** $\{-3\}$ **11.** $\{5\}$ **13.** $\{-17\}$ **15.** $\left\{-\frac{16}{3}\right\}$ **17.** $\{-3\}$ **19.** $\left\{\frac{9}{5}\right\}$ **21.** $\{-2\}$

23. $\left\{\frac{59}{20}\right\}$ **25.** $\{0\}$ **27.** $\{12\}$ **29.** $\left\{\frac{9}{4}\right\}$ **31.** $\Re$ **33.** $\varnothing$ **35.** $\Re$ **37.** $\{-2, 2\}$ **39.** $\{-929, 929\}$ **41.** $\{1, 15\}$ **43.** $\left\{-4, \frac{4}{3}\right\}$

45. $\varnothing$ **47.** $\{14, 2\}$ **49.** $\{-2, 8\}$ **51.** $\left\{0, \frac{11}{2}\right\}$ **53.** $\left\{-5, \frac{11}{3}\right\}$ **55.** $\{-9, 3\}$ **57.** $\{-18, 4\}$ **59.** $\{-1, 21\}$ **61.** $\{-7, 3\}$

(Answers will vary for exercises 63–65.)

63. $|x| = 2$ **65.** $|x+1| = 4$ **67.** Answers will vary. **69.** Answers will vary. **71.** Answers will vary.

Quick Review **1.** -32 **2.** -120 **3.** $\frac{17}{13}$ **4.** 10

Quick Check 8.2 **1.** $x \le 2$, ![number line], $(-\infty, 2]$ **2.** $x \ge -2$, ![number line], $[-2, \infty)$

3. $x < 5$, ![number line], $(-\infty, 5)$ **4.** $-5 \le x \le -2$, ![number line], $[-5, -2]$ **5.** $x < -6$ or $x > \frac{9}{2}$,

![number line], $(-\infty, -6) \cup \left(\frac{9}{2}, \infty\right)$ **6.** $-4 \le x \le -2$, ![number line], $[-4, -2]$ **7.** $-2 < x < 3$,

![number line], $(-2, 3)$ **8.** $\varnothing$ **9.** $x < -2$ or $x > 6$, ![number line], $(-\infty, -2) \cup (6, \infty)$

10. $x < -6$ or $x > -1$, ![number line], $(-\infty, -6) \cup (-1, \infty)$ **11.** $\Re$, ![number line], $(-\infty, \infty)$

Section 8.2 **1.** $(-\infty, 7)$![number line] **3.** $(-\infty, -5]$![number line] **5.** $[-4, \infty)$![number line]

7. $\left(-\frac{7}{4}, \infty\right)$![number line] **9.** $[18, \infty)$![number line] **11.** $(-33, \infty)$![number line]

13. $[8, 14]$![number line] **15.** $[-3, 7]$![number line] **17.** $\left(-7, -\frac{9}{2}\right)$![number line]

19. $\left(-\frac{32}{3}, -4\right)$![number line] **21.** $(-\infty, 2) \cup (8, \infty)$![number line] **23.** $(-\infty, -7] \cup [9, \infty)$

![number line] **25.** $(-\infty, 4) \cup (6, \infty)$![number line] **27.** $(-5, 1)$![number line]

29. $(-\infty, -5) \cup (-1, \infty)$![number line] **31.** $(-5, 5)$![number line] **33.** $(-\infty, \infty)$

![number line] **35.** $\left(-\infty, \frac{1}{4}\right] \cup [1, \infty)$![number line] **37.** $\left[-\frac{13}{4}, -\frac{5}{4}\right]$![number line]

39. $(-\infty, 4] \cup [10, \infty)$![number line] **41.** $\varnothing$![number line] **43.** $\left[-\frac{19}{5}, 5\right]$![number line]

45. $[3, 13]$![number line] **47.** $\left(-\infty, -\frac{8}{3}\right) \cup \left(\frac{4}{3}, \infty\right)$![number line] **49.** $(-\infty, -12) \cup (2, \infty)$

![number line] **51.** $|x| < 2$ **53.** $|x-5| > 4$ **55.** Answers will vary. **57.** Answers will vary.

Quick Check 8.3 **1.** $(-2, 0), (0, -5)$ **2.** $\left(\frac{8}{7}, 0\right), (0, -4)$ **3.** $(6, 0), (0, 6)$ **4.**

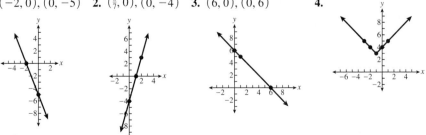

5. Domain: $(-\infty, \infty)$
Range: $[-4, \infty)$

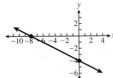

6. Domain: $(-\infty, \infty)$
Range: $(-\infty, -5]$

7. $f(x) = |x + 6| - 3$

Section 8.3 **1.** $(-8, 0), (0, -4)$ **3.** $(-4, 0), (0, 3)$ **5.** $(9, 0), (0, 6)$ **7.** $\left(-\frac{9}{5}, 0\right), (0, 3)$ **9.** $\left(\frac{4}{3}, 0\right), (0, 4)$

11.   **13.**   **15.**   **17.**   **19.**

21.   **23.**   **25.**   **27.**   **29.**

31. Domain: $(-\infty, \infty)$
Range: $[0, \infty)$

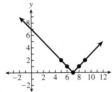

33. Domain: $(-\infty, \infty)$
Range: $[0, \infty)$

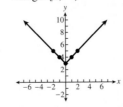

35. Domain: $(-\infty, \infty)$
Range: $[3, \infty)$

37. Domain: $(-\infty, \infty)$
Range: $[-6, \infty)$

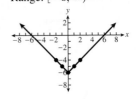

39. Domain: $(-\infty, \infty)$
Range: $[-4, \infty)$

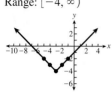

41. Domain: $(-\infty, \infty)$
Range: $[3, \infty)$

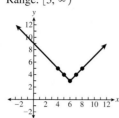

43. Domain: $(-\infty, \infty)$
Range: $[0, \infty)$

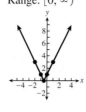

45. Domain: $(-\infty, \infty)$
Range: $[-8, \infty)$

47. Domain: $(-\infty, \infty)$
Range: $[3, \infty)$

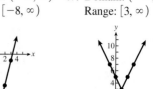

49. Domain: $(-\infty, \infty)$
Range: $(-\infty, 0]$

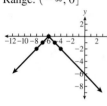

51. Domain: $(-\infty, \infty)$
Range: $(-\infty, -4]$

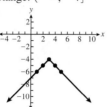

53. Domain: $(-\infty, \infty)$
Range: $(-\infty, 9]$

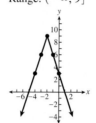

	55.	**57.**	**59.**
x-int.	$(-1, 0), (-9, 0)$	$(8, 0)$	None
y-int.	$(0, 1)$	$(0, 8)$	$(0, 4)$

61. $[2, 6]$ **63.** $(-\infty, -3) \cup (3, \infty)$ **65.** $f(x) = |x + 1| + 3$ **67.** $f(x) = |x + 6|$ **69.** Answers will vary.

Quick Check 8.4 1. $12x^3(3x^3 + 5x^2 - 1)$ **2.** $(x - 8)(x^2 + 4)$ **3. a)** $(x + 4)(x + 7)$ **b)** $(x - 3)(x - 10)$
4. $2(x - 9)(x + 2)$ **5.** $(2x - 5)(3x + 4)$ **6.** $(3x + 10)(3x - 10)$ **7.** $\{-4, 8\}$ **8.** $\{-2, \frac{5}{2}\}$ **9.** $\{5\}$ **10.** $\{1, 3\}$

Section 8.4 1. $2x^3(18x^5 - 8x + 25)$ **3.** $4a^2b(5a - 7b^3 - 4a^3b)$ **5.** $(n + 4)(n^2 + 5)$ **7.** $(x + 7)(x^2 - 3)$ **9.** $(x - 5)(x - 8)$
11. $2(x - 8)(x + 4)$ **13.** $(a + 4b)(a + 9b)$ **15.** $3(x - 9)(x + 7)$ **17.** $(5x + 1)(x - 3)$ **19.** $(3t + 4)(2t + 3)$
21. $(x + 10)(x - 10)$ **23.** $3(a + 5)(a - 5)$ **25.** Prime **27.** $(x - 2)(x^2 + 2x + 4)$ **29.** $(3x + 5)(9x^2 - 15x + 25)$
31. $\{-4, 1\}$ **33.** $\{-9, -8\}$ **35.** $\{-6, 8\}$ **37.** $\{-1, 5\}$ **39.** $\{-\frac{1}{2}, -2\}$ **41.** $\{-4, 4\}$ **43.** $\{-\frac{5}{3}, \frac{5}{3}\}$ **45.** $\{0, 64\}$ **47.** $\{-7, 3\}$
49. $\{-6, -1\}$ **51.** $\{-3, -\frac{1}{2}\}$ **53.** $\{-2, 15, 3, 10\}$ **55.** $\{-12, 2, -4, -6\}$ **57.** $x^2 - x - 12 = 0$ **59.** $2x^2 - 7x + 3 = 0$
61. $x^2 - 64 = 0$ **63.** $\{16\}$ **65.** $\{-2, 5\}$ **67.** $\{-11, 4\}$ **69.** $\{-3, 2\}$ **71.** $\{-4\}$ **73.** $\{-5, 5\}$ **75.** $\varnothing$ **77.** $\{-15, 3\}$
79. $\{-8, 6, -12, 4\}$ **81.** Answers will vary.

Quick Check 8.5 1. $(-2, 4)$ **2.** $(3, 2)$ **3.** No **4.** $(1, 2, -3)$ **5.** $(\frac{1}{2}, -4, 5)$ **6.** 16 $1's, 9 $5's, 5 $10's

Section 8.5 1. $(7, -2)$ **3.** $(3, -5)$ **5.** $(2, 5)$ **7.** $(3, 2)$ **9.** $(8, 2)$ **11.** $(5, -\frac{1}{2})$ **13.** $(-4, -4)$ **15.** $\varnothing$ **17.** $(4, -6)$ **19.** No
21. Yes **23.** $(3, -2, 0)$ **25.** $(2, 2, -3)$ **27.** $(4, \frac{1}{2}, -1)$ **29.** $(-30, 20, 5)$ **31.** $(-7, 4, 7)$ **33.** $(1, 2, 3)$ **35.** 19 **37.** 5
39. A: 60°, B: 40°, C: 80° **41.** $8000 at 3%, $20,000 at 5%, $12,000 at 6% **43.** Answers will vary.

Chapter 8 Review 1. $\{6\}$ **2.** $\{\frac{13}{4}\}$ **3.** $\{\frac{5}{2}\}$ **4.** $\{7\}$ **5.** $\{-6, 6\}$ **6.** $\varnothing$ **7.** $\{-15, 1\}$ **8.** $\{-7, -\frac{5}{3}\}$ **9.** $\{-3, -5\}$

10. $\{-6, 3\}$ **11.** $(-\infty, -2)$ **12.** $(-\infty, 7]$ **13.** $[9, \infty)$

14. $[17, \frac{39}{2}]$ **15.** $(-8, 9)$ **16.** $(-\infty, 2) \cup [6, \infty)$

17. $[-10, 10]$ **18.** $(\frac{4}{3}, 6)$

19. $(-8, 7)$ **20.** $(-\infty, -9) \cup (9, \infty)$

21. $(-\infty, -6] \cup [-1, \infty)$ **22.** $(-\infty, \infty)$

23. $(6, 0), (0, 5)$ **24.** $\left(-\frac{3}{2}, 0\right), (0, 4)$ **25.**   **26.**   **27.**   **28.**

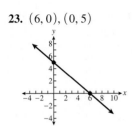 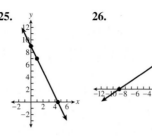 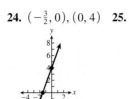 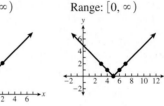 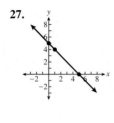 

29.   **30.**   **31.** Domain: $(-\infty, \infty)$ **32.** Domain: $(-\infty, \infty)$ **33.** Domain: $(-\infty, \infty)$
Range: $[3, \infty)$ Range: $[0, \infty)$ Range: $[-4, \infty)$

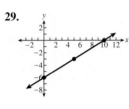  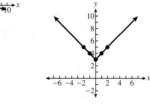 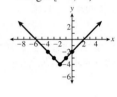

34. Domain: $(-\infty, \infty)$
Range: $(-\infty, 2]$

35. $2x(3x^8 - 2x^6 + 5x^2 - 9x + 1)$ **36.** $5a^2b(a^2b^2 - 2b^4 - 5a)$ **37.** $(x - 8)(x^2 + 11)$
38. $(x + 9)(x + 2)(x - 2)$ **39.** $(x - 7)(x + 2)$ **40.** $3(x - 3)(x - 8)$ **41.** $(x + 8y)(x + 10y)$
42. $(x + 15)(x - 2)$ **43.** $4(x + 6)(x - 1)$ **44.** $(x - 9)(x + 6)$ **45.** $(2x - 9)(3x - 1)$
46. $(2x + 3)(2x - 7)$ **47.** $(x + 6)(x - 6)$ **48.** $(3x + 1)(3x - 1)$ **49.** $(2x + 7)(2x - 7)$
50. $(5x + 8y)(5x - 8y)$ **51.** $(x + 3)(x^2 - 3x + 9)$ **52.** $(x - 9)(x^2 + 9x + 81)$ **53.** $\{-2, 3\}$
54. $\{-4, -6\}$ **55.** $\{-9, 2\}$ **56.** $\{-12, 12\}$ **57.** $\{13\}$ **58.** $\{-3, 8\}$ **59.** $\varnothing$ **60.** $\{5\}$
61. $(4, -1)$ **62.** $(-5, 3)$ **63.** $(1, 7)$ **64.** $(-7, 8)$ **65.** $(5, 4, -2)$ **66.** $(3, -8, 4)$
67. $(0, 6, -6)$ **68.** $\left(6, -\frac{5}{3}, \frac{7}{2}\right)$ **69.** 50   **70.** A: 72°, B: 48°, C: 60°

Chapter 8 Test   **1.** $\{-16\}$ [8.1]   **2.** $\{-16, -2\}$ [8.1]   **3.** $\{-3, 13\}$ [8.1]   **4.** $(-4, \infty)$ [8.2]
5. $[-6, 5]$ [8.2]   **6.** $(-\infty, 3) \cup (5, \infty)$ [8.2]
7. $[-9, 1]$ [8.2]   **8.** $(-\infty, 0) \cup (5, \infty)$ [8.2]
9. [8.3]   **10.** [8.3]   **11.** Domain: $(-\infty, \infty)$ [8.3]   **12.** $(x + 8)(x - 5)$ [8.4]   **13.** $4(x - 2)(x - 5)$ [8.4]
Range: $[3, \infty)$ **14.** $(2x + 5)(3x + 2)$ [8.4]   **15.** $(3x + 10)(3x - 10)$ [8.4]
16. $\{-9, 3\}$ [8.4]   **17.** $\{-7, 1\}$ [8.4]   **18.** $(-3, 4)$ [8.5]
19. $(8, 2, 1)$ [8.5]   **20.** A: 115°, B: 45°, C: 20° [8.5]

Chapter 9

Quick Check 9.1   **1. a)** 7 **b)** $\frac{2}{5}$ **c)** -6 **2. a)** $|a^7|$ **b)** $5|b^{11}|$ **3.** 10.440   **4. a)** 3 **b)** 4 **c)** -7 **5.** x^7 **6.** $-2x^7y^8$ **7.** $x + 5$
8. $30b^4$ **9.** 144   **10. a)** 5 **b)** $3a^3$ **11.** 35   **12.** $[-18, \infty)$ **13.** $(-\infty, \infty)$

Section 9.1   **1.** 6   **3.** 2   **5.** $\frac{1}{4}$ **7.** $\frac{6}{5}$ **9.** -4 **11.** Not a real number   **13.** a^8 **15.** x^{10} **17.** $3|x^3|$ **19.** $\frac{1}{7}x^2$ **21.** $|m^7n^5|$ **23.** $x^2y^4z^{16}$
25. 81   **27.** $9x^2$ **29.** 7.416   **31.** 18.055   **33.** 0.728   **35.** 2   **37.** 5   **39.** a^6 **41.** $7m^7$ **43.** $3x^3y^5$ **45.** $x + 3$ **47.** 9   **49.** 30   **51.** 6
53. 10   **55.** 6   **57.** 6   **59.** 3   **61.** 200   **63.** 600   **65.** 5   **67.** 392   **69.** 4   **71.** 3   **73.** 8.828   **75.** 5   **77.** $[8, \infty)$ **79.** $\left[\frac{15}{2}, \infty\right)$
81. $\left[-\frac{5}{2}, \infty\right)$ **83.** $(-\infty, \infty)$ **85.** Real: $-\sqrt{64}, -\sqrt[3]{64}, \sqrt[3]{-64}$; not real: $\sqrt{-64}$. Explanations will vary.   **87.** Answers will vary.

Quick Check 9.2 **1. a)** $\sqrt{36} = 6$ **b)** $\sqrt[5]{-32x^{15}} = -2x^3$ **2.** $\sqrt[4]{x^{32}y^4z^{24}} = x^8yz^6$ **3.** $(\sqrt[6]{4096x^{18}})^5 = 1024x^{15}$ **4.** $(\sqrt[24]{x})^{29}$

5. $(\sqrt[45]{x})^{14}$ **6.** $\dfrac{x^{15}}{y^5z^{20}}$ **7.** $\dfrac{1}{729}$ **8.** $\sqrt[12]{x^7}$

Section 9.2 **1.** $x^{1/4}$ **3.** $7^{1/2}$ **5.** $9d^{1/4}$ **7.** 8 **9.** 4 **11.** -7 **13.** x^3 **15.** x^8y^7 **17.** $2x^8y^{15}$ **19.** $-6x^6y^5z^7$ **21.** $x^{3/5}$ **23.** $y^{8/9}$ **25.** $x^{5/8}$
27. $(3x^2)^{7/3}$ **29.** $(10x^4y^5)^{3/8}$ **31.** 125 **33.** 4 **35.** x^8 **37.** $64a^6b^{18}$ **39.** $625x^{12}y^{20}z^4$ **41.** $\sqrt[5]{x^4}$ **43.** $\sqrt[20]{x^{13}}$ **45.** $\sqrt[6]{x^{11}}$ **47.** $\sqrt[12]{a^{31}}$
49. $\sqrt{xy}$ **51.** $\sqrt[10]{x}$ **53.** $\sqrt[9]{b^4}$ **55.** $\sqrt{x^7}$ **57.** $\sqrt{x}$ **59.** $\sqrt[24]{x^7}$ **61.** 1 **63.** 1 **65.** $\frac{1}{4}$ **67.** $\frac{1}{7}$ **69.** $\frac{1}{128}$ **71.** $\frac{1}{6}$ **73.** $\frac{1}{3125}$ **75.** $\frac{1}{6}$
77. $\sqrt[20]{x^9}$ **79.** $\sqrt[3]{a}$ **81.** $\sqrt[6]{m}$ **83.** $\dfrac{1}{\sqrt[6]{x}}$ **85.** Yes; explanations will vary. **87.** Answers will vary.

Quick Check 9.3 **1.** $x^3\sqrt[5]{x^2}$ **2.** $a^4b^7c^3\sqrt{bcd}$ **3.** $3\sqrt{10}$ **4.** $2\sqrt[3]{35}$ **5.** $x^6yz^{10}w^3\sqrt[5]{x^3yw^3}$ **6.** $21\sqrt[5]{2x^2}$ **7.** $15\sqrt{10} - 5\sqrt{5}$
8. $11\sqrt[5]{a^4b^3} - 11\sqrt[7]{a^4b^3}$ **9.** $4\sqrt{7}$ **10.** $6\sqrt{2}$

Section 9.3 **1.** $2\sqrt{3}$ **3.** $4\sqrt{5}$ **5.** $11\sqrt{3}$ **7.** $3\sqrt[3]{3}$ **9.** $5\sqrt[3]{6}$ **11.** $2\sqrt[4]{75}$ **13.** $x^4\sqrt{x}$ **15.** $m^4\sqrt[3]{m}$ **17.** $x^{14}\sqrt[5]{x^4}$ **19.** $x^7y^6\sqrt{x}$
21. $y^3\sqrt{xy}$ **23.** $x^{16}y^8z^8\sqrt{xy}$ **25.** $a^4c^6\sqrt[3]{b^2c}$ **27.** $a^{11}b^5\sqrt[6]{a^3b^5c^5}$ **29.** $2a^4b^4\sqrt{2a}$ **31.** $2b^3c^5\sqrt[4]{7a^3bc^3}$ **33.** $27\sqrt{5}$ **35.** $-6\sqrt[3]{4}$
37. $-5\sqrt{21}$ **39.** $16\sqrt{10} - 6\sqrt{15}$ **41.** $3\sqrt{2} + 10\sqrt{3}$ **43.** $-2\sqrt{x}$ **45.** $11\sqrt[5]{a}$ **47.** $22\sqrt{x}$ **49.** $5\sqrt{x} - 17\sqrt{y}$
51. $21\sqrt[3]{x} - 3\sqrt[4]{x}$ **53.** $13\sqrt{2}$ **55.** $21\sqrt{7}$ **57.** $-24\sqrt[3]{3}$ **59.** $162\sqrt{3} + 6\sqrt{6}$ **61.** $7\sqrt{3} - 11\sqrt{2}$ **63.** $-6\sqrt{2} + 17\sqrt{5}$
65. Answers will vary. **67.** Answers will vary.

Quick Check 9.4 **1. a)** 60 **b)** $x^4yz\sqrt[3]{z^2}$ **c)** $144\sqrt{3}$ **2.** $6 + 6\sqrt{5}$ **3.** $x^6y^6 - x^8y^6\sqrt{xy}$ **4. a)** $57 - 2\sqrt{6}$ **b)** $17 - 2\sqrt{70}$
5. a) 9 **b)** 529 **6.** $\dfrac{\sqrt{210}}{3}$ **7.** $\dfrac{\sqrt{5}}{5}$ **8.** $\dfrac{2y^2\sqrt{xy}}{5x^2z^5}$ **9.** $\dfrac{5\sqrt{3} - 3\sqrt{5}}{2}$ **10.** $\dfrac{33 - 20\sqrt{3}}{3}$

Section 9.4 **1.** 6 **3.** 6 **5.** 5 **7.** 60 **9.** x^8 **11.** $m^9n^8\sqrt{m}$ **13.** $x^{10}y^7\sqrt[4]{x}$ **15.** $4 - 2\sqrt{3}$ **17.** $42 + 14\sqrt{5}$ **19.** $6\sqrt{15} - 40\sqrt{6}$
21. $m^4n^6\sqrt{n} + m^6n^8\sqrt{m}$ **23.** $a^6b^{15}c^4\sqrt[3]{c} + a^{10}b^3c^7\sqrt[3]{b}$ **25.** $2 - \sqrt{6} + \sqrt{10} - \sqrt{15}$ **27.** 5 **29.** $19\sqrt{6} - 153$ **31.** $9\sqrt{6} - 315\sqrt{2} +$
$12\sqrt{3} - 420$ **33.** $98 - 24\sqrt{10}$ **35.** -3 **37.** 414 **39.** -178 **41.** $\dfrac{6}{5}$ **43.** $\dfrac{2\sqrt{3}}{x^2y^3}$ **45.** $3\sqrt{2}$ **47.** $\dfrac{3\sqrt{3}}{2}$ **49.** $\dfrac{2\sqrt{6}}{3}$
51. $\dfrac{\sqrt{2}}{2}$ **53.** $\dfrac{3\sqrt{33a}}{11a^4}$ **55.** $\dfrac{s^2\sqrt[3]{4rs^2t^2}}{2rt}$ **57.** $\dfrac{5\sqrt{2}}{2}$ **59.** $\dfrac{b^4\sqrt{ac}}{a^2c^4}$ **61.** $\dfrac{3\sqrt{13} + 3\sqrt{7}}{2}$ **63.** $-\dfrac{3 - \sqrt{33}}{2}$ **65.** $\dfrac{3\sqrt{39} + 9\sqrt{3}}{2}$
67. $\dfrac{3\sqrt{2} + \sqrt{6} + 4\sqrt{3} + 4}{4}$ **69.** -3 **71.** $-\dfrac{7 + 10\sqrt{6}}{19}$ **73.** $\dfrac{3\sqrt{5}}{7}$ **75.** $573 + 80\sqrt{35}$ **77.** $5a^3b^2c^4\sqrt[3]{5c^2}$ **79.** $-\dfrac{99 - 17\sqrt{35}}{2}$
81. $120 + 135\sqrt{2} - 56\sqrt{5} - 63\sqrt{10}$ **83.** $-14x^5\sqrt{2} + 7x^{11}$ **85.** $\dfrac{a^2b^4\sqrt{3ac}}{6c^5}$ **87.** $\dfrac{5\sqrt[3]{6x}}{3x}$ **89.** 2257 **91.** $a + 2\sqrt{ab} + b$
93. Answers will vary. **95.** Answers will vary.

Quick Review **1.** $2^4 \cdot 3^2$ **2.** $2^3 \cdot 5^2 \cdot 7$ **3.** 2^{10} **4.** $2 \cdot 3^3 \cdot 7^2 \cdot 11$
Quick Check 9.5 **1. a)** $\{47\}$ **b)** $\{-73\}$ **2.** $\varnothing$ **3.** 11 **4. a)** $\{9\}$ **b)** $\{-220\}$ **5.** $\{2, 10\}$ **6.** $\{8\}$ **7.** $\{-22\}$ **8.** $\{6\}$
Section 9.5 **1.** $\{23\}$ **3.** $\{16\}$ **5.** $\varnothing$ **7.** $\{7\}$ **9.** $\varnothing$ **11.** $\{-7, 4\}$ **13.** 45 **15.** 13 **17.** $-2, 7$ **19.** $\{4\}$ **21.** $\{39\}$ **23.** $\{42\}$
25. $\{8\}$ **27.** $\{9\}$ **29.** $\{7\}$ **31.** $\varnothing$ **33.** $\{3\}$ **35.** $\{2\}$ **37.** $\{26\}$ **39.** $\{9\}$ **41.** $\{-5, \frac{9}{2}\}$ **43.** $\{5\}$ **45.** $\{10\}$ **47.** $\{263\}$
49. Answers will vary.
Quick Check 9.6 **1.** 2.72 seconds **2.** 7.30 feet **3.** 70 miles per hour
Section 9.6 **1.** 2.22 sec **3.** 1.76 sec **5.** 0.3 ft **7.** 1.32 sec **9.** 3.48 sec **11.** 40 mph **13.** 73 mph **15.** 138.6 gal/min
17. 71.4 gal/min **19.** 2.3 ft **21.** 8.04 knots **23.** 50.3 ft **25.** 534.0 mph **27.** 3.8 mi **29.** Answers will vary.
Quick Check 9.7 **1. a)** $6i$ **b)** $3\sqrt{7}i$ **c)** $-3\sqrt{6}i$ **2. a)** $-5 + 27i$ **b)** $11 - 28i$ **3.** -60 **4.** $-10\sqrt{6}$ **5.** $48 - 42i$ **6.** $26 - 13i$
7. $65 - 72i$ **8.** 65 **9.** 137 **10.** $\frac{12 - 9i}{10}$ or $\frac{6}{5} - \frac{9}{10}i$ **11.** $\frac{57 + 41i}{58}$ or $\frac{57}{58} + \frac{41}{58}i$ **12.** $-\frac{8 + 15i}{12}$ or $-\frac{2}{3} - \frac{5}{4}i$ **13.** -1

Section 9.7 **1.** $2i$ **3.** $9i$ **5.** $-11i$ **7.** $3\sqrt{3}\,i$ **9.** $4\sqrt{2}\,i$ **11.** $11 + 14i$ **13.** $4 - 17i$ **15.** $4i$ **17.** $26i$ **19.** $-2 - 24i$
21. $6 + 9i$ **23.** $2 + 9i$ **25.** -98 **27.** 20 **29.** -56 **31.** $-6\sqrt{6}$ **33.** -30

(Answers may vary for Exercises 35–37.)

35. $5i \cdot 13i$ **37.** $-20i \cdot 2i$ **39.** $18 + 48i$ **41.** $-49 - 28i$ **43.** $29 - 37i$ **45.** $20 - 48i$ **47.** 82 **49.** $-48 + 14i$ **51.** 29 **53.** 208
55. $2 - i$ **57.** $\frac{-12 + 21i}{13}$ or $-\frac{12}{13} + \frac{12}{13}i$ **59.** $\frac{-6 - 35i}{13}$ or $-\frac{6}{13} - \frac{35}{13}i$ **61.** $\frac{-11 + 60i}{61}$ or $-\frac{11}{61} + \frac{60}{61}i$ **63.** $-8i$ **65.** $\frac{-7 - 5i}{3}$ or $-\frac{7}{3} - \frac{5}{3}i$
67. $\frac{-1 + 6i}{4}$ or $-\frac{1}{4} + \frac{3}{2}i$ **69.** $\frac{6 + 14i}{29}$ or $\frac{6}{29} + \frac{14}{29}i$ **71.** $\frac{14 - 27i}{25}$ or $\frac{14}{25} - \frac{27}{25}i$ **73.** $3 + 8i$ **75.** $31 + 33i$ **77.** $-i$ **79.** 1 **81.** -1 **83.** 0
85. $48 - 9i$ **87.** $\frac{4 - 9i}{5}$ or $\frac{4}{5} - \frac{9}{5}i$ **89.** $\frac{-6 + 12i}{5}$ or $-\frac{6}{5} + \frac{12}{5}i$ **91.** 29 **93.** $-18\sqrt{2}$ **95.** $\frac{-6 + 27i}{17}$ or $-\frac{6}{17} + \frac{27}{17}i$ **97.** $6\sqrt{7}\,i$
99. $72 - 54i$ **101.** $23 + 10i$ **103.** 126 **105.** $-i$ **107.** $-24 + 16i$ **109.** Answers will vary. **111.** Answers will vary.

Chapter 9 Review **1.** 5 **2.** 13 **3.** -4 **4.** x^3 **5.** $9x^6$ **6.** $3x^3y^2z^5$ **7.** 4.123 **8.** 6.403 **9.** 9 **10.** 2.828 **11.** $[-3, \infty)$
12. $\left[-\frac{25}{3}, \infty\right)$ **13.** $a^{1/5}$ **14.** $x^{3/4}$ **15.** $x^{13/2}$ **16.** $n^{8/5}$ **17.** 2 **18.** $49x^{14}$ **19.** $x\sqrt[3]{x^2}$ **20.** $\sqrt[16]{x^7}$ **21.** $\sqrt[4]{x}$ **22.** $\frac{1}{7}$ **23.** $\frac{1}{1024}$ **24.** $\frac{1}{81}$
25. $2\sqrt{7}$ **26.** $x^3\sqrt[3]{x}$ **27.** $x^2yz\sqrt[3]{xy^3}$ **28.** $3a^8b^4c^5\sqrt[3]{15ac}$ **29.** $6rs^5t^6\sqrt{5rs}$ **30.** $5r^3s^4t^2\sqrt{7}$ **31.** $12\sqrt{2}$ **32.** $6\sqrt[3]{x}$ **33.** $25\sqrt{5}$
34. $57\sqrt{3} - 40\sqrt{2}$ **35.** $2x^2\sqrt[3]{9}$ **36.** $30\sqrt{2} - 60$ **37.** $3\sqrt{2} - \sqrt{6} + 4\sqrt{3} - 4$ **38.** $296 - 43\sqrt{55}$ **39.** $303 - 108\sqrt{5}$ **40.** -648
41. $4x^2\sqrt{7}$ **42.** $\frac{5a^3b^3\sqrt{b}}{2c^5}$ **43.** $\frac{2\sqrt[3]{50}}{5}$ **44.** $\frac{4\sqrt{14}}{7}$ **45.** $\frac{\sqrt{22}}{6}$ **46.** $\frac{a^3c^2\sqrt{bc}}{b^2}$ **47.** $\frac{4(\sqrt{11} + \sqrt{5})}{3}$ **48.** $\frac{5\sqrt{5} + 3}{116}$ **49.** $\frac{13\sqrt{14} + 44}{10}$
50. $\frac{27\sqrt{5} - 53}{209}$ **51.** $\{12\}$ **52.** $\{20\}$ **53.** $\{9\}$ **54.** $\{2\}$ **55.** $\{9\}$ **56.** $\varnothing$ **57.** 4 **58.** $-2, 8$ **59.** 2.48 sec **60.** 67 mph **61.** $7i$
62. $2\sqrt{10}\,i$ **63.** $6\sqrt{7}\,i$ **64.** $-15\sqrt{3}\,i$ **65.** $11 + 5i$ **66.** $16 - 2i$ **67.** $1 + 5i$ **68.** $1 + 7i$ **69.** -16 **70.** $-10\sqrt{3}$ **71.** $28 + 21i$
72. $114 + 106i$ **73.** $72 - 54i$ **74.** 205 **75.** $\frac{21 + 9i}{29}$ or $\frac{21}{29} + \frac{9}{29}i$ **76.** $\frac{12 + 9i}{10}$ or $\frac{6}{5} + \frac{9}{10}i$ **77.** $\frac{13 + 19i}{106}$ or $\frac{13}{106} + \frac{19}{106}i$ **78.** $-16i$ **79.** i **80.** -1

Chapter 9 Test **1.** $5x^5y^7$ [9.1] **2.** 8.718 [9.1] **3.** $\sqrt{a}$ [9.2] **4.** $\sqrt[24]{b^{11}}$ [9.2] **5.** $\frac{1}{25}$ [9.2] **6.** $6\sqrt{2}$ [9.3] **7.** $a^5b^{10}c\sqrt[4]{ab^2}$ [9.3]
8. $5\sqrt{2}$ [9.3] **9.** $140\sqrt{2} - 40\sqrt{7}$ [9.4] **10.** $226 - 168\sqrt{3}$ [9.4] **11.** $\frac{x\sqrt{xy}}{y^4}$ [9.4] **12.** $\frac{-564 + 167\sqrt{14}}{50}$ [9.4] **13.** $\{20\}$ [9.5]
14. $\{8\}$ [9.5] **15.** 13.0 ft [9.6] **16.** $3\sqrt{11}\,i$ [9.7] **17.** $-18 - 11i$ [9.7] **18.** 85 [9.7] **19.** $\frac{-33 + 47i}{34}$ or $-\frac{33}{34} + \frac{47}{34}i$ [9.7]

Chapter 10

Quick Check 10.1 **1. a)** $\{-3, 5\}$ **b)** $\{-11, 11\}$ **2. a)** $\{-2\sqrt{7}, 2\sqrt{7}\}$ **b)** $\{-4\sqrt{2}\,i, 4\sqrt{2}\,i\}$ **3. a)** $\left\{-\frac{11}{3}, 3\right\}$ **b)** $\left\{\frac{3 - 3i}{2}, \frac{3 + 3i}{2}\right\}$
4. a) $\{-6, -2\}$ **b)** $\{3 - i, 3 + i\}$ **5.** $\{-9, 2\}$

Section 10.1 **1.** $\{-2, 5\}$ **3.** $\{-3, -8\}$ **5.** $\{-2, -4\}$ **7.** $\{-4, 6\}$ **9.** $\{-8, 8\}$ **11.** $\{-2, 13\}$ **13.** $\{-6, 6\}$ **15.** $\{\pm 7\sqrt{2}\}$ **17.** $\{\pm 5i\}$
19. $\{\pm 2\sqrt{7}\}$ **21.** $\{-1, 11\}$ **23.** $\{-3 \pm 3\sqrt{3}\,i\}$ **25.** $\{-5 \pm 3\sqrt{2}\,i\}$ **27.** $\{9 \pm 5\sqrt{2}\}$ **29.** $\left\{-1, \frac{9}{5}\right\}$ **31.** $x^2 + 12x + 36 = (x + 6)^2$
33. $x^2 - 5x + \frac{25}{4} = \left(x - \frac{5}{2}\right)^2$ **35.** $x^2 - 8x + 16 = (x - 4)^2$ **37.** $x^2 + \frac{1}{4}x + \frac{1}{64} = \left(x + \frac{1}{8}\right)^2$ **39.** $\{-3, 11\}$ **41.** $\{-2 \pm \sqrt{3}\,i\}$
43. $\{-9, 1\}$ **45.** $\{-8, 2\}$ **47.** $\{-1, -6\}$ **49.** $\left\{\frac{-3 \pm \sqrt{29}}{2}\right\}$ **51.** $\left\{-\frac{1}{2}, -2\right\}$ **53.** $x^2 + 3x - 10 = 0$ **55.** $x^2 - 14x + 48 = 0$
57. $4x^2 + 5x - 6 = 0$ **59.** $x^2 - 9 = 0$ **61.** $x^2 + 25 = 0$ **63.** $\{-3, -10\}$ **65.** $\{-3 \pm \sqrt{26}\}$ **67.** $\{-1, 7\}$ **69.** $\{4 \pm 3\sqrt{3}\}$
71. $\{2 \pm 2\sqrt{6}\}$ **73.** $\{-1, 6\}$ **75.** $\{4 \pm \sqrt{3}\,i\}$ **77.** $\{-3 \pm 2\sqrt{2}\,i\}$ **79.** $\left\{\frac{-3 \pm 3\sqrt{3}\,i}{2}\right\}$ **81.** $\{-5, 7\}$ **83.** $\{-1, 5\}$
85. $\{-5, 1\}$ **87.** $3, 4$ **89.** $\pm 2\sqrt{10}$ **91.** $1, 9$ **93.** Answers will vary. **95.** Answers will vary.

Quick Check 10.2 **1. a)** $a = 1, b = 11$, and $c = -13$ **b)** $a = 5, b = -9$, and $c = -11$ **c)** $a = 1, b = 0$, and $c = -30$
2. a) $\{-10, 3\}$ **b)** $\left\{\frac{-21 + 3\sqrt{21}}{14}, \frac{-21 - 3\sqrt{21}}{14}\right\}$ **3.** $\left\{\frac{7 + 3\sqrt{3}\,i}{2}, \frac{7 - 3\sqrt{3}\,i}{2}\right\}$ **4.** $\left\{\frac{-2 + \sqrt{14}\,i}{3}, \frac{-2 - \sqrt{14}\,i}{3}\right\}$ **5.** $\left\{-\frac{5}{2}, \frac{4}{3}\right\}$
6. a) Two real solutions **b)** Two nonreal complex solutions **c)** One real solution **7. a)** Not factorable **b)** Factorable

Section 10.2 **1.** $\{-4, 9\}$ **3.** $\{2 \pm \sqrt{2}\}$ **5.** $\left\{\dfrac{-1 \pm 3\sqrt{3}\,i}{2}\right\}$ **7.** $\{-6, 0\}$ **9.** $\{-5, 2\}$ **11.** $\{-2, -3\}$ **13.** $\{-2, 2\}$

15. $\{-3, 5\}$ **17.** $\left\{\dfrac{3 \pm 3\sqrt{5}}{2}\right\}$ **19.** $\left\{-\dfrac{1}{5}, 4\right\}$ **21.** $\{3\}$ **23.** $\{\pm 2\sqrt{6}\}$ **25.** $\{-4, 5\}$ **27.** $\left\{\dfrac{1 \pm \sqrt{74}\,i}{10}\right\}$ **29.** $\{3, 4\}$

31. Two real **33.** Two nonreal complex **35.** Two nonreal complex **37.** One real **39.** Prime **41.** Factorable **43.** Factorable

45. Factorable

47. $\left\{\dfrac{5 \pm \sqrt{85}}{2}\right\}$ **49.** $\left\{-1, \dfrac{1}{3}\right\}$ **51.** $\left\{-\dfrac{2}{5}, 2\right\}$ **53.** $\left\{\dfrac{-4 \pm \sqrt{2}\,i}{2}\right\}$ **55.** $\{\pm 18i\}$ **57.** $\left\{\dfrac{4}{3}, \dfrac{7}{2}\right\}$ **59.** $\left\{\dfrac{-1 \pm \sqrt{10}}{2}\right\}$

61. $\left\{\dfrac{9 \pm \sqrt{165}}{2}\right\}$ **63.** $\left\{\dfrac{3}{4}\right\}$ **65.** $\left\{\dfrac{-1 \pm \sqrt{79}\,i}{2}\right\}$ **67.** $\{2 \pm \sqrt{6}\}$ **69.** $\{3 \pm i\}$ **71.** $\left\{\dfrac{-4 \pm \sqrt{70}}{9}\right\}$ **73.** $\left\{-\dfrac{7}{3}, \dfrac{5}{2}\right\}$

75. $\left\{\dfrac{9 \pm 4\sqrt{2}}{2}\right\}$ **77.** $\{-4, 1\}$ **79.** $\{5 \pm \sqrt{7}\}$ **81.** $\{7, 9\}$ **83.** $x = -\dfrac{b}{a}$ **85.** Answers will vary.

Quick Check 10.3 **1.** Base: 20 in., Height: 6 in. **2.** 9.80 in. **3.** 7.7 ft **4.** 250 ft **5. a)** 5 sec **b)** 2 sec, 3 sec **6. a)** 0.32 sec, 1.93 sec **b)** No

Section 10.3 **1.** 9 m **3.** 5.7 ft **5.** 5.6 in. **7.** Length: 8 in., Width: 10 in. **9.** Length: 7 in., Width: 14 in. **11.** Length: 17.9 in., Width: 2.7 in. **13.** Base: 12 in., Height: 7 in. **15.** Base: 3.6 in., Height: 8.2 in. **17.** Length: 8 in., Height: 5 in. **19.** Length: 5 ft, Width: 2.5 ft **21.** Length: 16.6 in., Width: 32.6 in. **23.** Base: 32 in., Height: 16 in. **25.** Base: 6.4 ft, Height: 14.4 ft **27.** 25 in. **29.** 8.7 cm **31.** 100 mi **33.** 6 ft **35.** 30.7 ft **37.** 16.4 in. **39.** Length: 5.1 ft, Width: 4.1 ft **41. a)** Base: 12 ft, Height: 9 ft **b)** 15 ft **43. a)** 69.3 cm **b)** 2772 sq cm **45.** 8 sec **47.** 5.4 sec **49.** 1 sec, 3 sec **51.** 1.6 sec, 4.6 sec **53.** No **55.** Answers will vary. **57.** Answers will vary.

Quick Check 10.4 **1.** $\{-2, 2, -\sqrt{3}\,i, \sqrt{3}\,i\}$ **2. a)** $u = \sqrt{x}$ **b)** $u = x^{1/3}$ **c)** $u = x^2 - 4x$ **3. a)** $\{1, 25\}$ **b)** $\{1, 27\}$ **4.** $\{-5\}$ **5.** $\{0, 3\}$ **6.** 71.5 min

Section 10.4 **1.** $\{1, -1, 2, -2\}$ **3.** $\{\pm\sqrt{3}\}$ **5.** $\{\pm 3i, \pm\sqrt{2}\}$ **7.** $\{-2, 2, -3, 3\}$ **9.** $\{1, 64\}$ **11.** $\{16, 256\}$ **13.** $\{9\}$ **15.** $\{4, 25\}$ **17.** $\{1, 3\}$ **19.** $\{-2\}$ **21.** $\{-1, -10\}$ **23.** $\{-216, 1\}$ **25.** $\{-64, -125\}$ **27.** $\{-1, 2\}$ **29.** $\{3, 9\}$ **31.** $\{-8, 2\}$ **33.** $\{2, 5\}$ **35.** $\{1\}$ **37.** $\{2, 10\}$ **39.** $\{-2, 5\}$ **41.** $\{-1, -6\}$ **43.** $\{1\}$ **45.** $\left\{-\dfrac{13}{5}, 5\right\}$ **47.** Small: 120 min, Large 60 min **49.** 4.6 hr

51. 71.6 min **53.** $\{4 \pm \sqrt{29}\}$ **55.** $\{64\}$ **57.** $\left\{\dfrac{2 \pm 4\sqrt{2}}{3}\right\}$ **59.** $\left\{\dfrac{5 \pm \sqrt{31}\,i}{2}\right\}$ **61.** $\{-7, 13\}$ **63.** $\{-1, 2\}$ **65.** $\{6\}$ **67.** $\left\{-\dfrac{8}{5}, 0\right\}$ **69.** $\{-8 \pm 12i\}$ **71.** $\{16\}$ **73.** $\{-125, -343\}$ **75.** $\{-5\}$ **77.** Answers will vary. **79.** Answers will vary.

Quick Review **1.** $\{-4, 10\}$ **2.** $\{-3, 12\}$ **3.** $\{3 \pm \sqrt{22}\}$ **4.** $\left\{\dfrac{-11 \pm \sqrt{39}\,i}{2}\right\}$

Quick Check 10.5 **1.** 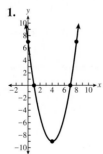 **2.** 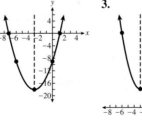 **3.**   **4.** 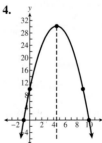 **5.**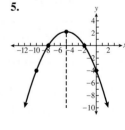

6. a) $(-2, -8)$, $x = -2$ **b)** $(8, 7)$, $x = 8$ **7. a)** $y = (x + 3)^2 - 49$, Vertex: $(-3, -49)$ **b)** $y = -(x - 6)^2 + 81$, Vertex: $(6, 81)$

8.

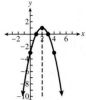

Section 10.5   **1.** $(3, -30)$, $x = 3$ **3.** $(-7, -9)$, $x = -7$ **5.** $(-5, 57)$, $x = -5$ **7.** $\left(\frac{9}{2}, -\frac{25}{4}\right)$, $x = \frac{9}{2}$ **9.** $(-2, -32)$, $x = -2$
11. $(0, -7)$, $x = 0$ **13.** $(3, -4)$, $x = 3$ **15.** $(-4, 4)$, $x = -4$

	17.	**19.**	**21.**	**23.**	**25.**	**27.**	**29.**
x-int.	$(-1, 0)$	None	$(-1.3, 0)$	$(-4, 0)$	$(3, 0)$	$(-1, 0)$	None
	$(-5, 0)$		$(4.8, 0)$	$(0, 0)$		$(-3, 0)$	
y-int.	$(0, 5)$	$(0, -7)$	$(0, -13)$	$(0, 0)$	$(0, 9)$	$(0, 3)$	$(0, -41)$

31.      **33.**   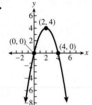   **35.**   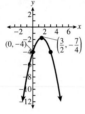   **37.**   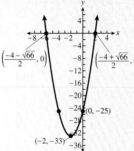   **39.**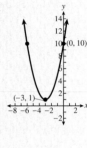

41.      **43.**      **45.**   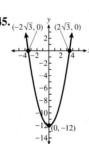   **47.**   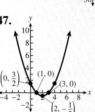   **49.**   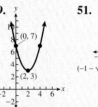   **51.**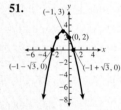

53. Answers will vary.   **55.** Answers will vary.

Quick Check 10.6   **1.** $[6, 8]$,   **2.** $\left(-\infty, \dfrac{-3 - \sqrt{69}}{2}\right] \cup \left[\dfrac{-3 + \sqrt{69}}{2}, \infty\right)$,

3. $(-\infty, \infty)$,   **4.** $(-\infty, -2) \cup (7, \infty)$ **5.** $(-7, -2] \cup [6, 8)$,

6. $(-\infty, -4) \cup (-1, 0) \cup (6, \infty)$,   **7.** $(-7, -3)$,

8. From 5 seconds after launch until 12 seconds after launch

Section 10.6 1. $(1, 5)$ **3.** $(-\infty, -3] \cup [6, \infty)$

5. $[3, 10]$ **7.** $(-\infty, -5) \cup (-3, \infty)$ **9.** $(0, 7)$

11. $(-\infty, -3 - \sqrt{6}) \cup (-3 + \sqrt{6}, \infty)$ **13.** $\left(-\infty, \dfrac{5 - \sqrt{37}}{6}\right] \cup \left[\dfrac{5 + \sqrt{37}}{6}, \infty\right)$

15. $(-\infty, \infty)$ **17.** $(5 - 2\sqrt{5}, 5 + 2\sqrt{5})$

19. $(-\infty, -5] \cup [9, \infty)$ **21.** $[-2, 1] \cup [4, \infty)$

23. $(-\infty, -6) \cup (-3, -2)$ **25.** $[-9, -5] \cup \{7\}$

27. $(-4, 3)$ **29.** $(-\infty, -2] \cup (-1, 2]$

31. $(-\infty, -2) \cup (-1, 5)$ **33.** $(-\infty, -8] \cup (-7, 1] \cup (7, \infty)$

35. $(-8, 2) \cup [5, 7]$ **37.** $(-8, 1) \cup (3, 4)$

39. 1 sec to 3 sec **41.** 0.5 sec to 1.5 sec **43.** 2 sec to 7 sec **45.** $[2, 6]$ **47.** $(-\infty, 2] \cup [3, \infty)$ **49.** $\varnothing$ **51.** $[-9, 6)$

53. $(-\infty, -7) \cup (-6, -3) \cup (11, \infty)$ **55.** $x^2 - 2x - 3 < 0$ **57.** $x^2 - 4 \geq 0$ **59.** $\dfrac{x - 5}{x + 4} \leq 0$ **61.** $\dfrac{x^2 - 2x - 3}{x^2 + 10x + 24} \leq 0$

63. $(-7, 1)$ **65.** $\varnothing$ **67.** Answers will vary.

Chapter 10 Review 1. $\{-7, -4\}$ **2.** $\{-10, 3\}$ **3.** $\{-4, 0\}$ **4.** $\{\frac{3}{2}, 5\}$ **5.** $\{-5, 1\}$ **6.** $\{7, 11\}$ **7.** $\{-7, 7\}$

8. $\{\pm 6\sqrt{2}\}$ **9.** $\{\pm 2i\}$ **10.** $\{6 \pm 6i\}$ **11.** $\{8 \pm 3i\}$ **12.** $\{-9, 2\}$ **13.** $\{-2, 4\}$ **14.** $\{3 \pm \sqrt{6}\}$ **15.** $\{-7 \pm \sqrt{2}\}$

16. $\left\{\dfrac{-5 \pm \sqrt{15}\,i}{2}\right\}$ **17.** $\left\{\dfrac{-7 \pm 3\sqrt{5}}{2}\right\}$ **18.** $\{-2 \pm \sqrt{6}\}$ **19.** $\left\{\dfrac{3 \pm 3\sqrt{3}\,i}{2}\right\}$ **20.** $\left\{\dfrac{5 \pm 2i}{2}\right\}$ **21.** $\{-7, 3\}$ **22.** $\{6 \pm i\}$

23. $\{-1, 1, -2, 2\}$ **24.** $\{64\}$ **25.** $\{-2, 1\}$ **26.** $\{-27, 343\}$ **27.** $\{-3, 2\}$ **28.** $\{-6\}$ **29.** $\{-6, -5\}$ **30.** $\{6\}$ **31.** $\{9, 10\}$

32. $\{-2, 2, \pm 3i\}$ **33.** $\{-6 \pm 2\sqrt{2}\,i\}$ **34.** $\{-2 \pm 2\sqrt{3}\,i\}$ **35.** $\{-10, 0\}$ **36.** $\{3 \pm \sqrt{29}\}$ **37.** $\{5\}$ **38.** $\{20\}$ **39.** $\{-4, 5\}$

40. $\left\{\dfrac{4 \pm \sqrt{10}}{3}\right\}$ **41.** $\{9\}$ **42.** $\{-9 \pm 8i\}$ **43.** $\{-16, -8\}$ **44.** $\left\{\dfrac{7 \pm \sqrt{129}}{2}\right\}$ **45.** $x^2 + 5x - 36 = 0$ **46.** $x^2 - 8x + 16 = 0$

47. $6x^2 - 47x + 52 = 0$ **48.** $x^2 - 8 = 0$ **49.** $x^2 + 25 = 0$ **50.** $x^2 + 64 = 0$ **51.** $(4, -36), x = 4$ **52.** $\left(-\frac{5}{2}, -\frac{89}{4}\right), x = -\frac{5}{2}$

53. $(-1, 36), x = -1$ **54.** $(3, -4), x = 3$

	55.	**56.**	**57.**	**58.**
x-int.	None	$(2 \pm \sqrt{7}, 0)$	$(5, 0)$	$(-4 \pm 2\sqrt{2}, 0)$
y-int.	$(0, 11)$	$(0, -3)$	$(0, -25)$	$(0, 8)$

59.

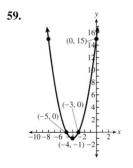

60.

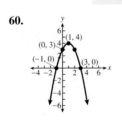

61.

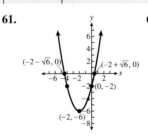

62.

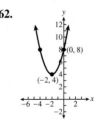

63.

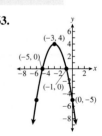

64.
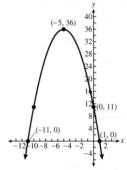

65. $[2, 7]$ 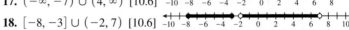   **66.** $(-\infty, -8) \cup (-1, \infty)$

67. $(-\infty, -2 - \sqrt{15}) \cup (-2 + \sqrt{15}, \infty)$

68. $(-\infty, \infty)$ **69.** $(-\infty, 6] \cup (7, 9]$

70. $(-\infty, -10) \cup [-2, 2] \cup (8, \infty)$

71. $(-\infty, -10] \cup [-5, \infty)$

72. $(-\infty, -4] \cup (1, 4] \cup (9, \infty)$

73. Length: 16 cm, Width: 18 cm   **74.** Length: 29.8 ft, Width: 16.8 ft   **75.** 1280.6 mi
76. Length: 9.9 ft, Width: 6.9 ft   **77.** $6\frac{1}{2}$ sec   **78.** 4.54 sec   **79.** 9.1 h   **80.** 61.1 h

Chapter 10 Test   **1.** $\{-6, -3\}$ [10.1]   **2.** $\{0, \frac{16}{3}\}$ [10.1]   **3.** $\{7 \pm 2\sqrt{13}\}$ [10.1]   **4.** $\{3 \pm \sqrt{3}\,i\}$ [10.2]   **5.** $\{\pm 2\sqrt{3}, \pm 2i\}$ [10.4]
6. $\{1\}$ [10.4]   **7.** $\{-1, \frac{3}{2}\}$ [10.1]   **8.** $\{-\frac{3}{2}, 9\}$ [10.2]   **9.** $\{4, 10\}$ [10.1]   **10.** $\{-4, 14\}$ [10.4]   **11.** $\{5 \pm 3i\}$ [10.2]
12. $7x^2 - 39x + 20 = 0$ [10.1]   **13.** $x^2 + 16 = 0$ [10.1]   **14.** x-int.: $(2.8, 0), (7.2, 0)$, y-int.: $(0, 20)$ [10.5]

15.
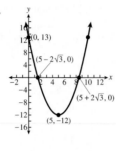

16.
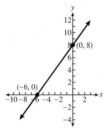

17. $(-\infty, -7) \cup (4, \infty)$ [10.6]

18. $[-8, -3] \cup (-2, 7)$ [10.6]

19. Length: 97.1 yd, Width: 37.1 yd [10.3]   **20.** 1700 mi [10.3]

Chapter 11

Quick Check 11.1   **1. a)** Yes **b)** No   **2.** No   **3.** No   **4.** 20   **5.** $-12m + 57$

Section 11.1   **1.** Yes   **3.** Yes   **5.** No   **7.** Function, Domain: {Bruce Springsteen, Bob Dylan, Madonna, Nirvana, Tone Loc, TLC},
Range: {"Born to Run," "Tangled Up in Blue," "Material Girl," "Smells Like Teen Spirit," "Wild Thing," "Waterfalls"}   **9.** Function,
Domain: $\{5, 3, 1, -1, -3, -5\}$, Range: $\{-5, -3, -1, 1, 3, 5\}$ **11.** Function, Domain: $\{-6, -3, 0, 3, 6\}$, Range: $\{5\}$ **13.** Yes   **15.** Yes
17. No   **19.** No   **21.** -8 **23.** 6   **25.** 3   **27.** $3a - 30$ **29.** $-28n + 40$ **31.** 116   **33.** -38 **35.** 66   **37.** 23   **39.** Domain:
$(-\infty, \infty)$, Range: $[-4, \infty)$ **41.** Domain: $(-\infty, 4]$, Range: $[2, \infty)$ **43.** x-int.: $(-2, 0)$ $(3, 0)$, y-int.: $(0, -6)$ **45.** 4   **47.** 8   **49.** 5
51. $-1, 9$ **53.** Answers will vary.   **55.** Answers will vary.

Quick Review   **1.**
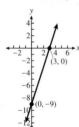

2.
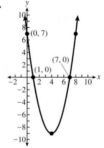

3.
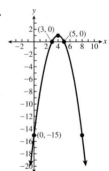

4.
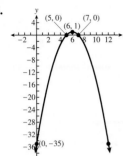

Quick Check 11.2 1. 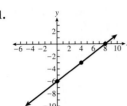 **2.** 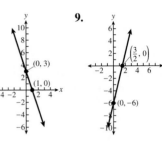 **3.** $f(x) = 2x - 6$ **4.** $f(x) = 20x + 500$

5. a) $C(x) = 50 + 1.25x$, $R(x) = 9.95x$, $P(x) = 8.70x - 50$ **b)** 121

Section 11.2 1. 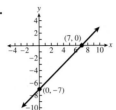 **3.** 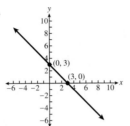 **5.** 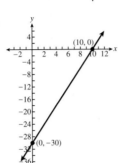 **7.** 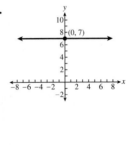 **9.**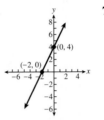

11. 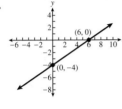 **13.** 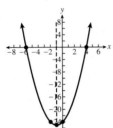 **15.** 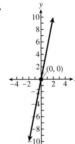 **17.**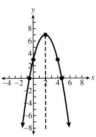

19. $f(x) = 3x + 7$ **21.** $f(x) = -\frac{4}{3}x + \frac{7}{3}$ **23.** $f(x) = 3x$ **25.** $f(x) = -6$ **27. a)** $f(x) = 0.03x + 0.39$ **b)** Slope: \$0.03/minute, y-int.: \$0.39 connection fee **c)** \$1.71 **29. a)** $f(x) = 0.15x + 20$ **b)** Slope: \$0.15/mile, y-int.: \$20 base fee **c)** \$59

31. a) $C(x) = 125$, $R(x) = 4x$, $P(x) = 4x - 125$ **b)** \$175 **33. a)** $C(x) = 0.40x + 12$, $R(x) = 3x$, $P(x) = 2.6x - 12$ **b)** \$144 **35. a)** $C(x) = 350x + 2000$, $R(x) = 600x$, $P(x) = 250x - 2000$ **b)** 48 **37.** $m = 3$ **39.** $m = 2$ **41.** Answers will vary. **43.** Answers will vary.

Quick Check 11.3 1. 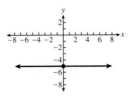 **2.** 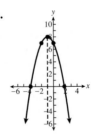 **3.** **4. a)** Maximum: -10 **b)** Minimum: 47

5. 961 **6.** 4356 square meters

7. $[-284, \infty)$ **8.** $2x + h + 3$

Section 11.3   **1.** 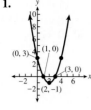   **3.** 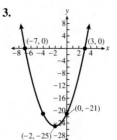   **5.** 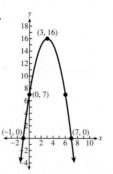   **7.** 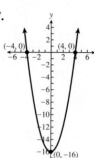   **9.**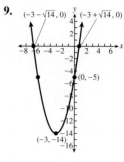

11. 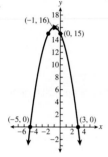   **13.** 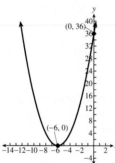   **15.** 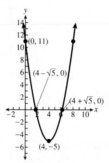   **17.** 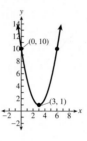   **19.**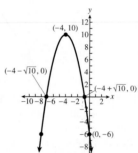

21. 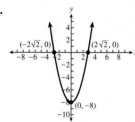   **23.** 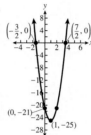   **25.** 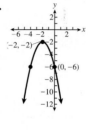   **27.**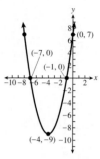

29. 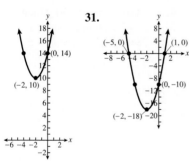   **31.**   **33.** Minimum, 4   **35.** Minimum, -78 **37.** Maximum, -26 **39.** Maximum, 43
41. Minimum, 18   **43.** Maximum, -33 **45.** Minimum, 46   **47.** Minimum, $\frac{263}{3}$
49. Maximum, 66,704   **51.** 21 ft   **53.** 1.725 m   **55.** 144   **57.** 75 ft by 75 ft, 5625 sq ft
59. 70 pans, \$3.4375   **61.** $[7, \infty)$ **63.** $(-\infty, 14]$ **65.** $(-\infty, -42]$ **67.** $2x + h$
69. $2x + h - 9$ **71.** $6x + 3h - 11$ **73.** Answers will vary.   **75.** Answers will vary.

Quick Check 11.4   **1.** Domain: $[-6, \infty)$,
Range: $[5, \infty)$

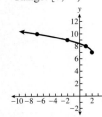

2. Domain: $(-\infty, 2]$,
Range: $[7, \infty)$

3. a) Domain: $\Re$
Range: $\Re$

b) Domain: $\Re$
Range: $\Re$

4. a) $f(x) = \sqrt{-x} + 3$

b) $f(x) = -(x + 2)^3$

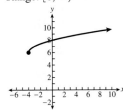

Section 11.4   **1.** Domain: $[-2, \infty)$
Range: $[0, \infty)$

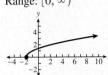

3. Domain: $[0, \infty)$
Range: $[2, \infty)$

5. Domain: $[-4, \infty)$
Range: $[6, \infty)$

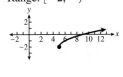

7. Domain: $[5, \infty)$
Range: $[-2, \infty)$

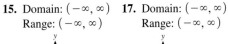

9. Domain: $[-4, \infty)$
Range: $(-\infty, 0]$

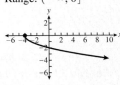

11. Domain: $[2, \infty)$
Range: $(-\infty, 3]$

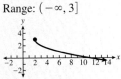

13. Domain: $(-\infty, 3]$
Range: $[1, \infty)$

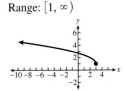

15. Domain: $(-\infty, \infty)$
Range: $(-\infty, \infty)$

17. Domain: $(-\infty, \infty)$
Range: $(-\infty, \infty)$

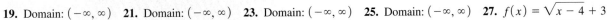

19. Domain: $(-\infty, \infty)$
Range: $(-\infty, \infty)$

21. Domain: $(-\infty, \infty)$
Range: $(-\infty, \infty)$

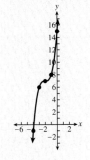

23. Domain: $(-\infty, \infty)$
Range: $(-\infty, \infty)$

25. Domain: $(-\infty, \infty)$
Range: $(-\infty, \infty)$

27. $f(x) = \sqrt{x - 4} + 3$

29. $f(x) = \sqrt{3 - x} - 1$

31. $f(x) = x^3 + 5$

33. a)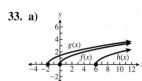

b) $f(x) = \sqrt{x + 19}$ would be 19 units to the left of $f(x) = \sqrt{x}$, $f(x) = \sqrt{x - 27}$ would be 27 units to the right of $f(x) = \sqrt{x}$.

35.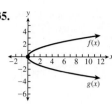

The graph of $g(x) = -\sqrt{x}$ is the reflection of $f(x) = \sqrt{x}$ about the x-axis. **37.** Answers will vary. **39.** Answers will vary.

Quick Check 11.5 **1. a)** $x^2 - 6x - 24$ **b)** $x^2 - 12x + 64$ **c)** -17 **d)** 92 **2.** $12x^2 - 5x - 28$ **3.** $\dfrac{x - 3}{x + 7}$

4. 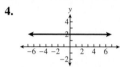 **5. a)** $155{,}889x + 12{,}119{,}000$: Total number of students enrolled in a public or private college x years after 1990. **c)** $67{,}943x + 6{,}937{,}000$: This function tells us how many more students there are enrolled at public colleges than at private colleges x years after 1990. **6.** $8x + 19$ **7. a)** $x^2 - 3x - 25$ **b)** $x^2 + 3x - 28$

8. $\dfrac{2x - 3}{x + 3}$, Domain: All real numbers except $x = -3$

Section 11.5 **1. a)** $7x - 6$ **b)** 50 **c)** -27 **3. a)** $x^2 + 4x - 11$ **b)** 85 **c)** -14 **5. a)** $8x - 6$ **b)** 42 **c)** -46 **7. a)** $-x^2 + 15x + 49$

b) 103 **c)** -51 **9. a)** $2x^2 + 4x - 70$ **b)** -22 **c)** 90 **11. a)** $4x^3 + 35x^2 - 69x - 540$ **b)** 0 **c)** -350 **13. a)** $\dfrac{3x + 1}{x + 7}$ **b)** $\dfrac{11}{7}$

c) -1 **15. a)** $\dfrac{x + 5}{x - 4}$ **b)** 4 **c)** $-\dfrac{1}{2}$ **17.** 37 **19.** -25 **21.** 377 **23.** $-\dfrac{5}{17}$ **25.** $7x - 26$ **27.** $10x^2 - 103x + 153$

29. $-x^2 + 9x + 10$ **31.** $\dfrac{1}{x + 2}$ **33.** $g(x) = 4x + 9$ **35.** $g(x) = -4x + 6$ **37.** 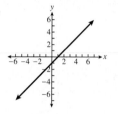 **39.**

41. a) $(f + g)(x) = 19.5x + 447$; this is the total number of doctors, in thousands, in the United States x years after 1980. **b)** 1032; there will be 1,032,000 doctors in the United States in the year 2010. **43. a)** $(f - g)(x) = 5.8x + 20.9$; this tells how many more females than males, in thousands, will earn a master's degree in the United States x years after 1990. **b)** 194.9; there will be 194,900 more master's degrees earned by females than by males in the United States in the year 2020. **45.** 15 **47.** 0 **49.** 11 **51.** 550 **53. a)** $6x + 17$ **b)** $6x + 14$

c) 35 **d)** -10 **55. a)** $x^2 + 3x - 36$ **b)** $x^2 + 11x - 12$ **c)** -18 **d)** -40 **57.** $\dfrac{10x + 65}{2x + 20}$, All real numbers except -10

59. $\sqrt{2x + 12}, [-6, \infty)$ **61.** $9x - 40$ **63.** $x^4 + 12x^3 + 66x^2 + 180x + 228$ **65.** $x = -4$ **67.** $x = -3$ **69.** $f(x) = x + 8$

71. $f(x) = 5x - 16$ **73.** $f(x) = x - 8$ **75.** $f(x) = \dfrac{x - 8}{3}$ **77.** Yes, explanations will vary.

Quick Check 11.6 **1.** Not one-to-one **2.** Not one-to-one **3.** Inverse functions **4.** Inverse functions **5.** $f^{-1}(x) = 2x - 20$

6. $f^{-1}(x) = \dfrac{3}{x - 5}$ **7.** $f^{-1}(x) = x^2 + 16x + 73$, Domain: $[8, \infty)$ **8.**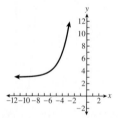

Section 11.6 **1.** Yes **3.** No **5.** No **7.** No **9.** No **11.** Yes **13.** No **15.** Yes **17.** No **19.** Inverses **21.** Not inverses

23. Not inverses **25.** Not inverses **27.** $f^{-1}(x) = x - 5$ **29.** $f^{-1}(x) = \dfrac{x}{4}$ **31.** $f^{-1}(x) = -x + 9$ **33.** $f^{-1}(x) = \dfrac{x - 17}{2}$

35. $f^{-1}(x) = \dfrac{-3x - 24}{2}$ **37.** $f^{-1}(x) = \dfrac{x}{m}$ **39.** $f^{-1}(x) = \dfrac{1}{x - 3}$ **41.** $f^{-1}(x) = \dfrac{1 - 3x}{x}$ **43.** $f^{-1}(x) = \dfrac{5}{x + 6}$ **45.** $f^{-1}(x) = \dfrac{8}{3x - 4}$

47. $f^{-1}(x) = x^2, x \geq 0$ **49.** $f^{-1}(x) = x^2 - 10x + 25, x \geq 5$ **51.** $f^{-1}(x) = x^2 - 8x + 8, x \geq 4$ **53.** $f^{-1}(x) = \sqrt{x + 9}, x \geq -9$

55. $f^{-1}(x) = \sqrt{x + 1} + 3, x \geq -1$ **57.** $\{(-17, -5), (-9, -1), (-5, 1), (1, 4), (7, 7)\}$ **59.** Not one-to-one

61.
63.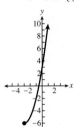
65.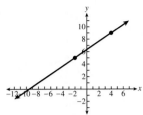
67.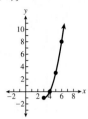
69. False, Answers will vary.
71. Answers will vary.

Chapter 11 Review   **1.** Yes   **2.** No   **3.** Yes   **4.** No   **5.** -10 **6.** 5   **7.** $21b - 67$ **8.** 32   **9.** 0   **10.** $4n^2 + 40n + 50$ **11.** -10
12. 33   **13.** Domain: $(-\infty, \infty)$, Range: $[-7, \infty)$ **14.** Domain: $[1, \infty)$, Range: $[5, \infty)$ **15.** -3 **16.** 8   **17.** $-4, 2$ **18.** 2
19.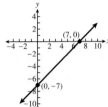
20.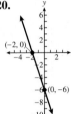
21.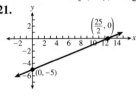
22.
23. $f(x) = 2x + 6$ **24.** $f(x) = -\frac{1}{4}x$
25. a) $f(x) = 0.05x + 30$ **b)** \$33.75
26. a) $C(x) = 0.10x + 100$
   b) $R(x) = 1.50x$
   c) $P(x) = 1.40x - 100$ **d)** \$460

27.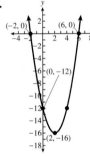
28.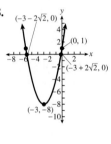
29.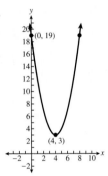

30.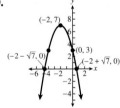
31.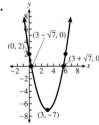
32.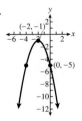

33. Minimum, 24   **34.** Minimum, $-\frac{121}{4}$ **35.** Maximum, -6 **36.** Maximum, -39 **37.** 526 ft

38. 36 ft by 36 ft, 1296 sq ft   **39.** $\left[\frac{143}{4}, \infty\right)$ **40.** $(-\infty, 81]$ **41.** $2x + h$ **42.** $2x + h - 6$

43. Domain: $[-4, \infty)$ **44.** Domain: $[2, \infty)$ **45.** Domain: $[-1, \infty)$ **46.** Domain: $[-5, \infty)$
Range: $[0, \infty)$ Range: $[7, \infty)$ Range: $(-\infty, 2]$ Range: $[-3, \infty)$

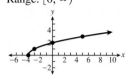   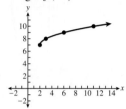   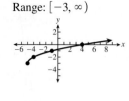

47. Domain: $(-\infty, \infty)$ **48.** Domain: $(-\infty, \infty)$ **49.** $f(x) = \sqrt{x - 1} + 6$ **50.** $f(x) = (x - 6)^3$ **51.** -36
Range: $(-\infty, \infty)$ Range: $(-\infty, \infty)$ **52.** 312   **53.** 2166   **54.** -3 **55.** $10x + 24$ **56.** $16x - 88$

57. $x^2 - 5x - 36$ **58.** $\dfrac{x + 2}{x - 7}$ **59.** 4   **60.** 35

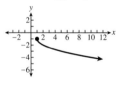   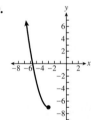

	61.	**62.**	**63.**	**64.**
$(f \circ g)(x)$	$18x + 103$	$-8x + 33$	$x^2 + 7x - 64$	$3x^2 - 27x + 130$
$(g \circ f)(x)$	$18x - 31$	$-8x - 51$	$x^2 - 9x - 48$	$9x^2 + 15x + 27$

65. No   **66.** Yes   **67.** Yes   **68.** No   **69.** No   **70.** Yes   **71.** No   **72.** Yes   **73.** $f^{-1}(x) = x + 10, (-\infty, \infty)$

74. $f^{-1}(x) = \dfrac{x + 15}{2}, (-\infty, \infty)$ **75.** $f^{-1}(x) = -x + 19, (-\infty, \infty)$ **76.** $f^{-1}(x) = \dfrac{-x + 27}{8}, (-\infty, \infty)$

77. $f^{-1}(x) = \dfrac{2}{3x - 9}$, All real numbers except 3   **78.** $f^{-1}(x) = \dfrac{21x}{7 - 4x}$, All real numbers except $\frac{7}{4}$

79.   **80.**

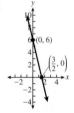   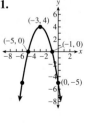

Chapter 11 Test   **1.** Yes [11.1]   **2.** No [11.1]   **3.** 51 [11.1]   **4.** $30a - 9$ [11.1]   **5.** 270 [11.1]   **6.** Domain: $(-\infty, \infty)$,
Range: $(-\infty, 5]$ [11.1]   **7.**   [11.2]   **8.**   [11.2]   **9.** $f(x) = 7x + 35$ [11.2]

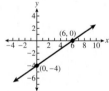

10.   [11.3]   **11.**   [11.3]

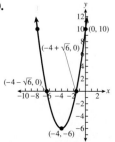

12. 9.25 ft [11.3]   **13.** Domain: $[-1, \infty)$ [11.4]   **14.** Domain: $(-\infty, \infty)$ [11.4]
  Range: $[-3, \infty)$ Range: $(-\infty, \infty)$

15. $f(x) = \sqrt{x + 6}$ [11.4]   **16.** -13 [11.5]   **17.** $x^3 - 216$ [11.5]   **18.** 11 [11.5]   **19.** 18 [11.5]   **20.** $-8x + 105$ [11.5]
21. $x^2 + 15x + 6$ [11.5]   **22.** No [11.6]   **23.** Yes [11.6]   **24.** $f^{-1}(x) = \dfrac{-x + 20}{4}$ [11.6]   **25.** $f^{-1}(x) = \dfrac{-9}{4x - 1}, x \neq \dfrac{1}{4}$ [11.6]

Cumulative Review Answers, Chapter 8–11   **1.** $\{-9, 3\}$ **2.** $(-\infty, 4]$    **3.** $[\frac{1}{2}, 13]$
4. $(-\infty, -3) \cup (2, \infty)$ **5.** $(-5, 5)$ **6.** $(x - 9)(x + 6)$
7. $4(x - 1)(x - 8)$ **8.** $(x + 5)(x + 6)$ **9.** $(7x + 8)(7x - 8)$ **10.** $(6, -2)$ **11.** $(3, 5, -2)$ **12.** 7.483   **13.** $64x^{15}$
14. $\sqrt[14]{x^9}$ **15.** $5r^6s^5t^4\sqrt{3t}$ **16.** $34\sqrt{3}$ **17.** $20 + 6\sqrt{15}$ **18.** $\dfrac{c\sqrt{ac}}{ab}$ **19.** $\dfrac{136 + 39\sqrt{10}}{62}$ **20.** 17   **21.** $6\sqrt{10}i$ **22.** $28 + 41i$
23. $\dfrac{4 + 3i}{5}$ or $\dfrac{4}{5} + \dfrac{3}{5}i$ **24.** $\{-2 \pm 3\sqrt{2}i\}$ **25.** $\left\{\dfrac{5 \pm 2\sqrt{22}}{3}\right\}$ **26.** $\{\pm\sqrt{2}, \pm 3\}$ **27.** $\{4\}$ **28.** $\{2\}$ **29.** $\{3, 9\}$

30. 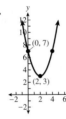   **31.** 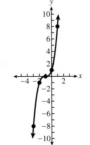   **32.** $[4, 6]$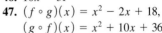
33. $(-\infty, -6] \cup (-5, 8]$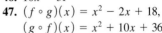
34. Length: 28.9 ft, Width: 33.9 ft   **35.** 5.15 sec
36. 11.72 hours   **37.** $12n - 43$ **38.** 126   **39.**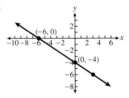

40. a) $f(x) = 0.15x + 20$ **b)** \$65   **41.** Minimum, -96 **42.** 374 ft
43. Domain: $(-\infty, \infty)$ **44.** Domain: $[-4, \infty)$ **45.** Domain: $(-\infty, \infty)$ **46.** $10x - 31$
  Range: $[-5, \infty)$ Range: $[-1, \infty)$ Range: $(-\infty, \infty)$ **47.** $(f \circ g)(x) = x^2 - 2x + 18$,
                                                                                                    $(g \circ f)(x) = x^2 + 10x + 36$
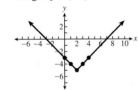      **48.** Inverses
                                                                                          49. $f^{-1}(x) = \dfrac{x + 19}{3}$
                                                                                          50. $f^{-1}(x) = \dfrac{3 + 9x}{x}, x \neq 0$

Chapter 12

Quick Check 12.1   **1.** $f(5) = 1024$, $f(-3) = \frac{1}{64}$ **2.** 79   **3.** Domain: $(-\infty, \infty)$ **4.** Domain: $(-\infty, \infty)$ **5.** Domain: $(-\infty, \infty)$
Range: $(0, \infty)$ Range: $(0, \infty)$ Range: $(3, \infty)$

 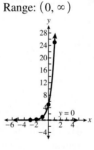 

6. Domain: $(-\infty, \infty)$ **7.** $f(-1) \approx 0.368$,   **8.** Domain: $(-\infty, \infty)$,   **9. a)** 4 **b)** 9   **10.** Approximately \$34,197.4 million
Range: $(-2, \infty)$ $f(4.3) \approx 73.700$ Range: $(5, \infty)$

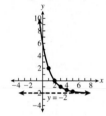 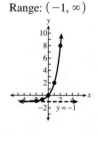

Section 12.1   **1.** 16   **3.** 1   **5.** $\frac{4}{25}$ **7.** $\frac{625}{16}$ **9.** 11   **11.** 251   **13.** 10   **15.** $\frac{1}{64}$ **17.** 2087   **19.** $\frac{1}{100}, \frac{1}{10}, 1, 10, 100$
21. $-8\frac{15}{16}, -8\frac{3}{4}, -8, -5, 7$ **23.** Domain: $(-\infty, \infty)$ **25.** Domain: $(-\infty, \infty)$ **27.** Domain: $(-\infty, \infty)$ **29.** Domain: $(-\infty, \infty)$
Range: $(0, \infty)$ Range: $(0, \infty)$ Range: $(0, \infty)$ Range: $(-1, \infty)$

31. Domain: $(-\infty, \infty)$ **33.** Domain: $(-\infty, \infty)$ **35.** 7.389   **37.** 0.368   **39.** 0.006   **41.** 20.086   **43.** 0.001
Range: $(-8, \infty)$ Range: $(2, \infty)$

 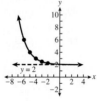 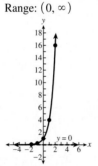

45. Domain: $(-\infty, \infty)$ **47.** Domain: $(-\infty, \infty)$ **49.** Domain: $(-\infty, \infty)$
Range: $(0, \infty)$ Range: $(2, \infty)$ Range: $(4, \infty)$

51. $\{4\}$ **53.** $\{3\}$ **55.** $\{-1\}$ **57.** $\{12\}$ **59.** $\{1\}$
61. $\{3\}$ **63.** $\{3\}$ **65.** $\{3\}$ **67.** $\{5\}$ **69.** $663,907.76
71. $15,605.09 **73.** $99.33 **75.** $13,548 million
77. $36.8°$ C **79.** $1.20 **81.** No **83.** Yes
85. Answers will vary. **87.** Answers will vary.

Practice Quiz **1.** -20 **2.** 4 **3.** Domain: $(-\infty, \infty)$, **4.** Domain: $(-\infty, \infty)$, **5.** 8512.084 **6.** Domain: $(-\infty, \infty)$, **7.** $\{9\}$
Range: $(0, \infty)$ Range: $(5, \infty)$ Range: $(-7, \infty)$ **8.** $\{7, -2\}$
9. $2288.50
10. $7414.80

Quick Check 12.2 **1. a)** 2 **b)** -3 **c)** 1 **d)** 0 **2.** -4 **3. a)** 4 **b)** 1.875 **4. a)** -3 **b)** 6.405 **5.** $\log_7 20 = x$ **6.** $3^6 = x$
7. $\{1024\}$ **8.** $\{3\}$ **9. a)** **b)** **10.** Approximately $43.47 billion **11.** 3.4

Section 12.2 **1.** 2 **3.** 3 **5.** 3 **7.** 0 **9.** -2 **11.** -3 **13. a)** 2 **b)** 5 **15. a)** 0 **b)** 7 **17.** 3 **19.** -2 **21.** 2.049 **23.** 0.870
25. 1 **27.** -6 **29.** 3.466 **31.** 2.787 **33.** $\log_2 8 = 3$ **35.** $\log_9 729 = 3$ **37.** $\log_6 30 = x$ **39.** $\log 500 = x$ **41.** $4^5 = 1024$
43. $16^{3/4} = 8$ **45.** $10^3 = x$ **47.** $e^7 = x$ **49.** $\{32\}$ **51.** $\{\frac{1}{81}\}$ **53.** $\{10,000\}$ **55.** $\{7.389\}$ **57.** $\{9\}$ **59.** $\{5.135\}$ **61.** $\{216\}$
63. $\{110\}$ **65.** $\{3\}$ **67.** $\{5\}$ **69.** $\{\frac{3}{2}\}$ **71.** **73.** **75.**

77. **79.**

81. $f(x) = \log_2 x$ **83.** $f(x) = \log_{1/3} x$ **85.** 19,934 **87.** 33.8 in.
89. 4.2 **91.** 630,957,345 times larger **93.** Answers will vary.
95. Answers will vary. **97.** Answers will vary.

Quick Review **1.** x^{16} **2.** b^{12} **3.** n^{56} **4.** 1
Quick Check 12.3 **1.** $\log(35x)$ **2.** $\log_2 5 + \log_2 b$ **3.** 2 **4.** $\log_b x - 1$ **5.** $2 + 2\log_3 x + 4\log_3 y$ **6.** $\frac{1}{7}\ln x$ **7.** $\log_8 x^8$

8. -4 **9.** 16 **10.** $\log_2\left(\frac{acd}{b}\right)$ **11.** $\log\left(\dfrac{a^6\sqrt[5]{b}}{c^9 d^{12}}\right)$ **12.** $3\log_5 z - 9\log_5 x - 4\log_5 y$ **13.** 4.395

Section 12.3   1. $\log_2(5x)$ **3.** $\log_4(bc)$ **5.** $\log_4 256 = 4$ **7.** $\log_3 10 + \log_3 x$ **9.** $\ln 3 + 1$ **11.** $\log_4 a + \log_4 b + \log_4 c$
13. $\log_5 6$ **15.** $\log\left(\frac{39}{x}\right)$ **17.** $\log_4 64 = 3$ **19.** $\log_8 10 + \log_8 a - \log_8 b$ **21.** $\log_4 6 + \log_4 x + \log_4 y + \log_4 z - \log_4 w$
23. $\log_5 x + \log_5 y - 2 - \log_5 z$ **25.** $10 \log a$ **27.** $2 \log_7 x + 3 \log_7 y$ **29.** $\frac{1}{2} \log_3 x$ **31.** $4 + 5 \log_2 x + 7 \log_2 y$
33. $\log_5 x^8$ **35.** $\ln \sqrt[3]{x}$ **37.** $\log_3 x^{10}$ **39.** 12   **41.** 9   **43.** 7   **45.** b **47.** $\log_3 \frac{10}{7}$ **49.** $\log_5 5832$ **51.** $\ln \frac{4}{5}$ **53.** $\log_2 \frac{x}{yzw}$
55. $\ln \dfrac{x^4 y^9}{z^{13}}$ **57.** $\log(x^2 - 4x - 12)$ **59.** $\log_3(2x^2 + 17x + 35)$ **61.** $\ln \dfrac{a^3 \sqrt[3]{b}}{c^{10}}$ **63.** $A + B$ **65.** $C + D$ **67.** $3A$
69. $3C$ **71.** $\log_8 a - \log_8 b - \log_8 c - \log_8 d$ **73.** $5 + 4 \log_2 x + 8 \log_2 y + 5 \log_2 z$ **75.** $6 \ln a + \frac{1}{2} \ln b - 7 \ln c$
77. $-3 \ln a - 2 \ln d + \ln b + 7 \ln c$ **79.** 2.807   **81.** 1.549   **83.** 6.481   **85.** 0.528   **87.** -1.521 **89.** 1.829   **91.** 1.701
93. Answers will vary.

Quick Check 12.4   1. a) $\{2\}$ **b)** $\{12\}$ **2.** $\left\{ \dfrac{\log 12}{\log 4} + 5 \right\}, \dfrac{\log 12}{\log 4} + 5 \approx 6.7925$ **3.** $\left\{ \dfrac{\log 11}{\log 9} - 3 \right\}, \dfrac{\log 11}{\log 9} - 3 \approx -1.909$
4. $\{\ln 28 - 2\}, \ln 28 - 2 \approx 1.332$ **5.** $\{2\}$ **6.** $\varnothing$ **7.** $\{-2, 5\}$ **8.** $\{333, 338\}$ **9. a)** $\{4\}$ **b)** $\{3\}$ **10.** $\{11\}$
11. $f^{-1}(x) = \ln(x - 9) - 6$ **12.** $f^{-1}(x) = e^{x+18} - 10$

Section 12.4   1. $\{3\}$ **3.** $\{6\}$ **5.** $\{5\}$ **7.** $\{\frac{3}{2}\}$ **9.** $\{\frac{9}{2}\}$ **11.** $\{-12\}$ **13.** $\{2.044\}$ **15.** $\{2.111\}$ **17.** $\{9.755\}$ **19.** $\{-0.820\}$
21. $\{1.922\}$ **23.** $\{3.305\}$ **25.** $\{-0.726\}$ **27.** $\{4\}$ **29.** $\{9\}$ **31.** $\varnothing$ **33.** $\{2, 12\}$ **35.** $\{-6\}$ **37.** $\{243\}$ **39.** $\{\frac{1}{625}\}$ **41.** $\{407.429\}$
43. $\{24\}$ **45.** $\{-12, 8\}$ **47.** $\{-5, 8\}$ **49.** $\{8\}$ **51.** $\{34\}$ **53.** $\{2\}$ **55.** $\{3\}$ **57.** $\{8\}$ **59.** $\{3\}$ **61.** $\{13\}$ **63.** $\{9\}$
65. $\{-10, 7\}$ **67.** $\{-2, 15\}$ **69.** $\{-5.267\}$ **71.** $\{5.577\}$ **73.** $\{-5\}$ **75.** $\{347\}$ **77.** $\{-3\}$ **79.** $\{10.069\}$ **81.** $\{\frac{5}{3}\}$ **83.** $\{4\}$
85. $\{37\}$ **87.** $\{111\}$ **89.** $\{-7\}$ **91.** $\{12\}$ **93.** $\{-2.432\}$ **95.** $\{-6\}$ **97.** $f^{-1}(x) = \ln x - 3$ **99.** $f^{-1}(x) = \ln(x - 4) + 5$
101. $f^{-1}(x) = e^x + 6$ **103.** $f^{-1}(x) = e^{x-5} - 2$ **105.** Answers will vary.   **107.** Answers will vary.

Quick Check 12.5   1. \$464.65   **2.** Approximately 14 years   **3.** Approximately 15.35 years   **4.** Approximately \$1.45
5. Approximately 117.5 days   **6.** In 2013   **7.** 2.8, acid   **8.** 5.01×10^{-1} moles/liter   **9.** 60 db   **10.** 10^{-4} watts per square meter

Section 12.5   1. \$5634.13   **3.** \$27,198.41   **5.** 10.2 yr   **7.** 23.1 yr   **9.** \$1343.13   **11.** 5.8 yr   **13.** 3250   **15.** 13.7 billion
17. In 2004 (4.7 yr)   **19.** In 2011 (12.5 yr)   **21.** 86.4 min   **23.** 358   **25.** 22.6 mg   **27.** 167 yr   **29.** 229,919.7 yr   **31.** 2094
33. 2009   **35.** 120,243   **37.** 178   **39.** In 2020 (25.9 yr)   **41.** 2047   **43.** 7 yr   **45. a)** $90.3°$ F **b)** 2.5 hr   **47.** In 2018 (29.2 yr)
49. 12.6, base   **51.** 2.3, acid   **53.** 1.0×10^{-5} moles per liter   **55.** 6.31×10^{-11} moles per liter   **57.** 80 db   **59.** 110 db   **61.** 85 db
63. 10^{-1} watts per square meter   **65.** $10^{-2.5}$ watts per square meter   **67.** 80   **69.** In 2016 (21.1 yr)   **71.** Answers will vary.
73. Answers will vary.

Quick Check 12.6   1. Domain: $(-\infty, \infty)$ **2.** Domain: $(-\infty, \infty)$ **3.** Domain: $(-2, \infty)$ **4.** Domain: $(6, \infty)$
Range: $(-6, \infty)$ Range: $(-9, \infty)$ Range: $(-\infty, \infty)$ Range: $(-\infty, \infty)$

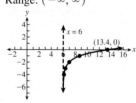

Section 12.6

	1.	**3.**	**5.**	**7.**	**9.**
x-int.	None	$(-0.3, 0)$	$(-0.7, 0)$	$(13, 0)$	$(51.6, 0)$
y-int.	$(0, 22)$	$(0, 15)$	$(0, 10.1)$	None	$(0, -2.9)$

	11.	**13.**	**15.**	**17.**
Asymptote	$y = 8$	$y = -9$	$x = 5$	$x = \frac{1}{2}$
Domain	$(-\infty, \infty)$	$(-\infty, \infty)$	$(5, \infty)$	$(\frac{1}{2}, \infty)$
Range	$(8, \infty)$	$(-9, \infty)$	$(-\infty, \infty)$	$(-\infty, \infty)$

19. Domain: $(-\infty, \infty)$
Range: $(4, \infty)$

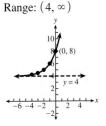

21. Domain: $(-\infty, \infty)$
Range: $(-15, \infty)$

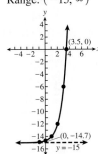

23. Domain: $(-\infty, \infty)$
Range: $(6, \infty)$

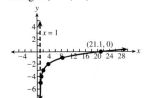

25. Domain: $(2, \infty)$
Range: $(-\infty, \infty)$

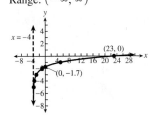

27. Domain: $(-2, \infty)$
Range: $(-\infty, \infty)$

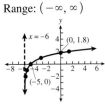

29. Domain: $(1, \infty)$
Range: $(-\infty, \infty)$

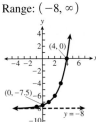

31. Domain: $(-\infty, \infty)$
Range: $(-2, \infty)$

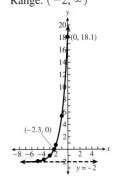

33. Domain: $(-4, \infty)$
Range: $(-\infty, \infty)$

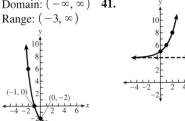

35. Domain: $(-6, \infty)$
Range: $(-\infty, \infty)$

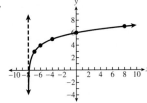

37. Domain: $(-\infty, \infty)$
Range: $(-8, \infty)$

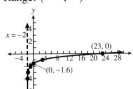

39. Domain: $(-\infty, \infty)$
Range: $(-3, \infty)$

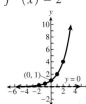

41.

43.

45. $f^{-1}(x) = 2^x$

47. $f^{-1}(x) = \log_2(x + 8) + 4$ **49.** Answers will vary.

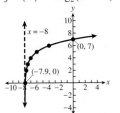

Chapter 12 Review **1.** 1015 **2.** 7.086 **3.** -2 **4.** 5 **5.** 0 **6.** 12 **7.** 23.761 **8.** -2.756 **9.** $\log_4 1024 = x$ **10.** $\ln 20 = x$

11. $10^{3.2} = x$ **12.** $e^{11} = x$ **13.** $\log_2 104$ **14.** $\log\left(\dfrac{a^2}{b^5}\right)$ **15.** $\log_2\left(\dfrac{x^4 z^9}{y^3}\right)$ **16.** $\log_b \frac{1}{972}$ **17.** $2 + \log_8 x$ **18.** $\log_3 n - 3$

19. $6 \log_b a + 3 \log_b c - 2$ **20.** $3 + 9 \log_2 x - \log_2 y - 7 \log_2 z$ **21.** 2.322 **22.** -8.205 **23.** 10.641 **24.** 4.340 **25.** $\{3\}$
26. $\{7\}$ **27.** $\{\frac{5}{4}\}$ **28.** $\{12\}$ **29.** $\{1.232\}$ **30.** $\{0.468\}$ **31.** $\{12.127\}$ **32.** $\{5.806\}$ **33.** $\{14\}$ **34.** $\{1\}$ **35.** $\{148.413\}$ **36.** $\{3, 10\}$
37. $\{2\}$ **38.** $\{10\}$ **39.** $\{4\}$ **40.** $\{2\}$ **41.** $\{3.107\}$ **42.** $\{54.598\}$ **43.** $f^{-1}(x) = \ln(x - 4)$ **44.** $f^{-1}(x) = \log_2(x + 1) + 5$
45. $f^{-1}(x) = e^x - 8$ **46.** $f^{-1}(x) = 3^{x+10} - 6$ **47.** 13.3 yr **48.** 334,133 **49.** 6.25 g **50.** 167 yr **51.** In 2007 (17.3 yr)
52. 78.5° F **53.** 4.3 **54.** 3.16×10^{-4} moles/liter **55.** Domain: $(-\infty, \infty)$ **56.** Domain: $(-\infty, \infty)$ **57.** Domain: $(-\infty, \infty)$
Range: $(2, \infty)$ Range: $(-8, \infty)$ Range: $(-6, \infty)$

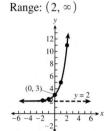
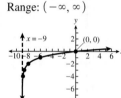
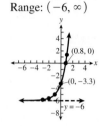

58. Domain: $(5, \infty)$ **59.** Domain: $(-9, \infty)$ **60.** Domain: $(-8, \infty)$
Range: $(-\infty, \infty)$ Range: $(-\infty, \infty)$ Range: $(-\infty, \infty)$

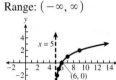
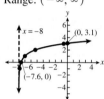

Chapter 12 Test **1.** -2.611 **2.** 6.890 **3.** 2 **4.** -3 **5.** 3 **6.** $\log_b\left(\dfrac{x^2 y^5}{z^3}\right)$ **7.** $4 \ln a + 6 \ln b - 8 \ln c - \ln d$ **8.** 1.878 **9.** $\{-6\}$

10. $\{12.124\}$ **11.** $\{4.207\}$ **12.** $\{8\}$ **13.** $\{60\}$ **14.** $\{8\}$ **15.** $f^{-1}(x) = \ln(x + 17) - 9$ **16.** 6 yr **17.** $80,946.56 **18.** 870.7 yr
19. Domain: $(-\infty, \infty)$ **20.** Domain: $(5, \infty)$
Range: $(4, \infty)$ Range: $(-\infty, \infty)$

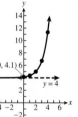

Chapter 13

Quick Check 13.1 **1.**

2.

3.

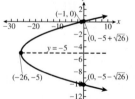

4.

5.
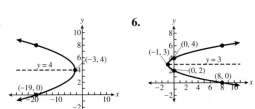

6.
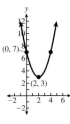

7. a) $y = (x + 2)^2 + 16$, Vertex: $(-2, 16)$
b) $x = -(y + 3)^2 + 26$, Vertex: $(26, -3)$ **8.** $y = x^2 - 6x + 8$

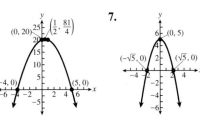

Section 13.1   1.
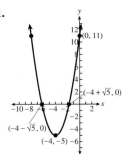

3.
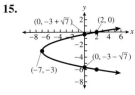

5.
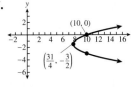

7.

9.
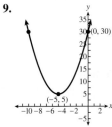

11.
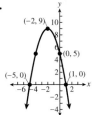

13.
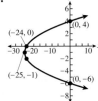

15.
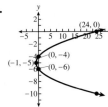

17.
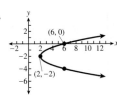

19.
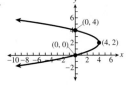

21.
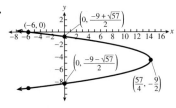

23.

25.

27.

29.
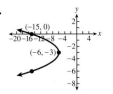

31.
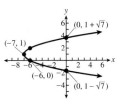

33.
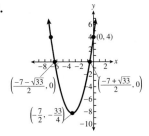

35.
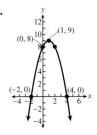

37.
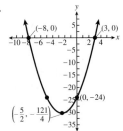

39.
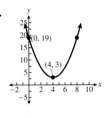

41. $(1, 14)$ **43.** $(7, 125)$ **45.** $(-49, -3)$ **47.** $(-47, 2)$
49. $y = x^2 - 6x + 5$ **51.** $x = y^2 - 4y - 5$ **53. a)** $y = \frac{1}{500}x^2 - 5$
b) 4.2 ft   **55.** Answers will vary.   **57.** Answers will vary.
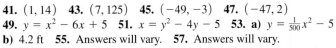

Quick Check 13.2 **1. a)** 10 **b)** 12.6 **2.** $\left(\frac{1}{2}, 2\right)$
3. $(0, 0), r = 2\sqrt{3}$ **4.** $(6, 5), r = 3$ **5. a)** $(1, -6), r = 4$ **b)** $(-2, 3), r = 8$ **6.** $(-7, 3), r = 1$

  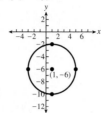 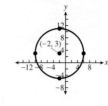 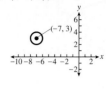

7. $(x + 3)^2 + (y - 5)^2 = 16$ **8.** $(x - 1)^2 + (y + 2)^2 = 34$

Section 13.2 **1.** 17 **3.** 89 **5.** 13 **7.** 4.2 **9.** 6.4 **11.** $(3, 2)$ **13.** $(7, -17)$
15. $(0, 0), r = 1$ **17.** $(0, 0), r = 4$ **19.** $(0, 0), r = 10$ **21.** $(0, 0), r = \sqrt{6}$ **23.** $(5, 9), r = 4$

    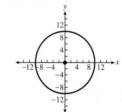  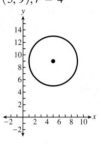

25. $(3, 2), r = 2$ **27.** $(-4, 5), r = 3$ **29.** $(-2, -1), r = 7$ **31.** $(-5, 0), r = 5$ **33.** $(2, -6), r = 3\sqrt{2}$

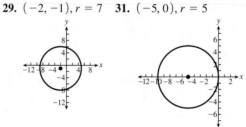

35. $(-3, 7), r = 3\sqrt{5}$

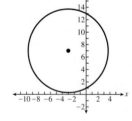

37. $(-2, -6), 9$ **39.** $(5, -1), 7$ **41.** $(-8, 0), 11$ **43.** $(-2, -7), 4\sqrt{5}$ **45.** $x^2 + y^2 = 36$
47. $(x - 4)^2 + (y - 2)^2 = 49$ **49.** $(x + 9)^2 + (y + 10)^2 = 25$ **51.** $x^2 + (y + 8)^2 = 15$
53. $x^2 + y^2 = 64$ **55.** $(x - 1)^2 + (y - 5)^2 = 16$ **57.** $(x + 1)^2 + (y - 1)^2 = 49$
59. $(x - 30)^2 + (y + 40)^2 = 2500$ **61.** $(x - 5)^2 + (y - 7)^2 = 16$
63. $(x - 2)^2 + (y - 7)^2 = 25$ **65.** $(x - 8)^2 + (y + 1)^2 = 20$ **67.**

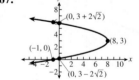

69. **71.** **73.** **75.** **77.**

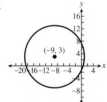 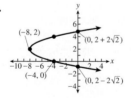 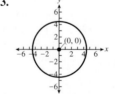 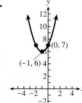 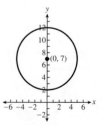

79. a) $x^2 + y^2 = 725$ **b)** 1277.7 sq ft **81.** Answers will vary. **83.** Answers will vary.

Quick Check 13.3   1. a)

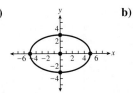

b)

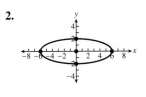

2.

3.

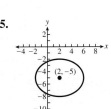

4.

5.

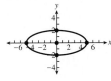

6. $\dfrac{x^2}{49} + \dfrac{(y-5)^2}{9} = 1$

Section 13.3   1. $(0,0), a = 2, b = 3$ **3.** $(0,0), a = 5, b = 2$ **5.** $(0,0), a = 1, b = 5$ **7.** $(0,0), a = 6, b = 2\sqrt{3}$

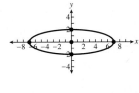

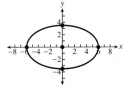

9. $(0,0), a = 3, b = 5$ **11.** $(0,0), a = 7, b = 2$ **13.** $(6,8), a = 5, b = 6$ **15.** $(9,-4), a = 9, b = 3$

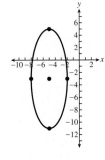

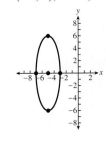

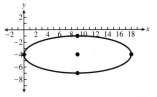

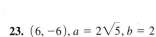

17. $(-5,-3), a = 3, b = 8$ **19.** $(-5,0), a = 2, b = 6$ **21.** $(-2,4), a = 7, b = 1$ **23.** $(6,-6), a = 2\sqrt{5}, b = 2$

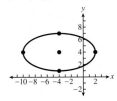

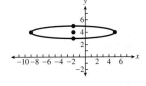

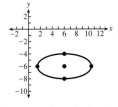

25. $(3,5), a = 3, b = 2$ **27.** $(-4,4), a = 6, b = 3$

29. $\dfrac{(x+2)^2}{4} + \dfrac{(y-8)^2}{100} = 1$ **31.** $\dfrac{(x-6)^2}{64} + \dfrac{(y+3)^2}{49} = 1$

33. $\dfrac{x^2}{9} + \dfrac{(y+6)^2}{25} = 1$ **35.** $\dfrac{x^2}{25} + \dfrac{y^2}{9} = 1$ **37.** $\dfrac{(x-4)^2}{25} + \dfrac{y^2}{4} = 1$

39. $\dfrac{x^2}{9} + \dfrac{(y+4)^2}{25} = 1$ **41.** $\dfrac{(x-3)^2}{36} + \dfrac{(y+4)^2}{25} = 1$

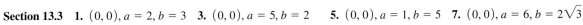

43.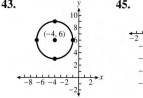

45.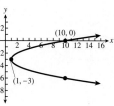

47.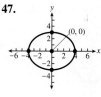

49.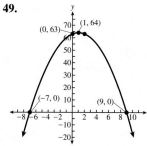

51.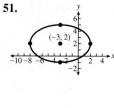

53.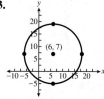

55.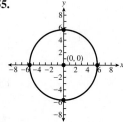

57.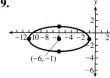

59.

61. a) 100 ft **b)** 66 ft
63. 471.2 sq ft
65. Answers will vary.

Quick Check 13.4 **1.** Center: $(0, 0)$
Vertices: $(-4, 0), (4, 0)$
Asymptotes: $y = \frac{3}{4}x,\ y = -\frac{3}{4}x$

2. Center: $(0, 0)$
Vertices: $(0, -2), (0, 2)$
Asymptotes: $y = \frac{1}{3}x,\ y = -\frac{1}{3}x$

 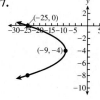 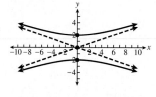

3. Center: $(3, 5)$
Vertices: $(1, 5), (5, 5)$
Asymptotes: $y = 2(x - 3) + 5,$
$\qquad\qquad y = -2(x - 3) + 5$

4. Center: $(-3, 2)$
Vertices: $(-3, 9), (-3, -5)$
Asymptotes: $y = \frac{7}{3}(x + 3) + 2,$
$\qquad\qquad y = -\frac{7}{3}(x + 3) + 2$

5. Center: $(2, -3)$
Vertices: $(-1, -3), (5, -3)$
Asymptotes: $y = \frac{4}{3}(x - 2) - 3,$
$\qquad\qquad y = -\frac{4}{3}(x - 2) - 3$

6. $\dfrac{(y - 4)^2}{4} - \dfrac{(x + 3)^2}{25} = 1$

 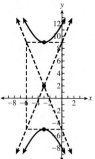 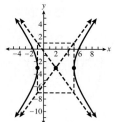

Section 13.4   **1.** $(0,0), a = 3, b = 2$ **3.** $(0,0), a = 3, b = 7$ **5.** $(0,0), a = 3, b = 3$ **7.** $(0,0), a = 1, b = 5$

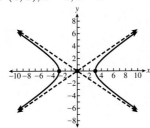 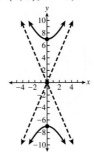 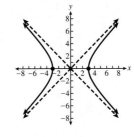 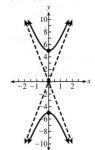

9. $(0,0), a = 2\sqrt{3}, b = 3$ **11.** $(0,0), a = 2, b = 7$ **13.** $(0,0), a = 1, b = 3$ **15.** $(2,5), a = 3, b = 4$

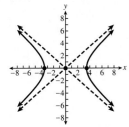 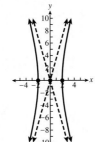 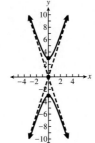 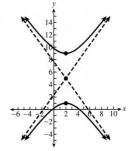

17. $(-1,-4), a = 3, b = 2$ **19.** $(0,1), a = 2, b = 2$ **21.** $(4,-2), a = 7, b = 4$ **23.** $(-3,5), a = 3, b = 1$

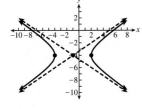 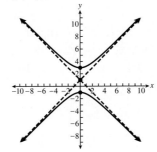 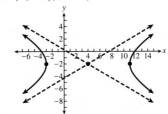 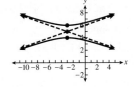

25. $(-2,-3), a = 5, b = 2\sqrt{7}$ **27.** $(-1,-3), a = 2, b = 3$ **29.** $(-2,7), a = 4, b = 5$

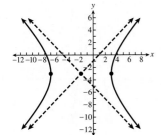 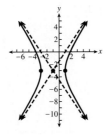 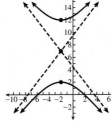

31. $\dfrac{(y-2)^2}{9} - \dfrac{(x-4)^2}{25} = 1$

33. $\dfrac{(x-6)^2}{16} - \dfrac{(y+2)^2}{16} = 1$

35. $\dfrac{y^2}{25} - \dfrac{x^2}{9} = 1$

37. $\dfrac{(y-2)^2}{25} - \dfrac{(x-3)^2}{4} = 1$

39. $\dfrac{(x+1)^2}{25} - \dfrac{(y+5)^2}{9} = 1$

41.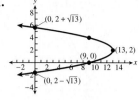

43.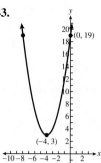

45.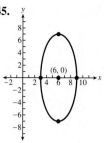

47.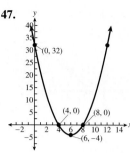

49.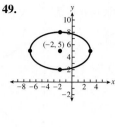

51.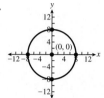

53.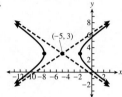

55.

57.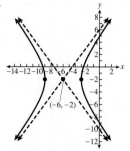

59. Answers will vary.

Quick Review **1.** $(3, 2)$ **2.** $(-3, -9)$ **3.** $(4, -1)$ **4.** $\left(\frac{3}{2}, -\frac{9}{5}\right)$

Quick Check 13.5 **1.** $(2, -5), \left(-\frac{50}{17}, -\frac{43}{17}\right)$ **2.** $(-\sqrt{3}, 3), (\sqrt{3}, 3)$ **3.** $(3, 5), (-3, 5), (3, -5), (-3, -5)$ **4.** 3 ft by 12 ft
5. 9 in. by 12 in. **6.** 2 sec

Section 13.5 **1.** $(8, -6), (-6, -8)$ **3.** $(-9, -2), \left(-\frac{42}{5}, -\frac{19}{5}\right)$ **5.** $(5, 0), (0, 2)$ **7.** $(0, -4), (2, 0)$ **9.** $(1, 6), (4, 12)$
11. $(2, -1), (4, 3)$ **13.** $(3, 1)$ **15.** $(0, -5), (3, 4), (-3, 4)$ **17.** $(5, \sqrt{5}), (5, -\sqrt{5})$ **19.** $(0, 1)$ **21.** $(1, 0)$ **23.** $\varnothing$
25. $(10, \sqrt{21}), (10, -\sqrt{21}), (-6, \sqrt{5}), (-6, -\sqrt{5})$ **27.** $(3, 2), (-3, 2)$ **29.** $(14, -2)$ **31.** $(2, 1), (2, -1), (-2, 1), (-2, -1)$
33. $(2, \sqrt{2}), (2, -\sqrt{2}), (-2, \sqrt{2}), (-2, -\sqrt{2})$ **35.** $(\sqrt{2}, 2\sqrt{2}), (\sqrt{2}, -2\sqrt{2}), (-\sqrt{2}, 2\sqrt{2}), (-\sqrt{2}, -2\sqrt{2})$
37. $\left(3, \frac{\sqrt{6}}{2}\right), \left(3, -\frac{\sqrt{6}}{2}\right), \left(-3, \frac{\sqrt{6}}{2}\right), \left(-3, -\frac{\sqrt{6}}{2}\right)$ **39.** $(\sqrt{15}, 3), (\sqrt{15}, -3), (-\sqrt{15}, 3), (-\sqrt{15}, -3)$
41. $(\sqrt{3}, \sqrt{3}), (\sqrt{3}, -\sqrt{3}), (-\sqrt{3}, \sqrt{3}), (-\sqrt{3}, -\sqrt{3})$ **43.** $(3, 3), (3, -3), (-3, 3), (-3, -3)$
45. $(1, 3), (1, -3), (-1, 3), (-1, -3)$ **47.** 7 ft by 15 ft **49.** 16 ft by 25 ft **51.** 8 ft by 24 ft **53.** 45 ft by 60 ft
55. 40 ft by 20 ft **57.** 4.5 sec **59. a)** 3 sec **b)** 36 ft **61.** Answers will vary.

Chapter 13 Review **1.** 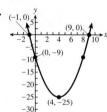 **2.** 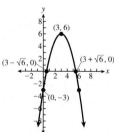 **3.** 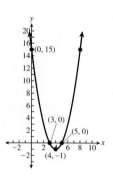 **4.**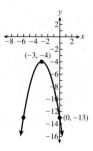

5.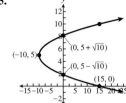

6.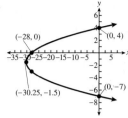

7.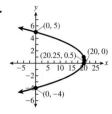

8.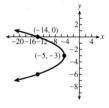

9.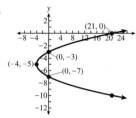

10.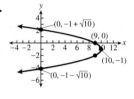

11.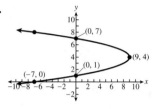

12.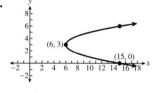

13. 17 **14.** 9.5 **15.** $r = 6$

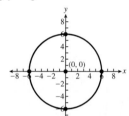

16. $r = \sqrt{20}$

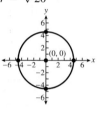

17. $r = 2$

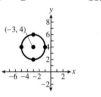

18. $r = 3$

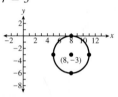

19. $r = 5$

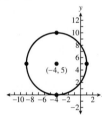

20. $r = 4$

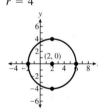

21. $r = 7$

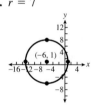

22. $(-3, 9), r = 10$ **23.** $(4, 10), r = \sqrt{17}$
24. $x^2 + y^2 = 9$ **25.** $(x + 2)^2 + (y - 4)^2 = 25$

26. $a = 5, b = 3$

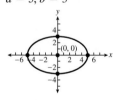

27. $a = 1, b = 4$

28. $a = 3, b = 4$

29. $a = 5, b = 2$

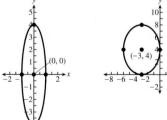

30. $a = 3, b = 6$

31. $a = 4, b = 3$

32. $(6, 1), a = 7, b = 2$

33. $(-2, -4), a = 4, b = 9$ **34.** $\dfrac{x^2}{4} + \dfrac{y^2}{25} = 1$

35. $\dfrac{(x + 1)^2}{64} + \dfrac{(y - 4)^2}{9} = 1$

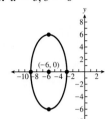

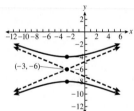

36. $a = 3, b = 5$

37. $a = 4, b = 4$

38. $a = 2, b = 3$

39. $a = 5, b = 6$

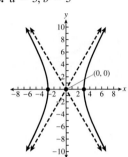
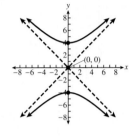
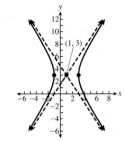
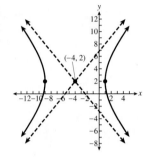

40. $a = 9, b = 2$

41. $a = 5, b = 2$

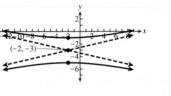
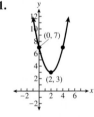

	42.	**43.**
Center	$(5, 7)$	$(-8, 4)$
Transverse Axis	$2b = 6$	$2a = 12$
Conjugate Axis	$2a = 4$	$2b = 14$

44. $\dfrac{x^2}{4} - \dfrac{y^2}{25} = 1$ **45.** $\dfrac{(y - 2)^2}{9} - \dfrac{(x + 1)^2}{16} = 1$ **46.** $(5, 5), (1, 7)$ **47.** $(3, 2), (0, -4)$ **48.** $(6, 11), (4, 7)$ **49.** $(10, 7), (-2, 5)$

50. $(1, 4), (-1, 4)$ **51.** $\left(\dfrac{2\sqrt{15}}{5}, -\dfrac{18}{5}\right), \left(-\dfrac{2\sqrt{15}}{5}, -\dfrac{18}{5}\right), (3, 3), (-3, 3)$ **52.** $(\sqrt{10}, 9), (-\sqrt{10}, 9), (1, 0), (-1, 0)$ **53.** $(3, 9), (-2, 4)$

54. $(\sqrt{19}, 9), (-\sqrt{19}, 9), (\sqrt{19}, -9), (-\sqrt{19}, -9)$ **55.** $(3, 1), (-3, 1), (3, -1), (-3, -1)$ **56.** $(2, 2), (-2, 2), (2, -2), (-2, -2)$

57. $(5, 3), (-5, 3), (5, -3), (-5, -3)$ **58.** 15 ft by 20 ft   **59.** 9 in. by 12 in.   **60. a)** 4 sec **b)** 48 ft

Chapter 13 Test   1. [13.1]   **2.** [13.1]   **3.** [13.1]

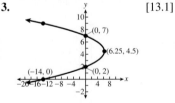

4. 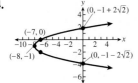 [13.1]   **5.** 8.9 [13.2]   **6.** $r = 5$ [13.2]   **7.** $(6, 3), r = 4\sqrt{2}$ [13.2]

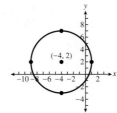

8. $(x - 3)^2 + (y + 8)^2 = 256$ [13.2]   **9.** $a = 4\ b = 3$ [13.3]   **10.** $(8, -2), a = 5, b = 2$ [13.3]   **11.** $a = 4, b = 1$ [13.4]

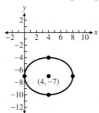

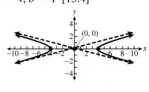

12. $(x - 3)^2 + (y - 2)^2 = 25$ [13.2]   **13.** $\frac{(x - 5)^2}{4} + \frac{(y + 1)^2}{36} = 1$ [13.3]   **14.** $(10, 0), (-8, 6)$ [13.5]   **15.** $(5, 2), \left(-\frac{19}{5}, -\frac{12}{5}\right)$ [13.5]
16. $(\sqrt{39}, 7), (-\sqrt{39}, 7), (\sqrt{26}, -6), (-\sqrt{26}, -6)$ [13.5]   **17.** $(7, 3), (-7, 3), (7, -3), (-7, -3)$ [13.5]   **18.** 50 ft by 120 ft [13.5]

Chapter 14

Quick Check 14.1   **1.** $\frac{1}{7}, \frac{1}{10}, \frac{1}{13}, \frac{1}{16}$ **2.** $\frac{1}{2}, -\frac{1}{4}, \frac{1}{6}, -\frac{1}{8}$ **3.** $-37, 43, -49, a_n = (-1)^{n+1}(6n + 1)$ **4.** $2, 5, 26, 677, 458330$ **5.** 60   **6.** 112
7. 60   **8.** \$6400, \$5120, \$4096, \$3276.80

Section 14.1   **1.** 11   **3.** 2187   **5.** $-1, -2, -3, -4, -5$ **7.** $1, 5, 9, 13, 17$ **9.** $-5, 7, -9, 11, -13$ **11.** $-2, -2, 0, 4, 10$
13. $30, 35, 40, a_n = 5n$ **15.** $59,049, 531,441, 4,782,969, a_n = 9^{n-1}$ **17.** $\frac{1}{20}, \frac{1}{23}, \frac{1}{26}, a_n = \frac{1}{3n+2}$ **19.** $216, 343, 512, a_n = n^3$
21. $6, 9, 12, 15, 18$ **23.** $-5, -7, -13, -31, -85$ **25.** $16, -48, 144, -432, 1296$ **27.** $-35, 18, -35, 18, -35$ **29.** 300
31. -399 **33.** 3066   **35.** -45 **37.** 95   **39.** -147 **41.** 55   **43.** 25   **45.** 117   **47.** -212 **49.** 25,263   **51.** 43,053,375
53. 120   **55.** 64   **57.** 385   **59.** $\frac{15}{16}$ **61.** $\sum_{i=1}^{4} 7i$ **63.** $\sum_{i=1}^{6} (-1)^{i+1} \cdot 5i$ **65.** $\sum_{i=1}^{7} i^2$ **67.** \$1500, \$900, \$540, \$324
69. a) \$32,000, \$32,800, \$33,600, \$34,400, \$35,200, \$36,000, \$36,800, \$37,600 **b)** \$278,400   **71.** Answers will vary.

Quick Check 14.2   **1.** $-8, -5, -2, 1, 4, 7$ **2.** $a_n = 12n - 10$ **3.** 177   **4.** 396   **5.** 12,870   **6.** 3906

Section 14.2   **1.** 9   **3.** $\frac{5}{2}$ **5.** 3   **7.** 8   **9.** $1, 6, 11, 16, 21$ **11.** $-9, 3, 15, 27, 39$ **13.** $-10, -31, -52, -73, -94$ **15.** $a_n = 13n - 4$
17. $a_n = 5n - 13$ **19.** $a_n = -35n + 122$ **21.** $a_n = 4n - \frac{11}{2}$ **23.** 106   **25.** 62nd   **27.** 123rd   **29.** 533   **31.** 816   **33.** -1900
35. 910   **37.** 1887   **39.** -5690 **41.** 52,592   **43.** 500,500   **45.** 4489   **47.** 40,200   **49. a)** \$10, \$11, \$12, \$13, \$14 **b)** $a_n = n + 9$
51. a) \$38,500, \$39,250, \$40,000, \$40,750, \$41,500, \$42,250 **b)** $a_n = 750n + 37,750$ **c)** \$656,250   **53.** Answers will vary.
55. Answers will vary.

Quick Check 14.3   **1.** $\frac{1}{4}$ **2.** $-\frac{1}{2}$ **3.** $5, -\frac{15}{4}, \frac{45}{16}, -\frac{135}{64}, \frac{405}{256}$ **4.** $a_n = 9 \cdot 2^{n-1}$ **5.** $a_n = 11 \cdot (-5)^{n-1}$ **6.** 39,360   **7.** $-14,762$ **8.** Yes, 9
9. No   **10. a)** \$17,500, \$12,250, \$8575, \$6002.50, \$4201.75, \$2941.23 **b)** \$706.19

Section 14.3   **1.** 3   **3.** $\frac{1}{5}$ **5.** -2 **7.** $-\frac{4}{3}$ **9.** $1, 10, 100, 1000, 10,000$ **11.** $4, -8, 16, -32, 64$ **13.** $8, 6, \frac{9}{2}, \frac{27}{8}, \frac{81}{32}$
15. $-20, 35, -\frac{245}{4}, \frac{1715}{16}, -\frac{12,005}{64}$ **17.** $a_n = 8^{n-1}$ **19.** $a_n = -1 \cdot 6^{n-1}$ **21.** $a_n = 9 \cdot 2^{n-1}$ **23.** $a_n = 16(-\frac{3}{4})^{n-1}$ **25.** 27,305
27. $-16,777,200$ **29.** 511.5   **31.** $-29,127$ **33.** 4372   **35.** 3,528,889   **37.** 2047.875   **39.** Yes, 45   **41.** Yes, $\frac{5}{9}$ **43.** No limit
45. Yes, $\frac{1}{3}$ **47.** Yes, 30   **49.** Yes, $-\frac{80}{7}$ **51.** No limit   **53. a)** \$6400, \$5120, \$4096 **b)** \$1342.18   **55.** $a_n = 40,000(0.5)^{n-1}$
57. \$159,374,246   **59.** Answers will vary.   **61.** Answers will vary.   **63.** Answers will vary.

Quick Review   1. $x^2 + 10x + 25$ **2.** $x^2 - 2xy + y^2$ **3.** $4x^2 + 36xy + 81y^2$ **4.** $25a^6 - 40a^3b^4 + 16b^8$

Quick Check 14.4   1. 362,880   **2.** 15   **3.** 1   **4.** $_9C_0 = 1, _9C_1 = 9, _9C_3 = 84, _9C_5 = 126, _9C_7 = 36, _9C_8 = 9, _9C_9 = 1$

5. $x^3 + 3x^2y + 3xy^2 + y^3$ **6.** $x^4 - 16x^3 + 96x^2 - 256x + 256$ **7.** $x^7 + 7x^6y + 21x^5y^2 + 35x^4y^3 + 35x^3y^4 + 21x^2y^5 + 7xy^6 + y^7$

Section 14.4   1. 720   **3.** 40,320   **5.** 6   **7.** 1   **9.** 15!   **11.** 45!   **13.** 1,663,200   **15.** 84   **17.** 45,360   **19.** 20   **21.** 36   **23.** 6   **25.** 5

27. 1   **29.**

$_{11}C_0 = 1$	$_{11}C_1 = 11$	$_{11}C_2 = 55$	$_{11}C_3 = 165$	$_{11}C_4 = 330$	$_{11}C_5 = 462$
$_{11}C_6 = 462$	$_{11}C_7 = 330$	$_{11}C_8 = 165$	$_{11}C_9 = 55$	$_{11}C_{10} = 11$	$_{11}C_{11} = 1$

31. $x^3 + 3x^2y + 3xy^2 + y^3$

33. $x^3 + 6x^2 + 12x + 8$ **35.** $x^4 + 24x^3 + 216x^2 + 864x + 1296$ **37.** $x^5 - 5x^4y + 10x^3y^2 - 10x^2y^3 + 5xy^4 - y^5$

39. $x^4 - 12x^3 + 54x^2 - 108x + 81$ **41.** $x^3 + 9x^2y + 27xy^2 + 27y^3$ **43.** $64x^3 - 48x^2y + 12xy^2 - y^3$

45.

$n = 8$		1	8	28	56	70	56	28	8	1		
$n = 9$		1	9	36	84	126	126	84	36	9	1	
$n = 10$	1	10	45	120	210	252	210	120	45	10	1	
$n = 11$	1	11	55	165	330	462	462	330	165	55	11	1
$n = 12$	1	12	66	220	495	792	924	792	495	220	66	12	1

47. $x^5 + 5x^4y + 10x^3y^2 + 10x^2y^3 + 5xy^4 + y^5$ **49.** $x^7 + 14x^6 + 84x^5 + 280x^4 + 560x^3 + 672x^2 + 448x + 128$

51. $x^3 - 3x^2y + 3xy^2 - y^3$ **53.** $x^4 - 12x^3 + 54x^2 - 108x + 81$ **55.** Answers will vary.

Chapter 14 Review   1. 8, 9, 10, 11, 12   **2.** $-14, -9, -4, 1, 6$ **3.** $78, 91, 104, a_n = 13n$ **4.** $47, 54, 61, a_n = 7n + 5$ **5.** $-9, 8, 25,$
$42, 59$ **6.** $-4, -31, -85, -193, -409$ **7.** $a_n = a_{n-1} + 14$ **8.** $a_n = -2(a_{n-1})$ **9.** 174   **10.** 4,031,078   **11.** 21   **12.** 228   **13.** 182
14. 7380   **15.** 78   **16.** 396   **17.** 24   **18.** -6 **19.** $16, 7, -2, -11, -20$ **20.** $-18, -7, 4, 15, 26$ **21.** $a_n = 7n - 3$ **22.** $-7n + 15$
23. $2n - 17$ **24.** $\frac{3}{4}n + \frac{1}{12}$ **25.** 823   **26.** -607 **27.** 236th   **28.** 495th   **29.** 276   **30.** 360   **31.** -444 **32.** 2745   **33.** 57,970
34. a) \$20,000,000, \$22,000,000, \$24,000,000, \$26,000,000, \$28,000,000   **b)** $a_n = 2,000,000n + 18,000,000$ **c)** \$290,000,000   **35.** 10
36. 6   **37.** $a_n = 1 \cdot 12^{n-1}$ **38.** $a_n = 8(\frac{3}{4})^{n-1}$ **39.** $a_n = -4 \cdot 3^{n-1}$ **40.** $a_n = -24(-\frac{1}{3})^{n-1}$ **41.** 1152   **42.** 9841   **43.** 6830
44. 55,987   **45.** 3,595,117.5   **46.** 639.375   **47.** 1,835,015   **48.** No limit   **49.** $\frac{108}{7}$ **50.** $-\frac{147}{5}$ **51.** 24   **52.** \$1,715,383.22
53. 362,880   **54.** 5040   **55.** 56   **56.** 165   **57.** $x^6 + 6x^5y + 15x^4y^2 + 20x^3y^3 + 15x^2y^4 + 6xy^5 + y^6$
58. $x^4 + 24x^3 + 216x^2 + 864x + 1296$ **59.** $x^3 - 27x^2 + 243x - 729$
60. $x^9 + 9x^8y + 36x^7y^2 + 84x^6y^3 + 126x^5y^4 + 126x^4y^5 + 84x^3y^6 + 36x^2y^7 + 9xy^8 + y^9$

Chapter 14 Test   1. $86, 101, 116, a_n = 15n - 4$ [14.1]   **2.** 279 [14.1]   **3.** $9, 17, 25, 33, 41$ [14.2]   **4.** $a_n = -7n + 27$ [14.2]
5. 577.5 [14.2]   **6.** 40,200 [14.2]   **7.** $5, 40, 320, 2560, 20,480$ [14.3]   **8.** $a_n = 4 \cdot 7^{n-1}$ [14.3]   **9.** 7,324,218 [14.3]   **10.** Yes, $\frac{25}{4}$
[14.3]   **11. a)** \$2500, \$3000, \$3500, \$4000, \$4500, \$5000   **b)** $500n + 2000$ **c)** \$121,500 [14.3]   **12. a)** \$13,500, \$10,125, \$7593.75,
\$5695.31   **b)** $a_n = 18,000(0.75)^{n-1}$ [14.3]   **13.** 792 [14.4]   **14.** $x^4 + 4x^3 + 6x^2 + 4x + 1$ [14.4]
15. $16,384x^7 + 28,672x^6y + 21,504x^5y^2 + 8960x^4y^3 + 2240x^3y^4 + 336x^2y^5 + 28xy^6 + y^7$ [14.4]

Cumulative Review: Chapters 12–14   1. 127.413   **2.** 35.466   **3.** 4   **4.** 8   **5.** $\ln\left(\dfrac{x^2z^5}{y^4}\right)$ **6.** $4\log b + 5\log c - \log a$ **7.** 2.066

8. 3.677   **9.** $\{8\}$ **10.** $\{2.769\}$ **11.** $\{1.733\}$ **12.** $\{4\}$ **13.** $\{90\}$ **14.** $\{-1\}$ **15.** $f^{-1}(x) = \ln(x - 3) + 2$ **16.** $f^{-1}(x) = e^{x+16} - 9$
17. 27 years   **18.** Approximately 234,373   **19.** Domain: $(-\infty, \infty)$,   **20.** Domain: $(-5, \infty)$,   **21.**
Range: $(-1, \infty)$ Range: $(-\infty, \infty)$

22. 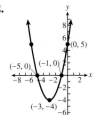 **23.** 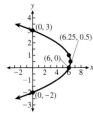 **24.** 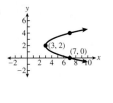 **25.** Center: $(0, 0)$, $r = 8$ **26.** Center: $(-3, 2)$, $r = 5$

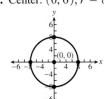 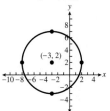

27. Center: $(0, 0)$, $a = 3$, $b = 7$ **28.** Center: $(4, 6)$, $a = 5$, $b = 6$ **29.** Center: $(0, 0)$, $a = 5$, $b = 2$ **30.** Center: $(-4, 2)$, $a = 3$, $b = 1$

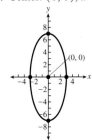 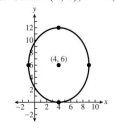 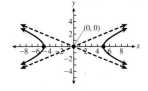 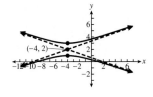

31. $(x - 3)^2 + (y + 2)^2 = 36$ **32.** $\dfrac{x^2}{16} + \dfrac{(y - 1)^2}{49} = 1$ **33.** $(2, 6)$, $(6, -2)$ **34.** $(0, -3)$, $(4, 1)$ **35.** $(8, 6)$, $(-8, 6)$, $(8, -6)$, $(-8, -6)$

36. $(3, 2)$, $(-3, 2)$, $(3, -2)$, $(-3, -2)$ **37.** 30 ft by 20 ft **38.** $61, 74, 87, a_n = 13n - 17$ **39.** $13, 31, 67, 139, 283$ **40.** $a_n = 9n - 6$
41. 258 **42.** 20,100 **43.** $a_n = 12 \cdot \left(\frac{3}{4}\right)^{n-1}$ **44.** 21,845 **45.** 274,514 **46.** Yes, $\frac{48}{5}$ **47.** 210 **48.** $x^4 + 4x^3y + 6x^2y^2 + 4xy^3 + y^4$
49. $x^6 - 6x^5y + 15x^4y^2 - 20x^3y^3 + 15x^2y^4 - 6xy^5 + y^6$ **50.** $x^8 + 8x^7y + 28x^6y^2 + 56x^5y^3 + 70x^4y^4 + 56x^3y^5 + 28x^2y^6 + 8xy^7 + y^8$

Appendix A

1. $(2, 3)$ **3.** $(-3, 7)$ **5.** $(0, 6)$ **7.** $(3, 2, 1)$ **9.** $(-1, -2, 6)$ **11.** $(0, 4, -4)$

Appendix B

1. $x - 11$ **3.** $x + 16$ **5.** $7x - 25 + \dfrac{38}{x + 2}$ **7.** $12x - 11 - \dfrac{308}{x - 14}$ **9.** $x^2 + 5x - 9 + \dfrac{13}{x - 5}$ **11.** $x^3 - 8x^2 + 72x - 576 + \dfrac{4575}{x + 8}$

Page 14 © Thinkstock **Page 15** © Rubberball Productions **Page 26** (infant twins) © PhotoDisc Red **Page 26** (female chemist) © Digital Vision **Page 27** © PhotoDisc **Page 33** © PhotoDisc **Page 43** © Digital Vision **Page 56** (Tom Selleck as Magnum, P.I.) © The Kobal Collection **Page 56** (James Garner as Jim Rockford) © The Kobal Collection **Page 63** © Murleen Ray **Page 72** © Philip Gould/Corbis **Page 74** © Rubberball Productions **Page 84** © Comstock **Page 85** © Stockbyte Platinum **Page 91** © Brand X Pictures **Page 92** © The Kobal Collection **Page 96** © Corbis **Page 100** © Peter Harholdt/Corbis **Page 103** © Rubberball **Page 105** © Guy Motil/Corbis **Page 119** (glassy-winged sharpshooter) © George D. Lepp/Corbis **Page 119** (young woman in a formal dress) © PhotoDisc Red **Page 152** © Digital Vision **Page 163** © Comstock **Page 168** © PhotoDisc Red **Page 171** © PhotoDisc Blue **Page 177** © S. Carmona/Corbis **Page 197** © PhotoDisc Blue **Page 205** © PhotoDisc Red **Page 219** © Murleen Ray **Page 236** © Robbie Jack/Corbis **Page 239** (college basketball game) © David Bergman/Corbis **Page 239** (female doctor) © PhotoDisc Red **Page 249** © PhotoDisc Blue **Page 254** © PhotoDisc PP **Page 257** © Reuters/Corbis **Page 275** © PhotoDisc Blue **Page 278** © Murleen Ray **Page 299** © NASA **Page 301** (Milky Way galaxy) © NASA **Page 301** (college students) © PhotoDisc **Page 330** © Anthony Reynolds; Cordaiy Photo Library Ltd./Corbis **Page 373** © Getty Images Inc.—Stone Allstock **Page 375** © Thinkstock **Page 377** © Image Source **Page 389** © PhotoDisc Red **Page 391** © Library of Congress **Page 440** © Stockbyte Platinum **Page 444** © PhotoDisc Blue **Page 448** © Brand X Pictures **Page 449** © Randy Wells/Corbis **Page 456** © Najlah Feanny/Corbis SABA **Page 508** © Robert Maass/Corbis **Page 516** © PhotoDisc **Page 518** © St. Andrew's University MacTutor Archive **Page 562** © PhotoEdit **Page 563** © PhotoDisc Blue **Page 579** © SAU **Page 605** (quilt) © PhotoDisc **Page 605** (hang glider) © Brand X Pictures **Page 607** © Taxi **Page 612** © Brand X Pictures **Page 614** © PhotoDisc **Page 649** © Corbis/Bettmann **Page 653** © Yannis Behrakis/Reuters/Corbis **Page 660** © Mike Blake/Reuters/Corbis **Page 675** © Digital Vision **Page 717** © PhotoDisc Red **Page 744** © The Granger Collection **Page 760** © PhotoDisc **Page 763** © PhotoDisc Red **Page 775** © Comstock **Page 779** © Roger Ressmeyer/Corbis **Page 802** © PhotoDisc Red **Page 806** © PhotoDisc Blue **Page 810** © Corbis **Page 811** © Brand X Pictures **Page 831** © Lucien Aigner/Corbis **Page 865** © Thinkstock **Page 909** © PhotoDisc **Page 910** © Getty Images, Inc.—Stone Allstock **Page 919** © Institut Mittag Leffler **Page 922** © PhotoDisc **Page 926** © PhotoDisc **Page 941** © Digital Vision **Page 944** © Comstock **Page 957** © SAU

Index of Applications

Index

Formulas and Equations

Distance Formula (see page 91) $d = r \cdot t$

Basic Percent Equation (see page 98)
Amount = Percent $\cdot$ Base

Simple Interest (see page 101) $I = P \cdot r \cdot t$

Horizontal Line (see page 145) $y = b$

Vertical Line (see page 146) $x = a$

Slope Formula (see page 156) $m = \dfrac{y_2 - y_1}{x_2 - x_1}$

Slope-Intercept Form of a Line (see page 159)
$y = mx + b$

Point-Slope Form of a Line (see page 190)
$y - y_1 = m(x - x_1)$

Rules for Exponents (see pages 290, 293)
$$x^m \cdot x^n = x^{m+n}$$
$$(x^m)^n = x^{mn}$$
$$(xy)^n = x^n y^n$$
$$\frac{x^m}{x^n} = x^{m-n} \ (x \neq 0)$$
$$x^0 = 1 \ (x \neq 0)$$
$$\left(\frac{x}{y}\right)^n = \frac{x^n}{y^n} \ (y \neq 0)$$
$$x^{-n} = \frac{1}{x^n} \ (x \neq 0)$$

Special Products (see page 313)
$$(a + b)(a - b) = a^2 - b^2$$
$$(a + b)^2 = a^2 + 2ab + b^2$$
$$(a - b)^2 = a^2 - 2ab + b^2$$

Factoring Formulas
Difference of Squares:
$a^2 - b^2 = (a + b)(a - b)$ (see page 352)
Difference of Cubes:
$a^3 - b^3 = (a - b)(a^2 + ab + b^2)$ (see page 354)
Sum of Cubes:
$a^3 + b^3 = (a + b)(a^2 - ab + b^2)$ (see page 355)

Height, in feet, of a projectile after t seconds
(see page 382) $h(t) = -16t^2 + v_0 t + s$

Direct Variation (see page 445) $y = kx$

Inverse Variation (see page 446) $y = \dfrac{k}{x}$

Properties of Radicals (see pages 525, 526)
$$\sqrt[n]{a} \cdot \sqrt[n]{b} = \sqrt[n]{ab}$$
$$\frac{\sqrt[n]{a}}{\sqrt[n]{b}} = \sqrt[n]{\frac{a}{b}}$$

Fractional Exponents (see page 531) $a^{m/n} = (\sqrt[n]{a})^m$

Period of a Pendulum (see page 560) $T = 2\pi\sqrt{\dfrac{L}{32}}$

Imaginary Numbers (see page 564) $i = \sqrt{-1}, i^2 = -1$

Quadratic Formula (see page 591) $x = \dfrac{-b \pm \sqrt{b^2 - 4ac}}{2a}$

Pythagorean Theorem (see page 600) $a^2 + b^2 = c^2$

Properties of Logarithms (see pages 781, 782, 783)
$$\log_b x + \log_b y = \log_b(xy) \ (x > 0, y > 0, b > 0, b \neq 1)$$
$$\log_b x - \log_b y = \log_b\left(\frac{x}{y}\right) \ (x > 0, y > 0, b > 0, b \neq 1)$$
$$\log_b x^r = r \cdot \log_b x \ (x > 0, b > 0, b \neq 1)$$

Change-of-Base Formula (see page 787)
$$\log_b x = \frac{\log_a x}{\log_a b} \ (x > 0, a > 0, b > 0, a \neq 1, b \neq 1)$$

Compound Interest Formula (see page 802)
$$A = P\left(1 + \frac{r}{n}\right)^{nt}$$

Continuous Compound Interest (see page 804)
$A = Pe^{rt}$

Exponential Growth and Decay (see pages 805, 806)
$P = P_0 e^{kt}$

pH of a Liquid (see page 808) $pH = -\log[H^+]$

Volume of a Sound (see page 809) $L = 10\log\left(\dfrac{I}{10^{-12}}\right)$

Distance Formula (see page 851)
$$d = \sqrt{(x_2 - x_1)^2 + (y_2 - y_1)^2}$$

General Term of an Arithmetic Sequence (see page 930)
$a_n = a_1 + (n - 1)d$

Partial Sum of an Arithmetic Sequence (see page 931)
$$s_n = \frac{n}{2}(a_1 + a_n)$$

General Term of a Geometric Sequence (see page 937)
$a_n = a_1 \cdot r^{n-1}$

Partial Sum of a Geometric Sequence (see page 938)
$$s_n = a_1 \cdot \frac{1 - r^n}{1 - r}$$

Limit of an Infinite Geometric Series (If It Exists)
(see page 940)
$$s = \frac{a_1}{1 - r}$$

Binomial Theorem (see page 947)
$$(x + y)^n = \sum_{r=0}^{n} {}_nC_r \cdot x^{n-r} y^r$$

Functions, Equations, and Their Graphs

Linear Equations and Functions (see page 174)
$$Ax + By = C, \; y = mx + b, \; f(x) = mx + b$$

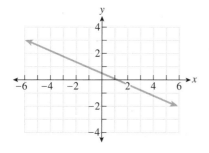

Square Root Functions (see page 700)
$$f(x) = a\sqrt{x - h} + k$$

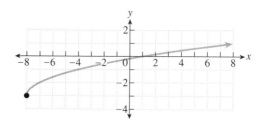

Absolute Value Functions (see page 485)
$$f(x) = a|x - h| + k$$

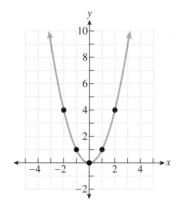

Cubic Functions (see page 700)
$$f(x) = a(x - h)^3 + k$$

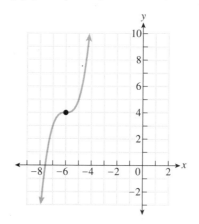

Quadratic Equations and Functions (see page 616)
$$y = ax^2 + bx + c, \quad y = a(x - h)^2 + k,$$
$$f(x) = ax^2 + bx + c, \quad f(x) = a(x - h)^2 + k$$

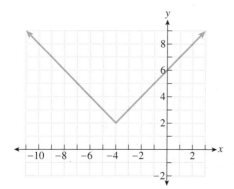

Exponential Functions (see page 753)
$$f(x) = b^{x-h} + k$$

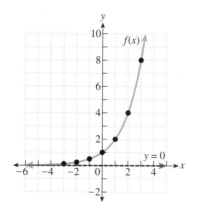